Introductory DC/AC Circuits

Sixth Edition

Nigel P. Cook

Upper Saddle River, New Jersey
Columbus, Ohio

Library of Congress Cataloging-in-Publication Data
Cook, Nigel P.
Introductory DC/AC circuits / Nigel P. Cook.—6th ed.
p. cm.
Includes index.
ISBN 0-13-114006-X
1. Electronics. I. Title.

TK7816.C656 2005
621.381—dc21

00-053757

Acquistions Editor: Dennis Williams
Development Editor: Kate Linsner
Production Editor: Rex Davidson
Design Coordinator: Diane Ernsberger
Cover Designer: Kristina D. Holmes
Cover Photo: Digital Vision
Illustrations: Rolin Graphics
Production Manager: Pat Tonneman
Project Management: Holly Henjum, Carlisle Publishers Services

This book was set in Times Roman by Carlisle Communiations, Ltd. It was printed and bound by R. R. Donnelley & Sons Company. The cover was printed by Coral Graphic Services, Inc.

Pearson Education Ltd.
Pearson Education Singapore Pte. Ltd.
Pearson Education Canada, Ltd.
Pearson Education—Japan
Pearson Education Australia Pty. Limited
Pearson Education North Asia Ltd.
Pearson Educación de Mexico, S.A. de C.V.
Pearson Education Malaysia Pte. Ltd.

10 9 8 7 6 5 4 3 2 1
ISBN 0-13-114006-X

To Dawn, Candy, and Jon

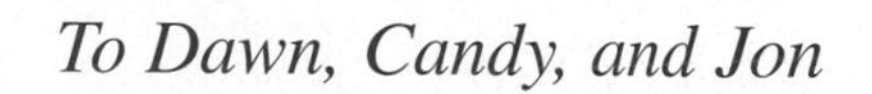

Books by Nigel P. Cook

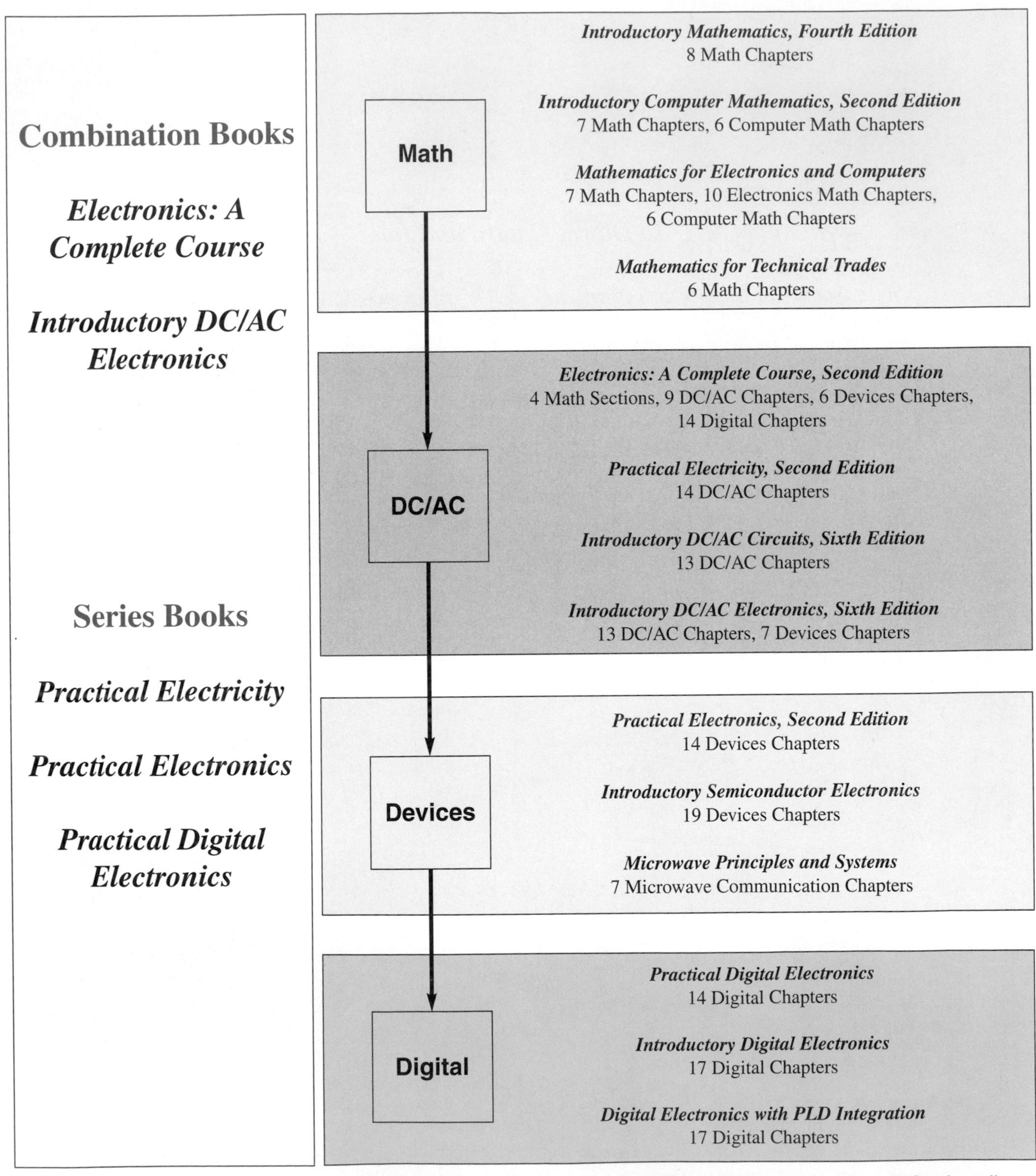

For more information and a desk copy of any of these textbooks by Nigel P. Cook, call 1-800-228-7854, visit the Prentice Hall website at www.prenhall.com, or ask your local Prentice Hall representative.

Preface

TO THE STUDENT

Since World War II, no branch of science has contributed more to the development of the modern world than *electronics*. It has stimulated dramatic advances in the fields of communication, computing, consumer products, industrial automation, test and measurement, and health care. It has now become the largest single industry in the world, exceeding the automobile and oil industries, with annual sales of electronic systems greater than $2 trillion.

The early pioneers in electronics were intrigued by the mystery and wonder of a newly discovered science, whereas people today are attracted by its utility in any application that allows them to accomplish almost anything imaginable. If you analyze exactly how you feel at this stage, you will probably discover that you have mixed emotions about the journey ahead. On the one hand, imagination, curiosity, and excitement are driving you on, while apprehension and reservations may be slowing you down. Your enthusiasm will overcome any indecision you have once you become actively involved in electronics and realize that it is as exciting as you ever expected it to be.

ORGANIZATION OF THE TEXTBOOK

This textbook has been divided into three basic parts. Chapters 1 through 3 introduce you to the world of electronics and the fundamentals of electricity. Chapters 4 through 7 cover direct current, or dc, circuits. Chapters 8 through 13 cover alternating current, or ac, circuits.

Introduction: Careers in Electronics

Part A **The Fundamentals of Electricity**
Chapter 1 Current and Voltage
Chapter 2 Resistance and Power
Chapter 3 Conductors, Insulators, and Resistors

Part B **Direct-Current Electronics**
Chapter 4 Direct Current (DC)
Chapter 5 Series Circuits
Chapter 6 Parallel Circuits
Chapter 7 Series–Parallel Circuits

Part C	Alternating-Current Electronics
Chapter 8	Alternating Current (AC)
Chapter 9	Capacitors
Chapter 10	Electromagnetic Devices
Chapter 11	Inductors
Chapter 12	Transformers
Chapter 13	Resistive, Inductive, and Capacitive *(RLC)* Circuits

The material covered in this book has been logically divided and sequenced to provide a gradual progression from the known to the unknown, and from the simple to the complex.

ANCILLARIES ACCOMPANYING THIS TEXT

The following ancillaries accompanying this text provide extensive opportunity for further study and support:

- ***Electronics Workbench Multisim® CD-ROM.*** Packaged with each copy of this text, the CD-ROM contains over 100 circuits from the text, created in Multisim Version 7.
- ***Laboratory Manual.*** Coauthored by Nigel Cook and Gary Lancaster, the lab manual offers numerous experiments designed to translate all of the textbook's theory into practical experimentation.
- ***Companion Website,*** located at http://www.prenhall.com/cook. Numerous interactive study questions are provided on this site to reinforce the concepts covered in the book.

To complete the ancillary package, the following supplements are essential elements for any instructor using this text for a course:

- ***Instructor's Resource Manual,*** containing solutions for the text and lab manual as well as a Test Item File.
- ***PowerPoint® Slides.*** This CD-ROM includes a full set of **lecture presentations** as well as slides for all schematics appearing in the text.
- ***PH TestGen,*** a computerized test bank.

CIRCUIT SIMULATION CD-ROM USING MULTISIM VERSION 7

Multisim® is a schematic capture, simulation, and programmable logic tool used by college and university students in their course study of Electronics and Electrical Engineering. The circuits on the CD in this text were created for use with Multisim software.

Multisim is widely regarded as an excellent tool for classroom and laboratory learning. However, no part of this textbook is dependent upon the Multisim software or the files provided with this text. These files are provided at no extra cost to the consumer and are for use by anyone who chooses to utilize Multisim software.

The first 25% of the circuits on the CD included with this text are already rendered "live" for you by Electronics Workbench in the Textbook Edition of Multisim 7, enabling you to do the following:

- Manipulate the interactive components and adjust the value of any virtual components.
- Run interactive simulation on the active circuits and use any pre-placed virtual instruments.

- Run analyses.
- Run/print/save simulation results for the pre-defined viewable circuits.
- Create your own circuits up to a maximum of 15 components.

The balance of the circuits requires that you have access to Multisim 7 in your school lab (the Lab Edition) or on your computer (Electronics Workbench Student Suite). If you do not currently have access to this software and wish to purchase it, *please call Prentice Hall Customer Service at 1-800-282-0693 or send a fax request to 1-800-835-5327.*

If you need *technical assistance or have questions concerning the Multisim software,* contact Electronics Workbench directly for support at 416-977-5550 or via the EWB website located at **www.electronicsworkbench.com.**

DEVELOPMENT, CLASS TESTING, AND REVIEWING

The first phase of the development of this text was conducted in the classroom with students and instructors as critics. Each topic was class-tested by videotaping each lesson, evaluating the results, and then implementing recommended changes. This invaluable feedback enabled me to fine-tune my presentation of topics and instill understanding and confidence in the students.

The second phase of development involved sending a copy of the text to several instructors at schools throughout the country. Their technical and topical critiques helped to mold the text into a more accurate and effective form.

The third and final phase was to class-test the text and then commission the last technical review in the final stages of production.

ACKNOWLEDGMENTS

My appreciation and thanks are extended to the following instructors who have reviewed and contributed greatly to the development of this sixth edition: Gary House, DeVry University—Atlanta; Lili Muljadi, DeVry University—Denver; David Skeen, Bates Technical College; Roger Wendler, Southeast College of Technology; and Kathy Zarrinkoub, DeVry University—Crystal City.

Nigel P. Cook

ILLUSTRATED TOUR OF TEXTBOOK FEATURES

NEW Math Review Section

■ **EXAMPLE:**

The voltage at the wall outlet in your home was measured at different times in the day and equaled 108.6, 110.4, 115.5, and 123.6 volts (V). Find the average voltage.

■ *Solution:*

Add all the voltages.

$$108.6 + 110.4 + 115.5 + 123.6 = 458.1 \text{ V}$$

Divide the result by the number of readings.

$$\frac{458.1 \text{ V}}{4} = 114.525 \text{ V}$$

The average voltage present at your home was therefore 114.525 volts.

3-1-6 *Graphing Data*

In many areas of science and technology, you will often see data shown visually in the form of a **graph** (Figure 3-1). To obtain this data, a *practical analysis* is performed on a device, in which the input is physically changed in increments and the output results are recorded in a table. Read the notes in Figure 3-1 to see how a line graph is plotted from the example table of device data.

Figure 3-1 would accommodate any data in which the input and output data are positive, but what sort of graph could be used if positive and negative values had to be plotted? The answer is a **Cartesian coordinate system.** With this system, the previous positive vertical *x*-axis and horizontal *y*-axis are still used, but are extended beyond the zero point in the opposite direction to accommodate negative values, as in Figure 3-2. Again, read the notes in Figure 3-1 to see how the table of results can be used to plot a graph.

Graph
A diagram (as a series of one or more points, lines, line segments, curves, or areas) that represents the variation of a variable in comparison with that of one or more other variables.

Cartesian Coordinate System
Two perpendicular number lines used to place points in a plane.

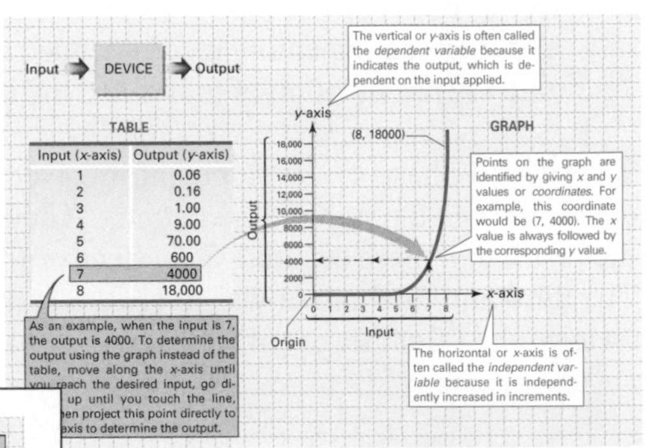

3 / CONDUCTORS, INSULATORS, AND RESISTORS

Type	Construction	Characteristics
Carbon Composition	Solder Coated Leads; Color Coding; Solid Resistive Element; Leads Solidly Embedded	Resistive material is a combination of powdered carbon and carbon insulator, the ratio of which determines the resistance value. Low cost, but inherently noisy; noise is generated by small arcs within the resistive material.
Carbon Film	Helixing; Ceramic Substrate; Color Bands; Insulation; Carbon Film; End Cap and Lead Wire	Smaller tolerances (±5% to ±2%) and better temperature stability than carbon composition.
Metal Film	Metal Film; Epoxy Coating; End Cap; Ceramic; Leads	Good tolerances (±1% to ±0.1%) and temperature stability; however, high cost.
Metal Oxide	Epoxy Coating; End Cap; Leads; Film Resistor Material	Has excellent temperature stability; however, high cost.
Wirewound (100 MΩ)		High power rating and good tolerance (±1%); however, very high cost.
Thick Film Networks Chips (SIP; DIP; Chip; Actual Size)	SIP Single In-line Package; DIP Double In-line Package; A single row of connecting pins; A dual row of connecting pins; Surface Mount Chip Resistor; Glass Coat; Substrate; Resistive Film; Termination	SIP and DIP Resistor Networks: ±2% tolerance, 1/8 W heat dissipation, 10Ω to 3.3MΩ. Chip Resistor: Surface Mount Technology (SMT) device uses less printed circuit board (PCB) space than through-hole resistor, as shown here. Conductive end terminations are soldered onto pads on the PCB. Thick film networks and chip resistors have a very high cost. (Through-Hole Resistor; Surface Mount Resistor; Printed Circuit Board; Conductive Strips Printed on Insulating Board)

FIGURE 3-20 **Fixed-Value Resistor Types.**

NEW Component Type Tables

NEW Topic Coverage

switch and connects the current source to the load. For momentary small overcurrent surges, the electromagnet is activated but does not have enough force to trip the breaker. If overcurrent continues, however, the bimetallic strip will have heated and the combined forces of the bimetallic strip and electromagnet will trip the breaker and protect the load by disconnecting the source. Momentary large overcurrent surges will supply enough current to the electromagnet to cause it to trip the breaker independently of the bimetallic strip and so disconnect the source from the load.

4-4-3 *Ground Fault Interrupters*

Gounding
Connecting a point in a circuit to Earth or some connecting body that serves in place of Earth.

Before we discuss the operation of ground fault interrupters, or GFIs, let us first discuss the need for grounding. **Grounding** is not normally discussed to any great extent, and yet an understanding of why it is necessary to have it is very important from a safety standpoint. In Figure 4-31(a) you can see a toaster (with a 10 Ω load resistance) connected to the 120 V outlet. In nearly all cases, appliances are grounded, which means that the chassis (the unit's metal frame) is connected to the ground pin in the plug, which connects back through the circuit breaker box to an 8 foot copper bar that is in the ground. If the chassis is not grounded, as shown in Figure 4-31(b), an accidental internal short within the appliance could connect

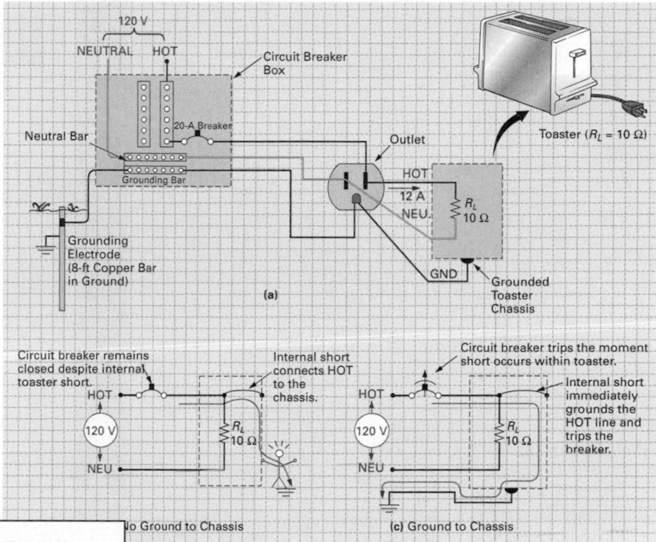

son for Grounding.

CHAPTER 4 / DIRECT CURRENT (DC)

tant after the fuse has blown, prevents arcing across the blown fuse contacts, as illustrated in Figure 4-25(e). Once the fuse has blown, the circuit's positive and negative voltages are now connected across the fuse contacts. If the voltage is too great, an arc can jump across the gap, causing a sudden surge of current, damaging the equipment connected.

Fuses are mounted within fuse holders and normally placed at the back of the equipment for easy access, as is shown in Figure 4-25(c). When replacing blown fuses, you must be sure that the equipment is turned OFF and disconnected from the source. If you do not remove power you could easily get a shock, since the entire source voltage appears across the fuse, as is shown in Figure 4-25(e). Also, make sure that a fuse with the correct current and voltage rating is used.

A good fuse should have a resistance of 0 Ω when it is checked with an ohmmeter. In the circuit and with power ON, the fuse should have no voltage drop across it because it is just a piece of wire. A blown or burned-out fuse will, when removed, read infinite ohms, and when in its holder and power ON, will have the full applied voltage across its two terminals.

EXAMPLE

Figure 4-26 details how a variable resistor is used to dim car lights.

a. Is a rheostat or potentiometer being used to adjust current?
b. When the switch is in position 2, what lights are ON and what are OFF?
c. Based on the values given, what is the resistance of the ashtray light?
d. What is the fuse rating for the dashboard lights and the headlights?

Solution

a. Rheostat
b. Position 2: Parking lights, side lights, licence plate lights, and dashboard lights.

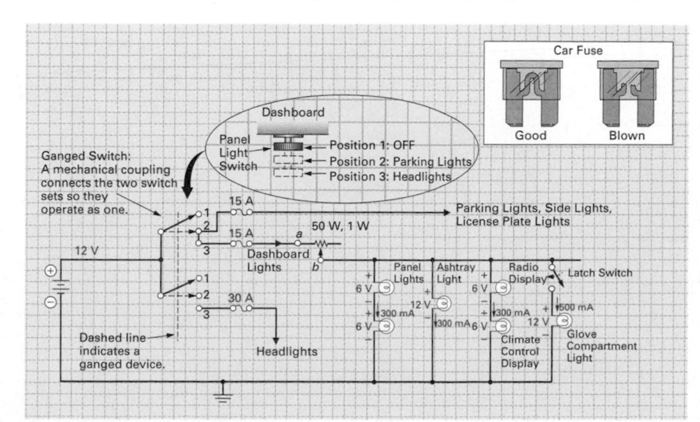

FIGURE 4-26 **Automobile Dimmer Control.**

158 CHAPTER 4 / DIRECT CURRENT (DC)

NEW Application Circuit Examples

NEW Historical Vignettes and Time Lines

PART B DIRECT-CURRENT ELECTRONICS

Direct Current (DC)

The First Computer Bug

Mathematician Grace Murray Hopper, an extremely independent U.S. naval officer, was assigned to the Bureau of Ordnance Computation Project at Harvard during World War II. As Hopper recalled, "We were not programmers in those days, the word had not yet come over from England. We were 'coders,'" and with her colleagues she was assigned to compute ballistic firing tables on the Harvard Mark 1 computer. In carrying out this task, Hopper developed programming method fundamentals that are still in use.

On a less important note, Hopper is also credited with creating a term frequently used today falling under the category of computer jargon. During the hot summer of 1945, the computer developed a mysterious problem. Upon investigation, Hopper discovered that a moth had somehow strayed into the computer and prevented the operation of one of the thousands of electromechanical relay switches. In her usual meticulous manner, Hopper removed the remains and taped and entered it into the logbook. In her own words, "From then on, when an officer came in to ask if we were accomplishing anything, we told him we were 'debugging' the computer," a term that is still used to describe the process of finding problems in a computer program.

Multisim and Companion Website

No matter how many resistors are connected in series, the total resistance or opposition to current flow is always equal to the sum of all the resistor values. This formula can be stated mathematically as

$$R_T = R_1 + R_2 + R_3 + \cdots$$

Total resistance = value of R_1 + value of R_2 + value of R_3, and so on.

Equivalent Resistance (R_{eq}) Total resistance of all the individual resistances in a circuit.

Total resistance (R_T) is the only opposition a source can sense. It does not see the individual separate resistors, but one **equivalent resistance.** Based on the source's voltage and the circuit's total resistance, a value of current will be produced to flow through the circuit (Ohm's law, $I = V/R$).

EXAMPLE:

Referring to Figure 5-8(a) and (b), calculate:

a. The circuit's total resistance

b. The current flowing through R_2

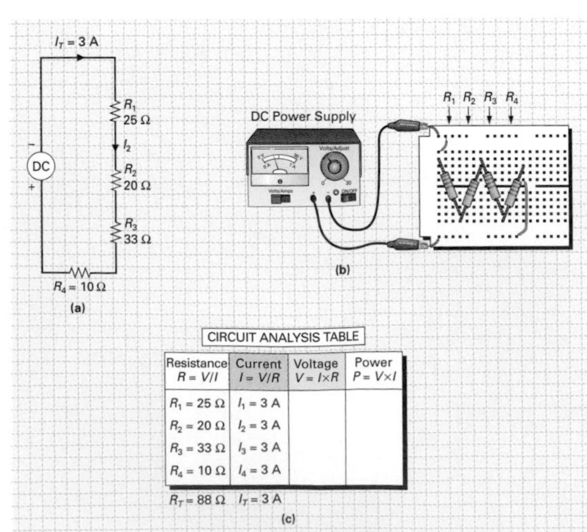

CIRCUIT ANALYSIS TABLE

Resistance $R = V/I$	Current $I = V/R$	Voltage $V = I \times R$	Power $P = V \times I$
$R_1 = 25\ \Omega$	$I_1 = 3$ A		
$R_2 = 20\ \Omega$	$I_2 = 3$ A		
$R_3 = 33\ \Omega$	$I_3 = 3$ A		
$R_4 = 10\ \Omega$	$I_4 = 3$ A		
$R_T = 88\ \Omega$	$I_T = 3$ A		

FIGURE 5-8 **Total Resistance Example. (a) Schematic. (b) Protoboard Circuit. (c) Circuit Analysis Table.**

Accessible Writing Style

From this example you can see that the current source delivered an almost constant output current regardless of the large load resistance change.

A Two-Source Circuit Example

In some instances, as seen in Figure 7-58, a circuit can contain more than one voltage source. In this example, the schematic diagram shows how car one's battery can be connected across car two's dead battery to boost current to the starter motor.

In the following sections, you will be introduced to some of the theorems that enable us to analyze these complex networks and reduce them to a simple equivalent circuit.

7-9-2 *Superposition Theorem*

The **superposition theorem** is used not only in electronics but also in physics and even economics. It is used to determine the net effect in a circuit that has two or more sources connected. The basic idea behind this theorem is that if two voltage sources are both producing a current through the same circuit, the net current can be determined by first finding the individual currents and then adding them together. Stated formally: *In a network containing two or more voltage sources, the current at any point is equal to the algebraic sum of the individual source currents produced by each source acting separately.*

Superposition Theorem
In a network or circuit containing two or more voltage sources, the current at any point is equal to the algebraic sum of the individual source currents produced by each source acting separately.

The best way to fully understand the theorem is to apply it to a few examples. Figure 7-59(a) illustrates a simple series circuit with two resistors and two voltage sources. The 12 V source (V_1) is trying to produce a current in a clockwise direction, while the 24 V source (V_2) is trying to force current in a counterclockwise direction. What will be the resulting net current in this circuit?

STEP 1 To begin, let's consider what current would be produced in this circuit if only V_1 were connected, as shown in Figure 7-59(b).

$$I_1 = \frac{V_1}{R_T} = \frac{12 \text{ V}}{9 \ \Omega} = 1.33 \text{ A}$$

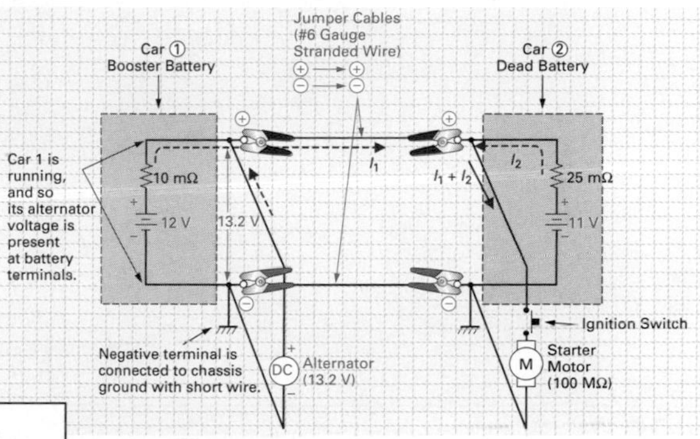

...ping a Car's Dead Battery.

Circuit Analysis Tables

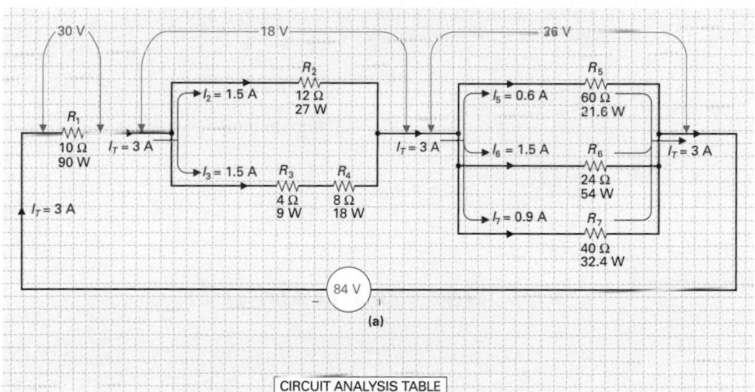

(a)

CIRCUIT ANALYSIS TABLE

Resistance $R = V/I$	Voltage $V = I \times R$	Current $I = V/R$	Power $P = V \times I$
$R_1 = 10\ \Omega$	$V_{R1} = 30$ V	$I_1 = 3$ A	$P_1 = 90$ W
$R_2 = 12\ \Omega$	$V_{R2} = 18$ V	$I_2 = 1.5$ A	$P_2 = 27$ W
$R_3 = 4\ \Omega$	$V_{R3} = 6$ V	$I_3 = 1.5$ A	$P_3 = 9$ W
$R_4 = 8\ \Omega$	$V_{R4} = 12$ V	$I_4 = 1.5$ A	$P_4 = 18$ W
$R_5 = 60\ \Omega$	$V_{R5} = 36$ V	$I_5 = 0.6$ A	$P_5 = 21.6$ W
$R_6 = 24\ \Omega$	$V_{R6} = 36$ V	$I_6 = 1.5$ A	$P_6 = 54$ W
$R_7 = 40\ \Omega$	$V_{R7} = 36$ V	$I_7 = 0.9$ A	$P_7 = 32.4$ W
$R_T = 28\ \Omega$	$V_T = V_S = 84$ V	$I_T = 3$ A	$P_T = 252$ W

(b)

FIGURE 7-18 **Series–Parallel Circuit Example with All Information Inserted. (a) Schematic. (b) Circuit Analysis Table.**

7-6 FIVE-STEP METHOD FOR SERIES–PARALLEL CIRCUIT ANALYSIS

Let's now combine and summarize all the steps for calculating resistance, voltage, current, and power in a series–parallel circuit by solving another problem. Before we begin, however, let us review the five-step procedure as shown at the top of the next page.

EXAMPLE:

Referring to Figure 7-19, calculate:

a. Total resistance

b. Total current

Descriptive Illustrations

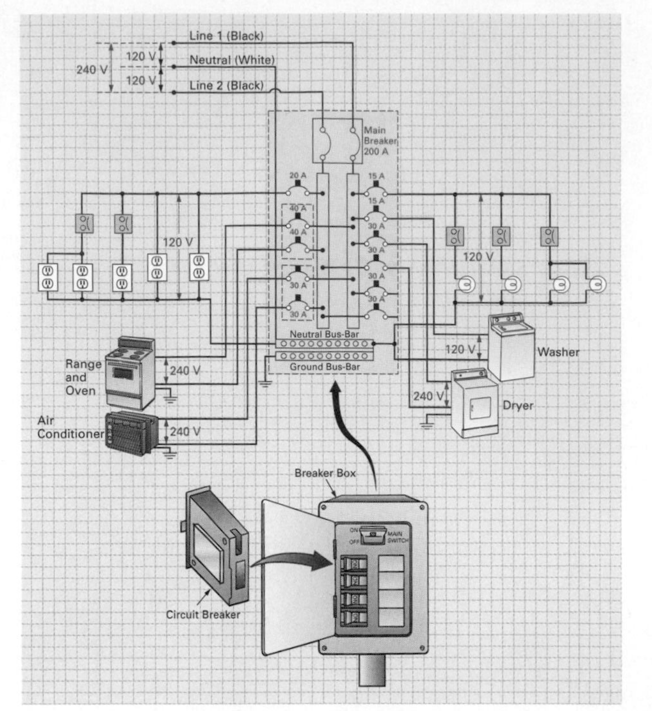

FIGURE 4-30 **Thermomagnetic Circuit Breakers in the Home Breaker Box.**

Thermomagnetic Circuit Breakers

Figure 4-30 illustrates the typical residential-type thermomagnetic circuit breaker and the schematic of a typical home breaker box. The thermal circuit breaker is similar to a slow-blow fuse in that it is ideal for passing momentary surges without tripping because of the delay caused by heating of the bimetallic strip. The magnetic circuit breaker, on the other hand, is most similar to a fast-blow fuse in that it is tripped immediately when an increase of current occurs. ...agnetic circuit breaker combines the advantages of both previously men- ...ers by incorporating both a bimetallic strip and an electromagnet in its ac- ... For currents at and below the current rating, the circuit breaker is a closed

Solution:

a. $R_T = R_1 + R_2 + R_3 + R_4$
$= 25\ \Omega + 20\ \Omega + 33\ \Omega + 10\ \Omega$
$= 88\ \Omega$

b. $I_T = I_1 = I_2 = I_3 = I_4$. Therefore, $I_2 = I_T = 3$ A.

The details of this circuit are summarized in the analysis table shown in Figure 5-8(c).

EXAMPLE:

Figure 5-9(a) and (b) shows how a single-pole three-position switch is being used to provide three different lamp brightness levels. In position ① R_1 is placed in series with the lamp, in position ② R_2 is placed in series with the lamp, and in position ③ R_3 is placed in series with the lamp. If the lamp has a resistance of 75 Ω, calculate the three values of current for each switch position.

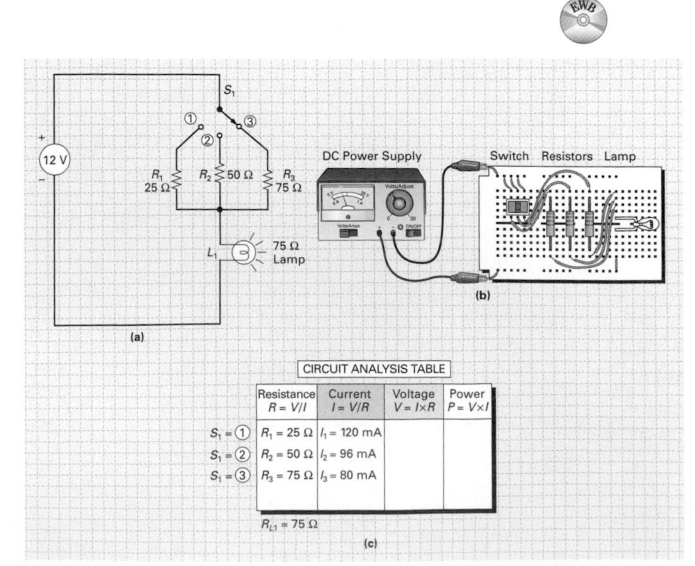

FIGURE 5-9 **Three-Position Switch Controlling Lamp Brightness. (a) Schematic. (b) Protoboard Circuit. (c) Circuit Analysis Table.**

Protoboard Pictorials

Contents in Brief

Contents

Introduction: Careers in Electronics

Your Course in Electronics

Your future in the electronics industry begins with this text. To give you an idea of where you are going and what we will be covering, Figure I-1 acts as a sort of road map, breaking up your study of electronics into four basic steps.

Step 1: Basics of Electricity
Step 2: Electronic Components
Step 3: Electronic Circuits
Step 4: Electronic Systems

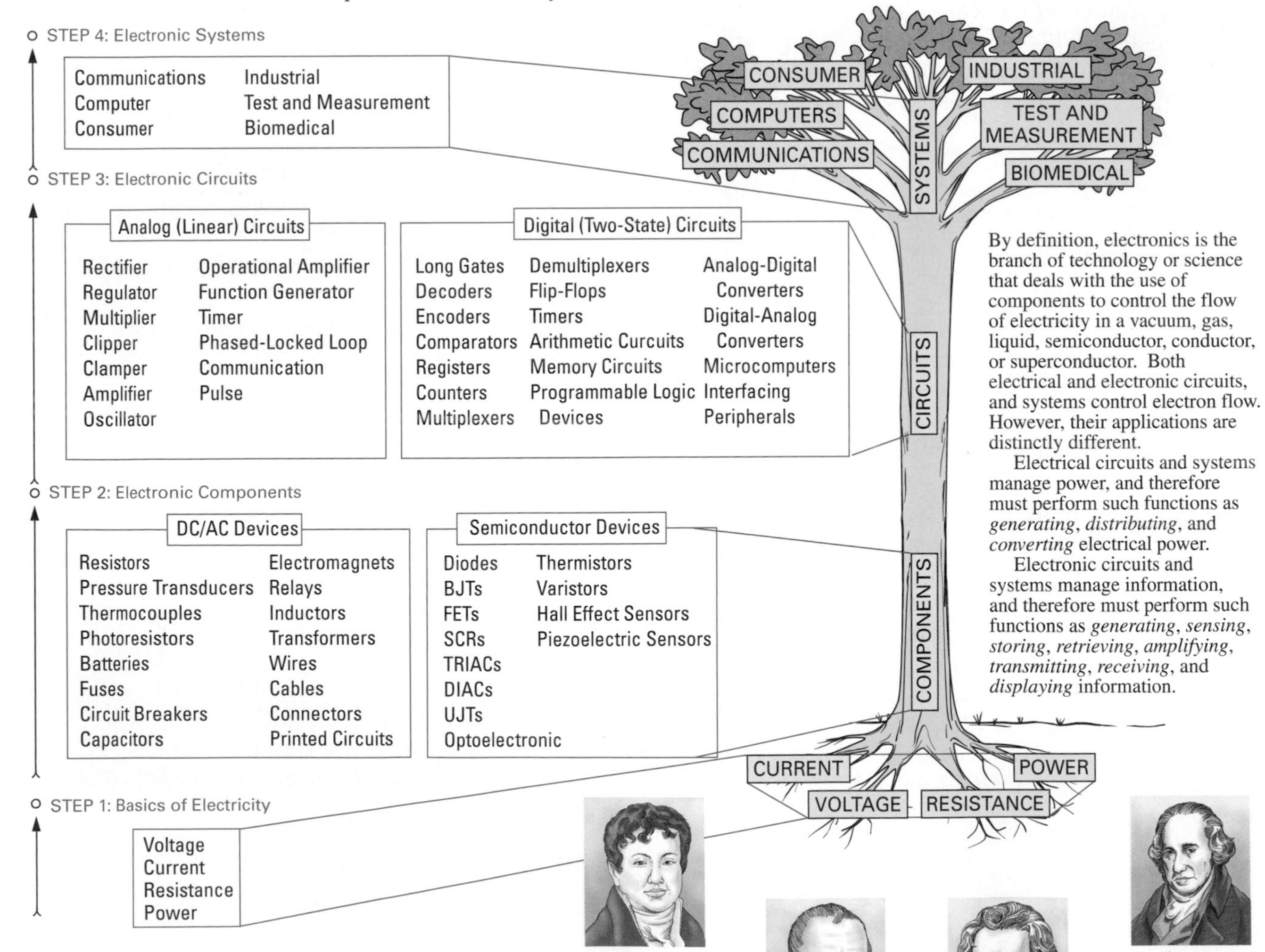

FIGURE I-1 The Steps Involved in Studying Electronics.

The main purpose of this introduction is not only to introduce you to the terms of the industry but also to show you why the first two chapters in this text begin at the very beginning with "current and voltage," and then "resistance and power." **Components,** which are the basic electronic building blocks, were developed to control these four roots or properties, and when these devices are combined they form **circuits.** Moving up the tree to the six different branches of electronics, you will notice that just as components are the building blocks for circuits, circuits are in turn the building blocks for **systems.**

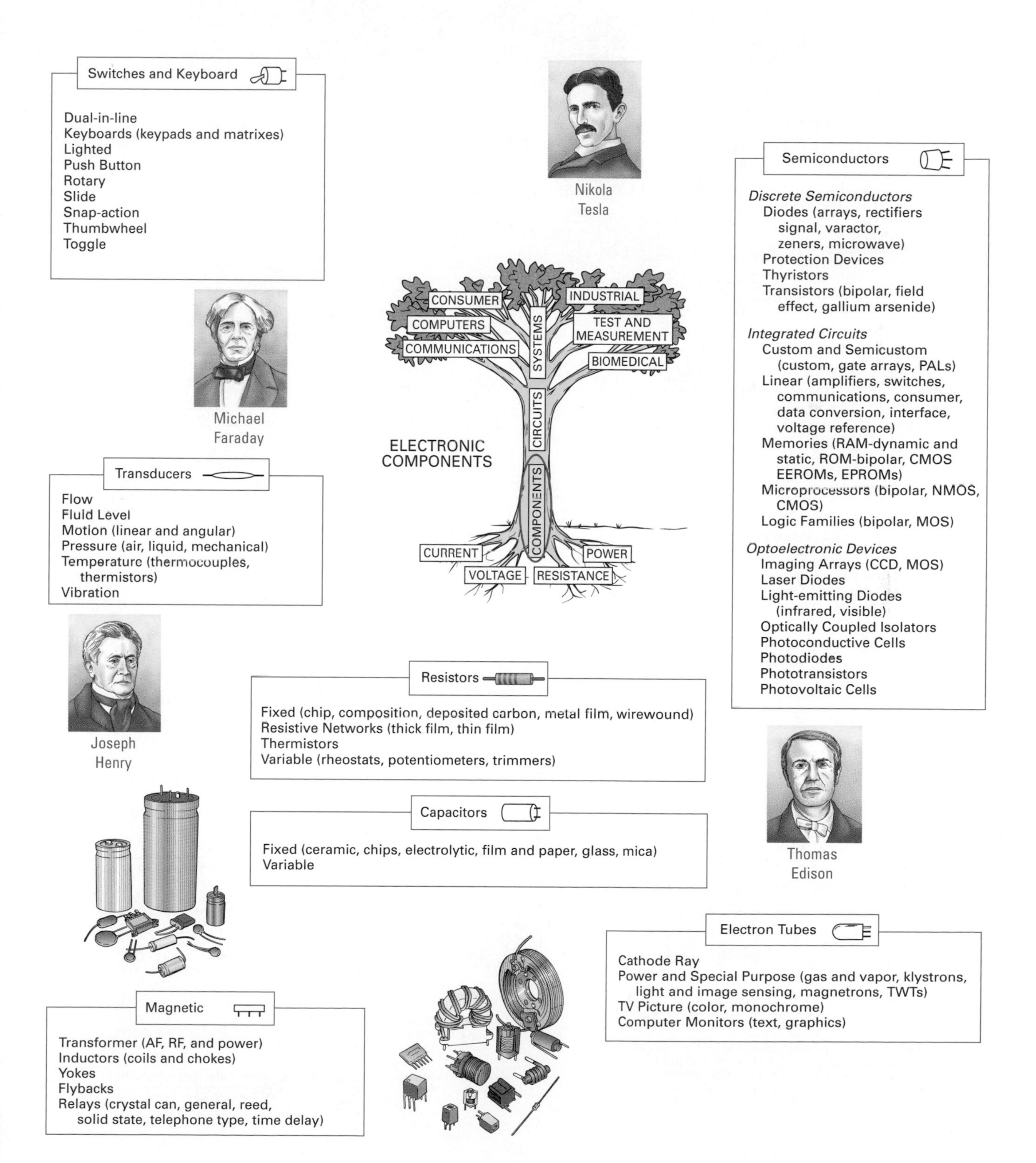

Charles
Wheatstone

Signal Generators and Timers

- Sine-wave Generators (oscillators)
- Square- and Pulse-wave Generators
- Ramp- and Triangular-wave Generators

Digital Circuits

- Oscillators and Generators
- Gates and Flip-flops
- Display Drivers
- Counters and Dividers
- Encoders and Decoders
- Memories
- Input/Output
- Microprocessors
- Latches
- Registers
- Multiplexers
- Demultiplexers
- Gate Arrays

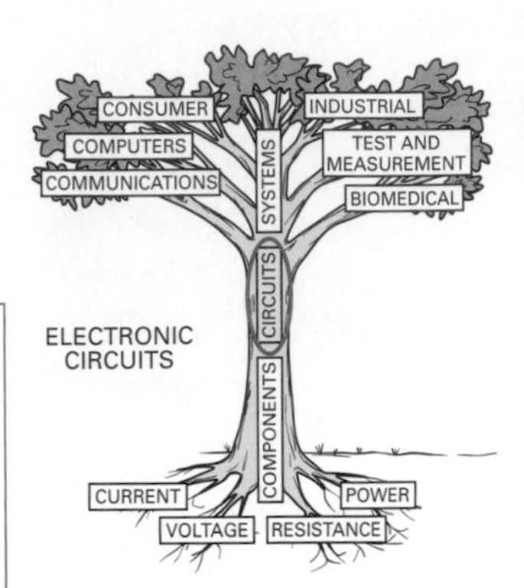

ELECTRONIC
CIRCUITS

René
Descartes

Amplifiers

- Bipolar Transistor
- Field-effect Transistor
- Operational Amplifier

Power Supplies

- Switching
- Linear
- Uninterruptible

Robert
Noyce

Miscellaneous

- Detectors and Mixers
- Filters
- Phase-locked Loops
- Converters
- Data Acquisition
- Synthesizers

Gustav
Kirchoff

Radio

- Amateur (mobile and base stations)
- Aviation Mobile and Ground Support Stations
- Broadcast Equipment
- Land Mobile (mobile and base stations)
- Marine Mobile (ship and shore stations)
- Microwave Systems
- Satellite Systems
- Radar and Sonar Systems

Guglielmo
Marconi

Lee
DeForest

Telecommunications

- Switching Systems
- Data and Voice Switching
- Voice-only Switching
- Cellular Systems
- Telephones
- Corded
- Cordless
- Telephone/Video Equipment
- Facsimile Terminals
- Fiber Optic Communication Systems

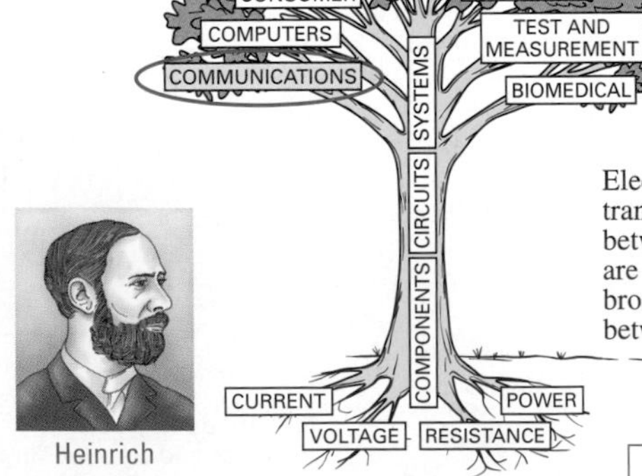

Heinrich
Hertz

COMMUNICATION
SYSTEMS

Electronic communications allow the transmission and reception of information between two points. Radio and television are obvious communication devices, broadcasting data or entertainment between two points.

John
Baird

Data Communications

- Concentrators
- Front-end Communications Processors
- Message-switching Systems
- Modems
- Multiplexers
- Network Controllers
- Mixed Service (combining voice, data, video, imaging)

Television

- Broadcast Equipment
- CATV Equipment
- CCTV Equipment
- Satellite TV Equipment
- HDTV (High Definition TV) Equipment

COMPUTER SYSTEMS

The computer is proving to be one of the most useful of all systems. Its ability to process, store, and manipulate large groups of information at an extremely fast rate makes it ideal for almost any and every application. Systems vary in complexity and capability, ranging from the Cray supercomputer to the home personal computer. The applications of word processing, record keeping, inventory, analysis, and accounting are but a few examples of data processing systems.

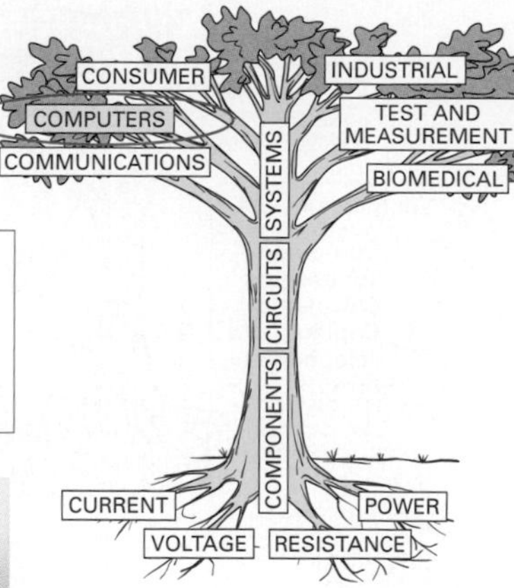

Data Terminals

CRT Terminals
- ASCII Terminals
- Graphics Terminals (color, monochrome)

Remote Batch Job Entry Terminals

Computer Systems

Microcomputers and Supermicrocomputers
Minicomputers (personal computers) and
- Superminicomputers (technical workstations, multiuser)

Mainframe Computers
Supercomputers

Charles Babbage

John von Neuman

George Boole

Alan Turing

Data Storage Devices

Fixed Disk (14, 8, $5\frac{1}{4}$, and $3\frac{1}{2}$ in.)
Flexible Disk (8, $5\frac{1}{4}$, and $3\frac{1}{2}$ in.)
Optical Disk Drives (read-only, write once, erasable)
Cassette
Cartridge Magnetic Tape ($\frac{1}{4}$ in.)
Cartridge Tape Drives ($\frac{1}{2}$ in.)
Reel-type Magnetic Tape Drives

I/O Peripherals

Computer Microfilm
Digitizers
Graphics Tablets
Light Pens
Trackball and Mice
Optical Scanning Devices
Plotters
Printers
- Impact
- Nonimpact (laser, thermal, electrostatic, inkjet)

CONSUMER SYSTEMS

From the smart computer-controlled automobiles, which provide navigational information and monitor engine functions and braking, to the compact disc players, video camcorders, satellite TV receivers, and wide-screen stereo TVs, this branch of electronics provides us with entertainment, information, safety, and, in the case of the pacemaker, life.

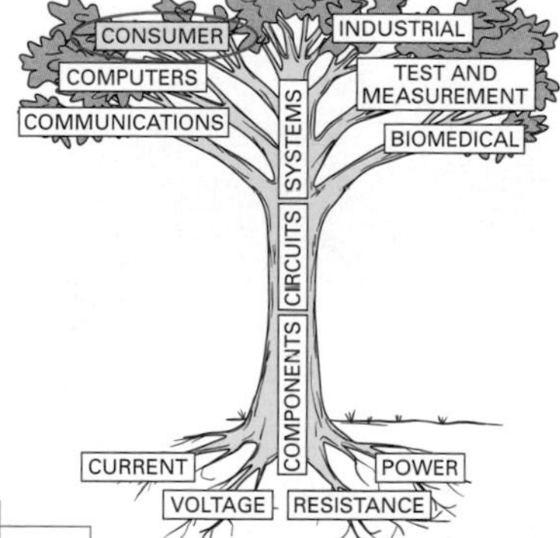

Jack Kilby

Automobile Electronics

Dashboard
Engine Monitoring and Analysis
Computer Navigation Systems
Alarms
Telephones

William Shockley

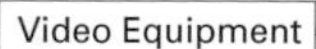

Video Equipment

TV Receiver (color, monochrome)
Projection TV Receivers
Video Cassette Recorders (VCRs)
Video Disk Players
Camcorders (8 mm, $\frac{1}{2}$ in.)
Home Satellite Receivers

Nolan Bushnell

Personal

Calculators, Cameras, Watches
Telephone Answering Equipment
Personal Computers
Microwave Ovens
Musical Equipment and Instruments
Pacemakers and Hearing Aids
Alarms and Smoke Detectors

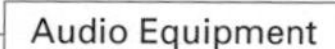

Audio Equipment

Car
Stereo Equipment
- Compact Systems (miniature components)
- Components (speakers, amps, turntables, tuners, tape decks)

Phonographs and Radio Phonographs
Radios (table, clock, portable)
Tape Players/Recorders
Compact Disk Players
Digital Tape Players

Chester Carlson

Manufacturing Equipment

- Energy Management Equipment
- Inspection Systems
- Motor Controllers (speed, torque)
- Numerical-control Systems
- Process Control Equipment (data-acquisition systems, process instrumentation, programmable controllers)
- Robot Systems
- Vision Systems

Computer-aided Design and Engineering CAD/CAE

- Hardware Equipment
  - Design Work Stations (PC based, 32 microprocessor based platform, host based)
  - Application Specific Hardware
- Design Software
  - Design Capture (schematic capture, logic fault and timing simulators, model libraries)
  - IC Design (design rule checkers, logic synthesizers, floor planners–place and route, layout editors)
  - Printed Circuit Board Design Software
  - Project Management Software
  - Test Equipment

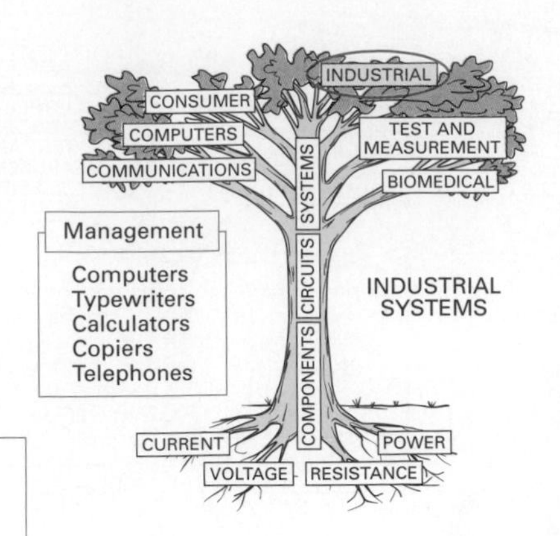

Management

- Computers
- Typewriters
- Calculators
- Copiers
- Telephones

Almost any industrial company can be divided into three basic sections, all of which utilize electronic equipment to perform their functions. The manufacturing section will typically use power, motor, and process control equipment, along with automatic insertion, inspection, and vision systems, for the fabrication of a product. The engineering section uses computers and test equipment for the design and testing of a product, while the management section uses electronic equipment such as computers, copiers, telephones, and so on.

James Joule

Carl Gauss

Charles Steinmetz

Werner Von Siemens

CONSUMER
INDUSTRIAL
COMPUTERS
TEST AND MEASUREMENT
COMMUNICATIONS
BIOMEDICAL
SYSTEMS
CIRCUITS
COMPONENTS
TEST AND MEASUREMENT SYSTEMS
CURRENT
POWER
VOLTAGE
RESISTANCE

Isaac Newton

Charles Napier

The rising complexity of electronic components, circuits, and equipment is causing a demand for sophisticated automatic test equipment for both the manufacturer and the customer to test their products.

Benjamin Franklin

Grace Hopper

General Test and Measurement Equipment

- Amplifiers (lab)
- Arbitrary Waveform Generators
- Analog Voltmeters, Ammeters, and Multimeters
- Audio Oscillators
- Audio Waveform Analyzers and Distortion Meters
- Calibrators and Standards
- Dedicated IEEE–488 Bus Controllers
- Digital Multimeters
- Electronic Counters (RF, Microwave, Universal)
- Frequency Synthesizers
- Function Generators
- Pulse/Timing Generators
- Signal Generators (RF, Microwave)
- Logic Analyzers
- Microprocessor Development Systems
- Modulation Analyzers
- Noise-measuring Equipment
- Oscilloscopes (Analog, Digital)
- Panel Meters
- Personal Computer (PC) Based Instruments
- Recorders and Plotters
- RF/Microwave Network Analyzers
- RF/Microwave Power-measuring Equipment
- Spectrum Analyzers
- Stand-alone In-circuit Emulators
- Temperature-measuring Instruments

Automated Test Systems

- Active and Discrete Component Test Systems
- Automated Field Service Testers
- IC Testers (benchtop, general purpose, specialized)
- Interconnect and Bare Printed Circuit-Board Testers
- Loaded Printed Circuit-Board Testers (in-circuit, functional, combined)

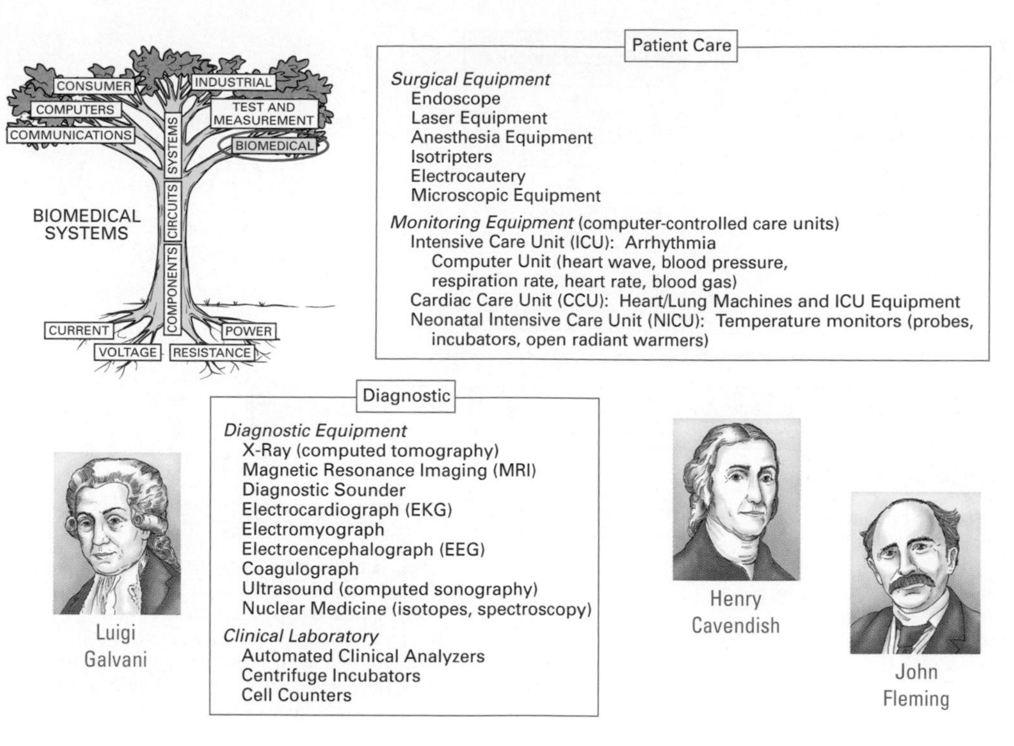

Electronic equipment is used more and more within the biological and medical fields, which can be categorized simply as being either patient care or diagnostic equipment. In the operating room, the endoscope, which is an instrument used to examine the interior of a canal or hollow organ, and the laser, which is used to coagulate, cut or vaporize tissue with extremely intense light, both reduce the use of invasive surgery. A large amount of monitoring equipment is used both in and out of operating rooms, and the equipment consists of generally large computer-controlled systems that can have a variety of modules inserted (based on the application) to monitor, on a continuous basis, body temperature, blood pressure, pulse rate, and so on. In the diagnostic group of equipment, the clinical laboratory test results are used as diagnostic tools. With the advances in automation and computerized information systems, multiple tests can be carried out at increased speeds. Diagnostic imaging, in which a computer constructs an image of a cross-sectional plane of the body, is probably one of the most interesting equipment areas.

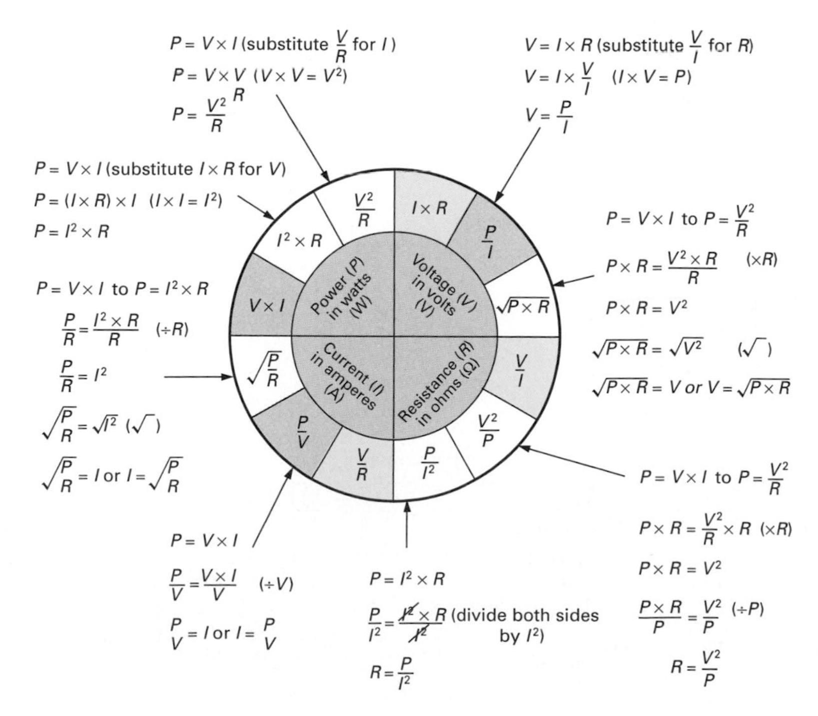

The formula circle shows how the four basic properties—voltage (V), current (I), resistance (R), and power (P)—are all related. This text will start at the very beginning with these basic roots of electronics. Voltage and current will be discussed in Ch. 1, resistance and power in Ch. 2, and then subsequent chapters will proceed to components, circuits, and systems.

OHM'S LAW $V = I \times R$

WATT'S LAW $P = V \times I$

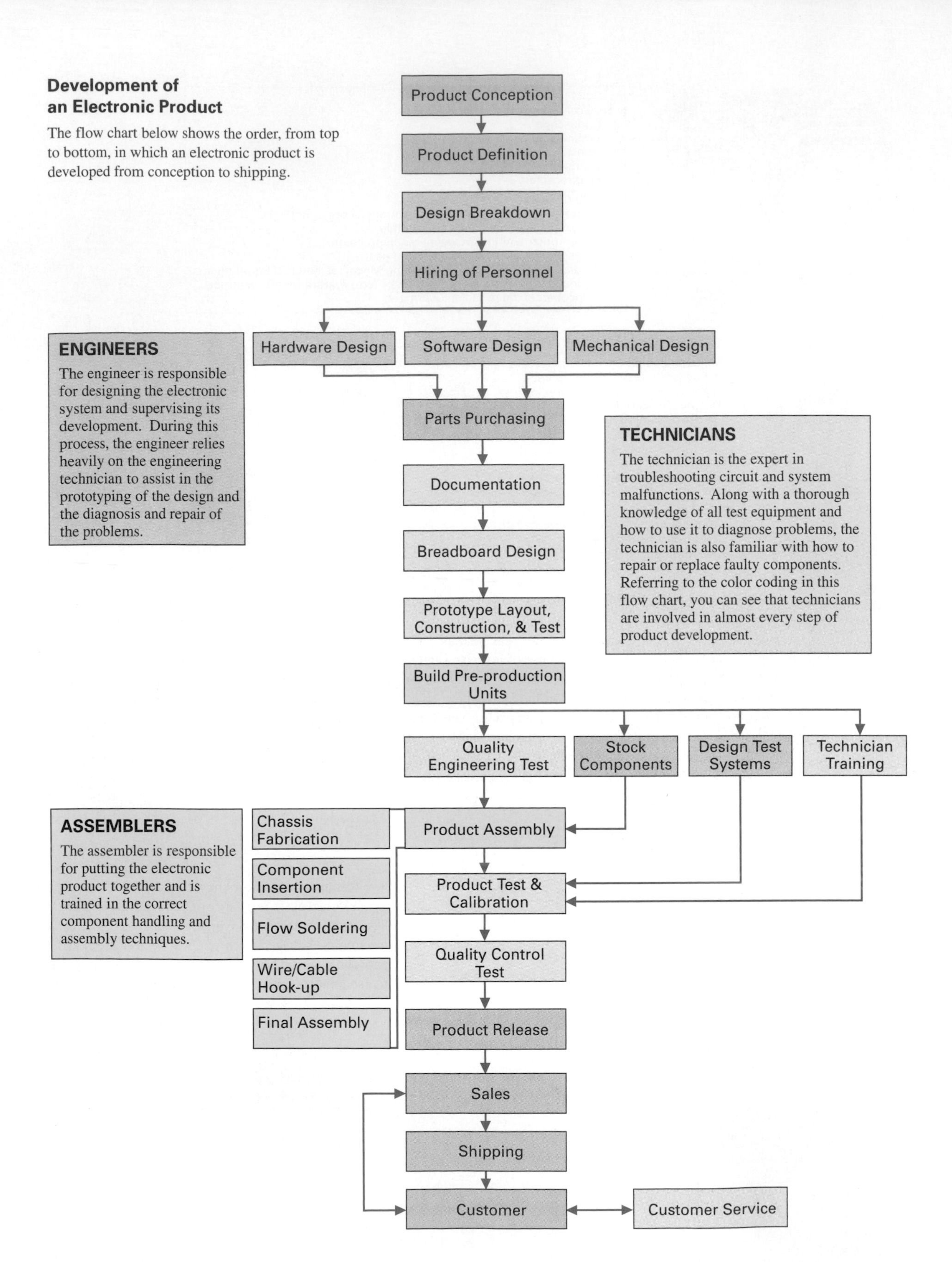
Development of an Electronic Product
The flow chart below shows the order, from top to bottom, in which an electronic product is developed from conception to shipping.
Product Conception
Product Definition
Design Breakdown
Hiring of Personnel
Hardware Design
Software Design
Mechanical Design
ENGINEERS
The engineer is responsible for designing the electronic system and supervising its development. During this process, the engineer relies heavily on the engineering technician to assist in the prototyping of the design and the diagnosis and repair of the problems.
Parts Purchasing
TECHNICIANS
The technician is the expert in troubleshooting circuit and system malfunctions. Along with a thorough knowledge of all test equipment and how to use it to diagnose problems, the technician is also familiar with how to repair or replace faulty components. Referring to the color coding in this flow chart, you can see that technicians are involved in almost every step of product development.
Documentation
Breadboard Design
Prototype Layout, Construction, & Test
Build Pre-production Units
Quality Engineering Test
Stock Components
Design Test Systems
Technician Training
ASSEMBLERS
The assembler is responsible for putting the electronic product together and is trained in the correct component handling and assembly techniques.
Chassis Fabrication
Component Insertion
Flow Soldering
Wire/Cable Hook-up
Final Assembly
Product Assembly
Product Test & Calibration
Quality Control Test
Product Release
Sales
Shipping
Customer
Customer Service

Electronic Technicians in the System Development Process

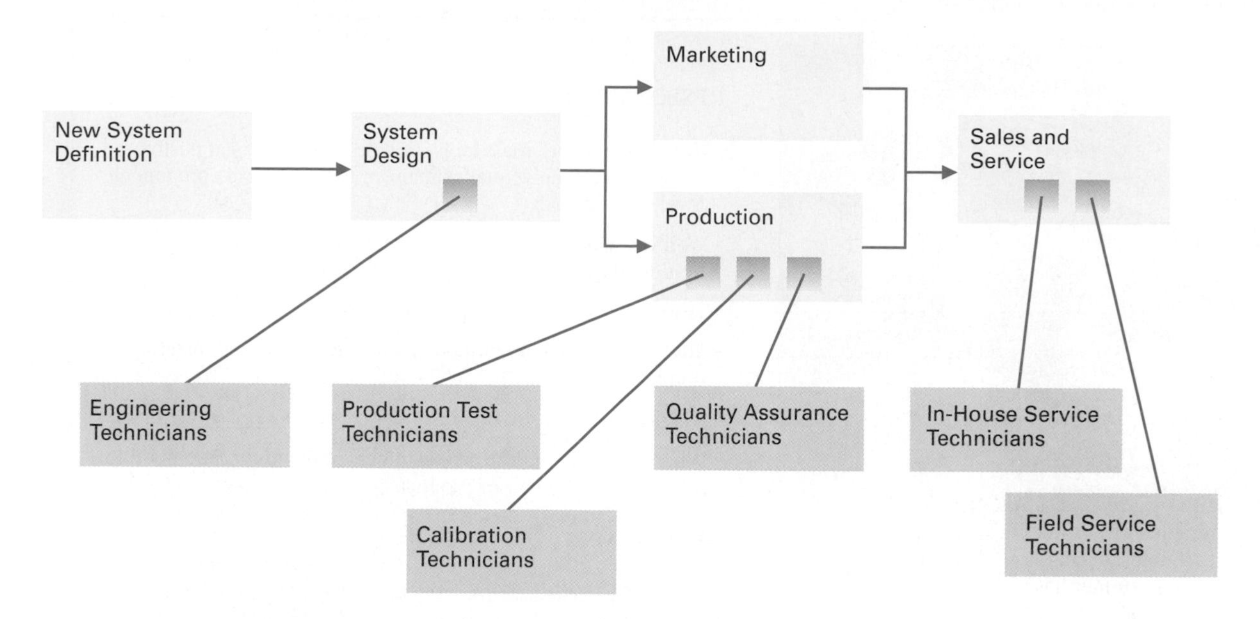

ENGINEERING TECHNICIAN

Here you can see an engineering technician breadboarding the design. From sketches supplied by the engineers, a breadboard model of the design is constructed. The breadboard model is an experimental arrangement of a circuit in which the components are temporarily attached to a flat board. In this arrangement, the components can be tested to prove the feasability of the circuit. A breadboard facilitates making easy changes when they are necessary.

Working under close supervision, the engineering technician performs all work assignments as given by all levels of engineers.

RESPONSIBILITIES:

- Breadboard electronic circuits from schematics.
- Test, evaluate, and document circuits and system performance under the engineer's direction.
- Check out, evaluate, and take data for the engineering prototypes including mechanical assembly of prototype circuits.
- Help to generate and maintain preliminary engineering documentations. Assist engineers and senior Engineering Technicians in ERN documentation.
- Maintain the working station equipment and tools in orderly fashion.
- Support the engineers in all aspects of the development of new products.

REQUIREMENTS:

AS/AAS Degree in Electronics or equivalent plus 1–2 years of technician experience. Ability to read color code, to solder properly, and to bond wires where the skill is required to complete breadboards and prototypes. Ability to use common machinery required to build prototype circuits. Working knowledge of common electronic components: TTL logic circuits, op-amps, capacitors, resistors, inductors, semiconductor devices.

IN-HOUSE SERVICE TECHNICIAN

This photograph shows some in-house service technicians troubleshooting problems on returned units. Once the customer has received the electronic equipment, customer service provides assistance in maintenance and repair of the unit through direct in-house service or at service centers throughout the world.

Working under moderate supervision, the in-house service technician performs all work assignments given by lead tech or direct supervisor.

RESPONSIBILITIES:

- Utilizing all appropriate tools, troubleshoot and repair customer systems in a timely, quality manner, to the general component level.
- Working with basic test equipment, perform timely quality calibration of customer systems to specifications.
- Timely repair of QA rejects.
- Solder and desolder components where appropriate, meeting company standards.
- Aid marketing in solving customer problems via the telephone.
- When appropriate, instruct customers in the proper methods of calibration and repair of products.

REQUIREMENTS:

AS/AAS Degree in Electronics or equivalent plus 2–3 years of experience troubleshooting analog and/or digital systems, at least 6–12 months of which should be in a service environment. Must be able to read and understand flow charts, block diagrams, schematics, and truth tables. Must be able to operate and utilize test equipment such as oscilloscopes, counters, voltmeters, and analyzers.

Ability to effectively communicate and work with customers.

FIELD SERVICE TECHNICIAN

The field service technician seen here has been requested by the customer to make a service call on a malfunctioning unit that currently is under test.

Working under moderate supervision, the field service technician performs all work assignments given by the lead tech or direct supervisor.

RESPONSIBILITIES:

- Utilizing all appropriate tools, troubleshoot and repair customer systems in a timely, quality manner, to the general component level.
- Working with basic test equipment, perform timely quality calibration of customer systems to specifications.
- When appropriate, instruct others in proper soldering techniques meeting company standards.
- Timely repair of QA rejects.
- Aid marketing in solving customer problems via the telephone or at the customer's facility at marketing's discretion.
- When appropriate, aid marketing with sales applications.
- Help QA, Production, and Engineering in solving field problems.
- Evaluate manuals and other customer documents for errors or omissions.

REQUIREMENTS:

AS/AAS Degree in Electronics or equivalent plus 4–5 years of experience troubleshooting analog and/or digital systems, at least 6–12 months of which should be in a service environment. Must be able to read and understand flow charts, block diagrams, schematics, and truth tables. A demonstrated ability to effectively communicate and work with customers and suggest alternative applications for product utilization are also required.

CALIBRATION TECHNICIAN

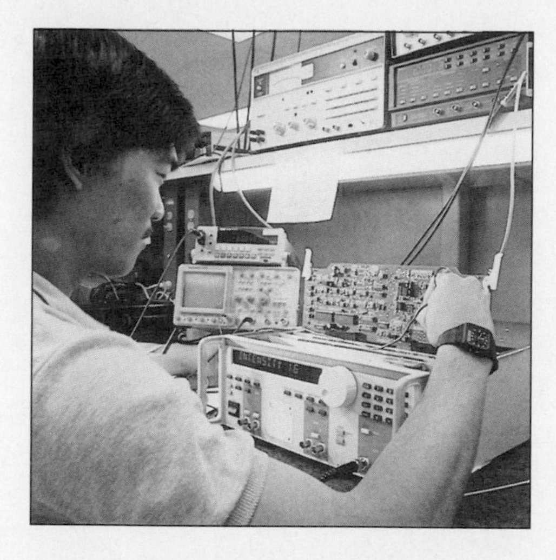

The calibration technician shown here is undertaking a complete evaluation of the newly constructed breadboard's mechanical and electrical form, design, and performance.

Working under general supervision, the calibration technician interfaces with test equipment in a system environment requiring limited decision-making.

RESPONSIBILITIES:

- Work in an interactive mode with test station. Test, align, and calibrate products to defined specifications.
- May set up own test stations and those of other operators.
- Perform multiple alignments to get products to meet specifications.
- May perform other manufacturing-related tasks as required.

REQUIREMENTS:

1–2 years of experience with test and measurement equipment, experience with multiple alignment and calibration of assemblies including test station setups. Able to follow written instructions and write clearly.

QUALITY ASSURANCE TECHNICIANS

The quality assurance (QA) technician takes one of the pre-production units through an extensive series of tests to determine whether it meets the standards listed. This technician is evaluating the new product as it is put through an extensive series of tests.

Working under direct supervision, the quality assurance technician performs functional tests on completed instruments.

RESPONSIBILITIES:

- Using established Acceptance Test Procedures, perform operational tests of all completed systems to ensure that all functional and electrical parameters are within specified limits.
- Perform visual inspection of all completed systems for cleanliness and absence of cosmetic defects.
- Reject all systems that do not meet specifications and/or established parameters of function and appearance.
- Make appropriate notations on the system history sheet.
- Maintain the QA Acceptance Log in accordance with current instructions.
- Refer questionable characteristics to supervisor.

REQUIREMENTS:

AS/AAS Degree in Electronics or equivalent including the use of test equipment. Must know color code and be able to distinguish between colors. Must have a working knowledge of related test equipment. Must know how to read and interpret drawings.

PRODUCTION TEST TECHNICIAN

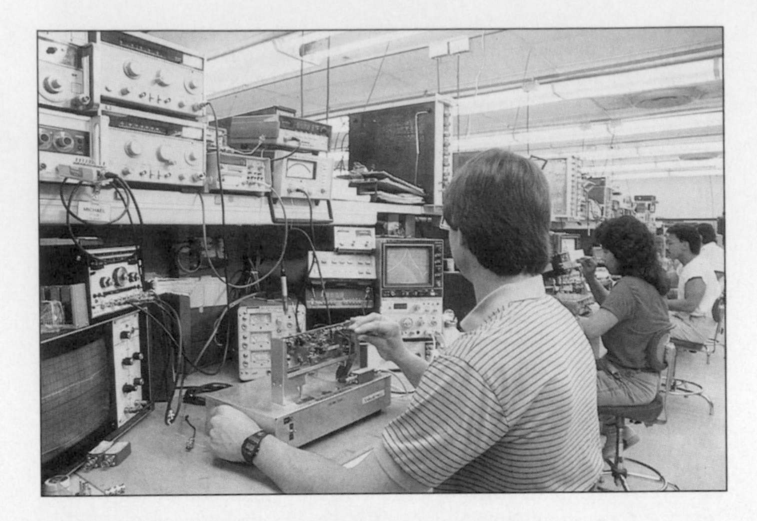

Once the system is fully operational, it is calibrated by a calibration technician. The more complex problems are handled by the production test technicians seen in this photograph.

TECHNICIAN I

Working under close supervision, the production test technician performs all work assignments, and exercises limited decision-making.

RESPONSIBILITIES:

- Perform routine, simple operational tests and fault isolation on simple components, circuits, and systems for verification of product performance to well-defined specifications.
- May perform standard assembly operations and simple alignment of electronic components and assemblies.
- May set up simple test equipment to test performance of products to specifications.

REQUIREMENTS:

AS/AAS Degree in Electronic Technology or equivalent work experience.

TECHNICIAN II

Working under moderate supervision, the production test technician exercises general decision-making involving simple cause and effect relationships to identify trends and common problems.

RESPONSIBILITIES:

- Perform moderately complex operational tests and fault isolation on components, circuits, and systems for verification of product performance to well-defined specifications.
- Perform simple mathematical calculations to verify test measurements and product performance to well-defined specifications.
- Set up general test stations, utilizing varied test equipment, including some sophisticated equipment.
- Perform standard assembly operations.

REQUIREMENTS:

AS/AAS Degree in Electronic Technology or equivalent plus 2–4 years of directly related experience. Demonstrated experience working mathematical formulas and equations. Working knowledge of counters, scopes, spectrum analyzers, and related industry standard test equipment.

Introductory DC/AC Circuits

Current and Voltage

Magnetic Attraction

Andre Ampere was born near Lyon, France, in 1775. His father was a rich silk merchant who tutored him privately; however, by the time the boy had reached his teens he had read the works of many of the great mathematicians. Ampere possessed a photographic memory which, in conjunction with a precocious ability in mathematics, made him a challenge for any professor of science.

In 1796 he began giving private lessons in mathematics, chemistry, and languages. It was in this capacity that he met his wife and was married in 1799. In 1801 he was offered a position as the professor of physics at Bourg, and because his wife was sick at the time, he traveled on ahead. His wife died a few days later and he never recovered from the blow. In fact, in later life he confided to a friend that he realized at the time of his wife's death that he could love nothing else except his work. In 1809 he became professor of mathematics at Ecole Polytechnique in Paris, a position he held for the remainder of his life.

In 1820 Ampere witnessed Hans Christian Oersted's discovery that a compass needle could be deflected by a current-carrying wire. Inspired by this first basic step in electromagnetism, Ampere began to experiment with characteristic industry and care. In only a few weeks he advanced Oersted's discovery by leaps and bounds, developing several mathematical laws of electromagnetism. He also discovered that a coil of wire carrying a current would act like a magnet, and if an iron bar were placed in its center, it would become magnetized. He called this device a solenoid, a name that is still given to electromagnets containing a movable element.

In honor of his achievements in electricity the basic unit of electric current, the ampere, is named after him. On June 10th, 1836, Ampere died in Marseille of what he told a friend just moments before his death was a "broken heart."

1

Outline and Objectives

MAGNETIC ATTRACTION
INTRODUCTION

1-1 MINI-MATH REVIEW

1-1-1 Calculating in Decimal

Objective 1: Describe the relationships among the four basic arithmetic operations: addition, subtraction, multiplication, and division.

1-1-2 Positive and Negative Numbers

Objective 2: Define the difference between a positive number and a negative number.

Objective 3: Explain how to express positive and negative numbers.

1-1-3 Exponents

Objective 4: Define the term *exponent.*

Objective 5: Describe what is meant by raising a number to a higher power.

Objective 6: Explain how to find the square and root of a number.

1-1-4 Powers of 10

Objective 7: Explain the powers-of-10 method and how to convert to powers of 10.

1-1-5 Scientific and Engineering Notation

Objective 8: Describe the two following floating-point number systems:
a. Scientific notation
b. Engineering notation

1-1-6 Metric Prefixes

Objective 9: Define and explain the purpose of the metric system.

Objective 10: List the metric prefixes and describe the purpose of each.

1-2 THE STRUCTURE OF MATTER

1-2-1 The Atom

Objective 11: Explain the atom's subatomic particles.

Objective 12: State the difference between an atomic number and atomic weight.

Objective 13: Understand the term *natural element.*

Objective 14: Describe what is meant by and state how many shells or bonds exist around an atom.

Objective 15: Explain the difference between
a. An atom and a molecule
b. An element and a compound
c. A proton and an electron

1-2-2 Laws of Attraction and Repulsion

Objective 16: State the laws of attraction and repulsion.

Objective 17: Define the terms:
a. Free electron
b. Compound
c. Molecule

1-3 CURRENT

1-3-1 Coulombs per Second

Objective 18: Describe the terms:
a. Neutral atom
b. Negative ion
c. Positive ion

Objective 19: Define *electrical current.*

1-3-2 The Ampere

Objective 20: Describe the ampere in relation to coulombs per second.

1-3-3 Units of Current

Objective 21: Demonstrate how to convert between different current prefix values.

1-3-4 Conventional versus Electron Flow

Objective 22: Describe the difference between conventional current flow and electron flow.

1-3-5 How is Current Measured?

Objective 23: Describe the steps that should be followed when using an ammeter to make a current measurement.

1-4 VOLTAGE

Objective 24: Define the term *electrical voltage,* and explain the difference between a large and small potential difference.

1-4-1 A Simple Voltage Source

Objective 25: Describe the basic operation of a battery as a voltage source.

1-4-2 Components and Circuits

Objective 26: List and describe the function of some simple circuit components.

1-4-3 Basic Circuit Conditions

Objective 27: Explain the following circuit conditions:

a. Open circuit
b. Closed circuit
c. Short circuit

1-4-4 Power Supply Unit

Objective 28: Describe the basic operation of a power supply unit as a voltage source.

1-4-5 Units of Voltage

Objective 29: Describe how to convert between different voltage prefix values.

1-4-6 How is Voltage Measured?

Objective 30: Describe the steps that should be followed when using a voltmeter to make voltage measurements.

1-4-7 Summary: Fluid Analogy of Current and Voltage

Objective 31: Compare electrical current and voltage by means of a fluid analogy equivalent.

Objective 32: Explain why current is directly proportional to voltage.

Introduction

Before we begin, let me try to put you in the right frame of mind. As you proceed through this chapter and the succeeding chapters, it is imperative that you study every section, example, self-test evaluation point, and end-of-chapter question. If you cannot understand a particular section or example, go back and review the material that led up to the problem and make sure that you fully understand all the basics before you continue. Since each chapter builds on previous chapters, you may find that you need to return to an earlier chapter to refresh your understanding before moving on with the current chapter. This process of moving forward and then backtracking to refresh your understanding is very necessary and helps to engrave the material in your mind. Try never to skip a section or chapter because you feel that you already have a good understanding of the subject matter. If it is a basic topic that you have no problem with, read it anyway to refresh your understanding about the steps involved and the terminology because these may be used in a more complex operation in a later chapter.

In this chapter, we will examine the smallest and most significant part of electricity and electronics—the **electron.** By having a good understanding of the electron and the atom, we will be able to obtain a clearer understanding of three basic electrical quantities: **voltage, current,** and **resistance.** In this chapter we will be discussing the basic building blocks of matter and the electrical quantities of voltage and current. To begin with, though, let's review the math skills you will need for this chapter's material.

1-1 MINI-MATH REVIEW—CALCULATION, SIGNED NUMBERS, EXPONENTS, POWERS OF 10, AND METRIC PREFIXES

This first "Mini-Math Review" is included to overview the mathematical details you need for the electronic concepts covered in this chapter. In this review, we will be examining basic calculation, positive and negative numbers, exponents, powers of 10, and metric prefixes.

1-1-1 *Calculating in Decimal*

As you will discover, there are really only *two basic mathematical operations: addition and subtraction.* If you can perform an addition, you can perform a multiplication because *multiplication is simply repeated addition.* For example, a problem such as 4×2 (4 multiplied by 2) is asking you to calculate the sum of four 2s ($2 + 2 + 2 + 2$). Similarly, *division is simply repeated subtraction.* For example, a problem such as $9 \div 3$ (9 divided by 3) is asking you to calculate how many times 3 can be subtracted from 9 ($9 - 3 = 6, 6 - 3 = 3, 3 - 3 = 0$; therefore, there are three 3s in 9). In summary, therefore, the two basic mathematical operations (addition and subtraction) and their associated mathematical operations (multiplication and division) are as follows:

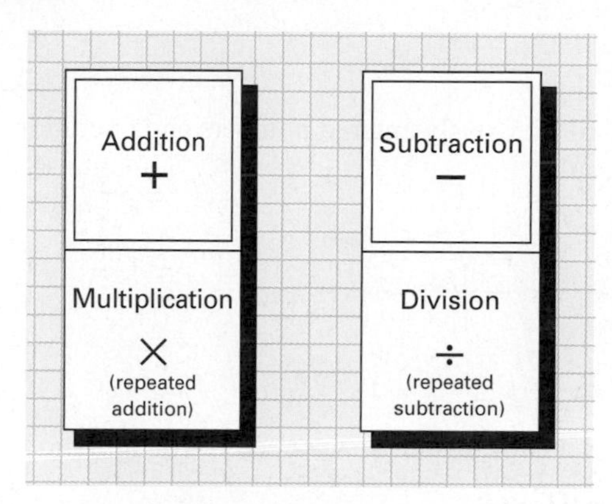

There is another relationship that needs to be discussed, and that is the inverse or opposite nature between addition and subtraction and between multiplication and division. For example, *subtraction is the opposite arithmetic operation of addition.* This can be proved by starting with a number such as 10, then adding 5, then subtracting 5.

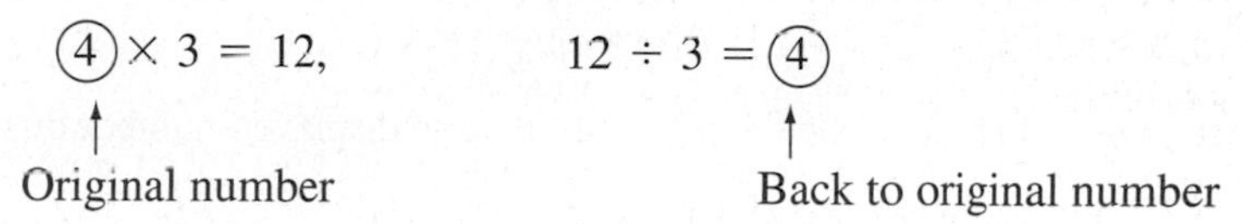

Similarly, *multiplication is the opposite arithmetic operation of division.* This can be proved by starting with a number such as 4, then multiplying by 3, then dividing by 3.

$$④ \times 3 = 12, \qquad 12 \div 3 = ④$$

Original number — Back to original number

In summary, therefore, addition is the opposite mathematical operation of subtraction, and multiplication is the opposite mathematical operation of division. This opposite relationship will be made use of in a subsequent chapter.

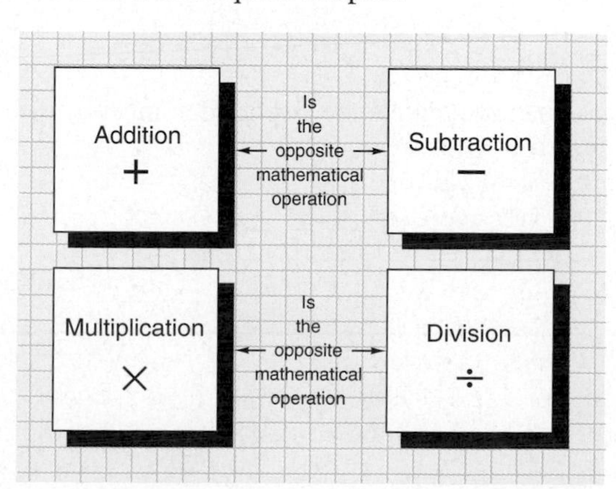

Every calculator has keys for performing these four basic arithmetic operations. The name, function, and procedure for using these keys are as follows.

CALCULATOR KEYS

Name: Decimal point key [.]

Function: Used in conjunction with the digit keys 0 through 9 to enter a decimal point. The position of the decimal point separates the whole number on the left from the decimal fraction on the right.

Name: Equals key [=]

Function: Combines all previously entered numbers and operations. Used to obtain intermediate and final results.

Example: $3 \times 4 + 6 = ?$

Press keys: [3] [×] [4] [=] [+] [6] [=]

Display shows: 18.0

Name: Add key [+]

Function: Instructs calculator to add the next entered quantity to the displayed number.

Example: Add $31.6 + 2.9 = ?$

Press keys: [3] [1] [.] [6] [+] [2] [.] [9] [=]

Displays shows: 34.5

Name: Subtract key [−]

Function: Instructs calculator to subtract the next entered quantity from the displayed number.

Example: $266.3 - 5.35 = ?$

Press keys: [2] [6] [6] [.] [3] [−] [5] [.] [3] [5] [=]

Display shows: 260.95

Name: Multiply key [×]

Function: Instructs calculator to multiply the displayed number by the next entered quantity.

Example: $3.85 \times 2.9 = ?$

Press keys: [3] [.] [8] [5] [×] [2] [.] [9] [=]

Display shows: 11.165

Name: Divide key [÷]

Function: Instructs calculator to divide the displayed number by the next entered quantity.

Example: $28.98 \div 2.3 = ?$

Press keys: [2] [8] [.] [9] [8] [÷] [2] [.] [3] [=]

Display shows: 12.6

1-1-2 *Positive and Negative Numbers*

Positive and negative numbers, or signed numbers, are needed to indicate values that are above and below a reference point. As an example, let us assume that the rainfall over the last 2 years has been 3 and 9 graduations (marks on the scale) above the average reference point of zero. Because these two values of 3 and 9 are above the zero reference, they are positive numbers and can therefore be written as

$$3 \text{ and } 9 \qquad \text{or} \qquad +3 \text{ and } +9$$

To calculate the total amount of rainfall over the last 2 years, therefore, we must add these two values, as follows:

$$3 + 9 = 12 \qquad \text{or} \qquad +3 + (+9) = +12$$

If each graduation or mark on the scale was equivalent to 1 inch, over the past 2 years we would have had 12 inches of rain (+12) above the average of zero. Both preceding statements or expressions are identical, although the positive signs precede the values or numbers in the statement on the right. You can also see how the parentheses were used to separate the *plus sign* (indicating mathematical addition) from the *positive sign* (indicating that 9 is a positive number). If the parentheses had not been included, you might have found the statement confusing, because it would have appeared as follows:

$$+3 + +9 = +12$$

Plus sign ⤴ ⤴ Positive sign

The parentheses are included to isolate the sign of a number from a mathematical operation sign and therefore to prevent confusion.

As another example, let us imagine that last year the rainfall was 7 inches or graduations above the average, and the year before that there was a bit of a drought and the rainfall was 5 graduations or inches below the average. The total amount of rainfall for the last 2 years was therefore

$$+7 + (-5) = +2$$ ← Expression or meaningful combination of symbols

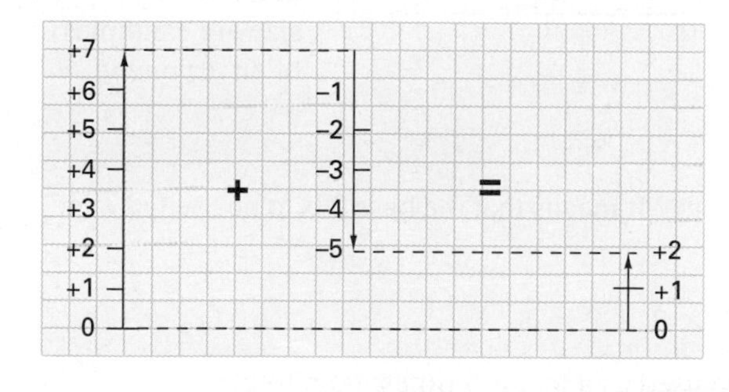

← Graphic representation of expression

Once again, notice how the parentheses were used to separate the addition symbol (plus sign) from the sign of the number (negative 5). This example is probably best understood by referring to the graphic representation below the expression, which shows how the −5 arrow cancels 5 graduations of the +7 arrow, resulting in a sum of 2 positive graduations (+2).

A *positive/negative* or *change-sign key* is available on most calculators and can be used to change the sign of an inputted number as shown at the top of the next page.

1-1-3 *Exponents*

Exponent
A symbol written above and to the right of a mathematical expression to indicate the operation of raising to a power.

Like many terms used in mathematics, the word **exponent** sounds as if it will be complicated; however, once you find out that exponents are simply a sort of "math shorthand," the topic

CALCULATOR KEYS

Name: Change-sign, positive/negative, or negation key

Function: Generally used after a number has been entered to change the sign of the number. Numbers with no sign indicated are always positive. Normally, you press the positive/negative key [+/−] or the negation key [(−)] before you enter the value.

Example: [+/−] $+7 + (-5) = ?$

Press keys: [7] [+] [5] [+/−] [=]

Answer: 2 or +2

Example: [(−)] $-3 + (-15) = ?$

Press keys: [(−)] [3] [+] [(−)] [1] [5] [=]

Answer: −18

Note: If the subtraction key [−] is used in place of the negation key [(−)], or vice versa, an error occurs.

loses its intimidation. Many of the values used in science and technology contain numbers that have exponents. An exponent is a number in a smaller type size that appears to the right of and slightly higher than another number, for example:

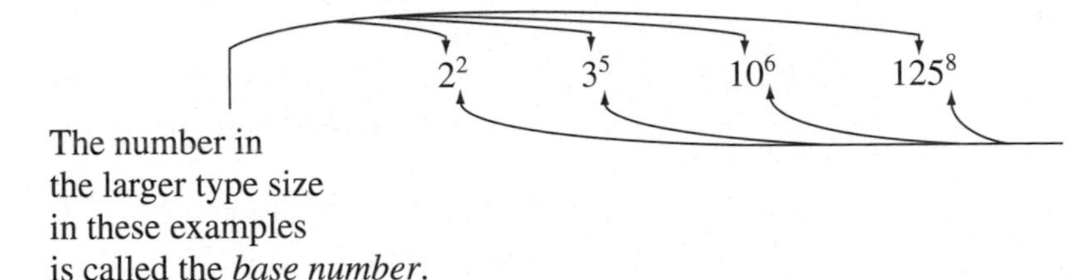

The number in the larger type size in these examples is called the *base number*.

All these numbers in the smaller type size are examples of an *exponent*.

However, what does a term like 2^3 mean? It means that the base 2 is to be used as a factor 3 times; therefore,

$$2^3 = 2 \times 2 \times 2 = 8$$

Similarly, 3^5 means that the base 3 is to be used as a factor 5 times; therefore,

$$3^5 = 3 \times 3 \times 3 \times 3 \times 3 = 243$$

As you can see, it is much easier to write 3^5 than to write $3 \times 3 \times 3 \times 3 \times 3$, and since both mean the same thing and equal the same amount (which is 243), exponents are a quick and easy math shorthand.

Square of a Number

The *square* of a base number means that the base number is to be multiplied by itself. Most calculators also have a key for raising a base number of any value to any power. This is called the *y to the x power key* or *power key,* and it operates as follows:

CALCULATOR KEYS

Name: *y* to the *x* power key [y^x] or power key [∧]

Function: Raises the displayed value *y* to the *x*th power.

Example: $5^4 = ?$

Press keys: [5] [y^x] [4] [=]

Display shows: 625

Example: $2^3 = ?$

Press keys: [2] [∧] [3] [=]

Display shows: 8

Raising a Base Number to Any Power

What would we do if we had the result 64 and we didn't know the value of the number that was multiplied by itself to get 64? In other words, we wanted to *find the source or root number that was squared to give us the result*. This process, called **square root,** uses a special symbol called a *radical sign* and a smaller number called the *index*.

Square Root
A factor of a number that when squared gives the number.

$\sqrt[2]{64} = ?$

Smaller number is called the *index*. It indicates how many times a number was multiplied by itself to get the value shown inside the radical sign. A 2 index is called the *square root*.

Radical sign ($\sqrt{\ }$) indicates that the value inside is the result of a multiplication of a number 2 or more times.

In this example the index is 2, which indicates that the number we are trying to find was multiplied by itself 2 times. Of course, in this example we already know that the answer is 8 because $8 \times 8 = 64$.

If squaring a number takes us forward ($8^2 = 8 \times 8 = 64$), taking the square root of a number must take us backward ($\sqrt[2]{64} = 8$). Almost nobody extracts the squares from square root problems by hand because calculators make this process more efficient. Most calculators have a special key just for determining the square root of a number. Called the *square root key,* it operates as follows.

CALCULATOR KEYS

Name: Square root key

Function: Calculates the square root of the number in the display.

Example: $\sqrt{81} = ?$

Press keys: [8] [1] [$\sqrt{x}$]

Display shows: 9

1-1-4 *Powers of 10*

Many of the sciences deal with numbers that contain a large number of zeros, for example:

14,000

0.000032

By using exponents, we can eliminate the large number of zeros to obtain a shorthand version of the same number. This method is called *powers of 10.*

As an example, let us remove all the zeros from the number 14,000 until we are left with simply 14; however, this number (14) is not equal to the original number (14,000), and therefore simply removing the zeros is not an accurate shorthand. Another number needs to be written with the 14 to indicate what has been taken away—this number is called a *multiplier*. The multiplier must indicate what you have to multiply 14 by to get back to 14,000; therefore,

$$14{,}000 = 14 \times 1000$$

As you know from our discussion on exponents, we can replace the 1000 with 10^3 because $1000 = 10 \times 10 \times 10$. Therefore, the powers-of-10 notation for 14,000 is 14×10^3. To convert 14×10^3 back to its original form (14,000), simply remember that each time a number is multiplied by 10, the decimal place is moved one position to the right. In this example 14 is multiplied by 10 three times (10^3, or $10 \times 10 \times 10$), and therefore the decimal point will have to be moved three positions to the right.

$$14 \times 10^3 = 14 \times 10 \times 10 \times 10 = 14_{\circ}0_{\circ}0_{\circ}0. = 14{,}000$$

As another example, what is the powers-of-10 notation for the number 0.000032? If we once again remove all the zeros to obtain the number 32, we will again have to include a multiplier with 32 to indicate what 32 has to be multiplied by to return it to its original form. In this case 32 will have to be multiplied by 1/1,000,000 (one millionth) to return it to 0.000032.

$$32 \times \frac{1}{1{,}000{,}000} = 0.000032$$

This can be verified because when you divide any number by 10, you move the decimal point one position to the left. Therefore, to divide any number by 1,000,000, you simply move the decimal point six positions to the left.

$$32 \times \frac{1}{1{,}000{,}000} = \frac{32}{1{,}000{,}000} = \frac{32}{10 \times 10 \times 10 \times 10 \times 10 \times 10} = 0.0_{\circ}0_{\circ}0_{\circ}0_{\circ}3_{\circ}2_{\circ}$$

Once again an exponential expression can be used in place of the 1/1,000,000 multiplier, namely,

$$\frac{1}{1{,}000{,}000} = \frac{1}{10^6} = 0.000001 = 10^{-6}$$

Whenever you divide a number into 1, you get the *reciprocal* of that number. In this example, when you divide 1,000,000 into 1, you get 0.000001, which is equal to power-of-10 notation with a negative exponent of 10^{-6}. The multiplier 10^{-6} indicates that the decimal point must be moved back (to the left) by six places, and therefore

$$32 \times \frac{1}{1{,}000{,}000} = 32 \times 0.000001 = 32 \times 10^{-6} = 0.0_{\circ}0_{\circ}0_{\circ}0_{\circ}3_{\circ}2_{\circ}$$

Now that you know exactly what a multiplier is, you have only to remember these simple rules:

1. A *negative exponent* tells you how many places *to the left* to move the decimal point.
2. A *positive exponent* tells you how many places *to the right* to move the decimal point.

Most calculators have a key specifically for entering powers of 10. It is called the *exponent key* and operates as follows.

CALCULATOR KEYS

Name: Exponent entry key [EXP] or [EE] or [^]

Function: Prepares calculator to accept the next digits entered as a power-of-10 exponent. The sign of the exponent can be changed by using the change-sign key [+/−] or the negation key [(−)].

Example: 24×10^3

Press keys: [2] [4] [×] [1] [0] [^] [3]

Display shows: 24000

Example: 85×10^{-5}

Press keys: [8] [5] [EXP] [5] [+/−]

Display shows: 85. −05

1-1-5 *Scientific and Engineering Notation*

As mentioned previously, powers of 10 are used in science and technology as a shorthand owing to the large number of zeros in many values. There are basically two systems or notations used, involving values that have exponents that are a power of 10. They are called **scientific notation** and **engineering notation.**

A number in *scientific notation* is expressed as a base number between 1 and 10 multiplied by a power of 10. In the following examples, the values on the left have been converted to scientific notation.

Scientific Notation
A widely used floating-point system in which numbers are expressed as products consisting of a number between 1 and 10 multiplied by an appropriate power of 10.

Engineering Notation
A widely used floating-point system in which numbers are expressed as products consisting of a number that is greater than 1 multiplied by a power of 10 that is some multiple of 3.

EXAMPLE A

$$32{,}000 = 3.2\,0\,0\,0_{\circ} = 3.2 \times 10^4$$

Scientific notation

Decimal point is moved to a position that results in a base number between 1 and 10. If decimal point is moved left, exponent is positive. If decimal point is moved right, exponent is negative.

EXAMPLE B

$$0.0019 = 0_{\circ}0\,0\,1.9 = 1.9 \times 10^{-3}$$

Scientific notation

As you can see from the preceding examples, the decimal point floats backward and forward, which explains why scientific notation is called a **floating-point number system.**

Floating-Point Number System
A system in which numbers are expressed as products consisting of a number and a power-of-10 multiplier.

Although scientific notation is used in science and technology, the engineering notation system, discussed next, is used more frequently.

In *engineering notation* a number is represented as a base number that is greater than 1 multiplied by a power of 10 that is some multiple of 3. In the following examples, the values on the left have been converted to engineering notation.

EXAMPLE A

$$32{,}000 = 32.000_{0} = 32 \times 10^{3}$$

Engineering notation

Decimal point is moved to a position that results in a base number that is greater than 1 and a power-of-10 exponent that is some multiple of 3.

EXAMPLE B

$$0.0019 = 0_{0}001.9 = 1.9 \times 10^{-3}$$

Engineering notation

CALCULATOR KEYS

Name: Normal, scientific, engineering modes

Function: Most calculators have different notation modes that affect the way an answer is displayed on the calculator's screen. Numeric answers can be displayed with up to 10 digits and a two-digit exponent. You can enter a number in any format.

Example: Normal notation mode is the usual way we express numbers, with digits to the left and right of the decimal, as in 12345.67.

Sci (scientific) notation mode expresses numbers in two parts. The significant digits display with one digit to the left of the decimal. The appropriate power of 10 displays to the right of E, as in 1.234567E4.

Eng (engineering) notation mode is similar to scientific notation; however, the number can have one, two, or three digits before the decimal, and the power-of-10 exponent is a multiple of 3, as in 12.34567E3.

Metric system
A decimal system of weights and measures based on the meter and the gram.

Unit
A determinate quantity adopted as a standard of measurement.

1-1-6 *Metric Prefixes*

The **metric system of weights and measures** was developed to make working with values easier by having only a few units (meter, liter, and gram) and prefixes that are multiples of 10 (milli = 1/1000, kilo = 1000).

To help explain what we mean by units and prefixes, let us begin by examining how we measure length using the metric system. The standard **unit** of length in the metric system is the *meter* (abbreviated m). You have probably not seen the unit "meter" used a lot on its own. More frequently, you have heard and seen the terms *centimeter, millimeter,* and *kilometer*. All these words have two parts: a *prefix name* and *unit*. For example, with the name *centimeter*, *centi* is the prefix, and *meter* is the unit. Similarly, with the names *millimeter* and *kilometer*, *milli* and *kilo* are the prefixes, and *meter* is the unit. The next question, therefore, is: What are these prefixes? A **prefix** *is simply a power of 10 or multiplier that precedes the unit*. Figure 1-1 shows the names, symbols, and values of the most frequently used metric prefixes.

Prefix
An affix attached to the beginning of a word, base, or phrase.

What does *centimeter* mean? If you look up the prefix *centi* in Figure 1-1 , you can see that it is a prefix indicating a fraction (less than 1). Its value or power of 10 is 10^{-2}, or one hundredth $(\frac{1}{100})$. This means that 1 meter has been divided up into 100 pieces, and each of these pieces is a centimeter or a hundredth of a meter. Millimeters have been used in photography for many years because the width of film negatives is measured in millimeters (for example, 35 mm film). Looking up the prefix *milli* in Figure 1-1, you see that its power of 10 is 10^{-3}, or one thousandth $(\frac{1}{1000})$. This prefix is used to indicate smaller fractions because *milli* measures length in thousandths of a meter, whereas *centi* measures length in hundredths of a meter. When the length of some object is less than 1 meter, therefore, the unit *meter* will have a prefix indicating the fractional multiplier, and this power of 10 will have a negative exponent value. Remember that these negative exponents do not indicate negative numbers; they simply indicate a fractional multiplier.

Looking at the number scale in Figure 1-1, you can see that when values are between 1 and 99, we do not need to use a prefix; for example, 1 meter, 62 meters, 84 meters, and so on. A value like 4500 meters, however, would be shortened to include the $1000(10^3)$ prefix *kilo*, as follows:

$$4500 \text{ m} = 4.5 \times 1000 \text{ m} \leftarrow (4.5\,0\,0.)$$
$$= 4.5 \times 10^3 \text{ m} \leftarrow \text{(Because the 1000 or } 10^3 \text{ multiplier is } kilo\text{, we can substitute } kilo\text{, or k, for } 10^3.)$$
$$= 4.5 \text{ km}$$

The value 4500 is shortened to 4.5 with the prefix *kilo*.

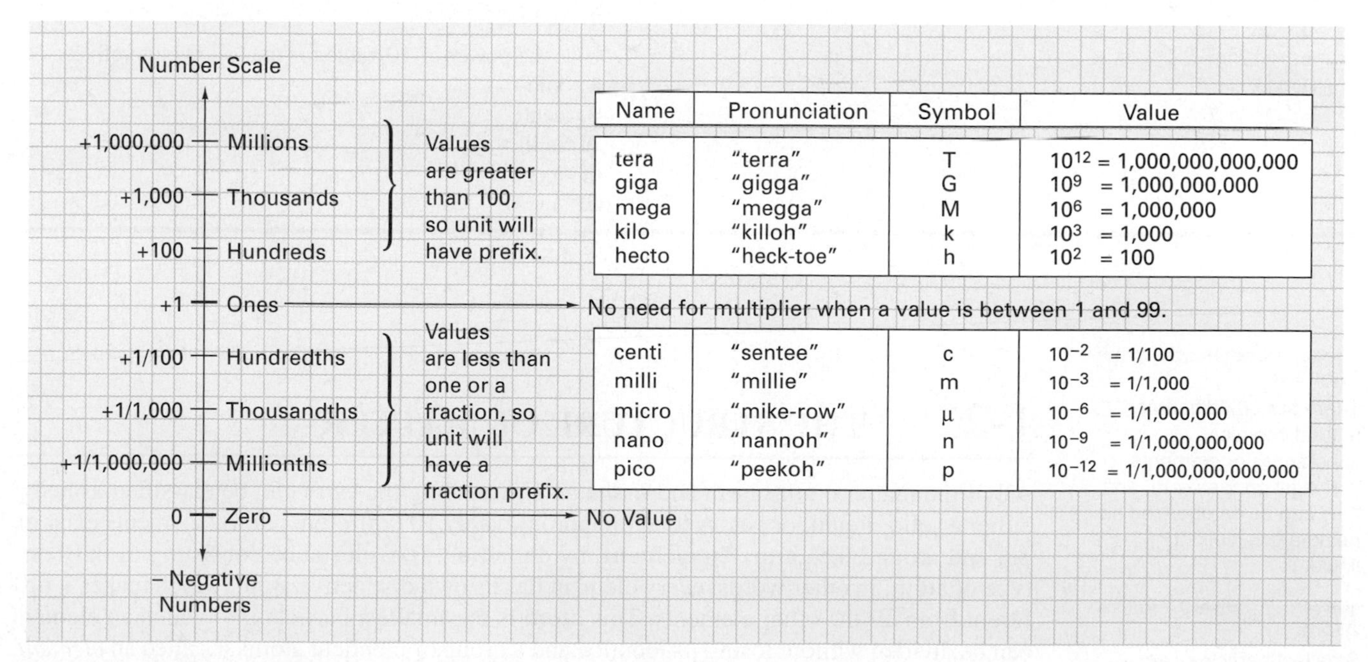

Name	Pronunciation	Symbol	Value
tera	"terra"	T	10^{12} = 1,000,000,000,000
giga	"gigga"	G	10^{9} = 1,000,000,000
mega	"megga"	M	10^{6} = 1,000,000
kilo	"killoh"	k	10^{3} = 1,000
hecto	"heck-toe"	h	10^{2} = 100
centi	"sentee"	c	10^{-2} = 1/100
milli	"millie"	m	10^{-3} = 1/1,000
micro	"mike-row"	μ	10^{-6} = 1/1,000,000
nano	"nannoh"	n	10^{-9} = 1/1,000,000,000
pico	"peekoh"	p	10^{-12} = 1/1,000,000,000,000

FIGURE 1-1 Metric Prefixes.

SELF-TEST EVALUATION POINT FOR SECTION 1-1

Now that you have completed this section, you should be able to:

Objective 1. Describe the relationships among the four-basic arithmetic operations: addition, subtraction, multiplication, and division.

Objective 2. Define the difference between a positive number and a negative number.

Objective 3. Explain how to express positive and negative numbers.

Objective 4. Define the term *exponent.*

Objective 5. Describe what is meant by raising a number to a higher power.

Objective 6. Explain how to find the square and root of a number.

Objective 7. Explain the powers-of-10 method and how to convert to powers of 10.

Objective 8. Describe the two following floating-point number systems:

a. Scientific notation
b. Engineering notation

Objective 9. Define and explain the purpose of the metric system.

Objective 10. List the metric prefixes and describe the purpose of each.

Use the following questions to test your understanding of Section 1-1:

1. Raise the following base numbers to the power indicated by the exponent, and give the answer:
   a. $16^4 = ?$
   b. $32^3 = ?$
   c. $112^2 = ?$
   d. $15^6 = ?$
   e. $2^3 = ?$
   f. $3^{12} = ?$
2. Give the following roots:
   a. $\sqrt[2]{144} = ?$
   b. $\sqrt[3]{3375} = ?$
   c. $\sqrt{20} = ?$
   d. $\sqrt[3]{9} = ?$
3. Convert the following to powers of 10:
   a. $100 = ?$
   b. $1 = ?$
   c. $10 = ?$
   d. $1{,}000{,}000 = ?$
   e. $\frac{1}{1{,}000} = ?$
   f. $\frac{1}{1{,}000{,}000} = ?$
4. Convert the following to common numbers without exponents:
   a. $6.3 \times 10^3 = ?$
   b. $114{,}000 \times 10^{-3} = ?$
   c. $7{,}114{,}632 \times 10^{-6} = ?$
   d. $6624 \times 10^6 = ?$
5. Perform the indicated operation on the following:
   a. $\sqrt{3 \times 10^6} = ?$
   b. $(2.6 \times 10^{-6}) - (9.7 \times 10^{-9}) = ?$
   c. $\frac{(4.7 \times 10^3)^2}{3.6 \times 10^6} = ?$
6. Convert the following common numbers to engineering notation:
   a. $47{,}000 = ?$
   b. $0.00000025 = ?$
   c. $250{,}000{,}000 = ?$
   d. $0.0042 = ?$
7. Give the power-of-10 value for the following prefixes:
   a. kilo
   b. centi
   c. milli
   d. mega
   e. micro

Element
There are 107 different natural chemical substances or elements that exist on the earth, and they can be categorized as being a gas, solid, or liquid.

Atom
Smallest particle of an element.

1-2 THE STRUCTURE OF MATTER

All of the matter on the earth and in the air surrounding the earth can be classified as being either a solid, liquid, or gas. A total of approximately 107 different natural elements exist in, on, and around the earth. An **element,** by definition, is a substance consisting of only one type of atom; in other words, every element has its own distinctive atom, which makes it different from all the other elements. This **atom** is the smallest particle into which an element can be divided without losing its identity, and a group of identical atoms is called an *element,* as shown in Figure 1-2(a).

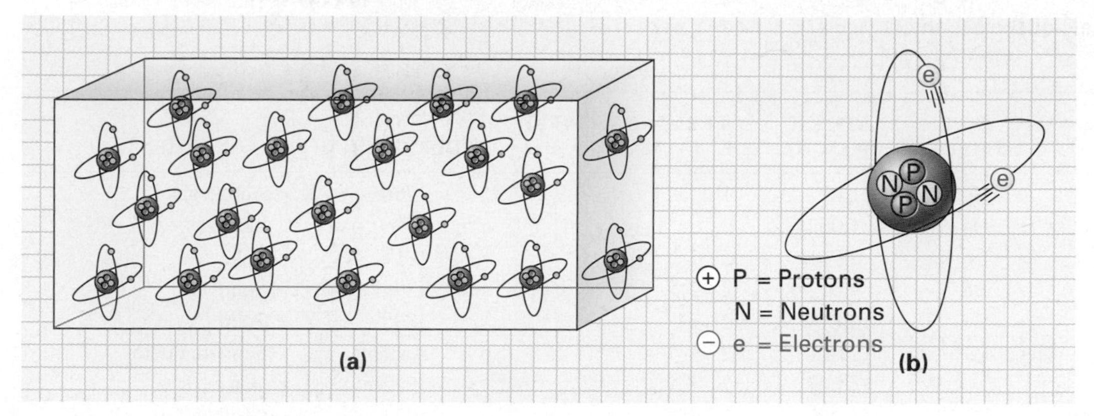

FIGURE 1-2 **(a) Element: Many Similar Atoms. (b) Atom: Smallest Unit.**

1-2-1 *The Atom*

The word *atom* is a Greek word meaning a particle that is too small to be subdivided. At present, we cannot clearly see the atom; however, physicists and researchers do have the ability to record a picture as small as 12 billionths of an inch (about the diameter of one atom), and this image displays the atom as a white fuzzy ball.

In 1913, a Danish physicist, Neils Bohr, put forward a theory about the atom, and his basic model outlining the **subatomic** particles that make up the atom is still in use today and is illustrated in Figure 1-2(b). Bohr actually combined the ideas of Lord Rutherford's (1871–1937) nuclear atom with Max Planck's (1858–1947) and Albert Einstein's (1879–1955) quantum theory of radiation.

The three important particles of the atom are the *proton,* which has a positive charge, the *neutron,* which is neutral or has no charge, and the **electron,** which has a negative charge. Referring to Figure 1-2(b), you can see that the atom consists of a positively charged central mass called the *nucleus,* which is made up of protons and neutrons surrounded by a quantity of negatively charged orbiting electrons.

Table 1-1 lists the periodic table of the elements in order of their atomic number. The **atomic number** of an atom describes the number of protons that exist within the nucleus.

The proton and the neutron are almost 2000 times heavier than the very small electron, so if we ignore the weight of the electron, we can use the fourth column in Table 1-1 (weight of an atom) to give us a clearer picture of the protons and neutrons within the atom's nucleus. For example, a hydrogen atom, shown in Figure 1-3(a), is the smallest of all atoms and has an atomic number of 1, which means that hydrogen has a one-proton nucleus. Helium, however [Figure 1-3(b)], is second on the table and has an atomic number of 2, indicating that two protons are within the nucleus. The **atomic weight** of helium, however, is 4, meaning that two protons and two neutrons make up the atom's nucleus.

Subatomic
Particles such as electrons, protons, and neutrons that are smaller than atoms.

Electron
Smallest subatomic particle of negative charge that orbits the nucleus of the atom.

Atomic Number
Number of positive charges or protons in the nucleus of an atom.

Atomic Weight
The relative weight of a neutral atom of an element, based on a neutral oxygen atom having an atomic weight of 16.

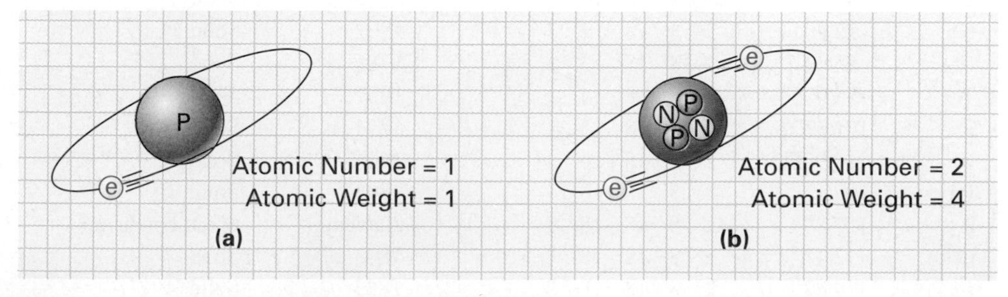

FIGURE 1-3 **(a) Hydrogen Atom. (b) Helium Atom.**

TABLE 1-1 Periodic Table of the Elements

ATOMIC NUMBER	ELEMENT NAME	SYMBOL	ATOMIC WEIGHT	ELECTRONS/SHELL K	L	M	N	O	P	Q	DISCOVERED	COMMENT
1	Hydrogen	H	1.007	1							1766	Active gas
2	Helium	He	4.002	2							1895	Inert gas
3	Lithium	Li	6.941	2	1						1817	Solid
4	Beryllium	Be	9.01218	2	2						1798	Solid
5	Boron	B	10.81	2	3						1808	Solid
6	Carbon	C	12.011	2	4						Ancient	Semiconductor
7	Nitrogen	N	14.0067	2	5						1772	Gas
8	Oxygen	O	15.9994	2	6						1774	Gas
9	Fluorine	F	18.998403	2	7						1771	Active gas
10	Neon	Ne	20.179	2	8						1898	Inert gas
11	Sodium	Na	22.98977	2	8	1					1807	Solid
12	Magnesium	Mg	24.305	2	8	2					1755	Solid
13	Aluminum	Al	26.98154	2	8	3					1825	Metal conductor
14	Silicon	Si	28.0855	2	8	4					1823	Semiconductor
15	Phosphorus	P	30.97376	2	8	5					1669	Solid
16	Sulfur	S	32.06	2	8	6					Ancient	Solid
17	Chlorine	Cl	35.453	2	8	7					1774	Active gas
18	Argon	Ar	39.948	2	8	8					1894	Inert gas
19	Potassium	K	39.0983	2	8	8	1				1807	Solid
20	Calcium	Ca	40.08	2	8	8	2				1808	Solid
21	Scandium	Sc	44.9559	2	8	9	2				1879	Solid
22	Titanium	Ti	47.90	2	8	10	2				1791	Solid
23	Vanadium	V	50.9415	2	8	11	2				1831	Solid
24	Chromium	Cr	51.996	2	8	13	1				1798	Solid
25	Manganese	Mn	54.9380	2	8	13	2				1774	Solid
26	Iron	Fe	55.847	2	8	14	2				Ancient	Solid (magnetic)
27	Cobalt	Co	58.9332	2	8	15	2				1735	Solid
28	Nickel	Ni	58.70	2	8	16	2				1751	Solid
29	Copper	Cu	63.546	2	8	18	1				Ancient	Metal conductor
30	Zinc	Zn	65.38	2	8	18	2				1746	Solid
31	Gallium	Ga	69.72	2	8	18	3				1875	Liquid
32	Germanium	Ge	72.59	2	8	18	4				1886	Semiconductor
33	Arsenic	As	74.9216	2	8	18	5				1649	Solid
34	Selenium	Se	78.96	2	8	18	6				1818	Photosensitive
35	Bromine	Br	79.904	2	8	18	7				1898	Liquid
36	Krypton	Kr	83.80	2	8	18	8				1898	Inert gas
37	Rubidium	Rb	85.4678	2	8	18	8	1			1861	Solid
38	Strontium	Sr	87.62	2	8	18	8	2			1790	Solid
39	Yttrium	Y	88.9059	2	8	18	9	2			1843	Solid
40	Zirconium	Zr	91.22	2	8	18	10	2			1789	Solid
41	Niobium	Nb	92.9064	2	8	18	12	1			1801	Solid
42	Molybdenum	Mo	95.94	2	8	18	13	1			1781	Solid
43	Technetium	Tc	98.0	2	8	18	14	1			1937	Solid
44	Ruthenium	Ru	101.07	2	8	18	15	1			1844	Solid
45	Rhodium	Rh	102.9055	2	8	18	16	1			1803	Solid
46	Palladium	Pd	106.4	2	8	18	18				1803	Solid
47	Silver	Ag	107.868	2	8	18	18	1			Ancient	Metal conductor
48	Cadmium	Cd	112.41	2	8	18	18	2			1803	Solid
49	Indium	In	114.82	2	8	18	18	3			1863	Solid
50	Tin	Sn	118.69	2	8	18	18	4			Ancient	Solid
51	Antimony	Sb	121.75	2	8	18	18	5			Ancient	Solid

(continued)

TABLE 1-1 Periodic Table of the Elements *(continued)*

ATOMIC NUMBER[a]	ELEMENT NAME	SYMBOL	ATOMIC WEIGHT	ELECTRONS/SHELL K	L	M	N	O	P	Q	DISCOVERED	COMMENT
52	Tellurium	Te	127.60	2	8	18	18	6			1783	Solid
53	Iodine	I	126.9045	2	8	18	18	7			1811	Solid
54	Xenon	Xe	131.30	2	8	18	18	8			1898	Inert gas
55	Cesium	Cs	132.9054	2	8	18	18	8	1		1803	Liquid
56	Barium	Ba	137.33	2	8	18	18	8	2		1808	Solid
57	Lanthanum	La	138.9055	2	8	18	18	9	2		1839	Solid
72	Hafnium	Hf	178.49	2	8	18	32	10	2		1923	Solid
73	Tantalum	Ta	180.9479	2	8	18	32	11	2		1802	Solid
74	Tungsten	W	183.85	2	8	18	32	12	2		1783	Solid
75	Rhenium	Re	186.207	2	8	18	32	13	2		1925	Solid
76	Osmium	Os	190.2	2	8	18	32	14	2		1804	Solid
77	Iridium	Ir	192.22	2	8	18	32	15	2		1804	Solid
78	Platinum	Pt	195.09	2	8	18	32	16	2		1735	Solid
79	Gold	Au	196.9665	2	8	18	32	18	1		Ancient	Solid
80	Mercury	Hg	200.59	2	8	18	32	18	2		Ancient	Liquid
81	Thallium	Tl	204.37	2	8	18	32	18	3		1861	Solid
82	Lead	Pb	207.2	2	8	18	32	18	4		Ancient	Solid
83	Bismuth	Bi	208.9804	2	8	18	32	18	5		1753	Solid
84	Polonium	Po	209.0	2	8	18	32	18	6		1898	Solid
85	Astatine	At	210.0	2	8	18	32	18	7		1945	Solid
86	Radon	Rn	222.0	2	8	18	32	18	8		1900	Inert gas
87	Francium	Fr	223.0	2	8	18	32	18	8	1	1945	Liquid
88	Radium	Ra	226.0254	2	8	18	32	18	8	2	1898	Solid
89	Actinium	Ac	227.0278	2	8	18	32	18	9	2	1899	Solid

[a]Rare earth series 58–71 and 90–107 have been omitted.

In most instances, atoms like the beryllium atom will not be drawn in the three-dimensional way shown in Figure 1-4(a). Figure 1-4(b) shows how a beryllium atom could be more easily drawn as a two-dimensional figure.

A **neutral atom** or *balanced atom* is one that has an equal number of protons and orbiting electrons, so the net positive proton charge is equal but opposite to the net negative electron charge, resulting in a balanced or neutral state. For example, Figure 1-5 illustrates a copper atom, which is the most commonly used metal in the field of electronics. It has an atomic number of 29, meaning that 29 protons and 29 electrons exist within the atom when it is in its neutral state.

Neutral Atom
An atom in which the number of positive charges in the nucleus (protons) is equal to the number of negative charges (electrons) that surround the nucleus.

The number of neutrons within an atom's nucleus can therefore be calculated by subtracting the atomic number (protons) from the atomic weight (protons and neutrons). For example, Figure 1-4(a) illustrates a beryllium atom, which has the following atomic number and weight:

Beryllium
Atomic number: 4 (protons)
Atomic weight: 9 (protons and neutrons)

Subtracting the beryllium atom's weight from the beryllium atomic number, we can determine the number of neutrons in the beryllium atom's nucleus, as shown in Figure 1-4(a).

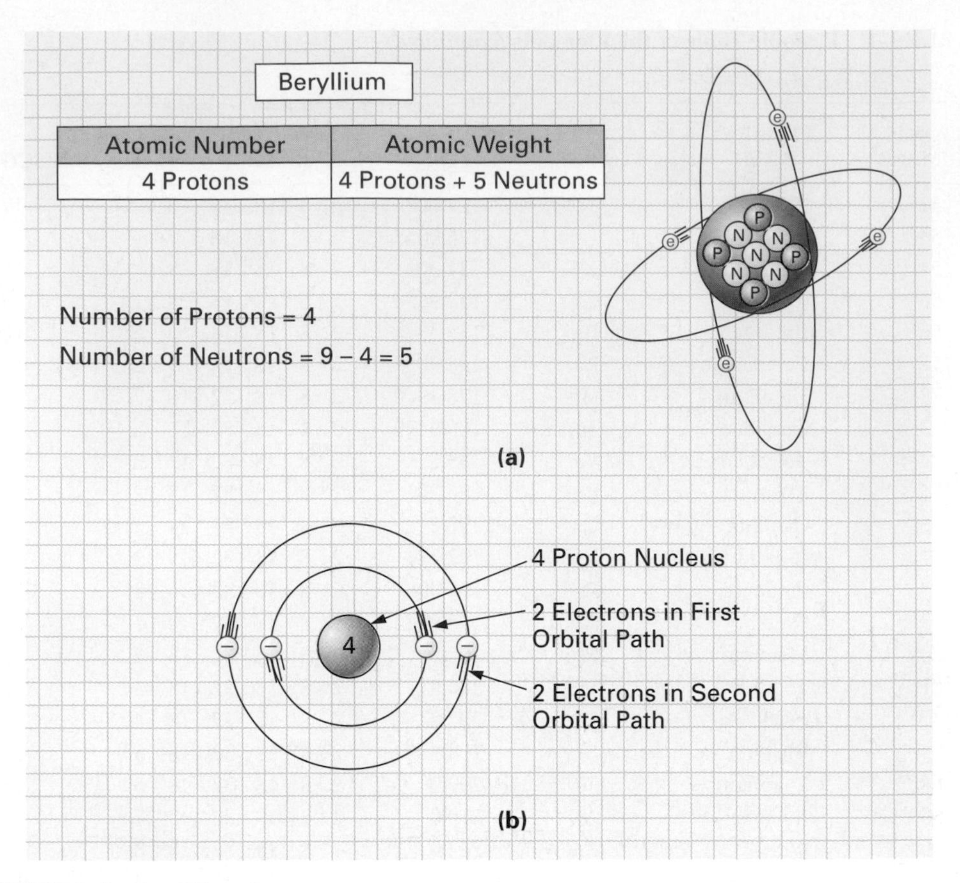

FIGURE 1-4 **Beryllium Atom.**

Shells or Bands
An orbital path containing a group of electrons that have a common energy level.

Orbiting electrons travel around the nucleus at varying distances from the nucleus, and these orbital paths are known as **shells** or **bands.** The orbital shell nearest the nucleus is referred to as the first or K shell. The second is known as the L, the third is M, the fourth is N, the fifth is O, the sixth is P, and the seventh is referred to as the Q shell. There are seven shells available for electrons (K, L, M, N, O, P, and Q) around the nucleus, and each of these seven shells can only hold a certain number of electrons, as shown in Figure 1-6.

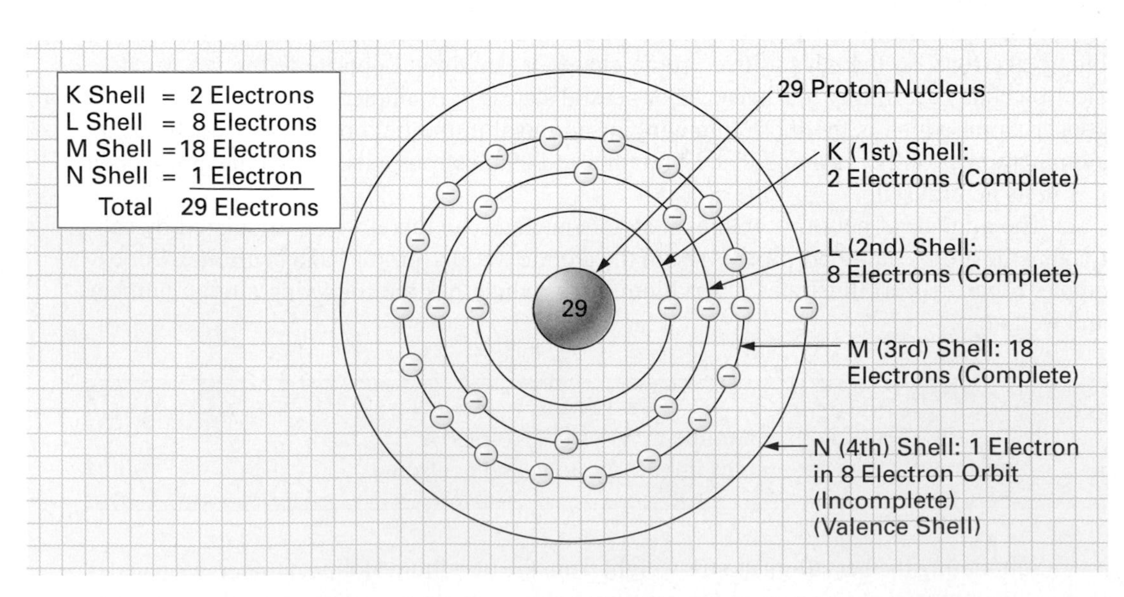

FIGURE 1-5 **Copper Atom.**

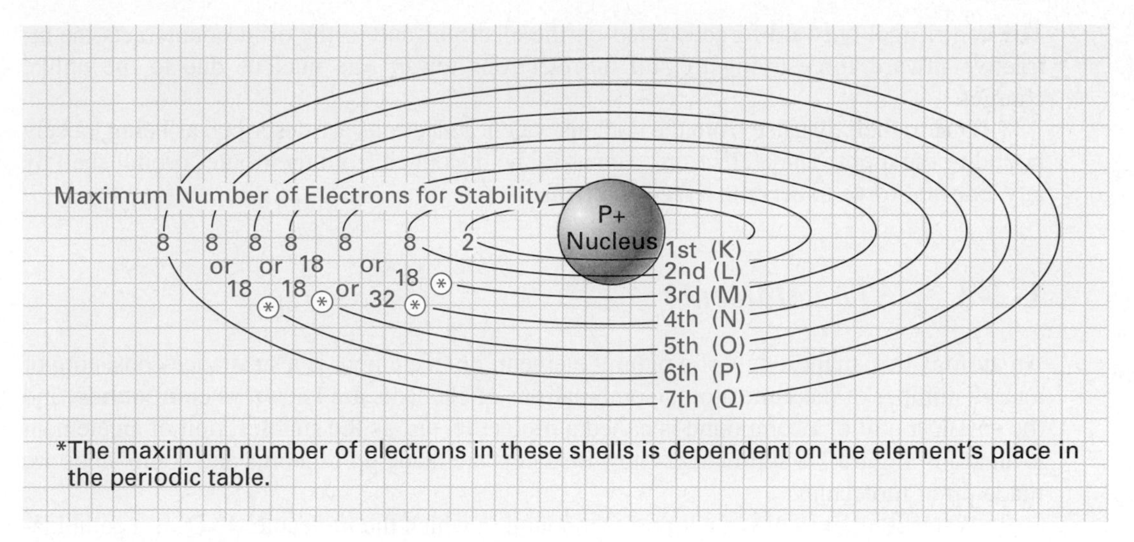

FIGURE 1-6 Electrons and Shells.

An atom's outermost electron-occupied shell is referred to as the **valence shell** or **ring,** and electrons in this shell are termed *valence electrons*. In the case of the copper atom, a single valence electron exists in the valence N shell.

Valence Shell or Ring
Outermost shell formed by electrons.

All matter exists in one of three states: solid, liquid, or gas. The atoms of a solid are fixed in relation to one another but vibrate in a back-and-forth motion, unlike liquid atoms, which can flow over each other. The atoms of a gas move rapidly in all directions and collide with one another. The far-right column of Table 1-1 indicates whether the element is a gas, a solid, or a liquid.

1-2-2 *Laws of Attraction and Repulsion*

For the sake of discussion and understanding, let us theoretically imagine that we are able to separate some positive and negative subatomic particles. Using these separated protons and electrons, let us carry out a few experiments, the results of which are illustrated in Figure 1-7. Studying Figure 1-7, you will notice that

1. *Like charges* (positive and positive or negative and negative) repel one another.
2. *Unlike charges* (positive and negative or negative and positive) attract one another.

Orbiting negative electrons are therefore attracted toward the positive nucleus, which leads us to the question of why the electrons do not fly into the atom's nucleus. The answer is that the orbiting electrons remain in their stable orbit due to two equal but opposite forces.

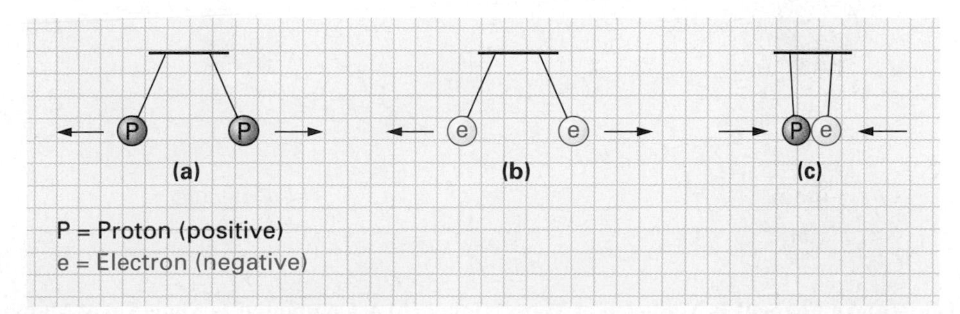

FIGURE 1-7 Attraction and Repulsion. (a) Positive Repels Positive. (b) Negative Repels Negative. (c) Unlike Charges Attract.

The centrifugal outward force exerted on the electrons due to the orbit counteracts the attractive inward force trying to pull the electrons toward the nucleus due to the unlike charges.

Due to their distance from the nucleus, valence electrons are described as being loosely bound to the atom. These electrons can easily be dislodged from their outer orbital shell by any external force to become a **free electron.**

Free Electron
An electron that is not in any orbit around a nucleus.

1-2-3 *The Molecule*

Compound
A material composed of united separate elements.

Molecule
Smallest particle of a compound that still retains its chemical characteristics.

An atom is the smallest unit of a natural element, or an element is a substance consisting of a large number of the same atom. Combinations of elements are known as **compounds,** and the smallest unit of a compound is called a **molecule,** just as the smallest unit of an element is an atom. Figure 1-8 summarizes how elements are made up of atoms and compounds are made up of molecules.

Water is an example of a liquid compound in which the molecule (H_2O) is a combination of an explosive gas (hydrogen) and a very vital gas (oxygen). Table salt is another example of a compound; here the molecule is made up of a highly poisonous gas atom (chlorine) and a potentially explosive solid atom (sodium). These examples of compounds each contain atoms that, when alone, are both poisonous and explosive, yet when combined the resulting substance is as ordinary and basic as water and salt.

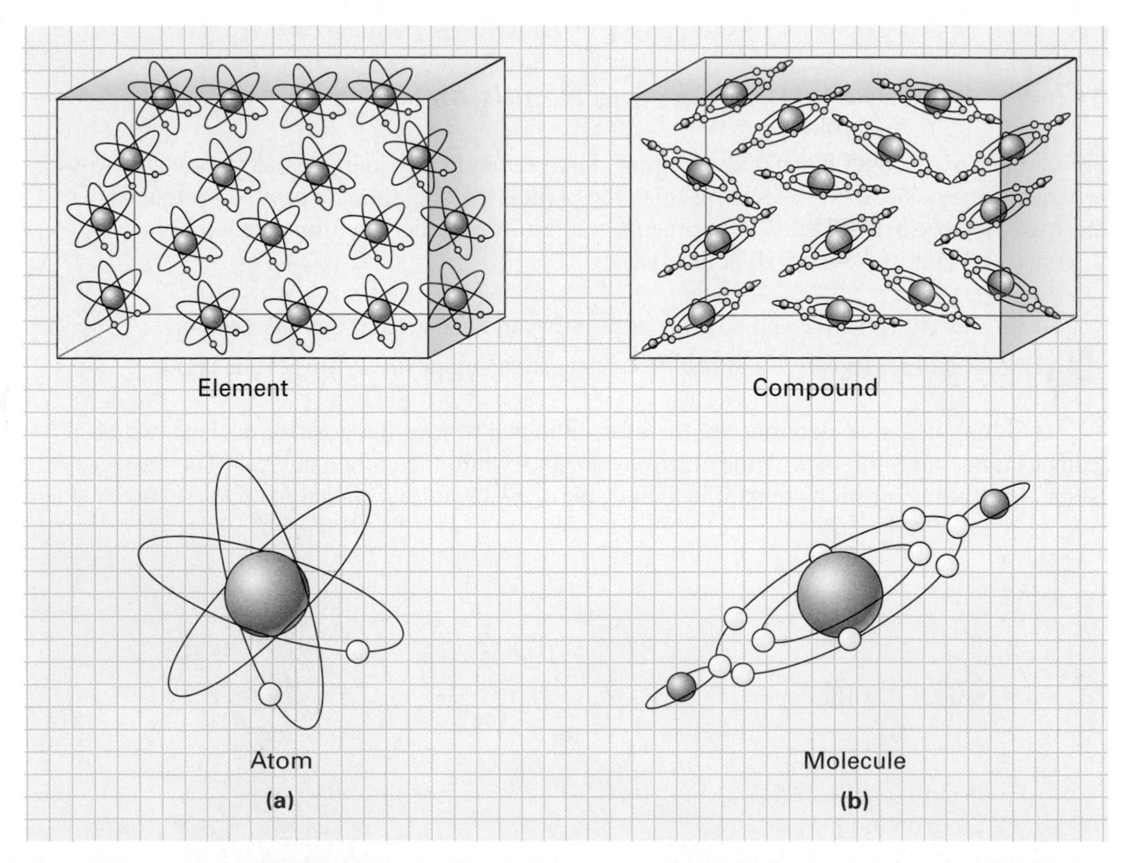

FIGURE 1-8 **(a) An Element Is Made Up of Many Atoms. (b) A Compound Is Made Up of Many Molecules.**

SELF-TEST EVALUATION POINT FOR SECTION 1-2

Now that you have completed this section, you should be able to:

Objective 11. Explain the atom's subatomic particles.

Objective 12. State the difference between an atomic number and atomic weight.

Objective 13. Understand the term *natural element.*

Objective 14. Describe what is meant by and state how many shells or bonds exist around an atom.

Objective 15. Explain the difference between

a. An atom and a molecule
b. An element and a compound
c. A proton and an electron

Objective 16. State the laws of attraction and repulsion.

Objective 17. Define the terms:

a. Free electron
b. Compound
c. Molecule

Use the following questions to test your understanding of Section 1-2:

1. Define the difference between an element and a compound.
2. Name the three subatomic particles that make up an atom.
3. What is the most commonly used metal in the field of electronics?
4. State the laws of attraction and repulsion.

1-3 CURRENT

The movement of electrons from one point to another is known as *electrical* **current.** Energy in the form of heat or light can cause an outer-shell electron to be released from the valence shell of an atom. Once an electron is released, the atom is no longer electrically neutral and is called a **positive ion,** as it now has a net positive charge (more protons than electrons). The released electron tends to jump into a nearby atom, which will then have more electrons than protons and is referred to as a **negative ion.**

Let us now take an example and see how electrons move from one point to another. Figure 1-9 shows a broken metal conductor between two charged objects. The metal conductor could be either gold, silver, or copper, but whichever it is, one common trait can be noted: The valence electrons in the outermost shell are very loosely bound and can easily be pulled from their parent atom.

In Figure 1-10, the conductor between the two charges has been joined so that a path now exists for current flow. The negative ions on the right in Figure 1-10 have more electrons than protons, while the positive ions on the left in Figure 1-10 have fewer electrons than protons and so display a **positive charge.** The metal joining the two charges has its own atoms, which begin in the neutral condition.

Current (I)
Measured in amperes or amps, it is the flow of electrons through a conductor.

Positive Ion
Atom that has lost one or more of its electrons and therefore has more protons than electrons, resulting in a net positive charge.

Negative Ion
Atom that has more than the normal neutral amount of electrons.

Positive Charge
The charge that exists in a body which has fewer electrons than normal.

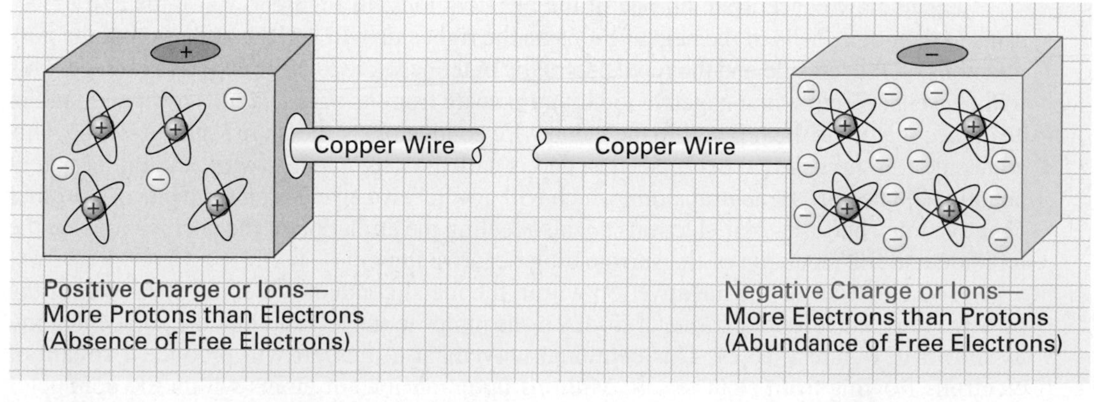

FIGURE 1-9 **Positive and Negative Charges.**

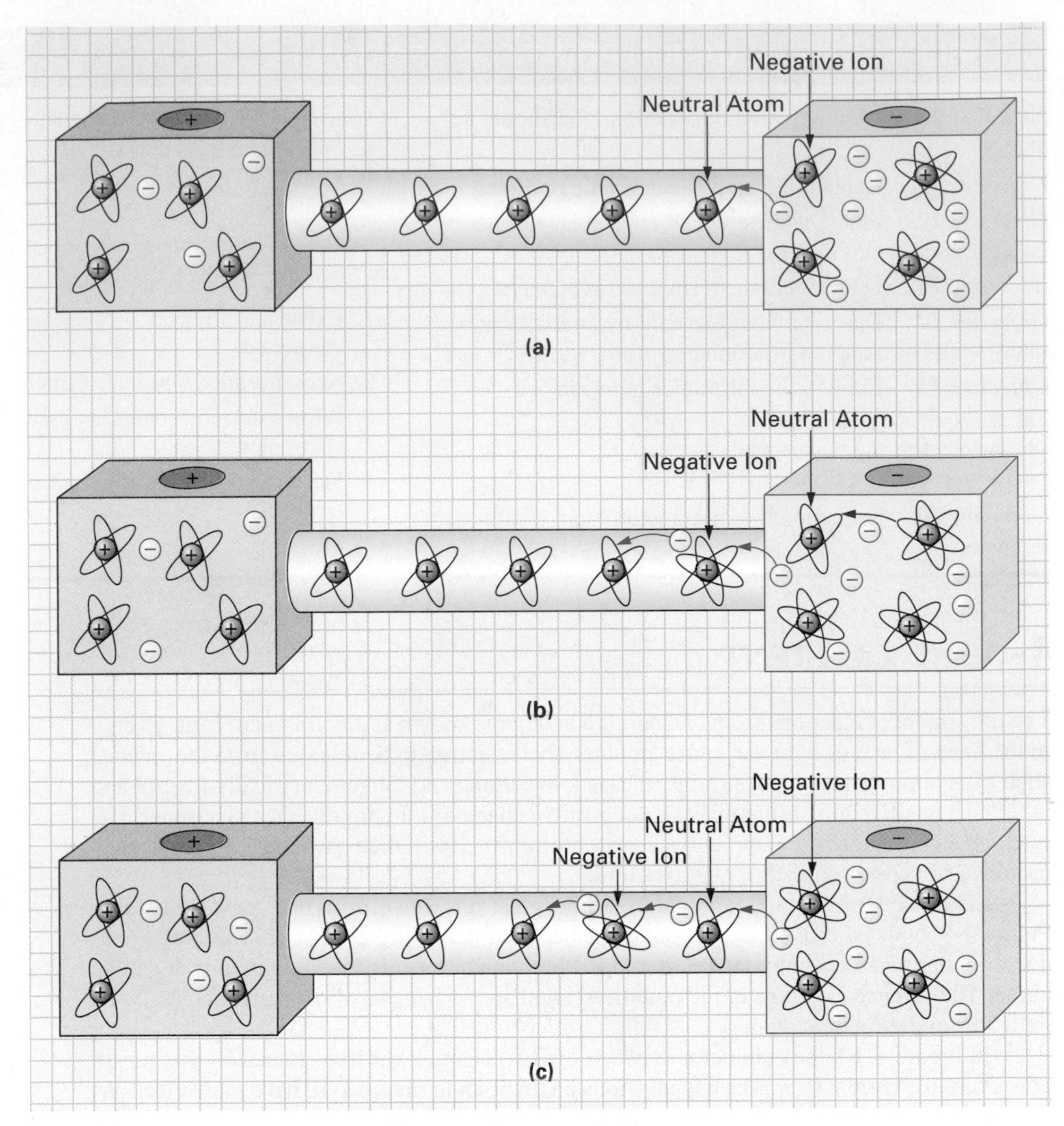

FIGURE 1-10 Electron Migration Due to Forces of Positive Attraction and Negative Repulsion on Electrons.

Negative Charge
An electric charge that has more electrons than protons.

Electric Current (*I*)
Measured in amperes or amps, it is the flow of electrons through a conductor.

Let us now concentrate on one of the negative ions. In Figure 1-10(a), the extra electrons in the outer shells of the negative ions on the right side will feel the attraction of the positive ions on the left side and the repulsion of the outer negative ions, or **negative charge.** This will cause an electron in a negative ion to jump away from its parent atom's orbit and land in an adjacent atom to the left within the metal wire conductor, as shown in Figure 1-10(b). This adjacent atom now has an extra electron and is called a *negative ion,* while the initial parent negative ion becomes a neutral atom, which will now receive an electron from one of the other negative ions, because their electrons are also feeling the attraction of the positive ions on the left side and the repulsion of the surrounding negative ions.

The electrons of the negative ion within the metal conductor feel the attraction of the positive ions, and eventually one of its electrons jumps to the left and into the adjacent atom, as shown in Figure 1-10(c). This continual movement to the left will produce a stream of electrons flowing from right to left. Millions upon millions of atoms within the conductor pass a continuous movement of billions upon billions of electrons from right to left. This electron flow is known as **electric current.**

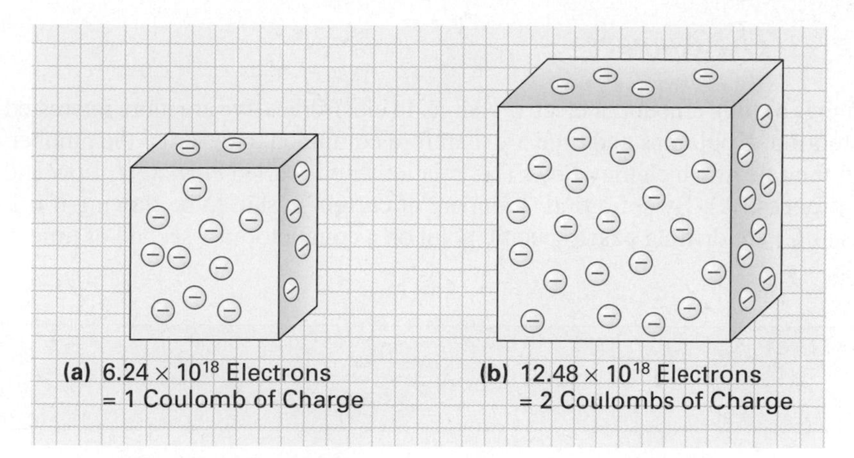

FIGURE 1-11 Coulomb of Charge. (a) 1 C of Charge. (b) 2 C of Charge.

To summarize, we could say that as long as a force or pressure, produced by the positive charge and negative charge, exists it will cause electrons to flow from the negative to the positive terminal. The positive side has a deficiency of electrons and the negative side has an abundance, and so a continuous flow or migration of electrons takes place between the negative and positive terminal through our metal conducting wire. This electric current or electron flow is a measurable quantity, as will now be explained.

1-3-1 *Coulombs per Second*

There are 6.24×10^{18} electrons in 1 **coulomb of charge,** as illustrated in Figure 1-11. To calculate coulombs of charge (designated Q), we can use the formula

Coulomb of Charge
Unit of electric charge. One coulomb equals 6.24×10^{18} electrons.

$$\text{charge, } Q = \frac{\text{total number of electrons } (n)}{6.24 \times 10^{18}}$$

where Q is the electric charge in coulombs.

EXAMPLE:

If a total of 3.75×10^{19} free electrons exist within a piece of metal conductor, how many coulombs (C) of charge would be within this conductor?

Solution:

By using the charge formula (Q) we can calculate the number of coulombs (C) in the conductor.

$$Q = \frac{n}{6.24 \times 10^{18}}$$
$$= \frac{3.75 \times 10^{19}}{6.24 \times 10^{18}}$$
$$= 6 \text{ C}$$

A total of 6 C of charge exists within the conductor.

In the calculator sequence you will see how the exponent key (E, EE, EXP, or ∧) on your calculator can be used.

Calculator Sequence

STEP	KEYPAD ENTRY	DISPLAY RESPONSE
1.	3 . 7 5 E (Exponent) 1 9	3.75E19
2.	÷	
3.	6 . 2 4 E 1 8	6.24E18
4.	=	6.0096

1-3-2 *The Ampere*

Ampere (A)
Unit of electric current.

A coulomb is a static amount of electric charge. In electronics, we are more interested in electrons in motion. Coulombs and time are therefore combined to describe the number of electrons and the rate at which they flow. This relationship is called *current (I)* flow and has the unit of **amperes (A).** By definition, 1 ampere of current is said to be flowing if 6.24×10^{18} electrons (1 C) are drifting past a specific point on a conductor in 1 second of time. Stated as a formula,

$$\text{current } (I) = \frac{\text{coulombs } (Q)}{\text{time } (t)}$$

$$1 \text{ ampere} = 1 \text{ coulomb per 1 second}$$

$$1 \text{ A} = \frac{1 \text{ C}}{1 \text{ s}}$$

In summary, 1 ampere equals a flow rate of 1 coulomb per second, and current is measured in amperes.

EXAMPLE:

If 5×10^{19} electrons pass a point in a conductor in 4 s, what is the amount of current flow in amperes?

Solution:

Current (I) is equal to Q/t. We must first convert electrons to coulombs.

$$Q = \frac{n}{6.24 \times 10^{18}}$$

$$= \frac{5 \times 10^{19}}{6.24 \times 10^{18}}$$

$$= 8 \text{ C}$$

Now, to calculate the amount of current, we use the formula

$$I = \frac{Q}{t}$$

$$= \frac{8 \text{ C}}{4 \text{ s}}$$

$$= 2 \text{ A}$$

This means that 2 A or 1.248×10^{19} electrons (2 C) are passing a specific point in the conductor every second.

Calculator Sequence

STEP	KEYPAD ENTRY	DISPLAY RESPONSE
1.	[5] [E] (exponent) [1] [9]	5E19
2.	[÷]	
3.	[6] [.] [2] [4] [E] [1] [8]	6.24E18
4.	[=]	8.012
5.	[÷]	
6.	[4]	2.003
7.	[=]	2.003

1-3-3 *Units of Current*

Current within electronic equipment is normally a value in milliamperes or microamperes and very rarely exceeds 1 ampere. Table 1-2 lists all the prefixes related to current. For example, 1 milliampere is one-thousandth of an ampere, which means that if 1 ampere were divided into 1000 parts, 1 part of the 1000 parts would be flowing through the circuit.

TABLE 1-2 Current Units

NAME	SYMBOL	VALUE
Picoampere	pA	$10^{-12} = \frac{1}{1{,}000{,}000{,}000{,}000}$
Nanoampere	nA	$10^{-9} = \frac{1}{1{,}000{,}000{,}000}$
Microampere	μA	$10^{-6} = \frac{1}{1{,}000{,}000}$
Milliampere	mA	$10^{-3} = \frac{1}{1000}$
Ampere	A	$10^{0} = 1$
Kiloampere	kA	$10^{3} = 1000$
Megaampere	MA	$10^{6} = 1{,}000{,}000$
Gigaampere	GA	$10^{9} = 1{,}000{,}000{,}000$
Teraampere	TA	$10^{12} = 1{,}000{,}000{,}000{,}000$

EXAMPLE:

Convert the following:

a. 0.003 A = ___________ mA (milliamperes)
b. 0.07 mA = ___________ μA (microamperes)
c. 7333 mA = ___________ A (amperes)
d. 1275 μA = ___________ mA (milliamperes)

Solution:

a. 0.003A = ___________ mA. In this example, 0.003 A has to be converted so that it is represented in milliamperes (10^{-3} or 1/1000 of an ampere). The basic algebraic rule to be remembered is that both expressions on either side of the equal sign must be equal.

LEFT			RIGHT		
Base		Multiplier	Base		Multiplier
0.003	×	10^{0}	= _______	×	10^{-3}

The multiplier on the right in this example is going to be decreased 1000 times (10^{0} to 10^{-3}), so for the statement to balance, the number on the right will have to be increased 1000 times; that is, the decimal point will have to be moved to the right three places (0.003 or 3). Therefore,

$$0.003 \times 10^{0} = 3 \times 10^{-3}$$

or

$$0.003 \text{ A} = 3 \times 10^{-3} \text{ A or } 3 \text{ mA}$$

b. 0.07 mA = __________ μA. In this example the unit is going from milliamperes to microamperes (10^{-3} to 10^{-6}) or 1000 times smaller, so the number must be made 1000 times greater.

0.070 or 70.0

Therefore, 0.07 mA = 70 μA.

c. 7333 mA = __________ A. The unit is going from milliamperes to amperes, increasing 1000 times, so the number must decrease 1000 times.

7333 or 7.333

Therefore, 7333 mA = 7.333 A.

d. 1275 μA = __________ mA. The unit is changing from microamperes to milliamperes, an increase of 1000 times, so the number must decrease by the same factor.

1275.0 or 1.275

Therefore, 1275 μA = 1.275 mA.

1-3-4 *Conventional versus Electron Flow*

Electron Flow
A current produced by the movement of free electrons toward a positive terminal.

Conventional Current Flow
A current produced by the movement of positive charges toward a negative terminal.

Electrons drift from a negative to a positive charge, as illustrated in Figure 1-12. As already stated, this current is known as **electron flow.**

In the eighteenth and nineteenth centuries, when very little was known about the atom, researchers believed that current was a flow of positive charges. Although this has now been proved incorrect, many texts still use **conventional current flow,** which is shown in Figure 1-13.

Whether conventional flow or electron flow is used, the same answers to problems, measurements, and designs are obtained. The key point to remember is that direction is not important, but the amount of current flow is.

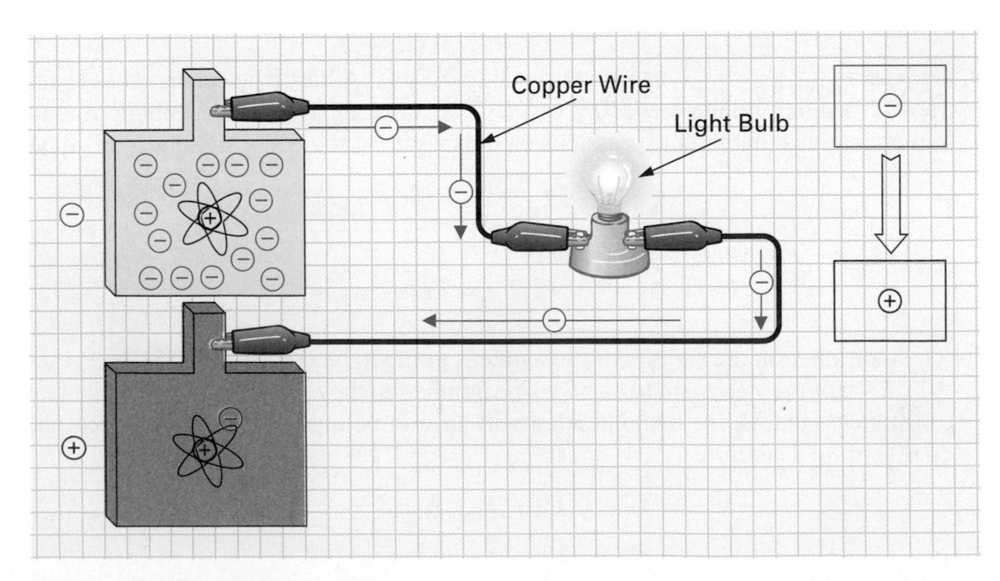

FIGURE 1-12 Electron Current Flow.

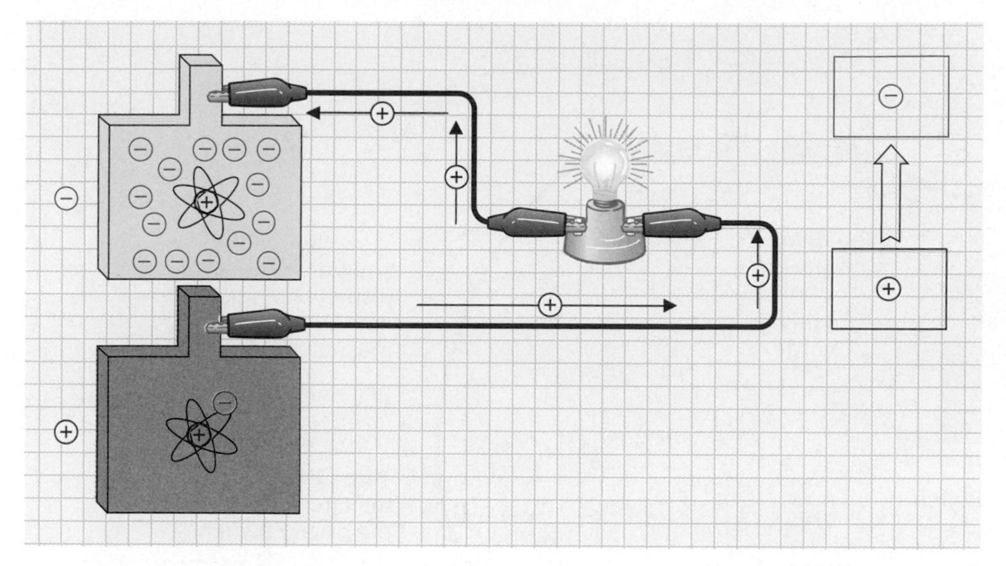

FIGURE 1-13 **Conventional Current Flow.**

TIME LINE

The galvanometer, which is used to measure electrical current, was named after Luigi Galvani (1737–1798), who conducted many experiments with electrical current, or, as it was known at the time, "galvanism."

Throughout this book we will be using electron flow so that we can relate back to the atom when necessary. If you wish to use conventional flow, just reverse the direction of the arrows. To avoid confusion, be consistent with your choice of flow.

1-3-5 *How Is Current Measured?*

Ammeter

Meter placed in the path of current flow to measure the amount.

Ammeters (ampere meters) are used to measure the current flow within a circuit. Stepping through the sequence detailed in Figure 1-14, you will see how an ammeter is used to measure the value of current within a circuit.

In the simple circuit shown in Figure 1-14, an ON/OFF switch is being used to turn ON or OFF a small lightbulb. One of the key points to remember with ammeters is that if you wish to measure the value of current flowing within a wire, the current path must be opened and the ammeter placed in the path so that it can sense and display the value.

SELF-TEST EVALUATION POINT FOR SECTION 1-3

Now that you have completed this section, you should be able to:

Objective 18. Describe the terms:

a. Neutral atom
b. Negative ion
c. Positive ion

Objective 19. Define *electrical current.*

Objective 20. Describe the ampere in relation to coulombs per second.

Objective 21. Demonstrate how to convert between different current prefix values.

Objective 22. Describe the difference between conventional current flow and electron flow.

Objective 23. Describe the steps that should be followed when using an ammeter to make a current measurement.

Use the following questions to test your understanding of Section 1-3:

1. What is the unit of current?
2. Define current in relation to coulombs and time.
3. What is the difference between conventional and electron current flow?
4. What test instrument is used to measure current?

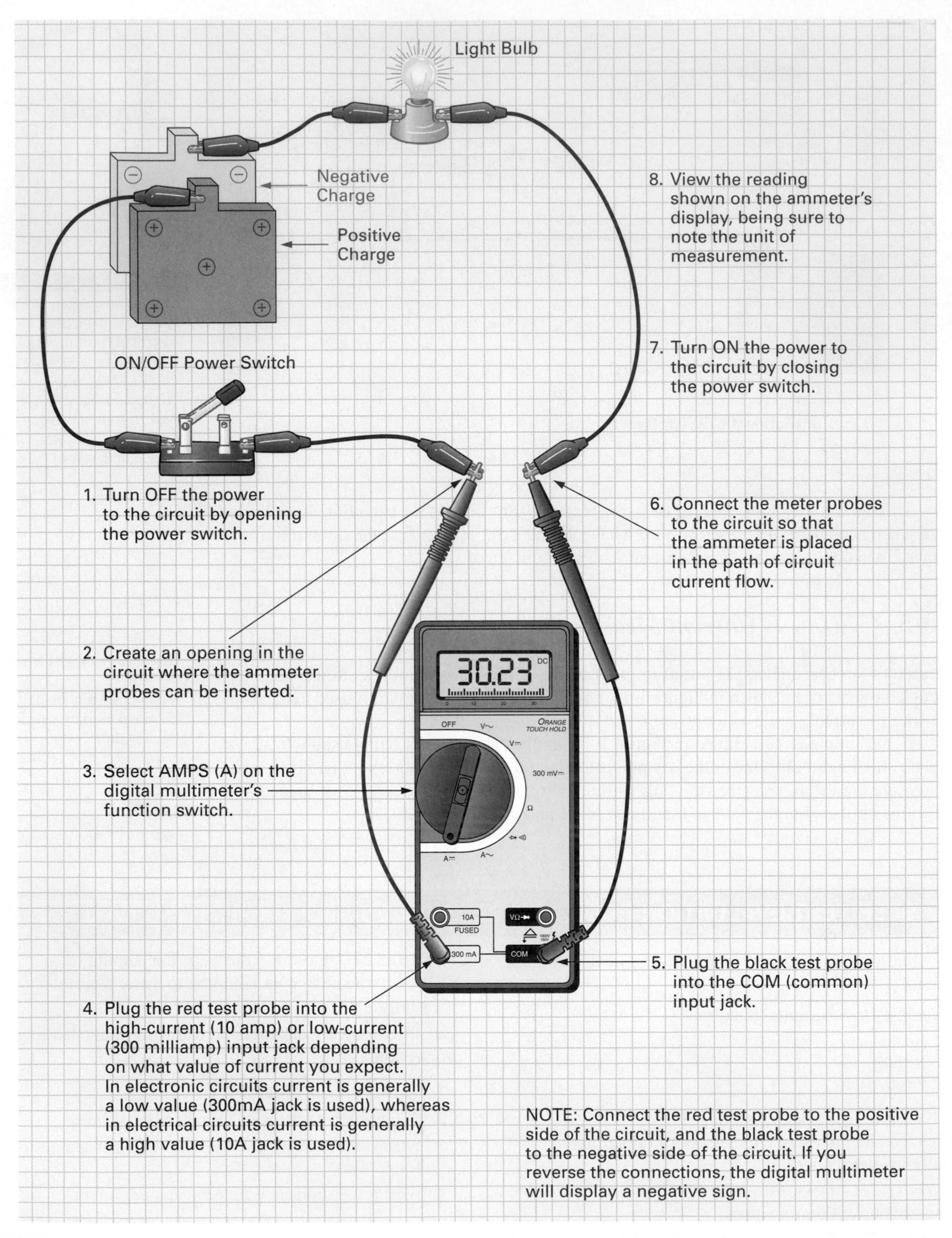

FIGURE 1-14 Measuring Current with the Ammeter.

1-4 VOLTAGE

Voltage is the force or pressure exerted on electrons. Referring to Figure 1-15(a) and (b), you will notice two situations. Figure 1-15(a) shows highly concentrated positive and negative charges or potentials connected to one another by a copper wire. In this situation, a large potential difference or voltage is being applied across the copper atom's electrons. This force or voltage causes a large amount of copper atom electrons to move from right to left. On the other hand, Figure 1-15(b) illustrates a low concentration of positive and negative potentials, so a small voltage or pressure is being applied across the conductor, causing a small amount of force, and therefore current, to move from right to left.

In summary, we could say that a highly concentrated charge produces a high voltage, whereas a low concentrated charge produces a low voltage. Voltage is also appropriately known as the "electron moving force" or **electromotive force (emf),** and since two opposite

Voltage (*V* or *E*)
Term used to designate electrical pressure or the force that causes current to flow.

Electromotive Force (emf)
Force that causes the motion of electrons due to a potential difference between two points.

TIME LINE
The unit of voltage, the volt, was named in honor of Alessandro Volta (1745–1827), an Italian physicist who is famous for his invention of the electric battery. In 1801, he was called to Paris by Napoleon to show his experiment on the generation of electric current.

Positive Charge
(Strong Attraction of Electrons)
Negative Charge
(Strong Repulsion of Electrons)
High Concentration of Positive Charge
High Concentration of Negative Charge
Large Positive Potential or Charge
Large Potential Difference (PD) or Voltage
Zero
Large Negative Potential or Charge
(a)

Positive Charge
(Weak Attraction of Electrons)
Negative Charge
(Weak Repulsion of Electrons)
Low Concentration of Positive Charge
Low Concentration of Negative Charge
Small Positive Potential or Charge
Small Potential Difference (PD) or Voltage
Zero
Small Negative Potential or Charge
(b)

FIGURE 1-15 (a) Large Potential Difference or Voltage. (b) Small Potential Difference or Voltage.

Potential Difference (PD)
Voltage difference between two points, which will cause current to flow in a closed circuit.

Battery
DC voltage source containing two or more cells that converts chemical energy into electrical energy.

potentials exist (one negative and one positive), the strength of the voltage can also be referred to as the amount of **potential difference (PD)** applied across the circuit. To compare, we can say that a large voltage, electromotive force, or potential difference exists across the copper conductor in Figure 1-15(a), and a small voltage, potential difference, or electromotive force is exerted across the conductor in Figure 1-15(b).

Voltage is the force, pressure, potential difference (PD), or electromotive force (emf) that causes electron flow or current and is symbolized by italic uppercase V. The unit for voltage is the volt, symbolized by roman uppercase V. This can become a bit confusing. For example, when the voltage applied to a circuit equals 5 volts, the circuit notation would appear as

$$V = 5\ \text{V}$$

You know the first V represents "voltage," not "volt," because 1 volt cannot equal 5 volts. To avoid confusion, some texts and circuits use *E,* symbolizing electromotive force, to represent voltage; for example,

$$E = 5\ \text{V}$$

In this text, we will maintain the original designation for voltage (V).

1-4-1 *A Simple Voltage Source*

A **battery,** like the one shown in Figure 1-16 (a), converts chemical energy into electrical energy. At the positive terminal of the battery, positive charges or ions (atoms with more protons than electrons) are present, and at the negative terminal, negative charges or ions (atoms with more electrons than protons) are available to supply electrons for current flow within a circuit. A battery, therefore, chemically generates negative and positive ions at its respective terminals. The symbol for the battery is shown in Figure 1-16(b).

1-4-2 *Components and Circuits*

Just as a complete path exists for an athlete around a track, a complete path also exists around an electric circuit for current. Components are individual devices such as wires, batteries, lamps, and switches that serve as the building blocks for these circuits. In Figure 1-17(a) you

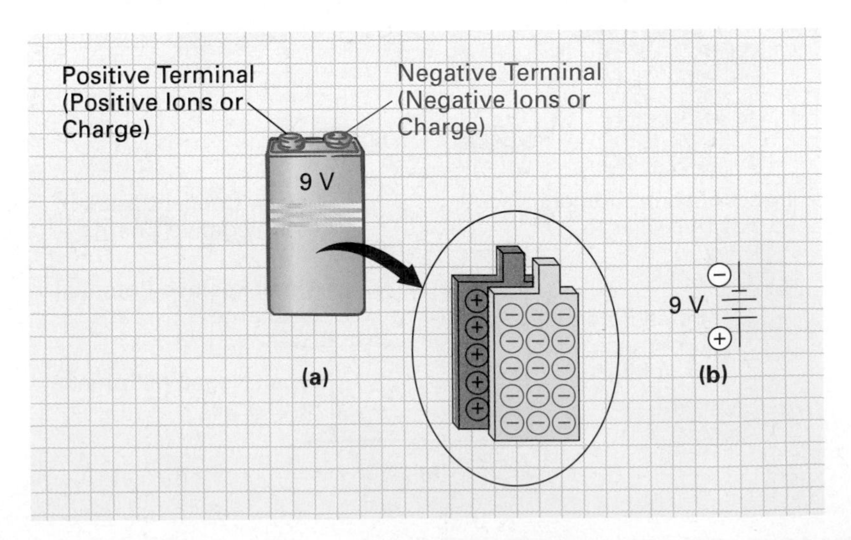

FIGURE 1-16 The Battery—A Source of Voltage. (a) Physical Appearance. (b) Schematic Symbol.

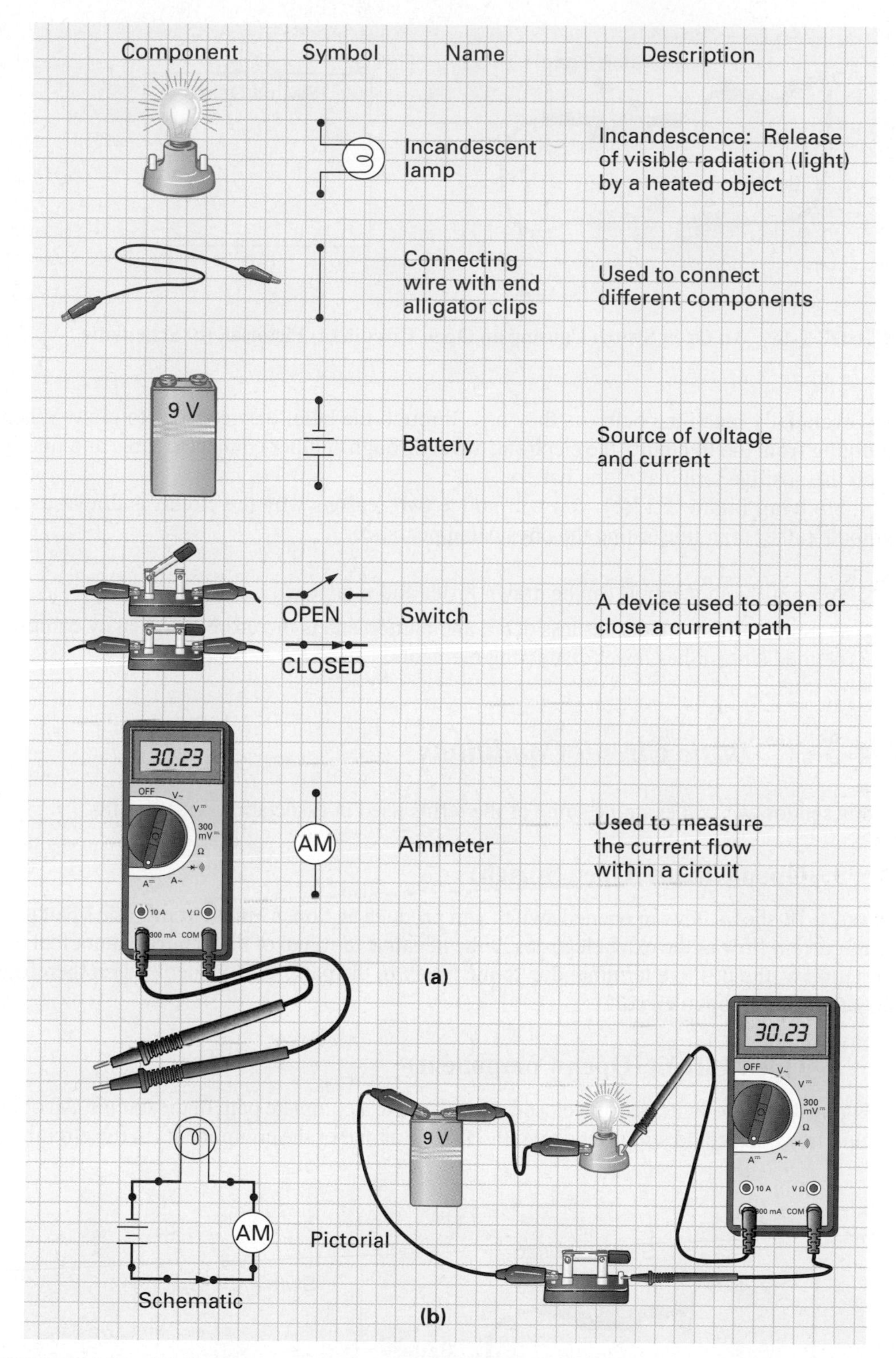

FIGURE 1-17 (a) Components. (b) Example Circuit.

can see the schematic symbols for many of the devices discussed so far, along with their physical appearance.

In the circuit shown in Figure 1-17(b), a 9 V battery chemically generates positive and negative ions. The negative ions at the negative terminal force away the negative electrons, which are attracted by the positive charge or absence of electrons at the positive terminal. As the electrons proceed through the copper conductor wire, jumping from one atom to the next,

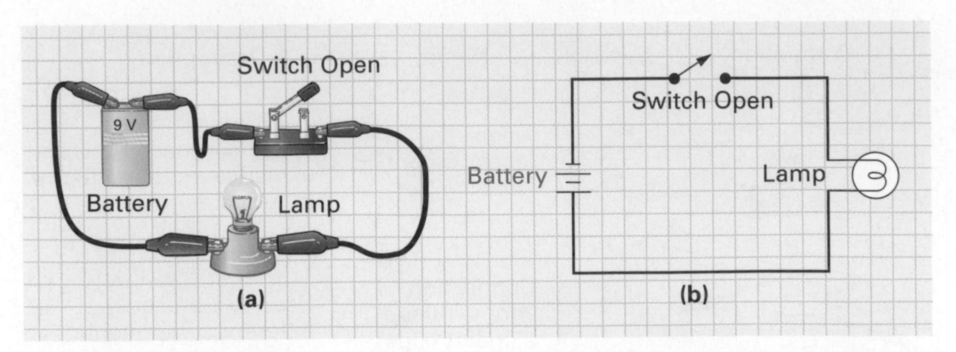

FIGURE 1-18 **An Open Switch Causing an Open Circuit. (a) Pictorial. (b) Schematic.**

they eventually reach the bulb. As they pass through the bulb, they cause it to glow. When emerging from the lightbulb, the electrons travel through another connector cable and finally reach the positive terminal of the battery.

Studying Figure 1-17(b), you will notice two reasons why the circuit is drawn using symbols rather than illustrating the physical appearance:

1. A circuit with symbols can be drawn faster and more easily.
2. A circuit with symbols has less detail and clutter, and is therefore more easily comprehended since it has fewer distracting elements.

1-4-3 *Basic Circuit Conditions*

In this section we will examine some of the terms used in association with circuits.

1. Open Circuit (Open Switch)

Figure 1-18 shows how an opened switch can produce an "open" in a circuit. The open prevents current flow as the extra electrons in the negative terminal of the battery cannot feel the attraction of the positive terminal due to the break in the path. The opened switch therefore has produced an **open circuit.**

Open Circuit
Break in the path of current flow.

2. Open Circuit (Open Component)

In Figure 1-19, the switch is now closed so as to make a complete path in the circuit. An open circuit, however, could still exist due to the failure of one of the components in the circuit. In

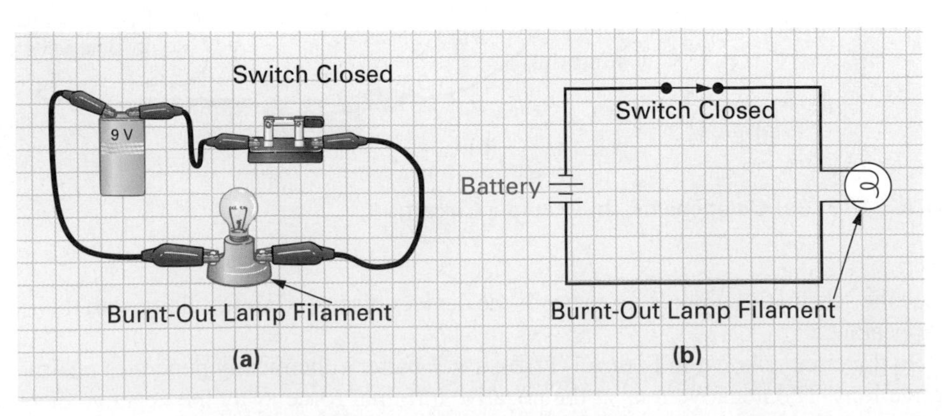

FIGURE 1-19 **An Open Lamp Filament Causing an Open Circuit. (a) Pictorial. (b) Schematic.**

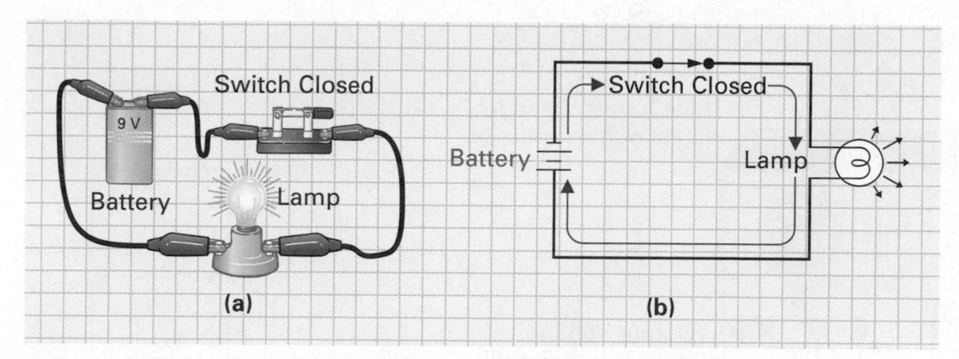

FIGURE 1-20 Closed Switch Causing a Closed Circuit. (a) Pictorial. (b) Schematic.

the example in Figure 1-19, the lightbulb filament has burned out, creating an open in the circuit, which prevents current flow. An open component therefore can also produce an open circuit.

3. Closed Circuit (Closed Switch)

If all of the devices in a circuit are operating properly and connected correctly, a closed switch produces a **closed circuit,** as shown in Figure 1-20. A closed circuit provides a complete path from the negative terminal of the battery to the positive, and so current will flow through the circuit.

Closed Circuit
Circuit having a complete path for current to flow.

4. Short Circuit (Shorted Component)

A **short circuit** normally occurs when one point in a circuit is accidentally connected to another. Figure 1-21(a) illustrates how a short circuit could occur. In this example, a set of pliers was accidentally laid across the two contacts of the lightbulb. Figure 1-21(b) shows the circuit's schematic diagram, with the effect of the short across the lightbulb drawn in. In this instance, nearly all the current will flow through the metal of the pliers, which offers no resistance or opposition to the current flow. Since the lightbulb does have some resistance or opposition, very little current will flow through the bulb, and therefore no light will be produced.

Short Circuit
Also called a short, it is a low-resistance connection between two points in a circuit, typically causing a large amount of current flow.

1-4-4 *Power Supply Unit*

A battery could be used to apply a voltage across a component or circuit, but batteries have a fixed output voltage and will, in time, run down and lose their potential.

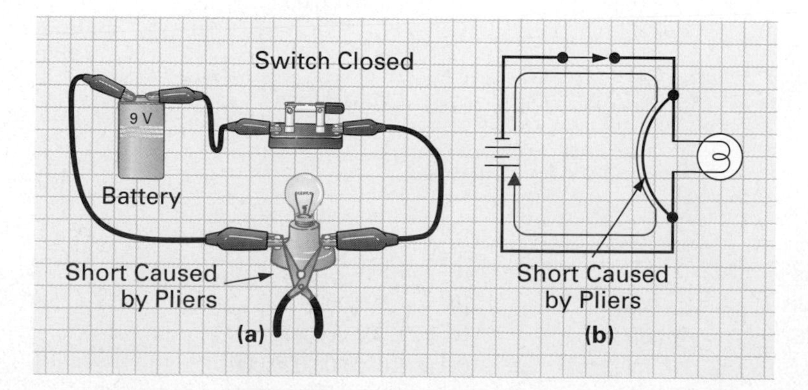

FIGURE 1-21 Short Circuit Due to Pliers across a Lamp. (a) Pictorial. (b) Schematic.

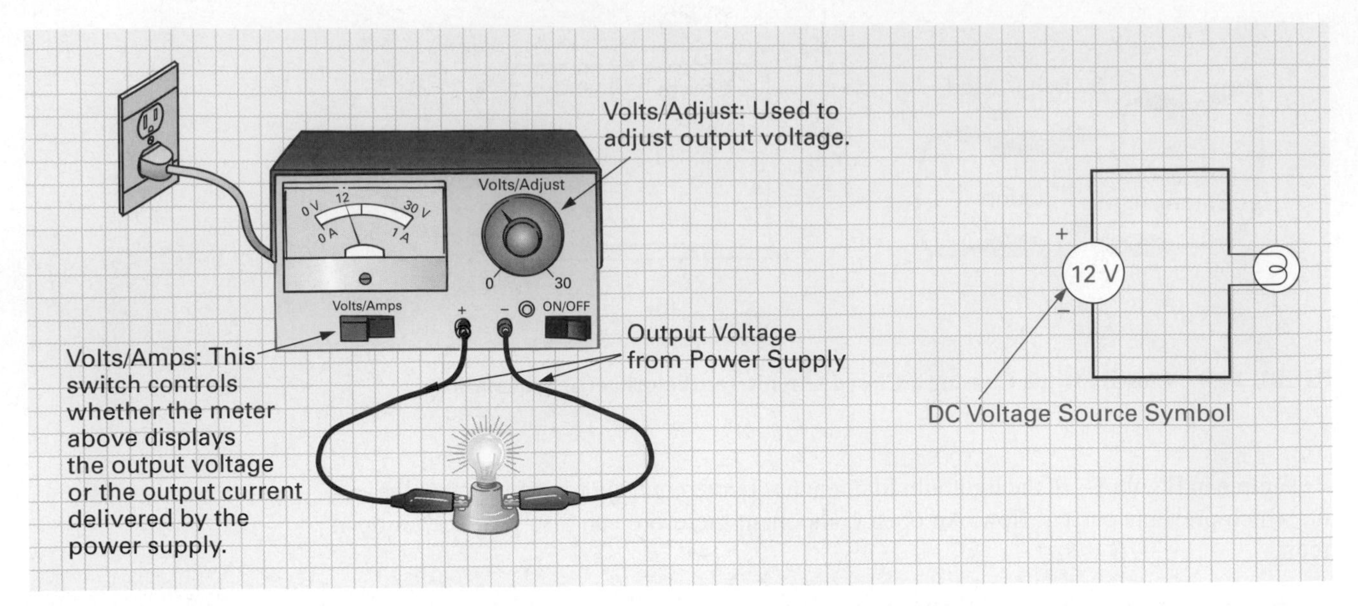

FIGURE 1-22 **Power Supply Unit.**

Power Supply Unit
Piece of electrical equipment used to deliver either ac or dc voltage.

The **power supply unit,** shown in Figure 1-22, can provide an accurate voltage at its output terminal that can be increased or decreased using the "volts/adjust" control. In the example shown in Figure 1-22, the power supply is being used to supply 12 volts across a lamp.

1-4-5 *Units of Voltage*

The unit for voltage is the volt (V). Voltage within electronic equipment is normally measured in volts, whereas heavy-duty industrial equipment normally requires high voltages that are generally measured in kilovolts (kV). Table 1-3 lists all the prefixes and values related to volts.

TABLE 1-3 **Voltage Units**

NAME	SYMBOL	VALUE
Picovolts	pV	$10^{-12} = \frac{1}{1,000,000,000,000}$
Nanovolts	nV	$10^{-9} = \frac{1}{1,000,000,000}$
Microvolts	μV	$10^{-6} = \frac{1}{1,000,000}$
Millivolts	mV	$10^{-3} = \frac{1}{1000}$
Volts	V	$10^{0} = 1$
Kilovolts	kV	$10^{3} = 1000$
Megavolts	MV	$10^{6} = 1,000,000$
Gigavolts	GV	$10^{9} = 1,000,000,000$
Teravolts	TV	$10^{12} = 1,000,000,000,000$

EXAMPLE:

Convert the following:

a. 3000 V = ___________ kV (kilovolts)
b. 0.14 V = ___________ mV (millivolts)
c. 1500 kV = ___________ MV (megavolts)

Solution:

a. 3000 V = 3 kV or 3×10^3 volts (multiplier ↑ 1000, number ↓ 1000)
b. 0.14 V = 140 mV or 140×10^{-3} volt (multiplier ↓ 1000, number ↑ 1000)
c. 1500 kV = 1.5 MV or 1.5×10^6 volts (multiplier ↑ 1000, number ↓ 1000)

1-4-6 *How Is Voltage Measured?*

Voltmeters (voltage meters) are used to measure electrical pressure or voltage. Stepping through the sequence detailed in Figure 1-23, you will see how a voltmeter can be used to measure a battery's voltage.

In many instances, the voltmeter is used to measure the potential difference or voltage drop across a device, as shown in Figure 1-24(a) and (b). Figure 1-24(a) shows how to measure the voltage across lightbulb 1 (L1), and Figure 1-24(b) shows how to measure the voltage across lightbulb 2 (L2).

Voltmeter
Instrument designed to measure the voltage or potential difference. Its scale can be graduated in kilovolts, volts, or millivolts.

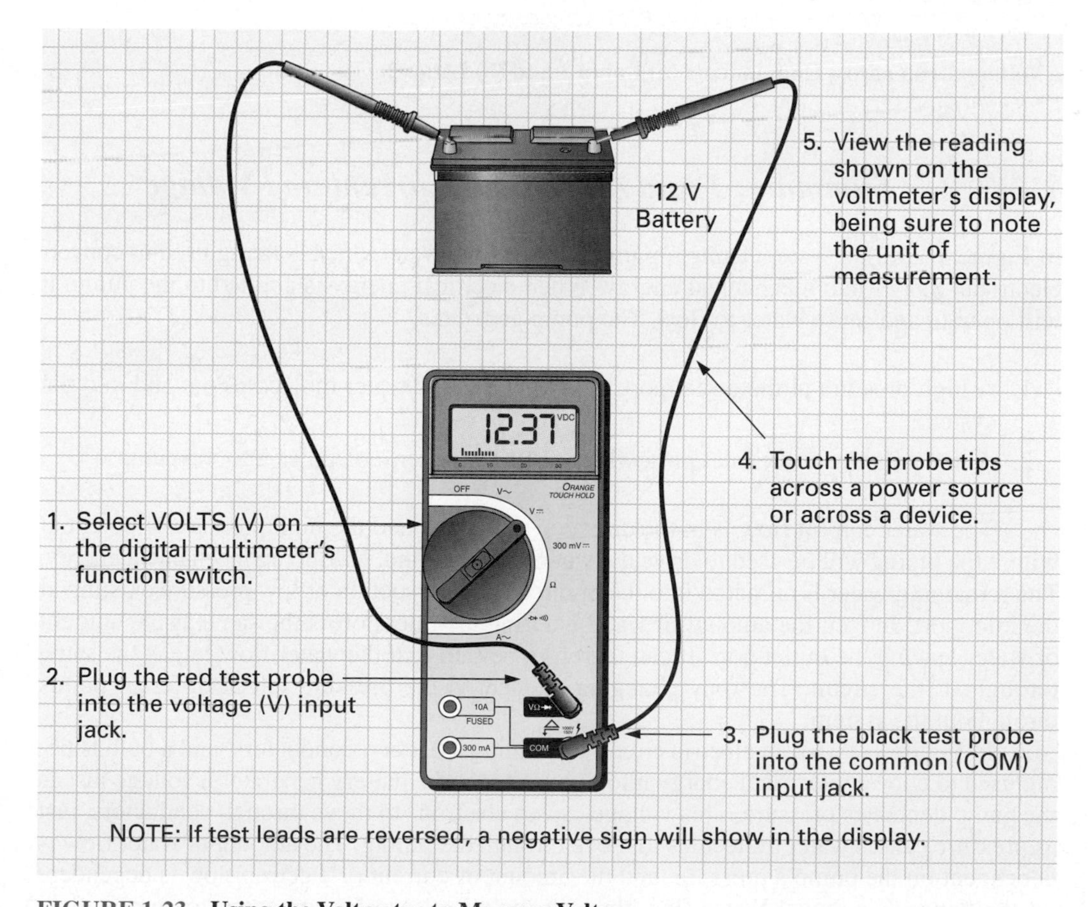

FIGURE 1-23 Using the Voltmeter to Measure Voltage.

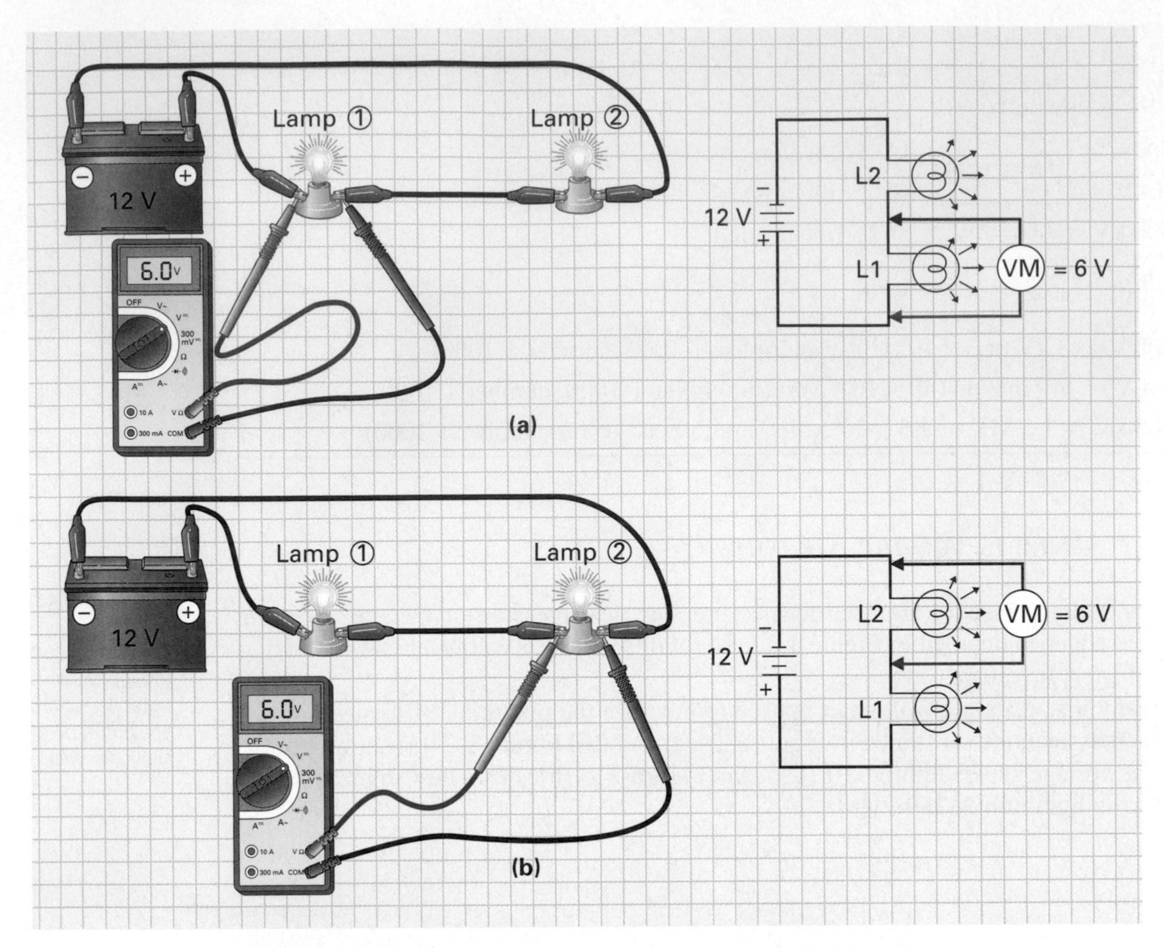

FIGURE 1-24 Measuring the Voltage Drop across Components (a) Lamp 1 and (b) Lamp 2.

1-4-7 *Summary: Fluid Analogy of Current and Voltage*

In Figure 1-25(a), a system using a pump, pipes, and a waterwheel is being used to convert electrical energy into mechanical energy. When electrical energy is applied to the pump, it will operate and cause water to flow. The pump generates

1. A high pressure at the outlet port, which pushes the water molecules out and into the system.
2. A low pressure at the inlet port, which pulls the water molecules into the pump.

The water current flow is in the direction indicated, and the high pressure or potential within the piping will be used to drive the waterwheel around, producing mechanical energy. The remaining water is attracted into the pump due to the suction or low pressure existing at the inlet port. In fact, the amount of water entering the inlet port is the same as the amount of water leaving the outlet port. It can therefore be said that the water flow rate is the same throughout the circuit. The only changing element is the pressure felt at different points throughout the system.

In Figure 1-25(b), an electric circuit containing a battery, conductors, and a bulb is being used to convert electrical energy into light energy. The battery generates a voltage just as the pump generates pressure. This voltage causes electrons to move through conductors, just as pressure causes water molecules to move through the piping. The amount of water flow is dependent on the pump's pressure, and the amount of current or electron flow is dependent on the battery's voltage. Water flow through the wheel can be compared to current flow

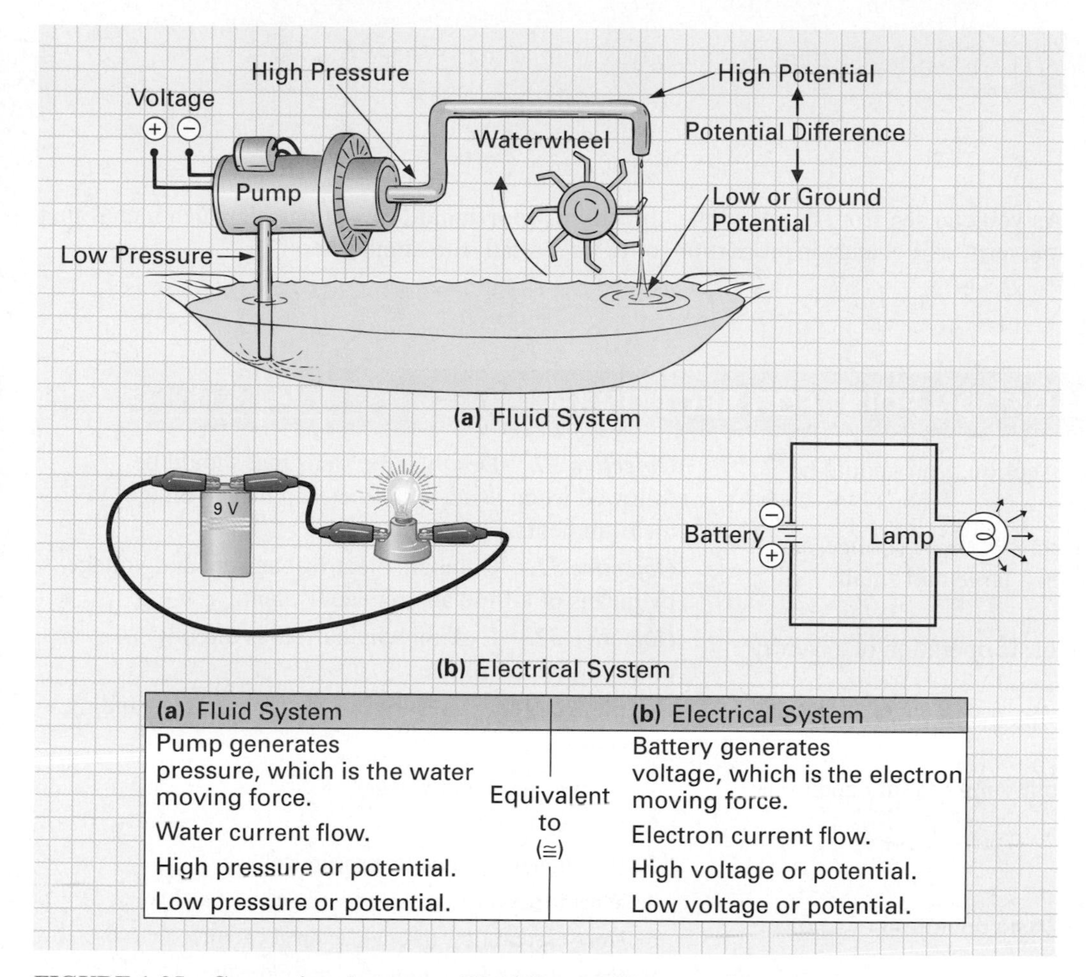

(a) Fluid System		(b) Electrical System
Pump generates pressure, which is the water moving force.	Equivalent to (≅)	Battery generates voltage, which is the electron moving force.
Water current flow.		Electron current flow.
High pressure or potential.		High voltage or potential.
Low pressure or potential.		Low voltage or potential.

FIGURE 1-25 Comparison between a Fluid System (a) and an Electrical System (b).

through the bulb. The high pressure is lost in turning the wheel and producing mechanical energy, just as voltage is lost in producing light energy out of the bulb. We cannot say that pressure or voltage flows: Pressure and voltage are applied and cause water or current to flow, and it is this flow that is converted to mechanical energy in our fluid system and light energy in our electrical system. Voltage is the force of repulsion and attraction needed to cause current to flow through a circuit, and without this potential difference or pressure there cannot be current.

Current (I) is therefore said to be *directly proportional* ($\propto$) *to voltage (V),* as a voltage decrease ($V\downarrow$) results in a current decrease ($I\downarrow$), and similarly, a voltage increase ($V\uparrow$) causes a current increase ($I\uparrow$), as in Figure 1-26.

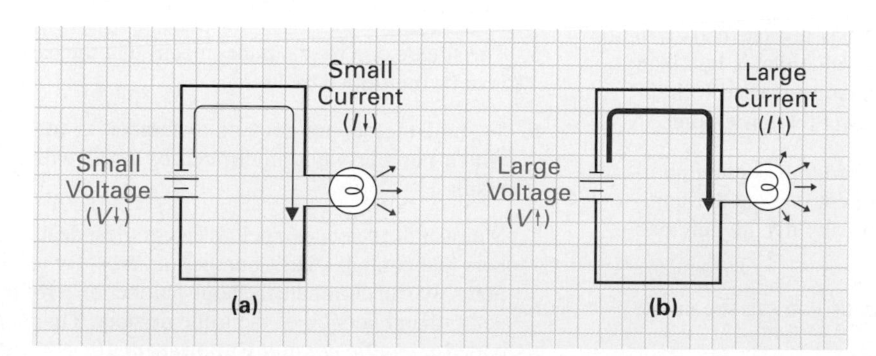

FIGURE 1-26 Current Is Directly Proportional to Voltage. (a) Small Voltage Produces a Small Current. (b) Large Voltage Produces a Large Current.

Current is directly proportional to voltage ($I \propto V$)

$V\downarrow$ causes a $I\downarrow$
$V\uparrow$ causes a $I\uparrow$

Proportional
A term used to describe the relationship between two quantities that have the same ratio.

As you can see from this example, directly **proportional** ($\propto$) is a phrase which means that one term will change in proportion, or in size, relative to another term.

SELF-TEST EVALUATION POINT FOR SECTION 1-4

Now that you have read this section, you should be able to:

Objective 24. Define the term *electrical voltage,* and explain the difference between a large and small potential difference.

Objective 25. Describe the basic operation of a battery as a voltage source.

Objective 26. List and describe the function of some simple circuit components.

Objective 27. Explain the following circuit conditions:

a. Open circuit
b. Closed circuit
c. Short circuit

Objective 28. Describe the basic operation of a power supply unit as a voltage source.

Objective 29. Describe how to convert between different voltage prefix values.

Objective 30. Describe the steps that should be followed when using a voltmeter to make voltage measurements.

Objective 31. Compare electrical current and voltage by means of a fluid analogy equivalent.

Objective 32. Explain why current is directly proportional to voltage.

Use the following questions to test your understanding of Section 1-4:

1. What is the unit of voltage?
2. Convert 3 MV to kilovolts.
3. Which meter is used to measure voltage?
4. What is the relationship between current and voltage?
5. Does current flow through an open circuit?
6. Does current flow through a short circuit?
7. Describe the difference between a closed circuit and a short circuit.

SUMMARY

Mini-Math Review (Section 1-1)

1. There are really only two basic mathematical operations: addition and subtraction. Multiplication is simply repeated addition, and division is repeated subtraction.
2. A positive number is any value that is greater than zero, whereas a negative number is any value that is less than zero. Positive and negative numbers are needed to indicate values that are above and below a reference point.
3. Many of the values inserted into formulas contain numbers that have *exponents*. An exponent is a smaller number that appears to the right and slightly higher than another number.
4. *Exponents allow us to save space,* as you can see in this example:

$$5 \times 5 \times 5 \times 5 = 5^4$$

Here the exponent (^{4}) is used to indicate how many times its base (5) is used as a *factor*. You can also see from this example that the other advantage is that *exponents are a sort of math shorthand*. It is much easier to write 5^4 than to write $5 \times 5 \times 5 \times 5$. The "$y^x$" key on your calculator allows you to enter in any base number (y) and any exponent (^{x}).

5. The square of a base number means that the base number is to be multiplied by itself.
6. To find the square root of a number is to determine which number when multiplied by itself will equal the result.
7. Many of the sciences, such as electronics, deal with numbers that contain a large number of zeros. By using exponents we can eliminate the large number of zeros to obtain a shorthand version of the same number. This method is called *scientific notation or powers of 10.*

8. A *negative exponent* tells you how many places *to the left* to move the decimal point. A *positive exponent* tells you how many places *to the right* to move the decimal point. Remember one important point: that a negative exponent does not indicate a negative number; it simply indicates a fraction.
9. *Scientific notation* is expressed as a base number between 1 and 10 with an exponent that is a power of 10. *Engineering notation,* however, is expressed as a base number that is greater than 1 with an exponent that is some multiple of 3.
10. A prefix is a power of 10, or multiplier, that precedes a unit.

The Structure of Matter (Section 1-2)

11. An *element,* by definition, is a substance consisting of only one type of atom.
12. The *atom* is the smallest particle into which an element can be divided without losing its identity, and a group of identical atoms is called an element.
13. The word *atom* is a Greek word meaning a particle that is too small to be subdivided.
14. In 1913, a Danish physicist, Neils Bohr, put forward a theory about the atom, and his basic model outlining the *subatomic* particles that make up the atom is still in use today.
15. The three important particles of the atom are the *proton,* which has a positive charge, the *neutron,* which is neutral or has no charge, and the *electron,* which has a negative charge.
16. The atom consists of a positively charged central mass called the *nucleus,* which is made up of protons and neutrons surrounded by a quantity of negatively charged orbiting electrons.
17. The *atomic number* of an atom describes the number of protons that exist within the nucleus.
18. The number of neutrons within an atom's nucleus can be calculated by subtracting the atomic number (protons) from the atomic weight (protons and neutrons).
19. A *neutral atom* or *balanced atom* is one that has an equal number of protons and orbiting electrons, so the net positive proton charge is equal but opposite to the net negative electron charge, resulting in a balanced or neutral state.
20. Orbiting electrons travel around the nucleus at varying distances from the nucleus, and these orbital paths are known as *shells* or *bands*. The orbital shell nearest the nucleus is referred to as the first or K shell. The second is known as the L, the third is M, the fourth is N, the fifth is O, the sixth is P, and the seventh is referred to as the Q shell. There are seven shells available for electrons (K, L, M, N, O, P, and Q) around the nucleus, and each of these seven shells can only hold a certain number of electrons. The outermost electron-occupied shell is referred to as the *valence shell* or *ring,* and its electrons are termed *valence electrons*.
21. All matter exists in one of three states: solid, liquid, or gas. The atoms of a solid are fixed in relation to one another but vibrate in a back-and-forth motion, unlike liquid atoms, which can flow over each other. The atoms of a gas move rapidly in all directions and collide with one another.
22. *Like charges* (positive and positive or negative and negative) repel one another. *Unlike charges* (positive and negative or negative and positive) attract one another.
23. Due to their distance from the nucleus, valence electrons are described as being loosely bound to the atom. The electrons can, therefore, easily be dislodged from their outer orbital shell by any external force, to become *free electrons*.
24. Combinations of elements are known as *compounds,* and the smallest unit of a compound is called a *molecule,* just as the smallest unit of an element is an atom.

Current (Section 1-3)

25. Once an electron is released, the atom is no longer electrically neutral and is called a *positive ion,* as it now has a net positive charge (more protons than electrons). The released electron tends to jump into a nearby atom, which will then have more electrons than protons and is referred to as a *negative ion.*
26. Millions upon millions of atoms within a conductor pass a continuous movement of billions upon billions of electrons from negative to positive. This electron flow is known as *electric current.*
27. A force or pressure produced by a positive charge and negative charge will cause electrons to flow from the negative to the positive terminal. The positive side has a deficiency of electrons and the negative side has an abundance, and so a continuous flow or migration of electrons takes place through a metal conducting wire. This electric current or electron flow is a measurable quantity.
28. There are 6.24×10^{18} electrons in 1 *coulomb of charge*.
29. Coulombs and time are therefore combined to describe the number of electrons and the rate at which they flow. This relationship is called *current (I)* flow and has the unit of *amperes* (A). If 6.24×10^{18} electrons (1 C) were to drift past a specific point on a conductor in 1 second of time, 1 ampere of current is said to be flowing.
30. Current within electronic equipment is normally a value in milliamperes or microamperes and very rarely exceeds 1 ampere.
31. Electrons drift from a negative to a positive charge. This current is known as *electron flow*. In the eighteenth and nineteenth centuries when very little was known about the atom, researchers believed that current was a flow of positive charges. Although this has now been proved incorrect, many texts still use *conventional current flow*. Whether conventional current flow or electron flow is used, the same answers to problems, measurements, or designs are

obtained. The key point to remember is that direction is not important, but the amount of current flow is.

32. *Ammeters* (ampere meters) are used to measure the current within a circuit. The current path must be opened and the ammeter placed in the path of current so that the instrument can sense and display the value.

Voltage (Section 1-4)

33. *Voltage* is the force or pressure exerted on electrons.

34. A highly concentrated charge produces a high voltage, whereas a low concentrated charge produces a low voltage. Voltage is also appropriately known as the electron moving force or *electromotive force (emf),* and since two opposite potentials exist, one negative and one positive, the strength of the voltage can also be referred to as the amount of *potential difference (PD)* applied across the circuit.

35. Voltage is the force, pressure, potential difference (PD), or electromotive force (emf) that causes electron flow or current and is symbolized by italic uppercase V. The unit for voltage is the volt, symbolized by roman uppercase V.

36. A *battery* converts chemical energy into electrical energy. Positive charges or ions (atoms with more protons than electrons) are present at the positive terminal of a battery, while negative charges or ions (atoms with more electrons than protons) are present at the negative terminal of a battery.

37. An opened switch produces an "open" in the circuit, preventing circuit current. If the switch is now closed so as to make a complete path in the circuit, an open could still exist due to the failure of one of the components. An open component can also produce an open circuit.

38. A closed switch produces a *closed circuit.* If all devices are operating correctly and all connections are correct, current will be present in a closed circuit.

39. A *short circuit* normally occurs when one point is accidentally connected to another.

40. A power supply unit can provide an accurate, adjustable voltage at its output terminals. The volt/amps Switch controls whether the meter displays the output voltage or the output current being supplied by the unit.

41. Voltage within electronic equipment is normally measured in volts, whereas heavy-duty industrial equipment normally requires high voltages that are generally measured in kilovolts (kV).

42. *Voltmeters* (voltage meters) are used to measure electrical pressure or voltage. To measure the voltage drop across a device, a voltmeter is connected across the device with the two leads of the voltmeter touching either side of the component.

43. If a larger voltage were applied to the electrical circuit, this larger electron moving force (emf) would cause more electrons or current to flow through the circuit. *Current (I)* is therefore said to be *directly proportional* ($\propto$) to *voltage (V),* as a voltage decrease ($V\downarrow$) results in a current decrease ($I\downarrow$), and a voltage increase ($V\uparrow$) causes a current increase ($I\uparrow$).

44. Directly proportional ($\propto$) is a phrase which means that one term will change in proportion, or in size, relative to another term.

REVIEW QUESTIONS

Multiple-Choice Questions

1. If $x^2 = 16$, then $\sqrt[2]{16} = x$. What is the value of x?
   a. 4
   b. 2
   c. 16
   d. 8

2. Calculate $\sqrt[5]{32}$.
   a. 2
   b. 1.3
   c. 4
   d. 6.4

3. Raise the following number to the power of 10 indicated by the exponent, and give the answer: 3.6×10^2.
   a. 3600
   b. 36
   c. 0.036
   d. 360

4. Convert the following number to powers of 10: 0.00029.
   a. 0.29×10^{-3}
   b. 2.9×10^{-4}
   c. 29×10^{-5}
   d. All the above

5. What is the power-of-10 value for the metric prefix *milli?*
   a. 10^{-3}
   b. 10^{-2}
   c. 10^3
   d. All of the above

6. Hydrogen is a
   a. Gas
   b. Solid
   c. Liquid
   d. Other

7. Which of the following is a liquid?
   a. Calcium
   b. Magnesium
   c. Mercury
   d. Helium

8. The atomic number of an atom describes
   a. The number of neutrons
   b. The number of electrons
   c. The number of nuclei
   d. The number of protons
9. The most commonly used metal in the field of electronics is:
   a. Silver
   b. Copper
   c. Mica
   d. Gold
10. The smallest unit of an element is
    a. A compound
    b. An atom
    c. A molecule
    d. A proton
11. The smallest unit of a compound is
    a. An element
    b. A neutron
    c. An electron
    d. A molecule
12. A water molecule is made up of
    a. 2 parts hydrogen and 1 part oxygen
    b. Chlorine and sodium
    c. 1 part oxygen and 2 parts sodium
    d. 3 parts chlorine and 1 part hydrogen
13. A negative ion has
    a. More protons than electrons
    b. More electrons than protons
    c. More neutrons than protons
    d. More neutrons than electrons
14. A positive ion has
    a. Lost some of its electrons
    b. Gained extra protons
    c. Lost neutrons
    d. Gained more electrons
15. One coulomb of charge is equal to:
    a. 6.24×10^{18} electrons
    b. $10^{18} \times 10^{12}$ electrons
    c. 6.24×10^{8} electrons
    d. 6.24×10^{81} electrons
16. If 14 C of charge passes by a point in 7 seconds, the current flow is said to be
    a. 2 A
    b. 98 A
    c. 21 A
    d. 7 A
17. How many electrons are there within 16 C of charge?
    a. 9.98×10^{19}
    b. 14
    c. 16
    d. 10.73×10^{19}
18. Current is measured in
    a. Volts
    b. Coulombs/second
    c. Ohms
    d. Siemens
19. Voltage is measured in units of
    a. Amperes
    b. Ohms
    c. Siemens
    d. Volts
20. Another word used to describe voltage is
    a. Potential difference
    b. Pressure
    c. Electromotive force (emf)
    d. All of the above

Communication Skill Questions

21. What is an exponent? (1-1-3)
22. Define the following terms: (1-1-3)
    a. Square of a number
    b. Square root of a number
23. What is a power-of-10 exponent? (1-1-4)
24. Describe the scientific and engineering notation systems. (1-1-5)
25. List the metric prefixes and their value. (1-1-6)
26. Define the following terms: (1-4-2)
    a. Components
    b. Circuits
27. Why is current directly proportional to voltage? (1-4-7)
28. List the four most commonly used current prefixes and their values in terms of the basic unit. (1-3-3)
29. What is a power supply unit, and what advantages does it have over a battery as a power source? (1-4-4)
30. List the four most commonly used voltage units and their values in terms of the basic unit. (1-4-5)
31. Describe what is meant by: (1-4-3)
    a. An open circuit
    b. A closed circuit
    c. A short circuit
32. What is the difference between conventional and electron current flow? (1-3-4)
33. Give the unit and symbol for the following:
    a. Voltage (V) is measured in __________ (1-4)
    b. Current (I) is measured in __________ (1-3-2)
34. In relation to the structure of matter and the atom, describe
    a. The atom's subatomic particles (1-2)
    b. An element (1-2)
    c. A compound (1-2)
    d. A molecule (1-2)
    e. A neutral atom (1-3)
    f. A positive ion (1-3)
    g. A negative ion (1-3)
35. Describe how to use a voltmeter to measure voltage and how to use an ammeter to measure current. (1-3-5, 1-4-6)

Practice Problems

36. Determine the square of the following values:
a. 9^2
b. 6^2
c. 2^2
d. 0^2
e. 1^2
f. 12^2

37. Determine the square root of the following values:
a. $\sqrt{81}$
b. $\sqrt{4}$
c. $\sqrt{0}$
d. $\sqrt{36}$
e. $\sqrt{144}$
f. $\sqrt{1}$

38. Raise the following base numbers to the power indicated by the exponent, and give the answer:
a. 9^3
b. 10^4
c. 4^6
d. 2.5^3

39. Determine the roots of the following values:
a. $\sqrt[3]{343} = ?$
b. $\sqrt{1296} = ?$
c. $\sqrt[4]{760}$
d. $\sqrt[6]{46{,}656}$

40. Convert the following values to powers of 10:
a. $\frac{1}{100}$
b. 1,000,000,000
c. $\frac{1}{1000}$
d. 1000

41. Convert the following to proper fractions and decimal fractions:
a. 10^{-4}
b. 10^{-2}
c. 10^{-6}
d. 10^{-3}

42. Convert the following values to powers of 10 in both scientific and engineering notation:
a. 475
b. 8200
c. 0.07
d. 0.00045

43. Convert the following to whole-number values with a metric prefix:
a. 0.005 A
b. 8000 m
c. 15,000,000 Ω
d. 0.000016 s

44. Calculate the total number of electrons in 6.5 C of charge.

45. Calculate the amount of current in amperes passing through a conductor if 3 C of charge passes by a point in 4 s.

46. Convert the following:
a. 0.014 A = __________ mA
b. 1374 A = __________ kA
c. 0.776 μA = __________ nA
d. 0.91 mA = __________ μA

47. Convert the following:
a. 1473 mV = __________ V
b. 7143 V = __________ kV
c. 0.139 kV = __________ V
d. 0.390 MV = __________ kV

48. Referring to Table 1-1, how many electrons are in the valence shell of the metals copper, silver, and gold.

49. An ammeter displays a current reading of 470 μA. What is this value in mA?

50. A voltmeter displays a voltage reading of 0.37 V. What is this value in mV?

Job Interview Test

These tests at the end of each chapter will challenge your knowledge up to this point, and give you the practice you need for a job interview. To make this more realistic, the test will be comprised of both technical and personal questions. In order to take full advantage of this exercise, you may want to set up a simulation of the interview environment, have a friend read the questions to you, and record your responses for later analysis.

Company Name: Multi-Meters, Inc.

Industry Branch: Test and Measurement.

Function: Design and Manufacture Multimeters.

Job Title: QA Technician.

51. What do you know about our company?

52. What steps would you follow to use a multimeter to measure current and voltage?

53. Have you had much experience working in a team environment?

54. Could you define, simply, the difference between current and voltage?

55. What is your understanding of the responsibilities of a QA technician?

56. If your meter showed a reading of 10 milliamps, what does the "milli" mean?

57. Could you use the multimeter to measure potential difference?

58. What would you say are your faults?

59. If you discovered an error in a circuit design, would you report it to your supervisor, or go find the engineer responsible?

60. Why would you want to work for our company?

Answers

51. Visit the company's web site before the interview to get an overall understanding of the company's ownership, product line, service, and support.

52. Sections 1-3-5 and 1-4-6

53. Discuss how you have worked with a partner in the lab on experiments and any other related work history.

54. Chapter 1 margin definitions.

55. See introduction, QA Technician.

56. One-thousandth.

57. Yes, PD is voltage.

58. Make sure your faults are positive; for example, you always like to finish what you start, and so on.

59. Your supervisor.

60. Refer back to the company's positive attributes listed in its web site.

Web Site Questions

Go to the web site http://www.prenhall.com/cook, select the textbook *Introductory DC/AC Electronics* or *Introductory DC/AC Circuits,* this chapter, and then follow the instructions when answering the multiple-choice practice problems.

Resistance and Power

Problem Solver

Charles Proteus Steinmetz (1865–1923) was an outstanding electrical genius who specialized in mathematics, electrical engineering, and chemistry. His three greatest electrical contributions were his investigation and discovery of the law of hysteresis, his investigations in lightning, which resulted in his theory on traveling waves, and his discovery that complex numbers could be used to solve ac circuit problems. Solving problems was in fact his specialty, and on one occasion he was commissioned to troubleshoot a failure on a large company system that no one else had been able to repair. After studying the symptoms and schematics for a short time, he chalked an X on one of the metal cabinets, saying that this was where they would find the problem, and left. He was right, and the problem was remedied to the relief of the company executives; however, they were not pleased when they received a bill for $1000. When they demanded that Steinmetz itemize the charges, he replied—$1 for making the mark and $999 for knowing where to make the mark.

2

Outline and Objectives

PROBLEM SOLVER INTRODUCTION

2-1 MINI-MATH REVIEW—ALGEBRA, EQUATIONS, AND FORMULAS

2-1-1 The Basics of Algebra

Objective 1: Define the following terms: *algebra, equation, formula,* and *literal numbers.*

Objective 2: Describe the rules regarding the equality on both sides of the equal sign.

Objective 3: Show how the equality of an equation is preserved if the same arithmetic operation is performed on both sides of an equation or formula.

Objective 4: Explain how equations can be rearranged and still maintain their equality.

2-1-2 Transposition

Objective 5: Demonstrate the process of transposing a formula.

Objective 6: Describe how the terms *directly proportional* and *inversely proportional* are used to show the relationship between quantities in a formula.

2-1-3 Substitution

Objective 7: Demonstrate how to use substitution to develop alternative formulas.

2-2 RESISTANCE

2-2-1 The Ohm

Objective 8: Define *resistance* and *ohm.*

2-2-2 Ohm's Law

Objective 9: Explain Ohm's law and describe its function.

Objective 10: Describe why:

a. Current is proportional to voltage.
b. Current is inversely proportional to resistance.

Objective 11: Show how to transpose Ohm's law into its three forms.

Objective 12: Define the terms:

a. Source
b. Load
c. Load resistance
d. Load current

2-2-3 How Is Resistance Measured?

Objective 13: Explain how an ohmmeter can be used to measure resistance.

2-2-4 Protoboards

Objective 14: Describe the purpose and function of the protoboard to construct circuits.

2-3 POWER

Objective 15: List the six basic forms of energy, and define the terms:

a. Energy
b. Work
c. Joule

2-3-1 What Is Power?

Objective 16: Describe the term *power* and define its unit, the watt.

2-3-2 Calculating Energy

Objective 17: Explain how to determine the amount of energy stored, when coulombs and voltage are known.

2-3-3 Calculating Power

Objective 18: List Watt's law, or the power formula, and transpose it to obtain its three forms.

2-3-4 Measuring Power

Objective 19: Explain how the multimeter can be used to measure power.

2-3-5 The Kilowatt-Hour

Objective 20: Define the term *kilowatt-hour,* and describe how it is calculated.

2-3-6 Body Shock

Objective 21: Describe the physiological effects that occur when a current passes through your body.

Introduction

Voltage, current, resistance, and power are the four basic properties of prime importance in our study of electronics. In Chapter 1, voltage and current were introduced, and in this chapter we will examine resistance and power.

In the previous chapter you discovered that current is the movement of free electrons. As these electrons move through a material, they occasionally collide with atoms, and it is these collisions that cause the electrons to lose some of their energy as their movement is restricted. This restriction or opposition to free-electron flow (or current) is called *resistance,* and whenever a current passes through a material that has resistance, heat will be produced. *Power* is the measure of how fast energy is being used, and if electrons are colliding into atoms and losing some of their energy as they give off heat, power indicates how much heat energy is being dissipated or given off for every 1 second of time.

To begin with, let's review the math skills you will need for this chapter's material.

2-1 MINI-MATH REVIEW—ALGEBRA, EQUATIONS, AND FORMULAS

This first section, "Mini-Math Review," is included to overview the mathematical details you need for the electronic concepts covered in this chapter. In this review, we will be examining algebra, equations, and formulas.

I contemplated calling this section "using letters in mathematics" because the word *algebra* seems to make many people back away. If you look up *algebra* in the dictionary, it states that it is "a branch of mathematics in which letters representing numbers are combined according to the rules of mathematics." In this chapter you will discover how easy algebra is to understand and then see how we can put it to some practical use, such as rearranging formulas. As you study this section you will find that algebra is quite useful, and because it is used in conjunction with formulas, an understanding is essential for anyone entering a technical field.

2-1-1 *The Basics of Algebra*

Algebra
A generalization of arithmetic in which letters representing numbers are combined according to the rules of arithmetic.

Formula
A general fact, rule, or principle expressed usually in mathematical symbols.

Equation
A formal statement of the equality or equivalence of mathematical or logical expressions.

Algebra by definition is a branch of mathematics in which letters representing numbers are combined according to the rules of mathematics. The purpose of using letters instead of values is to develop a general statement or **formula** that can be used for any values. For example, distance (d) equals velocity (v) multiplied by time (t), or

$$d = v \times t$$

where d = distance in miles
v = velocity or speed in miles per hour
t = time in hours

Using this formula, we can calculate how much distance was traveled if we know the speed or velocity at which we were traveling and the time for which we traveled at that speed. Therefore, if I were to travel at a speed of 20 miles per hour (20 mph) for 2 hours, how far, or how much distance, would I travel? Replacing the letters in the formula with values converts the problem from a formula to an **equation**, as shown:

$$d = v \times t \leftarrow \text{(Formula)}$$
$$d = 20 \text{ mph} \times 2 \text{ h} \leftarrow \text{(Equation)}$$
$$d = 40 \text{ mi} \leftarrow \text{(Answer)}$$

Literal Number
A number expressed as a letter.

Letters such as *d, v, t* and *a, b, c* are called **literal numbers** (letter numbers) and are used, as we have just seen, in general statements showing the relationship between quantities. They are also used in equations to signify an *unknown quantity,* as will be discussed in the following section.

The Equality on Both Sides of the Equal Sign

All equations or formulas can basically be divided into two sections that exist on either side of an equal sign, as shown:

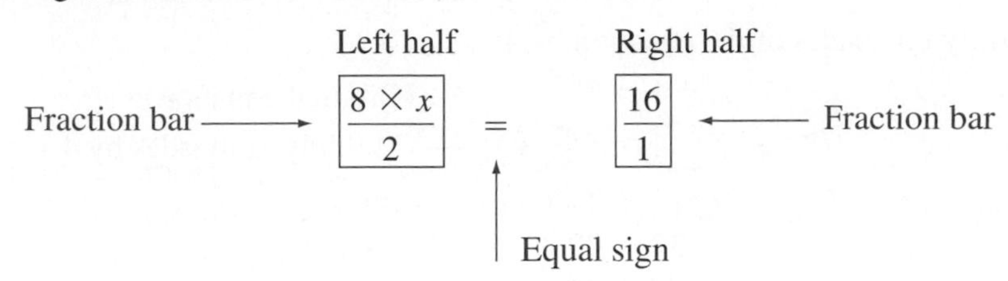

Everything in the left half of the equation is equal to everything in the right half of the equation. This means that 8 times x (which is an unknown value) divided by 2 equals 16 divided by 1. Generally, it is not necessary to put 1 under the 16 in the right section because any number divided by 1 equals the same number ($^{16}/_1 = 16 \div 1 = 16$). However, it was included in this introduction to show that each section has both a top and a bottom. If the equation is written without the fraction bar or 1 in the denominator position, it appears as follows:

$$\frac{8 \times x}{2} = 16$$

Although not visible, you should always assume that a number on its own on either side of the equal sign is above the fraction bar, as shown:

$$\frac{8 \times x}{2} = \frac{16}{}$$

Now that we understand the basics of an equation, let us see how we can manipulate it yet keep both sides equal to one another.

Treating Both Sides Equally

If you do exactly the same thing to both sides of an equation or formula, the two halves remain exactly equal, or in balance. This means that as long as you add, subtract, multiply, or divide both sides of the equation by the same number, the equality of the equation is preserved. For example, let us try adding, subtracting, multiplying, and dividing both sides of the following equation by 4 and see if both sides of the equation are still equal.

$2 \times 4 = 8$ ←Original equation

1. Add 4 to both sides of the equation:

$$2 \times 4 = 8 \quad \leftarrow \text{(Original equation)}$$
$$(2 \times 4) + 4 = 8 + 4 \quad \leftarrow \text{(Add 4 to both sides.)}$$
$$8 + 4 = 8 + 4$$
$$12 = 12$$

Both sides of the equation remain equal.

2. Subtract 4 from both sides of the equation:

$$2 \times 4 = 8 \quad \leftarrow \text{(Original equation)}$$
$$(2 \times 4) - 4 = 8 - 4 \quad \leftarrow \text{(Subtract 4 from both sides.)}$$
$$8 - 4 = 8 - 4$$
$$4 = 4$$

Both sides of the equation remain equal.

3. Multiply both sides of the equation by 4:

$$2 \times 4 = 8 \quad \leftarrow \text{(Original equation)}$$
$$(2 \times 4) \times 4 = 8 \times 4 \quad \leftarrow \text{(Multiply both sides by 4.)}$$
$$8 \times 4 = 8 \times 4$$
$$32 = 32$$

Both sides of the equation remain equal.

4. Divide both sides of the equation by 4:

$$2 \times 4 = 8 \quad \leftarrow \text{(Original equation)}$$
$$(2 \times 4) \div 4 = 8 \div 4 \quad \leftarrow \text{(Divide both sides by 4.)}$$
$$\frac{2 \times 4}{4} = \frac{8}{4}$$
$$\frac{8}{4} = \frac{8}{4}$$
$$2 = 2$$

Both sides of the equation remain equal.

As you can see from the four preceding procedures, if you add, subtract, multiply, or divide both halves of an equation by the same number, the equality of the equation is preserved. In the next section we will see how these operations can serve some practical purpose.

2-1-2 *Transposition*

Transposition
The transfer of any term of an equation from one side over to the other side with a corresponding change of the sign.

It is important to know how to transpose, or rearrange, equations and formulas so that you can determine the unknown quantity. This process of rearranging, called **transposition,** is discussed in this section.

Transposing Equations

As an example, let us return to the original problem introduced at the beginning of this chapter and try to determine the value of the unknown quantity x.

$$\frac{8 \times x}{2} = 16$$

To transpose the equation we must follow two steps:

Step 1: Move the unknown quantity so that it is above the fraction bar on one side of the equal sign.

Step 2: Isolate the unknown quantity so that it stands by itself on one side of the equal sign.

Looking at the first step, let us see if our unknown quantity is above the fraction bar on either side of the equal sign.

The unknown quantity x is above the fraction bar. $\quad \frac{8 \times \textcircled{x}}{2} = 16$

Since step 1 is done, we can move on to step 2. Looking at the equation, you can see that x does not stand by itself on one side of the equal sign. To satisfy this step, we must somehow move the 8 above the fraction bar and the 2 below the fraction bar away from the left side of the equation, so that x is on its own.

Let us begin by removing the 2. To remove a letter or number from one side of a formula or equation, simply remember this rule: *To move a quantity, simply apply to both sides the arithmetic opposite of that quantity*. Multiplication is the opposite of division, so to remove a "divide by 2" (÷ 2), simply multiply both sides by 2 (× 2), as follows:

$$\frac{8 \times x}{2} = 16 \quad \leftarrow \text{(Original equation)}$$

$$\frac{8 \times x}{2} \textcircled{\times 2} = 16 \textcircled{\times 2} \quad \leftarrow \text{(Multiply both sides by 2.)}$$

$$\frac{8 \times x}{\cancel{2}} \times \frac{\cancel{2}}{1} = 16 \times 2 \quad \leftarrow \text{(Because } 2/2 = 1\text{, the two 2s on the left side of the equation cancel.)}$$

$$(8 \times x) \times 1 = 16 \times 2$$

$$8 \times x = 16 \times 2$$

(Because anything multiplied by 1 equals the same number, the 1 on the left side of the equation can be removed: $8x \times 1 = 8x$.)

Looking at the result so far, you can see that by multiplying both sides by 2, we effectively moved the 2 that was under the fraction bar on the left side to the right side of the equation above the fraction bar.

$$\frac{8 \times x}{2} = 16 \quad \leftarrow \text{(Original equation)}$$

$$8 \times x = 16 \times 2 \quad \leftarrow \text{(Result after both sides of equation were multiplied by 2)}$$

However, we have still not completed step 2, which was to isolate x on one side of the equal sign. To achieve this we need to remove the 8 from the left side of the equation. Once again we will do the opposite: Because the opposite of multiply is divide, to remove a "multiply by 8" we must divide both sides by 8 (÷ 8), as follows:

$$\frac{8 \times x}{2} = 16 \quad \leftarrow \text{(Original equation)}$$

$$8 \times x = 16 \times 2 \quad \leftarrow \text{(Equation after both sides were multiplied by 2)}$$

$$\frac{8 \times x}{8} = \frac{16 \times x}{8} \quad \leftarrow \text{(Divide both sides by 8.)}$$

$$\frac{\cancel{8} \times x}{\cancel{8}} = \frac{16 \times 2}{8} \quad \text{(Because } 8/8 = 1\text{, the two 8s on the left side of the equation cancel.)}$$

$$1 \times x = \frac{16 \times 2}{8}$$

$$x = \frac{16 \times 2}{8}$$

(Anything multiplied by 1 equals the same number, so the 1 on the left side of the equation can be removed: $1 \times x = x$.)

Now that we have completed step 2, which was to isolate the unknown quantity so that it stands by itself on one side of the equation, we can calculate the value of the unknown x by performing the arithmetic operations indicated on the right side of the equation.

$$x = \frac{16 \times 2}{8} \qquad (16 \times 2 = 32)$$

$$x = \frac{32}{8} \qquad (32 \div 8 = 4)$$

$$x = 4$$

To double-check this answer, let us insert this value into the original equation to see if it works.

$$\frac{8 \times x}{2} = 16 \qquad \text{(Replace } x \text{ with 4, or substitute 4 for } x\text{.)}$$

$$\frac{8 \times 4}{2} = 16 \qquad (8 \times 4 = 32)$$

$$\frac{32}{2} = 6 \qquad (32 \div 2 = 16)$$

$$16 = 16 \qquad \text{(Answer checks out because } 16 = 16.\text{)}$$

Transposing Formulas

As mentioned previously, *a formula is a general statement using letters and sometimes numbers that enables us to calculate the value of an unknown quantity*. Some of the simplest relationships are formulas. For example, to calculate the distance traveled, you can use the following formula, which was discussed earlier.

$$d = v \times t$$

where d = distance in miles
v = velocity in miles per hour
t = time in hours

Using this formula we can calculate distance if we know the speed and time. If we need only to calculate distance, this formula is fine. But what if we want to know what speed to go (v = ?) to travel a certain distance in a certain amount of time, or what if we want to calculate how long it will take for a trip (t = ?) of a certain distance at a certain speed? The answer is: Transpose or rearrange the formula so that we can solve for any of the quantities in the formula.

EXAMPLE:

As an example, calculate how long it will take to travel 250 miles when traveling at a speed of 50 miles per hour (mph). To solve this problem, let us place the values in their appropriate positions in the formula.

$$d = v \times t$$

$$250 \text{ mi} = 50 \text{ mph} \times t$$

The next step is to transpose just as we did before with equations to determine the value of the unknown quantity t.

Steps:

$$250 = 50 \times t \qquad \text{To isolate } t\text{, divide both sides by 50.}$$

$$\frac{250}{50} = \frac{\cancel{50} \times t}{\cancel{50}} \qquad 50 \div 50 = 1,\ 1 \times t = t$$

$$\frac{250}{50} = t \qquad 250 \div 50 = 5$$

$$t = 5$$

It will therefore take 5 hours to travel 250 miles at 50 miles per hour (250 mi = 50 mph × 5 h).

All we have to do, therefore, is transpose the original formula so that we can obtain formulas for calculating any of the three quantities (distance, speed, or time) if two values are known. This time, however, we will transpose the formula instead of the inserted values.

$$d = v \times t$$

Solve for v:

$$d = v \times t$$

$$\frac{d}{t} = \frac{v \times \not{t}}{\not{t}}$$

$$\frac{d}{t} = v$$

Steps:
To isolate v, divide both sides by t.
$t \div t = 1$,
$1 \times v = v$

$$\text{velocity} = \frac{\text{distance}}{\text{time}} \qquad v = \frac{d}{t}$$

Solve for t

$$d = v \times t$$

$$\frac{d}{v} = \frac{\not{v} \times t}{\not{v}}$$

$$\frac{d}{v} = t$$

Steps:
To isolate t, divide both sides by v.
$v \div v = 1$,
$1 \times t = t$

$$t = \frac{d}{t} \qquad \text{time} = \frac{\text{distance}}{\text{velocity}}$$

To test if all three formulas are correct, we can replace each letter with the values from the preceding example (also called "plugging in the values"), as follows:

$$d = v \times t$$

$$250 \text{ mi} = 50 \text{ mph} \times 5 \text{ h}$$

$$(250 = 250)$$

$$v = \frac{d}{t}$$

$$50 \text{ mph} = \frac{250 \text{ mi}}{5 \text{ h}}$$

$$(50 = 50)$$

$$t = \frac{d}{v}$$

$$5 \text{ h} = \frac{250 \text{ mi}}{50 \text{ mph}}$$

$$(5 - 5)$$

Directly Proportional and Inversely Proportional

The word **proportional** is used to describe a relationship between two quantities. For example, reconsider the formula

$$\text{time } (t) = \frac{\text{distance } (d)}{\text{velocity } (v)}$$

With this formula we can say that the time (t) it takes for a trip is directly proportional (symbolized by $\propto$) to the amount of distance (d) that needs to be traveled. In symbols, the relationship appears as follows:

$$t \propto d \qquad \text{(Time is directly proportional to distance.)}$$

The term **directly proportional** can therefore be used to describe the relationship between time and distance because an increase in distance will cause a corresponding increase in time ($d\uparrow$ causes $t\uparrow$), and similarly, a decrease in distance will cause a corresponding decrease in time ($d\downarrow$ causes $t\downarrow$).

Proportional
The relation of one part to another or to the whole with respect to magnitude, quantity, or degree.

Directly Proportional
The relation of one part to one or more other parts in which a change in one causes a similar change in the other.

EXAMPLE:

Using the previous example,

$$5\text{ h }(t) = \frac{250\text{ mi }(d)}{50\text{ mph }(v)}$$

let us try doubling the distance and then halving the distance to see what effect it has on time. If time and distance are directly proportional to each other, they should change in proportion.

Original example $\rightarrow$ $\text{time} = \frac{\text{distance}}{\text{velocity}} = \frac{250\text{ mi}}{50\text{ mph}} = 5\text{ h}$

Doubling the distance from 250 mi to 500 mi should double the time it takes for the trip, from 5 h to 10 h.

$$t = \frac{d}{v} = \frac{500\text{ mi}}{50\text{ mph}} = 10\text{ h}$$

Doubling the distance did double from the time ($d\uparrow$, $t\uparrow$) 5 h to 10 h.

Halving the distance from 250 mi to 125 mi should halve the time it takes for the trip, from 5 h to 2½ h.

$$t = \frac{d}{v} = \frac{125\text{ mi}}{50\text{ mph}} = 2.5\text{ h}$$

Halving the distance did halve the time ($d\downarrow$, $t\downarrow$).

As you can see from this exercise, time is directly proportional to distance ($t \propto d$). Therefore, *formula quantities are always directly proportional to one another when both quantities are above the fraction bars on opposite sides of the equal sign.*

Inversely Proportional

The relation of one part to one or more other parts in which a change in one causes an opposite change in the other.

On the other hand, **inversely proportional** means that the two quantities compared are opposite in effect. For example, consider again our time formula:

$$t = \frac{d}{v}$$

Observing the relationship between these quantities again, we can say that the time (t) it takes for a trip is *inversely proportional* (symbolized by $1/\propto$) to the speed or velocity (v) we travel. In symbols, the relationship appears as follows:

$$t \frac{1}{\propto} v \qquad \text{(Time is inversely proportional to velocity.)}$$

The term *inversely proportional* can therefore be used to describe the relationship between time and velocity because an increase in velocity will cause a corresponding decrease in time ($v\uparrow$ causes $t\downarrow$), and similarly, a decrease in velocity will cause a corresponding increase in time ($v\downarrow$ causes $t\uparrow$).

EXAMPLE:

Using the previous example,

$$5\text{ h }(t) = \frac{250\text{ mi }(d)}{50\text{ mph }(v)}$$

let us try doubling the speed we travel, or velocity, and then halving the velocity to see what effect it has on time. If time and velocity are inversely proportional to each other, a velocity change should have the opposite effect on time.

Original example → $\text{time} = \dfrac{\text{distance}}{\text{velocity}} = \dfrac{250 \text{ mi}}{50 \text{ mph}} = 5 \text{ h}$

Doubling the speed from 50 mph to 100 mph should halve the time it takes to complete the trip, from 5 h to 2½ h. → $t = \dfrac{d}{v} = \dfrac{250 \text{ mi}}{100 \text{ mph}} = 2.5 \text{ h}$

Doubling the speed did halve the time ($v \uparrow, t \downarrow$).

Halving the speed from 50 mph to 25 mph should double the time it takes to complete the trip, from 5 h to 10 h. → $t = \dfrac{d}{v} = \dfrac{250 \text{ mi}}{25 \text{ mph}} = 10 \text{ h}$

Halving the speed did double the time ($v \downarrow, t \uparrow$).

As you can see from this exercise, time is inversely proportional to velocity ($t \propto 1/v$). Therefore, *formula quantities are always inversely proportional to one another when one of the quantities is above the fraction bar and the other is below the fraction bar on opposite sides of the equal sign.*

2-1-3 *Substitution*

Substitution is a mathematical process used to develop alternative formulas by replacing or substituting one mathematical term with an equivalent mathematical term. To explain this process, let us try an example.

A **circle** is a perfectly round figure in which every point is at an equal distance from the center. Some of the elements of a circle are as follows.

a. The **circumference** (C) of a circle is the distance around the perimeter of the circle.
b. The **radius** (r) of a circle is a straight line drawn from the center of the circle to any point of the circumference.
c. The **diameter** (d) of a circle is a straight line drawn through the center of the circle to opposite sides of the circle. The diameter of a circle is therefore equal to twice the circle's radius ($d = 2 \times r$).

To review, *pi* (symbolized by π) is the name given to the ratio, or comparison, of a circle's circumference to its diameter. This constant will always be equal to approximately 3.14 because the circumference of any circle will always be about 3 times larger than the same circle's diameter. Stated mathematically:

$$\pi = \frac{\text{circumference } (C)}{\text{diameter } (d)} = 3.14$$

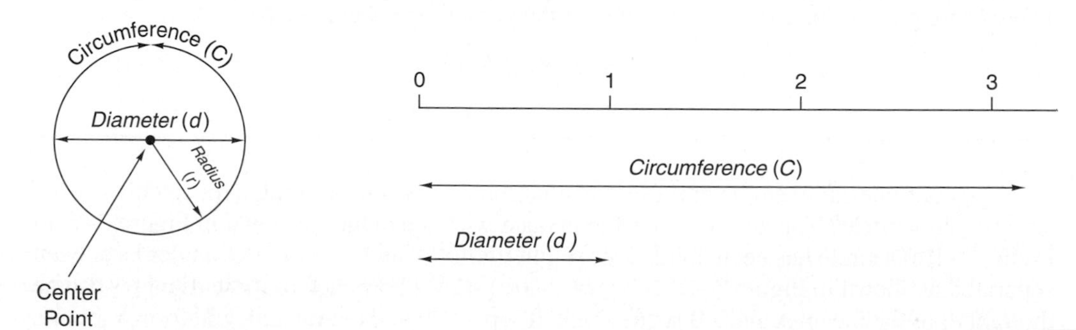

Substitution
The replacement of one mathematical entity by another of equal value.

Circle
A closed plane curve, every point of which is equidistant from a fixed point within the curve.

Circumference
The perimeter of a circle.

Radius
A line segment extending from the center of a circle or sphere to the circumference.

Diameter
The length of a straight line passing through the center of an object.

EXAMPLE:

Determine the circumference of a bicycle wheel that is 26 inches in diameter.

Solution:

To calculate the wheel's circumference, we will have to transpose the following formula to isolate C:

$$\pi = \frac{C}{d} \qquad \text{Multiply both sides by } d.$$

$$\pi \times d = \frac{C}{d} \times d \qquad d \div d = 1, \text{ and } C \times 1 = C.$$

$$\pi \times d = C$$

The circumference of the bicycle wheel is therefore

$$C = \pi \times d$$

$$C = 3.14 \times 26 \text{ in.} = 81.64 \text{ in.}$$

In some examples, only the circle's radius will be known. In situations like this, we can substitute the expression $\pi = C/d$ to arrive at an equivalent expression comparing a circle's circumference to its radius.

$$\pi = \frac{C}{d} \qquad \text{Because } d = 2 \times r, \text{ we can substitute } 2 \times r \text{ for } d.$$

$$\pi = \frac{C}{d} \qquad \text{Therefore, this is equivalent to } \pi = \frac{C}{2 \times r}.$$

EXAMPLE:

How much decorative edging will you need to encircle a garden that measures 6 feet from the center to the edge?

Solution:

In this example r is known, so to calculate the garden's circumference, we will have to transpose the following formula to isolate C:

$$\pi = \frac{C}{2 \times r} \qquad \text{Multiply both sides by } 2 \times r.$$

$$\pi \times (2 \times r) = \frac{C}{2 \times r} \times (2 \times r) \qquad (2 \times r) \div (2 \times r) = 1, \text{ and } C \times 1 = C.$$

$$\pi \times (2 \times r) = C$$

The edging needed for the circumference of the garden will therefore be

$$C = \pi \times (2 \times r)$$

$$C = 3.14 \times (2 \times 6 \text{ ft}) = 37.68 \text{ ft}$$

To continue our example of the circle, how could we obtain a formula for calculating the area within a circle? The answer is best explained with the diagram shown in Figure 2–1. In Figure 2–1(a) a circle has been divided into eight triangles, and then these triangles have been separated as shown in Figure 2–1(b). Figure 2–1(c) shows how each of these triangles can be thought of as having an A and a B section that, if separated and rearranged, will form a square.

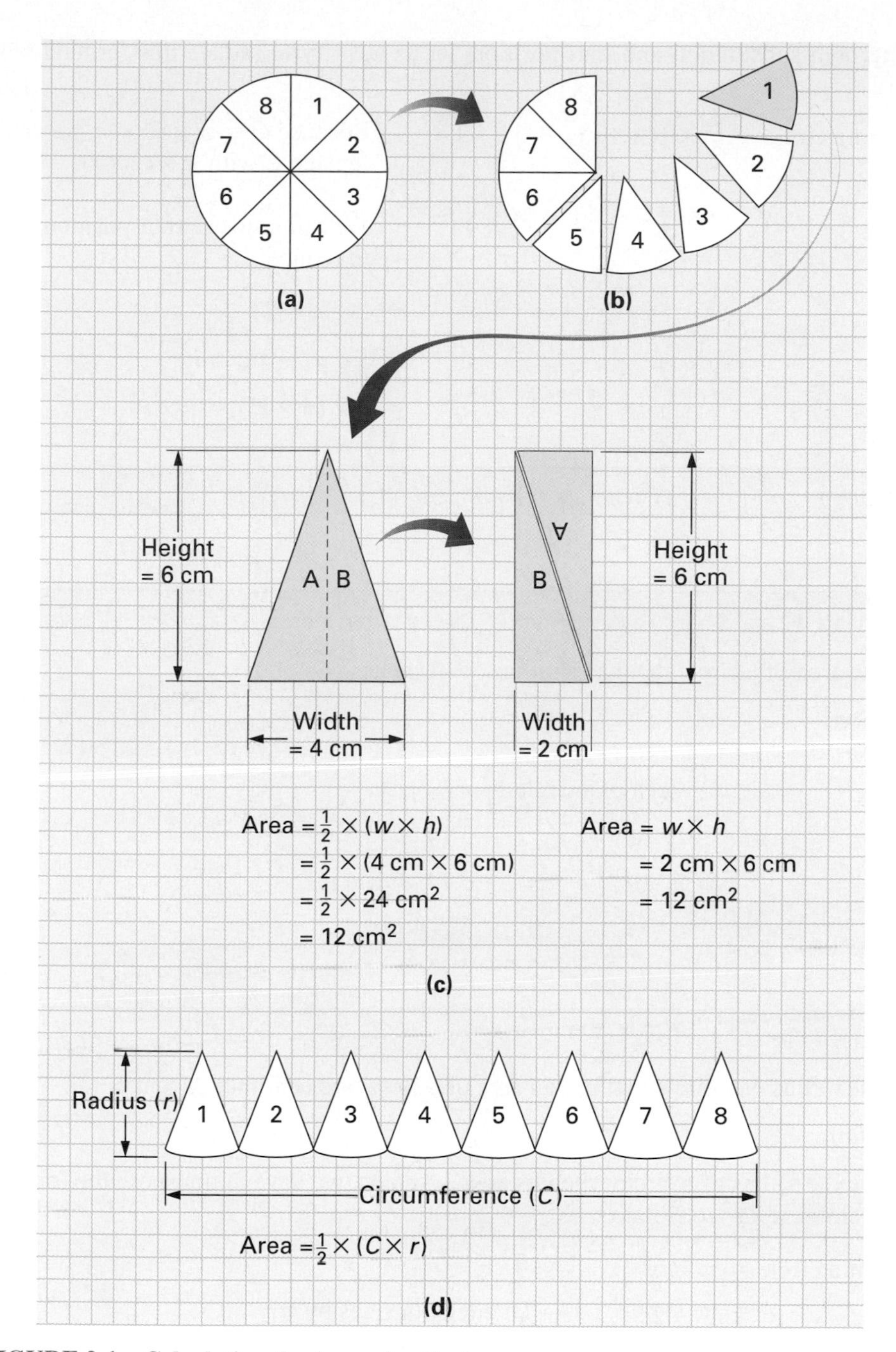

FIGURE 2-1 Calculating the Area of a Circle.

The area of a square is easy to calculate because it is simply the product of the square's width and height (area = $w \times h$). The area of the triangle shown in Figure 2–1(c), therefore, will be

$$\text{area} = \frac{1}{2} \times (w \times h)$$

Referring to Figure 2–1(d), you can see that the width of all eight triangles within a circle is equal to the circle's circumference ($w = C$), and the height of the triangles is equal to the circle's radius ($h = r$). The area of the circle, therefore, can be calculated with the following formula:

$$\text{area of a circle} = \frac{1}{2} \times (C \times r)$$

Using both substitution and transposition, let us now try to reduce this formula with $C = \pi \times (2 \times r)$.

$$\text{area of a circle} = \frac{1}{2} \times (C \times r)$$ ← Because $C = \pi \times (2 \times r)$, we can replace C with $\pi \times (2 \times r)$.

$$= \frac{1}{2} \times [\pi \times (2 \times r) \times r]$$ ← We can now use transposition to reduce.

$$= \frac{1 \times \pi \times 2 \times r \times r}{2}$$ ← $2 \div 2 = 1$

$$= 1 \times \pi \times r \times r$$ ← $r \times r = r^2$

$$= 1 \times \pi \times r^2$$ ← $1 \times (\pi \times r^2) = \pi \times r^2$

Area of a circle $= \pi \times r^2$

EXAMPLE:

Calculate the area of a circle that has a radius of 14 inches.

Solution:

$$\begin{aligned}\text{area of a circle} &= \pi \times r^2 \\ &= 3.14 \times (14 \text{ in.})^2 \\ &= 3.14 \times 196 \text{ in.}^2 \\ &= 615.4 \text{ in.}^2\end{aligned}$$

EXAMPLE:

Calculate the radius of a circle that has an area of 624 square centimeters (cm^2).

Solution:

To determine the radius of a circle when its area is known, we have to transpose the formula area of a circle $= \pi \times r^2$ to isolate r.

$$\text{area of a circle} = \pi \times r^2$$

$$A = \pi \times r^2$$ ← Divide both sides by π.

$$\frac{A}{\pi} = \frac{\pi \times r^2}{\pi}$$ ← $\pi \div \pi = 1$, $1 \times r^2 = r^2$

$$\frac{A}{\pi} = r^2$$ ← Take the square root of both sides.

$$\sqrt{\frac{A}{\pi}} = \sqrt{r^2}$$ $\sqrt{r^2} = r$

$$\sqrt{\frac{A}{\pi}} = r$$

Now we can calculate the radius of a circle that has an area of 624 cm^2:

$$r = \sqrt{\frac{A}{\pi}} = \sqrt{\frac{624}{3.14}} = \sqrt{198.7} = 14.1 \text{ cm}$$

SELF-TEST EVALUATION POINT FOR SECTION 2–1

Now that you have completed this section, you should be able to:

Objective 1. Define the terms *algebra, equation, formula,* and *literal numbers.*

Objective 2. Describe the rules regarding the equality on both sides of the equal sign.

Objective 3. Show how the equality of an equation is preserved if the same arithmetic operation is performed on both sides of an equation or formula.

Objective 4. Explain how equations can be rearranged and still maintain their equality.

Objective 5. Demonstrate the process of transposing a formula.

Objective 6. Describe how the terms *directly proportional* and *inversely proportional* are used to show the relationship between quantitites in a formula.

Objective 7. Demonstrate how to use substitution to develop alternative fomulas.

Use the following questions to test your understanding of Section 2–1:

1. Is the following equation true?

$$\frac{\frac{56}{14}}{2} = 2 \quad \text{or} \quad \frac{56 \div 14}{2} = 2$$

2. Fill in the missing values.

$$\frac{144}{12} \times \square = \frac{36}{6} \times 2 \times \square$$
$$60 = 60$$

3. Is the following equation balanced?

$$3.2 \text{ km} = 3200 \text{ m}$$

4. Transpose the following equations to determine the unknown quantity:

a. $x + 14 = 30$

b. $8 \times x = \frac{80 - 40}{10} \times 12$

c. $y - 4 = 8$

d. $(x \times 3) - 2 = \frac{26}{2}$

e. $x^2 + 5 = 14$

f. $2(3 + 4x) = (2x + 13)$

5. Transpose the following formulas:

a. $x + y = z, y = ?$

b. $Q = C \times V, C = ?$

c. $X_L = 2 \times \pi \times f \times L, L = ?$

d. $V = I \times R, R = ?$

6. Determine the unknown in the following equations by using transposition:

a. $I^2 = 9$ **b.** $\sqrt{z} = 8$

7. Use substitution to calculate the value of the unknown.

$$x = y \times z \quad \text{and} \quad a = x \times y$$

Calculate a if $y = 14$ and $z = 5$.

2-2 RESISTANCE

By definition, *resistance* is the opposition to current flow accompanied by the dissipation of heat. To help explain the concept of resistance, Figure 2–2 compares fluid opposition to electrical opposition.

In the fluid circuit in Figure 2–2(a), a valve has been opened almost completely, so a very small opposition to the water flow exists within the pipe. This small or low resistance within the pipe will not offer much opposition to water flow, so a large amount of water will flow through the pipe and gush from the outlet. In the electrical circuit in Figure 2–2(b), a small-value **resistor** (which is symbolized by the zigzag line) has been placed in the circuit. The resistor is a device that is included within electrical and electronic circuits to oppose current flow by introducing a certain value of circuit **resistance**. In this circuit, a small-value resistor has been included to provide very little resistance to the passage of current flow. This low resistance or small opposition will allow a large amount of current to flow through the conductor, as illustrated by the thick current line.

Resistor
Device constructed of a material that opposes the flow of current by having some value of resistance.

Resistance
Symbolized R and measured in ohms (Ω), it is the opposition to current flow with the dissipation of energy in the form of heat.

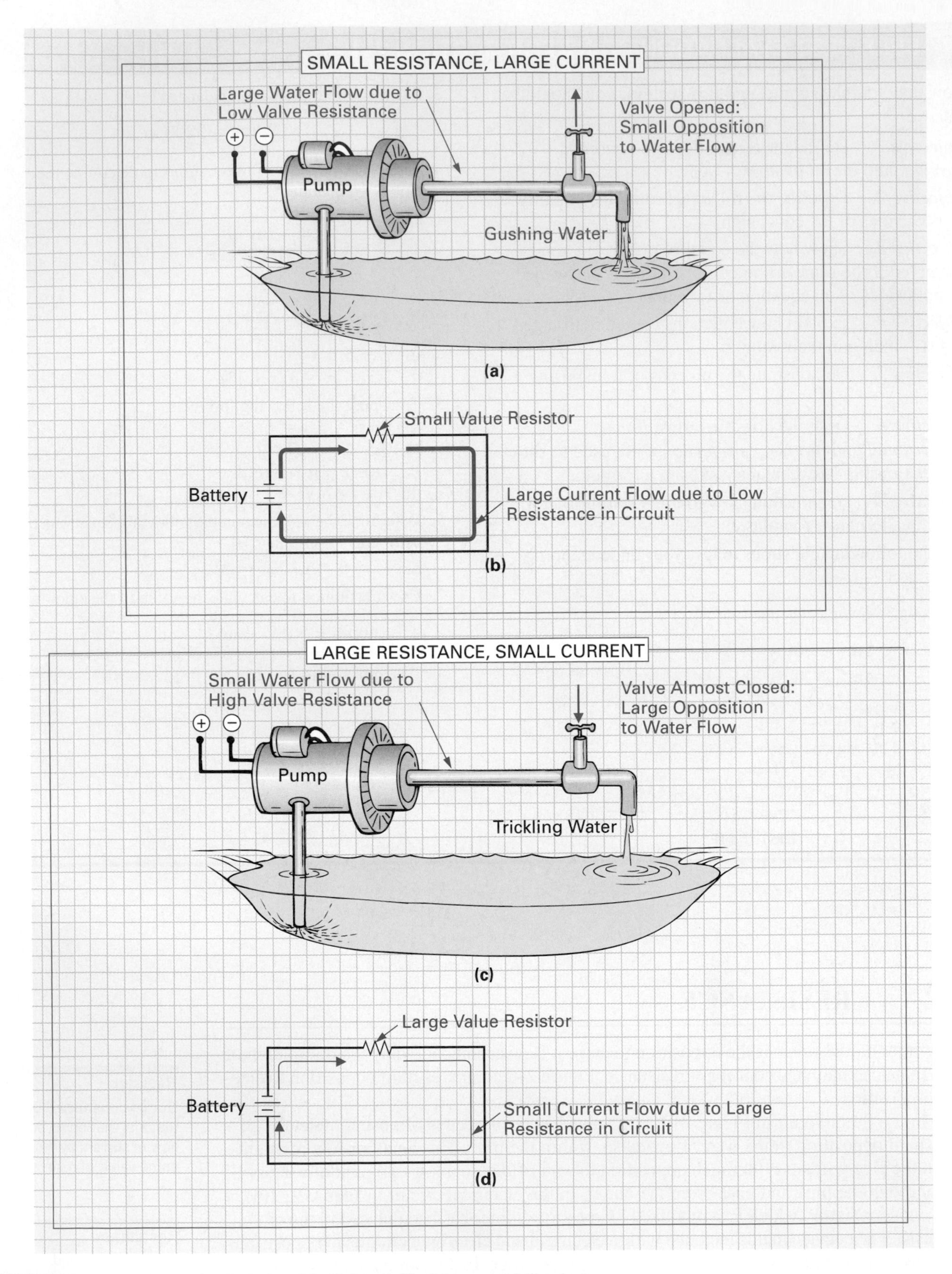

FIGURE 2-2 The Opposite Relationship Between Resistance and Current.

In the fluid circuit in Figure 2–2(c), the valve is almost completely closed, resulting in a high resistance or opposition to water flow, so only a trickle of water passes through the pipe and out from the outlet. In the electrical circuit in Figure 2–2(d), a large-value resistor has been placed in the circuit, causing a large resistance to the passage of current. This high resistance allows only a small amount of current to flow through the connecting wire, as illustrated by the thin current line.

In the previous examples of low resistance and high resistance shown in Figure 2–2, you may have noticed that resistance and current are inversely proportional ($1/\propto$) to one another. This means that if resistance is high the current is low, and if resistance is low, the current is high.

$$R \frac{1}{\propto} I$$

Resistance (R) is Inversely Proportional $\left(\frac{1}{\propto}\right)$ to Current (I)

This relationship between resistance and current means that if resistance is increased by some value, current will be decreased by the same value (assuming a constant electrical pressure or voltage). For example, if resistance is doubled current is halved and similarly, if resistance is halved current is doubled.

$$R\uparrow \frac{1}{\propto} I\downarrow \qquad R\downarrow \frac{1}{\propto} I\uparrow$$

In Figure 2–2, the fluid analogy was used alongside the electrical circuit to help you understand the idea of low resistance and high resistance. In the following section, we will examine resistance in more detail and define clearly exactly how much resistance exists within a circuit.

TIME LINE

Ohm's law, the best known law in electrical circuits, was formulated by Georg S. Ohm (1787–1854), a German physicist. His law was so coldly received that his feelings were hurt and he resigned his teaching post. When his law was finally recognized, he was reinstated. In honor of his accomplishments, the unit of resistance is called the ohm.

2-2-1 *The Ohm*

As discussed in the previous chapter, current is measured in amperes and voltage is measured in volts. Resistance is measured in **ohms**, in honor of Georg Simon Ohm, who was the first to formulate the relationship between current, voltage, and resistance.

Ohm
Unit of resistance, symbolized by the Greek capital letter omega (Ω).

The larger the resistance, the larger the value of ohms and the more the resistor will oppose current flow. The ohm is given the symbol Ω, which is the Greek capital letter omega. By definition, 1 ohm is the value of resistance that will allow 1 ampere of current to flow through a circuit when a voltage of 1 volt is applied, as shown in Figure 2–3(a), where the resistor is drawn as a zigzag. In some schematics or circuit diagrams the resistor is drawn as a rectangular block, as shown in Figure 2–3(b). The pictorial of this circuit is shown in Figure 2–3(c).

The circuit in Figure 2–4 reinforces our understanding of the ohm. In this circuit, a 1 V battery is connected across a resistor, whose resistance can be either increased or decreased. As the resistance in the circuit is increased the current will decrease and, conversely, as the resistance of the resistor is decreased the circuit current will increase. So how much is 1 ohm of resistance? The answer to this can be explained by adjusting the resistance of the variable resistor. If the resistor is adjusted until exactly 1 amp of current is flowing around the circuit, the value of resistance offered by the resistor is referred to as 1 ohm (Ω).

2-2-2 *Ohm's Law*

Current flows in a circuit due to the electrical force or voltage applied. The amount of current flow in a circuit, however, is limited by the amount of resistance in the circuit. It can be said,

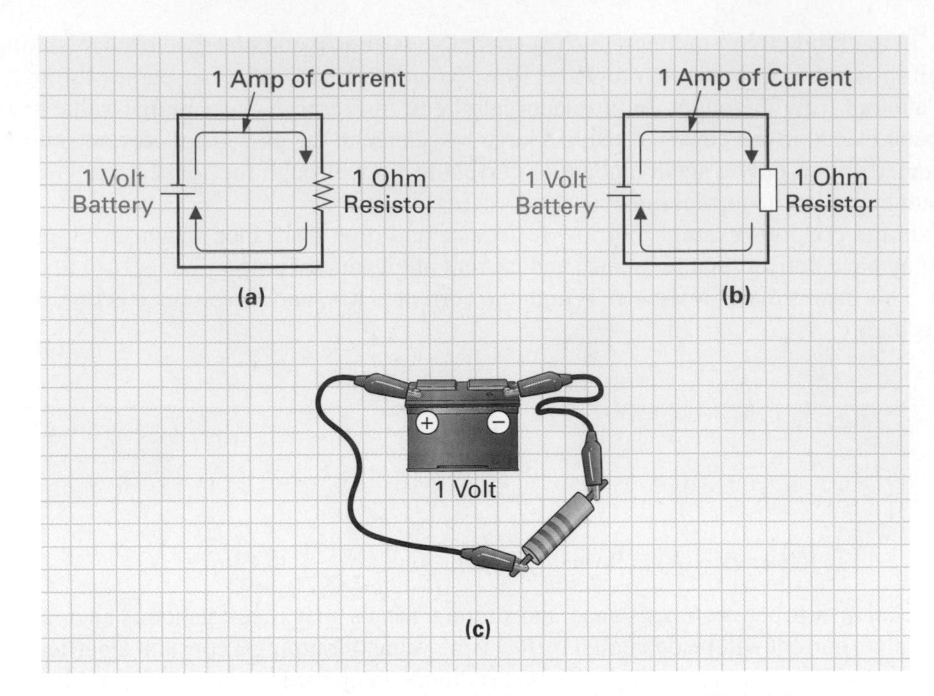

FIGURE 2-3 **One Ohm. (a) New and (b) Old Resistor Symbols. (c) Pictorial.**

therefore, that the amount of current flow around a circuit is dependent on both voltage and resistance. This relationship among the three electrical properties of current, voltage, and resistance was discovered by Georg Simon Ohm, a German physicist, in 1827. Published originally in 1826, **Ohm's law** states that the current flow in a circuit is directly proportional ($\propto$) to the source voltage applied and inversely proportional ($1/\propto$) to the resistance of the circuit.

Ohm's Law
Relationship among the three electrical properties of voltage, current, and resistance. Ohm's law states that the current flow within a circuit is directly proportional to the voltage applied across the circuit and inversely proportional to its resistance.

Stated in mathematical form, Ohm arrived at this formula:

$$\text{current } (I) = \frac{\text{voltage } (V)}{\text{resistance } (R)}$$

where current $(I) \propto$ voltage (V)

$$\text{current } (I) \frac{1}{\propto} \text{ resistance } (R)$$

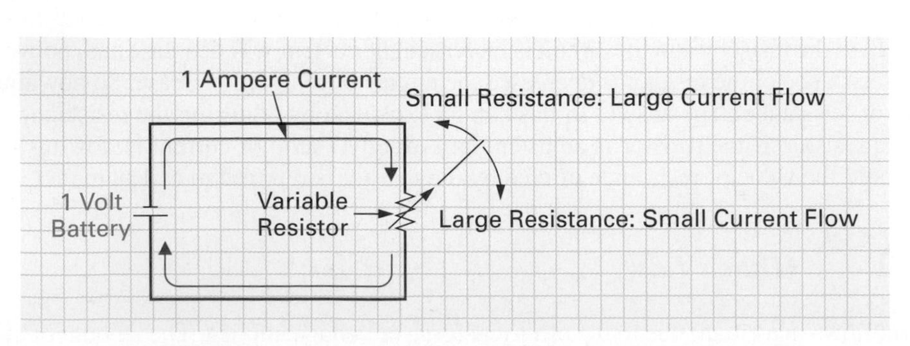

FIGURE 2-4 **1 Ohm Allows 1 Amp to Flow When 1 Volt Is Applied.**

1. Current Is Proportional to Voltage ($I \propto V$)

An increase in voltage (electron moving force), therefore, will exert a greater pressure on the circuit electrons and cause an increase in current flow. This point is reinforced with the Ohm's law formula:

$$I\uparrow = \frac{V\uparrow}{R}$$

Both properties are above the line and are therefore proportional.

In the two following examples, you will see how an increase in the voltage applied to a circuit results in a proportional increase in the circuit current.

EXAMPLE:

If the resistance in the circuit in Figure 2–5(a) remains constant at 1 Ω and the applied voltage equals 2 V, what is the value of current flowing through the circuit?

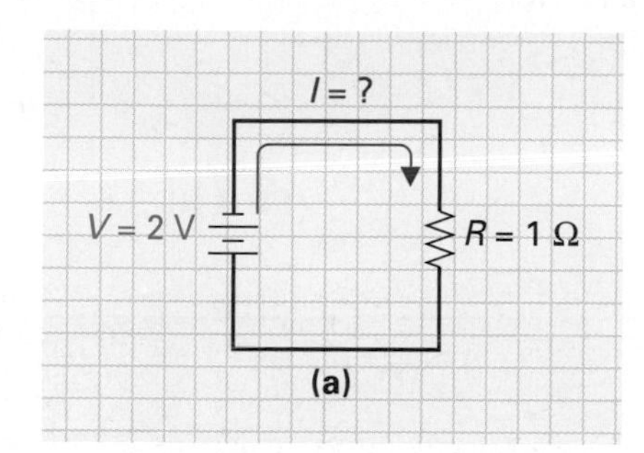

FIGURE 2-5

Solution:

$$\text{current } (I) = \frac{\text{voltage } (V)}{\text{resistance } (R)} = \frac{2\text{ V}}{1\ \Omega} = 2\text{ A}$$

Calculator Sequence

STEP	KEYPAD ENTRY	DISPLAY RESPONSE
1.	2	2
2.	÷	
3.	1	1
4.	=	2

EXAMPLE:

If the voltage from the previous example is now doubled to 4 V, as shown in Figure 2–5(b), what would be the change in current?

Solution:

$$\text{current } (I) = \frac{\text{voltage } (V)}{\text{resistance } (R)} = \frac{4\text{ V}}{1\ \Omega} = 4\text{ A}$$

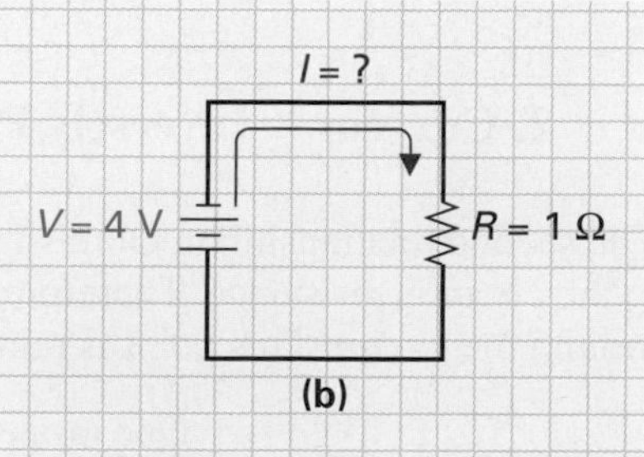

On the other hand, a decrease in the circuit's applied voltage will cause a decrease in circuit current. This is again reinforced by Ohm's law.

$$I\downarrow = \frac{V\downarrow}{R}$$

In the following two examples, you will see how a decrease in the applied voltage will result in a proportional decrease in circuit current.

EXAMPLE:

If the circuit shown in Figure 2–6 has a resistance of 2 Ω and an applied voltage of 8 V, what would be the circuit current?

Solution:

$$\text{current } (I) = \frac{\text{voltage } (V)}{\text{resistance } (R)} = \frac{8\text{ V}}{2\ \Omega} = 4\text{ A}$$

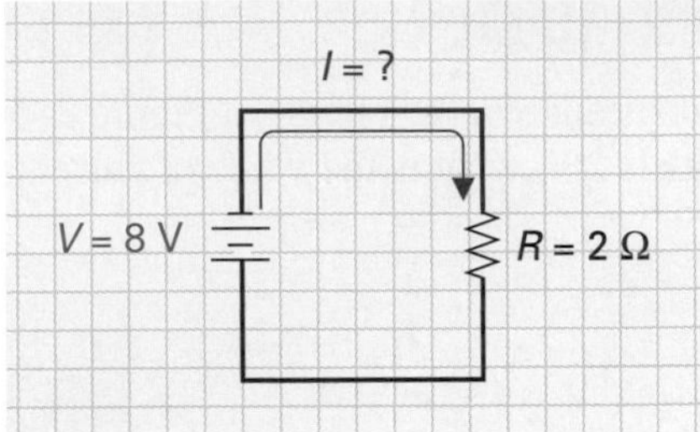

FIGURE 2-6

EXAMPLE:

If the circuit described in the previous example were to have its applied voltage halved to 4 V, what would be the change in circuit current?

Solution:

$$\text{current } (I) = \frac{\text{voltage } (V)}{\text{resistance } (R)} = \frac{4\text{ V}}{2\ \Omega} = 2\text{ A}$$

To summarize the relationship between circuit voltage and current, we can say that if the voltage were to double, the current within the circuit would also double (assuming the circuit's resistance were to remain at the same value). Similarly, if the voltage were halved, the current would also halve, making the two properties directly proportional to one another.

2. Current Is Inversely Proportional to Resistance $\left(\mathbf{I} \frac{1}{\propto} R\right)$

The second relationship in Ohm's law states that the circuit current is inversely proportional to the circuit's resistance. Therefore, a small resistance ($R\downarrow$) allows a large current ($I\uparrow$) flow around the circuit. This point is reinforced with Ohm's law:

$$I\uparrow = \frac{V}{R\downarrow}$$

One property is above the line, the other is below the line, and therefore the two are inversely proportional.

Conversely, a large resistance ($R\uparrow$) will always result in a small current ($I\downarrow$).

$$I\downarrow = \frac{V}{R\uparrow}$$

To help reinforce this relationship, let us look at a couple of examples.

EXAMPLE:

Figure 2–7 shows a circuit that has a constant 8 volts applied. If the resistance in this circuit were doubled from 2 ohms to 4 ohms, what would happen to the circuit current?

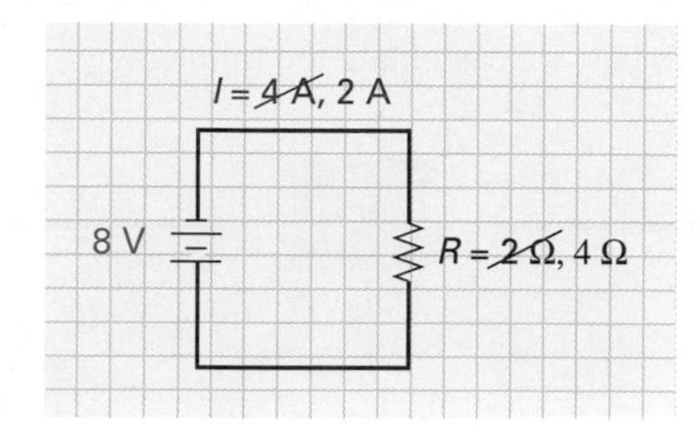

FIGURE 2-7

Solution:

If the resistance in Figure 2–7 were doubled from 2 Ω to 4 Ω, the circuit current would halve from 4 A to

$$\text{current } (I) = \frac{\text{voltage } (V)}{\text{resistance } (R)} = \frac{8 \text{ V}}{4\ \Omega} = 2 \text{ A}$$

EXAMPLE:

If the resistance in Figure 2–7 were returned to its original value of 2 Ω, what would happen to the circuit current?

Solution:

If the resistance in Figure 2–7 were halved from 4 Ω to 2 Ω, the circuit current would double from 2 A to

$$\text{current } (I) = \frac{\text{voltage } (V)}{\text{resistance } (R)} = \frac{8 \text{ V}}{2\ \Omega} = 4 \text{ A}$$

To summarize the relationship between circuit current and resistance, we can say that if the resistance were to double, the current within the circuit would be halved (assuming the circuit's voltage were to remain at the same value). Similarly, if the circuit resistance were halved, the circuit current would double, confirming that current is inversely proportional to resistance.

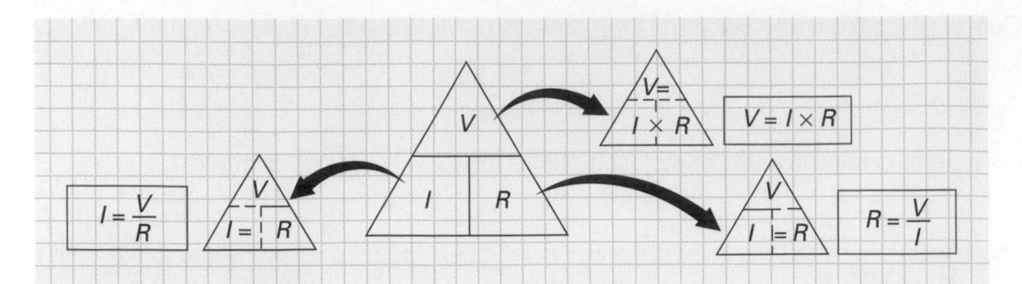

FIGURE 2-8 **Ohm's Law Triangle.**

3. The Three Forms of Ohm's Law

By transposing Ohm's law, we can obtain the following:

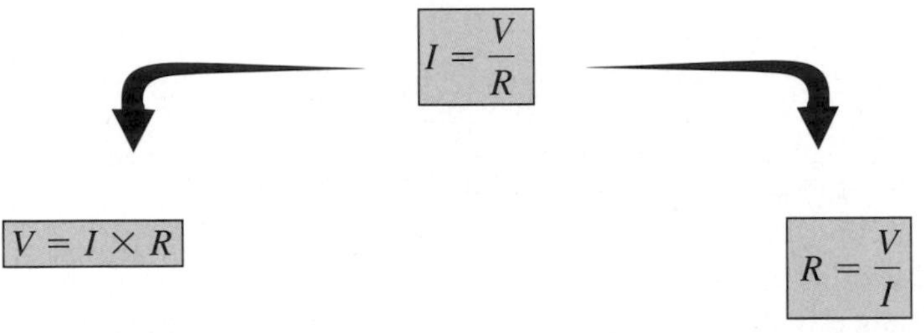

$$I = \frac{V}{R}$$

$$V = I \times R$$

This formula can be used to calculate V when I and R are known.

$$R = \frac{V}{I}$$

This formula can be used to calculate R when V and I are known.

Figure 2–8 shows how these three Ohm's law formulas can be placed within a triangle so that their relative proximity to one another can be used for easy memory recall.

EXAMPLE:

List the formula to use, and determine the unknown for the three examples shown in Figure 2–9.

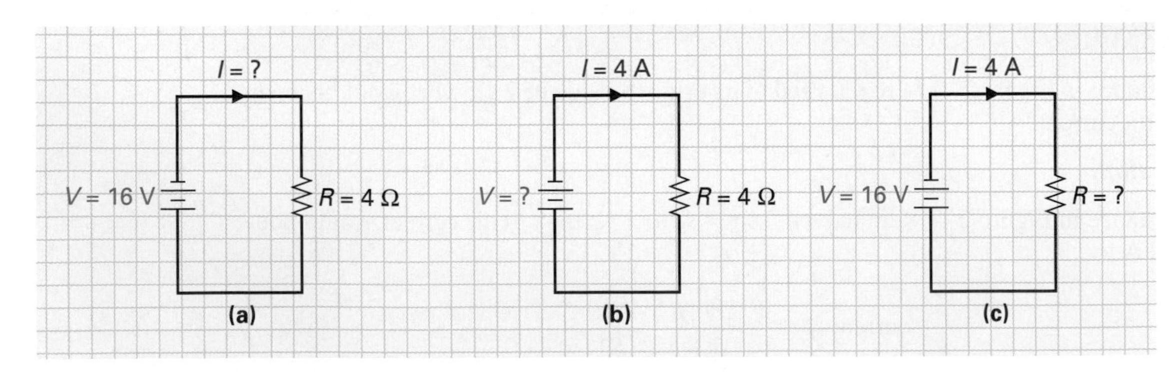

FIGURE 2-9 **Using Ohm's Law to Determine the Unknown.**

Solution:

a. In Figure 2–9(a), current (I) has to be determined when voltage (V) and resistance (R) are known. The Ohm's law formula for current should therefore be used as follows:

$$I = \frac{V}{R} = \frac{16\ V}{4\ \Omega} = 4\ \text{A}$$

b. In Figure 2–9(b), voltage (V) has to be determined when current (I) and resistance (R) are known. The Ohm's law formula for voltage should therefore be used as follows:

$$V = I \times R = 4\text{ A} \times 4\ \Omega = 16\text{ V}$$

c. In Figure 2–9(c), resistance (R) has to be determined when current (I) and voltage (V) are known. The Ohm's law formula for resistance should therefore be used as follows:

$$R = \frac{V}{I} = \frac{16\text{ V}}{4\text{ A}} = 4\ \Omega$$

4. Source and Load

Figure 2–10(a) illustrates a circuit with a battery connected across a lightbulb. In this circuit, the potential (static) energy of the battery (voltage) produces kinetic (dynamic or motion) energy (current) that is used to produce light energy from the bulb. The battery is the **source** in this circuit and the bulb is the **load**. By definition, a load is a device that absorbs the energy being supplied and converts it into the desired form. The **load resistance** will determine how hard the voltage source has to work. For example, the bulb has a resistance of 300 Ω. Consequently, a 300 Ω load resistance will permit 10 mA of **load current** to flow when a 3 V source is connected across the bulb ($I = V/R = 3\text{ V}/300\ \Omega = 10\text{ mA}$). If you were to change the load resistance by using a different bulb, the circuit current would be changed. To be specific, if the load resistance is increased, the load current decreases ($I\downarrow = V/R\uparrow$). Conversely, if the load resistance is decreased, the load current increases ($I\uparrow = V/R\downarrow$), assuming, of course, that the voltage source remains constant.

Figure 2–11 illustrates a heater, lightbulb, motor, computer, robot, television, and microwave oven. All these and all other devices or pieces of equipment are connected to some source, such as a battery or power supply. This supply or source does not see all the circuitry and internal workings of the equipment; it simply sees the whole device or piece of equipment as having some value of resistance. The resistance of the equipment is referred to as the device's *load resistance,* and it is this value of load resistance and the value of *source voltage* that determines the value of *load current.*

Source
Device that supplies the signal power or electric energy to a load.

Load
A component, circuit, or piece of equipment connected to the source. The load resistance will determine the load current.

Load Resistance
The resistance of the load.

Load Current
The current that is present in the load.

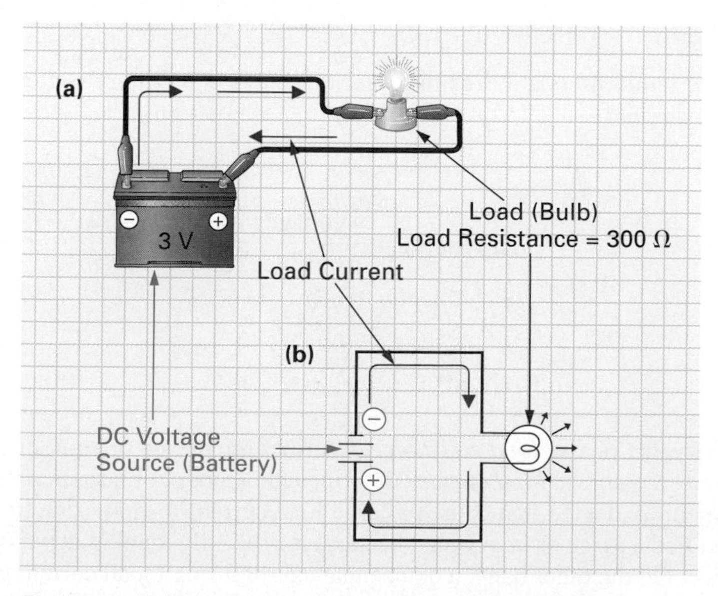

FIGURE 2-10 **Load on a Voltage Source. (a) Pictorial. (b) Schematic.**

FIGURE 2-11 **(a) Load Resistance of Equipment. (b) Schematic Symbols and Terms.**

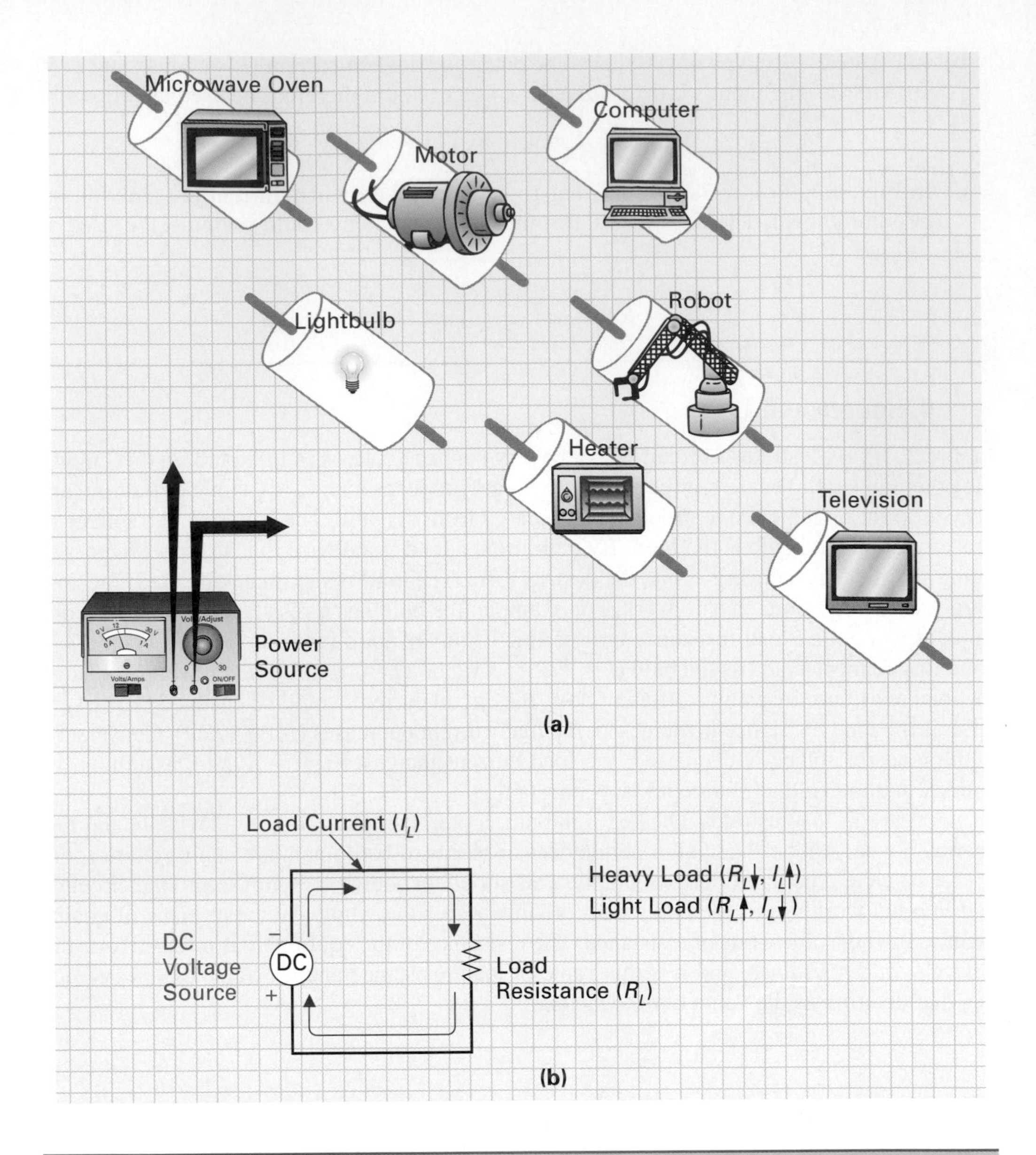

EXAMPLE:

A small portable radio offers a 390 Ω load resistance to a 9 V battery source. What current will flow in this circuit?

Solution:

$$\text{current } (I) = \frac{\text{voltage } (V)}{\text{resistance } (R)} = \frac{9\text{ V}}{390\ \Omega} = 23\text{ mA}$$

In summary, the phrase *load resistance* describes the device or equipment's circuit resistance, whereas the phrase *load current* describes the amount of current drawn by the device or equipment. A device that causes a large load current to flow (due to its small load resistance) is called a *large* or *heavy* load because it is heavily loading down or working the supply or source. On the other hand, a device that causes a small load current to flow (due to its large load resistance) is referred to as a *small* or *light* load because it is only lightly loading down or working the supply or source. It can be said, therefore, that load current and load resistance are inversely proportional to one another.

2-2-3 *How Is Resistance Measured?*

Resistance is measured with an **ohmmeter**. Stepping through the procedure described in Figure 2–12, you will see how the ohmmeter is used to measure the resistance of a resistor.

Ohmmeter
Measurement device used to measure electric resistance.

2-2-4 *Protoboards*

The solderless prototyping board (**protoboard**) or breadboard is designed to accommodate experiments. This protoboard will hold and interconnect resistors, lamps, switches and many other components, as well as provide a connecting point for electrical power. Figure 2–13(a) shows an experimental circuit wired up on a protoboard.

Protoboard
An experimental arrangement of a circuit on a board. Also called a breadboard.

Figure 2–13(b) shows the top view of a basic protoboard. As you can see by the cross section on the right side, electrical connector strips are within the protoboard. These conductive strips make a connection between the five hole groups. The bus strips have an electrical connector strip running from end to end, as shown in the cross section. They are usually connected to a power supply, as seen in the example circuit in Figure 2–13(a). In this circuit, you can see that the positive supply voltage is connected to the upper bus strip, and the negative supply (ground) is connected to the lower bus strip. These power supply "rails" can then be connected to a circuit formed on the protoboard with hookup wire, as shown in Figure 2–14(a). In this example, three resistors are connected end-to-end, as shown in the schematic diagram in Figure 2–14(b).

Figure 2–15 illustrates how the multimeter could be used to make voltage, current, and resistance measurements of a circuit constructed on the protoboard.

Other protoboards may vary slightly as far as layout, but you should be able to determine the pattern of conductive strips by making a few resistance checks with an ohmmeter.

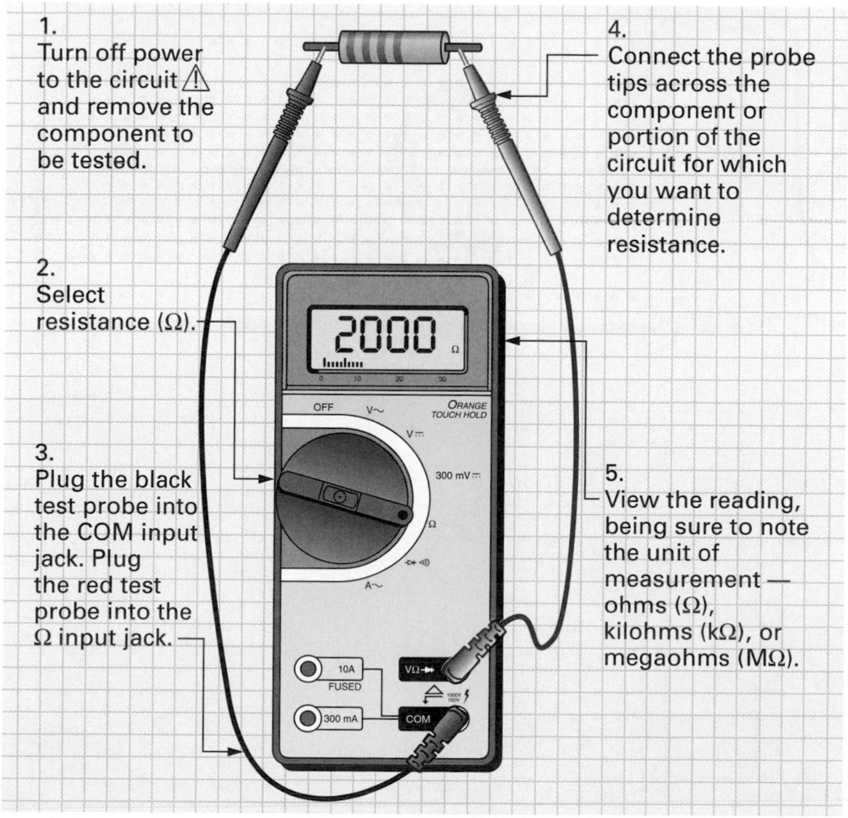

1. Turn off power to the circuit ⚠ and remove the component to be tested.

2. Select resistance (Ω).

3. Plug the black test probe into the COM input jack. Plug the red test probe into the Ω input jack.

4. Connect the probe tips across the component or portion of the circuit for which you want to determine resistance.

5. View the reading, being sure to note the unit of measurement—ohms (Ω), kilohms (kΩ), or megaohms (MΩ).

FIGURE 2-12
Measuring Resistance with the Ohmmeter.

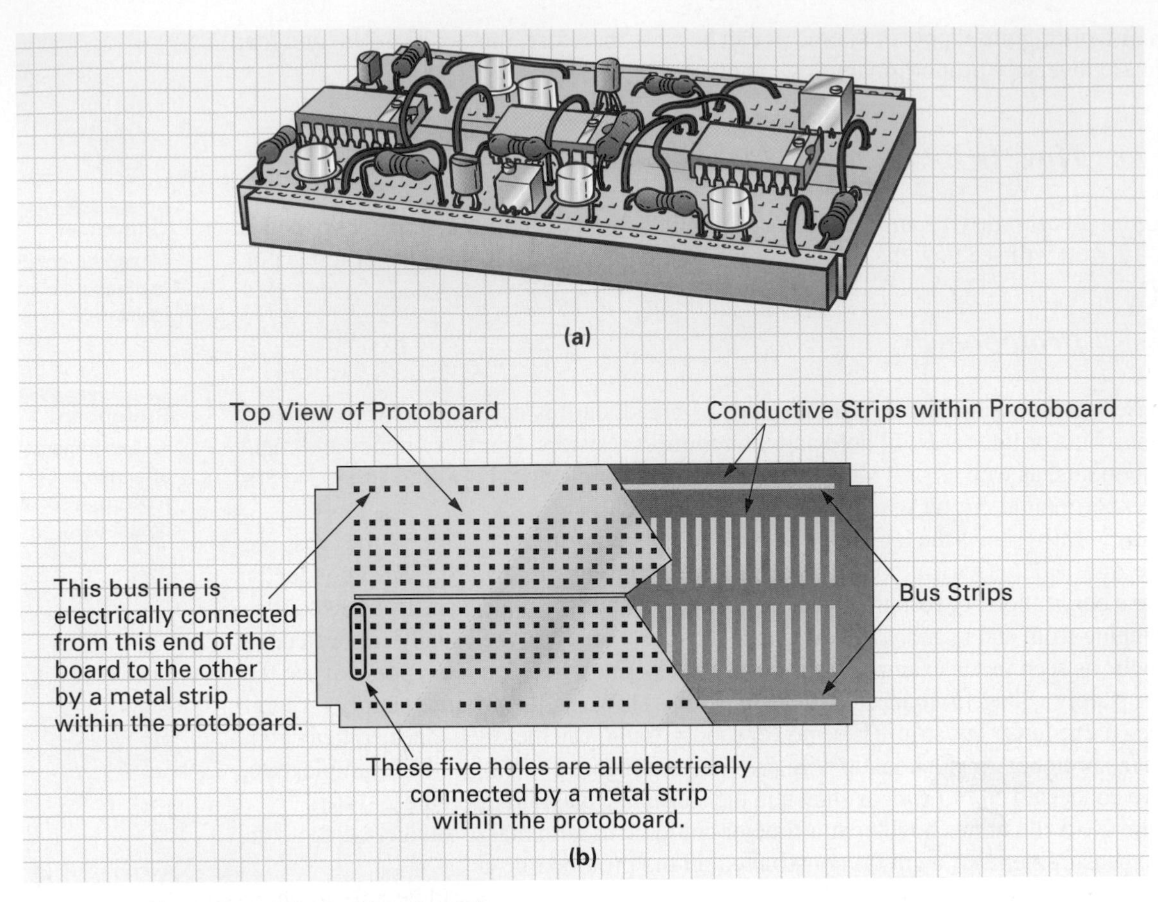

FIGURE 2-13 Experimenting with the Protoboard.

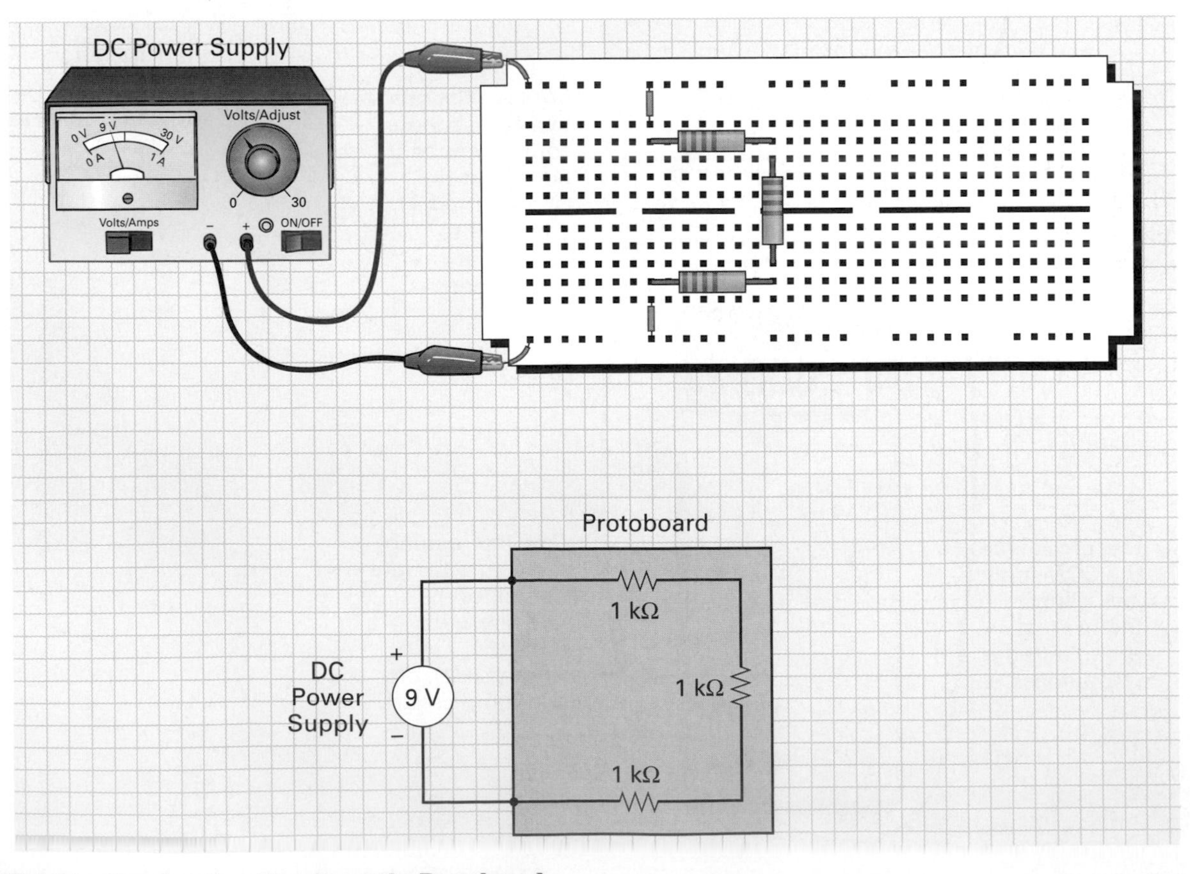

FIGURE 2-14 Constructing Circuits on the Protoboard.

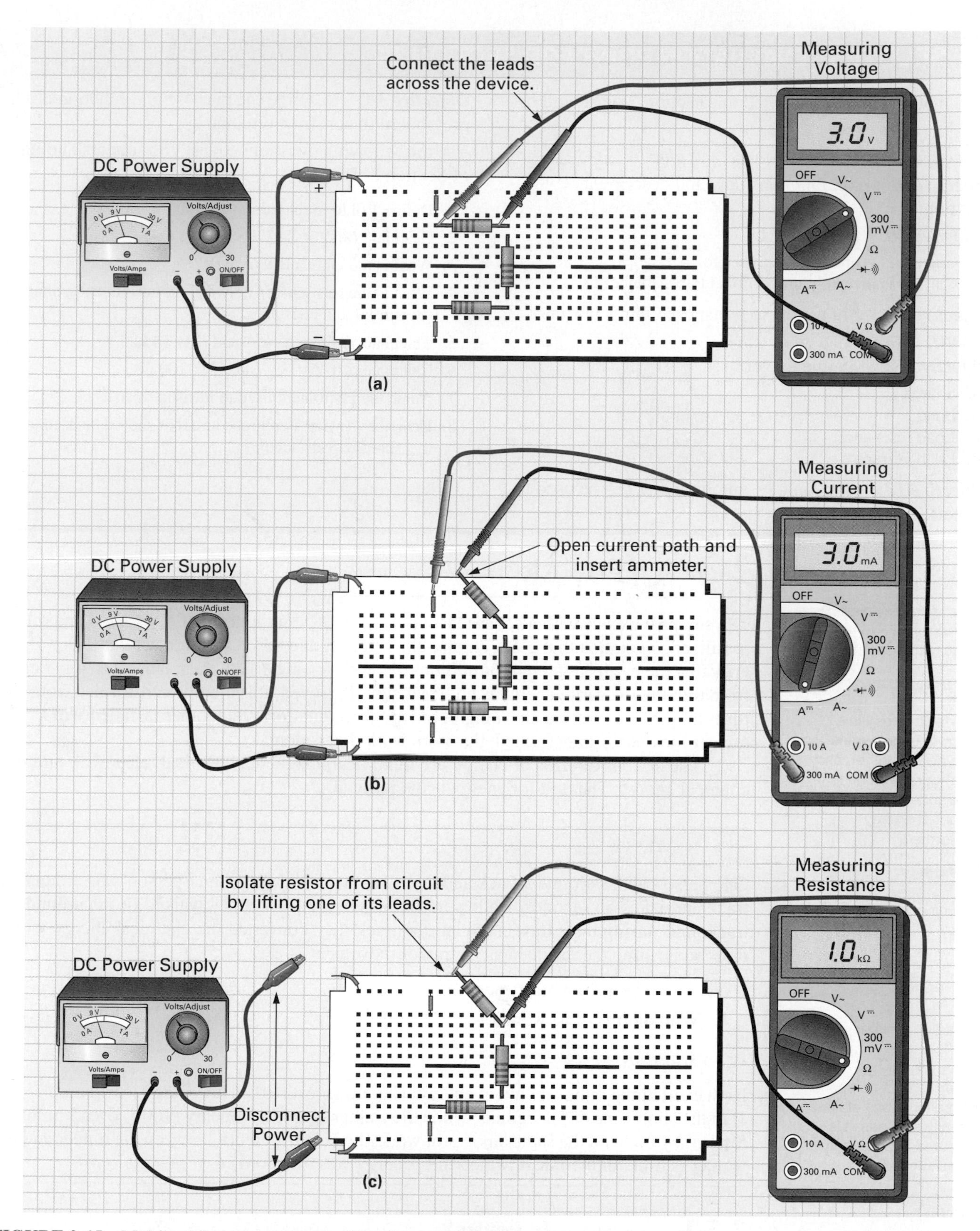

FIGURE 2-15 Making Measurements of a Circuit on the Protoboard.

SELF-TEST EVALUATION POINT FOR SECTION 2-2

Now that you have completed this section, you should be able to:

Objective 8. Define *resistance* and *ohm.*

Objective 9. Explain Ohm's law and describe its function.

Objective 10. Describe why:

a. Current is proportional to voltage.
b. Current is inversely proportional to resistance.

Objective 11. Show how to transpose Ohm's law into its three forms.

Objective 12. Define the terms:

a. Source
b. Load
c. Load resistance
d. Load current

Objective 13. Explain how an ohmmeter can be used to measure resistance.

Objective 14. Describe the purpose and function of the protoboard to construct circuits.

Use the following questions to test your understanding of Section 2–2:

1. Define 1 ohm in relation to current and voltage.
2. Calculate I if $V = 24$ V and $R = 6\ \Omega$.
3. What is the Ohm's law triangle?
4. What is the relationship between (a) current and voltage; (b) current and resistance?
5. Calculate V if $I = 25$ mA and $R = 1$ kΩ.
6. Calculate R if $V = 12$ V and $I = 100\ \mu$A.
7. Name the instrument used to measure resistance.
8. Describe the five-step procedure that should be used to measure the resistance.

2-3 POWER

Energy
Capacity to do work.

Work
Work is done anytime energy is transformed from one type to another, and the amount of work done is dependent on the amount of energy transformed.

Joule
The unit of work and energy.

Power
Amount of energy converted by a component or circuit in a unit of time, normally seconds. It is measured in units of watts (joules/second).

Watt (W)
Unit of electric power required to do work at a rate of 1 joule/second. One watt of power is expended when 1 ampere of direct current flows through a resistance of 1 ohm.

The sun provides us with a consistent supply of energy in the form of light. Coal and oil are fossilized vegetation that grew, among other things, due to the sun, and are examples of **energy** that the earth has stored for millions of years. It can be said, then, that all energy begins from the sun. On earth, energy is not created or destroyed; it is merely transformed from one form to another. The transforming of energy from one form to another is called **work**. The greater the energy transformed, the more work that is done.

The six basic forms of energy are light, heat, magnetic, chemical, electrical, and mechanical energy. The unit for energy is the **joule** (J).

EXAMPLE:

One person walks around a track and takes 5 minutes, while another person runs around the track and takes 50 seconds. Both were full of energy before they walked or ran around the track, and during their travels around the track they converted the chemical energy within their bodies into the mechanical energy of movement.

a. Who exerted the most energy?
b. Who did the most work?

Solution:

Both exerted the same amount of energy. The runner exerted all his energy (for example, 100 J) in the short time of 50 seconds, while the walker spaced his energy (100 J) over 5 minutes. Since they both did the same amount of work, the only difference between the runner and the walker is time, or the rate at which their energy was transformed.

2-3-1 *What Is Power?*

Power (P) is the rate at which work is performed and is given the unit of **watt** (W), which is joules per second (J/s).

Returning to the example involving the two persons walking and running around the track, we could say that the number of joules of energy exerted in 1 second by the runner was far greater than the number of joules of energy exerted in 1 second by the walker, although the total energy exerted by both persons around the entire track was equal and therefore the same amount of work was done. Even though the same amount of energy was used, and therefore the same amount of work was done by the runner and the walker, the output power of each was different. The runner exerted a large value of joules/second or watts (high power output) in a short space of time, while the walker exerted only a small value of joules/second or watts (low power output) over a longer period of time.

Whether discussing a runner, walker, electric motor, heater, refrigerator, lightbulb, or compact disc player—power is power. The output power, or power ratings, of electrical, electronic, or mechanical devices can be expressed in watts and describes the number of joules of energy converted every second. The output power of rotating machines is given in the unit *horsepower* (hp), the output power of heaters is given in the unit British thermal unit per hour (Btu/h), and the output power of cooling units is given in the unit *ton of refrigeration*. Despite the different names, they can all be expressed quite simply in the unit of watts. The conversions are as follows:

$$\text{1 horsepower (hp)} = 746\ \text{W}$$
$$\text{1 British thermal unit per hour (Btu/h)} = 0.293\ \text{W}$$
$$\text{1 ton of refrigeration} = 3.52\ \text{kW}\ (3520\ \text{W})$$

Now we have an understanding of power, work, and energy. Let's reinforce our knowledge by introducing the energy formula and try some examples relating to electronics.

TIME LINE

James P. Joule (1818–1889), an English physicist and self-taught scientist, conducted extensive research into the relationships among electrical, chemical, and mechanical effects, which led him to the discovery that one energy form can be converted into another. For this achievement, his name was given to the unit of energy, the joule.

2-3-2 *Calculating Energy*

The amount of energy stored (W) is dependent on the coulombs of charge stored (Q) and the voltage (V).

$$W = Q \times V$$

where W = energy stored, in joules (J)
Q = coulombs of charge (1 coulomb = 6.24×10^{18} electrons)
V = voltage, in volts (V)

If you consider a battery as an example, you can probably better understand this formula. The battery's energy stored is dependent on how many coulombs of electrons it holds (current) and how much electrical pressure it is able to apply to these electrons (voltage).

EXAMPLE:

How many coulombs of electrons would a 9 V battery have to store to have 63 J of energy?

Solution:

If $W = Q \times V$, then by transposition

$$\frac{W}{Q\ V} \qquad Q = \frac{W}{V}$$

$$\text{coulombs of electrons } (Q) = \frac{\text{energy in joules } (W)}{\text{battery voltage } (V)} = \frac{63\ \text{J}}{9\ \text{V}} = 7\ \text{C of electrons}$$

or

$$7 \times 6.24 \times 10^{18} = 4.36 \times 10^{19}\ \text{electrons}$$

Calculator Sequence

STEP	KEYPAD ENTRY	DISPLAY RESPONSE
1.	6 3	63
2.	÷	
3.	9	9
4.	=	7
5.	×	7
6.	6 . 2 4 E 1 8	6.24E18
7.	=	4.36E19

2-3-3 *Calculating Power*

Watt's Law
An electrical relationship developed by James Watt that states the amount of power delivered to a device is dependent on the electrical pressure (or voltage applied across the device) and the current flowing through the device.

Power, in connection with electricity and electronics, is the rate at which electric energy (W) is converted into some other form. A relationship between current, voltage, and power was first discovered by James Watt and is referred to as **Watt's law** or the power formula.

$$P = I \times V$$

where P = power, in watts (W)
I = current, in amperes (A)
V = voltage, in volts (V)

Watt's power formula states that the amount of power delivered to a device is dependent on the electrical pressure (or voltage applied across the device) and the current flowing through the device.

TIME LINE
The unit of power, the watt, was named after Scottish engineer and inventor James Watt (1736–1819) in honor of his advances in the field of science.

EXAMPLE:

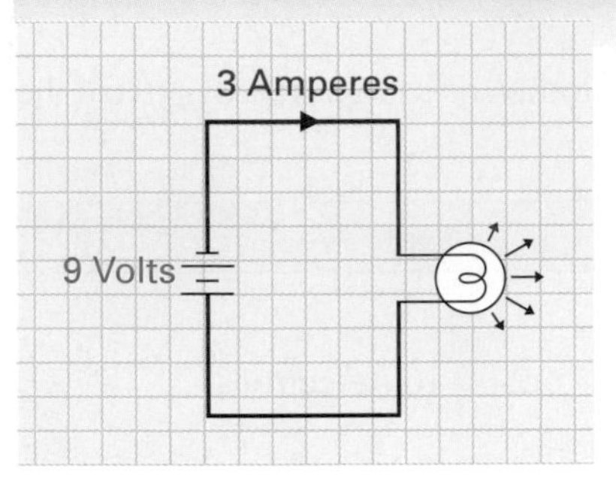

FIGURE 2-16 **Calculating Power.**

In regard to electrical and electronic circuits, power is the rate at which electric energy is converted into some other form. In the example in Figure 2–16, electric energy will be transformed into light and heat energy by the lightbulb. Power has the unit of watts, which is the number of joules of energy transformed per second (J/s). If 27 J of electric energy is being transformed into light and heat per second, how many watts of power does the lightbulb convert?

Solution

$$\text{power} = \frac{\text{joules}}{\text{second}} = \frac{27\text{ J}}{1\text{ s}} = 27\text{ W}$$

The power output in the previous example could have easily been calculated by merely multiplying current by voltage to arrive at the same result.

$$\text{power} = I \times V = 3\text{ A} \times 9\text{ V} = 27\text{ W}$$

We could say, therefore, that the lightbulb dissipates 27 watts of power, or 27 joules of energy per second.

Studying the power formula, $P = I \times V$, we can also say that 1 watt of power is expended when 1 ampere of current flows through a circuit that has 1 volt applied.

Like Ohm's law, we can transpose the power formula as follows:

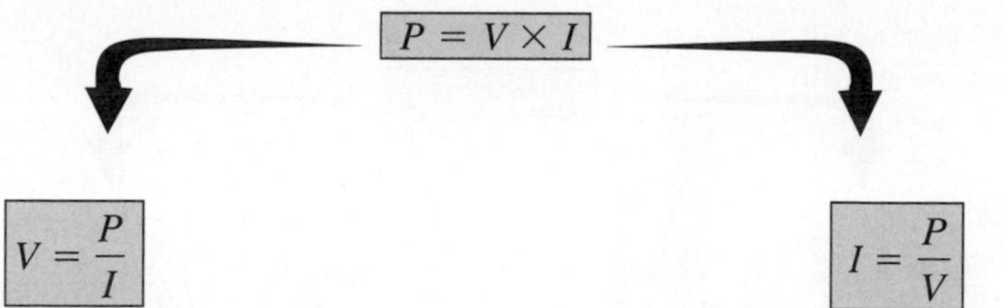

$$V = \frac{P}{I}$$

This formula can be used to calculate V when P and I are known.

$$I = \frac{P}{V}$$

This formula can be used to calculate I when P and V are known.

EXAMPLE:

A coffee cup warming plate has a power rating of 120 V, 23 W. How much current is flowing through the heating element, and what is the heating element's resistance?

Solution:

$$P = V \times I \qquad I = \frac{P}{V}$$

$$I = \frac{23\text{ W}}{120\text{ V}} = 191.7\text{ mA}$$

$$R = \frac{V}{I} = \frac{120\text{ V}}{191.7\text{ mA}} = 626\ \Omega$$

EXAMPLE:

To illustrate a point, let us work through a hypothetical example. Calculate the current drawn by a 100 W lightbulb if it is first used in the home (and therefore connected to a 120 V source) and then used in the car (and therefore connected across a 12 V source).

Solution:

$$P = V \times I \quad \text{therefore,} \quad I = \frac{P}{V}$$

In the home,

$$I = \frac{P}{V} = \frac{100\text{ W}}{120\text{ V}} = 0.83\text{ A}$$

In the car,

$$I = \frac{P}{V} = \frac{100\text{ W}}{12\text{ V}} = 8.33\text{ A}$$

This example brings out a very important point in relation to voltage, current, and power. In the home, only a small current needs to be supplied when the applied voltage is large ($P = V\uparrow \times I\downarrow$). However, when the source voltage is small, as in the case of the car, a large current must be supplied in order to deliver the same amount of power ($P = V\downarrow \times I\uparrow$).

When trying to calculate power we may not always have the values of V and I available. For example, we may know only I and R, or V and R. The following shows how we can substitute terms in the $P = V \times I$ formula to arrive at alternative power formulas for wattage calculations:

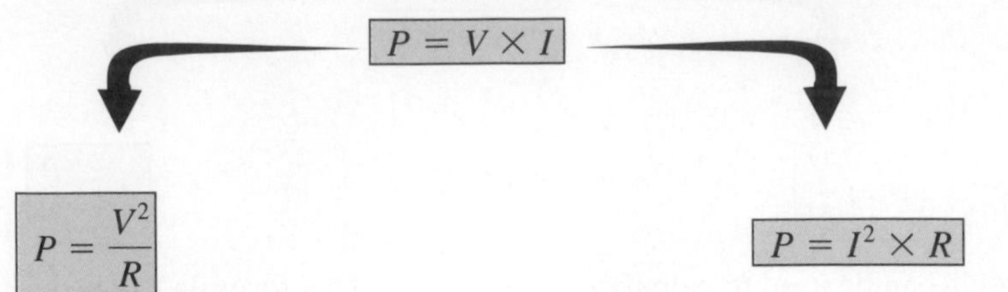

This formula can be used to calculate P when V and R are known.

This formula can be used to calculate P when I and R are known.

EXAMPLE:

In Figure 2–17 a 12 V battery is connected across a 36 Ω resistor. How much power does the resistor dissipate?

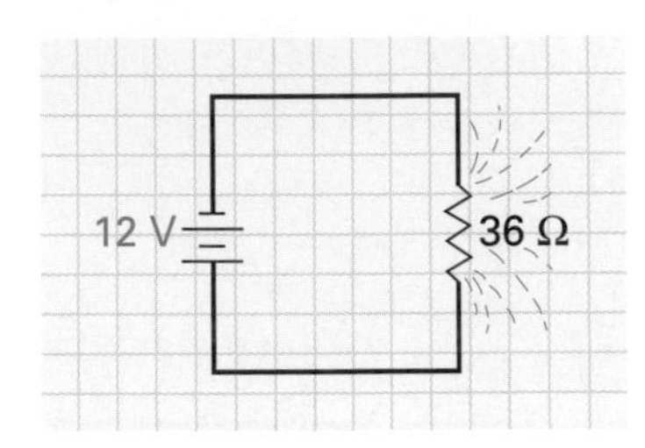

FIGURE 2-17

Solution:

Since V and R are known, the V^2/R power formula should be used.

$$\text{power} = \frac{\text{voltage}^2}{\text{resistance}}$$

$$= \frac{(12\text{ V})^2}{36\ \Omega} = \frac{144}{36}$$

$$= 4\text{ W}$$

Four joules of heat energy are being dissipated every second.

Calculator Sequence

STEP	KEYPAD ENTRY	DISPLAY RESPONSE
1.	[1] [2]	12
2.	[X^2] (square key)	144
3.	[÷]	
4.	[3] [6]	36
5.	[=]	4

The atoms of a resistor obstruct the flow of moving electrons (current), and this friction between the two causes heat to be generated. Current flow through any resistance therefore will always be accompanied by heat. In some instances, such as a heater or oven, the device has been designed specifically to generate heat. In other applications, it may be necessary to include a resistor to reduce current flow, and the heat generated will be an unwanted side effect.

Any one of the three power formulas can be used to calculate the power dissipated by a resistance. However, since power is determined by the friction between current (I) and the circuit's resistance (R), the $P = I^2 \times R$ formula is more commonly used to calculate the heat generated.

EXAMPLE:

If 100 mA of current is passing through a 33 Ω resistor, how much heat will it dissipate?

Solution:

$$
\begin{aligned}
P &= I^2 \times R \\
&= (100 \text{ mA})^2 \times 33\ \Omega \\
&= 0.33 \text{ W}
\end{aligned}
$$

2-3-4 *Measuring Power*

The multimeter can be used to measure power by following the procedure described in Figure 2–18. In this example, the power consumed by a car's music system is being determined.

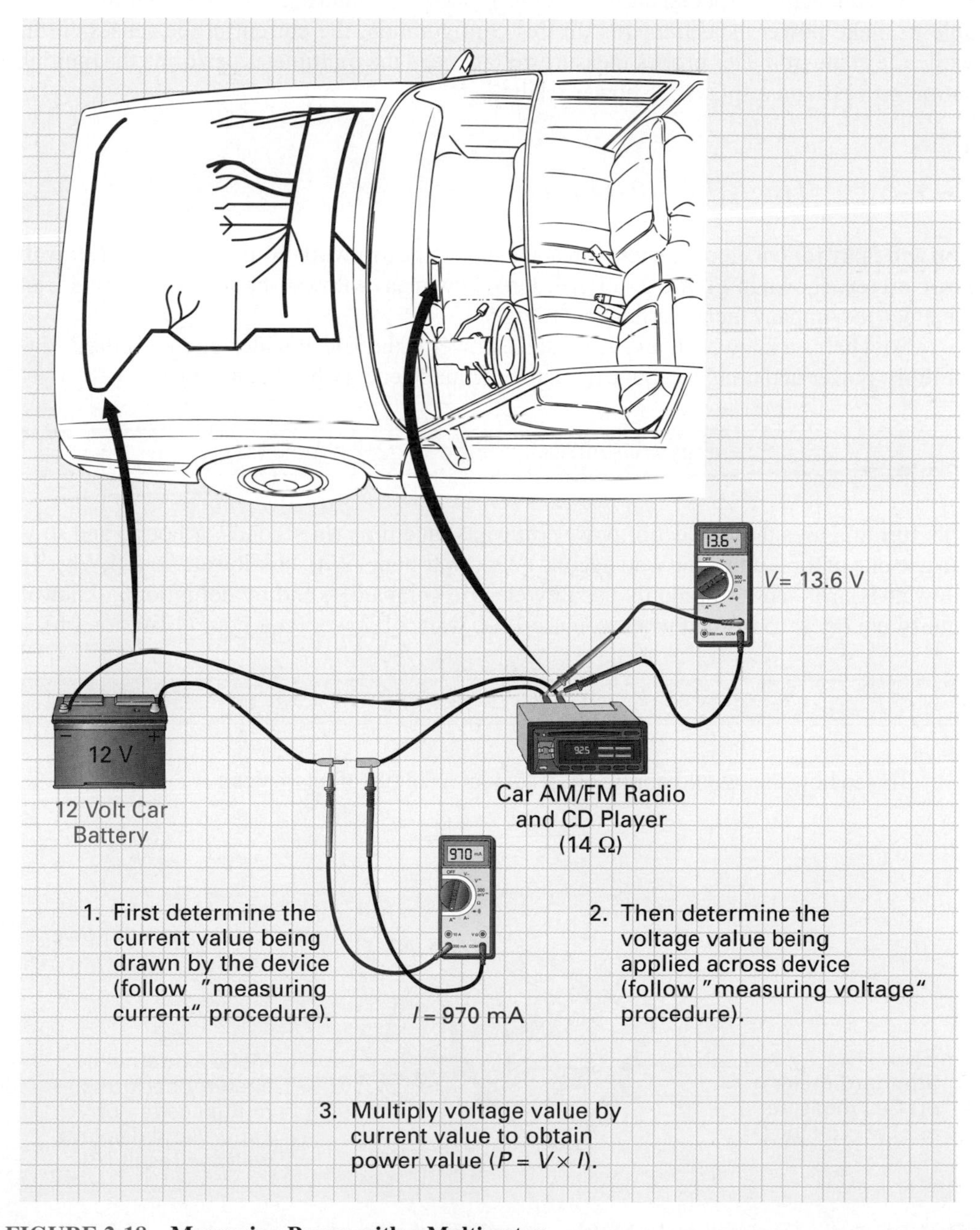

FIGURE 2-18 Measuring Power with a Multimeter.

After the current measurement and voltage measurement have been taken, the product of the two values will have to be calculated, since power = current × voltage ($P = I \times V$).

EXAMPLE:

Calculate the power consumed by the car music system shown in Figure 2–18.

Solution:

Current drawn from battery by music system = 970 mA. Voltage applied to music system power input = 13.6 V.

$$P = I \times V = 970 \text{ mA} \times 13.6 \text{ V} = 13.2 \text{ W}$$

By connecting a special current probe, as shown in Figure 2–19, some multimeters are able to make power measurements. In this configuration, the current probe senses current while the standard meter probes measure voltage, and the multimeter performs the multiplication process and displays the power reading.

2–3–5 *The Kilowatt-Hour*

Kilowatt-Hour
1000 watts for 1 hour.

Kilowatt-Hour Meter
A meter used by electric companies to measure a customer's electric power use in kilowatt-hours.

You and I pay for our electric energy in a unit called the **kilowatt-hour** (kWh). The **kilowatt-hour meter**, shown in Figure 2–20, measures how many kilowatt-hours are consumed, and the electric company then charges accordingly.

So what is a kilowatt-hour? Well, since power is the rate at which energy is used, if we multiply power and time, we can calculate how much energy has been consumed.

$$\text{energy consumed } (W) = \text{power } (P) \times \text{time } (t)$$

This formula uses the product of power (in watts) and time (in seconds or hours) and so we can use one of three units: the watt-second (Ws), watt-hour (Wh), or kilowatt-hour (kWh). The kilowatt-hour is most commonly used by electric companies, and by definition, a kilowatt-hour of energy is consumed when you use 1000 watts of power for 1 hour (1 kW for 1 h).

$$\text{energy consumed (kWh)} = \text{power (kW)} \times \text{time (h)}$$

To see how this would apply, let us look at a couple of examples.

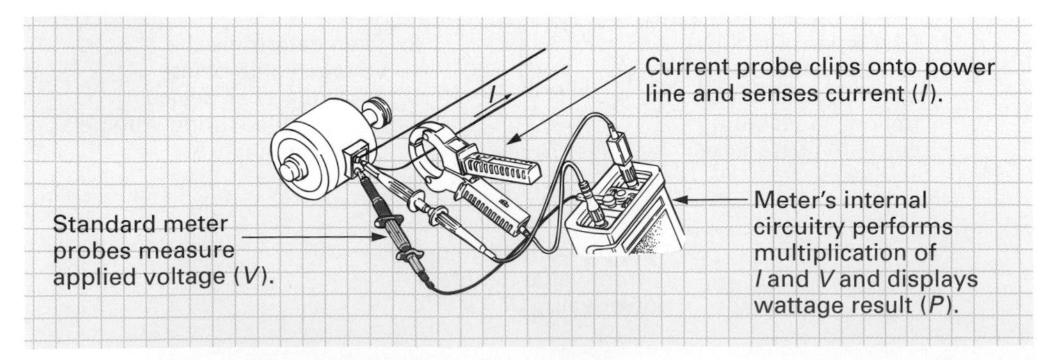

FIGURE 2-19 Configuring the Multimeter to Measure Power.

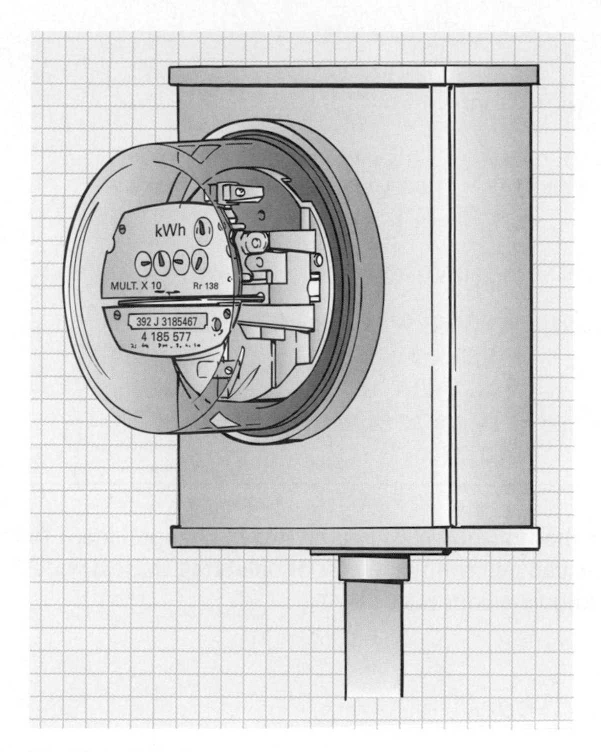

FIGURE 2-20 **Kilowatt-Hour Meter.**

EXAMPLE:

If a 100 W lightbulb is left on for 10 hours, how many kilowatt-hours will we be charged for?

Solution:

$$\begin{aligned}\text{power consumed (kWh)} &= \text{power (kW)} \times \text{time (hours)}\\ &= 0.1 \text{ kW} \times 10 \text{ hours}\\ &\quad (100 \text{ W} = 0.1 \text{ kW})\\ &= 1 \text{ kWh}\end{aligned}$$

Calculator Sequence

STEP	KEYPAD ENTRY	DISPLAY RESPONSE
1.	[0] [.] [1] [E] [3]	0.1E3
2.	[×]	
3.	[1] [0]	10
4.	[=]	1E3

EXAMPLE:

Figure 2–21 illustrates a typical household electric heater and an equivalent electrical circuit. The heater has a resistance of 7 Ω and the electric company is charging 12 cents/kWh. Calculate:

a. The energy consumed by the heater

b. The cost of running the heater for 7 hours

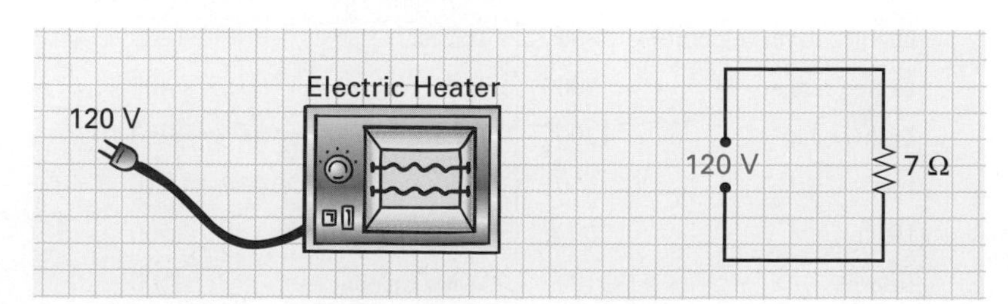

FIGURE 2-21 **An Electric Heater.**

■ ***Solution:***

$$\textbf{a.}\ \text{power}\ (P) = \frac{V^2}{R} = \frac{120^2}{7}$$

$$= \frac{14400\text{ V}}{7} = \frac{14.4\text{ kV}}{7}$$

$$= 2057\text{ W (approximately 2 kilowatts or 2 kW)}$$

$$\textbf{b.}\ \text{energy consumed} = \text{power (kW)} \times \text{time (hours)}$$

$$= 2.057 \times 7$$

$$= 14.4\text{ kWh}$$

$$\text{cost} = \text{kWh} \times \text{rate} = 14.4 \times 12\text{ cents}$$

$$= \$1.73$$

Table 2–1 lists the typical wattage rating for many appliances found in the home. Using these ratings, you can determine the cost of operating these devices for a certain amount of time at a certain kilowatt-hour rate.

2-3-6 *Body Shock*

Shock
The sudden pain, convulsion, unconsciousness, or death produced by the passage of electric current through the body.

Safety precautions should always be your first priority when working on electronic equipment, as there is always the possibility of receiving an electric shock. An electric **shock** is a sudden, uncontrollable reaction as current passes through your body and causes your muscles to contract, which delivers a certain amount of pain.

Figure 2–22 lists the physiological effects of different amounts of current. As you can see, even a current as small as 10 mA can be fatal. Any shock is dangerous, since even the mildest could surprise you and cause an involuntary action that could injure you or someone else. For example, a muscular spasm could throw you against a sharp object or move your arm to a point of higher voltage.

As you know, the value of current in a circuit is determined by the voltage applied and the circuit's resistance. When you come in contact with a high-voltage point, therefore, the value of current passing through your body will depend on the value of voltage at that point and your

TABLE 2-1 Typical Wattage Ratings for Household Appliances.

Appliance	Wattage Rating	Appliance	Wattage Rating
Air conditioner (small)	1600	Microwave oven	1200
Hair dryer	1300	Range	12,200
Portable CD player	5	Refrigerator	1800
Clock	2	Shaver	15
Clothes dryer (electric)	4800	Toaster	1200
Coffee maker	900	Trash compactor	400
Dishwasher	1200	TV	250
Fan	150	Videocassette recorder	110
Heater	2000	Washing machine	500
Iron	1100	Water heater	4500

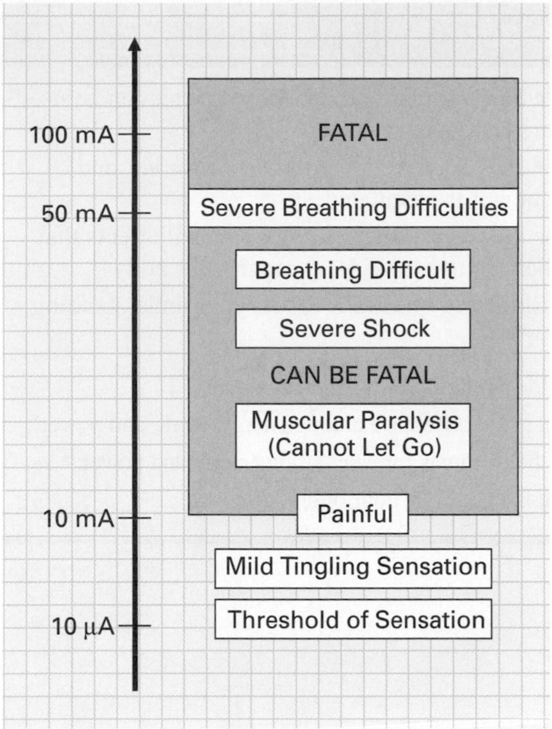

FIGURE 2-22 **Physiological Effects of Electric Current.**

body's resistance. The human body has a resistance of about 10,000 to 50,000 Ω, depending on how good a contact you make with the **"live"** (power present) conductor. Skin resistance is generally quite high and will subsequently oppose current flow; however, your resistance is lowered if your skin is wet due to perspiration or if your skin has a cut or an abrasion.

Live
Term used to describe a circuit or piece of equipment that is ON and has current flow within it.

Body Resistance
The resistance of the human body.

EXAMPLE:

Calculate the current through a **body resistance** of 10 kΩ if the body came in contact with 100 V. How much power is the body dissipating?

Solution:

$$I = \frac{V}{R} = \frac{100\text{ V}}{10\text{ k}\Omega} = 10\text{ mA}$$

$$P = V \times I = 100\text{ V} \times 10\text{ mA} = 1\text{ W}$$

This value of current will be painful and possibly fatal.

Power is a point to consider. For example, if an infrared laser's 1 mW power supply generates a 50,000 V output, would this be dangerous if a person was to come in contact with it? The answer is probably not, since even though the power supply generates an intimidating voltage, it cannot deliver a dangerous value of current:

$$I = \frac{P}{V} = \frac{1\text{ mW}}{50\text{ kV}}$$

$$= \frac{1 \times 10^{-3}\text{ W}}{50 \times 10^{3}\text{ V}}$$

$$= 0.02\ \mu\text{A}$$

This also works in reverse. For example, a car's battery may only be 12 V, but it can supply a very deadly value of current.

SELF-TEST EVALUATION POINT FOR SECTION 2–3

Now that you have completed this section, you should be able to:

Objective 15. List the six basic forms of energy, and define the terms:

a. Energy
b. Work
c. Joule

Objective 16. Describe the term *power* and define its unit, the watt.

Objective 17. Explain how to determine the amount of energy stored, when coulombs and voltage are known.

Objective 18. List Watt's law, or the power formula, and transpose it to obtain its three forms.

Objective 19. Explain how the multimeter can be used to measure power.

Objective 20. Define the term *kilowatt-hour*, and describe how it is calculated.

Objective 21. Describe the physiological effects that occur when a current passes through your body.

Use the following questions to test your understanding of Section 2–3:

1. List the six basic forms of energy.
2. What is the difference among energy, work, and power?
3. List the formulas for calculating energy and power.
4. What is 1 kilowatt-hour of energy?

SUMMARY

Mini-Math Review (Section 2–1)

Monomial (One Term): $2a$
Binomial (Two Terms): $2a + 3b$
Trinomial (Three Terms): $2a + 3b + c$
Polynomial (Any Number of Terms)
Order of Operations: Parentheses, Exponents, Multiplication, Division, Addition, Subtraction.
(Memory Aid: Please, Excuse, My, Dear, Aunt, Sally)

Basic Rules: $a + a = 2a$ $\quad a - a = 0$ $\quad a \cdot a = a^2$ $\quad a \div a = 1$ $\quad \sqrt{a^2} = a$
$a + 1 = a + 1$ $\quad a - 1 = a - 1$ $\quad a \cdot 1 = a$ $\quad a \div 1 = a$
$a + 0 = a$ $\quad a - 0 = a$ $\quad a \cdot 0 = 0$ $\quad a \div 0 =$ Impossible $\quad 0 \div a = 0$

Parentheses First: $(3 + 6) \cdot 2 = 9 \cdot 2 = 18$ $\quad 9 - (2 \cdot 3) = 9 - 6 = 3$ $\quad 6 + (9 - 7) = 6 + 2 = 8$

No Parentheses—Step 1: Multiplications and Divisions, left to right. $3 + 6 \cdot 4 = 3 + 24 = 27$
Step 2: Additions and Subtractions, left to right. $4 \cdot 2 + 3 - 4 \div 2 = 8 + 3 - 2 = 9$

Combine Only Like Terms: $2x + x + y = 3x + y$ $\quad 3a + 4b - a = 2a + 4b$

Commutative Property of Addition—order of terms makes no difference: $a + b = b + a$

Associative Property of Addition—grouping of terms makes no difference: $(a + b) + c = a + (b + c)$

Coefficients: $x + x + x = 3x$ $\quad x + y + x + y = 2x + 2y$

Exponents: $x \cdot x \cdot x = x^3$ $\quad x \cdot y \cdot x \cdot y = x^2 \cdot y^2 = x^2y^2$

$a^2 \cdot a^3 = a^{2+3} = a^5$ $\quad \dfrac{a^3}{a^2} = a^{3-2} = a^1 = a$

$a^{-2} = \dfrac{1}{a^2}$ $\quad (a^3)^2 = a^{3\times2} = a^6$

$\sqrt[3]{a^2} = \left(\sqrt[3]{a}\right)^2$ $\quad a^{\frac{2}{3}} = \sqrt[3]{a^2}$

Commutative Property of Multiplication—order of factors makes no difference: $a \cdot b = b \cdot a$

Associative Property of Multiplication—grouping of factors makes no difference: $(ab)c = a(bc)$

Distributive Property of Multiplication—$a(b + c) = ab + ac$ $\quad 2(a - 3b) = 2 \cdot a - 2 \cdot 3b = 2a - 6b$

Transposition—Step 1: Ensure unknown is above fraction bar. $x \cdot 8 = 16$ (divide both sides by 8, $\div 8$)
Step 2: Isolate unknown. $(x \cdot 8) \div 8 = (16) \div 8,$ $\quad x = 2$

Unknown Above Fraction Bar

$$\begin{aligned} (2 \cdot a) + 5 &= 23 \quad (-5) \\ 2 \cdot a &= 18 \quad (\div 2) \\ a &= 9 \end{aligned}$$

Unknown Below Fraction Bar

$$\begin{aligned} \frac{72}{x} &= 12 \quad (\cdot x) \\ 72 &= 12 \cdot x \quad (\div 12) \\ 6 &= x \end{aligned}$$

Unknown on Both Sides

$$\begin{aligned} (6 \cdot y) + 7 &= y + 27 \quad (-7) \\ 6y &= y + 20 \quad (-y) \\ 5y &= 20 \quad (\div 5) \\ y &= 4 \end{aligned}$$

Combining Unknowns

$$\begin{aligned} x + 2x + 4(x + 2x) &= 100 \\ x + 2x + 4x + 8x &= 100 \\ 15x &= 100 \\ x &= 6.67 \end{aligned}$$

Positive and Negative Number Rules

Addition ⊕	$(+) + (+) = (+), (-) + (-) = (-)$ $(+) + (-)$ or $(-) + (+) =$ Difference
Subtraction ⊖	Change sign of second number, and then add.
Multiplication ⊗	$(+) \cdot (+) = (+), (-) \cdot (-) = (+)$ $(+) \cdot (-)$ or $(-) \cdot (+) = (-)$
Division ⊕	$(+) \div (+) = (+), (-) \div (-) = (+)$ $(+) \div (-)$ or $(-) \div (+) = (-)$

Positive and Negative Numbers

Addition ⊕	$2x + 5x = 7x \quad -6a + (+3a) = -3a$
Subtraction ⊖	$-3y^2 - (-2y^2) = -3y^2 + (+2y^2) = -y^2$
Multiplication ⊗	$-3(6a) = -3 \cdot (+6a) = -18a$
Division ⊕	$14a \div (-7) = -2a$

Resistance (Section 2–2)

1. Resistance is the opposition to current flow accompanied by the dissipation of heat.

2. Resistance and current are inversely proportional ($1/\propto$) to one another. For example, if resistance is doubled, current is halved (assuming a constant voltage).

3. Current is measured in amperes, voltage is measured in volts, and resistance is measured in ohms, in honor of Georg Simon Ohm and his work with current, voltage, and resistance.

4. The larger the resistance, the larger the value of ohms and the more the resistor will oppose current flow.

5. The ohm is given the symbol Ω, which is the Greek capital letter omega. By definition, 1 ohm is the value of resistance that will allow 1 ampere of current to flow through a circuit when a voltage of 1 volt is applied.

6. Published originally in 1826, *Ohm's law* states that the current flow in a circuit is directly proportional ($\propto$) to the source voltage applied and inversely proportional ($1/\propto$) to the resistance of the circuit.

7. Like all formulas, the Ohm's law formula provides a relationship between quantities, or values. It is important to know how to rearrange or transpose a formula so that you can solve for any of the formula's quantities.

8. To transpose a formula, follow two steps. *First step*: Move the unknown quantity so that it is above the line. *Second step*: Move the unknown quantity so that it stands by itself on either side of the equal sign. *If you do exactly the same thing to both sides of the equation or formula, nothing is changed.*

9. If resistance were to remain constant and the voltage were to double, the current within the circuit would also double. Similarly, if the voltage were halved, the current would also halve, proving that current and voltage are directly proportional to one another.

10. If voltage were to remain constant and the resistance were to double, the current within the circuit would be halved. On the other hand, if the circuit resistance were halved, the circuit current would double, confirming that current is inversely proportional to resistance.

Power (Section 2–3)

11. The six basic forms of energy are light, heat, magnetic, chemical, electrical, and mechanical energy. The unit for energy is the *joule* (J).

12. Work is being done every time one form of energy is transformed to another.

13. Energy and work have the same symbol (W), the same formula, and the same unit (the joule). Energy is merely the capacity, potential, or ability to do work, and work is done when a transformation of the potential, capacity, or ability takes place.

14. *Power (P)* is the rate at which work is performed and is given the unit of watt (W), which is joules per second.

15. Whether discussing a runner, walker, electric motor, heater, refrigerator, lightbulb, or compact disc player—power is power. The output power, or power ratings, of electrical, electronic, or mechanical devices can be expressed in watts and describes the number of joules of energy converted every second.

The output power of rotating machines is given in the unit *horsepower* (hp), the output power of heaters is given in the unit *British thermal units per hour* (Btu/h), and the output power of cooling units is given in the unit *ton of refrigeration*. Despite the different names, they can all be expressed in the unit of watts.

16. The amount of energy stored (W) is dependent on the coulombs of charge stored (Q) and the voltage (V). The battery's stored energy is dependent on how many coulombs of electrons it holds (current) and how much electrical pressure it is able to apply to these electrons (voltage).

17. Power, in relation to electricity and electronics, is the rate (t) at which electric energy (W) is converted into some other form.

18. The $P = I \times V$ formula, or Watt's law, states that the amount of power delivered to a device is dependent on the electrical pressure or voltage applied across the device and the current flowing through the device. Looking at the power formula, $P = I \times V$, we can say that 1 watt of power is expended when 1 ampere of current flows through a circuit that has 1 volt applied.

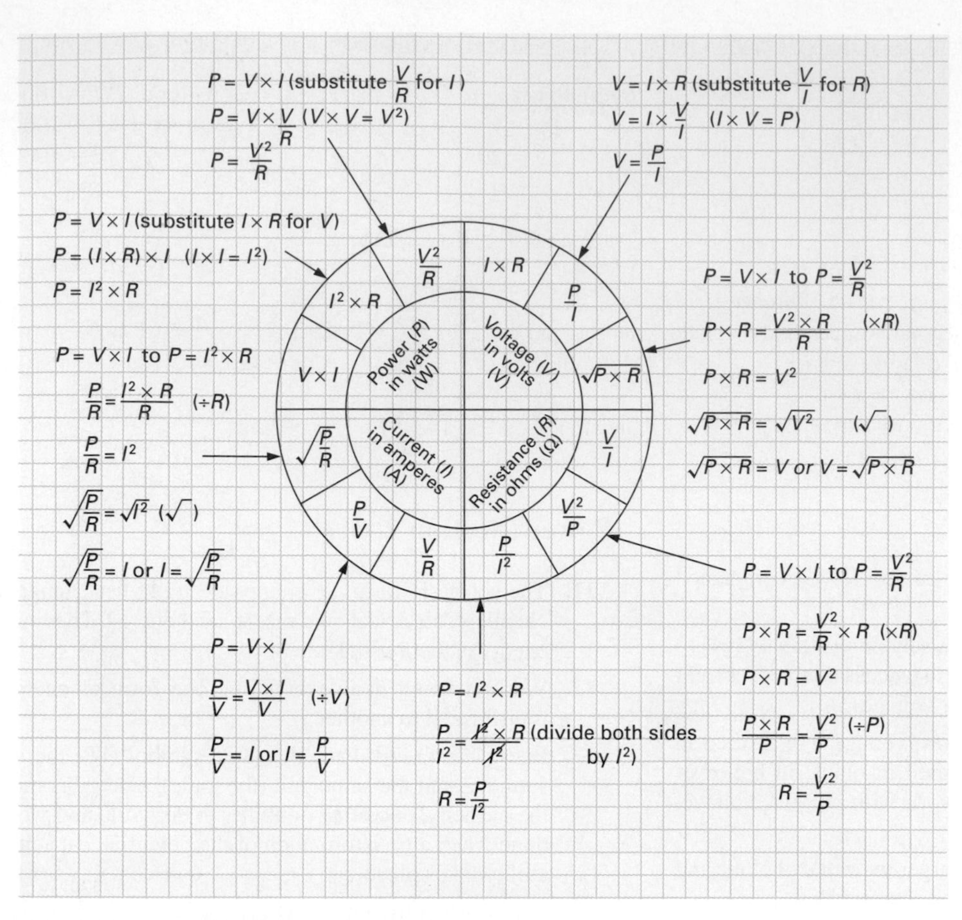

The Ohm's Law and Watt's Law Formula Circle

19. *Substitution* enables us to obtain alternative power formulas by replacing or substituting one mathematical term with an equivalent mathematical term.
20. The multimeter can be used to measure power by first making a current measurement, then a voltage measurement, and then calculating the product of the measured current and voltage values.
21. The *kilowatt-hour* meter measures how many kilowatt-hours are consumed, and the electric company then charges accordingly. Since power is the rate at which energy is used, if we multiply power and time, we can calculate how much energy has been consumed.
22. The kilowatt-hour is most commonly used by electric companies. A *kilowatt-hour* of energy is consumed when you use 1000 watts (1 kW) of power in 1 hour.

REVIEW QUESTIONS

Multiple-Choice Questions

1. What is the value of the unknown, s, in the following equation: $\sqrt{s^2} + 14 = 26$?
   a. 12 **c.** 4
   b. 26 **d.** 5
2. What is the correct transposition for the following formula: $a = \dfrac{b - c}{x}$, $x = ?$
   a. $x = \dfrac{b - a}{c}$ **c.** $x = \dfrac{b - c}{a}$
   b. $x = (b - a) \times c$ **d.** $x = x(b - c)$
3. If $V = I \times R$ and $P = V \times I$, develop a formula for P when I and R are known.
   a. $P = \dfrac{I}{R}$ **c.** $P = I^2 \times R$
   b. $P = I^2 \times R^2$ **d.** $P = \sqrt{\dfrac{I}{R}}$

4. If $E = M \times C^2$, $M = ?$ and $C = ?$
   a. $\sqrt{\frac{E}{C}}, \frac{E^2}{M}$
   b. $\frac{E}{C^2}, \sqrt{\frac{E}{M}}$
   c. $E \times C^2, \sqrt{C^2 \times E}$
   d. $\frac{C^2}{E}, \sqrt{\frac{M}{E}}$
5. If $P = \frac{V^2}{R}$, transpose to obtain a formula to solve for V.
   a. $V = \sqrt{P \times R}$
   b. $V = \frac{P^2}{R}$
   c. $V = P \times R^2$
   d. $V = \sqrt{P} \times R$
6. Resistance is measured in
   a. Ohms c. Amperes
   b. Volts d. Siemens
7. Current is proportional to
   a. Resistance c. Both (a) and (b)
   b. Voltage d. None of the above
8. Current is inversely proportional to
   a. Resistance c. Both (a) and (b)
   b. Voltage d. None of the above
9. If the applied voltage is equal to 15 V and the circuit resistance equals 5 Ω, the total circuit current would be equal to
   a. 4 A c. 3 A
   b. 5 A d. 75 A
10. Calculate the applied voltage if 3 mA flows through a circuit resistance of 25 kΩ.
    a. 63 mV c. 77 μV
    b. 25 V d. 75 V
11. Energy is measured in
    a. Volts c. Amperes
    b. Joules d. Watts
12. Chemical energy within a battery is converted into
    a. Electrical energy c. Magnetic energy
    b. Mechanical energy d. Heat energy
13. The water pump has the potential to cause water flow, just as the battery has the potential to cause
    a. Voltage c. Current
    b. Electron flow d. Both (b) and (c)
14. Work is measured in
    a. Joules c. Amperes
    b. Volts d. Watts
15. A portable CD player will have a typical voltage rating of:
    a. 5 W c. 500 W
    b. 50 W d. 5 kW
16. Power is the rate at which energy is transformed and is measured in
    a. Joules c. Volts
    b. Watts d. Amperes
17. Power is measured by using a (an)
    a. Ammeter c. Ohmmeter
    b. Voltmeter d. Wattmeter
18. Load current and load resistance are __________ to one another.
    a. Proportional
    b. Inversely proportional
19. When measuring __________, always open the circuit path to insert the multimeter.
    a. Current
    b. Voltage
    c. Resistance
20. When measuring __________, always disconnect the power and isolate the device from the circuit.
    a. Current
    b. Voltage
    c. Resistance

Communication Skill Questions

21. Define the following terms (2–1–1):
    a. Algebra c. Equation
    b. Formula d. Literal number
22. Why is it important to treat both sides of an equation equally? (2–1–1)
23. What two steps must be followed to transpose an equation? (2–1–2)
24. Explain how current (I), voltage (V), and resistance (R) can be combined in a formula. (2–1–2)
25. If $P = V \times I$, describe how you would solve for V and I. (2–1–2)
26. Briefly describe why
    a. Current is proportional to voltage. (2–2–2)
    b. Current is inversely proportional to resistance. (2–2–2)
27. State Ohm's law. (2–2–2)
28. List the three forms of Ohm's law. (2–2–2)
29. How is an ohmmeter used to measure resistance? (2–2–3)
30. Define power. (2–3–1)
31. Give three formulas for electric power. (2–3–3)
32. How is energy stored calculated, if coulombs and voltage are known? (2–3–2)
33. How can you measure electrical power? (2–3–4)
34. State the formula used by electric companies to determine the amount of power consumed. (2–3–5)
35. What is 1 kilowatt-hour? (2–3–5)

Practice Problems

36. Calculate the result of the following arithmetic operations involving literal numbers.
    a. $x + x + x = ?$ g. $4y \times 3y^2 = ?$
    b. $5x + 2x = ?$ h. $a \times a = ?$
    c. $y - y = ?$ i. $2y \div y = ?$
    d. $2x - x = ?$ j. $6b \div 3b = ?$
    e. $y \times y = ?$ k. $\sqrt{a^2} = ?$
    f. $2x \times 4x = ?$ l. $(x^3)^2 = ?$

37. Transpose the following equations to determine the unknown value:

a. $4x = 11$

b. $6a + 4a = 70$

c. $5b - 4b = \frac{7.5}{1.25}$

d. $\frac{2z \times 3z}{4.5} = 2z$

38. Apply transposition and substitution to Newton's second law of motion.

a. $F = ma$ [Force (F) = mass (m) × acceleration (a)]
If $F = m \times a$, $m = ?$ and $a = ?$

b. If acceleration (a) equals change in speed (ΔV) divided by time (t), how would the $F = ma$ formula appear if acceleration (a) was substituted for $\frac{\Delta V}{t}$?

39. a. Using the formula power (P) = voltage (V) × current (I), calculate the value of electric current through a hairdryer that is rated as follows: power = 1500 watts and voltage = 120 volts.

$$\boxed{P = V \times I}$$

where P = power in watts
V = voltage in volts
I = current in amperes

b. Using the Ohm's formula voltage (V) = current (I) × resistance (R), calculate the resistance or opposition to current flow offered by the hairdryer in Question 39a.

40. An electric heater with a resistance of 6 Ω is connected across a 120 V wall outlet.

a. Calculate the current flow.

b. Draw the schematic diagram.

41. What source voltage would be needed to produce a current flow of 8 mA through a 16 kΩ resistor?

42. If an electric toaster draws 10 A when connected to a power outlet of 120 V, what is its resistance?

43. Calculate the power used in Problems 40, 41, and 42.

44. Calculate the current flowing through the following lightbulbs when they are connected across 120 V:

a. 300 W **c.** 60 W

b. 100 W **d.** 25 W

45. If an electric company charges 9 cents/kWh, calculate the cost for each lightbulb in Problem 44 if on for 10 hours.

46. Indicate which of the following unit pairs is larger:

a. Millivolts or volts

b. Microamperes or milliamperes

c. Kilowatts or watts

d. Kilohms or megohms

47. Calculate the resistance of a lightbulb that passes 500 mA of current when 120 V is applied. What is the bulb's wattage?

48. Which of the following circuits has the largest resistance and which has the smallest?

a. $V = 120$ V, $I = 20$ mA **c.** $V = 9$ V, $I = 100$ μA

b. $V = 12$ V, $I = 2$ A **d.** $V = 1.5$ V, $I = 4$ mA

49. Calculate the power dissipated in each circuit in Problem 48.

50. How many watts are dissipated if 5000 J of energy is consumed in 25 s?

51. Convert the following:

a. 1000 W = __________ kW

b. 0.345 W = __________ mW

c. 1250 × 103 W = __________ MW

d. 0.00125 W = __________ μW

52. What is the value of the resistor when a current of 4 A is causing 100 W to be dissipated?

53. How many kilowatt-hours of energy are consumed in each of the following:

a. 7500 W in 1 hour

b. 25 W for 6 hours

c. 127,000 W for half an hour

54. What is the maximum output power of a 12 V, 300 mA power supply?

55. What power is being delivered to a load if load current equals 12 mA and load resistance is 1 kΩ?

Job Interview Test

These tests at the end of each chapter will challenge your knowledge up to this point and give you the practice you need for a job interview. To make this more realistic, the test will comprise both technical and personal questions. In order to take full advantage of this exercise, you may want to set up a simulation of the interview environment, have a friend read the questions to you, and record your responses for later analysis.

Company Name: Power Service, Inc.

Industry Branch: Test and Measurement.

Function: Maintain and Service Kilowatt-Hour Meters.

Job Title: Field Service Technician.

56. What do you know about us?

57. Could you tell me what a kilowatt-hour meter does?

58. How would you feel about working on your own in the field?

59. What is the difference between energy and power?

60. If you couldn't determine the answer to a problem, what would you do?

61. What is the difference between resistance and power?

62. What would you say is the job function of a field service technician?

63. Without supervision in the field, could you discipline yourself to get through a heavy schedule of stops?

64. Tell me what you know about Ohm's law.

65. What makes you think you are the right person for this job?

Answers

56. Visit the company's web site before the interview to get an overall understanding of the company's ownership, production line, service, and support.

57. Section 2–3–5.
58. Discuss how you have worked alone on project assignments and other related work history.
59. Section 2–3.
60. Say that you would report the problem to your supervisor and ask for help.
61. Sections 2–2 and 2–3.
62. See introduction, Field Service Technician.
63. Discuss how you have juggled school, studying, and a job to get to the position you currently hold, and other related work experience.
64. Section 2–2–2.
65. Describe why you were first attracted to the company's advertisement, why it felt compatible with your career goals, and how the company's positive attributes listed in its web site matched your expectations.

Web Site Questions

Go to the web site http://www.prenhall.com/cook, select the textbook *Introductory DC/AC Electronics* or *Introductory DC/AC Circuits*, this chapter, and then follow the instructions when answering the multiple-choice practice problems.

Conductors, Insulators, and Resistors

Communication Skills

Josiah Williard Gibbs was born in New Haven, Connecticut, in 1839. He graduated from Yale in 1858 and was appointed professor of mathematical physics at Yale in 1869, a position he held until his death in 1903. In his lifetime Gibbs wrote many important papers on the equilibrium of heterogeneous substances, the elements of vector analysis, and the electromagnetic theory of light.

Many historians rank Gibbs along with Newton and Einstein; however, he remains generally unknown to the public. This fact is due largely to his inability to communicate clearly and effectively. Strangely, for all of his technical genius, he just could not explain himself, a frustration that plagued him throughout his life. It took scientists years to comprehend what he was trying to explain, and as one scientist joked, "It was easier to rediscover Gibbs than to read him."

3

Outline and Objectives

COMMUNICATION SKILLS
INTRODUCTION

3-1 MINI-MATH REVIEW—MATHEMATICAL OPERATIONS AND GRAPHS

3-1-1 Ratio and Proportion

3-1-2 Rounding Off

3-1-3 Significant Places

3-1-4 Percentages

3-1-5 Averages

Objective 1: Explain the following mathematical operations:
a. Ratios
b. Rounding off
c. Significant places
d. Percentages
e. Averages

3-1-6 Graphing Data

Objective 2: Describe how to generate the data for a graph.

Objective 3: Define the terms:
a. *x*-axis
b. *y*-axis
c. Coordinates
d. Dependent variable
e. Independent variable

Objective 4: Explain how positive and negative values can be plotted on a Cartesian coordinate system.

3-2 CONDUCTORS

Objective 5: List the three atomic properties that govern why a material is a good conductor.

3-2-1 Conductance

3-2-2 Conductors and Their Resistance

Objective 6: Define the following terms:
a. Conductance
b. Relative conductivity
c. Resistivity

Objective 7: Describe how the resistance of a conductor can be calculated based on either its physical factors or electrical performance.

3-2-3 Temperature Effects on Conductors

Objective 8: Explain why conductors have a positive temperature coefficient of resistance.

Objective 9: Define the following terms:
a. Filament resistor
b. Cold resistance
c. Hot resistance
d. Ballast resistor

3-2-4 Maximum Conductor Current

Objective 10: Explain the purpose of the American Wire Gauge (AWG) standard.

3-2-5 Conductor and Connector Types

3-2-6 Switches

Objective 11: List some of the different types of conductors, connectors, and switches.

3-2-7 Superconductivity

Objective 12: Define the term *superconductor* and explain its advantages.

3-3 INSULATORS

Objective 13: Explain the difference between
a. A conductor
b. An insulator

Objective 14: Define the term *breakdown voltage* and describe how to determine dielectric thickness.

3-4 RESISTORS

3-4-1 Resistor Wattage, Tolerance, and Color Coding

Objective 15: Describe resistor characteristics such as wattage rating and tolerance.

Objective 16: Explain the color coding system used to encode value and tolerance for general-purpose and precision resistors.

3-4-2 Fixed-Value Resistor Types

3-4-3 Variable-Value Resistor Types

Objective 17: Describe the difference between a fixed- and a variable-value resistor.

Objective 18: Explain the difference between the six basic types of fixed-value resistors: carbon composition, carbon film, metal film, wirewound, metal oxide, and thick film.

Objective 19: Describe the SIP, DIP, and chip thick-film resistor packages.

Objective 20: Explain the following types of variable resistors:

a. Rheostat
b. Potentiometer

Objective 21: Describe the difference between a linear and tapered resistive track.

3-4-4 Testing Resistors

Objective 22: List and describe some of the resistor testing precautions and techniques.

Introduction

This chapter continues our discussion of resistance, examining the two extremes and everything in between. To explain this statement further, we begin this chapter with **conductors,** a group of metals that are used in electronics because they offer very little resistance to current. The next group of materials, **insulators,** are chosen for their ability to have extremely high resistance values. These are our two extremes—with conductors, the lower the resistance, the better they are at conducting; with insulators, the higher the resistance, the better they are at nonconducting.

So what is in between? Between the ideal conductor value of 0 Ω (zero ohms) and the ideal insulator value of ∞ Ω (infinite or maximum ohms), we have every value of resistance possible. **Resistors** are used to introduce desired values of resistance into a circuit. Resistance would seem to be an undesirable effect, as it reduces current flow and wastes energy as it dissipates heat. Resistors, however, are probably used more than any other component in electronics.

To begin with, let's review the math skills you will need for this chapter's material.

3-1 MINI-MATH REVIEW—MATHEMATICAL OPERATIONS AND GRAPHS

This first section, "Mini-Math Review," is included to overview the mathematical details you need for the electronic concepts covered in this chapter. In this review, we will be examining certain mathematical operations and graphs.

3-1-1 *Ratio and Proportion*

Ratio
The relationship in quantity, amount, or size between two or more things.

A **ratio** is a comparison of one number with another number. For example, a ratio such as 7 : 4 ("seven to four") is comparing the number 7 with the number 4. This ratio could be written in one of the following three ways:

$$\frac{7}{4} \leftarrow \text{Ratio is written as a fraction.}$$

$$7 : 4 \leftarrow \text{Ratio is written using the ratio sign (:).}$$

$$1.75 \text{ to } 1 \leftarrow \text{Ratio is written as a decimal.}$$

To express the ratio as a decimal, the number 7 was divided by the number 4, giving a result of 1.75 ($7 \div 4 = 1.75$). This result indicates that the number 7 is 1.75 times (or one and three-quarter times) larger than the number 4.

In the previous chapter, you were introduced to the terms *proportional* and *inversely proportional.* Let us now see how ratio and proportion are related by examining two simple examples.

EXAMPLE:

The electrical resistance of a wire is directly proportional to its length, as shown in the example illustration. Wire 1 has a small length and therefore small resistance ($L_1 \downarrow, R_1 \downarrow$). Conversely, wire 2 has a large length and therefore large resistance ($L_2 \uparrow, R_2 \uparrow$). Referring to the following equation, you can see this proportional relationship. The length ratio has wire 1 above the fraction bar, and similarly, the resistance ratio also has wire 1 above the fraction bar—indicating the proportional relationship.

a. Insert the values given in the illustration and determine whether both sides of the equation remain equal.

b. If you double the length of wire 1 (36 feet) and so double its resistance (6 ohms), will the equation remain equal?

$$\frac{\text{length of wire 1}}{\text{length of wire 2}} = \frac{\text{resistance of wire 1}}{\text{resistance of wire 2}}$$

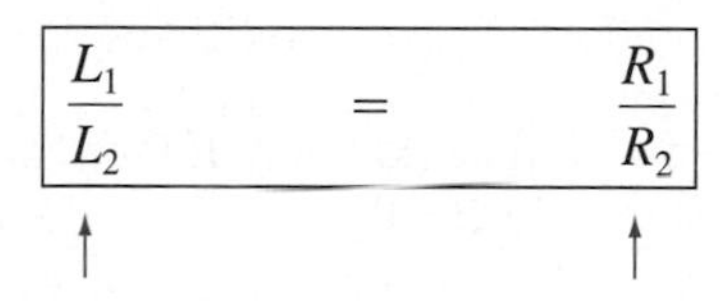

$$\boxed{\frac{L_1}{L_2} = \frac{R_1}{R_2}}$$

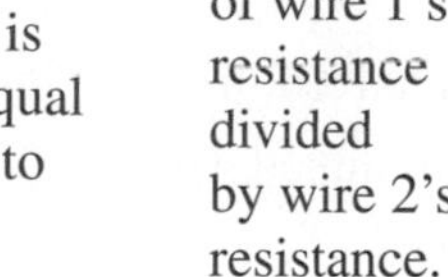

The ratio of wire 1's length divided by wire 2's length — is equal to — The ratio of wire 1's resistance divided by wire 2's resistance.

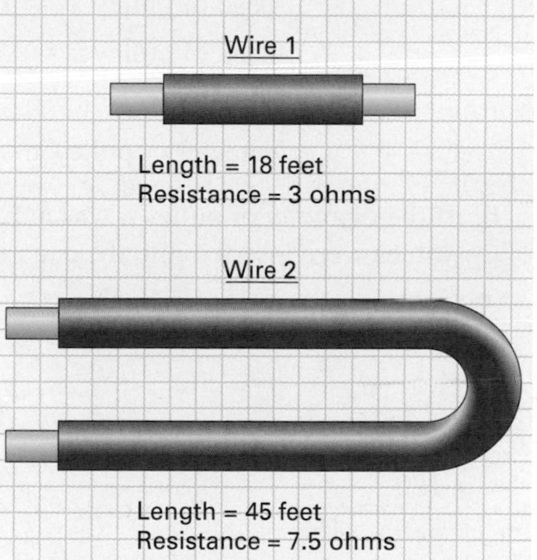

Solution:

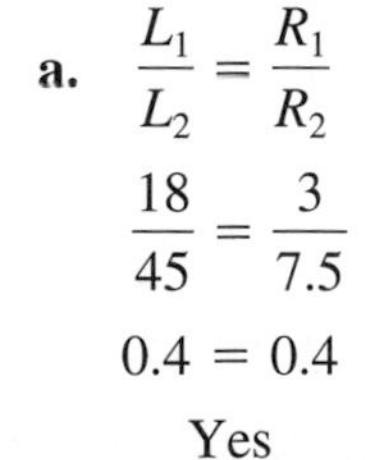

a. $\frac{L_1}{L_2} = \frac{R_1}{R_2}$

$\frac{18}{45} = \frac{3}{7.5}$

$0.4 = 0.4$

Yes

b. L_1 is doubled from 18 feet to 36 feet.

R_1 is doubled from 3 ohms to 6 ohms.

$\frac{36}{45} = \frac{6}{7.5}$

$0.8 = 0.8$

Yes

EXAMPLE:

The pulley system shown details the power transfer from an engine to an air-conditioning unit. With gears, the speed of a pulley is inversely proportional to its diameter (the larger the diameter, $d \uparrow$, the slower the speed, $s \downarrow$); therefore, the speed of pulley x is inversely proportional to its diameter $\left(s_x = \frac{1}{d_x}\right)$, and the speed of pulley y is inversely proportional to its diameter $\left(s_y = \frac{1}{d_y}\right)$.

The following equation describes this inverse relationship. The speed ratio has pulley x above the fraction bar; conversely, the diameter ratio has pulley x below the fraction bar indicating the inversely proportional relationship.

$$\frac{\text{speed of pulley } x}{\text{speed of pulley } y} = \frac{\text{diameter of pulley } y}{\text{diameter of pulley } x}$$

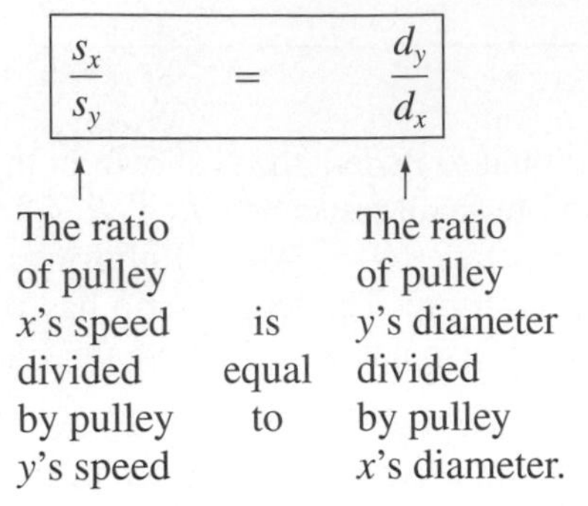

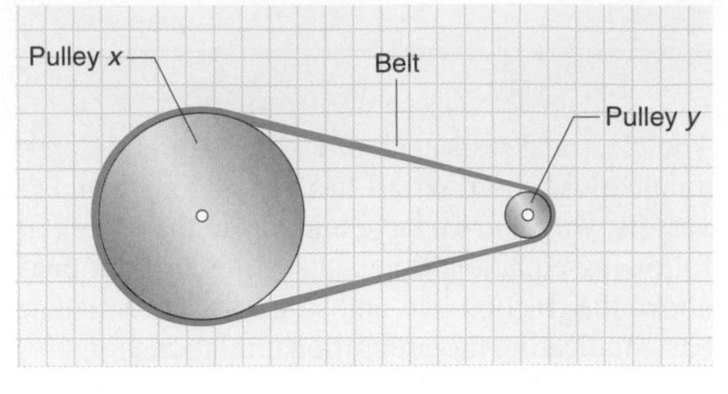

a. Insert the following values in the equation and determine whether both sides of the equation remain equal:

speed of pulley x = 120 revolutions per minute (rpm)
speed of pulley y = 480 rpm
diameter of pulley y = 4 inches
diameter of pulley x = 16 inches

b. If the diameter of pulley x was halved, its speed would be doubled. If these new values were inserted into the equation, would it remain equal?

Solution:

a. $\frac{s_x}{s_y} = \frac{d_y}{d_x}$

$\frac{120}{480} = \frac{4}{16}$

$0.25 = 0.25$

Yes

b. d_x is halved, from 16 inches to 8 inches.

s_x is doubled, from 120 rpm to 240 rpm.

$\frac{240}{480} = \frac{4}{8}$

$0.5 = 0.5$

Yes

3-1-2 *Rounding Off*

Rounding Off
An operation in which a value is abbreviated by applying the following rule: If the first digit to be dropped is 5 or more, increase the previous digit by 1. If the first digit to be dropped is less than 5, do not change the previous digit.

In many cases we will **round off** *decimal numbers because we do not need the accuracy indicated by a large number of decimal digits.* For example, it is usually unnecessary to have so many decimal digits in a value such as 74.139896428. If we were to round off this value to the nearest hundredths place, we would include two digits after the decimal point, which is 74.13. This is not accurate, however, since the digit following 74.13 was a 9 and therefore one count away from causing a reset and carry action into the hundredths column. To take into account the digit that is to be dropped when rounding off, therefore, we follow this basic three-step procedure:

Step 1: To round a decimal to a particular decimal place, first locate the digit in that place and call it the *rounding digit.*

Step 2: Look at the *test digit* to the right of the rounding digit.

Step 3: If the test digit is 5 or greater, add 1 to the rounding digit and drop all digits to the right of the rounding digit. If the test digit is less than 5, do not change the rounding digit, and drop all digits to the right of the rounding digit.

Therefore, the value 74.139896428, rounded off to the nearest hundredths place, would equal

Rounding digit: hundredths place

[74.13] 9896428 = 74.14

Test digit: First digit to be dropped is greater than 5, so rounding digit should be increased by 1.

EXAMPLE:

Round off the value 74.139896428 to the nearest ten, whole number, tenth, hundredth, thousandth, and ten-thousandth.

Solution:

[7] 4.139896428 rounded off to the nearest ten = 70
↑ Test digit to be dropped is less than 5; therefore, do not change rounding digit.

[74.] 139896428 rounded off to the nearest whole number = 74
↑ Test digit to be dropped is less than 5; therefore, do not change rounding digit.

[74.1] 39896428 rounded off to the nearest tenth = 74.1
↑ Test digit to be dropped is less than 5; therefore, do not change rounding digit.

[74.13] 9896428 rounded off to the nearest hundredth = 74.14
↑ Test digit to be dropped is 5 or greater; therefore, increase rounding digit by 1 (3 to 4).

[74.139] 896428 rounded off to the nearest thousandth = 74.140
↑ Test digit to be dropped is 5 or greater; therefore, increase rounding digit by 1. Since rounding digit is 9, allow reset and carry action to occur. The steps are

74.1398 ← 8 carries 1 into thousandths column.
74.130 ← 9 resets to 0, and carries 1 into hundredths column.
74.140 ← Hundredths digit is increased by 1.

[74.1398] 96428 rounded off to the nearest ten-thousandth = 74.1399
↑ Test digit to be dropped is 5 or greater; therefore, increase rounding digit by 1.

CALCULATOR KEYS

Name: Round function
Round (*value* [,*#decimals*])

Function: Value is rounded to the number of decimals specified.

Example:
round (π, 4)

3.1416

3-1-3 *Significant Places*

The number of significant places describes how many digits are in the value and how many digits in the value are accurate after rounding off. For example, a number such as 347.63 is a five-significant-place (or significant-figure) value because it has five digits in five columns. If we were to round off this value to a whole number, we would get 348.00. This value would still be a five-significant-place number; however, it would now be accurate to only three significant places. Let us examine a few problems to practice using the terms *significant places* and *accurate to significant places.*

EXAMPLE:

On a calculator, the value of π will come up as 3.141592654.

a. Write the number π to six significant places.

b. Give π to five significant places, and also round off π to ten-thousandths.

c. 3.14159000 is the value of π to ___________ significant places; however, it is accurate to only ___________ significant places.

Solution:

a. The value π to six significant places or figures is [3.14159] 27

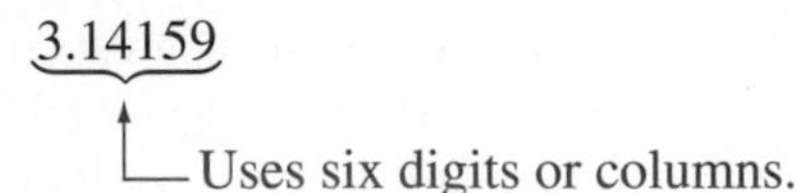

b. The value π to five significant places.

[3.1415] 92 rounded off to ten-thousandths = 3.1416

c. 3.14159000 is the value of π to nine significant places (has nine digits); however, it is accurate to only six significant places (because the zeros to the right are just extra, and the value 3.14159 has only six digits).

3-1-4 *Percentages*

Percent
In the hundred; of each hundred.

The **percent** sign (%) means hundredths, which as a proper fraction is $\frac{1}{100}$ or as a decimal fraction is 0.01. For example, 50% means 50 hundredths $\left(50 \times \frac{1}{100} = \frac{50}{100}\right)$, which is one-half. In decimal, 50% means 50 hundredths ($50 \times 0.01 = 0.5$), which is also one-half. The following shows how some of the more frequently used percentages can be expressed in decimal:

$$1\% = 1 \times 0.01 = 0.01$$
$$5\% = 5 \times 0.01 = 0.05$$
$$10\% = 10 \times 0.01 = 0.10$$
$$25\% = 25 \times 0.01 = 0.25$$
$$50\% = 50 \times 0.01 = 0.50$$
$$75\% = 75 \times 0.01 = 0.75$$

(For example, 1% is expressed mathematically in decimal as 0.01.)

Most calculators have a percent key that automatically makes this conversion to decimal hundredths. It operates as follows.

CALCULATOR KEYS

Name: Percent key

Function: Converts the displayed number from a percentage to a decimal fraction.

Example: 18.6% = ?

Press keys: [1] [8] [.] [6] [%]

Display shows: 0.186

Now that we understand that the percent sign stands for hundredths, what does the following question actually mean: What is 50% of 12? We now know that 50% is 50 hundredths ($\frac{50}{100}$), which is one-half. This question is actually asking: What is half of 12? Expressed mathematically as both proper fractions and decimal fractions, the question would appear as follows:

Proper Fractions	**Decimal Fractions**
50% of 12 = ?	50% of 12 = ?
$= \frac{50}{100} \times 12$	$= (50 \times 0.01) \times 12$
$= \frac{1}{2} \times 12$	$= 0.5 \times 12$
$= 6$	$= 6$

Explaining this another way, we could say that percentages are fractions in hundredths. In the preceding example, therefore, the question is actually asking us to divide 12 into 100 parts and then to determine what value we would have if we had 50 of those 100 parts. Therefore, if we were to split 12 into 100 parts, each part would have a value of 0.12 ($12 \div 100 = 0.12$). Having 50 of these 100 parts would give us a total of $50 \times 0.12 = 6$.

EXAMPLE:

A 330 ohm (330 Ω) resistor is listed as having a ±10% (plus or minus ten percent) tolerance, which means the manufacturer guarantees the value will not be greater than 330 ohms plus 10%, or less than 330 ohms minus 10%. List the resistor's possible range of values.

Solution:

$$330\ \Omega \times 10\% = 33\ \Omega$$
$$330\ \Omega + 33\ \Omega = 363\ \Omega$$
$$330\ \Omega - 33\ \Omega = 297\ \Omega$$

3-1-5 *Averages*

A *mean* **average** *is a value that summarizes a set of unequal values.* This value is equal to the sum of all the values divided by the number of values. For example, if a team has scores of 5, 10, and 15, what is their average score? The answer is obtained by adding all the scores ($5 + 10 + 15 = 30$) and then dividing the result by the number of scores ($30 \div 3 = 10$).

Average
A single value that summarizes or represents the general significance of a set of unequal values.

EXAMPLE:

The voltage at the wall outlet in your home was measured at different times in the day and equaled 108.6, 110.4, 115.5, and 123.6 volts (V). Find the average voltage.

Solution:

Add all the voltages.

$$108.6 + 110.4 + 115.5 + 123.6 = 458.1 \text{ V}$$

Divide the result by the number of readings.

$$\frac{458.1 \text{ V}}{4} = 114.525 \text{ V}$$

The average voltage present at your home was therefore 114.525 volts.

3-1-6 *Graphing Data*

Graph
A diagram (as a series of one or more points, lines, line segments, curves, or areas) that represents the variation of a variable in comparison with that of one or more other variables.

Cartesian Coordinate System
Two perpendicular number lines used to place points in a plane.

In many areas of science and technology, you will often see data shown visually in the form of a **graph** (Figure 3-1). To obtain this data, a *practical analysis* is performed on a device, in which the input is physically changed in increments and the output results are recorded in a table. Read the notes in Figure 3-1 to see how a line graph is plotted from the example table of device data.

Figure 3-1 would accommodate any data in which the input and output data are positive, but what sort of graph could be used if positive and negative values had to be plotted? The answer is a **Cartesian coordinate system.** With this system, the previous positive vertical x-axis and horizontal y-axis are still used, but are extended beyond the zero point in the opposite direction to accommodate negative values, as in Figure 3-2. Again, read the notes in Figure 3-1 to see how the table of results can be used to plot a graph.

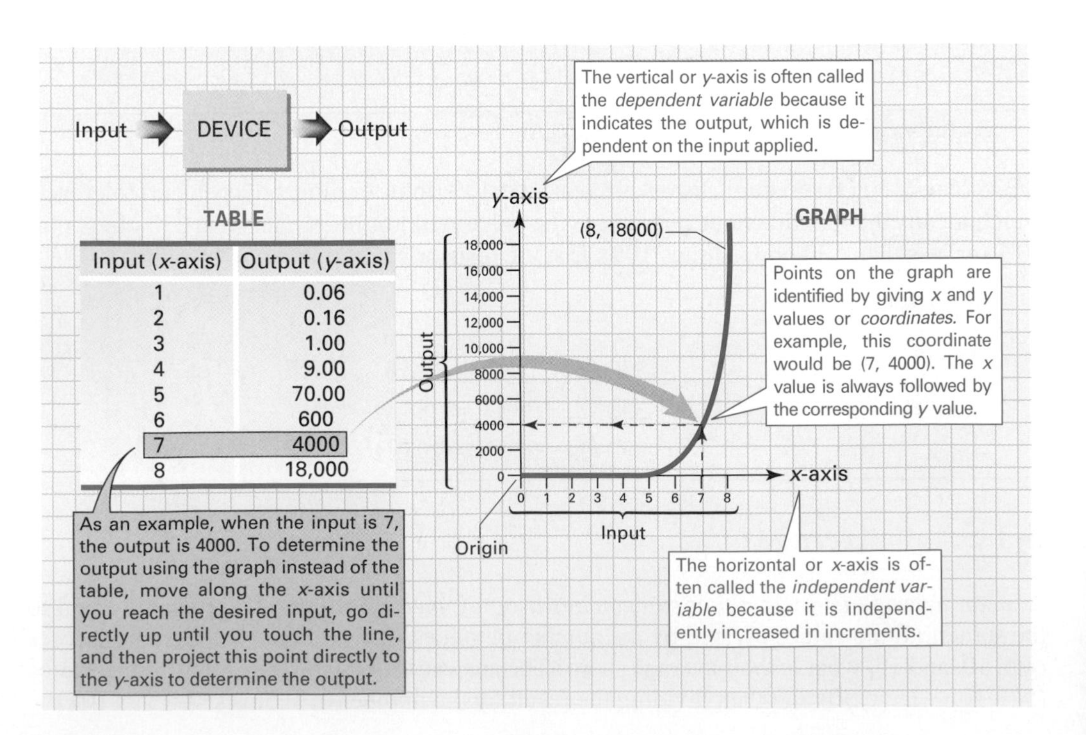

FIGURE 3-1

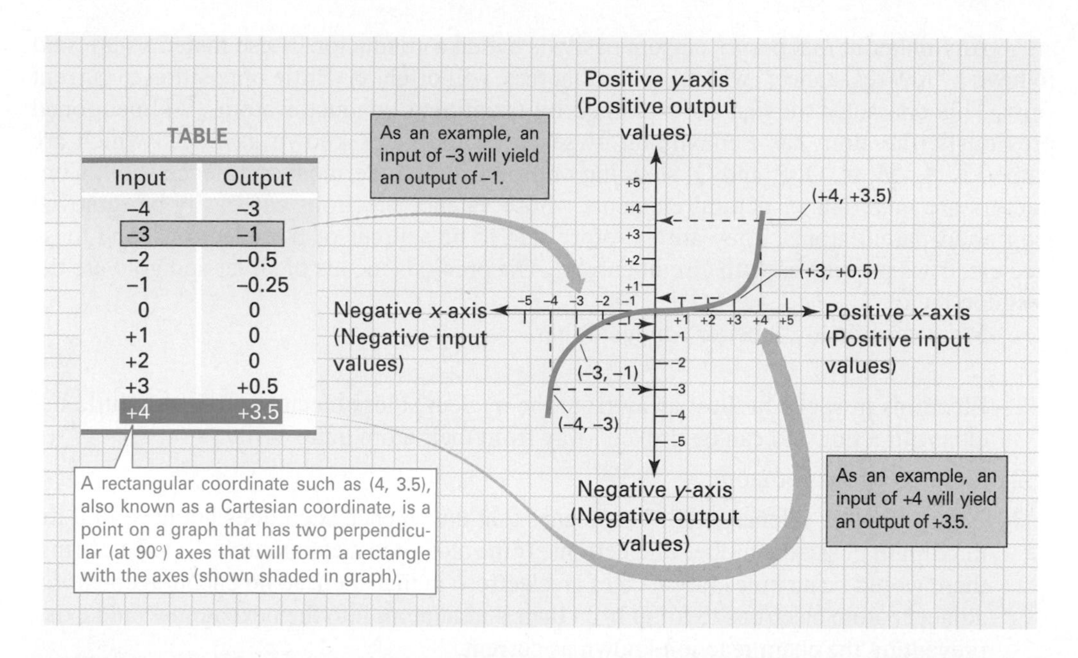

FIGURE 3-2

SELF-TEST EVALUATION POINT FOR SECTION 3-1

Now that you have completed this section, you should be able to:

Objective 1. Explain the following mathematical operations:

a. Ratios
b. Rounding off
c. Significant places
d. Percentages
e. Averages

Objective 2. Describe how to generate the data for a graph.

Objective 3. Define the terms:

a. *x*-axis
b. *y*-axis
c. Coordinates
d. Dependent variable
e. Independent variable

Objective 4. Explain how positive and negative values can be plotted on a Cartesian coordinate system.

Use the following questions to test your understanding of Section 3-1:

1. Express a 176 meter to 8 meter ratio as a fraction, using the ratio sign, and as a decimal.
2. Round off the following values to the nearest hundredth:
   a. 86.43760
   b. 12,263,415.00510
   c. 0.176600
3. Referring to the values in Question 2, describe
   a. Their number of significant places
   b. To how many significant places they are accurate
4. Calculate the following:
   a. 15% of 0.5 = ?
   b. 22% of 1000 = ?
   c. 2.35% of 10 = ?
   d. 96% of 20 = ?
5. Calculate the average of the following values:
   a. 20, 30, 40, and 50
   b. 4000, 4010, 4008, and 3998
6. Plot two lines on a graph to display the following data:

THE NUMBER OF INDOOR AND OUTDOOR MOVIE SCREENS

YEAR	INDOOR	OUTDOOR
1975	12,000	4000
1980	14,000	4000
1985	18,000	3000
1990	23,000	2000

3-2 CONDUCTORS

A lightning bolt that sets fire to a tree and the operation of your calculator are both electrical results achieved by the flow of electrons. The only difference is that your calculator's circuits control the flow of electrons, whereas the lightning bolt is the uncontrolled flow of electrons. In electronics, a **conductor** is used to channel or control the path in which electrons flow.

Conductor
Length of wire whose properties are such that it will carry an electric current.

Any material that passes current easily is called a conductor. These materials are said to have a "low resistance," which means that they will offer very little opposition to current flow. This characteristic can be explained by examining conductor atoms. As mentioned previously, the atom has a maximum of seven orbital paths known as shells, which are named K, L, M, N, O, P, and Q, stepping out toward the outermost or valence shell. Conductors are materials or natural elements whose valence electrons can easily be removed from their parent atoms. They are therefore said to be sources of free electrons, and these free electrons provide us with circuit current. The precious metals of silver and gold are the best conductors.

More specifically, a better conductor has

1. Electrons in shells farthest away from the nucleus; these electrons feel very little nucleus attraction and can be broken away from their atom quite easily.
2. More electrons per atom.
3. An incomplete valence shell. This means that the valence shell does not have in it the maximum possible number of electrons. If the atom had a complete (full) valence ring, there would be no holes (absence of an electron) in that shell, so no encouragement for adjacent atom electrons to jump from their parent atom into the next atom would exist, preventing the chain reaction known as current.

TIME LINE

Stephen Gray (1693–1736), an Englishman, discovered that certain substances would conduct electricity.

Economy must be considered when choosing a conductor. Large quantities of conductors using precious metals are obviously going to send the cost of equipment beyond reach. The conductor must also satisfy some physical requirements, in that we must be able to shape it into wires of different sizes and easily bend it to allow us to connect one circuit to the next.

Copper is the most commonly used conductor, as it meets the following three requirements:

1. It is a good source of electrons.
2. It is inexpensive.
3. It is physically pliable.

Aluminum is also a very popular conductor, and although it does not possess as many free electrons as copper, it is less expensive and lighter than copper.

3-2-1 *Conductance*

Conductance (*G*)
Measure of how well a circuit or path conducts or passes current. It is equal to the reciprocal of resistance.

Conductance is the measure of how good a conductor is at carrying current. Conductance (symbolized G) is equal to the reciprocal of resistance (or opposition) and is measured in the unit siemens (S):

$$\text{conductance } (G) = \frac{1}{\text{resistance } (R)}$$

Conductance (G) values are measured in siemens (S) and resistance (R) in ohms (Ω).

This means that conductance is inversely proportional to resistance. For example, if the opposition to current flow (resistance) is low, the conductance is high and the material is said to have a good conductance.

$$\text{high conductance } G\uparrow = \frac{1}{R\downarrow \text{ (low resistance)}}$$

MATERIAL	CONDUCTIVITY (RELATIVE)
Silver	1.000
Copper	0.945
Gold	0.652
Aluminum	0.575
Tungsten	0.297
Nichrome	0.015

TABLE 3-1 Relative Conductivity of Conductors

On the other hand, if the resistance of a conducting wire is high ($R\uparrow$), its conductance value is low ($G\downarrow$) and it is called a poor conductor. A good conductor therefore has a high conductance value and a very small resistance to current flow.

EXAMPLE:

A household electric blanket offers 25 ohms of resistance to current flow. Calculate the conductance of the electric blanket's heating element.

Solution:

$$\text{conductance} = \frac{1}{\text{resistance}}$$

$$G = \frac{1}{R} = \frac{1}{25 \text{ ohms}} = 0.04 \text{ siemens}$$

or 40 mS (millisiemens)

TIME LINE

Stephen Gray's lead was picked up in 1730 by Charles du Fay, a French experimenter, who believed that there were two types of electricity, which he called *vitreous* and *resinous* electricity.

To express how good or poor a conductor is, we must have some sort of reference point. The reference point we use is the best conductor, silver, which is assigned a relative conductivity value of 1.0. Table 3-1 lists other conductors and their relative conductivity values with respect to the best, silver. The formula for calculating relative conductivity is

$$\text{relative conductivity} = \frac{\text{conductor's relative conductivity}}{\text{reference conductor's relative conductivity}}$$

EXAMPLE:

What is the relative conductivity of tungsten if copper is used as the reference conductor?

Solution:

$$\text{tungsten} = 0.297$$

$$\text{copper} = 0.945$$

$$\text{relative conductivity} = \frac{\text{conductivity of conductor}}{\text{conductivity of reference}} = \frac{0.297}{0.945} = 0.314$$

Calculator Sequence

STEP	KEYPAD ENTRY	DISPLAY RESPONSE
1.	0 . 2 9 7	0.297
2.	÷	
3.	0 . 9 4 5	0.945
4.	=	0.314

EXAMPLE:

What is the relative conductivity of silver if copper is used as the reference?

Solution:

$$\text{relative conductivity} = \frac{\text{silver}}{\text{copper}}$$

$$= \frac{1.000}{0.945} = 1.058$$

3-2-2 *Conductors and Their Resistance*

If the current within a circuit needs to be reduced, a resistor is placed in the current path. Resistors are constructed using materials that are known to oppose current flow. Conductors, on the other hand, are not meant to offer any resistance or opposition to current flow. This, however, is not always the case since some are good conductors having a low resistance, and others are poor conductors having a high resistance. Conductance is the measure of how good a conductor is, and even the best conductors have some value of resistance. Up until this time, we have determined a conductor's resistance based on the circuit's electrical characteristics,

$$R = \frac{V}{I}$$

With this formula we could determine the conductor's resistance based on the voltage applied and circuit current. Another way to determine the resistance of a conductor is to examine the physical factors of the conductor.

The total resistance of a typical conductor, like the one shown in Figure 3-3, is determined by four main physical factors:

1. The type of conducting material used
2. The conductor's cross-sectional area
3. The total length of the conductor
4. The temperature of the conductor

By combining all of the physical factors of a conductor, we can arrive at a formula for its resistance.

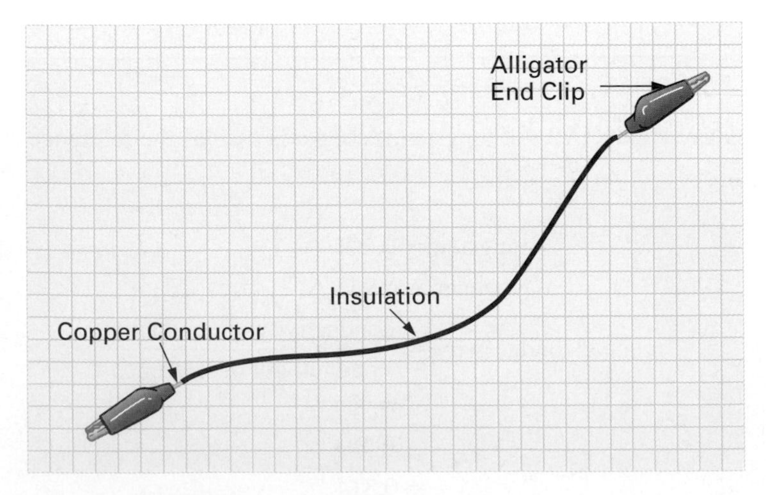

FIGURE 3-3 Copper Conductor.

MATERIAL	RESISTIVITY (cmil/ft) IN OHMS
Silver	9.9
Copper	10.7
Gold	16.7
Aluminum	17.0
Tungsten	33.2
Zinc	37.4
Brass	42.0
Nickel	47.0
Platinum	60.2
Iron	70.0

TABLE 3-2 **Material Resistivity**

$$R = \frac{\rho \times l}{a}$$

where R = resistance of conductor, in ohms
ρ = resistivity of conducting material
l = length of conductor, in feet
a = area of conductor, in circular mils

Resistivity, by definition, is the resistance (in ohms) that a certain length of conductive material (in feet) will offer to the flow of current. Table 3-2 lists the resistivity of the more commonly used conductors. To explain the area of a conductor, a unit of measure was needed to compare one thickness of wire conductor to another. Figure 3-4 illustrates a conductor with a diameter of 0.001 inch (1/1000 of an inch) or 1 **mil.**

Resistivity
Measure of a material's resistance to current flow.

Mil
One thousandth of an inch (0.001 in.).

The diameters of all conductors are measured in mils. As nearly all conductors are circular in cross section, the area of a conductor is normally given in **circular mils** (cmil), which can be calculated by squaring the diameter.

Circular Mil (cmil)
A unit of area equal to the area of a circle whose diameter is 1 mil.

$$\text{circular mils (cmil)} = \text{diameter}^2$$

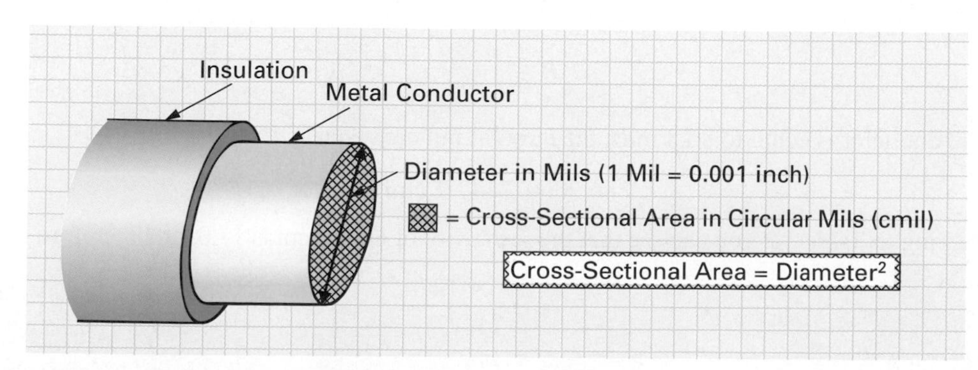

FIGURE 3-4 **Measuring a Conductor's Cross-Sectional Area.**

EXAMPLE:

If a conductor has a diameter of 71.96 mils, what is the circular mil area?

Solution:

$$\text{cmil} = \text{diameter}^2$$
$$= 71.96^2$$
$$= 5178 \text{ cmil}$$

Calculator Sequence

STEP	KEYPAD ENTRY	DISPLAY RESPONSE
1.	7 1 . 9 6	71.96
2.	x^2	5178.24

The resistance of a conductor, therefore, can be calculated based either on its physical factors or on electrical performance.

$$\frac{\rho \times l}{a} = R = \frac{V}{I}$$

Physical ⇔ Electrical

EXAMPLE:

Calculate the resistance of 333 ft of copper conductor with a conductor area of 3257 cmil.

Solution:

$$R = \frac{\rho \times l}{a}$$
$$= \frac{10.7 \times 333}{3257}$$
$$= 1.09\ \Omega$$

Calculator Sequence

STEP	KEYPAD ENTRY	DISPLAY RESPONSE
1.	1 0 . 7	10.7
2	×	
3.	3 3 3	333
4.	÷	3563.1
5.	3 2 5 7	3257
6.	=	1.09

EXAMPLE:

Calculate the resistance of 1274 ft of aluminum conductor with a diameter of 86.3 mils.

Solution:

$$R = \frac{\rho \times l}{a}$$

If the diameter is equal to 86.3 mils, the circular mil area equals d^2:

$$86.3^2 = 7447.7 \text{ cmil}$$

Referring to Table 3-2 you can see that the resistivity of aluminum is 17.0, and therefore

$$R = \frac{17 \times 1274}{7447.7}$$
$$= 2.9\ \Omega$$

3-2-3 *Temperature Effects on Conductors*

When heat is applied to a conductor, the atoms within the conductor convert this thermal energy into mechanical energy or movement. These random moving atoms cause collisions between the directed electrons (current flow) and the adjacent atoms, resulting in an opposition to the current flow (resistance).

Metallic conductors are said to have a **positive temperature coefficient of resistance** (+ Temp. Coe. of R). This means that the greater the heat applied to the conductor, the greater the atom movement, causing more collisions of atoms to occur and consequently, greater conductor resistance.

Positive Temperature Coefficient of Resistance
The rate of increase in resistance relative to an increase in temperature.

heat ↑ resistance ↑

To explain further the effects temperature has on a conductor, Figure 3-5(a) illustrates the **filament resistor** within a glass bulb. This component is more commonly known as the household lightbulb. The filament resistor is just a coil of wire that glows white hot when current is passed through it and, in so doing, dissipates both heat and light energy. This **incandescent lamp** is an electric lamp in which electric current flowing through a filament of resistive material heats the filament until it glows and emits light. Figure 3-5(b) shows a test circuit for varying current through a small incandescent lamp. Potentiometer R_1 is used to vary the current flow through the lamp. The lamp is rated at 6 V, 60 mA. Using Ohm's law, we can calculate the lamp's filament resistance at this rated voltage and current:

Filament Resistor
The resistor in a lightbulb or electron tube.

Incandescent Lamp
An electric lamp that generates light when an electric current is passed through its filament of resistance, causing it to heat to incandescence.

$$\text{filament resistance, } R = \frac{V}{I} = \frac{6\ V}{60\text{ mA}} = 100\ \Omega$$

The filament material (tungsten, for example) is like all other conductors in that it has a positive temperature coefficient of resistance. Therefore, as the current through the filament increases, so does the temperature and so does the filament's resistance ($I\uparrow$, temperature $\uparrow$, $R\uparrow$). Consequently, when variable resistor R_1's wiper is moved to the right so that it produces a high resistance ($R_1\uparrow$), the circuit current will be small ($I\downarrow$) and the lamp will glow dimly. Since the circuit current is small, the filament temperature will be small and so will the lamp's resistance ($I\downarrow$ temperature $\downarrow$, lamp resistance $\downarrow$). This small value of resistance is called the lamp's **cold resistance.** On the other hand, when R_1's wiper is moved to the left so that it produces a small resistance ($R\downarrow$), the circuit current will be large ($I\uparrow$), and the lamp will glow brightly. With the circuit current high, the filament temperature will be high and so will the lamp's resistance ($I\uparrow$, temperature $\uparrow$, lamp resistance $\uparrow$). This large value of resistance is called the lamp's **hot resistance.**

Cold Resistance
The resistance of a device when cold.

Hot Resistance
The resistance of a device when hot due to the generation of heat by electric current.

Figure 3-5(c) plots the filament voltage, which is being measured by the voltmeter, against the filament current, which is being measured by the ammeter. As you can see, an increase in current will cause a corresponding increase in filament resistance. Studying this graph, you may have also noticed that the lamp has been operated beyond its rated value of 6 V, 60 mA. Although the lamp can be operated beyond its rated value (for example, 10 V, 80 mA), its life expectancy will be decreased dramatically from several hundred hours to only a few hours.

A coil of wire similar to the one found in a lightbulb, called a **ballast resistor,** can be used to maintain a constant current despite a variation in voltage. Since the coil of wire is a conductor and conductors have a positive temperature coefficient, an increase in voltage will cause a corresponding increase in current and therefore in the heat generated, which will result in an increase in the wire's resistance. This increase in resistance will decrease the initial current rise. Similarly, a decrease in voltage and therefore current ($I \propto V$) will result in a decrease in heat and wire resistance. This decrease in resistance will permit an increase in current to counteract the original decrease. Current is therefore regulated or maintained constant by the ballast resistor despite variations in voltage.

Ballast Resistor
A resistor that increases in resistance when voltage increases. It can therefore maintain a constant current despite variations in line voltage.

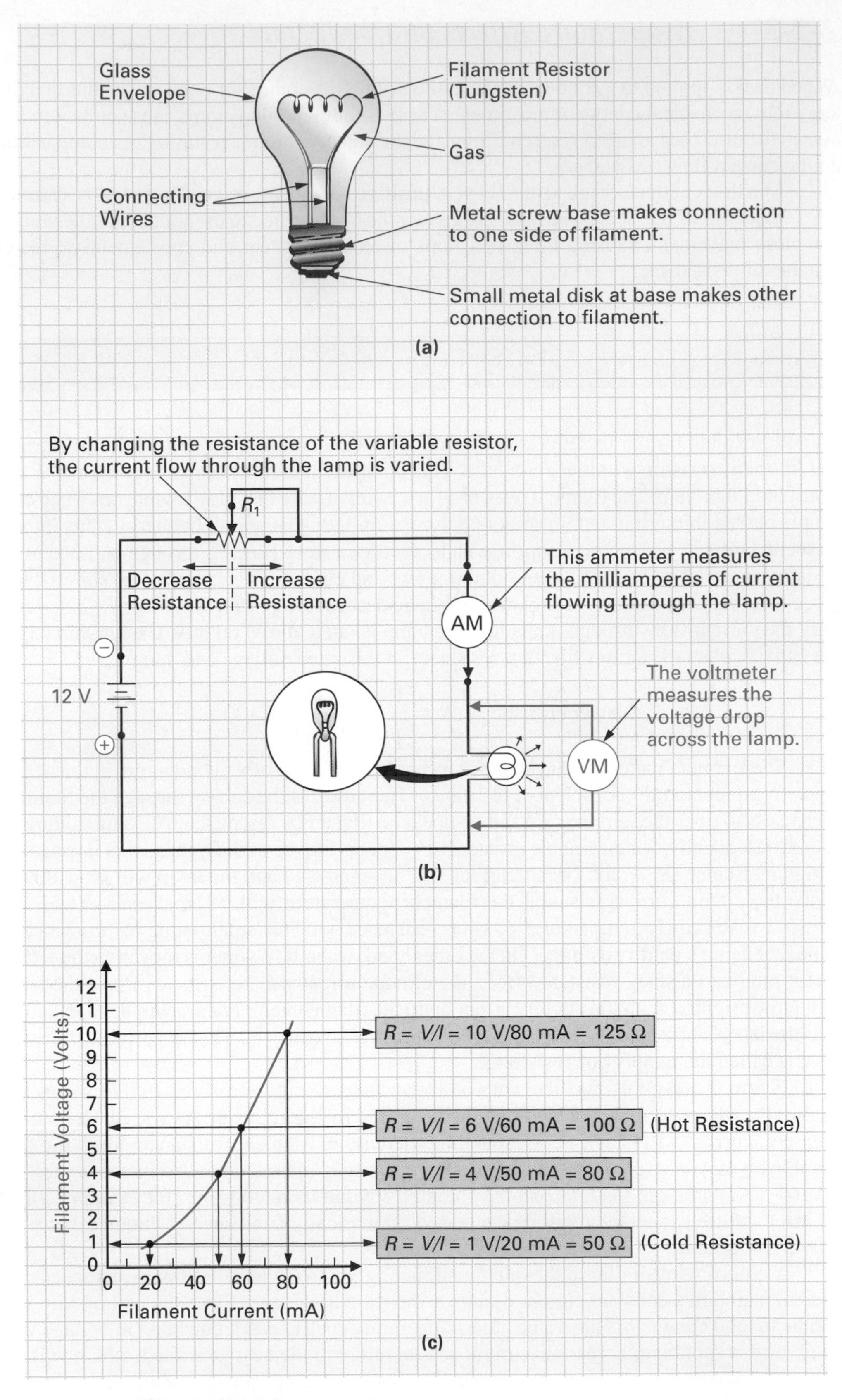

FIGURE 3-5 **Filament Resistor.**

3-2-4 *Maximum Conductor Current*

Any time that current flows through any conductor, a certain resistance or opposition is inherent in that conductor. This resistance will convert current to heat, and the heat further increases the conductor's resistance (+ Temp. Coe. of R), causing more heat to be generated due to the opposition. As a result, a conductor must be chosen carefully for each application so that it can carry the current without developing excessive heat. This is achieved by selecting a conductor with a greater cross-sectional area to decrease its resistance. The National Fire Protection Association has developed a set of standards known as the **American Wire Gauge** for all copper conductors, which lists their diameter, resistance, and maximum safe current in amperes. This data is given in Table 3-3. A rough guide for measuring wire size is shown in Figure 3-6.

Conducting wires are normally covered with a plastic or rubber type of material, known as insulation, as shown in Figure 3-7. This insulation is used to protect users and technicians from electrical shock and also to keep the conductor from physically contacting other conductors within the equipment. If the current through the conductor is too high, this insulation will burn due to the heat and may cause a fire hazard. This is why a conductor's maximum safe current value should not be exceeded.

American Wire Gauge
American Wire Gauge (AWG) is a system of numerical designations of wire sizes, with the first being 0000 (the largest size) and then going to 000, 00, 0, 1, 2, 3, and so on up to the smallest sizes of 40 and above.

AWG NUMBER	DIAMETER (mils)	MAXIMUM CURRENT (A)	Ω/1000 ft
0000	460.0	230	0.0490
000	409.6	200	0.0618
00	364.8	175	0.0780
0	324.9	150	0.0983
1	289.3	130	0.1240
2	257.6	115	0.1563
3	229.4	100	0.1970
4	204.3	85	0.2485
5	181.9	75	0.3133
6	162.0	65	0.3951
7	144.3	55	0.4982
8	128.5	45	0.6282
9	114.4	40	0.7921
10	101.9	30	0.9981
11	90.74	25	1.260
12	80.81	20	1.588
13	71.96	17	2.003
14	64.08	15	2.525
15	57.07		3.184
16	50.82	6	4.016
17	45.26		5.064
18	40.30	3	6.385
19	35.89	Wires of	8.051
20	31.96	this size	10.15
22	25.35	have	16.14
26	15.94	current	40.81
30	10.03	measured	103.21
40	3.145	in mA	1049.0

Note: The larger the AWG number, the smaller the size of the conductor.

TABLE 3-3 The American Wire Gauge (AWG) for Copper Conductor

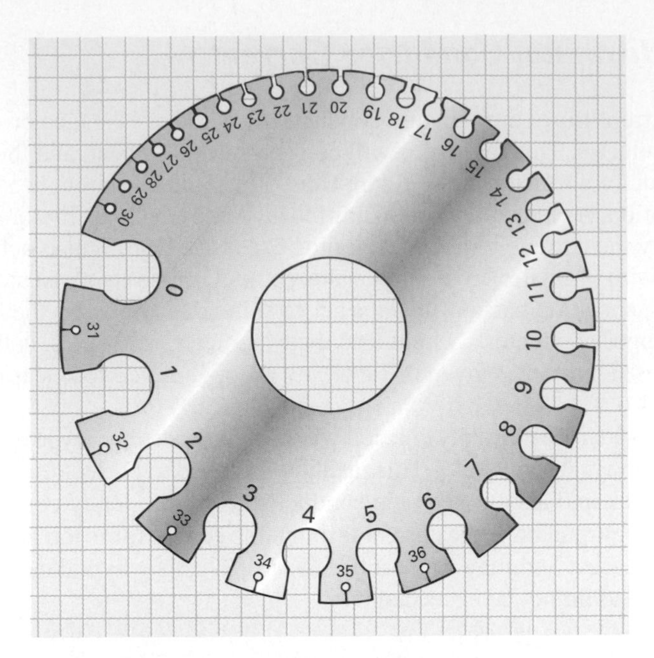

FIGURE 3-6 **Wire Gauge Size.**

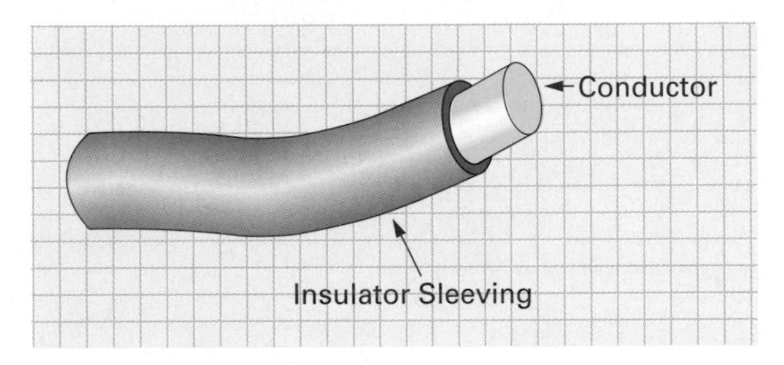

FIGURE 3-7 **Conductor with Insulator.**

3-2-5 *Conductor and Connector Types*

Wire
Single solid or stranded group of conductors having a low resistance to current flow.

Cable
Group of two or more insulated wires.

A cable is made up of two or more wires. Figure 3-8 illustrates some of the different types of **wires** and **cables.** In Figure 3-8(a), (b), and (c), only one conductor exists within the insulation, so they are classified as wires, whereas the cables seen in Figure 3-8(e) and (f) have two conductors. The coaxial and twin-lead cables are most commonly used to connect TV signals into television sets. Figure 3-8(d) illustrates how conducting copper, silver, or gold strips are printed on a plastic insulating board and are used to connect components such as resistors that are mounted on the other side of the board.

Wires and cables have to connect from one point to another. Some are soldered directly, whereas others are attached to plugs that plug into sockets. A sample of connectors is shown in Figure 3-9.

3-2-6 *Switches*

As we have seen previously, a *switch* is a device that completes (closes) or breaks (opens) the path of current, as seen in Figure 3-10. All mechanical switches can be classified into one of the eight categories shown in Figure 3-11. Many different variations of these eight classifications exist, and throughout this text you will see many of them incorporated into a variety of circuit applications.

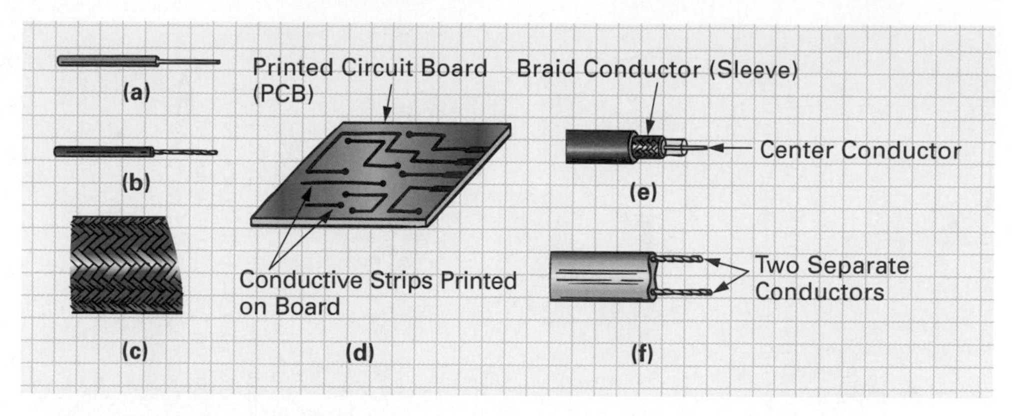

FIGURE 3-8 **Wires and Cables. (a) Solid Wire. (b) Stranded Wire. (c) Braided Wire. (d) Printed Wire. (e) Coaxial Cable. (f) Twin Lead.**

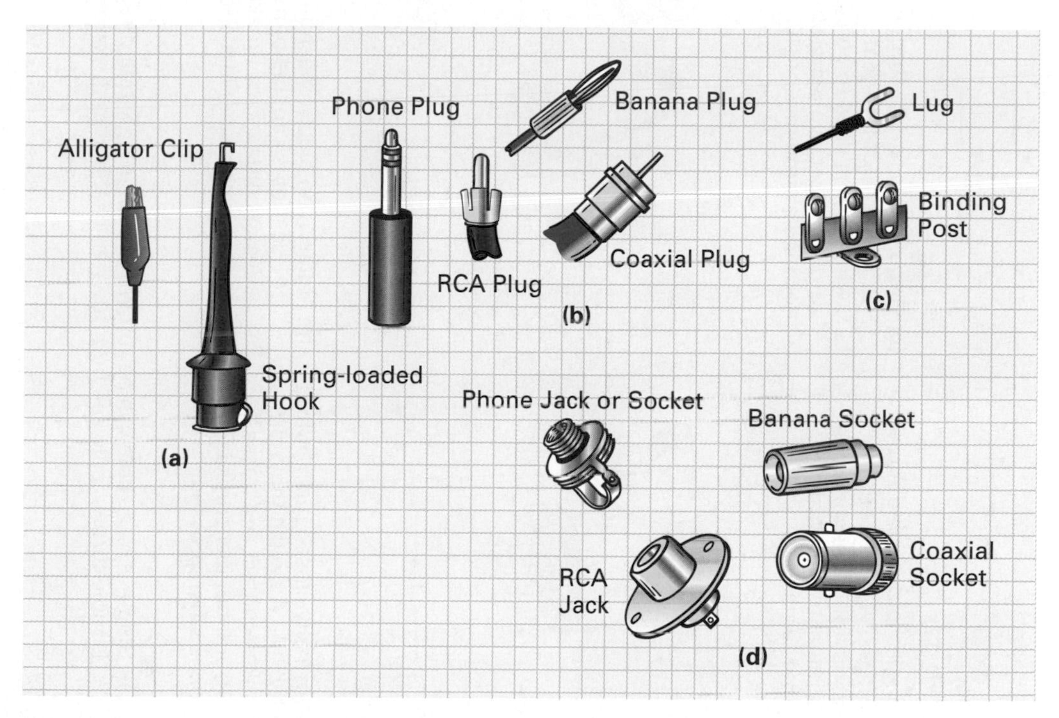

FIGURE 3-9 **Connectors. (a) Temporary Connectors. (b) Plugs. (c) Lug and Binding Post. (d) Sockets.**

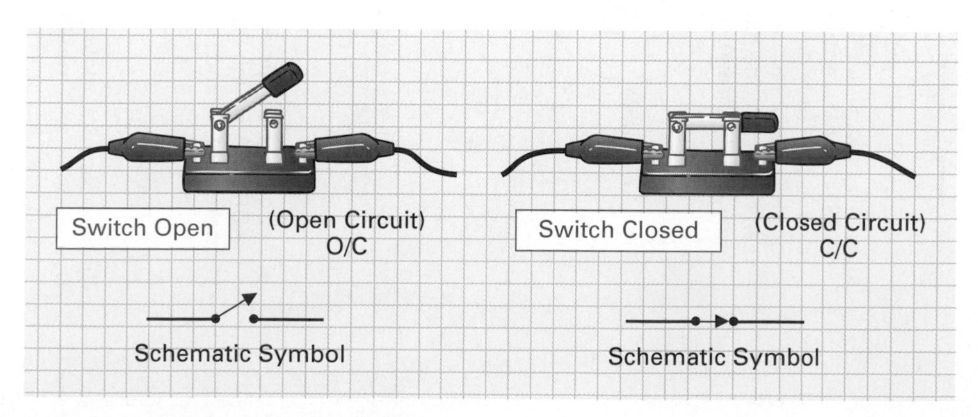

FIGURE 3-10 **Open and Closed Switches.**

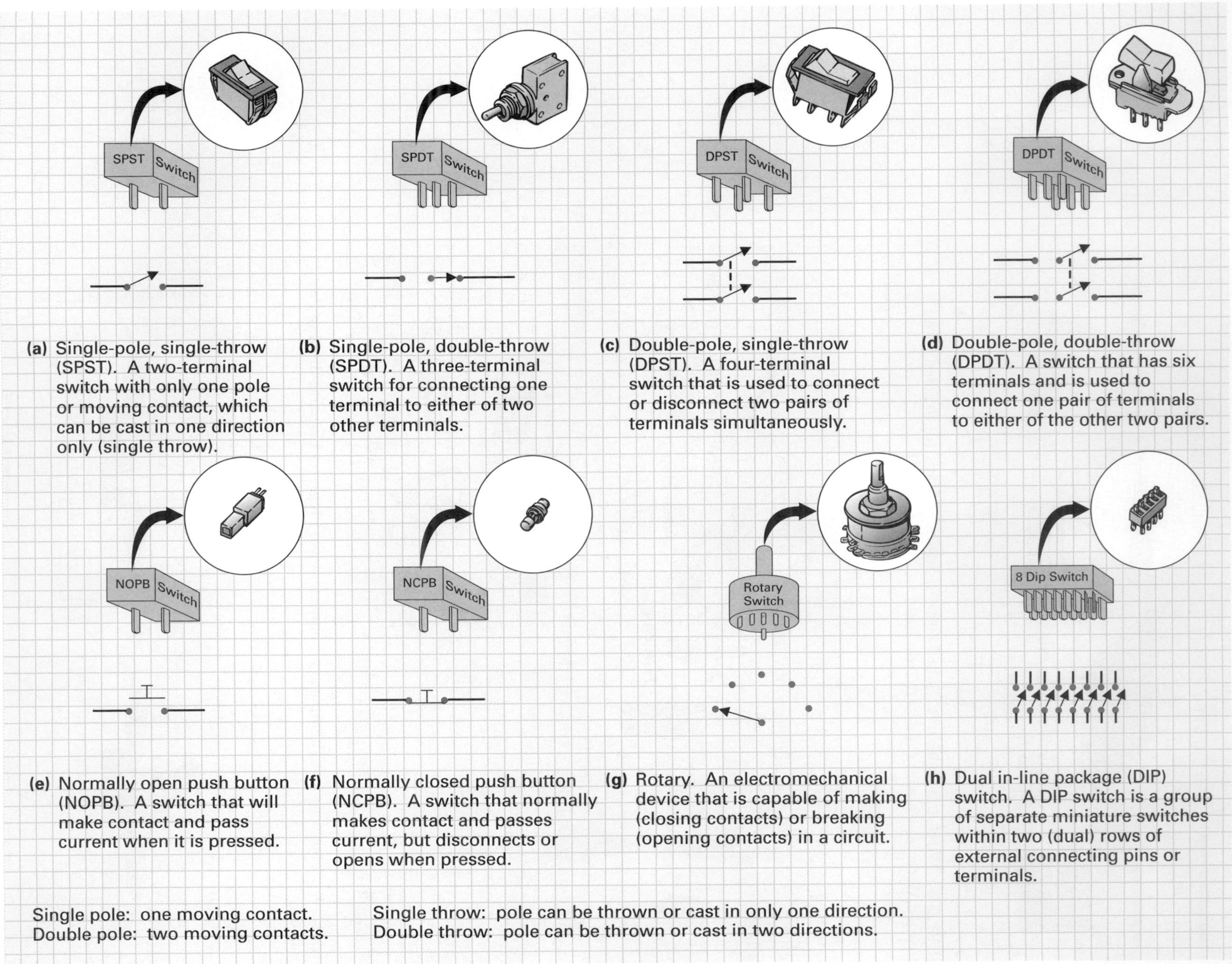
SPST Switch
SPDT Switch
DPST Switch
DPDT Switch
(a) Single-pole, single-throw (SPST). A two-terminal switch with only one pole or moving contact, which can be cast in one direction only (single throw).
(b) Single-pole, double-throw (SPDT). A three-terminal switch for connecting one terminal to either of two other terminals.
(c) Double-pole, single-throw (DPST). A four-terminal switch that is used to connect or disconnect two pairs of terminals simultaneously.
(d) Double-pole, double-throw (DPDT). A switch that has six terminals and is used to connect one pair of terminals to either of the other two pairs.
NOPB Switch
NCPB Switch
Rotary Switch
8 Dip Switch
(e) Normally open push button (NOPB). A switch that will make contact and pass current when it is pressed.
(f) Normally closed push button (NCPB). A switch that normally makes contact and passes current, but disconnects or opens when pressed.
(g) Rotary. An electromechanical device that is capable of making (closing contacts) or breaking (opening contacts) in a circuit.
(h) Dual in-line package (DIP) switch. A DIP switch is a group of separate miniature switches within two (dual) rows of external connecting pins or terminals.
Single pole: one moving contact.
Double pole: two moving contacts.
Single throw: pole can be thrown or cast in only one direction.
Double throw: pole can be thrown or cast in two directions.

FIGURE 3-11 Eight Basic Types of Switches.

3-2-7 *Superconductivity*

Conductors have a positive temperature coefficient, which means that if temperature increases, so does resistance. But what happens if the temperature is decreased? In 1911, a Dutch physicist, Heike Onnes, discovered that mercury (a liquid conductor) lost its resistance to electrical current when the temperature was decreased to −459.7 Fahrenheit (0 on the Kelvin temperature scale). Mercury actually became a **superconductor,** allowing a supercurrent to flow and not encounter any resistance and therefore not generate any heat. (Heat dissipated by any resistance can be calculated by the power formula $P = I^2 \times R$. If $R = 0$, the power dissipated by the conductor is 0 watts.)

Superconductor
Metal such as lead or niobium that when cooled to within a few degrees of absolute zero can conduct current with no electrical resistance.

In the spring of 1986, two IBM scientists discovered that a conductor compound made up of barium, lanthanum, copper, and oxygen would superconduct (have no resistance to current flow) at −406°F. Since then other scientists have increased the temperature to a point that now conductor compounds can be made to superconduct at −300°F. Some scientific projects that need to make use of the advantages of superconductivity immerse these conductor compounds in a bath of liquid nitrogen so that they can achieve superconductivity. Liquid nitrogen, however, is difficult to handle, so the real scientific achievement will be to find a conductor compound that will superconduct at room temperature.

When discussing the AWG table (Table 3-3), a maximum value of current was stated for a given thickness of wire. This is because a large current will generate a large amount of heat if the resistance is too large, so to decrease the resistance and therefore heat generated ($P\uparrow = I^2R\uparrow$), a thicker conductor with a lower resistance is used. Superconductivity allows a standard 1000 A conductor, which would be approximately 2 in. thick, to be replaced by a superconductor about the thickness of a human hair. Not only will conductors be reduced dramatically in size but the increased efficiency (as no power is being wasted in the form of heat) will result in high energy savings.

SELF-TEST EVALUATION POINT FOR SECTION 3-2

Now that you have completed this section, you should be able to:

Objective 5. List the three atomic properties that govern why a material is a good conductor.

Objective 6. Define the following terms:

- **a.** Conductance
- **b.** Relative conductivity
- **c.** Resistivity

Objective 7. Describe how the resistance of a conductor can be calculated based on either its physical factors or electrical performance.

Objective 8. Explain why conductors have a positive temperature coefficient of resistance.

Objective 9. Define the following terms:

- **a.** Filament resistor
- **b.** Cold resistance
- **c.** Hot resistance
- **d.** Ballast resistor

Objective 10. Explain the purpose of the American Wire Gauge (AWG) standard.

Objective 11. List some of the different types of conductors, connectors, and switches.

Objective 12. Define the term *superconductor,* and explain its advantages.

Use the following questions to test your understanding of Section 3-2:

1. True or false: A conductor is a material used to block the flow of current.
2. List the three atomic properties that make a better conductor.
3. What is the most commonly used conductor in the field of electronics?
4. Calculate the conductance of a 35-ohm heater element.
5. List the four physical factors that determine the total resistance of a conductor.
6. What is the relationship between the resistance of a conductor and its cross-sectional area, length, and resistivity?
7. True or false: Conductors are said to have a positive temperature coefficient.
8. True or false: The smaller the AWG number, the larger the size of the conductor.

3-3 INSULATORS

Insulator
A material that has few electrons per atom and electrons that are close to the nucleus and cannot be easily removed.

Breakdown Voltage
The voltage at which breakdown of an insulator occurs.

Any material that offers a high resistance or opposition to current flow is called an **insulator.** Conductors permit the easy flow of current and so have good conductivity, but insulators allow small to almost no amount of free electrons to flow. Insulators can, with sufficient pressure or voltage applied across them, "break down" and conduct current. This **breakdown voltage** must be great enough to dislodge the electrons from their close orbital shells (K, L shells) and release them as free electrons.

The best insulator should have the maximum possible resistance and conduct no current at all. The strength of an insulator indicates how good or bad an insulator is by indicating the voltage that will cause the insulating material to break down and conduct a large current. Table 3-4 lists some of the more popular insulators and the value of kilovolts that will cause a centimeter of insulator to break down. In Figure 3-12, for example, if 1 centimeter of paper is connected to a variable voltage source, a voltage of 500 kilovolts is needed to break down the paper and cause current to flow. Insulators are also referred to as *dielectrics,* because when a voltage is applied across them, as shown in Figure 3-12, an electric field is established within them. Between the two ends (*di*) of the insulator, therefore, an electric field exists when a voltage is applied, hence the term *dielectric.*

The following formula can be used to calculate the insulator or dielectric thickness needed to withstand a certain voltage.

$$\text{dielectric thickness} = \frac{\text{voltage to insulate}}{\text{insulator's breakdown voltage}}$$

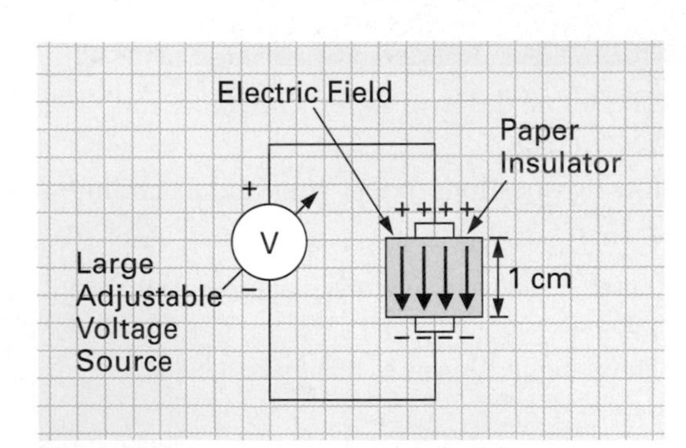

FIGURE 3-12 **Dielectric Thickness Test.**

MATERIAL	BREAKDOWN STRENGTH (kV/cm)
Mica	2000
Glass	900
Teflon	600
Paper	500
Rubber	275
Bakelite	151
Oil	145
Porcelain	70
Air	30

TABLE 3-4 **Breakdown Voltages of Certain Insulators**

EXAMPLE:

What thickness of mica would be needed to withstand 16,000 V?

Solution:

$$\text{mica strength} = 2000 \text{ kV/cm}$$

$$\text{dielectric thickness} = \frac{16{,}000 \text{ V}}{2000 \text{ kV/cm}} = 0.008 \text{ cm}$$

Calculator Sequence

STEP	KEYPAD ENTRY	DISPLAY RESPONSE
1.	[1] [6] [E] (exponent)[3]	16E3
2.	[÷]	
3.	[2] [E] [6] (or 2000 E3)	2E6
4.	[=]	0.008

EXAMPLE:

What maximum voltage could 1 mm of air withstand?

Solution:

There are 10 mm in 1 cm. If air can withstand 30,000 V/cm, it can withstand 3000 V/mm.

SELF-TEST EVALUATION POINT FOR SECTION 3-3

Now that you have completed this section, you should be able to:

Objective 13. Explain the difference between

a. A conductor
b. An insulator

Objective 14. Define the term *breakdown voltage,* and describe how to determine dielectric thickness.

Use the following questions to test your understanding of Section 3-3:

1. True or false: An insulator is a material used to block the flow of current.
2. What is considered to be the best insulator material?
3. Define *breakdown voltage.*
4. Would the conductance figure of a good insulator be large or small?

3-4 RESISTORS

Conductors are used to connect one device to another, and although they offer a small amount of resistance, this resistance is not normally enough. In electronic circuits, additional resistance is normally needed to control the amount of current flow, and the component used to supply this additional resistance is called a **resistor.** Resistors come in a variety of shapes and sizes. Some have a fixed value of resistance; others can be adjusted manually to change their resistance. To begin, let us examine some of the basic resistor characteristics.

Resistor
Component made of a material that opposes the flow of current and therefore has some value of resistance.

3-4-1 *Resistor Wattage, Tolerance, and Color Coding*

Figure 3-13 illustrates four resistors that range in value from 2 Ω to 10 MΩ (10 million Ω). Resistors are constructed by placing a piece of resistive material with embedded connecting leads at each end within an insulating cylindrical molded case, as seen in Figure 3-14. The resistance of the resistor is changed by varying the ratio of conductor to insulator within the resistive material. The 2 Ω resistor in Figure 3-13, therefore, has more conductive elements

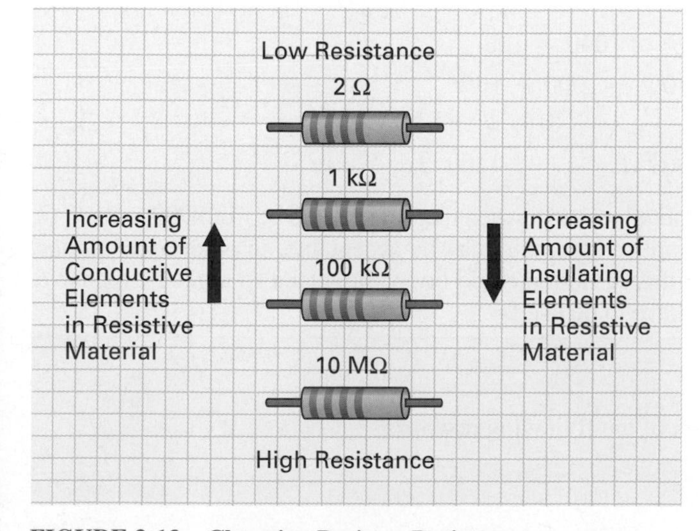

FIGURE 3-13 **Changing Resistor Resistance.**

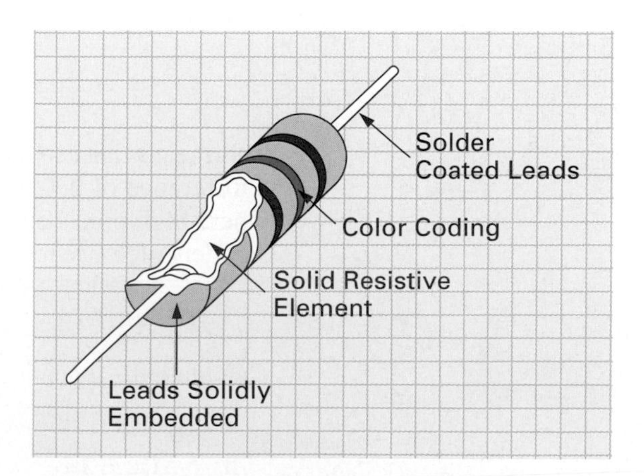

FIGURE 3-14 **Resistor Construction.**

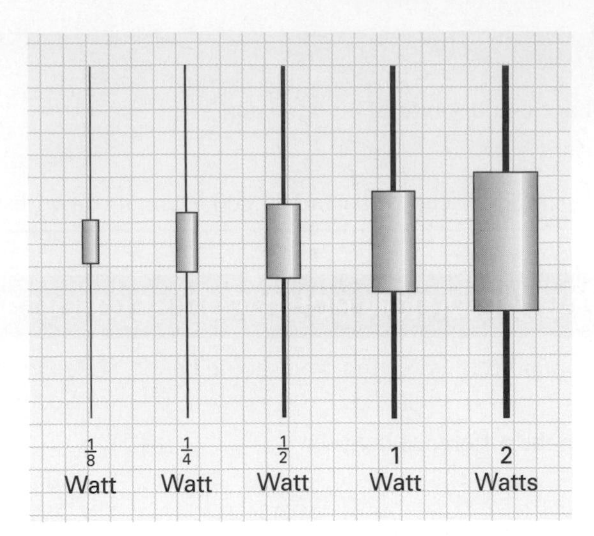

FIGURE 3-15 Resistor Wattage Rating Guide. All Resistor Silhouettes Are Drawn to Scale.

than insulating elements in its resistive material, whereas the 10 MΩ resistor has fewer conducting elements than insulating elements in its resistive material.

1. Resistor Wattage Rating

Dissipation
Release of electrical energy in the form of heat.

Wattage Rating
Maximum power a device can safely handle continuously.

The physical size of the resistors lets the user know how much power in the form of heat can be **dissipated,** as shown in Figure 3-15. As you already know, resistance is the opposition to current flow, and this opposition causes heat to be generated whenever current is passing through a resistor. The amount of heat dissipated each second is measured in watts and each resistor has its own **wattage rating.** For example, a 2-watt-size resistor can dissipate up to 2 joules of heat per second, whereas a 1/8-W-size resistor can only dissipate up to 1/8 joule of heat per second. The key point to remember is that resistors in high-current circuits should have a surface area that is large enough to dissipate the heat faster than it is being generated. If the current passing through a resistor generates heat faster than the resistor can dissipate it, the resistor will burn up and no longer perform its function.

2. Resistor Tolerance

Tolerance
Permissible deviation from a specified value, normally expressed as a percentage.

Another factor to consider when discussing resistors is their **tolerance.** Tolerance is the amount of deviation or error from the specified value. For example, a 1000 Ω (1 kΩ) resistor with a ±10% (plus and minus 10%) tolerance when manufactured could have a resistance anywhere between 900 and 1100 Ω.

$$10\% \text{ of } 1000 = 100$$

$$\begin{array}{ccc} 10\% & - \; 1000 \; + & 10\% \\ \downarrow & & \downarrow \\ 900\ \Omega & & 1100\ \Omega \end{array}$$

This means that two identically marked resistors when measured could be from 900 to 1100 Ω, a difference of 200 Ω. In some applications, this may be acceptable. In other applications, where high precision is required, this deviation could be too large and so a more expensive, smaller-tolerance resistor would have to be used.

EXAMPLE:

Calculate the amount of deviation of the following resistors:

a. 2.2 kΩ ± 10%
b. 5 MΩ ± 2%
c. 3 Ω ± 1%

■ *Solution:*

a. 10% of 2.2 kΩ = 220 Ω. For +10%, the value is

$$2200 + 220\ \Omega = 2420\ \Omega$$
$$= 2.42\ \text{k}\Omega$$

For −10%, the value is

$$2200 - 220\ \Omega = 1980\ \Omega$$
$$= 1.98\ \text{k}\Omega$$

The resistor will measure anywhere from 1.98 kΩ to 2.42 kΩ.

Calculator Sequence

STEP	KEYPAD ENTRY	DISPLAY RESPONSE
1.	1 0	10
2.	%	10
3.	×	0.10
4.	2 . 2 E 3	2.2E3
5.	=	220

b. 2% of 5 MΩ = 100 kΩ

$$5\ \text{M}\Omega + 100\ \text{k}\Omega = 5.1\ \text{M}\Omega$$
$$5\ \text{M}\Omega - 100\ \text{k}\Omega = 4.9\ \text{M}\Omega$$

Deviation = 4.9 MΩ to 5.1 MΩ

c. 1% of 3 Ω = 0.03 Ω or 30 milliohms (mΩ)

$$3\ \Omega + 0.03\ \Omega = 3.03\ \Omega$$
$$3\ \Omega - 0.03\ \Omega = 2.97\ \Omega$$

Deviation = 2.97 Ω to 3.03 Ω

3. Resistor Color Code

Manufacturers indicate the value and tolerance of resistors on the body of the component using either a color code (colored rings or bands) or printed alphanumerics (alphabet and numerals).

Resistors that are smaller in size tend to have the value and tolerance of the resistor encoded using colored rings, as shown in Figure 3-16(a).To determine the specifications of a color-coded resistor, therefore, you will have to decode the rings. Referring to Figure 3-16(b), you can see that when the value, tolerance, and wattage of a resistor are printed on the body, no further explanation is needed.

There are basically two different types of fixed-value resistors: general purpose and precision. Resistors with tolerances of ±2% or less are classified as *precision resistors* and have five bands. Resistors with tolerances of ±5% or greater have four bands and are referred to as *general-purpose resistors*. The color code and differences between precision and general-purpose resistors are explained in Figure 3-17.

When you pick up a resistor, look for the bands that are nearer to one end. This end should be held in your left hand. If there are four bands on the resistor, follow the general-purpose resistor code. If five bands are present, follow the precision resistor code.

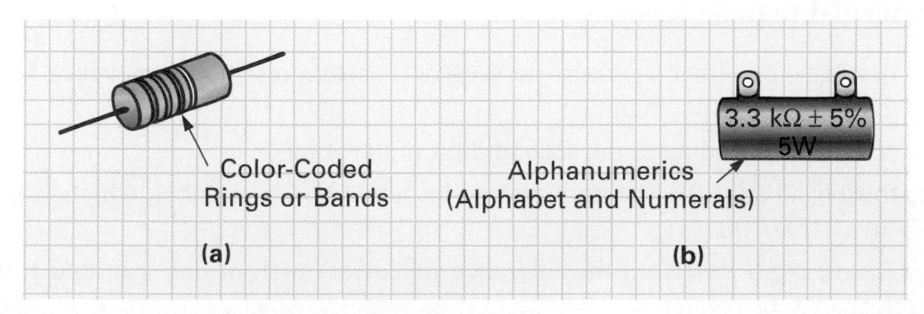

FIGURE 3-16 **Resistor Value and Tolerance Markings.**

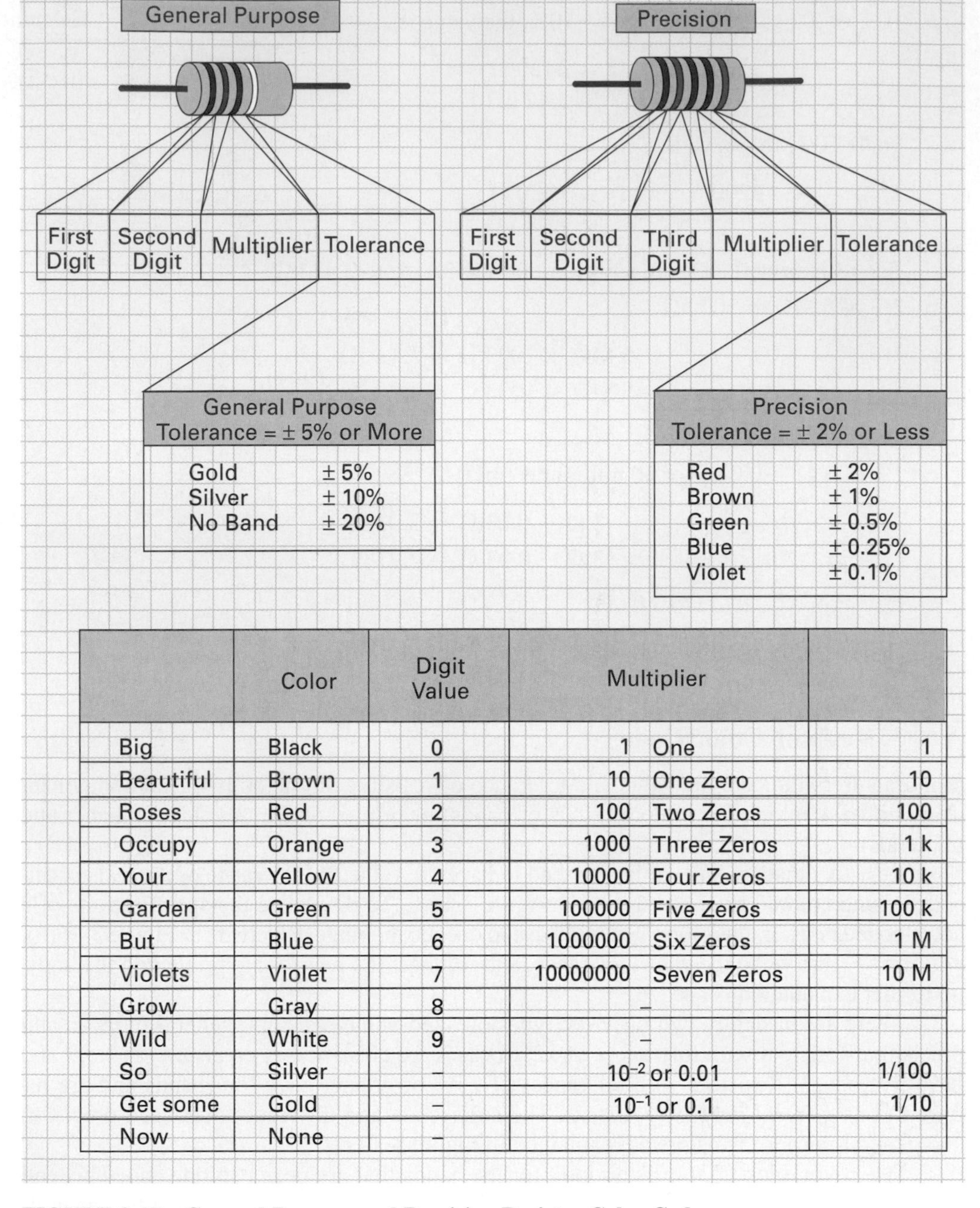

	Color	Digit Value	Multiplier		
Big	Black	0	1	One	1
Beautiful	Brown	1	10	One Zero	10
Roses	Red	2	100	Two Zeros	100
Occupy	Orange	3	1000	Three Zeros	1 k
Your	Yellow	4	10000	Four Zeros	10 k
Garden	Green	5	100000	Five Zeros	100 k
But	Blue	6	1000000	Six Zeros	1 M
Violets	Violet	7	10000000	Seven Zeros	10 M
Grow	Gray	8	–		
Wild	White	9	–		
So	Silver	–	10^{-2} or 0.01		1/100
Get some	Gold	–	10^{-1} or 0.1		1/10
Now	None	–			

FIGURE 3-17 General-Purpose and Precision Resistor Color Code.

General-Purpose Resistor Code

1. The first band on either a general-purpose or precision resistor can never be black, and it is the first digit of the number.
2. The second band indicates the second digit of the number.
3. The third band specifies the multiplier to be applied to the number, which ranges from × 1/100 to 10,000,000.
4. The fourth band describes the tolerance or deviation from the specified resistance, which is ±5% or greater.

Precision Resistor Code

1. The first band, like the general-purpose resistor, is never black and is the first digit of the three-digit number.
2. The second band provides the second digit.
3. The third band indicates the third and final digit of the number.
4. The fourth band specifies the multiplier to be applied to the number.
5. The fifth and final band indicates the tolerance figure of the precision resistor, which is always less than ±2%, which is why precision resistors are more expensive than general-purpose resistors.

EXAMPLE:

Figure 3-18 illustrates a ½ W resistor.

a. Is this a general-purpose or a precision resistor?
b. What is the resistor's value of resistance?
c. What tolerance does this resistor have, and what deviation plus and minus could occur from this value?

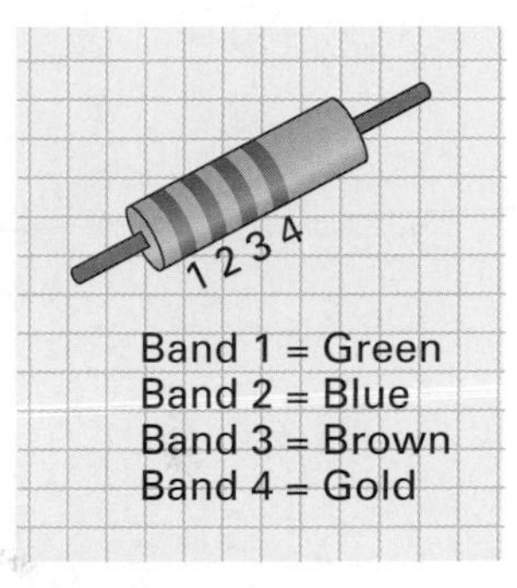

FIGURE 3-18

Solution:

a. General purpose (four bands)
b. green blue × brown

$$5 \quad 6 \times 10 = 560\ \Omega$$

c. Tolerance band is gold, which is ± 5%.

$$\text{deviation} = 5\% \text{ of } 560 = 28$$
$$560 + 28 = 588\ \Omega$$
$$560 - 28 = 532\ \Omega$$

The resistor could, when measured, be anywhere from 532 to 588 Ω.

EXAMPLE:

State the resistor's value and tolerance and whether it is general purpose or precision for the examples shown in Figure 3-19.

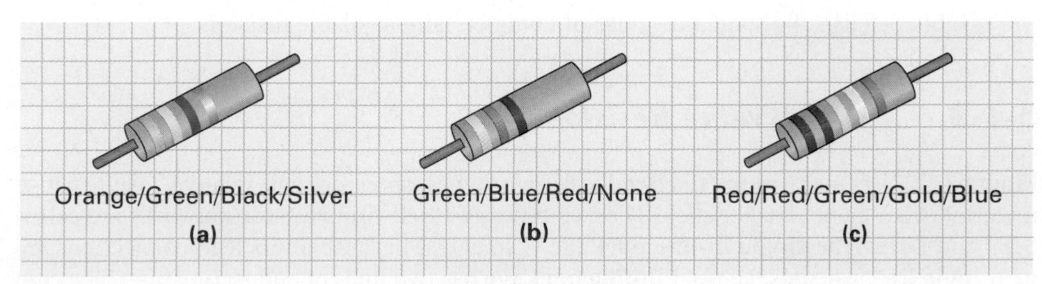

FIGURE 3-19

Ohms (Ω)					Kilohms ($k\Omega$)		Megohms ($M\Omega$)	
0.10	1.0	10	100	1000	10	100	1.0	10.0
0.11	1.1	11	110	1100	11	110	1.1	11.0
0.12	**1.2**	**12**	**120**	**1200**	**12**	**120**	**1.2**	**12.0**
0.13	1.3	13	130	1300	13	130	1.3	13.0
0.15	1.5	15	150	1500	15	150	1.5	15.0
0.16	1.6	16	160	1600	16	160	1.6	16.0
0.18	**1.8**	**18**	**180**	**1800**	**18**	**180**	**1.8**	**18.0**
0.20	2.0	20	200	2000	20	200	2.0	20.0
0.22	2.2	22	220	2200	22	220	2.2	22.0
0.24	2.4	24	240	2400	24	240	2.4	
0.27	**2.7**	**27**	**270**	**2700**	**27**	**270**	**2.7**	
0.30	3.0	30	300	3000	30	300	3.0	
0.33	3.3	33	330	3300	33	330	3.3	
0.36	3.6	36	360	3600	36	360	3.6	
0.39	**3.9**	**39**	**390**	**3900**	**39**	**390**	**3.9**	
0.43	4.3	43	430	4300	43	430	4.3	
0.47	4.7	47	470	4700	47	470	4.7	
0.51	5.1	51	510	5100	51	510	5.1	
0.56	**5.6**	**56**	**560**	**5600**	**56**	**560**	**5.6**	
0.62	6.2	62	620	6200	62	620	6.2	
0.68	6.8	68	680	6800	68	680	6.8	
0.75	7.5	75	750	7500	75	750	7.5	
0.82	**8.2**	**82**	**820**	**8200**	**82**	**820**	**8.2**	
0.91	9.1	91	910	9100	91	910	9.1	

TABLE 3-5 Standard Values of Commercially Available Resistors

Solution:

a. orange green black silver (four bands = general purpose)

3 5 × 1 10% $= 35\ \Omega \pm 10\%$

b. green blue red none (four bands = general purpose)

5 6 × 100 20% $= 5.6\ k\Omega \pm 20\%$

c. red red green gold blue (five bands = precision)

2 2 5 × 0.1 0.25% $= 22.5\ \Omega \pm 0.25\%$

Table 3-5 lists the standard values of commercially available resistors.

3-4-2 *Fixed-Value Resistor Types*

Figure 3-20 lists many of the fixed-value resistor types, showing their relative cost, resistive material used, and characteristics. The two types of **thick-film resistor** networks are called SIPs and DIPs. The **single in-line package** is so called because all its lead connections are in a single line, whereas the **dual in-line package** has two lines of connecting pins. The chip resistors are small thick-film resistors that are approximately the size of the tip of a pencil lead.

Thick-Film Resistors
Fixed-value resistor consisting of a thick-film resistive element made from metal particles and glass powder.

Single In-Line Package (SIP)
Package containing several electronic components (generally resistors) with a single row of external connecting pins.

Dual In-Line Package (DIP)
Package that has two (dual) sets or lines of connecting pins.

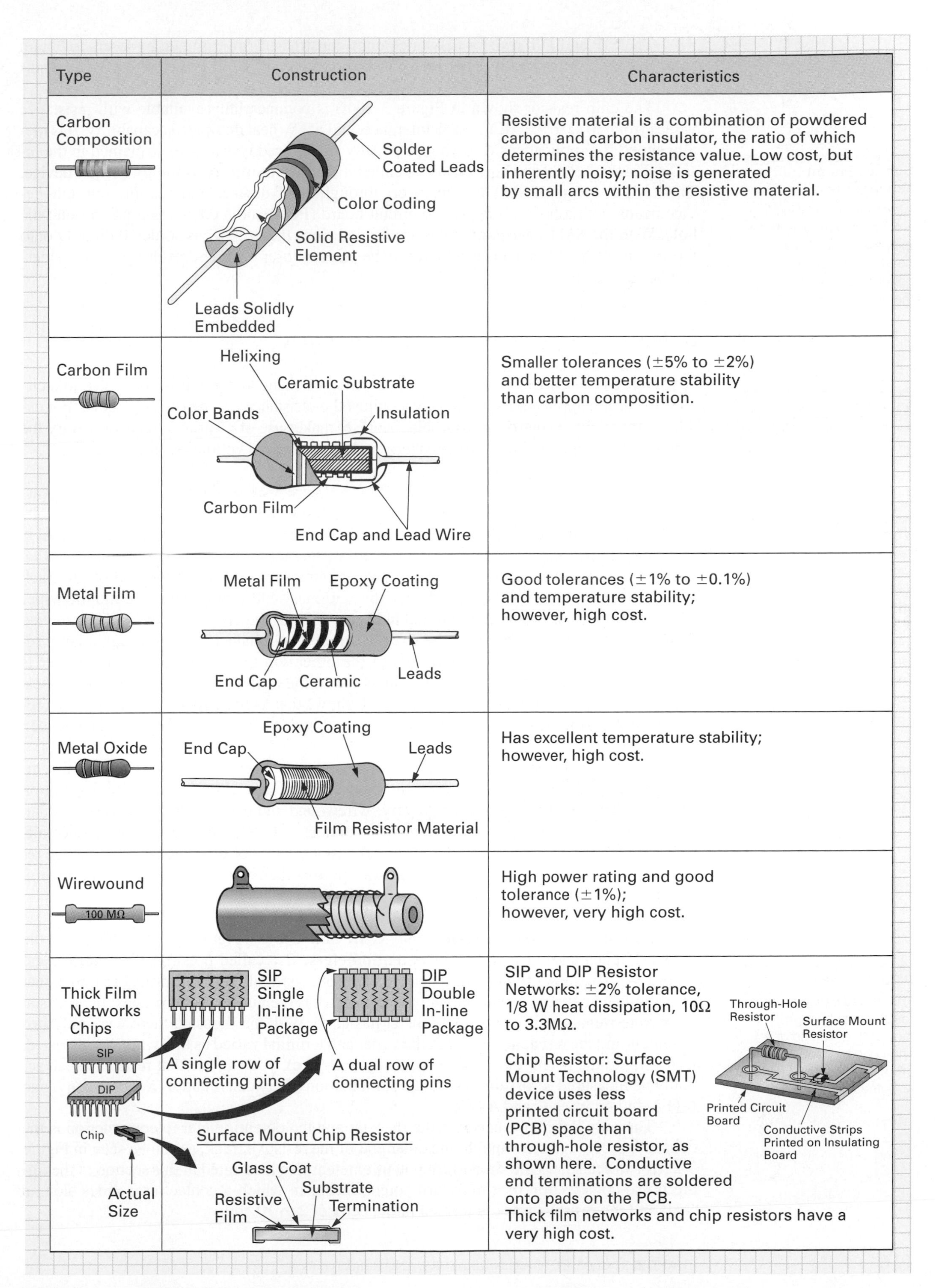

Type	Construction	Characteristics
Carbon Composition	Solder Coated Leads; Color Coding; Solid Resistive Element; Leads Solidly Embedded	Resistive material is a combination of powdered carbon and carbon insulator, the ratio of which determines the resistance value. Low cost, but inherently noisy; noise is generated by small arcs within the resistive material.
Carbon Film	Helixing; Ceramic Substrate; Color Bands; Insulation; Carbon Film; End Cap and Lead Wire	Smaller tolerances (±5% to ±2%) and better temperature stability than carbon composition.
Metal Film	Metal Film; Epoxy Coating; End Cap; Ceramic; Leads	Good tolerances (±1% to ±0.1%) and temperature stability; however, high cost.
Metal Oxide	Epoxy Coating; End Cap; Leads; Film Resistor Material	Has excellent temperature stability; however, high cost.
Wirewound (100 MΩ)		High power rating and good tolerance (±1%); however, very high cost.
Thick Film Networks Chips (SIP; DIP; Chip; Actual Size)	SIP Single In-line Package; A single row of connecting pins; DIP Double In-line Package; A dual row of connecting pins; Surface Mount Chip Resistor; Glass Coat; Substrate; Resistive Film; Termination	SIP and DIP Resistor Networks: ±2% tolerance, 1/8 W heat dissipation, 10Ω to 3.3MΩ. Chip Resistor: Surface Mount Technology (SMT) device uses less printed circuit board (PCB) space than through-hole resistor, as shown here. Conductive end terminations are soldered onto pads on the PCB. Thick film networks and chip resistors have a very high cost. (Through-Hole Resistor; Surface Mount Resistor; Printed Circuit Board; Conductive Strips Printed on Insulating Board)

FIGURE 3-20 Fixed-Value Resistor Types.

The SIP and DIP resistor networks, once constructed, are trimmed by lasers to obtain close tolerances of typically ±2%. Resistance values ranging from 10 Ω to 3.3 MΩ are available, with a power rating of ⅛ W.

The chip resistor shown in Figure 3-20(c) is commercially available with resistance values from 10 Ω to 3.3 MΩ, a ±2% tolerance, and a ⅛ W heat dissipation capability. It is ideally suited for applications requiring physically small-sized resistors, as explained in the inset in Figure 3-20. The chip resistor is called a **surface mount technology** (SMT) device. The key advantage of SMT devices over "through-hole" devices is that a through-hole device needs both a hole in the printed circuit board (PCB) and a connecting pad around the hole. With the SMT device, no holes are needed since the package is soldered directly onto the surface of the PCB. Pads can therefore be placed closer together, resulting in a considerable space saving.

Surface Mount Technology
A method of installing tiny electronic components on the same side of a circuit board as the printed wiring pattern that interconnects them.

3-4-3 *Variable-Value Resistor Types*

A **variable resistor** can have its resistance varied or changed while it is connected in a circuit. In certain applications, the ability to adjust the resistance of a resistor is needed. For example, the volume control on your television set makes use of a variable resistor to vary the amount of current passing to the speakers, and so change the volume of the sound.

Variable Resistor
A resistor whose value can be changed.

Rheostat (two terminals: A and B). Figure 3-21(a) shows the physical appearance of different **rheostats,** and Figure 3-21(b) shows the rheostat's schematic symbols. As can be seen in the construction of a circular rheostat in Figure 3-21(c), one terminal is connected to one side of a resistive track and the other terminal of this two-terminal device is connected to a movable wiper. As the wiper is moved away from the end of the track with the terminal, the resistance between the stationary end terminal and the mobile wiper terminal increases. This is summarized in Figure 3-21(d), where the wiper has been moved down by a clockwise rotation of the shaft. Current would have to flow through a large resistance as it travels from one terminal to the other. On the other hand, as the wiper is moved closer to the end of the track connected to the terminal, the resistance decreases. This is summarized in Figure 3–21(e), which shows that as the wiper is moved up, as a result of turning the shaft counterclockwise, current will see only a small resistance between the two terminals.

Rheostat
Two-terminal variable resistor that through mechanical turning of a shaft can be used to vary its resistance and therefore its value of terminal-to-terminal current.

Rheostats come in many shapes and sizes, as can be seen in Figure 3-21(a). Some employ a straight-line motion to vary resistance; others are classified as circular-motion rheostats. The resistive elements also vary; wirewound and carbon tracks are very popular. Cermet rheostats mix the ratio of ceramic (insulator) and metal (conductor) to produce different values of resistive tracks. A trimming rheostat is a miniature device used to change resistance by a small amount. Other circular-motion rheostats are available that require between 2 to 10 turns to cover the full resistance range.

Potentiometer (three terminals: A, B, and C). Figure 3-22(a) illustrates the physical appearance of a variety of **potentiometers,** also called pots (the slang term), and Figure 3-22(b) shows the potentiometer's schematic symbol. You will probably notice that the difference between a rheostat and potentiometer is the number of terminals; the rheostat has two terminals while the potentiometer has three. With the rheostat, there were only two terminals and the resistance between the wiper and terminal varied as the wiper was adjusted. Referring to the potentiometer shown in Figure 3-22(c), you can see that resistance can actually be measured across three separate combinations: between A and B (X), between B and C (Y), and between C and A (Z).

Potentiometer
Three-lead variable resistor that through mechanical turning of a shaft can be used to produce a variable voltage or potential.

The only difference between the rheostat and the potentiometer in construction is the connection of a third terminal to the other end of the resistive track, as can be seen in Figure 3-22(d), which shows the single-turn potentiometer. Also illustrated in this section of the figure is the construction of a multiturn potentiometer in which a contact arm slides along a shaft and the resistive track is formed into a helix of 2 to 10 coils.

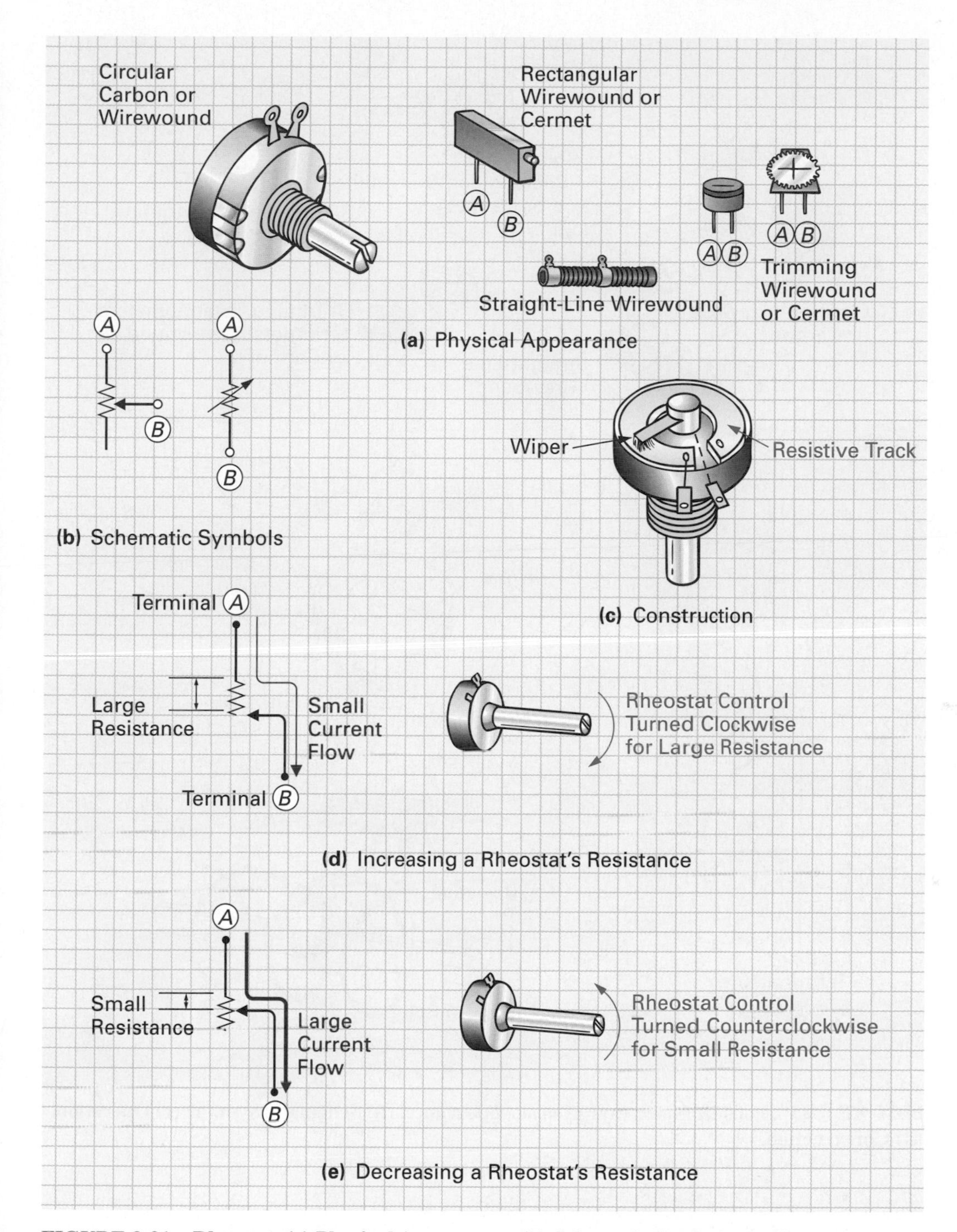

FIGURE 3-21 Rheostat. (a) Physical Appearance. (b) Schematic Symbols. (c) Construction. (d) Increasing a Rheostat's Resistance. (e) Decreasing a Rheostat's Resistance.

As shown in Figure 3-23, the resistance between *A* and *B* (*X*) and *B* and *C* (*Y*) will vary as the wiper's position is moved, as illustrated in Figure 3-23(a). If the user physically turns the shaft in a clockwise direction, the resistance between *A* and *B* increases while the resistance between *B* and *C* decreases. Similarly, if the user mechanically turns the shaft counterclockwise, a decrease occurs between *A* and *B* and there is a resulting increase in resistance between *B* and *C*. This point is summarized in Figure 3-23(b).

Whether rheostat or potentiometer, the resistive track can be classified as having either a **linear** or a **tapered** (nonlinear) resistance. In Figure 3-24(a) we have taken a 1 kΩ rheostat and illustrated the resistance value changes between *A* and *B* for a linear and a tapered one-

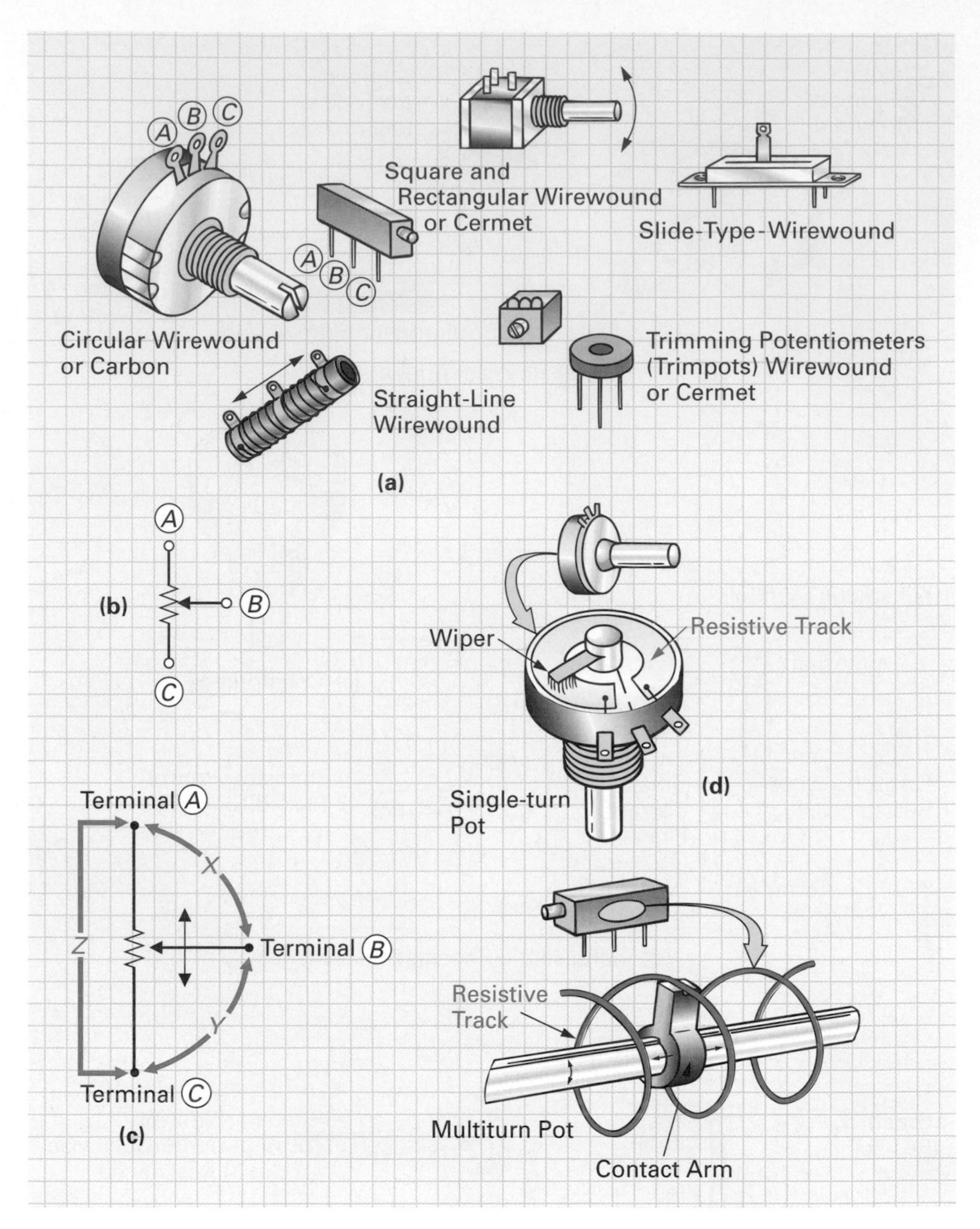

FIGURE 3-22 **Potentiometer. (a) Physical Appearance. (b) Schematic Symbol. (c) Operation. (d) Construction.**

Linear

Relationship between input and output in which the output varies in direct proportion to the input.

Tapered

Nonuniform distribution of resistance per unit length throughout the element.

turn rheostat. The definition of *linear* is having an output that varies in direct proportion to the input. The input in this case is the user turning the shaft, and the output is the linearly increasing resistance between *A* and *B*.

With a tapered rheostat or potentiometer, the resistance varies nonuniformly along its resistor element, sometimes being greater or less for equal shaft movement at various points along the resistance element, as shown in the table in Figure 3-24(a). Figure 3-24(b) plots the position of the variable resistor's wiper against the resistance between the two output terminals, showing the difference between a linear increase and a nonlinear or tapered increase.

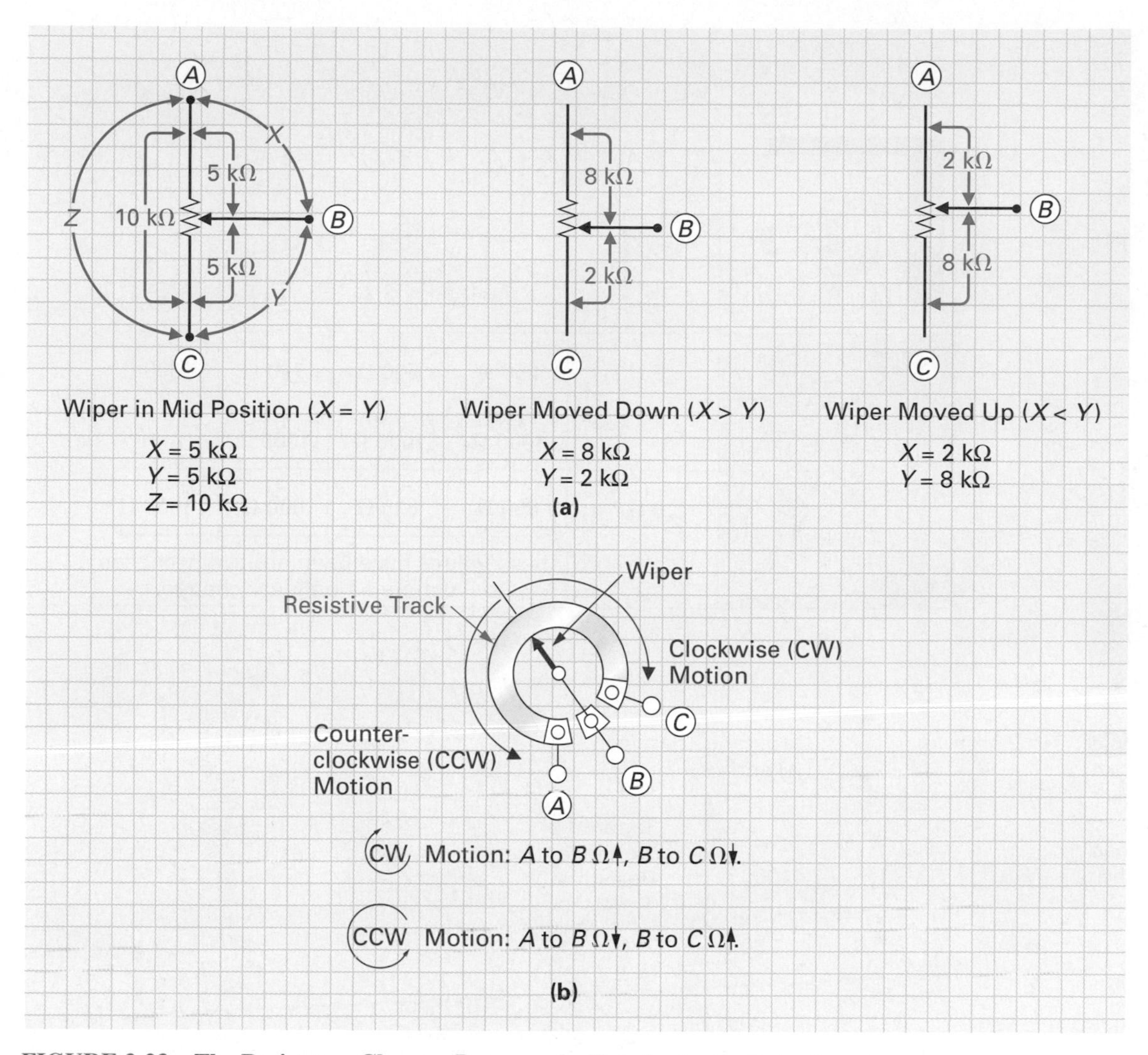

FIGURE 3-23 The Resistance Changes Between the Terminals of a Potentiometer.

3-4-4 *Testing Resistors*

It is virtually impossible for fixed-value resistors to internally short-circuit (no resistance or 0 ohms). Generally, the resistor's internal elements will begin to develop a higher resistance than its specified value (due to a partial internal open) or in some cases go completely open circuit (maximum resistance or infinite ohms).

Variable-value resistors have problems with the wiper making contact with the resistive track at all points. Faulty variable-value resistors in sound systems can normally be detected because they generate a scratchy noise whenever you adjust the volume control.

The ohmmeter is the ideal instrument for verifying whether or not a resistor is functioning correctly. When checking a resistor's resistance, here are some points that should be remembered.

1. The ohmmeter has its own internal power source (a battery), so always turn off the circuit power and remove from the circuit the resistor to be measured. If this is not done, you will not only obtain inaccurate readings, but you can damage the ohmmeter.
2. With an autoranging digital multimeter, the suitable ohms range is automatically selected by the ohmmeter. The range scales on a non-autoranging digital multimeter have to be selected and often confuse people. For example, if a 100 kΩ range is chosen the

TIME LINE

Stephen Wozniak (top) and Steven Jobs (bottom) met at their Los Altos, California, high school, and became friends due to their common interest in electronics. By selling a Volkswagen van and a programmable calculator, Wozniak and Jobs raised the initial capital of $1300 to start a business in April 1976, which they named Apple Computer as Jobs had a passion for the Beatles, who recorded under the Apple record lable. In just five years, Apple Computer grew faster than any other company in history, from a two-man asset-less partnership building computers in a family home, to a publicly traded corporation earning its youthful entrepreneurs sizeable fortunes.

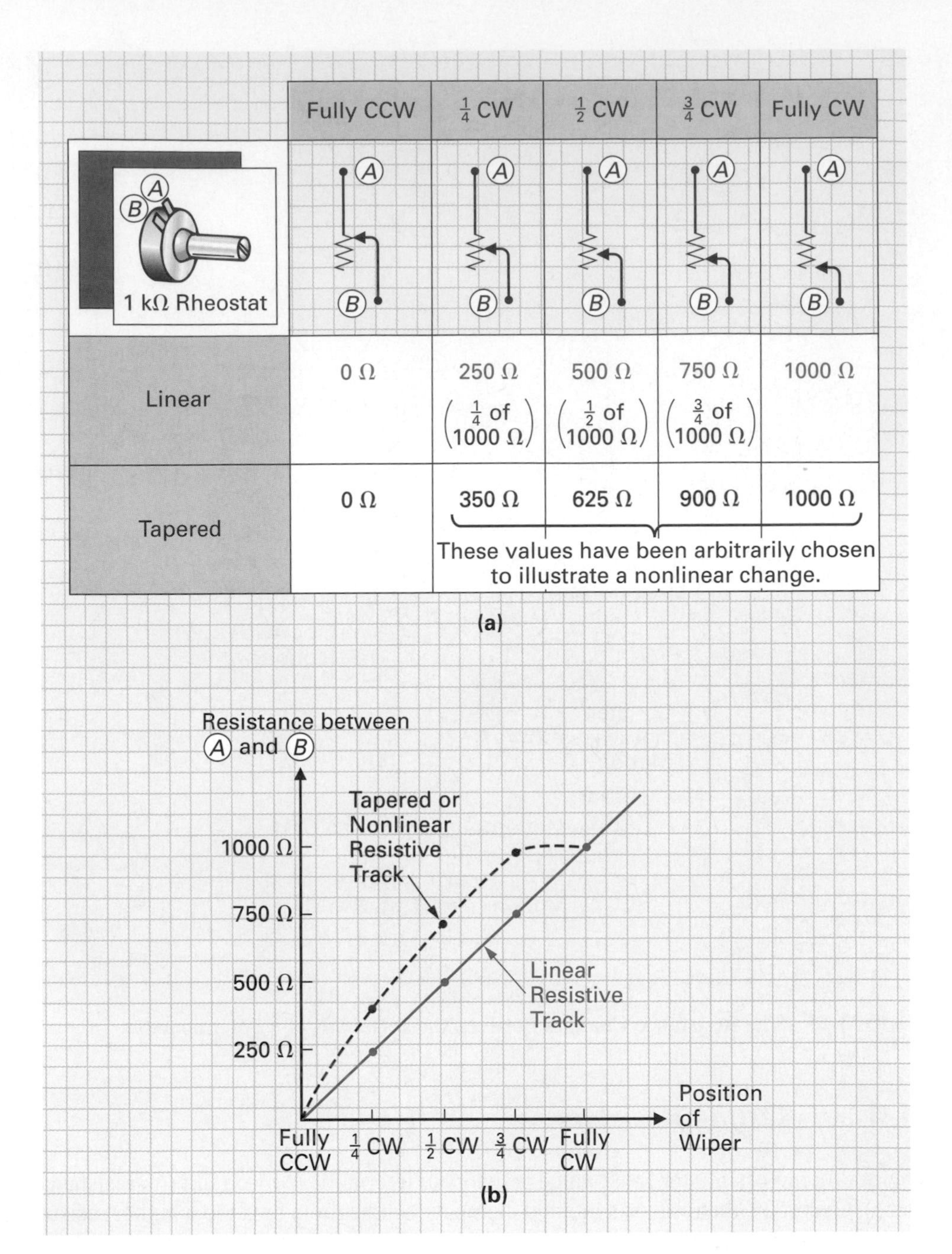

FIGURE 3-24 Linear versus Tapered Resistive Track.

highest reading that can be measured is 100 kΩ. If a 1 MΩ resistor is measured, the meter will indicate an infinite-ohms reading and you may be misled into believing that you have found the problem (an open resistor). To overcome this problem, always start on the highest range and then work down to a lower range for a more accurate reading.

3. Another point to keep in mind is tolerance. For example, if a suspected faulty 1 kΩ (1000 Ω) resistor is measured with an ohmmeter and reads 1.2 kΩ (1200 Ω), it could be within tolerance if no tolerance band is present on the resistor's body (±20%). A 1 kΩ resistor with ±20% tolerance could measure anywhere between 800 and 1200 Ω.
4. The ohmmeter's internal battery voltage is really too small to deliver an electrical shock. You should, however, avoid touching the bare metal parts of the probes or resistor leads, as your body resistance of approximately 50 kΩ will affect your meter reading.

SELF-TEST EVALUATION POINT FOR SECTION 3-4

Now that you have completed this section, you should be able to:

Objective 15. Describe resistor characteristics such as wattage rating and tolerance.

Objective 16. Explain the color coding system used to encode value and tolerance for general-purpose and precision resistors.

Objective 17. Describe the difference between a fixed- and a variable-value resistor.

Objective 18. Explain the difference between the six basic types of fixed-value resistors: carbon composition, carbon film, metal film, wirewound, metal oxide, and thick film.

Objective 19. Describe the SIP, DIP, and chip thick-film resistor packages.

Objective 20. Explain the following types of variable resistors:

a. Rheostat
b. Potentiometer

Objective 21. Describe the difference between a linear and tapered resistive track.

Objective 22. List and describe some of the resistor testing precautions and techniques.

Use the following questions to test your understanding of Section 3-4:

1. List the six types of fixed-value resistors.
2. What is the difference between SIPs and DIPs?
3. Name the two types of mechanically adjustable variable resistors, and state the difference between the two.
4. Describe the difference between a linear and a tapered resistive track.
5. What tolerance differences occur between a general-purpose and a precision resistor?
6. A red/red/red/gold/blue resistor has what value and tolerance figure?
7. What color code would appear on a 4.7 kΩ, ±10% tolerance resistor?
8. Would a three-band resistor be considered a general-purpose type?

SUMMARY

Mini-Math Review (Section 3-1)

1. A ratio is a comparison of one number with another number.
2. Rounding off is an operation in which a value is abbreviated by applying the following rule: If the first digit to be dropped is 5 or more, increase the previous digit by 1; if the first digit to be dropped is less than 5, do not change the previous digit.
3. The number of significant places describes how many digits are in a value and how many digits in the value are accurate after rounding off.
4. Percentages are fractions in hundredths.
5. A mean average is a value that summarizes a set of unequal values.
6. A graph is a diagram that represents the variation of a variable in comparison to that of one or more other variables.
7. The vertical or y-axis is often called the *dependent variable* because it indicates the output, which is dependent on the input applied.
8. The horizontal or x-axis is often called the *independent variable* because it is independently increased in increments.
9. Points on the graph are identified by giving x and y values or *coordinates*; for example, (7, 4000). The x value is always followed by the corresponding y value.
10. With the Cartesian coordinate system, the vertical and horizontal axes are extended beyond the zero point to accommodate negative values as well as positive.

Conductors (Section 3-2)

11. Materials that pass current easily are called *conductors*. Conductors are materials or natural elements whose valence electrons can easily be removed from their parent atoms. The precious metals of silver and gold are the best conductors.
12. Copper is the most commonly used conductor, as it meets the following three requirements:
    a. It is a good source of electrons.
    b. It is inexpensive.
    c. It is physically pliable.
13. *Conductance* is the measure of how good a conductor is at carrying current. Conductance (symbolized G) is equal to the reciprocal of resistance and is measured in the unit siemens (S).
14. Relative conductivity describes how good or poor a conductor is by comparing it to a reference point. The reference point we use is the best conductor, silver, which has a conductivity value of 1.0.
15. Resistors are normally made out of materials that cause an opposition to current flow. Conductance is the measure of how good a conductor is, and even the best conductors have some value of resistance.

16. The resistance of a conductor is inversely proportional to the conductor's cross-sectional area.
17. As nearly all conductors are circular in cross section, the area of a conductor is measured in *circular mils* (cmil).
18. Increasing the length of the conductor used increases the amount of resistance within the circuit.
19. By combining all of a conductor's physical factors, we can arrive at a formula for resistance.
20. *Resistivity*, by definition, is the resistance (in ohms) that a certain length of material (in centimeters) will offer to the flow of current.
21. Metallic conductors are said to have a *positive temperature coefficient of resistance* (+ Temp. Coe. of R), because the greater the heat applied to the conductor, the greater the atom movement, causing more collisions of atoms to occur, and consequently the greater the conductor's resistance.
22. Any time that current flows through a conductor, a certain resistance or opposition is inherent in that conductor. This resistance will convert current to heat, and the heat further increases the conductor's resistance, causing more heat to be generated due to the opposition. Consequently, a conductor must be chosen carefully for each application so that it can carry the current without developing excessive heat.
23. The atoms of a resistor obstruct the flow of moving electrons (current), and this friction between the two causes heat to be generated. Current flow through any resistance therefore will always be accompanied by heat. In some instances, such as a heater or oven, the device has been designed specifically to generate heat. In other applications, it may be necessary to include a resistor to reduce current flow, and the heat generated will be an unwanted side effect.
24. Any one of the three power formulas can be used to calculate the power dissipated by a resistance. However, since power is determined by the friction between current (I) and the circuit's resistance (R), the $P = I^2 \times R$ formula is more commonly used to calculate the heat generated.
25. The *filament resistor* is more commonly known as the household lightbulb. The filament resistor is just a coil of wire that glows white-hot when current is passed through it and, in so doing, dissipates both heat and light energy.
26. The *incandescent lamp* is an electric lamp in which electric current flowing through a filament of resistive material heats the filament until it glows and emits light.
27. The filament material (tungsten, for example) is like all other conductors in that it has a positive temperature coefficient of resistance. Therefore, as the current through the filament increases, so does the temperature and so does the filament's resistance ($I\uparrow$, temperature $\uparrow$, $R\uparrow$).
28. A coil of wire similar to the one found in a lightbulb, called a *ballast resistor,* is used to maintain a constant current despite variation in voltage. Current is regulated or maintained constant by the ballast resistor despite variations in voltage.
29. Conducting wires are normally covered with a plastic or rubber type of material, known as *insulation*. This insulation is used to protect users and technicians from electrical shock and also to keep the conductor from physically contacting other conductors within the equipment. If the current through the conductor is too high, insulation will burn due to the heat and may cause a fire hazard. The National Fire Protection Association has developed a set of standards known as the *American Wire Gauge* for all copper conductors, which lists their diameter, resistance, and maximum safe current in amperes.
30. In 1911, a Dutch physicist, Heike Onnes, discovered that mercury (a liquid conductor) lost its resistance to electrical current when the temperature was decreased to −459.7 Fahrenheit (0 on the Kelvin temperature scale). Mercury actually became a *superconductor,* allowing a supercurrent to flow and not encounter any resistance and therefore not generate any heat.
31. In the spring of 1986, two IBM scientists discovered that a conductor compound made up of barium, lanthanum, copper, and oxygen would superconduct (have no resistance to current flow) at −406°F. The real scientific achievement will be to find a conductor compound that will superconduct at room temperature.
32. A cable is made up of two or more wires.
33. Wires and cables have to connect from one point to another. Some are soldered directly, whereas others are attached to plugs that plug into sockets.

Insulators (Section 3-3)

34. Materials that are used to block current are called *insulators*. Insulators can, with sufficient pressure or voltage applied across them, "break down" and conduct current. The dielectric strength describes how good or poor an insulator is by listing what voltage will cause a small section to break down and conduct a large current.

Resistors (Section 3-4)

35. There are two basic types of resistors: fixed value and variable value. The fixed-value resistor has a value of resistance that cannot be changed; the variable-value resistor has a range of values that can be selected generally by mechanically adjusting a control.
36. By changing the ratio of powdered insulator to conductor, the value of resistance can be changed within the same area.
37. The physical size of a resistor lets the user know how much power in the form of heat can be *dissipated.* The amount of heat dissipated per unit of time is measured in watts, and each resistor has its own *wattage rating.* The bigger the resistor, the more heat that can be dissipated.
38. Another factor to consider when discussing resistors is their *tolerance.* Tolerance is the amount of deviation or error from the specified value. In some applications where high precision is required, smaller-tolerance resistors are used.
39. There are basically two different types of fixed resistors: general purpose and precision. Resistors with tolerances of ±2% or less are *precision resistors* and have five bands. Resistors with tolerances of ±5% or greater have four bands and are referred to as *general-purpose resistors.*

40. For the general-purpose resistor code,
    a. The first band on either a general-purpose or precision resistor can never be black, and it is the first digit of the number.
    b. The second band indicates the second digit of the number.
    c. The third band specifies the multiplier to be applied to the number, which ranges from × 1⁄100 to 10,000,000.
    d. The fourth band describes the tolerance or deviation from the specified resistance, which is ±5% or greater.
41. For the precision resistor code,
    a. The first band, like the general-purpose resistor, is never black and is the first digit of the three-digit number.
    b. The second band provides the second digit.
    c. The third band indicates the third and final digit of the number.
    d. The fourth band specifies the multiplier to be applied to the number.
    e. The fifth and final band indicates the tolerance figure of the precision resistor, which is always less than ±2%, which is why precision resistors are more expensive than general-purpose resistors.
42. Carbon film resistors have smaller tolerance figures (±5% to ±2%), have good temperature stability (maintain same resistance value over a wide range of temperatures), and have less internally generated noise (random small bursts of voltage) than do carbon composition resistors.
43. Metal film resistors have possibly the best tolerances commercially available of ±1% to ±0.1%. They also maintain a very stable resistance over a wide range of temperatures (good stability) and generate very little internal noise compared to any carbon resistor.
44. Wirewound resistors are constructed by wrapping a length of wire uniformly around a ceramic insulating core, with terminals making the connections at each end. The length and thickness of the wire are varied to change the resistance, which ranges from 1 Ω to 150 kΩ. Wirewound resistors are generally used in applications requiring low resistance values, which means that the current and therefore power dissipated are high.
45. Metal oxide resistors have excellent temperature stability, which is the ability of the resistor to maintain its value of resistance without change even when temperature is changed.
46. The two different types of *thick-film resistor* networks are called SIPs and DIPs. The *single in-line package* is so called because all its lead connections are in a single line, whereas the *dual in-line package* has two lines of connecting pins. The chip resistors are small thick-film resistors that are approximately the size of a pencil lead.
47. The SIP and DIP resistor networks, once constructed, are trimmed by lasers to obtain close tolerances of typically ±2%. Resistance values ranging from 22 Ω to 2.2 MΩ are available, with a power rating of ½ W.
48. The chip resistor is commercially available with resistance values from 10 Ω to 3.3 MΩ, a ±2% tolerance, and a $\frac{1}{8}$ W heat dissipation capability.
49. The *rheostat* is a two-terminal variable resistor that through mechanical turning of a shaft can be used to vary its resistance and therefore its value of terminal-to-terminal current.
50. The difference between a rheostat and potentiometer is the number of terminals; the rheostat has two terminals and the potentiometer has three.
51. Like the rheostat, the potentiometer comes in many different shapes and sizes. Wirewound, carbon, and cermet resistive tracks, circular or straight-line motion, 2 to 10 multiturn, and other variations are available for different applications.
52. Whether rheostat or potentiometer, the resistive track can be classified as having either a *linear* or a *tapered* (nonlinear) resistance. The definition of *linear* is having an output that varies in direct proportion to the input. The input, in this case, is the user turning the shaft, and the output is the linearly increasing resistance between *A* and *B*. With a tapered rheostat or potentiometer, the resistance varies nonuniformly along its resistor element, sometimes being greater or less for equal shaft movement at various points along the resistance element.
53. It is virtually impossible for fixed-value resistors to short-circuit internally. Generally, the resistor's internal resistive elements will begin to develop a higher resistance than its specified value or in some cases go completely open circuit.
54. Variable-value resistors have problems with the wiper making contact with the resistive track at all times.
55. The ohmmeter is the ideal instrument for verifying whether or not a resistor is functioning correctly.

REVIEW QUESTIONS

Multiple-Choice Questions

1. The Greek letter π represents a constant value that describes how much bigger a circle's __________ is compared with its __________, and it is frequently used in many formulas involved with the analysis of circular motion.
    a. Circumference, radius
    b. Circumference, diameter
    c. Diameter, radius
    d. Diameter, circumference

2. Is π a ratio?
   a. Yes
   b. No
3. Round off the following number to the nearest hundredth: 74.8552.
   a. 74.85
   b. 74.86
   c. 74.84
   d. 74.855
4. The number 27.0003 is a __________ -significant-place number and is accurate to __________ significant places.
   a. 6, 2
   b. 2, 6
   c. 6, 6
   d. 4, 6
5. Express 29.63% mathematically as a decimal.
   a. 29.63
   b. 2.963
   c. 0.02963
   d. 0.2963
6. Conductors offer a __________ resistance to current flow.
   a. High
   b. Low
   c. Medium
   d. Maximum
7. Conductors have
   a. Electrons in shells farthest away from the nucleus
   b. Relatively more electrons per atom
   c. An incomplete valence shell
   d. All of the above
   e. None of the above
8. Conductance is the measure of how good a conductor is at passing current and is measured in
   a. Siemens
   b. Volts
   c. Current
   d. Ohms
9. Insulators have
   a. Electrons close to the nucleus
   b. Relatively few electrons per atom
   c. An almost complete valence shell
   d. All of the above
   e. None of the above
10. A good conductor has a
    a. Large conductance figure
    b. Small resistance figure
    c. Both (a) and (b)
    d. None of the above
11. The resistance of a conductor is
    a. Proportional to the length of the conductor
    b. Inversely proportional to the area of the conductor
    c. Both (a) and (b)
    d. None of the above
12. AWG is an abbreviation for
    a. Alternate Wire Gauge
    b. Alternating Wire Gauge
    c. American Wave Guide
    d. American Wire Gauge
13. A SIP package has
    a. A single line of connectors
    b. A double line of connectors
    c. No connectors
    d. None of the above
14. A __________ band resistor is called a general-purpose resistor, and a __________ band resistor is known as a precision resistor.
    a. Four, five
    b. Three, four
    c. One, four
    d. Two, seven
15. The most commonly used mechanically adjustable variable-value resistor is the
    a. Rheostat
    b. Thermistor
    c. Potentiometer
    d. Photoresistor
16. What would be the power dissipated by a 2 kΩ carbon composition resistor when a current of 20 mA is flowing through it?
    a. 40 W
    b. 0.8 W
    c. 1.25 W
    d. None of the above
17. A rheostat is a __________ terminal device, and the potentiometer is a __________ terminal device.
    a. 2, 3
    b. 1, 2
    c. 2, 4
    d. 3, 2
18. Resistance is measured with a (an)
    a. Wattmeter
    b. Milliammeter
    c. Ohmmeter
    d. 100 meter
19. Which of the following is *not* true?
    a. Precision resistors have a tolerance of > 5%.
    b. General-purpose resistors can be recognized because they have either three or four bands.
    c. The fifth band indicates the tolerance of a precision resistor.
    d. The third band of a general-purpose resistor specifies the multiplier.
20. The first band on either a general-purpose or a precision resistor can never be
    a. Brown
    b. Red or black stripe
    c. Black
    d. Red
21. What is the name of the resistor used to maintain a constant current despite variations in voltage by changing its resistance?
    a. Precision resistor
    b. Rheostat
    c. Ballast resistor
    d. Potentiometer

22. The term *infinite ohms* describes
  a. A small finite resistance
  b. A resistance so large that a value cannot be placed on it
  c. A resistance between maximum and minimum
  d. None of the above
23. Resistance is always measured when
  a. Circuit power is ON
  b. Circuit power is OFF
  c. The ohmmeter selector switch is in the A position
  d. The ohmmeter is connected in the path of current
24. The typical problem with resistors is that they will
  a. Develop a partial internal open
  b. Completely open the circuit
  c. Develop a higher resistance than specified
  d. All of the above
  e. Both (a) and (c)
25. A 10% tolerance, 2.7 MΩ carbon composition resistor measures 2.99 MΩ when checked with an ohmmeter. It is
  a. Within tolerance
  b. Outside tolerance
  c. Faulty
  d. Both (a) and (c)
  e. Both (a) and (b)

Communication Skill Questions

26. What is a ratio, and how is it expressed? (3-1)
27. Briefly describe the rules for rounding off. (3-1)
28. What is a percentage? (3-1)
29. How is an average calculated? (3-1)
30. What is the approximate value of the constant pi (π), and what does it describe? (3-1)
31. List the four factors that determine a conductor's resistance. (3-2-2)
32. What is a circular mil? (3-2-2)
33. Define the resistivity of a conducting material. (3-2-2)
34. Describe why conductors have a positive temperature coefficient of resistance. (3-2-3)
35. What is the purpose(s) of placing an insulating sheath over conducting wires? (3-2-4)
36. What is the American Wire Gauge? (3-2-4)
37. What is a superconductor? (3-2-7)
38. List some of the advantages of superconductivity. (3-2-7)
39. What is the difference between a wire and a cable? (3-2-5)
40. Give some examples of different wires and cables. (3-2-5)
41. List examples of different conductor connectors. (3-2-5)
42. Describe the difference between a fixed- and a variable-value resistor. (3-4)
43. List the six basic categories of fixed-value resistors. (3-4-2)
44. Explain why a resistor's size determines its wattage rating. (3-4-1)
45. Why are resistors given a tolerance figure, and which is best, a small or a large tolerance? (3-4-1)
46. Briefly describe the construction of
  a. Carbon composition fixed-value resistors (3-4-2)
  b. Rheostats (3-4-3)
  c. Potentiometers (3-4-3)
  d. Multiturn precision potentiometers (3-4-3)
47. Draw the schematic symbol for a rheostat and potentiometer and describe the difference. Also show how a potentiometer can be used as a rheostat. (3-4-3)
48. Define linear and tapered resistive tracks. (3-4-3)
49. List the rules that should be applied when using an ohmmeter to measure resistance. (3-4-4)
50. Give the color code for the following resistor values (3-4):
  a. 1.2 MΩ, ±10%
  b. 10 Ω, ±5%
  c. 27 kΩ, ±20%
  d. 273 kΩ, ±0.5%

Practice Problems

51. Express the ratio of the following in their lowest terms. (Remember to compare like quantities.)
  a. The ratio of 20 feet to 5 feet
  b. The ratio of 2½ minutes to 30 seconds
52. Round off as indicated
  a. 48.36 to the nearest whole number
  b. 156.3625 to the nearest tenth
  c. 0.9254 to the nearest hundredth
53. Light travels in a vacuum at a speed of 186,282.3970 miles per second. This __________-significant-place value is accurate to __________ significant places.
54. Calculate the following percentages:
  a. 23% of 50 = ?
  b. 78% of 10 = ?
  c. 3% of 1.5 = ?
  d. 20% of 3300 = ?
  e. 5% of 10,000 = ?
55. Calculate the average value of the following groups of values (include units).
  a. 4, 5, 6, 7, and 8
  b. 15 seconds, 20 seconds, 18 seconds, and 16 seconds
  c. 110 volts, 115 volts, 121 volts, 117 volts, 128 volts, and 105 volts
  d. 33 ohms, 17 ohms, 1000 ohms, and 973 ohms
  e. 150 meters, 160 meters, and 155 meters
56. If a 1 kΩ rheostat has 50 V across it, calculate the different values of current if the rheostat is varied in 100 Ω steps. Insert the values into the table in Figure 3-25(a), and then plot these values in the graph in Figure 3-25(b).
57. What is the value of conductance in siemens for a 100 ohm resistor?
58. Calculate the resistance of 200 feet of copper having a diameter of 80 mils.
59. What AWG size wire should be used to safely carry just over 15 A?
60. Calculate the voltage dropped across 1000 ft of No. 4 copper conductor when a current of 7.5 A is flowing through it.

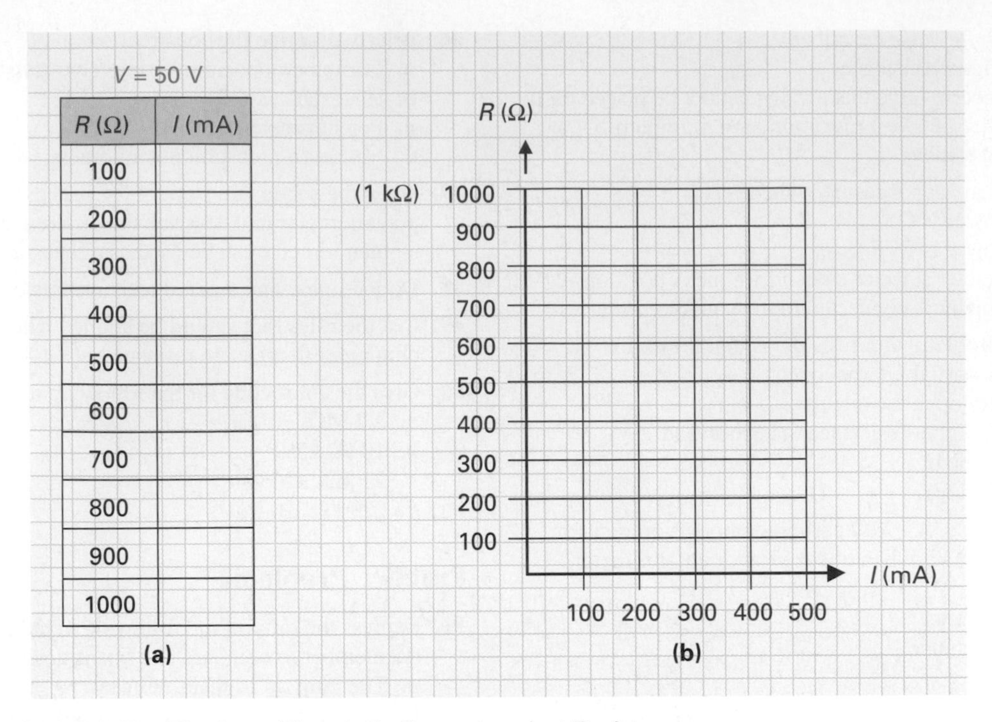

FIGURE 3-25 **Plotting a Rheostat's Current against Resistance.**

61. What is the relative conductivity of copper if silver is used as the reference conductor?

62. What minimum thickness of porcelain will withstand 24,000 V?

63. To insulate a circuit from 10 V, what insulator thickness would be needed if the insulator is rated at 750 kV/cm?

64. Convert the following:
a. 2000 kV/cm to __________ m
b. 250 kV/cm to __________ mm

65. What maximum voltage may be placed across 35 mm of mica without it breaking down?

66. If a 5.6 kΩ resistor has a tolerance of ±10%, what would be the allowable deviation in resistance above and below 5.6 kΩ?

67. (a) If a current of 50 mA is flowing through a 10 kΩ, 25 W resistor, how much power is the resistor dissipating? (b) Can the current be increased, and if so by how much?

68. What minimum wattage size could be used if a 12 Ω wirewound resistor were connected across a 12 V supply?

69. If a 20 kΩ, 1 W, ±10% tolerance resistor measures 20.39 kΩ, is it in or out of tolerance?

70. If 33 mA is flowing through a carbon film, 4.7 kΩ, ¼ W, ±5% resistor, will it burn up?

Job Interview Test

These tests at the end of each chapter will challenge your knowledge up to this point and give you the practice you need for a job interview. To make this more realistic, the test will be comprised of both technical and personal questions. In order to take full advantage of this exercise, you may want to set up a simulation of the interview environment, have a friend read the questions to you, and record your responses for later analysis.

Company Name: TRI, Inc.

Industry Branch: Industrial.

Function: Design and Manufacture Robotic Systems.

Job Title: Engineering Technician.

71. Would you say that your electronics education was complete?

72. What do you know about surface mount technology?

73. What is a tapered pot?

74. What is your attitude toward documentation?

75. In an automated manufacturing environment, ambient temperature can fluctuate quite dramatically. What effect does temperature have on conductors?

76. Can you tell me the difference between hot resistance and cold resistance?

77. What would you say is the job function of an engineering technician?

78. Do you know anything about the Altera or Xilinx PLD application software?

79. What can you tell me about linear variable resistors?

80. Can you list the resistor color code in order?

Answers

71. The answer must be yes, but go on to explain that you see your education as an ongoing process and, like industry, you will always be in a learning mode.

72. Section 3-4-1.

73. Section 3-4-3.

74. Documenting the engineering process is vitally important. As an engineering technician you will be called on to document design successess, and failures, in explicit detail. They are testing here to see if you have a talent for technical writing, since this is a must in this position.

75. Section 3-2-2.

76. Section 3-6.

77. See introduction, Field Service Technician.

78. Always be prepared for the interviewer to test your limits. Be honest and say no if you have no idea, but show an interest to understand and learn. A good response would be, "No, are they electronic circuit simulation programs?"

79. Section 3-2-2.

80. Section 3-4. If you are not completely proficient at this, take a moment to quickly jot down the first letters of the memory aid so that when you deliver your answer it will be smooth and accurate. Remember—less haste, more speed.

Web Site Questions

Go to the web site http://www.prenhall.com/cook, select the textbook *Introductory DC/AC Electronics* or *Introductory DC/AC Circuits,* this chapter, and then follow the instructions when answering the multiple-choice practice problems.

Direct Current (DC)

The First Computer Bug

Mathematician Grace Murray Hopper, an extremely independent U.S. naval officer, was assigned to the Bureau of Ordnance Computation Project at Harvard during World War II. As Hopper recalled, "We were not programmers in those days, the word had not yet come over from England. We were 'coders,'" and with her colleagues she was assigned to compute ballistic firing tables on the Harvard Mark 1 computer. In carrying out this task, Hopper developed programming method fundamentals that are still in use.

On a less important note, Hopper is also credited with creating a term frequently used today falling under the category of computer jargon. During the hot summer of 1945, the computer developed a mysterious problem. Upon investigation, Hopper discovered that a moth had somehow strayed into the computer and prevented the operation of one of the thousands of electromechanical relay switches. In her usual meticulous manner, Hopper removed the remains and taped and entered it into the logbook. In her own words, "From then on, when an officer came in to ask if we were accomplishing anything, we told him we were 'debugging' the computer," a term that is still used to describe the process of finding problems in a computer program.

4

Outline and Objectives

THE FIRST COMPUTER BUG
INTRODUCTION

4-1 DC (DIRECT-CURRENT) SOURCES

4-1-1 Batteries

4-1-2 Power Supplies

4-1-3 Electric Generators

4-1-4 Solar Cells

4-1-5 Thermocouples

Objective 1: Describe how direct current can be generated:

a. Chemically with a battery
b. Electrically with a power supply
c. Magnetically with a generator
d. Optically with a solar cell
e. Thermally with a thermocouple

Objective 2: Describe the operation of both a primary and a secondary cell.

Objective 3: State the difference between a primary and a secondary cell.

Objective 4: Describe the features, applications, and construction of all the popular primary and secondary cells.

Objective 5: Explain the advantages of connecting batteries in series and in parallel.

4-2 DC (DIRECT-CURRENT) SENSORS

Objective 6: Define the term *transducer*.

4-2-1 Temperature Sensors

Objective 7: Demonstrate how to convert among the four temperature scales of Fahrenheit, Celsius, Kelvin, and Rankine.

4-2-2 Light Sensors

4-2-3 Pressure Sensors

4-2-4 Magnetic Sensors

Objective 8: Describe the operation, characteristics, and applications for the following sensors:

a. Resistive temperature detector (RTD) and thin-film detector (TFD)
b. Thermistor
c. Light-dependent resistor (LDR)
d. Piezoelectric and piezoresistive sensor
e. Hall effect sensor

4-3 DC (DIRECT-CURRENT) ELECTROMAGNETISM

Objective 9: Describe what is meant by the word *electromagnetism*.

Objective 10: Explain the atomic theory of electromagnetism.

Objective 11: Describe how a magnetic field is generated by current flow through a

a. Conductor
b. Coil

4-3-1 Relays

4-3-2 Reed Switches

4-3-3 Solenoids

Objective 12: Describe the following applications of electromagnetism:

a. Relay
b. Reed switch
c. Solenoid-type electromagnet

4-4 EQUIPMENT PROTECTION

4-4-1 Fuses

Objective 13: Describe the function, operation, and characteristics of fuses.

4-4-2 Circuit Breakers

Objective 14: Describe the function, operation, and characteristics of the following circuit breaker types:

a. Thermal
b. Magnetic
c. Thermomagnetic

4-4-3 Ground Fault Interrupters

Objective 15: Explain the reason for grounding.

Objective 16: Describe the function, operation, and characteristics of ground fault interrupters.

Introduction

Direct current, abbreviated dc, is the flow of continuous current in only one DIRECTion. This should not be a new concept to you, as the battery and dc power supply are both sources of dc. It is probably difficult to imagine any other type of current flow. You have not been introduced to **alternating current,** or ac, yet, but later, you will find that ac voltage sources switch the current in a circuit first in one direction and then in the other. This makes dc very simple, since the positive terminal of the source voltage remains positive at all times, and, similarly, the negative terminal of the source always remains negative.

In this chapter, the first section will examine dc voltage sources, the second section will present several types of dc sensors, the third will introduce dc electromagnetic devices, and the final section will cover equipment protection.

4-1 DC (DIRECT-CURRENT) SOURCES

You have already been introduced to the most basic dc voltage source, the battery, but there are some additional details we should address. And so, in this first section, let us take a closer look at battery types and characteristics, and in the following sections we will discuss other sources of dc voltage.

4-1-1 *Batteries*

Voltaic Cell
A battery cell having two unlike metal electrodes immersed in a solution that chemically interacts with the plates to produce an emf.

Electrolyte
Electrically conducting liquid (wet) or paste (dry).

The battery is a chemical voltage source that contains one or more voltaic cells. The **voltaic cell** was discovered by Alessandro Volta, an Italian physicist, in 1800. Figure 4-1 shows the three basic components within a battery's cell, which are a negative plate, a positive plate, and an electrolyte.

Cell Operation

Two dissimilar, separated metal plates such as copper and zinc are placed within a container that is filled with a substance known as the **electrolyte,** usually an acid. A chemical reaction causes electrons to pass through the electrolyte as they are repelled from one plate and at-

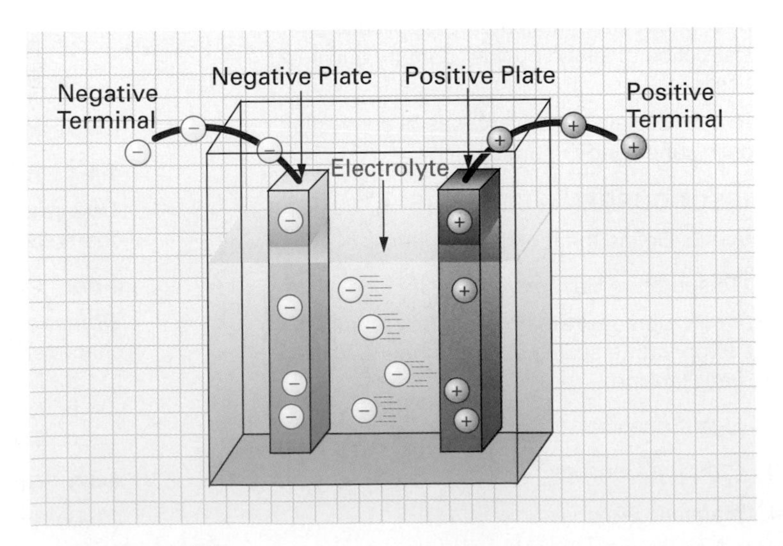

FIGURE 4-1 **A Battery's Cell.**

tracted to the other. This action causes a large number of electrons to collect on one plate (negative plate), and an absence or deficiency of electrons to exist on the opposite plate (positive plate). The electrolyte therefore acts on the two plates and transforms chemical energy into electrical energy, which can be taken from the cell at its two output terminals as an electrical current flow.

If nothing is connected across the battery, as shown in Figure 4-2(a), a chemical reaction between the electrolyte and the negative electrode produces free electrons that travel from atom to atom but are held in the negative electrode. On the other hand, if a lightbulb is connected between the negative and positive electrodes, as shown in Figure 4-2(b), the mutual repulsion of the free electrons at the negative electrode combined with the attraction of the positive electrode will cause a migration of free electrons (current flow) through the lightbulb, causing its filament to produce light.

Primary Cells

The chemical reaction within the cell will eventually dissolve the negative plate. This discharging process also results in hydrogen gas bubbles forming around the positive plate, causing a resistance between the two plates, known as the battery's internal resistance, to increase (called *polarization*). To counteract this problem, all dry cells have a chemical within them known as a depolarizer agent, which reduces the buildup of these gas bubbles. With time, however, the depolarizer's effectiveness will be reduced, and the battery's internal resistance will increase as the battery reaches a completely discharged condition. These types of batteries are known as **primary cells** (first time, last time) because they discharge once and must then be discarded. Almost all primary cells have their electrolyte in paste form, which is why they are also referred to as **dry cells.** Wet cells have an electrolyte that is in liquid form.

Primary Cell
Cell that produces electrical energy through an internal electrochemical action; once discharged, it cannot be reused.

Dry Cell
DC voltage-generating chemical cell using a nonliquid (paste) type of electrolyte.

Primary cell types. The shelf life of a battery is the length of time a battery can remain on the shelf (in storage) and still retain its usability. Most primary cells will deteriorate in a three-year period to approximately 80% of their original capacity. Of all the types of primary cells, five dominate the market as far as applications and sales are concerned. These are the carbon–zinc, alkaline–manganese, mercury, silver oxide, and lithium types, all of which are described and illustrated in Table 4-1. The cell voltage in Table 4-1 describes the voltage produced by one *cell* (one set of plates). If two or more sets of plates are installed in one package, the component is called a *battery*.

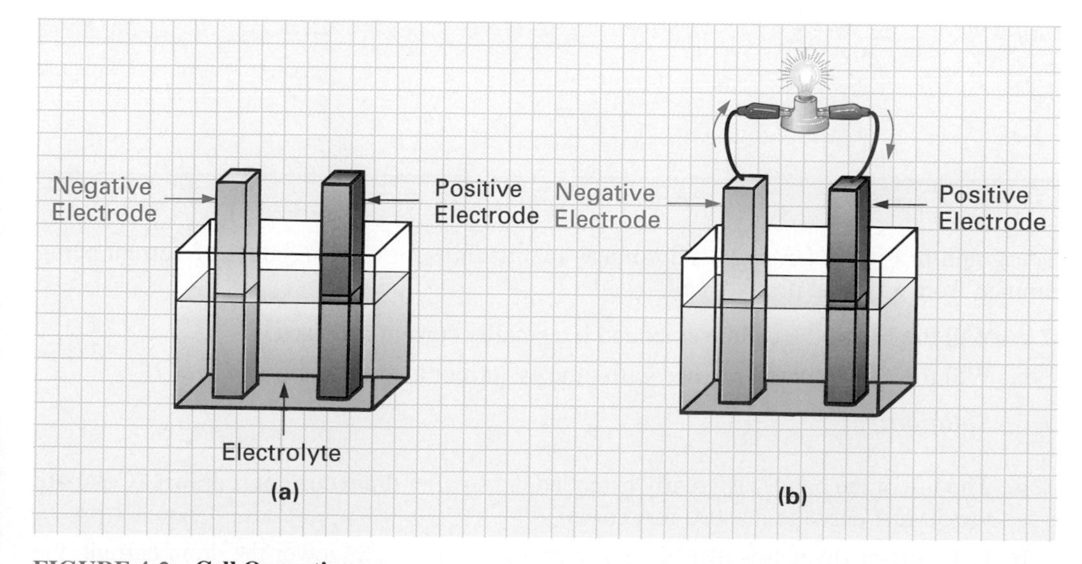

FIGURE 4-2 Cell Operation.

TABLE 4-1 Primary Cell Types

Battery Type	Features	Cell Voltage (V)	Applications	Construction
Carbon–zinc (also called Leclanché cell in honor of its inventor)	Most popular due to its low cost. Cylindrical D and C cells are most commonly used. Flat cells are stacked in series to obtain voltages greater than 1.5 V, as in the case of the 9 V battery.	1.5	Portable radios, tape players, televisions, toys.	Has a zinc case (negative), a carbon center rod (positive), with ammonium chloride as the electrolyte and manganese dioxide as the depolarizing agent. 9 V flat cell has 6 stacked 1.5 V cells
Alkaline–maganese (also called alkaline cell)	Has three times the shelf life and capacity of carbon zinc. Cylindrical or miniature cell sizes are available.	1.4	Portable radios, tape players, televisions, toys; large capacity is worth investment	The negative plate is granular zinc mixed with an electrolyte (alkaline); the positive plate is a polarizer in contact with the outer metal can. Alkaline Miniature Cell
Mercury (also called mercuric–oxide)	Higher energy density than preceding two with good shelf life and small size; used in low-power applications. Flat or cylindrical cells are available.	1.35 and 1.4	Watches, hearing aids, pacemakers, cameras, test equipment	A zinc anode and a mercuric oxide cathode in a potassium hydroxide electrolyte
Silver–oxide	High capacity; however, the material used (silver) makes it the most costly; can supply high currents for short periods of time; used in low-power applications.	1.5	Watches, hearing aids, pacemakers, cameras	Contains a cathode of silver oxide, an alkaline electrolyte, and zinc anode
Lithium	High discharge rate and long shelf life; light weight and higher output voltage; the familiar cylindrical and flat cell are available.	1.9	Liquid crystal watches, semi-conductor memories, hand-held calculators, sensor circuits	Contains a lithium anode, porous carbon cathode, and sulfur dioxide electrolyte

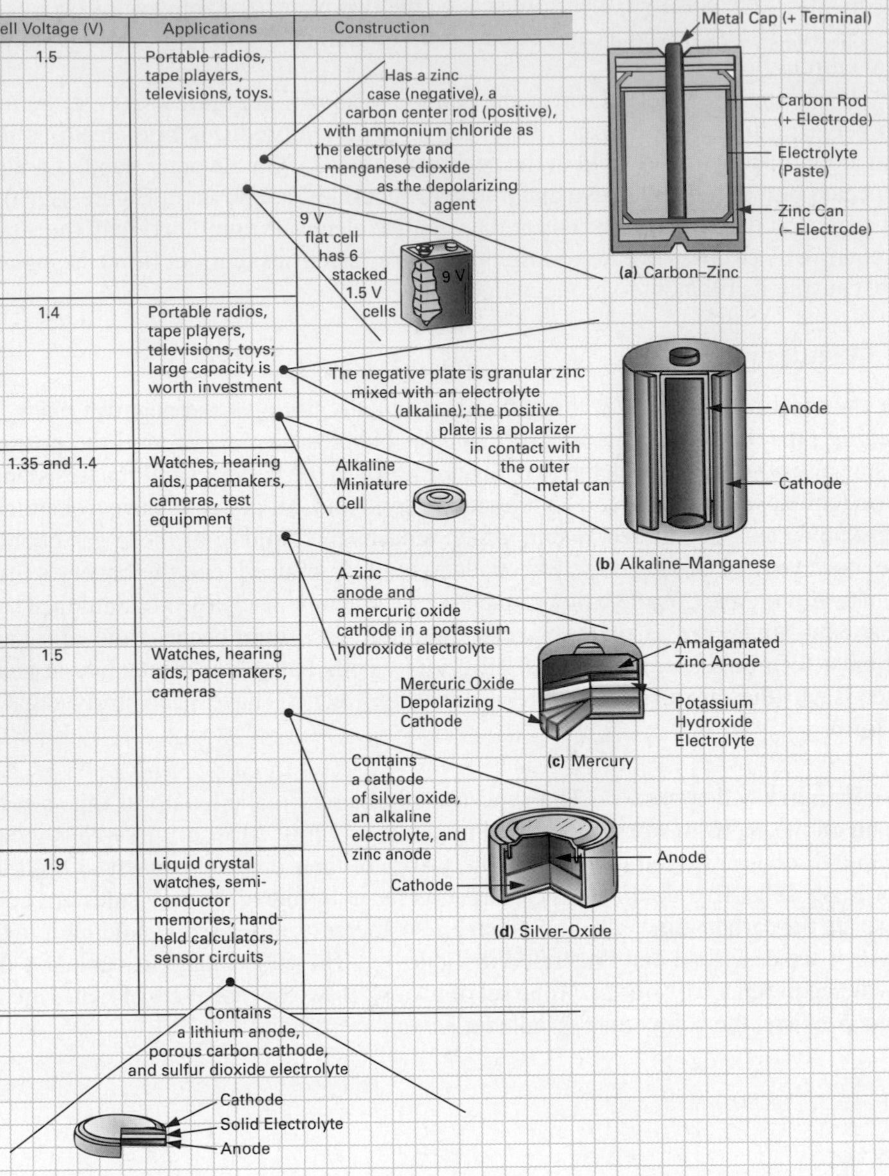

Metal Cap (+ Terminal)
Carbon Rod (+ Electrode)
Electrolyte (Paste)
Zinc Can (– Electrode)
9 V
(a) Carbon–Zinc

Anode
Cathode
(b) Alkaline–Manganese

Amalgamated Zinc Anode
Mercuric Oxide Depolarizing Cathode
Potassium Hydroxide Electrolyte
(c) Mercury

Anode
Cathode
(d) Silver-Oxide

Cathode
Solid Electrolyte
Anode

EXAMPLE:

The graph in Figure 4-3 plots cell voltage against time for certain values of current being drained from a typical primary cell.

a. Will the life of the cell be extended if a smaller current is drawn?

b. Will the cell voltage decrease with time as current is drained from the cell?

Solution:

a. The life of the battery is greatly extended if a smaller drain current is drawn. (Compare the 50 mA curve to the 100 mA curve).

b. Cell voltage decreases over time as current is drawn. The lower the drain current, the longer the cell will hold its cell voltage.

FIGURE 4-3

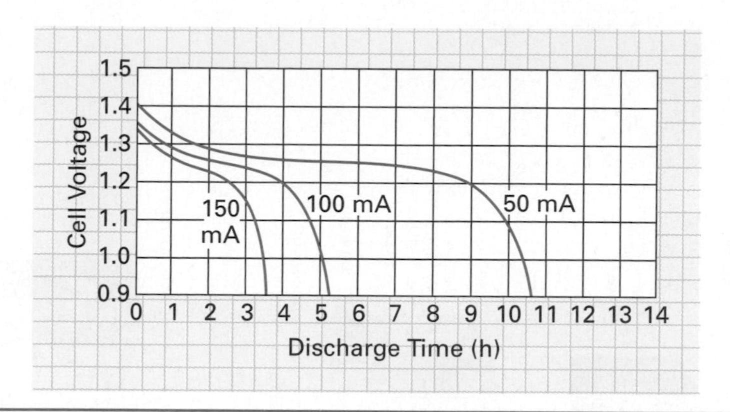

Secondary Cells

Secondary cells operate on the same basic principle as that of primary cells. In this case, however, the plates are not eaten away or dissolved, they only undergo a chemical change during discharge. Once discharged, the secondary cell can have the chemical change that occurred during discharge reversed by recharging, resulting once more in a fully charged cell. Restoring a secondary cell to the charged condition is achieved by sending a current through the cell in the direction opposite to that of the discharge current, as illustrated in Figure 4-4.

Secondary Cells
An electrolytic cell for generating electric energy. Once discharged, the cell may be restored or recharged by sending an electric current through the cell in the opposite direction to that of the discharge current.

In Figure 4-4(a), we have first connected a lightbulb across the battery, and free electrons are being supplied by the negative electrode and are moving through the lightbulb or load and back to the positive electrode. After a certain amount of time, any secondary cell will run down or will have discharged to such an extent that no usable value of current can be produced. The surfaces of the plates are changed, and the electrolyte lacks the necessary chemicals.

In Figure 4-4(b), during the recharging process, we use a battery charger, which reverses the chemical process of discharge by forcing electrons back into the cell and restoring the battery to its charged condition. The battery charger voltage is normally set to about 115% of the battery voltage. To use an example, a battery charger connected across a 12 V battery would be set to 115% of 12 V, or 13.8 V. If this voltage is set too high, an excessive current can result, causing the battery to overheat.

Secondary cells are often referred to as **wet cells** because the electrolyte is not normally in paste form (dry) as in the primary cell, but is in liquid (wet) form and is free to move and flow within the container.

Wet Cell
Cell using a liquid electrolyte.

Secondary cell types. Of all the secondary cells on the market today, the lead–acid and nickel–cadmium are the two most popular, and are illustrated and described in Table 4-2. Other secondary cell types include the nickel–metal hydride and lithium ion, which are used in cellular telephones, camcorders, and other portable electronic systems.

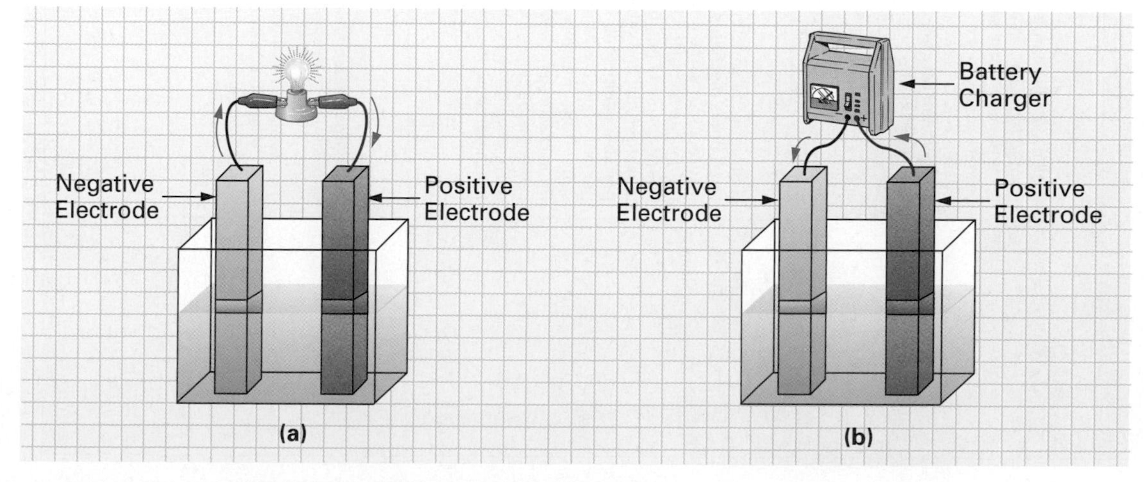

FIGURE 4-4 **Secondary Cell Operation. (a) Discharge. (b) Charge.**

TABLE 4-2 Secondary Cell Types

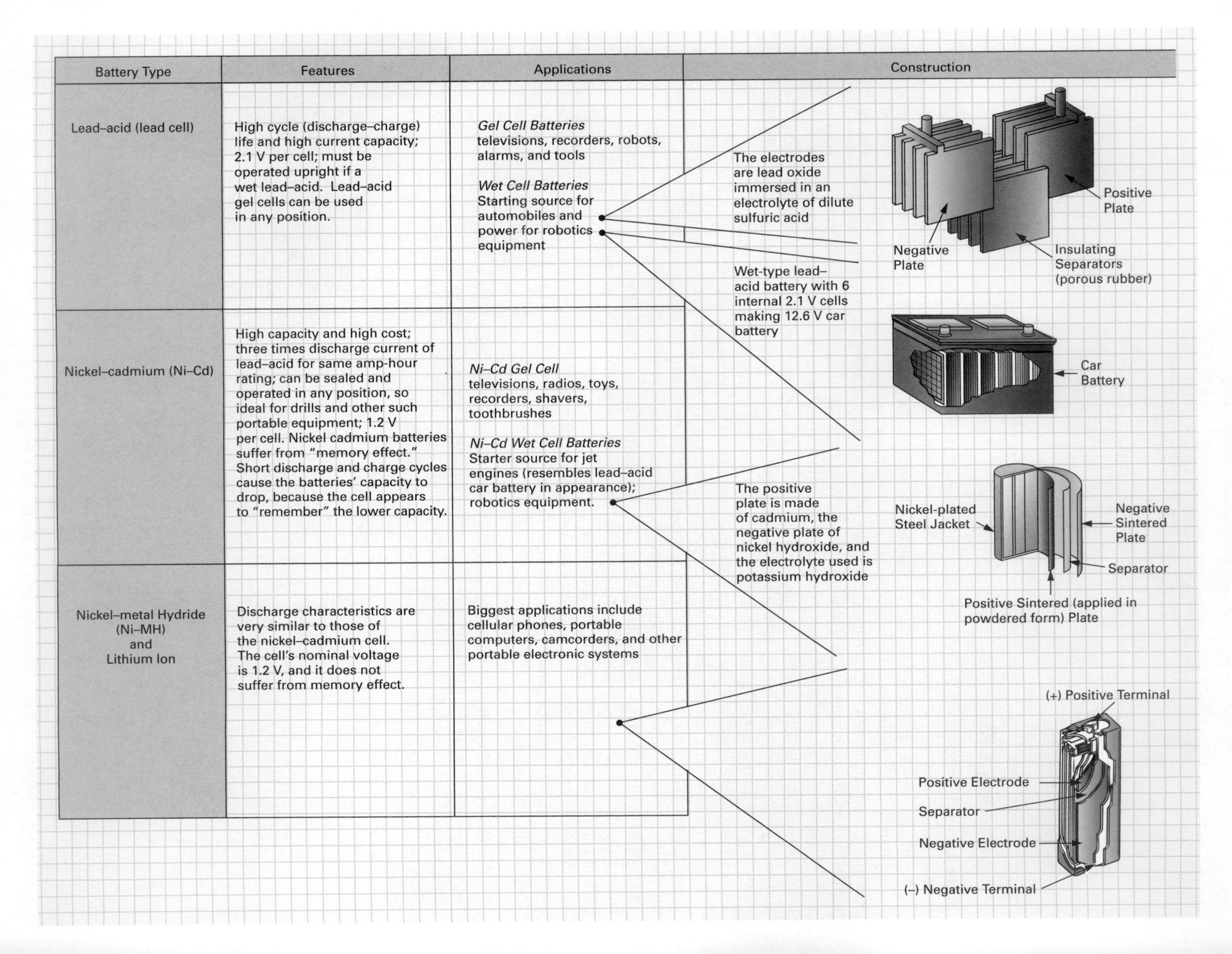

Battery Type	Features	Applications	Construction
Lead–acid (lead cell)	High cycle (discharge–charge) life and high current capacity; 2.1 V per cell; must be operated upright if a wet lead–acid. Lead–acid gel cells can be used in any position.	*Gel Cell Batteries* televisions, recorders, robots, alarms, and tools *Wet Cell Batteries* Starting source for automobiles and power for robotics equipment	The electrodes are lead oxide immersed in an electrolyte of dilute sulfuric acid Wet-type lead–acid battery with 6 internal 2.1 V cells making 12.6 V car battery
Nickel–cadmium (Ni–Cd)	High capacity and high cost; three times discharge current of lead–acid for same amp-hour rating; can be sealed and operated in any position, so ideal for drills and other such portable equipment; 1.2 V per cell. Nickel cadmium batteries suffer from "memory effect." Short discharge and charge cycles cause the batteries' capacity to drop, because the cell appears to "remember" the lower capacity.	*Ni–Cd Gel Cell* televisions, radios, toys, recorders, shavers, toothbrushes *Ni–Cd Wet Cell Batteries* Starter source for jet engines (resembles lead–acid car battery in appearance); robotics equipment.	The positive plate is made of cadmium, the negative plate of nickel hydroxide, and the electrolyte used is potassium hydroxide
Nickel–metal Hydride (Ni–MH) and Lithium Ion	Discharge characteristics are very similar to those of the nickel–cadmium cell. The cell's nominal voltage is 1.2 V, and it does not suffer from memory effect.	Biggest applications include cellular phones, portable computers, camcorders, and other portable electronic systems	

Secondary cell capacity. Capacity (C) is measured by the amount of **ampere-hours** (Ah) a battery can supply during discharge. For example, if a battery has a discharge capacity of 10 ampere-hours, the battery could supply 1 ampere for 10 hours, 10 amperes for 1 hour, 5 amperes for 2 hours, and so on, during discharge. Automobile batteries (12 V) will typically have an ampere-hour rating of between 100 and 300 Ah.

Ampere-hour units are actually specifying the coulombs of charge in the battery. For example, if a lead-acid car battery was rated at 150 Ah, this value would have to be converted to ampere-seconds to determine coulombs of charge, as 1 ampere-second (As) is equal to 1 coulomb. Ampere-hours are easily converted to ampere-seconds simply by multiplying ampere-hours by 3600, which is the number of seconds in 1 hour. In our example, a 150 Ah battery will have a charge of 54×10^4 coulombs.

Ampere-Hours
If you multiply the amount of current, in amperes, that can be supplied for a given time frame in hours, a value of ampere-hours is obtained.

EXAMPLE:

A Ni–Cd battery is rated at 300 Ah and is fully discharged.

a. How many coulombs of charge must the battery charger put into the battery to restore it to full charge?

b. If the charger is supplying a charging current of 3 A, how long will it take the battery to fully charge?

c. Once fully charged, the battery is connected across a load that is pulling a current of 30 A. How long will it take until the battery is fully discharged?

Solution:

a. The same amount that was taken out: $300 \text{ Ah} \times 3600 = 1080 \times 10^3 \text{ C}$

b. $\dfrac{300 \text{ Ah}}{3 \text{ A}} = 100$ hours until charged

c. $\dfrac{300 \text{ Ah}}{30 \text{ A}} = 10$ hours until discharged

Batteries in Series and Parallel

Batteries are often connected in combination to gain a higher voltage or current than can be obtained from one cell. Referring to Figure 4-5, you can see that when batteries are connected in series, the negative terminal of A is connected to the positive terminal of B. The total voltage across the bulb in this configuration will be the sum of the two cell voltages, which in this case will be 18 V.

Referring to Figure 4-6, you can see that when batteries are connected in parallel, the negative terminal connects to the other negative terminal and the positive terminal connects to the positive. The voltage in this configuration remains the same as for one cell. However, the current demand is now shared, and so the combined parallel batteries have the ability to supply a higher value of current if the load demands it.

TIME LINE
The electron was first discovered by Jean Baptiste Perrin (1870-1942), a French physicist who was awarded the Nobel prize for physics.

EXAMPLE:

Referring to the flashlight shown in Figure 4-7,

a. Are the two D-size batteries connected in series or parallel?

b. What voltage is applied across the bulb when the switch is closed?

Solution

a. series

b. 3 V

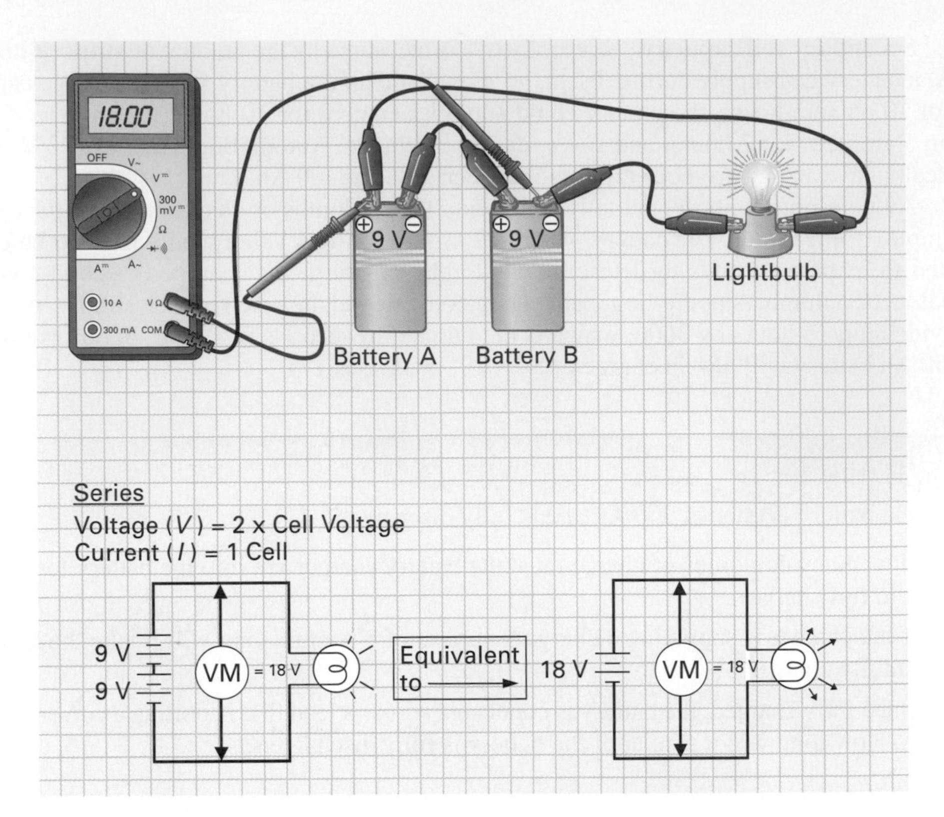

FIGURE 4-5 **Batteries in Series.**

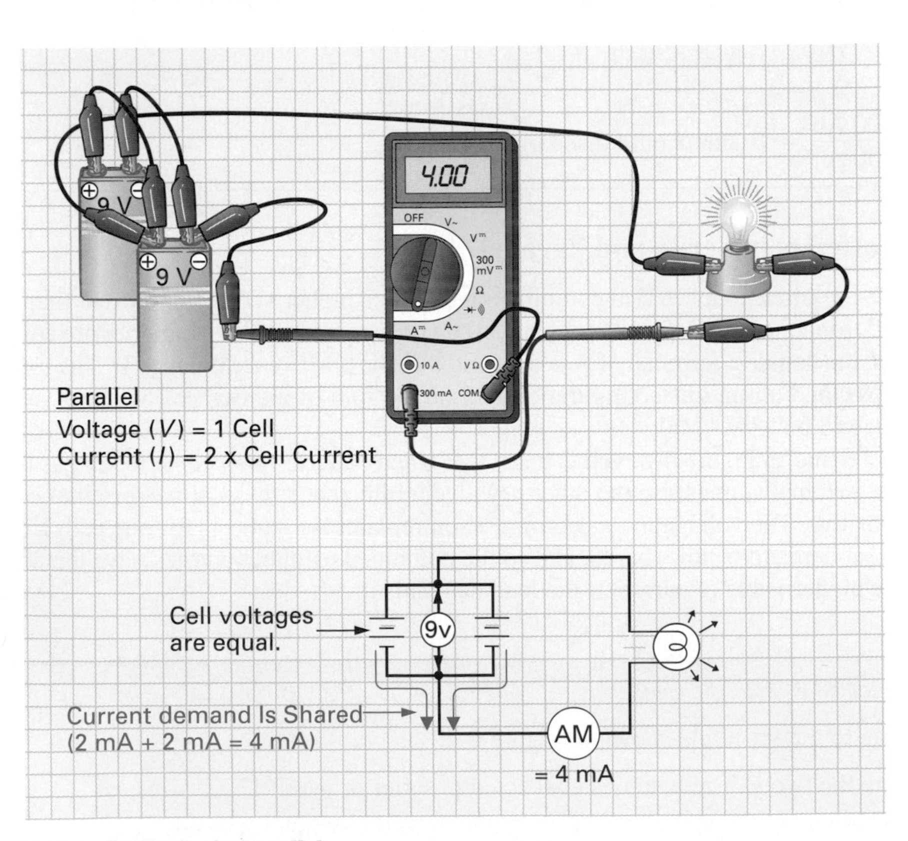

FIGURE 4-6 **Batteries in Parallel.**

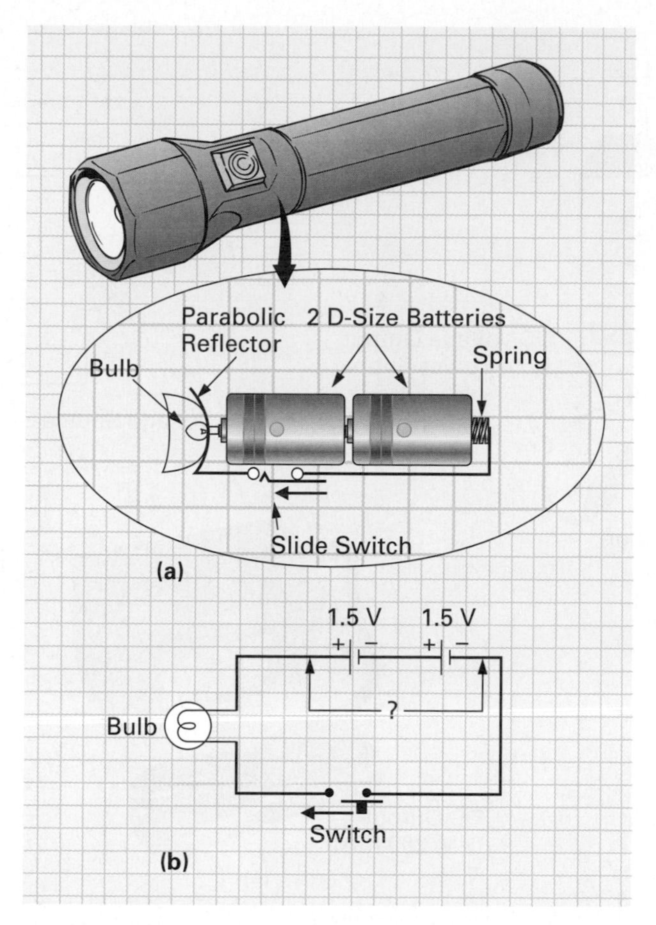

FIGURE 4-7 **Flashlight. (a) Physical Appearance. (b) Schematic Diagram.**

4-1-2 *Power Supplies*

Up until now we have only discussed electrical energy in the form of direct current (dc). However, as we will discover in Chapter 8, electrical energy is more readily available in the form of alternating current (ac). In fact, that is what arrives at the wall receptacle at your home and workplace. We can use a dc power supply, like the one seen in Figure 4-8(a), to convert the ac electrical energy at the wall outlet to dc electrical energy for experiments.

Power in the laboratory can be obtained from a battery, but since ac power is so accessible, dc power supplies are used. The power supply has advantages over the battery as a source of dc voltage in that it can quickly provide an accurate voltage that can easily be varied by a control on the front panel, and it never runs down. During your electronic studies, you will use a power supply in the lab to provide different voltages for various experiments. In Figure 4-8(a), you can see how a dc power supply is being used to supply 12 V to a 12 kΩ resistor, and in Figure 4-8(b), you can see how the schematic symbol for a dc power supply differs from the battery symbol.

Many electronic systems such as answering machines, cell phones, and CD players receive dc power from an adapter, like the one shown in Figure 4-8(c). Like the power supply, this unit is plugged into a wall outlet and will convert an ac input into a dc output.

4-1-3 *Electric Generators*

A magnet like the one in Figure 4-9(a) can be used to generate electrical current by movement of the magnet past a conductor, as shown in Figure 4-9(b). When the magnet is stationary, no voltage is induced into the wire and therefore the lighbulb is OFF. If the magnet

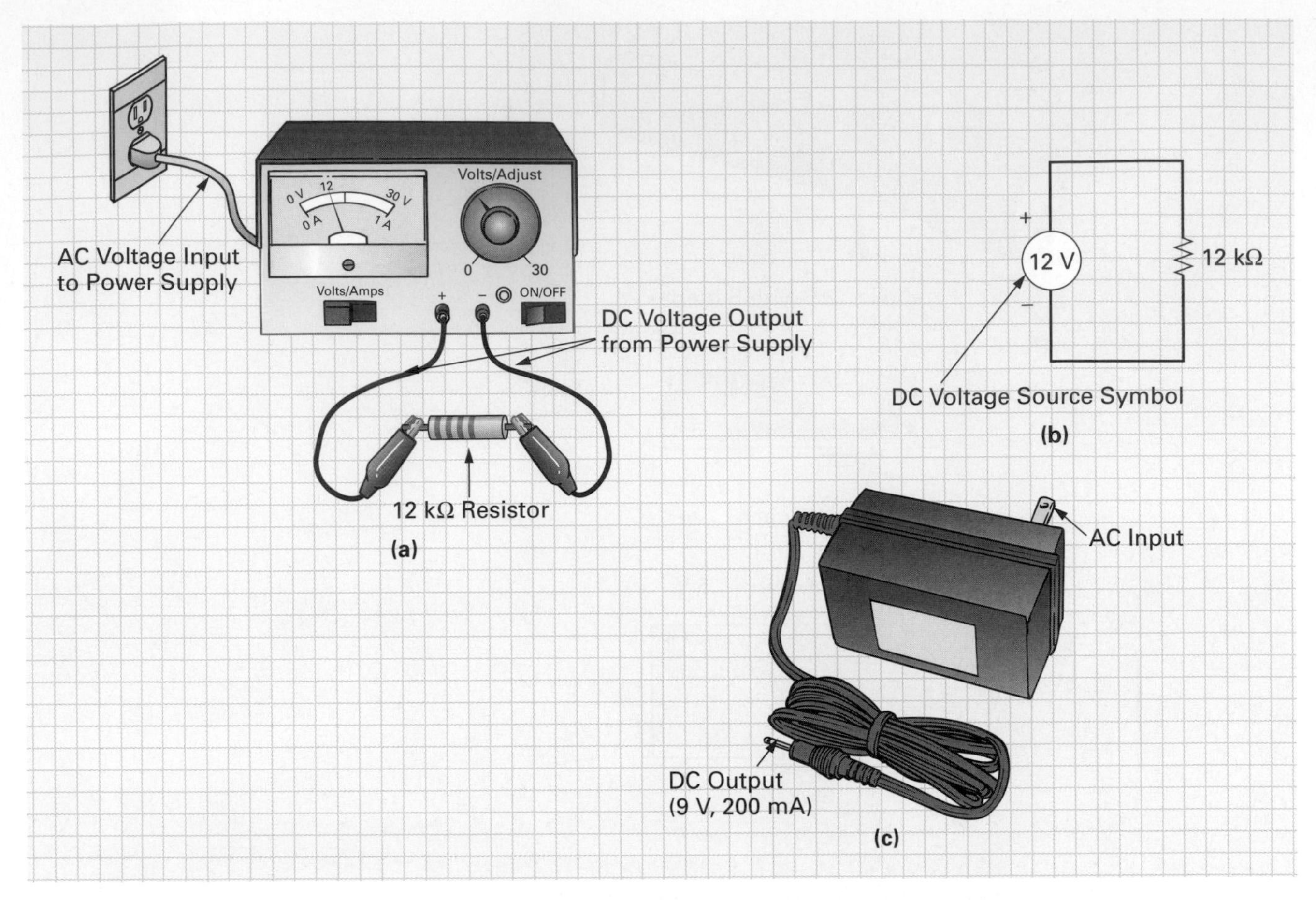

FIGURE 4-8 **Direct Current from Alternating Current. (a) Power Supply. (b) DC Voltage Source Symbol. (c) Adapter.**

is moved so that the invisible magnetic lines of force cut through the wire conductor, a voltage is induced in the conductor, and this voltage produces a current in the circuit causing the lightbulb to emit light. An energy conversion from magnetic to electric can be achieved by moving either the magnet past the wire or the wire past the magnet.

To produce a continuous supply of electric current, the magnet or wire must be constantly in motion. The device that achieves this is the dc electric generator, illustrated in Figure 4-9(c).

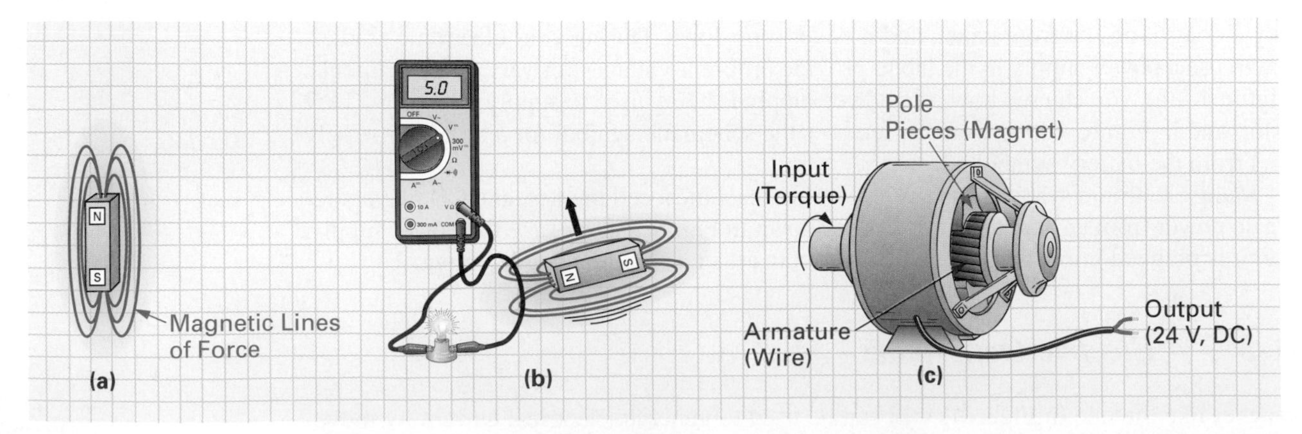

FIGURE 4-9 **Direct Current from Magnetism. (a) Stationary Magnet. (b) Moving Magnet. (c) Electric Generator.**

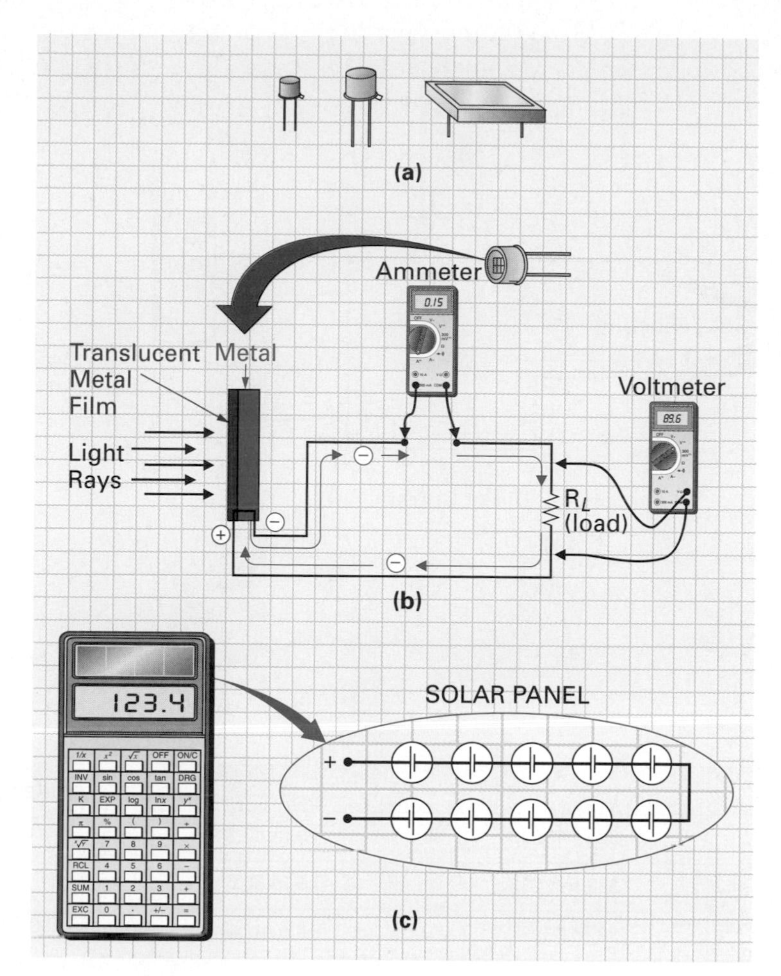

FIGURE 4-10 Direct Current from Light Using the Solar Cell. (a) Physical Appearance. (b) Construction and Operation. (c) Application: Solar Panel for a Calculator.

4-1-4 *Solar Cells*

A **photovoltaic cell,** shown in Figure 4-10(a), makes use of two dissimilar materials in its transformation of light to electrical energy. Figure 4-10(b) illustrates the construction and operation of a photovoltaic cell, which is also called a photoelectric cell or a **solar cell.**

A light-sensitive metal is placed behind a transparent piece of dissimilar metal. When the light illuminates the light-sensitive material, a charge is generated, causing current to flow and the meter to indicate current flow. This phenomenon is known as **photovoltaic action,** meaning *photo* (light) to *voltaic* (voltage) action.

To obtain a reasonable amount of power, solar cells are interconnected so that they work collectively on solar panels. These solar panels are used to power calculators, emergency freeway phones, and satellites, like the one shown in Figure 4-10(c). The power obtained on average is about 100 milliwatts per square centimeter.

Photovoltaic Cell
A solid-state device that generates electricity when light is applied. Also called a solar cell.

Solar Cell
Photovoltaic cell that converts light into electric energy. They are especially useful as a power source for space vehicles.

Photovoltaic Action
The development of a voltage across the junction of two dissimilar materials when light is applied.

4-1-5 *Thermocouples*

Heat can be used to generate an electric charge, as illustrated in Figure 4-11(a). When two dissimilar metals, such as copper and iron, are welded together and heated, an electrical charge is produced. This junction is called a **thermocouple,** and the heat causes the release of electrons from their parent atoms, resulting in a meter deflection indicating the generation of a charge. The size of the charge is proportional to the temperature difference between the

Thermocouple
Temperature transducer consisting of two dissimilar metals welded together at one end to form a junction that when heated will generate a voltage.

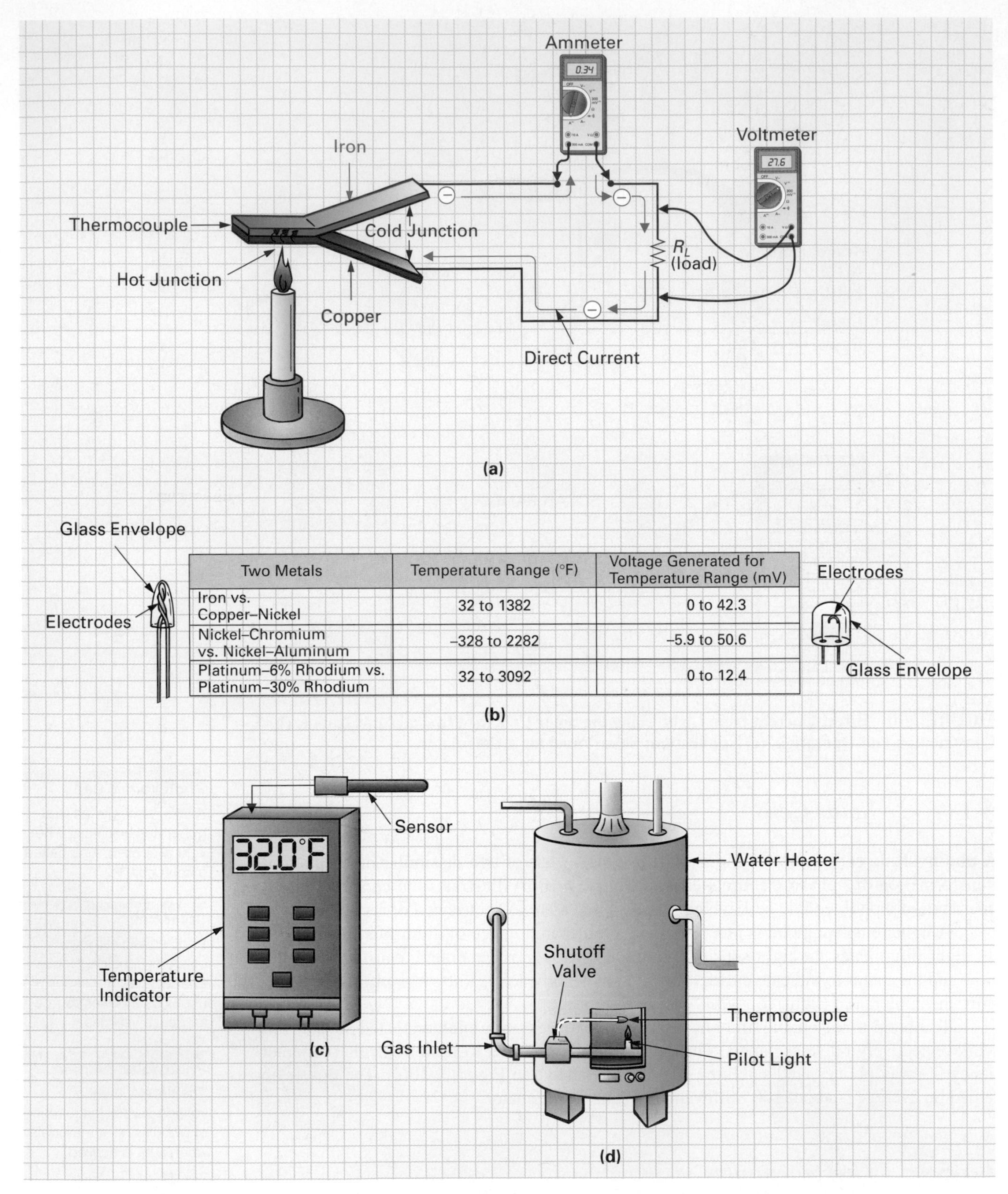

Two Metals	Temperature Range (°F)	Voltage Generated for Temperature Range (mV)
Iron vs. Copper–Nickel	32 to 1382	0 to 42.3
Nickel–Chromium vs. Nickel–Aluminum	–328 to 2282	–5.9 to 50.6
Platinum–6% Rhodium vs. Platinum–30% Rhodium	32 to 3092	0 to 12.4

FIGURE 4-11 Direct Current from Heat Using the Thermocouple. (a) Operation. (b) Typical Thermocouples. (c) Indicating Temperature. (d) Application: Water Heater.

two metals. Figure 4-11(b) lists some typical thermocouples and their voltages generated when heat is applied and shows the physical appearance of two types.

Thermocouples like the one in Figure 4-11(c) are normally used to indicate temperature on a display calibrated in degrees. The thermocouple seen in Figure 4-11(d) is being heated by the gas pilot light, and the electrical energy generated will allow a valve to open and let gas through to the water heater. If, by accident, the pilot is extinguished by a gust of wind, the thermocouple will not receive any more heat and therefore will not generate electrical energy, which will consequently close the electrically operated valve and prevent a large explosion.

SELF-TEST EVALUATION POINT FOR SECTION 4–1

Now that you have completed this section, you should be able to:

Objective 1. Describe how direct current can be generated:

- **a.** Chemically with a battery
- **b.** Electrically with a power supply
- **c.** Magnetically with a generator
- **d.** Optically with a solar cell
- **e.** Thermally with a thermocouple

Objective 2. Describe the operation of both a primary and a secondary cell.

Objective 3. State the difference between a primary and a secondary cell.

Objective 4. Describe the features, applications, and construction of all the popular primary and secondary cells.

Objective 5. Explain the advantages of connecting batteries in series and in parallel.

Use the following questions to test your understanding of Section 4–1:

1. What is a thermocouple?
2. How is light converted into a direct current?
3. What device is used to convert magnetic energy into electrical energy?
4. List the three basic components in a battery.
5. True or false: If two 1.5 V cells were connected in series, the total source voltage would be 1.5 V.

4-2 DC (DIRECT-CURRENT) SENSORS

A **transducer** is an electronic device that converts one form of energy to another. Input transducers are sensors that convert thermal, optical, mechanical, and magnetic energy variations into an equivalent signal voltage.

Transducer
Any device that converts energy from one form to another.

4-2-1 *Temperature Sensors*

When first discussing variable-value resistors, we talked about the rheostat and potentiometer, both of which need a mechanical input (the user turning the shaft) to produce a change in resistance. A **bolometer** is a device that changes its resistance when heat energy is applied. Before discussing the three basic types of bolometers, let's first consider the four units of temperature measurement detailed in Table 4-3.

Bolometer
Device whose resistance changes when heated.

The measurement of temperature (**thermometry**) is probably the most common type of measurement used in industry today. Commercially, temperature is normally expressed in degrees Celsius (°C) or degrees Fahrenheit (°F); however, although not commonly known, kelvin (K) and degrees Rankine (°R) are often used in industry, the kelvin being the international unit of temperature.

Thermometry
Relating to the measurement of temperature.

TABLE 4-3 Four Temperature Scales and Conversion Formulas

	FAHRENHEIT	CELSIUS	KELVIN	RANKINE
Absolute zero	–459.69°F	–273.16°C	0 K	0°R
Melting point of ice (x)	32°F	0°C	273.16 K	491.69°R
	↓	↓	↓	↓
(Division between x and y)	(180°F)	(100°C)	(100 K)	(180°R)
	↓	↓	↓	↓
Boiling point of water (y)	212°F	100°C	373.16 K	671.69°R

$F = (^9/_5 \times C) + 32$
$C = {^5/_9} \times (F - 32)$
$R = F + 460$
$F = R - 460$
$K = C + 273$
$C = K - 273$

EXAMPLE:

Convert the following:

a. 74°F = __________ °C
b. 45°C = __________ °F
c. 25°C = __________ K
d. 10°F = __________ °R

Solution:

a. $C = \frac{5}{9} \times (F - 32)$
$= \frac{5}{9} \times (74 - 32)$
$= 0.555 \times 42 = 23.3°C$

b. $F = (\frac{9}{5} \times C) + 32$
$= (\frac{9}{5} \times 45) + 32$
$= (1.8 \times 45) + 32 = 113°F$

c. $K = C + 273)$
$= 25 + 273 = 298$ K

d. $R = F + 260$
$= 10 + 460 = 470°R$

Calculator Sequence

STEP	KEYPAD ENTRY	DISPLAY RESPONSE
1.	[7] [4]	74
2.	[−]	
3.	[3] [2]	32
4.	[=]	42
5.	[STO] (store result in memory)	
6.	[C/CE] (cancel display)	0
7.	[5]	5
8.	[÷]	
9.	[9]	9
10.	[=]	0.55555
11.	[×]	
12.	[RCL] (recall value from memory)	42
13.	[=]	23.3

Resistive Temperature Detector (RTD)
A temperature detector consisting of a fine coil of conducting wire (such as platinum) that will produce a relatively linear increase in resistance as temperature increases.

Thin-Film Detector (TFD)
A temperature detector containing a thin layer of platinum and used for very precise temperature readings.

Thermistor
Temperature-sensitive semiconductor that has a negative temperature coefficient of resistance (as temperature increases, resistance decreases).

Bolometers are normally used as temperature sensors or detectors. Both the **resistive temperature detector** (RTD) and **thin-film detector** (TFD) shown in Figure 4-12(a) and (b) are temperature sensors that make use of a copper, nickel, or platinum conducting element and therefore have a positive temperature coefficient of resistance (resistance increases as temperature increases). Referring to the construction of the RTD in Figure 4-12(a), you can see that the sensing element consists of a coil of fine wire generally made of platinum, which gives a relatively linear increase in resistance as temperature increases, as indicated in the table in Figure 4-12(a).

The thin-film detector (TFD) shown in Figure 4-12(b) is used for very precise temperature readings and is constructed by placing a thin layer of platinum on a ceramic substrate. Because of its small size, the TFD responds rapidly to temperature change and is ideally suited for surface-temperature sensing.

Unlike the RTD and TFD, the **thermistor** contains a semiconductor material that has a negative temperature coefficient of resistance, which means that its resistance decreases as temperature increases. Semiconductor materials have characteristics that are midway between those of conductors and insulators. In addition to thermistors, semiconductor materials are also used to construct other electronic devices such as diodes and transistors, which will be discussed later.

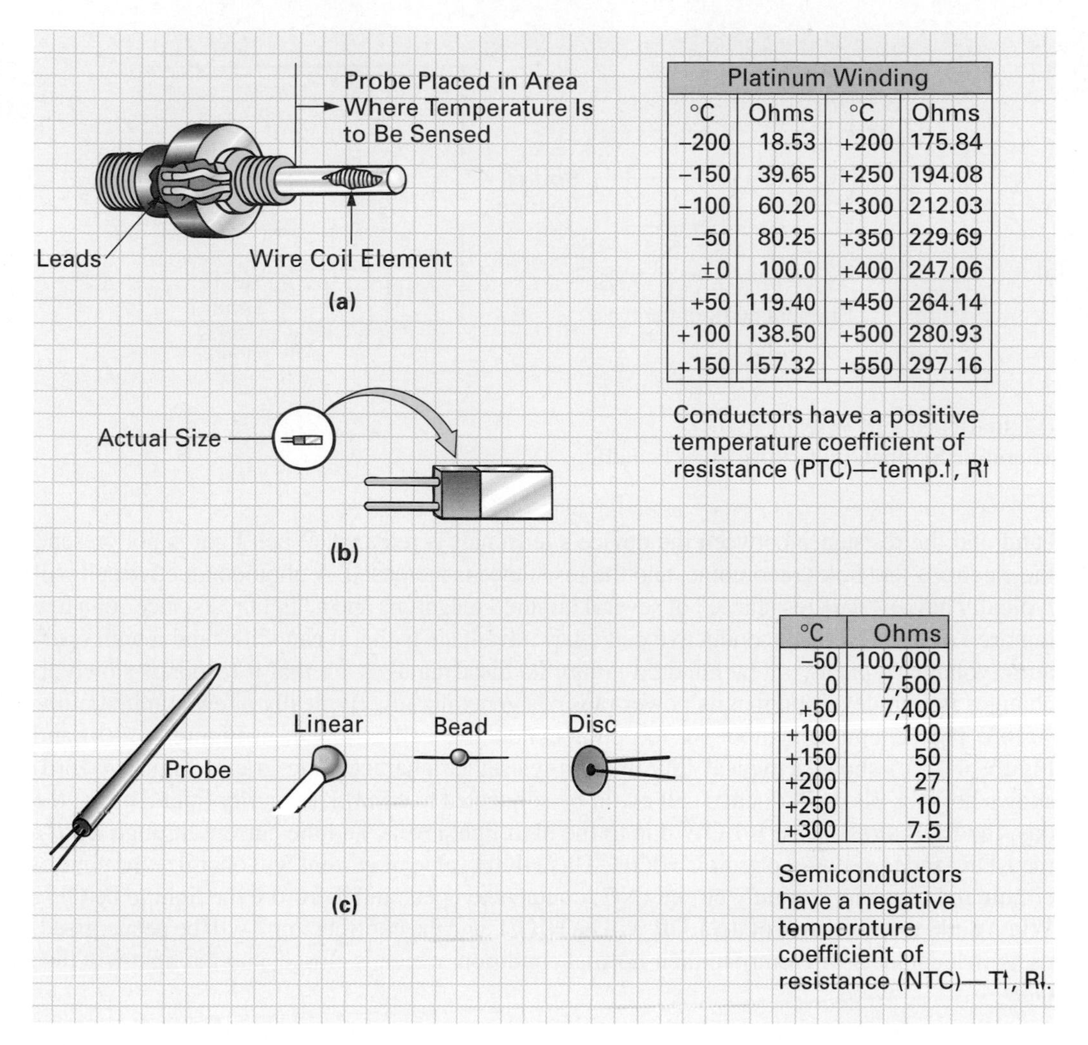

Platinum Winding			
°C	Ohms	°C	Ohms
−200	18.53	+200	175.84
−150	39.65	+250	194.08
−100	60.20	+300	212.03
−50	80.25	+350	229.69
±0	100.0	+400	247.06
+50	119.40	+450	264.14
+100	138.50	+500	280.93
+150	157.32	+550	297.16

°C	Ohms
−50	100,000
0	7,500
+50	7,400
+100	100
+150	50
+200	27
+250	10
+300	7.5

FIGURE 4-12

A variety of thermistors can be seen in Figure 4-12(c). The thermistor is the most common type of temperature sensor and produces rapid and extremely large changes in resistance for very small changes in temperature, as shown in the associated table.

An application of the thermistor could be as a temperature sensor inside an oven. As the oven heats up, the thermistor's resistance decreases, and at a certain decreased resistance the thermistor turns off the oven. Once the temperature starts to drop, the thermistor's resistance increases, and this increase of resistance turns the oven back on.

$$\text{temperature} \uparrow \quad \text{thermistor's } R \downarrow \quad \text{oven OFF}$$
$$\text{temperature} \downarrow \quad \text{thermistor's } R \uparrow \quad \text{oven ON}$$

4-2-2 *Light Sensors*

Photo means illumination, and the **photoresistor** is a resistor that is photoconductive. This means that as the material is exposed to light, it will become more conductive and less resistive. The photoresistor or **light-dependent resistor (LDR)** shown in Figure 4-13(a) is a two-terminal device that changes its resistance (conductance) when light (photo) is applied. The photoconductive cell is normally mounted in a metal or plastic case with a glass window that allows the sensed light to strike the S-shaped light-sensitive material (typically cadmium sulfide). When light strikes the photoconductive atoms, electrons are released into the conduction

Photoresistor
Also known as a photoconductive cell or light-dependent resistor, it is a device whose resistance varies with the illumination of the cell.

Light-Dependent Resistor (LDR)
A two-terminal device that changes its resistance when light is applied.

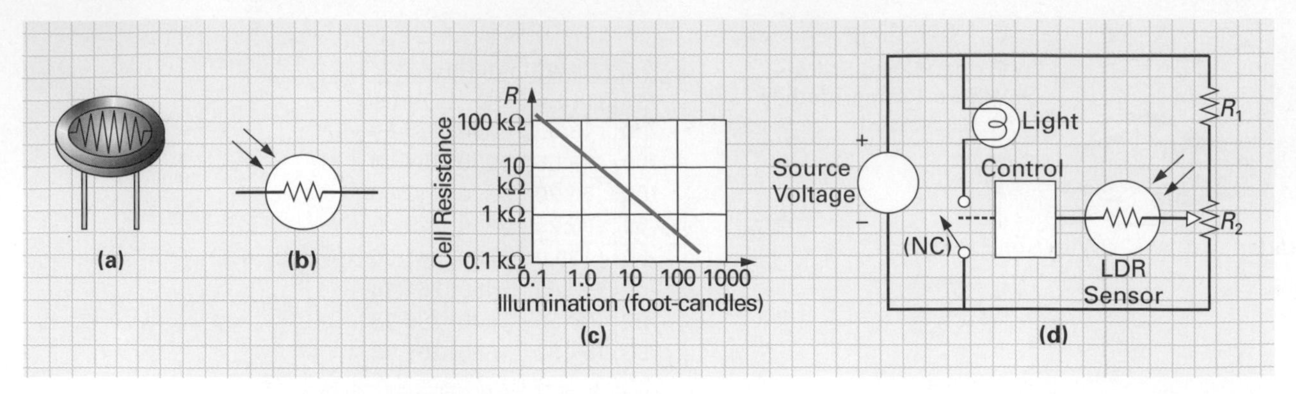

FIGURE 4-13 **Light Sensor–Light-Dependent Resistor (LDR).**
(a) Physical Appearance. (b) Schematic Symbol. (c) Graph. (d) Schematic.

band and the resistance between the device's terminals is reduced. When light is not present, the electrons and holes recombine, and the resistance is increased. A photoconductive cell will typically have a "dark resistance" of several hundred megohms and a "light resistance" of a few hundred ohms. The photoconductive cell's key advantage is that it can withstand a high operating voltage (typically a few hundred volts). Its disadvantages are that it responds slowly to changes in light level and that its power rating is generally low (typically a few hundred milliwatts). The schematic symbol for the photoconductive cell is shown in Figure 4-13(b) and Figure 4-13(c) shows a typical illumination-resistance characteristic graph. Figure 4-13(d) shows how the photoconductive cell could be connected to control a security light. During the day, the LDR's resistance is LOW due to the high light levels, and the current through R_1, R_2, the LDR, and the control circuit is HIGH. This HIGH value of current will energize the control circuit and cause its normally closed (NC) contacts to open, and therefore the light to be OFF. When dark, the resistance of the LDR will be HIGH, and the control circuit will be deenergized, its switch contacts will return to their normal condition, which is closed, and the light will turn ON.

4-2-3 *Pressure Sensors*

The quartz crystal is made of silicon dioxide and is naturally a six-sided (hexagonal) compound with pyramids at either end, as seen in Figure 4-14(a). To construct an electronic component, a thin slice or slab of crystal is cut from the mother stone, mounted between two metal plates that make electrical contact, then placed in a protective holder, as seen in Figure 4-14(b). On a schematic diagram, a crystal is generally labeled either *XTAL* or *Y*, and has the symbol shown in Figure 4-14(c).

The crystal is basically operated as a transducer, or energy converter, transforming mechanical energy or pressure to an equivalent electrical signal voltage, as shown in Figure 4-13(d), (e), and (f).

In Figure 4-14(d), you can see that a crystal normally has its internal charges evenly distributed throughout and the potential difference between its two plates is zero. If the crystal is compressed by applying pressure to either side, as shown in Figure 4-14(e), opposite charges accumulate on either side of the crystal and a potential difference is generated. Similarly, if the crystal is expanded by applying pressure to the top and bottom, as shown in Figure 4-14(f), opposite charges accumulate on either side of the crystal and a potential difference of the opposite polarity is generated. This action is described as **piezoelectric effect,** which is the generation of a voltage between the opposite faces of a crystal as a result of pressure being applied. This effect is made use of in most modern cigarette lighters. A voltage is produced between the opposite sides of a crystal (usually a crystalline lead compound) when it is either struck, pressured, or twisted. The resulting voltage is discharged across a gap to ignite the gas, which is released simultaneously by the operating button. The flame then burns for as long as the button is depressed.

Piezoelectric Effect
The generation of a voltage between the opposite sides of a piezoelectric crystal as a result of pressure or twisting. Also, the reverse effect in which the application of voltage to opposite sides causes deformation to occur at the frequency of the applied voltage.

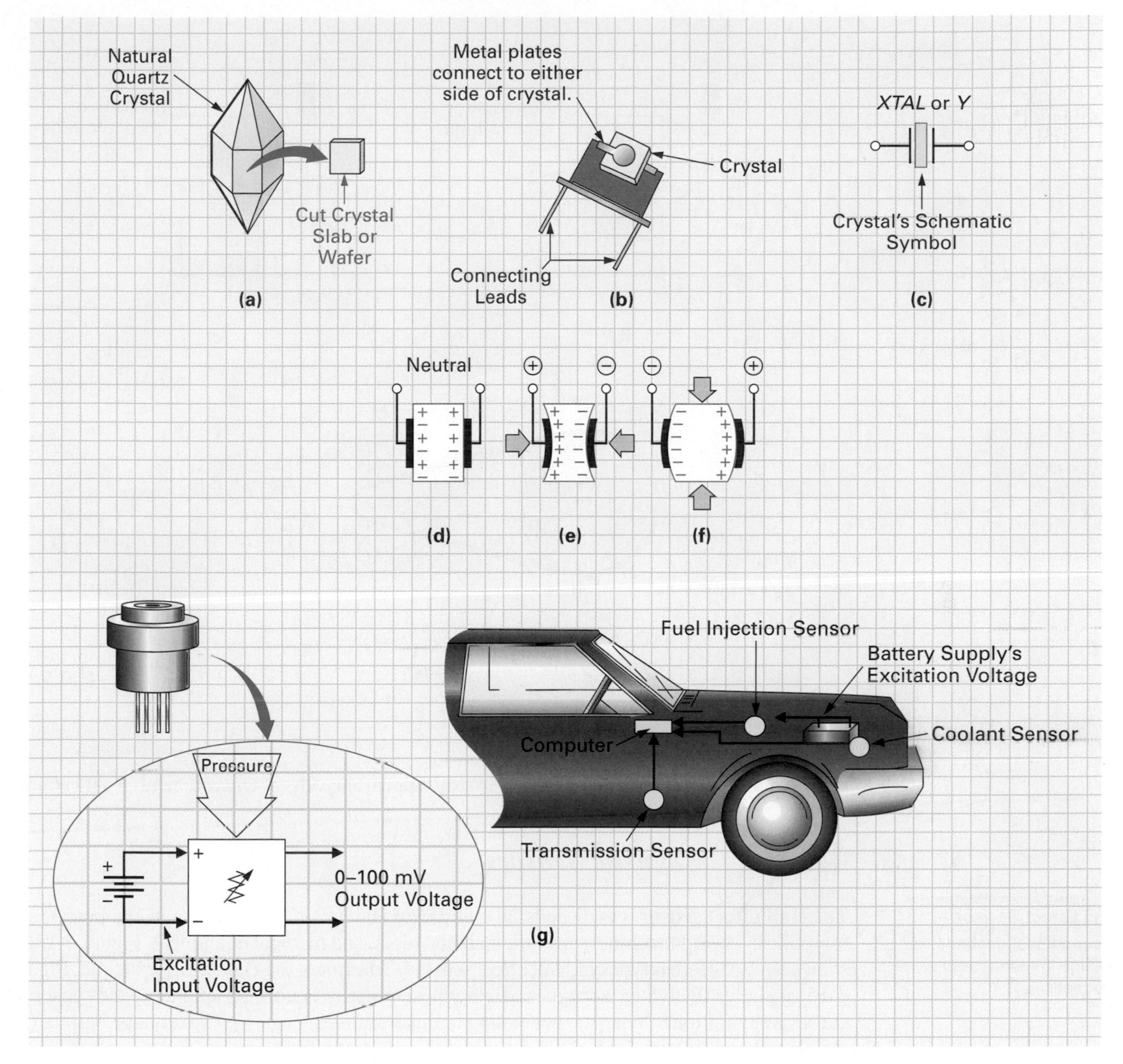

FIGURE 4-14 Pressure Sensors.
(a) Quartz Crystal. (b) Construction. (c) Schematic Symbol. (d–f) Operation. (g) Application.

Figure 4-14(g) illustrates the physical appearance and operation of a piezoresistive pressure sensor. Piezoresistance is described as a change in resistance due to a change in the applied pressure. Piezoresistive sensors need an excitation voltage applied, and it is this value that will determine the output voltage range. As an application, this sensor could be used to measure either the cooling system, hydraulic transmission, or fuel injection pressure in an automobile, which all vary in the 0 to 100 psi (pounds per square inch) range. The piezoresistive transducer would receive its excitation voltage from the car's battery (12 V) and produce a 0 to 100 mV output voltage, which would be sent to the car's computer, where the data would be analyzed and, if necessary, acted upon.

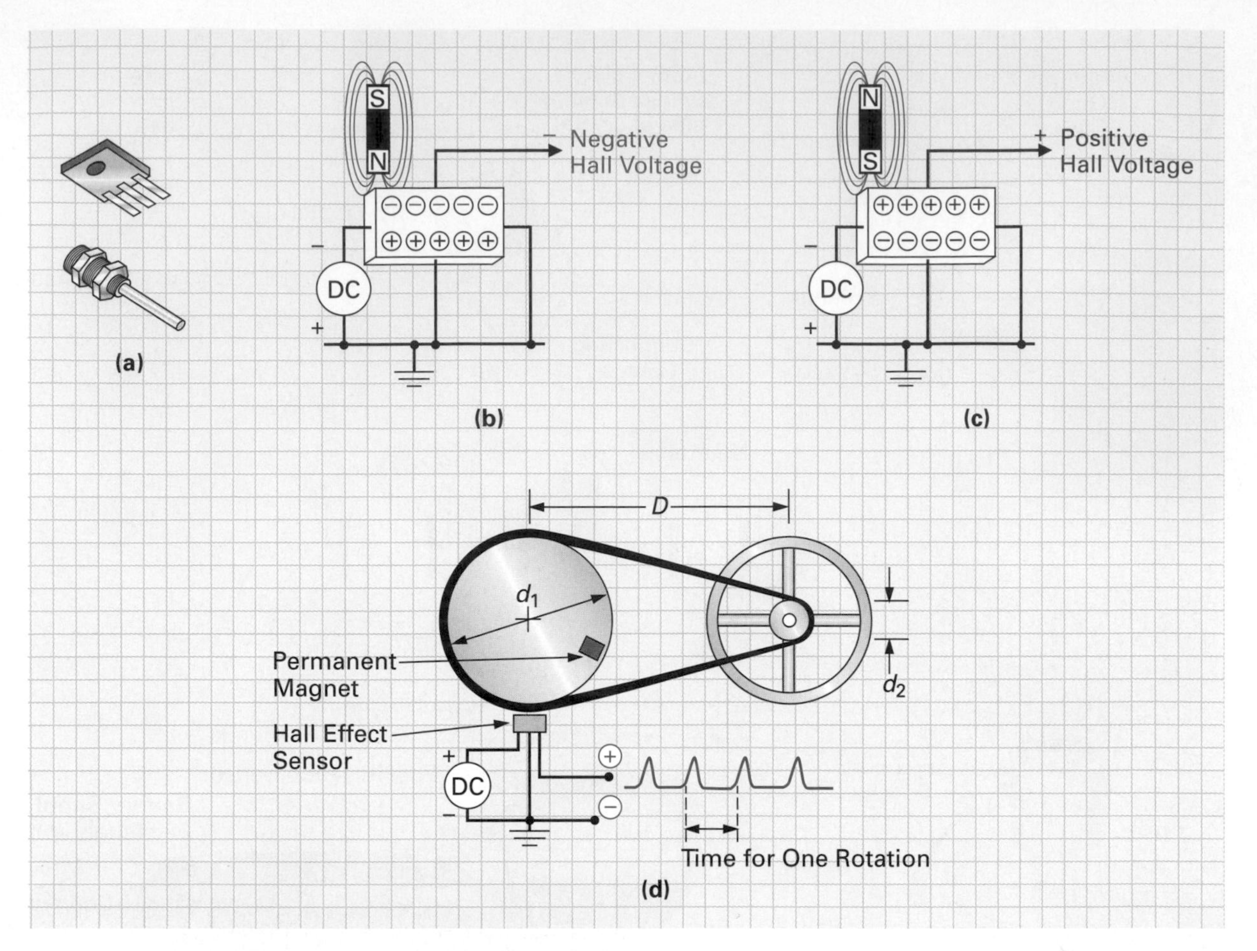

FIGURE 4-15 **Magnetic Sensors. (a) Physical Appearance. (b)(c) Operation. (d) Application.**

4-2-4 *Magnetic Sensors*

Hall Effect Sensor
A sensor that generates a voltage in response to a magnetic field.

The **Hall effect sensor** was discovered by Edward Hall in 1879 and is used in computers, automobiles, sewing machines, aircraft, machine tools, and medical equipment. Figure 4-15(a) shows a few different types of Hall effect sensors. To explain the Hall effect principle, Figure 4-15(b) shows that when a magnetic field whose polarity is north is applied to the sensor, it causes a separation of charges, generating in this example a negative Hall voltage output. On the other hand, Figure 4-15(c) shows that if the magnetic field polarity is reversed so that a south pole is applied to the sensor, it will cause a polarity separation of charges within the sensor that will result in a positive Hall voltage output. The amplitude of the generated positive or negative output voltage from the Hall effect sensor is directly dependent on the strength of the magnetic field.

Their small size, light weight, and ruggedness make the Hall effect sensors ideal in a variety of commercial and industrial applications. For example, Hall effect sensors are embedded in the human heart to serve as timing elements. They are also used to sense shaft rotation, as shown in Figure 4-15(d), camera shutter positioning, rotary position, flow rate, and so on.

SELF-TEST EVALUATION POINT FOR SECTION 4–2

Now that you have completed this section, you should be able to:

Objective 6. Define the term *transducer.*

Objective 7. Demonstrate how to convert among the four temperature scales of Fahrenheit, Celsius, Kelvin, and Rankine.

Objective 8. Describe the operation, characteristics, and applications for the following sensors:

- **a.** Resistive temperature detector (RTD) and thin-film detector (TFD)
- **b.** Thermistor
- **c.** Light-dependent resistor (LDR)
- **d.** Piezoelectric and piezoresistive sensor
- **e.** Hall effect sensor

Use the following questions to test your understanding of Section 4–2:

1. A __________ has a negative temperature coefficient of resistance.
   - **a.** RTD
   - **b.** Thermistor
   - **c.** TFD
   - **d.** None of the above
2. With the photoresistor, an increase in illumination will cause a/an __________ (increase/decrease) in resistance.
3. The generation of a voltage between the opposite sides of a crystal as a result of pressure is known as __________.
   - **a.** Piezoresistive effect
   - **b.** Hall effect
   - **c.** Piezoelectric effect
   - **d.** Thermometry
4. A __________ sensor generates a voltage in response to a magnetic field.
   - **a.** Piezoresistive
   - **b.** Hall effect
   - **c.** LDR
   - **d.** RTD
5. What is the difference between a piezoresistive device and a piezoelectric device?

4-3 DC (DIRECT-CURRENT) ELECTROMAGNETISM

The electron plays a very important role in magnetism. However, the real key or link between electricity and a magnetic field is motion. Anytime a charged particle moves, a magnetic field is generated. As a result, current flow, which is the movement of electrons, produces a magnetic field. To explain this further, let us examine the atomic theory of **electromagnetism**.

Electromagnetism
Relates to the magnetic field generated around a conductor when current is passed through it.

Every orbiting electron in motion around its nucleus generates a magnetic field. When electrons are forced to leave their parent atom by voltage and flow toward the positive polarity, they are all moving in the same direction and each electron's magnetic field will add to the next. The accumulation of all these electron fields will create a magnetic field around the conductor, as shown in Figure 4-16.

A simple experiment can be performed to prove that this invisible magnetic field does in fact exist around a conductor, the setup of which is illustrated in Figure 4-17. With the switch open, as shown in Figure 4-17(a), no current will flow, and therefore no magnetic field is generated around the conductor and the iron filings on the cardboard will be disorganized.

With the switch closed, a current of 3 mA ($I = V/R = 9\text{ V}/3\text{ k}\Omega = 3\text{ mA}$) will flow through the circuit and a magnetic field will be set up around the conductor. This magnetic field will cause the iron filings to become organized in circles, as shown in Figure 4-17(b).

A magnetic field results whenever a current flows through any piece of conductor or wire. If an insulated conductor is wound to form a spiral, as illustrated in Figure 4-18(a), the conductor, which is now referred to as a **coil**, will, as a result of current flow, develop a mag-

Coil
Number of turns of wire wound around a core to produce magnetic flux (an electromagnet) or to react to a changing magnetic flux (an inductor).

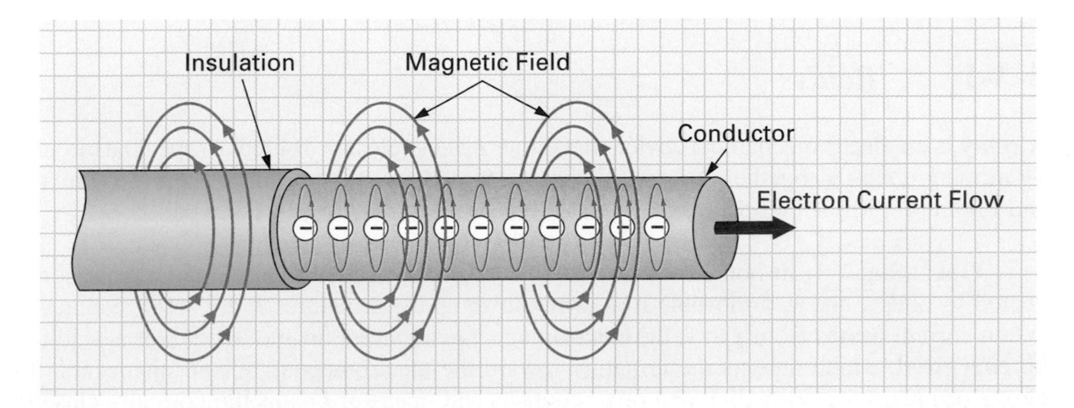

FIGURE 4-16 **Electron Magnetic Fields.**

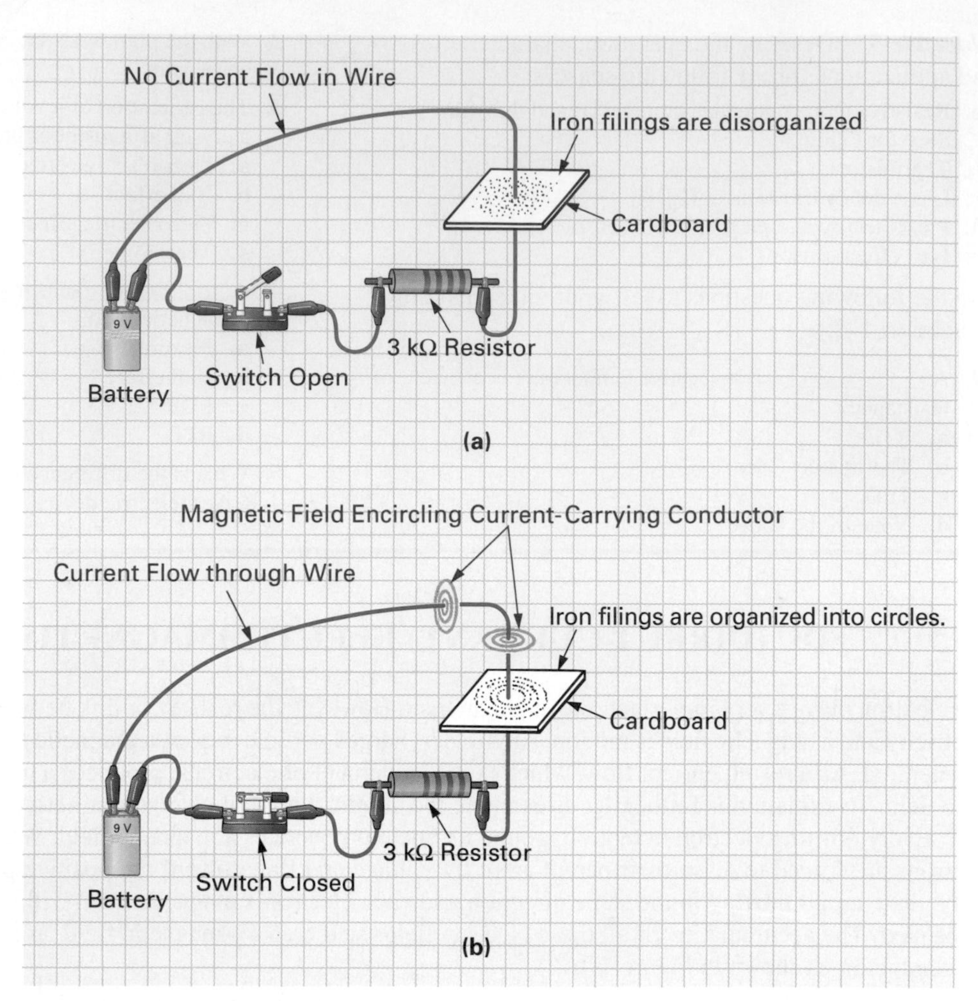

FIGURE 4-17 Magnetic Field Experiment. (a) Switch Open, No Current Flow, No Magnetic Field. (b) Switch Closed, Current Flow, Magnetic Field.

Electromagnet
A magnet consisting of a coil wound on a soft iron or steel core. When current is passed through the coil a magnetic field is generated and the core is strongly magnetized to concentrate the magnetic field.

netic field, which will sum or intensify within the coil. This device is known as an **electromagnet,** and it will produce a concentrated magnetic field whenever a current is passed through its coils.

The magnetic field strength of an electromagnet can be further increased if an iron core is placed within the electromagnet, as shown in Figure 4-18(b). This iron core has less opposition to magnetic lines of flux than air, and so the magnetic field is more concentrated.

We will be discussing electromagnetism in a little more detail in a later chapter, but for now let us see how this characteristic is employed by certain devices.

4-3-1 *Relays*

Relay
Electromechanical device that opens or closes contacts when a current is passed through a coil.

Energized
Being electrically connected to a voltage source so that the device is activated.

A **relay** is an electromechanical device that either makes (closes) or breaks (opens) a circuit by moving contacts together or apart. Figure 4-19(a) shows the normally open (NO) relay and Figure 4-19(b) shows the normally closed (NC) relay.

Operation. In both cases, the relay consists basically of an electromagnet connected to lines x and y, a movable iron arm known as the armature, and a set of contacts. When current passes from x to y, the electromagnet generates a magnetic field, or is said to be **energized,** which attracts the armature toward the electromagnet. When this occurs, it closes or makes the normally open relay contacts and opens or breaks the normally closed relay contacts.

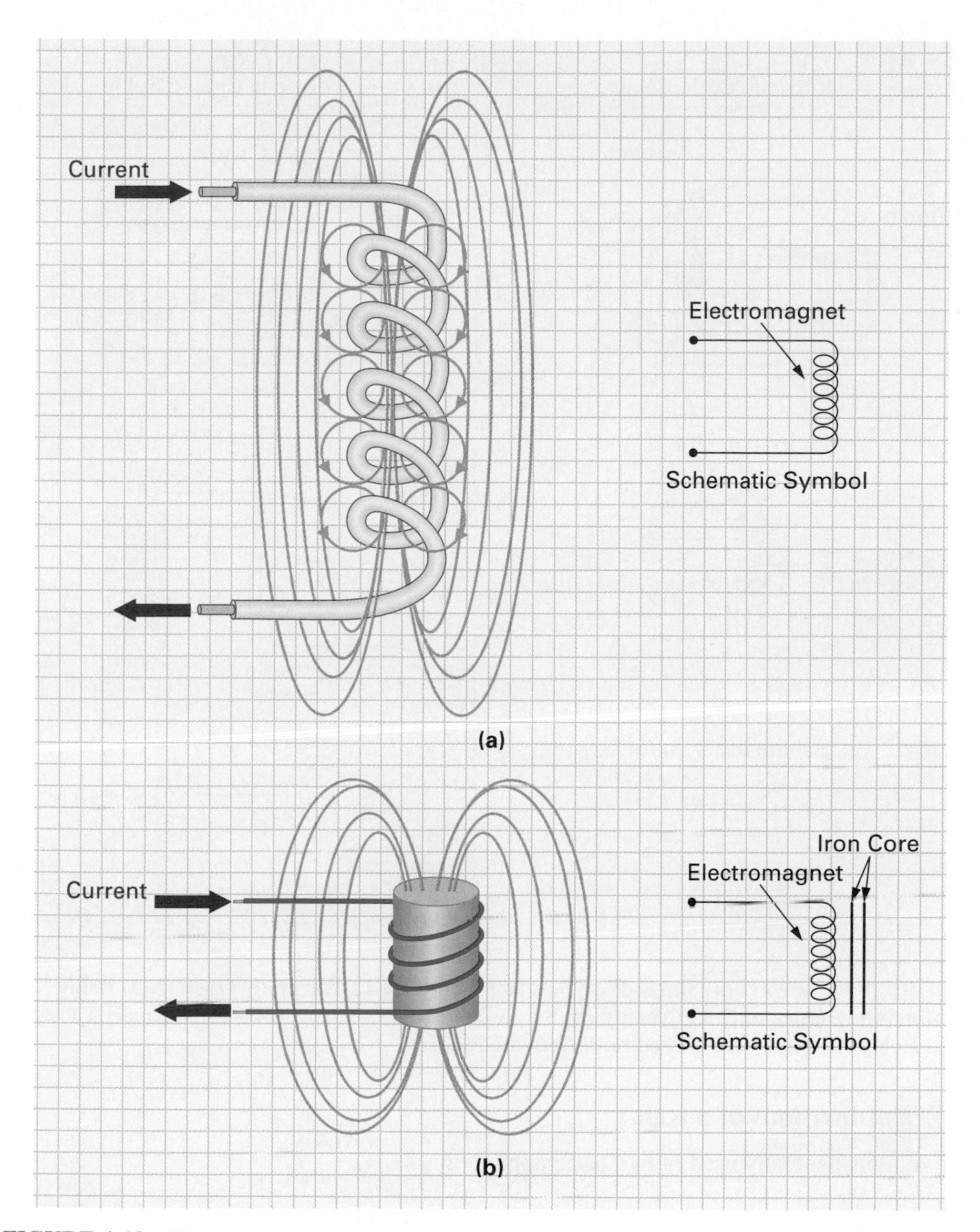

FIGURE 4-18 **Electromagnets. (a) Air Core Electromagnet. (b) Iron Core Electromagnet.**

If the electromagnet is deenergized by discontinuing the current through the coil, the spring will pull back the armature to open the NO relay contacts or close the NC relay contacts between *A* and *B*. The "normal" condition for the contacts between *A* and *B* is when the electromagnet is deenergized. In the deenergized condition, the normally open relay contacts are open and the normally closed relay contacts are closed.

The two relays discussed so far are actually single-pole, single-throw relays, as they have one movable contact (single pole) and one stationary contact that the pole can be thrown to. There are actually four basic configurations for relays and all are illustrated in Table 4-4(a). Variations on these basic four can come in all shapes and sizes, with one relay controlling sometimes several sets of contacts. Table 4-4(b) shows several different styles and packages of relays that are available.

TIME LINE

Ironically, Coulomb's law was not first discovered by Charles A. de Coulomb (1736–1806), but by Henry Cavendish, a wealthy scientist and philosopher, Cavendish did not publish his discovery, which he made several years before Coulomb discovered the law independently. James Clerk Maxwell published the scientific notebooks of Cavendish in 1879, describing his experiements and conclusions, however, about 100 years had passed, and Coulomb's name was firmly associated with the law. Many scientists demanded that the law be called Cavendish's law, while other scientists refused to change, stating that Coulomb was the discoverer because he made the law known promptly to the scientific community.

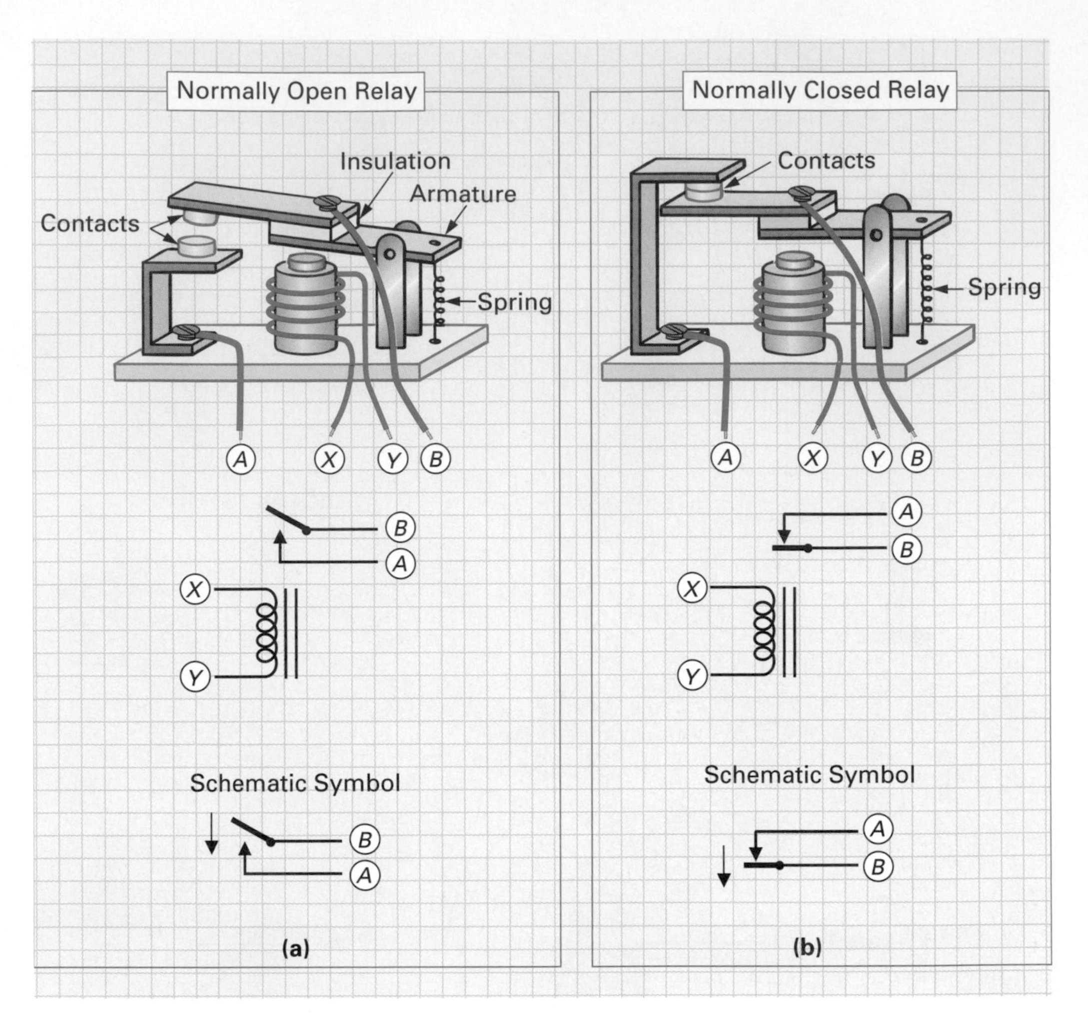

FIGURE 4-19 **Relays (a) Single-Pole, Single-Throw (SPST), Normally Open (NO) Relay (Contacts Are Open until Activated). (b) Single-Pole, Single-Throw (SPST), Normally Closed (NC) Relay (Contacts Are Closed until Activated).**

Applications of a Relay

The relay is generally used in two basic applications:

1. To enable one master switch to operate several remote or difficultly placed contact switches, as illustrated in Figure 4-20. When the master switch is closed, the relay is energized, closing all its contacts and turning on all the lights. The advantage of this is twofold in that, first, the master switch can turn on three lights at one time, which saves time for the operator, and second, only one set of wires need be taken from the master switch to the lights, rather than three sets for all three lights.
2. The second basic application of the relay is to enable a switch in a low-voltage circuit to operate relay contacts in a high-voltage circuit, as shown in Figure 4-21. The operator activates the switch in the safer low-voltage circuit, which will energize the relay, closing its contacts and connecting the more dangerous high voltage to the motor.

As an application, Figure 4-22 shows an automobile starter circuit. In this circuit, a relay is being used to supply the large dc current needed to activate a starter motor in an automobile. When the ignition switch is engaged in the passenger compartment by the driver, current flows through a light-gauge wire from the negative side of the battery, through the relay's electromagnet, through the ignition switch, and back to the positive side of the battery. This current flow through the electromagnet of the relay energizes the relay and closes the

TABLE 4-4 Relay Types

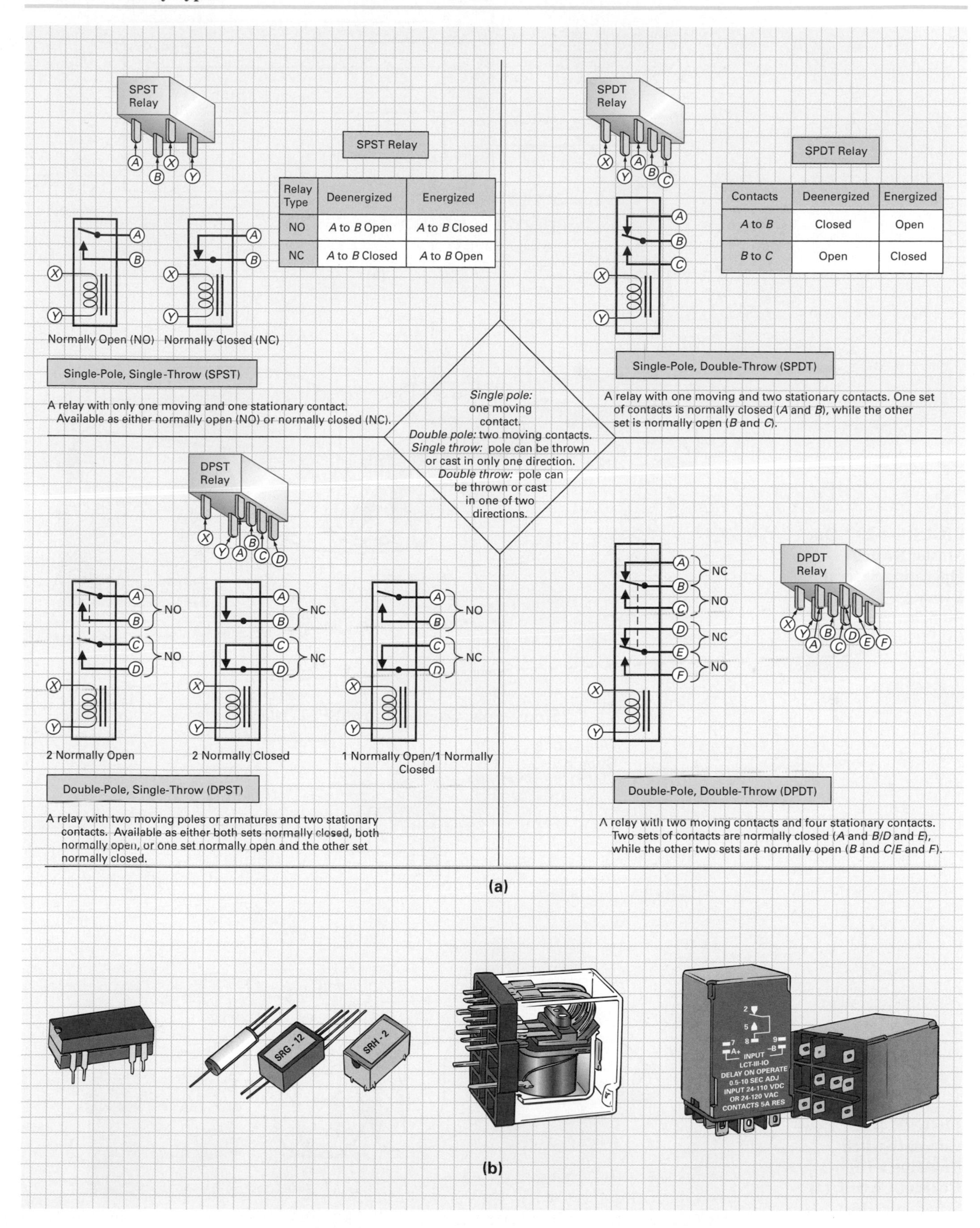

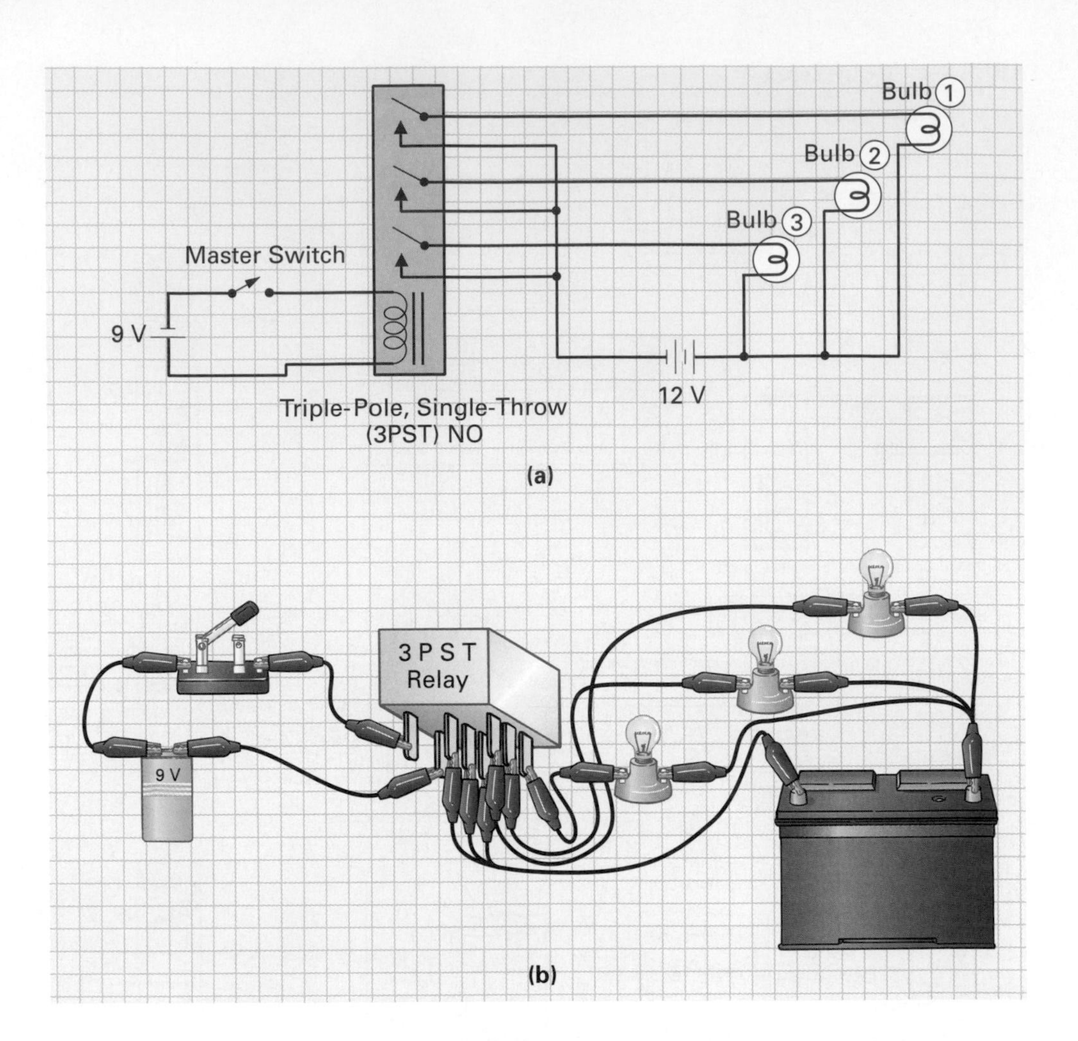

FIGURE 4-20 One Master Switch Operating Several Remotes. (a) Schematic. (b) Pictorial.

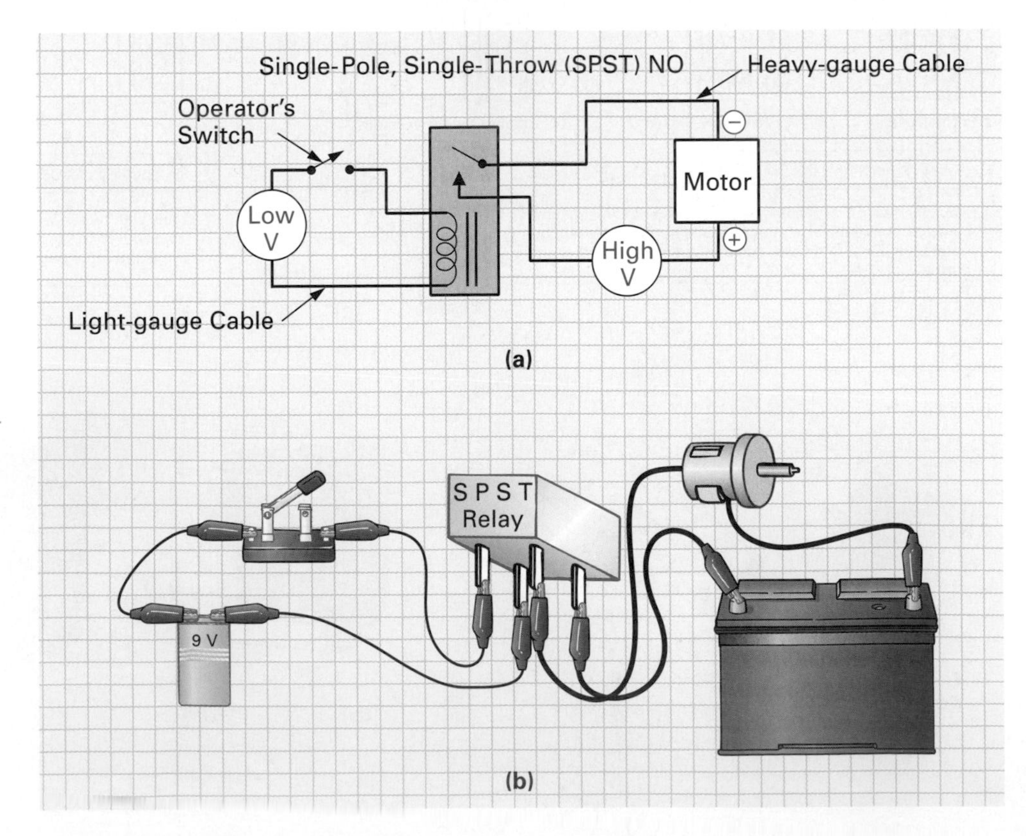

FIGURE 4-21 Low-Current Switch Enabling a High-Current Circuit. (a) Schematic. (b) Pictorial.

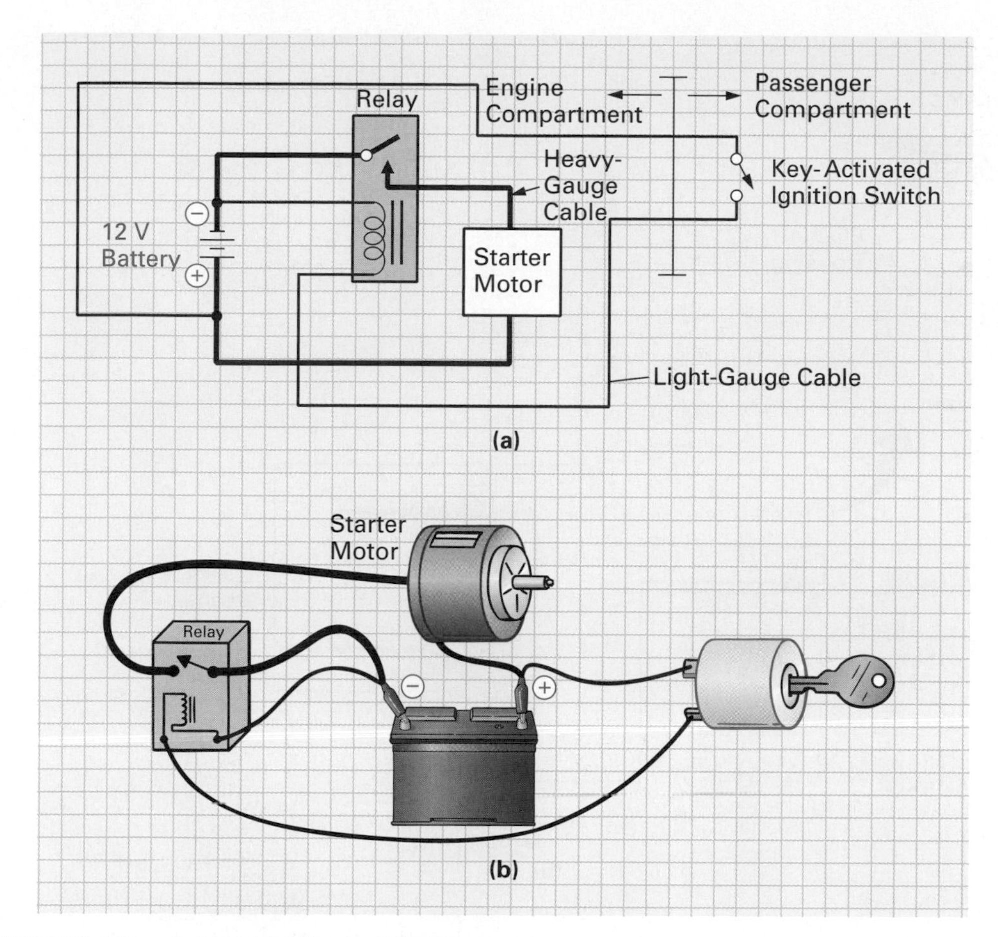

FIGURE 4-22 Automobile Starter Circuit.

relay's contacts. Closing the relay's contacts makes a path for the current to flow through the heavy-gauge cable from the negative side of the battery, through the relay contacts and starter motor, and back to the positive side of the battery. The starter motor's output shaft spins the engine, causing it to start.

This application is a perfect example of how a relay can be used to close contacts in a heavy-current (heavy-gauge cable) circuit, while the driver has only to close contacts in a small-current (light-gauge cable) circuit. If the relay were omitted, the driver's ignition switch would have to be used to connect the 12 V and large current to the starter motor. This would mean that

1. Heavy-gauge, expensive cable would need to be connected in a longer path between the starter motor and passenger compartment.
2. The ignition switch would need to be larger to handle the heavier current.
3. The driver would be in closer proximity to a more dangerous, high-current circuit.

4-3-2 *Reed Switches*

The **reed switch** consists of two flat magnetic strips mounted inside a capsule, which is normally made of plastic, as seen in Figure 4-23(a). The reed switch needs an external magnetic force to operate, and when the separate permanent magnet capsule is in close proximity, opposite magnetic polarities are induced in the overlapping reed blades, causing them to attract one another by induced magnetism and snap together, thus closing the circuit. When the magnetic force is removed, the blades spring apart due to spring tension and open the circuit. Although the reed switch does not employ an electromagnet, it is discussed here because it is used in similar applications to the relay.

Reed Switch
Relay that consists of two thin magnetic strips within a plastic envelope. When it is energized the contacts or strips will snap together, making a connection between the two leads.

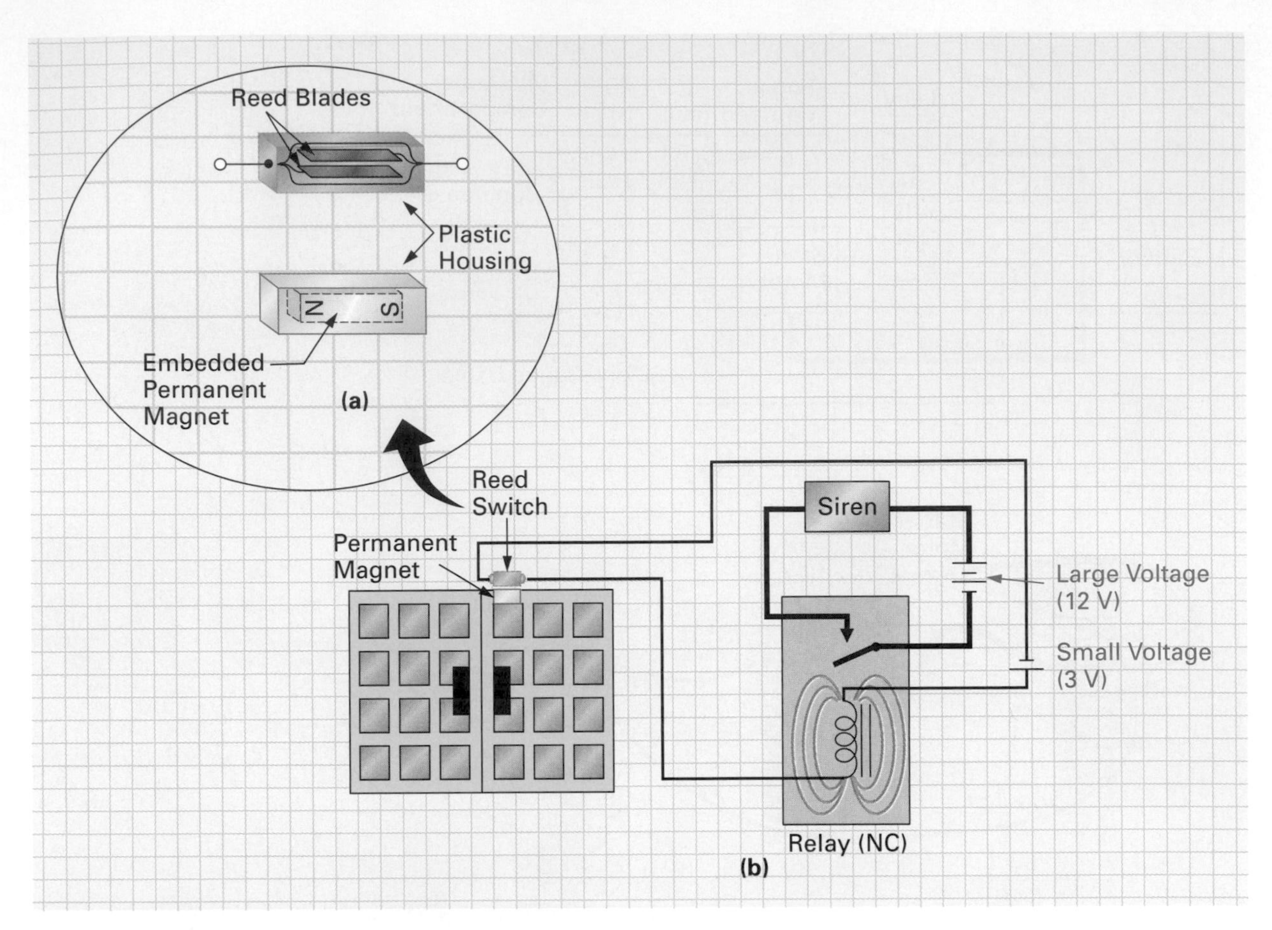

FIGURE 4-23 Reed Switch. (a) Construction. (b) Application: Security Sensor.

Referring to Figure 4-23(b), you will see a simple home security circuit. When the window is closed in the normal condition, the permanent magnet is directly adjacent to the reed switch, and therefore its contacts are closed, allowing current to flow from the negative side of the small 3 V battery, through the closed reed switch contacts, the relay's electromagnet, and back to the positive side of the 3 V battery. The relay is a normally closed (NC) SPST. As current is flowing through the coil of the relay, the contacts are open, preventing the large positive and negative voltage from the 12 V battery from reaching the siren.

If the window is forced and opened by an intruder, the permanent magnet will no longer be in close proximity to the reed switch and the reed switch's contacts will open. When the reed switch's contacts open, the relay's coil will no longer have current flowing through it, and therefore the relay will deenergize and its contacts will return to their normal closed condition. A large current is now permitted to flow from the negative side of the large 12 V battery, through the relay's contacts, the siren, and back to the positive side of the battery. The siren will sound, as it now has 12 V applied to it, and alert the occupant of the home.

4-3-3 *Solenoids*

Up to now, electromagnets have been used to close or open a set, or sets, of contacts to either make or break a current path. These electromagnets have used a stationary soft-iron core. Some electromagnets are constructed with movable iron cores, as shown in Figure 4-24, which can be used to open or block the passage of a gas or liquid through a valve. These are known as **solenoid**-type electromagnets.

Solenoid
Coil and movable iron core that when energized by an alternating or direct current will pull the core into a central position.

When no current is flowing through the solenoid coil, no magnetic field is generated, so no magnetic force is exerted on the movable iron core, as shown in Figure 4-24(a), and there-

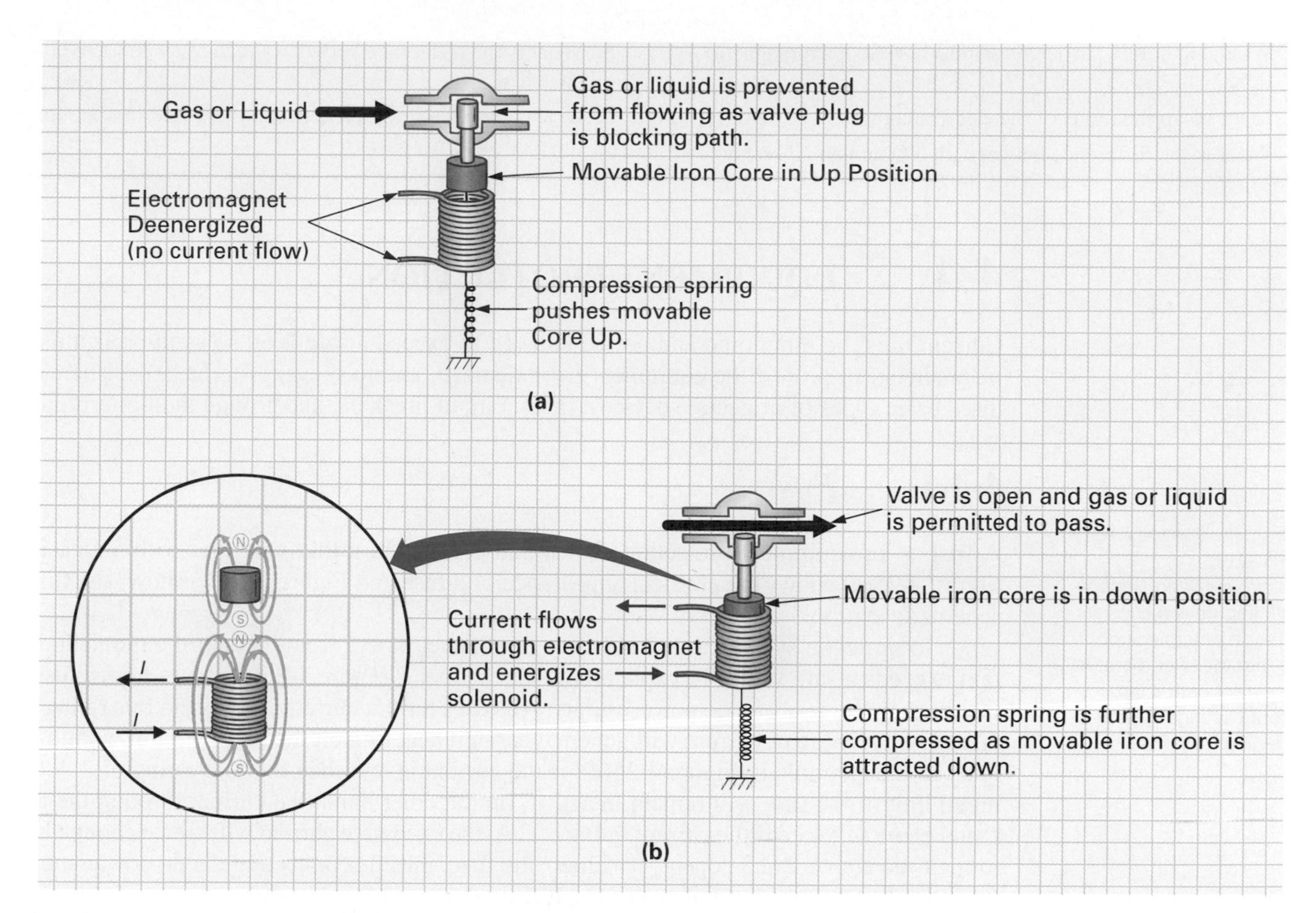

FIGURE 4-24 Solenoid-Type Electromagnet. (a) Deenergized. (b) Energized.

fore the compression spring maintains it in the up position, with the valve plug on the end of the core preventing the passage of either a liquid or gas through the valve (valve closed).

When a current flows through the electromagnet, the solenoid coil is energized, creating a magnetic field, as shown in Figure 4-24(b). Due to the influence of the coil's magnetic field, the movable soft-iron core will itself generate a magnetic field, as seen in the inset in Figure 4-24(b). This condition will create a north pole at the top of the solenoid coil and a south pole at the bottom of the movable core, and the resulting attraction will pull down the core (which is free to slide up and down), pulling with it the valve plug and opening the valve.

Solenoid-type electromagnets are actually constructed with the core partially in the coil and are used in washing machines to control water and in furnaces to control gas.

SELF-TEST EVALUATION POINT FOR SECTION 4-3

Now that you have completed this section, you should be able to:

Objective 9. Describe what is meant by the word *electromagnetism.*

Objective 10. Explain the atomic theory of electromagnetism.

Objective 11. Describe how a magnetic field is generated by current flow through a

a. Conductor
b. Coil

Objective 12. Describe the following applications of electromagnetism:

a. Relay
b. Reed switch
c. Solenoid-type electromagnet

Use the following questions to test your understanding of Section 4-3:

1. True or false: A magnetic field always encircles a current-carrying conductor.

2. What subatomic particle is said to have its own magnetic field?
3. List two applications of the electromagnet.
4. What is the difference between an NO and an NC relay?
5. What is the difference between a reed relay and a reed switch?
6. Give an application for a conventional relay and one for the reed switch.

4-4 EQUIPMENT PROTECTION

Current must be monitored and not be allowed to exceed a safe level so as to protect users from shock, to protect the equipment from damage, and to prevent fire hazards. There are three basic types of protective devices: fuses, circuit breakers, and ground fault interrupters.

4-4-1 *Fuses*

Fuse
This circuit- or equipment-protecting device consists of a short, thin piece of wire that melts and breaks the current path if the current exceeds a rated damaging level.

A **fuse** is an equipment protection device that consists of a thin wire link within a casing. Figure 4-25(a) shows the physical appearance of a fuse and Figure 4-25(b) shows the fuse's schematic symbol.

To protect a system, a fuse needs to disconnect power from the unit the moment current exceeds a safe value. To explain how this happens, Figure 4-25(c) shows how a fuse is mounted in the rear of a dc power supply. All fuses have a current rating, and this rating indicates what value of current will generate enough heat to melt the thin wire link within the fuse. In the example in Figure 4-25(c), if the current is less than the fuse rating of 2 A, the metal link of the fuse will remain intact. If, on the other hand, the current through the thin metal element exceeds the current rating of 2 A, the excessive current will create enough heat to melt the element and "open" or "blow" the fuse, thus disconnecting the dc power supply from the 120 volt source and protecting it from damage.

One important point to remember is that if the current increases to a damaging level, the fuse will open and protect the dc power supply. This implies that something went wrong with the 120 V source and it started to supply too much current. This is almost never the case. What actually happens is that, as with all equipment, eventually something internally breaks down. This can cause the overall load resistance of the piece of equipment to increase or decrease. An increase in load resistance ($R_L\uparrow$) means that the source sees a higher resistance in the current path and therefore a small current would flow from the source ($I\downarrow$) and the user would be aware of the problem due to the nonoperation of the equipment. If an internal equipment breakdown causes the equipment's load resistance to decrease ($R_L\downarrow$), the source will see a smaller circuit resistance and will supply a heavier circuit current ($I\uparrow$), which could severely damage the equipment before the user had time to turn it off. Fuse protection is needed to disconnect the current automatically in this situation, to protect the equipment from damage.

Fuse elements come in various shapes and sizes so as to produce either quick heating and then melting (fast blow) or delayed heating and then melting (slow blow), as illustrated in Figure 4-25(d). The reason for the variety is based in the application differences. When turned on, some pieces of equipment will be of such low resistance that a short momentary current surge, sometimes in the region of four times the fuse's current rating, will result for the first couple of seconds. If a fast-blow fuse were placed in the circuit, it would blow at the instant the equipment was turned on, even though the equipment did not need to be protected. The system merely needs a large amount of current initially to start up. A slow-blow fuse would be ideal in this application, as it would permit the initial heavy current and would begin to heat up and yet not blow due to the delay. Once the surge had ended, the current would decrease, and the fuse would still be intact (some slow-blow fuses will allow a 400% overload current to flow for a few seconds).

On the other hand, some equipment cannot take any increase in current without being damaged. The slow blow would not be good in this application, as it would allow an increase of current to pass to the equipment for too long a period of time. The fast blow would be

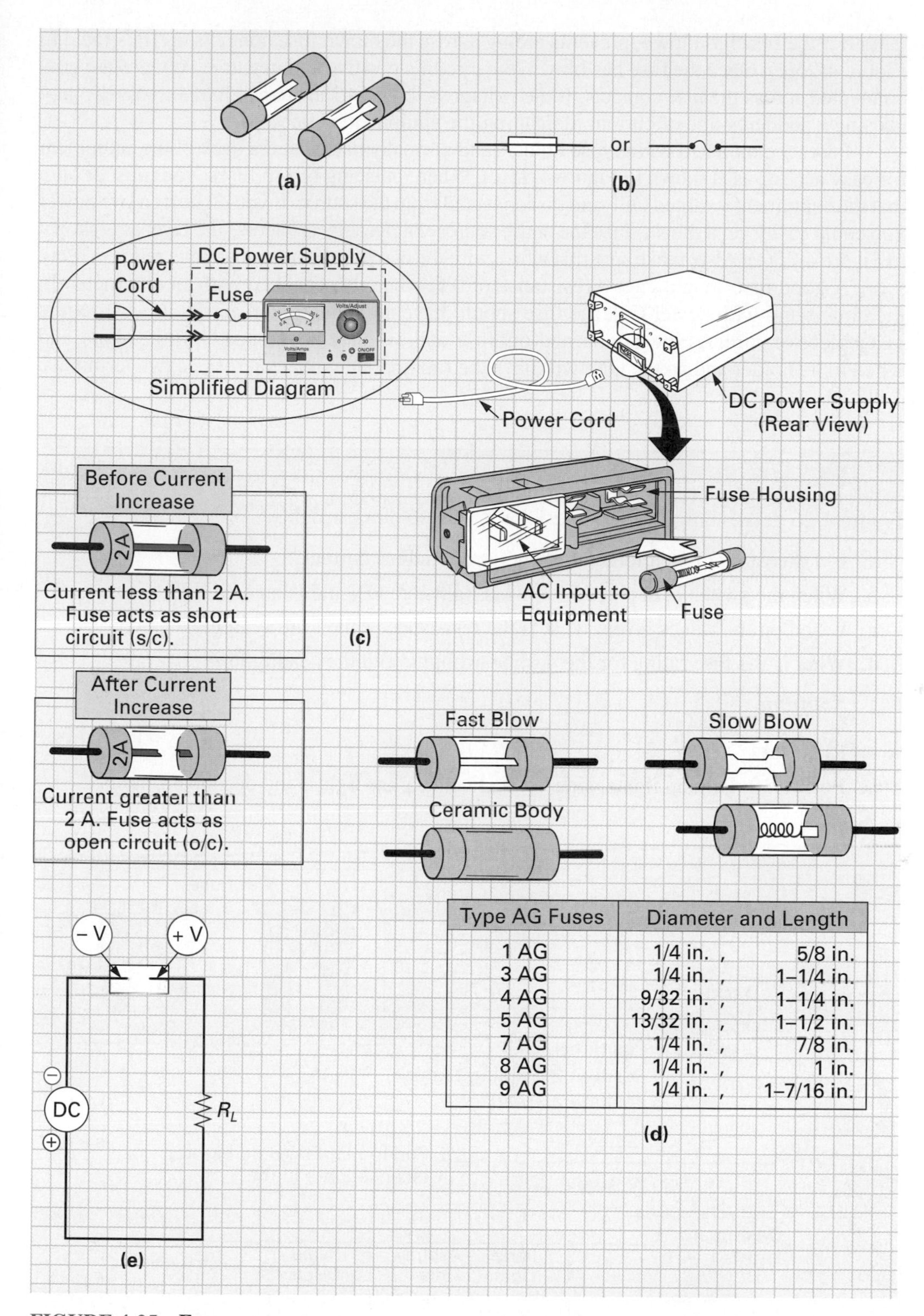

Type AG Fuses	Diameter and Length	
1 AG	1/4 in. ,	5/8 in.
3 AG	1/4 in. ,	1–1/4 in.
4 AG	9/32 in. ,	1–1/4 in.
5 AG	13/32 in. ,	1–1/2 in.
7 AG	1/4 in. ,	7/8 in.
8 AG	1/4 in. ,	1 in.
9 AG	1/4 in. ,	1–7/16 in.

FIGURE 4-25 Fuses.

ideal in this application, since it would disconnect the current instantly if the current rating of the fuse were exceeded and so prevent equipment damage. The automobile was the first application for a fuse in a glass holder, and so a size standard named AG (automobile glass) was established. This physical size standard is still used today and is listed in Figure 4-25(d).

Fuses also have a voltage rating that indicates the maximum circuit voltage that can be applied across the fuse by the circuit in which the fuse resides. This rating, which is important after the fuse has blown, prevents arcing across the blown fuse contacts, as illustrated in Figure 4-25(e). Once the fuse has blown, the circuit's positive and negative voltages are now connected across the fuse contacts. If the voltage is too great, an arc can jump across the gap, causing a sudden surge of current, damaging the equipment connected.

Fuses are mounted within fuse holders and normally placed at the back of the equipment for easy access, as is shown in Figure 4-25(c). When replacing blown fuses, you must be sure that the equipment is turned OFF and disconnected from the source. If you do not remove power you could easily get a shock, since the entire source voltage appears across the fuse, as is shown in Figure 4-25(e). Also, make sure that a fuse with the correct current and voltage rating is used.

A good fuse should have a resistance of 0 Ω when it is checked with an ohmmeter. In the circuit and with power ON, the fuse should have no voltage drop across it because it is just a piece of wire. A blown or burned-out fuse will, when removed, read infinite ohms, and when in its holder and power ON, will have the full applied voltage across its two terminals.

EXAMPLE

Figure 4-26 details how a variable resistor is used to dim car lights.

a. Is a rheostat or potentiometer being used to adjust current?

b. When the switch is in position 2, what lights are ON and what are OFF?

c. Based on the values given, what is the resistance of the ashtray light?

d. What is the fuse rating for the dashboard lights and the headlights?

Solution

a. Rheostat

b. Position 2: Parking lights, side lights, licence plate lights, and dashboard lights.

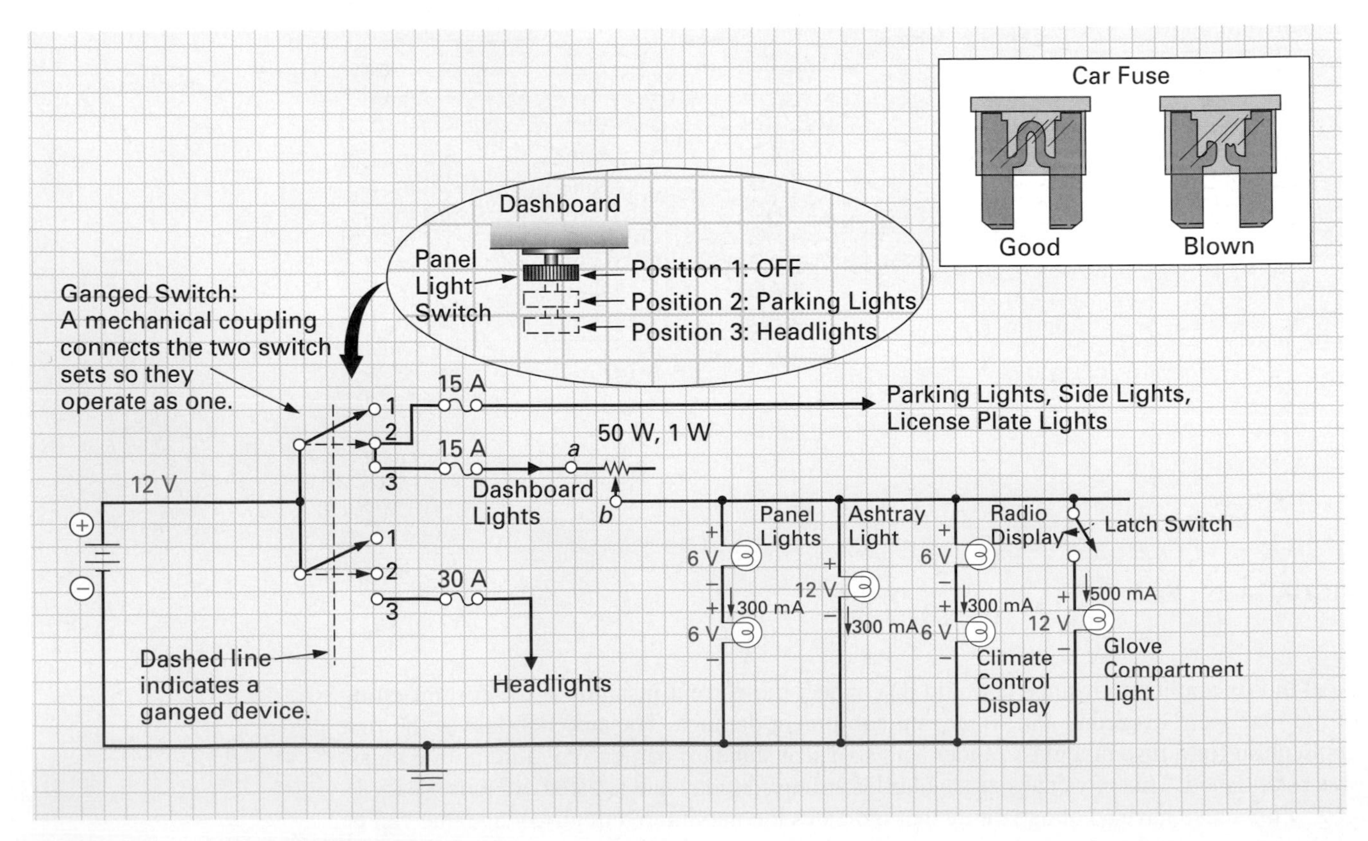

FIGURE 4-26 **Automobile Dimmer Control.**

c. $R = \frac{V}{I} = \frac{12\text{ V}}{300\text{ mA}} = 40\ \Omega$

d. Dashboard lights = 15 A
Headlights = 30 A

4-4-2 *Circuit Breakers*

In many appliances and in a home electrical system, **circuit breakers** are used in place of fuses to prevent damaging current. A circuit breaker can open a current-carrying circuit without damaging itself, then be manually reset and used repeatedly, unlike a fuse, which must be replaced when it blows. We can say, therefore, that a circuit breaker is a reusable fuse. Figure 4-27 illustrates a circuit breaker's schematic symbol, physical appearance, and a typical application in which it is used to protect a television.

Circuit Breaker
Reusable fuse. This device will open a current-carrying path without damaging itself once the current value exceeds its maximum current rating.

In this section, we will examine the three basic types of circuit breakers, which are thermal type, magnetic type, and thermomagnetic type.

Thermal Circuit Breakers

The operation of this type of circuit breaker depends on temperature expansion due to electrical heating. Figure 4-28 illustrates the construction of a thermal circuit breaker. A U-shaped bimetallic (two-metal) strip is attached to the housing of the circuit breaker. This strip is composed of a layer of brass on one side and a layer of steel on the other. The current arrives at terminal A, enters the right side of the bimetallic U-shaped strip, leaves the left side, and travels to the upper contact, which is engaging the bottom contact, where it leaves and exits the circuit breaker from terminal B.

If current were to begin to increase beyond the circuit breaker's current rating, heating of the bimetallic strip would occur. As with all metals, heat will cause the strip to expand. Some metals expand more than others, and this measurement is called the *thermal expansion coefficient.* In this situation, brass will expand more than steel, resulting in the lower end of the bimetallic strip bending to the right. This action allows a catch to release a pivoted arm. The arm's right side will be pulled down by the tension of a spring, lifting the left side of the arm and the attached top contact. The result is to separate the top contact surface from the bottom contact surface, thereby opening the current path and protecting the circuit from the excessive current. This action is also known as *tripping* the breaker. To reset the breaker, the reset button must be pressed to close the contacts. If the problem in the system still exists, however, the breaker will just trip once more, due to the excessive current.

Magnetic Circuit Breakers

The construction of a magnetic circuit breaker is shown in Figure 4-29; it will operate as follows. A small level of current flowing through the coil of the electromagnet will provide only

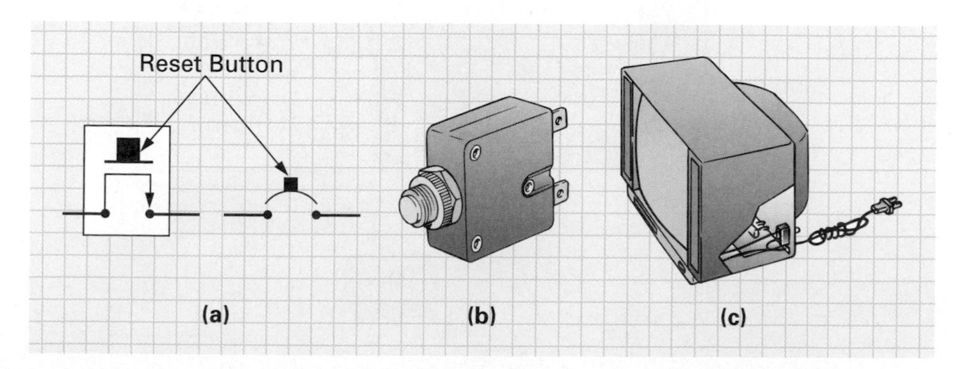

FIGURE 4-27 **Circuit Breaker. (a) Symbol. (b) Appearance. (c) Typical Application.**

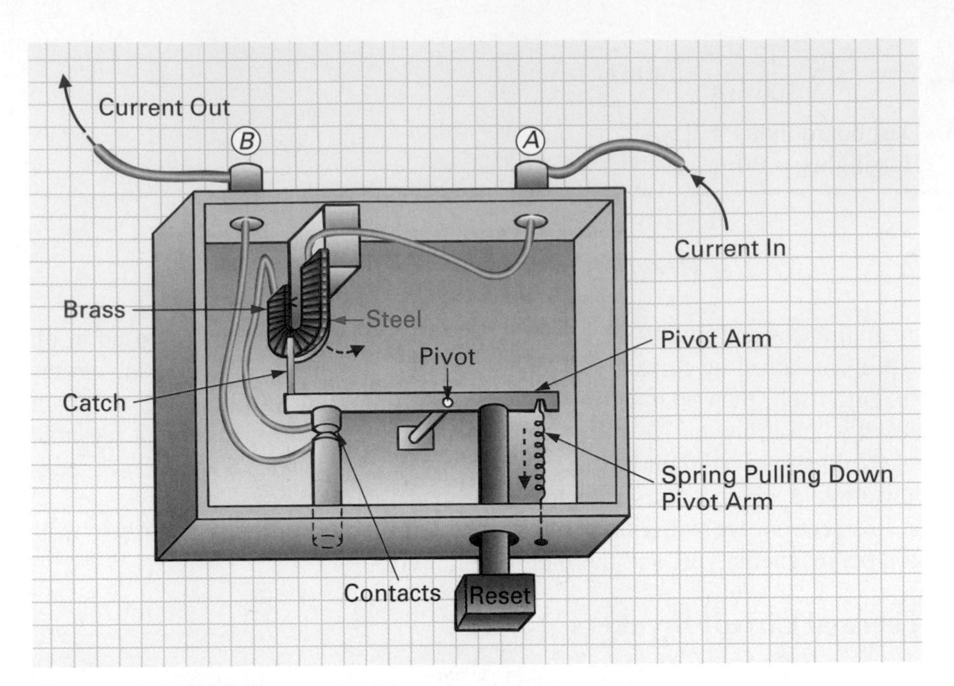

FIGURE 4-28 Thermal-Type Circuit Breaker.

a small amount of magnetic pull to the left on the iron arm. This magnetic force cannot overcome the pull to the right being generated by spring *A*. This safe value of current will therefore be allowed to pass from the *A* terminal, through the coil to the top contact, out of the bottom contact, and then exit the breaker at terminal *B*.

If the current exceeds the current rating of the circuit breaker, an increase in current through the coil of the electromagnet generates a greater magnetic force on the vertical arm, which pulls the top half of the vertical arm to the left and the lower half below the pivot to the right. This action releases the catch holding the horizontal arm, allowing spring *B* to pull the right side of the lower arm down, opening the contact and disconnecting or tripping the breaker. The reset button must now be pressed to close the contacts. Once again, however, if the problem still exists, the breaker will continue to trip, as the excessive circuit current still exists.

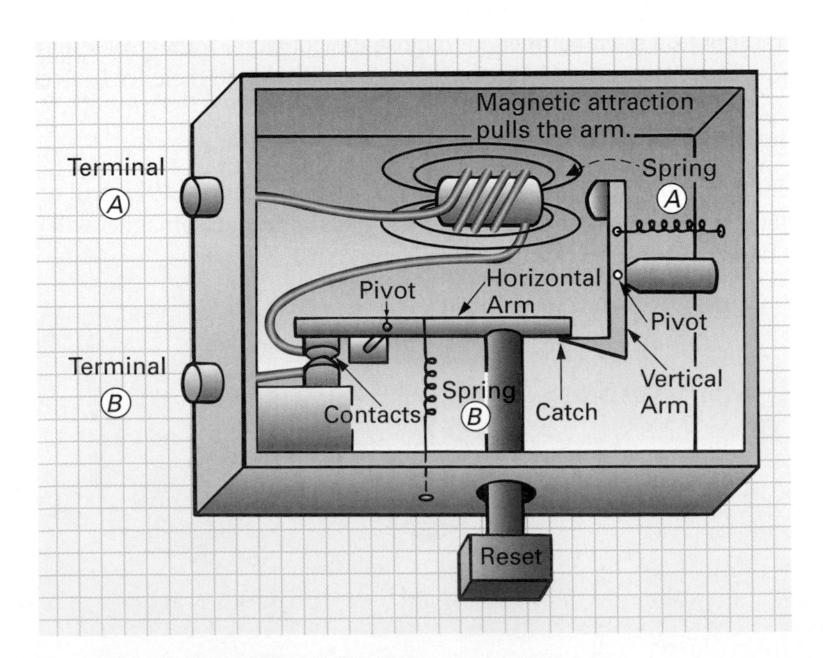

FIGURE 4-29 Magnetic-Type Circuit Breaker.

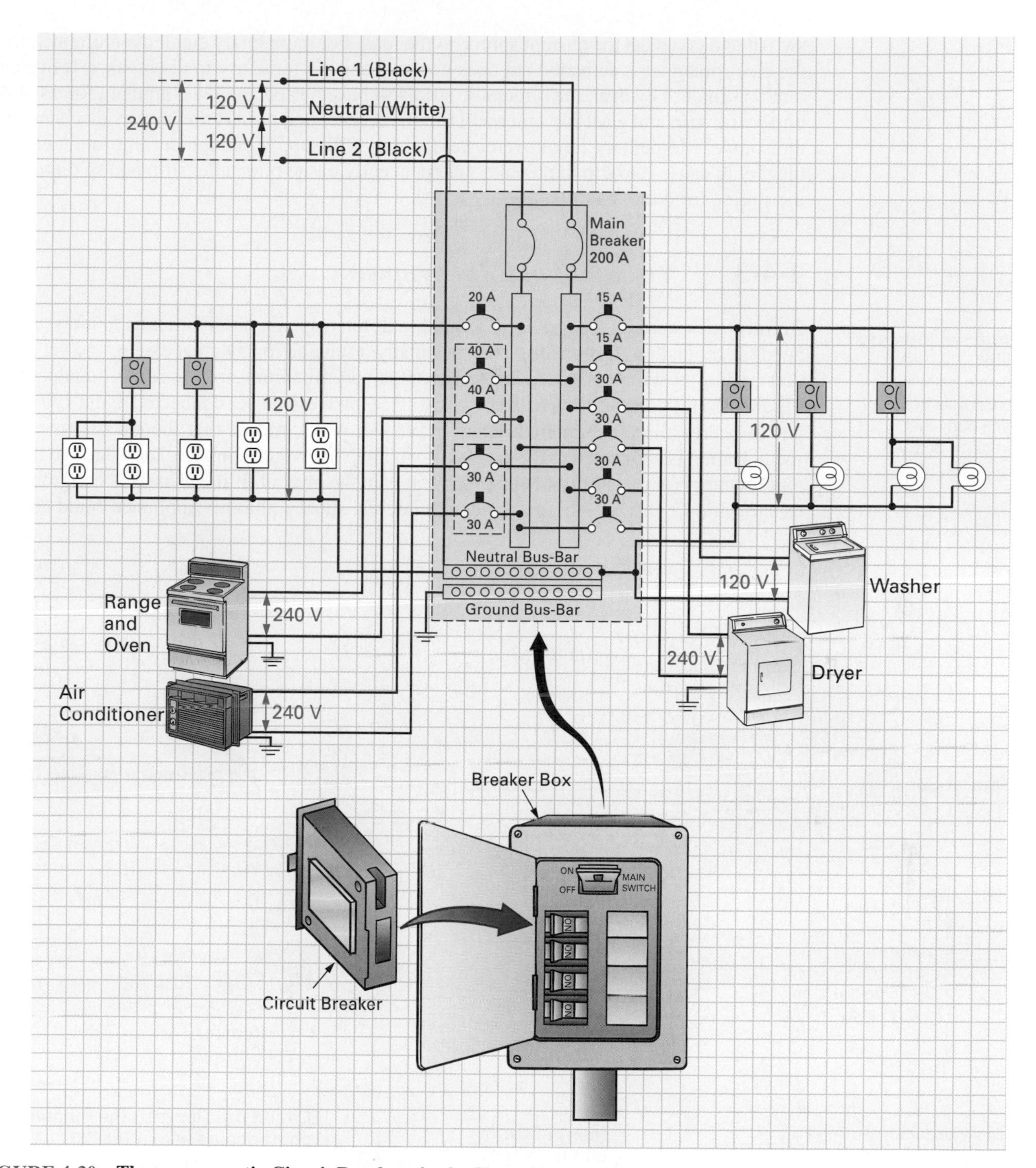

FIGURE 4-30 **Thermomagnetic Circuit Breakers in the Home Breaker Box.**

Thermomagnetic Circuit Breakers

Figure 4-30 illustrates the typical residential-type thermomagnetic circuit breaker and the schematic of a typical home breaker box. The thermal circuit breaker is similar to a slow-blow fuse in that it is ideal for passing momentary surges without tripping because of the delay caused by heating of the bimetallic strip. The magnetic circuit breaker, on the other hand, is most similar to a fast-blow fuse in that it is tripped immediately when an increase of current occurs.

The thermomagnetic circuit breaker combines the advantages of both previously mentioned circuit breakers by incorporating both a bimetallic strip and an electromagnet in its actuating mechanism. For currents at and below the current rating, the circuit breaker is a closed switch and connects the current source to the load. For momentary small overcurrent surges, the electromagnet is activated but does not have enough force to trip the breaker. If overcurrent continues, however, the bimetallic strip will have heated and the combined forces of the bimetallic strip and electromagnet will trip the breaker and protect the load by disconnecting the source. Momentary large overcurrent surges will supply enough current to the electromagnet to cause it to trip the breaker independently of the bimetallic strip and so disconnect the source from the load.

4-4-3 *Ground Fault Interrupters*

Gounding
Connecting a point in a circuit to Earth or some connecting body that serves in place of Earth.

Before we discuss the operation of ground fault interrupters, or GFIs, let us first discuss the need for grounding. **Grounding** is not normally discussed to any great extent, and yet an understanding of why it is necessary to have it is very important from a safety standpoint. In Figure 4-31(a) you can see a toaster (with a 10 Ω load resistance) connected to the 120 V outlet. In nearly all cases, appliances are grounded, which means that the chassis (the unit's metal frame) is connected to the ground pin in the plug, which connects back through the circuit breaker box to an 8 foot copper bar that is in the ground. If the chassis is not grounded, as shown in Figure 4-31(b), an accidental internal short within the appliance could connect

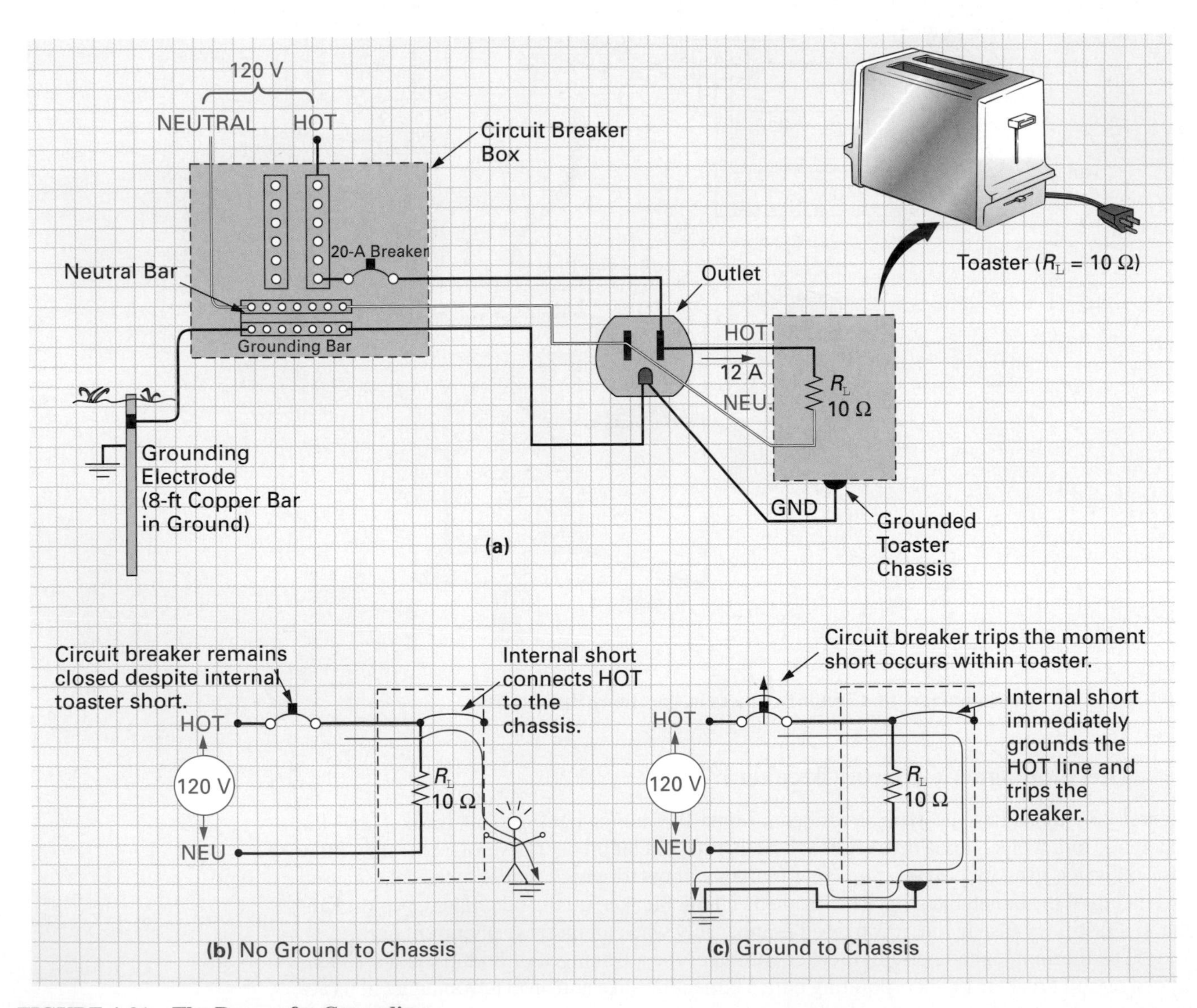

FIGURE 4-31 The Reason for Grounding.

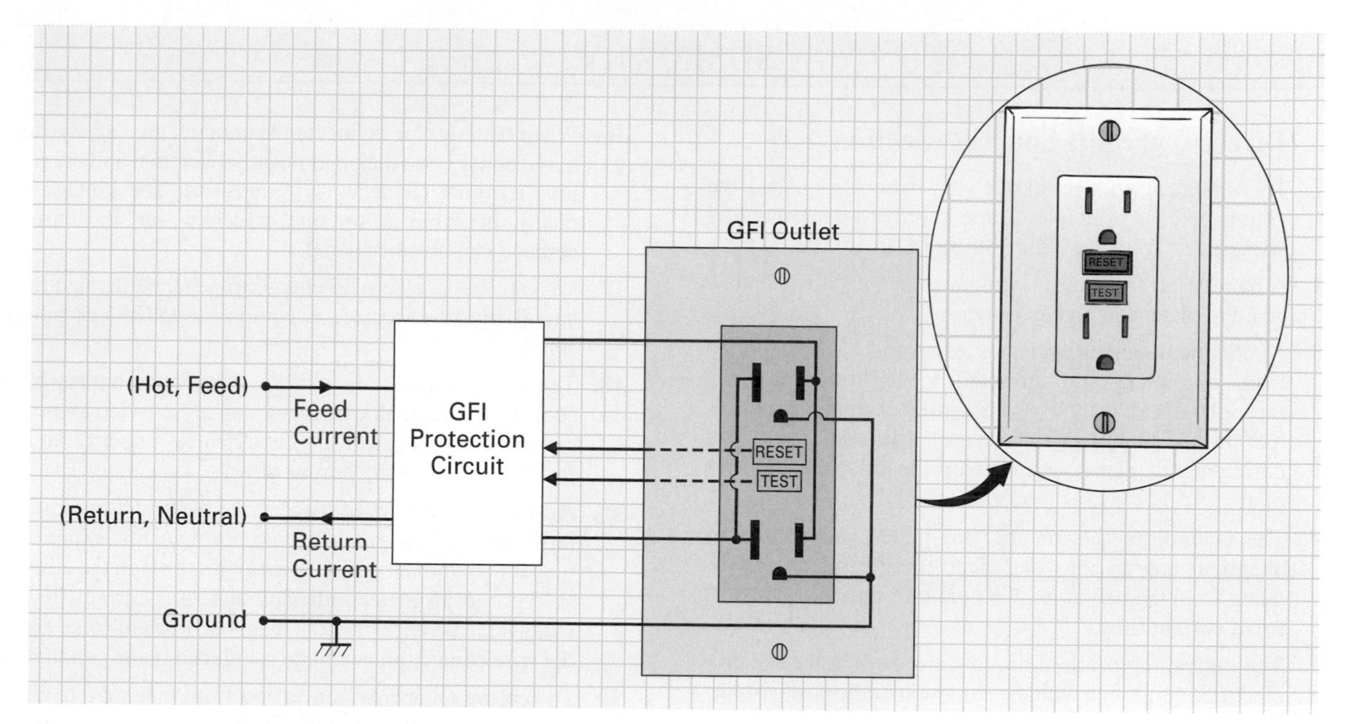

FIGURE 4-32 **Ground Fault Interrupters (GFIs).**

the HOT line to the chassis, and if a person were to then touch the chassis he would receive a shock. The circuit breaker in this instance would not trip until the person touches the chassis and provides a current path to ground.

On the other hand, if the appliance is grounded, as shown in Figure 4-31(c), an accidental short between the HOT line and the now grounded chassis will immediately trip the circuit breaker and disconnect the applied 120 V. If the breaker is reset, it will immediately trip again until the shorting appliance is unplugged from the outlet.

Ground fault interrupters will trip a lot faster than a standard circuit breaker and are generally installed where water and electricity are present. Any electrical current through a person to ground will be sensed, and the GFI will typically shut down the outlet in 500 ms (half a second). Figure 4-32 shows how the GFI provides this protection by monitoring the input and output current—these currents should normally be the same. If, however, a difference is sensed between the feed current and return current, the outlet will shut OFF and power will only be restored when the RESET button is pressed. The TEST button will simulate a "ground fault" and should trip the GFI to test whether it is operating correctly.

Ground Fault Interrupter
A fast-acting circuit breaker that senses small ground-fault currents that could occur if a person touches a hot ac line. This interrupter will trip the breaker in less than half a second, to protect the person from severe electrical shock.

SELF-TEST EVALUATION POINT FOR SECTION 4-4

Now that you have completed this section, you should be able to:

Objective 13. Describe the function, operation, and characteristics of fuses.

Objective 14. Describe the function, operation, and characteristics of the following circuit breaker types:

a. Thermal
b. Magnetic
c. Thermomagnetic

Objective 15. Explain the reason for grounding.

Objective 16. Describe the function, operation, and characteristics of ground fault interrupters.

Use the following questions to test your understanding of Section 4–4:

1. Why do fuses have a current and a voltage rating?
2. What is the difference between a slow-blow and a fast-blow fuse?
3. List the three types of circuit breakers.
4. What advantage does a circuit breaker have over a fuse?
5. Why is it important to ground appliances?
6. For what reason are GFIs installed instead of a standard outlet?

SUMMARY

DC (Direct-Current) Sources (Section 4-1)

1. The chemical voltage source (the battery) is called the voltaic cell. Its principle of operation was discovered by Alessandro Volta, an Italian physicist, in 1800.
2. A battery has three basic components: a negative plate, a positive plate, and an electrolyte.
3. A chemical reaction causes electrons to be repelled from one plate and attracted to the other, passing through the electrolyte. A large number of electrons collect on one plate (negative plate), while an absence or deficiency of electrons exists on the opposite, positive plate.
4. The electrolyte acts on the two plates and transforms chemical energy into electrical energy, which can be taken from the cell at its two output terminals as an electrical current flow.
5. The mutual repulsion of the free electrons at the negative electrode combined with the attraction of the positive electrode results in a migration of free electrons (current flow) through the lightbulb, causing its filament to produce light.
6. The chemical reaction within the cell will actually dissolve the negative plate until eventually it will be eaten away completely.
7. Primary (first time, last time) cells discharge once and must then be discarded. Almost all primary cells have their electrolyte in paste form and are therefore referred to as dry cells, as opposed to wet cells, whose electrolyte is in liquid form.
8. The shelf life of a battery is the length of time a battery can remain on the shelf (in storage) and still retain its usability.
9. The cell voltage describes the voltage produced by one cell (one set of plates). If two or more sets of plates are installed in one package, the component is called a *battery*.
10. Secondary cells operate on the same basic principle as that of primary cells. However, in this case the plates are not eaten away or dissolved. They only undergo a chemical change during discharge. Once discharged, the secondary cell can have the chemical change that occurred during discharge reversed by recharging, resulting once more in a fully charged cell.
11. The battery charger voltage is normally set to about 115% of the battery voltage.
12. Capacity (C) is measured by the amount of ampere-hours (Ah) a battery can supply during discharge.
13. Ampere-hour units are actually specifying the coulombs of charge in the battery.
14. Batteries are often connected together to gain a higher voltage or current than can be obtained from one cell.
15. When batteries are connected in series, the negative terminal of battery A is connected to the positive terminal of battery B. The total voltage across the load will be the sum of the two cell voltages.
16. When batteries are connected in parallel, the negative terminal connects to the other negative terminal and the positive terminal connects to the positive. The voltage remains the same as for one cell; however, the current demand can now be shared.
17. We can use ac electrical energy to generate dc electrical energy by use of a piece of equipment called a dc power supply.
18. The power supply has advantages over the battery as a source of voltage in that it can quickly provide an accurate voltage that can easily be varied by a control on the front panel, and it never runs down.
19. A magnet can be used to generate electrical current by movement of the magnet past a conductor. When the magnet is stationary, no current flows through the circuit, so the current meter indicates this zero current. If the magnet is moved so that the magnetic lines of force cross the wire conductor, a voltage is induced in the conductor.
20. To achieve an energy conversion from magnetic to electric, you can either move the magnet past the wire or move the wire past the magnet.
21. To produce a continuous supply of electric current, the magnet or wire must be constantly in motion. The device that achieves this is the dc electric generator.
22. A photoelectric cell is also called a solar cell.
23. When light illuminates light-sensitive material, a charge is generated, causing current to flow.
24. This phenomenon, known as photovoltaic action, finds application as a light meter for photographic purposes and as an electrical power source for use by a satellite.
25. Electric power companies make use of solar cells to convert sunlight into electrical energy for the consumer.
26. Heat can be used to generate an electric charge.
27. When two dissimilar metals, such as copper and iron, are welded together and heated, an electrical charge is produced. This junction is called a *thermocouple*, and the heat causes the release of electrons from their parent atoms.
28. The size of the charge is proportional to the temperature difference between the two metals.
29. Thermocouples are normally used to indicate temperature on a display calibrated in degrees.

DC (Direct-Current) Sensors (Section 4-2)

30. A semiconductor transducer is an electronic device that converts one form of energy to another.
31. A thermistor is a semiconductor device that acts as a temperature-sensitive resistor.
32. There are basically two different types of thermistors: positive temperature coefficient (PTC) thermistors and the more frequently used negative temperature coefficient (NTC) thermistors.

33. A temperature increase causes the resistance of an NTC thermistor to decrease. On the other hand, a temperature increase causes the resistance of a PTC thermistor to increase.

34. The photoconductive cell, or light-dependent resistor (LDR), is a two-terminal device that changes its resistance (conductance) when light (photo) is applied.

35. When light strikes the photoconductive atoms, electrons are released into the conduction band, and therefore the resistance between the device's terminals is reduced. When light is not present, the electrons and holes recombine, and therefore the resistance is increased.

36. A photoconductive cell will typically have a "dark resistance" of several hundred megohms and a "light resistance" of a few hundred ohms.

37. The photoconductive cell's key advantage is that it can withstand a high operating voltage (typically a few hundred volts). Its disadvantages are that it responds slowly to changes in light level and that its power rating is generally low (typically a few hundred milliwatts).

38. A semiconductor pressure transducer will change its resistance in accordance with changes in pressure.

39. Piezoresistance of a semiconductor is described as a change in resistance due to a change in the applied pressure. This device has a dc excitation voltage applied and will typically generate a 0 to 100 mV output voltage based on the pressure sensed.

40. The Hall effect sensor was discovered by Edward Hall in 1879 and is used in computers, automobiles, sewing machines, aircraft, machine tools, and medical equipment.

41. When a magnetic field whose polarity is south is applied to a Hall effect sensor, it causes a separation of charges, generating a negative Hall voltage output. On the other hand, if the magnetic field polarity is reversed so that a north pole is applied to the sensor, it will cause an opposite polarity separation of charges within the sensor, resulting in a positive Hall voltage output.

42. Their small size, light weight, and ruggedness make the Hall effect sensors ideal in a variety of commercial and industrial applications, such as sensing shaft rotation, camera shutter positioning, rotary position, flow rate, and so on.

DC (Direct-Current) Electromagnetism (Section 4-3)

43. Every orbiting electron in motion around its nucleus generates a magnetic field. When electrons are forced to leave their parent atom by voltage and flow toward the positive polarity, they are all moving in the same direction and each electron's magnetic field will add to the next. The accumulation of all these electron fields will create the magnetic field around the conductor.

44. If a conductor is wound to form a spiral, the conductor, which is now referred to as a *coil*, will, as a result of current flow, develop a magnetic field, which will sum or intensify within the coil. If many coils are wound in the same direction, an electromagnet is formed, which will produce a concentrated magnetic field whenever a current is passed through its coils.

45. A relay is an electromechanical device that either makes (closes) or breaks (opens) a circuit by moving contacts together or apart.

46. The relay consists of an electromagnet, a movable iron arm known as the armature, and some contacts. When current passes through the electromagnet, it is said to be energized. When the electromagnet is energized, it attracts the armature toward the electromagnet and either closes or makes the normally open relay contacts or opens or breaks the normally closed relay contacts.

    If the electromagnet is deenergized by stopping the current through the coil, the spring will pull back the armature to open the NO relay contacts or close the NC relay contacts.

47. The relay is generally used in two basic applications:
    a. To enable one master switch to operate several remote or difficulty placed contact switches.
    b. To enable a switch in a low-voltage circuit to operate relay contacts in a high-voltage circuit.

48. The reed switch consists of two flat magnetic strips mounted inside a capsule, which is normally made of plastic and needs an external magnetic force to operate.

49. Some electromagnets are constructed with movable iron cores that can be used to open or block the passage of a gas or liquid through a valve. These are known as solenoid-type electromagnets.

Equipment Protection (Section 4-4)

50. Current must be monitored and not be allowed to exceed a safe level so as to protect users from shock, to protect the equipment from damage, and to prevent fire hazards.

51. There are three basic types of protective devices: fuses, circuit breakers, and ground fault interrupters.

52. A fuse consists of a wire link or element of low melting point inside a casing. Fuses have a current rating that indicates the maximum amount of current they will allow to pass through the fuse to the equipment.

53. When current passes through the fuse, some of the electrical energy is transformed into heat. If the current through the thin metal element exceeds the current rating of the fuse, the excessive current will create enough heat to melt the element and "open" or "blow" the fuse, thus disconnecting the equipment and protecting it from damage.

54. An increase in load resistance ($R_L\uparrow$) means that the battery sees a higher resistance in the current path and therefore a small current would flow from the battery ($I\downarrow$) and the user would be aware of the problem due to the nonoperation of the equipment.

55. If an internal equipment breakdown causes the equipment's load resistance to decrease ($R_L\downarrow$), the battery will see a smaller circuit resistance and will supply a heavier

circuit current (I↑), which could severely damage the equipment before the user had time to turn it off. Fuse protection is needed to disconnect the current automatically in this situation, to protect the equipment from damage.

56. Fuse elements come in various shapes and sizes so as to produce either quick heating and then melting (fast blow) or delayed heating and then melting (slow blow).

57. Fuses also have a voltage rating that indicates the maximum circuit voltage that can be applied across the fuse by the circuit in which the fuse resides. This rating, which is important after the fuse has blown, prevents arcing across the blown fuse contacts.

58. Fuses are mounted within fuse holders and normally placed at the back of the equipment for easy access.

59. When replacing fuses, you must be sure that the power is off, because a potential is across it, and that a fuse with the correct current and voltage rating is used.

60. In many appliances and in a home electrical system, circuit breakers are used in place of fuses to prevent damaging current.

61. A circuit breaker can open a current-carrying circuit without damaging itself, then be manually reset and used repeatedly, unlike a fuse, which must be replaced when it blows. There are three types of circuit breakers: thermal type, magnetic type, and thermomagnetic type.

62. The operation of thermal circuit breakers depends on temperature expansion due to electrical heating.

63. Magnetic circuit breakers depend on the response of an electromagnet to break the circuit for protection.

64. The thermal circuit breaker is similar to a slow-blow fuse in that it is ideal for passing momentary surges without tripping because of the delay caused by heating of the bimetallic strip.

65. The magnetic circuit breaker is most similar to a fast-blow fuse in that it is tripped immediately when an increase of current occurs.

66. The thermomagnetic circuit breaker combines both a bimetallic strip and an electromagnet as an actuating mechanism. For currents at and below the current rating, the circuit breaker is a short circuit and connects the current source to the load. For small momentary overcurrent surges, the electromagnet is activated, but does not have enough force to trip the breaker. However, if overcurrent continues, the bimetallic strip will have heated and the combined forces of the bimetallic strip and electromagnet will trip the breaker and disconnect the current source from the load. For a very large surge of current, the electromagnet receives enough current to trip the breaker independently of the bimetallic strip and disconnect the current source from the load.

67. Grounding appliances is a preventative measure that will trip the circuit breaker if an accidental internal short to chassis occurs.

68. Ground fault interrupters or GFIs trip a lot faster than circuit breakers, shut down an outlet, and protect a user. They are generally installed if water is nearby the outlet.

REVIEW QUESTIONS

Multiple-Choice Questions

1. Direct current is
- **a.** A reversing of current continually in a circuit
- **b.** A flow of current in only one direction
- **c.** Produced by an ac voltage
- **d.** None of the above

2. If load resistance were doubled, load current will __________.
- **a.** Halve
- **b.** Double
- **c.** Triple
- **d.** Remain the same

3. Quartz is a solid compound that will produce electricity when __________ is applied.
- **a.** Friction
- **b.** An electrolyte
- **c.** A magnetic field
- **d.** Pressure

4. Direct current is generated thermally by use of a
- **a.** Crystal
- **b.** Thermocouple
- **c.** Thermistor
- **d.** None of the above

5. An application of a photovoltaic cell would be
- **a.** A satellite power source
- **b.** To turn on and off security lights
- **c.** Both (a) and (b)
- **d.** None of the above

6. Which type of battery can be used for only one discharge (and cannot be rejuvenated)?
- **a.** A lead–acid cell
- **b.** A secondary cell
- **c.** A nickel–cadmium cell
- **d.** A primary cell

7. A battery with a capacity of 12 ampere-hours could supply
- **a.** 12 A for 1 hour
- **b.** 3 A for 4 hours
- **c.** 6 A for 2 hours
- **d.** All of the above

8. Batteries are normally connected in series to obtain a higher total
- **a.** Voltage
- **b.** Current
- **c.** Resistance
- **d.** None of the above

9. The dc power supply converts
   a. ac to dc
   b. dc to ac
   c. High dc to a low dc
   d. Low ac to a high ac
10. A fuse's current rating states the
    a. Maximum amount of current allowed to pass
    b. Minimum amount of current allowed to pass
    c. Maximum permissible circuit voltage
    d. None of the above
11. A thermal-type circuit breaker is equivalent to a __________ fuse, whereas a magnetic-type circuit breaker is equivalent to a __________ fuse.
    a. Slow-blow, slow-blow
    b. Fast-blow, slow-blow
    c. Slow-blow, fast-blow
    d. Glass case, ceramic case
12. The type of circuit breaker found in the home is the
    a. Thermal type
    b. Thermomagnetic type
    c. Magnetic type
    d. Carbon–zinc type
13. A single-pole, double-throw switch would have __________ terminals.
    a. Two
    b. Three
    c. Four
    d. Five
14. A switch will very simply __________ (open) or __________ (close) a path of current.
    a. Break, make
    b. Make, break
    c. s/c, o/c
    d. o/c, break
15. With switches, the word *pole* describes a
    a. Stationary contact
    b. Moving contact
    c. Path of current
    d. None of the above
16. A thermistor has a __________ temperature coefficient of resistance.
    a. Negative
    b. Positive
17. A light-dependent resistor is a __________ device.
    a. Photoconductive
    b. Voltage source
    c. Photovoltaic
    d. Equipment protection
18. Which sensing device can be used to detect magnetic field strength?
    a. TFD
    b. Hall effect sensor
    c. Piezoresistive sensor
    d. LDR
19. Which sensing device can be used to detect temperature changes?
    a. TFD
    b. Hall effect sensor
    c. Piezoresistive sensor
    d. LDR
20. Which sensing device can be used to detect pressure changes?
    a. TFD
    b. Hall effect sensor
    c. Piezoresistive sensor
    d. LDR
21. Relays can be used to
    a. Allow one master switch to enable several others
    b. Allow several switches to enable one master
    c. Allow a switch in a low-current circuit to close contacts in a high-current circuit
    d. Both (a) and (b)
    e. Both (a) and (c)
22. The starter relay in an automobile is used to
    a. Allow one master switch to enable several others
    b. Allow several switches to enable one master
    c. Allow a switch in a low-current circuit to close contacts in a high-current circuit
    d. Both (a) and (b)
23. The __________ uses an electromagnet around two flat magnetic strips mounted inside a glass capsule.
    a. Reed switch
    b. Magnetic circuit breaker
    c. Reed relay
    d. Starter relay
24. The reed switch could be used in a/an
    a. Home security system
    b. Automobile starter
    c. Magnetic-type circuit breaker
    d. All of the above
25. A normally open relay (NO) will have
    a. Contacts closed until activated
    b. Contacts open until activated
    c. All contacts permanently open
    d. All contacts permanently closed

Communication Skill Questions

26. Describe briefly how direct current can be generated
    a. Chemically (4-1-1)
    b. Electrically (4-1-2)
    c. Magnetically (4-1-3)
    d. Optically (4-1-4)
    e. Thermally (4-1-5)
27. What is the difference between a cell and a battery? (4–1–1)
28. Briefly describe the operation of a battery. (4–1–1)
29. Simply state the difference between a primary and a secondary cell. (4–1–1)

30. Describe what can be gained by connecting batteries in series and parallel. (4-1-1)
31. List the four different temperature scales and describe the differences. (4-2-1)
32. What is the difference between a PTC and NTC temperature detector? (4-2-1)
33. Describe the operation of a light-dependent resistor. (4-2-2)
34. Define the following terms: (4-2)
    a. Transducer
    b. Bolometer
    c. Thermometry
    d. Piezoelectric effect
35. Describe the basic function of a Hall effect sensor. (4-2-4)
36. Why is it necessary to provide equipment protection? (4-4)
37. List three types of equipment protection devices. (4-4)
38. What is an electromagnet? (4-3)
39. Describe the operation and function of the relay. (4-3-1)
40. In what applications is the relay used? (4-3-1)
41. What is a reed switch? (4-3-2)
42. Describe the operation and purpose of a solenoid. (4-3-3)
43. Describe a fuse's (4-4-1)
    a. Current rating
    b. Voltage rating
44. Briefly describe the operation of the following types of circuit breakers: (4-4-2)
    a. Thermal type
    b. Magnetic type
45. What is the difference between a circuit breaker and a GFI? (4-4-3)

Practice Problems

46. If a 50 W lightbulb is connected across a 120 V source and the following fuses are available, which should be used to protect the circuit?
    a. 0.5 A/120 V (fast blow)
    b. ¾ A/120 V (fast blow)
    c. 1 A/12 V (slow blow)
    d. 0.5 A/120 V (slow blow)
47. Draw a diagram and indicate the total source voltage of
    a. Six 1.5 V cells connected in series with one another
    b. Six 1.5 V cells connected in parallel with one another
48. Draw a diagram and indicate the polarities of a 12 V lead–acid battery being charged by a 15 V battery charger. If this battery is rated at 150 Ah,
    a. How many coulombs of charge will be stored in the fully charged condition?
    b. How long will it take the battery to charge if a charging current of 5 A is flowing?
49. Show how to connect two 12 V batteries to increase (a) the voltage and (b) the current.
50. Referring to the alarm circuit in Figure 4-33,
    a. Briefly describe the operation of the circuit.
    b. Determine current (I), if all sensor switches are closed (inactive) and the alarm system is armed.

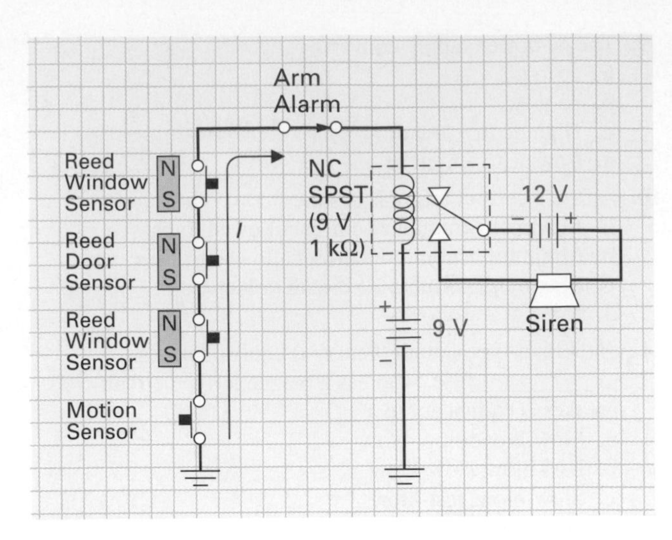

FIGURE 4-33 Alarm Circuit.

Job Interview Test

These tests at the end of each chapter will challenge your knowledge up to this point and give you the practice you need for a job interview. To make this more realistic, the test will be comprised of both technical and personal questions. In order to take full advantage of this exercise, you may want to set up a simulation of the interview environment, have a friend read the questions to you out loud, and record your responses for later analysis.

Company Name: LINK, Inc.

Industry Branch: Computers.

Function: Set up local area networks (LANs).

Job Title: Field Service Techician.

51. Is your interest in computers both personal and professional?
52. Could you define for me the difference between a source and a load?
53. Tell me what you know about nickel–metal hydride batteries for notebook computers.
54. What application software are you familiar with?
55. What is the difference between a photoconductive cell and a photovoltaic cell?
56. What is Ohm's law?
57. Would you say that you are personable?
58. What is the difference between a fuse and a circuit breaker?
59. What does the prefix "mega" mean, in, for example, a 500 megahertz computer system?
60. What does the abbreviation ISP stand for?

Answers

51. Answer must be yes. Go on to explain what computer system you have at home and school and in what capacity you have used them personally and professionally.
52. Chapter 3.

53. Section 4-1-1.
54. Discuss EWB's Multisym and/or Circuit Maker and all other application software you are familiar with.
55. Sections 4-1-4 and 4-2-2.
56. Chapter 2.
57. They're asking if you can get along well with customers and colleagues. Mention any work experience in which you had direct contact with customers, and discuss how you worked with fellow students in your class.
58. Section 4-3.
59. Chapter 2.
60. Internet Service Provider.

Web Site Questions

Go to the web site http://www.prenhall.com/cook, select the textbook *Introductory DC/AC Electronics* or *Introductory DC/AC Circuits*, this chapter, and then follow the instructions when answering the multiple-choice practice problems.

Series Circuits

The First Pocket Calculator

During the seventeenth century, European thinkers were obsessed with any device that could help them with mathematical calculation. Scottish mathematician John Napier decided to meet this need, and in 1614 he published his new discovery of logarithms. In this book, consisting mostly of tediously computed tables, Napier stated that a logarithm is the exponent of a base number. For example,

The common logarithm (base 10) of 100 is 2 ($100 = 10^2$).

The common logarithm of 10 is 1 ($10 = 10^1$).

The common logarithm of 27 is 1.43136 ($27 = 10^{1.43136}$).

The common logarithm of 6 is 0.77815 ($6 = 10^{0.77815}$).

Any number, no matter how large or small, can be represented by or converted to a logarithm. Napier also outlined how the multiplication of two numbers could be achieved by simply adding the numbers' logarithms. For example, if the logarithm of 2 (which is 0.30103) is added to the logarithm of 4 (which is 0.60206), the result will be 0.90309, which is the logarithm of the number 8 ($0.30103 + 0.60206 = 0.90309$, $2 \times 4 = 8$). Therefore, the multiplication of two large numbers can be achieved by looking up the logarithms of the two numbers in a log table, adding them together, and then finding the number that corresponds to the sum in an antilog (reverse log) table. In this example, the antilog of 0.90309 is 8.

Napier's table of logarithms was used by William Oughtred, who developed, just 10 years after Napier's death in 1617, a handy mechanical device that could be used for rapid calculation. This device, considered the first pocket calculator, was the slide rule.

5

Outline and Objectives

Introduction

Series Circuit
Circuit in which the components are connected end to end so that current has only one path to follow throughout the circuit.

A **series circuit,** by definition, is the connecting of components end to end in a circuit to provide a single path for the current. This is true not only for resistors, but also for other components that can be connected in series. In all cases, however, the components are connected in succession or strung together one after another so that only one path for current exists between the negative (−) and positive (+) terminals of the supply.

5-1 COMPONENTS IN SERIES

Figure 5-1 illustrates five examples of series resistive circuits. In all five examples, you will notice that the resistors are connected "in-line" with one another so that the current through the first resistor must pass through the second resistor, and the current through the second resistor must pass through the third, and so on.

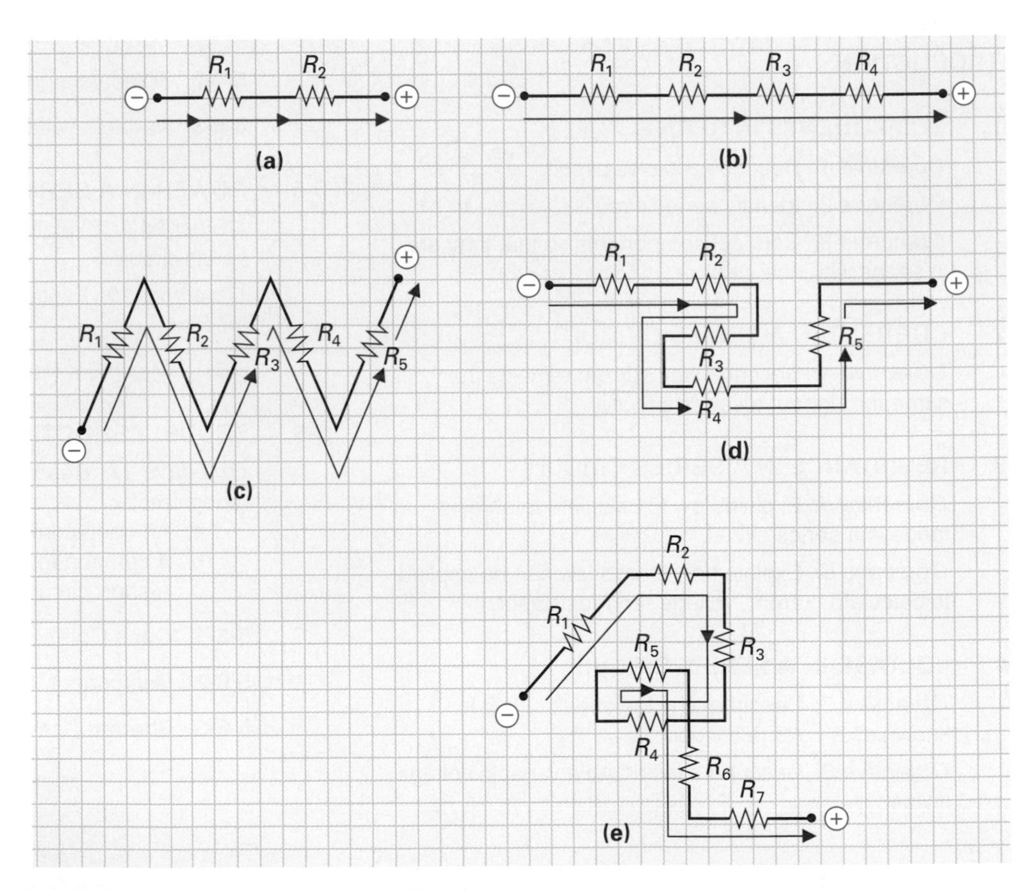

FIGURE 5-1 Five Series Resistive Circuits.

EXAMPLE:

In Figure 5-2(a), seven resistors are laid out on a tabletop. Using a protoboard, connect all the resistors in series, starting at R_1 and proceeding in numerical order through the resistors until reaching R_7. After completing the circuit, connect the series circuit to a dc power supply.

Solution:

In Figure 5-2(b), you can see that all the resistors are now connected in series (end-to-end) and the current has only one path to follow from negative to positive.

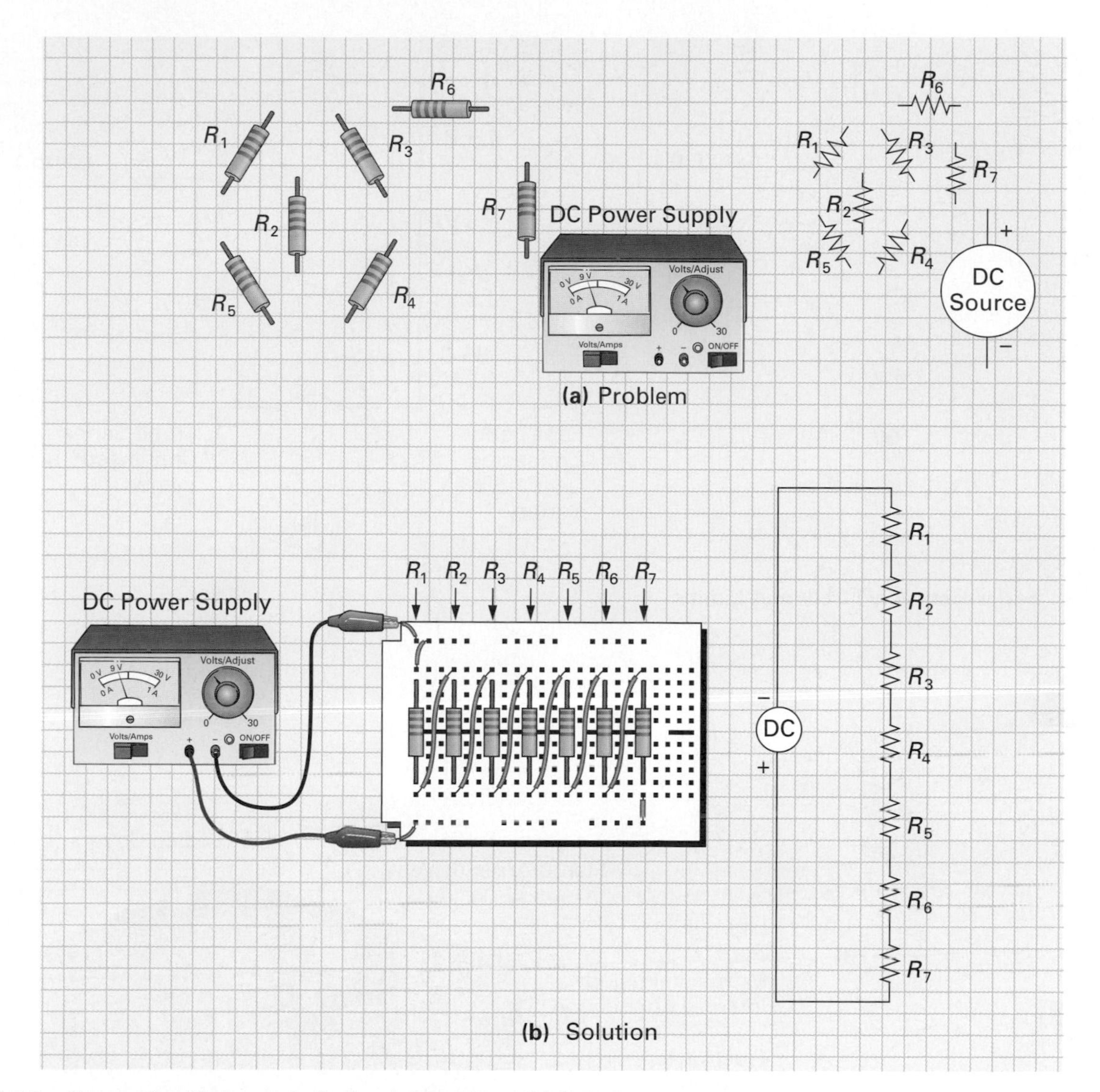

FIGURE 5-2 Connecting Resistors in Series. (a) Problem. (b) Solution.

EXAMPLE:

Figure 5-3(a) shows four 1.5 V cells and three lamps. Using wires, connect all of the cells in series to create a 6 V battery source. Then connect the three lamps in series with one another, and finally, connect the 6 V battery source across the three-series-connected-lamp load.

Solution:

In Figure 5-3(b) you can see the final circuit containing a source, made up of four series-connected 1.5 V cells, and a load, consisting of three series-connected lamps. As explained in Chapter 4, when cells are connected in series, the total voltage (V_T) will be equal to the sum of all the cell voltages:

$$V_T = V_1 + V_2 + V_3 + V_4 = 1.5\text{ V} + 1.5\text{ V} + 1.5\text{ V} + 1.5\text{ V} = 6\text{ V}$$

EXAMPLE:

The lead–acid car battery shown in Figure 5-4 has six series-connected 2.1 V cells. When charged, what voltage will be present across the output terminals?

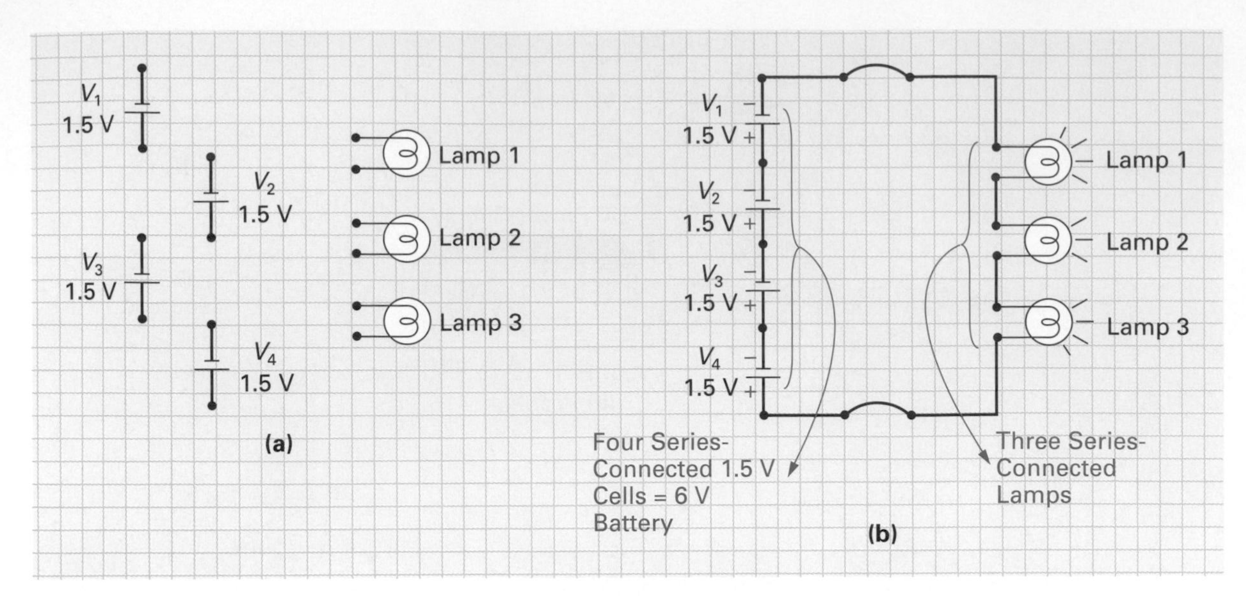

FIGURE 5-3 Series-Connected Cells and Lamps.

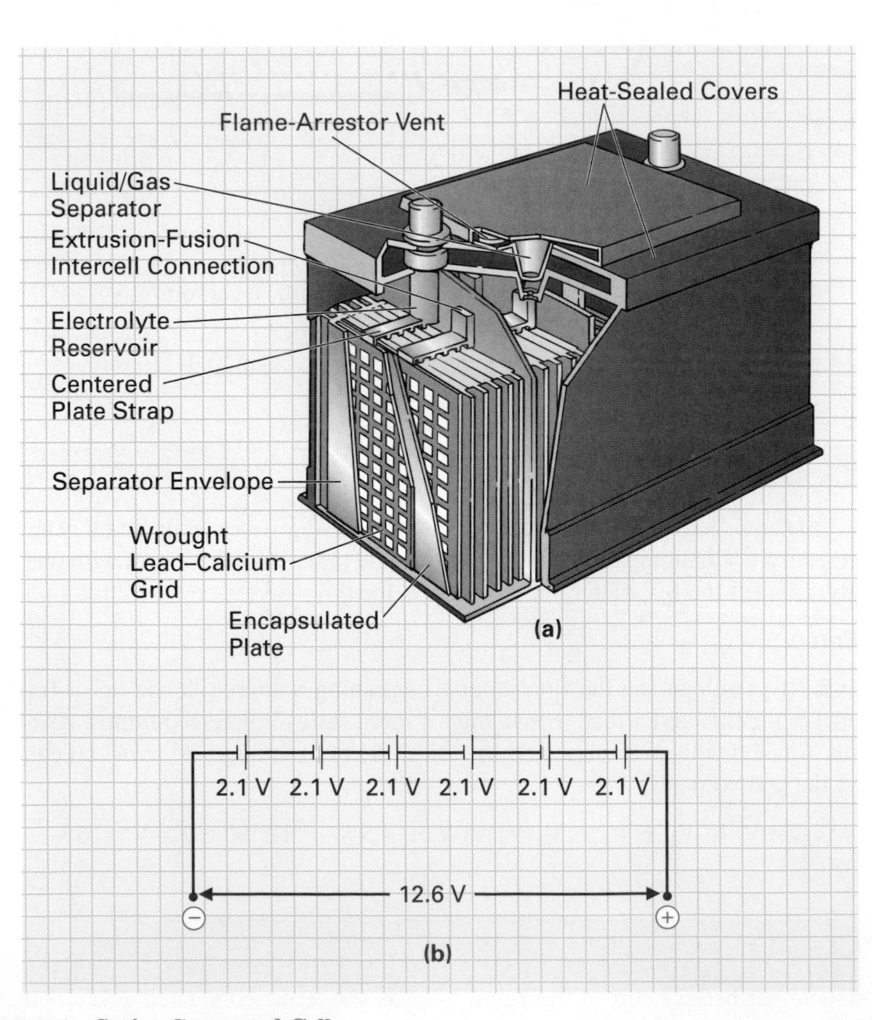

FIGURE 5-4 Series-Connected Cells.

■ *Solution:*

$$\begin{aligned} V_T &= V_1 + V_2 + V_3 + V_4 + V_5 + V_6 \\ &= 2.1\text{ V} + 2.1\text{ V} + 2.1\text{ V} + 2.1\text{ V} + 2.1\text{ V} + 2.1\text{ V} \\ &= 12.6\text{ V} \end{aligned}$$

5-2 CURRENT IN A SERIES CIRCUIT

The current in a series circuit has only one path to follow and cannot divert in any other direction. The current through a series circuit, therefore, is the same throughout that circuit.

Returning once again to the water analogy, you can see in Figure 5-5(a) that if 2 gallons of water per second are being supplied by the pump, 2 gallons per second must be pulled into the pump. If the rate at which water is leaving and arriving at the pump is the same, 2 gallons of water per second must be flowing throughout the circuit. It can be said, therefore, that the same value of water flow exists throughout a series-connected fluid system. This rule will always remain true, for if the valves were adjusted to double the opposition to flow, then half the flow, or 1 gallon of water per second, would be leaving the pump and flowing throughout the system.

Similarly, with the electronic series circuit shown in Figure 5-5(b), there is a total of 2 A leaving and 2 A arriving at the battery, and so the same value of current exists throughout the series-connected electronic circuit. If the circuit's resistance were changed, a new value of series circuit current would be present throughout the circuit. For example, if the resis-tance of the circuit was doubled, then half the current, or 1 A, will leave the battery, but that same value of 1 A will flow throughout the entire circuit. This series circuit current characteristic can be stated mathematically as

$$I_T = I_1 = I_2 = I_3 = \cdots$$

Total current = current through R_1 = current through R_2 = current through R_3, and so on.

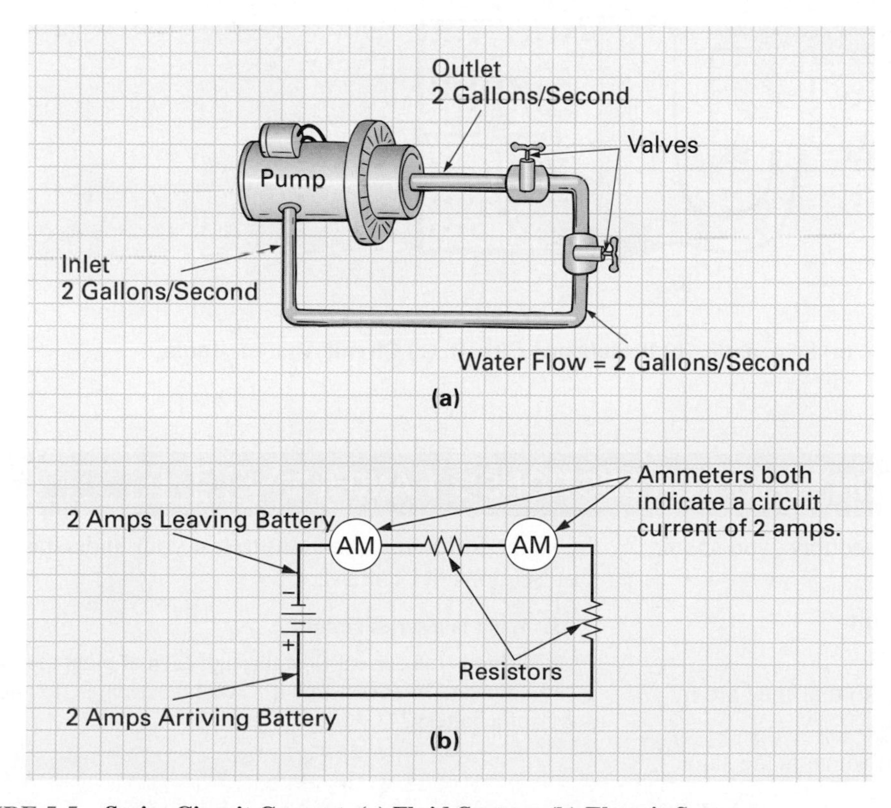

FIGURE 5-5 Series Circuit Current. (a) Fluid System. (b) Electric System.

EXAMPLE:

In Figure 5-6(a) and (b), a total current (I_T) of 1 A is flowing out of the negative terminal of the dc power supply, through two end-to-end resistors R_1 and R_2, and returning to the positive terminal of the dc power supply. Calculate:

a. The current through R_1 (I_1)
b. The current through R_2 (I_2)

Solution:

Since R_1 and R_2 are connected in series, the current through both will be the same as the circuit current, which is equal to 1 A.

$$I_T = I_1 = I_2$$
$$1\text{A} = 1\text{A} = 1\text{A}$$

The details of this circuit are summarized in the analysis table shown in Figure 5-6(c).

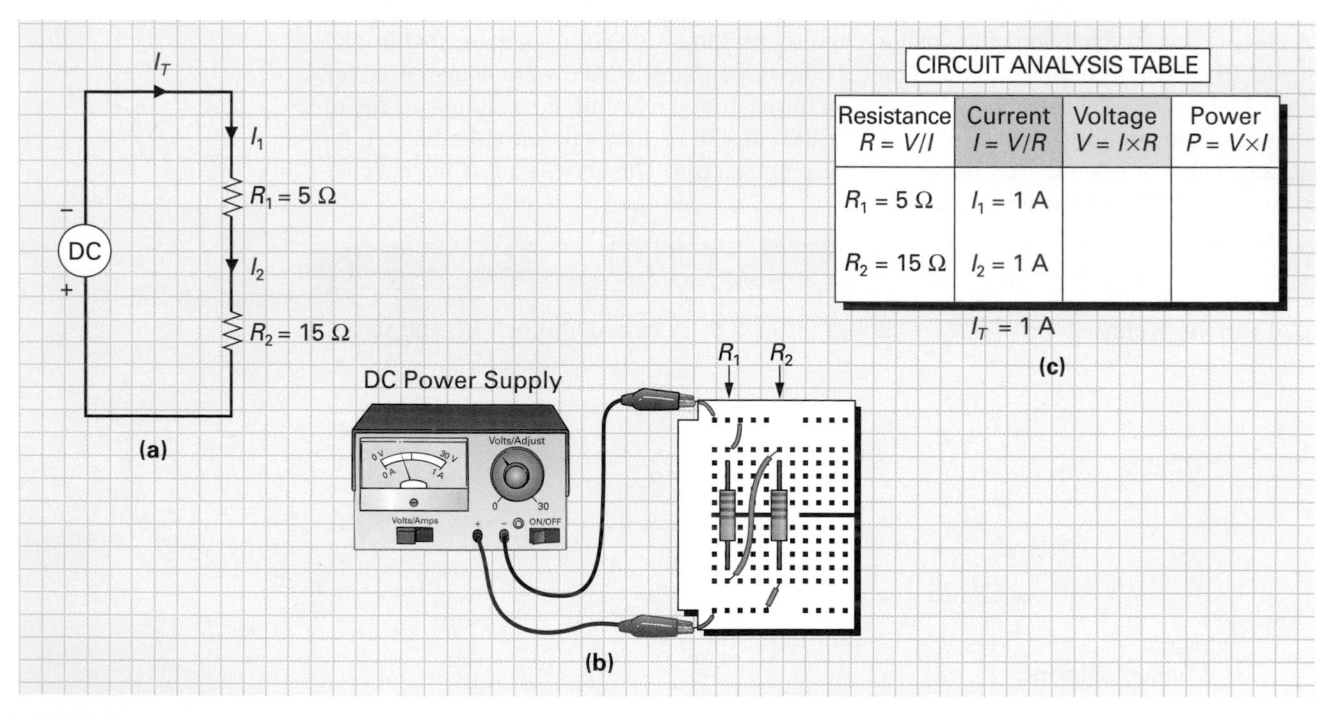

Resistance $R = V/I$	Current $I = V/R$	Voltage $V = I \times R$	Power $P = V \times I$
$R_1 = 5\ \Omega$	$I_1 = 1$ A		
$R_2 = 15\ \Omega$	$I_2 = 1$ A		

FIGURE 5-6 Total Current Example. (a) Schematic. (b) Protoboard Circuit. (c) Circuit Analysis Table.

SELF-TEST EVALUATION POINT FOR SECTIONS 5-1 AND 5-2

Now that you have completed these sections, you should be able to:

Objective 1. Describe a series circuit.

Objective 2. Identify series circuits.

Objective 3. Connect components so that they are in series with one another.

Objective 4. Describe why current remains the same throughout a series circuit.

Use the following questions to test your understanding of Sections 5-1 and 5-2:

1. What is a series circuit?
2. What is the current flow through each of eight series-connected 8 Ω resistors if 8 A total current is flowing out of a battery?

5-3 RESISTANCE IN A SERIES CIRCUIT

Resistance is the opposition to current flow, and in a series circuit every resistor in series offers opposition to the current flow. In the water analogy of Figure 5-5, the total resistance or opposition to water flow is the sum of the two individual valve opposition values. Like the battery, the pump senses the total opposition in the circuit offered by all the valves or resistors, and the amount of current that flows is dependent on this resistance or opposition.

The total resistance in a series-connected resistive circuit is thus equal to the sum of all the individual resistances, as shown in Figure 5-7(a) through (d). An equivalent circuit can be drawn for each of the circuits in Figure 5-7(b), (c), and (d) with one resistor of a value equal to the sum of all the series resistance values.

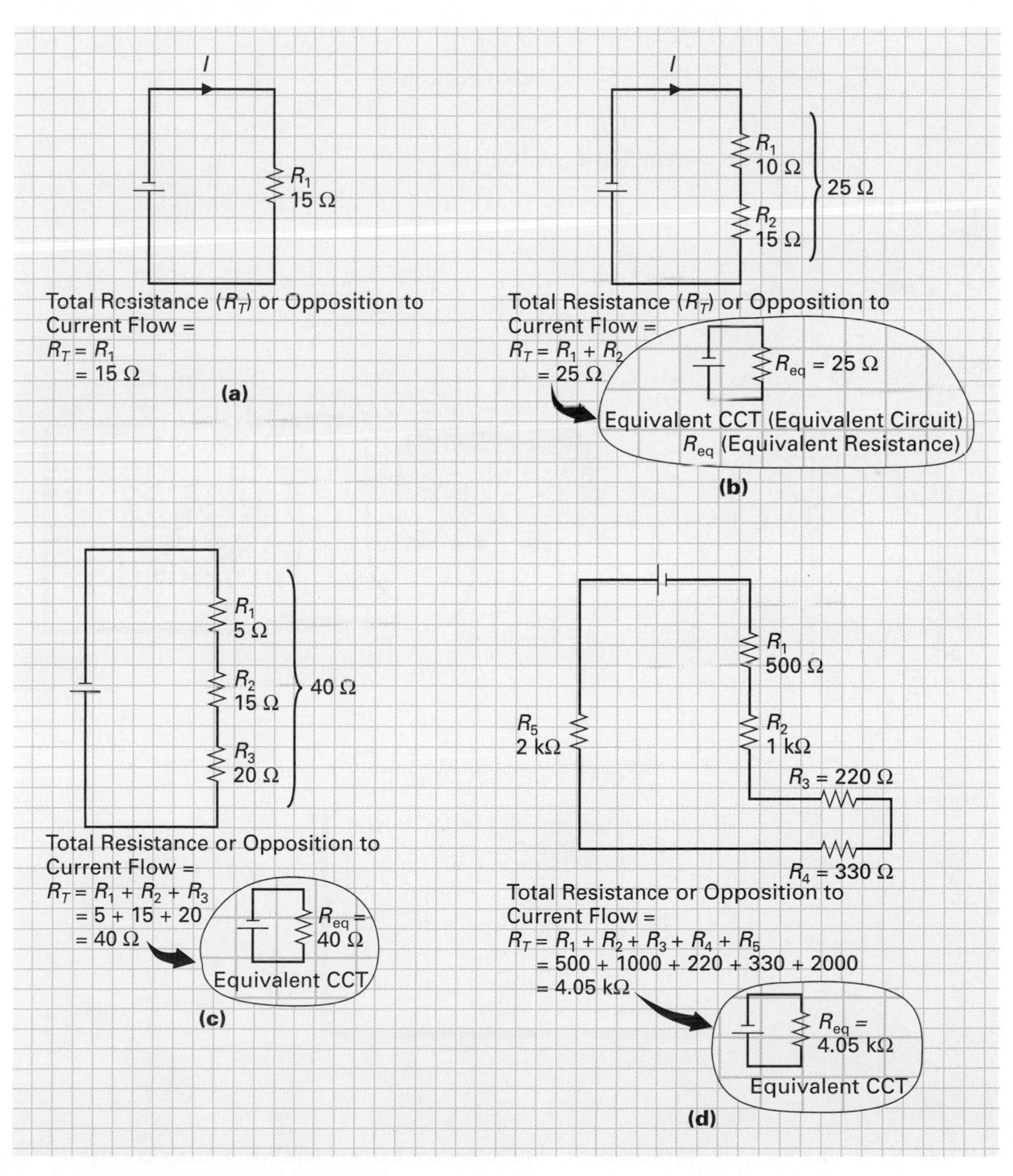

FIGURE 5-7 Total or Equivalent Resistance.

No matter how many resistors are connected in series, the total resistance or opposition to current flow is always equal to the sum of all the resistor values. This formula can be stated mathematically as

$$R_T = R_1 + R_2 + R_3 + \cdots$$

Total resistance = value of R_1 + value of R_2 + value of R_3, and so on.

Equivalent Resistance (R_{eq})
Total resistance of all the individual resistances in a circuit.

Total resistance (R_T) is the only opposition a source can sense. It does not see the individual separate resistors, but one **equivalent resistance.** Based on the source's voltage and the circuit's total resistance, a value of current will be produced to flow through the circuit (Ohm's law, $I = V/R$).

EXAMPLE:

Referring to Figure 5-8(a) and (b), calculate:

a. The circuit's total resistance

b. The current flowing through R_2

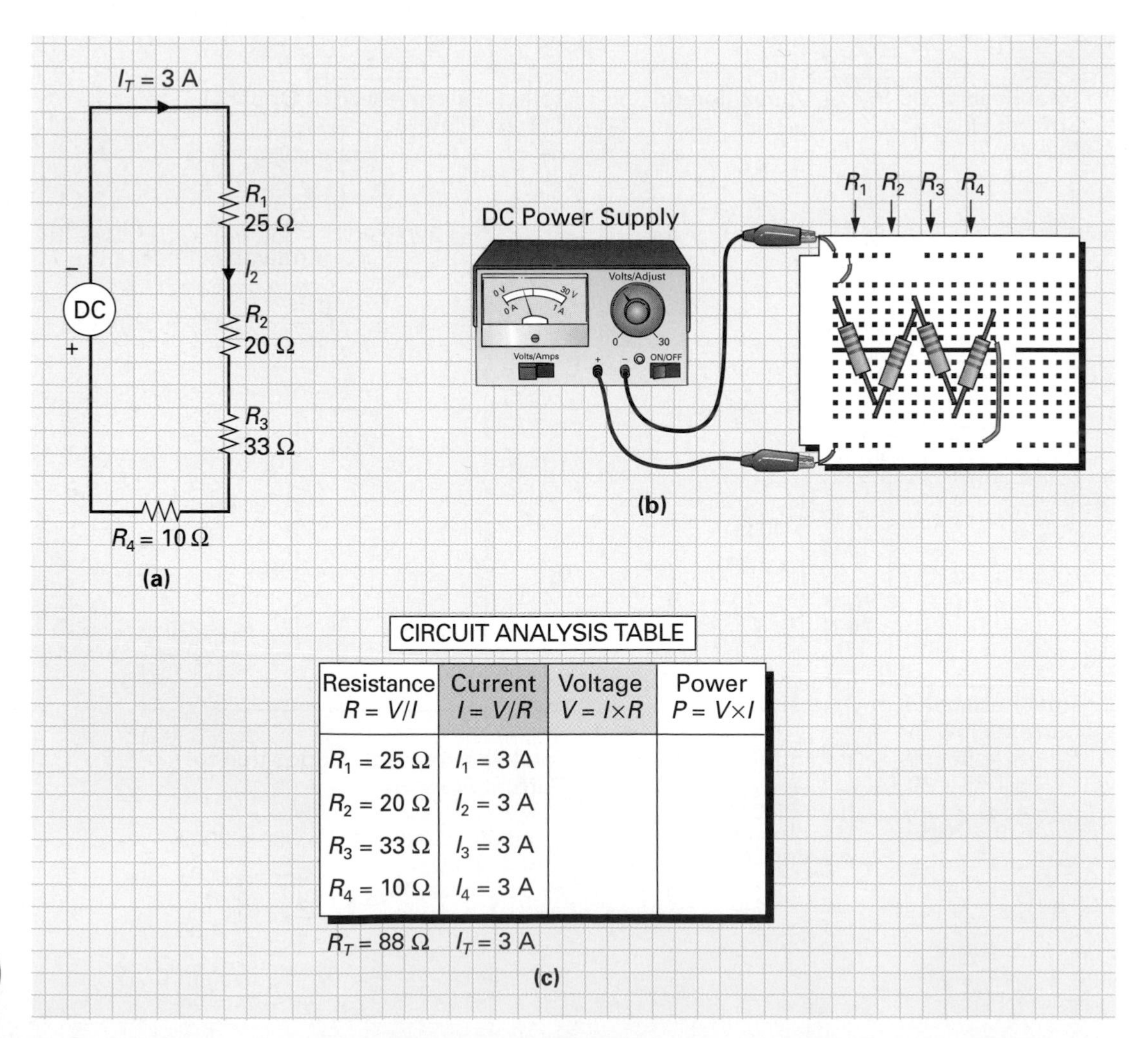

Resistance $R = V/I$	Current $I = V/R$	Voltage $V = I \times R$	Power $P = V \times I$
R_1 = 25 Ω	I_1 = 3 A		
R_2 = 20 Ω	I_2 = 3 A		
R_3 = 33 Ω	I_3 = 3 A		
R_4 = 10 Ω	I_4 = 3 A		
R_T = 88 Ω	I_T = 3 A		

FIGURE 5-8 **Total Resistance Example. (a) Schematic. (b) Protoboard Circuit. (c) Circuit Analysis Table.**

■ *Solution:*

a. $R_T = R_1 + R_2 + R_3 + R_4$
$= 25\ \Omega + 20\ \Omega + 33\ \Omega + 10\ \Omega$
$= 88\ \Omega$

b. $I_T = I_1 = I_2 = I_3 = I_4$. Therefore, $I_2 = I_T = 3$ A.

The details of this circuit are summarized in the analysis table shown in Figure 5-8(c).

EXAMPLE:

Figure 5-9(a) and (b) shows how a single-pole three-position switch is being used to provide three different lamp brightness levels. In position ① R_1 is placed in series with the lamp, in position ② R_2 is placed in series with the lamp, and in position ③ R_3 is placed in series with the lamp. If the lamp has a resistance of 75 Ω, calculate the three values of current for each switch position.

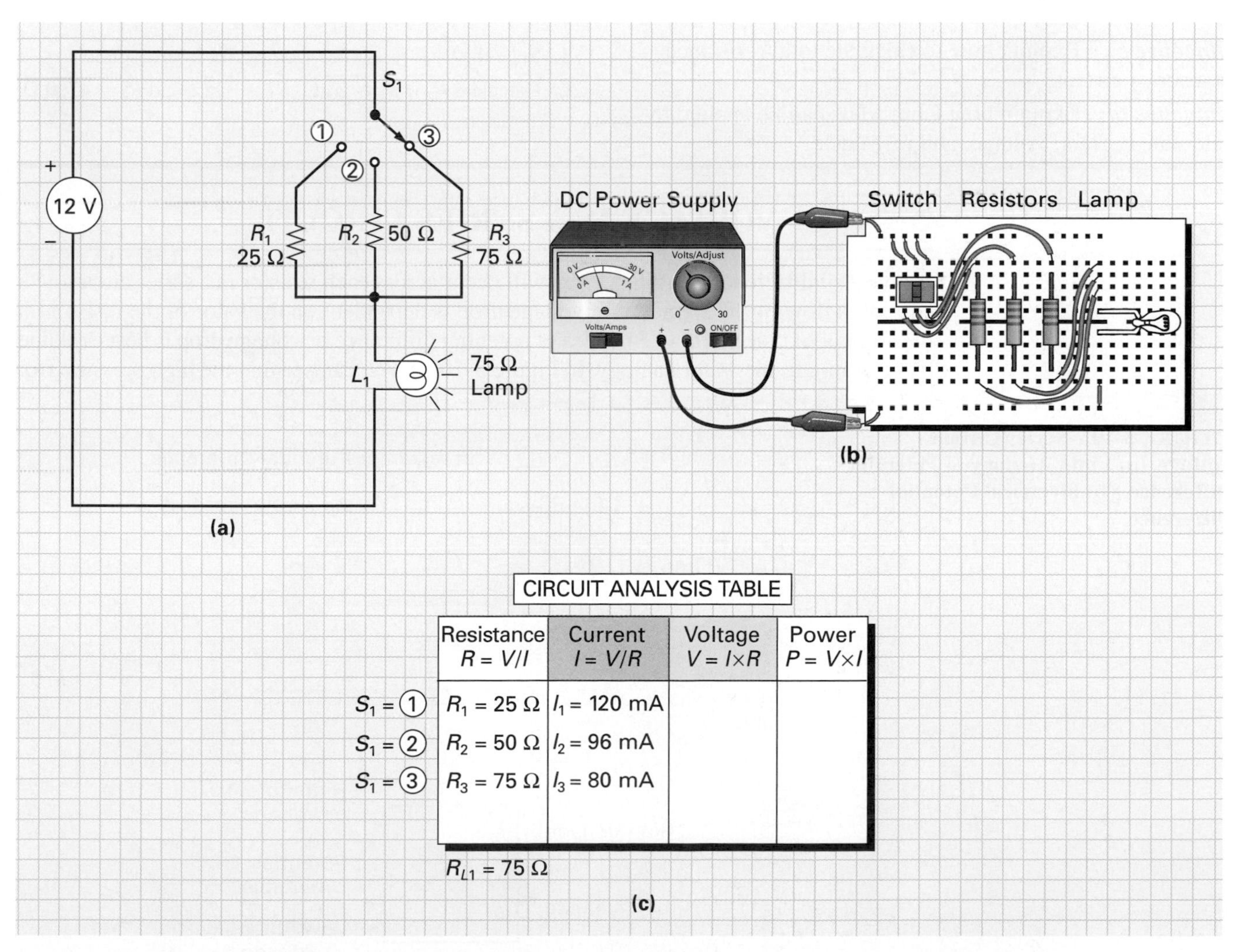

FIGURE 5-9 **Three-Position Switch Controlling Lamp Brightness. (a) Schematic. (b) Protoboard Circuit. (c) Circuit Analysis Table.**

■ *Solution:*

$$\text{Position 1: } R_T = R_1 + R_{\text{lamp}}$$
$$= 25\ \Omega + 75\ \Omega = 100\ \Omega$$
$$I_T = \frac{V_T}{R_T} = \frac{12\ \text{V}}{100\ \Omega} = 120\ \text{mA}$$
$$\text{Position 2: } R_T = R_2 + R_{\text{lamp}} = 50\ \Omega + 75\ \Omega = 125\ \Omega$$
$$I_T = \frac{V_T}{R_T} = \frac{12\ \text{V}}{125\ \Omega} = 96\ \text{mA}$$
$$\text{Position 3: } R_T = R_3 + R_{\text{lamp}} = 75\ \Omega + 75\ \Omega = 150\ \Omega$$
$$I_T = \frac{V_T}{R_T} = \frac{12\ \text{V}}{150\ \Omega} = 80\ \text{mA}$$

The details of this circuit are summarized in the analysis table shown in Figure 5-9(c).

SELF-TEST EVALUATION POINT FOR SECTION 5-3

Now that you have completed this section, you should be able to:

Objective 5. Explain how to calculate total resistance in a series circuit.

Objective 6. Explain how Ohm's law can be applied to calculate current, voltage, and resistance.

Use the following questions to test your understanding of Section 5-3:

1. State the total resistance formula for a series circuit.
2. Calculate R_T if $R_1 = 2\ \text{k}\Omega$, $R_2 = 3\ \text{k}\Omega$, and $R_3 = 4700\ \Omega$.

5-4 VOLTAGE IN A SERIES CIRCUIT

A potential difference or voltage drop will occur across each resistor in a series circuit when current is flowing. The amount of voltage drop is dependent on the value of the resistor and the amount of current flow. This idea of potential difference or voltage drop is best explained by returning to the water analogy. In Figure 5-10(a), you can see that the high pressure from

FIGURE 5-10 Series Circuit Voltage. (a) Fluid Analogy of Potential Difference. (b) Electrical Potential Difference.

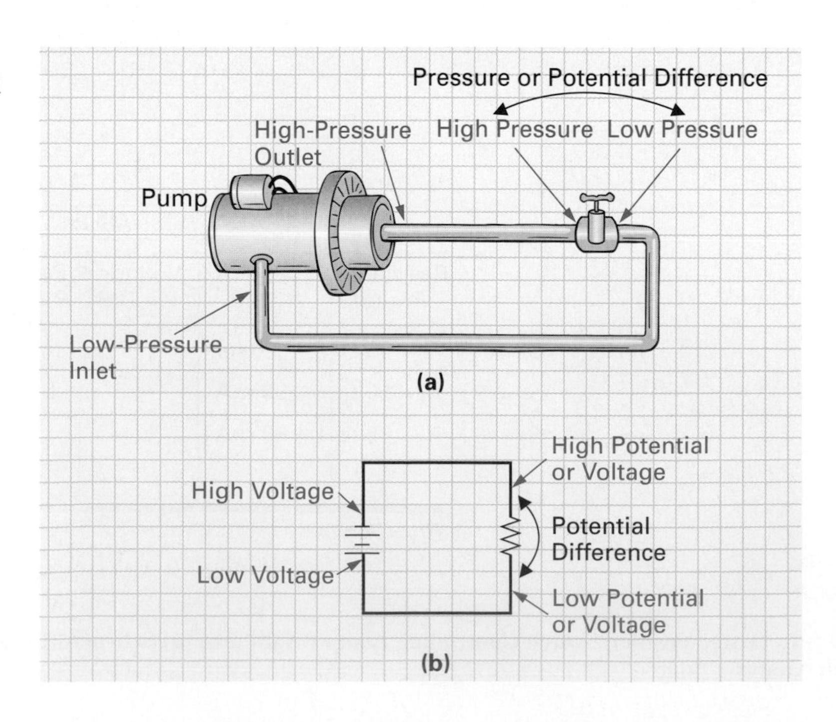

the pump's outlet is present on the left side of the valve. On the right side of the valve, however, the high pressure is no longer present. The high potential that exists on the left of the valve is not present on the right, so a potential or pressure difference is said to exist across the valve.

Similarly, with the electronic circuit shown in Figure 5-10(b), the battery produces a high voltage or potential that is present at the top of the resistor. The high voltage that exists at the top of the resistor, however, is not present at the bottom. Therefore, a potential difference or voltage drop is said to occur across the resistor. This voltage drop that exists across resistors can be found by utilizing Ohm's law: $V = I \times R$.

EXAMPLE:

Referring to Figure 5-11, calculate:

a. Total resistance (R_T)

b. Amount of series current flowing throughout the circuit (I_T)

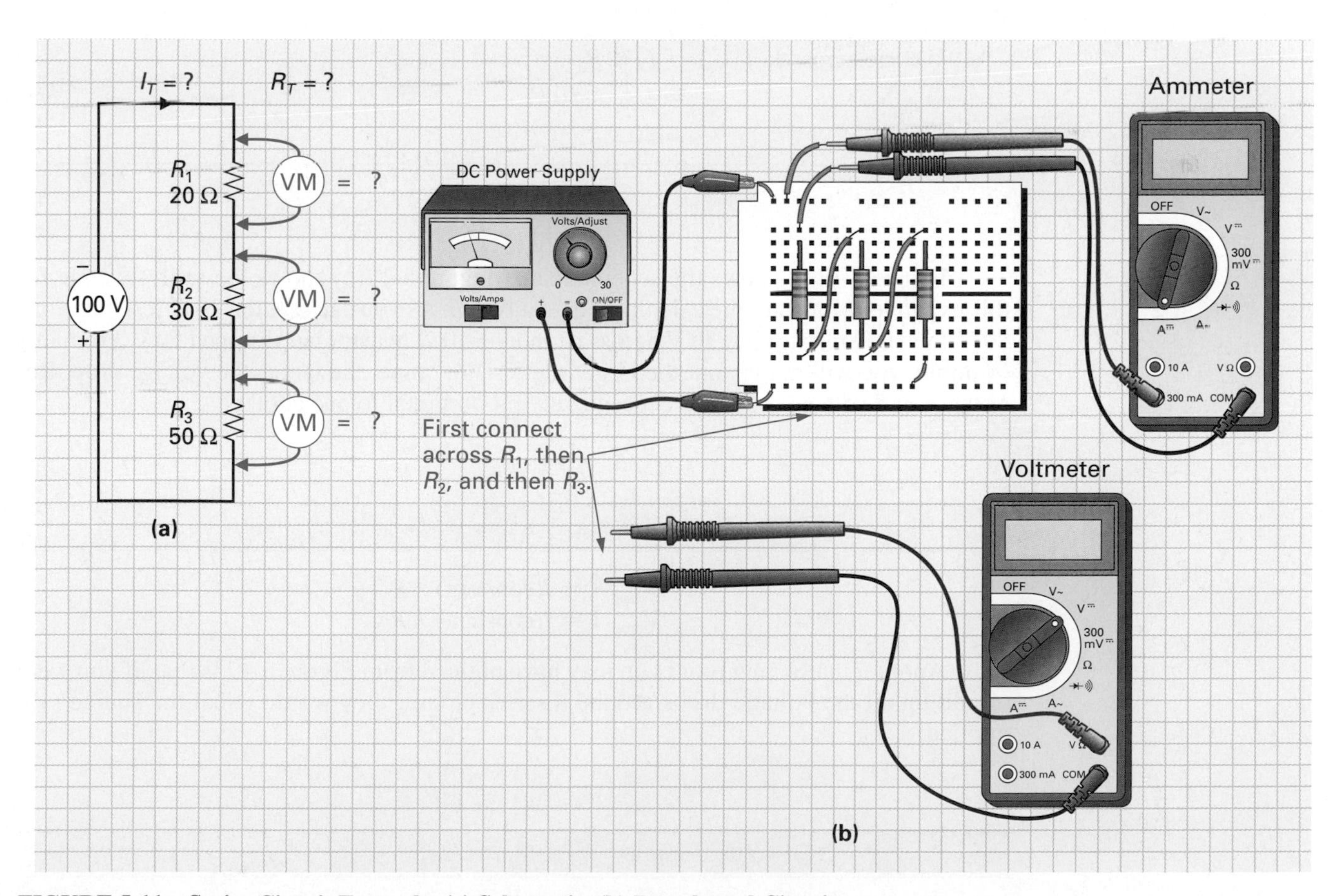

FIGURE 5-11 Series Circuit Example. (a) Schematic. (b) Protoboard Circuit.

c. Voltage drop across R_1

d. Voltage drop across R_2

e. Voltage drop across R_3

Solution:

a. Total resistance $(R_T) = R_1 + R_2 + R_3$

$$= 20\ \Omega + 30\ \Omega + 50\ \Omega$$
$$= 100\ \Omega$$

b. Total current $(I_T) = \dfrac{V_{source}}{R_T}$

$$= \frac{100\ V}{100\ \Omega}$$
$$= 1\ A$$

The same current will flow through the complete series circuit, so the current through R_1 will equal 1 A, the current through R_2 will equal 1 A, and the current through R_3 will equal 1 A.

c. Voltage across R_1 $(V_{R1}) = I_1 \times R_1$

$$= 1\ A \times 20\ \Omega$$
$$= 20\ V$$

d. Voltage across R_2 $(V_{R2}) = I_2 \times R_2$

$$= 1\ A \times 30\ \Omega$$
$$= 30\ V$$

e. Voltage across R_3 $(V_{R3}) = I_3 \times R_3$

$$= 1\ A \times 50\ \Omega$$
$$= 50\ V$$

Figure 5-12(a) shows the schematic and Figure 5-12(b) shows the analysis table for this example, with all of the calculated data inserted. As you can see, the 20 Ω resistor drops 20 V, the 30 Ω resistor has 30 V across it, and the 50 Ω resistor has dropped 50 V. From this example, you will notice that the larger the resistor value, the larger the voltage drop. Resistance and voltage drops are consequently proportional to one another.

The voltage drop across a device is ∝ to its resistance.

$$V_{drop} \uparrow = I(\text{constant}) \times R \uparrow$$
$$V_{drop} \downarrow = I(\text{constant}) \times R \downarrow$$

Another interesting point you may have noticed from Figure 5-12 is that, if you were to add up all the voltage drops around a series circuit, they would equal the source (V_S) applied:

Total voltage applied $(V_S$ or $V_T) = V_{R1} + V_{R2} + V_{R3} + \cdots$

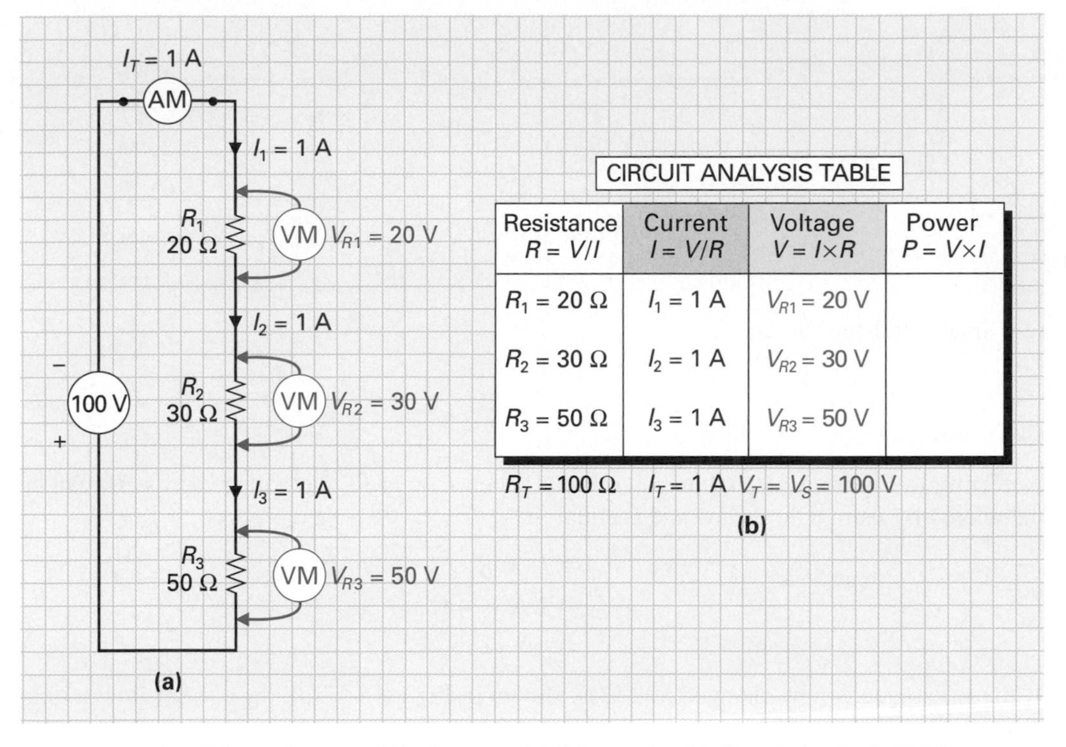

CIRCUIT ANALYSIS TABLE

Resistance $R = V/I$	Current $I = V/R$	Voltage $V = I \times R$	Power $P = V \times I$
$R_1 = 20\ \Omega$	$I_1 = 1$ A	$V_{R1} = 20$ V	
$R_2 = 30\ \Omega$	$I_2 = 1$ A	$V_{R2} = 30$ V	
$R_3 = 50\ \Omega$	$I_3 = 1$ A	$V_{R3} = 50$ V	
$R_T = 100\ \Omega$	$I_T = 1$ A	$V_T = V_S = 100$ V	

(b)

FIGURE 5-12 Voltage Drop and Resistance. (a) Schematic. (b) Circuit Analysis Table.

TIME LINE

German physicist Gustav Robert Kirchhoff (1824–1879) extended Ohm's law and developed two important laws of his own, known as Kirchhoff's voltage law and Kirchhoff's current law.

In the example in Figure 5-12, you can see that this is true, since

$$100\text{ V} = 20\text{ V} + 30\text{ V} + 50\text{ V}$$
$$100\text{ V} = 100\text{ V}$$

The series circuit has in fact divided up the applied voltage, and it appears proportionally across all the individual resistors. This characteristic was first observed by Gustav Kirchhoff in 1847. In honor of his discovery, this effect is known as **Kirchhoff's voltage law**, which states: The sum of the voltage drops in a series circuit is equal to the total voltage applied.

Kirchhoff's Voltage Law

The algebraic sum of the voltage drops in a closed-path circuit is equal to the algebraic sum of the source voltage applied.

To summarize the effects of current, resistance, and voltage in a series circuit so far, we can say that

1. The current in a series circuit has only one path to follow.
2. The value of current in a series circuit is the same throughout the entire circuit.
3. The total resistance in a series circuit is equal to the sum of all the resistances.
4. Resistance and voltage drops in a series circuit are proportional to one another, so a large resistance will have a large voltage drop and a small resistance will have a small voltage drop.
5. The sum of the voltage drops in a series circuit is equal to the total voltage applied.

EXAMPLE:

First, calculate the voltage drop across the resistor R_1 in the circuit in Figure 5-13(a) for a resistance of 4 Ω. Then, change the 4 Ω resistor to a 2 Ω resistor and recalculate the voltage drop across the new resistance value. Use a constant source of 4 V.

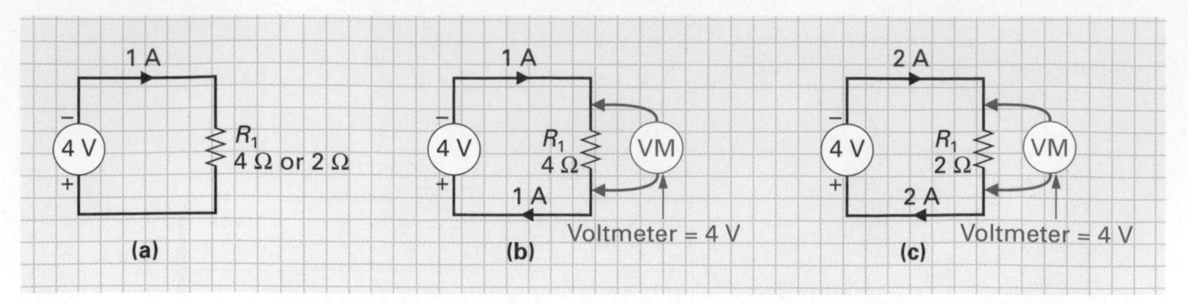

FIGURE 5-13 Single-Resistor Circuit.

Solution:

Referring to Figure 5-13(b), you can see that when $R_1 = 4\ \Omega$, the voltage across R_1 can be calculated by using Ohm's law and is equal to

$$V_{R1} = I_1 \times R_1$$
$$V_{R1} = 1\text{ A} \times 4\ \Omega$$
$$= 4\text{ V}$$

If the resistance is now changed to 2 Ω as shown in Figure 5-13(c), the current flow within the circuit will be equal to

$$I = \frac{V_S}{R} = \frac{4\text{ V}}{2\ \Omega} = 2\text{A}$$

The voltage dropped across the 2 Ω resistor will still be equal to

$$V_{R1} = I_1 \times R_1$$
$$= 2\text{ A} \times 2\ \Omega$$
$$= 4\text{ V}$$

As you can see from this example, if only one resistor is connected in a series circuit, the entire applied voltage appears across this resistor. The value of this single resistor will determine the amount of current flow through the circuit and this value of circuit current will remain the same throughout.

EXAMPLE:

Referring to Figure 5-14(a), calculate the following, and then draw the circuit schematic again with all of the new values inserted:

a. Total circuit resistance

b. Value of circuit current (I_T)

c. Voltage drop across each resistor

Solution:

a. $R_T = R_1 + R_2 + R_3 + R_4$
$= 2.2\text{ k}\Omega + 1.5\text{ k}\Omega + 3.8\text{ k}\Omega + 4.5\text{ k}\Omega$
$= (2.2 \times 10^3) + (1.5 \times 10^3) + (3.8 \times 10^3) + (4.5 \times 10^3)$
$= 12\text{ k}\Omega$ [Figure 5-14(b)]

b. $I_T = \frac{V_S}{R_T} = \frac{12\text{ V}}{12\text{ k}\Omega} = 1\text{ mA}$

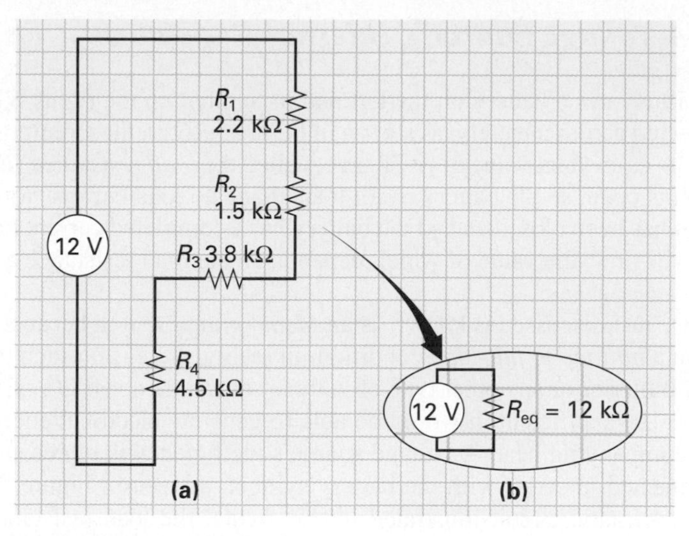

FIGURE 5-14 Series Circuit Example.

c. Voltage drop across each resistor:

$$\begin{aligned} V_{R1} &= I_T \times R_1 \\ &= 1 \text{ mA} \times 2.2 \text{ k}\Omega \\ &= 2.2 \text{ V} \\ V_{R2} &= I_T \times R_2 \\ &= 1 \text{ mA} \times 1.5 \text{ k}\Omega \\ &= 1.5 \text{ V} \\ V_{R3} &= I_T \times R_3 \\ &= 1 \text{ mA} \times 3.8 \text{ k}\Omega \\ &= 3.8 \text{ V} \\ V_{R4} &= I_T \times R_4 \\ &= 1 \text{ mA} \times 4.5 \text{ k}\Omega \\ &= 4.5 \text{ V} \end{aligned}$$

Figure 5-15(a) shows the schematic diagram for this example with all of the values inserted, and Figure 5-15(b) shows this circuit's analysis table.

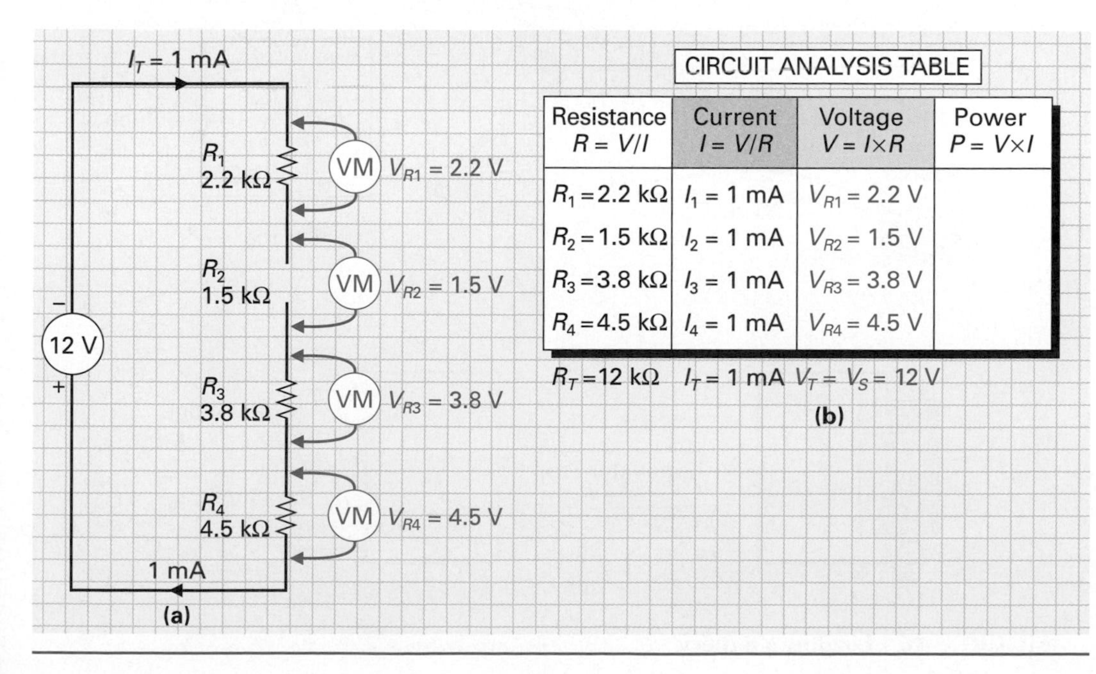

Resistance $R = V/I$	Current $I = V/R$	Voltage $V = I \times R$	Power $P = V \times I$
R_1 = 2.2 kΩ	I_1 = 1 mA	V_{R1} = 2.2 V	
R_2 = 1.5 kΩ	I_2 = 1 mA	V_{R2} = 1.5 V	
R_3 = 3.8 kΩ	I_3 = 1 mA	V_{R3} = 3.8 V	
R_4 = 4.5 kΩ	I_4 = 1 mA	V_{R4} = 4.5 V	
R_T = 12 kΩ	I_T = 1 mA	$V_T = V_S$ = 12 V	

(b)

FIGURE 5-15 Series Circuit Example with All Values Inserted. (a) Schematic. (b) Circuit Analysis Table.

5-4-1 *A Voltage Source's Internal Resistance*

Figure 5-16(a) illustrates a battery on the left and its symbol on the right. When a circuit or system is connected across a battery, as shown in Figure 5-16(a), the circuit or system can be represented by its equivalent value of resistance, called the *load resistance* (R_L). In Figure 5-16(a) the switch is open, so the battery is not loaded and no load current flows. In Figure 5-16(b) the switch has been closed and so the battery has a completed current path. As a result, a load current will flow, the value of which depends on the load resistance and battery voltage of 12 V.

The battery just discussed is known as an *ideal voltage source* because its voltage did not change from a *no-load* to *full-load* condition. In reality, there is no such thing as an ideal voltage source. Batteries or any other type of voltage source are not 100% efficient. If a voltage source were 100% efficient, it would be able to generate electrical energy at its output without any accompanying heat. A voltage source's inefficiency is represented by an internal resistor connected in series with the battery symbol, as seen in Figure 5-16(c). As you can see in the schematic circuit illustrated in this figure, the load resistance, R_L, and the source resistance, R_S (pronounced "R sub L", and "R sub S"), together form the total resistive load.

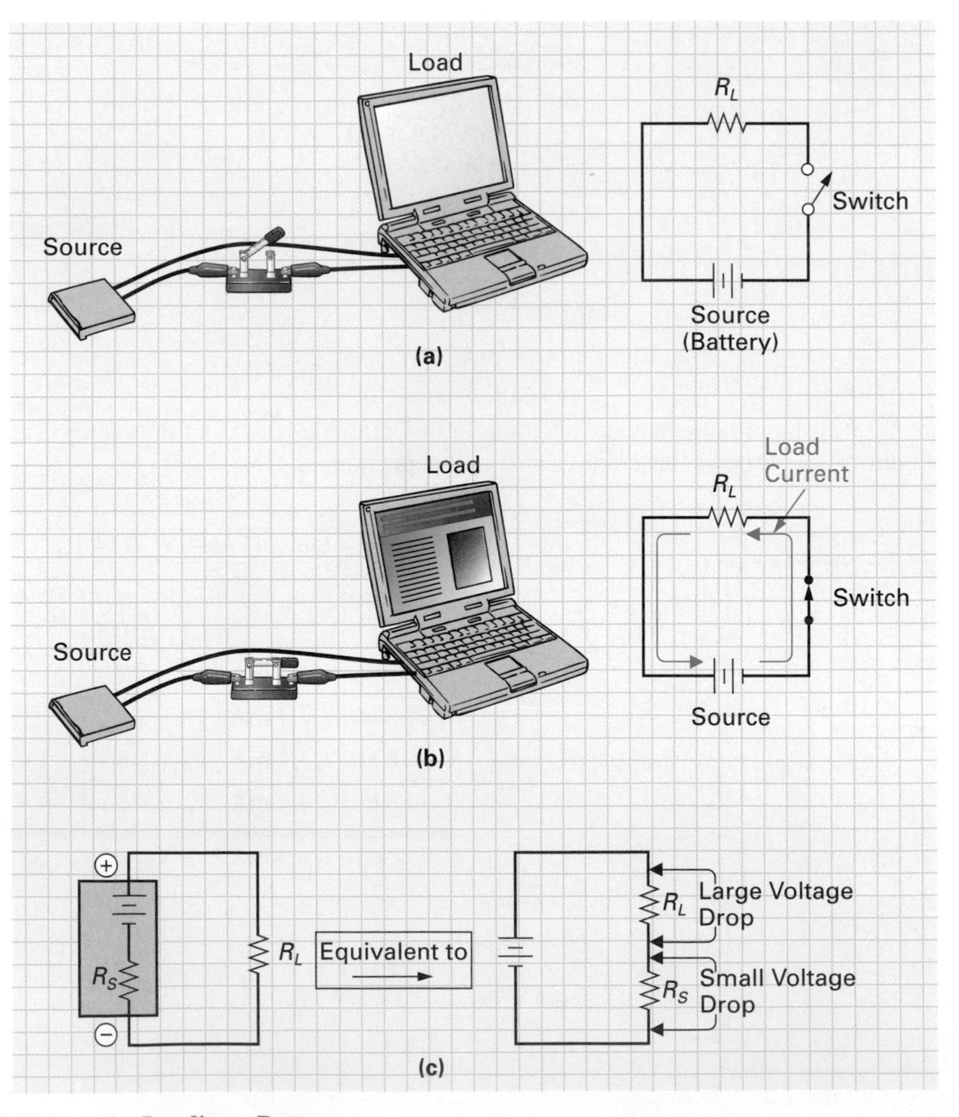

FIGURE 5-16 Loading a Battery.

Very little voltage will be dropped across R_S, as it is normally small compared to R_L. Internal inefficiencies of batteries and all other voltage sources must always be kept small because a large R_S would drop a greater amount of the supply voltage, resulting in less output voltage to the load and therefore a waste of electrical power. To give an example, a lead–acid cell will typically have an internal resistance of 0.01 Ω (10 mΩ).

EXAMPLE:

Figure 5-17 shows a 12 V battery that has an internal source resistance (R_S or R_{int}) of 0.5 Ω (500 mΩ). If the battery was to supply its maximum safe current, which in this example is 2.5 A, what would be the output terminal voltage of the battery?

Solution:

If the battery was supplying its maximum safe current of 2.5 A, the output voltage (V_{out}) would equal the source voltage ($V_S = 12$ V) minus the voltage drop across the internal battery resistance (V_{Rint}). First let us calculate V_{Rint}.

$$V_{Rint} = I \times R_{int}$$
$$= 2.5 \text{ A} \times 0.5 \ \Omega = 1.25 \text{ V}$$

The output voltage under full load (maximum safe current) will therefore be

$$V_{out} = V_S - V_{Rint} = 12 \text{ V} - 1.25 \text{ V} = 10.75 \text{ V}$$

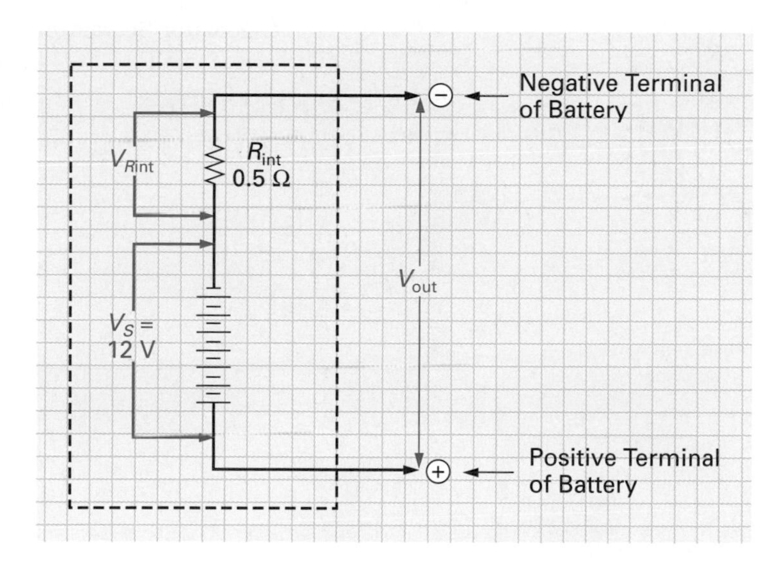

FIGURE 5-17 Voltage Drop across a Battery's Internal Series Resistance.

5-4-2 *Fixed Voltage Divider*

A series-connected circuit is often referred to as a *voltage-divider circuit,* because the total voltage applied (V_T) or source voltage (V_S) is divided and dropped proportionally across all the resistors in a series circuit. The amount of voltage dropped across a resistor is proportional to the value of resistance, and so a larger resistance will have a larger voltage drop, and a smaller resistance will have a smaller voltage drop.

The voltage dropped across a resistor is normally a factor that needs to be calculated. The voltage-divider formula allows you to calculate the voltage drop across any resistor without having to first calculate the value of circuit current. This formula is stated as

$$V_X = \frac{R_X}{R_T} \times V_S$$

where V_X = voltage dropped across selected resistor
R_X = selected resistor's value
R_T = total series circuit resistance
V_S = source or applied voltage

Figure 5-18 illustrates a circuit from a previous example. To calculate the voltage drop across each resistor, we would normally have to do the following:

1. Calculate the total resistance by adding up all the resistance values.
2. Once we have the total resistance and source voltage (V_S), we could then calculate current.
3. Having calculated the current flowing through each resistor, we could then use the current value to calculate the voltage dropped across any one of the four resistors merely by multiplying current by the individual resistance value.

The voltage-divider formula allows us to bypass the last two steps in this procedure. If we know total resistance, supply voltage, and the individual resistance values, we can calculate the voltage drop across the resistor without having to calculate steps 2 and 3. For example, what would be the voltage dropped across R_2 and R_4 for the circuit shown in Figure 5-18? The voltage dropped across R_2 is

$$V_{R2} = \frac{R_2}{R_T} \times V_S$$
$$= \frac{1.5\text{ k}\Omega}{12\text{ k}\Omega} \times 12\text{ V}$$
$$= 1.5\text{ V}$$

Calculator Sequence

STEP	KEYPAD ENTRY	DISPLAY RESPONSE
1.	1 . 5 E 3	1.5E3
2.	÷	
3.	1 2 E 3	12E3
4.	×	0.125
5.	1 2	12
6.	=	1.5

FIGURE 5-18 Series Circuit Example.

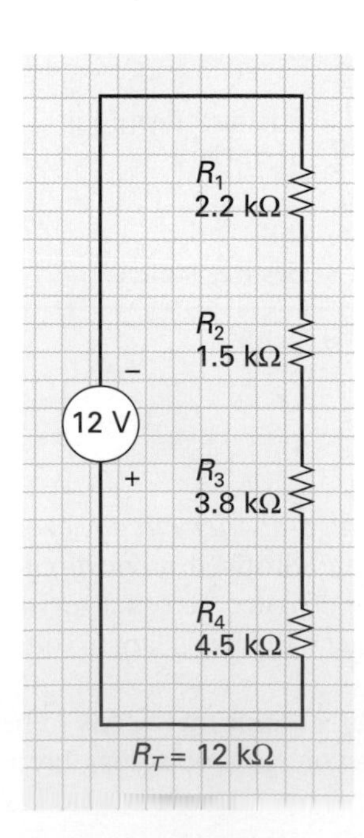

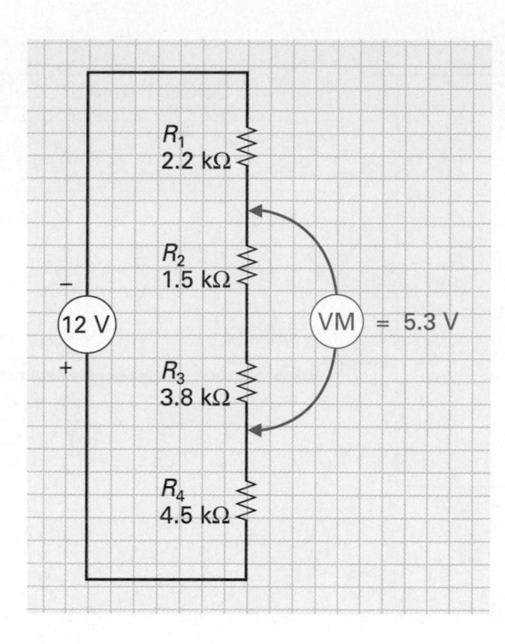

FIGURE 5-19 Series Circuit Example.

The voltage dropped across R_4 is

$$V_{R4} = \frac{R_4}{R_T} \times V_S$$
$$= \frac{4.5\text{ k}\Omega}{12\text{ k}\Omega} \times 12\text{ V}$$
$$= 4.5\text{ V}$$

The voltage-divider formula could also be used to find the voltage drop across two or more series-connected resistors. For example, referring again to the example circuit shown in Figure 5-18, what would be the voltage dropped across R_2 and R_3 combined? The voltage across $R_2 + R_3$ can be calculated using the voltage-divider formula as follows:

$$V_{R2} \text{ and } V_{R3} = \frac{R_2 + R_3}{R_T} \times V_S$$
$$= \frac{5.3\text{ k}\Omega}{12\text{ k}\Omega} \times 12\text{ V}$$
$$= 5.3\text{ V}$$

As can be seen in Figure 5-19, the voltage drop across R_2 and R_3 is 5.3 V.

EXAMPLE:

Referring to Figure 5-20(a), calculate the voltage drop across

a. R_1, R_2, and R_3 separately
b. R_2 and R_3 combined
c. R_1, R_2, and R_3 combined

FIGURE 5-20 **Series Circuit Example. (a) Schematic. (b) Circuit Analysis Table.**

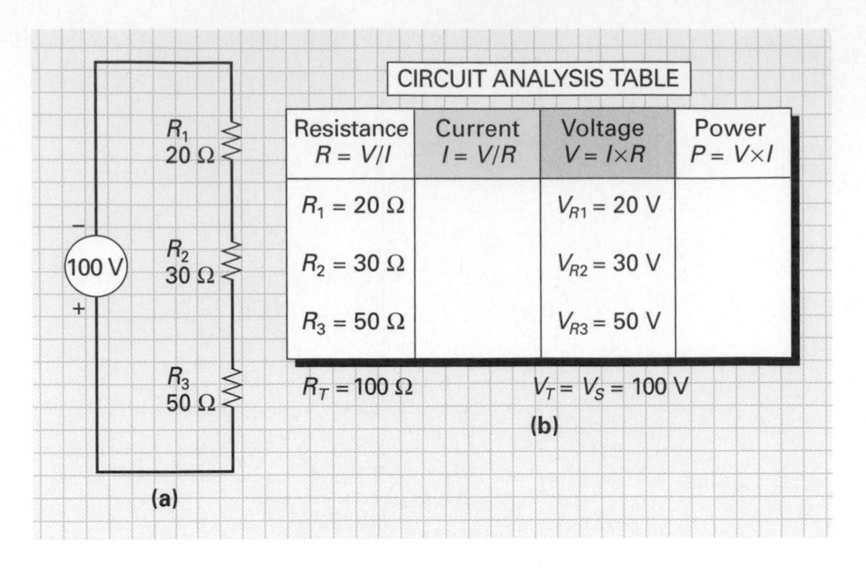

■ *Solution:*

a. The voltage drop across a resistor is proportional to the resistance value. The total resistance (R_T) in this circuit is 100 Ω or 100% of R_T. R_1 is 20% of the total resistance, so 20% of the source voltage will appear across R_1. R_2 is 30% of the total resistance, so 30% of the source voltage will appear across R_2. R_3 is 50% of the total resistance, so 50% of the source voltage will appear across R_3. This is a very simple problem in which the figures work out very neatly. The voltage-divider formula achieves the very same thing by calculating the ratio of the resistance value to the total resistance. This percentage is then multiplied by the source voltage in order to find the desired resistor's voltage drop:

$$\begin{aligned} V_{R1} &= \frac{R_1}{R_T} \times V_S \\ &= \frac{20\ \Omega}{100\ \Omega} \times V_S \\ &= 0.2 \times 100\ \text{V} \qquad (20\%\ \text{of}\ 100\ \text{V}) \\ &= 20\ \text{V} \end{aligned}$$

$$\begin{aligned} V_{R2} &= \frac{R_2}{R_T} \times V_S \\ &= \frac{30\ \Omega}{100\ \Omega} \times V_S \\ &= 0.3 \times 100\ \text{V} \qquad (30\%\ \text{of}\ 100\ \text{V}) \\ &= 30\ \text{V} \end{aligned}$$

$$\begin{aligned} V_{R3} &= \frac{R_3}{R_T} \times V_S \\ &= \frac{50\ \Omega}{100\ \Omega} \times V_S \\ &= 0.5 \times 100\ \text{V} \qquad (50\%\ \text{of}\ 100\ \text{V}) \\ &= 50\ \text{V} \end{aligned}$$

b. Voltage dropped across R_2 and $R_3 = 30 + 50 = 80$ V.

c. Voltage dropped across R_1, R_2, and $R_3 = 20 + 30 + 50 = 100$ V.

The details of this circuit are summarized in the analysis table shown in Figure 5-20(b).

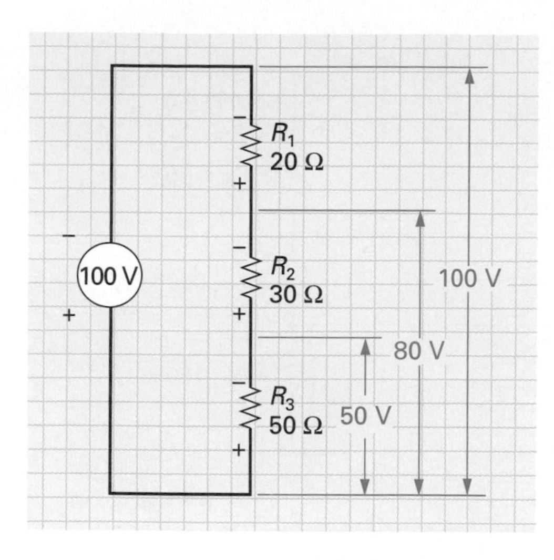

FIGURE 5-21 **Series Circuit Example.**

To summarize the voltage-divider formula, we can say: The voltage drop across a resistor or group of resistors in a series circuit is equal to the ratio of that resistance (R_X) to the total resistance (R_T), multiplied by the source voltage (V_S).

To show an application of a voltage divider, let us imagine that three voltages of 50, 80, and 100 V are required by an electronic system in order to make it operate. To meet this need, we could use three individual power sources, which would be very expensive, or use one 100 V voltage source connected across three resistors, as shown in Figure 5-21, to divide up the 100 V.

EXAMPLE:

Figure 5-22(a) shows a simplified circuit of an oscilloscope's cathode ray tube (CRT) or picture tube. In this example, a three-resistor series circuit and a −600 V supply voltage are being used to produce the needed three supply voltages for the CRT's three electrodes, called the focusing anode, the control grid, and the cathode. The heated cathode emits electrons that are collected and concentrated into a beam by the combined electrostatic effect of the control grid and focusing anode. This beam of electrons passes through the apertures of the control grid and focusing anode and strikes the inner surface of the CRT screen. This inner surface is coated with a phosphorescent material that emits light when it is struck by electrons. The voltage between the cathode (K) and grid (G) of the CRT (V_{KG}) determines the intensity of the electron beam and therefore the brightness of the trace seen on the screen. The voltage between the grid (G) and anode (A) of the CRT (V_{GA}) determines the sharpness of the electron beam and therefore the focus of the trace seen on the screen. For the resistance values given, calculate the voltages on each of the three CRT electrodes.

Solution:

Referring to the illustration and calculations in Figure 5-22(b), you can see how the voltage-divider formula can be used to calculate the voltage drop across R_1 ($V_{R1} = 300$ V), R_2 ($V_{R2} = 255$ V), and R_3 ($V_{R3} = 45$ V). Since the grid of the CRT is connected directly to the −600 V supply, the grid voltage (V_G) will be

$$V_G = V_T = -600 \text{ V}$$

The cathode voltage (V_K) will be equal to the total supply voltage (V_T) minus the voltage drop across R_3 (V_{R3}), so

$$V_K = V_T - V_{R3} = (-600) - (-45) = -555 \text{ V}$$

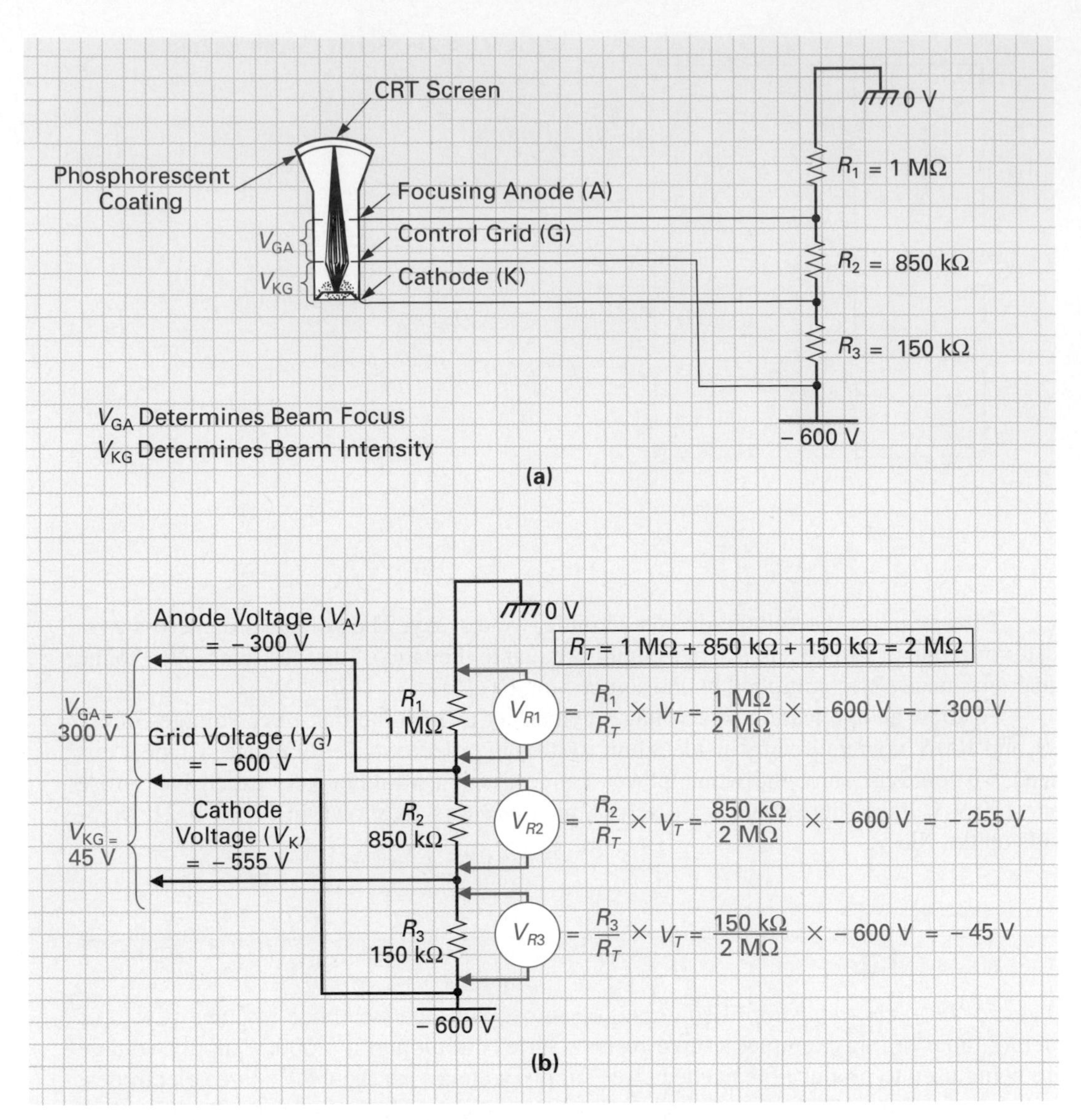

FIGURE 5-22 Fixed Voltage-Divider Circuit for Supplying Voltages to the Electrodes of a CRT.

The anode voltage (V_A) will be equal to the total supply voltage (V_T) minus the voltage drops across R_3 (V_{R3}) and R_2 (V_{R2}), and therefore

$$V_A = V_T - (V_{R3} + V_{R2}) = (-600) - (-45 + -255)$$
$$= -600 + 300\text{ V} = -300\text{ V}$$

The V_A, V_K, and V_G voltages are all negative with respect to 0 V. The voltages V_{GA} and V_{KG}, shown on the left of Figure 5-22(b), will be the potential difference between the two electrodes. Therefore, V_{KG} is equal to the difference between V_K and V_G (the difference between −600 V and −555 V) and is equal to 45 V ($V_{KG} = 45\text{ V} = V_{R3}$). The voltage between the CRT's grid and anode (V_{GA}) is equal to the difference between V_G and V_A (the difference between −600 V and −300 V), which will be 300 V ($V_{GA} = 300\text{ V} = V_{R1}$).

EXAMPLE:

Figure 5-23(a) shows a 24 V voltage source driving a 10 Ω resistor that is located 1000 ft from the battery. If two 1000 ft lengths of AWG No. 13 wire are used to connect the source to the load, what will be the voltage applied across the load?

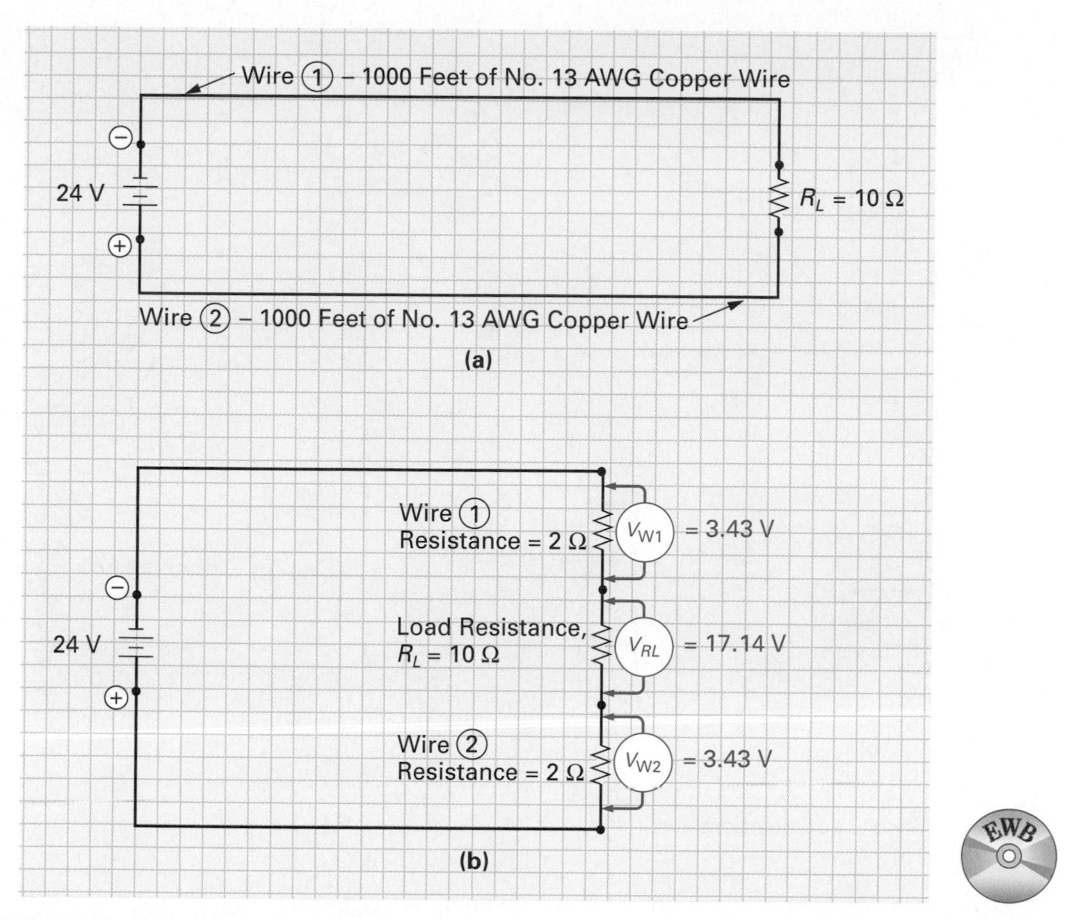

FIGURE 5-23 Series Wire Resistance.

Solution:

Referring back to Table 2-2, you can see that AWG No. 13 copper cable has a resistance of 2.003 Ω for every 1000 ft. To be more accurate, this means that our circuit should be redrawn as shown in Figure 5-23(b) to show the series resistances of wire 1 and wire 2. Using the voltage-divider formula, we can calculate the voltage drop across wire 1 and wire 2.

$$V_{W1} = \frac{R_{W1}}{R_T} \times V_T$$

$$= \frac{2\ \Omega}{14\ \Omega} \times 24\ \text{V} = 3.43\ \text{V}$$

Since the voltage drop across wire 2 will also be 3.43 V, the total voltage drop across both wires will be 6.86 V. The remainder, 17.14 V (24 V − 6.86 V = 17.14 V), will appear across the load resistor, R_L.

5-4-3 *Variable Voltage Divider*

When discussing variable-value resistors in Chapter 3, we talked about a potentiometer, or variable voltage divider, which consists of a fixed value of resistance between two terminals and a wiper that can be adjusted to vary resistance between its terminal and one of the other two.

To review, Figure 5-24(a) shows the potentiometer's schematic symbol and physical appearance. If the wiper of the potentiometer is moved down, as seen in Figure 5-24(b), the resistance between terminals *A* and *B* will increase, while the resistance between *B* and *C* will decrease. On the other hand, if the wiper is moved up, as seen in Figure 5-24(c), the resistance

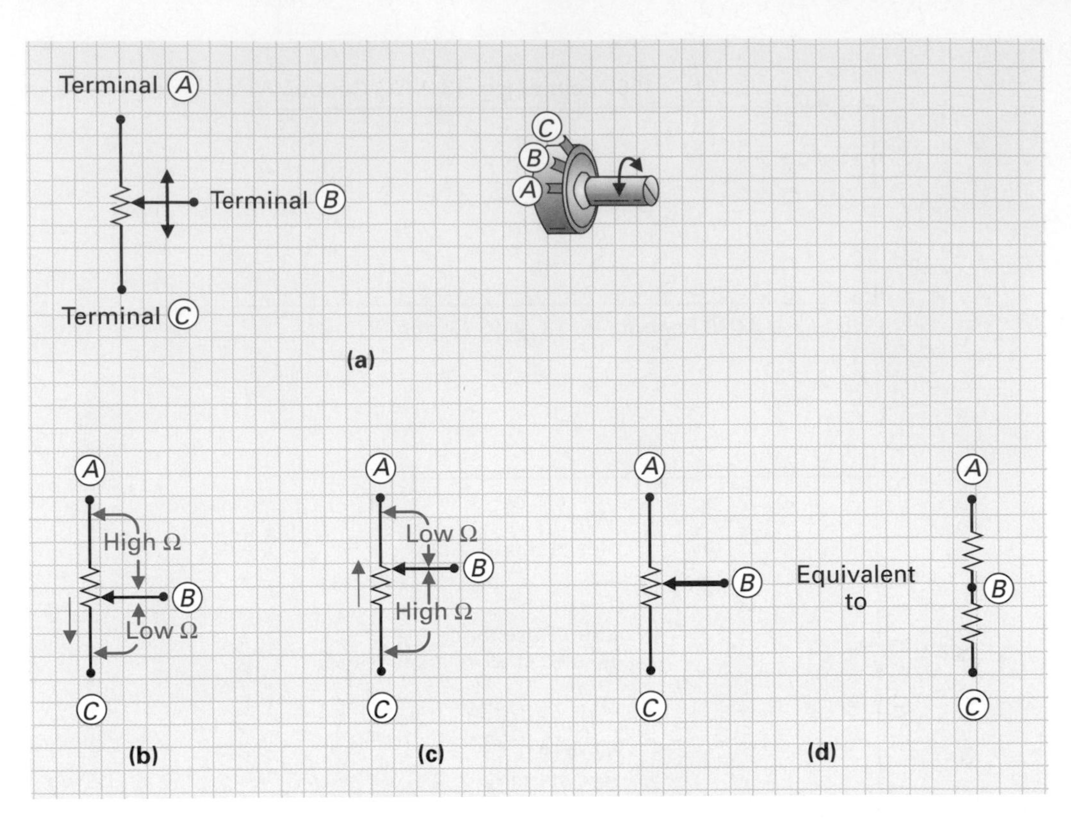

FIGURE 5-24 Potentiometer.

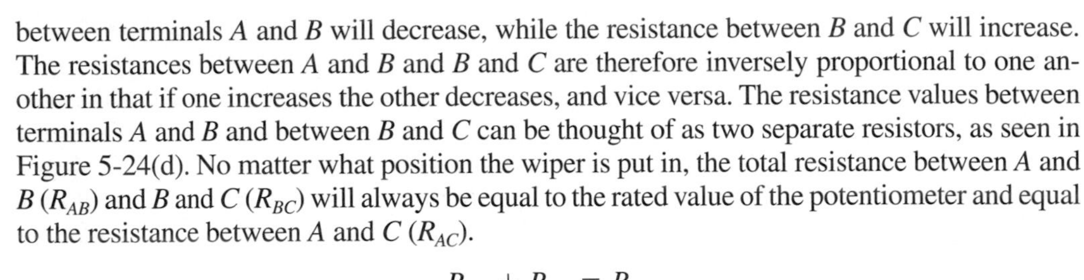

between terminals A and B will decrease, while the resistance between B and C will increase. The resistances between A and B and B and C are therefore inversely proportional to one another in that if one increases the other decreases, and vice versa. The resistance values between terminals A and B and between B and C can be thought of as two separate resistors, as seen in Figure 5-24(d). No matter what position the wiper is put in, the total resistance between A and B (R_{AB}) and B and C (R_{BC}) will always be equal to the rated value of the potentiometer and equal to the resistance between A and C (R_{AC}).

$$R_{AB} + R_{BC} = R_{AC}$$

As an example, Figure 5-25(a) illustrates a 10 kΩ potentiometer that has been hooked up across a 10 V dc source with a voltmeter between terminals B and C. If the wiper terminal is positioned midway between A and C, the voltmeter should read 5 V, and the potentiometer will be equivalent to two 5 kΩ resistors in series, as shown in Figure 5-25(b). Kirchhoff's voltage

TIME LINE

Samuel B. Morse (1791–1872) was an American painter and inventor who, independent of similar efforts in Europe, developed an electric telegraph (1832–35). In 1838 he developed the Morse Code.

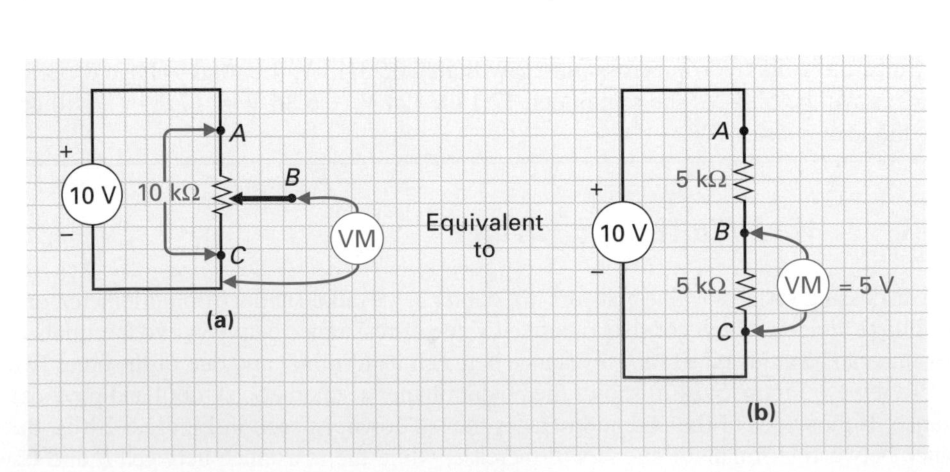

FIGURE 5-25 Potentiometer Wiper in Mid-Position.

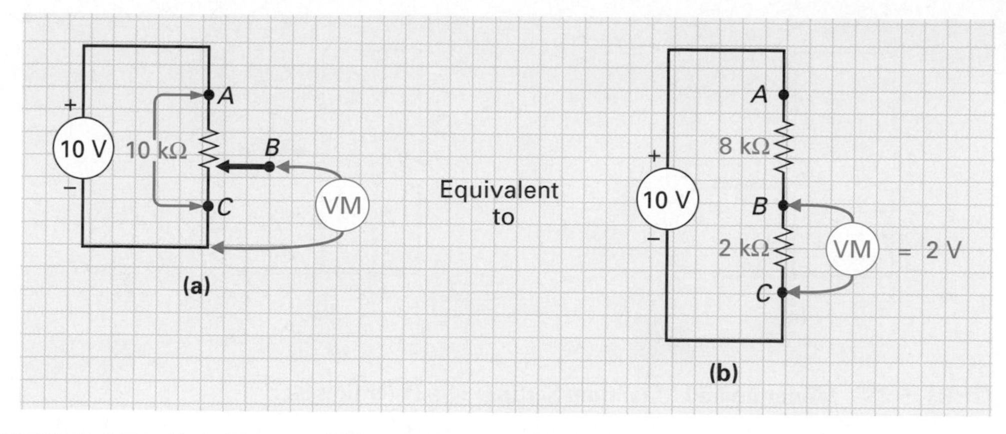

FIGURE 5-26 **Potentiometer Wiper in Lower Position.**

law states that the entire source voltage will be dropped across the resistances in the circuit, and since the resistance values are equal, each will drop half of the source voltage, that is, 5 V.

In Figure 5-26(a), the wiper has been moved down so that the resistance between *A* and *B* is equal to 8 kΩ and the resistance between *B* and *C* equals 2 kΩ. This will produce 2 V on the voltmeter, as shown in Figure 5-26(b). The amount of voltage drop is proportional to the resistance, so a larger voltage will be dropped across the larger resistance. Using the voltage-divider formula, you can calculate that the 8 kΩ resistance is 80% of the total resistance and therefore will drop 80% of the voltage:

$$V_{AB} = \frac{R_{AB}}{R_T} \times V_S = \frac{8\ \text{k}\Omega}{10\ \text{k}\Omega} \times 10\ \text{V} = 8\ \text{V}$$

The 2 kΩ resistance between *B* and *C* is 20% of the total resistance and consequently will drop 20% of the total voltage:

$$V_{BC} = \frac{R_{BC}}{R_T} \times V_S = \frac{2\ \text{k}\Omega}{10\ \text{k}\Omega} \times 10\ \text{V} = 2\ \text{V}$$

In Figure 5-27(a), the wiper has been moved up and now 2 kΩ exists between *A* and *B*, and 8 kΩ is present between *B* and *C*. In this situation, 2 V will be dropped across the 2 kΩ between *A* and *B*, and 8 V will be dropped across the 8 kΩ between *B* and *C*, as shown in Figure 5-27(b).

From this discussion, you can see that the potentiometer can be adjusted to supply different voltages on the wiper. This voltage can be decreased by moving the wiper down to

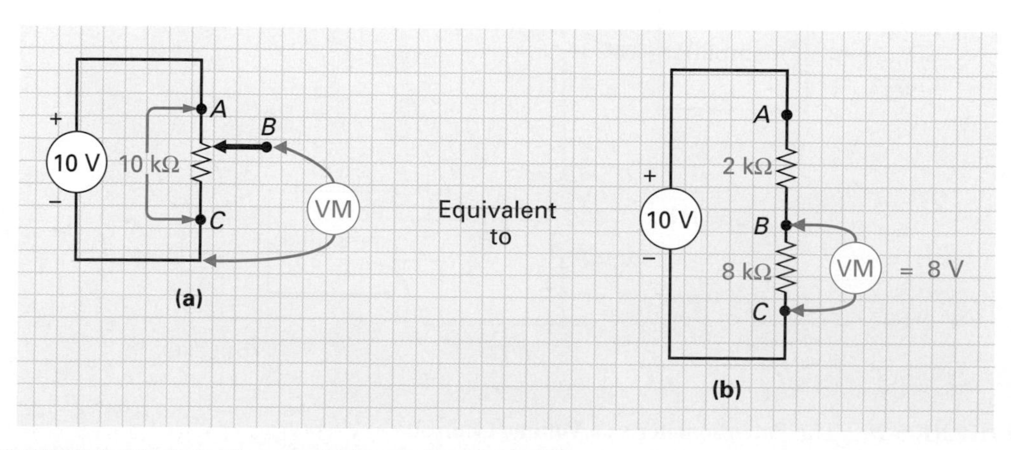

FIGURE 5-27 **Potentiometer Wiper in Upper Position.**

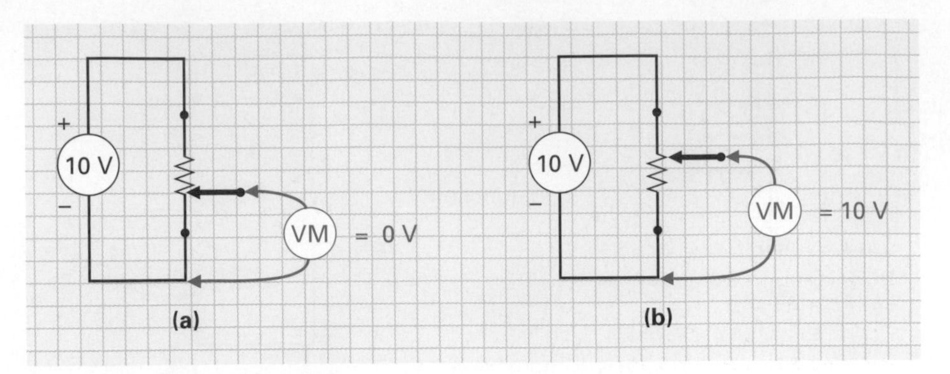

FIGURE 5-28 Minimum and Maximum Settings of a Potentiometer.

supply a minimum of 0 V, as shown in Figure 5-28(a), or the wiper can be moved up to supply a maximum of 10 V, as shown in Figure 5-28(b). By adjusting the wiper position, the potentiometer can be made to deliver any voltage between its maximum and minimum value, which is why the potentiometer is known as a variable voltage divider.

EXAMPLE:

Figure 5-29 illustrates how a potentiometer can be used to control the output volume of an amplifier that is being driven by a compact disc (CD) player. The preamplifier is producing an output of 2 V, which is developed across a 50 kΩ potentiometer. If the wiper of the potentiometer is in its upper position, the full 2 V from the preamp will be applied to the input of the power amplifier. The power amplifier has a fixed voltage gain (A_V) of 12, and therefore the power amplifier's output is always 12 times larger than the input voltage. An input of 2 V (V_{in}) will therefore produce an output voltage (V_{out}) of 24 V ($V_{out} = V_{in} \times A_V = 2\text{ V} \times 12 = 24\text{ V}$).

As the wiper is moved down, less of the 2 V from the preamp will be applied to the power amplifier, and therefore the output of the power amplifier and volume of the music heard will decrease. If the wiper of the potentiometer is adjusted so that a resistance of 20 kΩ exists between the wiper and the lower end of the potentiometer, what will be the input voltage to the power amplifier and output voltage to the speaker?

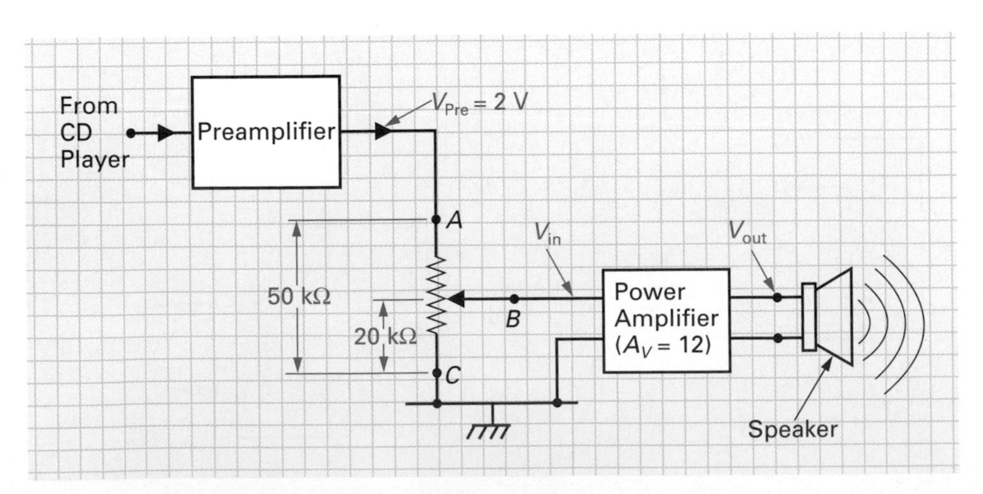

FIGURE 5-29 The Potentiometer as a Volume Control.

■ *Solution:*

By using the voltage-divider formula, we can determine the voltage developed across the potentiometer with 20 kΩ of resistance between the wiper (*B*) and lower end (*C*).

$$V_{in} = \frac{R_{AB}}{R_{AC}} \times V_{Pre}$$

$$= \frac{20\ \text{k}\Omega}{50\ \text{k}\Omega} \times 2\ \text{V} = 0.8\ \text{V}\ (800\ \text{mV})$$

The voltage at the input of the power amplifier (V_{in}) is applied to the input of the power amplifier, which has an amplification factor or gain of 12, and therefore the output voltage will be

$$V_{out} = V_{in} \times A_V$$

$$= 0.8\ \text{V} \times 12 = 9.6\ \text{V}$$

SELF-TEST EVALUATION POINT FOR SECTION 5-4

Now that you have completed this section, you should be able to:

Objective 7. Describe why the series circuit is known as a voltage divider.

Objective 8. Describe a fixed and a variable voltage divider.

Use the following questions to test your understanding of Section 5-4:

1. True or false: A series circuit is also known as a voltage-divider circuit.
2. True or false: The voltage drop across a series resistor is proportional to the value of the resistor.
3. If 6 Ω and 12 Ω resistors are connected across a dc 18 V supply, calculate I_T and the voltage drop across each.
4. State the voltage-divider formula.
5. Which component can be used as a variable voltage divider?
6. Could a rheostat be used in place of a potentiometer?

5-5 POWER IN A SERIES CIRCUIT

As discussed earlier in Chapter 3, power is the rate at which work is done, and work is said to have been done when energy is converted from one energy form to another. Resistors convert electrical energy into heat energy, and the rate at which they dissipate energy is called *power* and is measured in *watts* (joules per second). Resistors all have a resistive value, a tolerance, and a wattage rating. The wattage of a resistor is the amount of heat energy a resistor can safely dissipate per second, and this wattage is directly proportional to the resistor's size. Figure 5-30 reviews the size versus wattage rating of commercially available resistors. As you know, any of the power formulas can be used to calculate wattage. Resistors are manufactured in several different physical sizes, and if, for example, it is calculated that for a certain value of current

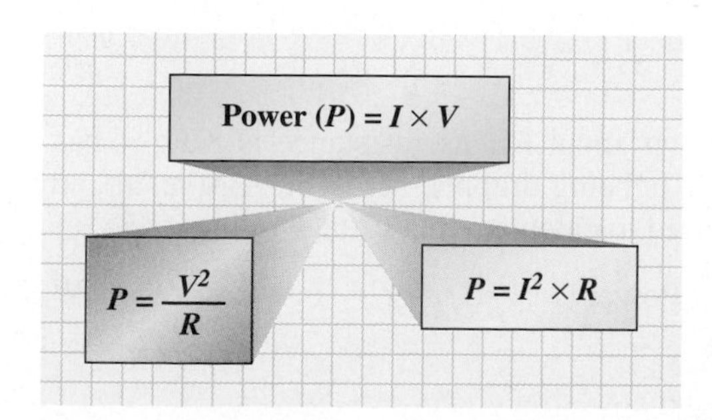

FIGURE 5-30 **Resistor Wattage Ratings.**

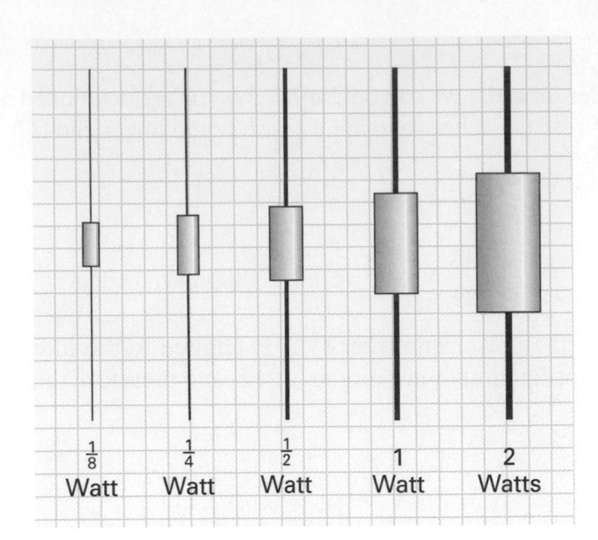

and voltage a 5 W resistor is needed and a ½ W resistor is put in its place, the ½ W resistor will burn out, because it is generating heat (5 W) faster than it can dissipate heat (½ W). A 10 W or 25 W resistor or greater could be used to replace a 5 W resistor, but anything less than a 5 W resistor will burn out. It is, therefore, necessary that we have some way of calculating which wattage rating is needed for each specific application.

A question you may be asking is why not just use large-wattage resistors everywhere. The two disadvantages with this are

1. The larger the wattage, the greater the cost.
2. The larger the wattage, the greater the size and area the resistor occupies within the equipment.

EXAMPLE:

Figure 5-31 shows a 20 V battery driving a 12 V/1 A television set. R_1 is in series with the TV set and is being used to drop 8 V of the 20 V supply, so 12 V will be developed across the television.

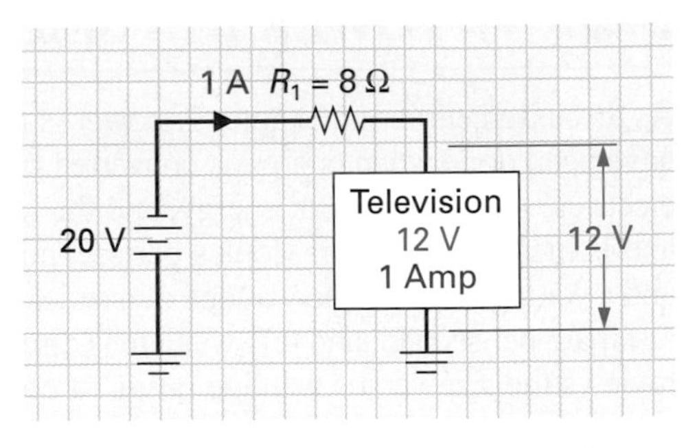

FIGURE 5-31 **Series Circuit Example.**

a. What is the wattage rating for R_1?

b. What is the series load resistance of the TV set?

c. What is the amount of power being consumed by the TV set?

d. Compile a circuit analysis table for this circuit.

Solution:

a. Everything is known about R_1. Its resistance is 8 Ω, it has 1 A of current flowing through it, and 8 V is being dropped across it. As a result, any one of the three power formulas can be used to calculate the wattage rating of R_1.

$$\text{Power } (P) = I \times V = 1\text{ A} \times 8\text{ V} = 8\text{ W}$$

or

$$P = I^2 \times R = 1^2 \times 8 = 8\text{ W}$$

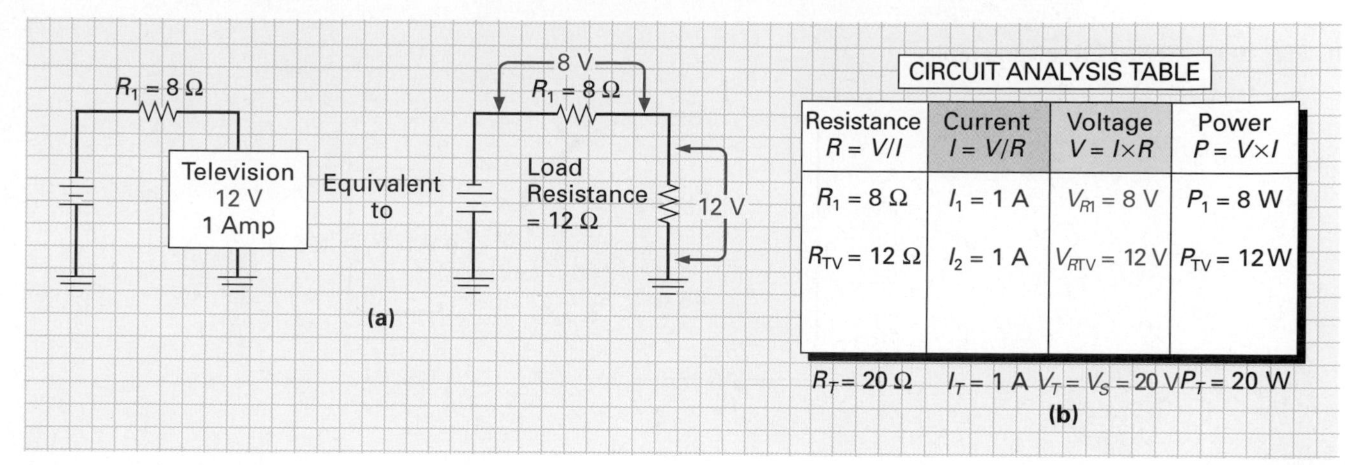

Resistance $R = V/I$	Current $I = V/R$	Voltage $V = I \times R$	Power $P = V \times I$
$R_1 = 8\ \Omega$	$I_1 = 1$ A	$V_{R1} = 8$ V	$P_1 = 8$ W
$R_{TV} = 12\ \Omega$	$I_2 = 1$ A	$V_{RTV} = 12$ V	$P_{TV} = 12$ W
$R_T = 20\ \Omega$	$I_T = 1$ A	$V_T = V_S = 20$ V	$P_T = 20$ W

FIGURE 5-32 Series Circuit Example with Values Inserted. (a) Schematic. (b) Circuit Analysis Table.

or

$$P = \frac{V^2}{R} = \frac{8^2}{8} = 8 \text{ W}$$

The nearest commercially available device would be a 10 W resistor. If size is not a consideration, it is ideal to double the wattage needed and use a 16 W resistor.

b. You may recall that any piece of equipment is equivalent to a load resistance. The TV set has 12 V across it and is pulling 1 A of current. Its load resistance can be calculated simply by using Ohm's law and deriving an equivalent circuit, as shown in Figure 5-32(a).

$$\begin{aligned} R_L \text{(load resistance)} &= \frac{V}{I} \\ &= \frac{12 \text{ V}}{1 \text{ A}} \\ &= 12\ \Omega \end{aligned}$$

c. The amount of power being consumed by the TV set is

$$P = V \times I = 12 \text{ V} \times 1 \text{ A} = 12 \text{ W}$$

d. Figure 5-32(b) shows the circuit analysis table for this example.

EXAMPLE:

Calculate the total amount of power dissipated in the series circuit in Figure 5-33, and insert any calculated values in a circuit analysis table.

Solution:

The total power dissipated in a series circuit is equal to the sum of all the power dissipated by all the resistors. The easiest way to calculate the total power is to simplify the circuit to one resistance, as shown in Figure 5-33.

$$\begin{aligned} R_T &= R_1 + R_2 + R_3 + R_4 \\ &= 5\ \Omega + 33\ \Omega + 45\ \Omega + 75\ \Omega \\ &= 158\ \Omega \end{aligned}$$

FIGURE 5-33 Series Circuit Example.

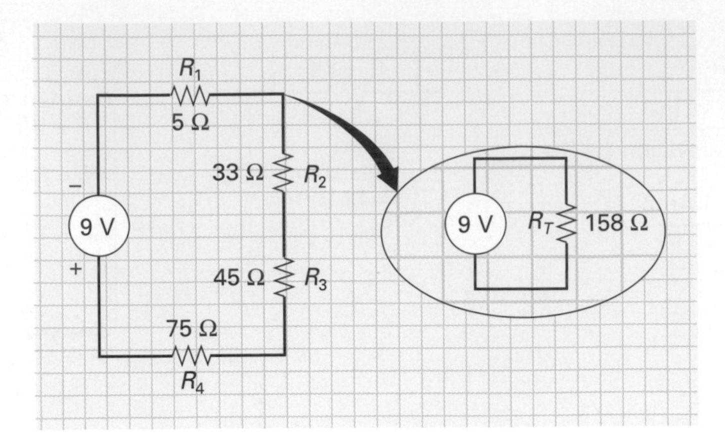

We now have total resistance and total voltage, so we can calculate the total power:

$$P_T = \frac{V_S^{\,2}}{R_T} = \frac{(9\text{ V})^2}{158\ \Omega} = \frac{81}{158} = 512.7\text{ milliwatts (mW)}$$

The longer method would have been to first calculate the current through the series:

$$I = \frac{V_S}{R_T} = \frac{9\text{ V}}{158\ \Omega} = 56.96\text{ mA} \quad \text{or} \quad 57\text{ mA}$$

We could then calculate the power dissipated by each separate resistor and add up all the individual values to gain a total power figure. This is illustrated in Figure 5-34(a).

$$P_T = P_1 + P_2 + P_3 + P_4 + \cdots$$

$$\text{Total power} = \text{addition of all the individual power losses}$$

$$P_T = 16\text{ mW} + 107\text{ mW} + 146\text{ mW} + 243\text{ mW}$$

$$= 512\text{ mW}$$

The calculated values for this example are shown in the circuit analysis table in Figure 5-34(b).

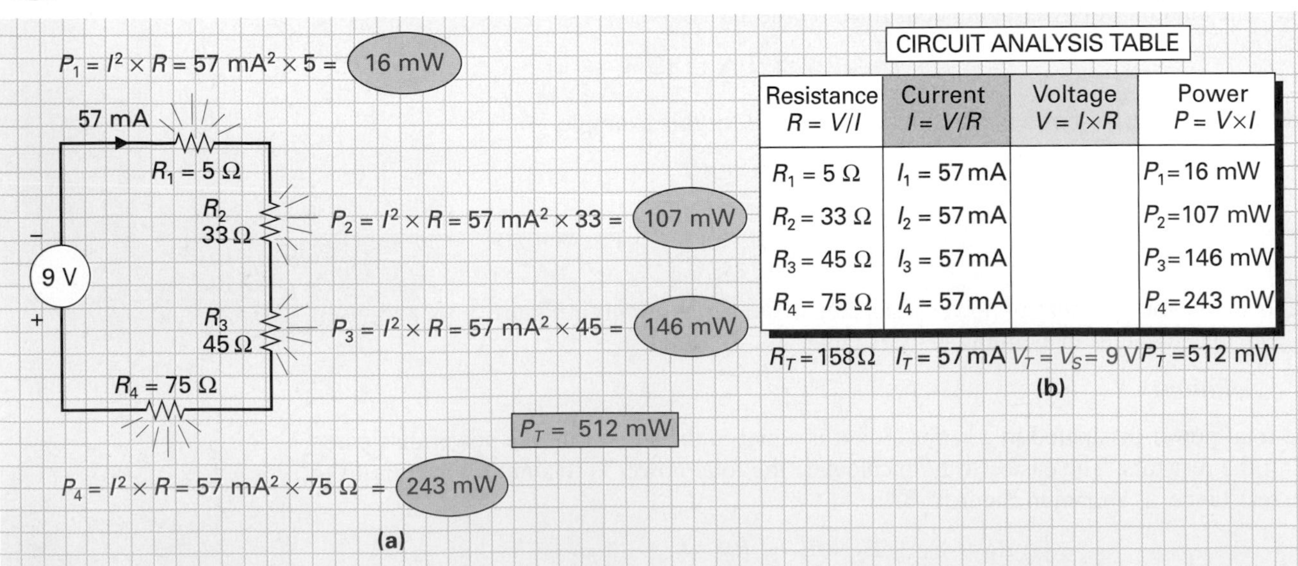

Resistance $R = V/I$	Current $I = V/R$	Voltage $V = I \times R$	Power $P = V \times I$
$R_1 = 5\ \Omega$	$I_1 = 57$ mA		$P_1 = 16$ mW
$R_2 = 33\ \Omega$	$I_2 = 57$ mA		$P_2 = 107$ mW
$R_3 = 45\ \Omega$	$I_3 = 57$ mA		$P_3 = 146$ mW
$R_4 = 75\ \Omega$	$I_4 = 57$ mA		$P_4 = 243$ mW
$R_T = 158\ \Omega$	$I_T = 57$ mA	$V_T = V_S = 9$ V	$P_T = 512$ mW

FIGURE 5-34 Series Circuit Example with Values Inserted. (a) Schematic. (b) Circuit Analysis Table.

5-5-1 *Maximum Power Transfer*

The **maximum power transfer theorem** states that maximum power will be delivered to a load when the resistance of the load (R_L) is equal to the resistance of the source (R_S). The best way to see if this theorem is correct is to apply it to a series of examples and then make a comparison.

Maximum Power Transfer Theorem
A theorem that states maximum power will be transferred from source to load when the source resistance is equal to the load resistance.

EXAMPLE:

Figure 5-35(a) illustrates a 10 V battery with a 5 Ω internal resistance connected across a load. The load in this case is a lightbulb, which has a load resistance of 1 Ω. Calculate the power delivered to this lightbulb or load when $R_L = 1\ \Omega$.

Solution:

$$I\,(\text{circuit current}) = \frac{V}{R} = \frac{10\text{ V}}{R_S + R_L} = \frac{10\text{ V}}{6\ \Omega} = 1.66\text{ A}$$

The power supplied to the load is $P = I^2 \times R = 1.66^2 \times 1\ \Omega = 2.8\text{ W}$

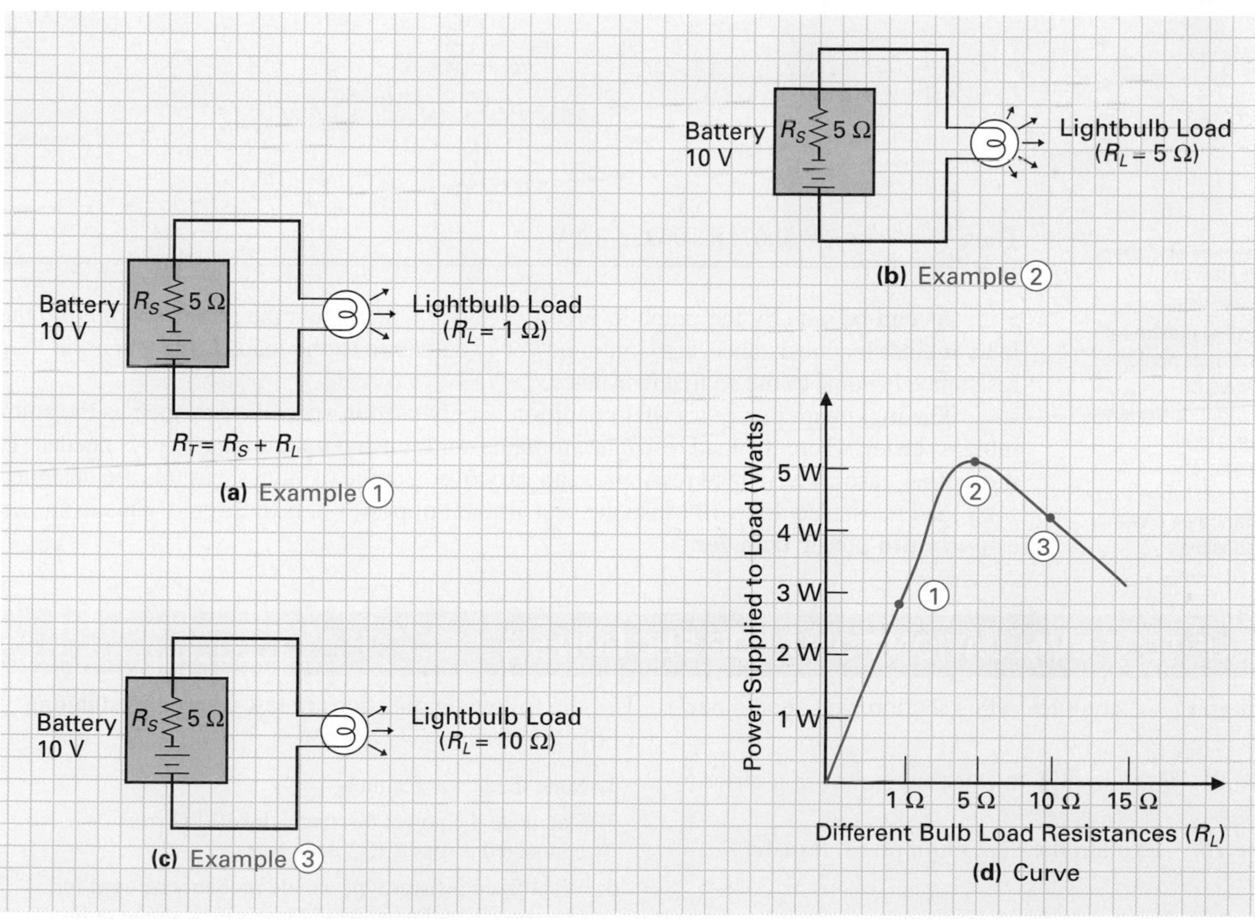

FIGURE 5-35 Maximum Power Transfer.

EXAMPLE:

Figure 5-35(b) illustrates the same battery and R_S, but in this case connected across a 5 Ω lightbulb. Calculate the power delivered to this lightbulb now that $R_L = 5\ \Omega$.

Solution:

$$I = \frac{V}{R}$$
$$= \frac{10\ \text{V}}{R_S + R_L}$$
$$= \frac{10\ \text{V}}{10\ \Omega}$$
$$= 1\ \text{A}$$

Power supplied is $P = I^2 \times R = 1^2 \times 5\ \Omega = 5\ \text{W}$.

EXAMPLE:

Figure 5-35(c) illustrates the same battery again, but in this case a 10 Ω lightbulb is connected in the circuit. Calculate the power delivered to this lightbulb now that $R_L = 10\ \Omega$.

Solution:

$$I = \frac{V}{R}$$
$$= \frac{10\ \text{V}}{R_S + R_L}$$
$$= \frac{10\ \text{V}}{15\ \Omega}$$
$$= 0.67\ \text{A}$$

Thus, $P = I^2 \times R = 0.67^2 \times 10\ \Omega = 4.5\ \text{W}$.

Optimum Power Transfer

Since the ideal maximum power transfer conditions cannot always be achieved, most designers try to achieve optimum power transfer and have the source resistance and load resistance as close in value as possible.

As can be seen by the graph in Figure 5-35(d), which plots the power supplied to the load against load resistance, maximum power is delivered to the load (5 W) when the load resistance is equal to the source resistance.

The maximum power transfer condition is only used in special cases such as the automobile starter, where the load resistance remains constant and maximum power is needed. In most other cases, where load resistance can vary over a range of values, circuits are designed for a load resistance that will cause the best amount of power to be delivered. This is known as **optimum power transfer**.

SELF-TEST EVALUATION POINT FOR SECTION 5-5

Now that you have completed this section, you should be able to:

Objective 9. Explain how to calculate power in a series circuit.

Objective 10. Explain the maximum power transfer theorem.

Use the following questions to test your understanding of Section 5-5:

1. State the power formula.
2. Calculate the power dissipated by a 12 Ω resistor connected across a 12 V supply.
3. What fixed resistor type should probably be used for Question 2, and what would be a safe wattage rating?
4. What would be the total power dissipated if R_1 dissipates 25 W and R_2 dissipates 3800 mW?

5-6 TROUBLESHOOTING A SERIES CIRCUIT

A resistor will usually burn out and cause an open between its two leads when an excessive current flow occurs. This can normally, but not always, be noticed by a visual check of the resistor, which will appear charred due to the excessive heat. In some cases, you will need to use your multimeter (combined ammeter, voltmeter, and ohmmeter) to check the circuit components to determine where a problem exists.

The two basic problems that normally exist in a series circuit are opens and shorts. In most instances, a problem is not always as drastic as a short or an open, but may be a variation in a component's value over a long period of time, which will eventually cause a problem.

To summarize, then, we can say that one of three problems can occur with components in a series circuit:

1. A component will open (infinite resistance).
2. A component's value will change over a period of time.
3. A component will short (zero resistance).

The voltmeter is the most useful tool for checking series circuits as it can be used to measure voltage drops by connecting the meter leads across the component or resistor. Let's now analyze a circuit problem and see if we can solve it by logically **troubleshooting** the circuit and isolating the faulty component. To begin with, let us take a look at the effects of an open component.

Troubleshooting
The process of locating and diagnosing malfunctions or breakdowns in equipment by means of systematic checking or analysis.

5-6-1 *Open Component in a Series Circuit*

A component is open when its resistance is the maximum possible (infinity).

EXAMPLE:

Figure 5-36(a) illustrates a TV set with a load resistance of 3 Ω. The TV set is off because R_2 has burned out and become an open circuit. How would you determine that the problem is R_2?

Solution:

If an open circuit ever occurs in a series circuit, due in this case to R_2 having burned out, there can be no current flow, because series circuits have only one path for current to flow and that path has been broken ($I = 0$ A). Using the voltmeter to check the amount of voltage drop across each resistor, two results will be obtained:

1. The voltage drop across a good resistor will be zero volts.
2. The voltage drop across an open resistor will be equal to the source voltage, V_S.

No voltage will be dropped across a good resistor because current is zero, and if $I = 0$, the voltage drop, which is the product of I and R, must be zero ($V = I \times R = 0 \times R = 0$ V). If no voltage is being dropped across the good resistor R_1 and the TV set resistance of 3 Ω, the entire source voltage will appear across the open resistor, R_2, in order that this series circuit comply with Kirchhoff's voltage law: V_S (9 V) $= V_{R1}$ (0 V) $+ V_{R2}$ (9 V) $+ V_L$ (0 V).

To explain this point further, refer to the fluid analogy in Figure 5-36(b). Like R_2, valve 2 has completely blocked any form of flow (water flow = 0). Looking at the pressure differences across all three valves, you can see that no pressure difference occurs across valves 1 and 3, but the entire pump pressure is appearing across valve 2, which is the component that has opened the circuit.

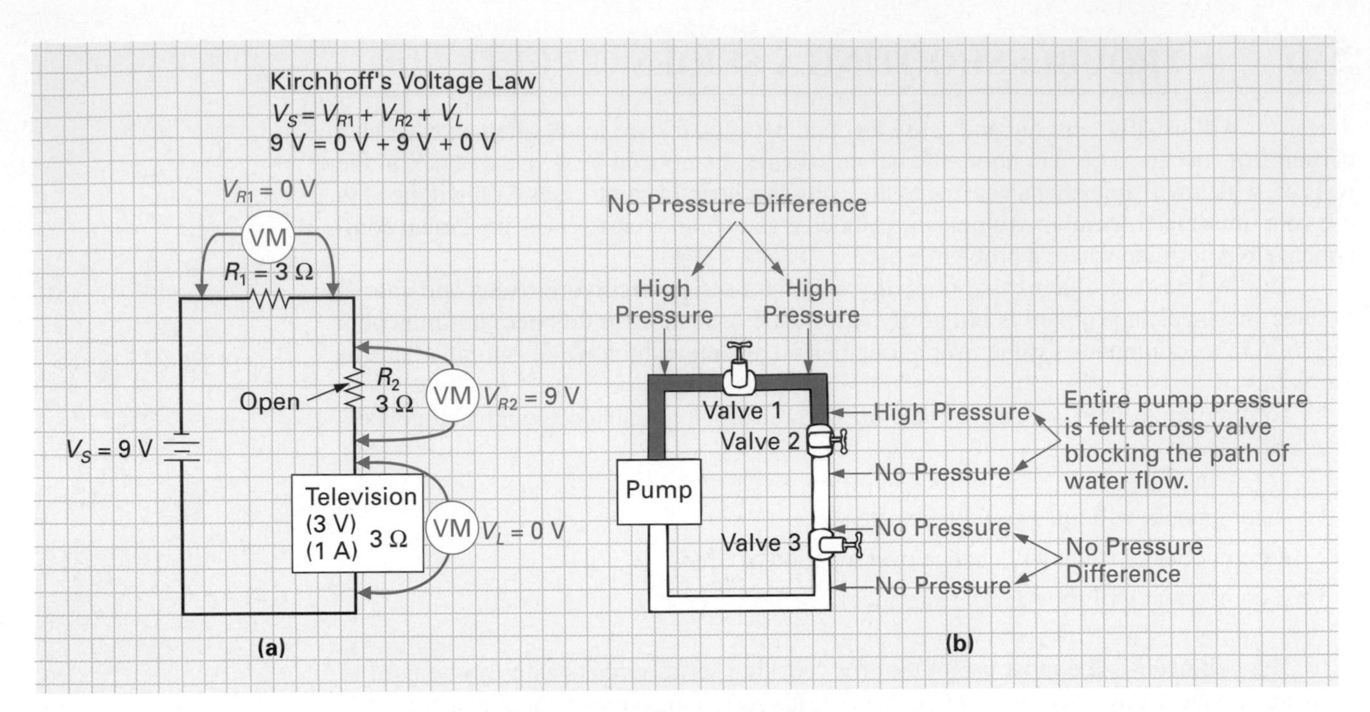

FIGURE 5-36 **Troubleshooting an Open in a Series Circuit.**

EXAMPLE:

Figure 5-37 illustrates a set of three lights connected across a 9 V battery. Bulb 3 is open and therefore there is no current flow. With no current flow, all three bulbs are off, and we need to isolate which of the three is faulty.

Solution:

Using the voltmeter, there will be

1. Zero volts dropped across bulb 1 (9 V appears on both sides of bulb 1, so the potential difference or voltage drop across bulb 1 is zero), and so bulb 1 is OK.

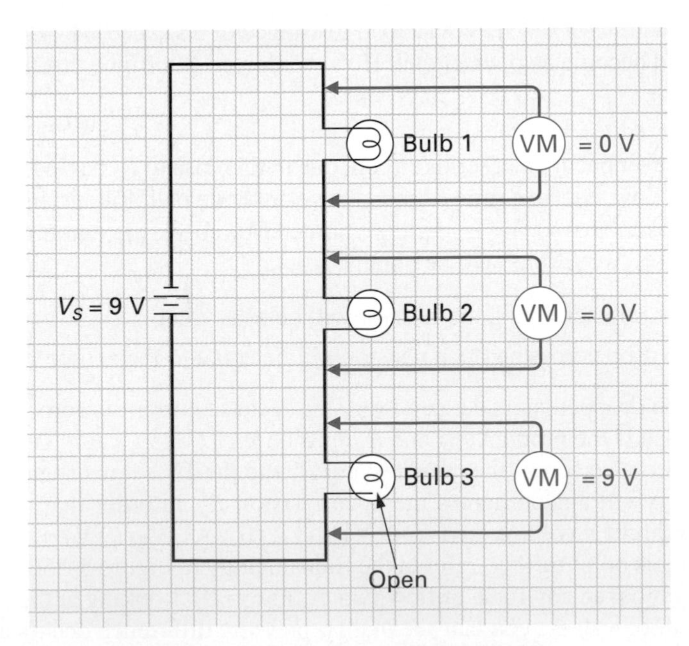

FIGURE 5-37 **Troubleshooting an Open Bulb in a Series Circuit.**

2. Zero volts dropped across bulb 2, and so bulb 2 is OK.
3. Nine volts dropped across bulb 3 (the entire source voltage is being dropped across bulb 3), and so bulb 3 is open and needs to be replaced.

The series-connected two-wire holiday light set has a unique way of overcoming the open circuit problem. Figure 5-38 shows a 50-string holiday light set. The flasher bulb turns the lights ON and OFF due to the heating and cooling action of a bimetal strip. The lights are ON as current passes through the bimetal strip, but in time the heated bimetal strip will move away from a center contact, break the current path, and turn OFF all the lights. As the bimetal strip cools, it returns to its original position, reestablishes contact, the lights turn ON, and the cycle repeats.

The standard bulb has a coil of fuse wire wrapped around the base of its terminals. This fuse wire is coated with an insulating varnish that prevents contact, and therefore will not short out the bulb filament. If the bulb burns out, the full 120 V will appear across the bad bulb, and this large potential difference from one terminal to the other will break down the insulating coating on the fuse wire, cause a current through the wire, which will melt, and provide a permanent connection between the posts and so bypass the bad bulb. The bad bulb is easily identified and should be replaced before other bulbs burn out and are bypassed. If this is not done, a large voltage will appear across each of the remaining good bulbs, and an increase in current due to the removal of filament resistors will, at some point, rapidly burn out all of the still-functioning bulbs.

5-6-2 *Component Value Variation in a Series Circuit*

Resistors will rarely go completely open unless severely stressed because of excessive current flow. With age, resistors will normally change their resistance value. This occurs slowly and will generally cause a decrease in the resistor's resistance and eventually cause a circuit

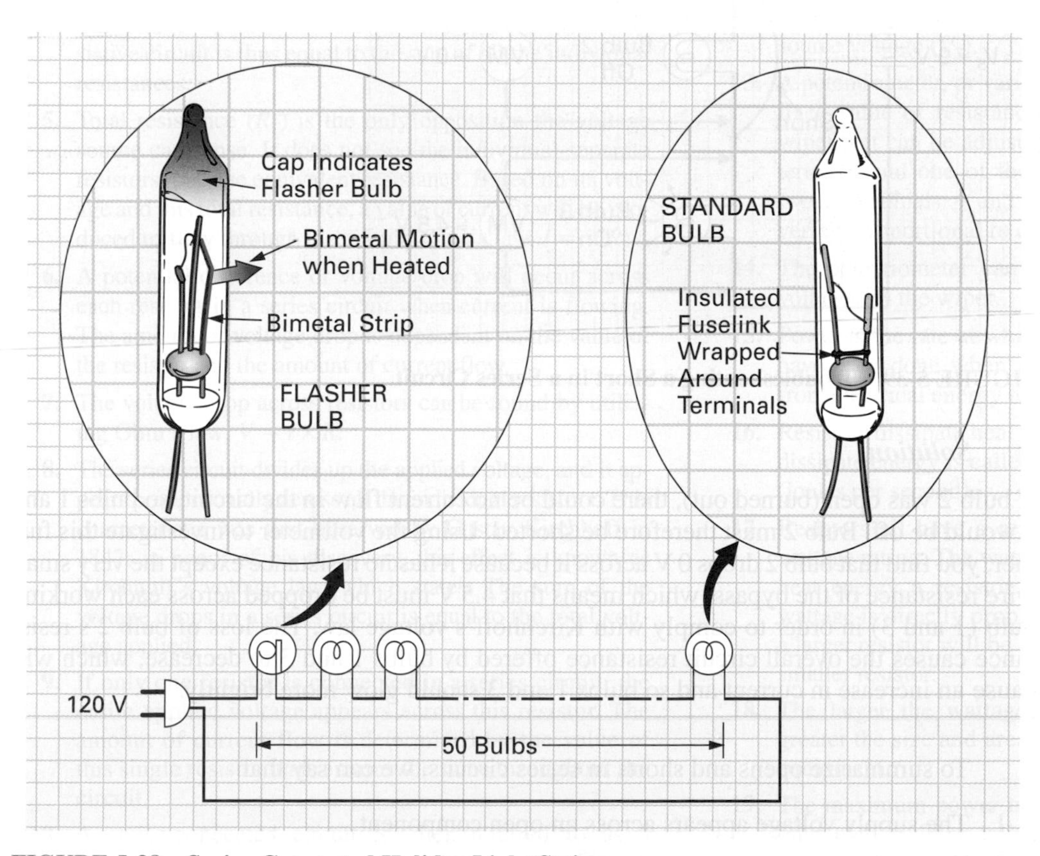

FIGURE 5-38 Series-Connected Holiday Light String.

20. A resistor will usually burn out and result in an open between its two leads when an excessive current flow occurs. This can normally, but not always, be noticed by a visual check of the resistor, which will appear charred due to the excessive heat. In some cases, you will need to use your multimeter to check the circuit components to determine where a problem exists.
21. The voltmeter is the most useful tool when checking series circuits as it can be used to measure voltage drops by connecting the meter leads across the component or resistors.
22. Using the voltmeter to check the amount of voltage drop across each resistor, two results will be obtained:
    - **a.** The voltage drop across a good resistor will be zero volts.
    - **b.** The voltage drop across an open resistor will be equal to the source voltage, V_S.
23. With age, resistors will normally change their resistance value. This occurs slowly and will generally always cause a decrease in the resistor's resistance and eventually cause a circuit problem. This lowering of resistance will cause an increase of current, which will cause an increase in the power dissipated; if the wattage of the resistor is exceeded, it can burn out. If they do not burn out but merely blow the circuit fuse due to this increase in current, the problem can be found by measuring the resistance values of each resistor or by measuring how much voltage is dropped across each resistor and comparing these to the calculated voltage, based on the parts list supplied by the manufacturer.
24. To summarize opens and shorts in series circuits, we can say that
    - **a.** The supply voltage appears across an open component.
    - **b.** Zero volts appears across a shorted component.

REVIEW QUESTIONS

Multiple-Choice Questions

1. A series circuit
a. Is the connecting of components end to end
b. Provides a single path for current
c. Functions as a voltage divider
d. All of the above

2. The total current in a series circuit is equal to
a. $I_1 + I_2 + I_3 + \cdots$ **c.** $I_1 = I_2 = I_3 = \cdots$
b. $I_1 - I_2$ **d.** All of the above

3. If R_1 and R_2 are connected in series with a total current of 2 A, what will be the current flowing through R_1 and R_2, respectively?
a. 1 A, 1 A
b. 2 A, 1 A
c. 2 A, 2 A
d. All of the above could be true on some occasions.

4. The total resistance in a series circuit is equal to
a. The total voltage divided by the total current
b. The sum of all the individual resistor values
c. $R_1 + R_2 + R_3 + \cdots$
d. All of the above
e. None of the above are even remotely true.

5. Which of Kirchhoff's laws applies to series circuits:
a. His voltage law
b. His current law
c. His power law
d. None of them apply to series circuits, only parallel circuits.

6. The amount of voltage dropped across a resistor is proportional to
a. The value of the resistor
b. The current flow in the circuit
c. Both (a) and (b)
d. None of the above

7. If three resistors of 6 kΩ, 4.7 kΩ, and 330 Ω are connected in series with one another, what total resistance will the battery sense?
a. 11.03 MΩ **c.** 6 kΩ
b. 11.03 Ω **d.** 11.03 kΩ

8. The voltage-divider formula states that the voltage drop across a resistor or multiple resistors in a series circuit is equal to the ratio of that __________ to the __________ multiplied by the __________.
a. Resistance, source voltage, total resistance
b. Resistance, total resistance, source voltage
c. Total current, resistance, total voltage
d. Total voltage, total current, resistance

9. The __________ can be used as a variable voltage divider.
a. Potentiometer **c.** SPDT switch
b. Fixed resistor **d.** None of the above

10. A resistor of larger physical size will be able to dissipate __________ heat than a small resistor.
a. More **c.** About the same
b. Less **d.** None of the above

11. The __________ is the most useful tool when checking series circuits.
a. Ammeter **c.** Voltmeter
b. Wattmeter **d.** Both (a) and (b)

12. When an open component occurs in a series circuit, it can be noticed because
a. Zero volts appears across it.
b. The supply voltage appears across it.
c. 1.3 V appears across it.
d. None of the above

13. Power can be calculated by
a. The addition of all the individual power figures
b. The product of the total current and the total voltage

c. The square of the total voltage divided by the total resistance
d. All of the above

14. A series circuit is known as a
    a. Current divider   c. Current subtractor
    b. Voltage divider   d. All of the above

15. In a series circuit only __________ path(s) exists for current flow, while the voltage applied is distributed across all the individual resistors.
    a. Three   c. Four
    b. Several   d. One

Communication Skill Questions

16. Describe a series-connected circuit. (5-1)
17. Describe and state mathematically what happens to current flow in a series circuit. (5-2)
18. Describe how total resistance can be calculated in a series circuit. (5-3)
19. Describe why voltage is dropped across a series circuit and how each voltage drop can be calculated. (5-4)
20. Briefly describe why resistance and voltage drops are proportional to one another. (5-4)
21. Describe a fixed and a variable voltage divider. (5-4-2 and 5-4-3)
22. How can individual and total power be calculated in a series circuit? (5-5)
23. How can you recognize shorts and opens in a series circuit when troubleshooting with a voltmeter? (5-6)
24. List the three problems that can occur in a series circuit. (5-6)
25. State Kirchhoff's voltage (series circuit) law. (5-4)

Practice Problems

26. If three resistors of 1.7 kΩ, 3.3 kΩ, and 14.4 kΩ are connected in series with one another across a 24 V source as shown in Figure 5-40, calculate:
    a. Total resistance (R_T)
    b. Circuit current
    c. Individual voltage drops
    d. Individual and total power dissipated
27. If 40 Ω and 35 Ω resistors are connected across a 24 V source, what would be the current flow through the resistors, and what resistance would cause half the current to flow?
28. Calculate the total resistance (R_T) of the following series-connected resistors: 2.7 kΩ, 3.4 MΩ, 370 Ω, and 4.6 MΩ.
29. Calculate the value of resistors needed to divide up a 90 V source to produce 45 V and 60 V outputs, with a divider circuit current of 1 A.
30. If R_1 = 4.7 kΩ and R_2 = 6.4 kΩ and both are connected across a 9 V source, how much voltage will be dropped across R_2?
31. What current would flow through R_1 if it were one-third the ohmic value of R_2 and R_3, and all were connected in series with a total current of 6.5 mA flowing out of V_S?

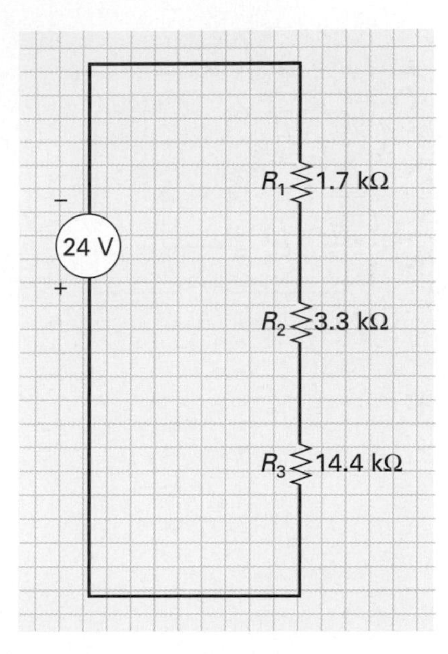

FIGURE 5-40

32. Draw a circuit showing R_1 = 2.7 kΩ, R_2 = 3.3 kΩ, and R_3 = 0.027 MΩ in series with one another across a 20 V source. Calculate:
    a. I_T d. P_2 g. V_{R2}
    b. P_T e. P_3 h. V_{R3}
    c. P_1 f. V_{R1} i. I_1
33. Calculate the current flowing through three lightbulbs that are dissipating 120 W, 60 W, and 200 W when they are connected in series across a 120 V source. How is the voltage divided around the series circuit?
34. If three equal-value resistors are series connected across a dc power supply adjusted to supply 10 V, what percentage of the source voltage will appear across R_1?
35. Refer to the following figures and calculate:
    a. I (Figure 5-41)
    b. R_T and P_T (Figure 5-42)
    c. V_S, V_{R1}, V_{R2}, V_{R3}, V_{R4}, P_1, P_2, P_3, and P_4 (Figure 5-43)
    d. P_T, I, R_1, R_2, R_3, and R_4 (Figure 5-44)

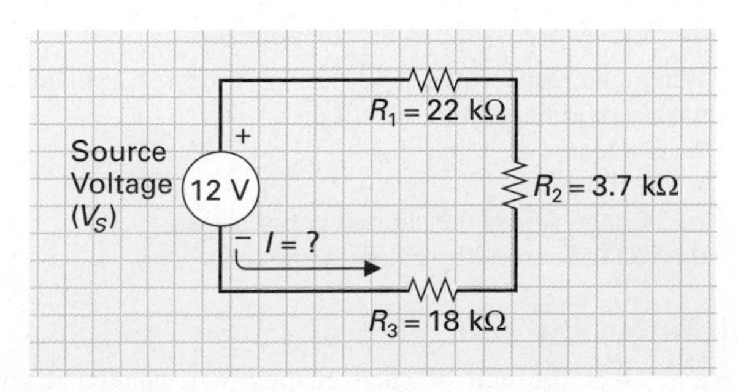

FIGURE 5-41

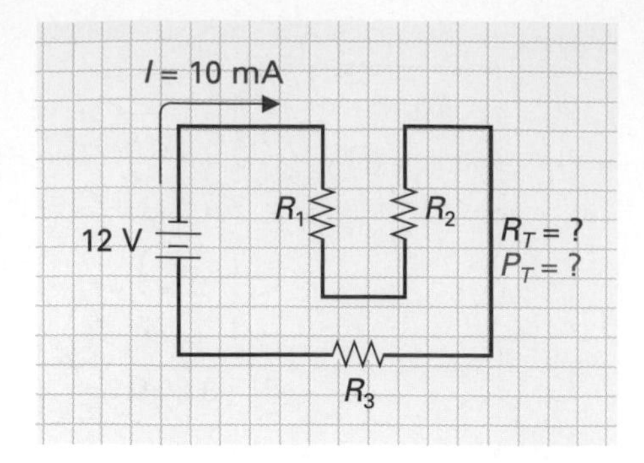

FIGURE 5-42

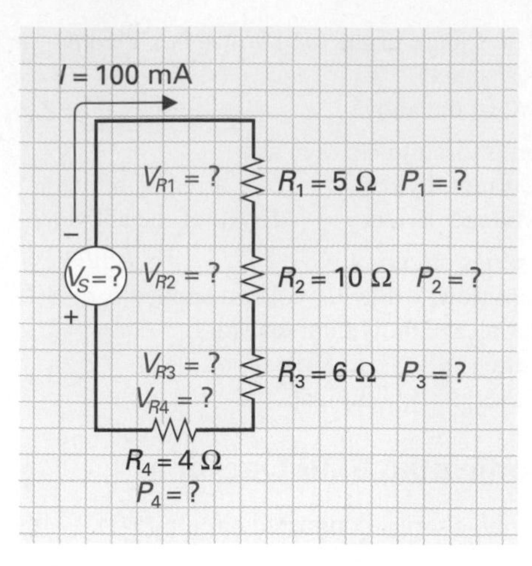

FIGURE 5-43

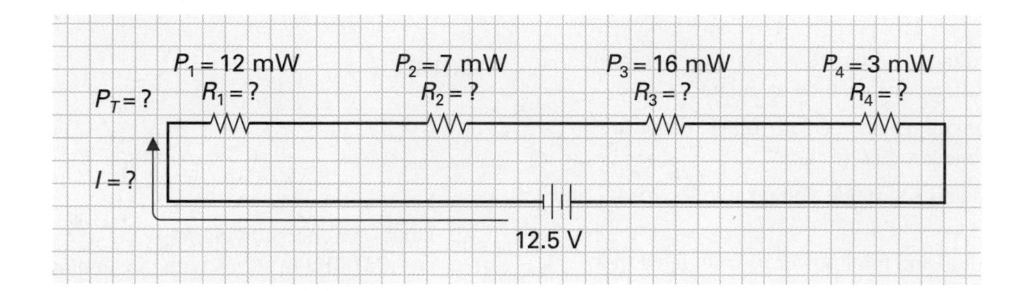

FIGURE 5-44

Troubleshooting Questions

36. If three bulbs are connected across a 9 V battery in series and the filament in one of the bulbs burned out, causing an open in the bulb, would the other lamps be on? Explain why.

37. Using a voltmeter, how would a short be recognized in a series circuit?

38. If one of three series-connected bulbs is shorted, will the other two bulbs be on? Explain why.

39. When one resistor in a series string is open, explain what would happen to the circuit's
  a. Current
  b. Resistance
  c. Voltage across the open component
  d. Voltage across the other components

40. When one resistor in a series string is shorted, explain what would happen to the circuit's
  a. Current
  b. Resistance
  c. Voltage across the shorted component
  d. Voltage across the other components

Job Interview Test

These tests at the end of each chapter will challenge your knowledge up to this point, and give you the practice you need for a job interview. To make this more realistic, the test will be comprised of both technical and personal questions. In order to take full advantage of this exercise, you may want to set up a simulation of the interview environment, have a friend read the questions to you, and record your responses for later analysis.

Company Name: NPC, Inc.

Industry Branch: Consumer Products.

Function: Audio/Video, Sales/Service.

Job Title: Customer Service Technician.

41. What do you know about our product line?

42. What is a series circuit?

43. How do you use a multimeter to measure resistance?

44. Would you object to attending a 6-week out-of-state training course?

45. What originally attracted you to the electronics industry?

46. What is meant by the expression "loading a battery"?

47. Would you say you're a self-starter?

48. What audio/video equipment do you own?

49. What would you say are the duties of a customer service technician?

50. Are you someone who enjoys interaction on a daily basis?

Answers

41. Visit the company's web site before the interview to get an overall understanding of the company's ownership, product line, service, and support.

42. Section 5-1.

43. Chapter 2.

44. Additional training only serves to increase your skills and marketability. If you are unsure, however, don't object to anything at the time. Later, if you decide that it is something you do not want to do, you can call and decline the offer.

45. Describe early interest and reasons for pursuing a career in electronics.

46. Section 5-4-1.

47. Discuss how you have disciplined yourself to manage school, studying, and a job, and how well you performed in all. Also, describe the lab environment, which was more than likely self-paced.

48. Discuss systems and mention any modifications or installations you have performed (such as installing your car's music system, etc.).

49. Quote introductory section in text and job description listed in newspaper.

50. They're asking if you can get along well with customers and colleagues. Mention any work experience in which you had direct contact with customers, and discuss how you worked with fellow students in your class.

Web Site Questions

Go to the web site http://www.prenhall.com/cook, select the textbook *Introductory DC/AC Electronics* or *Introductory DC/AC Circuits*, this chapter, and then follow the instructions when answering the multiple-choice practice problems.

Parallel Circuits

An Apple a Day

One of the early pioneers who laid the foundations of many branches of science was Isaac Newton. He was born in a small farmhouse near Woolsthorpe in Lincolnshire, England, on Christmas Day in 1642. He was an extremely small, premature baby, which worried the midwives who went off to get medicine and didn't expect to find him alive when they came back. Luckily for science, however, he did survive.

Newton's father was an illiterate farmer who died three months before Newton was born. His mother married the local vicar soon after Newton's birth and left him in the care of his grandmother. This parental absence while he was growing up had a traumatic effect on him and throughout his life affected his relationships with people.

At school in the nearby town of Grantham, Newton showed no interest in classical studies but rather in making working models and studying the world around him. When he was in his early teens, his stepfather died and Newton had to return to the farm to help his mother. Newton proved to be a hopeless farmer; in fact, on one occasion when he was tending sheep he became so engrossed with a stream that he followed it for miles and was missing for hours. Luckily, a schoolteacher recognized Newton's single-minded powers of concentration and convinced his mother to let him return to school, where he performed better and later went off to Cambridge University.

In 1665, in his graduation year, Newton left Cambridge to return home to escape an epidemic of bubonic plague that had spread throughout London. During this time, Newton reflected on his years of seclusion at his mother's cottage and called them the most significant time in his life. It was here on a warm summer's day that Newton saw the apple fall to the ground, leading him to develop his laws of motion and gravitation. It was here that he wondered about the nature of light and later built a prism and proved that white light contains all the colors in a rainbow.

Later, when Newton returned to Cambridge, he demonstrated many of his discoveries but was reluctant to publish the details and did so finally only at the insistence of others. Newton went on to build the first working reflecting astronomical telescope and wrote a paper on optics that was fiercely challenged by the physicist Robert Hooke. Hooke quarreled bitterly with Newton over the years, and there were also heated debates about whether Newton or the German mathematician Gottfried Leibniz invented calculus.

The truth is that many of Newton's discoveries roamed around with him in the English countryside, and even though many of these would, before long, have been put forward by others, it was Newton's genius and skill (and long walks) that tied together all the loose ends.

6

Outline and Objectives

Introduction

By tracing the path of current, we can determine whether a circuit has series-connected or parallel-connected components. In a series circuit, there is only one path for current, whereas in parallel circuits the current has two or more paths. These paths are known as *branches*. A parallel circuit, by definition, is the connection of two or more components to the same voltage source so that the current can branch out over two or more paths. In a parallel resistive circuit, two or more resistors are connected to the same source, so current splits to travel through each separate branch resistance.

6-1 COMPONENTS IN PARALLEL

Parallel Circuit
Also called shunt; circuit having two or more paths for current flow.

Many components, other than resistors, can be connected in parallel, and a **parallel circuit** can easily be identified because current is split into two or more paths. Being able to identify a parallel connection requires some practice, because they can come in many different shapes and sizes. If you can place your pencil at the negative terminal of the voltage source (battery) and follow the wire connections through components to the positive side of the battery and only have one path to follow, the circuit is connected in series. If, however, you can place your pencil at the negative terminal of the voltage source and follow the wire and at some point have a choice of two or more routes, the circuit is connected with two or more parallel branches. The number of routes determines the number of parallel branches. Figure 6-1 illustrates five examples of parallel resistive circuits.

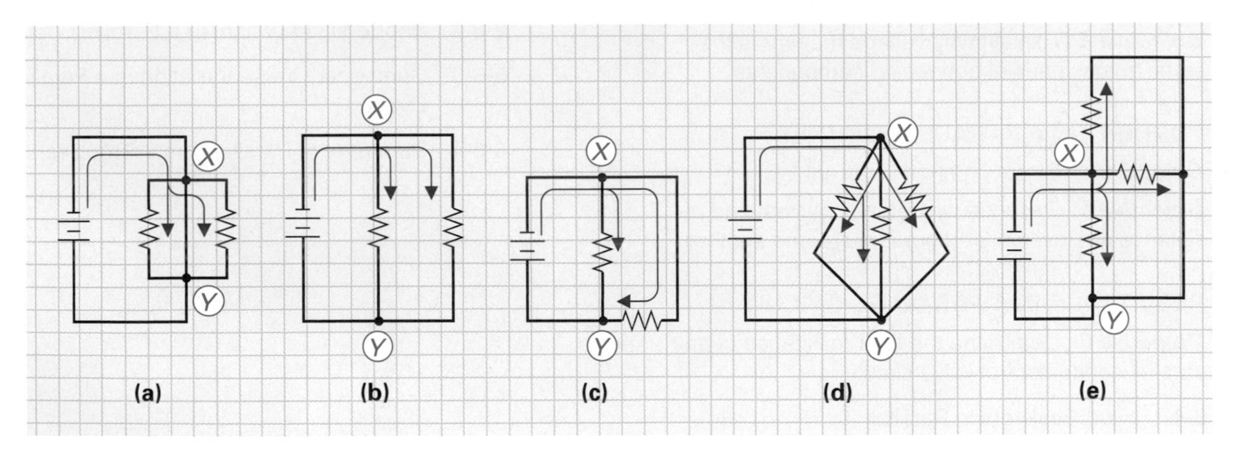

FIGURE 6-1 Parallel Circuits.

EXAMPLE:

Figure 6-2(a) illustrates four resistors laid out on a tabletop.

a. With wire leads, connect all four resistors in parallel on a protoboard, and then connect the circuit to a dc power supply.

b. Draw the schematic diagram of the parallel-connected circuit.

Solution:

Figure 6-2(b) shows how to connect the resistors in parallel and the circuit's schematic.

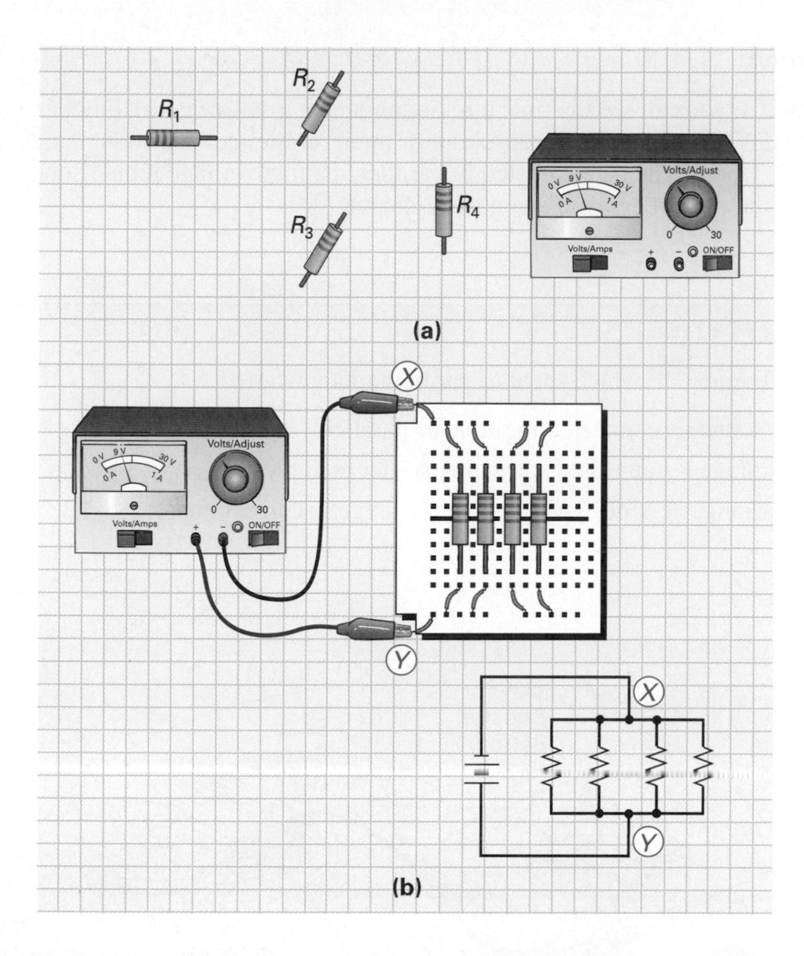

FIGURE 6-2 Connecting Resistors in Parallel. (a) Problem. (b) Solution.

EXAMPLE:

Figure 6-3(a) shows four 1.5 V cells and three lamps. Using wires, connect all of the cells in parallel to create a 1.5 V source. Then connect all of the three lamps in parallel with one another, and finally connect the 1.5 V source across the three parallel-connected lamp load.

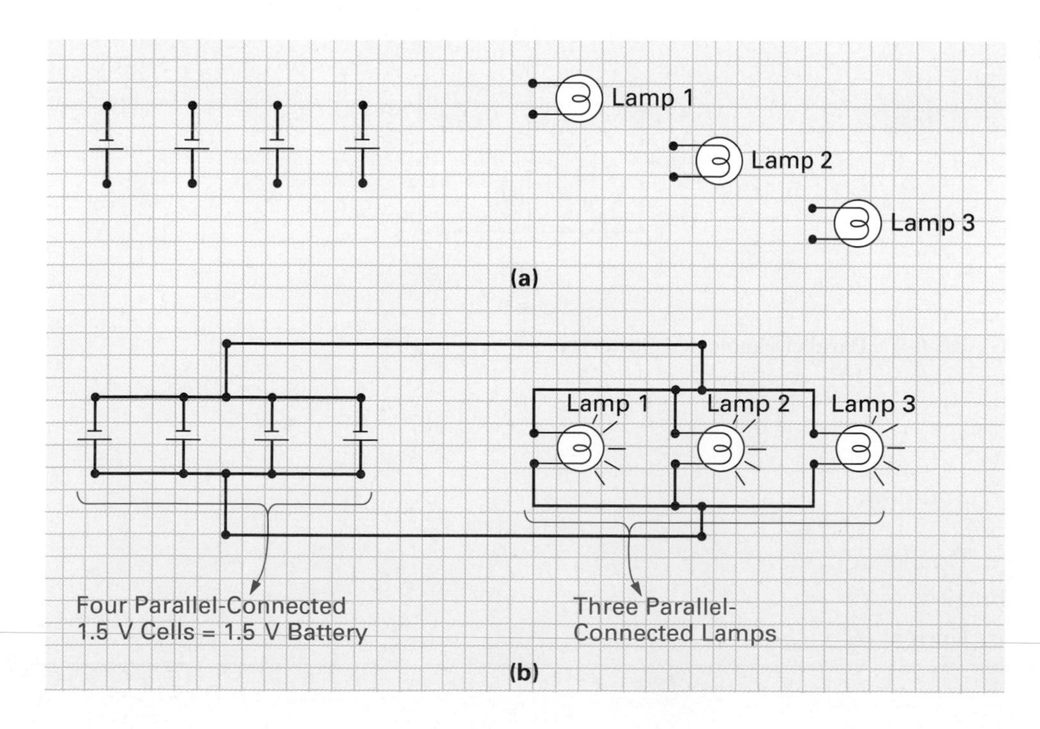

FIGURE 6-3 Parallel-Connected Cells and Lamps.

Solution:

In Figure 6-3(b) you can see the final circuit containing a source consisting of four parallel-connected 1.5 V cells, and a load consisting of three parallel-connected lamps. As explained in Chapter 4, when cells are connected in parallel, the total voltage remains the same as for one cell, but the current demand can now be shared.

EXAMPLE:

Referring to Figure 6-4, which shows a car being jump-started,

a. Are the batteries connected in series or parallel?

b. What voltage is connected across the starter motor?

c. Is this arrangement designed to give a voltage or current boost?

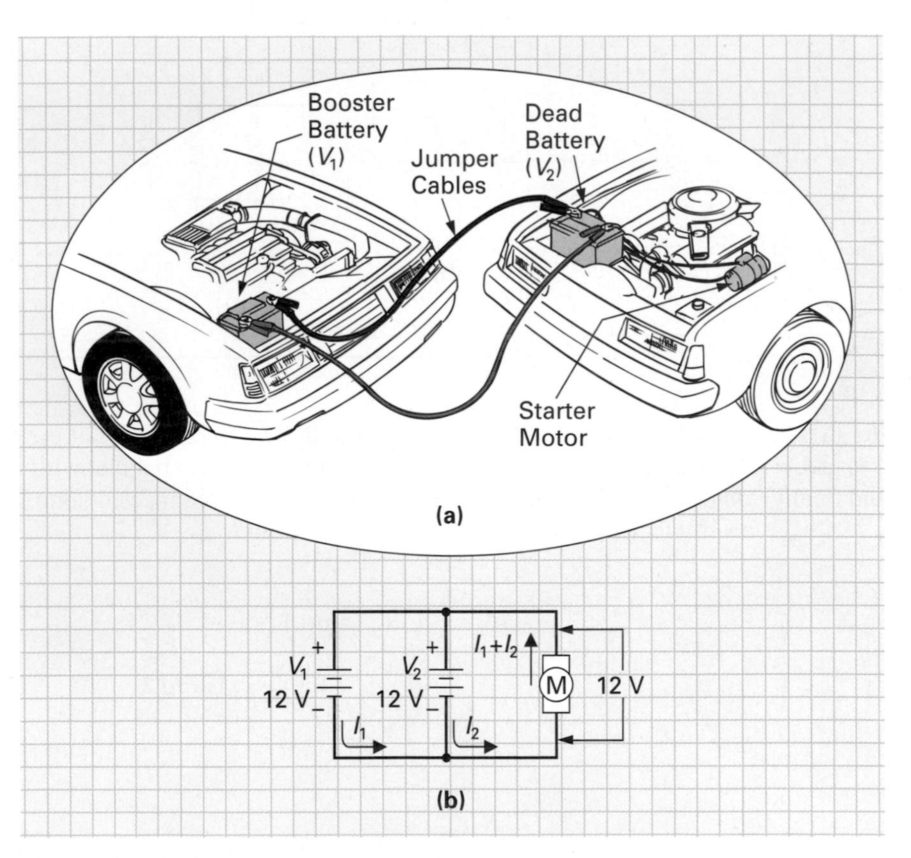

FIGURE 6-4 Parallel Connected Batteries.

Solution:

a. Parallel

b. 12 V

c. Current ($I_1 + I_2$)

6-2 VOLTAGE IN A PARALLEL CIRCUIT

Figure 6-5(a) shows a simple circuit with four resistors connected in parallel across the voltage source of a 9 V battery. The current from the negative side of the battery will split among the four different paths or branches, yet the voltage drop across each branch of a parallel circuit is equal to the voltage drop across all the other branches in parallel. This means that if the voltmeter were to measure the voltage across *A* and *B* or *C* and *D* or *E* and *F* or *G* and *H*, they would all be the same or, in this example, would all drop 9 V.

It is quite easy to imagine why there will be the same voltage drop across all the resistors, seeing that points *A, C, E,* and *G* are all one connection and points *B, D, F,* and *H* are all one connection. Measuring the voltage drop with the voltmeter across any of the resistors is the same as measuring the voltage across the battery, as shown in Figure 6-5(b). As long as the voltage source remains constant, the voltage drop will always be common (9 V) across the parallel resistors, no matter what value or how many resistors are connected in parallel. The voltmeter is therefore measuring the voltage between two common points that are directly connected to the battery, and the voltage dropped across all these parallel resistors will be equal to the source voltage.

In Figure 6-6(a) and (b), the same circuit is shown in two different ways, so you can see how the same circuit can look completely different. In both examples, the voltage drop

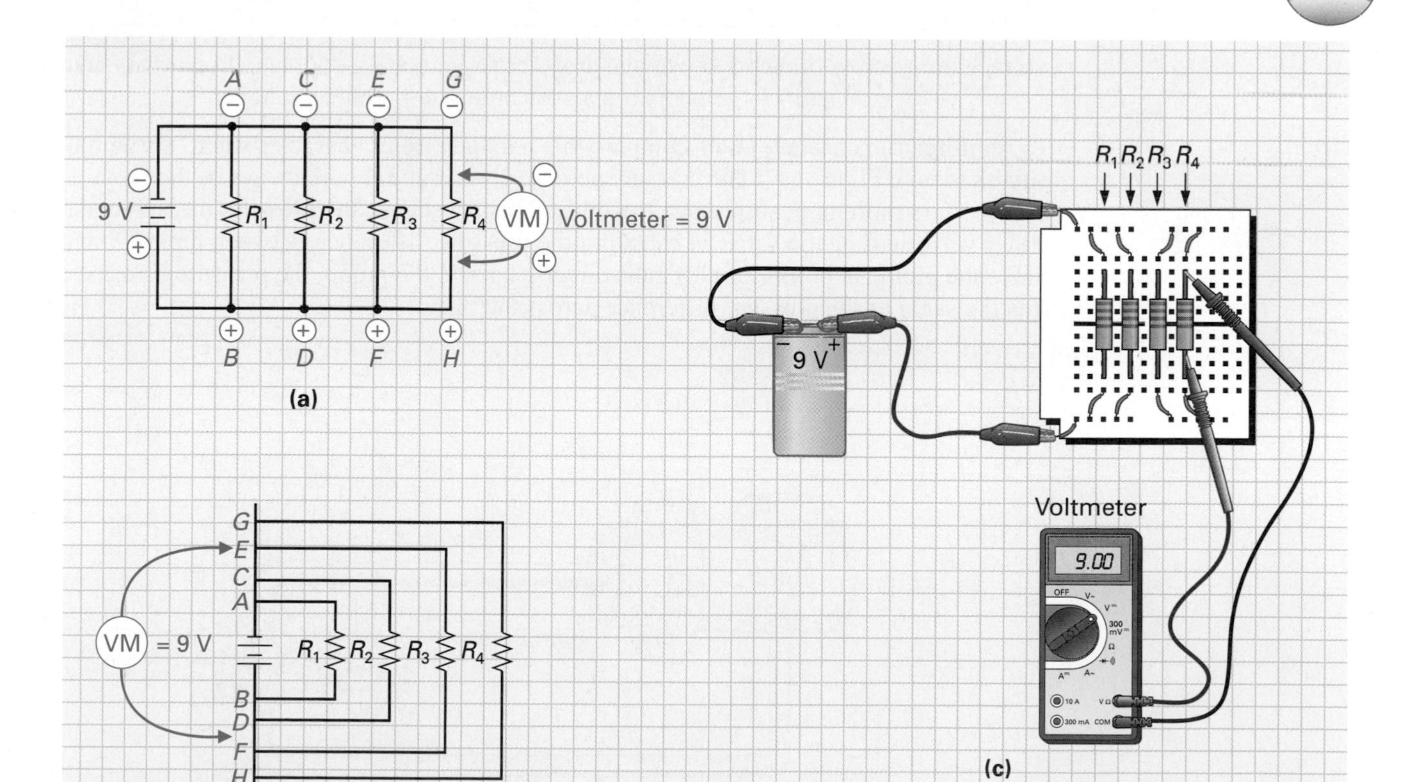

FIGURE 6-5 Voltage in a Parallel Circuit.

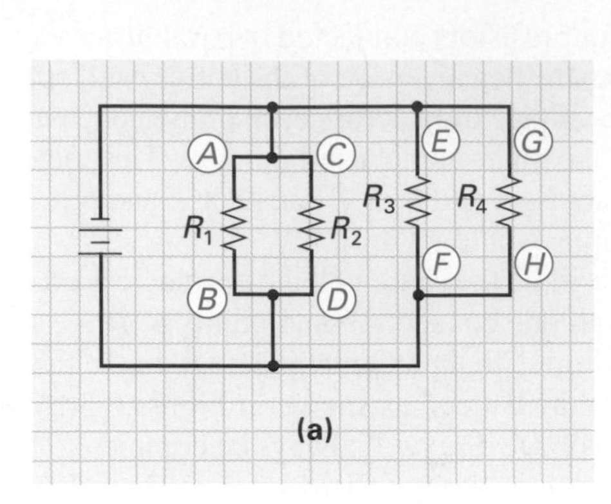

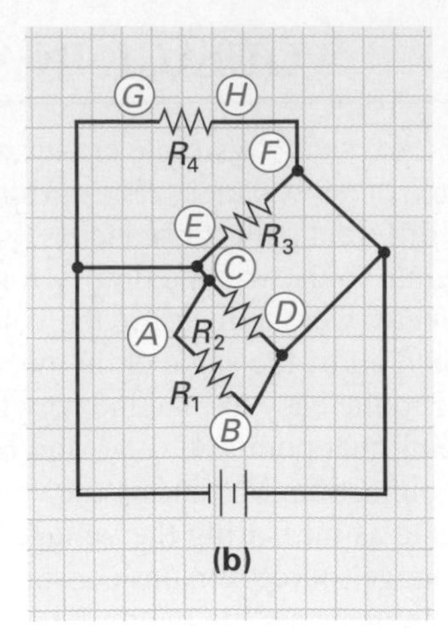

FIGURE 6-6 **Parallel-Circuit Voltage Drop.**

across any of the resistors will always be the same and, as long as the voltage source is not heavily loaded, equal to the source voltage. Just as you can trace the positive side of the battery to all four resistors, you can also trace the negative side to all four resistors.

Mathematically stated, we can say that in a parallel circuit

$$V_{B1} = V_{B2} = V_{B3} = V_{B4} = V_S$$

Voltage drop across branch 1 = voltage drop across branch 2 = voltage drop across branch 3 (etc.) = voltage drop across branch 4 = source voltage

To reinforce the concept, let us compare this parallel circuit characteristic to a water analogy, as seen in Figure 6-7. The pressure across valves *A* and *B* will always be the same, even if one offers more opposition than the other. This is because the pressure measured across either valve will be the same as checking the pressure difference between piping *X* and *Y*. Since the piping at points *X* and *Y* runs directly back to the pump, the pressure across *A* and *B* is the same as the pressure difference across the pump.

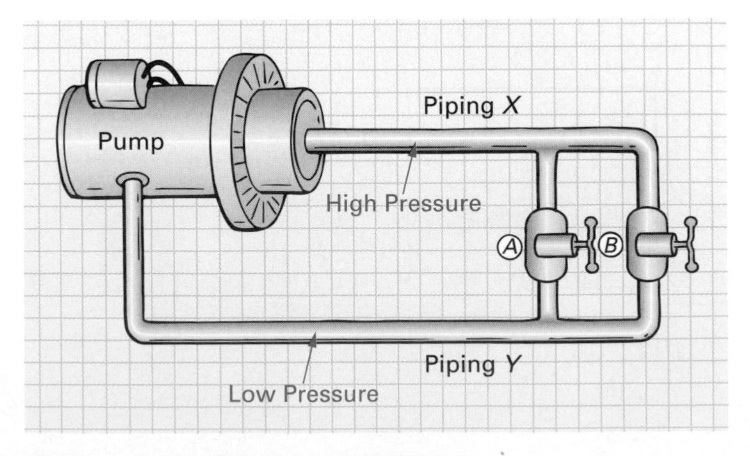

FIGURE 6-7 **Fluid Analogy of Parallel-Circuit Pressure.**

EXAMPLE:

Refer to Figure 6-8 and calculate:

a. Voltage drop across R_1

b. Voltage drop across R_2

c. Voltage drop across R_3

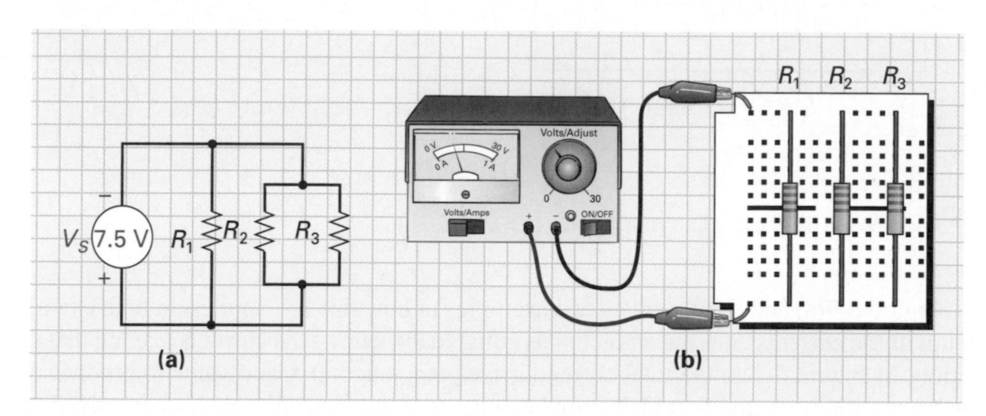

FIGURE 6-8 A Parallel Circuit Example (a) Schematic. (b) Protoboard Circuit.

Solution:

Since all these resistors are connected in parallel, the voltage across every branch will be the same and equal to the source voltage applied. Therefore,

$$V_{R1} = V_{R2} = V_{R3} = V_S$$
$$7.5 \text{ V} = 7.5 \text{ V} = 7.5 \text{ V} = 7.5 \text{ V}$$

SELF-TEST EVALUATION POINT FOR SECTIONS 6-1 AND 6-2

Now that you have completed these sections, you should be able to:

Objective 1. Describe the difference between a series and a parallel circuit.

Objective 2. Be able to recognize and determine whether circuit components are connected in series or parallel.

Objective 3. Explain why voltage measures the same across parallel-connected components.

Use the following questions to test your understanding of Sections 6-1 and 6-2:

1. Describe a parallel circuit.
2. True or false: A parallel circuit is also known as a voltage-divider circuit.
3. What would be the voltage drop across R_1 if $V_S = 12$ V and R_1 and R_2 are both in parallel with one another and equal to 24 Ω each?
4. Can Kirchhoff's voltage law be applied to parallel circuits?

6-3 CURRENT IN A PARALLEL CIRCUIT

In addition to providing the voltage law for series circuits, Gustav Kirchhoff was the first to observe (in 1847) and prove that the sum of all the **branch currents** in a parallel circuit ($I_1 + I_2 + I_3$, etc.) was equal to the total current (I_T). In honor of his second discovery, this phenomenon is known as **Kirchhoff's current law**, which states that the sum of all the currents entering a junction is equal to the sum of all the currents leaving that same junction.

Branch Current
A portion of the total current that is present in one path of a parallel circuit.

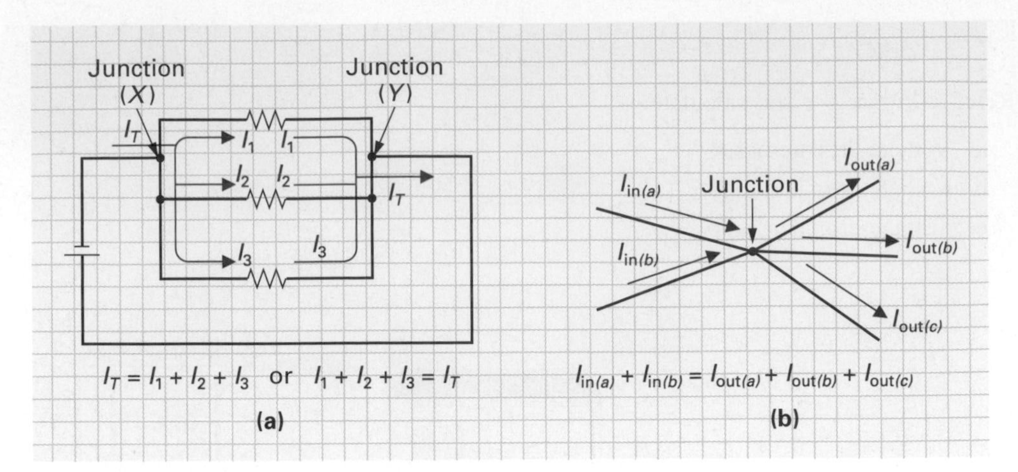

FIGURE 6-9 Kirchhoff's Current Law.

Kirchhoff's Current Law

The sum of the currents flowing into a point in a circuit is equal to the sum of the currents flowing out of that same point.

Figure 6-9(a) and (b) illustrates two examples of how this law applies. In both examples, the sum of the currents entering a junction is equal to the sum of the currents leaving that same junction. In Figure 6-9(a) the total current arrives at a junction X and splits to produce three branch currents, I_1, I_2, and I_3, which cumulatively equal the total current (I_T) that arrived at the junction X. The same three branch currents combine at junction Y, and the total current (I_T) leaving that junction is equal to the sum of the three branch currents arriving at junction Y.

$$I_T = I_1 + I_2 + I_3 + I_4 = \cdots$$

As another example, in Figure 6-9(b) you can see that there are two branch currents entering a junction [$I_{\text{in}(a)}$ and $I_{\text{in}(b)}$] and three branch currents leaving that same junction [$I_{\text{out}(a)}$, $I_{\text{out}(b)}$, and $I_{\text{out}(c)}$]. As stated below the illustration, the sum of the input currents will equal the sum of the output currents: $I_{\text{in}(a)} + I_{\text{in}(b)} = I_{\text{out}(a)} + I_{\text{out}(b)} + I_{\text{out}(c)}$.

EXAMPLE:

Refer to Figure 6-10 and calculate the value of I_T.

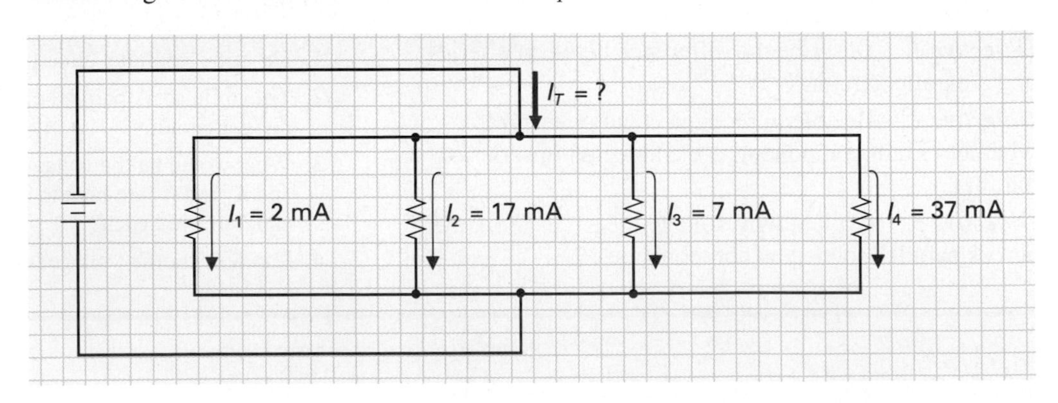

FIGURE 6-10 Calculating Total Current.

Solution:

By Kirchhoff's current law,

$$I_T = I_1 + I_2 + I_3 + I_4$$
$$? = 2\text{ mA} + 17\text{ mA} + 7\text{ mA} + 37\text{ mA}$$
$$I_T = 63\text{ mA}$$

EXAMPLE:

Refer to Figure 6-11 and calculate the value of I_1.

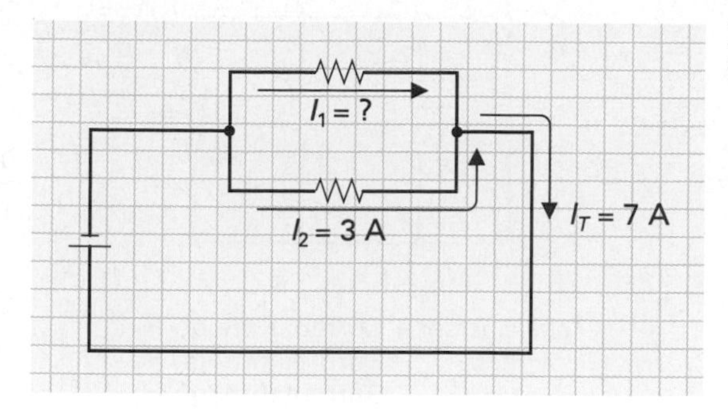

FIGURE 6-11 Calculating Branch Current.

Solution:

By transposing Kirchhoff's current law, we can determine the unknown value (I_1):

$$I_T = I_1 + I_2 \qquad I_T = I_1 + I_2 \quad (-I_2)$$
$$7\text{ A} = ? + 3\text{ A} \quad \text{or} \quad I_T - I_2 = I_1$$
$$I_1 = 4\text{ A} \qquad I_1 = I_T - I_2 = 7\text{ A} - 3\text{ A} = 4\text{ A}$$

As with series circuits, to find out how much current will flow through a parallel circuit, we need to find out how much opposition or resistance is being connected across the voltage source.

$$I_T = \frac{V_S}{R_T}$$

Total current equals source voltage divided by total resistance.

When we connect resistors in parallel, the total resistance in the circuit will actually decrease. In fact, the total resistance in a parallel circuit will always be less than the value of the smallest resistor in the circuit.

To prove this point, Figure 6-12 shows how two sets of identical resistors (R_1, R_2, and R_3) were used to build both a series and a parallel circuit. The total current flow in the parallel circuit would be larger than the total current in the series circuit, because the parallel circuit has two or more paths for current to flow, while the series circuit only has one.

To explain why the total current will be larger in a parallel circuit, let us take the analogy of a freeway with only one path for traffic to flow. A single-lane freeway is equivalent to a series circuit, and only a small amount of traffic is allowed to flow along this freeway. If the freeway is expanded to accommodate two lanes, a greater amount of traffic can flow along the freeway in the same amount of time. Having more lanes permits a greater total amount of traffic flow. In parallel circuits, more branches will allow a greater total amount of current flow because there is less resistance in more paths than there is with only one path. This concept is summarized in Figure 6-12.

Just as a series circuit is often referred to as a voltage-divider circuit, a parallel circuit is often referred to as a **current-divider** circuit, because the total current arriving at a junction will divide or split into branch currents (Kirchhoff's law), as shown in Figure 6-13. The current division is inversely proportional to the resistance in the branch, assuming that the voltage across both resistors is constant and equal to the source voltage (V_S). This means that a large branch resistance will cause a small branch current ($I\downarrow = V/R\uparrow$), and a small branch resistance will cause a large branch current ($I\uparrow = V/R\downarrow$).

Current Divider
A parallel network designed to proportionally divide the circuit's total current.

TIME LINE

Business in 1892 was not going at all well for Almon Strowger, a funeral home director in the small town of La Porte, Indiana. Strangely enough, the telephone, which promoted business for most shopkeepers, was Strowger's downfall. This was because the wife of the owner of a competing funeral parlor served as the town's telephone operator. When people called the operator and asked to be put through to the funeral parlor, they were naturally connected to the operator's husband.

As the old expression states, "necessity is the mother of invention," and for Strowger it became a case of do or die. Making up his mind to cure the problem, he began to study the details of telephone communications. He designed a mechanism that made telephone connection automatic and thus bypassed the town operator. The switch he invented consisted of two parts—a ten-by-ten array of terminals (called the bank) arranged in a cylindrical arc and a moveable switch (called the brush). The brush stepped along and around a cylinder and could be put in one of one hundred positions to connect to one of one hundred terminals. The position of the brush was driven by an electromagnet which responded to pulses produced by the telephone dial.

The "Strowger switch" brought about the first automatic telephone exchange system in the world, with improved versions still used extensively in offices up to the 1960s.

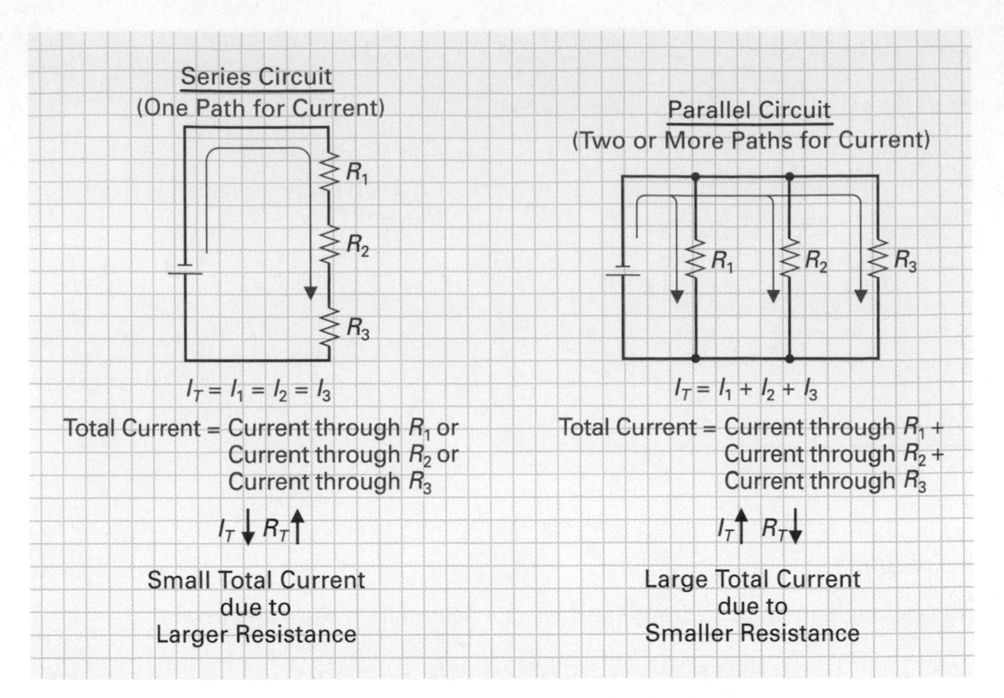

FIGURE 6-12 Series Circuit and Parallel Circuit Current Comparison.

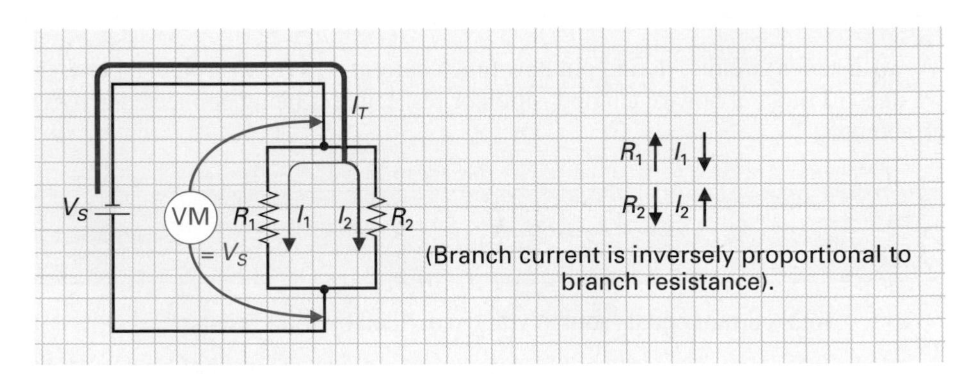

FIGURE 6-13 The Parallel Circuit Current Divider.

EXAMPLE:

Calculate the following for Figure 6-14(a), and then insert the values in a circuit analysis table:

a. I_1

b. I_2

c. I_T

Solution:

Since R_1 and R_2 are connected in parallel across the 10 V source, the voltage across both resistors will be 10 V.

a. $I_1 = \dfrac{V_{R_1}}{R_1} = \dfrac{10\text{ V}}{6\text{ k}\Omega} = 1.6\text{ mA}$ (smaller branch current through larger branch resistance)

b. $I_2 = \dfrac{V_{R_2}}{R_2} = \dfrac{10\text{ V}}{3\text{ k}\Omega} = 3.3\text{ mA}$ (larger branch current through smaller branch resistance)

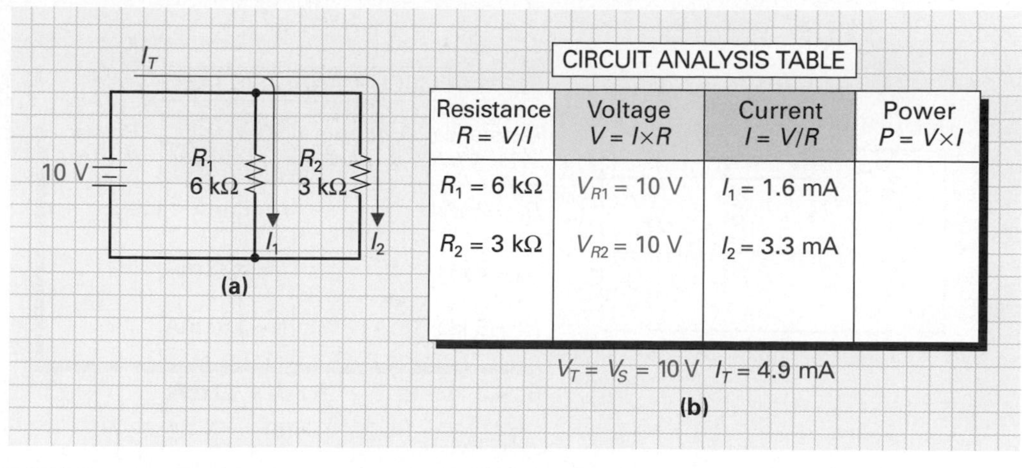

FIGURE 6-14 Parallel Circuit Example. (a) Schematic. (b) Circuit Analysis Table.

c. By Kirchhoff's current law,

$$I_T = I_1 + I_2$$
$$= 1.6 \text{ mA} + 3.3 \text{ mA}$$
$$= 4.9 \text{ mA}$$

The circuit analysis table for this example is shown in Figure 6-14(b).

By rearranging Ohm's law, we can arrive at another formula, called the *current-divider formula*, that can be used to calculate the current through any branch of a multiple-branch parallel circuit.

$$I_x = \frac{R_T}{R_x} \times I_T$$

where I_x = branch current desired
R_T = total resistance
R_x = resistance in branch
I_T = total current

EXAMPLE:

Refer to Figure 6-15 and calculate the following if the total circuit resistance (R_T) is equal to 1 kΩ:

a. $I_1 =$
b. $I_2 =$
c. $I_3 =$

Solution:

Since the source and therefore the voltage across each branch resistor are not known, we will use the current-divider formula to calculate I_1, I_2, and I_3.

a. $I_1 = \frac{R_T}{R_1} \times I_T = \frac{1 \text{ k}\Omega}{2 \text{ k}\Omega} \times 10 \text{ mA} = 5 \text{ mA}$

(smallest branch resistance has largest branch current)

Calculator Sequence

STEP	KEYPAD ENTRY	DISPLAY RESPONSE
1.	1 E 3	1.E3
2.	÷	
3.	2 E 3	2.E3
4.	×	0.5
5.	1 0 E 3 +/−	10.E − 3
6.	=	5.E − 03

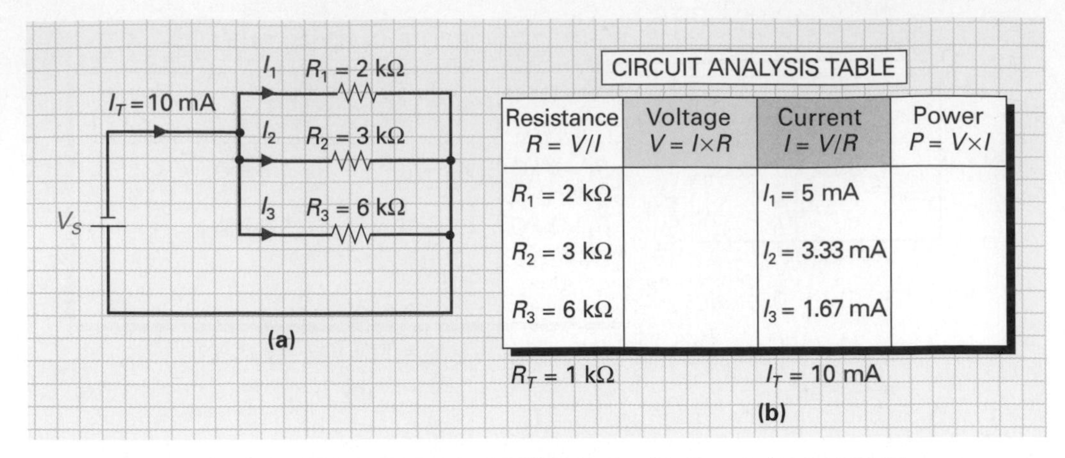

FIGURE 6-15 Parallel Circuit Example. (a) Schematic. (b) Circuit Analysis Table.

b. $I_2 = \dfrac{R_T}{R_2} \times I_T = \dfrac{1\text{ k}\Omega}{3\text{ k}\Omega} \times 10\text{ mA}$
$= 3.33\text{ mA}$

c. $I_3 = \dfrac{R_T}{R_3} \times I_T = \dfrac{1\text{ k}\Omega}{6\text{ k}\Omega} \times 10\text{ mA}$ (largest branch resistance has smallest branch current)
$= 1.67\text{ mA}$

To double-check that the values for I_1, I_2, and I_3 are correct, you can apply Kirchhoff's current law, which is

$$I_T = I_1 + I_2 + I_3$$
$$10\text{ mA} = 5\text{ mA} + 3.33\text{ mA} + 1.67\text{ mA}$$
$$= 10\text{ mA}$$

EXAMPLE:

A common use of parallel circuits is in the residential electrical system. All of the household lights and appliances are wired in parallel, as seen in the typical room wiring circuit in Figure 6-16(a). If it is a cold winter morning and lamp 1 is switched ON, together with the space heater and hair dryer, what will the individual branch currents be, and what will be the total current drawn from the source?

Solution:

Figure 6-16(b) shows the schematic of the pictorial in Figure 6-16(a). Since all resistances are connected in parallel across a 120 V source, the voltage across all devices will be 120 V. Using Ohm's law we can calculate the four branch currents:

$$I_1 = \frac{V_{\text{lamp1}}}{R_{\text{lamp1}}} = \frac{120\text{ V}}{125\ \Omega} = 960\text{ mA}$$

$$I_2 = \frac{V_{\text{hairdryer}}}{R_{\text{hairdryer}}} = \frac{120\text{ V}}{40\ \Omega} = 3\text{ A}$$

$$I_3 = \frac{V_{\text{heater}}}{R_{\text{heater}}} = \frac{120\text{ V}}{12\ \Omega} = 10\text{ A}$$

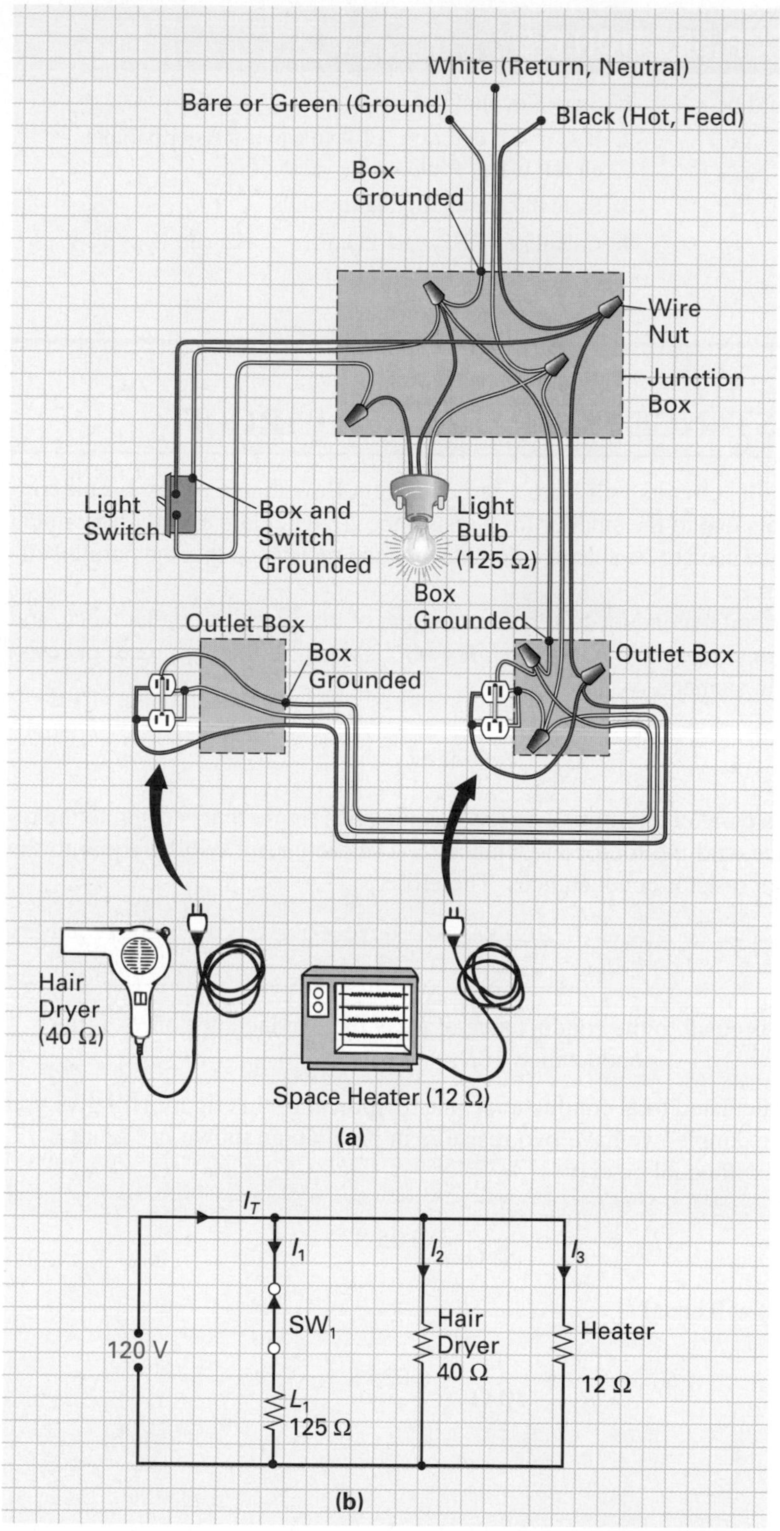

FIGURE 6-16 **The Parallel Home Electrical System. (a) Physical View. (b) Schematic Diagram.**

By Kirchhoff's current law,

$$I_T = I_1 + I_2 + I_3$$
$$= 960 \text{ mA} + 3 \text{ A} + 10 \text{ A}$$
$$= 13.96 \text{ A}$$

SELF-TEST EVALUATION POINT FOR SECTION 6-3

Now that you have completed this section, you should be able to:

Objective 4. State Kirchhoff's current law.

Objective 5. Describe why branch current and resistance are inversely proportional to one another.

Use the following questions to test your understanding of Section 6-3:

1. State Kirchhoff's current law.
2. If $I_T = 4$ A and $I_1 = 2.7$ A in a two-resistor parallel circuit, what would be the value of I_2?
3. State the current-divider formula.
4. Calculate I_1 if $R_T = 1$ kΩ, $R_1 = 2$ kΩ, and $V_T = 12$ V.

6-4 RESISTANCE IN A PARALLEL CIRCUIT

We now know that parallel circuits will have a larger current flow than a series circuit containing the same resistors due to the smaller total resistance. To calculate exactly how much total current will flow, we need to be able to calculate the total resistance that the parallel circuit presents to the source.

The ability of a circuit to conduct current is a measure of that circuit's conductance, and you will remember from Chapter 3 that conductance (G) is equal to the reciprocal of resistance and is measured in siemens.

$$G = \frac{1}{R} \quad \text{(siemens)}$$

Every resistor in a parallel circuit will have a conductance figure that is equal to the reciprocal of its resistance, and the total conductance (G_T) of the circuit will be equal to the sum of all the individual resistor conductances. Therefore,

$$G_T = G_{R1} + G_{R2} + G_{R3} + \cdots$$

Total conductance is equal to the conductance of R_1 + the conductance of R_2 + the conductance of R_3 +· · · .

Once you have calculated total conductance, the reciprocal of this figure will give you total resistance. If, for example, we have two resistors in parallel, as shown in Figure 6-17, the conductance for R_1 will equal

$$G_{R1} = \frac{1}{R_1} = \frac{1}{20\ \Omega} = 0.05\ \text{S}$$

The conductance for R_2 will equal

$$G_{R2} = \frac{1}{R_2} = \frac{1}{40\ \Omega} = 0.025\ \text{S}$$

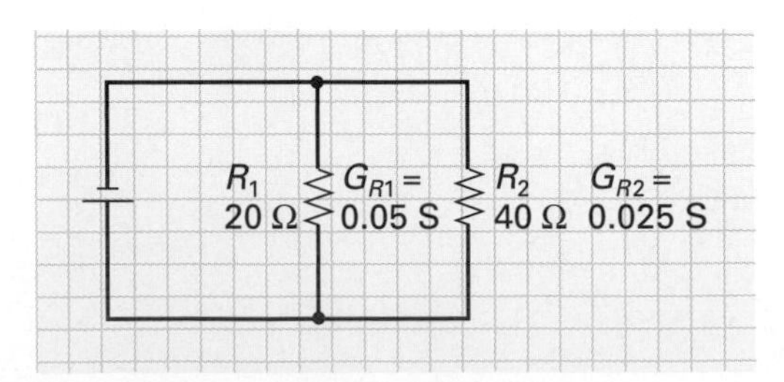

FIGURE 6-17 Parallel Circuit Conductance and Resistance.

The total conductance will therefore equal

$$G_T = G_{R1} + G_{R2}$$
$$= 0.05 + 0.025$$
$$= 0.075 \text{ S}$$

Since total resistance is equal to the reciprocal of total conductance, total resistance for the parallel circuit in Figure 6-17 will be

$$R_T = \frac{1}{G_T} = \frac{1}{0.075 \text{ S}} = 13.3\ \Omega$$

Combining these three steps (first calculate individual conductances, total conductance, and then total resistance), we can arrive at the following *reciprocal formula:*

$$R_T = \frac{1}{(1/R_1) + (1/R_2)}$$

This formula states that the conductance of R_1 (G_{R1})
+ conductance of R_2 (G_{R2})
= total conductance (G_T),
and the reciprocal of total conductance is equal to total resistance.

In the example in Figure 6-17, this combined general formula for total resistance can be verified by plugging in the example values.

$$R_T = \frac{1}{(1/R_1) + (1/R_2)}$$
$$= \frac{1}{(1/20) + (1/40)}$$
$$= \frac{1}{0.05 + 0.025}$$
$$= \frac{1}{0.075}$$
$$= 13.3\ \Omega$$

Calculator Sequence

STEP	KEYPAD ENTRY	DISPLAY RESPONSE
1.	(Clear Memory)	
2.	[2] [0]	20.
3.	[1/x]	5.E–2
4.	[+]	5.E–2
5.	[4] [0]	40
6.	[1/x]	2.5E–2
7.	[=]	7.5E–2
8.	[1/x]	13.33333

The *reciprocal formula* for calculating total parallel circuit resistance for any number of resistors is

$$R_T = \frac{1}{(1/R_1) + (1/R_2) + (1/R_3) + (1/R_4) + \cdots}$$

EXAMPLE:

Referring to Figure 6-18(a), calculate:

a. Total resistance

b. Voltage drop across R_2

c. Voltage drop across R_3

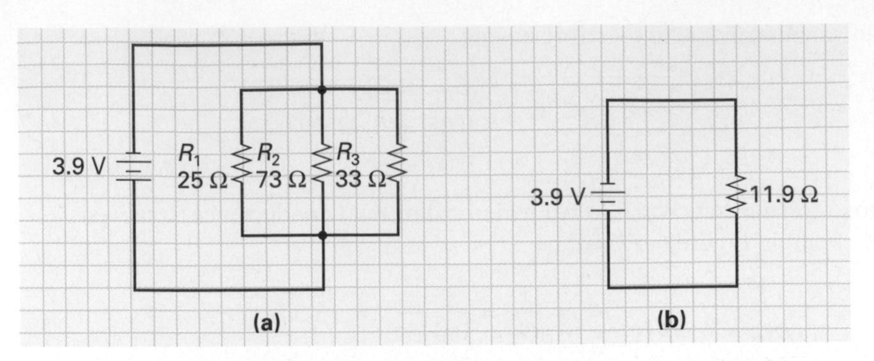

FIGURE 6-18 Parallel Circuit Example. (a) Schematic. (b) Equivalent Circuit.

Solution:

Total resistance can be calculated using the reciprocal formula:

$$R_T = \frac{1}{(1/R_1) + (1/R_2) + (1/R_3)}$$

$$= \frac{1}{(1/25\ \Omega) + (1/73\ \Omega) + (1/33\ \Omega)}$$

$$= \frac{1}{0.04 + 0.014 + 0.03}$$

$$= 11.9\ \Omega$$

With parallel resistance circuits, the total resistance is always smaller than the smallest branch resistance. In this example the total opposition of this circuit is equivalent to 11.9 Ω, as shown in Figure 6-18(b).

With parallel resistive circuits, the voltage drop across any branch is equal to the voltage drop across each of the other branches and is equal to the source voltage—in this example, 3.9 V.

6-4-1 *Two Resistors in Parallel*

If only two resistors are connected in parallel, a quick and easy formula called the *product-over-sum formula* can be used to calculate total resistance.

$$R_T = \frac{R_1 \times R_2}{R_1 + R_2}$$

$$\text{Total resistance} = \frac{\text{product of both resistance values}}{\text{sum of both resistance values}}$$

Using the example shown in Figure 6-19, let us compare the *product-over-sum* formula with the *reciprocal* formula.

FIGURE 6-19 Two Resistors in Parallel.

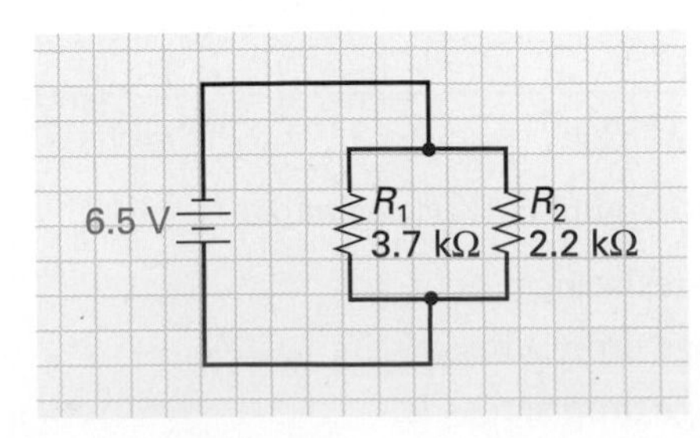

(a) Product-Over-Sum Formula

$$R_T = \frac{R_1 \times R_2}{R_1 + R_2}$$

$$= \frac{3.7\text{ k}\Omega \times 2.2\text{ k}\Omega}{3.7\text{ k}\Omega + 2.2\text{ k}\Omega}$$

$$= \frac{8.14\text{ k}\Omega^2}{5.9\text{ k}\Omega}$$

$$= 1.38\text{ k}\Omega$$

(b) Reciprocal Formula

$$R_T = \frac{1}{(1/R_1) + (1/R_2)}$$

$$= \frac{1}{(1/3.7\text{ k}\Omega) + (1/2.2\text{ k}\Omega)}$$

$$= \frac{1}{(270.2 \times 10^{-6}) + (454.5 \times 10^{-6})}$$

$$= \frac{1}{724.7 \times 10^{-6}}$$

$$= 1.38\text{ k}\Omega$$

As you can see from this example, the advantage of the product-over-sum parallel resistance formula (a) is its ease of use. Its disadvantage is that it can only be used for two resistors in parallel. The rule to adopt, therefore, is that if a circuit has two resistors in parallel use the *product-over-sum* formula, and in circuits containing more than two resistors, use the *reciprocal* formula.

6-4-2 *Equal-Value Resistors in Parallel*

If resistors of equal value are connected in parallel, a special-case *equal-value formula* can be used to calculate the total resistance.

$$R_T = \frac{\text{value of one resistor } (R)}{\text{number of parallel resistors } (n)}$$

EXAMPLE:

Figure 6-20(a) shows how a stereo music amplifier is connected to drive two 8 Ω speakers, which are connected in parallel with one another. What is the total resistance connected across the amplifier's output terminals?

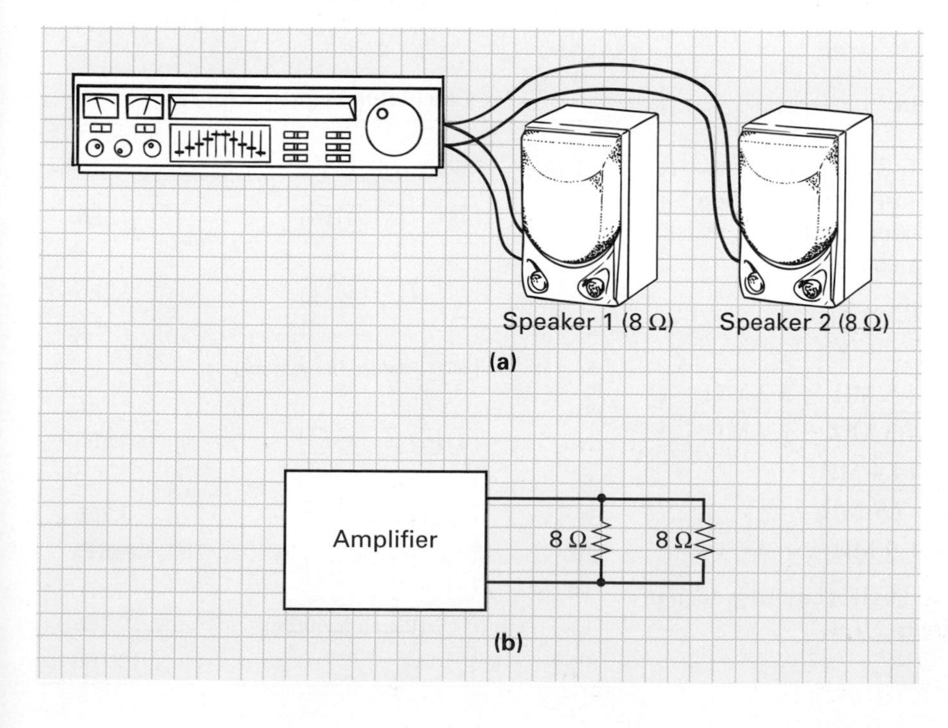

FIGURE 6-20 Parallel-Connected Speakers.

■ *Solution:*

Referring to the schematic in Figure 6-20(b), you can see that since both parallel-connected speakers have the same resistance, the total resistance is most easily calculated by using the equal-value formula:

$$R_T = \frac{R}{n} = \frac{8\ \Omega}{2} = 4\ \Omega$$

EXAMPLE:

Refer to Figure 6-21 and calculate:

a. Total resistance in part (a)

b. Total resistance in part (b)

FIGURE 6-21 Parallel Circuit Examples.

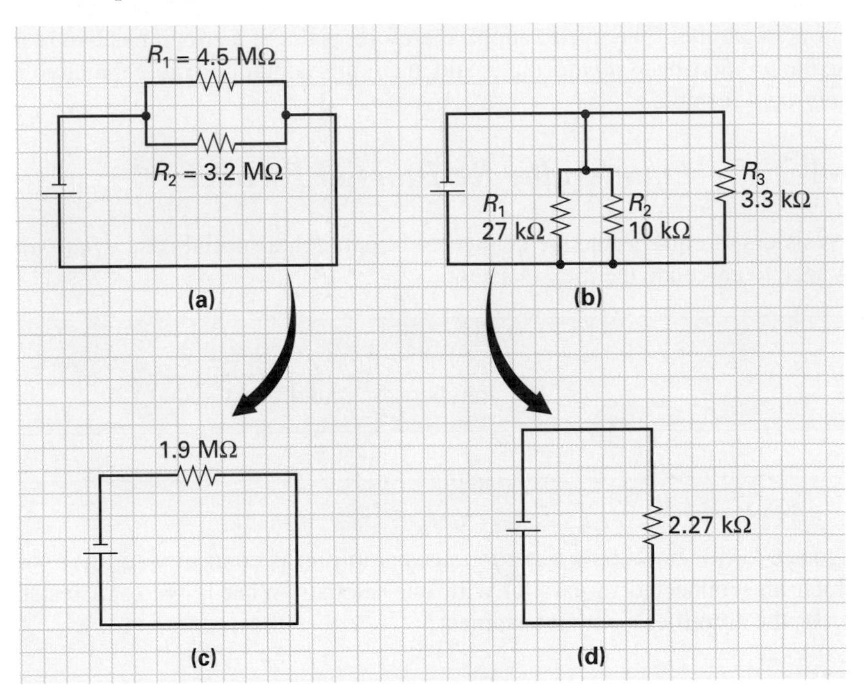

■ *Solution:*

a. Figure 6-21(a) has only two resistors in parallel and therefore the product-over-sum resistor formula can be used.

$$R_T = \frac{R_1 \times R_2}{R_1 + R_2}$$

$$= \frac{4.5\ \text{M}\Omega \times 3.2\ \text{M}\Omega}{4.5\ \text{M}\Omega + 3.2\ \text{M}\Omega}$$

$$= \frac{14.4\ \text{M}\Omega^2}{7.7\ \text{M}\Omega}$$

$$= 1.9\ \text{M}\Omega$$

The equivalent circuit is seen in Figure 6-21(c).

Calculator Sequence for (a)

STEP	KEYPAD ENTRY	DISPLAY RESPONSE
1.	4 . 5 E 6	4.5E6
2.	+	
3.	3 . 2 E 6	3.2E6
4.	=	7.7E6
5.	STO (store in memory)	
6.	C/CE	0.
7.	4 . 5 E 6	4.5E6
8.	×	
9.	3 . 2 E 6	3.2E6
10.	=	1.44E13
11.	÷	
12.	RM (Recall memory)	7.7E6
13.	=	1.87E6

b. Figure 6-21(b) has more than two resistors in parallel, and therefore the reciprocal formula must be used.

$$R_T = \frac{1}{(1/R_1) + (1/R_2) + (1/R_3)}$$

$$= \frac{1}{(1/27\ \text{k}\Omega) + (1/10\ \text{k}\Omega) + (1/3.3\ \text{k}\Omega)}$$

$$= \frac{1}{440.0 \times 10^{-6}}$$

$$= 2.27\ \text{k}\Omega$$

The equivalent circuit can be seen in Figure 6-21(d).

EXAMPLE:

Find the total resistance of the parallel circuit in Figure 6-22.

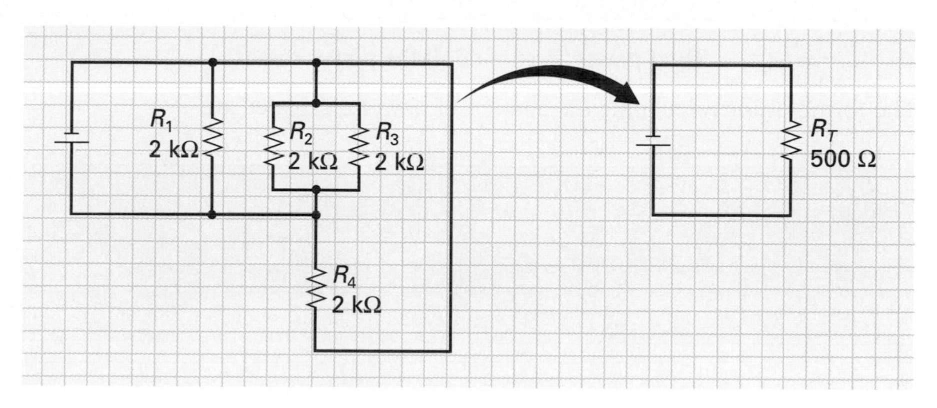

FIGURE 6-22 **Parallel Circuit Example.**

Solution:

Since all four resistors are connected in parallel and are all of the same value, the special-case equal-value formula can be used to calculate total resistance.

$$R_T = \frac{R}{n} = \frac{2\ \text{k}\Omega}{4} = 500\ \Omega$$

To summarize what we have learned so far about parallel circuits, we can say that

1. Components are said to be connected in parallel when the current has to travel two or more paths between the negative and positive sides of the voltage source.
2. The voltage across all the *parallel branches* is always the same.
3. The total current from the source is equal to the sum of all the branch currents (Kirchhoff's current law).
4. The amount of current flowing through each branch is inversely proportional to the resistance value in that branch.
5. The total resistance of a parallel circuit is always less than the value of the smallest branch *resistance*.

TIME LINE

Top-selling software programs for computers can originate from a diverse spectrum of sources. Some are developed by large research corporations in the center of major cities, while others are created by people in mountaintop cabins in the middle of nowhere.

Paul Lutus, who had created many best-selling software programs for the Apple computer, was fiercely independent. For a long period of time, Lutus lived and labored in his isolated, hand-built 12-foot by 16-foot cabin in the Oregon mountains, developing a word-processing program. Published in 1979, the highly acclaimed Apple Writer almost instantly became a best-seller, and the peace, tranquility, and hard work finally paid off in the form of royalty payments of more than $7500 per day.

SELF-TEST EVALUATION POINT FOR SECTION 6-4

Now that you have completed this section, you should be able to:

Objective 6. Determine the total resistance of any parallel-connected resistive circuit.

Use the following questions to test your understanding of Section 6-4:

State what parallel resistance formula should be used in questions 1 through 3 for calculating R_T.

1. Two resistors in parallel.
2. More than two resistors in parallel.
3. Equal-value resistors in parallel.
4. Calculate the total parallel resistance when $R_1 = 2.7\ \text{k}\Omega$, $R_2 = 24\ \text{k}\Omega$, and $R_3 = 1\ \text{M}\Omega$.

6-5 POWER IN A PARALLEL CIRCUIT

As with series circuits, the total power in a parallel resistive circuit is equal to the sum of all the power losses for each of the resistors in parallel.

$$P_T = P_1 + P_2 + P_3 + P_4 + \cdots$$

Total power = sum of all the power losses

The formulas for calculating the amount of power dissipated are

$$P = I \times V$$

$$P = \frac{V^2}{R}$$

$$P = I^2 \times R$$

EXAMPLE:

Calculate the total amount of power dissipated in Figure 6-23.

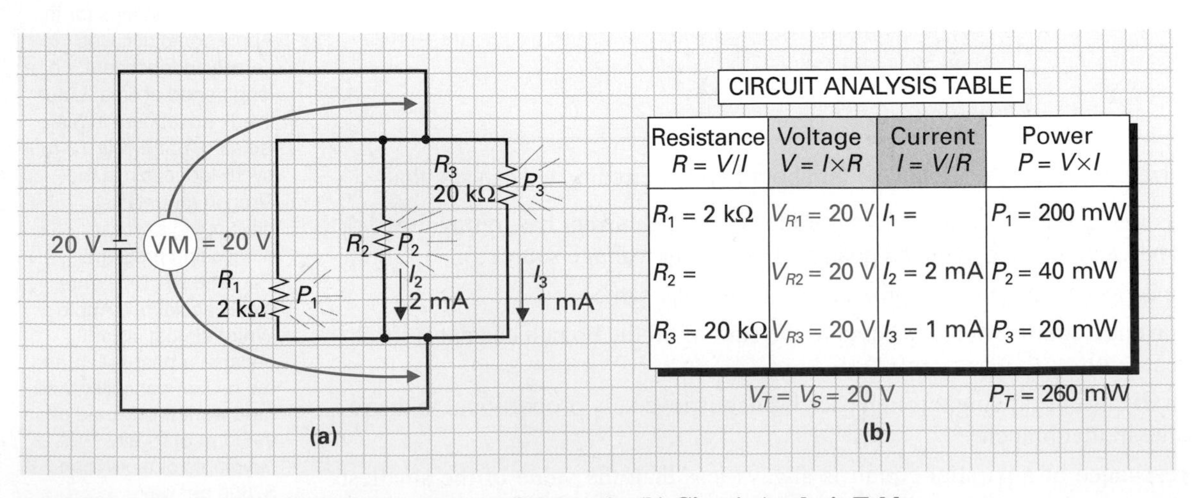

FIGURE 6-23 Parallel Circuit Example. (a) Schematic. (b) Circuit Analysis Table.

Solution:

The total power dissipated in a parallel circuit is equal to the sum of all the power dissipated by all the resistors. With P_1, we only know voltage and resistance and therefore we can use the formula

$$P_1 = \frac{V_{R1}^2}{R_1} = \frac{(20\text{ V})^2}{2\text{ k}\Omega} = 2.0\text{ W} \quad \text{or} \quad 200\text{ mW}$$

Calculator Sequence

STEP	KEYPAD ENTRY	DISPLAY RESPONSE
1.	2 0	20.
2.	x^2	400.
3.	÷	
4.	2 E 3	2.E3
5.	=	0.2

With P_2, we only know current and voltage, and therefore we can use the formula

$$P_2 = I_2 \times V_{R2} = 2\text{ mA} \times 20\text{ V} = 40\text{ mW}$$

Calculator Sequence

STEP	KEYPAD ENTRY	DISPLAY RESPONSE
1.	2 E 3 +/−	2.E–3
2.	×	
3.	2 0	20.
4.	=	40E–3

With P_3, we know V, I, and R; however, we will use the third power formula:

$$P_3 = I_3^2 \times R_3 = 1\text{ mA}^2 \times 20\text{ k}\Omega = 20\text{ mW}$$

Calculator Sequence

STEP	KEYPAD ENTRY	DISPLAY RESPONSE
1.	1 E 3 +/− x^2	1.E–6
2.	×	1.E–6
3.	2 0 E 3	20E3
4.	=	20.E–3

Total power (P_T) equals the sum of all the power or wattage losses for each resistor:

$$\begin{aligned} P_T &= P_1 + P_2 + P_3 \\ &= 200\text{ mW} + 40\text{ mW} + 20\text{ mW} \\ &= 260\text{ mW} \end{aligned}$$

EXAMPLE:

In Figure 6-24, there are two ½ W (0.5 W) resistors connected in parallel. Should the wattage ratings for each of these resistors be increased or decreased, or can they remain the same?

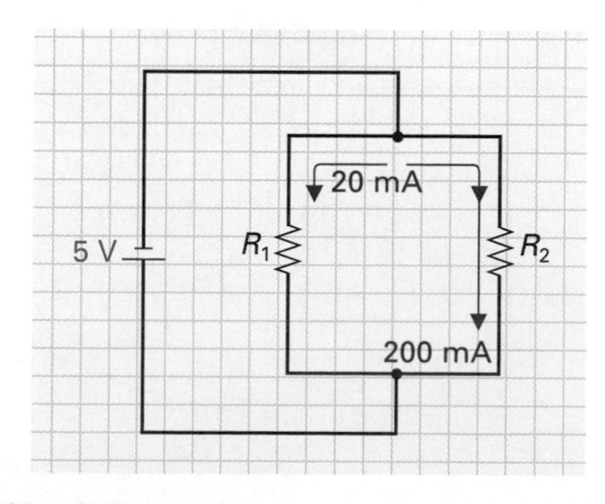

FIGURE 6-24 Parallel Circuit Example.

■ ***Solution:***

Since current and voltage are known for both branches, the power formula used in both cases can be $P = I \times V$.

$$P_1 = I \times V \qquad P_2 = I \times V$$
$$= 20 \text{ mA} \times 5 \text{ V} \qquad = 200 \text{ mA} \times 5 \text{ V}$$
$$= 0.1 \text{ W} \qquad = 1 \text{ W}$$

R_1 is dissipating 0.1 W and so a 0.5 W resistor will be fine in this application. On the other hand, R_2 is dissipating 1 W and is only designed to dissipate 0.5 W. R_2 will therefore overheat unless it is replaced with a resistor of the same ohmic value but with a 1 W or greater rating.

EXAMPLE:

Figure 6-25(a) shows a simplified diagram of an automobile external light system. A 12 V lead–acid battery is used as a source and is connected across eight parallel-connected lamps. The left and right brake lights are controlled by the brake switch, which is attached to the brake pedal. When the light switch is turned on, both the rear taillights and the low-beam headlights are brought into circuit and turned on. The high-beam set of headlights is activated

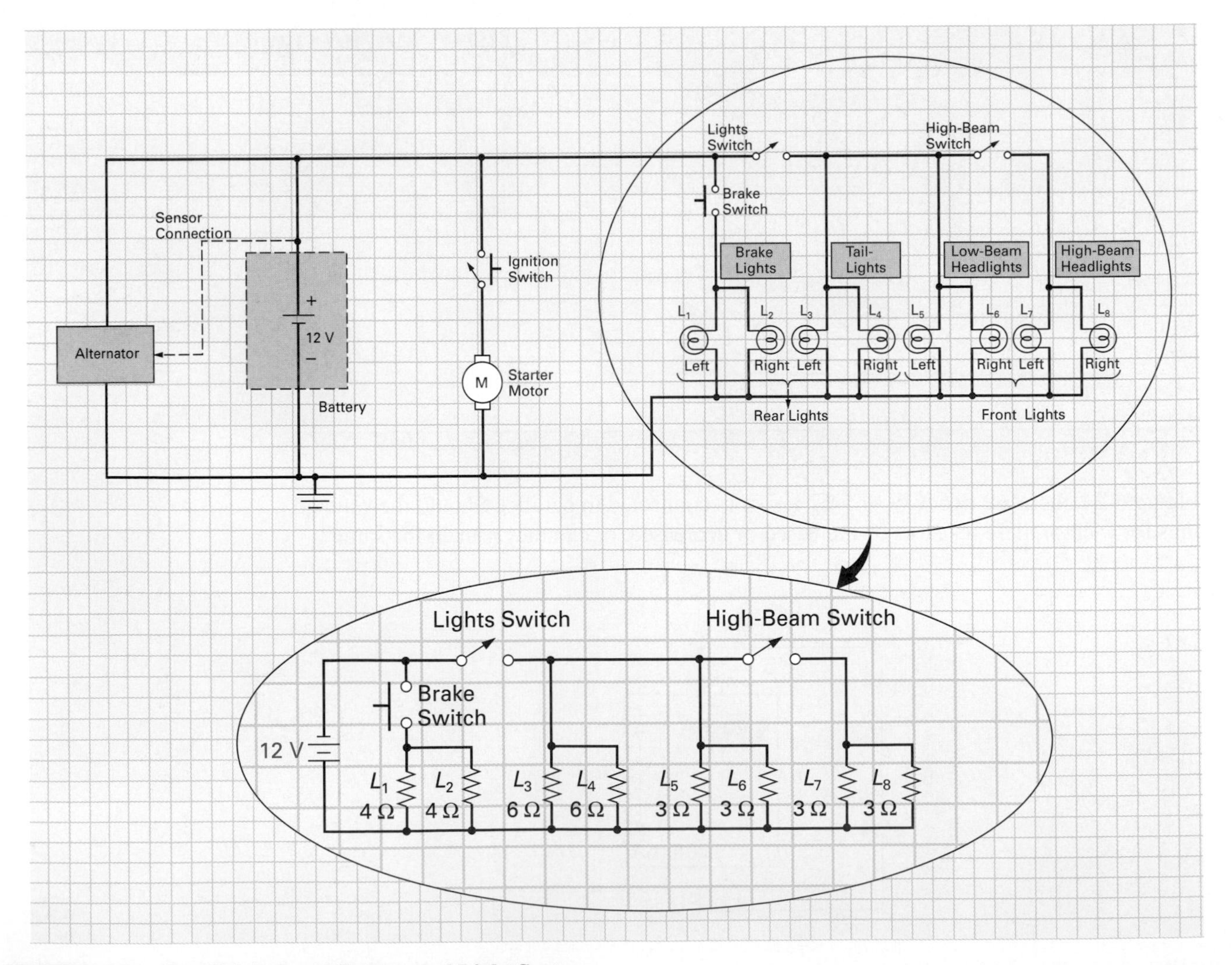

FIGURE 6-25 Parallel Automobile External Light System.

only if the high-beam switch is closed. For the lamp resistances given, calculate the output power of each lamp when in use.

Solution:

Figure 6-25(b) shows the schematic diagram of the pictorial in Figure 6-25(a). Since both V and R are known, we can use the V^2/R power formula.

Brake lights: Left lamp wattage $(P) = \frac{V^2}{R} = \frac{12\ \text{V}^2}{4\ \Omega} = 36\ \text{W}$

Right lamp wattage is the same as the left.

Taillights: Left lamp wattage $(P) = \frac{V^2}{R} = \frac{12\ \text{V}^2}{6\ \Omega} = 24\ \text{W}$

Right lamp wattage is the same as the left.

Low-beam headlights: $(P) = \frac{V^2}{R} = \frac{12\ \text{V}^2}{3\ \Omega} = 48\ \text{W}$

Each high-beam headlight is a 48 W lamp.

High-beam headlights: $(P) = \frac{V^2}{R} = \frac{12\ \text{V}^2}{3\ \Omega} = 48\ \text{W}$

Each high-beam headlight is a 48 W lamp.

SELF-TEST EVALUATION POINT FOR SECTION 6-5

Now that you have completed this section, you should be able to:

Objective 7. Describe and be able to apply all formulas associated with the calculation of voltage, current, resistance, and power in a parallel circuit.

Use the following questions to test your understanding of Section 6-5:

1. True or false: Total power in a parallel circuit can be obtained by using the same total power formula as for series circuits.
2. If $I_1 = 2$ mA and $V = 24$ V, calculate P_1.
3. If $P_1 = 22$ mW and $P_2 = 6400\ \mu$W, $P_T = ?$
4. Is it important to observe the correct wattage ratings of resistors when they are connected in parallel?

6-6 TROUBLESHOOTING A PARALLEL CIRCUIT

In the troubleshooting discussion on series circuits, we mentioned that one of three problems can occur, and these three also apply to parallel circuits:

1. A component will open.
2. A component will short.
3. A component's value will change over a period of time.

As a technician, it is important that you know how to recognize these circuit malfunctions and know how to isolate the cause and make the repair. Since you could be encountering opens, shorts, and component value changes in the parallel circuits you construct in lab, let us step through the troubleshooting procedure you should use.

As an example, Figure 6-26(a) indicates the normal readings that should be obtained from a typical parallel circuit if it is operating correctly. Keep these normal readings in mind, since they will change as circuit problems are introduced in the following section.

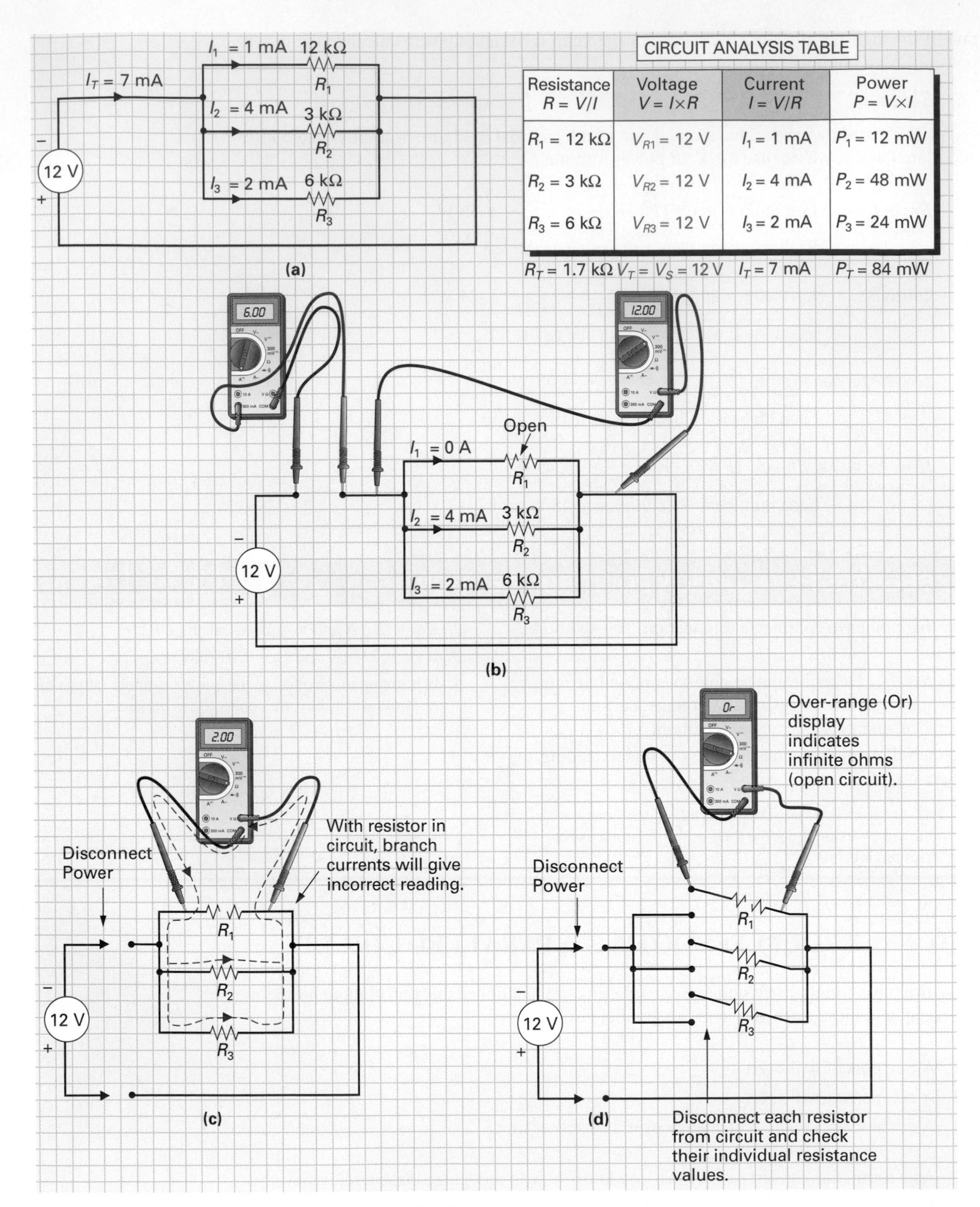

CIRCUIT ANALYSIS TABLE

Resistance $R = V/I$	Voltage $V = I \times R$	Current $I = V/R$	Power $P = V \times I$
$R_1 = 12\ k\Omega$	$V_{R1} = 12\ V$	$I_1 = 1\ mA$	$P_1 = 12\ mW$
$R_2 = 3\ k\Omega$	$V_{R2} = 12\ V$	$I_2 = 4\ mA$	$P_2 = 48\ mW$
$R_3 = 6\ k\Omega$	$V_{R3} = 12\ V$	$I_3 = 2\ mA$	$P_3 = 24\ mW$
$R_T = 1.7\ k\Omega$	$V_T = V_S = 12\ V$	$I_T = 7\ mA$	$P_T = 84\ mW$

FIGURE 6-26 Troubleshooting an Open in a Parallel Circuit.

6-6-1 *Open Component in a Parallel Circuit*

In Figure 6-26(b), R_1 has opened, so there can be no current flow through the R_1 branch. With one fewer branch for current, the parallel circuit's resistance will increase, causing the total current flow to decrease. The total current flow, which was 7 mA, will actually decrease by the amount that was flowing through the now open branch (1 mA) to a total current flow of 6 mA. If you had constructed this circuit in lab, you would probably not have noticed that there was anything wrong with the circuit until you started to make some multimeter measurements. A voltmeter measurement across the parallel circuit, as seen in Figure 6-26(b), would not have revealed any problem since the voltage measured across each resistor will always be the same and equal to the source voltage. A total current measurement, on the other hand, would indicate that there is something wrong because the measured reading does not equal the expected calculated value. Using your ammeter, you could isolate the branch fault by one of the following two methods:

1. If you measure each branch current, you will isolate the problem of R_1 because there will be no current flow through R_1. However, this could take three checks.
2. If you just measure total current (one check), you will notice that total current has decreased by 1 mA, and after making a few calculations you could determine that the only path that had a branch current of 1 mA was through R_1, so R_1 must have gone open.

We can summarize how method 2 could be used to isolate an open in this example circuit:

If R_1 opens, total current decreases by 1 mA to 6 mA ($I_T = I_1 + I_2 + I_3 = 0\text{ mA} + 4\text{ mA} + 2\text{ mA} = 6\text{ mA}$).
If R_2 opens, total current decreases by 4 mA to 3 mA ($I_T = I_1 + I_2 + I_3 = 1\text{ mA} + 0\text{ mA} + 2\text{ mA} = 3\text{ mA}$).
If R_3 opens, total current decreases by 2 mA to 5 mA ($I_T = I_1 + I_2 + I_3 = 1\text{ mA} + 4\text{ mA} + 0\text{ mA} = 5\text{ mA}$).

Method 2 can be used as long as the resistors are unequal. If they are equal, method 1 may be used to locate the 0 current branch.

In most cases, the ohmmeter is ideal for locating open circuits. To make a resistance measurement, you should disconnect power from the circuit, isolate the device from the circuit, and then connect the ohmmeter across the device. Figure 6-26(c) shows why it is necessary for you to isolate the device to be tested from the circuit. With R_1 in circuit, the ohmmeter will be measuring the parallel resistance of R_2 and R_3, even though the ohmmeter is connected across R_1. This will lead to a false reading and confuse the troubleshooting process. Disconnecting the device to be tested from the circuit, as shown in Figure 6-26(d), will give an accurate reading of R_1, then R_2, and then R_3's resistance and lead you to the cause of the problem.

EXAMPLE:

Figure 6-27 illustrates a set of three lightbulbs connected in parallel across a 9 V battery. The filament in bulb 2 has burned out, causing an open, so there is no current flow through that path. There will, however, be current flow through the other two, so bulbs 1 and 3 will be on. How would the faulty bulb be located?

Solution:

As always, a visual inspection would be your first approach. Since current is flowing through both B_1 and B_3, these two bulbs will be on. On the other hand, since B_2 is off, it would be easy for you to determine that bulb B_2 has blown.

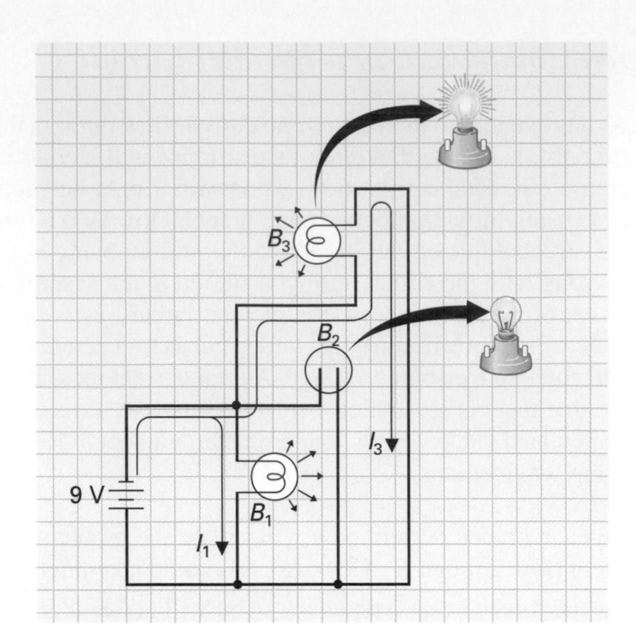

FIGURE 6-27 Open Bulb in a Parallel Circuit.

Opens in parallel paths are generally quite easy to isolate since the device in the parallel path will not operate because its power source has been disconnected.

The 3-wire holiday light string, shown in Figure 6-28, connects string sets in parallel, as can be seen in the equivalent schematic diagram. This arrangement has a distinct advantage over a series-connected one in that a failure in one string will not disable all of the strings that follow.

6-6-2 *Shorted Component in a Parallel Circuit*

In Figure 6-29(a), R_2, which has a resistance of 3 kΩ, has been shorted out. The only resistance in this center branch is the resistance of the wire, which would typically be a fraction of an ohm. In this example, let us assume that the resistance in the branch is 1 Ω, and therefore the current in this branch will attempt to increase to

$$I_2 = \frac{12 \text{ V}}{1 \ \Omega} = 12 \text{ A}$$

Before the current even gets close to 12 A, the fuse in the dc power supply will blow, as shown in the inset in Figure 6-29(a), and disconnect power from the circuit. This is typically the effect you will see from a circuit short. Replacing the fuse will be a waste of time and money since the fuse will simply blow again once power is connected across the circuit. As before, the tool to use to isolate the cause is the ohmmeter, as shown in Figure 6-29(b). To check the resistance of each branch, disconnect power from the circuit, disconnect the devices to be tested from the circuit to prevent false readings, and then test each branch until the shorted branch is located. Once you have corrected the fault, you will need to change the power supply fuse or reset its circuit breaker.

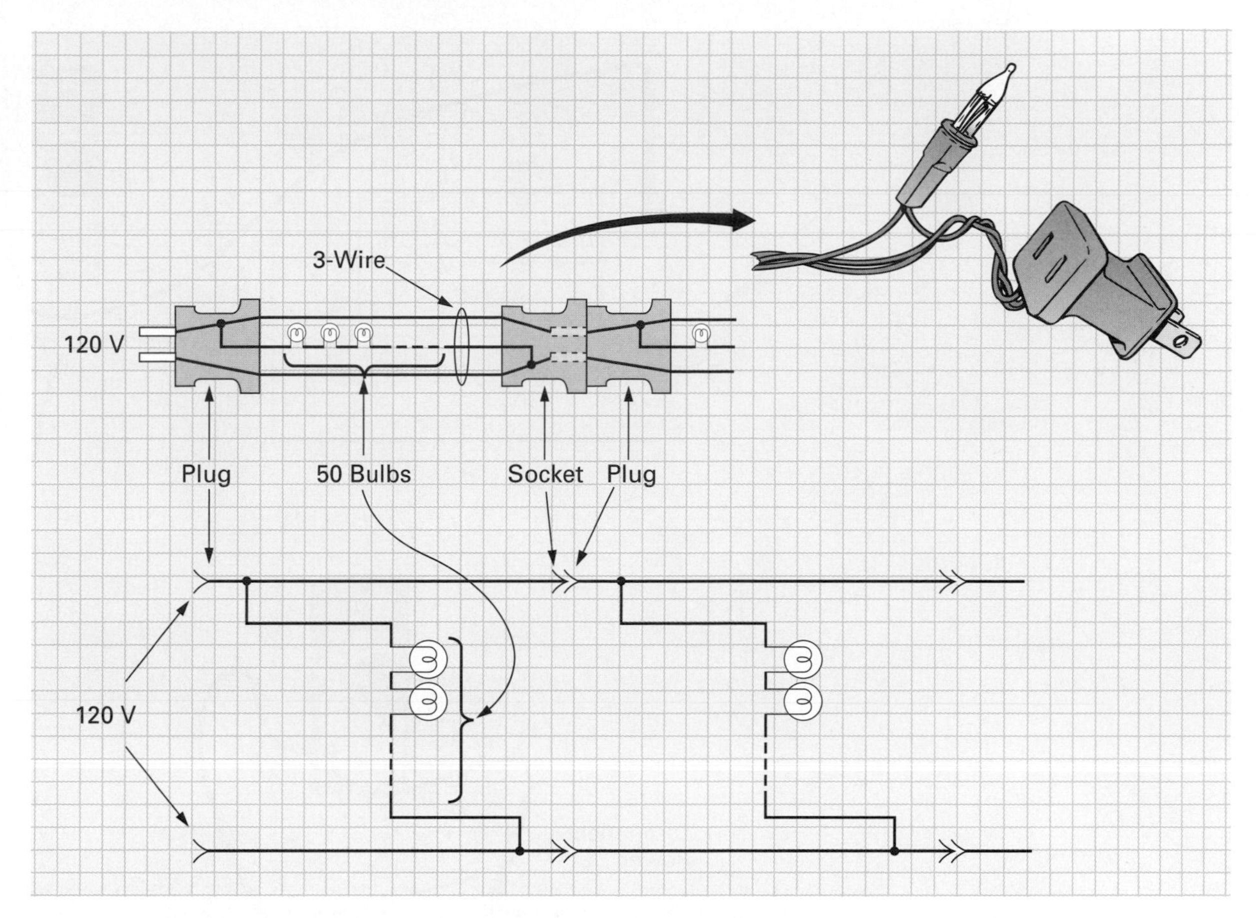

FIGURE 6-28 The 3-Wire Parallel-Connected Holiday Light String.

6-6-3 *Component Value Variation in a Parallel Circuit*

The resistive values of resistors will change with age. This increase or decrease in resistance will cause a corresponding decrease or increase in branch current and therefore total current. This deviation from the desired value can also be checked with the ohmmeter. Be careful to take into account the resistor's tolerance because this can make you believe that the resistor's value has changed when it is, in fact, within the tolerance rating.

6-6-4 *Summary of Parallel Circuit Troubleshooting*

1. An open component will prevent current flow within that branch. The total current will decrease, and the voltage across the component will be the same as the source voltage.
2. A shorted component will cause maximum current through that branch. The total current will increase and the voltage source fuse will normally blow.
3. A change in resistance value will cause a corresponding opposite change in branch current and total current.

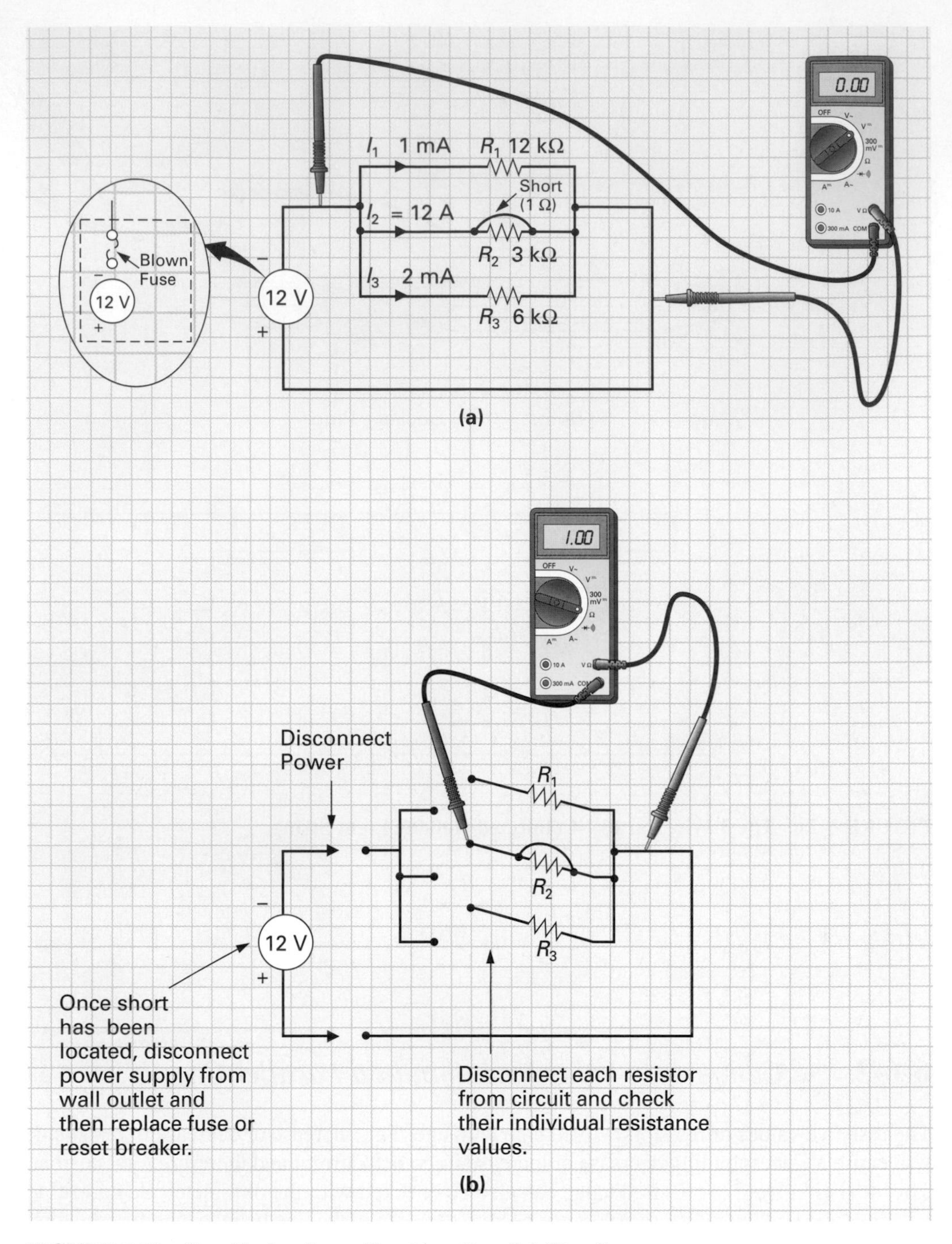

FIGURE 6-29 Troubleshooting a Short in a Parallel Circuit.

SELF-TEST EVALUATION POINT FOR SECTION 6-6

Now that you have completed this section, you should be able to:

Objective 8. Describe how a short, open, or component variation will affect a parallel circuit's operation and how it can be recognized.

Use the following questions to test your understanding of Section 6-6:

How would problems 1 through 3 be recognized in a parallel circuit?

1. An open component
2. A shorted component
3. A component's value variation
4. True or false: An ammeter is typically used to troubleshoot series circuits, whereas a voltmeter is typically used to troubleshoot parallel circuits.

SUMMARY

1. In a series circuit, there is only one path for current, whereas in parallel circuits the current has two or more paths. These paths are known as *branches*.
2. A parallel circuit is the connection of two or more components to the same voltage source so that the current can branch out over two or more paths.
3. In a parallel resistive circuit, two or more resistors are connected to the same source, so current splits to travel through each separate branch resistance.
4. Measuring the voltage drop with the voltmeter across any of the resistors in a parallel circuit is the same as measuring the voltage across the voltage source.
5. Gustav Kirchhoff was the first to observe and prove that the sum of all the branch currents ($I_1 + I_2 + I_3$, etc.) was equal to the total current (I_T).
6. Kirchhoff's current law states that the sum of all the currents entering a junction is equal to the sum of all the currents leaving that same junction.
7. When we connect resistors in parallel, the total resistance in the circuit will decrease. In fact, the total resistance in a parallel circuit will always be less than the value of the smallest resistor in the circuit.
8. Just as a series circuit is often referred to as a voltage-divider circuit, a parallel circuit is often referred to as a current-divider circuit, because the total current arriving at a junction will divide or split into branch currents (Kirchhoff's current law).
9. The current division is inversely proportional to the resistance in the branch given that the voltage across both resistors is constant and equal to the source voltage (V_S). This means that a large branch resistance will cause a small branch current ($I\downarrow = V/R\uparrow$), and a small branch resistance will cause a large branch current ($I\uparrow = V/R\downarrow$).
10. By rearranging Ohm's law, we can arrive at another formula, called the *current-divider formula,* that can be used to calculate the current through any branch of a multiple-branch parallel circuit.
11. Parallel circuits will have a larger current flow than a series circuit containing the same resistors due to the smaller total resistance.
12. To calculate exactly how much total current will flow, we need to calculate the total resistance that the parallel circuit develops or contains.
13. The reciprocal formula allows us to calculate total parallel circuit resistance for any number of resistors.
14. With parallel resistance circuits, the total resistance is always smaller than the smallest branch resistance.
15. With parallel resistive circuits, the voltage drop across any branch is equal to the voltage drop across each of the other branches and is equal to the source voltage.
16. If only two resistors are connected in parallel, a quick and easy formula called the *product-over-sum formula* can be used to calculate total resistance.
17. If resistors of equal value are connected in parallel, a special-case *equal-value formula* can be used to calculate the total resistance.
18. As with series circuits, the total power in a parallel resistive circuit is equal to the sum of all the power losses for each of the resistors in parallel.
19. An open component will prevent current flow within that branch. The total current will decrease, and the voltage across the component will be the same as the source voltage.
20. A shorted component will cause maximum current through that branch. The total current will increase and the voltage source fuse will normally blow.
21. A change in resistance value will cause a corresponding opposite change in branch current and total current.

REVIEW QUESTIONS

Multiple-Choice Questions

1. A parallel circuit has __________ path(s) for current to flow.
a. One **c.** Only three
b. Two or more **d.** None of the above

2. If a source voltage of 12 V is applied across four resistors of equal value in parallel, the voltage drops across each resistor would be equal to
a. 12 V **c.** 4 V
b. 3 V **d.** 48 V

3. What would be the voltage drop across two 25 Ω resistors in parallel if the source voltage were equal to 9 V?
a. 50 V **c.** 12 V
b. 25 V **d.** None of the above

4. If a four-branch parallel circuit has 15 mA flowing through each branch, the total current into the parallel circuit will be equal to
a. 15 mA **c.** 30 mA
b. 60 mA **d.** 45 mA

5. If the total three-branch parallel circuit current is equal to 500 mA and 207 mA is flowing through one branch and 153 mA through another, what would be the current flow through the third branch?
a. 707 mA **c.** 140 mA
b. 653 mA **d.** None of the above

6. A large branch resistance will cause a __________ branch current.
a. Large **c.** Medium
b. Small **d.** None of the above are true

7. What would be the conductance of a 1 kΩ resistor?
a. 10 mS **c.** 2 kΩ
b. 1 mS **d.** All of the above

8. If only two resistors are connected in parallel, the total resistance equals
a. The sum of the resistance values
b. Three times the value of one resistor
c. The product over the sum
d. All of the above

9. If resistors of equal value are connected in parallel, the total resistance can be calculated by
a. One resistor value divided by the number of parallel resistors
b. The sum of the resistor values
c. The number of parallel resistors divided by one resistor value
d. All of the above could be true.

10. The total power in a parallel circuit is equal to the
a. Product of total current and total voltage
b. Reciprocal of the individual power losses
c. Sum of the individual power losses
d. Both (a) and (b)
e. Both (a) and (c)

Communication Skill Questions

11. Describe the difference between a series and a parallel circuit. (6-1)

12. Explain and state mathematically the situation regarding voltage in a parallel circuit. (6-2)

13. State Kirchhoff's current law for parallel circuits. (6-3)

14. What is the current-divider formula? (6-3)

15. List the formulas for calculating the following total resistances: (6-4)
a. Two resistors of different values
b. More than two resistors of different values
c. Equal-value resistors

16. Describe the relationship between branch current and branch resistance. (6-3)

17. Briefly describe total and individual power in a parallel resistive circuit. (6-5)

18. Discuss troubleshooting parallel circuits as applied to: (6-6)
a. A shorted component
b. An open component

19. Explain why parallel circuits have a smaller total resistance and larger total current than series circuits. (6-3)

20. Briefly describe why Kirchhoff's voltage law applies to series circuits and why Kirchhoff's current law relates to parallel circuits. (6-3)

Practice Problems

21. Calculate the total resistance of four 30 kΩ resistors in parallel.

22. Find the total resistance for each of the following parallel circuits:
a. 330 Ω and 560 Ω
b. 47 kΩ, 33 kΩ, and 22 kΩ
c. 2.2 MΩ, 3 kΩ, and 220 Ω

23. If 10 V is connected across three 25 Ω resistors in parallel, what will be the total and individual branch currents?

24. If a four-branch parallel circuit has branch currents equal to 25 mA, 37 mA, 220 mA, and 0.2 A, what is the total circuit current?

25. If three resistors of equal value are connected across a 14 V supply and the total resistance is equal to 700 Ω, what is the value of each branch current?

26. If three 75 W lightbulbs are connected in parallel across a 110 V supply, what is the value of each branch current? What is the branch current through the other two lightbulbs if one burns out?

27. If 33 kΩ and 22 kΩ resistors are connected in parallel across a 20 V source, calculate:
    a. Total resistance
    b. Total current
    c. Branch currents
    d. Total power dissipated
    e. Individual resistor power dissipated
28. If four parallel-connected resistors are each dissipating 75 mW, what is the total power being dissipated?
29. Calculate the branch currents through the following parallel resistor circuits when they are connected across a 10 V supply:
    a. 22 kΩ and 33 kΩ
    b. 220 Ω, 330 Ω, and 470 Ω
30. If 30 Ω and 40 Ω resistors are connected in parallel, which resistor will generate the greatest amount of heat?
31. Calculate the total conductance and resistance of the following parallel circuits:
    a. Three 5 Ω resistors
    b. Two 200 Ω resistors
    c. 1 MΩ, 500 MΩ, 3.3 MΩ
    d. 5 Ω, 3 Ω, 2 Ω
32. Connect the three resistors in Figure 6-30 in parallel across a 12 V battery and then calculate the following:
    a. V_{R1}, V_{R2}, V_{R3}
    b. I_1, I_2, I_3
    c. I_T
    d. P_T
    e. P_1, P_2, P_3
    f. G_{R1}, G_{R2}, G_{R3}

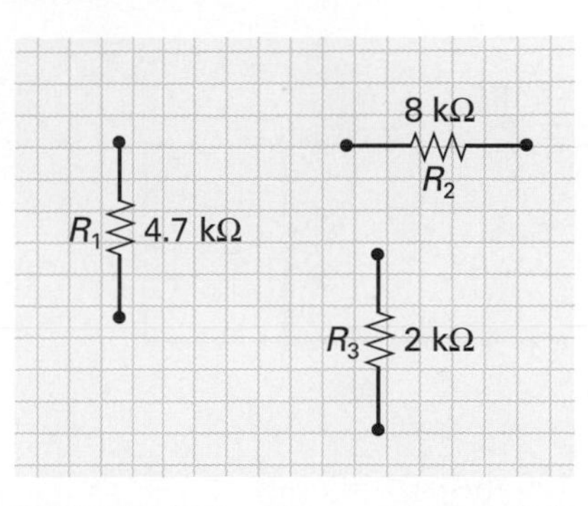

FIGURE 6-30 Connect in Parallel across a 12 Volt Source.

33. Calculate R_T in Figure 6-31(a), (b), (c), and (d).
34. Calculate the branch currents through four 60 W bulbs connected in parallel across 110 V. How much is the total current, and what would happen to the total current if one of the bulbs were to burn out? What change would occur in the remaining branch currents?
35. Calculate the following in Figure 6-32:
    a. I_2
    b. I_T
    c. V_S, I_1, I_2
    d. R_2, I_1, I_2, P_T

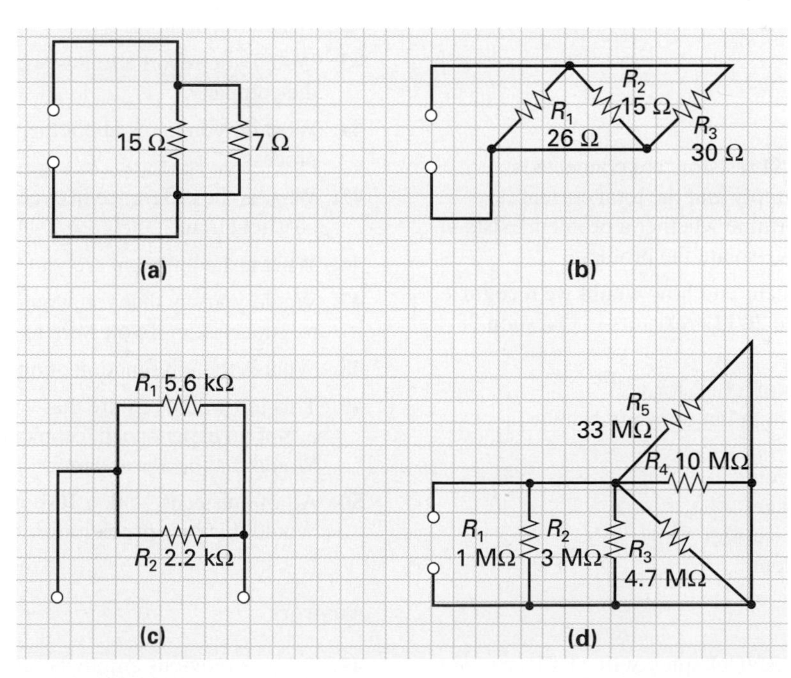

FIGURE 6-31 Calculate Total Resistance.

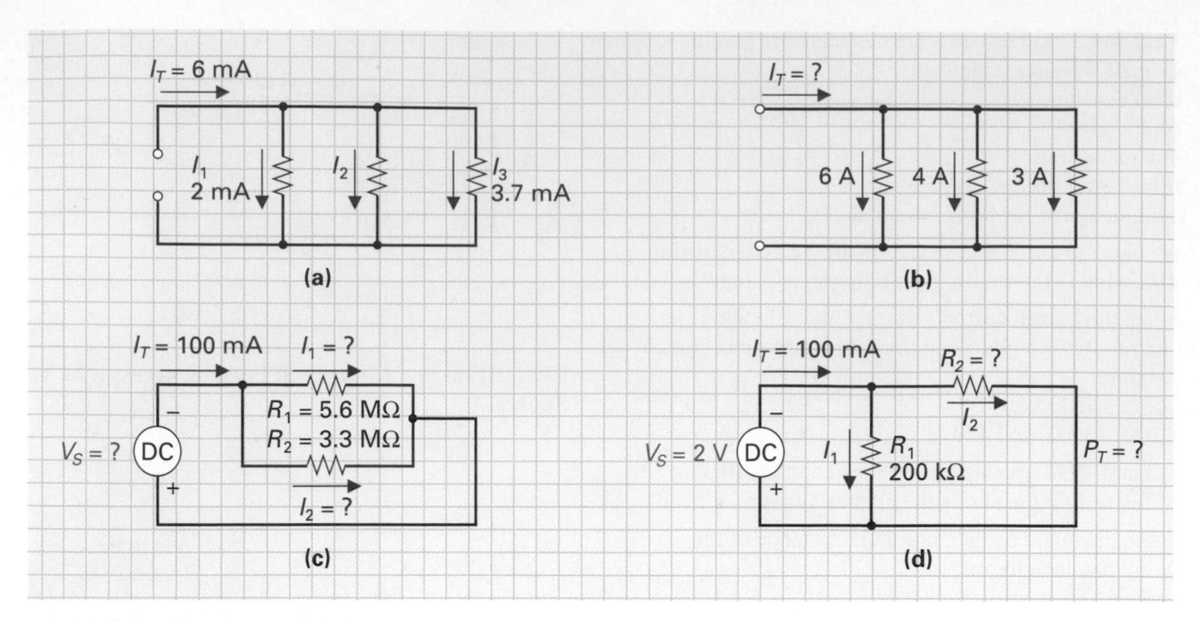

FIGURE 6-32 Calculate the Unknown.

Troubleshooting Questions

36. An open component in a parallel circuit will cause ___________ current flow within that branch, which will cause the total current to ___________.

a. Maximum, increase
b. Zero, decrease
c. Maximum, decrease
d. Zero, increase

37. A shorted component in a parallel circuit will cause ___________ current through a branch, and consequently the total current will ___________.

a. Maximum, increase
b. Zero, decrease
c. Maximum, decrease
d. Zero, increase

38. If a 10 kΩ and two 20 kΩ resistors are connected in parallel across a 20 V supply and the total current measured is 2 mA, determine whether a problem exists in the circuit and, if it does, isolate the problem.

39. What situation would occur and how would we recognize the problem if one of the 20 kΩ resistors in Question 38 were to short?

40. With age, the resistance of a resistor will ___________, resulting in a corresponding but opposite change in ___________.

a. Increase, branch current
b. Change, source voltage
c. Decrease, source resistance
d. Change, branch current

Job Interview Test

These tests at the end of each chapter will challenge your knowledge up to this point and give you the practice you need for a job interview. To make this more realistic, the test will be comprised of both technical and personal questions. In order to take full advantage of this exercise, you may want to set up a simulation of the interview environment, have a friend read the questions to you, and record your responses for later analysis.

Company Name: Pony, Inc.
Industry Branch: Medical Electronics.
Function: Manufacture ICU Equipment.
Job Title: Production Test Technician.

41. Who was your last employer?

42. What are Kirchhoff's laws?

43. Would you report a fellow worker who was stealing from the company?

44. What do you know about the product development process?

45. What is the difference between a series circuit and a parallel circuit?

46. What test equipment are you familiar with?

47. Would you say that you would need close, moderate, or no supervision if you were to get this job?

48. What is the job function of a production test technician?

49. This job would require that you work under the watchful eye of an experienced technician for at least 6 months. Would that be a problem for you?

50. All of our techs work a 10-hour day, 4 days a week. Would this be a problem for you?

Answers

41. Discuss previous employer and give a copy of letter of recommendation. If this job was not technical, refer back to how the lab work during training was structured like a work environment, and discuss other related topics.

42. Sections 5-4 and 6-3.
43. Yes, of course.
44. Describe the steps detailed in the Introduction section.
45. Chapter 6 introduction.
46. Up to this time, only the multimeter.
47. Say that you expect in the beginning you would have many questions, but as you became more familiar with company procedures and the product line, you would need less and less supervision.
48. Quote introductory section in text and job description listed in newspaper.
49. This would be an opportunity, since experienced technicians have a wealth of knowledge they can impart.
50. If you are unsure, don't object to anything at the time. Later, if you decide that it is something you do not want to do, you can call and decline the offer.

Web Site Questions

Go to the web site http://www.prenhall.com/cook, select the textbook *Introductory DC/AC Electronics* or *Introductory DC/AC Circuits,* this chapter, and then follow the instructions when answering the multiple-choice practice problems.

Series–Parallel Circuits

Charles Wheatstone

The Christie Bridge Circuit

Who invented the Wheatstone bridge circuit? It was obviously Sir Charles Wheatstone. Or was it?

The Wheatstone bridge was actually invented by S. H. Christie of the Royal Military Academy at Woolwich, England. He described the circuit in detail in a *Philosophical Transactions* paper dated February 28, 1833. Christie's name, however, was unknown and his invention was ignored.

Ten years later, Sir Charles Wheatstone called attention to Christie's circuit. Sir Charles was very well known, and from that point on, and even to this day, the circuit is known as a Wheatstone bridge. Later, Werner Siemens would modify Christie's circuit and invent the variable-resistance arm bridge circuit, which would also be called a Wheatstone bridge.

No one has given full credit to the real inventors of these bridge circuits, until now!

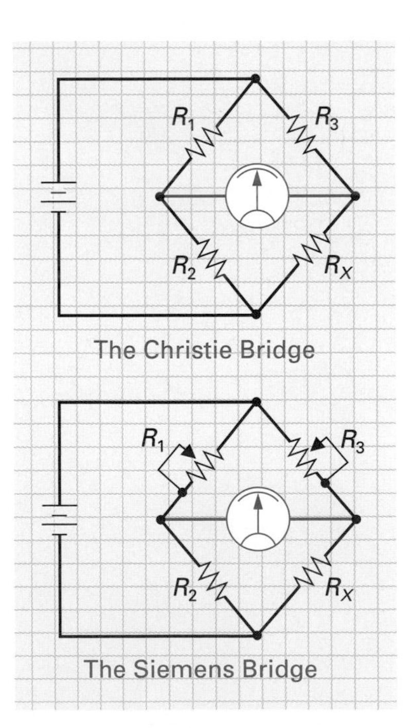

7

Outline and Objectives

THE CHRISTIE BRIDGE CIRCUIT
INTRODUCTION

7-1 SERIES- AND PARALLEL-CONNECTED COMPONENTS

Objective 1: Identify the differences among a series, a parallel, and a series–parallel circuit.

7-2 TOTAL RESISTANCE IN A SERIES–PARALLEL CIRCUIT

Objective 2: Describe how to use a three-step procedure to determine total resistance.

7-3 VOLTAGE DIVISION IN A SERIES–PARALLEL CIRCUIT

7-4 BRANCH CURRENTS IN A SERIES–PARALLEL CIRCUIT

7-5 POWER IN A SERIES–PARALLEL CIRCUIT

7-6 FIVE-STEP METHOD FOR SERIES–PARALLEL CIRCUIT ANALYSIS

Objective 3: Describe for the series–parallel circuit how to use a five-step procedure to calculate:

a. Total resistance.
b. Total current.
c. Voltage division.
d. Branch current.
e. Total power dissipated.

7-7 SERIES–PARALLEL CIRCUITS

7-7-1 Loading of Voltage-Divider Circuits

Objective 4: Explain what loading effect a piece of equipment will have when connected to a voltage divider.

7-7-2 The Wheatstone Bridge

Objective 5: Identify and describe the Wheatstone bridge circuit in both the balanced and unbalanced condition.

7-7-3 The R–$2R$ Ladder Circuit

Objective 6: Describe the R–$2R$ ladder circuit used for digital-to-analog conversion.

7-8 TROUBLESHOOTING SERIES–PARALLEL CIRCUITS

Objective 7: Explain how to identify the following problems in a series–parallel circuit:

a. Open series resistor.
b. Open parallel resistor.
c. Shorted series resistor.
d. Shorted parallel resistor.
e. Resistor value variation.

7-8-1 Open Component

7-8-2 Shorted Component

7-8-3 Resistor Value Variation

7-9 THEOREMS FOR DC CIRCUITS

Objective 8: Describe the differences between a voltage and a current source.

Objective 9: Analyze series–parallel networks using:

a. The superposition theorem.
b. Thévenin's theorem.
c. Norton's theorem.

7-9-1 Voltage and Current Sources

7-9-2 Superposition Theorem

7-9-3 Thévenin's Theorem

7-9-4 Norton's Theorem

Introduction

In the two previous chapters, you were introduced first to series circuits and then to parallel circuits. To review the key differences, let's compare these circuit types to a single-lane and multilane road. Cars traveling down a single-lane road have to travel end-on-end (in series), which naturally offers a comparatively high resistance to the passage of cars ($R\uparrow$), resulting in a low traffic flow value ($I\downarrow$). Conversely, cars traveling down a multilane road can travel side-by-side (in parallel), which naturally offers a comparatively low resistance to the passage of cars ($R\downarrow$), resulting in a high traffic flow value ($I\uparrow$).

In most applications, circuits are rarely as straightforward as series or parallel. In most instances, you will come across a combination circuit that has both series-connected and parallel-connected components.

7-1 SERIES- AND PARALLEL-CONNECTED COMPONENTS

Series–Parallel Circuit
Network or circuit that contains components that are connected in both series and parallel.

Figure 7-1(a) through (f) shows six examples of **series–parallel resistive circuits.** The most important point to learn is how to distinguish between the resistors that are connected in series and the resistors that are connected in parallel, which will take a little practice. One thing that you may not have noticed when examining Figure 7-1 is that

Circuit 7-1(a) is equivalent to 7-1(b)
Circuit 7-1(c) is equivalent to 7-1(d)
Circuit 7-1(e) is equivalent to 7-1(f)

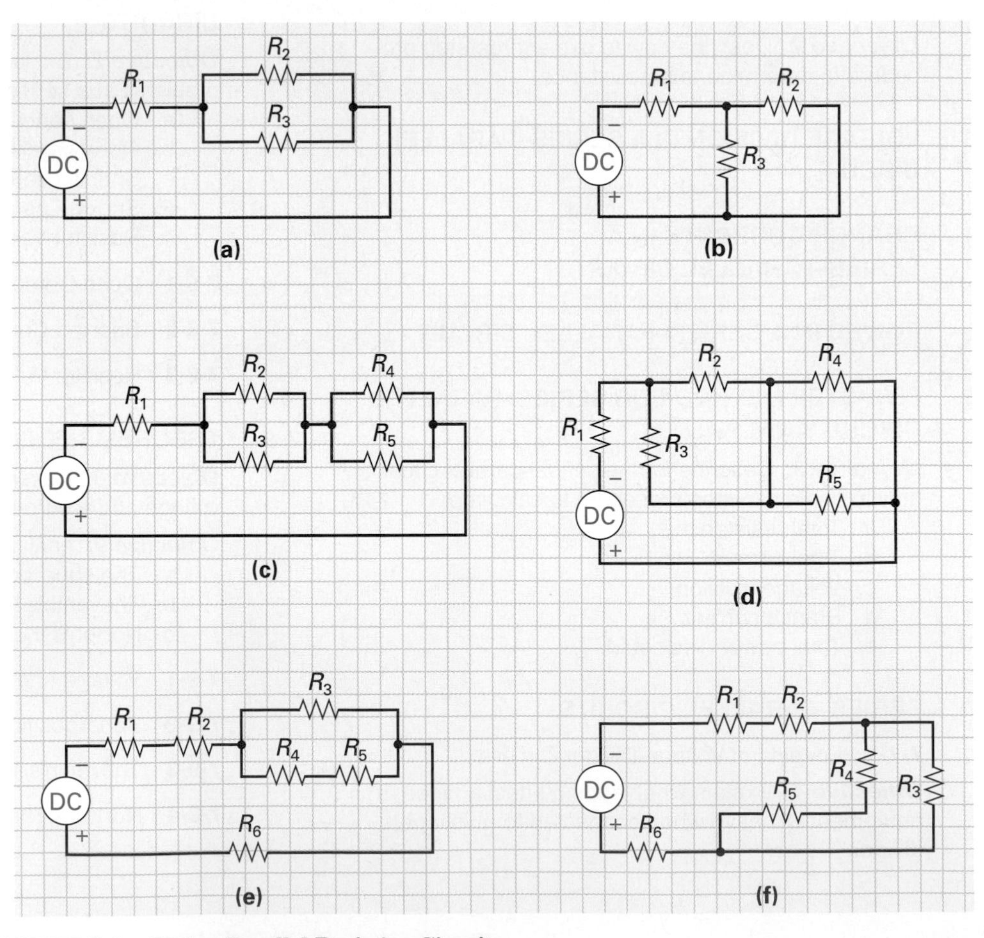

FIGURE 7-1 **Series–Parallel Resistive Circuits.**

When analyzing these series–parallel circuits, always remember that current flow determines whether the resistor is connected in series or parallel. Begin at the negative side of the battery and apply these two rules:

1. If the total current has only one path to follow through a component, that component is connected in series.
2. If the total current has two or more paths to follow through two or more components, those components are connected in parallel.

Referring again to Figure 7-1, you can see that series or parallel resistor networks are easier to identify in parts (a), (c), and (e) than in parts (b), (d), and (f). Redrawing the circuit so that the components are arranged from left to right or from top to bottom is your first line of attack in your quest to identify series- and parallel-connected components.

EXAMPLE:

Refer to Figure 7-2 and identify which resistors are connected in series and which are in parallel.

Solution:

First, let's redraw the circuit so that the components are aligned either from left to right as shown in Figure 7-3(a) or from top to bottom as shown in Figure 7-3(b). Placing your pencil at the negative terminal of the battery on whichever figure you prefer, either Figure 7-3(a) or (b), trace the current paths through the circuit toward the positive side of the battery, as illustrated in Figure 7-4.

The total current arrives first at R_1. There is only one path for current to flow, which is through R_1, and therefore R_1 is connected in series. The total current proceeds on past R_1 and arrives at a junction where current divides and travels through two branches, R_2 and R_3. Since current had to split into two paths, R_2 and R_3 are therefore connected in parallel. After the parallel connection of R_2 and R_3, total current combines and travels to the positive side of the battery.

In this example, therefore, R_1 is in series with the parallel combination of R_2 and R_3.

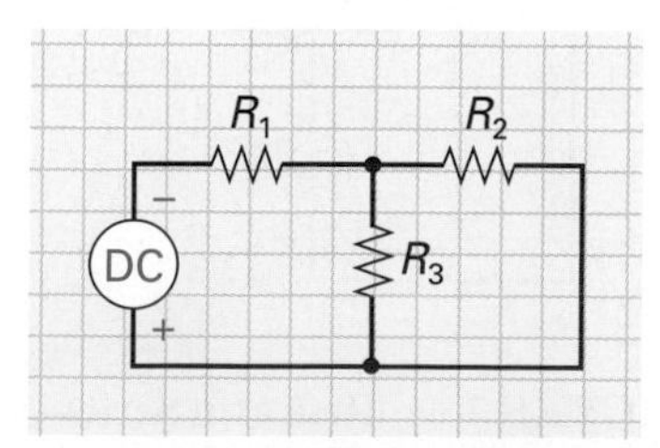

FIGURE 7-2 Series–Parallel Circuit Example.

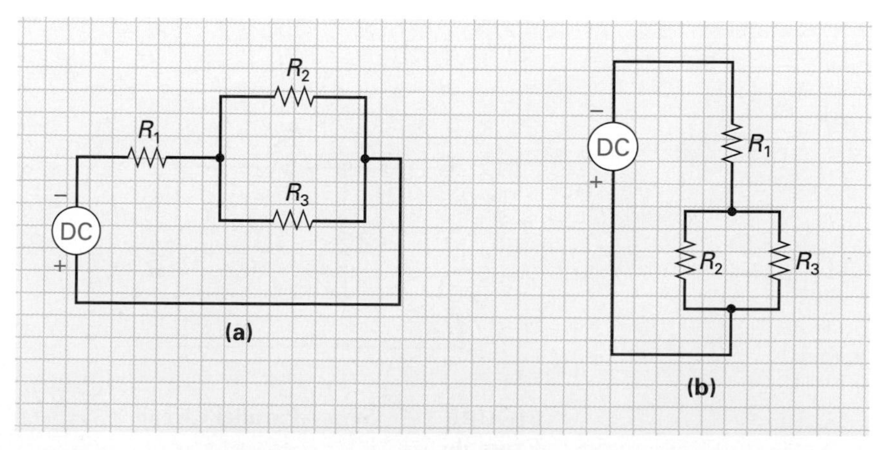

FIGURE 7-3 Redrawn Series–Parallel Circuit. (a) Left to Right. (b) Top to Bottom.

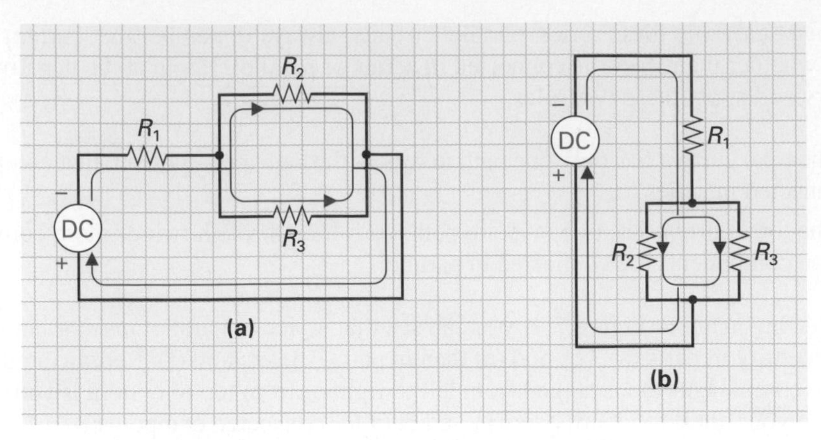

FIGURE 7-4 Tracing Current through a Series–Parallel Circuit.

EXAMPLE:

Refer to Figure 7-5 and identify which resistors are connected in series and which are connected in parallel.

Solution:

Figure 7-6 illustrates the simplified, redrawn schematic of Figure 7-5. Total current leaves the negative terminal of the battery, and all of this current has to travel through R_1, which is therefore a series-connected resistor. Total current will split at junction A, and consequently, R_3 and R_4 with R_2 make up a parallel combination. The current that flows through R_3 (I_2) will also flow through R_4 and therefore R_3 is in series with R_4. I_1 and I_2 branch currents combine at junction B to produce total current, which has only one path to follow through the series resistor R_5 and, finally, to the positive side of the battery.

In this example, therefore, R_3 and R_4 are in series with one another and both are in parallel with R_2, and this combination is in series with R_1 and R_5.

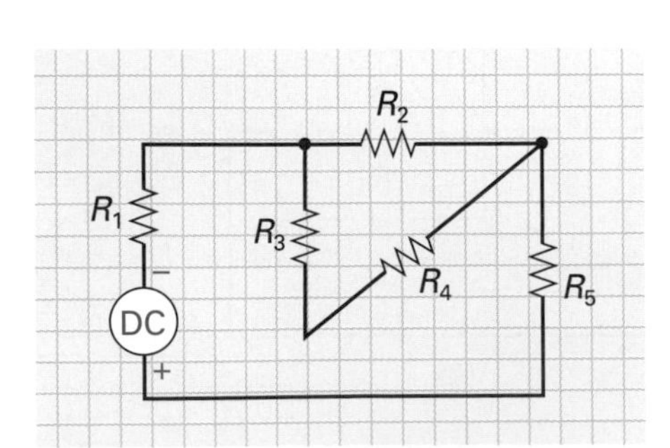

FIGURE 7-5 Series–Parallel Circuit Example.

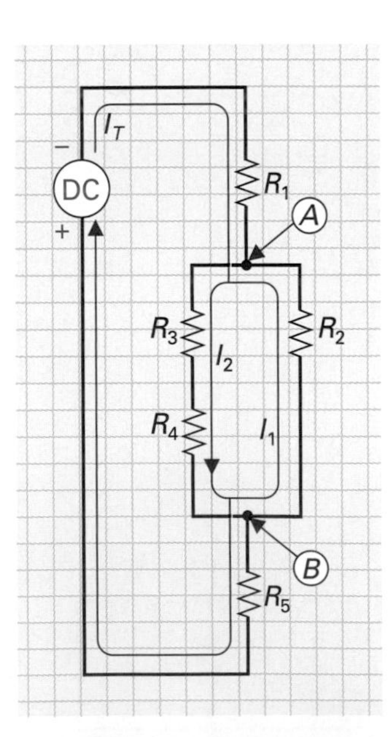

FIGURE 7-6 Redrawn Series–Parallel Circuit Example.

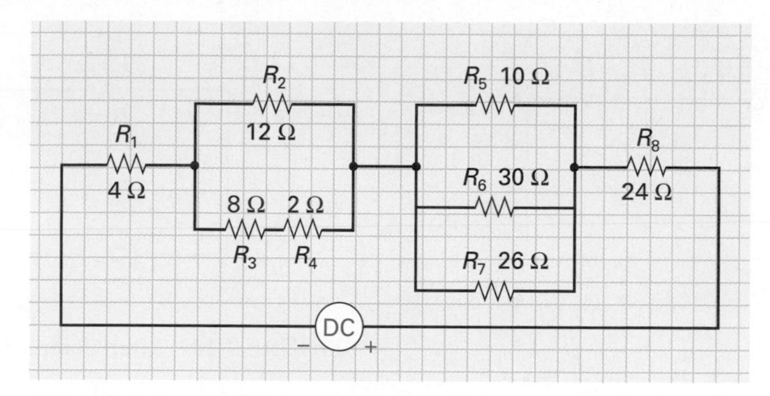

FIGURE 7-7 **Total Series–Parallel Circuit Resistance.**

7-2 TOTAL RESISTANCE IN A SERIES–PARALLEL CIRCUIT

No matter how complex or involved the series–parallel circuit, there is a simple three-step method to simplify the circuit to a single equivalent total resistance. Figure 7-7 illustrates an example of a series–parallel circuit. Once you have analyzed and determined the series–parallel relationship, we can proceed to solve for total resistance.

The three-step method is

Step A: Determine the equivalent resistances of all branch series-connected resistors.

Step B: Determine the equivalent resistances of all parallel-connected combinations.

Step C: Determine the equivalent resistance of the remaining series-connected resistances.

Let's now put this procedure to work with the example circuit in Figure 7-7.

STEP A Solve for all branch series-connected resistors. In our example, this applies only to R_3 and R_4, and since this is a series connection, we have to use the series resistance formula.

$$R_{3,4} = R_3 + R_4 = 8 + 2 = 10\ \Omega \qquad \text{(series resistance formula)}$$

With R_3 and R_4 solved, the circuit now appears as indicated in Figure 7-8.

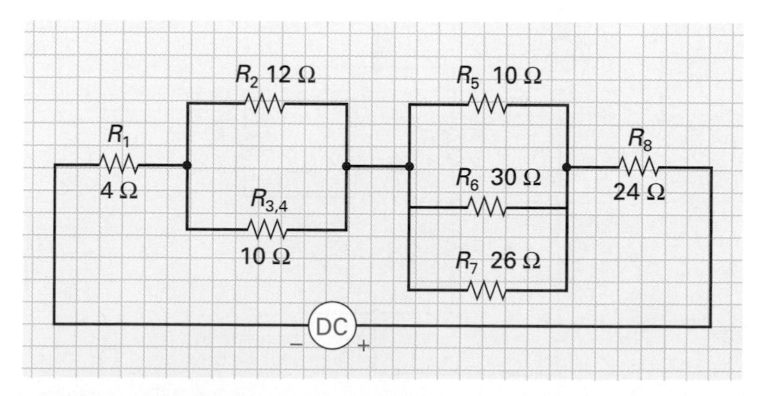

FIGURE 7-8 **After Completing Step A.**

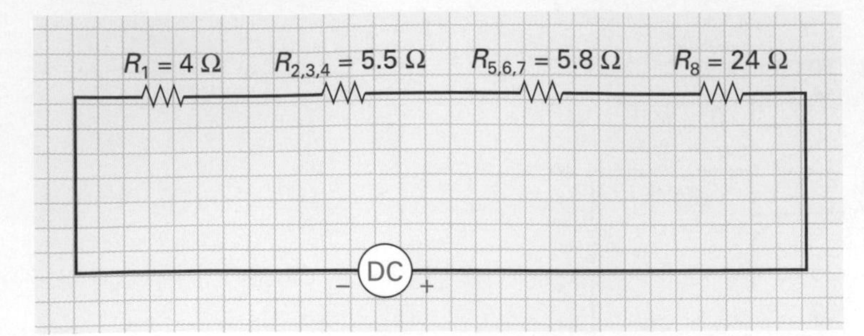

FIGURE 7-9 After Completing Step B.

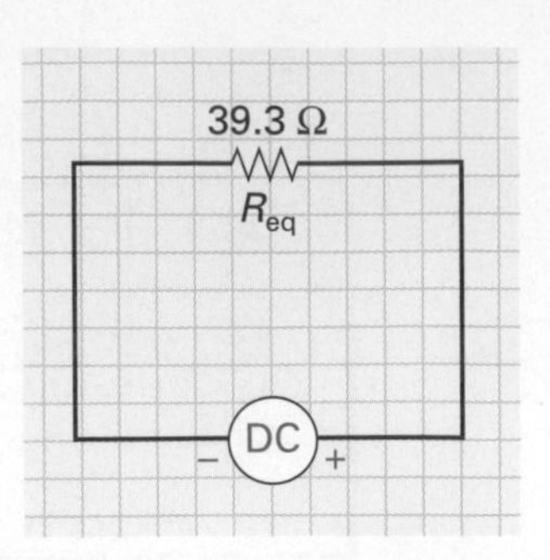

FIGURE 7-10 After Completing Step C.

STEP B Solve for all parallel combinations. In this example, they are the two parallel combinations of (a) R_2 and $R_{3,4}$ and (b) R_5 and R_6 and R_7. Since these are parallel connections, use the parallel resistance formulas.

$$R_{2,3,4} = \frac{R_2 \times R_{3,4}}{R_2 + R_{3,4}} = \frac{12 \times 10}{12 + 10} = 5.5\ \Omega \quad \text{(product-over-sum formula)}$$

$$R_{5,6,7} = \frac{1}{(1/R_5) + (1/R_6) + (1/R_7)} = 5.8\ \Omega \quad \text{(reciprocal formula)}$$

With $R_{2,3,4}$ and $R_{5,6,7}$ solved, the circuit now appears as illustrated in Figure 7-9.

STEP C Solve for the remaining series resistances. There are now four remaining series resistances, which can be reduced to one equivalent resistance (R_{eq}) or total resistance (R_T). As seen in Figure 7-10, by using the series resistance formula, the total equivalent resistance for this example circuit will be

$$\begin{aligned} R_{eq} &= R_1 + R_{2,3,4} + R_{5,6,7} + R_8 \\ &= 4\ \Omega + 5.5\ \Omega + 5.8\ \Omega + 24\ \Omega \\ &= 39.3\ \Omega \end{aligned}$$

EXAMPLE:

Find the total resistance of the circuit in Figure 7-11.

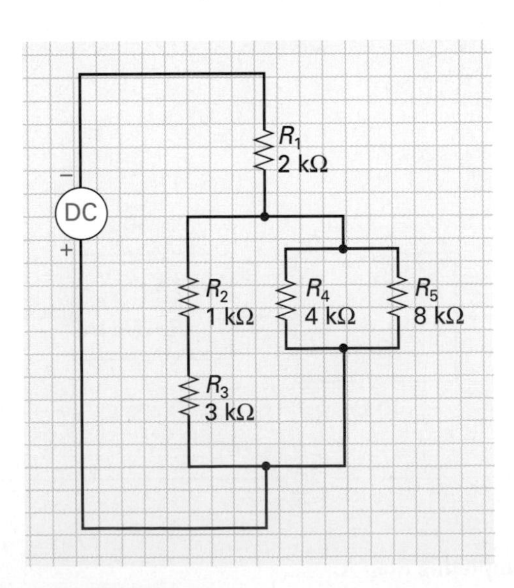

FIGURE 7-11 Calculate Total Resistance.

■ *Solution:*

STEP A Solve for all branch series-connected resistors. This applies to R_2 and R_3 (series connection):

$$R_{2,3} = R_2 + R_3 = 1\ \text{k}\Omega + 3\ \text{k}\Omega = 4\ \text{k}\Omega$$

The resulting circuit, after completing step A, is illustrated in Figure 7-12(a).

STEP B Solve for all parallel combinations. Looking at Figure 7-12(a), which shows the circuit resulting from step A, you can see that current branches into three paths, so the parallel reciprocal formula must be used for this step.

$$R_{2,3,4,5} = \frac{1}{(1/R_{2,3}) + (1/R_4) + (1/R_5)}$$

$$= \frac{1}{(1/4\ \text{k}\Omega) + (1/4\ \text{k}\Omega) + (1/8\ \text{k}\Omega)} = 1.6\ \text{k}\Omega$$

The resulting circuit, after completing step B, is illustrated in Figure 7-12(b).

STEP C Solve for the remaining series resistances. Looking at Figure 7-12(b), which shows the circuit resulting from step B, you can see that there are two remaining series resistances. The equivalent resistance (R_{eq}) is therefore equal to

$$R_{eq} = R_1 + R_{2,3,4,5} = 2\ \text{k}\Omega + 1.6\ \text{k}\Omega = 3.6\ \text{k}\Omega$$

The total equivalent resistance, after completing all three steps, is illustrated in Figure 7-12(c).

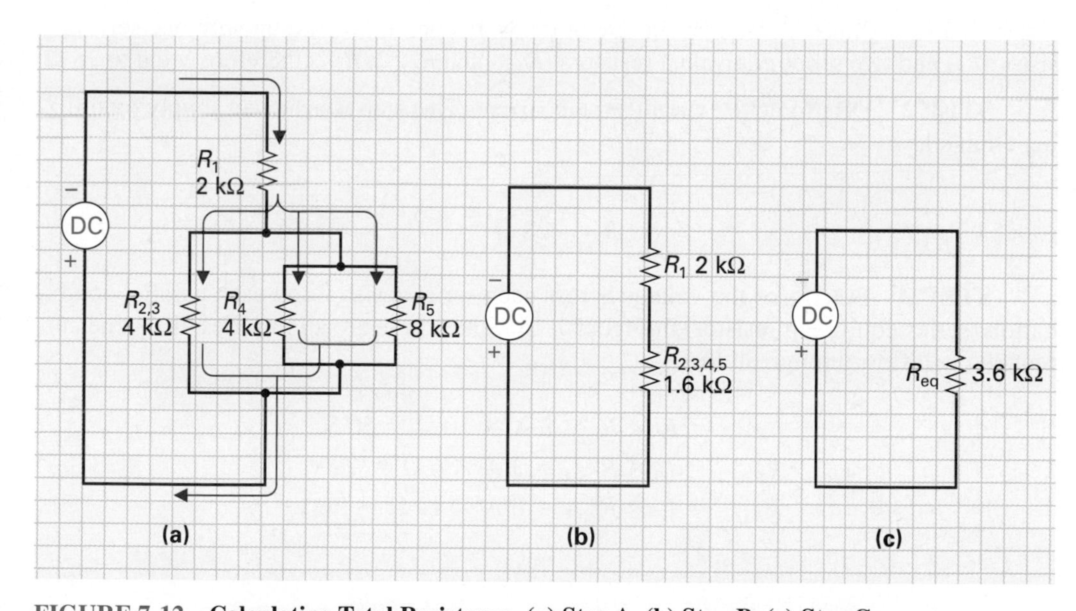

FIGURE 7-12 **Calculating Total Resistance. (a) Step A. (b) Step B. (c) Step C.**

SELF-TEST EVALUATION POINT FOR SECTIONS 7-1 AND 7-2

Now that you have completed these sections, you should be able to:

Objective 1. Identify the differences among a series, a parallel, and a series–parallel circuit.

Objective 2. Describe how to use a three-step procedure to determine total resistance.

Use the following questions to test your understanding of Sections 7-1 and 7-2:

1. How can we determine which resistors are connected in series and which are connected in parallel in a series–parallel circuit?
2. Calculate the total resistance if two series-connected 12 kΩ resistors are connected in parallel with a 6 kΩ resistor.
3. State the three-step procedure used to determine total resistance in a circuit made up of both series and parallel resistors.
4. Sketch the following series–parallel resistor network made up of three resistors. R_1 and R_2 are in series with each other and are connected in parallel with R_3. If R_1 = 470 Ω, R_2 = 330 Ω, and R_3 = 270 Ω, what is R_T?

7-3 VOLTAGE DIVISION IN A SERIES–PARALLEL CIRCUIT

There is a simple three-step procedure for finding the voltage drop across each part of the series–parallel circuit. Figure 7-13 illustrates an example of a series–parallel circuit to which we will apply the three-step method for determining voltage drop.

STEP 1 Determine the circuit's total resistance. This is achieved by following the three-step method used previously for calculating total resistance.

Step A: $R_{3,4} = 4 + 8 = 12\ \Omega$

Step B: $R_{2,3,4} = \dfrac{1}{(1/R_2) + (1/R_{3,4})} = 6\ \Omega$

$$R_{5,6,7} = \frac{1}{(1/R_5) + (1/R_6) + (1/R_7)} = 12\ \Omega$$

Figure 7-14 illustrates the equivalent circuit up to this point. We end up with one series resistor (R_1) and two series equivalent resistors ($R_{2,3,4}$ and $R_{5,6,7}$). R_T is therefore equal to 28 Ω.

STEP 2 Determine the circuit's total current. This step is achieved simply by utilizing Ohm's law.

$$I_T = \frac{V_T}{R_T} = \frac{84\text{ V}}{28\ \Omega} = 3\text{ A}$$

STEP 3 Determine the voltage across each series resistor and each parallel combination (series equivalent resistor) in Figure 7-14. Since these are all in series, the same current (I_T) will flow through all three.

$$V_{R1} = I_T \times R_1 = 3\text{ A} \times 10\ \Omega = 30\text{ V}$$
$$V_{R2,3,4} = I_T \times R_{2,3,4} = 3\text{ A} \times 6\ \Omega = 18\text{ V}$$
$$V_{R5,6,7} = I_T \times R_{5,6,7} = 3\text{ A} \times 12\ \Omega = 36\text{ V}$$

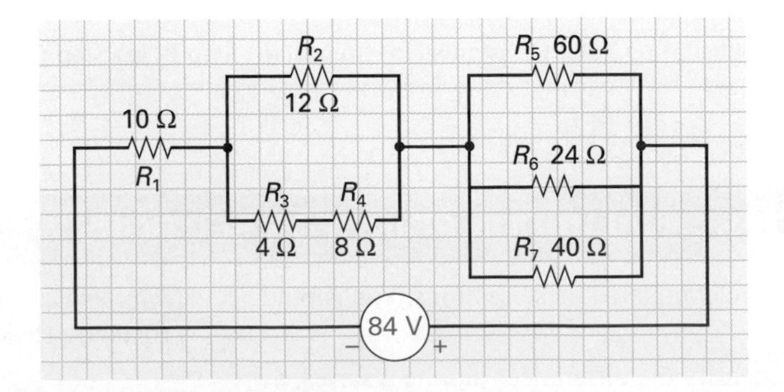

FIGURE 7-13 **Series–Parallel Circuit Example.**

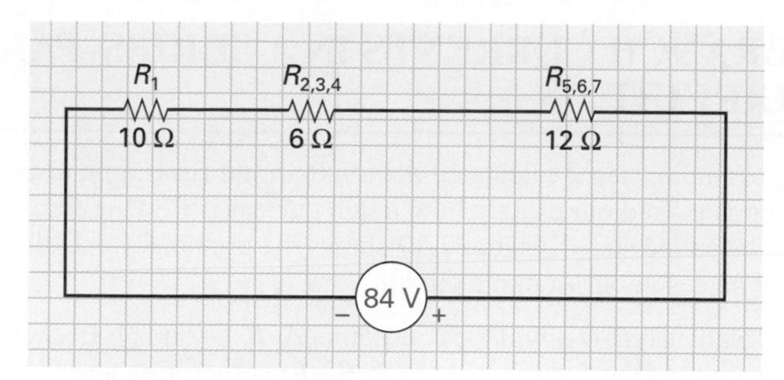

FIGURE 7-14 **After Completing Step 1.**

The voltage drops across the series resistor (R_1) and series equivalent resistors ($R_{2,3,4}$ and $R_{5,6,7}$) are illustrated in Figure 7-15.

Kirchhoff's voltage law states that the sum of all the voltage drops is equal to the source voltage applied. This law can be used to confirm that our calculations are all correct:

$$
\begin{aligned}
V_T &= V_{R1} + V_{R2,3,4} + V_{R5,6,7} \\
&= 30\text{ V} + 18\text{ V} + 36\text{ V} \\
&= 84\text{ V}
\end{aligned}
$$

To summarize, refer to Figure 7-16, which shows these voltage drops inserted into our original circuit. As you can see from this illustration,

30 V is dropped across R_1.
18 V is dropped across R_2.
18 V is dropped across both R_3 and R_4.
36 V is dropped across R_5.
36 V is dropped across R_6.
36 V is dropped across R_7.

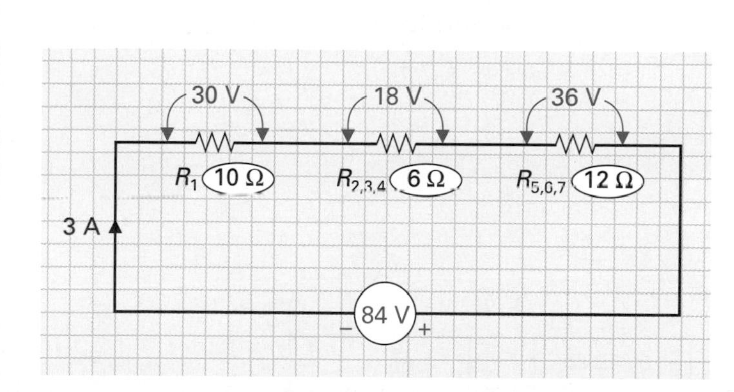

FIGURE 7-15 **After Completing Steps 2 and 3.**

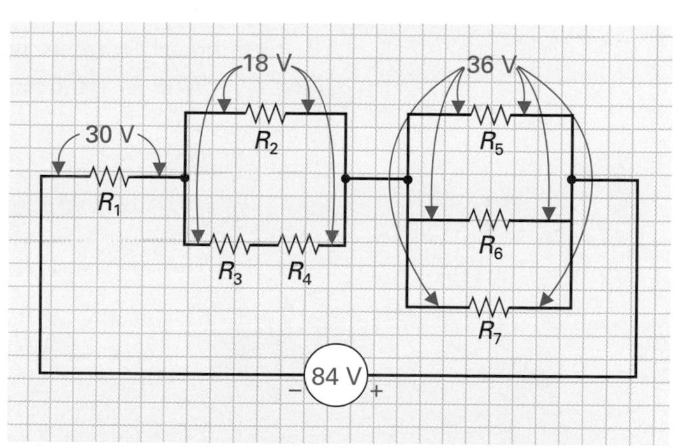

FIGURE 7-16 **Detail of Step 3.**

SELF-TEST EVALUATION POINT FOR SECTION 7-3

Use the following questions to test your understanding of Section 7-3:

1. State the three-step procedure used to calculate the voltage drop across each part of a series–parallel circuit.
2. Referring to Figure 7-13, double the values of all the resistors. Would the voltage drops calculated previously change, and if so, what would they be?

7-4 BRANCH CURRENTS IN A SERIES–PARALLEL CIRCUIT

In the preceding example, step 2 calculated the total current flowing in a series–parallel circuit. The next step is to find out exactly how much current is flowing through each parallel branch. This will be called step 4. Figure 7-17 shows the previously calculated data inserted in the appropriate places in our example circuit.

STEP 4 Total current (I_T) will exist at points *A, B, C,* and *D*. Between *A* and *B*, current has only one path to flow, which is through R_1. R_1 is therefore a series resistor, so $I_1 = I_T = 3$ A. Between points *B* and *C*, current has two paths: through R_2 (12 Ω) and through R_3 and R_4 (12 Ω).

$$I_2 = \frac{V_{R2}}{R_2} = \frac{18 \text{ V}}{12 \ \Omega} = 1.5 \text{ A}$$

$$I_{3,4} = \frac{V_{R3,4}}{R_{3,4}} = \frac{18 \text{ V}}{12 \ \Omega} = 1.5 \text{ A}$$

Not surprisingly, the total current of 3 A is split equally due to both branches having equal resistance.

The two 1.5 A branch currents will combine at point *C* to produce once again the total current of 3 A. Between points *C* and *D*, current has three paths to flow through, R_5, R_6, and R_7.

$$I_5 = \frac{V_{R5}}{R_5} = \frac{36 \text{ V}}{60 \ \Omega} = 0.6 \text{ A}$$

$$I_6 = \frac{V_{R6}}{R_6} = \frac{36 \text{ V}}{24 \ \Omega} = 1.5 \text{ A}$$

$$I_7 = \frac{V_{R7}}{R_7} = \frac{36 \text{ V}}{40 \ \Omega} = 0.9 \text{ A}$$

All three branch currents will combine at point *D* to produce the total current of 3 A ($I_T = I_5 + I_6 + I_7 = 0.6 + 1.5 + 0.9 = 3$ A), proving Kirchhoff's current law.

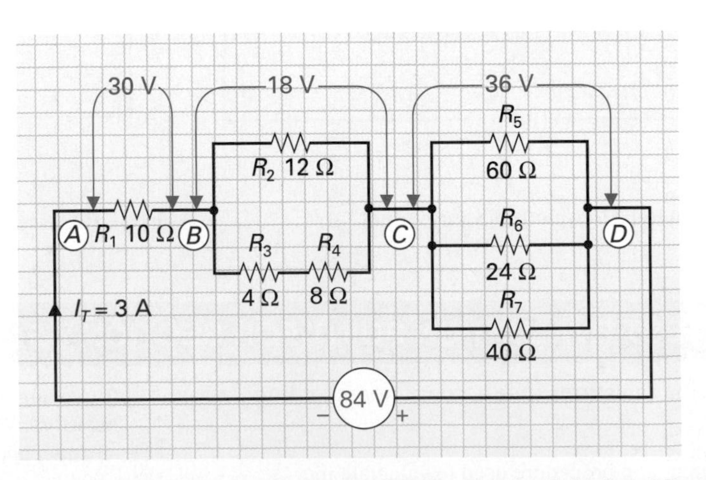

FIGURE 7-17 Series–Parallel Circuit Example with Previously Calculated Data.

7-5 POWER IN A SERIES–PARALLEL CIRCUIT

Whether resistors are connected in series or in parallel, the total power in a series–parallel circuit is equal to the sum of all the individual power losses.

$$P_T = P_1 + P_2 + P_3 + P_4 + \cdots$$

Total power = addition of all power losses.

The formulas for calculating the amount of power lost by each resistor are

$$P = \frac{V^2}{R} \qquad P = I \times V \qquad P = I^2 \times R$$

Let us calculate the power dissipated by each resistor. This final calculation will be called step 5.

STEP 5 Since resistance, voltage, and current are known, either of the three formulas for power can be used to determine power.

$$P_1 = \frac{V_{R1}{}^2}{R_1} = \frac{(30\ \text{V})^2}{10\ \Omega} = 90\ \text{W}$$
$$P_2 = \frac{V_{R2}{}^2}{R_2} = \frac{(18\ \text{V})^2}{12\ \Omega} = 27\ \text{W}$$
$$P_3 = I_{R3}{}^2 \times R_3 = (1.5\ \text{A})^2 \times 4\ \Omega = 9\ \text{W}$$
$$P_4 = I_{R4}{}^2 \times R_4 = (1.5\ \text{A})^2 \times 8\ \Omega = 18\ \text{W}$$
$$P_5 = \frac{V_{R5}{}^2}{R_5} = \frac{(36\ \text{V})^2}{60\ \Omega} = 21.6\ \text{W}$$
$$P_6 = \frac{V_{R6}{}^2}{R_6} = \frac{(36\ \text{V})^2}{24\ \Omega} = 54\ \text{W}$$
$$P_7 = \frac{V_{R7}{}^2}{R_7} = \frac{(36\ \text{V})^2}{40\ \Omega} = 32.4\ \text{W}$$

$$\begin{aligned} P_T &= P_1 + P_2 + P_3 + P_4 + P_5 + P_6 + P_7 \\ &= 90 + 27 + 9 + 18 + 21.6 + 54 + 32.4 \\ &= 252\ \text{W} \end{aligned}$$

or

$$P_T = \frac{V_T{}^2}{R_T} = \frac{(84\ \text{V})^2}{28\ \Omega}$$
$$= 252\ W$$

The total power dissipated in this example circuit is 252 W. All the information can now be inserted in a final diagram for the example, as shown in Figure 7-18.

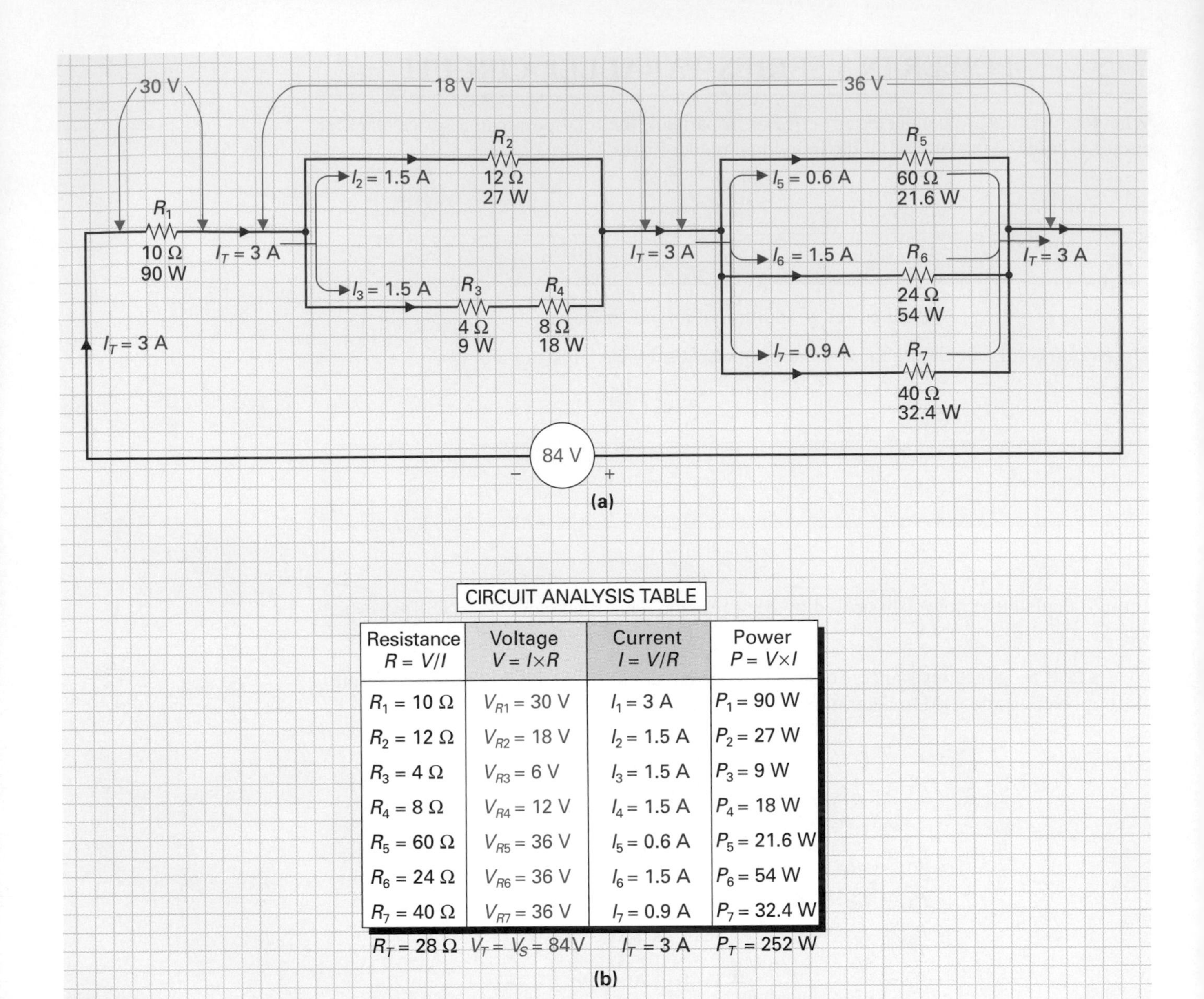

Resistance $R = V/I$	Voltage $V = I \times R$	Current $I = V/R$	Power $P = V \times I$
$R_1 = 10\ \Omega$	$V_{R1} = 30$ V	$I_1 = 3$ A	$P_1 = 90$ W
$R_2 = 12\ \Omega$	$V_{R2} = 18$ V	$I_2 = 1.5$ A	$P_2 = 27$ W
$R_3 = 4\ \Omega$	$V_{R3} = 6$ V	$I_3 = 1.5$ A	$P_3 = 9$ W
$R_4 = 8\ \Omega$	$V_{R4} = 12$ V	$I_4 = 1.5$ A	$P_4 = 18$ W
$R_5 = 60\ \Omega$	$V_{R5} = 36$ V	$I_5 = 0.6$ A	$P_5 = 21.6$ W
$R_6 = 24\ \Omega$	$V_{R6} = 36$ V	$I_6 = 1.5$ A	$P_6 = 54$ W
$R_7 = 40\ \Omega$	$V_{R7} = 36$ V	$I_7 = 0.9$ A	$P_7 = 32.4$ W
$R_T = 28\ \Omega$	$V_T = V_S = 84$ V	$I_T = 3$ A	$P_T = 252$ W

FIGURE 7-18 Series–Parallel Circuit Example with All Information Inserted. (a) Schematic. (b) Circuit Analysis Table.

7-6 FIVE-STEP METHOD FOR SERIES–PARALLEL CIRCUIT ANALYSIS

Let's now combine and summarize all the steps for calculating resistance, voltage, current, and power in a series–parallel circuit by solving another problem. Before we begin, however, let us review the five-step procedure as shown at the top of the next page.

EXAMPLE:

Referring to Figure 7-19, calculate:

a. Total resistance

b. Total current

SOLVING FOR RESISTANCE, VOLTAGE, CURRENT, AND POWER IN A SERIES–PARALLEL CIRCUIT

STEP 1 Determine the circuit's total resistance.

Step A Solve for series-connected resistors in all parallel combinations.

Step B Solve for all parallel combinations.

Step C Solve for remaining series resistances.

STEP 2 Determine the circuit's total current.

STEP 3 Determine the voltage across each series resistor and each parallel combination (series equivalent resistor).

STEP 4 Determine the value of current through each parallel resistor in every parallel combination.

STEP 5 Determine the total and individual power dissipated by the circuit.

c. Voltage drop across all resistors

d. Current through each resistor

e. Total power dissipated by the circuit

Solution:

This problem has asked us to calculate everything about the series–parallel circuit shown in Figure 7-19 and is an ideal application for our five-step series–parallel circuit analysis procedure.

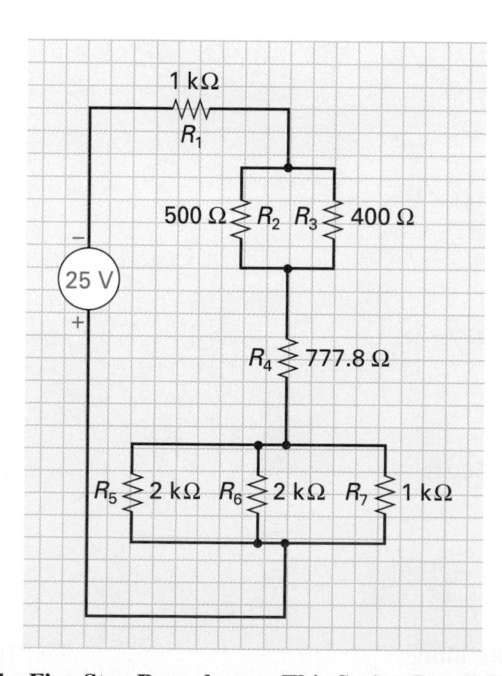

FIGURE 7-19 **Apply the Five-Step Procedure to This Series–Parallel Circuit Example.**

STEP 1 Determine the circuit's total resistance.

Step A: There are no series resistors within parallel combinations.

Step B: There are two-resistor (R_2, R_3) and three-resistor (R_5, R_6, R_7) parallel combinations in this circuit.

$$R_{2,3} = \frac{1}{(1/R_2) + (1/R_3)} = 222.2\ \Omega$$

$$R_{5,6,7} = \frac{1}{(1/R_5) + (1/R_6) + (1/R_7)} = 500\ \Omega$$

Figure 7-20 illustrates the circuit resulting after step B.

Step C: Solve for the remaining four resistances to gain the circuit's total resistance (R_T) or equivalent resistance (R_{eq}).

$$\begin{aligned} R_{eq} &= R_1 + R_{2,3} + R_4 + R_{5,6,7} \\ &= 1000\ \Omega + 222.2\ \Omega + 777.8\ \Omega + 500\ \Omega \\ &= 2500\ \Omega \quad \text{or} \quad 2.5\ \text{k}\Omega \end{aligned}$$

Figure 7-21 illustrates the circuit resulting after step C.

STEP 2 Determine the circuit's total current.

$$I_T = \frac{V_S}{R_T} = \frac{25\ \text{V}}{2.5\ \text{k}\Omega} = 10\ \text{mA}$$

STEP 3 Determine the voltage across each series resistor and each series equivalent resistor. To achieve this, we utilize the diagram obtained after completing step B (Figure 7-20):

$$\begin{aligned} V_{R1} &= I_T \times R_1 = 10\ \text{mA} \times 1\ \text{k}\Omega = 10\ \text{V} \\ V_{R2,3} &= I_T \times R_{2,3} = 10\ \text{mA} \times 222.2\ \Omega = 2.222\ \text{V} \\ V_{R4} &= I_T \times R_4 = 10\ \text{mA} \times 777.8\ \Omega = 7.778\ \text{V} \\ V_{R5,6,7} &= I_T \times R_{5,6,7} = 10\ \text{mA} \times 500\ \Omega = 5\ \text{V} \end{aligned}$$

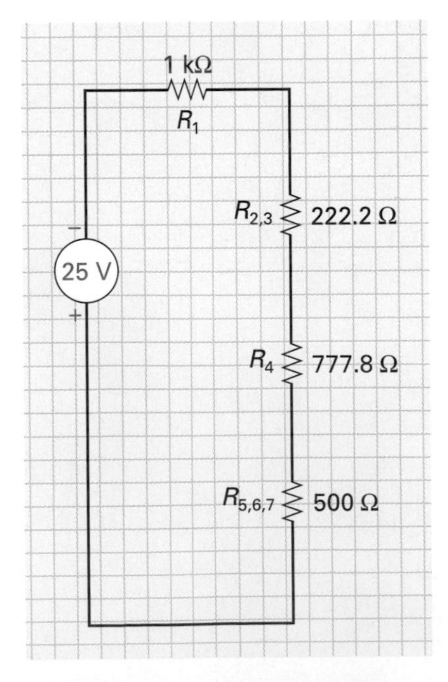

FIGURE 7-20 Circuit Resulting after Step 1B.

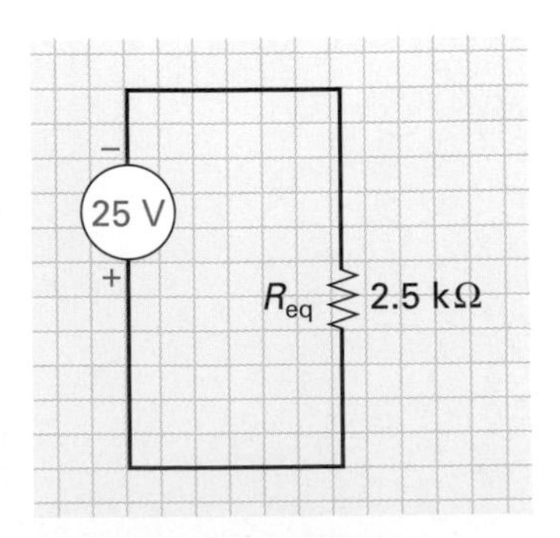

FIGURE 7-21 Circuit Resulting after Step 1C.

Figure 7-22 illustrates the results after step 3.

STEP 4 Determine the value of current through each parallel resistor (Figure 7-23). R_1 and R_4 are series-connected resistors, and therefore their current will equal 10 mA.

$$I_1 = 10 \text{ mA}$$
$$I_4 = 10 \text{ mA}$$

The current through the parallel resistors is calculated by Ohm's law.

$$I_2 = \frac{V_{R2}}{R_2} = \frac{2.222 \text{ V}}{500 \ \Omega} = 4.4 \text{ mA}$$

$$I_3 = \frac{V_{R3}}{R_3} = \frac{2.222 \text{ V}}{400 \ \Omega} = 5.6 \text{ mA}$$

$$\left.\begin{aligned} I_T &= I_2 + I_3 \\ 10 \text{ mA} &= 4.4 \text{ mA} + 5.6 \text{ mA} \end{aligned}\right\} \text{Kirchhoff's current law}$$

$$I_5 = \frac{V_{R5}}{R_5} = \frac{5 \text{ V}}{2 \text{ k}\Omega} = 2.5 \text{ mA}$$

$$I_6 = \frac{V_{R6}}{R_6} = \frac{5 \text{ V}}{2 \text{ k}\Omega} = 2.5 \text{ mA}$$

$$I_7 = \frac{V_{R7}}{R_7} = \frac{5 \text{ V}}{1 \text{ k}\Omega} = 5 \text{ mA}$$

$$\left.\begin{aligned} I_T &= I_5 + I_6 + I_7 \\ 10 \text{ mA} &= 2.5 \text{ mA} + 2.5 \text{ mA} + 5 \text{ mA} \end{aligned}\right\} \text{Kirchhoff's current law}$$

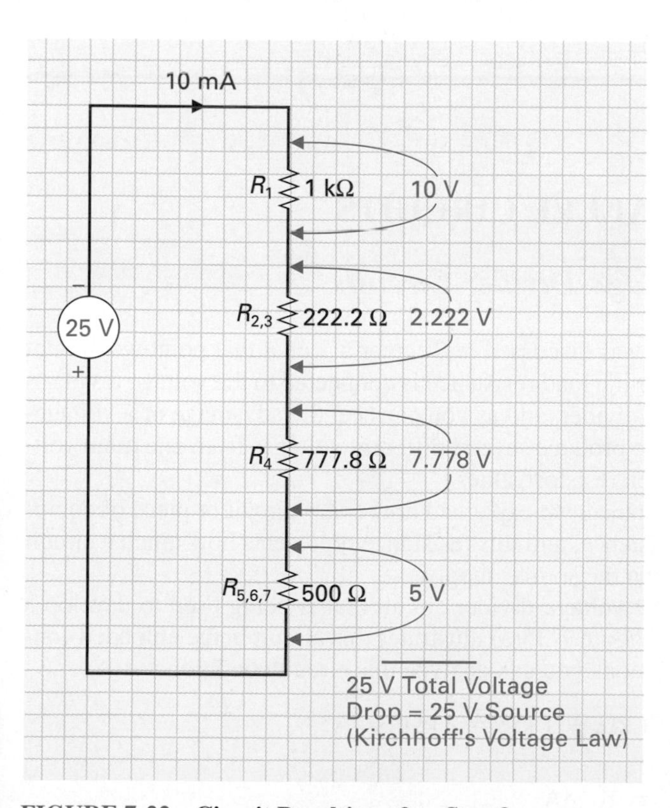

FIGURE 7-22 Circuit Resulting after Step 3.

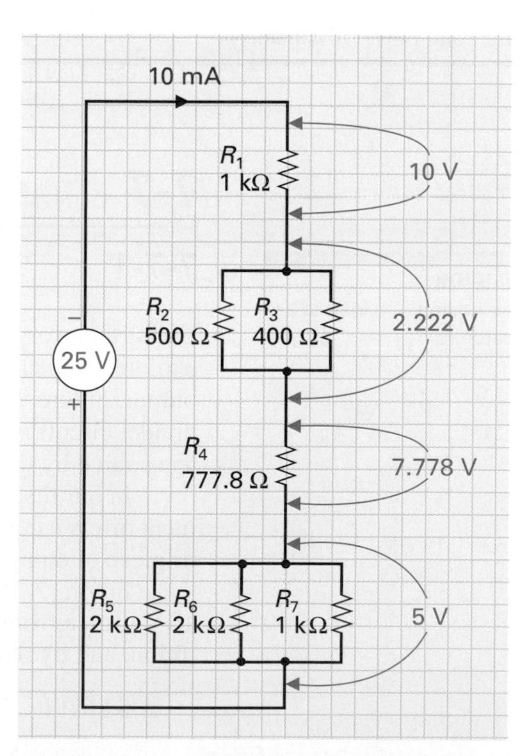

FIGURE 7-23 Series–Parallel Circuit Example with Steps 1, 2, and 3 Data Inserted.

STEP 5 Determine the total power dissipated by the circuit.

$$P_T = P_1 + P_2 + P_3 + P_4 + P_5 + P_6 + P_7$$

or

$$P_T = \frac{V_T^2}{R_T}$$

Each resistor's power figure can be calculated and the sum would be the total power dissipated by the circuit. Since the problem does not ask for the power dissipated by each individual resistor, but for the total power dissipated, it will be easier to use the formula:

$$P_T = \frac{V_T^2}{R_T} = \frac{(25\text{ V})^2}{2.5\text{ k}\Omega} = 0.25\text{ W}$$

SELF-TEST EVALUATION POINT FOR 7-4, 7-5, AND 7-6

Now that you have completed these sections, you should be able to:

Objective 3. Describe for the series–parallel circuit how to use a five-step procedure to calculate:

- **a.** Total resistance.
- **b.** Total current.
- **c.** Voltage division.
- **d.** Branch current.
- **e.** Total power dissipated.

Use the following questions to test your understanding of Sections 7-4, 7-5, and 7-6:

1. State the five-step method used for series–parallel circuit analysis.
2. Design your own five-resistor series–parallel circuit, assign resistor values and a source voltage, and then apply the five-step analysis method.

7-7 SERIES–PARALLEL CIRCUITS

7-7-1 *Loading of Voltage-Divider Circuits*

Loading
The adding of a load to a source.

The straightforward voltage divider was discussed in Chapter 5, but at that point we did not explore some changes that will occur if a load resistance is connected to the voltage divider's output. Figure 7-24 shows a voltage divider, and as you can see, the advantage of a voltage-divider circuit is that it can be used to produce several different voltages from one main voltage source by the use of a few chosen resistor values.

In our discussion on load resistance, we explained how every circuit or piece of equipment offers a certain amount of resistance, and this resistance represents how much a circuit or piece of equipment will load down the source supply.

Figure 7-25 shows an example voltage-divider circuit that is being used to develop a 10 V source from a 20 V dc supply. Figure 7-25(a) illustrates this circuit in the unloaded condition, and by making a few calculations you can analyze this circuit condition.

STEP 1 $R_T = R_1 + R_2 = 1\text{ k}\Omega + 1\text{ k}\Omega = 2\text{ k}\Omega$

STEP 2 $I_T = \frac{V_T}{R_T} = \frac{20\text{ V}}{2\text{ k}\Omega} = 10\text{ mA}$

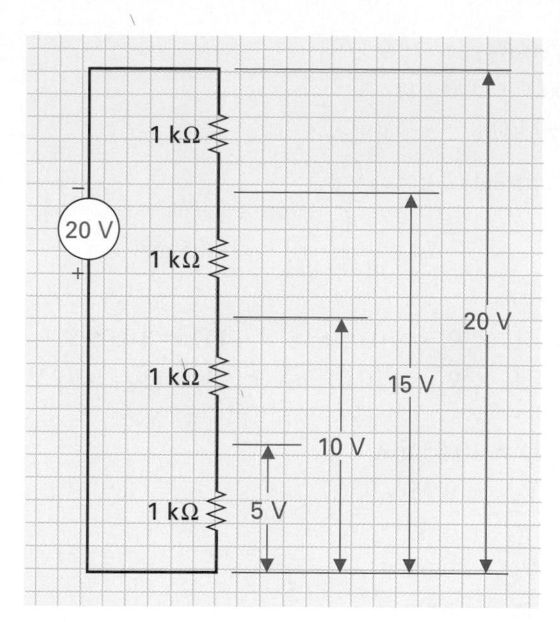

FIGURE 7-24 **Voltage-Divider Circuit.**

The current that flows through a voltage divider, without a load connected, is called the **bleeder current.** In this example, the bleeder current is equal to 10 mA. It is called the bleeder current because it is continually drawing or bleeding this current from the voltage source.

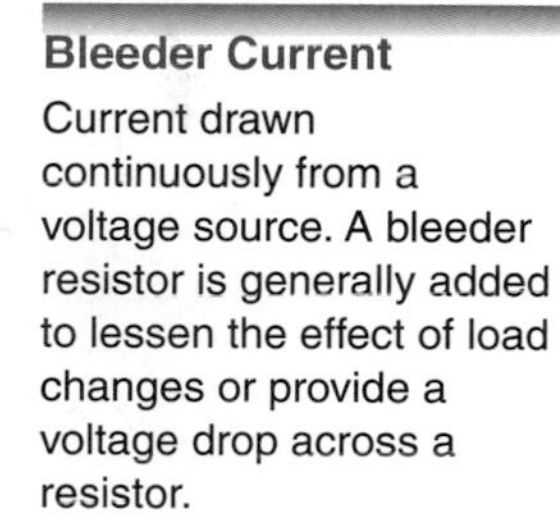

Bleeder Current
Current drawn continuously from a voltage source. A bleeder resistor is generally added to lessen the effect of load changes or provide a voltage drop across a resistor.

STEP 3 $V_{R1} = V_{R2}$ (as resistors are the same value)

$V_{R1} = 10\text{ V}$

$V_{R2} = 10\text{ V}$

In Figure 7-25(b) we have connected a piece of equipment, represented as a resistance (R_3), across the 10 V supply. This automatically turns the previous series circuit of R_1 and R_2 into a series–parallel circuit made up of R_1, R_2, and the 100 kΩ load resistance. By making a few more calculations, we can discover the changes that have occurred by connecting this load resistance.

STEP 1 Total resistance (R_T):

Step B:

$$R_{2,3} = \frac{R_2 \times R_3}{R_2 + R_3} = \frac{1\text{ k}\Omega \times 100\text{ k}\Omega}{1\text{ k}\Omega + 100\text{ k}\Omega} = 990.1\ \Omega$$

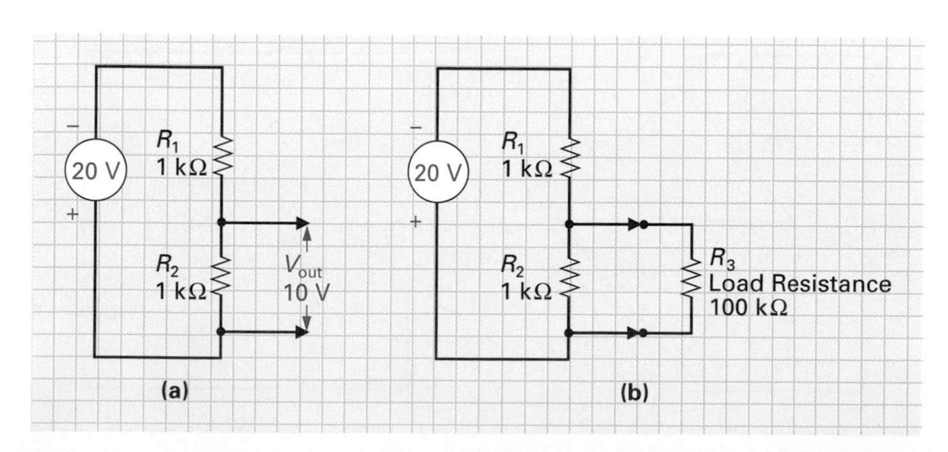

FIGURE 7-25 **Voltage-Divider Circuit. (a) Unloaded Output Voltage. (b) Loaded Output Voltage.**

$$Step\ C: \quad R_{1,2,3} = R_1 + R_{2,3}$$
$$= 1\ \text{k}\Omega + 990.1\ \Omega$$
$$= 1.99\ \text{k}\Omega$$

STEP 2 Total current (I_T):

$$I_T = \frac{20\ \text{V}}{1.99\ \text{k}\Omega} = 10.05\ \text{mA}$$

STEP 3 $V_{R1} = I_T \times R_1 = 10.05\ \text{mA} \times 1\ \text{k}\Omega = 10.05\ \text{V}$

$V_{R2,3} = I_T \times R_{2,3} = 10.05\ \text{mA} \times 990.1\ \Omega = 9.95\ \text{V}$

STEP 4 $I_1 = I_T = 10.05\ \text{mA}$

$$I_2 = \frac{V_{R2}}{R_2} = \frac{9.95\ \text{V}}{1\ \text{k}\Omega} = 9.95\ \text{mA}$$

$$I_3 = \frac{V_{R3}}{R_3} = \frac{9.95\ \text{V}}{100\ \text{k}\Omega} = 99.5\ \mu\text{A}$$

$$\left.\begin{array}{r} I_2 + I_3 = I_T \\ 9.95\ \text{mA} + 99.5\ \mu\text{A} = 10.05\ \text{mA} \end{array}\right\} \text{Kirchhoff's current law}$$

As you can see, the load resistance is pulling 99.5 μA from the source, and this pulls the voltage down to 9.95 V from the required 10 V that was desired and is normally present in the unloaded condition.

When designing a voltage divider, design engineers need to calculate how much current a particular load will pull and then alter the voltage-divider resistor values to offset the loading effect when the load is connected.

7-7-2 *The Wheatstone Bridge*

Wheatstone Bridge
A four-arm, generally resistive bridge that is used to measure resistance.

In 1850, Charles Wheatstone developed a circuit to measure resistance. This circuit, which is still widely used today, is called the **Wheatstone bridge** and is illustrated in Figure 7-26(a). In Figure 7-26(b), the same circuit has been redrawn so that the series and parallel resistor connections are easier to see.

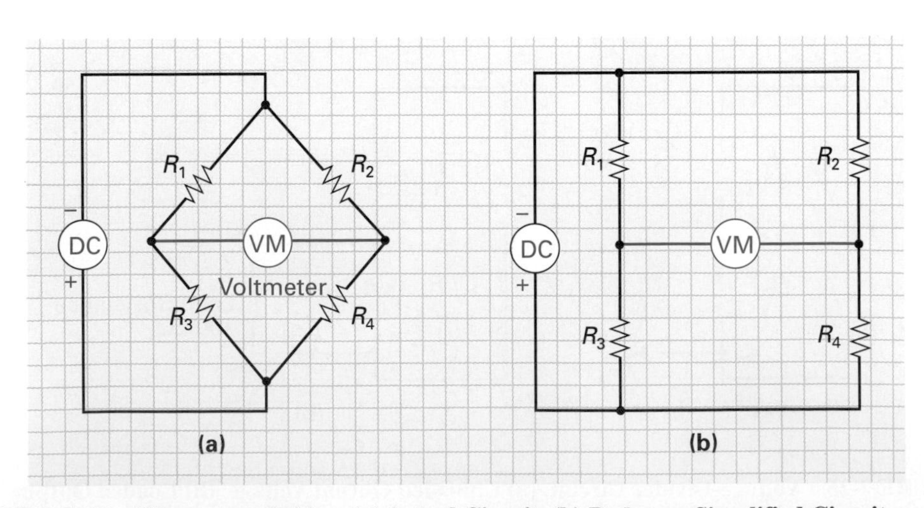

FIGURE 7-26 **Wheatstone Bridge. (a) Actual Circuit. (b) Redrawn Simplified Circuit.**

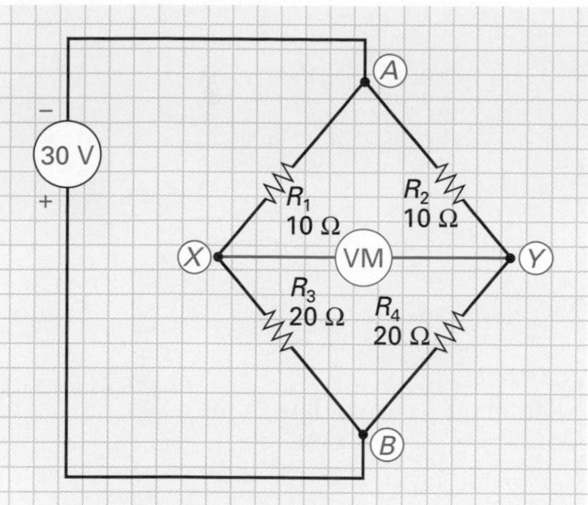

FIGURE 7-27 **Wheatstone Bridge Circuit Example.**

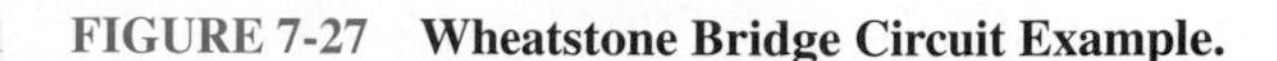

Balanced Bridge

Figure 7-27 illustrates an example circuit in which four resistors are connected to form a series–parallel arrangement. Let us now use the five-step procedure to find out exactly what resistance, current, voltage, and power values exist throughout the circuit.

STEP 1 Total resistance (R_T):

Step A: $R_{1,3} = R_1 + R_3 = 10 + 20 = 30\ \Omega$

$R_{2,4} = R_2 + R_4 = 10 + 20 = 30\ \Omega$

Step B:

$$R_T\text{: } (R_{1,2,3,4}) = \frac{R_{1,3} \times R_{2,4}}{R_{1,3} + R_{2,4}}$$

$$= \frac{30 \times 30}{30 + 30} = 15\ \Omega$$

Total resistance $= 15\ \Omega$

STEP 2 Total current (I_T):

$$I_T = \frac{V_T}{R_T} = \frac{30\text{ V}}{15\ \Omega} = 2\text{ A}$$

STEP 3 Since $R_{1,3}$ is in parallel with $R_{2,4}$, 30 V will appear across both $R_{1,3}$ and $R_{2,4}$.

$$V_T = V_{R1,3} = V_{R2,4} = 30\text{ V}$$

voltage-divider formula:

$$V_{R1} = \frac{R_1}{R_{1,3}} \times V_T$$

$$= \frac{10}{30} \times 30 = 10\text{ V}$$

$$V_{R3} = \frac{R_3}{R_{1,3}} \times V_T = 20\text{ V}$$

$$V_{R2} = \frac{R_2}{R_{2,4}} \times V_T = 10\text{ V}$$

$$V_{R4} = \frac{R_4}{R_{2,4}} \times V_T = 20\text{ V}$$

STEP 4 $I_{1,3} = \frac{V_{R1,3}}{R_{1,3}} = \frac{30\text{ V}}{30\ \Omega} = 1\text{ A}$

$$I_{2,4} = \frac{V_{R2,4}}{R_{2,4}} = \frac{30\text{ V}}{30\ \Omega} = 1\text{ A}$$

$$\left.\begin{array}{c} I_{1,3} + I_{2,4} = I_T \\ 1\text{ A} + 1\text{ A} = 2\text{ A} \end{array}\right\} \text{Kirchhoff's current law}$$

STEP 5 Total power dissipated (P_T):

$$\begin{aligned} P_T &= I_T^{\,2} \times R_T \\ &= (2\text{ A})^2 \times 15\ \Omega \\ &= 4\text{ A}^2 \times 15\ \Omega \\ &= 60\text{ W} \end{aligned}$$

Figure 7-28 shows all of the step results inserted in the Wheatstone bridge example schematic. The Wheatstone bridge is said to be in the balanced condition when the voltage at point X equals the voltage at point Y ($V_{R3} = V_{R4}$, 20 V = 20 V). This same voltage exists across R_3 and R_4, so the voltmeter, which is measuring the voltage difference between X and Y, will indicate 0 V potential difference, and the circuit is said to be a *balanced bridge*.

Unbalanced Bridge

In Figure 7-29 we have replaced R_3 with a variable resistor and set it to 10 Ω. The R_2 and R_4 resistor combination will not change its voltage drop. However, R_1 and R_3, which are now equal, will each split the 30 V supply, producing 15 V across R_3. The voltmeter will indicate the difference in potential (5 V) from the voltage across R_3 at point X(15 V) and across R_4 at point Y(20 V). The voltmeter is actually measuring the imbalance in the circuit, which is why this circuit in this condition is known as an *unbalanced bridge.*

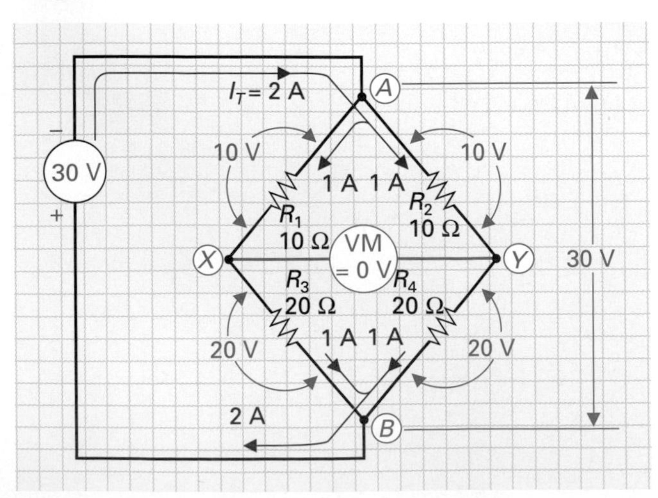

FIGURE 7-28 **Balanced Wheatstone Bridge.**

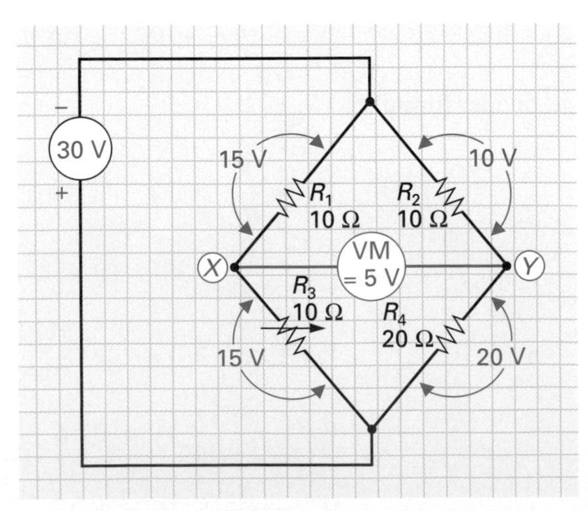

FIGURE 7-29 **Unbalanced Wheatstone Bridge.**

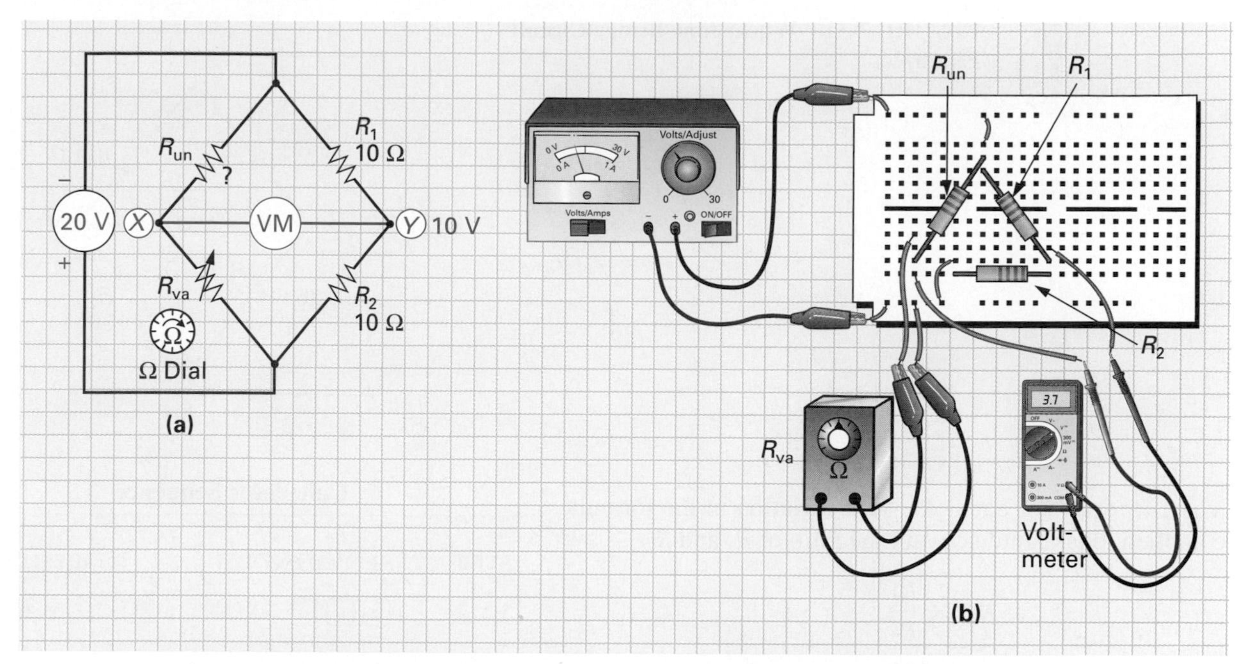

FIGURE 7-30 **Using a Wheatstone Bridge to Determine an Unknown Resistance. (a) Schematic. (b) Pictorial.**

Determining Unknown Resistance

Figure 7-30 shows how a Wheatstone bridge circuit can be used to find the value of an unknown resistor (R_{un}). The variable-value resistor (R_{va}) is a calibrated resistor, which means that its resistance has been checked against a known, accurate resistance and its value can be adjusted and read from a calibrated dial.

The procedure to follow to find the value of the unknown resistor is as follows:

1. Adjust the variable-value resistor until the voltmeter indicates that the Wheatstone bridge is balanced (voltmeter indicates 0 V).
2. Read the value of the variable-value resistor. As long as $R_1 = R_2$, the variable resistance value will be the same as the unknown resistance value.

$$R_{va} = R_{un}$$

Since R_1 and R_2 are equal to one another, the voltage will be split across the two resistors, producing 10 V at point *Y.* The variable-value resistor must therefore be adjusted so that it equals the unknown resistance, and therefore the same situation will occur, in that the 20 V source will be split, producing 10 V at point *X,* indicating a balanced 0 V condition on the voltmeter. For example, if the unknown resistance is equal to 5 Ω, then only when the variable-value resistor is adjusted and equal to 5 Ω would 10 V appear at point *X* and allow the circuit meter to read zero volts, indicating a balance. The variable-value resistor resistance could be read (5 Ω) and the unknown resistor resistance would be known (5 Ω).

EXAMPLE:

What is the unknown resistance in Figure 7-31?

Solution:

The bridge is in a balanced condition as the voltmeter is reading a 0 V difference between points *X* and *Y.* In the previous section we discovered that if $R_1 = R_2$, then

$$R_{va} = R_{un}$$

FIGURE 7-31 **Wheatstone Bridge Circuit Example.**

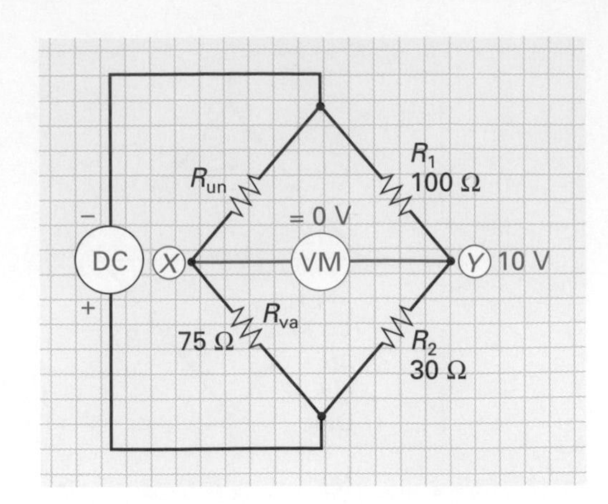

In this case, R_1 does not equal R_2, so a variation in the formula must be applied to take into account the ratio of R_1 and R_2.

$$R_{un} = R_{va} \times \frac{R_1}{R_2}$$

$$= 75\ \Omega \times \frac{100}{30}$$

$$= 75\ \Omega \times 3.33 = 250\ \Omega$$

Since R_1 is 3.33 times greater than R_2, then R_{un} must be 3.33 times greater than R_{va} if the Wheatstone bridge is in the balanced condition.

Calculator Sequence

STEP	KEYPAD ENTRY	DISPLAY RESPONSE
1.	1 0 0	100
2.	÷	
3.	3 0	30
4.	=	3.333
5.	×	
6.	7 5	75
7.	=	250

EXAMPLE:

Referring to Figure 7-32, answer the following questions:

a. When no smoke is present, the light level applied to LDR 1 and LDR 2 is ___________ (the same/not the same) and so the resistance of LDR 1 and LDR 2 is ___________ (the same/not the same), and the Wheatstone Bridge is ___________ (balanced/unbalanced). In this condition, the voltage at points *x* and *y* is ___________ (the same/not the same), and so ___________ (a/no) potential difference is applied across the alarm circuit, making it ___________ (active/inactive).

b. When smoke is present, ___________ (more/less) light is applied to the LDR in the sensing chamber, and the Wheatstone Bridge is ___________ (balanced/unbalanced). In this condition, the voltage at point *x* and *y* is ___________ (the same/not the same), and so ___________ (a/no) potential difference is applied across the alarm circuit, making it ___________ (active/inactive).

Solution:

a. the same, the same, balanced, the same, no, inactive

b. less, unbalanced, not the same, a, active

R–2R Ladder Circuit
A network or circuit composed of a sequence of *L* networks connected in tandem. This *R–2R* circuit is used in digital-to-analog converters.

7-7-3 *The R–2R Ladder Circuit*

Figure 7-33 illustrates an ***R–2R* ladder circuit,** which is a series–parallel circuit used within computer systems to convert digital information to analog information. To fully understand this circuit, our first step should be to find out exactly which branches will have which val-

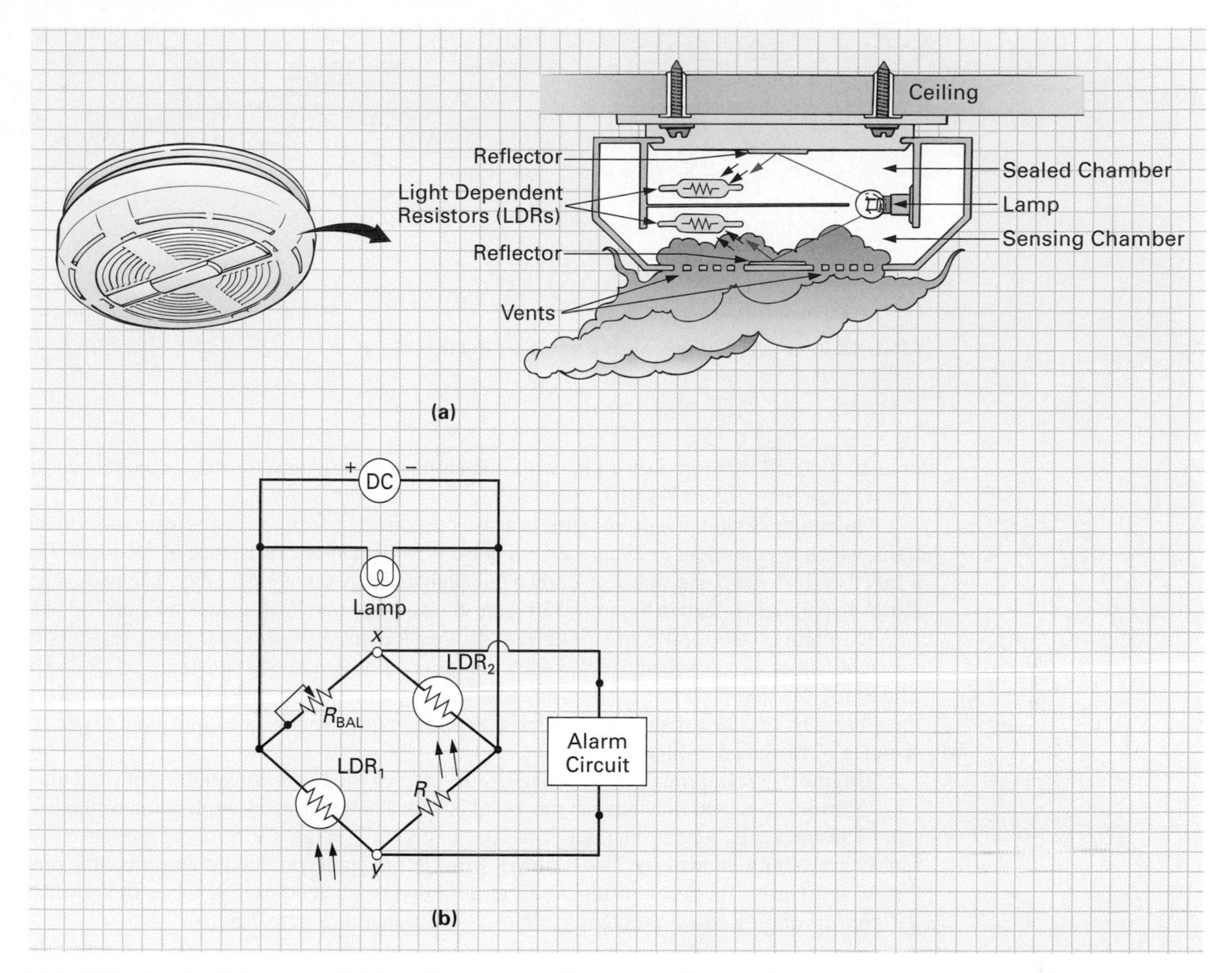

FIGURE 7-32 **The Wheatstone Bridge within a Smoke Detector. (a) Physical Elements. (b) Schematic Diagram.**

ues of current. This can be obtained by finding out what value of resistance the current sees when it arrives at the three junctions *A, B,* and *C.* Let us begin with point *C* first and simplify the circuit. This is illustrated in Figure 7-34(a). No specific resistance value has been chosen, but in all cases 2*R* resistors (2 × *R*) are twice the resistance of an *R* resistor.

In Figure 7-34(a), if 2 mA of current arrives at point *C,* it sees 2*R* of resistance in parallel with a 2*R* resistance, and so the 2 mA of current splits and 1 mA flows through each

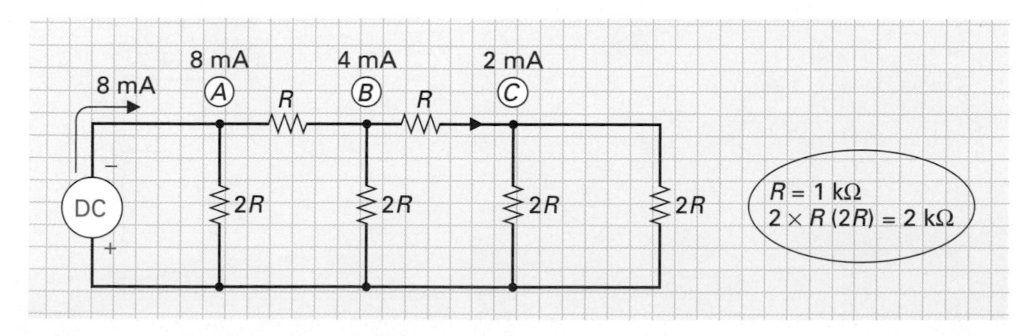

FIGURE 7-33 ***R–2R* Ladder Circuit.**

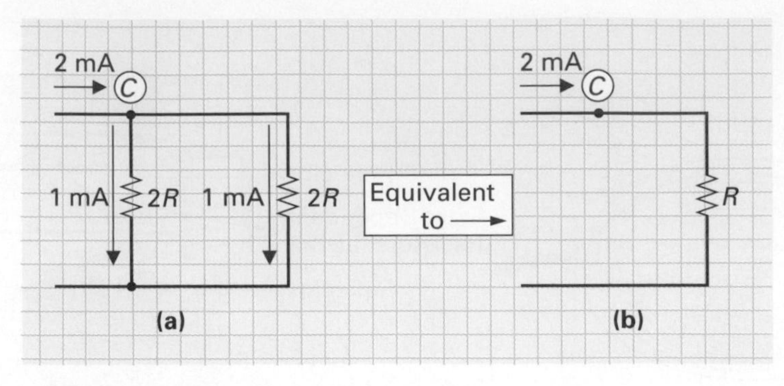

FIGURE 7-34 ***R*–2*R* Equivalent Circuit at Junction C.**

branch. Two 2*R* resistors in parallel with one another would consequently be equivalent to one *R,* as seen in Figure 7-34(b).

In Figure 7-35(a), if 4 mA of current arrives at point *B,* it sees two series resistances to the right, which is equivalent to 2*R* as seen in Figure 7-35(b), and one resistance of 2*R* down. The 4 mA therefore splits, causing 2 mA down one path and 2 mA down the other. The two 2*R* resistors in parallel, as seen in Figure 7-35(b), are equivalent to one *R,* as shown in Figure 7-35(c).

In Figure 7-36(a), if 8 mA of current arrives at point *A,* it sees two series resistors to the right, which is equivalent to 2*R* as seen in Figure 7-36(b), and one resistance of 2*R* down. The 8 mA of current therefore splits equally, causing 4 mA down one path and 4 mA down the other. The two 2*R* resistors in parallel, as seen in Figure 7-36(b), are equivalent to one resistance *R,* as shown in Figure 7-36(c).

Figure 7-37(a) through (f) summarizes the step-by-step simplification of this circuit.

The question you may have at this point is: What is the primary application of this circuit? The answer is as a current divider, as seen in Figure 7-38(a). The 8 mA of reference current is repeatedly divided by 2 as it moves from left to right, producing currents of 4 mA, 2 mA, and 1 mA. The result of this *R*–2*R* current division can be used in a circuit known as a *digital-to-analog converter* (DAC), which is illustrated in Figure 7-38(b).

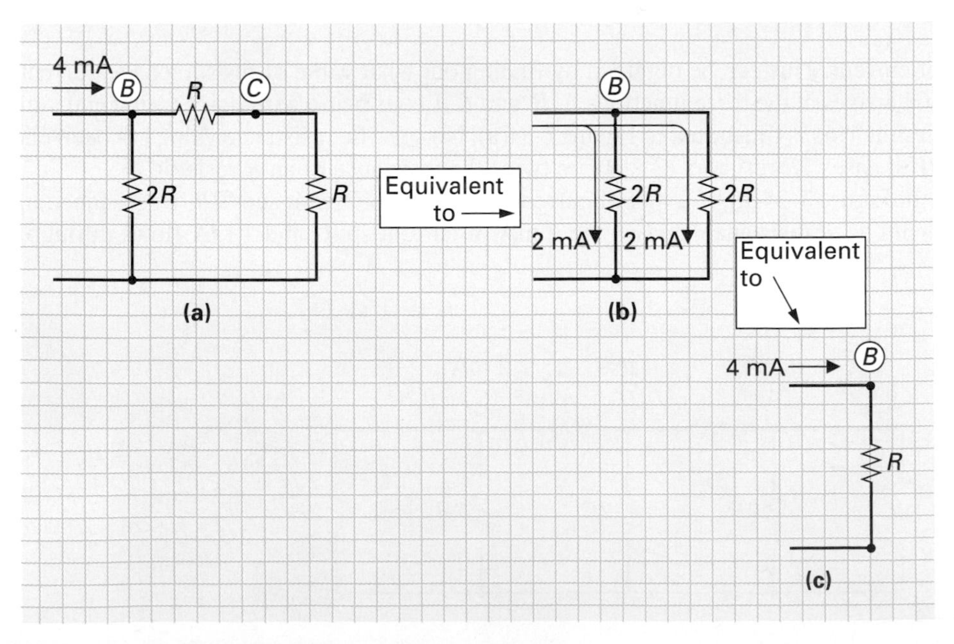

FIGURE 7-35 ***R*–2*R* Equivalent Circuit at Junction B.**

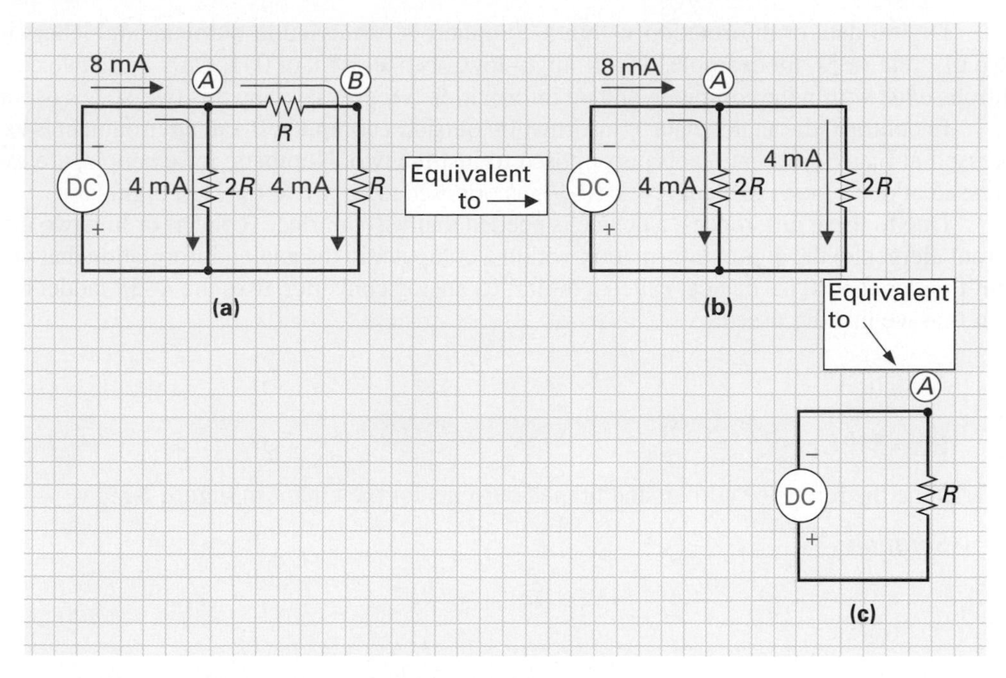

FIGURE 7-36 *R*–2*R* Equivalent Circuit at Junction A.

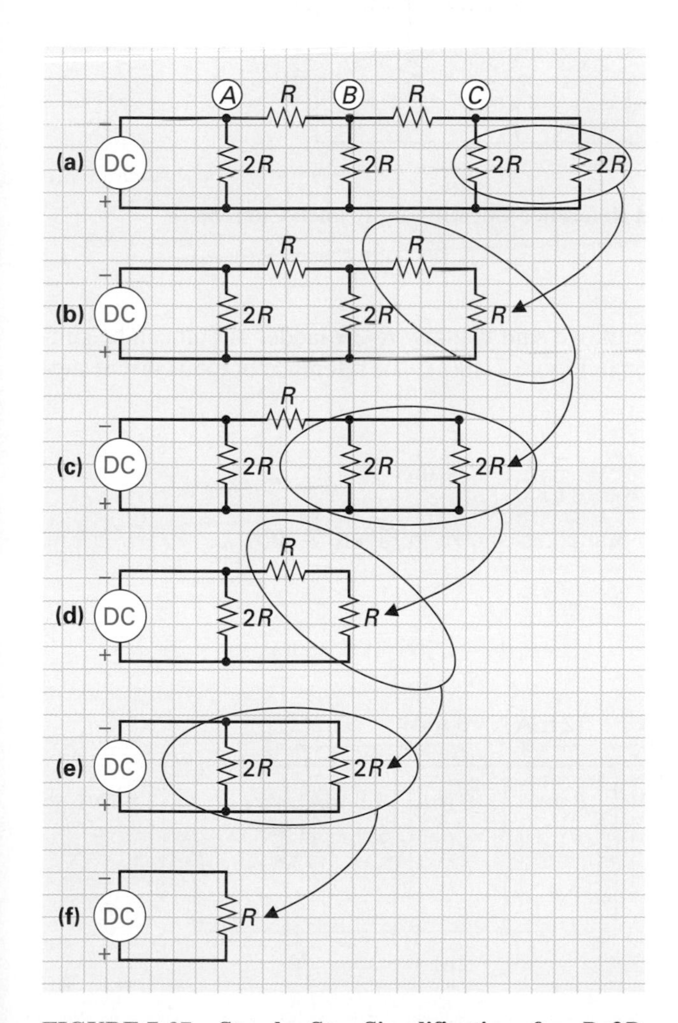

FIGURE 7-37 Step-by-Step Simplification of an *R*–2*R* Ladder Circuit.

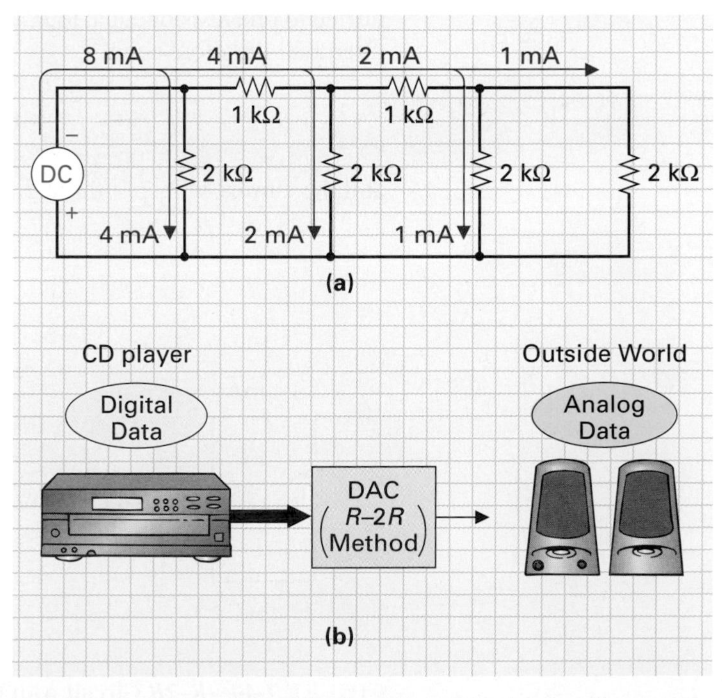

FIGURE 7-38 *R*–2*R* Ladder in a Digital-to-Analog Converter.

Digital data or information within a computer is expressed in numbers and letters in two discrete steps, for example, on–off, high–low, open–closed, or 0–1. Only these two conditions exist within the computer, and all information is represented by this two-state system.

In contrast, the analog data or information outside a computer in our environment is expressed at many different levels, as opposed to just the two. Numbers, for example, are expressed at one of ten different levels (0–9), as opposed to only two (0–1) in digital.

Due to these differences, a device is needed that will interface (convert or link two different elements) the digital information within a computer to the analog information that you and I understand. This device, called a digital-to-analog converter, uses the *R*–2*R* ladder circuit that we just discussed.

EXAMPLE:

Determine the reference current and branch currents for the circuit in Figure 7-39.

Solution:

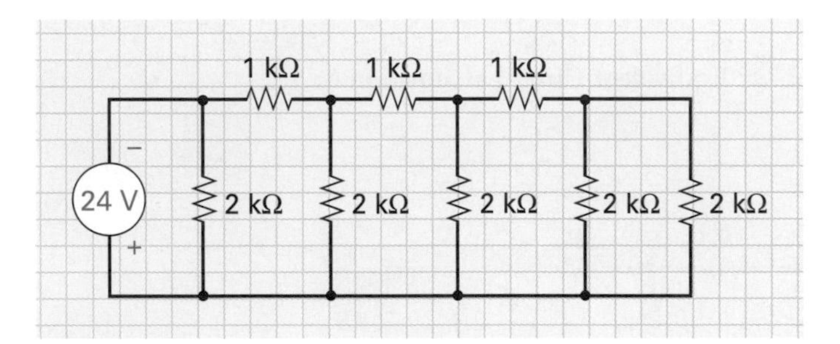

FIGURE 7-39 **Calculate *R*–2*R* Circuit Reference Current and Branch Currents.**

In our simplification of the ladder circuit, we said that any *R*–2*R* ladder circuit can be simplified to one resistor equal to *R*, as seen in Figure 7-40(a). The reference or total current supplied will therefore equal

$$I_T = \frac{V_T}{R_T} = \frac{24 \text{ V}}{1 \text{ k}\Omega} = 24 \text{ mA}$$

and the current will split through each branch, as shown in Figure 7-40(b).

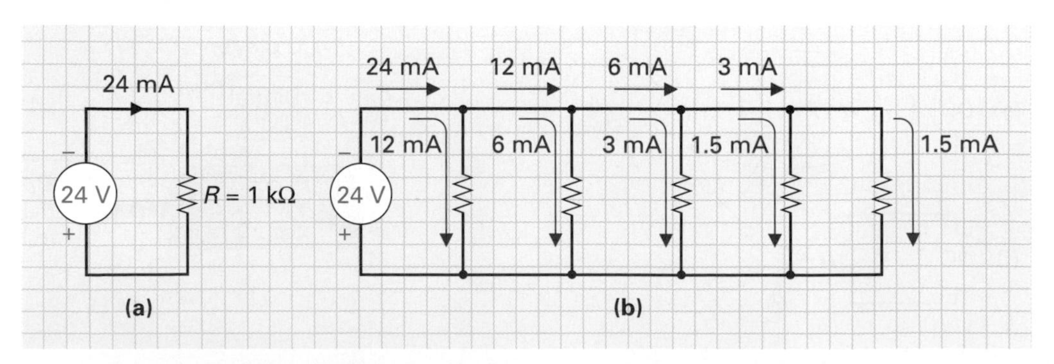

FIGURE 7-40 ***R*–2*R* Circuit with Data Inserted.**

SELF-TEST EVALUATION POINT FOR SECTION 7-7

Now that you have completed this section, you should be able to:

Objective 4. Explain what loading effect a piece of equipment will have when connected to a voltage divider.

Objective 5. Identify and describe the Wheatstone bridge circuit in both the balanced and unbalanced condition.

Objective 6. Describe the R–2R ladder circuit used for digital-to-analog conversion.

Use the following questions to test your understanding of Section 7-7:

1. What is meant by the loading of a voltage-divider circuit?
2. Sketch a Wheatstone bridge circuit and list an application of this circuit.
3. What value will the total resistance of an *R*–2*R* ladder circuit always equal?
4. For what application could the *R*–2*R* ladder be used?

7-8 TROUBLESHOOTING SERIES–PARALLEL CIRCUITS

Troubleshooting is defined as the process of locating and diagnosing malfunctions or breakdowns in equipment by means of systematic checking or analysis. As discussed in previous resistive-circuit troubleshooting procedures, there are basically only three problems that can occur:

1. A component will open. This usually occurs if a resistor burns out or a wire or switch contact breaks.
2. A component will short. This usually occurs if a conductor, such as solder, wire, or some other conducting material, is dropped or left in the equipment, making or connecting two points that should not be connected.
3. There is a variation in a component's value. This occurs with age in resistors over a long period of time and can eventually cause a malfunction of the equipment.

Using the example circuit in Figure 7-41, we will step through a few problems, beginning with an open component. Throughout the troubleshooting, we will use the voltmeter

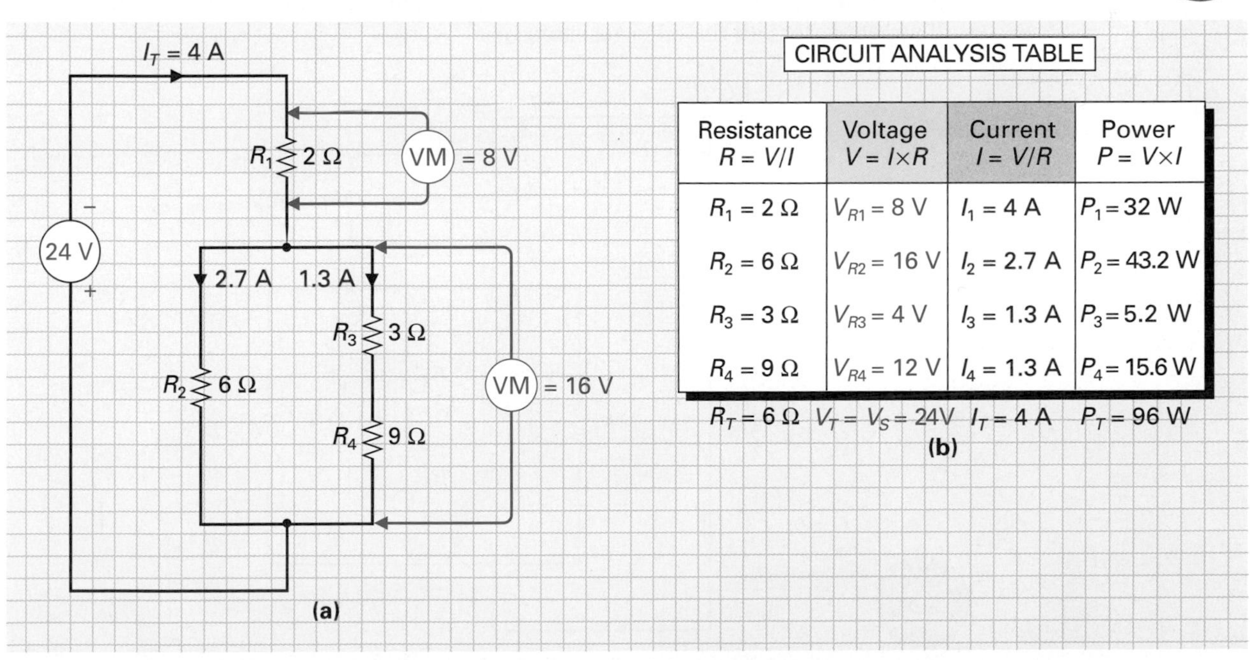

Resistance $R = V/I$	Voltage $V = I \times R$	Current $I = V/R$	Power $P = V \times I$
$R_1 = 2\ \Omega$	$V_{R1} = 8$ V	$I_1 = 4$ A	$P_1 = 32$ W
$R_2 = 6\ \Omega$	$V_{R2} = 16$ V	$I_2 = 2.7$ A	$P_2 = 43.2$ W
$R_3 = 3\ \Omega$	$V_{R3} = 4$ V	$I_3 = 1.3$ A	$P_3 = 5.2$ W
$R_4 = 9\ \Omega$	$V_{R4} = 12$ V	$I_4 = 1.3$ A	$P_4 = 15.6$ W
$R_T = 6\ \Omega$	$V_T = V_S = 24$V	$I_T = 4$ A	$P_T = 96$ W

FIGURE 7-41 Series–Parallel Circuit Example. (a) Schematic. (b) Circuit Analysis Table.

whenever possible, as it can measure voltage by just connecting the leads across the component, rather than the ammeter, which has to be placed in the circuit, in which case the circuit path has to be opened. In some situations, using an ammeter can be difficult.

To begin, let's calculate the voltage drops and branch current obtained when the circuit is operating normally.

STEP 1 (A) $R_{3,4} = R_3 + R_4 = 3\ \Omega + 9\ \Omega = 12\ \Omega$

(B) $$R_{2,3,4} = \frac{R_2 \times R_{3,4}}{R_2 + R_{3,4}} = \frac{6\ \Omega \times 12\ \Omega}{6\ \Omega + 12\ \Omega} = 4\ \Omega$$

(C) $$R_{1,2,3,4} = R_T = R_1 + R_{2,3,4} = 2\ \Omega + 4\ \Omega = 6\ \Omega$$

STEP 2 $$I_T = \frac{V_T}{R_T} = \frac{24\ \text{V}}{6\ \Omega} = 4\ \text{A}$$

STEP 3 $$V_{R1} = I_{R1} \times R_1 = 4\ \text{A} \times 2\ \Omega = 8\ \text{V}$$

$$V_{R2,3,4} = I_{R2,3,4} \times R_{2,3,4} = 4\ \text{A} \times 4\ \Omega = 16\ \text{V}\ \ \text{(Kirchoff's voltage law)}$$

STEP 4 $I_1 = 4\ \text{A}$ (series resistor)

$$I_2 = \frac{V_{R2}}{R_2} = \frac{16\ \text{V}}{6\ \Omega} = 2.7\ \text{A}$$

$$I_{3,4} = \frac{V_{R3,4}}{R_{3,4}} = \frac{16\ \text{V}}{12\ \Omega} = 1.3\ \text{A}$$

All these results have been inserted in the schematic in Figure 7-41(a) and in the circuit analysis table in Figure 7-41(b).

7-8-1 *Open Component*

R_1 Open (Figure 7-42)

With R_1 open, there cannot be any current flow through the circuit because there is no path from one side of the power supply to the other. This fault can be recognized easily because approximately all of the applied 24 V will be measured across the open resistor (R_1) and 0 V will appear across all the other resistors.

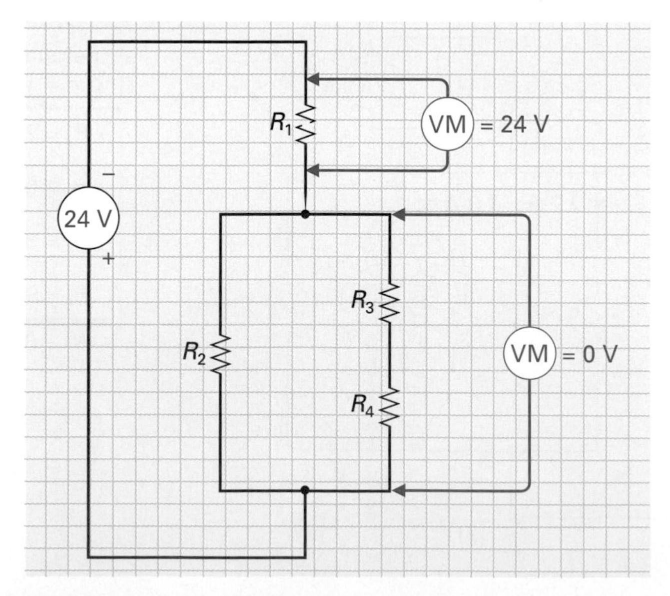

FIGURE 7-42 Open Series-Connected Resistor in a Series–Parallel Circuit.

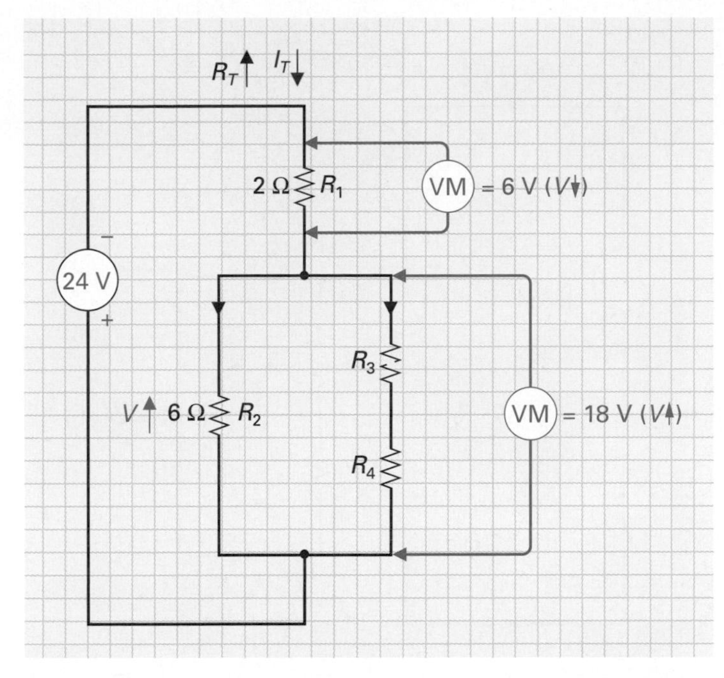

FIGURE 7-43 **Open Parallel-Connected Resistor in a Series–Parallel Circuit.**

R_3 Open (Figure 7-43)

With R_3 open, there will be no current through the branch made up of R_3 and R_4. The current path will be through R_1 and R_2, and therefore the total resistance will now increase ($R_T\uparrow$) from 6 Ω to

$$R_T = R_1 + R_2$$
$$= 2\ \Omega + 6\ \Omega = 8\ \Omega$$

This 8 Ω is an increase in circuit resistance from the normal resistance, which was 6 Ω, which implies that an open has occurred to increase resistance. The total current will decrease ($I_T\downarrow$) from 4 A to

$$I_T = \frac{V_T}{R_T} = \frac{24\text{ V}}{8\ \Omega} = 3\text{ A}$$

The voltage drop across the resistors will be

$$V_{R1} = I_T \times R_1 = 3\text{ A} \times 2\ \Omega = 6\text{ V}$$
$$V_{R2} = I_T \times R_2 = 3\text{ A} \times 6\ \Omega = 18\text{ V}$$

If one of the parallel branches is opened, the overall circuit resistance will always increase. This increase in the total resistance will cause an increase in the voltage dropped across the parallel branch (the greater the resistance, the greater the voltage drop), which enables the technician to localize the fault area and also to determine that the fault is an open.

The voltage measured with a voltmeter will be

$$V_{R1} = 6\text{ V}$$
$$V_{R2} = 18\text{ V}$$
$$V_{R3} = 18\text{ V (open)}$$
$$V_{R4} = 0\text{ V}$$

This identifies the problem as R_3 being open, as it drops the entire parallel circuit voltage (18 V) across itself, whereas normally the voltage would be dropped proportionally across R_3 and R_4, which are in series with one another.

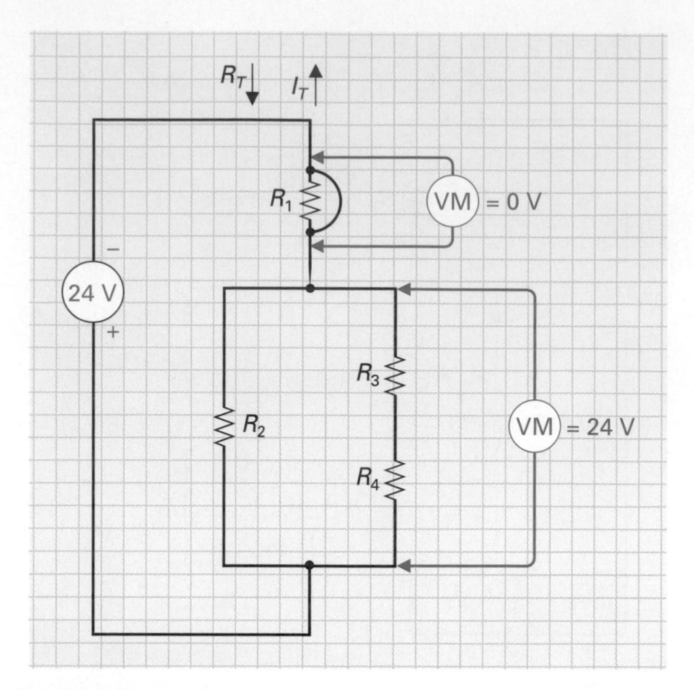

FIGURE 7-44 Shorted Series-Connected Resistor in a Series–Parallel Circuit.

7-8-2 *Shorted Component*

R_1 Shorted (Figure 7-44)

With R_1 shorted, the total circuit resistance will decrease ($R_T\downarrow$), causing an increase in circuit current ($I_T\uparrow$). This increase in current will cause an increase in the voltage dropped across the parallel branch. However, the fault can be located once you measure the voltage across R_1, which will read 0 V, indicating that this resistor has almost no resistance as it has no voltage drop across it.

R_3 Shorted (Figure 7-45)

With R_3 shorted, there will be a decrease in the circuit's total resistance from 6 Ω to

$$R_{2,3,4} = \frac{R_2 \times R_{3,4}}{R_2 + R_{3,4}} = \frac{6 \times 9}{6 + 9} = 3.6\ \Omega$$

$$\begin{aligned} R_{1,2,3,4} &= R_1 + R_{2,3,4} \\ &= 2\ \Omega + 3.6\ \Omega \\ &= 5.6\ \Omega \end{aligned}$$

This decrease in total resistance ($R_T\downarrow$) will cause an increase in total current ($I_T\uparrow$), which implies that a short has occurred to decrease resistance. The total current will now increase from 4 A to

$$I_T = \frac{V_T}{R_T} = \frac{24\ \text{V}}{5.6\ \Omega} = 4.3\ \text{A}$$

The voltage drops across the resistors will be

$$\begin{aligned} V_{R1} &= I_T \times R_1 = 4.3\ \text{A} \times 2\ \Omega = 8.6\ \text{V} \\ V_{R2,3,4} &= I_T \times R_{2,3,4} = 4.3\ \text{A} \times 3.6\ \Omega = 15.4\ \text{V} \end{aligned}$$

When one of the parallel branch resistors is shorted, the overall circuit resistance will always decrease. This decrease in total resistance will cause a decrease in the voltage dropped across the parallel branch (the smaller the resistance, the smaller the voltage drop), and this enables the technician to localize the faulty area and also to determine that the fault is a short.

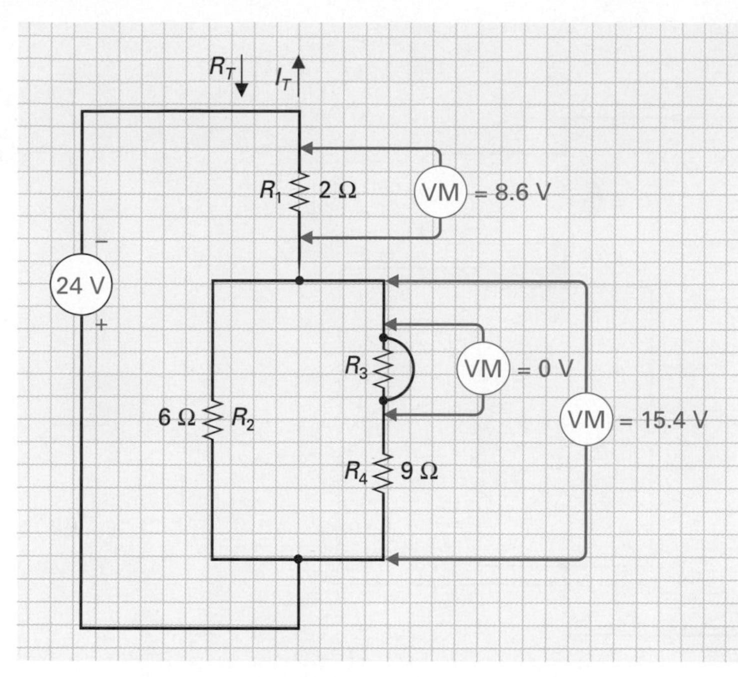

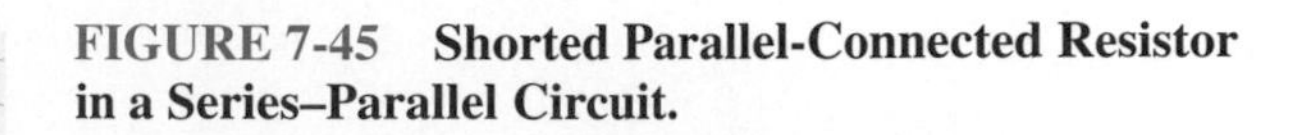
FIGURE 7-45 Shorted Parallel-Connected Resistor in a Series–Parallel Circuit.

The voltage measured with the voltmeter (VM) will be

$$V_{R1} = 8.6 \text{ V}$$
$$V_{R2} = 15.4 \text{ V}$$
$$V_{R3} = 0 \text{ V (short)}$$
$$V_{R4} = 15.4 \text{ V}$$

This identifies the problem as R_3 being a short, as 0 V is being dropped across it.

In summary, you may have noticed that (Figure 7-46):

1. An open component causes total resistance to increase ($R_T\uparrow$) and, therefore, total current to decrease ($I_T\downarrow$). The open component, if in series, has the supply voltage across it; if in a parallel branch, it has the parallel branch voltage across it.

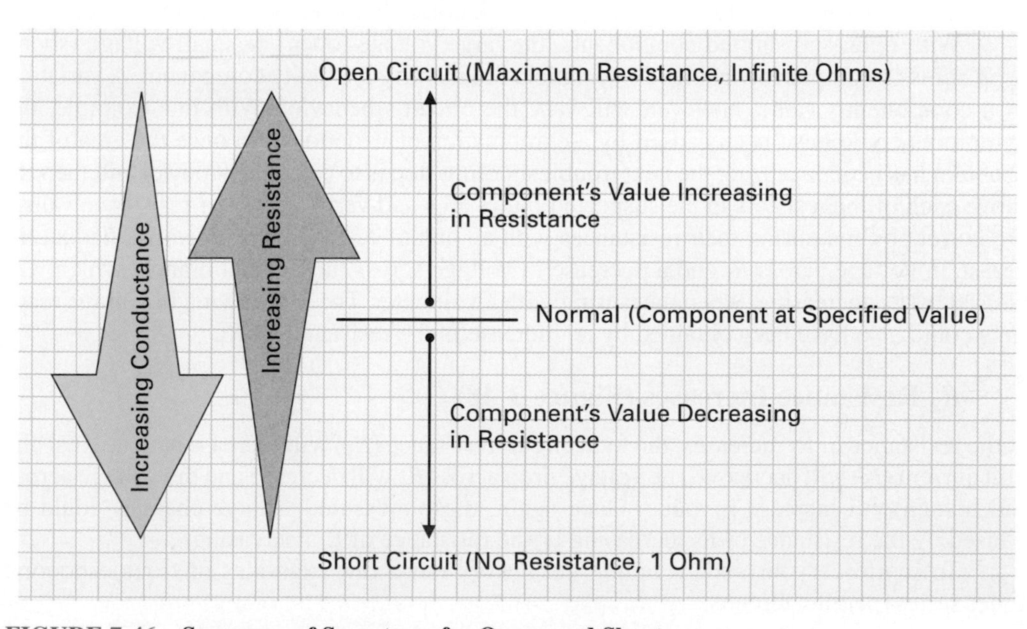

FIGURE 7-46 Summary of Symptoms for Opens and Shorts.

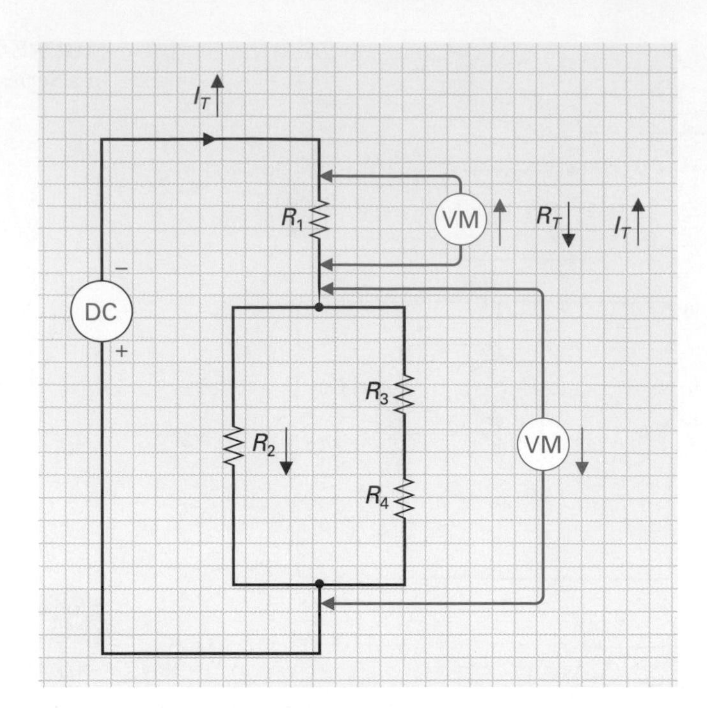

FIGURE 7-47 Parallel-Connected Resistor Value Decrease in Series–Parallel Circuit.

2. A shorted component causes total resistance to decrease ($R_T\downarrow$) and, therefore, total current to increase ($I_T\uparrow$). The shorted component, if in series or parallel branches, will have 0 V across it.

7-8-3 *Resistor Value Variation*

R_2 Resistance Decreases (Figure 7-47)

If the resistance of R_2 decreases, the total circuit resistance (R_T) will decrease and the total circuit current (I_T) will increase. The result of this problem and the way in which the fault can be located is that when the voltage drop across R_1 and the parallel branch is tested, there will be an increased voltage drop across R_1 due to the increased current flow and a decrease in the voltage drop across the parallel branch due to a decrease in the parallel branch resistance.

With open and shorted components, the large voltage (open) or small voltage (short) drop across a component enables the technician to identify the faulty component. A variation in a component's value, however, will vary the circuit's behavior. With this example, the symptoms could have been caused by a combination of variations. So once the area of the problem has been localized, the next troubleshooting step is to disconnect power and then remove each of the resistors in the suspected faulty area and verify that their resistance values are correct by measuring their resistances with an ohmmeter. In this problem we had an increase in voltage across R_1 and a decrease in voltage across the parallel branch, which was caused by R_2 decreasing, when measuring with a voltmeter. The same swing in voltage readings could also have been obtained by an increase in the resistance of R_1.

R_2 Resistance Increases (Figure 7-48)

If the resistance of R_2 increases, the total circuit resistance (R_T) will increase and the total circuit current (I_T) will decrease. The voltage drop across R_1 will decrease and the voltage across the branch will increase in value. Once again, these measured voltage changes could be caused by the resistance of R_2 increasing or the resistance of R_1 decreasing.

To reinforce your understanding, let's work out a few examples of troubleshooting other series–parallel circuits.

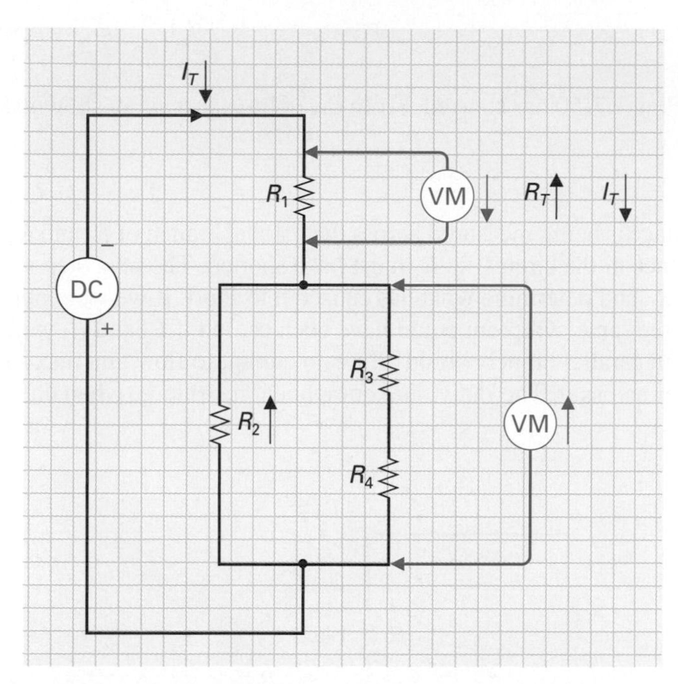

FIGURE 7-48 Parallel-Connected Resistor Value Increase in Series–Parallel Circuit.

EXAMPLE:

If bulb 1 in Figure 7-49 goes open, what effect will it have, and how will the fault be recognized?

Solution:

A visual inspection of the circuit shows all bulbs off, as there is no current path from one side of the battery to the other, since bulb 1 is connected in series and has opened the only path. Since all bulbs are off, the faulty bulb cannot be visually isolated. However, you can easily localize the faulty bulb by using one of two methods:

1. Use the voltmeter and check the voltage across each bulb. Bulb 1 would have 12 V across it, while 2 and 3 would have 0 V across them. This isolates the faulty component to bulb 1, since all the supply voltage is being measured across it.
2. By analyzing the circuit diagram, you can see that only one bulb can open and cause all the bulbs to go out, and that is bulb 1. If bulb 2 opens, 1 and 3 would still be on, and if bulb 3 opens, 1 and 2 would still remain on. With power off, the ohmmeter could verify this open.

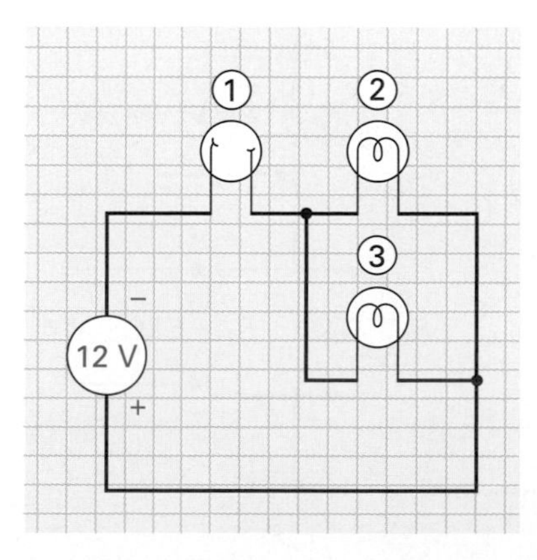

FIGURE 7-49 Series-Connected Open Bulb in Series–Parallel Circuit.

EXAMPLE:

One resistor in Figure 7-50 has shorted. From the voltmeter reading shown, determine which one.

■ ***Solution:***

If the supply voltage is being measured across the parallel branch of R_2 and R_3, there cannot be any other resistance in the circuit, so R_1 must have shorted. The next step would be to locate the component, R_1, and determine what has caused it to short. If we were not told that a resistor had shorted, the same symptom could have been caused if R_2 and R_3 were both open, and therefore the open parallel branch would allow no current to flow and maximum supply voltage would appear across it. The individual component resistance, when checked, will isolate the problem.

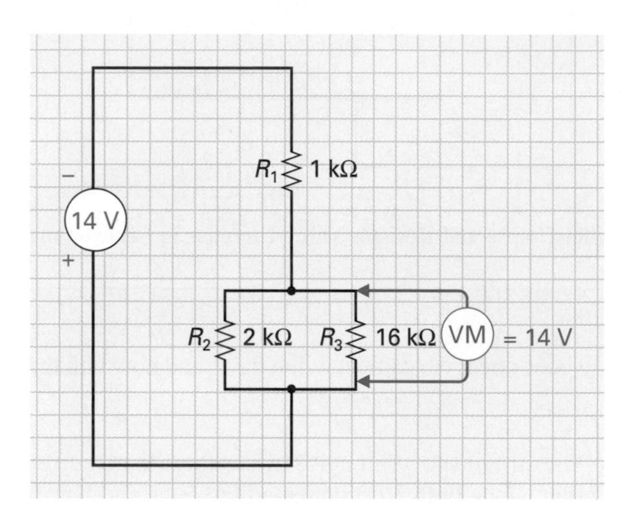

FIGURE 7-50 Find the Shorted Resistor.

EXAMPLE:

Determine if there is an open or short in Figure 7-51. If so, isolate it by the two voltage readings that are shown in the circuit diagram.

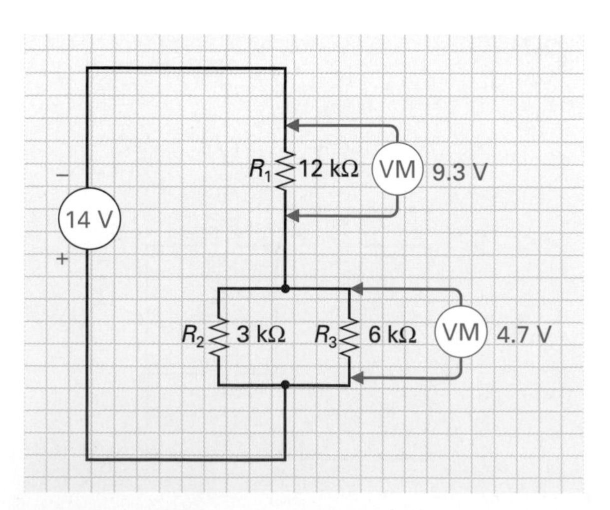

FIGURE 7-51 Does a Problem Exist?

■ *Solution:*

Performing a few calculations, you should come up with a normal total circuit resistance of 14 kΩ and a total circuit current of 1 mA. This should cause 12 V across R_1 and 2 V across the parallel branch under no-fault conditions. The decrease in the voltage drop across the series resistor R_1 leads you to believe that there has been a decrease in total circuit current, which must have been caused by a total resistance increase, which points to an open component (assuming that only an open or short can occur and not a component value variation).

If R_1 was open, all the 14 V would have been measured across R_1, which did not occur.

If R_3 opened,

$$\text{Total resistance} = 15\ \text{k}\Omega$$

$$\text{Total current} = 0.93\ \text{mA}\left(\frac{V_T}{R_T} = \frac{14\ \text{V}}{15\ \text{k}\Omega}\right)$$

$$V_{R1} = I_T \times R_1 = 11.16\ \text{V}$$

Since the voltage dropped across R_1 was 9.3 V, R_3 is not the open.

If R_2 opened,

$$\text{Total resistance} = 18\ \text{k}\Omega$$

$$\text{Total current} = 0.78\ \text{mA}\left(\frac{V_T}{R_T} = \frac{14\ \text{V}}{18\ \text{k}\Omega}\right)$$

$$V_{R1} = I_T \times R_1 = 9.36\ \text{V}$$

This circuit's problem is resistor R_2, which has opened.

SELF-TEST EVALUATION POINT FOR SECTION 7-8

Now that you have completed this section, you should be able to:

Objective 7. Explain how to identify the following problems in a series–parallel circuit:

a. Open series resistor.
b. Open parallel resistor.
c. Shorted series resistor.
d. Shorted parallel resistor.
e. Resistor value variation.

Use the following questions to test your understanding of Section 7-8:

Describe how to isolate the following problems in a series–parallel circuit:

1. An open component
2. A shorted component
3. A resistor value variation

7-9 THEOREMS FOR DC CIRCUITS

Series–parallel circuits can become very complex in some applications, and the more help you have in simplifying and analyzing these networks, the better. The following theorems can be used as powerful analytical tools for evaluating circuits. To begin, let's discuss the differences between voltage and current sources.

7-9-1 *Voltage and Current Sources*

The easiest way to understand a current source is to compare its features to a voltage source, so let's begin by discussing voltage sources.

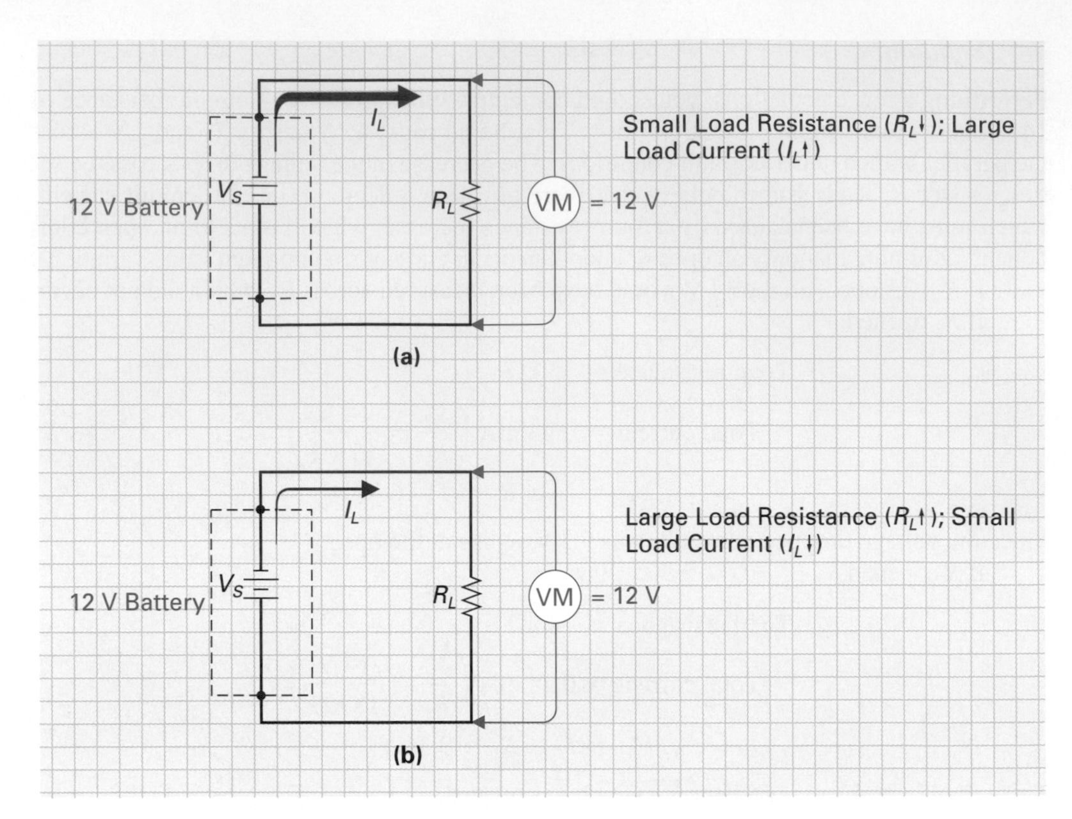

FIGURE 7-52 Ideal Voltage Source. (a) Heavy Load. (b) Light Load.

Voltage Source

Voltage Source
The circuit or device that supplies voltage to a load circuit.

The battery is an example of a **voltage source** that under ideal conditions will produce a fixed output voltage regardless of what load resistance is connected across its terminals. This means that even if a large load current is drawn from the battery (heavy load, due to a small load resistance) or if a small load current results (light load, due to a large load resistance), the battery will always produce a constant output voltage, as seen in Figure 7-52.

In reality, every voltage source, whether a battery, power supply, or generator, will have some level of inefficiency and generate not only an output electrical dc voltage but also heat. This inefficiency is represented as an internal resistance, as seen in Figure 7-53(a), and in most cases this internal source resistance (R_{int}) is very low (several ohms) compared to the load resistance (R_L). In Figure 7-53(a), no load has been connected, so the output or open-circuit voltage will be equal to the source voltage, V_S. When a load is connected across the battery, as shown in Figure 7-53(b), R_{int} and R_L form a series circuit, and some of the source

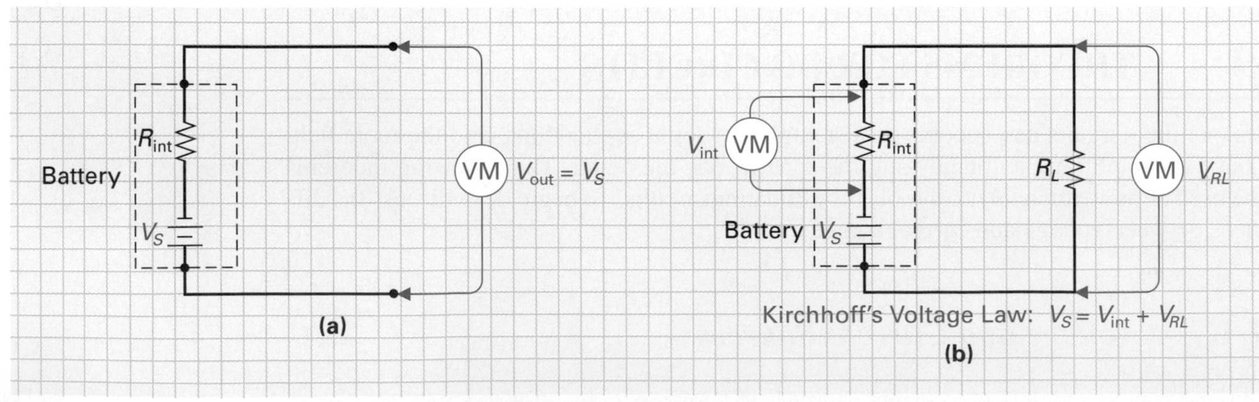

FIGURE 7-53 Realistic Voltage Source. (a) Unloaded. (b) Loaded.

voltage appears across R_{int}. As a result, the output or load voltage is always less than V_S. Since R_{int} is normally quite small compared to R_L, the voltage source approaches ideal, as almost all the source voltage (V_S) appears across R_L.

In conclusion, a voltage source should have the smallest possible internal resistance so that the output voltage (V_{out}) will remain constant and approximately equal to V_S independent of whether a light load (large R_L, small I_L) or heavy load (small R_L, large I_L) is connected across its output terminals.

EXAMPLE:

Calculate the output voltage in Figure 7-54 if R_L is equal to

a. 100 Ω
b. 1 kΩ
c. 100 kΩ

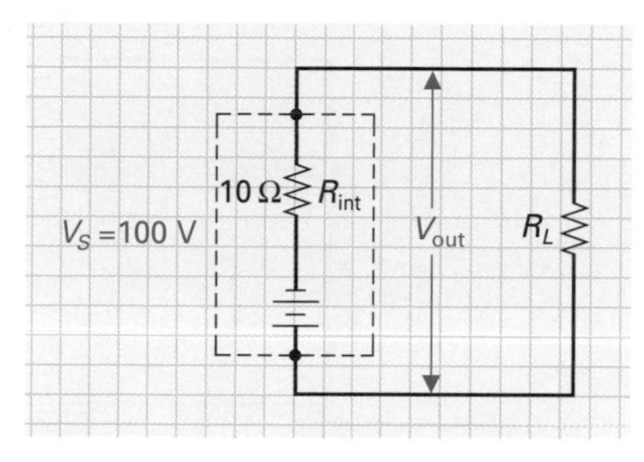

FIGURE 7-54 Voltage Source Circuit Example.

Solution (Voltage-divider formula):

a. $R_L = 100\ \Omega$:

$$V_{out} = \frac{R_L}{R_T} \times V_S$$

$$= \frac{100\ \Omega}{110\ \Omega} \times 100\ \text{V} = 90.9\ \text{V}$$

b. $R_L = 1\ \text{k}\Omega\ (1000\ \Omega)$:

$$V_{out} = \frac{1000\ \Omega}{1010\ \Omega} \times 100\ \text{V} = 99.0\ \text{V}$$

c. $R_L = 100\ \text{k}\Omega\ (100{,}000\ \Omega)$:

$$V_{out} = \frac{100{,}000\ \Omega}{100{,}010\ \Omega} \times 100\ \text{V} = 99.99\ \text{V}$$

From this example you can see that the larger the load resistance, the greater the output voltage (V_{out} or V_{RL}). To explain this in a little more detail, we can say that a large R_L is considered a light load for the voltage source, as it only has to produce a small load current ($R_L\uparrow, I_L\downarrow$), and consequently, the heat generated by the source is small ($P_{R_{int}}\downarrow = I^2\downarrow \times R$) and the voltage source is more efficient (V_{out} almost equals V_S), approaching ideal ($V_{out} = V_S$). The voltage source in this example, however, produced an almost constant output voltage (within 10% of V_S) despite the very large changes in R_L.

Current Source

Just as a voltage source has a voltage rating, a **current source** has a current rating. An ideal voltage source should deliver a constant output voltage, and similarly an ideal current source should deliver its constant rated current, regardless of what value of load resistance is connected across its output terminals, as seen in Figure 7-55.

Current Source
The circuit or device that supplies current to a load circuit.

A current source can be thought of as a voltage source with an extremely large internal resistance, as seen in Figure 7-56(a) and symbolized in Figure 7-56(b). The current source should have a large internal resistance so that, whatever the load resistance connected across the output, it will have very little effect on the total resistance and the load current will remain constant. The symbol for a constant current source has an arrow within a circle and this arrow points in the direction of current flow.

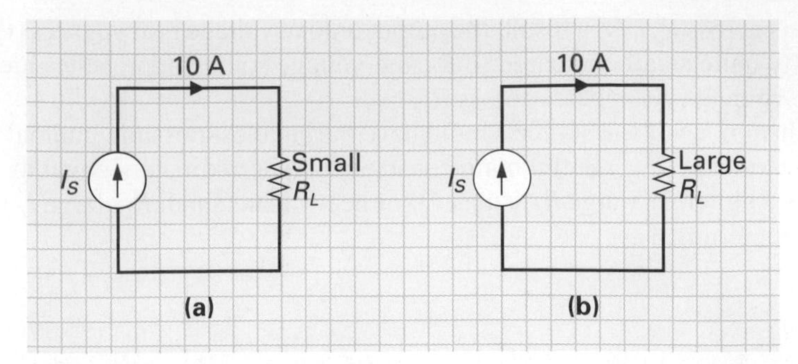

FIGURE 7-55 The Current Source Maintains a Constant Output Current Whether the Load is (a) Heavy (Small R_L) or (b) Light (Large R_L).

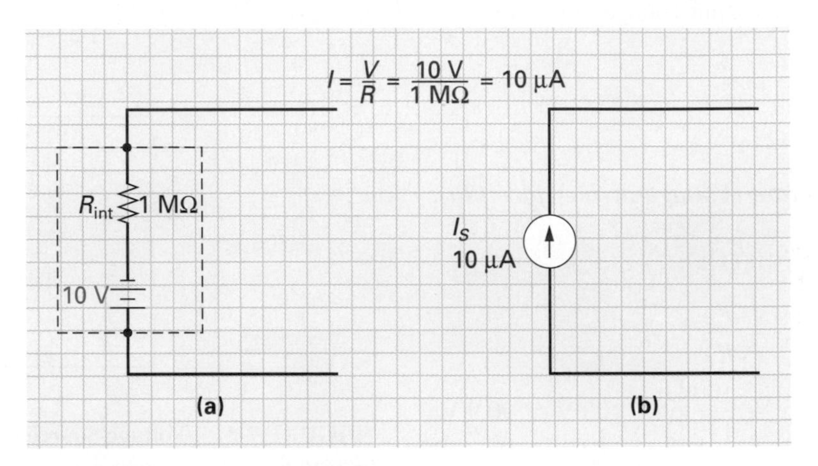

FIGURE 7-56 The Large Internal Resistance of the Current Source.

EXAMPLE:

Calculate the load current supplied by the current source in Figure 7-57 if the following values of R_L are connected across the output terminals:

a. 100 Ω
b. 1 kΩ
c. 100 kΩ

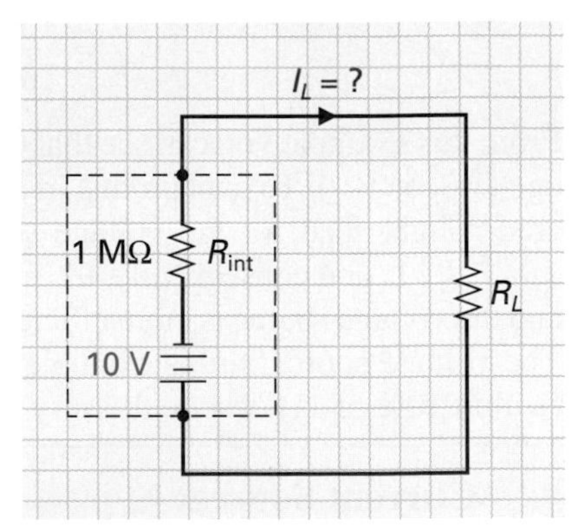

FIGURE 7-57 Current Source Circuit Example.

Solution:

a. $R_L = 100 \ \Omega \ (R_T = R_{int} + R_L = 1{,}000{,}000 \ \Omega + 100 \ \Omega = 1{,}000{,}100 \ \Omega)$

$$I_L = \frac{V_S}{R_T}$$

$$= \frac{10 \text{ V}}{1{,}000{,}100 \ \Omega} = 9.999 \ \mu\text{A}$$

b. $R_L = 1 \text{ k}\Omega \ (R_T = R_{int} + R_L = 1{,}001{,}000 \ \Omega)$

$$I_L = \frac{10 \text{ V}}{1{,}001{,}000 \ \Omega} = 9.99 \ \mu\text{A}$$

c. $R_L = 100 \text{ k}\Omega \ (R_T = R_{int} + R_L = 1{,}100{,}000 \ \Omega)$

$$I_L = \frac{10 \text{ V}}{1{,}100{,}000 \ \Omega} = 9.09 \ \mu\text{A}$$

From this example you can see that the current source delivered an almost constant output current regardless of the large load resistance change.

A Two-Source Circuit Example

In some instances, as seen in Figure 7-58, a circuit can contain more than one voltage source. In this example, the schematic diagram shows how car one's battery can be connected across car two's dead battery to boost current to the starter motor.

In the following sections, you will be introduced to some of the theorems that enable us to analyze these complex networks and reduce them to a simple equivalent circuit.

7-9-2 *Superposition Theorem*

The **superposition theorem** is used not only in electronics but also in physics and even economics. It is used to determine the net effect in a circuit that has two or more sources connected. The basic idea behind this theorem is that if two voltage sources are both producing a current through the same circuit, the net current can be determined by first finding the individual currents and then adding them together. Stated formally: *In a network containing two or more voltage sources, the current at any point is equal to the algebraic sum of the individual source currents produced by each source acting separately.*

Superposition Theorem
In a network or circuit containing two or more voltage sources, the current at any point is equal to the algebraic sum of the individual source currents produced by each source acting separately.

The best way to fully understand the theorem is to apply it to a few examples. Figure 7-59(a) illustrates a simple series circuit with two resistors and two voltage sources. The 12 V source (V_1) is trying to produce a current in a clockwise direction, while the 24 V source (V_2) is trying to force current in a counterclockwise direction. What will be the resulting net current in this circuit?

STEP 1 To begin, let's consider what current would be produced in this circuit if only V_1 were connected, as shown in Figure 7-59(b).

$$I_1 = \frac{V_1}{R_T} = \frac{12\text{ V}}{9\ \Omega} = 1.33\text{ A}$$

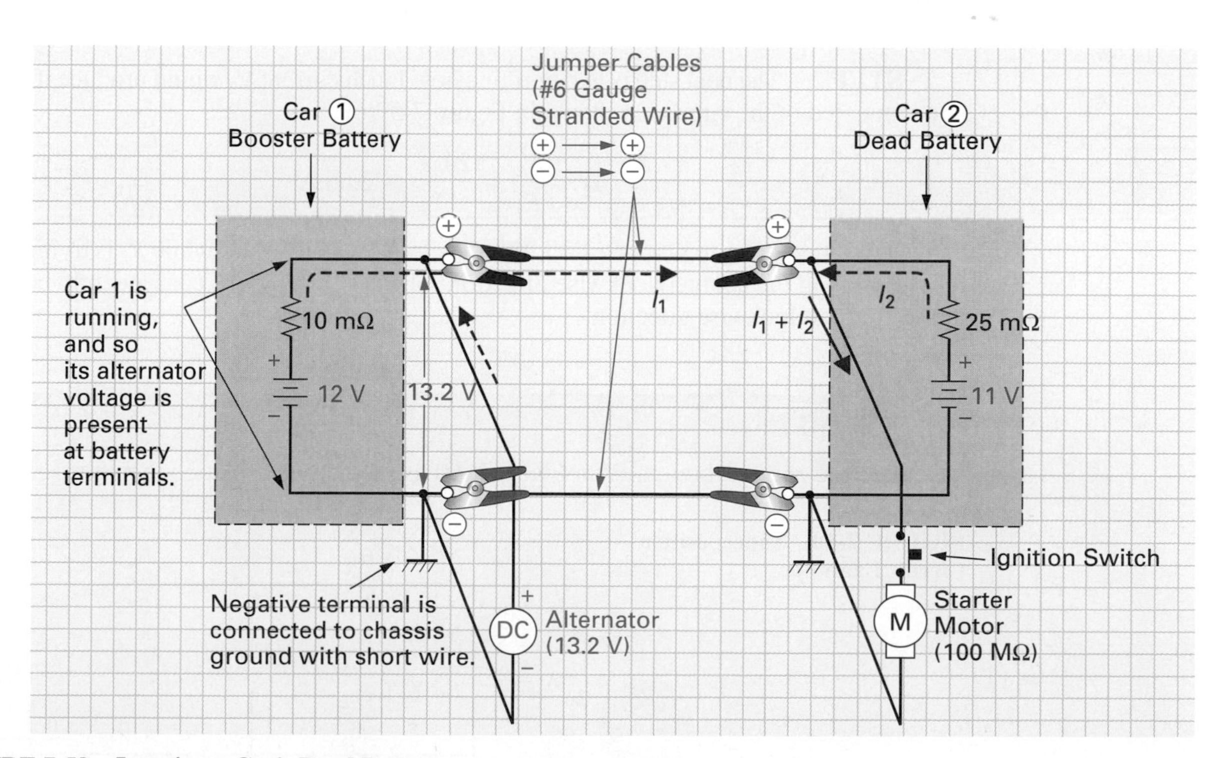

FIGURE 7-58 Jumping a Car's Dead Battery.

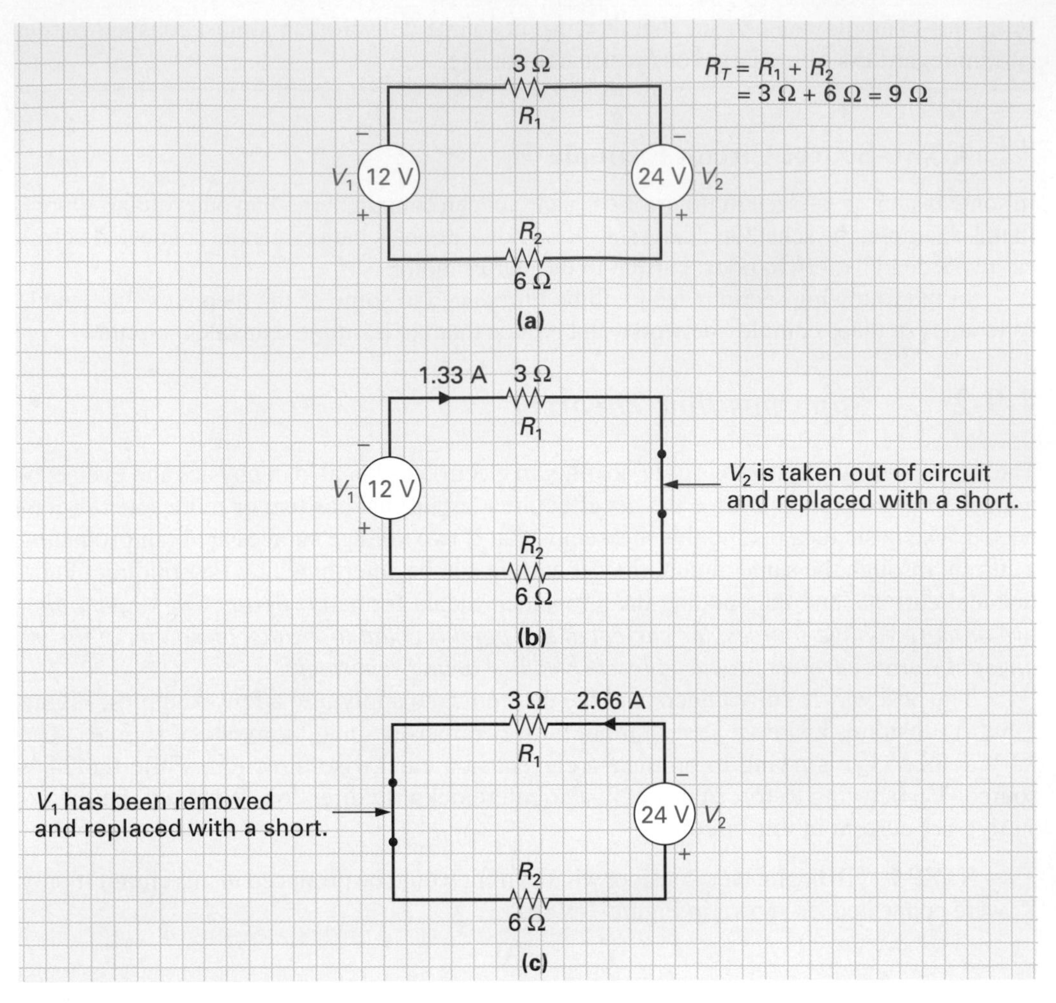

FIGURE 7-59 Superposition.

STEP 2 The next step is to determine how much current V_2 would produce if V_1 were not connected in the circuit, as shown in Figure 7-59(c).

$$I_2 = \frac{V_2}{R_T} = \frac{24\ \text{V}}{9\ \Omega} = 2.66\ \text{A}$$

V_1 is attempting to produce 1.33 A in the clockwise direction, while V_2 is trying to produce 2.66 A in the counterclockwise direction. The net current will consequently be 1.33 A in the counterclockwise direction.

EXAMPLE:

Calculate the current through R_2 in Figure 7-60 using the superposition theorem.

FIGURE 7-60 Superposition Circuit Example.

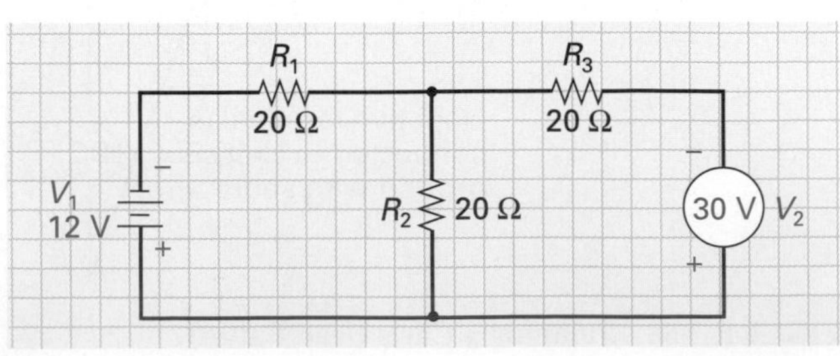

■ ***Solution:***

The first step is to calculate the current through R_2 due only to the voltage source V_1. This is shown in Figure 7-61(a). R_1 is a series-connected resistor, and R_2 and R_3 are connected in parallel with one another. So:

$$R_{2,3} = \frac{R_{ev}}{n} = \frac{20\ \Omega}{2} = 10\ \Omega \qquad \text{(equal-value parallel resistor formula)}$$

$$R_T = R_1 + R_{2,3} = 20\ \Omega + 10\ \Omega = 30\ \Omega \qquad \text{(total resistance)}$$

$$I_T = \frac{V_1}{R_T} = \frac{12\ \text{V}}{30\ \Omega} = 400\ \text{mA} \qquad \text{(total current)}$$

$$I_2 = \frac{R_{2,3}}{R_2} \times I_T = \frac{10\ \Omega}{20\ \Omega} \times 400\ \text{mA} = 200\ \text{mA} \qquad \text{(current-divider formula)}$$

This 200 mA of current is flowing down through R_2.

The next step is to find the current flow through R_2 due only to the voltage source V_2. This is shown in Figure 7-61(b). In this instance, R_3 is a series-connected resistor and R_1 and R_2 make up a parallel circuit. So:

$$R_{1,2} = 10\ \Omega$$

$$R_T = R_{1,2} + R_3 = 10\ \Omega + 20\ \Omega = 30\ \Omega$$

$$I_T = \frac{V_2}{R_T} = \frac{30\ \text{V}}{30\ \Omega} = 1\ \text{A} \quad (1000\ \text{mA})$$

$$I_2 = \frac{R_{1,2}}{R_2} \times I_T = \frac{10\ \Omega}{20\ \Omega} \times 1000\ \text{mA} = 500\ \text{mA}$$

This 500 mA of current will flow down through R_2.

Both V_1 and V_2 produce a current flow down through R_2, so I_{R2} and I_{R3} have the same algebraic sign, and the total current through R_2 is equal to the sum of the two currents produced by V_1 and V_2.

$$\begin{aligned} I_2\,(\text{total}) &= I_2 \text{ due to } V_1 + I_2 \text{ due to } V_2 \\ &= 200\ \text{mA} + 500\ \text{mA} \\ &= 700\ \text{mA} \end{aligned}$$

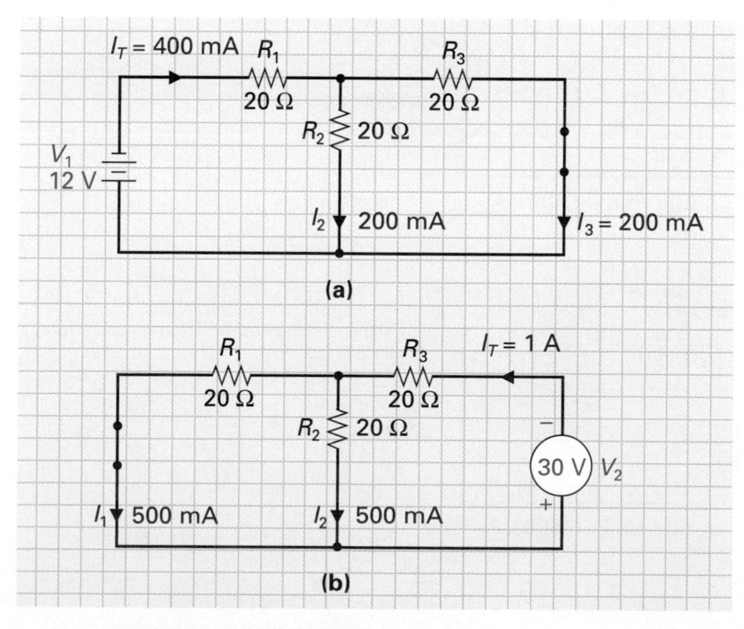

FIGURE 7-61 Superposition Circuit Solution.

EXAMPLE:

Calculate the current flow through R_1 in Figure 7-62.

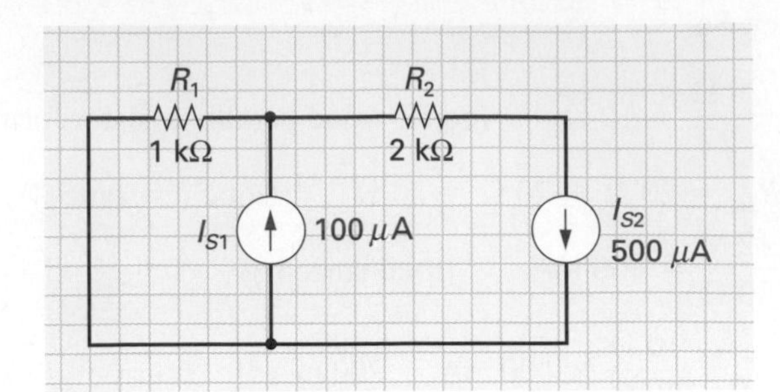

FIGURE 7-62 Superposition Circuit Example.

Solution:

With the superposition theorem, current sources are treated differently from voltage sources in that each *current source is removed from the circuit and replaced with an open,* as illustrated in the first step of the solution shown in Figure 7-63(a). In this instance you can see that current has only one path (series circuit), so the current through R_1 is counterclockwise and equal to the I_{S1} source current, 100 μA.

In Figure 7-63(b), the current source I_{S1} has been removed and replaced with an open. R_1 and R_2 form a series circuit, so the total source current from I_{S2} flows through R_1 ($I_1 = 500$ μA) in a clockwise direction.

Since I_{S1} is producing 100 μA of current through R_1 in a counterclockwise direction and I_{S2} is producing 500 μA of current in a clockwise direction, the resulting current through R_1 will equal

$$\begin{aligned} I_1 &= I_{S2}\,(\text{cw}) - I_{S1}\,(\text{ccw}) \\ &= 500\ \mu\text{A} - 100\ \mu\text{A} = 400\ \mu\text{A clockwise} \end{aligned}$$

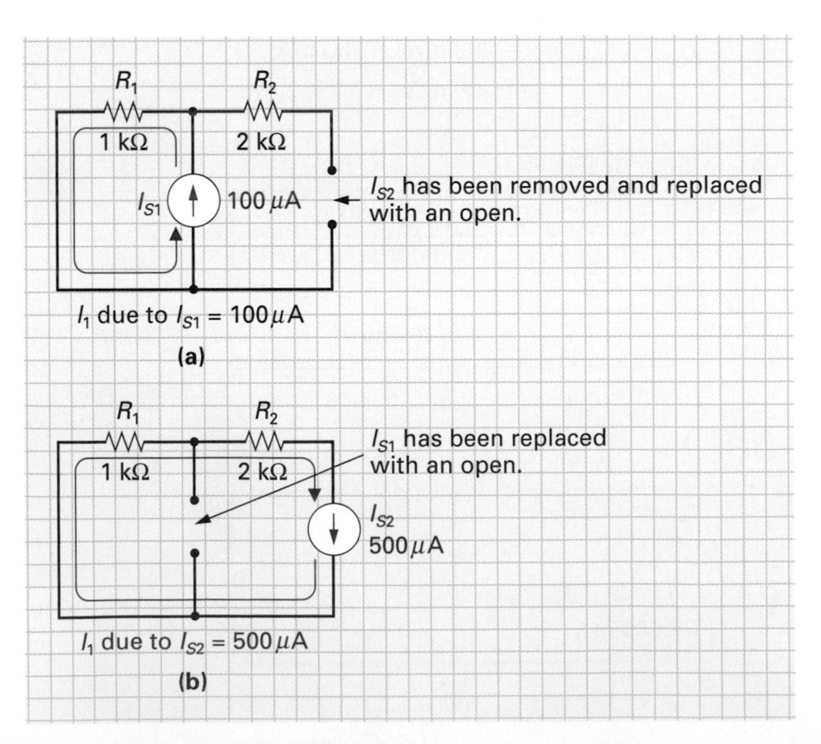

FIGURE 7-63 Superposition Circuit Solution.

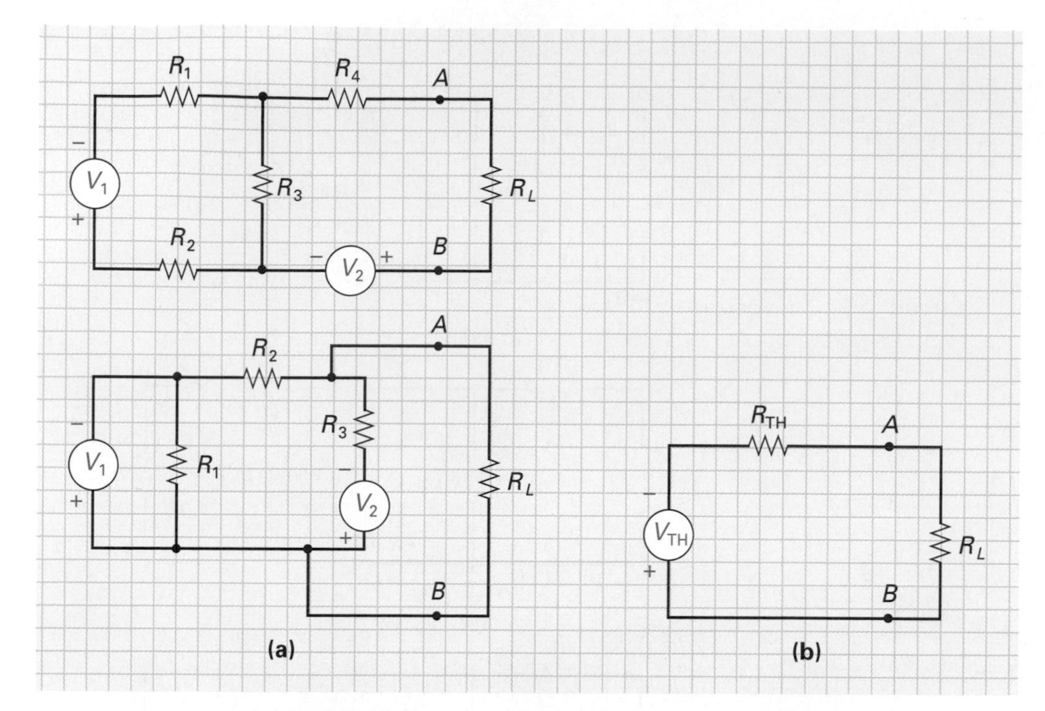

FIGURE 7-64 Thévenin's Theorem. (a) Complex Multiple Resistors and Source Networks Are Replaced by (b) One Source Voltage (V_{TH}) and One Series-Connected Resistance (R_{TH}).

TIME LINE
Leon-Charles Thévenin (1857–1927), a French telegraph engineer, educator, and accomplished snow skier, presented his theorem in the *Journal of Physics* in 1883. In the early 1920s AT&T used the theorem in the design of recording instrumentation and was the first to refer to it as Thévenin's theorem. Later, Edward L. Norton, an engineer at AT&T, would develop a current source equivalent to Thévenin's voltage source circuit.

7-9-3 *Thévenin's Theorem*

Thévenin's Theorem
Any network of voltage sources and resistors can be replaced by a single equivalent voltage source (V_{TH}) in series with a single equivalent resistance (R_{TH}).

Thévenin's theorem allows us to replace the complex networks in Figure 7-64(a) with an equivalent circuit containing just one source voltage (V_{TH}) and one series-connected resistance (R_{TH}), as in Figure 7-64(b). Stated formally: *Any network of voltage sources and resistors can be replaced by a single equivalent voltage source (V_{TH}) in series with a single equivalent resistance (R_{TH}).*

Figure 7-65(a) illustrates an example circuit. As with any theorem, a few rules must be followed to obtain an equivalent V_{TH} and R_{TH}.

STEP 1 The first step is to disconnect the load (R_L) and calculate the voltage that will appear across points A and B, as in Figure 7-65(b). This open-circuit voltage will be the same as the voltage drop across R_2 (V_{R2}) and is called the *Thévenin equivalent voltage* (V_{TH}). First, let's calculate current:

$$I_T = \frac{V_S}{R_T} = \frac{12\text{ V}}{9\ \Omega} = 1.333\text{ A}$$

Therefore, V_{R2} or V_{TH} will equal

$$V_{R2} = I_T \times R_2 = 1.333 \times 6\ \Omega = 8\text{ V}$$

so $V_{TH} = 8$ V.

STEP 2 Now that the Thévenin equivalent voltage has been calculated, the next step is to calculate the Thévenin equivalent resistance. In this step, the source voltage is removed and replaced with a short, as seen in Figure 7-65(c), and the Thévenin equivalent resistance is equal to whatever resistance exists between points A and B. In this example, R_1 and R_2 form a parallel circuit, the total resistance of which is equal to

$$R_T = \frac{R_1 \times R_2}{R_1 + R_2} = \frac{3 \times 6}{3 + 6} = \frac{18}{9} = 2\ \Omega$$

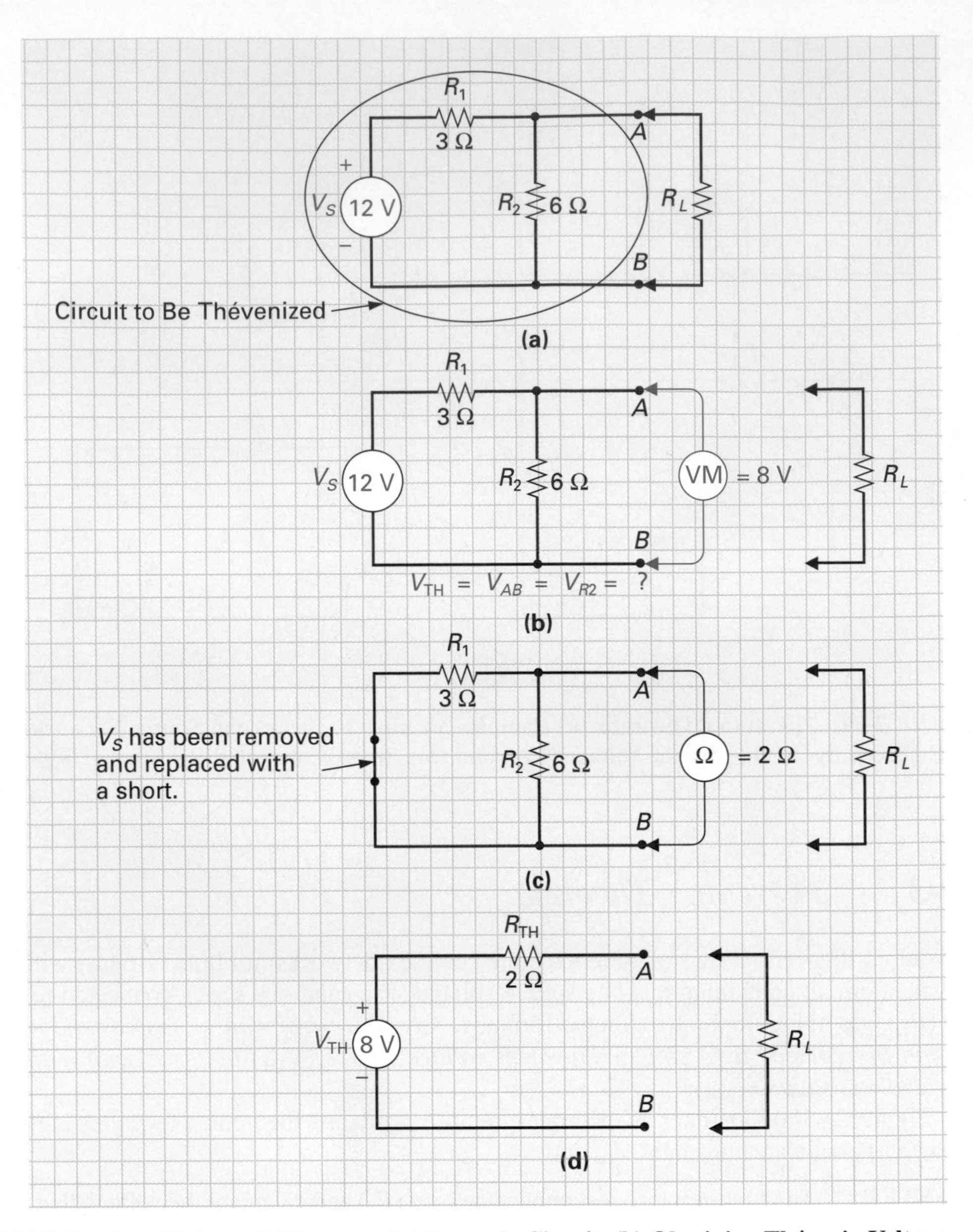

FIGURE 7-65 Thévenin's Theorem. (a) Example Circuit. (b) Obtaining Thévenin Voltage (V_{TH}). (c) Obtaining Thévenin Resistance. (d) Thévenin Equivalent Circuit.

so $R_{TH} = 2\ \Omega$. The circuit to be Thévenized in Figure 7-65(a) can be represented by the Thévenin equivalent circuit shown in Figure 7-65(d).

The question you may have at this time is why we would need to simplify such a basic circuit when Ohm's law could have been used just as easily to analyze the network. Thévenin's theorem has the following advantages:

1. If you had to calculate load current and load voltage (I_{RL} and V_{RL}) for 20 different values of R_L, it would be far easier to use the Thévenin equivalent circuit with the series-connected resistors R_{TH} and R_L, rather than applying Ohm's law to the series–parallel circuit made up of R_1, R_2, and R_L.
2. Thévenin's theorem permits you to solve complex circuits that could not easily be analyzed using Ohm's law.

EXAMPLE:

Determine V_{TH} and R_{TH} for the circuit in Figure 7-66.

Solution:

The first step is to remove the load resistor R_L and calculate what voltage will appear between points A and B, as shown in Figure 7-67(a). Removing R_L will open the path for current to flow through R_4, which will consequently have no voltage drop across it. The voltage between points A and B, therefore, will be equal to the voltage dropped across R_3, and since R_1, R_2, and R_3 form a series circuit, the voltage-divider formula can be used:

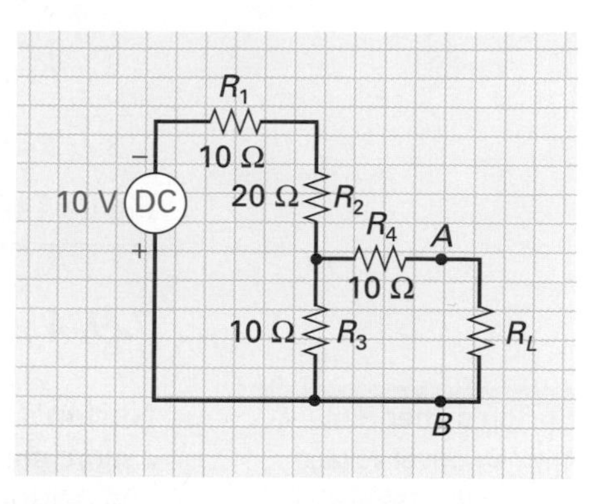

FIGURE 7-66 Thévenin Circuit Example.

$$V_{R3} = \frac{R_3}{R_T} \times V_S$$

$$= \frac{10\ \Omega}{40\ \Omega} \times 10\ \text{V} = 2.5\ \text{V}$$

Therefore,

$$V_{AB} = V_{R3} = V_{TH} = 2.5\ \text{V}$$

The next step is to calculate the Thévenin resistance, which will equal whatever resistance appears across the terminals A and B with the voltage source having been removed and replaced with a short, as shown in Figure 7-67(b). In Figure 7-67(c), the circuit has been redrawn so that the relationship between the resistors can be seen in more detail. As you can

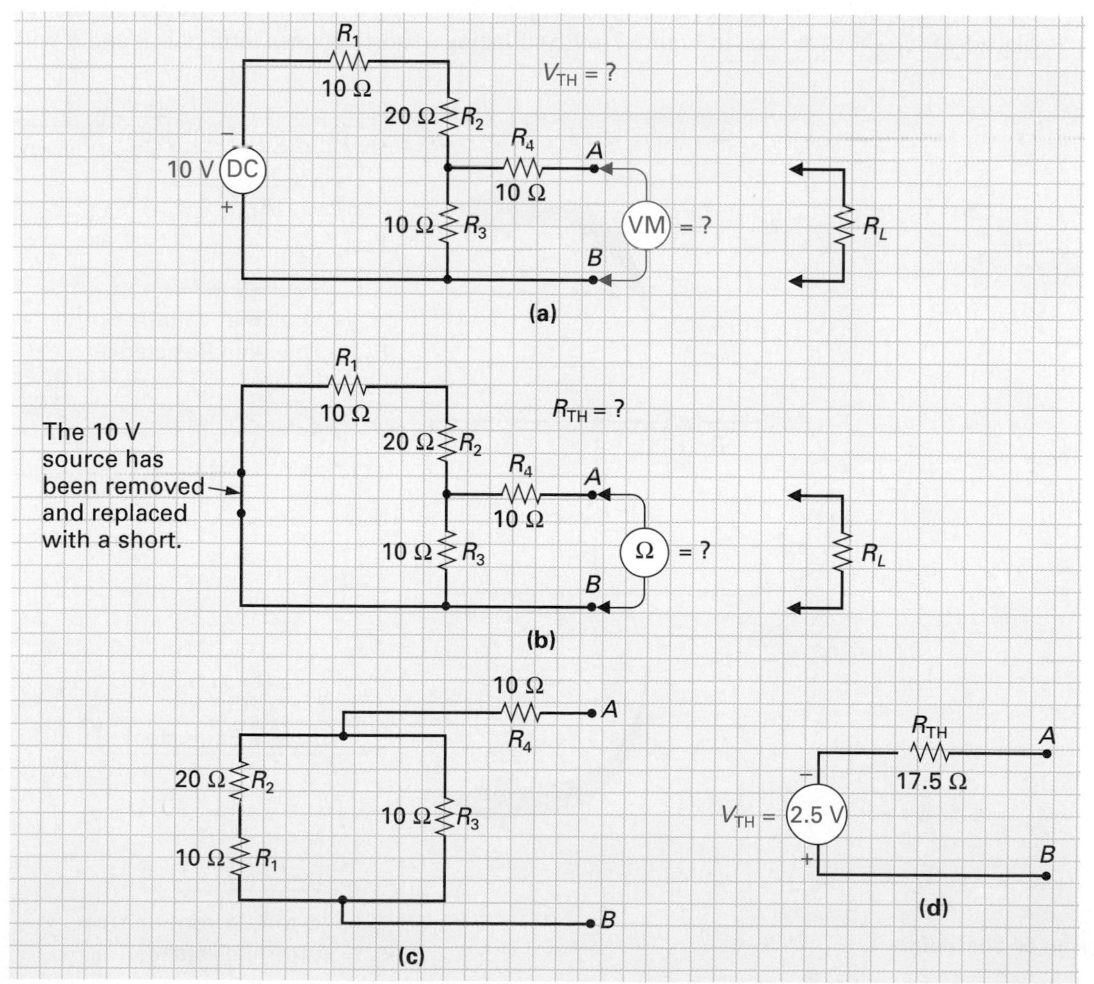

FIGURE 7-67 Thévenin Circuit Solution.

see, R_1 and R_2 are in series with one another and both are in parallel with R_3, and this combination is in series with R_4. Using our three-step procedure for calculating total resistance in a series–parallel circuit, the following results are obtained:

1. $R_{1,2} = R_1 + R_2 = 10\ \Omega + 20\ \Omega = 30\ \Omega$
2. $R_{1,2,3} = \dfrac{R_{1,2} \times R_3}{R_{1,2} + R_3} = \dfrac{30\ \Omega \times 10\ \Omega}{30\ \Omega + 10\ \Omega} = \dfrac{300\ \Omega}{40\ \Omega} = 7.5\ \Omega$
3. $R_T = R_{1,2,3} + R_4 = 7.5\ \Omega + 10\ \Omega = 17.5\ \Omega$

Figure 7-67(d) on the previous page illustrates the Thévenin equivalent circuit.

7-9-4 *Norton's Theorem*

Norton's Theorem

Any network of voltage sources and resistors can be replaced by a single equivalent current source (I_N) in parallel with a single equivalent resistance (R_N).

Norton's theorem, like Thévenin's theorem, is a tool for simplifying a complex circuit into a more manageable one. Figure 7-68 illustrates the difference between a Thévenin equivalent and Norton equivalent circuit. Thévenin's theorem simplifies a complex network and uses an equivalent voltage source (V_{TH}) and an equivalent series resistance (R_{TH}). Norton's theorem, on the other hand, simplifies a complex circuit and represents it with an equivalent current source (I_N) in parallel with an equivalent Norton resistance (R_N), as shown in Figure 7-68.

As with any theorem, a set of steps has to be carried out to arrive at an equivalent circuit. The example we will use is shown in Figure 7-69(a) and is the same example used in Figure 7-68.

TIME LINE

While working at AT&T in 1926, Edward L. Norton (1898–1983) proposed simplifying a circuit to an equivalent current source and parallel resistor, as opposed to Thévenin's theorem that simplified a circuit to an equivalent voltage source and series resistor.

STEP 1 Calculate the Norton equivalent current source, which will be equal to the current that would flow between terminals *A* and *B* if the load resistor were removed and replaced with a short, as seen in Figure 7-69(b). Placing a short between terminals *A* and *B* will

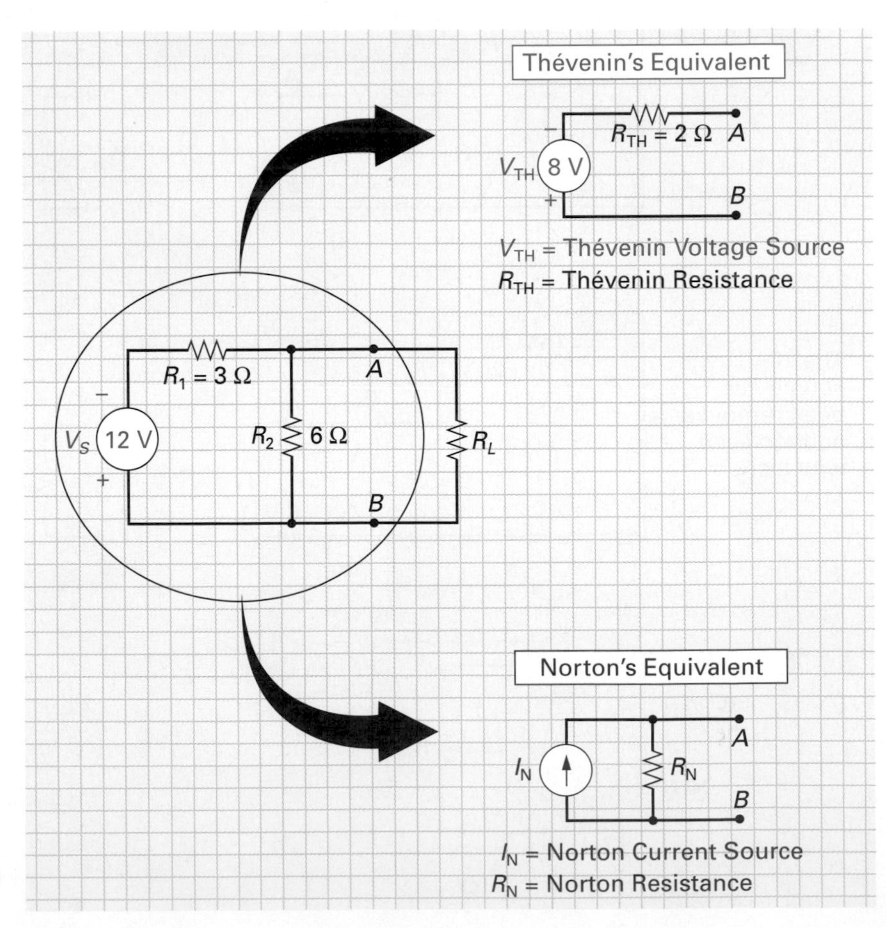

FIGURE 7-68 Comparison of Thévenin's and Norton's Circuits.

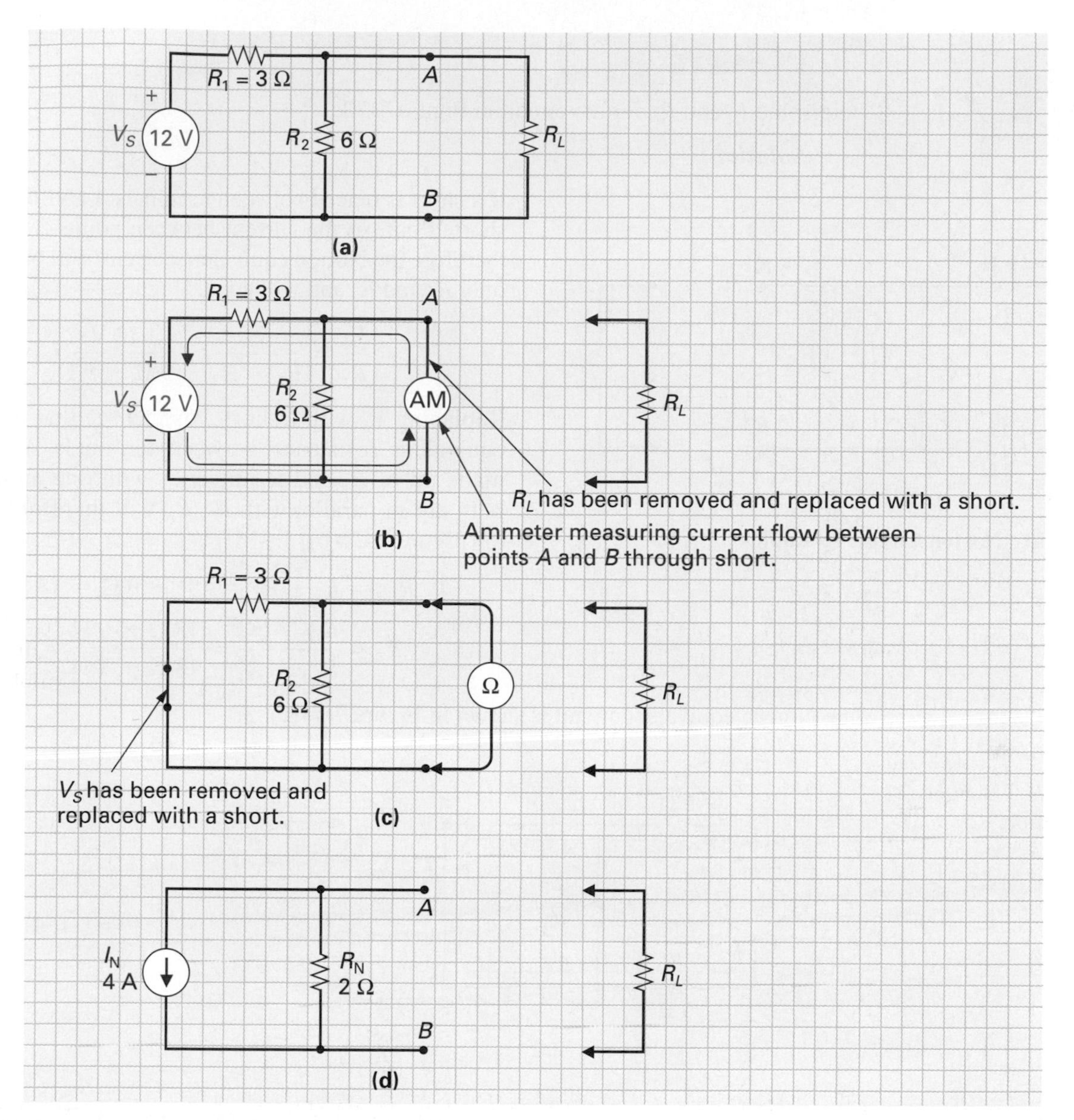

FIGURE 7-69 Norton's Theorem. (a) Example Circuit. (b) Obtaining Norton Current. (c) Obtaining Norton Resistance (R_N). (d) Norton Equivalent Circuit.

short out the resistor R_2, so the only resistance in the circuit will be R_1. The Norton equivalent current source in this example will therefore be equal to

$$I_N = \frac{V_S}{R_T} = \frac{12\text{ V}}{3\ \Omega} = 4\text{ A}$$

STEP 2 The next step is to determine the value of the Norton equivalent resistance that will be placed in parallel with the current source, unlike Thévenin's equivalent resistance, which was placed in series. Like Thévenin's theorem, though, Norton's equivalent resistance (R_N) is equal to the resistance between terminals A and B when the voltage source is removed and replaced with a short, as shown in Figure 7-69(c). Since R_1 and R_2 form a parallel circuit, the Norton equivalent resistance will be equal to

$$R_N = \frac{R_1 \times R_2}{R_1 + R_2} = \frac{3\ \Omega \times 6\ \Omega}{3\ \Omega + 6\ \Omega} = \frac{18\ \Omega^2}{9\ \Omega} = 2\ \Omega$$

The Norton equivalent circuit has been determined simply by carrying out these two steps and is illustrated in Figure 7-69(d).

EXAMPLE:

Determine I_N and R_N for the circuit in Figure 7-70.

Solution:

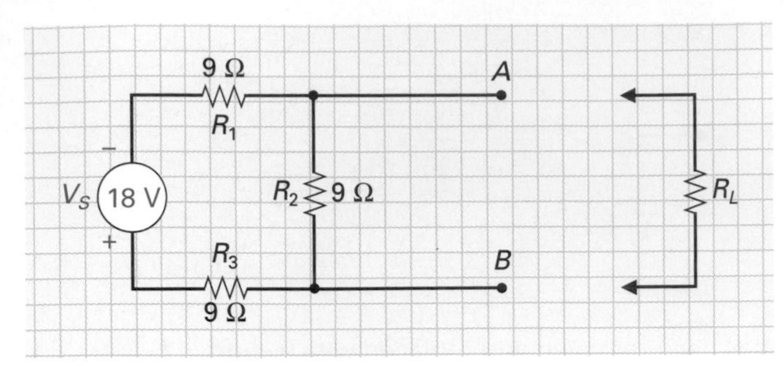

FIGURE 7-70 Norton Circuit Example.

If a short is placed between terminals *A* and *B*, R_2 will be shorted out, so the current between points *A* and *B*, and therefore the Norton equivalent current, will be limited by only R_1 and R_3 and will equal [Figure 7-71(a)]

$$I_N = \frac{V_S}{R_T} = \frac{V_S}{R_1 + R_3} = \frac{18\text{ V}}{9\ \Omega + 9\ \Omega} = 1\text{ A}$$

Replacing V_S with a short, you can see that our Norton equivalent resistance between terminals *A* and *B* is made up of R_1 and R_3 in series with one another, and both are in parallel with R_2. R_N will therefore be equal to [Figure 7-71(b)]

$$R_{1,3} = R_1 + R_3 = 9\ \Omega + 9\ \Omega = 18\ \Omega$$

$$R_T = \frac{R_{1,3} \times R_2}{R_{1,3} + R_2} = \frac{18\ \Omega \times 9\ \Omega}{18\ \Omega + 9\ \Omega} = 6\ \Omega$$

The Norton equivalent circuit is shown in Figure 7-71(c).

FIGURE 7-71 Norton Circuit Solution.

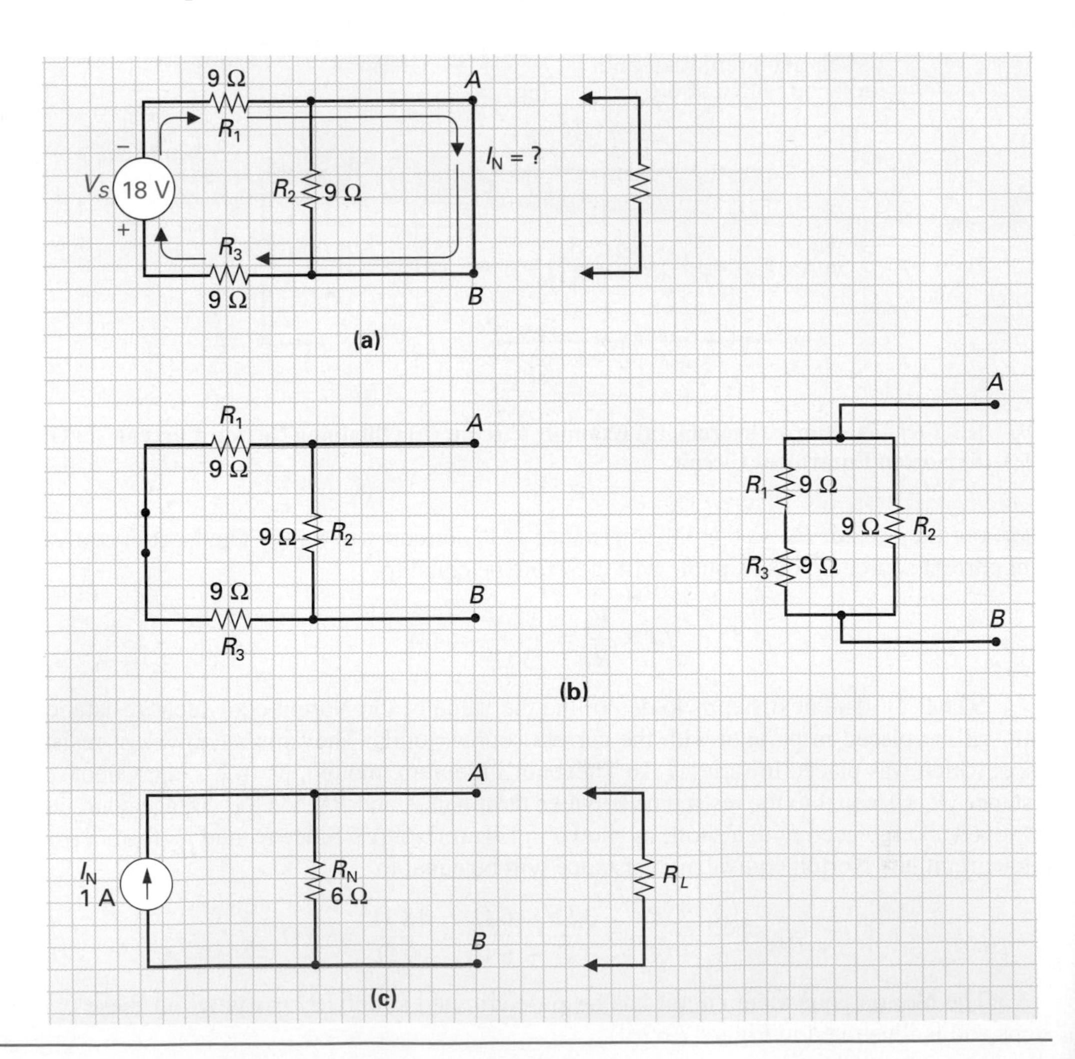

SELF-TEST EVALUATION POINT FOR SECTION 7-9

Now that you have completed this section, you should be able to:

Objective 8. Describe the differences between a voltage and a current source.

Objective 9. Analyze series–parallel networks using:

a. The superposition theorem.
b. Thévenin's theorem.
c. Norton's theorem.

Use the following questions to test your understanding of Section 7-9:

1. A constant voltage source will have a __________ internal resistance, and a constant current source will have a __________ internal resistance. (large or small)
2. The superposition theorem is a logical way of analyzing networks with more than one __________.
3. Thévenin's theorem represents a complex two-terminal network as a single __________ source with a series-connected single __________.
4. Norton's theorem also allows you to analyze complex two-terminal networks as a single __________ source in parallel with a single resistor.

SUMMARY

1. If current has only one path to follow through a component, that component is connected in series.
2. If the total current has two or more paths to follow, these components are connected in parallel.
3. All electronic equipment is composed of many components interconnected to form a combination of series and parallel (series–parallel) circuits.
4. The R–$2R$ ladder circuit is a series–parallel circuit used for digital-to-analog conversion.
5. Troubleshooting is the process of locating and diagnosing malfunctions or breakdowns in equipment by means of systematic checking or analysis.
6. **a.** If a series-connected resistor in a series–parallel circuit opens, there cannot be any current flow, and the source voltage will appear across the open and 0 V will appear across all the others.
   b. If a parallel branch in a series–parallel circuit opens, the overall circuit resistance will increase and the total current will decrease. A greater parallel resistance will cause a greater voltage drop across that parallel circuit.
7. **a.** If a series-connected resistor in a series–parallel circuit shorts, the total circuit resistance will decrease, resulting in an increase in total current. The fault can be located by measuring the voltage across the shorted resistor, which will be 0 V.
   b. If a parallel branch in a series–parallel circuit shorts, the total resistance will decrease, resulting in a total current increase. This smaller parallel circuit resistance will result in a smaller voltage drop across the parallel circuit.
8. A resistor's resistance will typically change with age, resulting in a total resistance change and therefore a total current change. The faulty component can be located due to its abnormal voltage drop.
9. An ideal voltage source will have an internal resistance of 0 Ω and provide a constant output voltage regardless of what load resistance is connected across its output.
10. In reality, every practical voltage source will have some level of inefficiency, and this is represented as an internal resistance, and the output voltage will remain relatively constant despite variations in load resistance.
11. An ideal current source will have an infinite internal resistance and provide a constant output current, regardless of what load resistance is connected across its output.
12. In reality, every practical current source has a relatively high internal resistance, and the output current will remain relatively constant despite variations in load resistance.
13. The superposition theorem is a useful tool when analyzing circuits with more than one voltage source.
14. Thévenin's theorem is another handy tool that can be used to represent complex series–parallel networks as a single voltage source (V_{TH}) in series with a single resistor (R_{TH}).
15. Norton's theorem can also simplify a complex series–parallel network to an equivalent form consisting of a single current source (I_N) in parallel with a single resistor (R_N).

REVIEW QUESTIONS

Multiple-Choice Questions

1. A series–parallel circuit is a combination of
a. Components connected end to end
b. Series (one current path) circuits
c. Both series and parallel circuits
d. Parallel (two or more current paths) circuits

2. Total resistance in a series–parallel circuit is calculated by applying the __________ resistance formula to series-connected resistors and the __________ resistance formula to resistors connected in parallel.
a. Series, parallel **c.** Series, series
b. Parallel, series **d.** Parallel, parallel

3. Total current in a series–parallel circuit is determined by dividing the total __________ by the total __________.
a. Power, current **c.** Current, resistance
b. Voltage, resistance **d.** Voltage, power

4. Branch current within series–parallel circuits can be calculated by
a. Ohm's law
b. The current-divider formula
c. Kirchhoff's current law
d. All of the above

5. All voltages on a circuit diagram are with respect to __________ unless otherwise stated.
a. The other side of the component **c.** Ground
b. The high-voltage source **d.** All of the above

6. A __________ ground has the negative side of the source voltage connected to ground, and a __________ ground has the positive side of the source voltage connected to ground.
a. Positive, negative **c.** Positive, earth
b. Chassis, earth **d.** Negative, positive

7. The output voltage will always __________ when a load or voltmeter is connected across a voltage divider.
a. Decrease
b. Remain the same
c. Increase
d. All of the above could be considered true.

8. A Wheatstone bridge was originally designed to measure
a. An unknown voltage **c.** An unknown power
b. An unknown current **d.** An unknown resistance

9. A balanced bridge has an output voltage
a. Equal to the supply voltage **c.** Of 0 V
b. Equal to half the supply voltage **d.** Of 5 V

10. The total resistance of a ladder circuit is best found by starting at the point __________ the source.
a. Nearest to **c.** Midway between
b. Farthest from **d.** All of the above

11. The *R*–2*R* ladder circuit finds its main application as a(an)
a. Analog-to-digital converter
b. Digital-to-analog converter
c. Device to determine unknown resistance values
d. All of the above

12. The Norton equivalent resistance (R_N) is always in __________ with the Norton equivalent current source (I_N).
a. Proportion **c.** Parallel
b. Series **d.** Either series or parallel

13. An ideal current source has __________ internal resistance, and an ideal voltage source has __________ internal resistance.
a. An infinite, 0 Ω of **c.** 0 Ω of, an infinite
b. No, a large amount **d.** No, an infinite

14. The Thévenin equivalent resistance (R_{TH}) is always in __________ with the Thévenin equivalent voltage (V_{TH}).
a. Proportion **c.** Parallel
b. Series **d.** Either series or parallel

15. The superposition theorem is useful for analyzing circuits with
a. Two or more voltage sources
b. A single voltage source
c. Only two voltage sources
d. A single current source

16. A resistor, when it burns out, will generally
a. Decrease slightly in value **c.** Short
b. Increase slightly in value **d.** Open

17. In a series–parallel resistive circuit, an open series-connected resistor will cause __________ current, whereas an open parallel-connected resistor will result in a total current __________.
a. An increase in, decrease **c.** Zero, decrease
b. A decrease in, increase **d.** None of the above

18. In a series–parallel resistive circuit, a shorted series-connected resistor will cause __________ current, whereas a shorted parallel-connected resistor will result in a total current __________.
a. An increase in, increase **c.** An increase in, decrease
b. A decrease in, decrease **d.** A decrease in, increase

19. A resistor's resistance will typically __________ with age, resulting in a total circuit current __________.
a. Decrease, decrease **c.** Decrease, increase
b. Increase, increase **d.** All of the above

20. The maximum power transfer theorem states that maximum power is transferred from the source to a load when load resistance is equal to source resistance.
a. True **b.** False

Communication Skill Questions

21. State the five-step method for determining a series–parallel circuit's resistance, voltage, current, and power values. (7-6)

22. Illustrate the following series–parallel circuits:
    a. R_1 in series with a parallel combination R_2, R_3, and R_4.
    b. R_1 in series with a two-branch parallel combination consisting of R_2 and R_3 in series and R_4 in parallel.
    c. R_1 in parallel with R_2, which is in series with a three-resistor parallel combination, R_3, R_4, and R_5.
23. Using the example in Question 22(c), apply values of your choice and apply the five-step procedure. (7-6)
24. Describe what is meant by "loading of a voltage-divider circuit." (7-7-1)
25. Illustrate and describe the Wheatstone bridge in the (7-7-2)
    a. Balanced condition
    b. Unbalanced condition
    c. Application of measuring unknown resistances
26. Describe how the ladder circuit acts as a current divider. (7-7-3)
27. Briefly describe the difference between a voltage source and a current source. (7-9-1)
28. Briefly describe the following theorems: (7-9)
    a. Superposition
    b. Thévenin's
    c. Norton's
    d. Maximum power transfer
29. What would be the advantages of Thévenin's and Norton's theorems to obtain an equivalent circuit? (7-9)
30. Draw the components that would exist in a (7-9)
    a. Thévenin equivalent circuit
    b. Norton equivalent circuit
31. Describe the steps involved in obtaining a (7-9)
    a. Thévenin equivalent circuit
    b. Norton equivalent circuit
32. When troubleshooting series–parallel circuits, describe what effect (7-8-1)
    a. An open series-connected
    b. An open parallel-connected

    resistor would have on total current and resistance, and how the opened resistor could be isolated.
33. When troubleshooting series–parallel circuits, describe what effect (7-8-2)
    a. A shorted series-connected
    b. A shorted parallel-connected

    resistor would have on total current and resistance, and how the shorted resistors could be isolated.
34. Describe what effect a resistor's value variation would have and how it could be recognized. (7-8-3)
35. Give the divider formula, Ohm's law, and Kirchhoff's laws used to determine (7-6)
    a. Branch currents
    b. Voltage drops in a series–parallel circuit

    (List all six.)

Practice Problems

36. R_3 and R_4 are in series with one another and are both in parallel with R_5. This parallel combination is in series with two series-connected resistors, R_1 and R_2. $R_1 = 2.5$ kΩ, $R_2 = 10$ kΩ, $R_3 = 7.5$ kΩ, $R_4 = 2.5$ kΩ, $R_5 = 2.5$ MΩ, and $V_S = 100$ V. For these values, calculate:
    a. Total resistance
    b. Total current
    c. Voltage across series resistors and parallel combinations
    d. Current through each resistor
    e. Total and individual power figures
37. Referring to the example in Question 36, calculate the voltage at every point of the circuit with respect to ground.
38. A 10 V source is connected across a series–parallel circuit made up of R_1 in parallel with a branch made up of R_2 in series with a parallel combination of R_3 and R_4. $R_1 = 100$ Ω, $R_2 = 100$ Ω, $R_3 = 200$ Ω, and $R_4 = 300$ Ω. For these values, apply the five-step procedure, and also determine the voltage at every point of the circuit with respect to ground.
39. Calculate the output voltage (V_{RL}) in Figure 7-72 if R_L is equal to
    a. 25 Ω b. 2.5 kΩ c. 2.5 MΩ
40. What load current will be supplied by the current source in Figure 7-73 if R_L is equal to
    a. 25 Ω b. 2.5 kΩ c. 2.5 MΩ
41. Use the superposition theorem to calculate total current
    a. Through R_2 in Figure 7-74(a)
    b. Through R_3 in Figure 7-74(b)

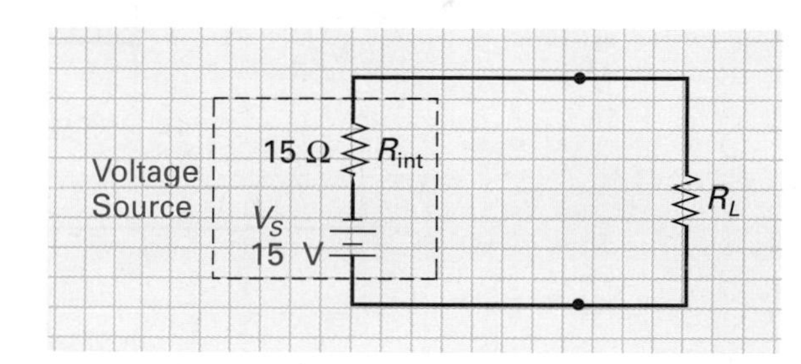

FIGURE 7-72

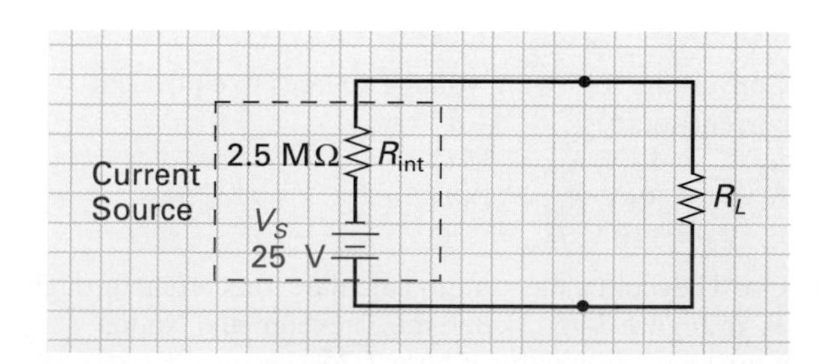

FIGURE 7-73

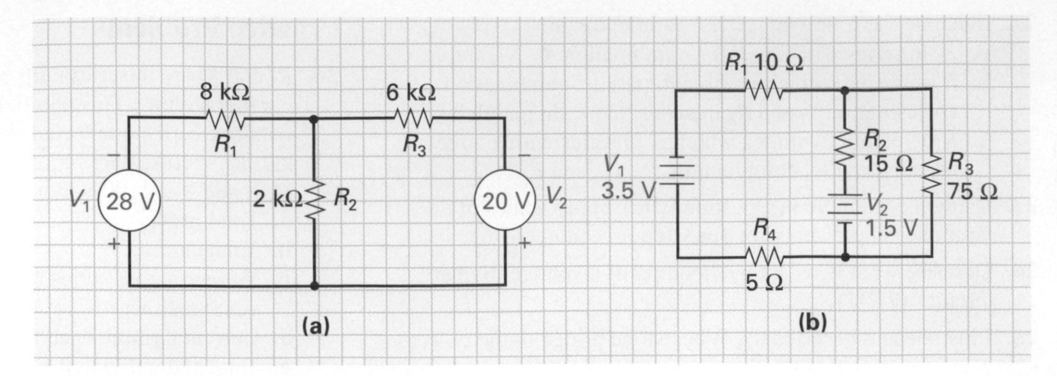

FIGURE 7-74

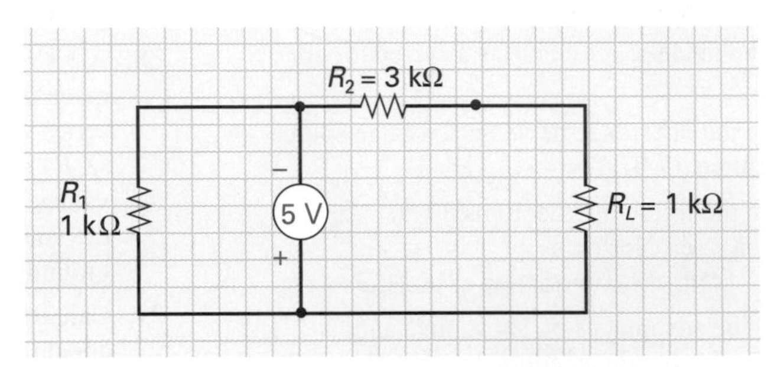

FIGURE 7-75

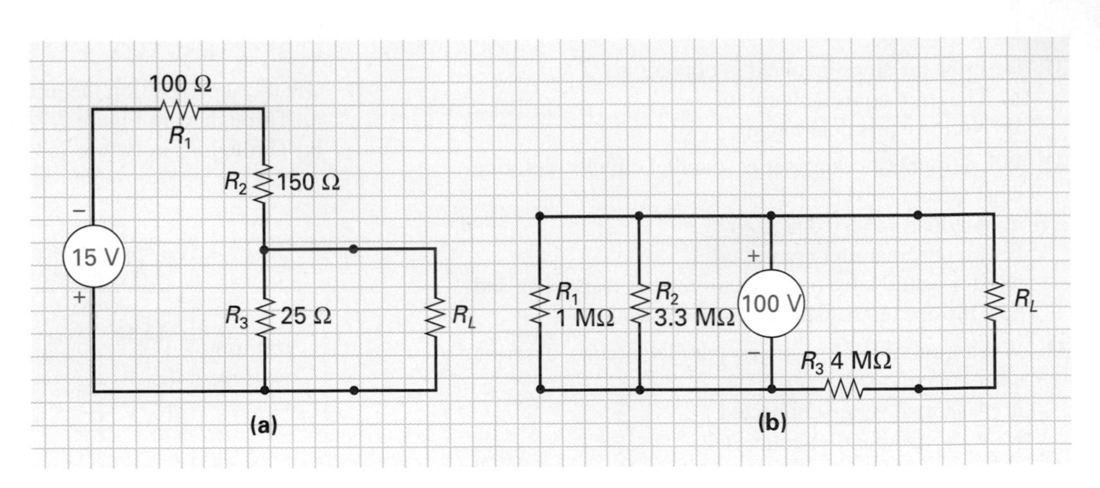

FIGURE 7-76

42. Convert the following voltage sources to equivalent current sources:

a. $V_S = 10$ V, $R_{int} = 15\ \Omega$
b. $V_S = 36$ V, $R_{int} = 18\ \Omega$
c. $V_S = 110$ V, $R_{int} = 7\ \Omega$

43. Use Thévenin's theorem to calculate the current through R_L in Figure 7-75. Sketch the Thévenin and Norton equivalent circuits for Figure 7-75.

44. Sketch the Thévenin and Norton equivalent circuits for the networks in Figure 7-76.

45. Convert the following current sources to equivalent voltage sources:

a. $I_S = 5$ mA, $R_{int} = 5$ MΩ
b. $I_S = 10$ A, $R_{int} = 10$ kΩ
c. $I_S = 0.0001$ A, $R_{int} = 2.5$ kΩ

Troubleshooting Questions

46. Referring to the example circuit in Figure 7-50, describe the effects you would get if a resistor were to short and if a resistor were to open.

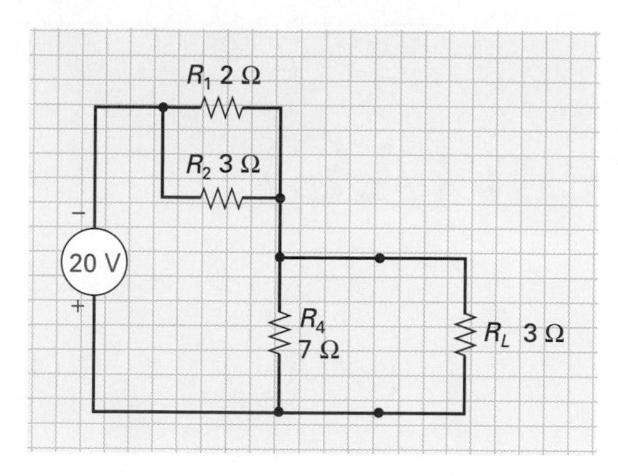

FIGURE 7-77

47. Design a simple five-resistor series–parallel circuit and insert a source voltage and resistance values. Apply the five-step series–parallel circuit procedure, and then theoretically open and short all the resistors and calculate what effect would occur and how you would recognize the problem.

48. Carbon composition resistors tend to increase in resistance with age, whereas most other types generally decrease in resistance. What effects would resistance changes have on their respective voltage drops?

49. Use Thévenin's theorem to simplify the circuit in Figure 7-77. What effect would the following faults have on the Thévenin equivalent circuit?
a. R_2 is shorted.
b. R_2 is open.

50. Calculate the Norton equivalent for Figure 7-77 and describe what circuit differences will occur for the same faults listed in Question 49.

Job Interview Test

These tests at the end of each chapter will challenge your knowledge up to this point and give you the practice you need for a job interview. To make this more realistic, the test will be comprised of both technical and personal questions. In order to take full advantage of this exercise, you may want to set up a simulation of the interview environment, have a friend read the questions to you, and record your responses for later analysis.

Company Name: DVD, Inc.

Industry Branch: Consumer Electronics.

Function: Assisting engineers with R&D.

Job Title: Engineering Technician.

51. How did you hear of the job opening?

52. What is a Wheatstone bridge circuit?

53. What quality advantages does a DVD have over a VHS tape?

54. What sort of responsibilities do you envision this job to have?

55. What could you tell me about Thévenin's theorem?

56. What symptoms would you expect from a series circuit open?

57. What symptoms would you expect from a parallel circuit short?

58. Would you be prepared to work a large number of extra hours for a short period of time if a project is behind schedule?

59. What is a series–parallel circuit?

60. Could you handle an emergency calmly and efficiently?

Answers

51. Describe source.

52. Section 7-7-2.

53. Knowing this company specializes in the design and development of DVD systems, you should have visited your local consumer electronics retailer prior to this interview and probed the sales staff for information on model types, advantages and features, etc.

54. Quote introductory section in text and job description listed in newspaper.

55. Section 7-9-3.

56. Chapter 5.

57. Chapter 6.

58. To be more competitive, most companies are bringing products to market in a much shorter space of time, and so this pattern has become more and more prevalent. If you are unsure, don't object to anything at the time. Later, if you decide that it is something you do not want to do, you can call and decline the offer.

59. Chapter 7.

60. The answer must be yes. Qualify this by discussing lab- and job-related safety practices and procedures.

Web Site Questions

Go to the web site http://www.prenhall.com/cook, select the textbook *Introductory DC/AC Electronics* or *Introductory DC/AC Circuits,* this chapter, and then follow the instructions when answering the multiple-choice practice problems.

Alternating Current (AC)

Charles Townes

The Laser

In 1898, H. G. Wells's famous book *The War of the Worlds* described Martian invaders with laserlike death rays blasting bricks, incinerating trees, and piercing iron as if it were paper. In 1917, Albert Einstein stated that, under certain conditions, atoms or molecules could absorb light and then be stimulated to release this borrowed energy. In 1954, Charles H. Townes, a professor at Columbia University, conceived and constructed with his students the first "maser" (acronym for "microwave amplification by stimulated emission of radiation"). In 1958, Townes and Arthur L. Shawlow wrote a paper showing how stimulated emission could be used to amplify light waves as well as microwaves, and the race was on to develop the first "laser." In 1960, Theodore H. Maiman, a scientist at Hughes Aircraft Company, directed a beam of light from a flash lamp into a rod of synthetic crystal, which responded with a burst of crimson light so bright that it outshone the sun.

An avalanche of new lasers emerged, some as large as football fields and others no bigger than a pinhead. They can be made to produce invisible infrared or ultraviolet light or any visible color in the rainbow. The high-power lasers can vaporize any material a million times faster and more intensely than a nuclear blast, but the low-power lasers are safe to use in children's toys.

At present, the laser is being used by the FBI to detect fingerprints that are 40 years old, in defense programs, in compact disc players, in underground fiber-optic communication to transmit hundreds of telephone conversations, to weld car bodies, to drill holes in baby-bottle nipples, to create three-dimensional images called holograms, and as a surgeon's scalpel in the operating room. Not a bad beginning for a device that when first developed was called "a solution looking for a problem."

8

Outline and Objectives

THE LASER
INTRODUCTION

8-1 MINI-MATH REVIEW—TRIGONOMETRY

8-1-1 The 3, 4, 5 Right-Angle Triangle

8-1-2 Opposite, Adjacent, Hypotenuse, and Theta

Objective 1: Explain how the Pythagorean theorem relates to right-angle triangles.

Objective 2: Determine the length of one side of a right-angle triangle when the lengths of the other two sides are known.

Objective 3: Describe how vectors are used to represent the magnitude and direction of physical quantities and how they are arranged in a vector diagram.

Objective 4: Explain how the Pythagorean theorem can be applied to a vector diagram to calculate the magnitude of a resultant vector.

Objective 5: Define the trigonometric terms:

a. Opposite
b. Adjacent
c. Hypotenuse
d. Theta
e. Sine
f. Cosine
g. Tangent

Objective 6: Demonstrate how the sine, cosine, and tangent trigonometric functions can be used to calculate:

a. The length of an unknown side of a right-angle triangle if the length of another side and the angle theta are known.
b. The angle theta of a right-angle triangle if the lengths of two sides of the triangle are known.

8-2 THE DIFFERENCE BETWEEN DC AND AC

Objective 7: Explain the difference between alternating current and direct current.

8-3 WHY ALTERNATING CURRENT?

Objective 8: Describe how ac is used to deliver power and to represent information.

Objective 9: Give the three advantages that ac has over dc from a power point of view.

8-3-1 Power Transfer

Objective 10: Describe basically the ac power distribution system from the electric power plant to the home or industry.

8-3-2 Information Transfer

Objective 11: Describe the three waves used to carry information between two points.

Objective 12: Explain how the three basic information carriers are used to carry many forms of information on different frequencies within the frequency spectrum.

8-3-3 Electrical Equipment or Electronic Equipment?

8-4 AC WAVE SHAPES

Objective 13: Explain the different characteristics of the five basic wave shapes.

Objective 14: Describe fundamental and harmonic frequencies.

8-4-1 The Sine Wave

8-4-2 The Square Wave

8-4-3 The Rectangular or Pulse Wave

8-4-4 The Triangular Wave

8-4-5 The Sawtooth Wave

8-4-6 Other Waveforms

8-5 MEASURING AND GENERATING AC SIGNALS

Objective 15: Explain the operation of the AC multimeter, and the following special attachments:

a. Current Clamp
b. Radio Frequency Probe
c. High Voltage Probe

Objective 16: Describe the operation of the oscilloscope and how it can be used to make measurements.

Objective 17: Describe the purpose and operation of the

a. Function generator.
b. Frequency counter.
c. Hand-held scopemeter.

Introduction

In this chapter you will be introduced to alternating current (ac). Direct current (dc) can be used in many instances to deliver power or represent information, but there are certain instances in which ac is preferred. For example, ac is easier to generate and is more efficiently transmitted as a source of electrical power for the home or business. Audio (speech and music) and video (picture) information are generally always represented in electronic equipment as an alternating current or alternating voltage signal.

In this chapter, we will begin by describing the difference between dc and ac and then examine where ac is used. Following this we will discuss all the characteristics of ac waveform shapes and, finally, ac measuring and generating instruments.

To begin with, though, let's review the math skills you will need for this chapter's material.

8-1 MINI-MATH REVIEW—TRIGONOMETRY

This "Mini-Math Review" is included to overview the mathematical details you need for the electronic concepts covered in this chapter. In this review we will be examining trigonometry.

Electronic and electrical technicians and engineers, mechanical engineers, carpenters, architects, navigators, and many other people in technical trades make use of **trigonometry** on a daily basis. Although volumes of theory are available about this subject, most of the material is not that useful to us for everyday applications. In this chapter we will concentrate on the more practical part of *trigonometry—the right-angle triangle* and its applications.

Trigonometry
The study of the properties of triangles, trigonometric functions, and their applications.

8-1-1 *The 3, 4, 5 Right-Angle Triangle*

Like all triangles, the **right-angle triangle,** or *right triangle,* has three sides and three corners. Its distinguishing feature, however, is that two of the sides of this triangle are at right angles (at 90°) to each other, as shown in Figure 8-1(a). The small square box within the triangle is placed in the corner to show that sides A and B are square, or at right angles to each other.

Right-Angle Triangle
A triangle having a 90° angle.

If you study this right triangle you may notice two interesting facts about the relative lengths of sides A, B, and C. These observations are as follows:

1. Side C is always longer than side A or side B.
2. The total length of side A and side B is always longer than side C.

The right triangle in Figure 8-1(b) has been drawn to scale to demonstrate another interesting fact about this triangle. If side A were to equal 3 cm and side B were to equal 4 cm, side C would equal 5 cm. This demonstrates a basic relationship among the three sides and accounts for why this right triangle is sometimes referred to as a *3, 4, 5 triangle*. As long as the relative lengths remain the same, it makes no difference whether the sides are 3 cm, 4 cm,

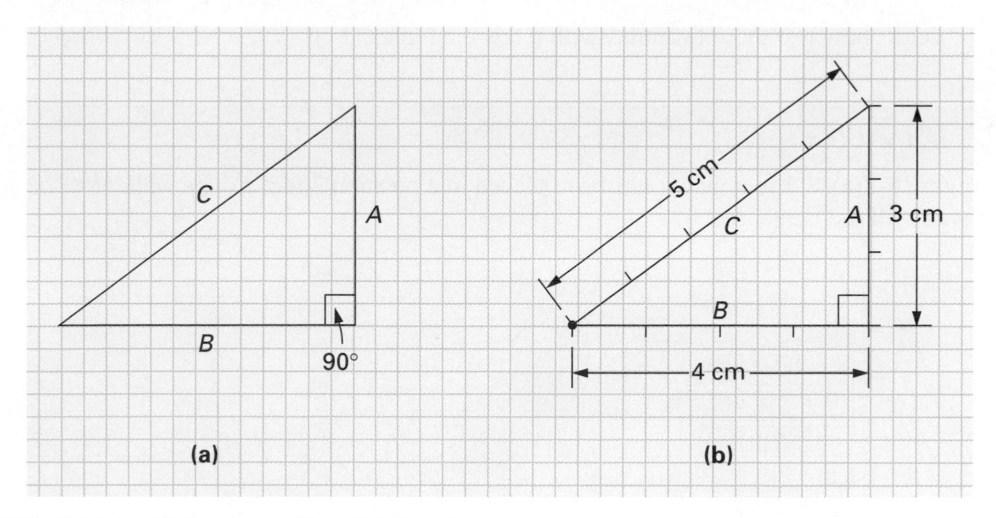

FIGURE 8-1 Right-Angle (3, 4, 5) Triangle.

and 5 cm or 30 km, 40 km, and 50 km. Unfortunately, in most applications our lengths of *A, B,* and *C* will not work out as easily as 3, 4, 5.

When a relationship exists among three quantities, we can develop a formula to calculate an unknown when two of the quantities are known. It was Pythagoras who first developed this basic formula or equation (known as the **Pythagorean theorem**), which states that *the square of the length of the hypotenuse (side C) of a right triangle equals the sum of the squares of the lengths of the other two sides*:

Pythagorean Theorem
A theorem in geometry: The square of the length of the hypotenuse of a right triangle equals the sum of the squares of the lengths of the other two sides.

$$\boxed{C^2 = A^2 + B^2} \qquad \begin{pmatrix} 5^2 = 3^2 + 4^2 \\ 25 = 9 + 16 \end{pmatrix}$$

By using the rules of algebra, we can transpose this formula to derive formulas for sides *C, B,* and *A*.

$$\boxed{C^2 = A^2 + B^2}$$

Solve for *B*

$$C^2 = A^2 + B^2$$
$$C^2 - A^2 = (A^2 + B^2) - A^2$$
$$C^2 - A^2 = B^2$$

Subtract A^2 from both sides.

$$\sqrt{C^2 - A^2} = \sqrt{B^2}$$
$$\sqrt{C^2 - A^2} = B$$

Take the square root of both sides.

$$\boxed{B = \sqrt{C^2 - A^2}}$$

$$4 = \sqrt{5^2 - 3^2}$$
$$= \sqrt{25 - 9}$$
$$= \sqrt{16}$$
$$= 4$$

Solve for C

$$C^2 = A^2 + B^2$$
$$\sqrt{C^2} = \sqrt{A^2 + B^2}$$

← Take the square root of both sides.

$$\boxed{C = \sqrt{A^2 + B^2}}$$

$$5 = \sqrt{3^2 + 4^2}$$
$$= \sqrt{9 + 16}$$
$$= \sqrt{25}$$
$$= 5$$

Solve for *A*

$$C^2 = A^2 + B^2$$
$$C^2 - B^2 = (A^2 + B^2) - B^2$$
$$C^2 - B^2 = A^2$$

Subtract B^2 from both sides.

$$\sqrt{C^2 - B^2} = \sqrt{A^2}$$
$$\sqrt{C^2 - B^2} = A$$

Take the square root of both sides.

$$\boxed{A = \sqrt{C^2 - B^2}}$$

$$3 = \sqrt{5^2 - 4^2}$$
$$= \sqrt{25 - 16}$$
$$= \sqrt{9}$$
$$= 3$$

Let us now test these formulas with a few examples.

EXAMPLE:

In Figure 8-2(a), a 4 foot ladder has been placed in a position 2 feet from a wall. How far up the wall will the ladder reach?

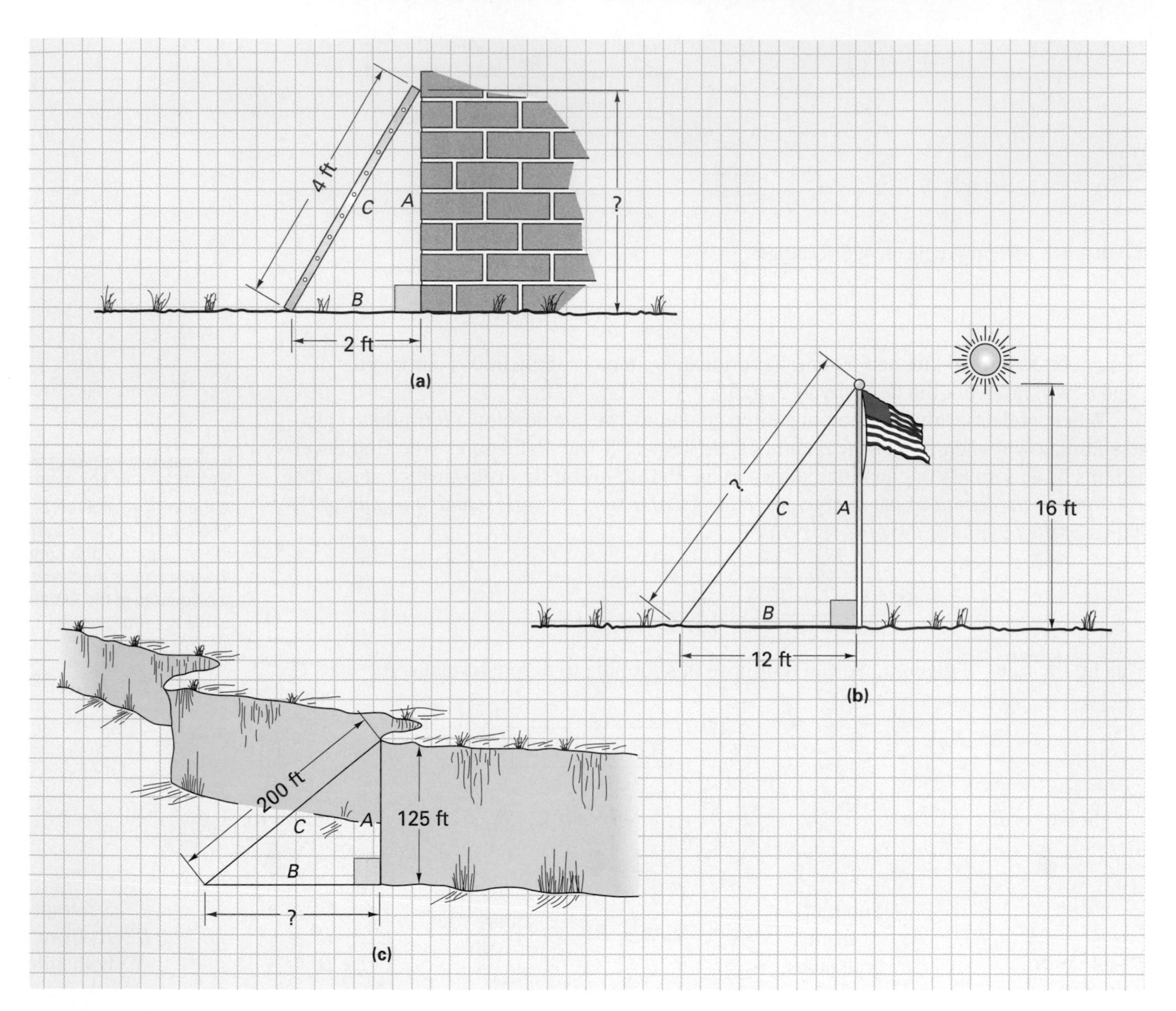

FIGURE 8-2 Right-Triangle Examples.

Solution:

In this example, A is unknown, and therefore

$$\begin{aligned} A &= \sqrt{C^2 - B^2} \\ &= \sqrt{4^2 - 2^2} \\ &= \sqrt{16 - 4} \\ &= \sqrt{12} \\ &= 3.46 \quad \text{or} \quad \approx 3.5 \text{ ft} \end{aligned}$$

EXAMPLE:

In Figure 8-2(b), a 16 foot flagpole is casting a 12 foot shadow. What is the distance from the end of the shadow to the top of the flagpole?

Solution:

In this example, C is unknown, and therefore

$$\begin{aligned} C &= \sqrt{A^2 + B^2} \\ &= \sqrt{16^2 + 12^2} \\ &= \sqrt{256 + 144} \\ &= \sqrt{400} \\ &= 20 \text{ ft} \end{aligned}$$

EXAMPLE:

In Figure 8-2(c), a 200 foot piece of string is stretched from the top of a 125 foot cliff to a point on the beach. What is the distance from this point to the cliff?

Solution:

In this example, B is unknown, and therefore

$$\begin{aligned} B &= \sqrt{C^2 - A^2} \\ &= \sqrt{200^2 - 125^2} \\ &= \sqrt{40{,}000 - 15{,}625} \\ &= \sqrt{24{,}375} \\ &= 156 \text{ ft} \end{aligned}$$

Vectors and Vector Diagrams

A **vector** or **phasor** is an arrow used to represent the magnitude and direction of a quantity. Vectors are generally used to represent a physical quantity that has two properties. For example, Figure 8-3(a) shows a motorboat heading north at 12 miles per hour. Figure 8-3(b) shows how vector **A** could be used to represent the vessel's direction and speed. The size of the vector represents the speed of 12 miles per hour by being 12 centimeters long, and because we have made the top of the page north, vector **A** should point straight up so that it represents the vessel's direction. Referring to Figure 8-3(a) you can see that there is another factor that also needs to be considered, a 12 mile per hour easterly tide. This tide is represented in our vector diagram in Figure 8-3(b) by vector **B,** which is 12 centimeters long and pointing east. Since

Vector or Phasor
A quantity that has magnitude and direction and that is commonly represented by a directed line segment whose length represents the magnitude and whose orientation in space represents the direction.

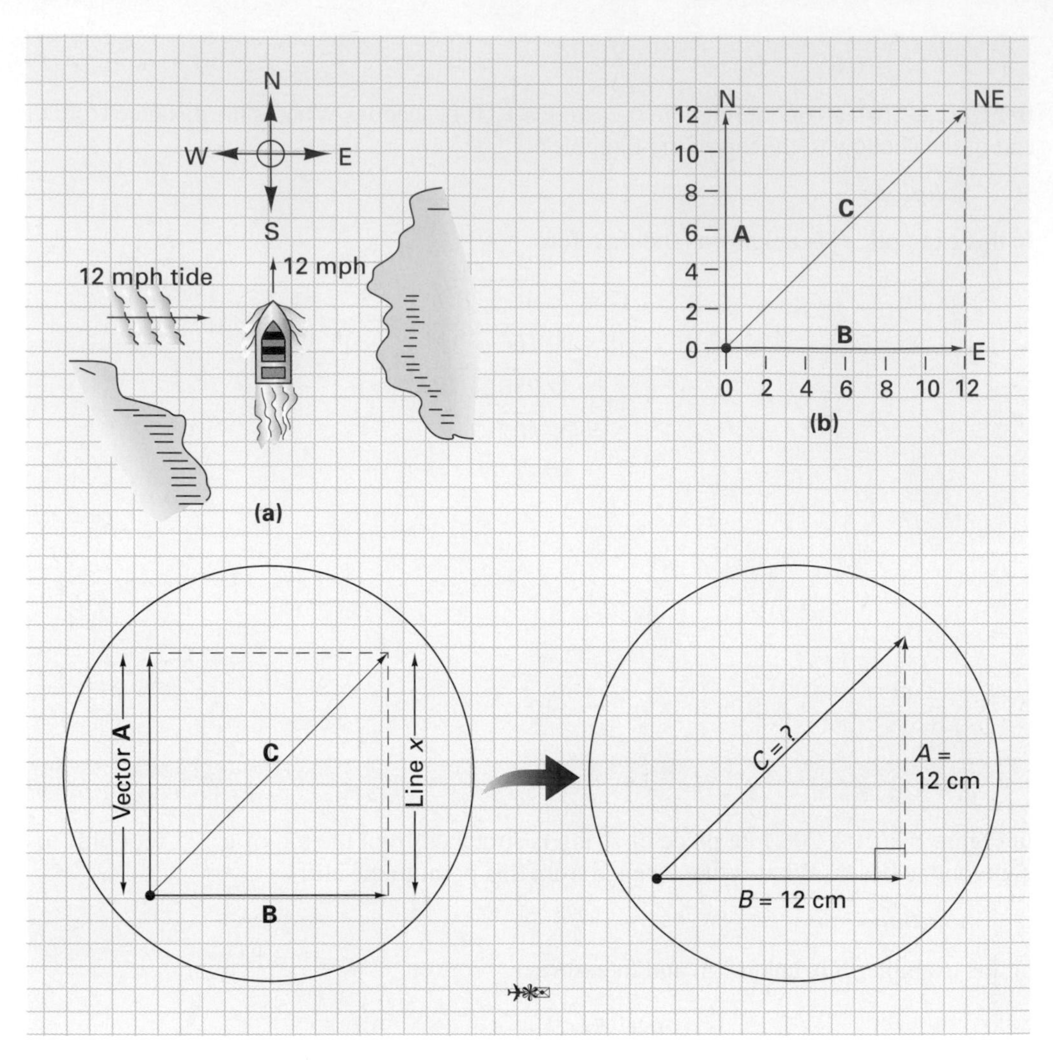

FIGURE 8-3 Vectors and Vector Diagrams.

Vector Addition
Determination of the sum of two out-of-phase vectors using the Pythagorean theorem.

Resultant Vector
A vector derived from or resulting from two or more other vectors.

Vector Diagram
A graphic drawing using vectors that shows arrangement and relations.

the motorboat is pushing north at 12 miles per hour and the tide is pushing east at 12 miles per hour, the resultant course will be northeast, as indicated by vector **C** in Figure 8-3(b). Vector **C** was determined by **vector addition** (as seen by the dashed lines) of vectors **A** and **B.** The result, vector **C,** is called the **resultant vector.** This resultant vector indicates the motorboat's course and speed. The direction or course, we can see, is northeast because vector **C** points in a direction midway between north and east. The speed of the motorboat, however, is indicated by the length or magnitude of vector **C.** This length, and therefore the motorboat's speed, can be calculated by using the Pythagorean theorem. Probably your next question is: How is the **vector diagram** in Figure 8-3(b) similar to a right-angle triangle? The answer is to redraw Figure 8-3(b), as shown in Figure 8-3(c). Because the dashed line x is equal in length to vector **A,** vector **A** can be put in the position of line x to form a right-angle triangle with vector **B** and vector **C.** Now that we have a right-angle triangle with two known lengths and one unknown length, we can calculate the unknown vector's length and therefore the motorboat's speed.

$$\begin{aligned} C &= \sqrt{A^2 + B^2} \\ &= \sqrt{12^2 + 12^2} \\ &= \sqrt{144 + 144} \\ &= \sqrt{288} = 16.97 \end{aligned}$$

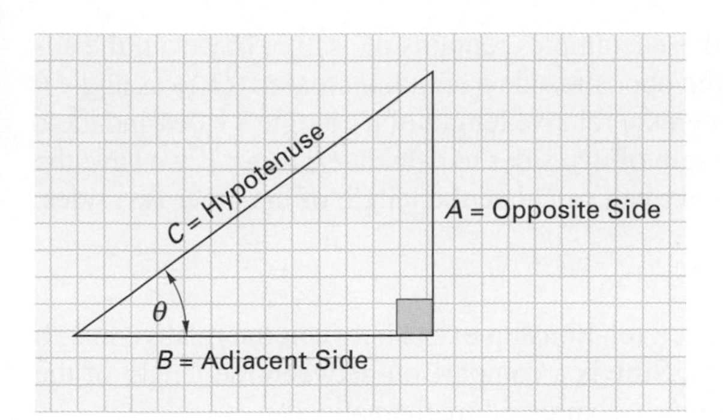

FIGURE 8-4 Names Given to the Sides of a Right Triangle.

The length of vector **C** is 16.97 centimeters, and because each 1 mile per hour of the motorboat was represented by 1 centimeter, the motorboat will travel at a speed of 16.97 miles per hour in a northeasterly direction.

8-1-2 *Opposite, Adjacent, Hypotenuse, and Theta*

Until this point we have called the three sides of our right triangle *A, B,* and *C*. Figure 8-4 shows the more common names given to the three sides of a right-angle triangle. Side *C*, called the **hypotenuse,** is always the longest of the three sides. Side *B*, called the **adjacent side,** always extends between the hypotenuse and the vertical side. An angle called **theta** (θ, a Greek letter) is formed between the hypotenuse and the adjacent side. The side that is always opposite the angle θ is called the **opposite side.**

Angles are always measured in degrees because all angles are part of a circle, like the one shown in Figure 8-5(a). A circle is divided into 360 small sections called **degrees.** In Figure 8-5(b) our right triangle has been placed within the circle. In this example the triangle occupies a 45° section of the circle and therefore theta equals 45 degrees ($\theta = 45°$). Moving from left to right in Figure 8-5(c) you will notice that angle θ increases from 5° to 85°.

Hypotenuse
The side of a right-angled triangle that is opposite the right angle.

Adjacent
The side of a right-angled triangle that has a common endpoint, in that it extends between the hypotenuse and the vertical side.

Theta
The eighth letter of the Greek alphabet, used to represent an angle.

Opposite
The side of a right-angle triangle that is opposite the angle theta.

Degree
A unit of measure for angles equal to an angle with its vertex at the center of a circle and its sides cutting off 1/360 of the circumference.

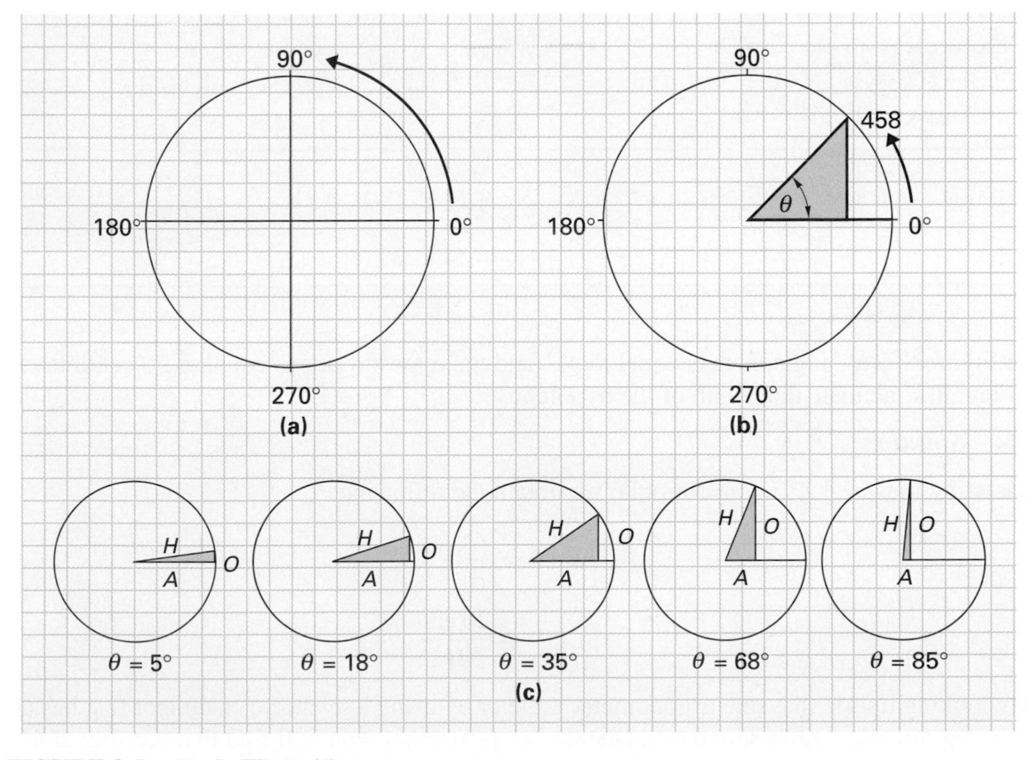

FIGURE 8-5 Angle Theta (θ).

The length of the hypotenuse (H) in all these examples remains the same; however, the adjacent side's length decreases ($A \downarrow$) and the opposite side's length increases ($O \uparrow$) as angle θ is increased ($\theta \uparrow$). This relationship between the relative length of a triangle's sides and theta means that we do not have to know the length of two sides to calculate a third. If we have the value of just one side and the angle theta, we can calculate the length of the other two sides.

Sine of Theta (Sin θ)

Sine
The trigonometric function that for an acute angle is the ratio between the leg opposite the angle when it is considered part of a right triangle, and the hypotenuse.

In the preceding section we discovered that a relationship exists between the relative length of a triangle's sides and the angle of theta. **Sine** is a comparison between the length of the opposite side and the length of the hypotenuse. Expressed mathematically,

$$\text{Sine of theta } (\sin\theta) = \frac{\text{opposite side } (O)}{\text{hypotenuse } (H)}$$

Because the hypotenuse is always larger than the opposite side, the result will always be less than 1 (a decimal fraction). Let us use this formula in a few examples to see how it works.

CALCULATOR KEYS

Name: Sine key

Function: Instructs the calculator to find the sine of the displayed value (angle → value).

Example: sin 37° = ?
Press keys: [3] [7] [SIN]
Answer: 0.601815

Name: Arcsine ($\sin^{-1}$) or inverse sine sequence.

Function: Calculates the smallest angle whose sine is in the display (value → angle).

Example: invsin 0.666 = ?
Press keys: [.] [6] [6] [6] [INV] [SIN]
Answer: 41.759°

EXAMPLE:

In Figure 8-6(a), angle theta is equal to 41° and the opposite side is equal to 20 centimeters in length. Calculate the length of the hypotenuse.

Solution:

Inserting these values in our formula, we obtain the following:

$$\sin\theta = \frac{O}{H}$$

$$\sin 41^\circ = \frac{20\text{ cm}}{H}$$

By looking up 41° in a sine trigonometry table or by using a scientific calculator that has all the trigonometry tables stored permanently in its memory, you will find that the sine of 41°

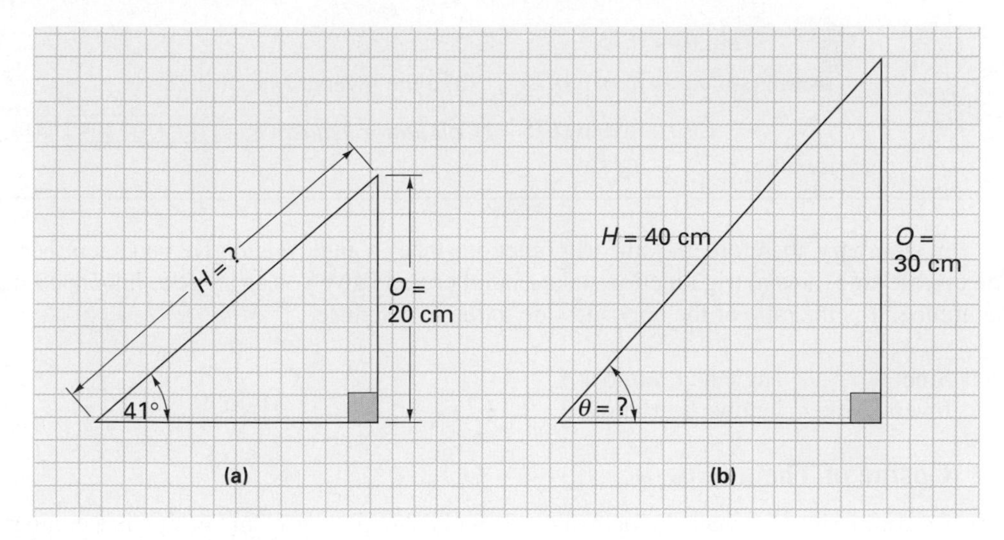

FIGURE 8-6 Sine of Theta.

is 0.656. This value describes the fact that when $\theta = 41°$, the opposite side will be 0.656, or 65.6%, as long as the hypotenuse side. By inserting this value into our formula and transposing the formula according to the rules of algebra, we can determine the length of the hypotenuse.

$$\sin 41° = \frac{20 \text{ cm}}{H}$$

Calculator sequence: [4] [1] [SIN]

$$0.656 = \frac{20 \text{ cm}}{H}$$

$$0.656 \times H = \frac{20 \text{ cm } \cancel{H}}{\cancel{H}}$$

Multiply both sides by H.

$$\frac{\cancel{0.656} \times H}{\cancel{0.656}} = \frac{20 \text{ cm}}{0.656}$$

Divide both sides by 0.656.

$$H = \frac{20 \text{ cm}}{0.656} = 30.5 \text{ cm}$$

EXAMPLE:

Figure 8-6(b) illustrates another example; however, in this case the lengths of sides H and O are known but θ is not.

Solution:

$$\sin \theta = \frac{O}{H}$$

$$\sin \theta = \frac{30 \text{ cm}}{40 \text{ cm}}$$

$$= 0.75$$

The ratio of side O to side H is 0.75, or 75%, which means that the opposite side is 75%, or 0.75, as long as the hypotenuse. To calculate angle θ we must isolate it on one side of the equation. To achieve this we must multiply both sides of the equation by **arcsine,** or **inverse sine** (invsin), which does the opposite of sine.

Arcsine or Inverse Sine
The inverse function to the sine. (If y is the sine of θ, then θ is the arcsine of y.)

$$\sin \theta = 0.75$$

$$\text{invsin} (\sin \theta) = \text{invsin } 0.75 \quad \text{Take the inverse sine of 0.75.}$$

$$\theta = \text{invsin } 0.75 \quad \textit{Calculator sequence: } \boxed{.}\ \boxed{7}\ \boxed{5}\ \boxed{\text{INV}}\ \boxed{\text{SIN}}$$

$$= 48.6°$$

In summary, therefore, the sine trig functions take an angle θ and give you a number x. The inverse sine (arcsin) trig functions take a number x and give you an angle θ. In both cases, the number x is the ratio of the opposite side to the hypotenuse.

Sine: angle $\theta \rightarrow$ number x
Inverse sine: number $x \rightarrow$ angle θ

Cosine of Theta (Cos θ)

Cosine

The trigonometric function that for an acute angle is the ratio between the leg adjacent to the angle when it is considered part of a right triangle, and the hypotenuse.

Sine is a comparison between the opposite side and the hypotenuse, and **cosine** is a comparison between the adjacent side and the hypotenuse.

$$\text{Cosine of theta } (\cos \theta) = \frac{\text{adjacent } (A)}{\text{hypotenuse } (H)}$$

CALCULATOR KEYS

Name: Cosine key

Function: Instructs the calculator to find the cosine of the displayed value (angle → value).

Example: cos 26° = ?
Press keys: [2] [6] [COS]
Answer: 0.89879

Name: Arccosine ($\cos^{-1}$) or inverse cosine sequence.

Function: Calculates the smallest angle whose cosine is in the display (value → angle).

Example: invcos 0.234 = ?
Press keys: [.] [2] [3] [4] [INV] [COS]
Answer: 76.467

EXAMPLE:

Figure 8-7(a) illustrates a right triangle in which the angle θ and the length of the hypotenuse are known. From this information, calculate the length of the adjacent side.

Solution:

$$\cos \theta = \frac{A}{H}$$

$$\cos 30° = \frac{A}{40 \text{ cm}} \quad \textit{Calculator sequence: } \boxed{3}\ \boxed{0}\ \boxed{\text{COS}}$$

$$0.866 = \frac{A}{40 \text{ cm}}$$

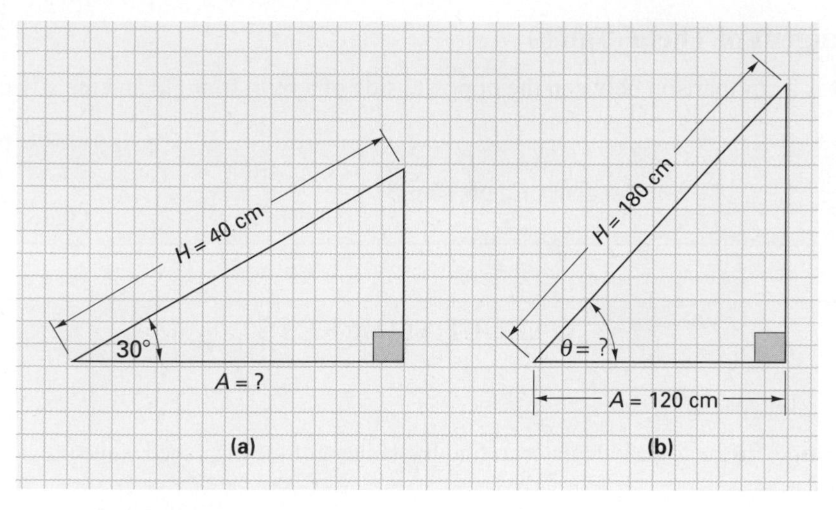

FIGURE 8-7 Cosine of Theta.

Looking up the cosine of 30°, you will obtain the fraction 0.866. This value states that when $\theta = 30°$, the adjacent side will always be 0.866, or 86.6%, as long as the hypotenuse. By transposing this equation we can calculate the length of the adjacent side:

$$0.866 = \frac{A}{40 \text{ cm}}$$

$$40 \times 0.866 = \frac{A}{40 \text{ cm}} \times 40 \qquad \text{Multiply both sides by 40.}$$

$$A = 0.866 \times 40$$

$$= 34.64 \text{ cm}$$

EXAMPLE:

Calculate the angle θ in Figure 8-7(b) if $A = 120$ centimeters and $H = 180$ centimeters.

Solution:

$$\cos\theta = \frac{A}{H}$$

$$= \frac{120 \text{ cm}}{180 \text{ cm}}$$

$$= 0.667 \qquad A \text{ is 66.7\% as long as } H.$$

$$\text{invcos}(\cos\theta) = \text{invcos } 0.667 \qquad \text{Multiply both sides by invcos.}$$

$$\theta = \text{invcos } 0.667$$

Calculator sequence: [0] [.] [6] [6] [7] [INV] [COS]

$$= 48.2°$$

The inverse cosine trig function does the reverse operation of the cosine function:

Cosine: angle $\theta \rightarrow$ number x
Inverse cosine: number $x \rightarrow$ angle θ

Tangent of Theta (Tan θ)

Tangent

The trigonometric function that for an acute angle is the ratio between the leg opposite the angle when it is considered part of a right triangle, and the leg adjacent.

Tangent is a comparison between the opposite side of a right triangle and the adjacent side.

$$\text{Tangent of theta } (\tan\theta) = \frac{\text{opposite } (O)}{\text{adjacent } (A)}$$

CALCULATOR KEYS

Name: Tangent key

Function: Instructs the calculator to find the tangent of the displayed value (angle → value).

Example: tan 73° = ?
Press keys: [7] [3] [TAN]
Answer: 3.2709

Name: Arctangent ($\tan^{-1}$) or inverse tangent sequence.

Function: Calculates the smallest angle whose tangent is in the display (value → angle).

Example: invtan 0.95 = ?
Press keys: [.] [9] [5] [INV] [TAN]
Answer: 43.5312

EXAMPLE:

Figure 8-8(a) illustrates a right triangle in which $\theta = 65°$ and the opposite side is 43 centimeters. Calculate the length of the adjacent side.

Solution:

$$\tan\theta = \frac{O}{A}$$

$$\tan 65° = \frac{43 \text{ cm}}{A}$$

$$2.14 = \frac{43 \text{ cm}}{A}$$ *Calculator sequence:* [6] [5] [TAN]

FIGURE 8-8 Tangent of Theta.

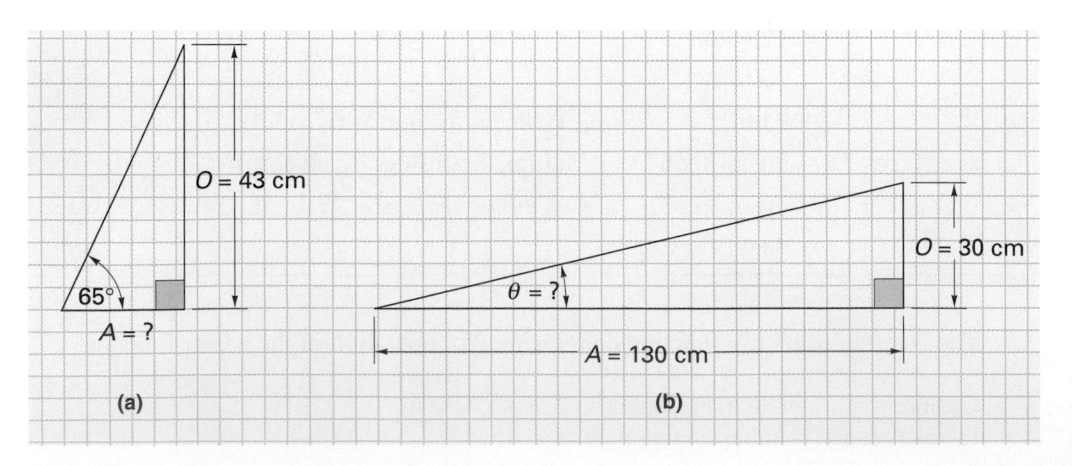

$$2.14 \times A = \frac{43\text{ cm}}{A} \times A$$

$$\frac{\cancel{2.14} \times A}{\cancel{2.14}} = \frac{43\text{ cm}}{2.14}$$

$$A = \frac{43\text{ cm}}{2.14}$$

$$= 20.1\text{ cm}$$

Whenever $\theta = 65°$, the opposite side will be 2.14 times longer than the adjacent side.
Multiply both sides by A.
Divide both sides by 2.14.

EXAMPLE:

Calculate the angle θ in Figure 8-8(b) if $O = 30$ centimeters and $A = 130$ centimeters.

Solution:

$$\tan\theta = \frac{O}{A}$$

$$= \frac{30\text{ cm}}{130\text{ cm}}$$

$$= 0.231$$ O is 0.231 or 23.1% as long as A.

$$\text{invtan}(\tan\theta) = \text{invtan } 0.231$$ Multiply both sides by invtan.

$$\theta = \text{invtan } 0.231$$ *Calculator sequence:* 0 . 2 3 1 INV TAN

$$= 12.99 \text{ or } 13°$$

Summary

As you have seen, trigonometry involves the study of the relationships among the three sides (O, H, A) of a right triangle and also the relationships among the sides of the right triangle and the number of degrees contained in the angle theta (θ).

If the lengths of two sides of a right triangle are known, and the length of the third side is needed, remember that

$$H^2 = O^2 + A^2$$

If the angle θ is known along with the length of one side, or if angle θ is needed and the lengths of the two sides are known, one of the three formulas can be chosen based on which variables are known and what is needed.

$$\sin\theta = \frac{O}{H} \qquad \cos\theta = \frac{A}{H} \qquad \tan\theta = \frac{O}{A}$$

When I was introduced to trigonometry, my mathematics professor spent 15 minutes having the whole class practice what he described as an old Asian war cry that went like this: "SOH CAH TOA." After he explained that it wasn't a war cry but in fact a memory aid to help us remember that SOH was in fact $\sin\theta = O/H$, CAH was $\cos\theta = A/H$, and TOA was $\tan\theta = O/A$, we understood the method in his madness.

SELF-TEST EVALUATION POINT FOR SECTION 8-1

Now that you have completed this section, you should be able to:

Objective 1. Explain how the Pythagorean theorem relates to right-angle triangles.

Objective 2. Determine the length of one side of a right-angle triangle when the lengths of the other two sides are known.

Objective 3. Describe how vectors are used to represent the magnitude and direction of physical quantities and how they are arranged in a vector diagram.

Objective 4. Explain how the Pythagorean theorem can be applied to a vector diagram to calculate the magnitude of a resultant vector.

Objective 5. Define the trigonometric terms:

a. Opposite
b. Adjacent
c. Hypotenuse
d. Theta
e. Sine
f. Cosine
g. Tangent

Objective 6. Demonstrate how the sine, cosine, and tangent trigonometric functions can be used to calculate:

a. The length of an unknown side of a right-angle triangle if the length of another side and the angle theta are known.
b. The angle theta of a right-angle triangle if the lengths of two sides of the triangles are known.

Use the following questions to test your understanding of Section 8-1:

1. Assume that the two shorter sides of a right-angle triangle are 15 meters and 22 meters. Calculate the length of the other side.

2. Calculate the lengths of the unknown sides in the triangles at the right.

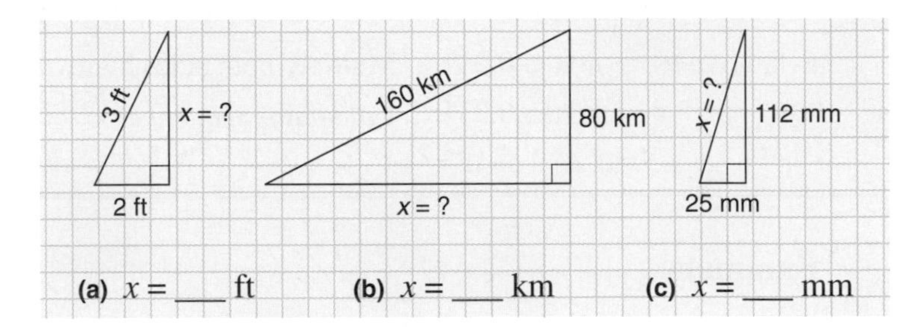

3. Calculate the length of the unknown side in the triangles at the right.

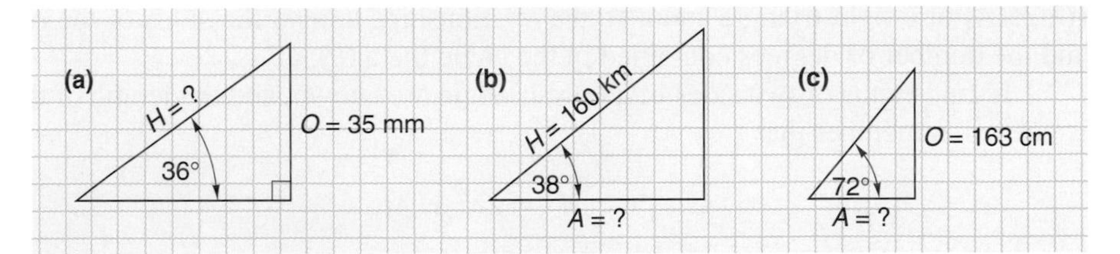

4. Calculate the angle θ for the right-angle triangles at the right.

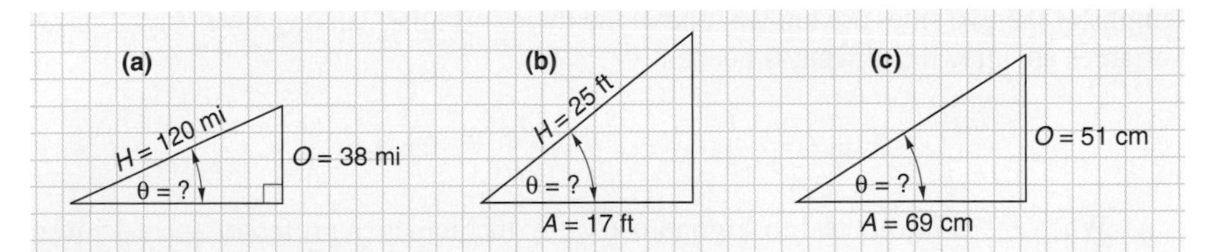

5. Calculate the magnitude of the resultant vectors in the vector diagrams at the right.

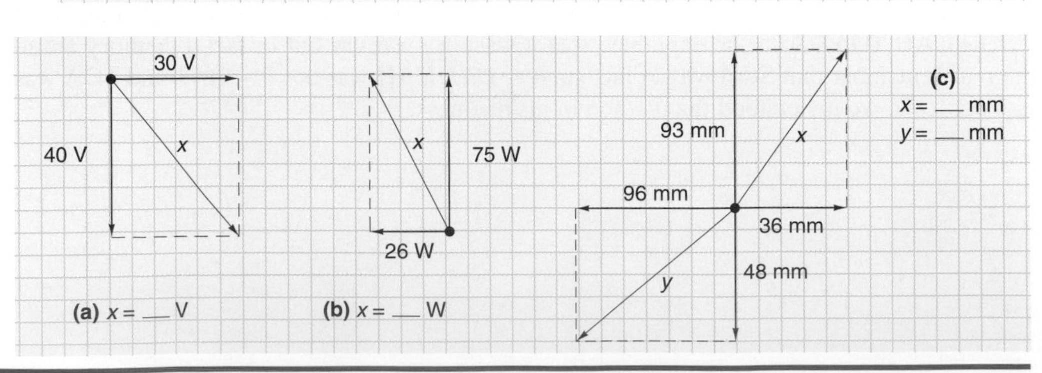

8-2 THE DIFFERENCE BETWEEN DC AND AC

One of the best ways to describe anything new is to begin by redescribing something known and then discuss the unknown. The known topic in this case is **direct current.** Direct current (dc) is the flow of electrons in one DIRECTion and one direction only. Direct current voltage is nonvarying and normally obtained from a battery or power supply unit, as seen in Figure 8-9(a). The only variation in voltage from a battery occurs due to the battery's discharge, but even then, the current will still flow in only one direction, as seen in Figure 8-9(b). A dc voltage of 9 or 6 V could be illustrated graphically as shown in Figure 8-9(c), which shows that whether 9 or 6 V, the voltage is constant or the same at any time.

Direct Current
Current flow in only one direction.

Some power supplies supply a form of dc known as *pulsating* dc, which varies periodically from zero to a maximum, back to zero, and then repeats. Figure 8-10(a) illustrates the physical appearance and schematic diagram of a battery charger that is connected across two series resistors. The battery charger is generating a waveform, shown in Figure 8-10(b), known as pulsating dc. At time 1 [Figure 8-10(c)], the power supply is generating 9 V and

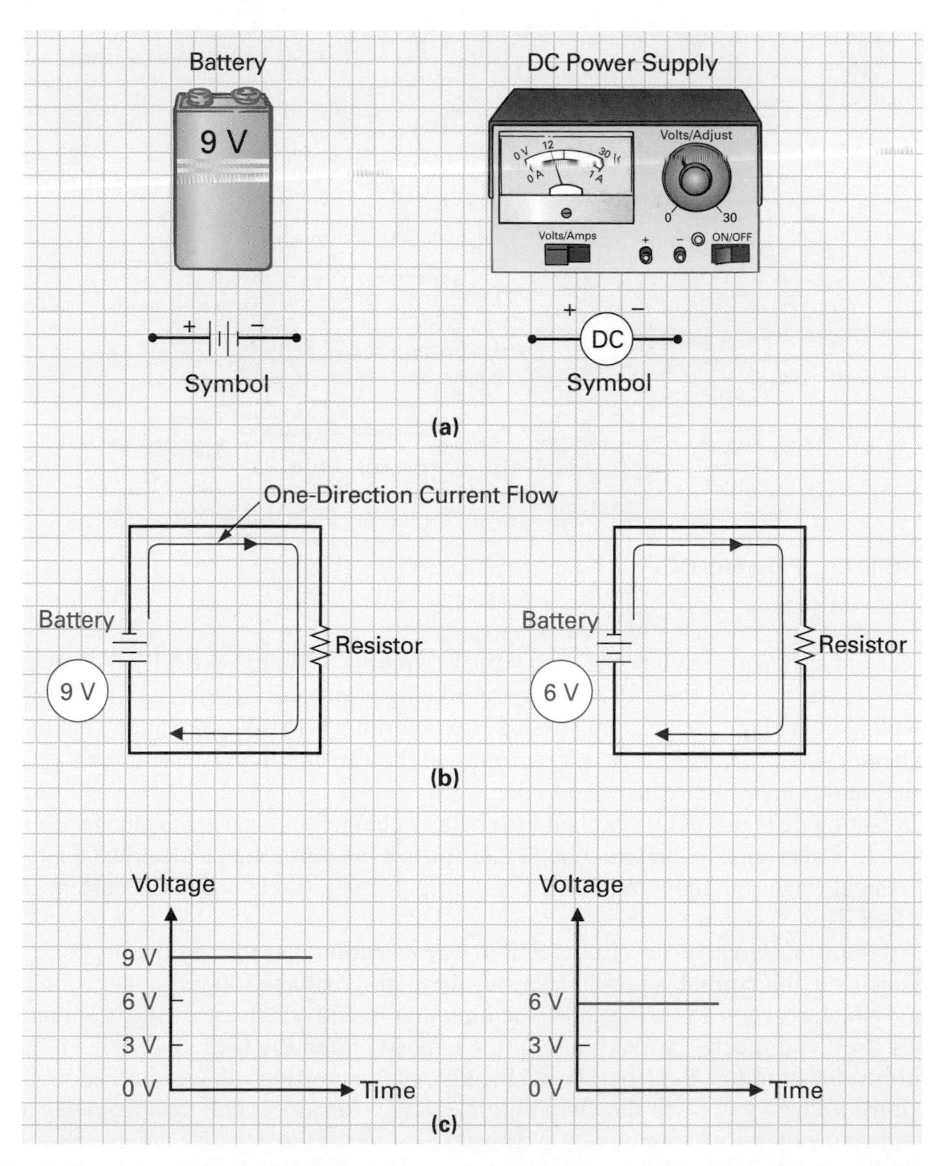

FIGURE 8-9 Direct Current. (a) DC Sources. (b) DC Flow. (c) Graphical Representation of DC.

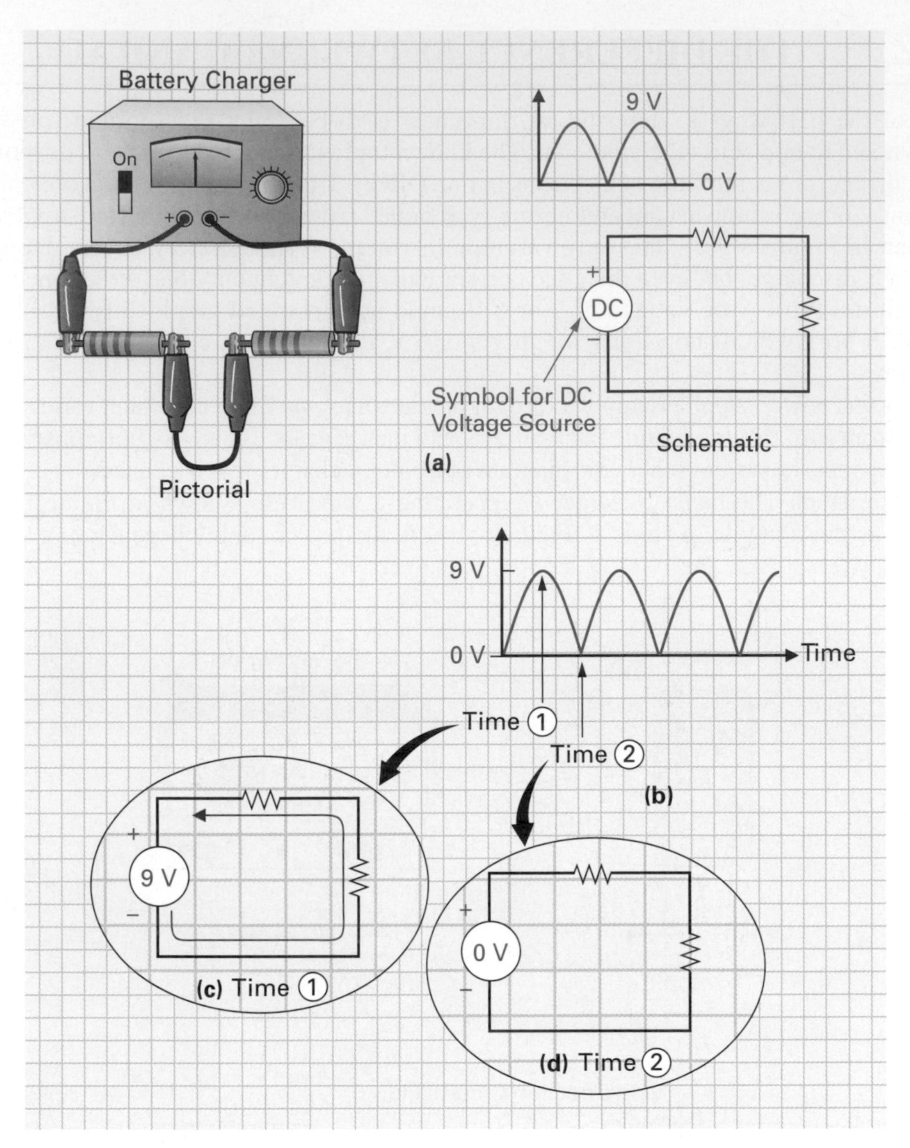

FIGURE 8-10 Pulsating DC.

direct current is flowing from negative to positive. At time 2 [Figure 8-10(d)], the power supply is producing 0 V, and therefore no current is being produced. In between time 1 and time 2, the voltage out of the power supply will decrease from 9 V to 0 V. No matter what the voltage, whether 8, 7, 6, 5, 4, 3, 2, or 1 V, current will be flowing in only one direction (unidirectional) and is therefore referred to as dc.

Pulsating dc is normally supplied by a battery charger and is used to charge secondary batteries. It is also used to operate motors that convert the pulsating dc electrical energy into a mechanical rotation output. Whether steady or pulsating, direct current is current in only one DIRECTion.

Alternating Current
Electric current that rises from zero to a maximum in one direction, falls to zero, rises to a maximum in the opposite direction, and then repeats another cycle, the positive and negative alternations being equal.

Alternating current (ac) flows first in one direction and then in the opposite direction. This reversing current is produced by an alternating voltage source, as shown in Figure 8-11(a), which reaches a maximum in one direction (positive), decreases to zero, and then reverses itself and reaches a maximum in the opposite direction (negative). This is graphically illustrated in Figure 8-11(b). During the time of the positive voltage alternation, the polarity of the voltage will be as shown in Figure 8-11(c), so current will flow from negative to positive in a counterclockwise direction. During the time of the negative voltage alternation, the polarity of the voltage will reverse, as shown in Figure 8-11(d), causing current to flow once again from negative to positive, but in this case, in the opposite, clockwise direction.

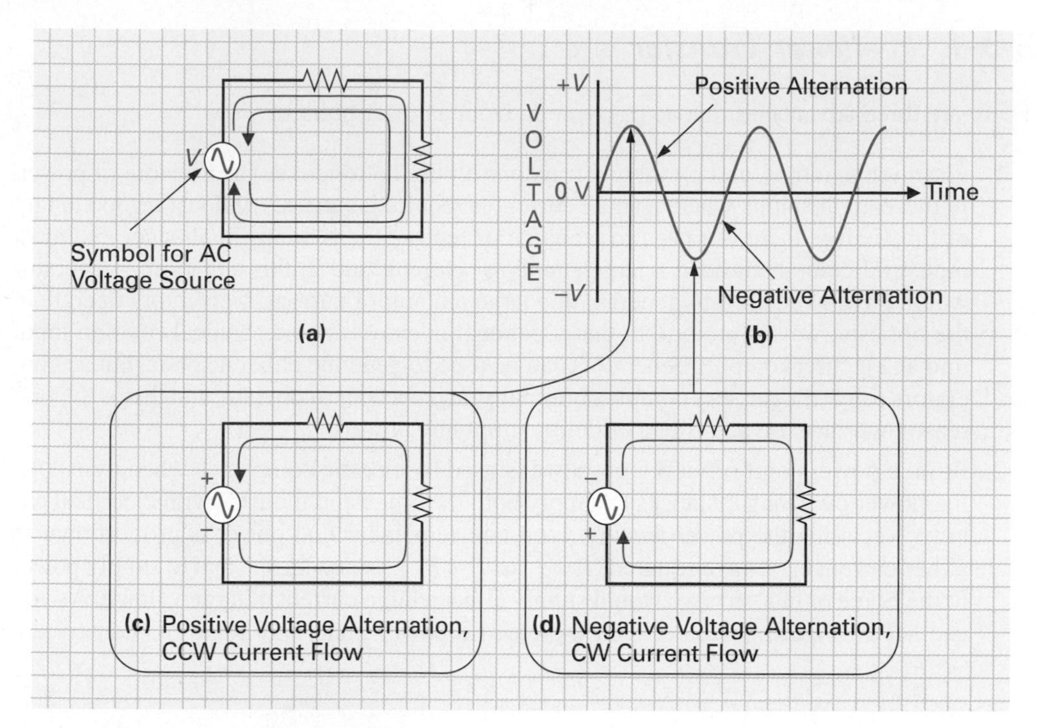

FIGURE 8-11 Alternating Current.

SELF-TEST EVALUATION POINT FOR SECTION 8-2

Now that you have completed this section, you should be able to:

Objective 7. Explain the difference between alternating current and direct current.

Use the following questions to test your understanding of Section 8-2:

1. Give the full names of the following abbreviations: (a) ac; (b) dc.
2. Is the pulsating waveform generated by a battery charger considered to be ac or dc? Why?
3. The polarity of a/an __________ voltage source will continually reverse, and therefore so will the circuit current.
4. The polarity of a/an __________ voltage source will remain constant, and therefore current will flow in only one direction.
5. List the two main applications of ac.
6. State briefly the difference between ac and dc.

8-3 WHY ALTERNATING CURRENT?

The question that may be troubling you at this point is: If we have been managing fine for the past chapters with dc, why do we need ac?

There are two main applications for ac:

1. *Power transfer:* to supply electrical power for lighting, heating, cooling, appliances, and machinery in both the home and industry
2. *Information transfer:* to communicate or carry information, such as radio music and television pictures, between two points

To begin with, let us discuss the first of these applications, power transfer.

8-3-1 *Power Transfer*

There are three advantages that ac has over dc from a power point of view.

Generator
Device used to convert a mechanical energy input into an electrical energy output.

1. Flashlights, radios, and portable televisions all use batteries (dc) as a source of power. In these applications where a small current is required, batteries will last a good length of time before there is a need to recharge or replace them. Many appliances and most industrial equipment need a large supply of current, and in this situation a generator would have to be used to generate this large amount of current. Generators operate in the opposite way to motors, in that a **generator** converts a mechanical rotation input into an electrical output. Generators can be used to generate either dc or ac, but ac generators can be larger, less complex internally, and cheaper to operate, and this is the first reason why we use ac instead of dc for supplying power.
2. From a power point of view, ac is always used by electric companies when transporting power over long distances to supply both the home and industry with electrical energy. Recalling the power formula, you will remember that power is proportional to either current or voltage squared ($P \propto I^2$ or $P \propto V^2$), which means that to supply power to the home or industry, we would supply either a large current or large voltage. As you can see in Figure 8-12, between the electric power plant and home or industry are

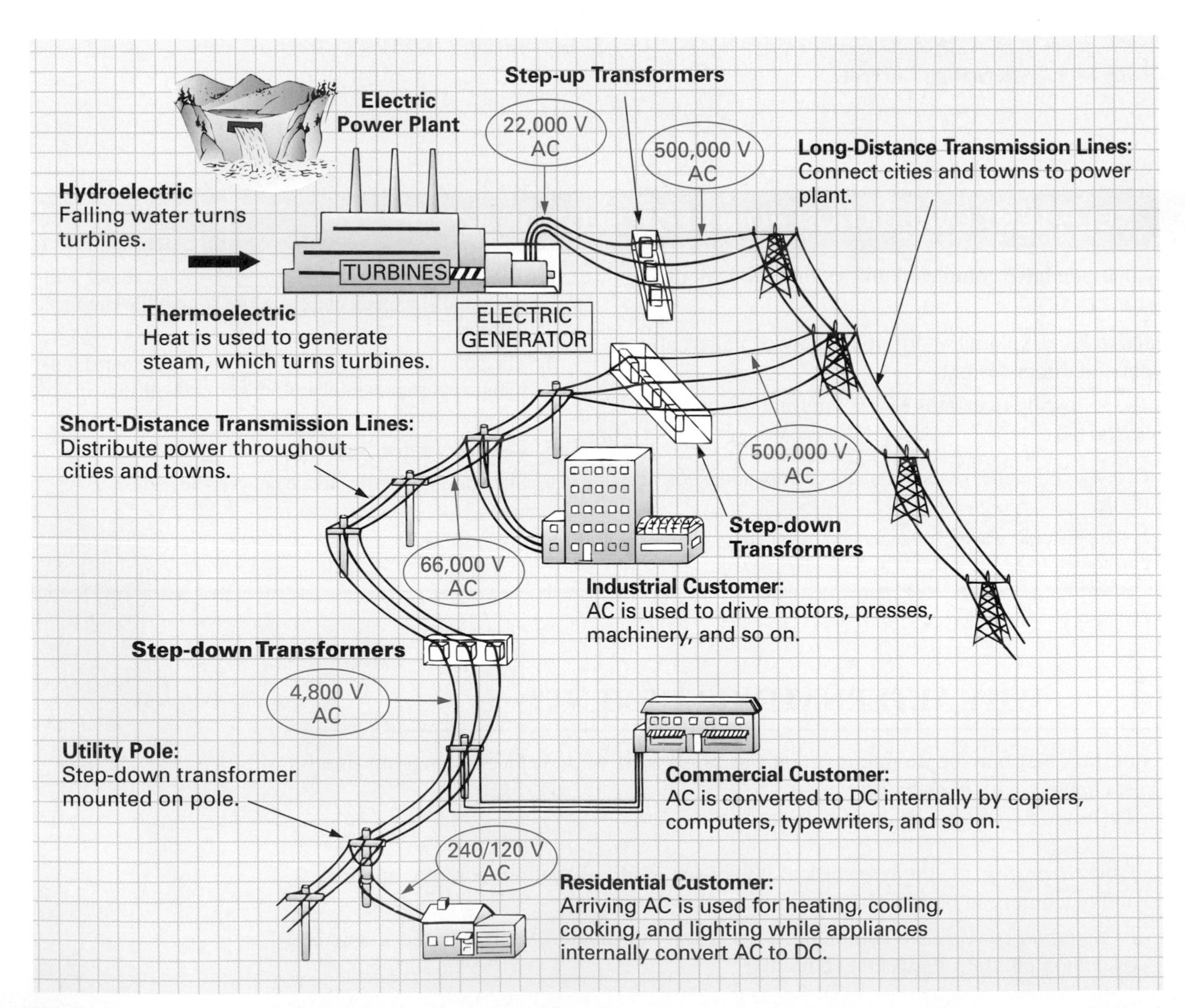

FIGURE 8-12 AC Power Distribution.

power lines. The amount of power lost (heat) in these power lines can be calculated by using the formula $P = I^2 \times R$, where I is the current flowing through the line and R is the resistance of the power lines. This means that the larger the current, the greater the amount of power lost in the lines in the form of heat and therefore the less the amount of power supplied. For this reason, power companies transport electric energy at a very high voltage, between 200,000 and 600,000 V. Since the voltage is high, the current can be low and still provide the same amount of power to the consumer ($P = V\uparrow \times I\downarrow$). Yet, by keeping the current low, the amount of heat loss generated in the power lines is minimal.

Now that we have discovered why it is more efficient to transport high voltages over a long distance than high current, what does this have to do with ac? An ac voltage can easily and efficiently be transformed up or down to a higher or lower voltage by utilizing a device known as a **transformer,** and even though dc voltages can be stepped up and down, the method is inefficient and more complex.

Transformer
Device consisting of two or more coils that are used to couple electric energy from one circuit to another, yet maintain electrical isolation between the two.

3. Nearly all electronic circuits and equipment are powered by dc voltages, which means that once the ac power arrives at the home or industry, in most cases it will have to be converted into dc power to operate electronic equipment. It is a relatively simple process to convert ac to dc, but conversion from dc to ac is a complex and comparatively inefficient process.

Figure 8-12 illustrates ac power distribution from the electric power plant. The distribution system begins at the electric power plant, which has powerful, large generators driven by turbines to generate large ac voltages. The turbines can be driven either by falling water (hydroelectric) or by steam, which is produced with intense heat by burning either coal, gas, or oil or from a nuclear reactor (thermoelectric). The turbine supplies the mechanical energy to the generator that transforms it into ac electrical energy.

The generator generates an ac voltage of approximately 22,000 V, which is stepped up by transformers to approximately 500,000 V. This voltage is applied to the long-distance transmission lines, which connect the power plant to the city or town. At each city or town, the voltage is tapped off the long-distance transmission lines and stepped down to approximately 66,000 V, and is distributed to large-scale industrial customers. The 66,000 V is stepped down again to approximately 4800 V and distributed throughout the city or town by short-distance transmission lines. This 4800 V is used by small-scale industrial customers and residential customers, who receive the ac power via step-down transformers on utility poles, which step down the 4800 V to 240 V and 120 V.

Most equipment and devices within industry and the home will run directly from the ac power, such as heating, lighting, and cooling. Some equipment that runs on dc, such as televisions and computers, has an internal dc power supply that will accept the 120 V ac from the wall outlet and convert it to the dc voltages required to power the system.

SELF-TEST EVALUATION POINT FOR SECTION 8-3-1

Now that you have completed this section, you should be able to:

Objective 8. Describe how ac is used to deliver power and to represent information.

Objective 9. Give the three advantages that ac has over dc from a power point of view.

Objective 10. Describe basically the ac power distribution system from the electric power plant to the home or industry.

Use the following questions to test your understanding of Section 8-3-1:

1. In relation to power transfer, what three advantages does ac have over dc?
2. True or false: A generator converts an electrical input into a mechanical output.
3. What formula is used to calculate the amount of power lost in a transmission line?
4. What is a transformer?
5. What voltage is provided to the wall outlet in the home?
6. Most appliances internally convert the __________ input voltage into a __________ voltage.

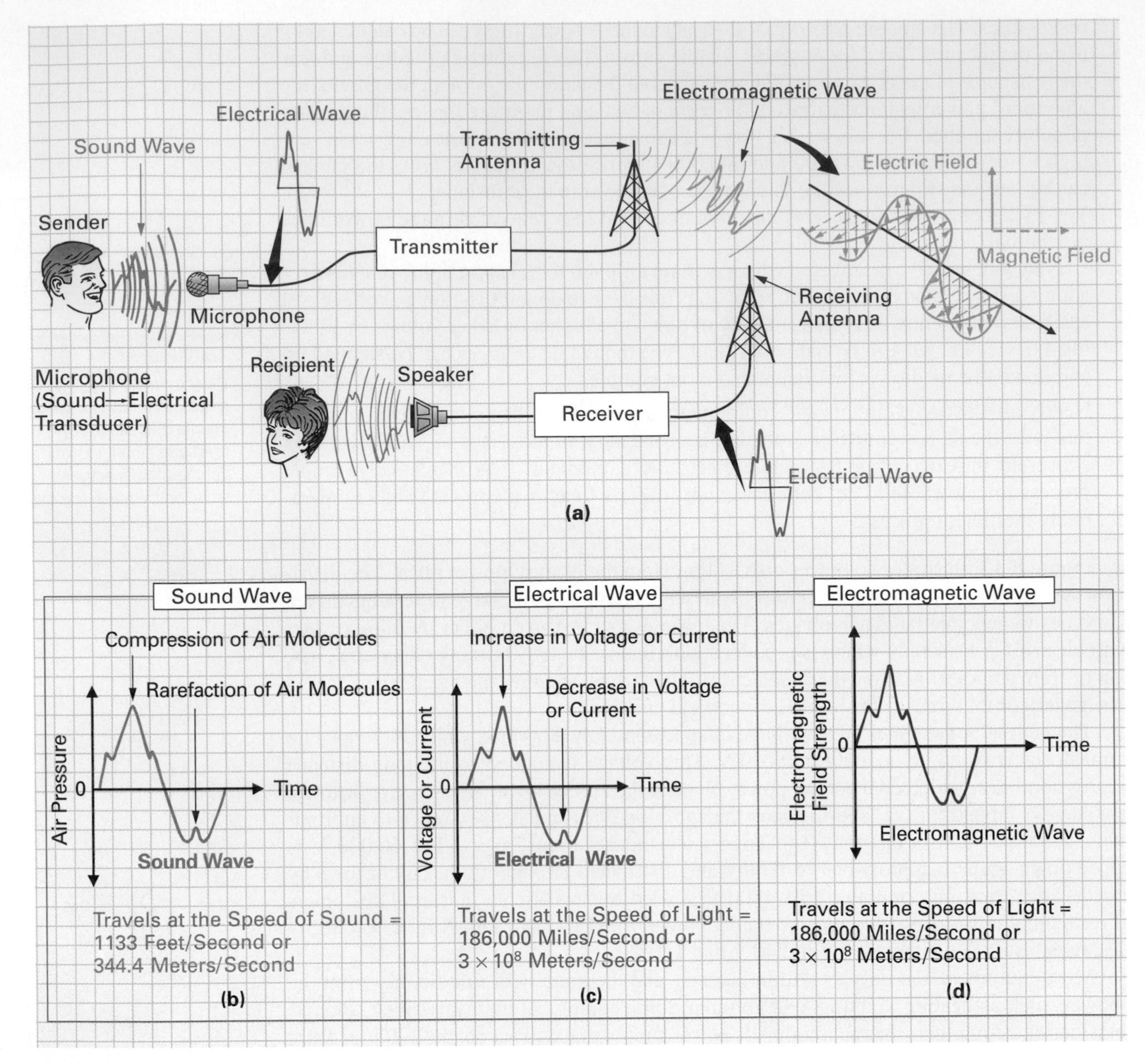

FIGURE 8-13 Information Transfer.

8-3-2 *Information Transfer*

Communication
Transmission of information between two points.

Information, by definition, is the property of a signal or message that conveys something meaningful to the recipient. **Communication,** which is the transfer of information between two points, began with speech and progressed to handwritten words in letters and printed words in newspapers and books. Face-to-face communications evolved into telephone and radio communications over greater distances.

A simple communication system can be seen in Figure 8-13(a). The voice information or sound wave produced by the sender is a variation in air pressure and travels at the speed of sound, as detailed in Figure 8-13. Sound waves or sounds, like those generated by a vibrating reed or plucked string in the case of musical instruments, are generated by the sender's vocal cords, which vibrate backward and forward, producing a rarefaction or decreased air pressure, where few air molecules exist, and a compression or increased air pressure, where many air molecules exist. Like the ripples produced by a stone falling in a pond, the sound waves produced by the sender are constantly expanding and traveling outward.

The microphone is in fact a **transducer** (energy converter), because it converts the sound wave (which is a form of mechanical energy) into electrical energy in the form of voltage and current, which varies in the same manner as the sound wave and therefore contains the sender's message or information.

The **electrical wave,** shown in Figure 8-13(c), is a variation in voltage or current and can only exist in a wire conductor or circuit. This electrical signal travels at the speed of light.

The speaker, like the microphone, is also an electroacoustical transducer that converts the electrical energy input into a mechanical sound-wave output. These sound waves strike the outer eardrum, causing the ear diaphragm to vibrate, and these mechanical vibrations actuate nerve endings in the ear, which convert the mechanical vibrations into electrochemical impulses that are sent to the brain. The brain decodes this information by comparing these impulses with a library of previous sounds and so provides the sensation of hearing.

To communicate between two distant points, a wire must be connected between the microphone and speaker. However, if an electrical wave is applied to an antenna, the electrical wave is converted into a radio or electromagnetic wave, as shown in the inset in Figure 8-13(a), and communication is established without the need of a connecting wire—hence the term **wireless communication.** Antennas are designed to radiate and receive electromagnetic waves, which vary in field strength, as shown in Figure 8-13(d), and can exist in either air or space. These radio waves, as they are also known, travel at the speed of light and allow us to communicate over great distances.

More specifically, radio waves have two basic components. The electrical voltage applied to the antenna is converted into an electric field and the electrical current into a magnetic field. This **electromagnetic** (electric–magnetic) **wave** is used to carry a variety of information, such as speech, radio broadcasts, television signals, and so on.

In summary, the sound wave is a variation in air pressure, the electrical wave is a variation of voltage or current, and the electromagnetic wave is a variation of electric and magnetic field strength.

Transducer
Any device that converts energy from one form to another.

Electrical Wave
Traveling wave propagated in a conductive medium that is a variation in voltage or current and travels at slightly less than the speed of light.

Wireless Communication
Term describing radio communication that requires no wires between the two communicating points.

Electromagnetic (Radio) Wave
Wave that consists of both an electric and magnetic variation and travels at the speed of light.

TIME LINE
Studying the experiments of Maxwell and Heinrich Hertz, Guglielmo Marconi (1874–1937) invented a practical system of telegraphy communication. In an evolutionary process, Marconi extended his distance of communication from 1 1/2 miles in 1896 to 6000 miles in 1902. In September 1899, Marconi equipped two U.S. ships with equipment and used them in the Atlantic Ocean to transmit to America the progress of the America's Cup yacht race.

EXAMPLE:

How long will it take the sound wave produced by a rifle shot to travel 9630.5 feet?

Solution:

This problem makes use of the following formula:

$$\text{Distance} = \text{velocity} \times \text{time}$$

or

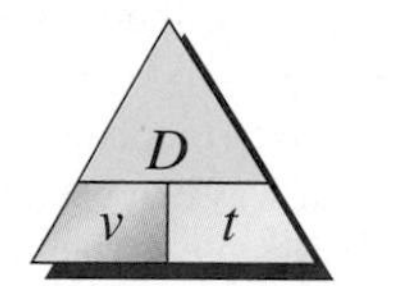

D = distance
v = velocity
t = time

If someone travels at 20 miles per hour for 2 hours, the person will travel 40 miles ($D = v \times t = 20 \text{ mph} \times 2 \text{ h} = 40 \text{ miles}$). In this problem, the distance (9630.5 ft) and the sound wave's velocity (1130 ft/s) are known, and so by rearranging the formula, we can find time:

$$\text{Time} = \frac{\text{distance}}{\text{velocity}} = \frac{9630.5 \text{ ft}}{1130 \text{ ft/s}} = 8.5 \text{ s}$$

EXAMPLE:

How long will it take an electromagnetic (radio) wave to reach a receiving antenna that is 2000 miles away from the transmitting antenna?

■ ***Solution:***

In this problem, both distance (2000 miles) and velocity (186,000 miles/s) are known, and time has to be calculated:

$$\text{Time} = \frac{\text{distance}}{\text{velocity}} = \frac{2000 \text{ miles}}{186{,}000 \text{ miles/s}}$$
$$= 1.075 \times 10^{-2} \quad \text{or} \quad 10.8 \text{ ms}$$

8-3-3 *Electrical Equipment or Electronic Equipment?*

In the beginning of this chapter, it was stated that ac is basically used in two applications: (1) power transfer and (2) information transfer. These two uses for ac help define the difference between electricity and electronics. Electronic equipment manages the flow of information, while electrical equipment manages the flow of power. In summary:

EQUIPMENT	MANAGES
Electrical	Power (large values of *V* and *I*)
Electronic	Information (small values of *V* and *I*)

To use an example, we can say that a dc power supply is a piece of electrical equipment since it is designed to manage the flow of power. A TV set, however, is an electronic system since its electronic circuits are designed to manage the flow of audio (sound) and video (picture) information.

Since most electronic systems include a dc power supply, we can also say that the electrical circuits manage the flow of power, and this power supply enables the electronic circuits to manage the flow of information.

SELF-TEST EVALUATION POINT FOR SECTIONS 8-3-2 AND 8-3-3

Now that you have completed these sections, you should be able to:

Objective 11. Describe the three waves used to carry information between two points.

Objective 12. Explain how the three basic information carriers are used to carry many forms of information.

Use the following questions to test your understanding of Section 8-3-2:

1. Define *information and communication.*
2. The ___________ wave is a variation in air pressure, the ___________ wave is a variation in field strength, and the ___________ wave is a variation of voltage or current.
3. Which of the three waves described in Question 2
   - **a.** Can only exist in the air?
   - **b.** Can exist in either air or a vacuum?
   - **c.** Exists in a wire conductor?
4. Sound waves travel at the speed of sound, which is ___________, while electrical and electromagnetic waves travel at the speed of light, which is ___________.
5. A human ear is designed to receive ___________ waves, an antenna is designed to transmit or receive ___________ waves, and an electronic circuit is designed to pass only ___________ waves.
6. Give the names of the following energy converters or transducers:
   - **a.** Sound wave (mechanical energy) to electrical wave
   - **b.** Electrical wave to sound wave
   - **c.** Electrical wave to electromagnetic wave
   - **d.** Sound wave to electrochemical impulses
7. ___________ equipment manages the flow of information, and these ac waveforms normally have small values of current and voltage.
8. ___________ equipment manages the flow of power, and these ac waveforms normally have large values of current and voltage.

8-4 AC WAVE SHAPES

In all fields of electronics, whether medical, industrial, consumer, or data processing, different types of information are being conveyed between two points and electronic equipment is managing the flow of this information. Let's now discuss the basic types of ac wave shapes. The way in which a wave varies in magnitude with respect to time describes its wave shape. All ac waves can be classified into one of the six types illustrated in Figure 8-14.

8-4-1 *The Sine Wave*

The **sine wave** is the most common type of waveform. It is the natural output of a generator that converts a mechanical input in the form of a rotating shaft into an electrical output in the form of a sine wave. In fact, for one cycle of the input shaft, the generator will produce one sinusoidal ac voltage waveform, as shown in Figure 8-15. When the input shaft of the generator is at 0°, the ac output is 0 V. As the shaft is rotated through 360°, the ac output voltage will rise to a maximum positive voltage at 90°, fall back to 0 V at 180°, and then reach a maximum negative voltage at 270°, and finally return to 0 V at 360°. If this ac voltage is applied across a closed circuit, it produces a current that continually reverses or alternates in each direction.

Sine Wave
Wave whose amplitude is the sine of a linear function of time. It is drawn on a graph that plots amplitude against time or radial degrees relative to the angular rotation of an alternator.

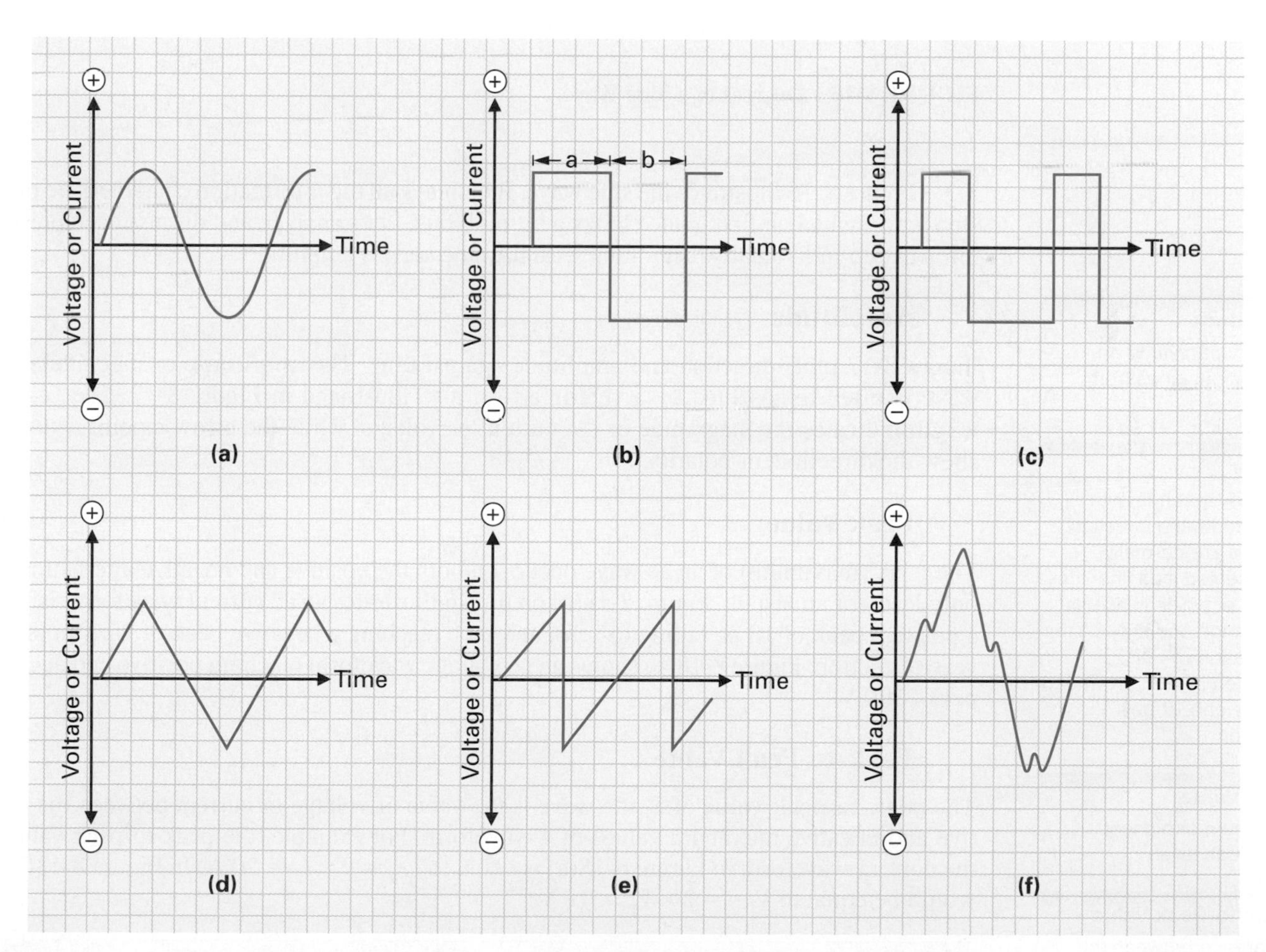

FIGURE 8-14 **AC Wave Shapes. (a) Sine Wave. (b) Square Wave. (c) Pulse Wave. (d) Triangular Wave. (e) Sawtooth Wave. (f) Irregular Wave.**

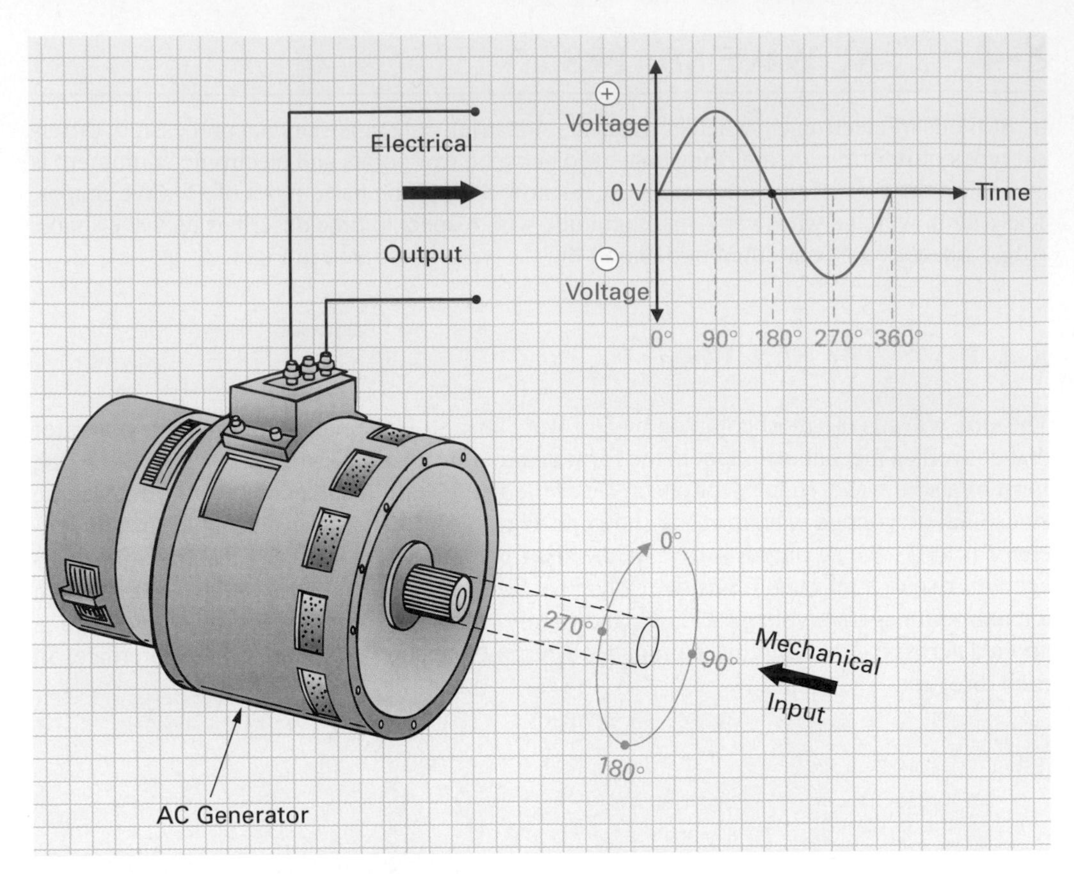

FIGURE 8-15 **Degrees of a Sine Wave.**

Figure 8-16 illustrates the sine wave. It displays all the information characteristic of the sine wave, which at first glance looks a bit ominous. Let's analyze and discuss each piece of information individually, beginning with the sine wave's amplitude.

Amplitude
Size or magnitude an alternation varies from zero.

Vector
Quantity that has both magnitude and direction. Vectors are normally represented as a line, the length of which indicates magnitude and the orientation of which, indicated by the arrowhead on one end, shows direction.

Peak Value
Maximum or highest-amplitude level.

Peak-to-Peak Value
Difference between the maximum positive and maximum negative values.

Amplitude

Figure 8-17 plots direction and amplitude against time. The **amplitude** or magnitude of a wave is often represented by a **vector** arrow, also illustrated in Figure 8-17. The vector's length indicates the magnitude of the current or voltage, while the arrow's point is used to show the direction, or polarity.

Peak Value

The peak (maximum) of an ac wave occurs on both the positive and negative alternation but only lasts for an instant. Figure 8-18(a) on p. 326 illustrates an ac current waveform rising to a positive peak of 10 A, falling to zero, and then reaching a negative peak of 10 A in the reverse direction. Figure 8-18(b) shows an ac voltage waveform reaching positive and negative peaks of 9 V.

Peak-to-Peak Value

The **peak-to-peak value** of a sine wave is the value of voltage or current between the positive and negative maximum values of a waveform. For example, the peak-to-peak value of the current waveform in Figure 8-18(a) is equal to $I_{p\text{-}p} = 2 \times I_p = 20$ A. In Figure 8-18(b), it would be equal to $V_{p\text{-}p} = 2 \times V_p = 18$ V.

$$\text{p-p} = 2 \times \text{peak}$$

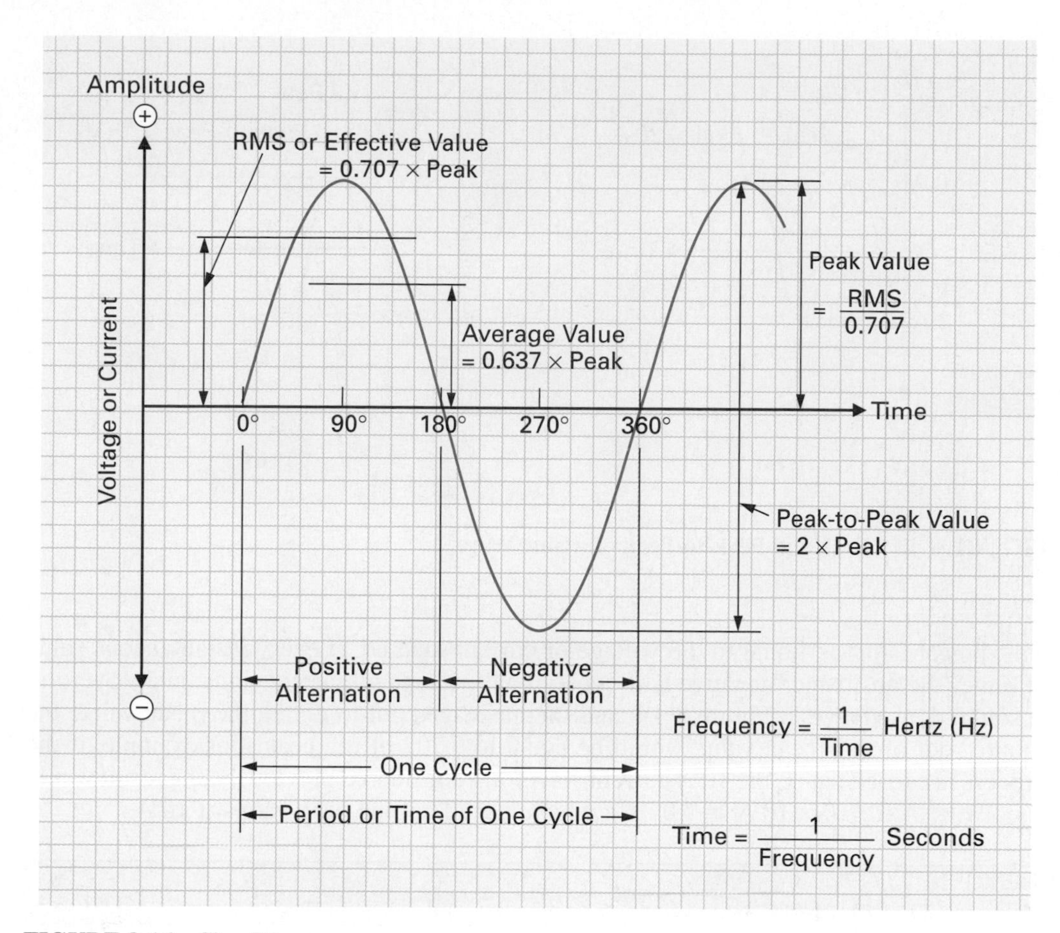

FIGURE 8-16 **Sine Wave.**

RMS or Effective Value

Both the positive and negative alternation of a sine wave can accomplish the same amount of work, but the ac waveform is only at its maximum value for an instant in time, spending most of its time between peak currents. Our examples in Figure 8-18(a) and (b), therefore, cannot supply the same amount of power as a dc value of 10 A or 9 V.

The effective value of a sine wave is equal to 0.707 of the peak valuc. Let's now see how this value was obtained. Power is equal to either $P = I^2 \times R$ or $P = V^2/R$. Said another

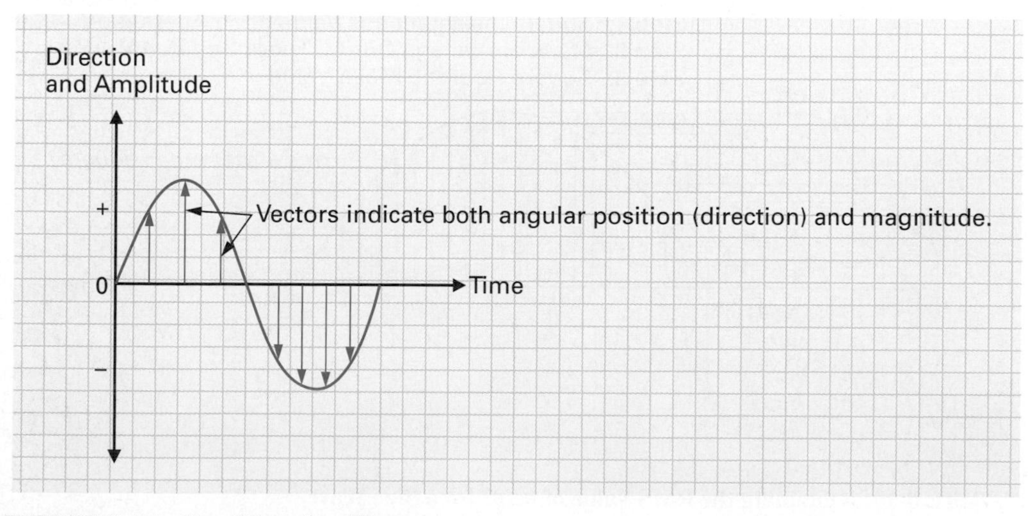

FIGURE 8-17 **Sine-Wave Amplitude.**

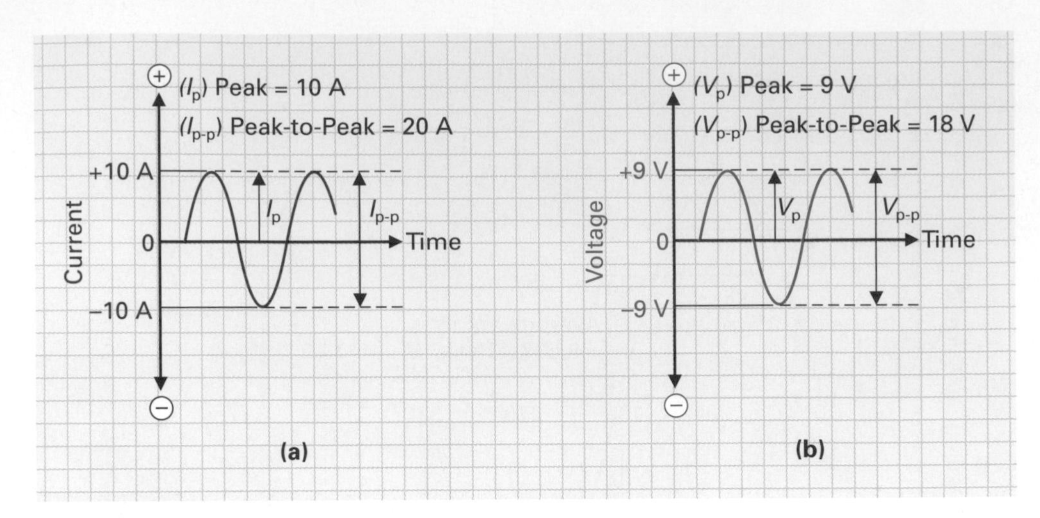

FIGURE 8-18 Peak and Peak-to-Peak of a Sine Wave.

way, power is proportional to the voltage or current squared. If every instantaneous value of either the positive or negative half-cycle of any voltage or current sinusoidal waveform is squared, as shown in Figure 8-19, and then averaged out to obtain the mean value, the square root of this mean value would be equal to 0.707 of the peak. For example, if the process is carried out on the 10 A current waveform considered previously, the result would equal 7.07 A (0.707 × 10 A = 7.07 A), which is 0.707 of the peak value of 10 A.

$$\text{rms} = 0.707 \times \text{peak}$$

RMS Value

RMS value of an ac voltage, current, or power waveform is equal to 0.707 times the peak value. The rms value is the effective or dc value equivalent of the ac wave.

This **root-mean-square (rms) value** of 0.707 can always be used to tell how effective an ac sine wave will be. For example, a 10 A dc source would be 10 A effective because it is continually at its peak value and always delivering power to the circuit to which it is connected, whereas a 10 A ac source would only be 7.07 A effective, as seen in Figure 8-20, because it is at 10 A for only a short period of time. As another example, a 10 V ac sine-wave

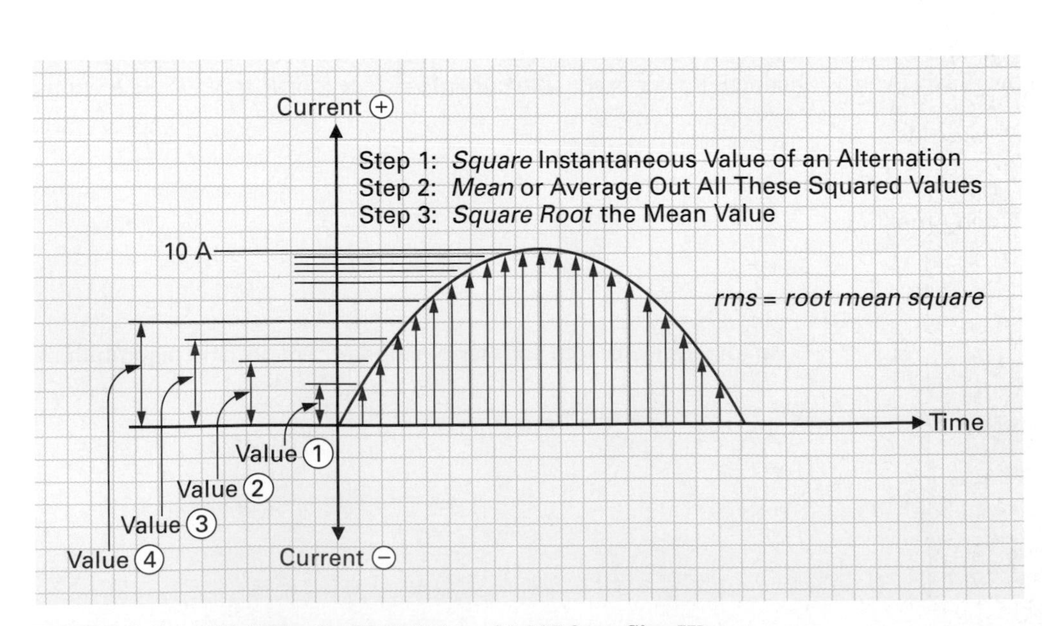

FIGURE 8-19 Obtaining the RMS Value of 0.707 for a Sine Wave.

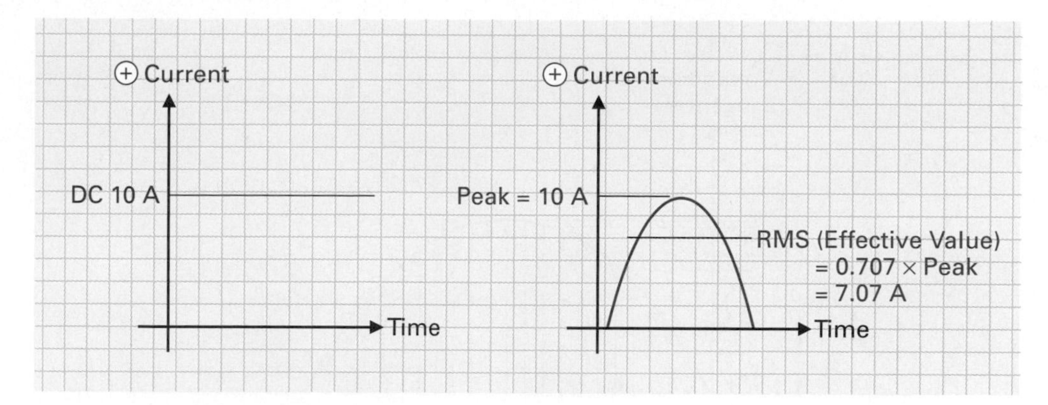

FIGURE 8-20 Effective Equivalent.

alternation would be as effective or supply the same amount of power to a circuit as a 7.07 V dc source.

Unless otherwise stated, ac values of voltage or current are always given in rms. The peak value can be calculated by transposing the original rms formula of rms = peak × 0.707, ending up with

$$\text{Peak} = \frac{\text{rms}}{0.707}$$

Since 1/0.707 = 1.414, the peak can also be calculated by

$$\text{Peak} = \text{rms} \times 1.414$$

Average Value

Average Value
Mean value found when the area of a wave above a line is equal to the area of the wave below the line.

The **average value** of either the positive or negative alternation is found by listing the amplitude or vector length of current or voltages at 1° intervals, as shown in Figure 8-21(a). The sum of all these values is then divided by the total number of values (averaging), which for all sine waves will calculate out to be 0.637 of the peak voltage or current. For example, the average value of the sine-wave alternation with a peak of 10 V seen in Figure 8-21(b) is equal to

$$\text{Average} = 0.637 \times \text{peak}$$

$$= 0.637 \times 10 \text{ V}$$
$$= 6.37 \text{ V}$$

Although the average of a positive or negative alternation (half-cycle) is equal to 0.637 × peak, the average of the complete cycle, including both the positive and negative half-cycles, is mathematically zero, as the amount of voltage or current above the zero line is equal but opposite to the amount of voltage or current below the zero line, as shown in Figure 8-22.

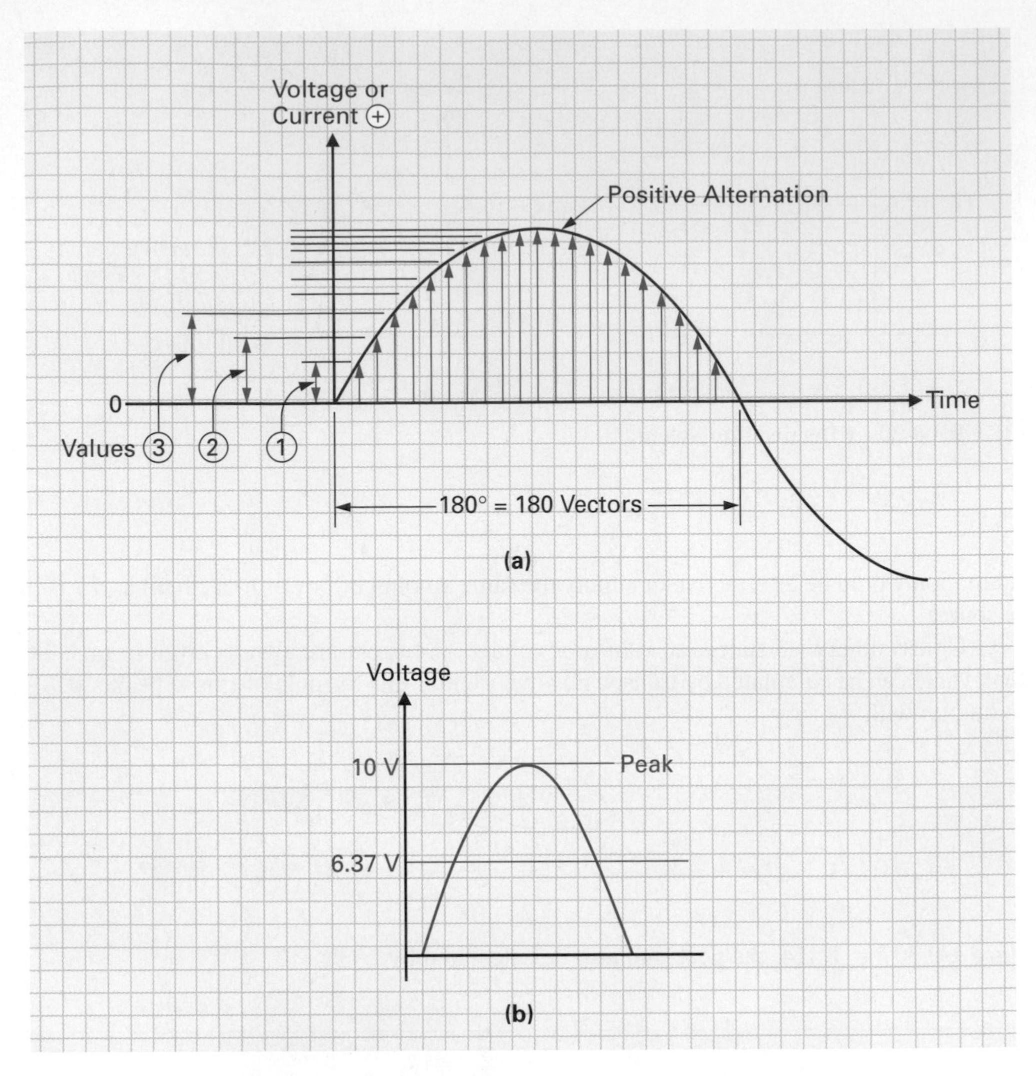

FIGURE 8-21 Average Value of a Sine-Wave Alternation = 0.637 × Peak.

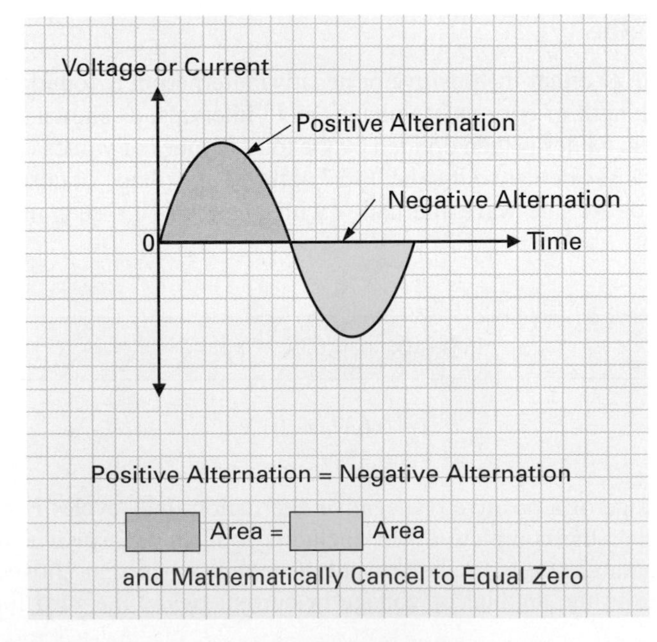

FIGURE 8-22 Average Value of a Complete Sine-Wave Cycle = 0.

EXAMPLE:

Calculate V_p, $V_{p\text{-}p}$, V_{rms}, and V_{avg} of a 16 V peak sine wave.

Solution:

$$V_p = 16 \text{ V}$$
$$V_{p\text{-}p} = 2 \times V_p = 2 \times 16 \text{ V} = 32 \text{ V}$$
$$V_{rms} = 0.707 \times V_p = 0.707 \times 16 \text{ V} = 11.3 \text{ V}$$
$$V_{avg} = 0.637 \times V_p = 0.637 \times 16 \text{ V} = 10.2 \text{ V}$$

EXAMPLE:

Calculate V_p, $V_{p\text{-}p}$, and V_{avg} of a 120 V (rms) ac main supply.

Solution:

$$V_p = \text{rms} \times 1.414 = 120 \text{ V} \times 1.414 = 169.68 \text{ V}$$
$$V_{p\text{-}p} = 2 \times V_p = 2 \times 169.68 \text{ V} = 339.36 \text{ V}$$
$$V_{avg} = 0.637 \times V_p = 0.637 \times 169.68 \text{ V} = 108.09 \text{ V}$$

The 120 V (rms) that is delivered to every home and business has a peak of 169.68 V. This ac value will deliver the same power as 120 V dc.

Frequency and Period

As shown in Figure 8-23, the **period** (t) is the time required for one complete cycle (positive and negative alternation) of the sinusoidal current or voltage waveform. A *cycle,* by definition, is the change of an alternating wave from zero to a positive peak, to zero, then to a negative peak, and finally back to zero.

Period
Time taken to complete one complete cycle of a periodic or repeating waveform.

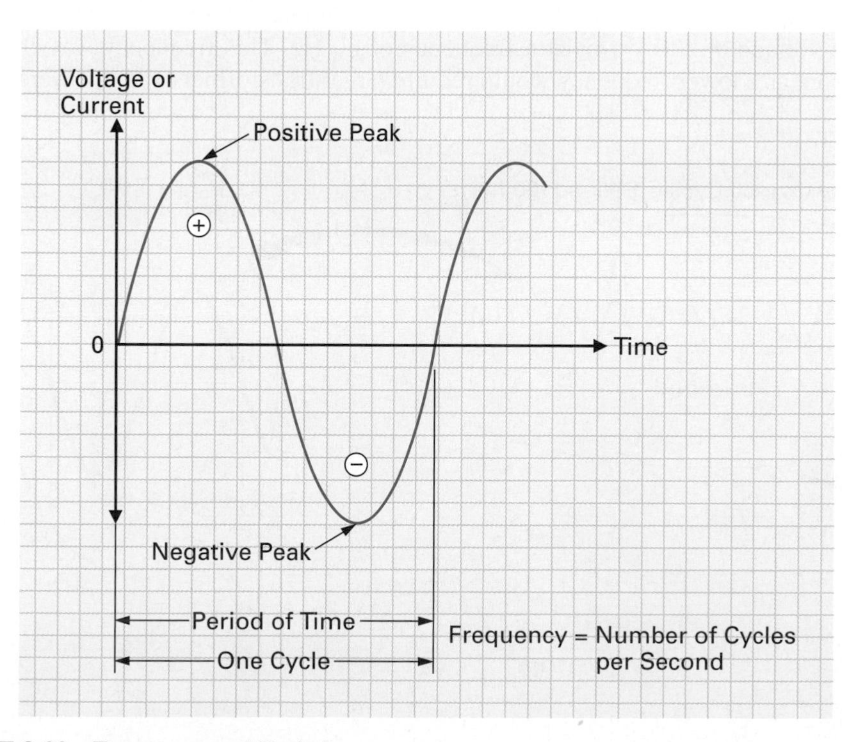

FIGURE 8-23 Frequency and Period.

Frequency
Rate of recurrences of a periodic wave, normally within a unit of 1 second, measured in hertz (cycles/second).

TIME LINE
Heinrich R. Hertz (1857–1894), a German physicist, was the first to demonstrate the production and reception of electromagnetic (radio) waves. In honor of his work in this field, the unit of frequency is called the hertz.

Frequency is the number of repetitions of a periodic wave in a unit of time. It is symbolized by f and is given the unit hertz (cycles per second), in honor of a German physicist, Heinrich Hertz.

Sinusoidal waves can take a long or a short amount of time to complete one cycle. This time is related to frequency in that period and is equal to the reciprocal of frequency, and vice versa.

$$f\,(\text{hertz}) = \frac{1}{t}$$

$$t\,(\text{seconds}) = \frac{1}{f}$$

where t = period, f = frequency

For example, the ac voltage of 120 V (rms) arrives at the household electrical outlet alternating at a frequency of 60 hertz (Hz). This means that 60 cycles arrive at the household electrical outlet in 1 second. If 60 cycles occur in 1 second, as seen in Figure 8-24(a), it is actually taking $^1/_{60}$ of a second for one of the 60 cycles to complete its cycle, which calculates out to be

$$\frac{1}{60} \text{ of } 1\text{ s} = \frac{1}{60 \text{ cycles}} \times 1\text{ s} = 16.67 \text{ milliseconds (ms)}$$

So the time or period of one cycle can be calculated by using the formula period $(t) = 1/f = 1/60$ Hz $= 16.67$ ms, as shown in Figure 8-24(b). If the period or time of a cycle is known, the frequency can be calculated. For example,

$$\text{frequency } (f) = \frac{1}{\text{period}} = 1/16.67 \text{ ms} = 60 \text{ Hz}$$

As illustrated in Figure 8-12, all homes in the United States receive at their wall outlets an ac voltage of 120 V rms at a frequency of 60 Hz. This frequency was chosen for good reason, as a lower frequency would require larger transformers and if it were too low,

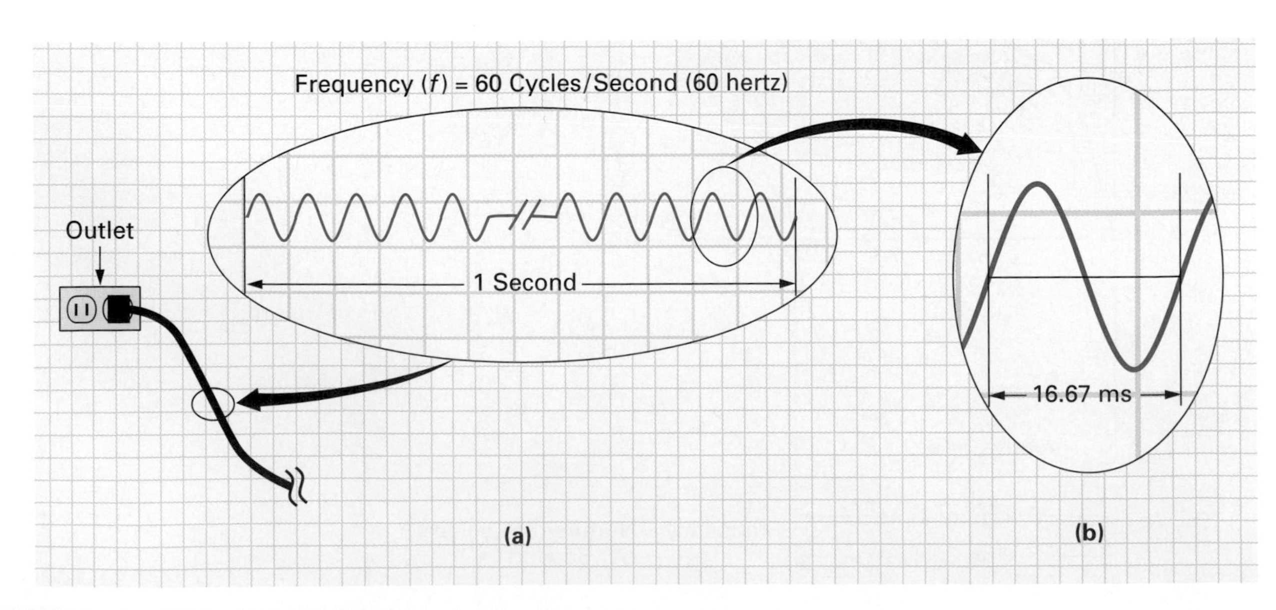

FIGURE 8-24 **120 V, 60 Hz AC Supply.**

the slow switching (alternating) of current through lightbulbs would cause the lights to flicker. A frequency higher than 60 Hz was found to cause an increase in the amount of heat generated in the core of all power distribution transformers due to eddy currents and hysteresis losses. As a result, a frequency of 60 Hz was chosen in the United States; however, other countries, such as England and most of Europe, use an ac power line frequency of 50 Hz (240 V).

EXAMPLE:

If a sine wave has a period of 400 μs, what is its frequency?

Solution:

$$\text{frequency } (f) = \frac{1}{\text{time } (t)} = \frac{1}{400\ \mu\text{s}} = 2.5 \text{ kHz or 2500 cycles per second}$$

EXAMPLE:

If it takes a sine wave 25 ms to complete two cycles, how many of the cycles will be received in 1 s?

Solution:

If the period of two cycles is 25 ms, one cycle period will equal 12.5 ms. The number of cycles per second or frequency will equal

$$f = \frac{1}{t} = \frac{1}{12.5 \text{ ms}} = 80 \text{ Hz} \quad \text{or} \quad 80 \text{ cycles/second}$$

EXAMPLE:

Calculate the period of the following:

a. 100 MHz
b. 40 cycles every 5 seconds
c. 4.2 kilocycles/second
d. 500 kHz

Solution:

$$f = \frac{1}{t} \quad \text{therefore,} \quad t = \frac{1}{f}$$

a. $t = \dfrac{1}{100 \text{ MHz}} = 10 \text{ nanoseconds (ns)}$

b. $40 \text{ cycles/5 s} = 8 \text{ cycles/second (8 Hz)}$

$t = \dfrac{1}{8 \text{ Hz}} = 125 \text{ ms}$

c. $t = \dfrac{1}{4.2 \text{ kHz}} = 238\ \mu\text{s}$

d. $t = \dfrac{1}{500 \text{ kHz}} = 2\ \mu\text{s}$

Wavelength

Wavelength
Distance between two corresponding points of a phase and is equal to waveform velocity or speed divided by frequency.

Wavelength, as its name states, is the physical length of one complete cycle and is generally measured in meters. The wavelength (λ, lambda) of a complete cycle is dependent on the frequency and velocity of the transmission:

$$\lambda = \frac{\text{velocity}}{\text{frequency}}$$

TIME LINE
During World War II, there was a need for microwave-frequency vacuum tubes. British inventor Henry Boot developed the magnetron in 1939.

Electromagnetic waves. Radio waves travel at the speed of light in air or a vacuum, which is 3×10^8 meters/second or 3×10^{10} centimeters/second.

$$\lambda\,(\text{m}) = \frac{3 \times 10^8\,\text{m/s}}{f(\text{Hz})}$$

or

$$\lambda\,(\text{cm}) = \frac{3 \times 10^{10}\,\text{cm/s}}{f(\text{Hz})}$$

There are 100 centimeters [cm] in 1 meter [m], therefore cm $= 10^{-2}$ and m $= 10^0$ or 1.

Subsequently, the higher the frequency, the shorter the wavelength, which is why a shortwave radio receiver is designed to receive high frequencies ($\lambda\downarrow = 3 \times 10^8/f\uparrow$).

EXAMPLE:

Calculate the wavelength of the electromagnetic waves illustrated in Figure 8-25.

Solution:

a. $\lambda = \dfrac{3 \times 10^8}{f(\text{Hz})}\,\text{m/s} = \dfrac{3 \times 10^8}{10\,\text{kHz}} = 30{,}000\,\text{m} \quad \text{or} \quad 30\,\text{km}$

b. $\lambda = \dfrac{3 \times 10^{10}}{f(\text{Hz})}\,\text{cm/s} = \dfrac{3 \times 10^{10}}{2182\,\text{kHz}} = 13{,}748.9\,\text{cm} \quad \text{or} \quad 137.489\,\text{m}$

c. $\lambda = \dfrac{3 \times 10^{10}}{f(\text{Hz})}\,\text{cm/s} = \dfrac{3 \times 10^{10}}{4.0\,\text{GHz}} = \dfrac{3 \times 10^{10}}{4 \times 10^9} = 7.5\,\text{cm} \quad \text{or} \quad 0.075\,\text{m}$

TIME LINE
In 1939, an American brother duo, Russel and Sigurd Varian, invented the klystron. In 1943, the traveling-wave tube amplifier was invented by Rudolf Komphner, and up to this day these three microwave tubes are still used extensively.

Sound waves. Sound waves travel at a slower speed than electromagnetic waves, as their mechanical vibrations depend on air molecules, which offer resistance to the traveling wave. For sound waves, the wavelength formula will be equal to

$$\lambda\,(\text{m}) = \frac{344.4\,\text{m/s}}{f(\text{Hz})}$$

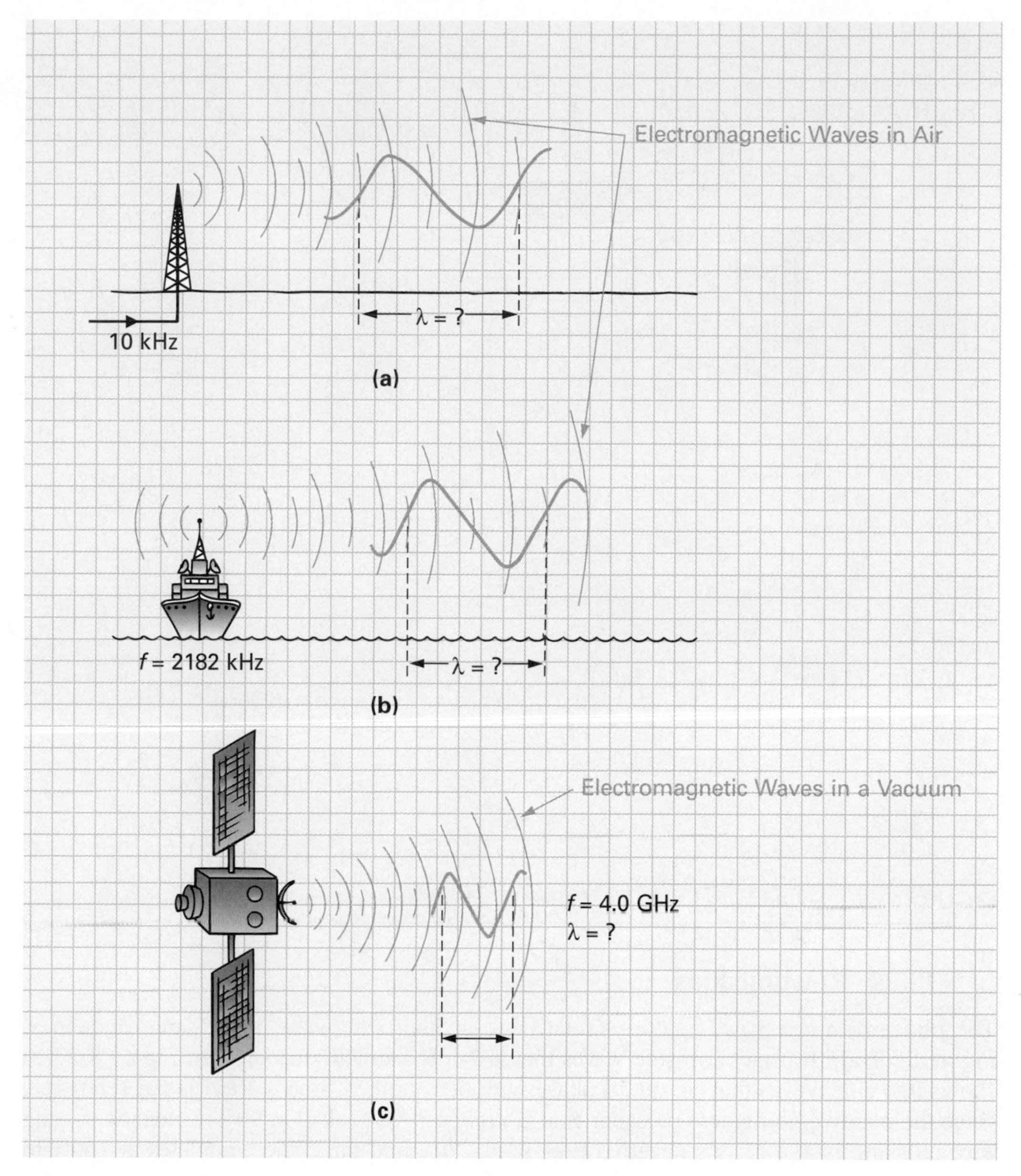

FIGURE 8-25 Electromagnetic Wavelength Examples.

EXAMPLE:

Calculate the wavelength of the sound waves illustrated in Figure 8-26.

Solution:

a. $\lambda\,(\text{m}) = \dfrac{344.4\ \text{m/s}}{f(\text{Hz})} = \dfrac{344.4\ \text{m/s}}{35\ \text{kHz}} = 9.8 \times 10^{-3}\ \text{m} = 9.8\ \text{mm}$

b. 300 Hz: $\lambda\,(\text{m}) = \dfrac{344.4\ \text{m/s}}{300\ \text{Hz}} = 1.15\ \text{m}$

3000 Hz: $\lambda\,(\text{m}) = \dfrac{344.4\ \text{m/s}}{3000\ \text{Hz}} = 0.115\ \text{m}$ or 11.5 cm

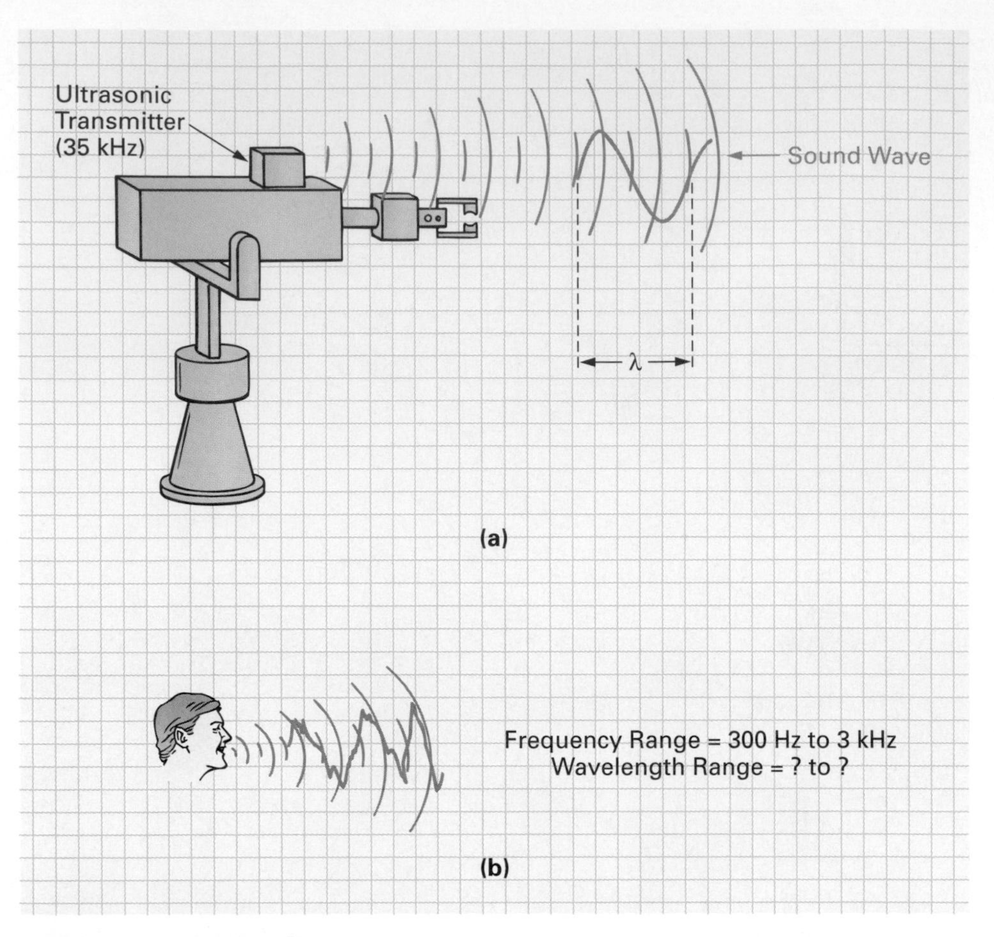

FIGURE 8-26 Sound Wavelength Examples.

Phase Relationships

Phase
Angular relationship between two waves, normally between current and voltage in an ac circuit.

The **phase** of a sine wave is always relative to another sine wave of the same frequency. Figure 8-27(a) illustrates two sine waves that are in phase with one another; Figure 8-27(b) shows two sine waves that are out of phase with one another. Sine wave *A* is our reference, since the positive-going zero crossing is at 0°, its positive peak is at 90°, its negative-going zero crossing is at 180°, its negative peak is at 270°, and the cycle completes at 360°. In Figure 8-27(a), sine wave *B* is in phase with *A* since its peaks and zero crossings occur at the same time as those of sine wave *A*.

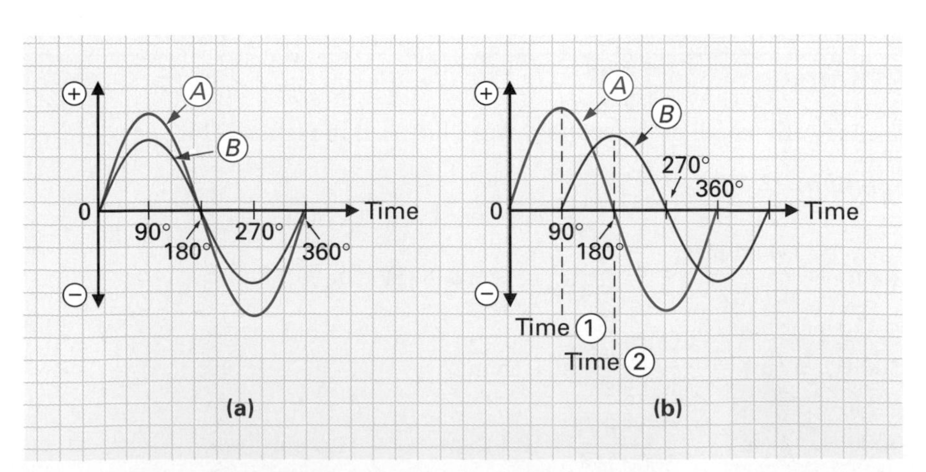

FIGURE 8-27 Phase Relationship. (a) In Phase. (b) Out of Phase.

In Figure 8-27(b), sine wave B has been shifted to the right by 90° with respect to the reference sine wave A. This **phase shift** or **phase angle** of 90° means that sine wave A *leads* B by 90°, or B *lags* A by 90°. Sine wave A is said to lead B, as, for example, its positive peak occurs first at time 1; the positive peak of B occurs later at time 2.

Phase Shift or Angle
Change in phase of a waveform between two points, given in degrees of lead or lag. Phase difference between two waves, normally expressed in degrees.

EXAMPLE:

What are the phase relationships between the two waveforms illustrated in Figure 8-28(a) and (b)?

Solution:

a. The phase shift or angle is 90°. Sine wave B leads sine wave A by 90°, or A lags B by 90°.

b. The phase shift or angle is 45°. Sine wave A leads sine wave B by 45°, or B lags A by 45°.

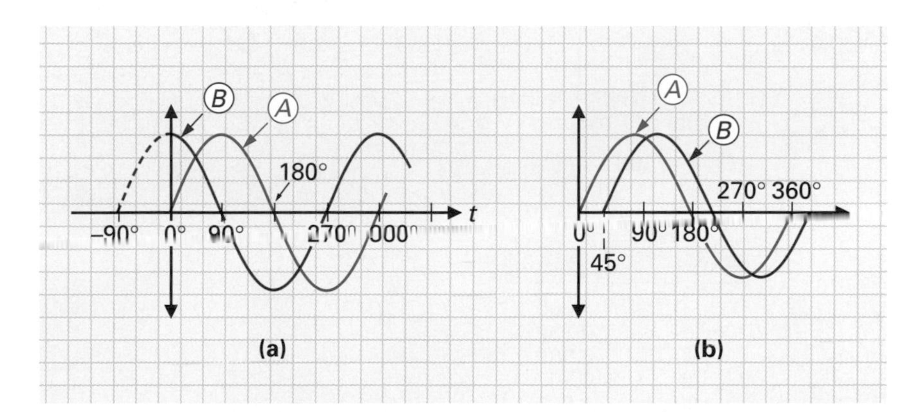

FIGURE 8-28 **Phase Relationship Examples.**

The Meaning of Sine

The names of the square, rectangular, triangular, and sawtooth waveforms reflect their waveform shapes. The sine wave is the most common type of waveform shape, and why the name sine was given to this wave needs to be explained further.

Figure 8-29(a) shows the correlation between the 360° of a circle and the 360° of a sine wave. A triangle has been drawn within the circle to represent the right-angle triangle formed at 30°. The hypotenuse side will always remain at the same length throughout 360° and will equal the peak of the sine wave. The opposite side of the triangle is equal to the amplitude vector of the sine wave at 30°. To calculate the amplitude of the opposite side, and therefore the amplitude of the sine wave at 30°, we use the sine of theta formula discussed previously in the mini-math review of trigonometry.

$$\sin \theta = \frac{\text{opposite}}{\text{hypotenuse}}$$

$$\sin 30^\circ = \frac{O}{H} \qquad \left(\textit{calculator sequence:} \quad \boxed{3}\boxed{0}\boxed{\text{SIN}} \right)$$

$$0.500 = \frac{O}{H}$$

This tells us that at 30°, the opposite side is 0.500, or half the size of the hypotenuse. At 30°, therefore, the amplitude of the sine wave will be 0.5 (50%) of the peak value.

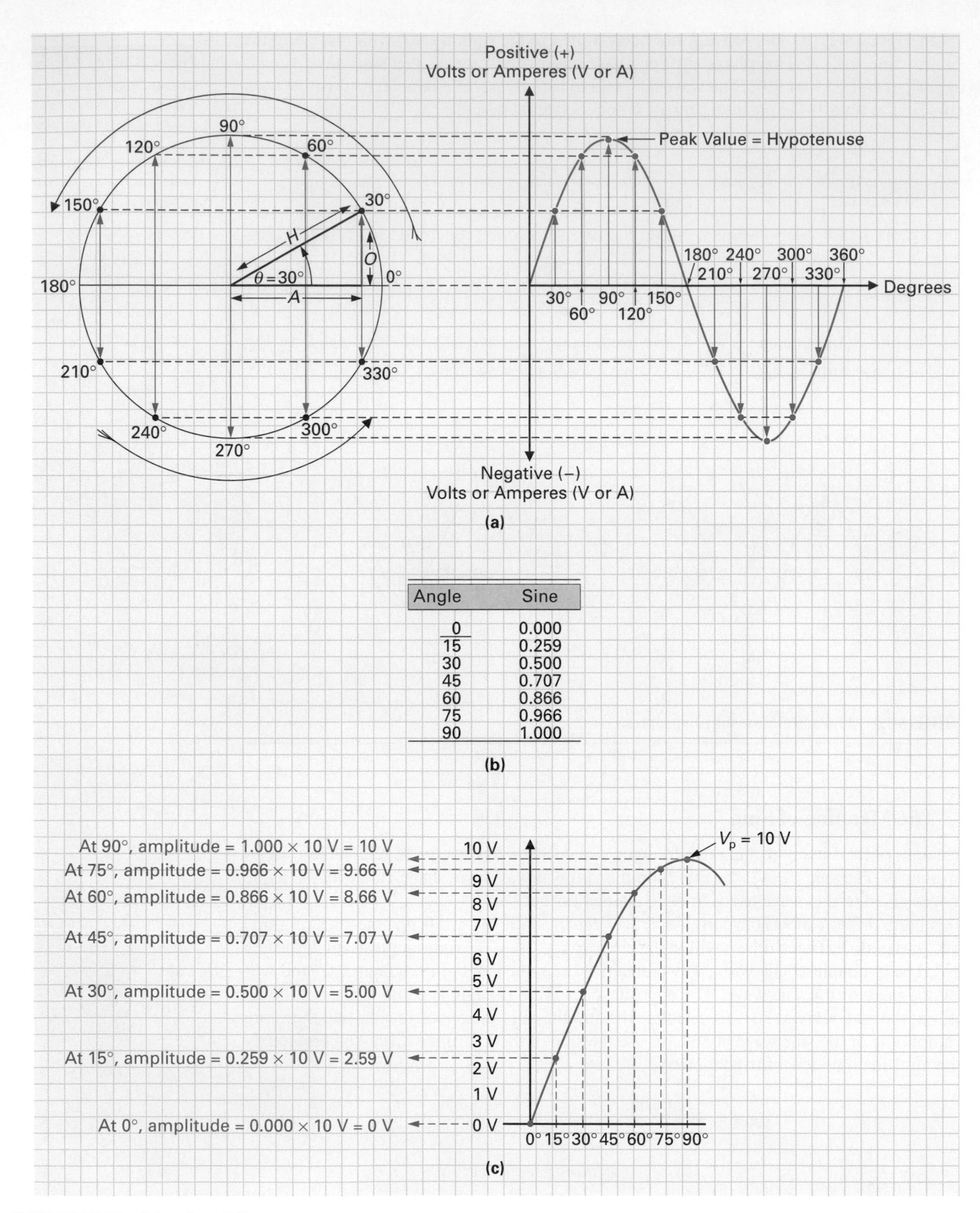

Angle	Sine
0	0.000
15	0.259
30	0.500
45	0.707
60	0.866
75	0.966
90	1.000

FIGURE 8-29 Meaning of Sine.

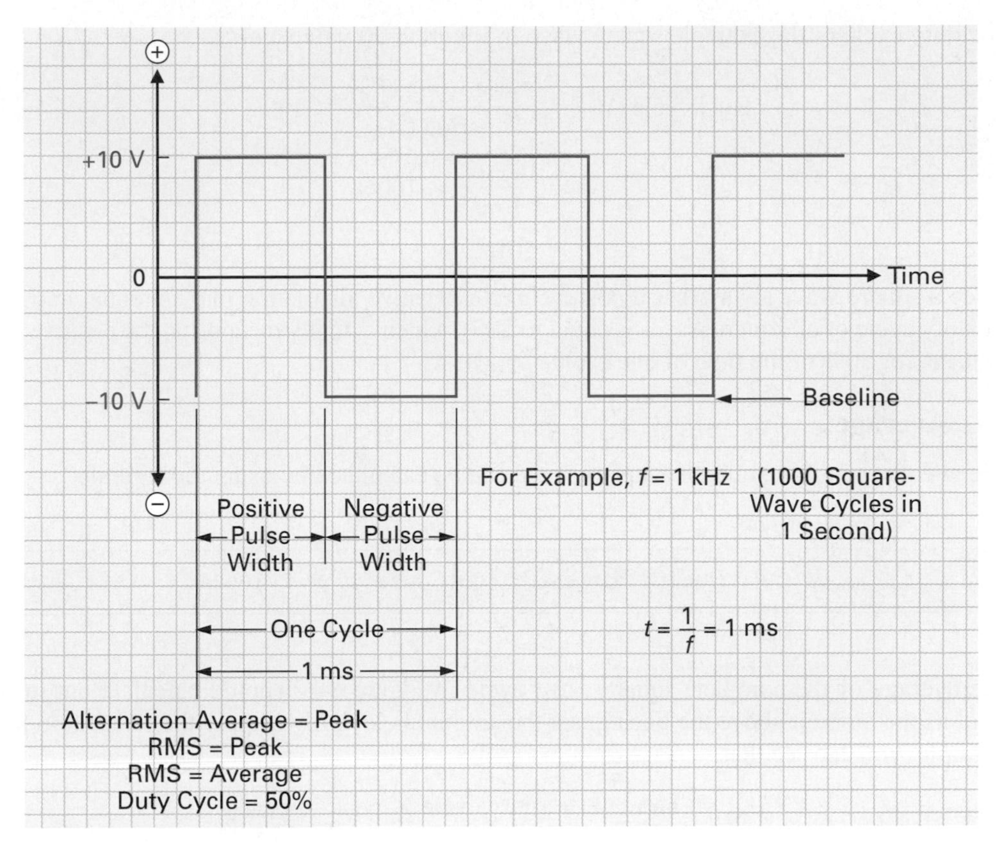

FIGURE 8-30 **Square Wave.**

Figure 8-29(b) lists the sine values at 15° increments. Figure 8-29(c) shows an example of a 10 V peak sine wave. At 15°, a sine wave will always be at 0.259 (sine of 15°) of the peak value, which for a 10 V sine wave will be 2.59 V (0.259 × 10 V = 2.59 V). At 30°, a sine wave will have increased to 0.500 (sine of 30°) of the peak value. At 45°, a sine wave will be at 0.707 of the peak, and so on. The sine wave is called a sine wave, therefore, because it changes in amplitude at the same rate as the sine trigonometric function.

8-4-2 *The Square Wave*

The **square wave** is a periodic (repeating) wave that alternates from a positive peak value to a negative peak value, and vice versa, for equal lengths of time.

Square Wave
Wave that alternates between two fixed values for an equal amount of time.

In Figure 8-30 you can see an example of a square wave that is at a frequency of 1 kHz and has a peak of 10 V. If the frequency of a wave is known, its period or time of one cycle can be calculated by using the formula $t = 1/f = 1/1\text{ kHz} = 1\text{ ms}$ or $\frac{1}{1000}$ of a second. One complete cycle will take 1 ms to complete, so the positive and negative alternations will each last for 0.5 ms. If the peak of the square wave is equal to 10 V, the peak-to-peak value of this square wave will equal $V_{p\text{-}p} = 2 \times V_p = 20$ V. To summarize, the square wave alternates from a positive peak value of +10 V to a negative peak value of −10 V for equal time lengths (half-cycles) of 0.5 ms.

Duty Cycle

Duty cycle is an important relationship that has to be considered when discussing square waveforms. The **duty cycle** is the ratio of a pulse width (positive or negative pulse or cycle) to the overall period or time of the wave and is normally given as a percentage.

Duty Cycle
A term used to describe the amount of ON time versus OFF time. ON time is usually expressed as a percentage.

$$\text{Duty cycle } (\%) = \frac{\text{pulse width } (P_w)}{\text{period } (t)} \times 100\%$$

The duty cycle of the example square wave in Figure 8-30 will equal

$$\text{Duty cycle } (\%) = \frac{\text{pulse width } (P_w)}{\text{period } (t)} \times 100\%$$

$$= \frac{0.5 \text{ ms}}{1 \text{ ms}} \times 100\%$$

$$= 50\%$$

Since a square wave always has a positive and a negative alternation that are equal in time, the duty cycle of all square waves is equal to 50%, which actually means that the positive cycle lasts for 50% of the time of one cycle.

Average

The average or mean value of a square wave can be calculated by using the formula

$$V \text{ or } I \text{ average} = \text{baseline} + (\text{duty cycle} \times \text{peak-to-peak})$$

The average of the complete square-wave cycle in Figure 8-30 should calculate out to be zero, as the amount above the line equals the amount below. If we apply the formula to this example, you can see that

$$V_{avg} = \text{baseline} + (\text{duty cycle} \times \text{peak-to-peak})$$

$$= -10 \text{ V} + (0.5 \times 20 \text{ V})$$

$$= -10 \text{ V} + 10 \text{ V}$$

$$= 0 \text{ V}$$

However, a square wave does not always alternate about 0. For example, Figure 8-31 illustrates a 16 $V_{p\text{-}p}$ square wave that rests on a baseline of 2 V. The average value of this square wave is equal to

$$V_{avg} = \text{baseline} + (\text{duty cycle} \times \text{peak-to-peak})$$

$$= 2 \text{ V} \times (0.5 \times 16 \text{ V})$$

$$= (+2) + (+8 \text{ V})$$

$$= 10 \text{ V}$$

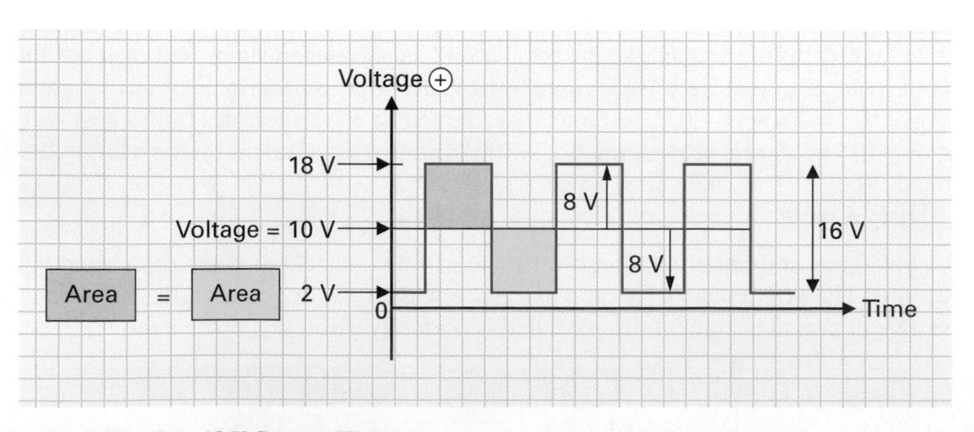

FIGURE 8-31 2 to 18 V Square Wave.

EXAMPLE:

Calculate the duty cycle and V_{avg} of a square wave of 0 to 5 V.

Solution:

The duty cycle of a square wave is always 0.5 or 50%. The average is

$$\begin{aligned} V_{avg} &= \text{baseline} + (\text{duty cycle} \times V_{p\text{-}p}) \\ &= 0\text{ V} + (0.5 \times 5\text{ V}) \\ &= 0\text{ V} + 2.5\text{ V} = 2.5\text{ V} \end{aligned}$$

Up to this point, we have examined the ideal square wave, which has instantaneous transition from the negative to the positive values, and vice versa, as shown in Figure 8-32(a). In fact, the transitions from negative to positive (positive or leading edge) and from positive to negative (negative or trailing edge) are not ideal. It takes a small amount of time for the wave to increase to its positive value (the **rise time**) and an equal amount of time for a wave to decrease to its negative value (the **fall time**). Rise time (T_R), by definition, is the time it takes for an edge to rise from 10% to 90% of its full amplitude; fall time (T_F) is the time it takes for an edge to fall from 90% to 10% of its full amplitude, as shown in Figure 8-32(b).

With a waveform such as that in Figure 8-32(b), it is difficult, unless a standard is used, to know exactly what points to use when measuring the width of either the positive or negative alternation. The standard width is always measured between the two 50% amplitude points, as shown in Figure 8-33.

Rise Time
Time it takes a positive edge of a pulse to rise from 10% to 90% of its peak value.

Fall Time
Time it takes a negative edge of a pulse to fall from 90% to 10% of its peak value.

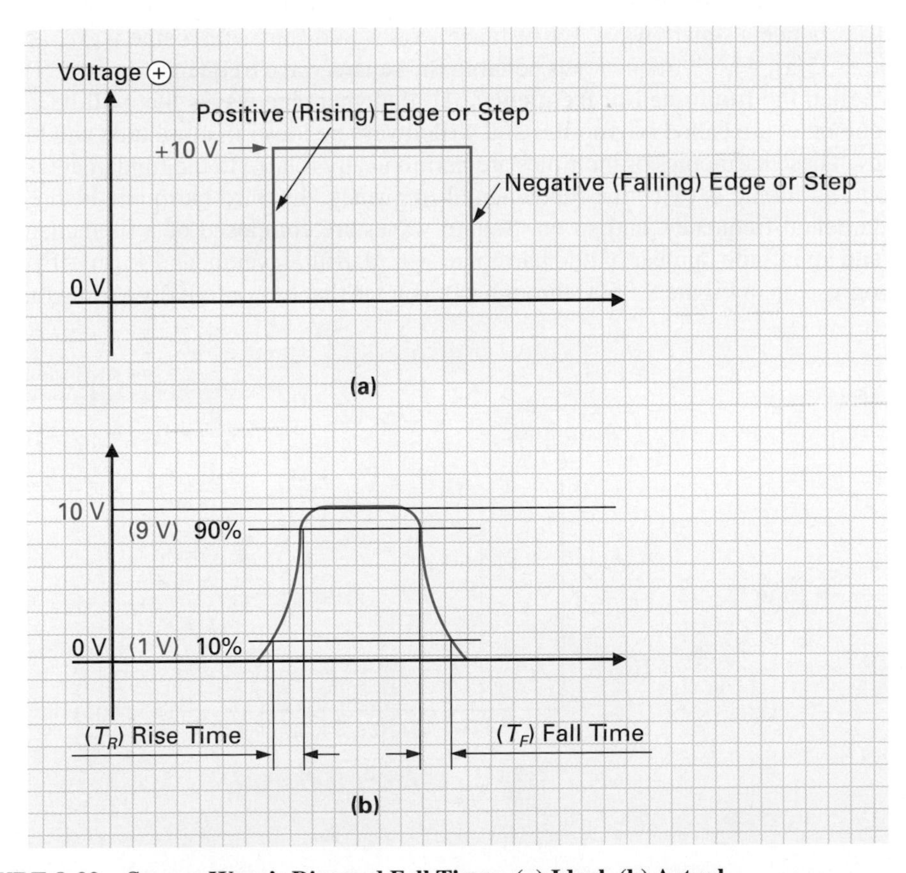

FIGURE 8-32 Square Wave's Rise and Fall Times. (a) Ideal. (b) Actual.

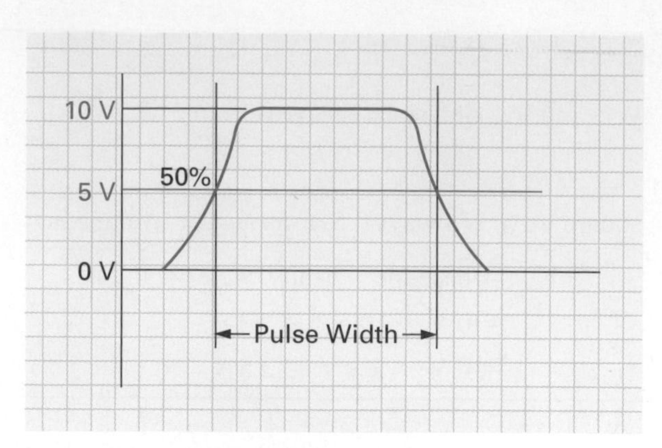

FIGURE 8-33 **Pulse Width of a Square Wave.**

Time-Domain Analysis
A method of representing a waveform by plotting its amplitude versus time.

Frequency-Domain Analysis
A method of representing a waveform by plotting its amplitude versus frequency.

Fundamental Frequency
This sine wave that is always the lowest-frequency and largest-amplitude component of any waveform shape and is used as a reference.

Harmonic Frequency
Sine wave that is smaller in amplitude and is some multiple of the fundamental frequency.

Frequency-Domain Analysis

A *periodic wave* is a wave that repeats the same wave shape from one cycle to the next. Figure 8-34(a) is a **time-domain** representation of a periodic sine wave as it would appear on an oscilloscope display plotting the sine wave's amplitude against time. Figure 8-34(b) is a **frequency-domain** representation of the same periodic sine wave. This graph, which shows the wave as it would appear on a spectrum analyzer, plots the sine wave's amplitude against frequency instead of time. The graph shows all the frequency components contained within a wave, and since, in this example, the sine wave has a period of 1 ms and therefore a frequency of 1 kHz, there is one bar at the 1 kHz point of the graph, its size representing the sine wave's amplitude. Pure sine waves have no other frequency components.

Other periodic wave shapes, such as square, pulse, triangular, sawtooth, or irregular, are actually made up of a number of sine waves having a particular frequency, amplitude, and phase. To produce a square wave, for instance, you would start with a sine wave, as shown in Figure 8-35(a), whose frequency is equal to the square-wave frequency desired. This sine wave is called the **fundamental frequency.** All the other sine waves that will be added to this fundamental are called **harmonics** and will always be lower in amplitude and higher in frequency. These harmonics or multiples are harmonically related to the fundamental, in that the second harmonic is twice the fundamental frequency, the third harmonic is three times the fundamental frequency, and so on. Square waves are composed of a fundamental frequency and an infinite number of odd harmonics (third, fifth, seventh, and so on). If you look at the progression in Figure 8-35(a) through (d), you see that by continually adding these odd

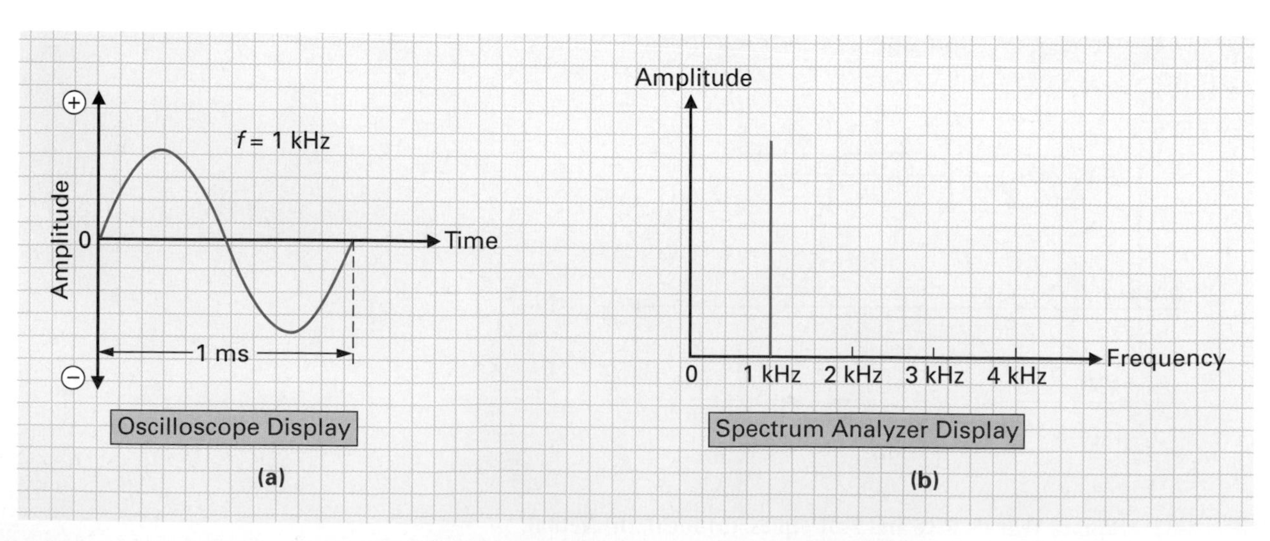

FIGURE 8-34 **Analysis of a 1 kHz Sine Wave. (a) Time Domain. (b) Frequency Domain.**

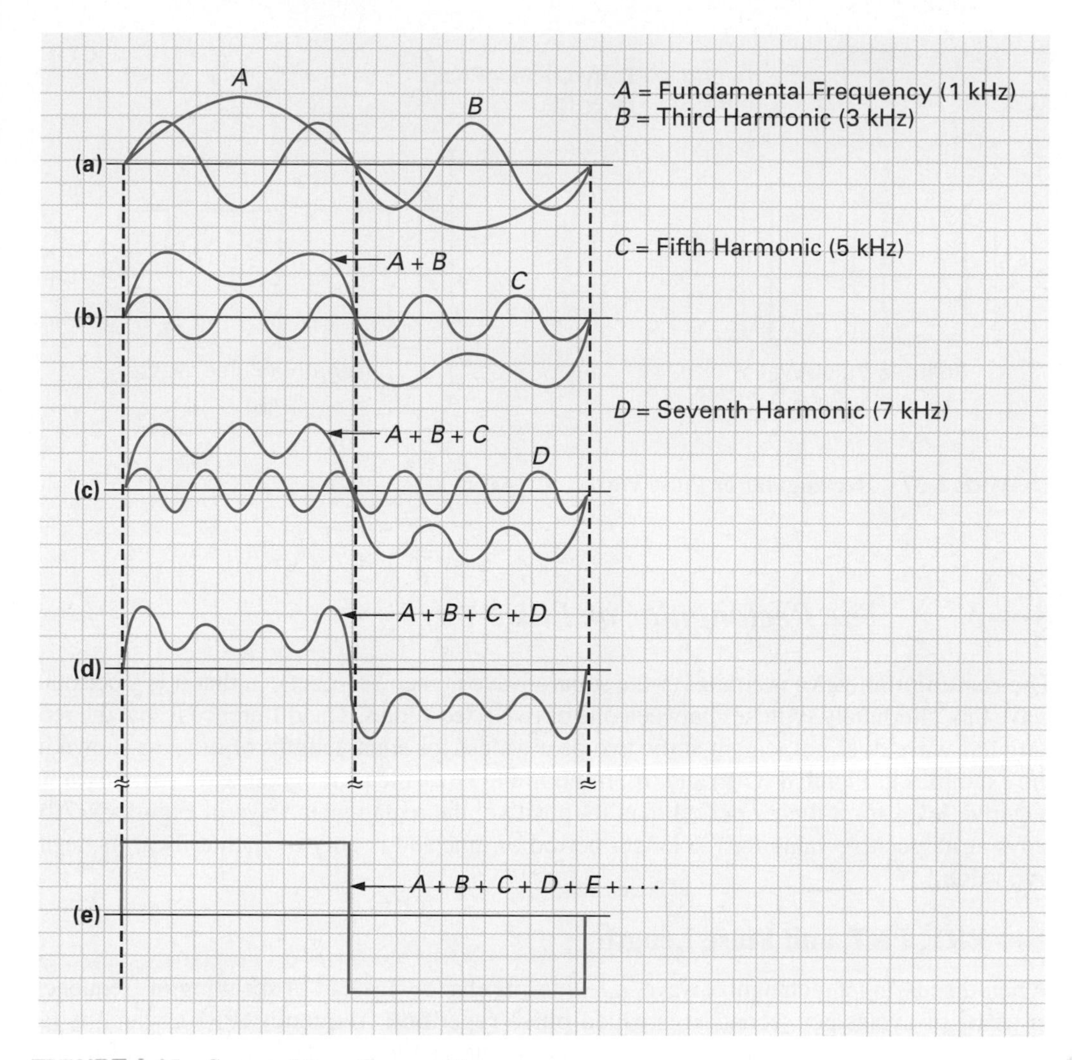

FIGURE 8-35 Square-Wave Composition.

harmonics the waveform comes closer to a perfect square wave, as shown in Figure 8-35(e). Figure 8-36 plots the frequency domain of a square wave, with the bars representing the odd harmonics of decreasing amplitude.

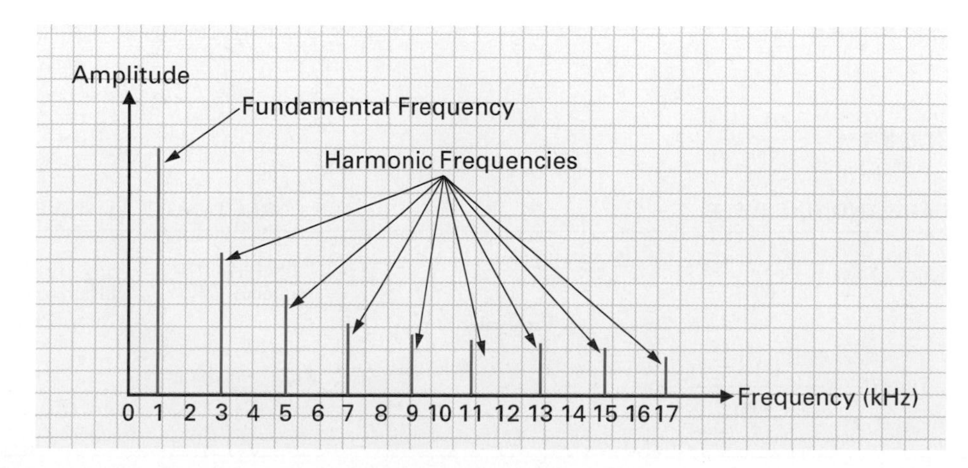

FIGURE 8-36 Frequency-Domain Analysis of a Square Wave.

TIME LINE

Jean Fourier (1768–1830) developed a series of terms, known as the *Fourier series,* that can be used to represent a nonsinusoidal periodic waveform. Primarily a mathematician, he applied much of his work to real-world physics. In 1806, in response to a request from Napoleon, he traveled to Egypt for three years to help research Egyptian antiquities.

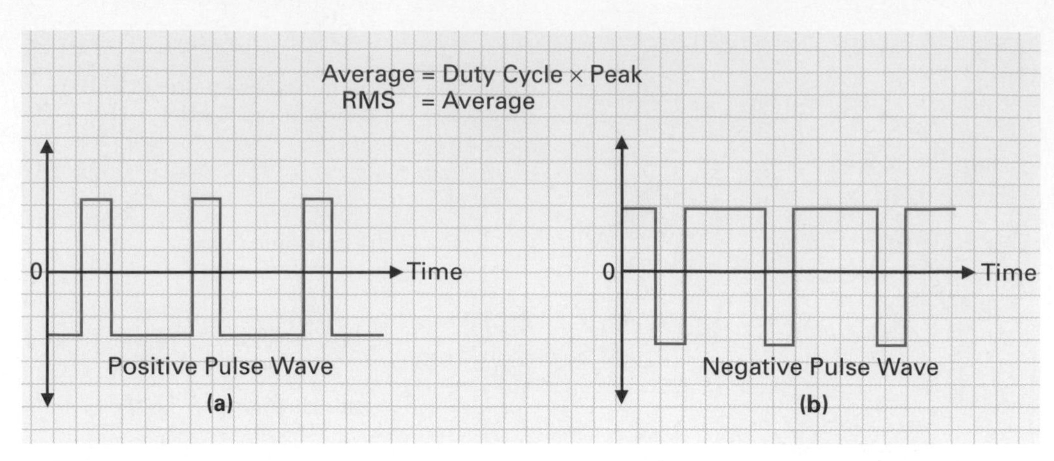

FIGURE 8-37 **Rectangular or Pulse Wave.**

Rectangular (Pulse) Wave
Also known as a pulse wave, it is a repeating wave that only alternates between two levels or values and remains at one of these values for a small amount of time relative to the other.

Pulse Repetition Frequency
The number of times per second that a pulse is transmitted.

Pulse Repetition Time
The time interval between the start of two consecutive pulses.

Pulse Width, Pulse Length, or Pulse Duration
The time interval between the leading edge and trailing edge of a pulse at which the amplitude reaches 50% of the peak pulse amplitude.

8-4-3 *The Rectangular or Pulse Wave*

The **rectangular wave** is similar to the square wave in many respects, in that it is a periodic wave that alternately switches between one of two fixed values. The difference is that the rectangular wave does not remain at the two peak values for equal lengths of time, as shown in the examples in Figure 8-37(a) and (b). In Figure 8-37(a), the rectangular wave remains at its negative level for a longer period than its positive; the rectangular wave in Figure 8-37(b) stays at its positive value for the longer period of time and is only momentarily at its negative value.

PRF, PRT, and Pulse Length

When discussing a rectangular wave, a few terms change. Instead of speaking of frequency in terms of cycles per second, it is called **pulse repetition frequency** (PRF), which is far more descriptive. The reciprocal of frequency is time, and with rectangular pulse waveforms the reciprocal of the PRF is **pulse repetition time** (PRT), as summarized in Table 8-1. With rectangular or pulse waves, therefore, frequency is equivalent to PRF, time to PRT, and the only difference is the name.

Let us look at the example in Figure 8-38 of a 5 V rectangular wave at a frequency of 1 kHz and a pulse width of 1 μs, and practice with these new terms. With a pulse repetition frequency of 1 kHz, the time between the leading edges of pulses (PRT) will be 1/1 kHz = 1 ms. **Pulse width** (P_w), **pulse duration** (P_d), or **pulse length** (P_l) are all terms that describe the length of time for which the pulse lasts, and in this example it is equal to 1 μs, which means that 999 μs exists between the end of one pulse and the beginning of the next.

TABLE 8-1

SQUARE AND SINE WAVE		RECTANGULAR WAVE
$\text{Frequency } (f) = \dfrac{1}{\text{time } (t)}$	**Equivalent to**	$\text{Pulse repetition frequency (PRF)} = \dfrac{1}{\text{pulse repetition time (PRT)}}$
$\text{Time } (t) = \dfrac{1}{\text{frequency } (f)}$		$\text{Pulse repetition time (PRT)} = \dfrac{1}{\text{pulse repetition frequency (PRF)}}$

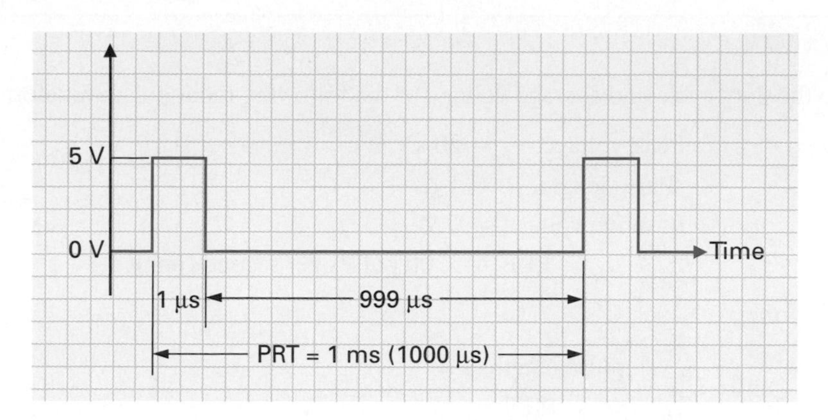

FIGURE 8-38 PRF and PRT of a Pulse Wave.

Duty Cycle

The duty cycle is calculated in exactly the same way as for the square wave and is a ratio of the pulse width to the overall time (PRT). In our example in Figure 8-38 the duty cycle will be equal to

$$\begin{aligned}\text{Duty cycle } (\%) &= \frac{\text{pulse width } (P_w)}{\text{PRT}} \times 100\% \\ &= \frac{1\ \mu\text{s}}{1000\ \mu\text{s}} \times 100\% \\ &= \text{duty cycle figure of } 0.001 \times 100\% \\ &= 0.001 \times 100\% \\ &= 0.1\%\end{aligned}$$

The result tells us that the positive pulse lasts for 0.1% of the total time (PRT).

Average

The average or mean value of this waveform is calculated by using the same square-wave formula. The average of the pulse wave in Figure 8-38 will be

$$\begin{aligned}V \text{ or } I \text{ average} &= \text{baseline} + (\text{duty cycle} \times \text{peak-to-peak}) \\ V_{\text{avg}} &= 0\ \text{V} + (0.001 \times 5\ \text{V}) \\ &= 0\ \text{V} + (5\ \text{mV}) \\ &= 5\ \text{mV}\end{aligned}$$

Figure 8-39 illustrates the average value of this rectangular waveform. If the voltage and width of the positive pulse are spread out over the entire PRT, they will have a mean level equal, in this example, to 5 mV.

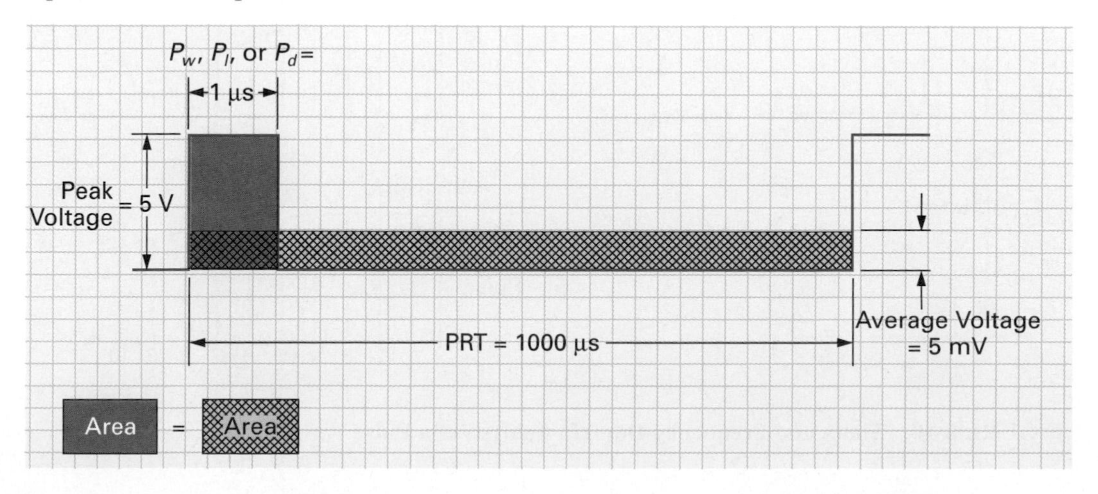

FIGURE 8-39 Average of a Pulse Wave.

EXAMPLE:

Calculate the duty cycle and average voltage of the following radar pulse waveform:

$$\text{Peak voltage, } V_p = 20 \text{ kV}$$
$$\text{Pulse length, } P_l = 1\ \mu\text{s}$$
$$\text{Baseline voltage} = 0 \text{ V}$$
$$\text{PRF} = 3300 \text{ pulses per second (pps)}$$

Solution:

$$\text{Duty cycle} = \frac{\text{pulse length } (P_l)}{\text{PRT}} \times 100\%$$
$$= \frac{1\ \mu\text{s}}{303\ \mu\text{s}} \times 100\% \left(\text{PRT} = \frac{1}{\text{PRF}} = \frac{1}{3300} = 303\ \mu\text{s} \right)$$
$$= (3.3 \times 10^{-3}) \times 100\%$$
$$= 0.33\%$$
$$V_{\text{avg}} = \text{baseline} + (\text{duty cycle} \times V_{\text{p-p}})$$
$$= 0 \text{ V} + [(3.3 \times 10^{-3}) \times 20 \times 10^3]$$
$$= 66 \text{ V}$$

Frequency-Domain Analysis

The pulse or rectangular wave is closely related to the square wave, as shown in Figure 8-40, but there are some differences in its harmonic content. One is that even-number har-

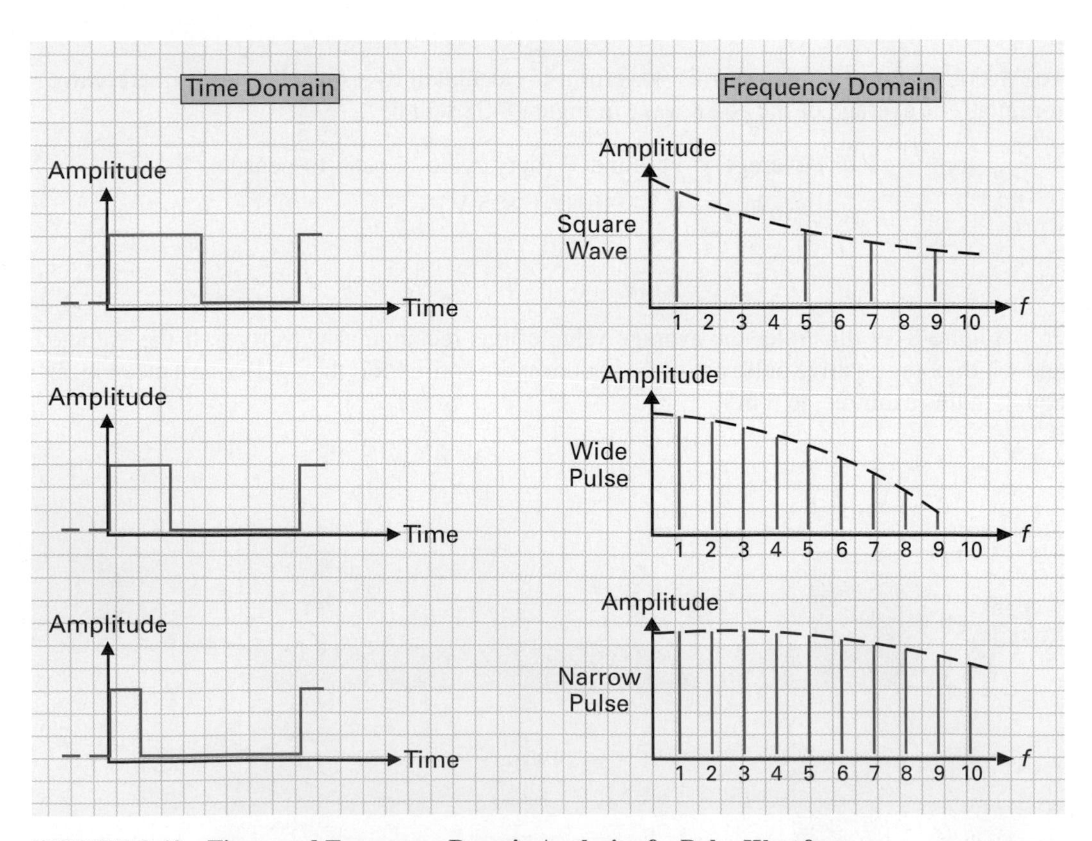

FIGURE 8-40 Time- and Frequency-Domain Analysis of a Pulse Waveform.

monics are present and their amplitudes do not fall off as quickly as do those of the square wave. The amplitude and phase of these sine-wave harmonics are determined by the pulse width and pulse repetition frequency; the narrower the pulse, the greater the number of harmonics present.

8-4-4 *The Triangular Wave*

A **triangular wave** consists of a positive and negative ramp of equal values, as shown in Figure 8-41. Both the positive and negative ramps have, respectively, a linear increase and decrease. Linear, by definition, describes a relationship between two quantities that exists when a change in a second quantity is directly proportional to a change in the first quantity. The two quantities in this case are voltage or current and time. As shown in Figure 8-41, if the increment of change in voltage, ΔV (pronounced "delta vee"), is changing at the same rate as the increment of time, Δt ("delta tee"), the ramp is said to be **linear.**

Triangular Wave
A repeating wave that has equal positive and negative ramps that have linear rates of change with time.

Linear
Relationship between input and output in which the output varies in direct proportion to the input.

With Figure 8-42(a), the voltage has risen 1 V in 1 second (time 1) and maintains that rise through to time 2; consequently, it is known as a linearly rising slope. In Figure 8-42(b), the voltage is falling first from 6 to 5 V, which is a 1 V drop in 1 second, and in time 2 from 6 V to 2 V, which is a 4 V drop in 4 seconds. The rate of fall still remains the same (1 volt per second), so the waveform is referred to as a linearly falling slope.

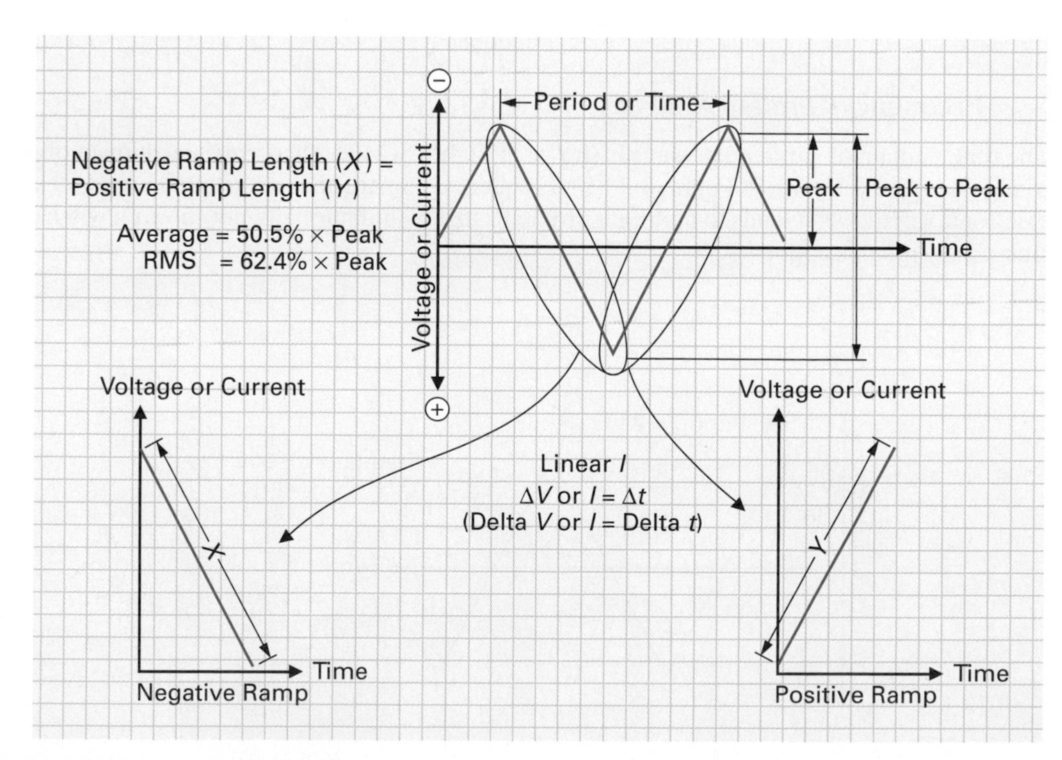

FIGURE 8-41 **Triangular Wave.**

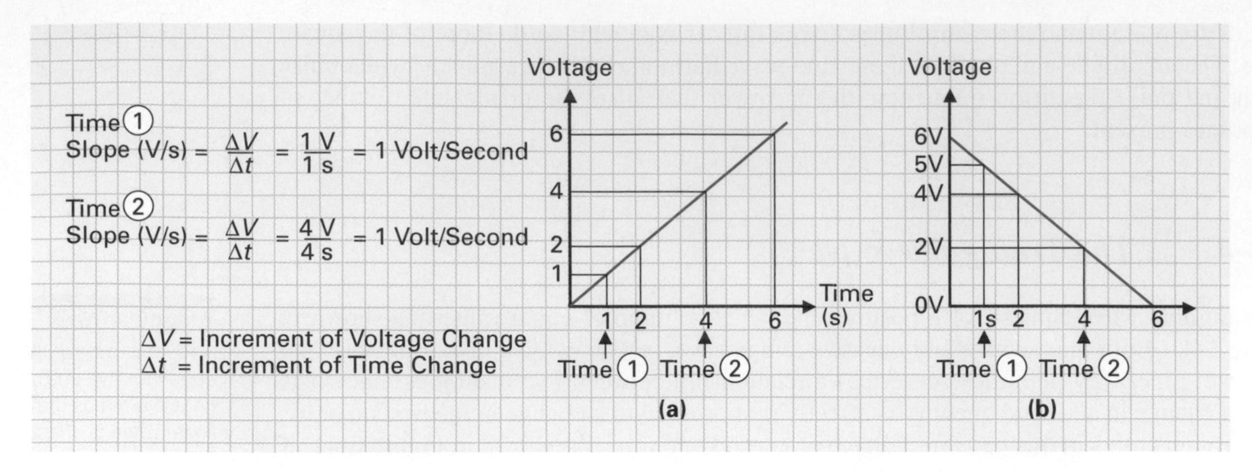

FIGURE 8-42 Linear Triangular Wave Rise and Fall.

This formula for slope will also apply to a current waveform and is

$$\text{Slope (A/s)} = \frac{\Delta I}{\Delta t}$$

where ΔI = increment of current change and Δt = increment of time change.

With triangular waves, frequency and time apply as usual, as seen in Figure 8-43, with

$$\text{Frequency} = \frac{1}{\text{time}} \quad (\text{Hz})$$

$$\text{Time} = \frac{1}{\text{frequency}} \quad (\text{s})$$

Frequency-Domain Analysis

The time domain of a triangular wave is shown in Figure 8-44(a); the frequency domain of a triangular wave is shown in Figure 8-44(b). Frequency-domain analysis is often used to test electronic circuits as it tends to highlight problems that would normally not show up when a sine-wave signal is applied.

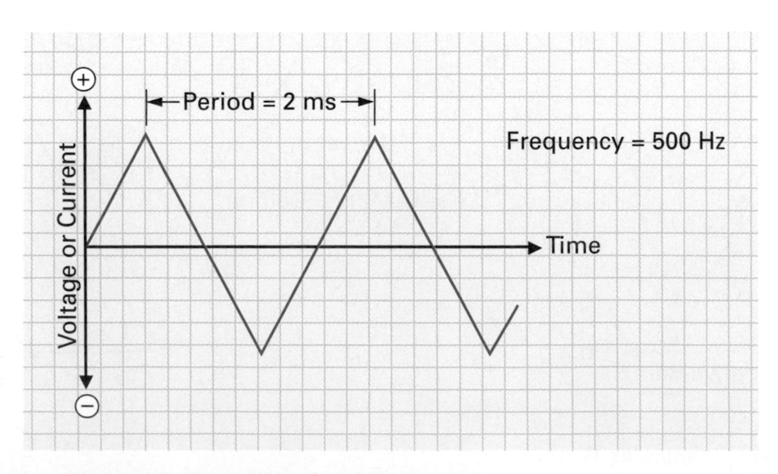

FIGURE 8-43 Triangular Wave Period and Frequency.

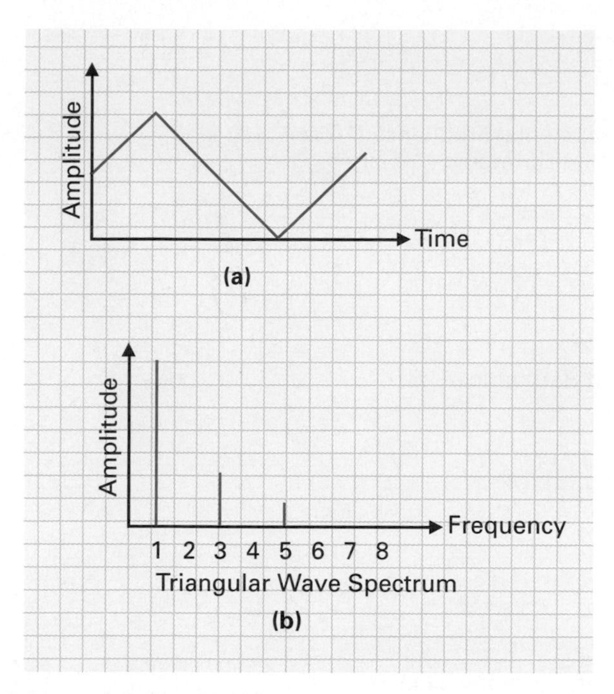

FIGURE 8-44 **Analysis of a Triangular Wave. (a) Time Domain. (b) Frequency Domain.**

8-4-5 *The Sawtooth Wave*

On an oscilloscope display (time-domain presentation), the **sawtooth wave** is very similar to a triangular wave, in that a sawtooth wave has a linear ramp. However, unlike the triangular wave, which reverses and has an equal but opposite ramp back to its starting level, the sawtooth "flies" back to its starting point immediately and then repeats the previous ramp, as seen in Figure 8-45, which shows both a positive and a negative ramp sawtooth. Figure 8-46 shows the odd and even harmonics contained in the frequency domain analysis of a negative-going ramp.

Sawtooth Wave
Repeating waveform that rises from zero to a maximum value linearly and then falls to zero and repeats.

8-4-6 *Other Waveforms*

The waveforms discussed so far are some of the more common types; however, since every waveform shape (except a pure sine wave) is composed of a large number of sine waves combined in an infinite number of ways, any waveform shape is possible. Figure 8-47 illustrates a variety of waveforms that can be found in all fields of electronics, and the appendix contains a detailed frequency spectrum chart detailing the applications of many electromagnetic frequencies.

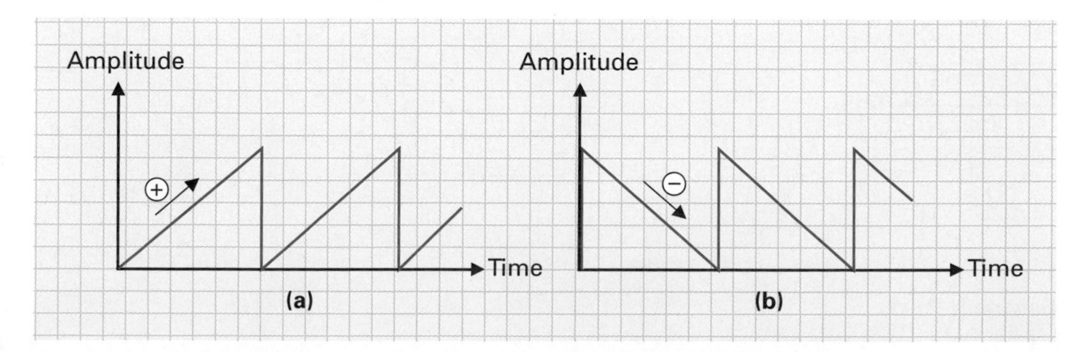

FIGURE 8-45 **Sawtooth Wave. (a) Positive Ramp. (b) Negative Ramp.**

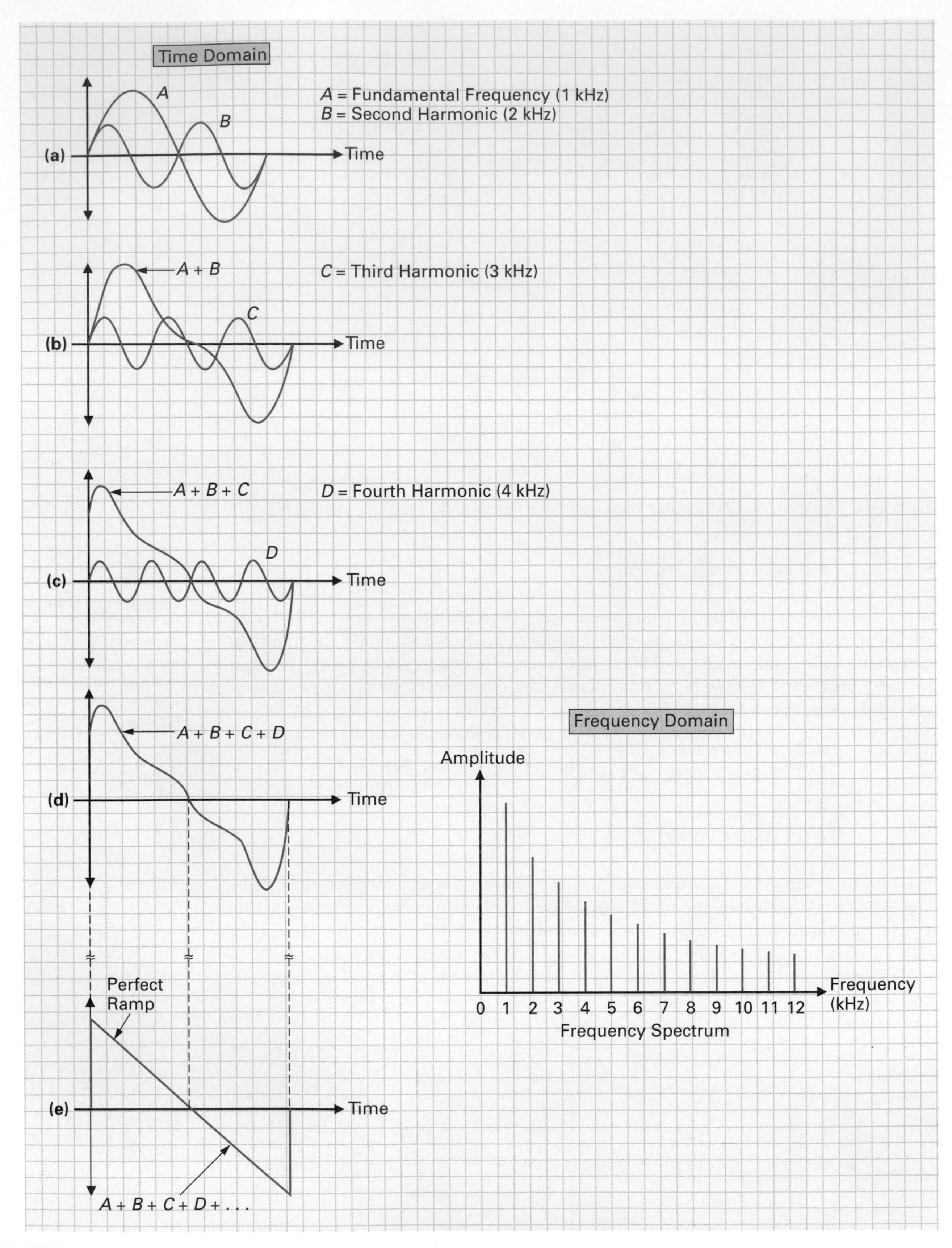

FIGURE 8-46 Analysis of a Sawtooth Wave.

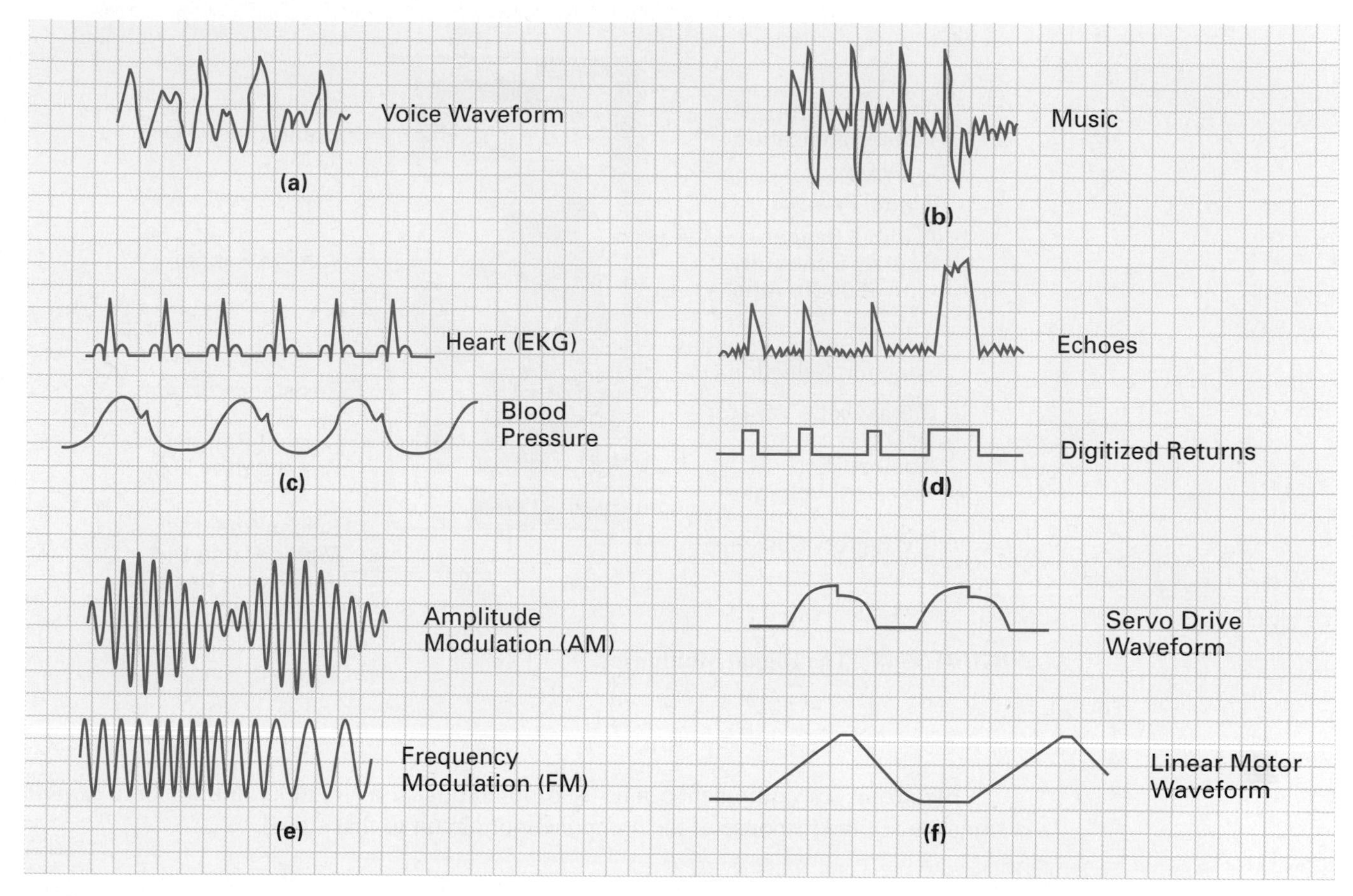

FIGURE 8-47 **Waveforms. (a) Telephone Communications. (b) Radio Broadcast. (c) Medical. (d) Radar/Sonar. (e) Communications. (f) Industrial.**

SELF-TEST EVALUATION POINT FOR SECTION 8-4

Now that you have completed this section, you should be able to:

Objective 13. Explain the different characteristics of the five basic wave shapes.

Objective 14. Describe fundamental and harmonic frequencies.

Use the following questions to test your understanding of Section 8-4:

1. Sketch the following waveforms:
   a. Sine wave
   b. Square wave
   c. Rectangular wave
   d. Triangular wave
   e. Sawtooth wave
2. An oscilloscope gives a __________ -domain representation of a periodic wave; a spectrum analyzer gives a __________ -domain representation.
3. What are the wavelength formulas for sound waves and electromagnetic waves, and why are they different?
4. A square wave is composed of an infinite number of __________ harmonics.

8-5 MEASURING AND GENERATING AC SIGNALS

As a technician or engineer, you are going to be required to diagnose and repair a problem in the shortest amount of time possible. To improve the efficiency of this fault-finding and repair process, you can make use of certain pieces of test equipment. Humans have five kinds of sensory systems: touch, taste, sight, sound, and smell. Four can be used for electronic troubleshooting: sight, sound, touch, and smell.

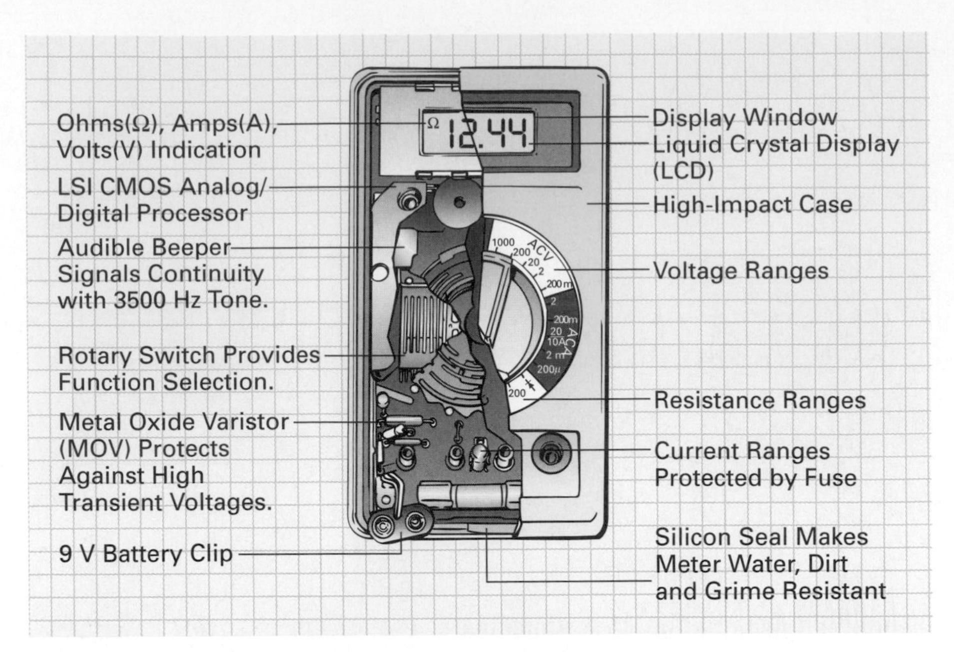

FIGURE 8-48 **The Digital Multimeter.**

Electronic test equipment can be used either to *sense* a circuit's condition or to *generate* a signal to see the response of the component or circuit to that signal.

8-5-1 *The AC Meter*

Rectifier
A device that converts alternating current into a unidirectional or dc current.

Up to this point, we have seen how a multimeter, like the one shown in Figure 8-48, can be used to measure direct current and voltage. Most multimeters can be used to measure either dc or ac. When the technician or engineer wishes to measure ac, the ac current or voltage is converted to dc internally by a circuit known as a **rectifier,** as shown in Figure 8-49, before passing on to the meter.

The dc produced by the rectification process is in fact pulsating, as shown in Figure 8-50, so the current through the meter is a series of pulses rising from zero to maximum (peak) and from maximum back to zero. Frequencies below 10 Hz (lower frequency limit) will cause the digits on a digital multimeter to continually increase in value and then decrease in value as the meter follows the pulsating dc. This makes it difficult to read the meter. This effect will also occur with an analog multimeter, which, from 10 Hz to approximately 2 kHz, will not be able to follow the fluctuation, causing the needle to remain in a position equal to the average value of the pulsating dc from the rectifier (0.637 of peak). Most meters are normally calibrated internally to indicate rms values (0.707 of peak) rather than average values, because this effective value is most commonly used when expressing ac voltage or current. The upper frequency limit of the ac meter is approximately 2 to 8 kHz. Beyond this limit the meter becomes progressively inaccurate due to the reactance of the capacitance in the rectifier. This reactance, which will be discussed later, will result in inaccurate indications due to the change in opposition at different ac input frequencies.

Current Clamp
A device used in conjunction with an ac ammeter, containing a magnetic core in the form of hinged jaws that can be snapped around the current-carrying wire.

Current Clamps

The voltmeter is probably the most frequently used setting on the multimeter. A meter reading can be obtained by just connecting the probes across the component or source to be measured, unlike the ammeter, which requires that the circuit current path be opened and the ammeter inserted in the path of current flow. If a current measurement is required, a clamp

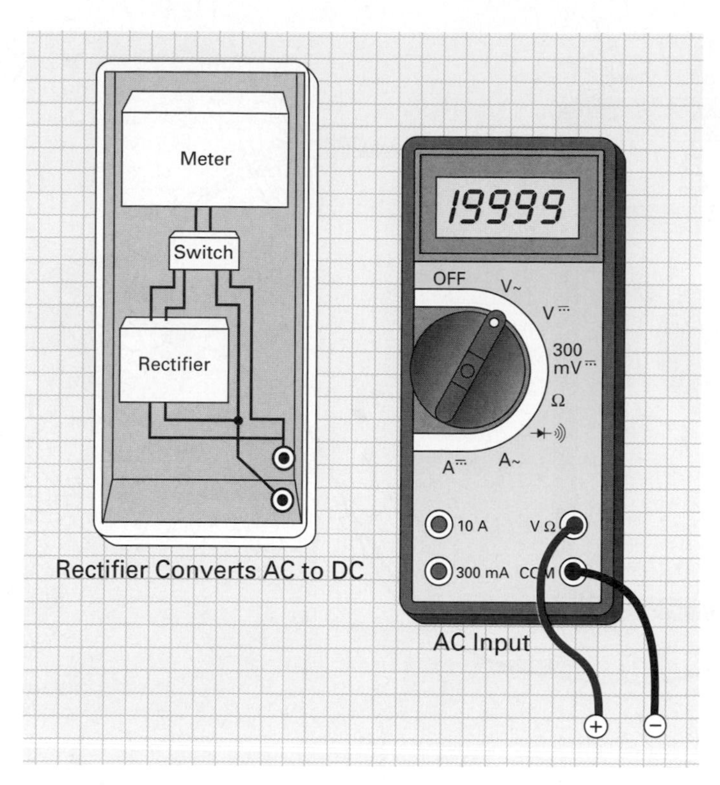

FIGURE 8-49 **AC Meter Uses a Rectifier to Produce DC.**

can be used, as seen in Figure 8-51, which allows us to sense the amount of current flow through the conductor without opening the current path.

The alternating current flowing through the conductor produces an expanding and collapsing magnetic field, which cuts across the coil of wire wound around the core of the clamp and induces an alternating voltage in the coil (1 mA induces 1 mV). The induced alternating voltage causes an alternating current to flow, which is converted to dc by the rectifier and used to operate the meter. The larger the current flowing in the conductor, the larger the magnetic field surrounding the conductor, which results in a greater induced voltage, current, and consequent meter readings. These clamps are generally ineffective at measuring smaller currents (microamperes) because the magnetic field produced by the current is too weak.

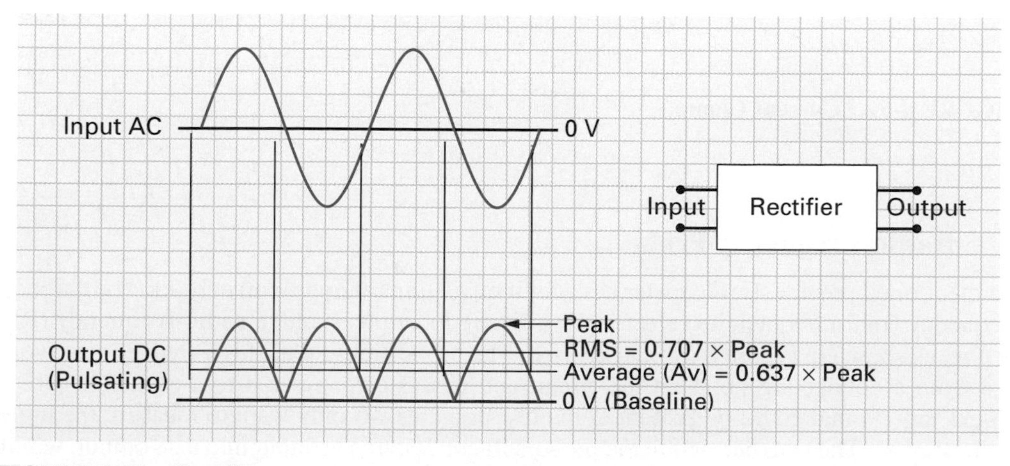

FIGURE 8-50 **Rectifier.**

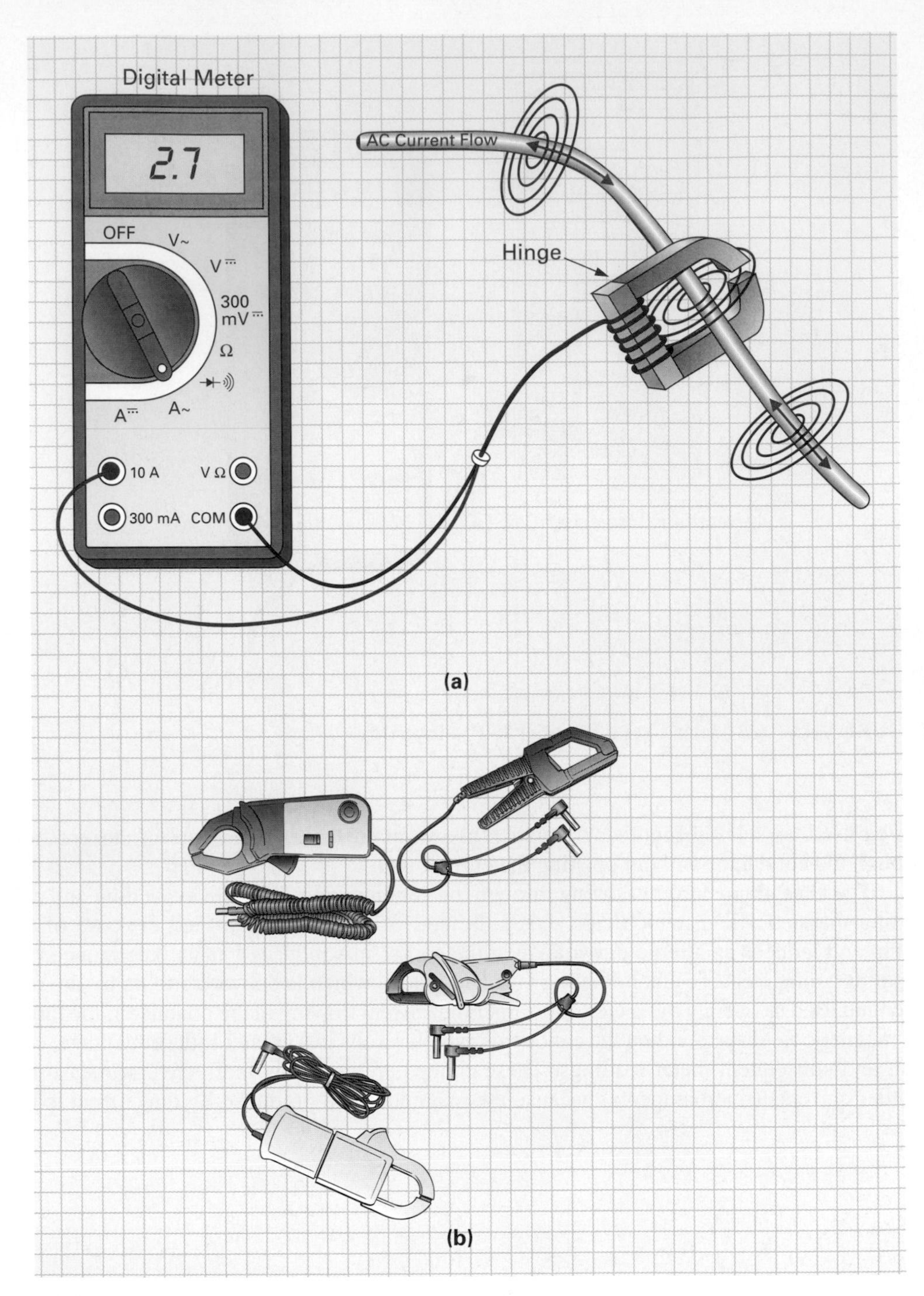

FIGURE 8-51 Current Clamp.

Radio-Frequency Probe

Radio-Frequency Probe
A probe used in conjunction with an ac meter to measure high-frequency RF signals.

As mentioned previously, the meter has a high range limit of approximately 2 kHz. If higher-frequency (radio frequencies) electrical waves are to be measured, a **radio-frequency** (RF) **probe,** as shown in Figure 8-52, can be used. The probe picks up the high-frequency ac voltage from a conducting point on the circuit and passes or couples it to a capacitor, which blocks any dc that could be present at the test point, as we only want to measure the high-frequency ac. The rectifier within the probe will convert this ac input into a dc output, which will be displayed as an rms value on the meter display.

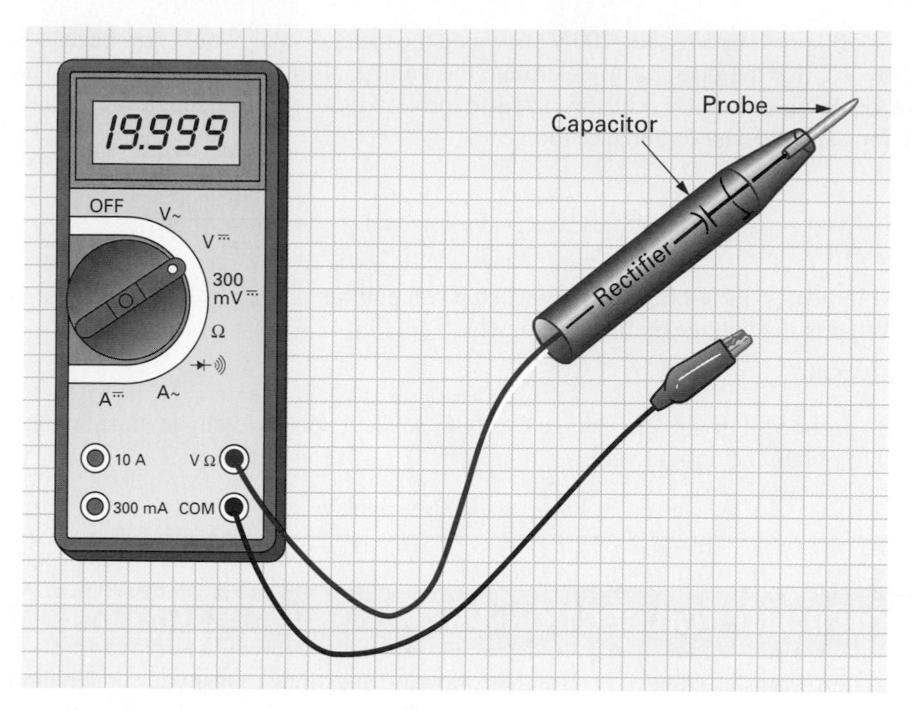

FIGURE 8-52 **Radio-Frequency Probe.**

High-Voltage Probe

The typical multimeter can handle voltages up to approximately 1000 V. If you wish to measure voltages higher than this, another component, known as a **high-voltage** (HV) **probe**, such as the one shown in Figure 8-53, can be used. The high-voltage probe has additional multiplier resistors to drop the extra voltage. Most high-voltage probes are designed so that $\frac{1}{100}$ of voltage at the probe tip from the test point will be applied to the meter. For example, if 10 kV is being measured, 100 V will be applied to the meter ($\frac{1}{100} \times 10{,}000\ \text{V} = 100\ \text{V}$). This probe would be

High-Voltage Probe
Accessory to the voltmeter that has added multiplier resistors within the probe to divide up the large potential being measured by the probe.

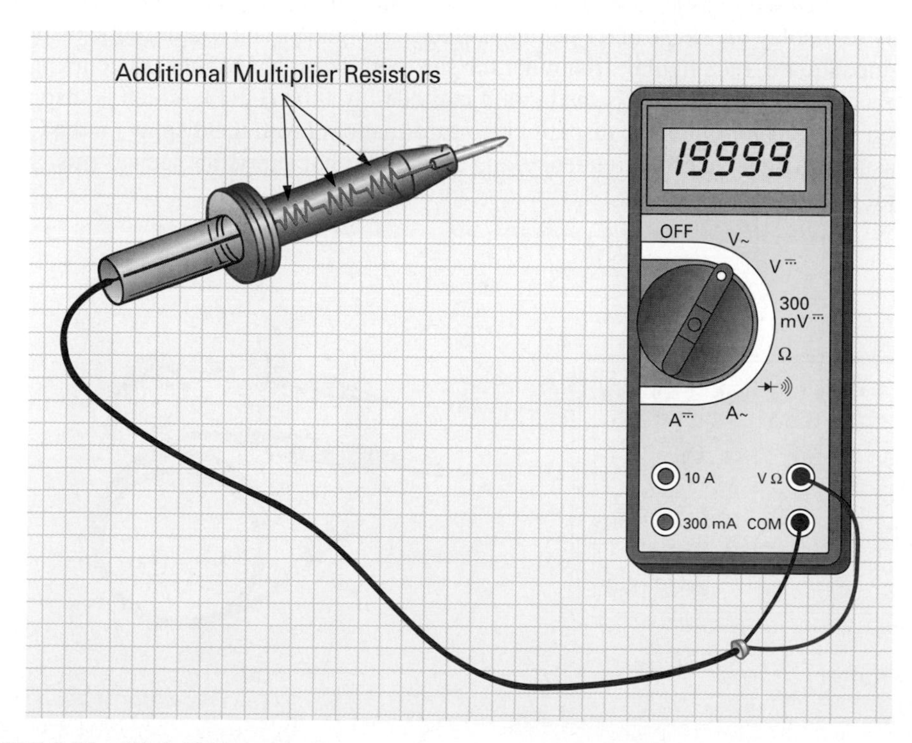

FIGURE 8-53 **High-Voltage Probe.**

called a × 100 probe because the 100 V shown on the meter display would now have to be multiplied by 100 for the operator to determine the voltage (100 V × 100 = 10 kV). The high-voltage probe is especially well insulated to protect its user, who should apply all safety precautions and exercise extreme caution.

EXAMPLE:

A DMM indicates 3.9 V on its display when a test point is probed by a × 1000 high-voltage probe. What is the voltage at this test point?

Solution:

A × 1000 probe will divide the voltage by 1000, so the displayed voltage must be multiplied by 1000 to obtain the correct value:

$$3.9\text{ V} \times 1000 = 3.9\text{ kV}$$

SELF-TEST EVALUATION POINT FOR SECTION 8-5-1

Now that you have completed this section, you should be able to:

Objective 15. Explain the operation of the ac multimeter, and the following special attachments:

a. Current clamp
b. Radio-frequency probe
c. High-voltage probe

Use the following questions to test your understanding of Section 8-5-1:

1. True or false: The current clamp does not require the circuit current path to be broken in order to measure current. Can it be used to measure dc amps?
2. What is the difference between an RF and an HV probe?

8-5-2 *The Oscilloscope*

Oscilloscope
Instrument used to view signal amplitude, frequency, and shape at different points throughout a circuit.

Figure 8-54 illustrates a typical **oscilloscope** (sometimes abbreviated to *scope*), which is used primarily to display the shape and spacing of electrical signals. The oscilloscope displays the actual sine, square, rectangular, triangular, or sawtooth wave shape that is occurring at any point in a circuit. This display is made on a cathode-ray tube (CRT), which is also used in television sets and computers to display video information. From the display on the CRT, we can measure or calculate time, frequency, and amplitude characteristics such as rms, average, peak, and peak-to-peak.

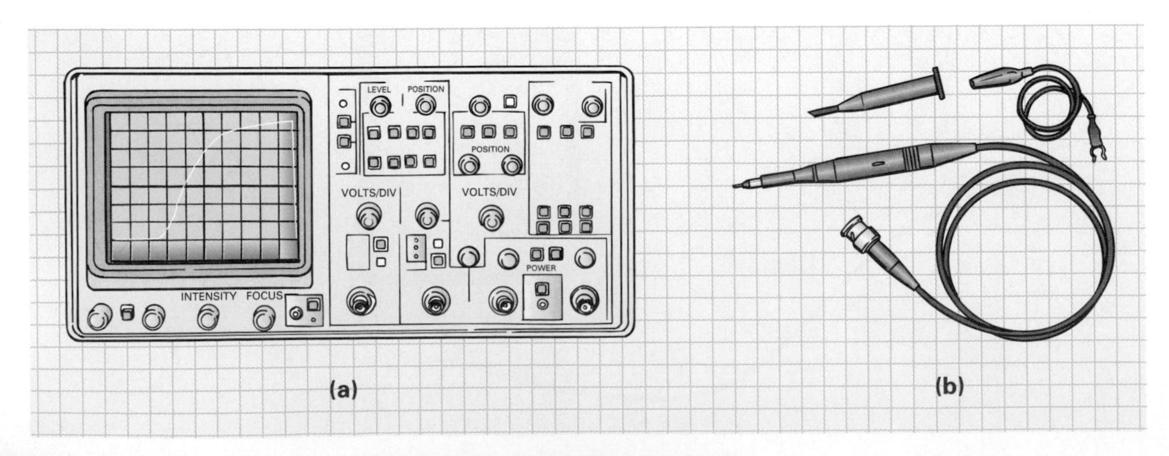

FIGURE 8-54 **Typical Oscilloscope. (a) Oscilloscope. (b) Oscilloscope Probe.**

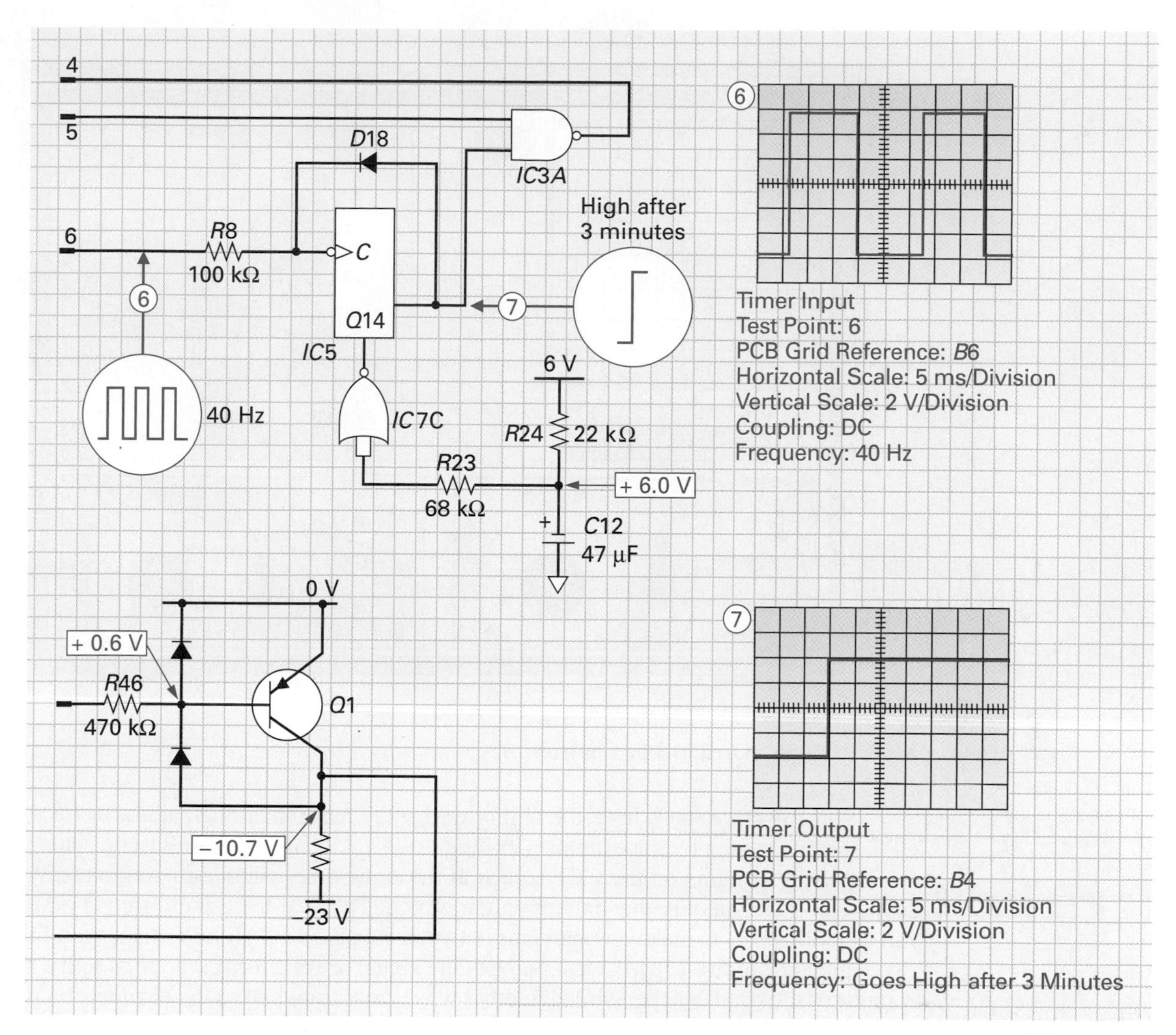

FIGURE 8-55 Schematic Diagram with Voltage and Waveform Test Points.

The oscilloscope allows us to see what is happening at every point through a circuit. In Figure 8-55 you can see the different waveforms at different points on the circuit. There are also voltage test points that can be tested with a voltmeter if a scope is not available. A voltmeter, as you can well imagine, does not supply the technician or engineer with as much information as the oscilloscope.

Controls

Oscilloscopes come with a wide variety of features and functions, but the basic operational features are almost identical. Figure 8-56 illustrates the front panel of a typical oscilloscope. Some of these controls are difficult to understand without practice and experience, so practical experimentation is essential if you hope to gain a clear understanding of how to operate an oscilloscope.

General controls (see Figure 8-56)
Intensity control: Controls the brightness of the trace, which is the pattern produced on the screen of a CRT.
Focus control: Used to focus the trace.
Power OFF/ON: Switch will turn on oscilloscope while indicator shows when oscilloscope is turned on.

TIME LINE

The father of television, Viadymir Zworykin (1889–1982), developed the first television picture tube, called the kinescope, in 1920.

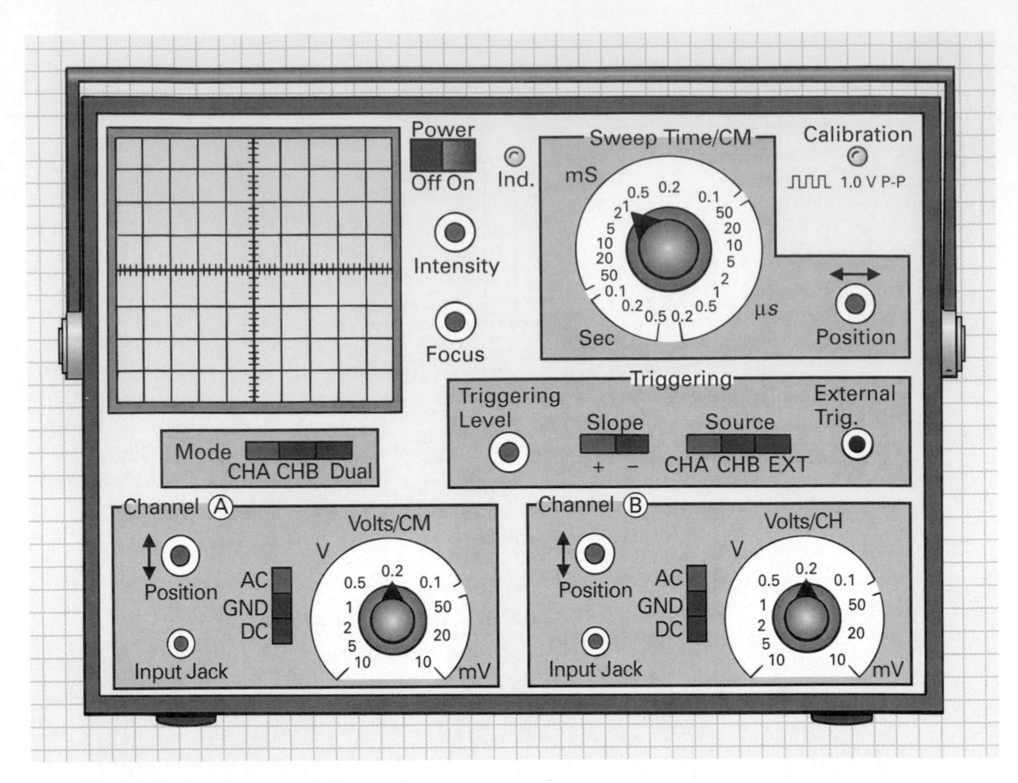

FIGURE 8-56 Oscilloscope Controls.

Some oscilloscopes have the ability to display more than one pattern or trace on the CRT screen, as seen by the examples in Figure 8-57. A dual-trace oscilloscope can produce two traces or patterns on the CRT screen at the same time, whereas a single-trace oscilloscope can trace out only one pattern on the screen. The dual-trace oscilloscope is very useful, as it allows us to make comparisons between the phase, amplitude, shape, and timing of two signals from two separate test points. One signal or waveform is applied to the channel *A* input of the oscilloscope, while the other waveform is applied to the channel *B* input.

Channel selection (see Figure 8-56)
Mode switch: This switch allows us to select which channel input should be displayed on the CRT screen.
CHA: The input arriving at channel *A*'s jack is displayed on the screen as a single trace.
CHB: The input arriving at channel *B*'s jack is displayed on the screen as a single trace.
Dual: The inputs arriving at jacks *A* and *B* are both displayed on the screen, as a dual trace.

Calibration
To determine, by measurement or comparison with a standard, the correct value of each scale reading.

(a) **Calibration Output** This output connection provides a point where a fixed 1 V peak-to-peak square-wave signal can be obtained at a frequency of 1 kHz. This signal is normally fed into either channel *A* or *B*'s input to test probes and the oscilloscope operation.

(b) **Channel *A* and *B* Horizontal Controls**

↔ *Position control:* This control will move the position of the one (single-trace) or two (dual-trace) waveforms horizontally (left or right) on the CRT screen.

Sweep time/cm switch: The oscilloscope contains circuits that produce a beam of light that is swept continually from the left to the right of the CRT screen. When no input signal is applied, this sweep will produce a straight horizontal line in the center of the screen. When an input signal is present, this horizontal sweep is influenced by the input signal, which moves it up and down to produce a pattern on the CRT screen that is the same as the input pattern (sine, square, sawtooth, and so on). This sweep time/cm switch selects the speed of the sweep from left to right, and it can be either fast (0.2 microseconds per centimeter; 0.2 μs/cm) or slow (0.5 second per centimeter; 0.5 s/cm). A

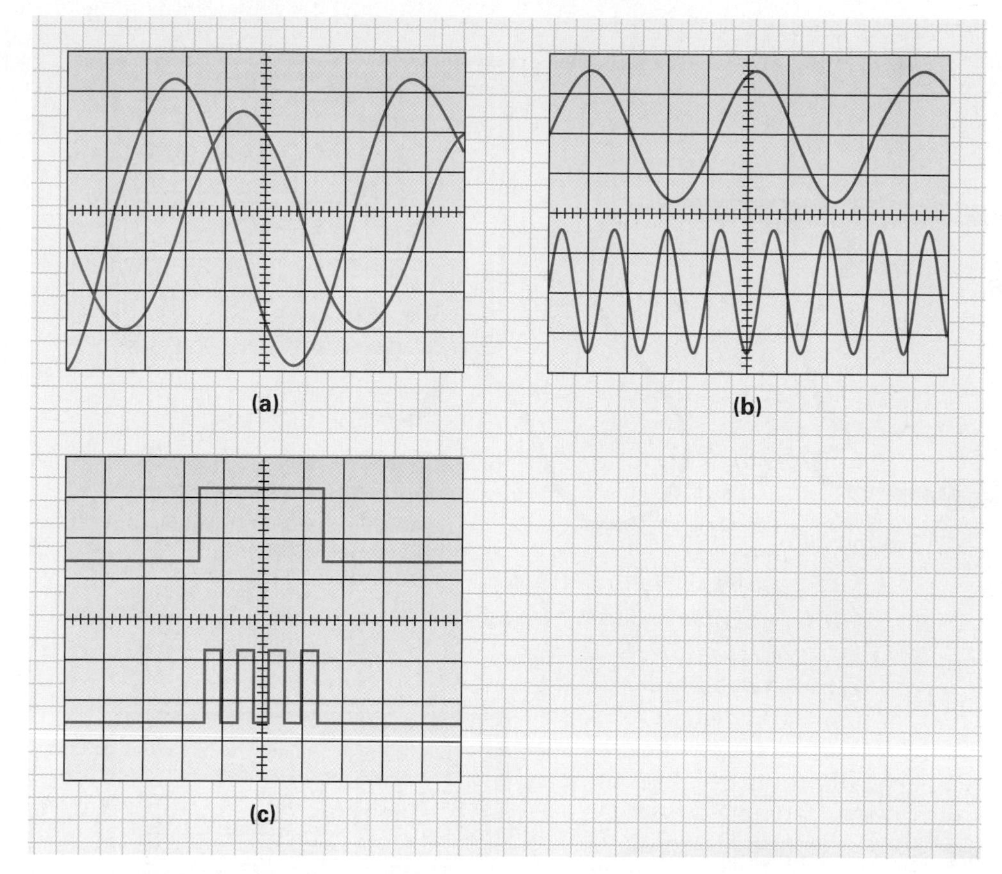

FIGURE 8-57 **Sample of Dual-Trace Oscilloscope Displays for Comparison.**

low-frequency input signal (long cycle time or period) will require a long time setting (0.5 s/cm) so that the sweep can capture and display one or more cycles of the input. A number of settings are available, with lower time settings displaying fewer cycles and higher time settings showing more cycles of an input.

(c) **Triggering Controls** These provide the internal timing control between the sweep across the screen and the input waveform.

Triggering
Initiation of an action in a circuit which then functions for a predetermined time, for example, the duration of one sweep in a cathode-ray tube.

Triggering level control: This determines the point where the sweep starts.

Slope switch (+): Sweep is triggered on positive-going slope.
(−): Sweep is triggered on negative-going slope.

Source switch, CHA: The input arriving at the channel *A* jack triggers the sweep.
CHB: The input arriving at the channel *B* jack triggers the sweep.
EXT: The signal arriving at the external trigger jack is used to trigger the sweep.

(d) **Channel *A* and *B* Vertical Controls** The *A* and *B* channel controls are identical.

Volts/cm switch: This switch sets the number of volts to be displayed by each major division on the vertical scale of the screen.

↕ *Position control:* Moves the trace up or down for easy measurement or viewing.

AC-DC-GND *switch:* In the AC position, a capacitor on the input will pass the ac component entering the input jack, but block any dc components.
In the GND position, the input is grounded (0 V) so that the operator can establish a reference.
In the DC position, both ac and dc components are allowed to pass on to and be displayed on the screen.

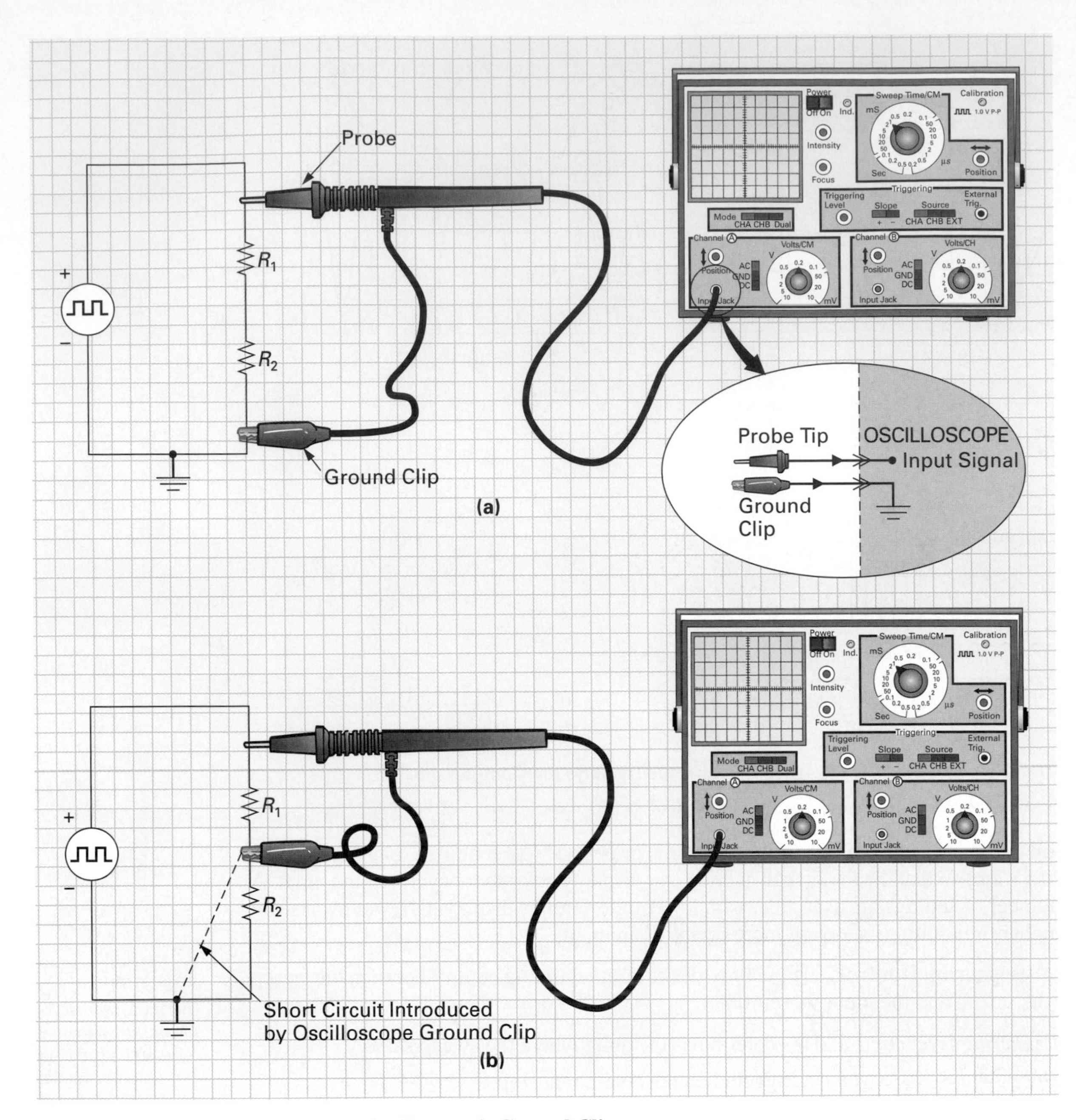

FIGURE 8-58 Correctly Connecting the Oscilloscope's Ground Clip.

The Oscilloscope's Ground Clip

Before you probe a circuit, you must first connect the probe's ground clip to a point on the circuit that you are sure is a circuit ground, as shown in Figure 8-58(a). It is important that you double-check your circuit ground point before you connect the ground clip, as this lead is connected to ground through the oscilloscope, as shown in the inset in Figure 8-58(a). Your measurements will now be accurate, and relative to the circuit ground.

If the ground clip is incorrectly connected to a point other than circuit ground, as shown in Figure 8-58(b), a short could be introduced. In this example, the ground connection within the oscilloscope has grounded the top of R_2 through the ground clip. Shorting out R_z will lower the circuit's resistance, and in most instances, a short circuit will cause a chain reaction of circuit damage.

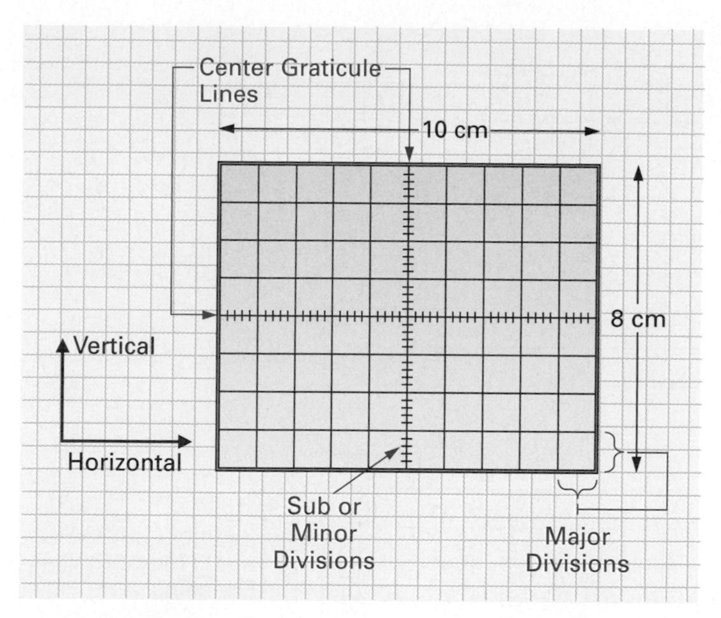

FIGURE 8-59 Oscilloscope Grid.

Measurements

The oscilloscope is probably the most versatile of test equipment, as it can be used to test dc voltage, ac voltage, waveform duration, waveform frequency, and waveform shape.

(a) **Voltage Measurement** The screen is divided into eight vertical and ten horizontal divisions, as shown in Figure 8-59. This 8 × 10 cm grid is called the *graticule*. Every vertical division has a value depending on the setting of the volts/cm control. For example, if the volts/cm control is set to 5 V/div or 5 V/cm, the waveform shown in Figure 8-60(a), which rises up four major divisions, will have a peak positive alternation value of 20 V (4 div × 5 V/div = 20 V).

As another example, look at the positive alternation in Figure 8-60(b). The positive alternation rises up three major divisions and then extends another three subdivisions, which are each equal to 1 V because five subdivisions exist within one major division, and one major division is, in this example, equal to 5 V. The positive alternation shown in Figure 8-60(b) therefore has a peak of three major divisions (3 × 5 V/cm = 15 V), plus three subdivisions (3 × 1 V = 3 V), which equals 18 volts peak.

In Figure 8-60(c), we have selected the 10 volt/cm position, which means that each major division is equal to 10 V and each subdivision is equal to 2 V. In this example, the waveform peak will be equal to two major divisions (2 × 10 V = 20 V), plus four subdivisions (4 × 2 V = 8 V), which is equal to 28 V. Once the peak value of a sine wave is known, the peak-to-peak, average, and rms can be calculated mathematically.

When measuring a dc voltage with the oscilloscope, the volts/cm is applied in the same way, as shown in Figure 8-61. A positive dc voltage in this situation will cause deflection toward the top of the screen, whereas a negative voltage will cause deflection toward the bottom of the screen. To determine the dc voltage, count the number of major divisions and to this add the number of minor divisions. In Figure 8-61, a major division equals 1 V/cm and, therefore, a minor division equals 0.2 V/cm, so the dc voltage being measured is interpreted as +2.6 V.

(b) **Time and Frequency Measurement** The frequency of an alternating wave, such as that seen in Figure 8-62(a), is inversely proportional to the amount of time it takes to complete one cycle ($f = 1/t$). Consequently, if time can be measured, frequency can be determined.

The time/cm control relates to the horizontal line on the oscilloscope graticule and is used to determine the period of a cycle so that frequency can be calculated. For example, in Figure 8-62(b), a cycle lasts five major horizontal divisions, and since the 20 μs/division setting has

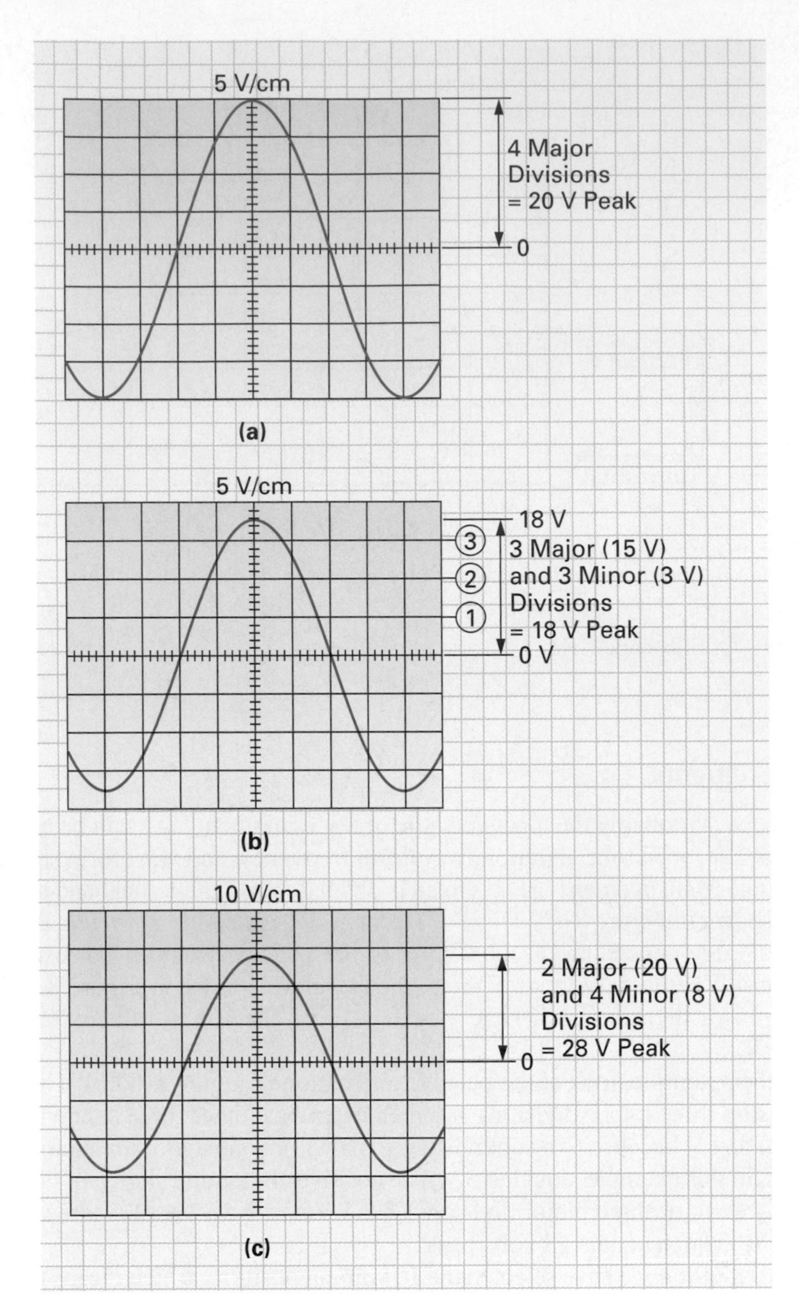

FIGURE 8-60 Measuring AC Peak Voltage.

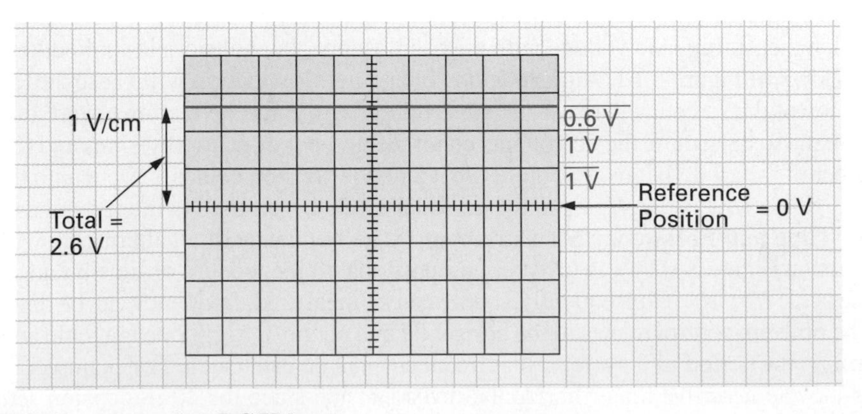

FIGURE 8-61 Measuring DC Voltage.

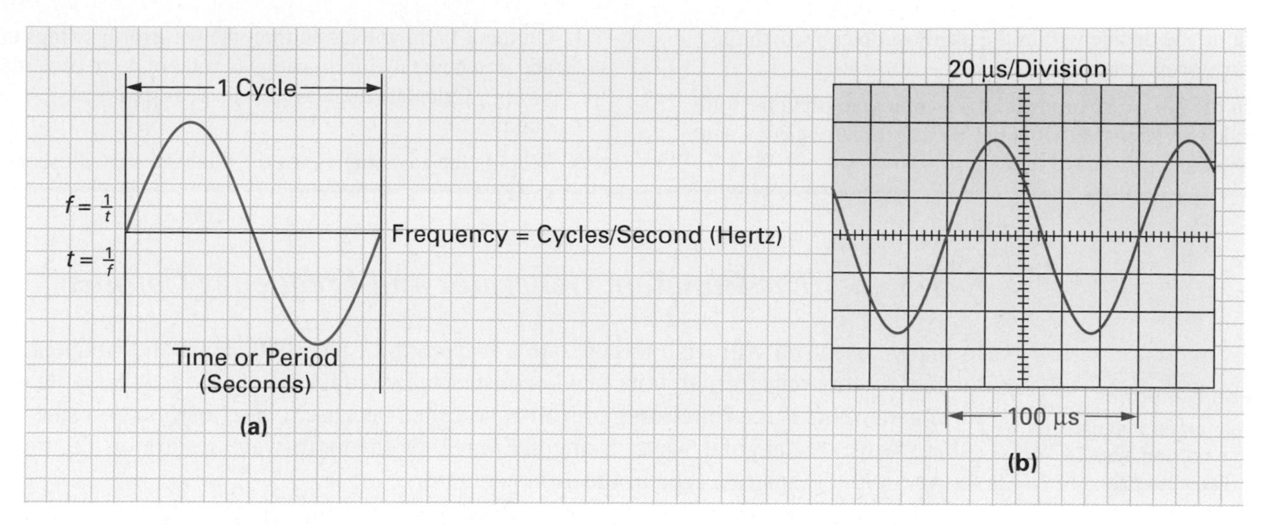

FIGURE 8-62 **Time and Frequency Measurement.**

been selected, the period of the cycle will equal $5 \times 20\ \mu s$/division $= 100\ \mu s$. If the period is equal to $100\ \mu s$, the frequency of the waveform will be equal to $f = 1/t = 1/100\ \mu s = 10$ kHz.

EXAMPLE:

A complete sine-wave cycle occupies four horizontal divisions and four vertical divisions from peak to peak. If the oscilloscope is set on 20 ms/cm and 500 mV/cm, calculate:

a. $V_{p\text{-}p}$
b. t
c. f
d. V_p
e. V_{avg} of peak
f. V_{rms}

Solution:

$$4 \text{ horizontal divisions} \times 20 \text{ ms/div} = 80 \text{ ms}$$
$$4 \text{ vertical divisions} \times 500 \text{ mV/div} = 2 \text{ V}$$

a. $V_{p\text{-}p} = 2$ V
b. $t = 80$ ms
c. $f = \dfrac{l}{t} = 12.5$ Hz
d. $V_p = 0.5 \times V_{p\text{-}p} = 1$ V
e. $V_{avg} = 0.637 \times V_p = 0.637$ V
f. $V_{rms} = 0.707 \times V_p = 0.707$ V

SELF-TEST EVALUATION POINT FOR SECTION 8-5-2

Now that you have completed this section, you should be able to:

Objective 16. Describe the operation of the oscilloscope and how it can be used to make measurements.

Use the following questions to test your understanding of Section 8-5-2:

1. The __________ is used primarily to display the shape and spacing of electrical signals.

2. Name the device within the oscilloscope on which the waveforms can be seen.
3. On the 20 μs/div time setting, a full cycle occupies four major divisions. What is the waveform's frequency and period?
4. On the 2 V/cm voltage setting, the waveform swings up and down two major divisions (a total of 4 cm of vertical swing). Calculate the waveform's peak and peak-to-peak voltage.
5. What is the advantage(s) of the dual-trace oscilloscope?

8-5-3 *The Function Generator and Frequency Counter*

Function Generator
Signal generator that can function as a sine, square, rectangular, triangular, or sawtooth waveform generator.

Frequency Counter
Meter used to measure the frequency or cycles per second of a periodic wave.

In many cases you will wish to generate a waveform of a certain shape and frequency and then apply this waveform to your newly constructed circuit. One of the most versatile waveform generators is the **function generator,** so called because it can function as a sine wave, square wave, rectangular wave, triangular wave, or sawtooth wave generator. Figure 8-63 shows a photograph of a typical function generator.

It is very important that anyone involved in the design, manufacture, and servicing of electronic equipment be able to accurately measure the frequency of a periodic wave. Without the ability to measure frequency accurately, there could be no communications, home entertainment, or a great number of other systems. A **frequency counter,** like the one shown in Figure 8-64, can be used to analyze the frequency of any periodic wave applied to its input jack and provide a readout on the display of its frequency.

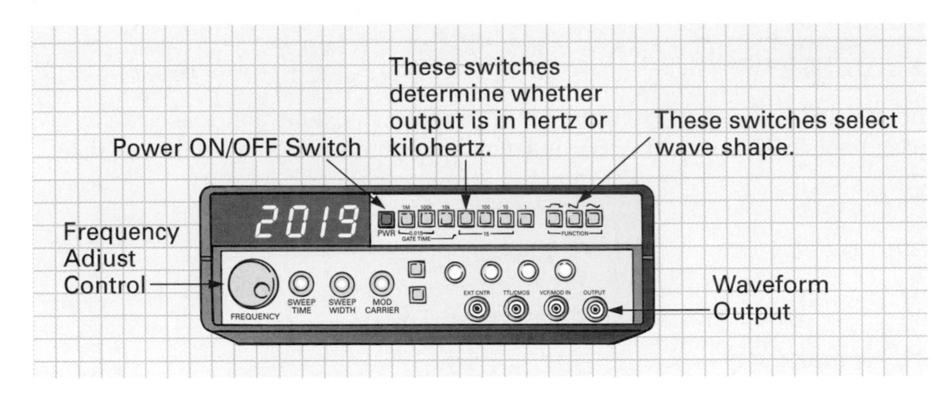

FIGURE 8-63 **Function Generator.**

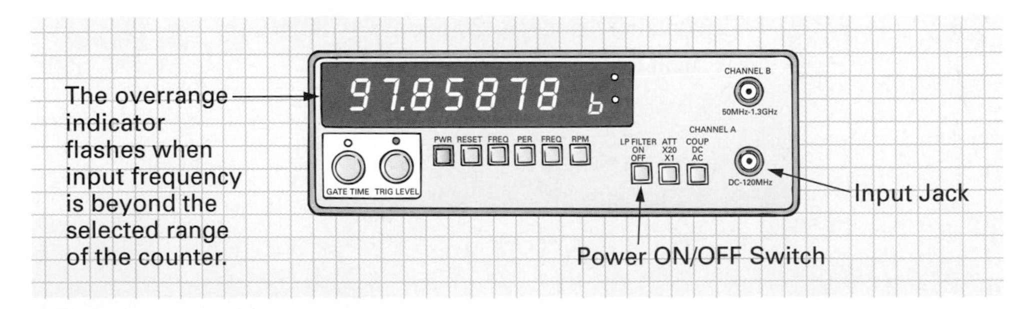

FIGURE 8-64 **Frequency Counter.**

SELF-TEST EVALUATION POINT FOR SECTION 8-5-3

Use the following questions to test your understanding of Section 8-5-3:

1. List some of the waveform shapes typically generated by a function generator.
2. What is the function of a frequency counter?

8-5-4 *Handheld Scopemeters*

The handheld **scopemeter,** shown in Figure 8-65, combines a multimeter, oscilloscope, frequency counter, and signal generator in one easy-to-carry battery-operated unit. For a technician, this test instrument is portable, easy to set up, and easy to use since it will automatically change its settings for the best operating mode and continue to adjust itself as the input changes. To explain this test instrument in more detail, let us examine some of its key functions.

Scopemeter
A handheld, battery-operated Instrument that combines a multimeter, oscilloscope, frequency counter, and signal generator in one.

Measure menu: The measurement menu button gives you direct access to a quick pop-up menu of more than 30 measurements, including: V_{rms}, V_{mean} (arithmetic average), $V_{peak\text{-}to\text{-}peak}$, frequency, time delay, rise time, phase, current, and so on. Once the measurement is selected, the scopemeter automatically sets itself up and takes the measurement, as seen in the example in Figure 8-66(a).

Auto set: As you move from test point to test point, the scopemeter will handle the changing inputs automatically in this continuous autoset mode. Each time the signal input changes, the scopemeter will automatically search for the best trigger level, time base (sweep time), and range scale to speed up measurements and reduce errors, as seen in Figure 8-66(b).

Min max trendplot: This function is used to display and record up to 40 days of signal trend, which is sometimes needed in order to pinpoint intermittent problems that occur randomly. Figure 8-66(c) shows an example of how the display shows minimum, maximum, and average readings.

Save: Most scopemeters have large internal memories for saving screen images, setups, and waveforms from the field. These stored measurements can be recalled at any time, downloaded or sent to a personal computer for reports, or sent directly to a printer. Figure 8-66(d) shows how these waveforms and setups would appear on a computer screen.

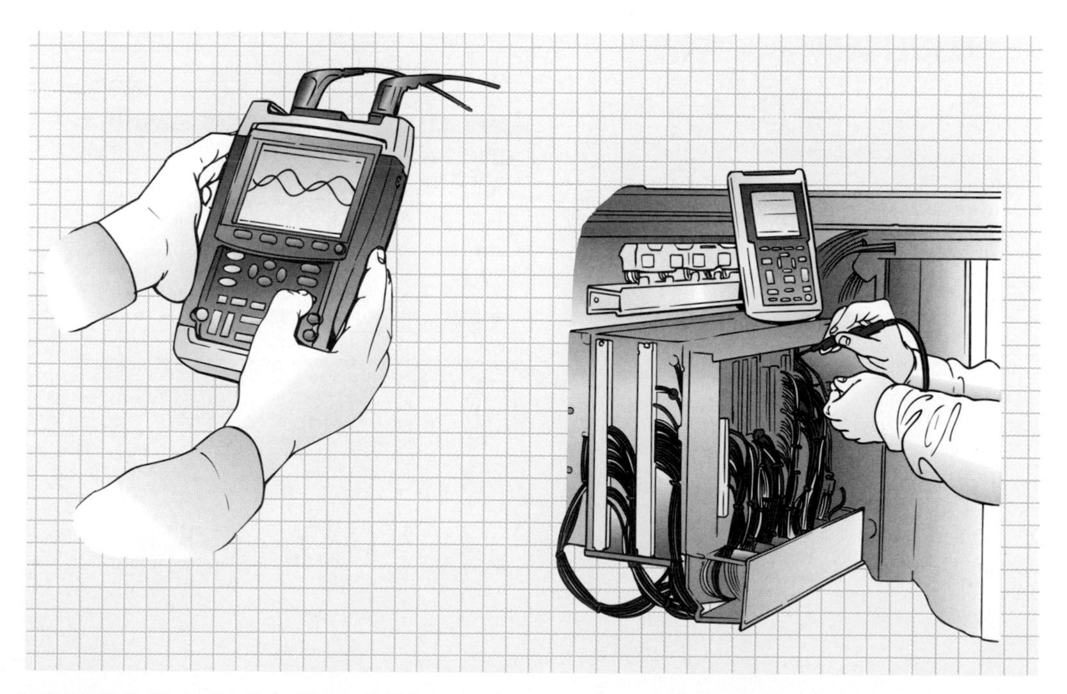

FIGURE 8-65 The Handheld Scopemeter.

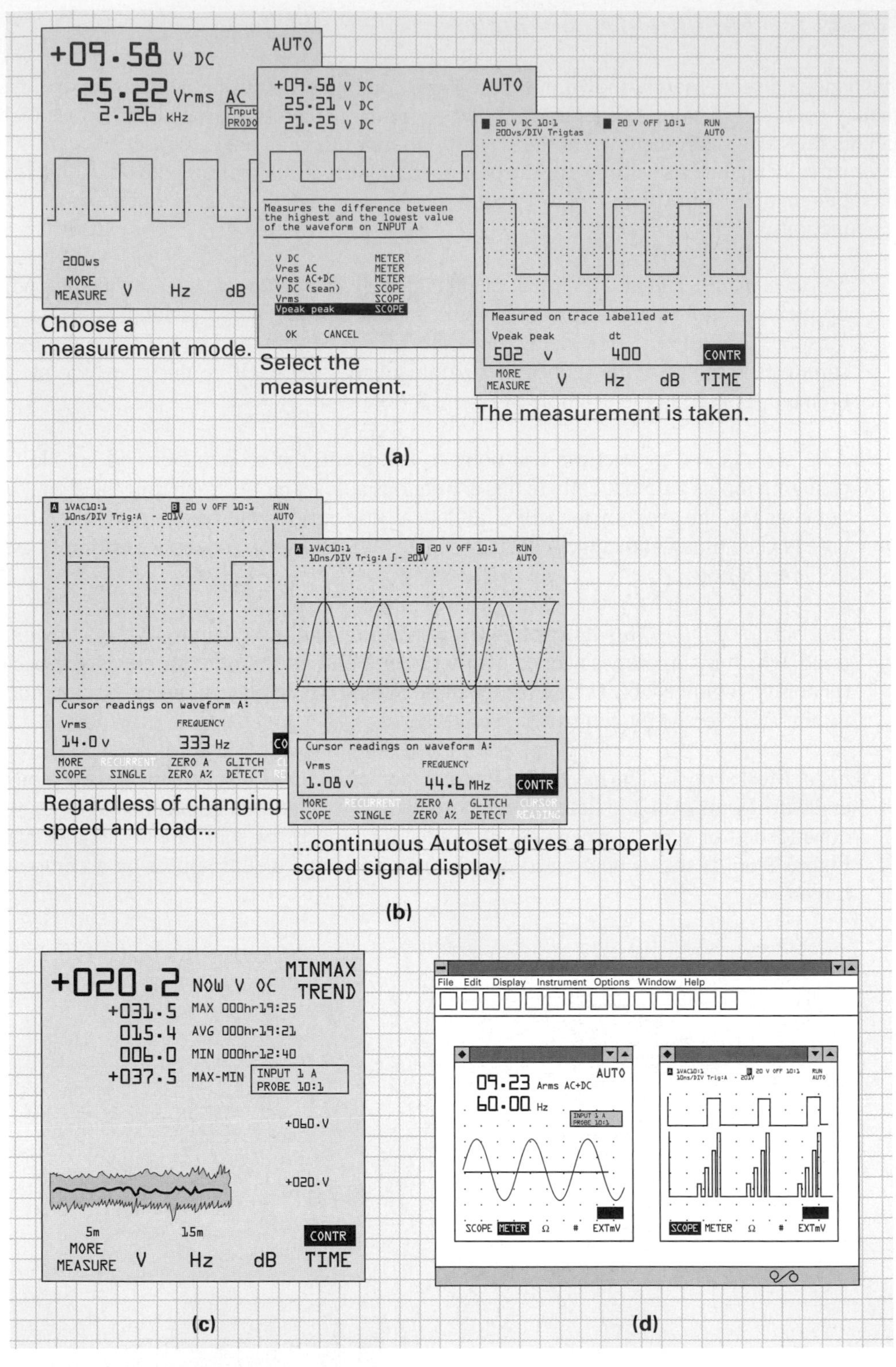

FIGURE 8-66 Scopemeter Operation.

SELF-TEST EVALUATION POINT FOR SECTION 8-5-4

Now that you have completed this section, you should be able to:

Objective 17. Describe the purpose and operation of the

a. Function generator
b. Frequency counter
c. Handheld scopemeter

Use the following questions to test your understanding of Section 8-5-4:

1. What is a scopemeter?
2. What advantages does the scopemeter have over the previously discussed instruments?

SUMMARY

Mini-Math Review—Trigonometry (Section 8-1)

1. Trigonometry is the study of the properties of triangles, trigonometric functions, and their applications.
2. A right-angle triangle is a three-sided figure having a 90° angle.
3. The Pythagorean theorem states that the square of the length of the hypotenuse of a right triangle equals the sum of the squares of the lengths of the other two sides.
4. A vector or phasor is an arrow used to represent the magnitude and direction of a quantity.
5. Side C of a right triangle is always the longest side. Side B, called the adjacent side, always extends between the hypotenuse and the vertical side. An angle called theta (θ) is formed between the hypotenuse and the adjacent side. The side opposite angle θ is called the opposite side.
6. If the angle θ is known along with the length of one side, or if angle θ is needed and the lengths of the two sides are known, one of three formulas can be chosen based on what variables are known and what is needed:

$$\sin\theta = \frac{O}{H} \qquad \cos\theta = \frac{A}{H} \qquad \tan\theta = \frac{O}{A}$$

The Difference Between DC and AC (Section 8-2)

7. Although direct current (dc) can be used in many instances to deliver power or represent information, there are certain instances in which ac is preferred. For example, ac is easier to generate and is more efficiently transmitted as a source of electrical power for the home or business. Audio (speech and music) and video (picture) information are generally always represented in electronic equipment as an alternating current or alternating voltage signal.
8. Direct current (dc) is the flow of electrons in one DIRECTion and one direction only. DC voltage is nonvarying and normally obtained from a battery or power supply unit. The only variation in voltage from a battery occurs due to the battery's discharge, but even then, the current will still flow in only one direction.
9. Some power supplies supply a form of dc known as pulsating dc, which varies periodically from zero to a maximum, back to zero, and then repeats. Pulsating dc is normally supplied by a battery charger and is used to charge secondary batteries. It is also used to operate motors that convert the pulsating dc electrical energy into a mechanical rotation output. Whether steady or pulsating, direct current is current in only one DIRECTion.
10. Alternating current (ac) flows first in one direction and then in the opposite direction. This reversing current is produced by an alternating voltage source, which reaches a maximum in one direction (positive), decreases to zero, and then reverses itself and reaches a maximum in the opposite direction (negative). During the time of the positive voltage alternation, the polarity of the voltage will cause current to flow from negative to positive in one direction. During the time of the negative voltage alternation, the polarity of the voltage will reverse, causing current to flow once again from negative to positive, but in this case, in the opposite direction.

Why Alternating Current? (Section 8-3)

11. There are two main applications for ac:
    - **a.** *Power transfer:* to supply electrical power for lighting, heating, cooling, appliances, and machinery in both home and industry.
    - **b.** *Information transfer:* to communicate or carry information, such as radio music and television pictures, between two points.
12. There are three advantages that ac has over dc from a power point of view:
    - **a.** Generators can be used to generate either dc or ac, but ac generators can be larger, less complex internally, and cheaper to operate, which is the first reason why we use ac instead of dc for supplying power.
    - **b.** Power companies transport electric energy at a very high voltage, between 200,000 and 600,000 V. Since the voltage is high, the current can be low and provide the same amount of power to the consumer. An ac voltage can easily and efficiently be transformed up or down to a higher or lower voltage by utilizing a device known as a transformer, and even though dc voltages can be stepped up and down, the method is inefficient and more complex.
    - **c.** It is a relatively simple process to convert ac to dc, but conversion from dc to ac is a complex and comparatively inefficient process.

13. The ac power distribution system begins at the electric power plant, which has powerful large generators driven by turbines to generate large ac voltages.
14. The turbines can be driven by either falling water (hydroelectric) or steam, which is produced with intense heat by burning either coal, gas, or oil or from a nuclear reactor (thermoelectric).
15. The turbine supplies the mechanical energy to the generator to be transformed into ac electrical energy.
16. The generator generates an ac voltage of approximately 22,000 V, which is stepped up by transformers to approximately 500,000 V.
17. At each city or town, the voltage is tapped off the long-distance transmission lines and stepped down to approximately 66,000 V and is distributed to large-scale industrial customers.
18. The 66,000 V is stepped down again to approximately 4800 V and distributed throughout the city or town by short-distance transmission lines. This 4800 V is used by small-scale industrial customers and residential customers, who receive the ac power via step-down transformers on utility poles, which step down the 4800 V to 240 V and 120 V.
19. A large amount of equipment and devices within industry and the home will run directly from the ac power, such as heating, lighting, and cooling.
20. Some equipment that runs on dc, such as televisions and computers, will accept the 120 V ac and internally convert it to the dc voltages required.
21. Information, by definition, is the property of a signal or message that conveys something meaningful to the recipient.
22. Communication, which is the transfer of information between two points, began with speech and progressed to handwritten words in letters and printed words in newspapers and books. To achieve communication at greater distances, face-to-face communications evolved into telephone and radio communications.
23. Sound waves or sounds are normally generated by a vibrating reed, a plucked string, or a person's vocal cords.
24. Like the ripples produced by a stone falling in a pond, sound waves are constantly expanding and traveling outward.
25. The microphone is in fact a transducer (energy converter), because it converts a sound wave (which is a form of mechanical energy) into electrical energy in the form of voltage and current, which varies in the same manner as the sound wave and therefore contains the sender's message or information.
26. The electrical wave is a variation in voltage or current and can only exist in a wire conductor or circuit. This electrical signal travels at the speed of light.
27. The speaker, like the microphone, is also an electroacoustical transducer that converts the electrical energy input into a mechanical sound-wave output.
28. To communicate between two distant points, a wire must be connected between the microphone and speaker. However, if an electrical wave is applied to an antenna, the electrical wave is converted into a radio or electromagnetic wave.
29. Antennas are designed to radiate and receive electromagnetic waves, which vary in field strength and can exist in either air or space.
30. Radio waves are composed of two basic components. The electrical voltage applied to the antenna is converted into an electric field and the electrical current into a magnetic field.
31. AC is basically used in two applications: (1) power transfer and (2) information transfer.
32. Electronic equipment manages the flow of information, while electrical equipment manages the flow of power.
33. A dc power supply is a piece of electrical equipment since it is designed to manage the flow of power. A TV set, however, is an electronic system since its electronic circuits are designed to manage the flow of audio (sound) and video (picture) information.

AC Wave Shapes (Section 8-4)

34. The sine wave is the most common type of waveform. It is the natural output of a generator. If this ac voltage is applied across a closed circuit, it produces a current that continually reverses or alternates in each direction.
35. The amplitude or magnitude of a wave is often represented by a vector arrow. The vector's length indicates the magnitude of the current or voltage, and the arrow's point is used to show the direction, or polarity.
36. The peak of an ac wave occurs on both the positive and negative alternation but is only at the peak (maximum) for an instant.
37. The peak-to-peak value of a sine wave is the value of voltage or current between the positive and negative maximum values of a waveform.
38. Both the positive and negative alternation of a sine wave can accomplish the same amount of work, but the ac waveform is only at its maximum value for an instant in time, spending most of its time between peak currents. The effective value of a sine wave is equal to 0.707 of the peak value.
39. This root-mean-square (rms) result of 0.707 can always be used to tell us how effective an ac sine wave will be.
40. Unless otherwise stated, ac values of voltage or current are always given in rms.
41. The average of a positive or negative alternation (half-cycle) is equal to $0.637 \times$ peak.
42. The period (t) is the time required for one complete cycle (positive and negative alternation) of the sinusoidal current or voltage waveform. A *cycle,* by definition, is the change of an alternating wave from zero to a positive peak, to zero, then to a negative peak, and finally back to zero.
43. Frequency is the number of repetitions of a periodic wave in a unit of time. It is symbolized by f and is given the unit hertz (cycles per second), in honor of a German physicist, Heinrich Hertz.
44. Sinusoidal waves can take a long or a short amount of time to complete one cycle. This time is related to frequency, in that period is equal to the reciprocal of frequency.

45. All homes in the United States receive at their wall outlets an ac voltage of 120 V rms at a frequency of 60 Hz. This frequency was chosen for convenience, as a lower frequency would require larger transformers; if the frequency were too low, the slow-switching (alternating) current through the lightbulb would cause it to flicker.
46. A higher frequency than 60 Hz was found to cause an increase in the amount of heat generated in the core of all power distribution transformers due to eddy currents and hysteresis losses.
47. Wavelength, as its name states, is the physical length of one complete cycle and is generally measured in meters. The wavelength (λ, lambda) of a complete cycle is dependent on the frequency and velocity of the transmission.
48. Electromagnetic waves or radio waves travel at the speed of light in air or a vacuum, which is 3×10^8 meters/second or 3×10^{10} centimeters/second.
49. Sound waves travel at a slower speed than electromagnetic waves, as their mechanical vibrations depend on air molecules, which offer resistance to the traveling wave.
50. The phase of a sine wave is always relative to another sine wave of the same frequency.
51. The sine wave derives its name from the fact that it changes in amplitude at the same rate as the sine trigonometric function.
52. The square wave is a periodic (repeating) wave that alternates from a positive peak value to a negative peak value, and vice versa, for equal lengths of time.
53. The duty cycle is the ratio of a pulse width (positive or negative pulse or cycle) to the overall period or time of the wave and is normally given as a percentage.
54. Since a square wave always has a positive and a negative alternation that are equal in time, the duty cycle of all square waves is equal to 50%, which actually means that the positive cycle lasts for 50% of the time of one cycle.
55. The average of the complete square wave cycle should calculate out to be zero, as the amount above the line equals the amount below.
56. It takes a small amount of time for the wave to increase to its positive value (the rise time) and consequently an equal amount of time for a wave to decrease to its negative value (the fall time). Rise time (T_R), by definition, is the time it takes for an edge to rise from 10% to 90% of its full amplitude; fall time (T_F) is the time it takes for an edge to fall from 90% to 10% of its full amplitude.
57. A *periodic wave* is a wave that repeats the same wave shape from one cycle to the next.
58. A time-domain representation of a periodic sine wave, which is the same way it would appear on an oscilloscope display, plots the sine wave's amplitude against time.
59. A frequency-domain representation of the same periodic sine wave, which shows the wave as it would appear on a spectrum analyzer, plots the sine wave's amplitude against frequency instead of time. Pure sine waves have no other frequency components.
60. Other periodic wave shapes, such as square, pulse, triangular, sawtooth, or irregular, are actually made up of a number of sine waves having a particular frequency, amplitude, and phase.
61. The fundamental-frequency sine wave is always the lowest frequency and largest amplitude component of any waveform shape and is used as a reference.
62. Sine-wave harmonic frequencies are smaller in amplitude and are some multiple of the fundamental frequency.
63. Square waves are composed of a fundamental frequency and an infinite number of odd harmonics (third, fifth, seventh, and so on).
64. The rectangular wave is similar to the square wave in many respects, in that it is a periodic wave that alternately switches between one of two fixed values. The difference is that the rectangular wave does not remain at the two peak values for equal lengths of time.
65. Pulse repetition frequency is the number of times per second that a pulse is transmitted.
66. Pulse repetition time is the time interval between the start of two consecutive pulses.
67. Pulse width, pulse length, or pulse duration is the time interval between the leading edge and trailing edge of a pulse at which the amplitude reaches 50% of the peak pulse amplitude.
68. The triangular wave is a repeating wave with equal positive and negative ramps that have linear rates of change with time.
69. The term *linear* describes a relationship between input and output in which the output varies in direct proportion to the input.
70. The sawtooth wave is a repeating waveform that rises from zero to a maximum value linearly and then falls to zero and repeats.

Measuring and Generating AC Signals (Section 8-5)

71. Test equipment can be used to either sense a circuit's condition or generate a signal to see the response of the component or circuit to that signal.
72. Most multimeters can be used to measure either ac or dc.
73. When multimeters are selected to measure ac, an internal rectifier is used to convert the ac input into a dc voltage.
74. Multimeters are normally calibrated to indicate rms values of the ac being measured.
75. A current clamp allows the technician or engineer to measure ac current without opening the current path.
76. The RF probe can be used with the multimeter to more accurately measure higher frequencies above 2 kHz.
77. The high-voltage probe can be used to measure voltages in the kilovolt range by connecting additional multiplier resistors.
78. The analog multimeter, unlike the DMM, can be used to measure low-frequency ac.
79. The digital multimeter is generally more popular because of its easy-to-read display and accuracy.
80. The oscilloscope displays the shape and spacing of electrical signals and can therefore be used to measure dc

voltage, ac voltage, waveform duration, waveform frequency, and waveform shape.

81. The oscilloscope can be used to display waveform shapes, and from this presentation we can calculate the waveform's time, frequency, and amplitude characteristics.

82. A dual-trace oscillocope can produce two traces or waveforms on the screen, which allows the technician or engineer to make comparisons between phase, amplitude, shape, or timing.

83. The function generator can produce a sine wave, square wave, rectangular wave, trangular wave, or sawtooth wave.

84. The frequency counter measures and displays the number of cycles per second (hertz) on a digital display.

REVIEW QUESTIONS

Multiple-Choice Questions

1. Trigonometry is the study of
   a. Triangles
   b. Angles
   c. Trigonometric functions
   d. All of the above

2. The Pythagorean theorem states that the __________ of the length of the hypotenuse of a right-angle triangle equals the sum of the __________ of the lengths of the other two sides.
   a. Square, square roots
   b. Square root, squares
   c. Square, squares
   d. Square root, square roots

3. A vector is an arrow whose length is used to represent the __________ of a quantity and whose point is used to indicate the same quantity's __________.
   a. Magnitude, size
   b. Magnitude, direction
   c. Direction, size
   d. Direction, magnitude

4. If two vectors in a vector diagram are not working together, or are out of phase with one another, they must be __________ to obtain a __________ vector.
   a. Added, resultant
   b. Multiplied by one another, parallel
   c. Vectorially added, resultant
   d. Both (a) and (b)

5. The __________ of a right-angle triangle is always the longest side.
   a. Opposite side c. Hypotenuse
   b. Adjacent side d. Angle theta

6. The __________ of a right-angle triangle is always across from the angle theta.
   a. Opposite side c. Hypotenuse
   b. Adjacent side d. Angle theta

7. The __________ of a right-angle triangle is always formed between the hypotenuse and the adjacent side.
   a. Opposite side c. Hypotenuse
   b. Adjacent side d. Angle theta

8. Which trigonometric function can be used to determine the length of the hypotenuse if the angle theta and the length of the opposite side are known?
   a. Tangent
   b. Sine
   c. Cosine

9. Which trigonometric function can be used to determine the angle theta when the lengths of the opposite and adjacent sides are known?
   a. Tangent
   b. Sine
   c. Cosine

10. Which trigonometric function can be used to determine the length of the adjacent side when the angle theta and the length of the hypotenuse are known?
    a. Tangent
    b. Sine
    c. Cosine

11. A current that rises from zero to maximum positive, returns to zero, and then repeats is known as
    a. Alternating current c. Pulsating dc
    b. AC d. Steady dc

12. A current that rises from zero to maximum positive, decreases to zero, and then reverses to reach a maximum in the opposite direction (negative) is known as
    a. Alternating current
    b. Pulsating direct current
    c. Steady direct current
    d. All of the above

13. The advantage(s) of ac over dc from a power distribution point of view is/are
    a. Generators can supply more power than batteries.
    b. AC can be transformed to a high or low voltage easily, minimizing power loss.
    c. AC can easily be converted into dc.
    d. All of the above.
    e. Only (a) and (c).

14. The approximate voltage carried by long-distance transmission lines in the ac distribution system is
    a. 250 V c. 500,000 V
    b. 2500 V d. 250,000 V

15. The most common type of alternating wave shape is the
    a. Square wave c. Rectangular wave
    b. Sine wave d. Triangular wave

16. __________ equipment manages the flow of information.
    a. Electronic c. Discrete
    b. Electrical d. Integrated

17. __________ equipment manages the flow of power.
    a. Electronic c. Discrete
    b. Electrical d. Integrated

18. The peak-to-peak value of a sine wave is equal to
   a. Twice the rms value
   b. 0.707 times the rms value
   c. Twice the peak value
   d. 1.14 × the average value

19. The rms value of a sine wave is also known as the
   a. Effective value
   b. Average value
   c. Peak value
   d. All of the above

20. The peak value of a 115 V (rms) sine wave is
   a. 115 V
   b. 230 V
   c. 162.7 V
   d. Two of the above could be true.

21. The mathematical average value of a sine wave cycle is
   a. 0.637 × peak
   b. 0.707 × peak
   c. 1.414 × rms
   d. Zero

22. The frequency of a sine wave is equal to the reciprocal of __________.
   a. The period
   b. One cycle time
   c. One alternation
   d. Both (a) and (b)
   e. None of the above

23. What is the period of a 1 MHz sine wave?
   a. 1 ms
   b. One-millionth of a second
   c. 10 ms
   d. 100 μs

24. The sine of 90° is
   a. 0
   b. 0.5
   c. 1
   d. Any of the above

25. What is the frequency of a sine wave that has a cycle time of 1 ms?
   a. 1 MHz
   b. 1 kHz
   c. 200 m
   d. 10 kHz

26. The pulse width (P_w) is the time between the __________ points on the positive and negative edges of a pulse.
   a. 10%
   b. 90%
   c. 50%
   d. All of the above

27. The duty cycle is the ratio of __________ to period.
   a. Peak
   b. Average power
   c. Pulse length
   d. Both (a) and (c)

28. With a pulse waveform, PRF can be calculated by taking the reciprocal of
   a. The duty cycle
   b. PRT
   c. P_d
   d. P_l

29. The sound wave exists in __________ and travels at approximately __________.
   a. Space, 1130 ft/s
   b. Wires, 186,282.397 miles/s
   c. Air, 3×10^6 m/s
   d. None of the above

30. Electrical and electromagnetic waves travel at a speed of
   a. 186,000 miles/s
   b. 3 × 108 meters/s
   c. 162,000 nautical miles/s
   d. All of the above

Communication Skill Questions

31. What is trigonometry? (Introduction)
32. Describe the Pythagorean theorem. (8-1)
33. Describe how $c^2 = a^2 + b^2$ can be transposed to solve for *a, b,* and *c*. (8-1)
34. What is a vector, and what does it represent? (8-1)
35. Sketch a right triangle and indicate the adjacent and opposite sides and the hypotenuse. (8-2)
36. State and explain the purpose of the sine of theta formula. (8-2)
37. How would you transpose the sine of theta formula to determine the angle theta if the lengths of the two sides are known? (8-2)
38. What is the difference between the sine and the inverse sine function? (8-2)
39. State the cosine of theta formula, and explain its function. (8-2)
40. State the tangent of theta formula, and explain its function. (8-2)
41. Describe the three advantages that ac has over dc from a power point of view. (8-3-1)
42. Describe briefly the ac power distribution system. (8-3-1)
43. What is the difference between electricity and electronics? (8-5)
44. What are the six basic ac information wave shapes? (8-4)
45. Describe briefly the following terms as they relate to the sine wave: (8–4–1)
   a. RMS
   b. Peak
   c. Peak-to-peak
   d. Average
   e. The name sine
   f. Frequency
   g. Period
   h. Wavelength
   i. Phase
46. Describe briefly the following terms as they relate to the square wave: (8-4-2)
   a. Duty cycle
   b. Average
47. Describe briefly the following terms as they relate to the rectangular wave: (8-4-3)
   a. PRT
   b. PRF
   c. Duty cycle
   d. Average
48. Briefly describe the meaning of the terms *fundamental frequency* and *harmonics*. (8-4-2)
49. List and describe all the pertinent information relating to the following information carriers: (8-3-2)
   a. Sound wave
   b. Electrical wave
   c. Electromagnetic wave
50. Describe the difference between frequency- and time-domain analysis. (8-4-2)

Practice Problems

51. Calculate the length of the hypotenuse (*C* side) of the following right triangles:
   a. $A = 20$ mi, $B = 53$ mi
   b. $A = 2$ km, $B = 3$ km
   c. $A = 4$ in., $B = 3$ in.
   d. $A = 12$ mm, $B = 12$ mm
52. Calculate the length or magnitude of the resultant vectors in the vector diagrams in Figure 8-67.
53. Calculate the value of the following trigonometric functions:
   a. sin 0°
   b. sin 30°
   c. sin 45°
   d. sin 60°
   e. sin 90°
   f. cos 0°
   g. cos 30°
   h. cos 45°
   i. cos 60°
   j. cos 90°
   k. tan 0°
   l. tan 30°
   m. tan 45°
   n. tan 60°
   o. tan 90°

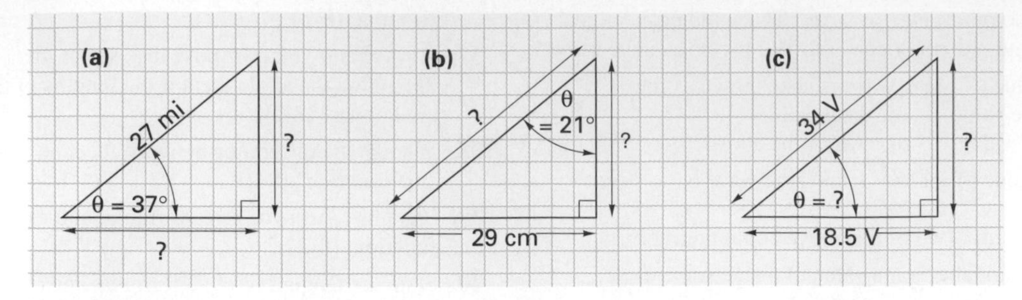

FIGURE 8-67

54. Calculate angle θ from the function given.
    a. $\sin \theta = 0.707, \theta = ?$
    b. $\sin \theta = 0.233, \theta = ?$
    c. $\cos \theta = 0.707, \theta = ?$
    d. $\cos \theta = 0.839, \theta = ?$
    e. $\tan \theta = 1.25, \theta = ?$
    f. $\tan \theta = 0.866, \theta = ?$
55. In Figure 8-67, name each of the triangle's sides, and calculate the unknown values.
56. Calculate the periods of the following sine-wave frequencies:
    a. 27 kHZ
    b. 3.4 MHz
    c. 25 Hz
    d. 365 Hz
    e. 60 Hz
    f. 200 kHz
57. Calculate the frequency for each of the following values of time:
    a. 16 ms
    b. 1 s
    c. 15 μs
    d. 0.05 s
    e. 200 μs
    f. 350 ms
58. A 22 V peak sine wave will have the following values:
    a. Rms voltage =
    b. Average voltage =
    c. Peak-to-peak voltage =
59. A 40 mA rms sine wave will have the following values:
    a. Peak current =
    b. Peak-to-peak current =
    c. Average current =
60. How long would it take an electromagnetic wave to travel 60 miles?
61. An 11 kHz rectangular pulse, with a pulse width of 10 μs, will have a duty cycle of ___________%.
62. Calculate the PRT of a 400 kHz pulse waveform.
63. Calculate the average current of the pulse waveform in Question 61 if its peak current is equal to 15 A.
64. What is the duty cycle of a 10 V peak square wave at a frequency of 1 kHz?
65. Considering a fundamental frequency of 1 kHz, calculate the frequency of its
    a. Third harmonic
    b. Second harmonic
    c. Seventh harmonic
66. If one cycle of a sine wave occupies 4 cm on the oscilloscope horizontal grid and 5 cm from peak-to-peak on the vertical grid, calculate frequency, period, rms, average, and peak for the following control settings:
    a. 0.5 V/cm, 20 μs/cm
    b. 10 V/cm, 10 ms/cm
    c. 50 mV/cm, 0.2 μs/cm
67. Assuming the same graticule and switch settings of the oscilloscope in Figure 8-9, what would be the lowest setting of the volts/cm and time/division switches to fully view a 6 V rms, 350 kHz sine wave?
68. If the volts/cm switch is positioned to 10 V/cm and the waveform extends 3.5 divisions from peak to peak, what is the peak-to-peak value of this wave?
69. If a square wave occupies 5.5 horizontal cm on the 1 μs/cm position, what is its frequency?
70. Which settings would you use on an autoset scopemeter to fully view an 8 V rms, 20 kHz triangular wave?

Job Interview Test

These tests at the end of each chapter will challenge your knowledge up to this point and give you the practice you need for a job interview. To make this more realistic, the test will comprise both technical and personal questions. In order to take full advantage of this exercise, you may like to set up a simulation of the interview environment, have a friend read the questions to you, and record your responses for later analysis.

Company Name: Free-2-Roam, Inc.

Industry Branch: Communications

Function: Quality Testing wireless products

Job Applying For: QA Technician

71. Would you say that you have good written and oral communication skills?
72. What is ac?
73. What is the difference between the ac waves we receive at the wall outlet and the ac waves we receive with our car radio?
74. How are your teamwork skills?
75. What is wireless communication?
76. Can you tell me the differences between a sound wave and an electromagnetic wave?
77. How easily do you feel you could adapt to new technologies?

78. What is a sine wave?
79. What is the duty cycle of a pulse wave?
80. What test instrument would you use to measure the frequency domain of a waveform?

Answers

71. Your resume and cover letter were heavily scrutinized in order to evaluate your written communication skills, and they must have passed the test, otherwise you would not have been granted an interview. Your oral communication skills are now being tested, and it is important that all of your answers are clear and concise. You more than likely practiced communication skills during your training, and so quote these examples and any previous job-related experience.
72. Section 8-2.
73. Section 8-3.
74. The reason for this question lies in industry's drive for a high-skill, high-quality, high-performance organization and reduced cost. In order to achieve this goal, the work environment has changed, resulting in the integration of electronic system design, manufacture, and service. There is a need, therefore, for a technician to have better teamwork skills. Teamwork is defined in the dictionary as "combined effort or organized co-operation." To operate in a team environment, therefore, a person will need good communication and interpersonal skills. Emphasize your lab-team training and any previous job-related experience.
75. Section 8-3-2.
76. Section 8-3-2.
77. Technology, especially electronics technology, seems to advance in leaps and bounds, and so there will always be a need for you to be able to independently learn and adapt to the new changes. Your math skills, logic and reasoning skills, learning skills, and relationship skills develop intelligence, which in turn develops better adaptability and organizational and problem-solving skills. Success in these areas will ensure career advancement. Discuss your training in each of these areas, and if you feel the same way I do, let them know that your interest in electronics is both personal and professional and so there is a strong desire on your part to understand and embrace the advances.
78. Section 8-4-1.
79. Section 8-4-3.
80. Section 8-4-2.

Web Site Questions

Go to the web site http://www.prenhall.com/cook, select the textbook *Introductory DC/AC Electronics* or *Introductory DC/AC Circuits,* this chapter, and then follow the instructions when answering the multiple-choice practice problems.

Capacitors

Back to the Future

Born in England in 1791, Charles Babbage became very well known for both his mathematical genius and eccentric personality. Babbage's ultimate pursuit was that of mathematical accuracy. He delighted in spotting errors in everything from log tables (used by astronomers, mathematicians, and navigators) to poetry. In fact, he once wrote to poet Alfred Lord Tennyson, pointing out an inaccuracy in his line "Every moment dies a man—every moment one is born." Babbage explained to Tennyson that since the world population was actually increasing and not, as he indicated, remaining constant, the line should be rewritten to read "Every moment dies a man—every moment one and one-sixteenth is born."

In 1822, Babbage described in a paper and built a model of what he called "a difference engine," which could be used to calculate mathematical tables. The Royal Society of Scientists described his machine as "highly deserving of public encouragement," and a year later the government awarded Babbage £1500 for his project. Babbage originally estimated that the project should take 3 years; however, the design had its complications, and after 10 years of frustrating labor, in which the government grants increased to £17,000, Babbage was still no closer to completion. Finally, the money stopped and Babbage reluctantly decided to let his brainchild go.

In 1833, Babbage developed an idea for a much more practical machine, which he named "the analytical engine." It was to be a more general machine that could be used to solve a variety of problems, depending on instructions supplied by the operator. It would include two units, called a "mill" and a "store," both of which would be made of cogs and wheels. The store, which was equivalent to a modern-day computer memory, could hold up to 100 forty-digit numbers. The mill, which was equivalent to a modern computer's arithmetic and logic unit (ALU), could perform both arithmetic and logic operations on variables or numbers retrieved from the store, and the result could be stored in the store and then acted upon again or printed out. The program of instructions directing these operations would be fed into the analytical engine in the form of punched cards.

The analytical engine was never built. All that remains are the volumes of descriptions and drawings and a section of the mill and printer built by Babbage's son, who also had to concede defeat. It was, unfortunately for Charles Babbage, a lifetime of frustration to have conceived the basic building blocks of the modern computer a century before the technology existed to build it.

9

Outline and Objectives

9-1 CAPACITOR CHARACTERISTICS

9-1-1 Charging a Capacitor

Objective 1: List the principal parts that make up a capacitor.

9-1-2 Discharging a Capacitor

9-1-3 Electrostatics

Objective 2: Explain the charging and discharging process and its relationship to electrostatics.

9-1-4 The Unit of Capacitance

Objective 3: State the unit of capacitance and explain how it relates to charge and voltage.

9-1-5 Capacitance Formula

Objective 4: List and explain the factors determining capacitance.

9-1-6 Capacitors in Combination

Objective 5: Calculate total capacitance in parallel and series capacitance circuits.

9-1-7 Dielectric Breakdown and Leakage

Objective 6: Describe capacitance breakdown and capacitor leakage.

9-2 CAPACITOR TYPES, CODING, AND TESTING

9-2-1 Capacitor Types

Objective 7: Describe the advantages of and differences between the five basic types of fixed capacitors.

Objective 8: Describe the advantages of and differences between the four basic types of variable capacitors.

9-2-2 Capacitor Coding

Objective 9: Describe the coding of capacitor values on the body by use of alphanumerics or color.

9-2-3 Capacitor Testing

Objective 10: Explain some of the more common capacitor failures and how to use an ohmmeter and capacitance analyzer to test them.

9-3 DC CAPACITIVE CIRCUITS

9-3-1 Charging

Objective 12: Explain the capacitor time constant as it relates to dc charging and discharging.

9-3-2 Discharging

9-4 AC CAPACITIVE CIRCUITS

Objective 13: Explain how the capacitor charges and discharges when ac is applied.

9-4-1 Phase Relationship Between Current and Voltage

Objective 14: Describe why there is a phase difference between circuit current and capacitor voltage.

9-4-2 Capacitive Reactance

Objective 15: Define and explain capacitive reactance.

9-4-3 Series *RC* Circuit

9-4-4 Parallel *RC* Circuit

Objective 16: Describe impedance, phase angle, power, and power factor as they relate to a series and parallel *RC* circuit.

9-5 APPLICATIONS OF CAPACITORS

9-5-1 Combining AC and DC

9-5-2 The Capacitive Voltage Divider

9-5-3 *RC* Filters

9-5-4 The *RC* Integrator

9-5-5 The *RC* Differentiator

Objective 17: Describe how the capacitor and resistor in combination can be used

a. To combine ac and dc.
b. To act as a voltage divider.
c. As a filter.
d. As an integrator.
e. As a differentiator.

Introduction

Up to this point we have concentrated on circuits containing only resistance, which opposes the flow of current and then converts or dissipates power in the form of heat. Capacitance and inductance are two circuit properties that act differently from resistance in that they will charge or store the supplied energy and then return almost all the stored energy back to the circuit, rather than lose it in wasted heat. Inductance will be discussed in Ch. 11.

Capacitance is the ability of a circuit or device to store electrical charge. A device or component specifically designed to have this capacity or capacitance is called a *capacitor.* A capacitor stores an electrical charge similar to a bucket holding water. Using the analogy throughout this chapter will help you gain a clear understanding of the capacitor's operation. For example, the capacitor holds charge in the same way that a bucket holds water. A larger capacitor will hold more charge and will take longer to charge, just as a larger bucket will hold more water and take longer to fill. A larger circuit resistance means a smaller circuit current, and therefore a longer capacitor charge time. Similarly, a smaller hose will have a greater water resistance, producing a smaller water flow, and therefore the bucket will take a longer time to fill. Unlike a bucket, capacitors store electrons, and basically, the amount of electrons stored is a measure of the capacitor's capacitance.

9-1 CAPACITOR CHARACTERISTICS

Capacitor
Previously called a condenser, it is a device that stores electric energy in the form of an electric field that exists within a dielectric (insulator) between two conducting plates each of which is connected to a lead.

Figure 9-1 illustrates the main parts and schematic symbol of the **capacitor.** Two leads are connected to two parallel metal conductive plates, which are separated by an insulating material known as a *dielectric.* The conductive plates are normally made of metal foil; the dielectric can be paper, air, glass, ceramic, mica, or some other form of insulator. In the following sections you will will see that, like a secondary battery, a capacitor can be made to charge or discharge, and when discharging it will return nearly all of the energy it consumed during charge.

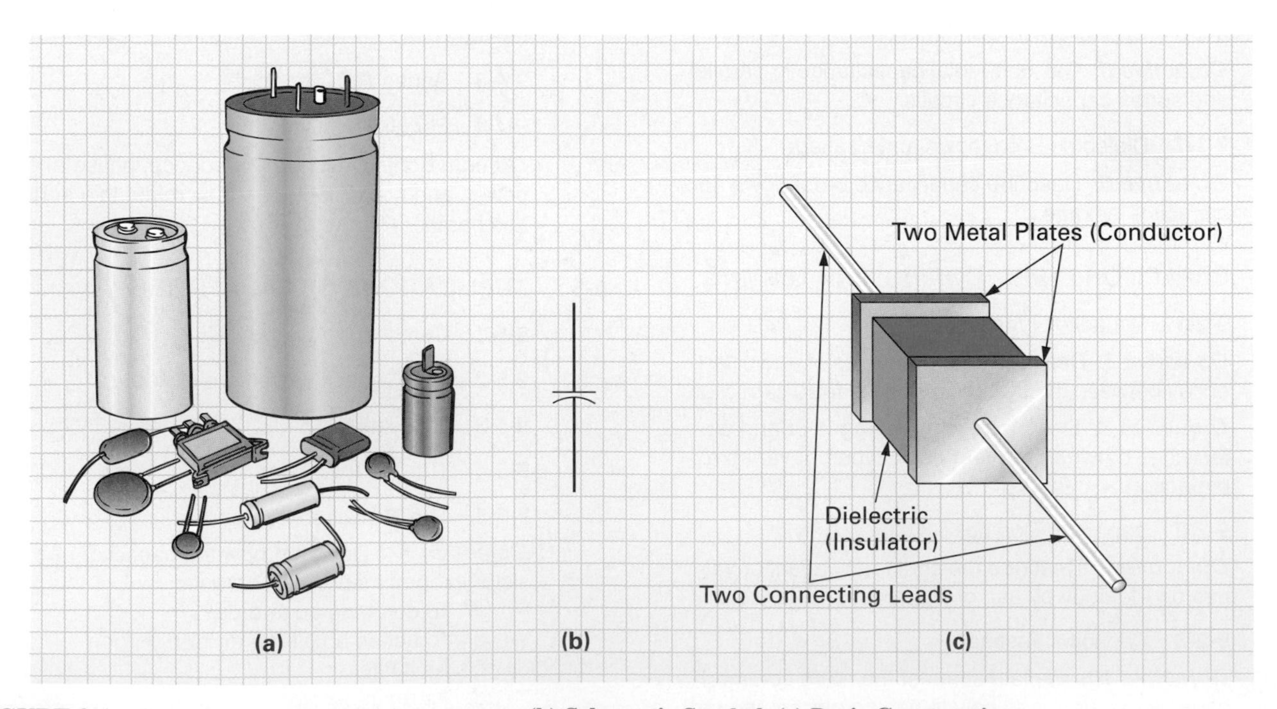

FIGURE 9-1 **Capacitor. (a) Physical Appearance. (b) Schematic Symbol. (c) Basic Construction.**

9-1-1 *Charging a Capacitor*

Capacitance is the ability of a capacitor to store an electrical charge. Figure 9-2 illustrates how the capacitor stores an electric charge. This capacitor is shown as two plates with air acting as the dielectric.

In Figure 9-2(a), the switch is open, so no circuit current results. An equal number of electrons exist on both plates, so the voltmeter (VM) indicates zero, which means no potential difference exists across the capacitor.

In Figure 9-2(b), the switch is now closed and electrons travel to the positive side of the battery, away from the right-hand plate of the capacitor. This creates a positive right-hand capacitor plate, which results in an attraction of free electrons from the negative side of the battery to the left-hand plate of the capacitor. In fact, for every electron that leaves the right-hand capacitor plate and is attracted into the positive battery terminal, another electron leaves the negative side of the battery and travels to the left-hand plate of the capacitor. Current appears to be flowing from the negative side of the battery, around to the positive, through the capacitor. This is not really the case, because no electrons can flow through the insulator or gap between the plates, and although there appears to be one current flowing throughout the circuit, there are in fact two separate currents—one from the battery to the capacitor and the other from the capacitor to the battery.

TIME LINE

Robert Moog (1934–   ), a self-employed engineer of Trumansburg, New York, built an analog music synthesizer that allowed the operator to control many tones simultaneously and rapidly alter tones during performances.

Synthesizers, devices used to generate a number of frequencies, had been around for a number of years, but electronic music did not find wide acceptance with the popular music audience until a young musician named Walter Carlos used a Moog instrument to produce a totally synthesized hit record. The album, *Switched-On Bach*, released in 1968, was an interpretation of music by eighteenth-century composer Johann Sebastian Bach, and both the album and instrument were such breakthroughs in electronic music that in no time at all composers were writing a wide variety of original music specifically for Moog synthesizers.

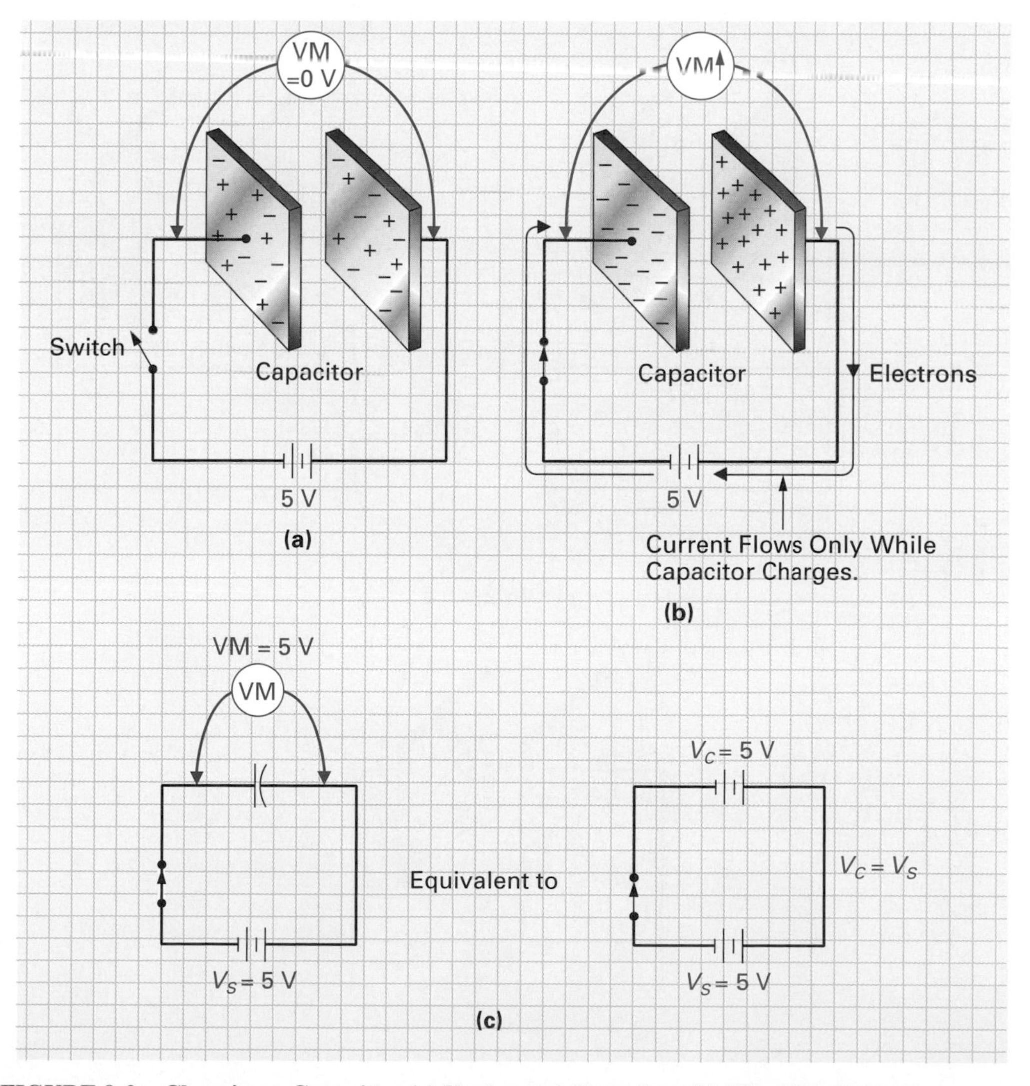

FIGURE 9-2 Charging a Capacitor. (a) Uncharged Capacitor. (b) Charging Capacitor. (c) Charged Capacitor.

A voltmeter across the capacitor will indicate an increase in the potential difference between the plates, and the capacitor is said to be charging toward, in this example, 5 V. This potential difference builds up across the two plates until the voltage across the capacitor is equal to the voltage of the battery. In this example, when the capacitor reaches a charge of 5 V, the capacitor will be equivalent to a 5 V battery, as seen in Figure 9-2(c). Once charged, there will be no potential difference between the battery and capacitor, and so circuit current flow will be zero when the capacitor is fully charged.

9-1-2 *Discharging a Capacitor*

In Figure 9-3(a), you can see that when switch 1 is closed and switch 2 is open, the capacitor will charge. If switch 1 was then opened, the capacitor will remain in its charged condition. If switch 2 is now closed, a path exists across the charged capacitor, as shown in Figure 9-3(b), and the excess of electrons on the left plate will flow through the conducting wire to the positive plate on the right side. The capacitor is now said to be *discharging*. When an equal number of electrons exist on both sides, the capacitor is said to be *discharged* since both plates have an equal charge and the potential difference across the capacitor will be zero, as shown in Figure 9-3(c).

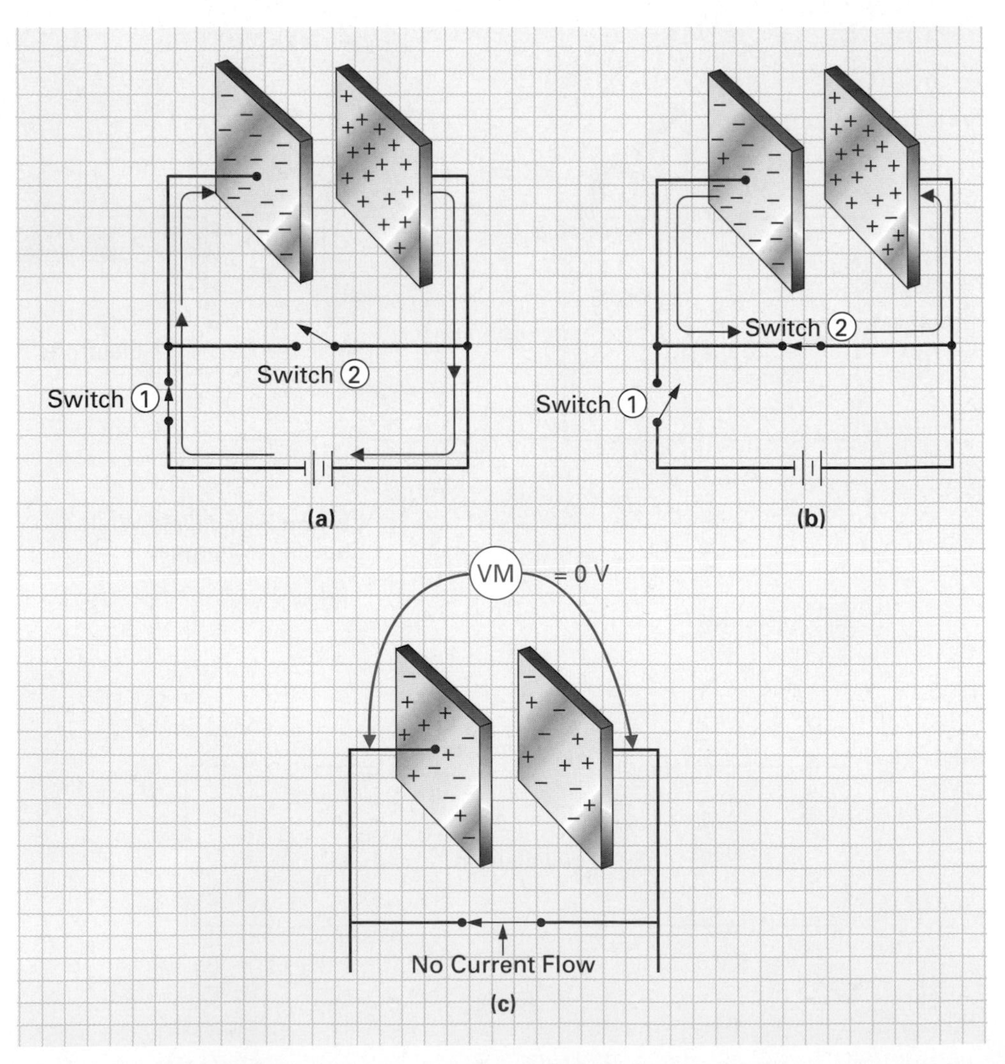

FIGURE 9-3 Discharging a Capacitor. (a) Charged Capacitor. (b) Discharging Capacitor. (c) Discharged Capacitor.

Figure 9-3(a) and (b) illustrate the capacitor's charge and discharge currents, which flow in opposite directions. In both cases the current flow is always from one plate to the other and never exists through the dielectric insulator.

9-1-3 *Electrostatics*

Just as a magnetic field is produced by the flow of current, an **electric field** is produced by voltage.

Electrostatic or Electric Field

Force field produced by static electrical charges. Also called a voltage field, it is a field or force that exists in the space between two different potentials or voltages.

Current generates a magnetic field

Voltage generates an electric field

Figure 9-4(a) illustrates an example capacitor circuit with the capacitor charged and the switch open. In this condition, the capacitor retains its charge and an invisible electric or electrostatic (voltage) field will be produced by nonmoving or static electrical (electrostatic) charges of different polarities. You will remember that like charges repel and unlike charges attract. Invisible electrostatic lines of flux or force can be illustrated to show this electrostatic force of attraction or repulsion. These lines are polarized away from a positive electrostatic (stationary electrical) charge and toward the negative electrostatic charge, as shown in Figure 9-4(b). If two like charges are in close proximity to one another, the electrostatic lines organize themselves into a pattern, as shown in Figure 9-4(c).

A charged capacitor has an electric or electrostatic field existing between the positively charged and negatively charged plates. The **strength of the electrostatic field** is proportional to the charge or potential difference on the plates and inversely proportional to the distance between the plates.

Field Strength

The strength of an electric, magnetic, or electromagnetic field at a given point.

$$\text{Field strength (V/m)} = \frac{\text{charge difference } (V)\text{, volts}}{\text{distance between plates } (d)\text{, meters}}$$

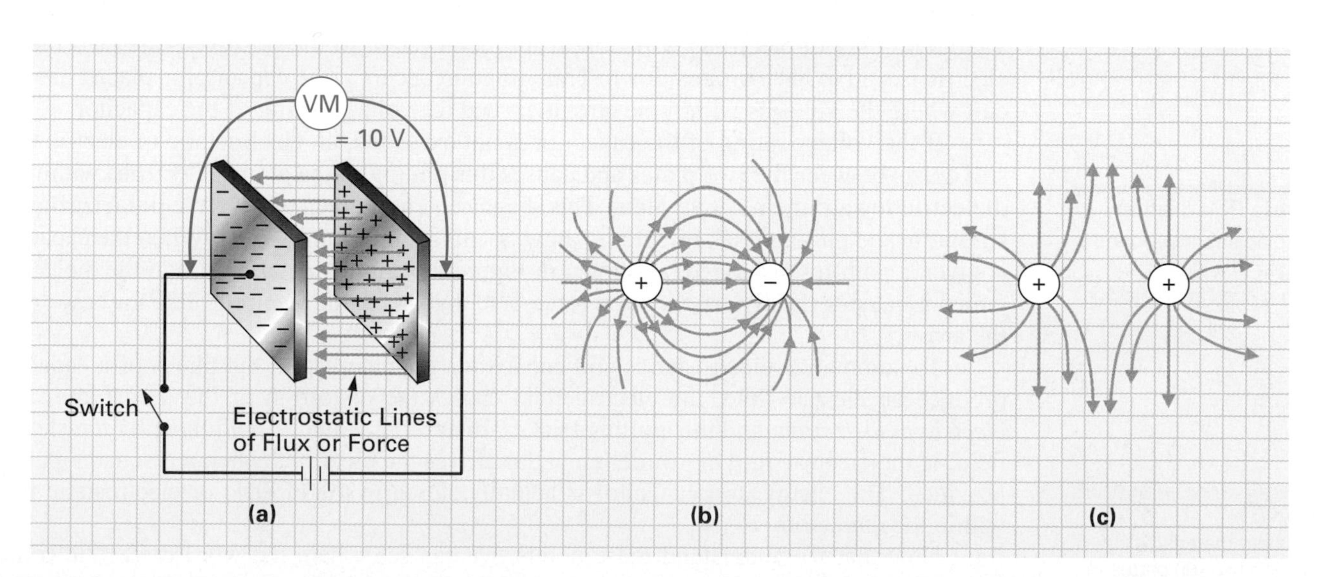

FIGURE 9-4 **Electrostatic (Electric) Field. (a) Between the Plates of a Charged Capacitor. (b) Electrostatic Lines of Attraction. (c) Electrostatic Lines of Repulsion.**

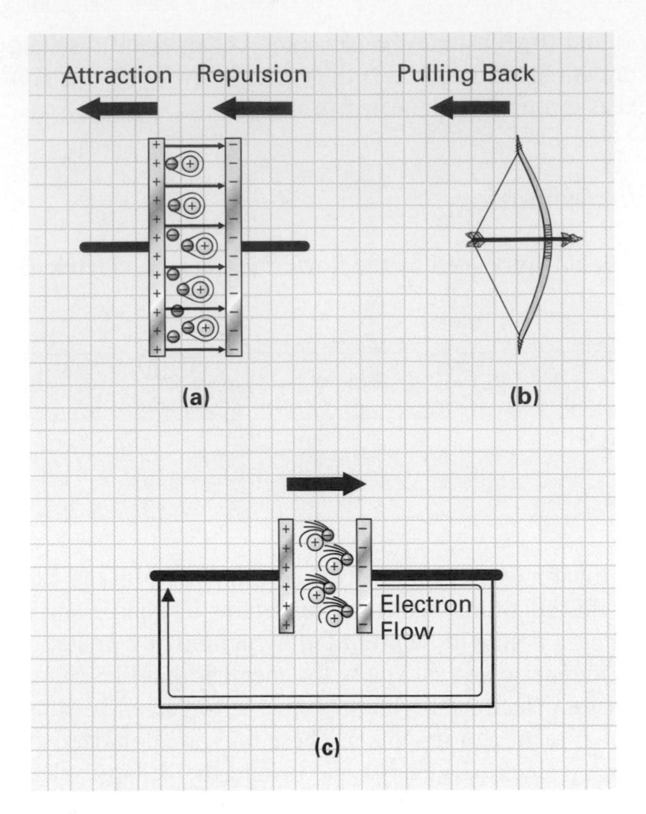

FIGURE 9-5 **Electric Polarization.**

The dielectric or insulator between the plates, like any other material, has its own individual atoms, and although the dielectric electrons are more tightly bound to their atoms than conductor electrons, stresses are placed on the atoms within the dielectric, as seen in Figure 9-5. The electrons in orbit around the dielectric atoms are displaced or distorted by the electric field existing between the positive and negative plates. If the charge potential across the capacitor is high enough and the distance between the plates is small enough, the attraction and repulsion exerted on the dielectric atom can be large enough to free the dielectric atom's electrons. The material then becomes ionized, and a chain reaction of electrons jumping from one atom to the next in a right-to-left movement occurs. If this occurs, a large number of electrons will flow from the negative to the positive plate, and the dielectric is said to have broken down. This situation occurs if the capacitor is placed in a circuit where the voltages within the circuit exceed the voltage rating of the capacitor.

If the voltage rating of the capacitor is not exceeded, an electrostatic or electric field still exists between the two plates and causes this pulling of the atom's electrons within the dielectric toward the positive plate. This displacement, known as **electric polarization**, is similar to the pulling-back effect on a bow, as shown in Figure 9-5(b). When the capacitor is given a path for discharge, as shown in Figure 9-5(c), the electric field in the dielectric, which is causing the distortion, is the force field that drives the electrons, like the bow drives the arrow.

Electric Polarization
A displacement of bound charges in a dielectric when placed in an electric field.

To summarize electrostatics and capacitors, the charges on the plates of a capacitor produce an electric field, the electric field causes the distortion of the atoms known as *electric polarization*, and this pulling back or distortion, which is held there by the electric field, is the electron moving force (emf) that drives the electrons when a discharge path is provided. The energy in a capacitor is actually stored in the electric or electrostatic field within the dielectric.

Dielectric
Insulating material between two (di) plates in which the electric field exists.

Now that we understand these points, we can see where the word **dielectric** comes from. The dielectric is the insulating material that exists between two (di) plates and undergoes electric polarization when an electric field exists within it (dielectric).

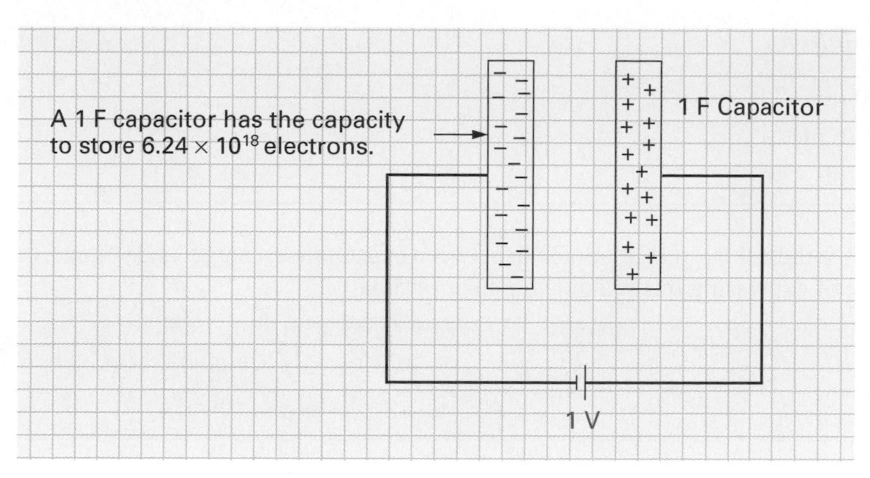

FIGURE 9-6 One Farad of Capacitance.

9-1-4 *The Unit of Capacitance*

Capacitance is the ability of a capacitor to store an electrical charge, and the unit of capacitance is the **farad** (F), named in honor of Michael Faraday's work in 1831 in the field of capacitance. A capacitor with the capacity of 1 farad (1 F) can store 1 coulomb of electrical charge (6.24×10^{18} electrons) if 1 volt is applied across the capacitor's plates, as seen in Figure 9-6.

Farad
Unit of capacitance.

A 1 F capacitor is a very large value and not frequently found in electronic equipment. Most values of capacitance found in electronic equipment are in the units between the microfarad ($\mu F = 10^{-6}$ F) and picofarad ($pF = 10^{-12}$ F). A microfarad is 1 millionth of a farad (10^{-6}). So if a 1 F capacitor can store 6.24×10^{18} electrons with 1 V applied, a 1 μF capacitor, which has 1-millionth the capacity of a 1 F capacitor, can store only 1 millionth of a coulomb, or $(6.24 \times 10^{18}) \times (1 \times 10^{-6}) = 6.24 \times 10^{12}$ electrons when 1 V is applied, as shown in Figure 9-7.

EXAMPLE:

Convert the following to either microfarads or picofarads (whichever is more appropriate):

a. 0.00002 F

b. 0.00000076 F

c. 0.00047×10^{-7} F

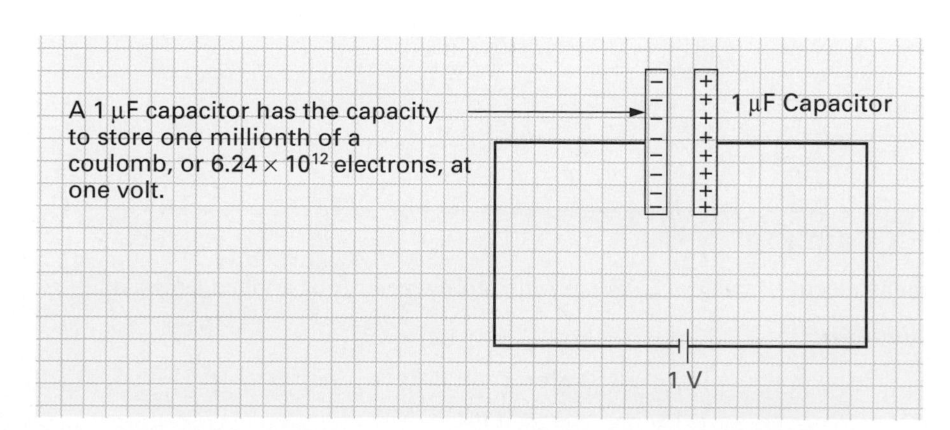

FIGURE 9-7 One-Millionth of a Farad.

Solution:

a. 20 μF
b. 0.76 μF
c. 47 pF

Since there is a direct relationship between capacitance, charge, and voltage, there must be a way of expressing this relationship in a formula.

$$\text{Capacitance, } C \text{ (farads)} = \frac{\text{charge, } Q \text{ (coulombs)}}{\text{voltage, } V \text{ (volts)}}$$

where C = capacitance, in farads
Q = charge, in coulombs
V = voltage, in volts

By transposition of the formula, we arrive at the following combinations for the same formula:

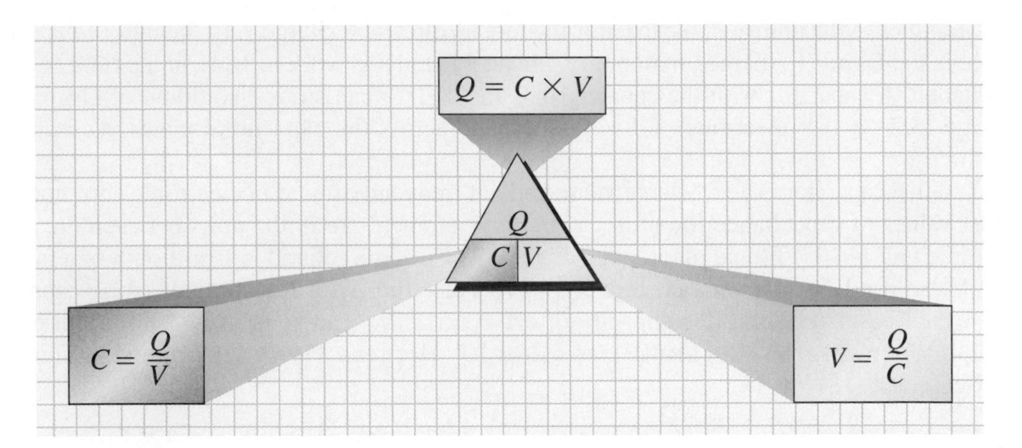

EXAMPLE:

If a capacitor has the capacity to hold 36 C ($36 \times 6.24 \times 10^{18} = 2.25 \times 10^{20}$ electrons) when 12 V is applied across its plates, what is the capacitance of the capacitor?

Solution:

$$C = \frac{Q}{V}$$
$$= \frac{36\text{ C}}{12\text{ V}}$$
$$= 3\text{ F}$$

EXAMPLE:

How many electrons could a 3 μF capacitor store when 5 V is applied across it?

Solution:

$$Q = C \times V$$
$$= 3\ \mu\text{F} \times 5\text{ V}$$
$$= 15\ \mu\text{C}$$

(15 microcoulombs is 15 millionths of a coulomb.) Since 1 C = 6.24×10^{18} electrons, 15 μC = $(15 \times 10^{-6}) \times 6.24 \times 10^{18} = 9.36 \times 10^{13}$ electrons.

EXAMPLE:

If a capacitor of 2 F has stored 42 C of charge (2.63×10^{20} electrons), what is the voltage across the capacitor?

Solution:

$$V = \frac{Q}{C}$$
$$= \frac{42\ \text{C}}{2\ \text{F}}$$
$$= 21\ \text{V}$$

9-1-5 *Capacitance Formula*

The capacitance of a capacitor is determined by three factors:

1. The plate area of the capacitor
2. The distance between the plates
3. The type of dielectric used

1. Plate Area (*A*)

The capacitance of a capacitor is directly proportional to the plate area. This area in square centimeters is the area of only one plate and is calculated by multiplying length by width. This is illustrated in Figure 9-8(a) and (b). In these two examples, the (b) capacitor plate is twice as large as the (a) capacitor plate, and since capacitance is proportional to plate area ($C \propto A$), the capacitor in example (b) will have double the capacity or capacitance of the capacitor in example (a). Since the energy of a charged capacitor is in the electric field between the plates and the plates of capacitor (b) are double those of (a), there is twice as much area for the electric field to exist in, and this doubles the capacitor's capacitance.

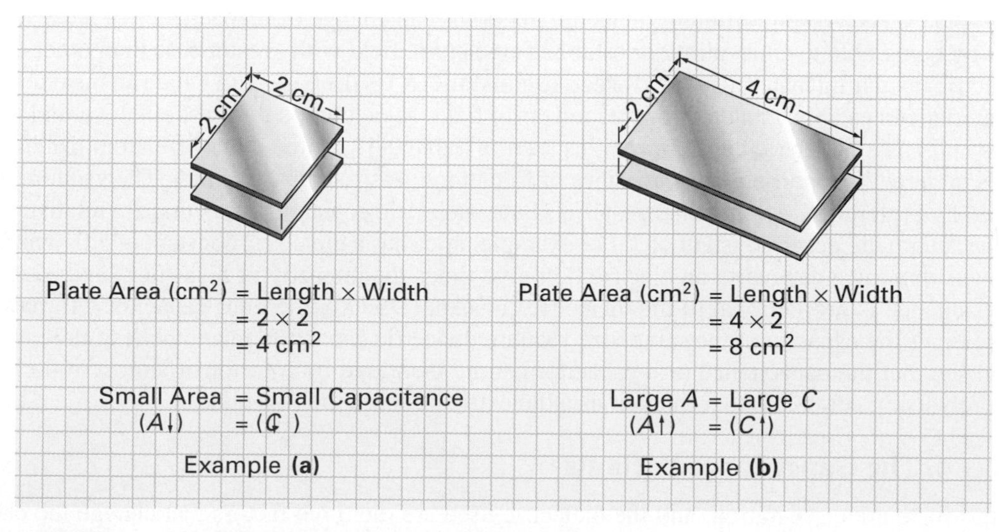

FIGURE 9-8 Capacitance Is Proportional to Plate Area.

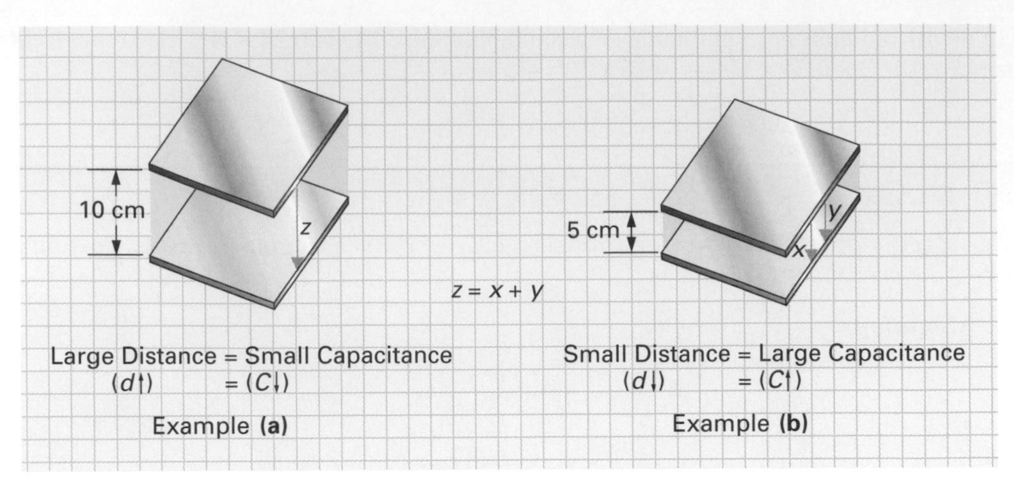

FIGURE 9-9 Capacitance Is Inversely Proportional to Plate Separation or Distance (d).

2. Distance between the Plates (d)

The distance or separation between the plates is dependent on the thickness of the dielectric used. The capacitance of a capacitor is inversely proportional to this distance between the plates, in that an increase in the distance ($d\uparrow$) causes a decrease in the capacitor's capacitance ($C\downarrow$). In Figure 9-9(a), a large distance between the capacitor plates results in a small capacitance, whereas in Figure 9-9(b), the dielectric thickness and the plate separation are half that of capacitor (a). This illustrates also how the capacitance of a capacitor can be doubled, in this case by halving the space between the plates. The gap across which the electric lines of force exist is halved in capacitor (b), and this doubles the strength of the electric field, which consequently doubles capacitance. Simply stated, an electric line of force (z) in Figure 9-9(a) can be used to produce two electric lines of force (x and y) in Figure 9-9(b) if the distance is half.

3. Dielectric Constant

Dielectric Constant (K)
The property of a material that determines how much electrostatic energy can be stored per unit volume when unit voltage is applied. Also called permittivity.

The insulating dielectric of a capacitor concentrates the electric lines of force between the two plates. As a result, different dielectric materials can change the capacitance of a capacitor by being able to concentrate or establish an electric field with greater ease than other dielectric insulating materials. The **dielectric constant (K)** is the ease with which an insulating material can establish an electrostatic (electric) field. A vacuum is the least effective dielectric and has a dielectric constant of 1, as seen in Table 9-1. All the other insulators listed in this table will support electrostatic lines of force more easily than a vacuum. The vacuum is used as a reference. All the other materials have dielectric constant values that are relative to the vacuum dielectric constant of 1. For example, mica has a dielectric constant of 5.0, which means that mica can cause an electric field five times the intensity of a vacuum; and, since capacitance is proportional to the dielectric constant ($C \propto K$), the mica capacitor will have five times the capacity of the same-size vacuum dielectric capacitor. In another example, we can see that the capacitance of a capacitor can be increased by a factor of almost 8000 by merely using ceramic rather than air as a dielectric between the two plates.

4. The Capacitance Formula

Thus plate area, separation, and the dielectric used are the three factors that change the capacitance of a capacitor. The formula that combines these three factors is

TABLE 9-1 Dielectric Constants

MATERIAL	DIELECTRIC CONSTANT (K)[a]
Vacuum	1.0
Air	1.0006
Teflon	2.0
Wax	2.25
Paper	2.5
Amber	2.65
Rubber	3.0
Oil	4.0
Mica	5.0
Ceramic (low)	6.0
Bakelite	7.0
Glass	7.5
Water	78.0
Ceramic (high)	8000.

[a]The different material compositions can cause different values of ***K***.

$$C = \frac{(8.85 \times 10^{-12}) \times K \times A}{d}$$

where C = capacitance, in farads (F)
8.85×10^{-12} is a constant
K = dielectric constant
A = plate area, in square meters (m^2)
d = distance between the plates, in meters (m)

This formula summarizes what has been said in relation to capacitance. The capacitance of a capacitor is directly proportional to the dielectric constant (K) and the area of the plates (A) and is inversely proportional to the dielectric thickness or distance between the plates (d).

EXAMPLE:

What is the capacitance of a ceramic capacitor with a 0.3 m^2 plate area and a dielectric thickness of 0.0003 m?

Solution:

$$C = \frac{(8.85 \times 10^{-12}) \times K \times A}{d}$$

$$= \frac{(8.85 \times 10^{-12}) \times 6 \times 0.3\ \text{m}^2}{0.0003\ \text{m}}$$

$$= 5.31 \times 10^{-8}$$

$$= 0.0531\ \mu\text{F}$$

Calculator Sequence

STEP	KEYPAD ENTRY	DISPLAY RESPONSE
1.	8 . 8 5 E 1 2 +/−	8.85E–12
2.	×	
3.	6	6
4.	×	5.31E–11
5.	0 . 3	
6.	÷	1.59E–11
7.	0 . 0 0 0 3	
8.	=	5.31E–8

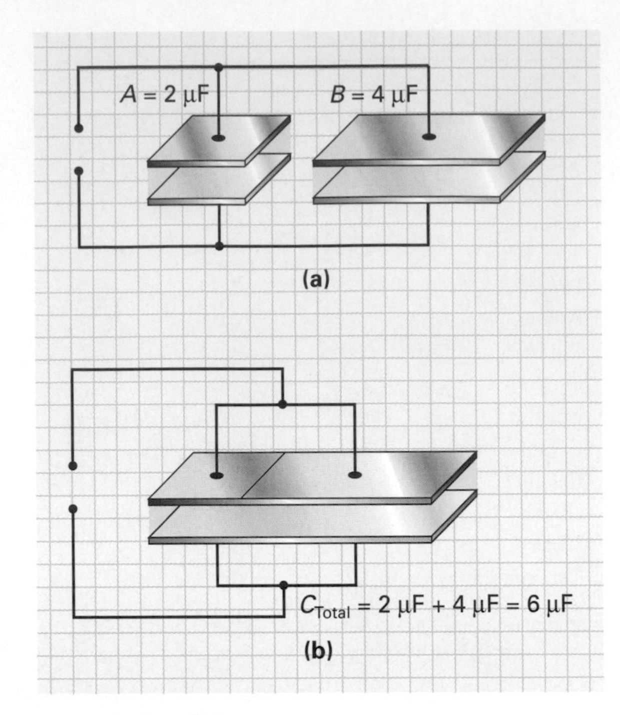

FIGURE 9-10 **Capacitors in Parallel.**

9-1-6 *Capacitors in Combination*

Like resistors, capacitors can be connected in either series or parallel. As you will see in this section, the rules for determining total capacitance for parallel- and series-connected capacitors are opposite to those for series- and parallel-connected resistors.

1. Capacitors in Parallel

In Figure 9-10(a), you can see a 2 μF and 4 μF capacitor connected in parallel with one another. As the top plate of capacitor *A* is connected to the top plate of capacitor *B* with a wire, and a similar situation occurs with the bottom plates, you can see that this is the same as if the top and bottom plates were touching one another, as shown in Figure 9-10(b). When drawn so that the respective plates are touching, the dielectric constant and plate separation is the same as shown in Figure 9-10(a), but now we can easily see that the plate area is actually increased. Consequently, if capacitors are connected in parallel, the effective plate area is increased; and since capacitance is proportional to plate area $[C\uparrow = (8.85 \times 10^{-12}) \times K \times A\uparrow/d]$, the capacitance will also increase. Total capacitance is actually calculated by adding the plate areas, so total capacitance is equal to the sum of all the individual capacitances in parallel.

$$C_T = C_1 + C_2 + C_3 + C_4 + \cdots$$

EXAMPLE:

Determine the total capacitance of the circuit in Figure 9-11(a). What will be the voltage drop across each capacitor?

Solution:

$$\begin{aligned} C_T &= C_1 + C_2 + C_3 \\ &= 1\ \mu\text{F} + 0.5\ \mu\text{F} + 0.75\ \mu\text{F} \\ &= 2.25\ \mu\text{F} \end{aligned}$$

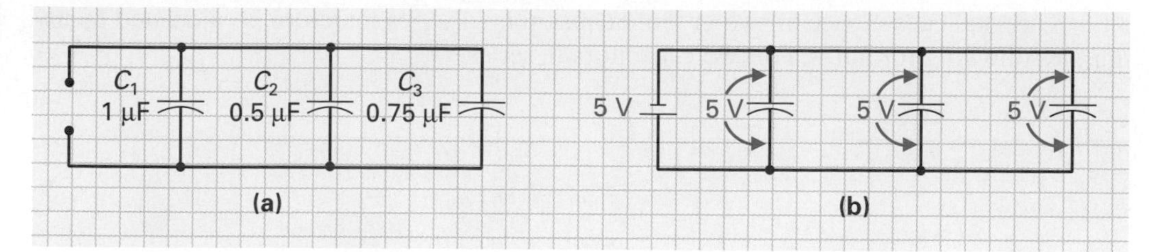

FIGURE 9-11 Example of Parallel-Connected Capacitors.

As with any parallel-connected circuit, the source voltage appears across all the components. If, for example, 5 V is connected to the circuit of Figure 9-11(b), all the capacitors will charge to the same voltage of 5 V because the same voltage always exists across each section of a parallel circuit.

2. Capacitors in Series

In Figure 9-12(a), we have connected the two capacitors of 2 μF and 4 μF in series. Since the bottom plate of the *A* capacitor is connected to the top plate of the *B* capacitor, they can be redrawn so that they are touching, as shown in Figure 9-12(b).

The top plate of the *A* capacitor is connected to a wire into the circuit, and the bottom plate of *B* is connected to a wire into the circuit. This connection creates two center plates that are isolated from the circuit and can therefore be disregarded, as shown in Figure 9-12(c). The first thing you will notice in this illustration is that the dielectric thickness ($d\uparrow$) has increased, causing a greater separation between the plates. The effective plate area of this capacitor has decreased, as it is just the area of the top plate only. Even though the bottom plate extends out further, the electric field can only exist between the two plates, so the surplus metal of the bottom plate has no metal plate opposite for the electric field to exist in. Consequently, when capacitors are connected in series the effective plate area is decreased ($A\downarrow$) and the dielectric thickness increased ($d\uparrow$), and both of these effects result in an overall capacitance decrease ($C\downarrow\downarrow = (8.85 \times 10^{-12}) \times K \times A\downarrow/d\uparrow$).

The plate area is actually decreased to the smallest individual capacitance connected in series, which in this example is the plate area of *A*. If the plate area were the only factor, then capacitance would always equal the smallest capacitor value. However, the dielectric thickness is always equal to the sum of all the capacitor dielectrics, and this factor always causes the total capacitance (C_T) to be less than the smallest individual capacitance when capacitors are connected in series.

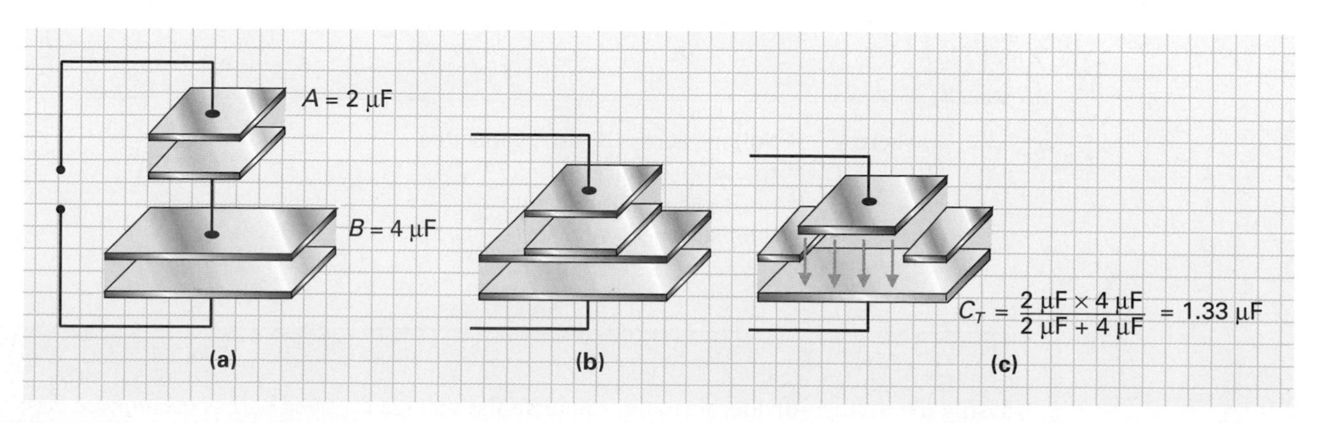

FIGURE 9-12 Capacitors in Series.

The total capacitance of two or more capacitors in series therefore is calculated by using the following formulas. For two capacitors in series,

$$C_T = \frac{C_1 \times C_2}{C_1 + C_2}$$

(product-over-sum formula)

For more than two capacitors in series,

$$C_T = \frac{1}{(1/C_1) + (1/C_2) + (1/C_3) + \cdots}$$

(reciprocal formula)

EXAMPLE:

Determine the total capacitance of the circuit in Figure 9-13.

Solution:

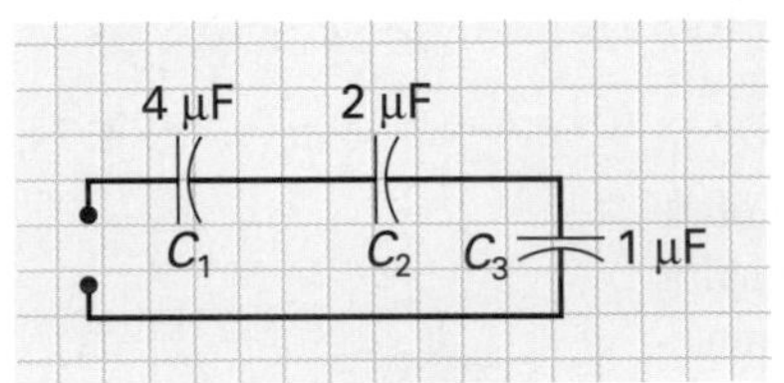

FIGURE 9-13 Example of Series-Connected Capacitors.

$$\begin{aligned} C_T &= \frac{1}{(1/C_1) + (1/C_2) + (1/C_3)} \\ &= \frac{1}{(1/4\ \mu\text{F}) + (1/2\ \mu\text{F}) + (1/1\ \mu\text{F})} \\ &= \frac{1}{1.75 \times 10^6} = 5.7143 \times 10^{-7} \\ &= 0.5714\ \mu\text{F} \quad \text{or} \quad 0.6\ \mu\text{F} \end{aligned}$$

The total capacitance for capacitors in series is calculated in the same way as total resistance when resistors are in parallel.

As with series-connected resistors, the sum of all of the voltage drops across the series-connected capacitors will equal the voltage applied (Kirchhoff's voltage law). With capacitors connected in series, the charged capacitors act as a voltage divider, and therefore the voltage-divider formula can be applied to capacitors in series.

$$V_{cx} = \frac{C_T}{C_x} \times V_T$$

where V_{cx} = voltage across desired capacitor
C_T = total capacitance
C_x = desired capacitor's value
V_T = total supplied voltage

EXAMPLE:

Using the voltage-divider formula, calculate the voltage dropped across each of the capacitors in Figure 9-13 if $V_T = 24$ V.

■ ***Solution:***

$$V_{C1} = \frac{C_T}{C_1} \times V_T = \frac{0.5714\ \mu\text{F}}{4\ \mu\text{F}} \times 24\text{ V} = 3.4\text{ V}$$

$$V_{C2} = \frac{C_T}{C_2} \times V_T = \frac{0.5714\ \mu\text{F}}{2\ \mu\text{F}} \times 24\text{ V} = 6.9\text{ V}$$

$$V_{C3} = \frac{C_T}{C_3} \times V_T = \frac{0.5714\ \mu\text{F}}{1\ \mu\text{F}} \times 24\text{ V} = 13.7\text{ V}$$

$$V_T = V_{C1} + V_{C2} + V_{C3} = 3.4 + 6.9 + 13.7 = 24\text{ V}$$

(Kirchoff's voltage law)

If the capacitor values are the same, as seen in Figure 9-14(a), the voltage is divided equally across each capacitor, as each capacitor has an equal amount of charge and therefore has half of the applied voltage (in this example, 3 V across each capacitor).

When the capacitor values are different, the smaller-value capacitor will actually charge to a higher voltage than the larger capacitor. In the example in Figure 9-14(b), the smaller capacitor is actually half the size of the other capacitor, and it has charged to twice the voltage. Since Kirchhoff's voltage law has to apply to this and every series circuit, you can easily calculate that the voltage across C_1 will equal 4 V and is twice that of C_2, which is 2 V. To understand this fully, we must first understand that although the capacitance is different, both capacitors have an equal value of coulomb charge held within them, which in this example is 8 μC.

$$Q_1 = C_1 \times V_1$$
$$= 2\ \mu\text{F} \times 4\text{ V} = 8\ \mu\text{C}$$
$$Q_2 = C_2 \times V_2$$
$$= 4\ \mu\text{F} \times 2\text{ V} = 8\ \mu\text{C}$$

This equal charge occurs because the same amount of current flow exists throughout a series circuit, so both capacitors are being supplied with the same number or quantity of electrons. The charge held by C_1 is large with respect to its small capacitance, whereas the same charge held by C_2 is small with respect to its larger capacitance.

6 V, C_1 2 µF, C_2 2 µF

$$V_{C1} = \frac{C_T}{C_1} \times V_T = \frac{1\ \mu\text{F}}{2\ \mu\text{F}} \times 6\text{ V} = 3\text{ V}$$

$$V_{C2} = \frac{C_T}{C_2} \times V_T = \frac{1\ \mu\text{F}}{2\ \mu\text{F}} \times 6\text{ V} = 3\text{ V}$$

(a)

6 V, C_1 2 µF, C_2 4 µF

$$V_{C1} = \frac{C_T}{C_1} \times V_T = \frac{1.33\ \mu\text{F}}{2\ \mu\text{F}} \times 6\text{ V} = 4\text{ V}$$

$$V_{C2} = \frac{C_T}{C_2} \times V_T = \frac{1.33\ \mu\text{F}}{4\ \mu\text{F}} \times 6\text{ V} = 2\text{ V}$$

(b)

FIGURE 9-14 Voltage Drops across Series-Connected Capacitors.

If the charge remains the same (Q is constant) and the capacitance is small, the voltage drop across the capacitor will be large, because the charge is large with respect to the capacitance:

$$V\uparrow = \frac{Q}{C\downarrow}$$

On the other hand, for a constant charge, a large capacitance will have a small charge voltage because the charge is small with respect to the capacitance:

$$V\downarrow = \frac{Q}{C\uparrow}$$

We can apply the water analogy once more and imagine two series-connected buckets, one of which is twice the size of the other. Both are being supplied by the same series pipe, which has an equal flow of water throughout, and are consequently each holding an equal amount of water, for example, 1 gallon. The 1 gallon of water in the small bucket is large with respect to the size of the bucket, and a large amount of pressure exists within that bucket. The 1 gallon of water in the large bucket is small with respect to the size of the bucket, so a small amount of pressure exists within this bucket. The pressure within a bucket is similar to the voltage across a capacitor, and therefore a small bucket or capacitor will have a greater pressure or voltage associated with it, while a large bucket or capacitor will develop a small pressure or voltage.

To summarize capacitors in series, all the series-connected components will have the same charging current throughout the circuit, and because of this, two or more capacitors in series will always have equal amounts of coulomb charge. If the charge (Q) is equal, the voltage across the capacitor is determined by the value of the capacitor. A small capacitance will charge to a larger voltage ($V\uparrow = Q/C\downarrow$), whereas a large value of capacitance will charge to a smaller voltage ($V\downarrow = Q/C\uparrow$).

9-1-7 *Dielectric Breakdown and Leakage*

Capacitors store a charge just as a container or tank stores water. The amount of charge stored by a capacitor is proportional to the capacitor's capacitance and the voltage applied across the capacitor ($Q = C \times V$). The charge stored by a fixed-value capacitor (C is fixed) therefore can be increased by increasing the voltage across the plates, as shown in Figure 9-15(a). If the voltage across the capacitor is increased further, the charge held by the capacitor will increase until the dielectric between the two plates of the capacitor breaks down and a spark jumps or arcs between the plates.

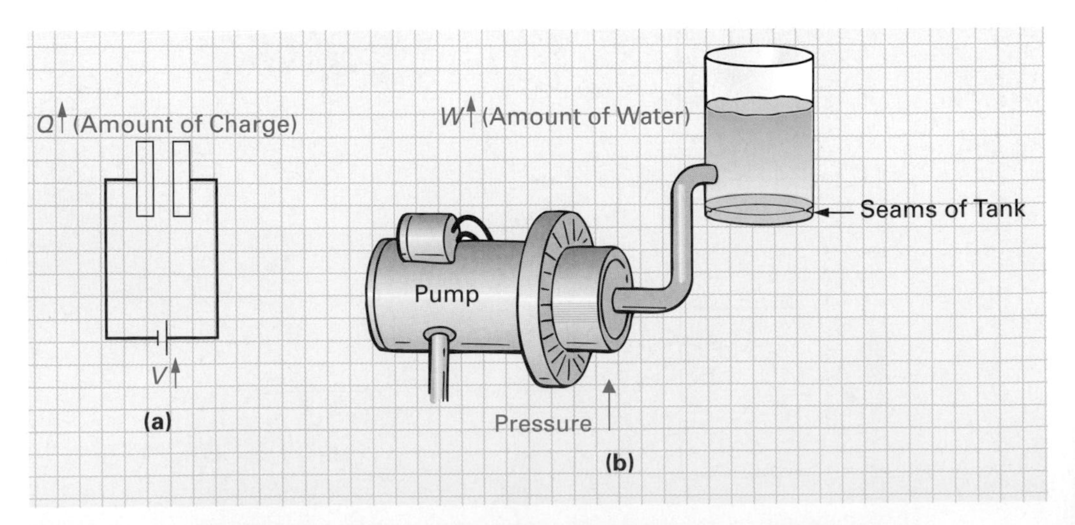

FIGURE 9-15 Breakdown. (a) Voltage Increase, Charge Increases until Dielectric Breakdown. (b) Pressure Increase, Water Increases until Seams Break Down.

TABLE 9-2 Dielectric Strengths

MATERIAL	DIELECTRIC STRENGTH (V/mm)
Air	787
Oil	12,764
Ceramic	39,370
Paper	49,213
Teflon	59,055
Mica	59,055
Glass	78,740

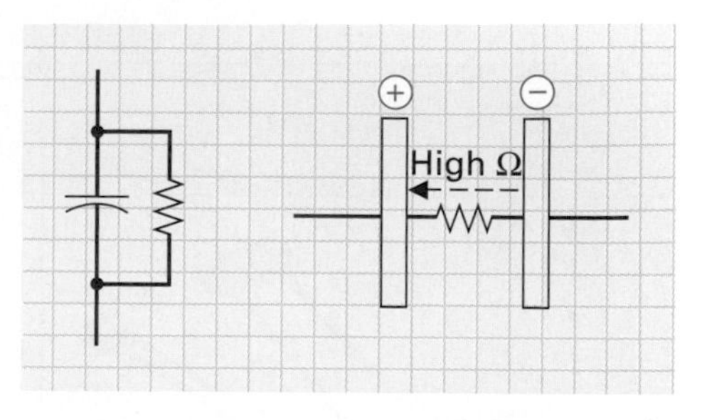

FIGURE 9-16 **Capacitor Leakage.**

Using the water analogy shown in Figure 9-15(b), if the pressure of the water being pumped in is increased, the amount of water stored in the tank will also increase until a time is reached when the tank's seams at the bottom of the tank cannot contain the large amount of pressure and break down under strain. The amount of water stored in the tank is proportional to the pressure applied, just as the amount of charge stored in a capacitor is proportional to the amount of voltage applied.

The **breakdown voltage** of a capacitor is determined by the strength of the dielectric used. Table 9–2 illustrates some of the different strengths of many of the common dielectrics. As an example, let's consider a capacitor that uses 1 mm of air as a dielectric between its two plates. This particular capacitor can withstand any voltage up to 787 V. If the voltage is increased further, the dielectric will break down and current will flow between the plates, destroying the capacitor (air capacitors, however, can recover from ionization).

Breakdown Voltage
The voltage at which breakdown occurs in a dielectric or insulation.

The ideal or perfect insulator should have a resistance equal to infinite ohms. Insulators or the dielectric used to isolate the two plates of a capacitor are not perfect and therefore have some very high values of resistance. This means that some value of resistance exists between the two plates, as shown in Figure 9-16; although this value of resistance is very large, it will still allow a small amount of current to flow between the two plates (in most applications, a few nanoamperes or picoamperes). This small current, referred to as **leakage current**, causes any charge in a capacitor to slowly, over a long period of time, discharge between the two plates. A capacitor should ideally have a large leakage resistance to ensure the smallest possible leakage current.

Leakage Current
Small, undesirable flow of current through an insulator or dielectric.

SELF-TEST EVALUATION POINT FOR SECTION 9-1

Now that you have completed this section, you should be able to:

Objective 1. List the principal parts that make up a capacitor.

Objective 2. Explain the charging and discharging process and its relationship to electrostatics.

Objective 3. State the unit of capacitance and explain how it relates to charge and voltage.

Objective 4. List and explain the factors determining capacitance.

Objective 5. Calculate total capacitance in parallel and series capacitance circuits.

Objective 6. Describe capacitance breakdown and capacitor leakage.

Use the following questions to test your understanding of Section 9-1:

1. True or false: When both plates of a capacitor have an equal charge, the capacitor is said to be charged.
2. If 2 μA of current flows into one plate of a capacitor and 2 μA flows out of the other plate, how much current is flowing through the dielectric of this working capacitor?
3. What is the unit of capacitance?
4. State the formula for capacitance in relation to charge and voltage.
5. Convert 30,000 μF to farads.
6. If a capacitor holds 17.5 C of charge when 9 V is applied, what is the capacitance of the capacitor?
7. State the formula for capacitance.
8. If 7 pF, 2 pF, and 14 pF capacitors are connected in parallel, what will be the total circuit capacitance?
9. State the voltage-divider formula as it applies to capacitance.
10. True or false: With resistors, the large-value resistor will drop a larger voltage, whereas with capacitors the smaller-value capacitor will actually charge to a higher voltage.

Name	Physical Appearance	Construction	Approximate Range of Values and Tolerances	Characteristics and Applications
Silver Mica	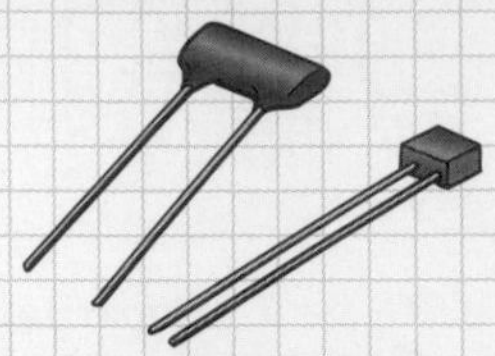	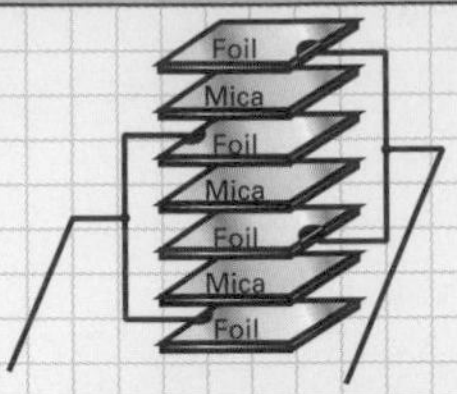 Foil, Mica, Foil, Mica, Foil, Mica, Foil	1 pF–0.1 μF ±1% to ±5% 50 V–500 V	Lower voltage rating than other capacitors of the same size. Used in oscillator circuits and for coupling.
Ceramic (Surface Mount Technology –SMT)	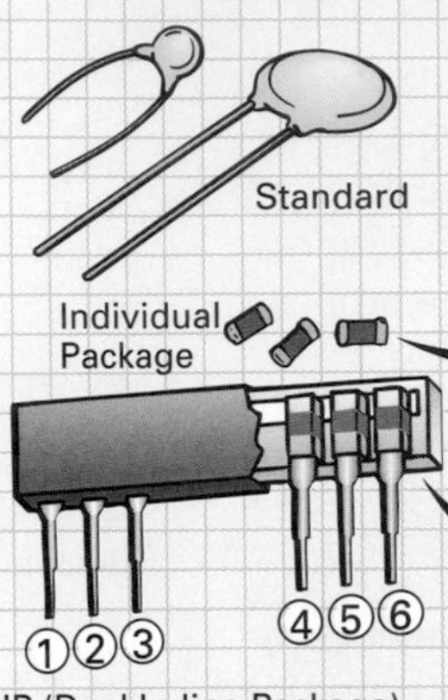 Standard Individual Package ①②③ ④⑤⑥ DIP (Dual In-line Package)	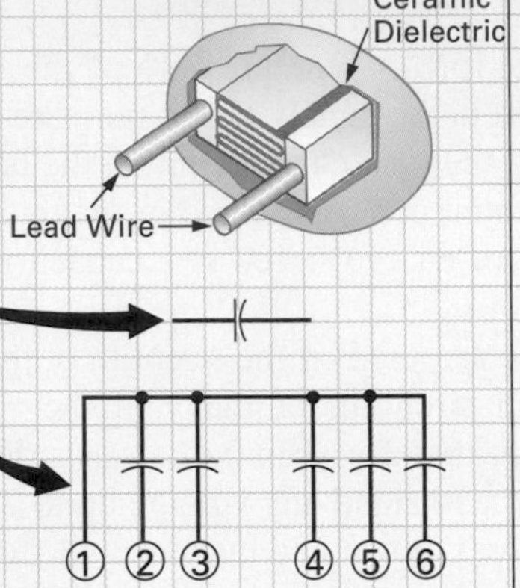 Ceramic Dielectric Lead Wire ① ② ③ ④ ⑤ ⑥	*Low Dielectric K*: 1 pF–0.01 μF ±0.5% to ±10% *High Dielectric K*: 1 pF–0.1 μF ±10% to ±80% Standard: 10 pF–0.047 μF 100 V–6 kV SMT: 10 pF–10 μF 6.3V–16V	Most popular small value capacitor due to lower cost and its ruggedness. Standard: oscillators, filters and coupling. SMT: High-density boards where space saving is crucial.
Mylar Paper	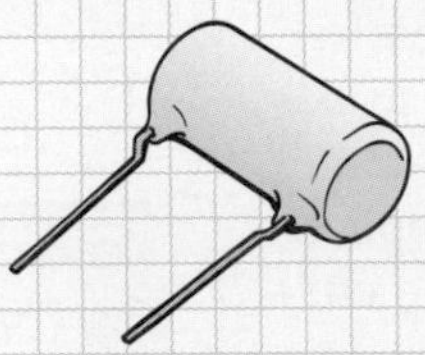	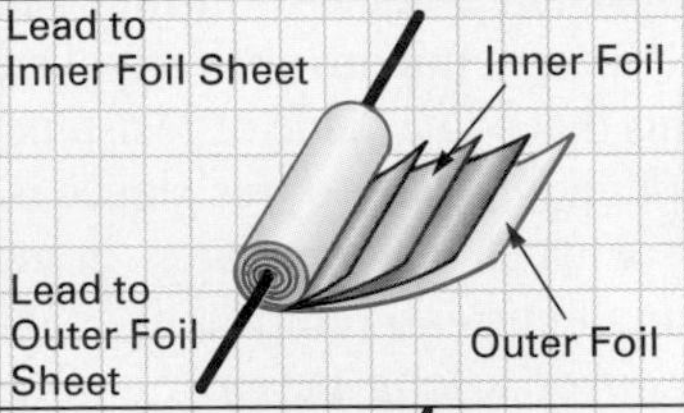 Lead to Inner Foil Sheet Inner Foil Lead to Outer Foil Sheet Outer Foil	1 pF–1 μF ± 10% 50 V–600 V	Has a large plate area and therefore large capacitance for a small size.
Plastic	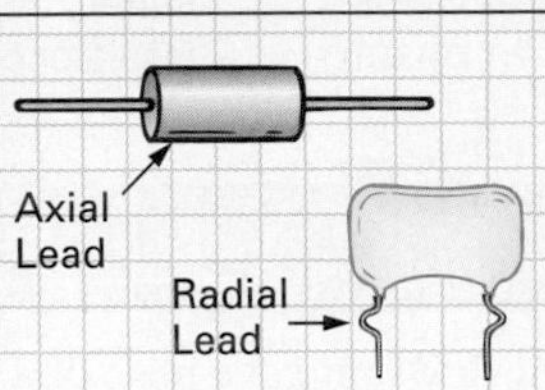 Axial Lead Radial Lead	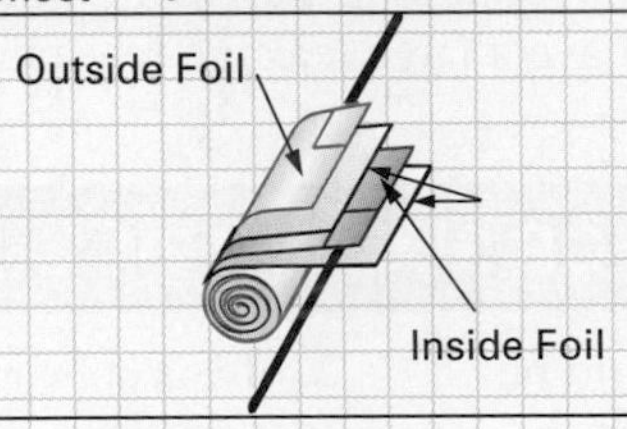 Outside Foil Inside Foil	1 pF–10 μF ± 5% to ±10% 50 V–600 V	Has almost completely replaced paper capacitors; has large capacitance values for small size and high voltage ratings.
Electrolytic (Aluminum and Tantalum)	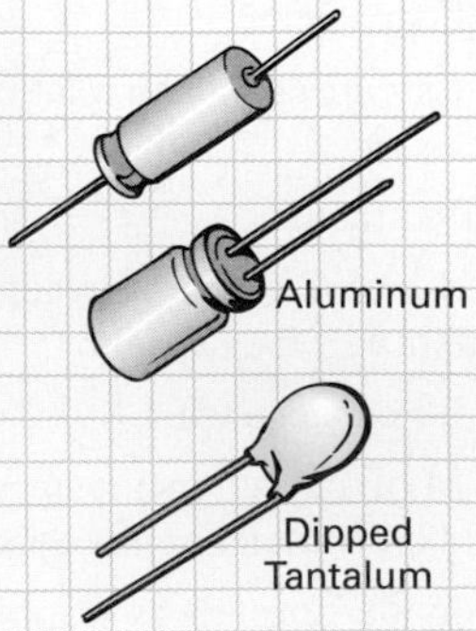 Aluminum Dipped Tantalum	(Tantalum has tantalum rather than aluminum foil plates) 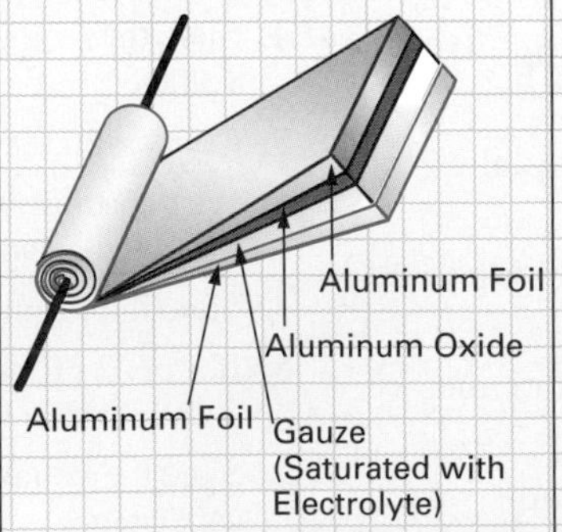 Aluminum Foil Aluminum Oxide Aluminum Foil Gauze (Saturated with Electrolyte)	Aluminum: 0.1 μF–15,000 μF 5 V–450 V ±20% Tantalum: 0.047 μF–470 μF 6.3 V–50 V ±10% to ±20%	Most popular large-value capacitor: large capacitance into small area, wide range of values. Used in power supplies. Disadvantages are: cannot be used in ac circuits as they are polarized; poor tolerances; low leakage resistance and so high leakage current. *Tantalum* advantages over aluminum include smaller size, longer life than aluminum, which has an approximate lifespan of 12 years. Disadvantages: 4 to 5 times the price.

Small-Capacitor Values (Silver Mica, Ceramic); Large-Capacitor Values (Mylar Paper, Plastic, Electrolytic)

FIGURE 9-17 Fixed-Value Capacitor Types.

9-2 CAPACITOR TYPES, CODING, AND TESTING

Capacitors come in a variety of shapes and sizes and can be either fixed or variable in their values of capacitance. Within these groups, capacitors are generally classified by the dielectric used between the plates.

9-2-1 *Capacitor Types*

A **fixed-value capacitor** is a capacitor whose capacitance value remains constant and cannot be altered. Fixed capacitors normally come in a disk or a tubular package and consist of metal foil plates separated by some type of insulator (dielectric), which is the means by which we classify them, as shown in Figure 9-17.

Fixed-Value Capacitor
A capacitor whose value is fixed and cannot be varied.

Variable-value capacitors are shown in Figure 9-18. The *trimmer variable* capacitor type generally has one stationary plate and one spring-metal moving plate. The screw forces

Variable-Value Capacitor
A capacitor whose value can be varied.

Name	Physical Appearance	Construction	Characteristics and Applications
Trimmer Variable (Mica, Ceramic and Plastic)		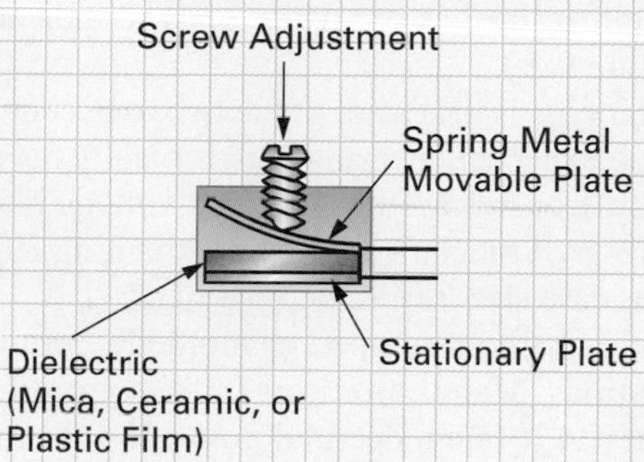 Screw Adjustment Spring Metal Movable Plate Dielectric (Mica, Ceramic, or Plastic Film) Stationary Plate	1.5 pF to 600 pF 5V – 100V ±10% Oscillators, Tuning Circuits and Filters
Tuning Variable (Air)	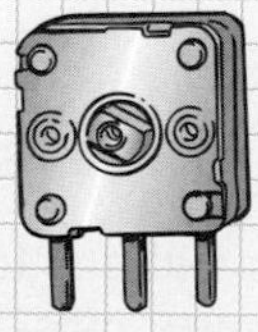	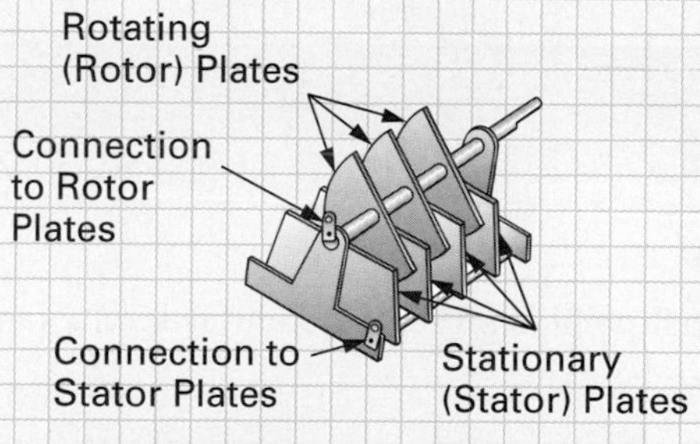 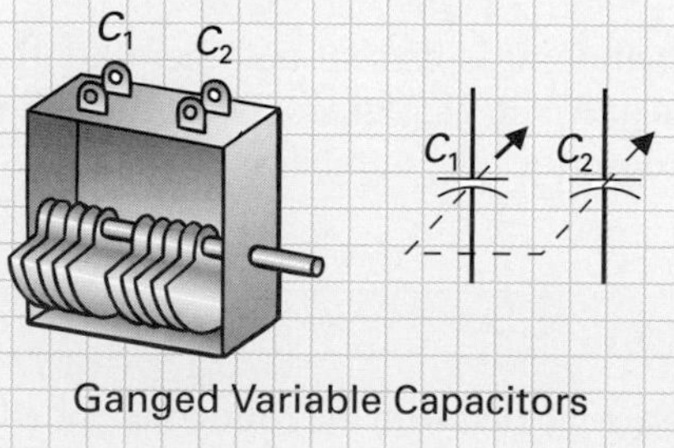 Rotating (Rotor) Plates Connection to Rotor Plates Connection to Stator Plates Stationary (Stator) Plates C_1 C_2 C_1 C_2 Ganged Variable Capacitors	10 pF to .2 μF 5V – 100V ±10% Radio Tuning Circuits, Oscillators and Filters

FIGURE 9-18 **Variable-Value Capacitor Types.**

the spring-metal plate closer or farther away from the stationary plate, varying the distance between the plates and so changing capacitance. The two plates are insulated from one another by either mica, ceramic, or plastic film, and the advantages of each are the same as for fixed-value capacitors. With the *tuning variable* capacitor type the effective plate area is adjusted to vary capacitance by causing a set of rotating plates (rotor) to mesh with a set of stationary plates (stator). When the rotor plates are fully out, the capacitance is minimum, and when the plates are fully in, the capacitance is maximum, because the maximum amount of rotor plate area is now opposite the stator plate, creating the maximum value of capacitance.

In radio equipment, it is sometimes necessary to have two or more variable-value capacitors that have been constructed in such a way that a common shaft (rotor) runs through all the capacitors and varies their capacitance simultaneously, as shown in Figure 9-18. If you mechanically couple (gang) two or more variable capacitors so that they can all be operated from a single control, the component is known as a **ganged capacitor**; the symbol for this arrangement is also shown in the figure.

Ganged Capacitor
Mechanical coupling of two or more capacitors, switches, potentiometers, or any other components so that the activation of one control will operate all sections.

Variable-value capacitors should only be adjusted with a plastic or nonmetallic alignment tool, because a metal screwdriver may affect the capacitance of the capacitor when nearby, making it very difficult to adjust for a specific value of capacitance.

9-2-2 *Capacitor Coding*

Manufacturers today most commonly use letters of the alphabet and numbers (alphanumerics) printed on either the disk or tubular body to indicate the capacitor's specifications, as illustrated in Figure 9-19.

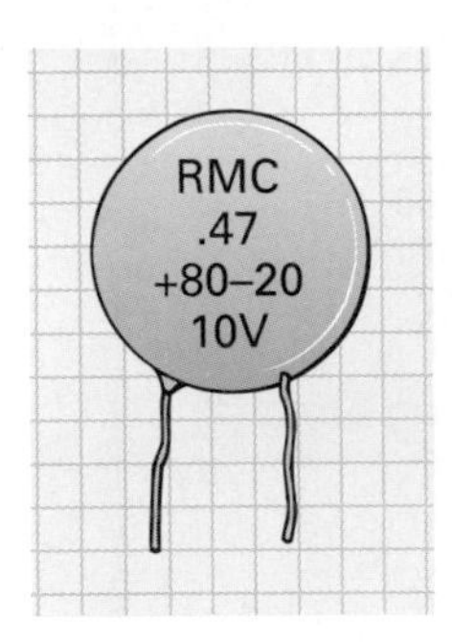

FIGURE 9-19 Alphanumeric Coding of Capacitors. 0.47 μF, +80%, −20% Tolerance, 10 V.

The tubular type is the easier of the two to decode since the information is basically uncoded. The value of capacitance and unit, typically the microfarad (μF or MF), tolerance figure (preceded by ± or followed by %), and voltage rating (followed by a V for voltage) are printed on all sizes of tubular cases. The remaining letters or numbers are merely manufacturers' codes for case size, series, and the like.

With disk capacitors (dipped or molded), certain rules have to be applied when decoding the notations. Many capacitors of this type do not define the unit of capacitance; in this situation, try to locate a decimal point. If a decimal point exists, for example 0.01 or 0.001, the value is in microfarads (10^{-6}). If no decimal point exists, for example 50 or 220, the value is in picofarads (10^{-12}) and you must analyze the number in a little more detail.

EXAMPLE:

What is the value of a capacitor if it is labeled 50, 50 V, ±5?

Solution:

Since no decimal point is present, the unit is in picofarads:

$$50 \text{ pF}, \quad 50 \text{ V}, \quad \pm 5\%$$

If no decimal point is present and there are three digits with the last digit a zero, the value is as stands and is in picofarads. If the third digit is a number other than 0 (1 to 9), it is a multiplier and describes the number of zeros to be added to the picofarad value.

EXAMPLE:

If two capacitors are labeled with the following coded values, how should they be interpreted?

a. 220
b. 104

Solution:

Since no decimal point is present, they are both in picofarads.

a. If the last of the three digits is a zero, the value is as it stands.

$$220 = 220 \text{ pF}$$

three-digit value

b. If the third digit is a number from 1 to 9, it is a multiplier.

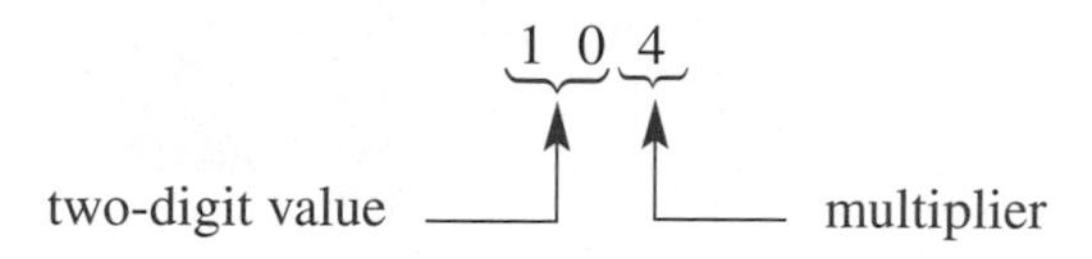

So,

$$10 \times 10^4 = 100{,}000 \text{ pF}$$

or

$$100{,}000 \times 10^{-12} = 0.1 \times 10^{-6} = 0.1 \; \mu\text{F}$$

The tolerance of the capacitor is sometimes clearly indicated, for example, ± 5 or 10%; in other cases, a letter designation is used, such as

F = ±1%
G = ±2%
J = ±5%
K = ±10%
M = ±20%
Z = −20%, +80%

Unfortunately, there does not seem to be a standard among capacitor manufacturers, which can cause confusion when trying to determine the value of capacitance. Therefore, if you are not completely sure, you should always measure the value or consult technical data or information sheets from the manufacturer.

9-2-3 *Capacitor Testing*

Now that you have a good understanding as to how a capacitor should function, let us investigate how to diagnose a capacitor malfunction.

1. The Ohmmeter

A faulty capacitor may have one of three basic problems:

1. A short, which is easy to detect and is caused by a contact from plate to plate.
2. An open, which is again quite easy to detect and is normally caused by one of the leads becoming disconnected from its respective plate.
3. A leaky dielectric or capacitor breakdown, which is quite difficult to detect, as it may only short at a certain voltage. This problem is usually caused by the deterioration of the dielectric, which starts displaying a much lower dielectric resistance than it was designed for. The capacitor with this type of problem is referred to as a *leaky capacitor*.

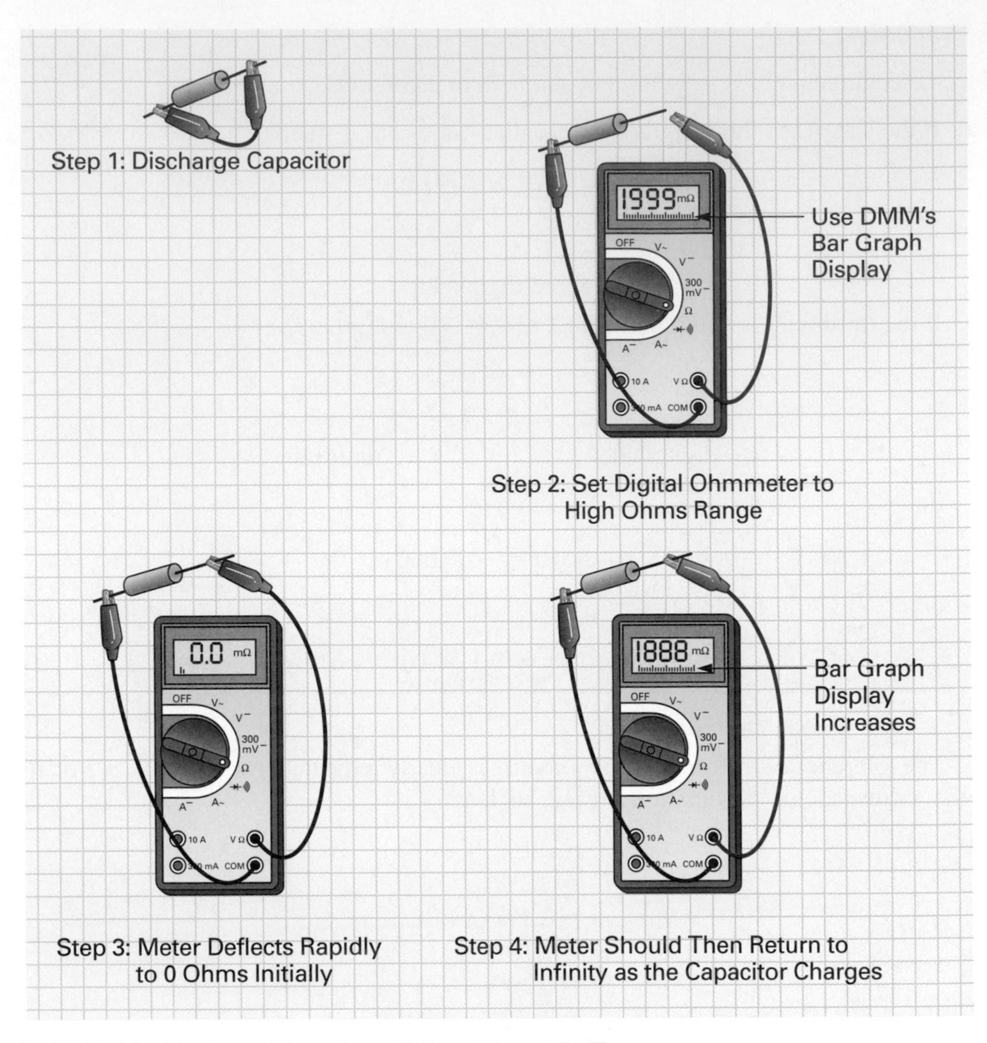

FIGURE 9-20 Testing a Capacitor of More Than 0.5 μF.

Capacitors of 0.5 μF and larger can be checked by using an analog ohmmeter or a digital multimeter with a bar graph display by using the procedure shown in Figure 9-20.

Step 1: Ensure that the capacitor is discharged by shorting the leads together.

Step 2: Set the multimeter to the ohms scale.

Step 3: Connect the meter to the capacitor, observing the correct polarity if an electrolytic is being tested, and observe the bar graph display. The capacitor will initially be discharged, and therefore maximum current will flow from the meter battery to the capacitor. Maximum current means low resistance, which is why the meter's display indicates 0 Ω.

Step 4: As the capacitor charges, it will cause current flow from the meter's battery to decrease, and consequently, the bar graph display will increase.

A good capacitor will cause the meter to react as just explained. A larger capacitance will cause the meter to increase slowly to infinity (∞), as it will take a longer time to charge; a smaller value of capacitance will charge at a much faster rate, causing the meter to increase rapidly toward ∞. For this reason, the ohmmeter cannot be reliably used to check capacitors with values of less than 0.5 μF, because the capacitor charges up too quickly and the meter does not have enough time to respond.

A shorted capacitor will cause the meter to show 0 ohms and remain in that position. An open capacitor will cause a maximum reading (infinite resistance) because there is no path for current to flow. A leaky capacitor will cause the bar graph to reduce its bars to the left, and then return almost all the way back to ∞ if only a small current is still flowing and the capacitor has a small dielectric leak. If the meter bars only come back to halfway or a large distance away from infinity, a large amount of current is still flowing and the capacitor has a large dielectric leak (defect).

When using the ohmmeter to test capacitors, there are some other points that you should be aware of:

1. Electrolytics are noted for having a small yet noticeable amount of inherent leakage, and so do not expect the meter's bar display to move all the way to the right (∞ ohms). Most electrolytic capacitors that are still functioning normally will show a resistance of 200 kΩ or more.
2. Some ohmmeters utilize internal batteries of up to 15 V, so be careful not to exceed the voltage rating of the capacitor.

2. The Capacitance Meter or Analyzer

The ohmmeter check tests the capacitor under a low-voltage condition. This may be adequate for some capacitor malfunctions; however, a problem that often occurs with capacitors is that they short or leak at a high voltage. The ohmmeter test is also adequate for capacitors of 0.5 μF or greater; however, a smaller capacitor cannot be tested because its charge time is too fast for the meter to respond. The ohmmeter cannot check for high-voltage failure, for small-value capacitance, or if the value of capacitance has changed through age or extreme thermal exposure.

A **capacitance meter** or analyzer, which is illustrated in Figure 9-21, can totally check all aspects of a capacitor in a range of values from approximately 1 pF to 20 F. The tests that are generally carried out include:

Capacitance Meter
Instrument used to measure the capacitance of a capacitor or circuit.

1. *Capacitor value change* [Figure 9-22(a)]: Capacitors will change their value over a period of time. Ceramic capacitors often change 10 to 15% within the first year as the ceramic material relaxes. Electrolytics change their value due to the electrolytic solution drying out. Some capacitors are simply labeled incorrectly by manufacturers or the technician cannot determine the correct value because of the labeling used. A value change accounts for approximately 25% of all defective capacitors.
2. *Capacitor leakage* [Figure 9-22(b)]: Leakage occurs due to an imperfection of the dielectric. Although the dielectric's resistance is very high, a small amount of leakage current will flow between the plates. This resistance, which is between the plates and therefore effectively in parallel with the capacitor, can become too low and cause the circuit that the capacitor is in to malfunction. Most capacitance meters will perform a leakage test with operating potentials up to 650 V. Leakage accounts for approximately 40% of all defective capacitors.

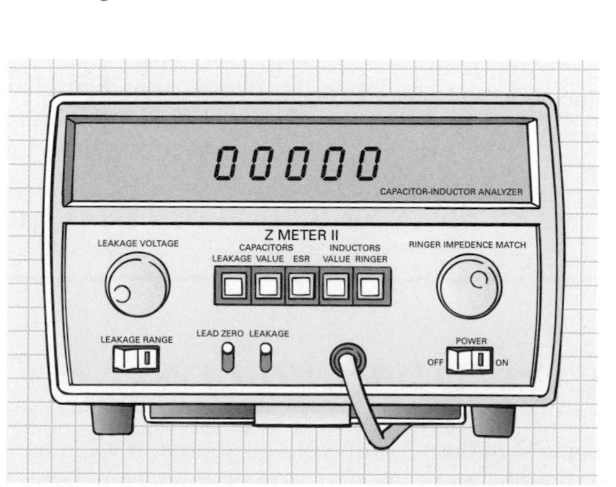

FIGURE 9-21 Capacitor Analyzer.

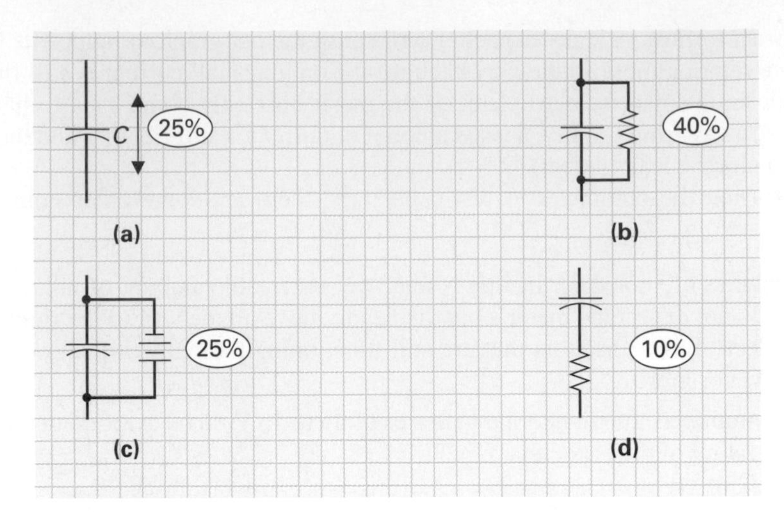

FIGURE 9-22 **Problems with Capacitors. (a) Value Change. (b) Leakage. (c) Dielectric Absorption. (d) Equivalent Series Resistance.**

3. *Dielectric absorption* [Figure 9-22(c)]: This occurs mainly in electrolytics when they take on a charge but will not fully discharge. This residual charge remains within the capacitor, similar to a small dc battery, and this changes the effective value of the capacitor once it is in circuit during operation. If this causes the value to change more than 15%, the capacitor should be rejected. Dielectric absorption accounts for approximately 25% of all defective capacitors.
4. *Equivalent series resistance* (ESR) [Figure 9-22(d)]: Series resistance is found in the capacitor leads, lead-to-plate connection, and electrolyte (almost always occurs in electrolytics) and causes the effective circuit capacitance value to change. Equivalent series resistance accounts for 10% of all defective capacitors.

SELF-TEST EVALUATION POINT FOR SECTION 9-2

Now that you have completed this section, you should be able to:

Objective 7. Describe the advantages of and differences between the five basic types of fixed capacitors.

Objective 8. Describe the advantages of and differences between the four basic types of variable capacitors.

Objective 9. Describe the coding of capacitor values on the body by use of alphanumerics or color.

Objective 10. Explain some of the more common capacitor failures and how to use an ohmmeter and capacitance analyzer to test them.

Use the following questions to test your understanding of Section 9-2:

1. List the five types of fixed-value capacitors.
2. Which capacitor type is said to be polarity conscious?

What are the following values of capacitance?

3. 470 ± 2
4. 0.47 ± 5
5. The analog ohmmeter or digital multimeter with bar graph can only be used to test capacitors that are ___________ or more in value.
6. A ___________ ___________ tests a capacitor for value change, leakage, dielectric absorption, and equivalent series resistance.

9-3 DC CAPACITIVE CIRCUITS

When a capacitor is connected across a dc voltage source, it will charge to a value equal to the voltage applied. If the charged capacitor is then connected across a load, the capacitor will then discharge through the load. The time it takes a capacitor to charge or discharge can

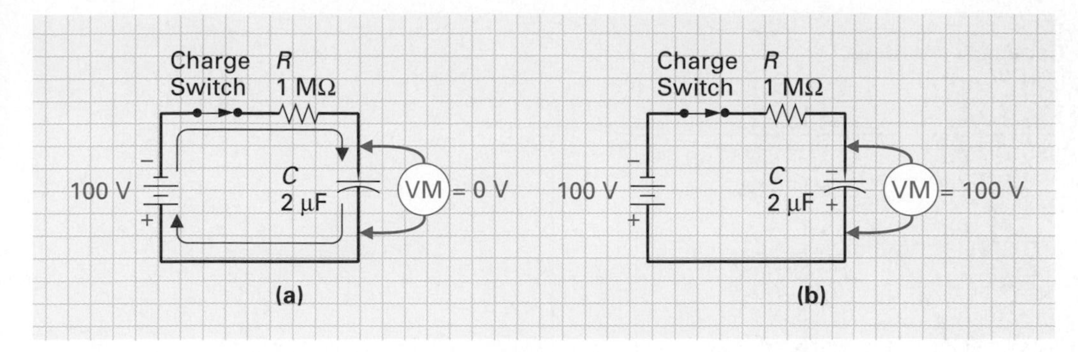

FIGURE 9-23 Capacitor Charging. (a) Switch Is Closed and Capacitor Begins to Charge. (b) Capacitor Charged.

be calculated if the circuit's resistance and capacitance are known. Let us now see how we can calculate a capacitor's charge time and discharge time.

9-3-1 *Charging*

When a capacitor is connected across a dc voltage source, such as a battery or power supply, current will flow and the capacitor will charge up to a value equal to the dc source voltage, as shown in Figure 9-23. When the charge switch is first closed, as seen in Figure 9-23(a), there is no voltage across the capacitor at that instant and therefore a potential difference exists between the battery and capacitor. This causes current to flow and begin charging the capacitor.

Once the capacitor begins to charge, the voltage across the capacitor does not instantaneously rise to 100 V. It takes a certain amount of time before the capacitor voltage is equal to the battery voltage. When the capacitor is fully charged no potential difference exists between the voltage source and the capacitor. Consequently, no more current flows in the circuit as the capacitor has reached its full charge, as seen in Figure 9-23(b). The amount of time it takes for a capacitor to charge to the supplied voltage (in this example, 100 V) is dependent on the circuit's resistance and capacitance value. If the circuit's resistance is increased, the opposition to current flow will be increased, and it will take the capacitor a longer period of time to obtain the same amount of charge because the circuit current available to charge the capacitor is less. If the value of capacitance is increased, it again takes a longer time to charge to 100 V because a greater amount of charge is required to build up the voltage across the capacitor to 100 V.

The circuit's resistance (R) and capacitance (C) are the two factors that determine the charge time (τ). Mathematically, this can be stated as

$$\tau = R \times C$$

where τ = **time constant** (s)
R = resistance (Ω)
C = capacitance (F)

Time Constant
Time needed for either a voltage or current to rise to 63.2% of the maximum or fall to 36.8% of the initial value. The time constant of an *RC* circuit is equal to the product of *R* and *C*.

In the following example, we use a resistance of 1 MΩ and a capacitance of 2 μF, which means that the time constant is equal to

$$\begin{aligned}\tau &= R \times C\\ &= 2\ \mu\text{F} \times 1\ \text{M}\Omega\\ &= (2 \times 10^{-6}) \times (1 \times 10^{6})\\ &= 2\ \text{s}\end{aligned}$$

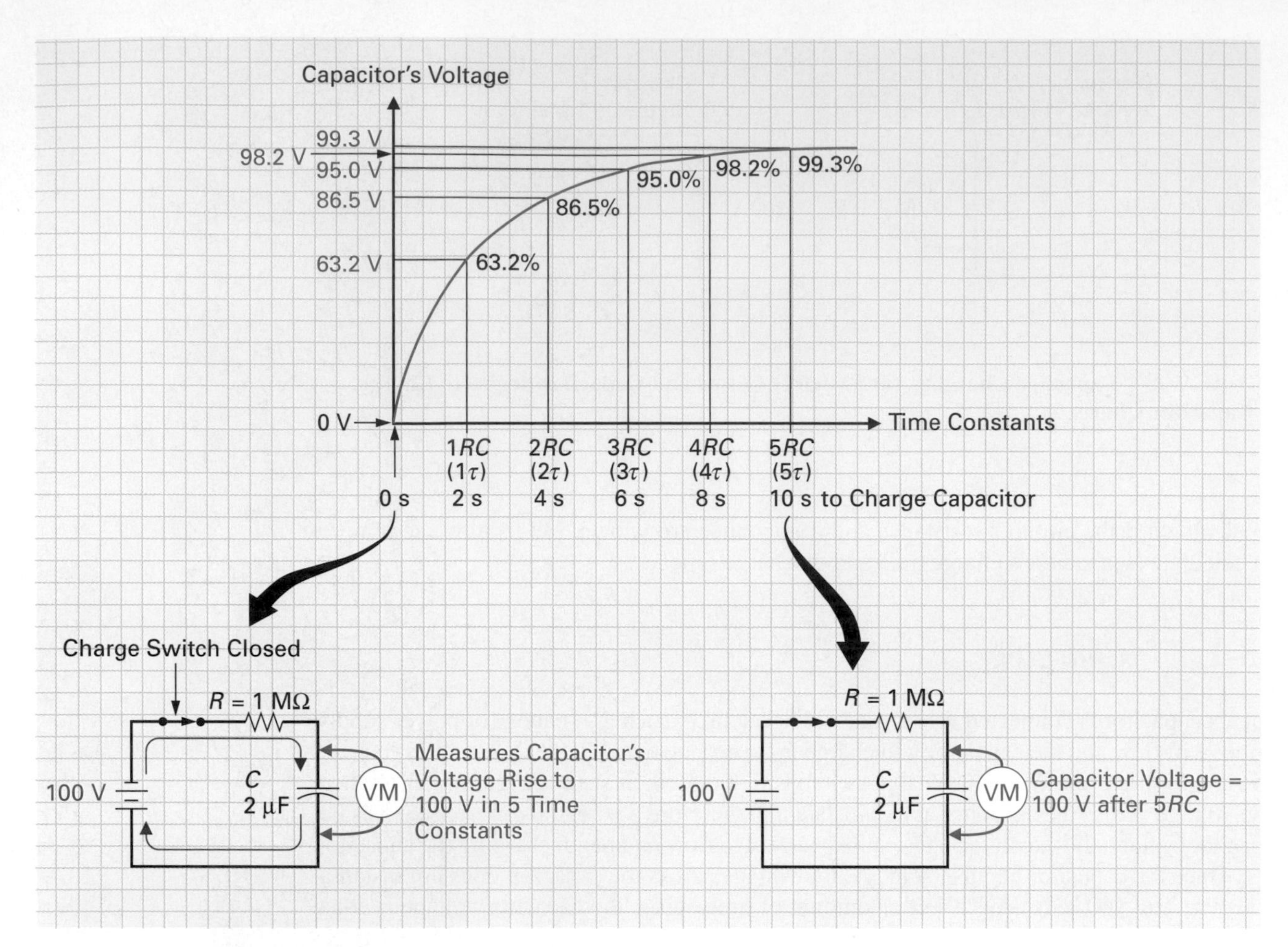

FIGURE 9-24 Charging Capacitor.

Two seconds is the time, so what is the constant? The constant value that should be remembered throughout this discussion is "**63.2**". Figure 9-24 illustrates the rise in voltage across the capacitor from 0 to a maximum of 100 V in five time constants ($5 \times 2\ \text{s} = 10\ \text{s}$). So where does 63.2 come into all this?

First time constant: In 1*RC* seconds ($1 \times R \times C = 2$ s), the capacitor will charge to 63.2% of the applied voltage (63.2% × 100 V = 63.2 V).

Second time constant: In 2*RC* seconds ($2 \times R \times C = 4$ s), the capacitor will charge to 63.2% of the remaining voltage. In the example, the capacitor will be charged to 63.2 V in one time constant, and therefore the voltage remaining is equal to 100 V − 63.2 V = 36.8 V. At the end of the second time constant, therefore, the capacitor will have charged to 63.2% of the remaining voltage (63.2% × 36.8 V = 23.3 V), which means that it will have reached 86.5 V (63.2 + 23.3 = 86.5 V) or 86.5% of the applied voltage.

Third time constant: In 3*RC* seconds (6 s), the capacitor will charge to 63.2% of the remaining voltage:

$$\begin{aligned}\text{Remaining voltage} &= 100\ \text{V} - 86.5\ \text{V} \\ &= 13.5\ \text{V} \\ 63.2\%\ \text{of}\ 13.5\ \text{V} &= 8.532\ \text{V}\end{aligned}$$

At the end of the third time constant, therefore, the capacitor will have charged to 86.5 V + 8.532 V = 95 V, or 95% of the applied voltage.

Fourth time constant: In $4RC$ seconds (8 s), the capacitor will have charged to 63.2% of the remaining voltage (100 V − 95 V = 5 V); therefore, 63.2% of 5 V = 3.2 V. So the capacitor will have charged to 95 V + 3.2 V = 98.2 V, or 98.2% of the applied voltage.

Fifth time constant: In $5RC$ seconds (10 s), the capacitor is considered to be fully charged since it will have reached 63.2% of the remaining voltage (100 V − 98.2 V = 1.8 V); therefore, 63.2% of 1.8 V = 1.1 V. So the capacitor will have charged to 98.2 V + 1.1 V = 99.3 V, or 99.3% of the applied voltage.

The voltage waveform produced by the capacitor acquiring a charge is known as an *exponential* waveform, and the voltage across the capacitor is said to rise exponentially. An exponential rise is also referred to as a *natural increase*. There are many factors that exponentially rise and fall. For example, we grow exponentially, in that there is quite a dramatic change in our height in the early years and then this increase levels off and reaches a maximum.

Before the switch is closed and even at the instant the switch is closed, the capacitor is not charged, which means that there is no capacitor voltage to oppose the supply voltage and, therefore, a maximum current of V/R, 100 V/1 MΩ = 100 μA flows. This current begins to charge the capacitor and a potential difference begins to build up across the plates of the capacitor, and this voltage opposes the supply voltage, causing a decrease in charging current. As the capacitor begins to charge, less of a potential difference exists between the supply voltage and capacitor voltage and so the current begins to decrease.

To calculate the current at any time, we can use the formula

$$i = \frac{V_S - V_C}{R}$$

where i = instantaneous current
V_S = source voltage
V_C = capacitor voltage
R = resistance

For example, the current flowing in the circuit after one time constant will equal the source voltage, 100 V, minus the capacitor's voltage, which in one time constant will be 63.2% of the source voltage or 63.2 V, divided by the resistance.

$$\begin{aligned} i &= \frac{V_S - V_C}{R} \\ &= \frac{100\text{ V} - 63.2\text{ V}}{1\text{ M}\Omega} \\ &= 36.8\ \mu\text{A} \end{aligned}$$

As the charging continues, the potential difference across the plates exponentially rises to equal the supply voltage, as seen in Figure 9-25(a), while the current exponentially falls to zero, as shown in Figure 9-25(b). The constant of 63.2 can be applied to the exponential fall of current from 100 μA to 0 μA in $5RC$ seconds.

When the switch was closed to start the charging of the capacitor, there was no charge on the capacitor; therefore, a maximum potential difference existed between the battery and capacitor, causing a maximum current flow of 100 μA ($I = V/R$).

First time constant: In $1RC$ seconds, the current will have exponentially decreased 63.2% (63.2% of 100 μA = 63.2 μA) to a value of 36.8 μA (100 μA − 63.2 μA). In the example of 2 μF and 1 MΩ, this occurs in 2 s.
Second time constant: In $2RC$ seconds ($2 \times R \times C$ = 4 s), the current will decrease 63.2% of the remaining current, which is

$$63.2\% \text{ of } 36.8\ \mu\text{A} = 23.26\ \mu\text{A}$$

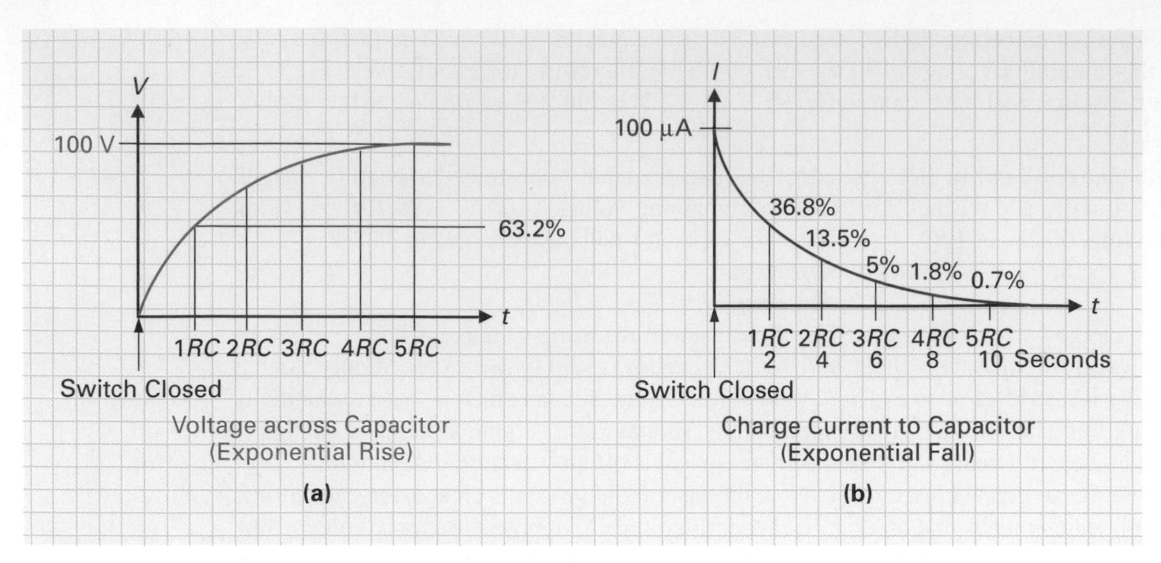

FIGURE 9-25 Exponential Rise in Voltage and Fall in Current in a Charging Capacitive Circuit.

The current will drop 23.26 μA from 36.8 μA and reach 13.5 μA or 13.5%.

> *Third time constant:* In 3*RC* seconds (6 s), the capacitor's charge current will decrease 63.2% of the remaining current (13.5 μA) to 5 μA or 5%.
> *Fourth time constant:* In 4*RC* seconds (8 s), the current will have decreased to 1.8 μA or 1.8%.
> *Fifth time constant:* In 5*RC* seconds (10 s), the charge current is now 0.7 μA or 0.7%. At this time, the charge current is assumed to be zero and the capacitor is now charged to a voltage equal to the applied voltage.

Studying the exponential rise of the voltage and the exponential decay of current in a capacitive circuit, you will notice an interesting relationship. In a pure resistive circuit, the current flow through a resistor will be in step with the voltage across that same resistor, in that an increased current will cause a corresponding increase in voltage drop across the resistor. Voltage and current are consequently said to be *in step* or *in phase* with one another. With the capacitive circuit, the current flow in the circuit and voltage across the capacitor are not in step or in phase with one another. When the switch is closed to charge the capacitor, the current is maximum (100 μA) and the voltage across the capacitor is zero. After five time constants (10 s), the capacitor's voltage is now maximum (100 V) and the circuit current is zero, as seen in Figure 9–26. The circuit current flow is out of phase with the capacitor voltage, and this difference is referred to as a *phase shift*. In any circuit containing capacitance, current will lead voltage.

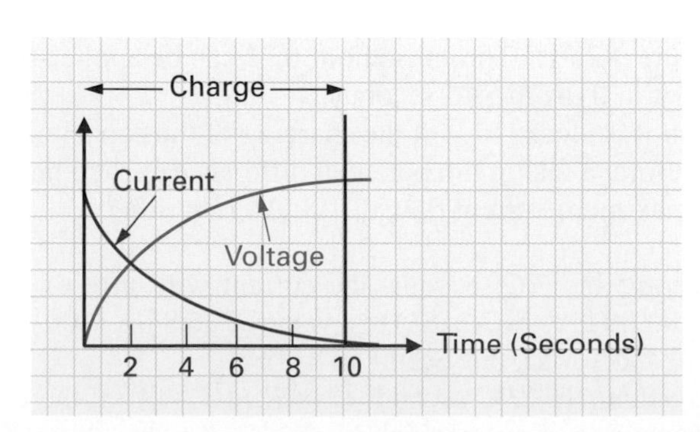

FIGURE 9-26 Phase Difference Between Voltage and Current in a Charging Capacitive Circuit.

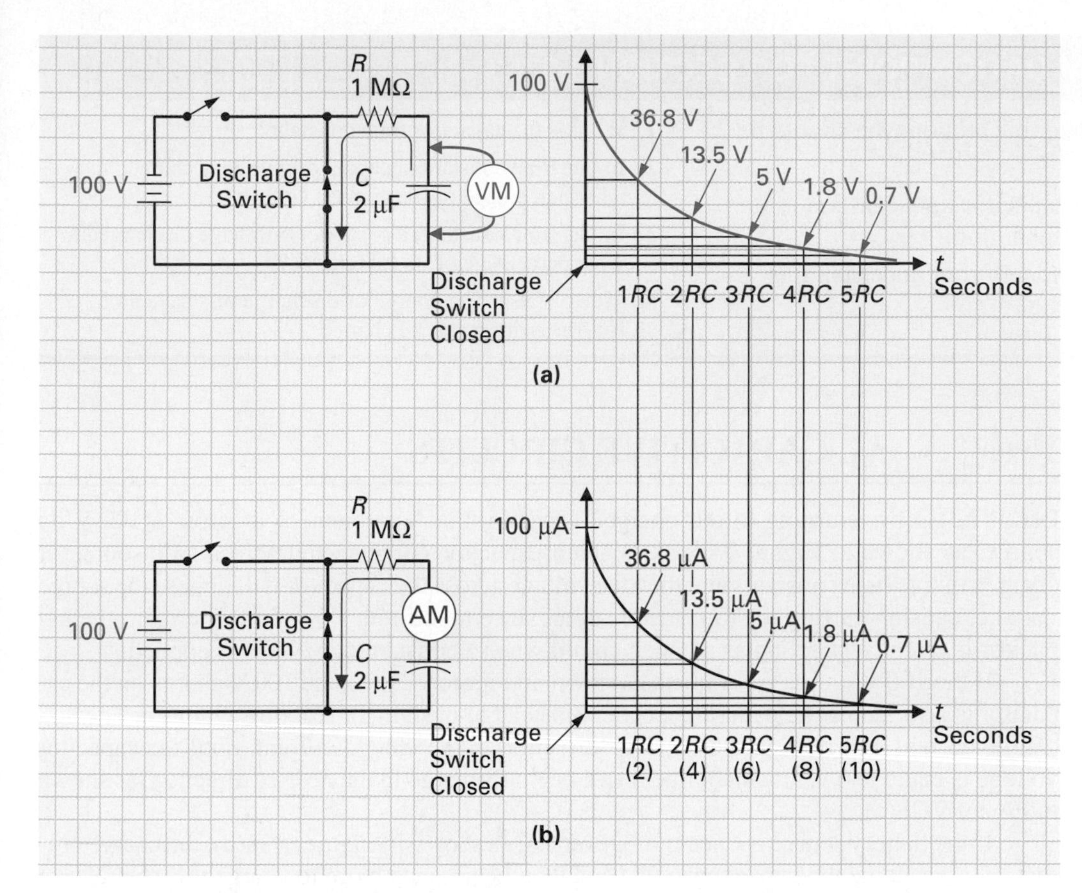

FIGURE 9-27 **Discharging Capacitor. (a) Voltage Waveform. (b) Current Waveform.**

9-3-2 *Discharging*

Figure 9-27 illustrates the circuit, voltage, and current waveforms that occur when a charged capacitor is discharged from 100 V to 0 V. The 2 μF capacitor, which was charged to 100 V in 10 s (5*RC*), is discharged from 100 to 0 V in the same amount of time.

Looking at the voltage curve, you can see that the voltage across the capacitor decreases exponentially, dropping 63.2% to 36.8 V in 1*RC* seconds, another 63.2% to 13.5 V in 2*RC* seconds, another 63.2% to 5 V in 3*RC* seconds, and so on, until zero.

The current flow within the circuit is dependent on the voltage in the circuit, which is across the 2 μF capacitor. As the voltage decreases, the current will also decrease by the same amount ($I\downarrow = V\downarrow/R$).

$$\text{Discharge switch closed: } I = \frac{V}{R} = \frac{100\ V}{1\ \text{M}\Omega} = 100\ \mu\text{A} \quad \text{maximum}$$

$$1RC\,(2)\ \text{seconds: } I = \frac{V}{R} = \frac{36.8\ \text{V}}{1\ \text{M}\Omega} = 36.8\ \mu\text{A}$$

$$2RC\,(4)\ \text{seconds: } I = \frac{V}{R} = \frac{13.5\ \text{V}}{1\ \text{M}\Omega} = 13.5\ \mu\text{A}$$

$$3RC\,(6)\ \text{seconds: } I = \frac{V}{R} = \frac{5\ \text{V}}{1\ \text{M}\Omega} = 5.0\ \mu\text{A}$$

$$4RC\,(8)\ \text{seconds: } I = \frac{V}{R} = \frac{1.8\ \text{V}}{1\ \text{M}\Omega} = 1.8\ \mu\text{A}$$

$$5RC\,(10)\ \text{seconds: } I = \frac{V}{R} = \frac{0.7\ \text{V}}{1\ \text{M}\Omega} = 0.7\ \mu\text{A} \quad \text{zero}$$

SELF-TEST EVALUATION POINT FOR SECTION 9-3

Now that you have completed this section, you should be able to:

Objective 12. Explain the capacitor time constant as it relates to dc charging and discharging.

Use the following questions to test your understanding of Section 9–3:

1. What is the capacitor time constant?
2. In one time constant, a capacitor will have charged to what percentage of the applied voltage?
3. In one time constant, a capacitor will have discharged to what percentage of its full charge?
4. True or false: The charge or discharge of a capacitor follows a linear rate of change.

9-4 AC CAPACITIVE CIRCUITS

Let us begin by returning to our charged capacitor that was connected across a 100 V dc source, as shown in Figure 9-28(a). When a capacitor is connected across a dc source, it charges in five time constants and then the voltage across the capacitor will oppose the source voltage, causing current to stop. At this time, since no current flows in this circuit, the capacitor is effectively acting as an open circuit when a constant dc voltage is applied.

If the 100 V source is reversed, as shown in Figure 9-28(b), the 100 V charge on the capacitor is now aiding the path of the battery instead of pushing or reacting against it. The discharge current will flow in the opposite direction to the charge current, until the capacitor voltage is equal to 0 V, at which time it will begin to charge in the reverse direction, as shown in Figure 9-28(c).

If the battery source is once more reversed, as shown in Figure 9-28(d), the charge current will discharge the capacitor to 0 V and then begin once more to charge it in the reverse direction, as shown in Figure 9-28(e).

The result of this switching or alternating procedure is that there is current flow in the circuit at all times (except for the instant when the battery source is removed and the polarity reversed). The switching of a dc voltage source in this way has almost the same effect as if we were applying an alternating voltage, as shown in Figure 9-28(f).

Figure 9-29 summarizes the capacitor's reaction to a dc source and an ac source. When a dc voltage is applied across a capacitive circuit, as shown in Figure 9-29(a), the capacitor will charge and then oppose the dc source, preventing current flow. Since no current can flow

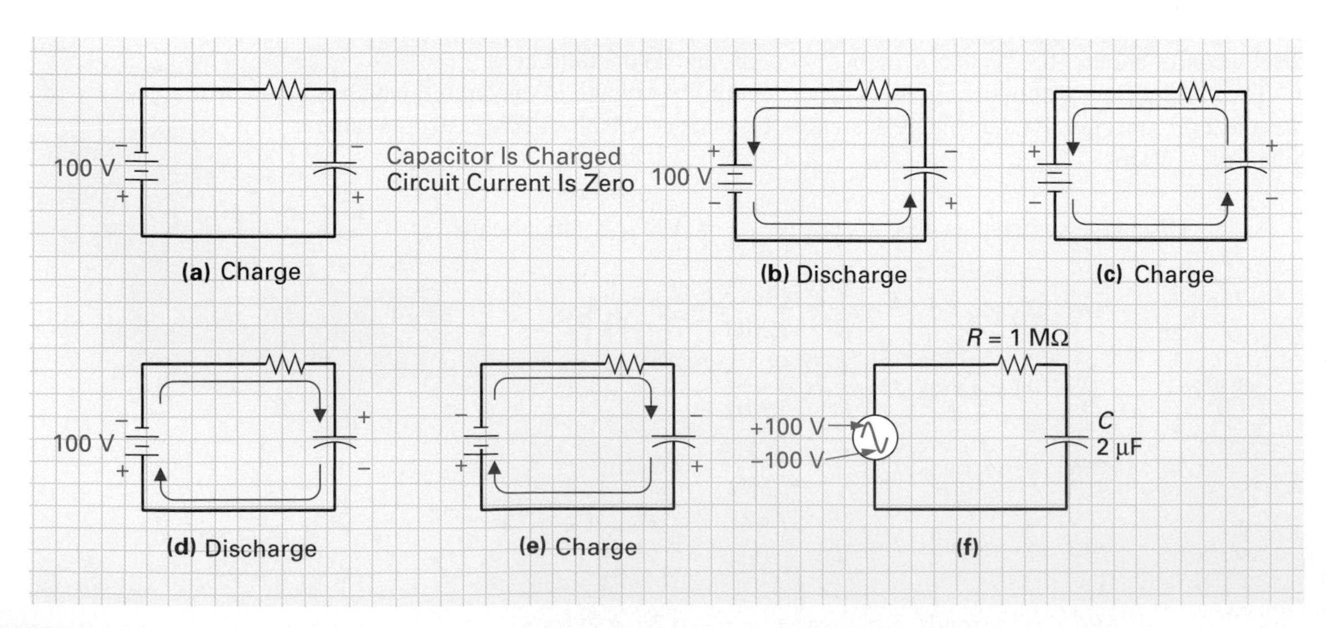

FIGURE 9-28 AC Charge and Discharge.

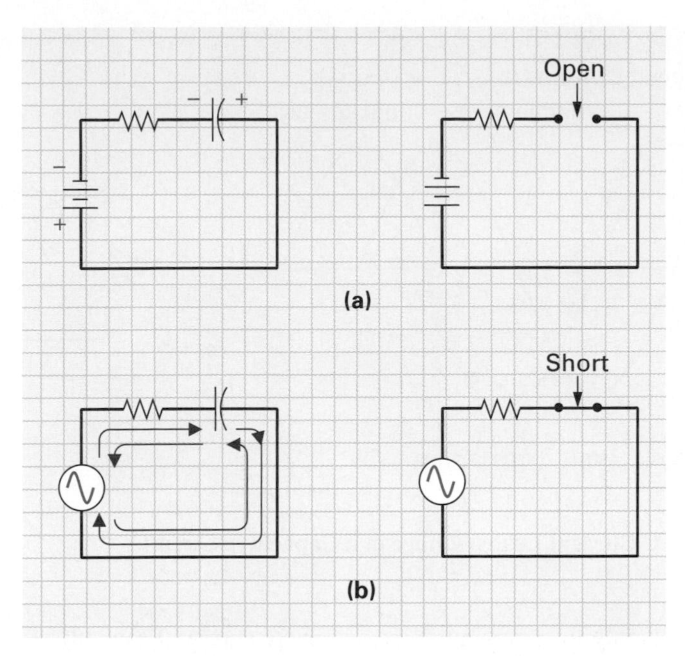

FIGURE 9-29 **Capacitors' Reaction to DC and AC. (a) Capacitor Will Block DC. (b) Capacitor Will Pass AC.**

when the capacitor is charged, the capacitor acts like an open circuit, and it is this action that accounts for why capacitors are referred to as an "open to dc."

When an ac voltage is applied across a capacitive circuit, as shown in Figure 9-29(b), the continual reversal of the applied voltage will cause a continual charging and discharging of the capacitor, and therefore circuit current will always be present. Any type of ac source or in fact any fluctuating or changing dc source will cause a circuit current, and if current is present, the opposition is low. It is this action that accounts for why capacitors are referred to as a "short to ac."

This ability of a capacitor to block dc and seem to pass ac will be exploited and explained later in applications of capacitors.

Capacitor is an open switch to dc, closed switch to ac

General Capacitor Rule

9-4-1 *Phase Relationship Between Current and Voltage*

Thinking of the applied alternating voltage as a dc source that is being continually reversed sometimes makes it easier to understand how a capacitor reacts to ac. Whether the applied voltage is dc or ac, the rule holds true in that a 90° phase shift or difference exists between circuit current and capacitor voltage. The exact relationship between ac current and voltage in a capacitive circuit is illustrated in Figure 9-30.

This phase shift that exists between the circuit current and the capacitor voltage is normally expressed in degrees. At 0°, the capacitor is fully discharged (0 V), and the source is supplying a maximum circuit charge current. From 0° to 90°, the capacitor will charge toward a maximum positive value, and this increase in capacitor voltage will oppose the source voltage, whose circuit charging current will slowly decrease to 0 A. At 90°, the capacitor is fully charged and the circuit current is at 0 A, so the capacitor will return to the circuit the energy it consumed during the positive (+) charge cycle. This discharge circuit current is in the opposite direction to the positive charge current. At 180°, the capacitor is fully discharged (0 V)

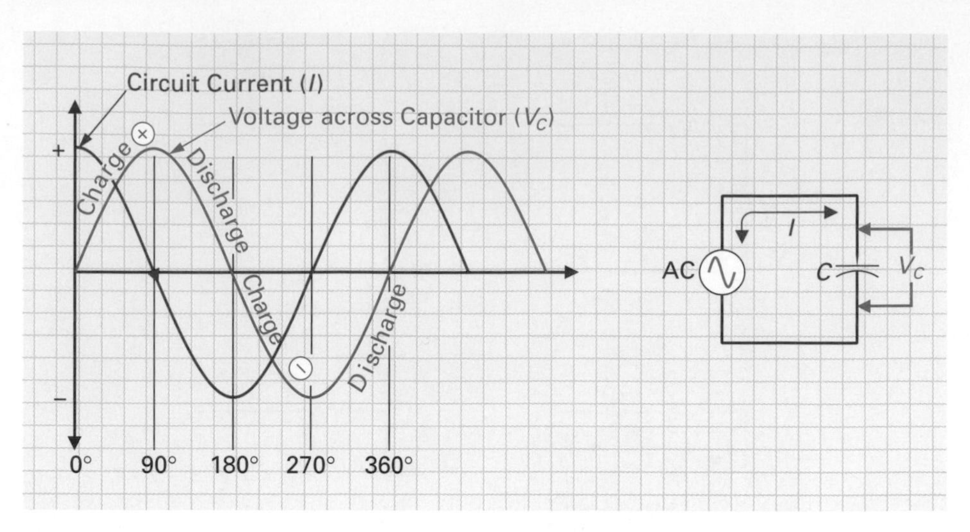

FIGURE 9-30 **AC Current Leads Voltage in a Capacitive Circuit.**

and the source is supplying a maximum circuit charge current. From 180° to 270°, the capacitor will charge toward a maximum negative value and this negative increase in capacitor voltage will oppose the source voltage, whose circuit-charging current will slowly decrease to 0 A. At 270°, the capacitor is fully charged and the circuit current is at 0 A, so the capacitor will discharge and return the energy it consumed during the negative (−) charge cycle. This discharge circuit current is in the opposite direction to the negative charge current.

Throughout this cycle, notice that the voltage across the capacitor follows the circuit current. This current, therefore, leads the voltage by 90°, and this 90° leading phase shift (current leading voltage) will occur only in a capacitive circuit.

Circuit current leads capacitor voltage in a purely capacitive circuit

9-4-2 *Capacitive Reactance*

Capacitive Reactance (X_C)
Measured in ohms, it is the ability of a capacitor to oppose current flow without the dissipation of energy.

Resistance (R), by definition, is the opposition to current flow with the dissipation of energy and is measured in ohms. Capacitors oppose current flow like a resistor. But a resistor dissipates energy, whereas a capacitor stores energy (when it charges) and then gives back its energy into the circuit (when it discharges). Because of this difference, a new term had to be used to describe the opposition offered by a capacitor. **Capacitive reactance (X_C),** by definition, is the opposition to current without the dissipation of energy and is also measured in ohms.

If capacitive reactance is basically opposition, it is inversely proportional to the amount of current flow. If a large current is within a circuit, the opposition or reactance must be low ($I\uparrow$, $X_C\downarrow$). Conversely, a small circuit current will be the result of a large opposition or reactance ($I\downarrow$, $X_C\uparrow$).

When a dc source is connected across a capacitor, current will flow only for a short period of time (5RC seconds) to charge the capacitor. After this time, there is no further current flow. Consequently, the capacitive reactance or opposition offered by a capacitor to dc is infinite (maximum).

Alternating current is continuously reversing in polarity, resulting in the capacitor continuously charging and discharging. This means that charge and discharge currents are always flowing around the circuit, and if we have a certain value of current, we must also have a certain value of reactance or opposition.

Initially, when the capacitor's plates are uncharged, they will not oppose or react against the charging current and therefore maximum current will flow ($I\uparrow$) and the reactance

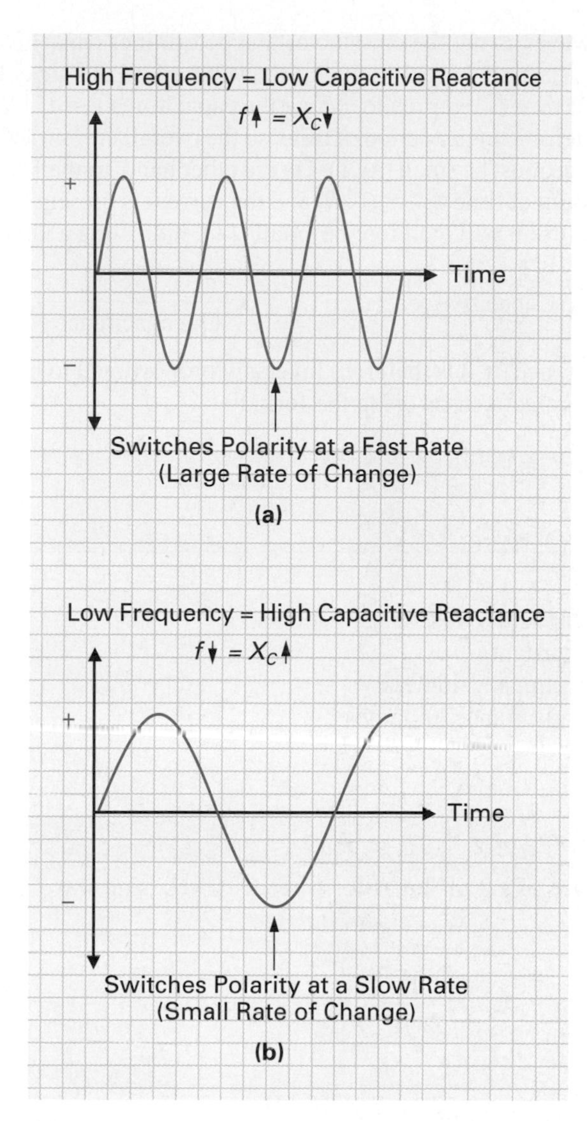

FIGURE 9-31 Capacitive Reactance Is Inversely Proportional to Frequency.

will be very low ($X_C\downarrow$). As the capacitor charges, it will oppose or react against the charge current, which will decrease ($I\downarrow$), so the reactance will increase ($X_C\uparrow$). The discharge current is also highest at the start of discharge ($I\uparrow$, $X_C\downarrow$) as the voltage of the charged capacitor is also high; but as the capacitor discharges, its voltage decreases and the discharge current will also decrease ($I\downarrow$, $X_C\uparrow$).

To summarize, at the start of a capacitor charge or discharge, the current is maximum, so the reactance is low. This value of current then begins to fall to zero, so the reactance increases.

If the applied alternating current is at a high frequency, as shown in Figure 9-31(a), it is switching polarity more rapidly than a lower frequency and there is very little time between the start of charge and discharge. As the charge and discharge currents are largest at the beginning of the charge and discharge of the capacitor, the reactance has very little time to build up and oppose the current, which is why the current is a high value and the capacitive reactance is small at higher frequencies. With lower frequencies, as shown in Figure 9-31(b), the applied alternating current is switching at a slower rate, and therefore the reactance, which is low at the beginning, has more time to build up and oppose the current. Capacitive reactance is therefore inversely proportional to frequency:

$$\text{Capacitive reactance } (X_C) \propto \frac{1}{f\,(\text{frequency})}$$

TIME LINE

Born in Buffalo, New York, Herman Hollerith was the son of German immigrants. After finishing his studies at Columbia University in 1879, he began work at the census office in Washington. Hundreds of clerks were needed to tabulate by hand the 1880 census of nearly 13 million people, which would take the next seven and a half years. Believing this task could be handled by a tabulating system, Hollerith worked for the next 10 years on a solution, which was unveiled in 1890 and took a third of the time to tabulate an increased population of more than 62.5 million people. The Hollerith tabulator used punched cards the size of dollar bills to hold data such as age, sex, country of birth, marital status, number of children, and so on. Hollerith formed a tabulating machine company to sell to government offices his tabulating inventions, one of which even found its way to Tsarist Russia, where they felt it was time for a modern census.

Over the years, the company merged with others and went through a number of name changes, the last of which occurred in 1924, five years before Hollerith died, when International Business Machines Corporation was created (now more commonly known as IBM).

Frequency, however, is not the only factor that determines capacitive reactance. Capacitive reactance is also inversely proportional to the value of capacitance. If a larger capacitor value is used, a longer time is required to charge the capacitor ($t\uparrow = C\uparrow R$), which means that current will be present for a longer period of time, so the overall current will be large ($I\uparrow$); consequently, the reactance must be small ($X_C\downarrow$). On the other hand, a small capacitance value will charge in a small amount of time ($t\downarrow = C\downarrow R$) and the current is present for only a short period of time. The overall current will therefore be small ($I\downarrow$), indicating a large reactance ($X_C\uparrow$).

$$\text{Capacitive reactance } (X_C) \propto \frac{1}{C\text{ (capacitance)}}$$

Capacitive reactance (X_C) is therefore inversely proportional to both frequency and capacitance and can be calculated by using the formula

$$X_C = \frac{1}{2\pi fC}$$

where X_C = capacitive reactance, in ohms
2π = constant
f = frequency, in hertz
C = capacitance, in farads

EXAMPLE:

Calculate the reactance of a 2 μF capacitor when a 10 kHz sine wave is applied.

Solution:

$$X_C = \frac{1}{2\pi fC}$$

$$= \frac{1}{2 \times \pi \times 10\text{ kHz} \times 2\ \mu\text{F}} = 8\ \Omega$$

Calculator Sequence

STEP	KEYPAD ENTRY	DISPLAY RESPONSE
1.	2	2.0
2.	×	
3.	π	3.1415927
4.	×	6.283185
5.	1 0 EE 3	10E3
6.	×	62831.8
7.	2 EE 6 +/−	2.−06
8.	=	0.1256637
9.	1/x	7.9577

9-4-3 *Series RC Circuit*

In a purely resistive circuit, as shown in Figure 9-32(a), the current flowing within the circuit and the voltage across the resistor are in phase with one another. In a purely capacitive circuit, as shown in Figure 9-32(b), the current flowing in the circuit leads the voltage across the capacitor by 90°.

Purely resistive: 0° phase shift (I is in phase with V)
Purely capacitive: 90° phase shift (I leads V by 90°)

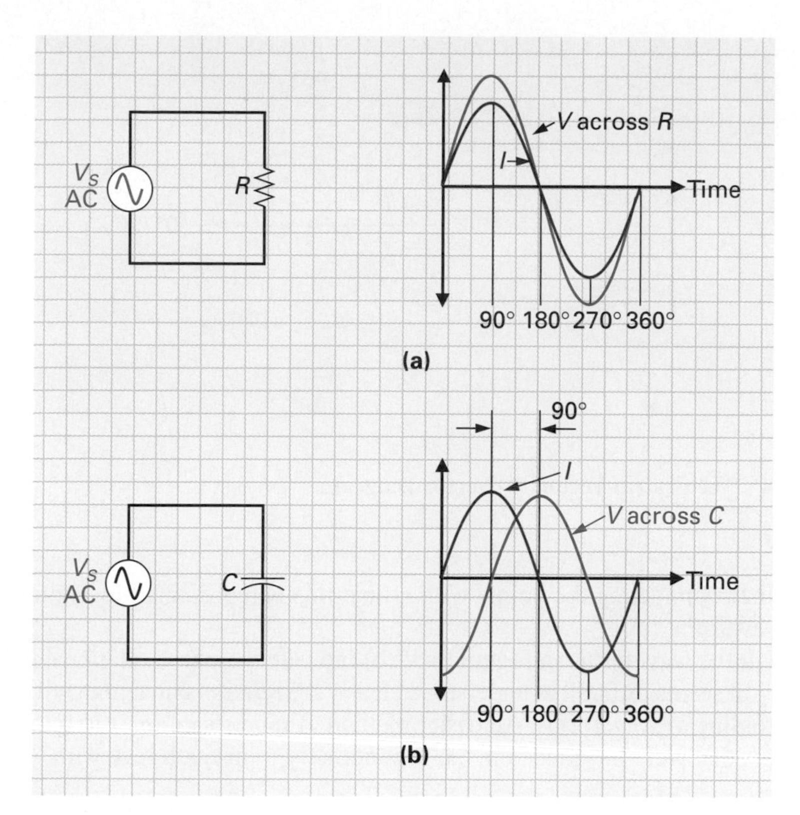

FIGURE 9-32 Phase Relationships Between *V* and *I*. (a) Resistive Circuit: Current and Voltage Are in Phase. (b) Capacitive Circuit: Current Leads Voltage by 90°.

If we connect a resistor and capacitor in series, as shown in Figure 9-33(a), we have probably the most commonly used electronic circuit, which has many applications. The voltage across the resistor (V_R) is always in phase with the circuit current (I), as can be seen in Figure 9-33(b), because maximum points and zero crossover points occur at the same time. The voltage across the capacitor (V_C) lags the circuit current by 90°.

Since the capacitor and resistor are in series, the same current is supplied to both components; Kirchhoff's voltage law can be applied, which states that the sum of the voltage drops around a series circuit is equal to the voltage applied (V_S). The voltage drop across the resistor (V_R) and the voltage drop across the capacitor (V_C) are out of phase with one another, which means that their peaks occur at different times. The signal for the applied voltage (V_S) is therefore obtained by adding the values of V_C and V_R at each instant in time, plotting the

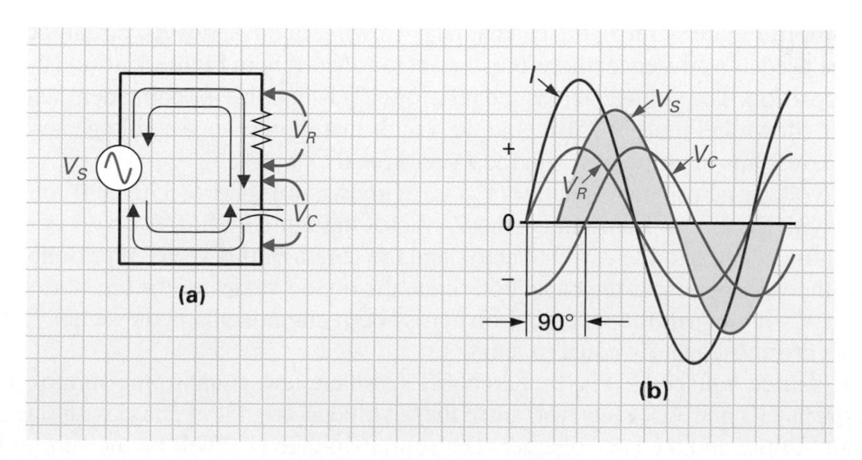

FIGURE 9-33 *RC* Series Circuit. (a) Circuit. (b) Waveforms.

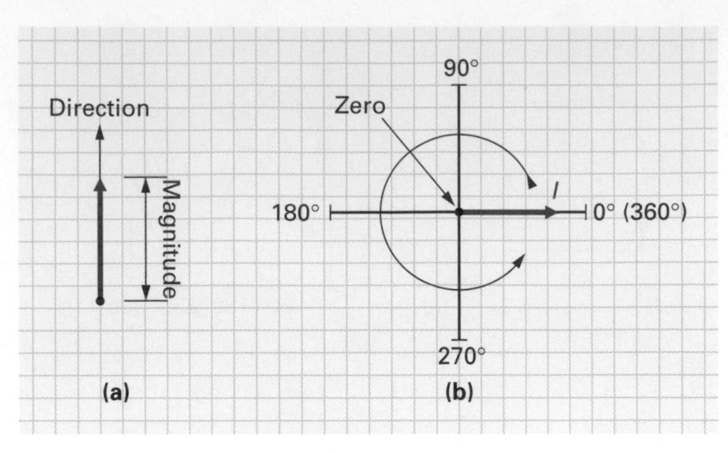

FIGURE 9-34 **Vectors. (a) Vector. (b) Vector Diagram.**

results, and then connecting the points with a line; this is represented in Figure 9-33(b) by the shaded waveform.

Although the waveforms in Figure 9–33(b) indicate the phase relationship between I, V_S, V_R, and V_C, it seems difficult to understand clearly the relationship among all four because of the crisscrossing of waveforms. An easier method of representation is to return to the circle and vectors that were introduced in the trigonometry mini-math review in Chapter 8.

1. Vector Diagram

A vector (or phasor) is a quantity that has both magnitude and direction and is represented by a line terminated at the end by an arrowhead, as seen in Figure 9-34(a). Vectors are generally always used to represent a physical quantity that has two properties. For example, if you are traveling 60 miles per hour in a southeast direction, the size or magnitude of the vector would represent 60 mph, while the direction of the vector would point southeast. In an ac circuit containing a reactive component such as a capacitor, the vector is used to represent a voltage or current. The magnitude of the vector represents the value of voltage or current, while the direction of the vector represents the phase of the voltage or current.

Vector Diagram
Arrangement of vectors showing the phase relationships between two or more ac quantities of the same frequency.

A **vector diagram** is an arrangement of vectors to illustrate the magnitude and phase relationships between two or more quantities of the same frequency within an ac circuit. Figure 9-34(b) illustrates the basic parts of a vector diagram. As an example, the current (I) vector is at the 0° position and the size of the arrow represents the peak value of alternating current.

2. Voltage

Figure 9-35(a), (b), and (c) repeats our previous RC series circuit with waveforms and a vector diagram. In Figure 9-35(b), the current peak flowing in the series RC circuit occurs at 0° and will be used as a reference; therefore, the vector of current in Figure 9-35(c) is to the right in the 0° position. The voltage across the resistor (V_R) is in phase or coincident with the current (I), as shown in Figure 9-35(b), and the vector that represents the voltage across the resistor (V_R) overlaps or coincides with the I vector, at 0°.

The voltage across the capacitor (V_C) is, as shown in Figure 9-35(b), 90° out of phase (lagging) with the circuit's current, so the V_C vector in Figure 9-35(c) is drawn at $-90°$ (minus sign indicates lag) to the current vector, and the length of this vector represents the magnitude of the voltage across the capacitor. Since the ohmic values of the resistor (R) and the capacitor (X_C) are equal, the voltage drop across both components is the same. The V_R and V_C vectors are subsequently equal in length.

The source voltage (V_S) is, by Kirchhoff's voltage law, equal to the sum of the series voltage drops (V_C and V_R). However, since these voltages are not in phase with one another, we cannot simply add the two together. The source voltage (V_S) will be the sum of both V_C and V_R at a particular time. Studying the waveforms in Figure 9-35(b), you will notice that

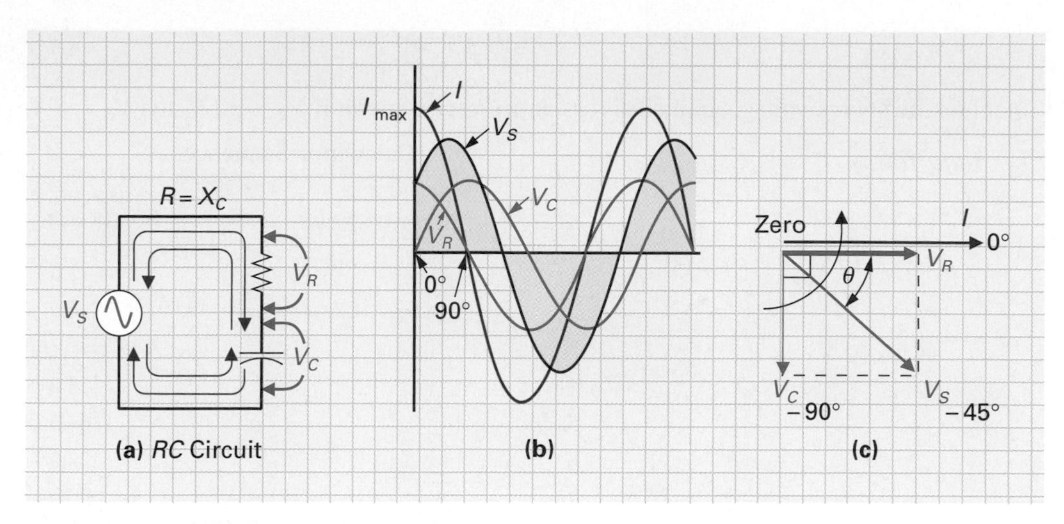

FIGURE 9-35 ***RC*** **Series Circuit Analysis.**

peak source voltage will occur at 45°. By vectorially adding the two voltages V_R and V_C in Figure 9-35(c), we obtain a resultant V_S vector that has both magnitude and phase. The angle theta (θ) formed between circuit current (I) and source voltage (V_S) will always be less than 90°, and in this example is equal to $-45°$ because the voltage drops across R and C are equal due to R and X_C being of the same ohmic value.

If V_C and V_R are drawn to scale, the peak source voltage (V_S) can be calculated by using the same scale, and a mathematical rather than graphical method can be used to save the drafting time.

In Figure 9–36(a), we have taken the three voltages (V_R, V_C, and V_S) and formed a right-angle triangle, as shown in Figure 9-36(b). The Pythagorean theorem for right-angle triangles states that if you take the square of a (V_R) and add it to the square of b (V_C), the square root of the result will equal c (the source voltage, V_S).

$$V_S = \sqrt{V_R^{\,2} + V_C^{\,2}}$$

By transposing the formula according to the rules of algebra we can calculate any unknown if two variables are known.

$$V_C = \sqrt{V_S^{\,2} - V_R^{\,2}}$$
$$V_R = \sqrt{V_S^{\,2} - V_C^{\,2}}$$

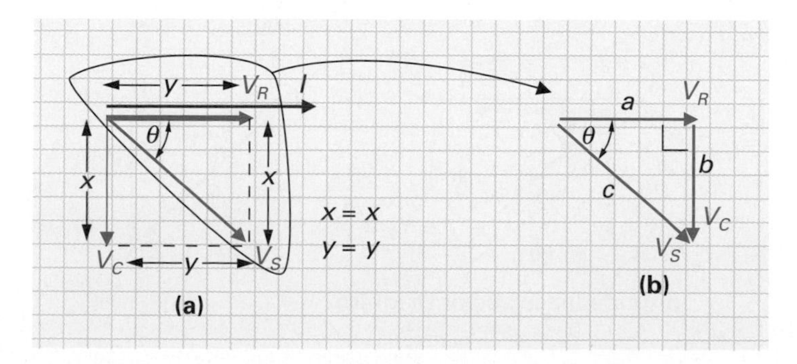

FIGURE 9-36 Voltage and Current Vector Diagram of a Series ***RC*** **Circuit.**

EXAMPLE:

Calculate the source voltage applied across an *RC* series circuit if $V_R = 12$ V and $V_C = 8$ V.

Solution:

$$V_S = \sqrt{V_R^2 + V_C^2}$$
$$= \sqrt{(12\text{ V})^2 + (8\text{ V})^2}$$
$$= \sqrt{144 + 64}$$
$$= 14.42\text{ V}$$

Calculator Sequence

STEP	KEYPAD ENTRY	DISPLAY RESPONSE
1.	[1] [2]	12.0
2.	[x²]	144.0
3.	[+]	
4.	[8]	8.0
5.	[x²]	64.0
6.	[=]	208.0
7.	[√x]	14.42220

3. Impedance

Impedance (*Z*)
Measured in ohms, it is the total opposition a circuit offers to current flow (reactive and resistive).

Since resistance is the opposition to current with the dissipation of heat, and reactance is the opposition to current without the dissipation of heat, a new term is needed to describe the total resistive and reactive opposition to current. **Impedance** (designated Z) is also measured in ohms and is the total circuit opposition to current flow. It is a combination of resistance (R) and reactance (X_C); however, in our capacitive and resistive circuit, a phase shift or difference exists, and just as V_C and V_R cannot be merely added together to obtain V_S, R and X_C cannot be simply summed to obtain Z.

If the current within a series circuit is constant (the same throughout the circuit), the resistance of a resistor (R) or reactance of a capacitor (X_C) will be directly proportional to the voltage across the resistor (V_R) or the capacitor (V_C).

$$V_R\updownarrow = I \times R\updownarrow, \qquad V_C\updownarrow = I \times X_C\updownarrow$$

A vector diagram can be drawn similarly to the voltage vector diagram to illustrate opposition, as shown in Figure 9-37(a). The current is used as a reference (0°); the resistance vector (R) is in phase with the current vector (I), since V_R is always in phase with I. The capacitive reactance (X_C) vector is at $-90°$ to the resistance vector, due to the 90° phase shift between a resistor and capacitor. The lengths of the resistance vector (R) and capacitive reactance vector (X_C) are equal in this example. By vectorially adding R and X_C, we have a resulting impedance (Z) vector.

By using the three variables, which have again formed a right-angle triangle [Figure 9-37(b)], we can apply the Pythagorean theorem to calculate the total opposition or impedance (Z) to current flow, taking into account both R and X_C.

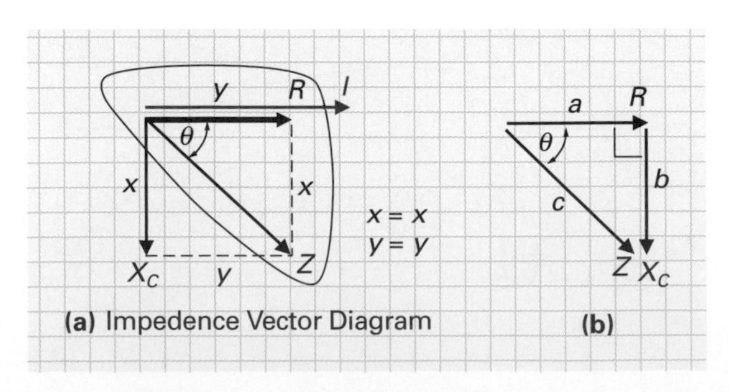

FIGURE 9-37 **Resistance, Reactance, and Impedance Vector Diagram of a Series *RC* Circuit.**

$$Z = \sqrt{R^2 + X_C^2}$$

$$R = \sqrt{Z^2 - X_C^2}$$
$$X_C = \sqrt{Z^2 - R^2}$$

EXAMPLE:

Calculate the total impedance of a series RC circuit if $R = 27\ \Omega$, $C = 0.005\ \mu\text{F}$, and the source frequency = 1 kHz.

Solution:

The total opposition (Z) or impedance is equal to

$$Z = \sqrt{R^2 + X_C^2}$$

R is known, but X_C will need to be calculated.

$$\begin{aligned} X_C &= \frac{1}{2\pi fC} \\ &= \frac{1}{2 \times \pi \times 1\text{ kHz} \times 0.005\ \mu\text{F}} \\ &= 31.8\text{ k}\Omega \end{aligned}$$

Since $R = 27\ \Omega$ and $X_C = 31.8\text{ k}\Omega$, then

$$\begin{aligned} Z &= \sqrt{R^2 + X_C^2} \\ &= \sqrt{(27\ \Omega)^2 + (31.8\text{ k}\ \Omega)^2} \\ &= \sqrt{729 + 1 \times 10^9} \\ &= 31.8\text{ k}\Omega \end{aligned}$$

As you can see in this example, the small resistance of 27 Ω has very little effect on the circuit's total opposition or impedance, due to the relatively large capacitive reactance of 31,800 Ω.

EXAMPLE:

Calculate the total impedance of a series RC circuit if $R = 45\text{ k}\Omega$ and $X_C = 45\ \Omega$.

Solution:

$$\begin{aligned} Z &= \sqrt{R^2 + X_C^2} \\ &= \sqrt{(45\text{ k}\Omega)^2 + (45\Omega)^2} \\ &= 45\text{ k}\Omega \end{aligned}$$

In this example, the relatively small value of X_C had very little effect on the circuit's opposition or impedance, due to the large circuit resistance.

EXAMPLE:

Calculate the total impedance of a series RC circuit if $X_C = 100\ \Omega$ and $R = 100\ \Omega$.

■ ***Solution:***

$$Z = \sqrt{R^2 + X_C^2}$$
$$= \sqrt{100^2 + 100^2}$$
$$= 141.4\ \Omega$$

In this example R was equal to X_C.

We can define the total opposition or impedance in terms of Ohm's law, in the same way as we defined resistance.

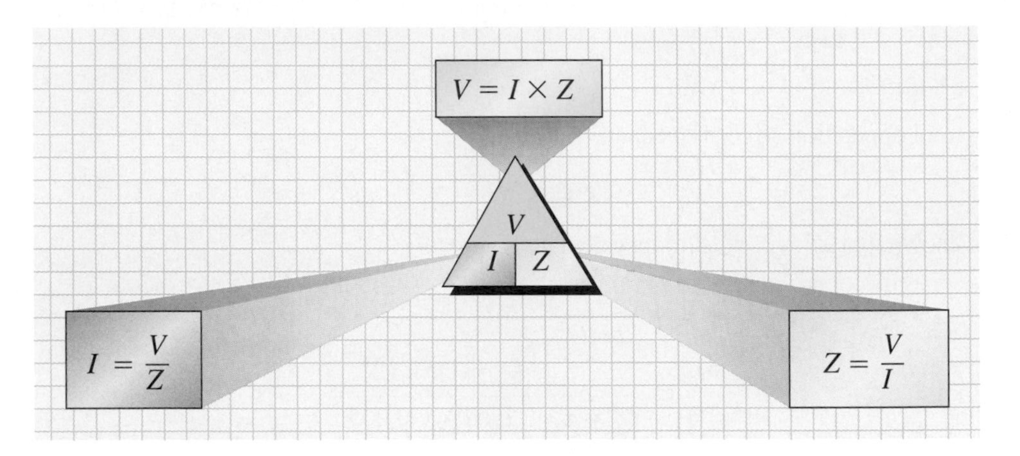

By transposition, we can arrive at the usual combinations of Ohm's law.

4. Phase Angle or Shift (θ)

In a purely resistive circuit, the total opposition (Z) is equal to the resistance of the resistor, so the phase shift (θ) is equal to 0° [Figure 9-38(a)]. In a purely capacitive circuit, the total opposition (Z) is equal to the capacitive reactance (X_C) of the capacitor, so the phase shift (θ) is equal to $-90°$ [Figure 9-38(b)].

When a circuit contains both resistance and capacitive reactance, the total opposition or impedance has a phase shift that is between 0 and 90°. Referring back to the impedance vector diagram in Figure 9-37 and the preceding example, you can see that we have used a simple example where R has equaled X_C, so the phase shift (θ) has always been $-45°$. If the resistance and reactance are different from one another, as was the case in the other two examples, the phase shift (θ) will change, as shown in Figure 9-39. As can be seen in this illustration, the phase shift (θ) is dependent on the ratio of capacitive reactance to resistance (X_C/R). A more resistive circuit will have a phase shift between 0 and 45°, and a more reactive circuit will have a phase shift between 45 and 90°. By the use of trigonometry (the science of triangles), we can derive a formula to calculate the degree of phase shift, since two quantities X_C and R are known.

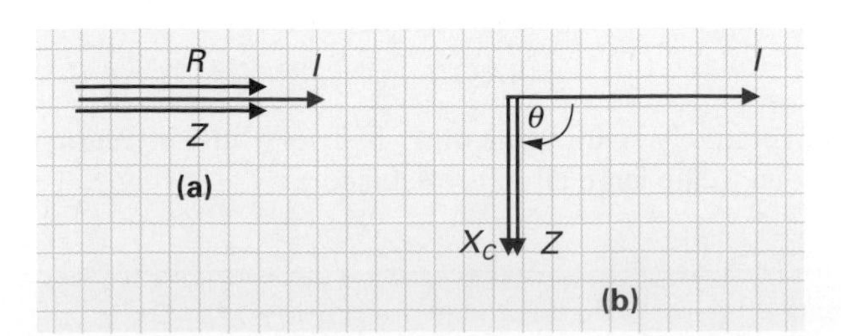

FIGURE 9-38 Phase Angles. (a) Purely Resistive Circuit. (b) Purely Capacitive Circuit.

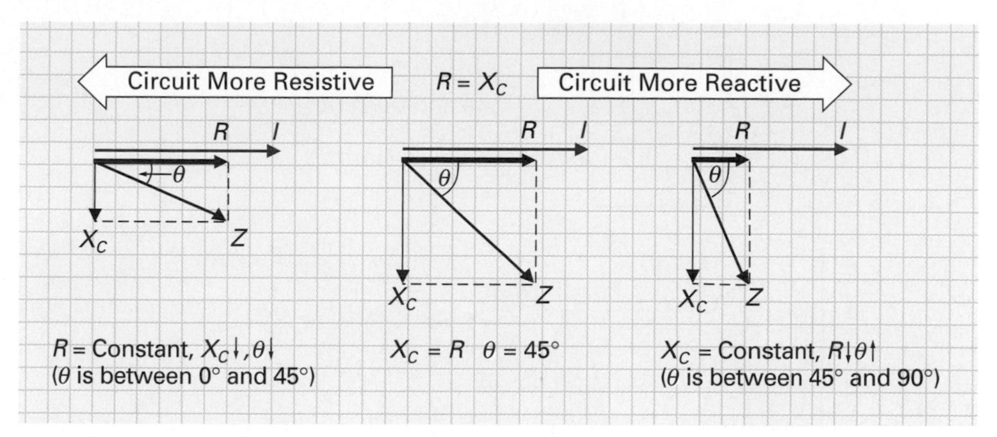

FIGURE 9-39 Phase Angle of a Series *RC* Circuit.

The phase angle, θ, is equal to

$$\theta = \text{invtan}\,\frac{X_C}{R}$$

This formula will determine by what angle Z lags R. Since X_C/R is equal to V_C/V_R, the phase angle can also be calculated if V_R and V_C are known.

$$\theta = \text{invtan}\,\frac{V_C}{V_R}$$

This formula will determine by what angle V_R leads V_S.

EXAMPLE:

Calculate the phase shift or angle in two different series *RC* circuits if

a. $V_R = 12\text{ V}, V_C = 8\text{ V}$
b. $R = 27\ \Omega, X_C = 31.8\text{ k}\Omega$

Solution:

a. $$\theta = \text{invtan}\,\frac{V_C}{V_R} = \text{invtan}\,\frac{8\text{ V}}{12\text{ V}} = 33.7^\circ \quad (V_R \text{ leads } V_S \text{ by } 33.7^\circ)$$

b. $$\theta = \text{invtan}\,\frac{X_C}{R} = \text{invtan}\,\frac{31.8\text{ k}\Omega}{27\ \Omega} = 89.95^\circ \quad (R \text{ leads } Z \text{ by } 89.95^\circ)$$

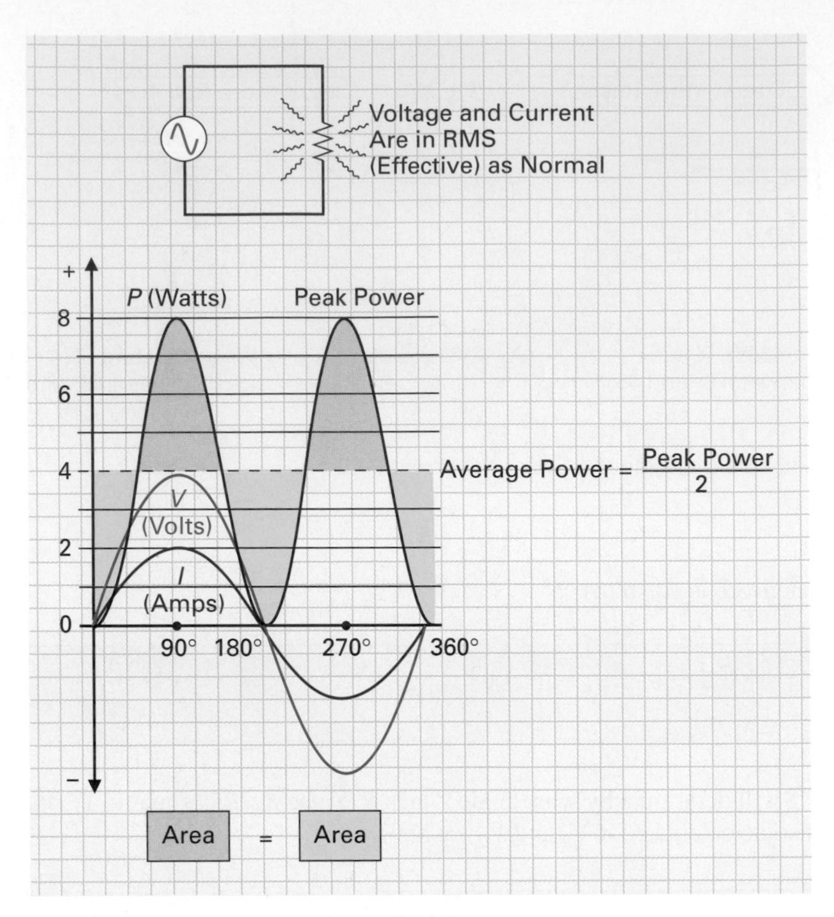

FIGURE 9-40 **Power in a Purely Resistive Circuit.**

5. Power

In this section we examine power in a series ac circuit. Let us begin with a simple resistive circuit and review the power formulas used previously.

Purely resistive circuit. In Figure 9-40 you can see the current, voltage, and power waveforms generated by applying an ac voltage across a purely resistive circuit. The applied voltage causes current to flow around the circuit, and the electrical energy is converted into heat energy. This heat or power is dissipated and lost and can be calculated by using the power formula.

$$P = V \times I$$

$$P = I^2 \times R$$

$$P = \frac{V^2}{R}$$

Voltage and current are in phase with one another in a resistive circuit, and instantaneous power is calculated by multiplying voltage by current at every instant through 360° ($P = V \times I$). The sinusoidal power waveform is totally positive, because a positive voltage multiplied by a positive current gives a positive value of power, and a negative voltage multiplied by a negative current will also produce a positive value of power. For these reasons, a resistor is said to generate a positive power waveform, which you may have noticed is twice the frequency of the voltage and current waveforms; two power cycles occur in the same time as one voltage and current cycle.

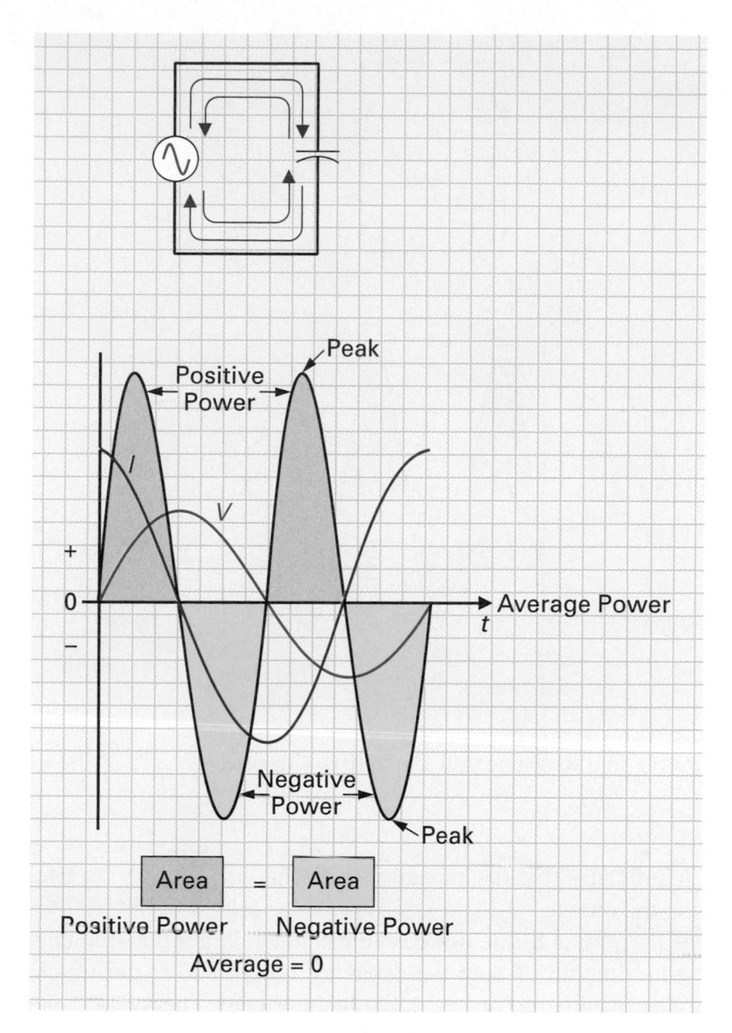

FIGURE 9-41 Power in a Purely Capacitive Circuit.

The power waveform has been split in half, and this line that exists between the maximum point (8 W) and zero point (0 W) is the average value of power (4 W) that is being dissipated by the resistor.

Purely capacitive circuit. In Figure 9-41, you can see the current, voltage, and power waveforms generated by applying an ac voltage source across a purely capacitive circuit. As expected, the current leads the voltage by 90°, and the power wave is calculated by multiplying voltage by current, as before, at every instant through 360°. The resulting power curve is both positive and negative. During the positive alternation of the power curve, the capacitor is taking power as the capacitor charges. When the power alternation is negative, the capacitor is giving back the power it took as it discharges back into the circuit.

The average power dissipated is once again the value that exists between the maximum positive and maximum negative points and causes the area above this line to equal the area below. This average power level calculates out to be zero, which means that no power is dissipated in a purely capacitive circuit.

Resistive and capacitive circuit. In Figure 9-42 you can see the current, voltage, and power waveforms generated by applying an ac voltage source across a series-connected *RC* circuit. The current leads the voltage by some phase angle less than 90°, and the power waveform is once again determined by the product of voltage and current. The negative alternation

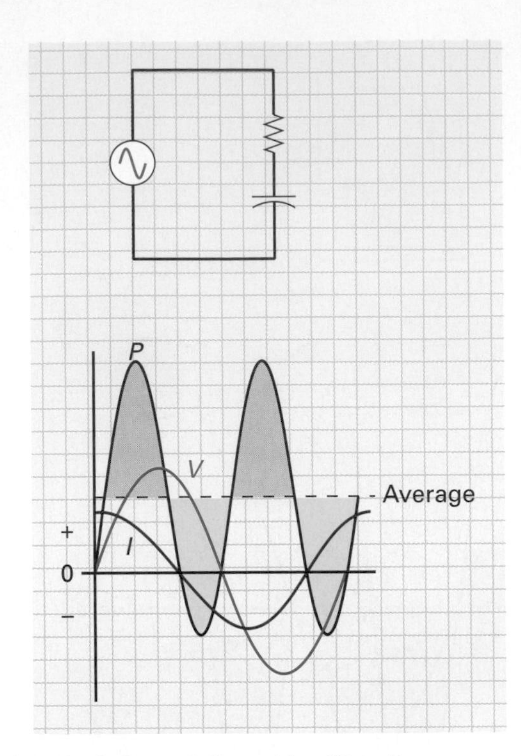

FIGURE 9-42 Power in a Resistive and Capacitive Circuit.

of the power cycle indicates that the capacitor is discharging and giving back the power that it consumed during the charge.

The positive alternation of the power cycle is much larger than the negative alternation because it is the combination of both the capacitor taking power during charge and the resistor consuming and dissipating power in the form of heat. The average power being dissipated will be some positive value, due to the heat being generated by the resistor.

Resistive Power or True Power
The average power consumed by a circuit during one complete cycle of alternating current.

Reactive Power or Imaginary Power
Also called wattless power, it is the power value obtained by multiplying the effective value of current by the effective value of voltage and the sine of the angular phase difference between current and voltage.

Apparent Power
The power value obtained in an ac circuit by multiplying together effective values of voltage and current, which reach their peaks at different times.

Power factor. In a purely resistive circuit, all the energy supplied to the resistor from the source is dissipated in the form of heat. This form of power is referred to as **resistive power** (P_R) or **true power**, and is calculated with the formula

$$P_R = I^2 \times R$$

In a purely capacitive circuit, all the energy supplied to the capacitor is stored from the source and then returned to the source, without energy loss. This form of power is referred to as **reactive power** (P_X) or **imaginary power**.

$$P_X = I^2 \times X_C$$

When a circuit contains both capacitance and resistance, some of the supply is stored and returned by the capacitor and some of the energy is dissipated and lost by the resistor.

Figure 9-43(a) illustrates another vector diagram. Just as the voltage across a resistor is 90° out of phase with the voltage across a capacitor, and resistance is 90° out of phase with reactance, resistive power will be 90° out of phase with reactive power.

If we take the three variables from Figure 9-43(b) to form a right-angle triangle as in Figure 9-43(c), we can vectorially add true power and imaginary power to produce a resultant **apparent power** vector. Apparent power is the power that appears to be supplied to the

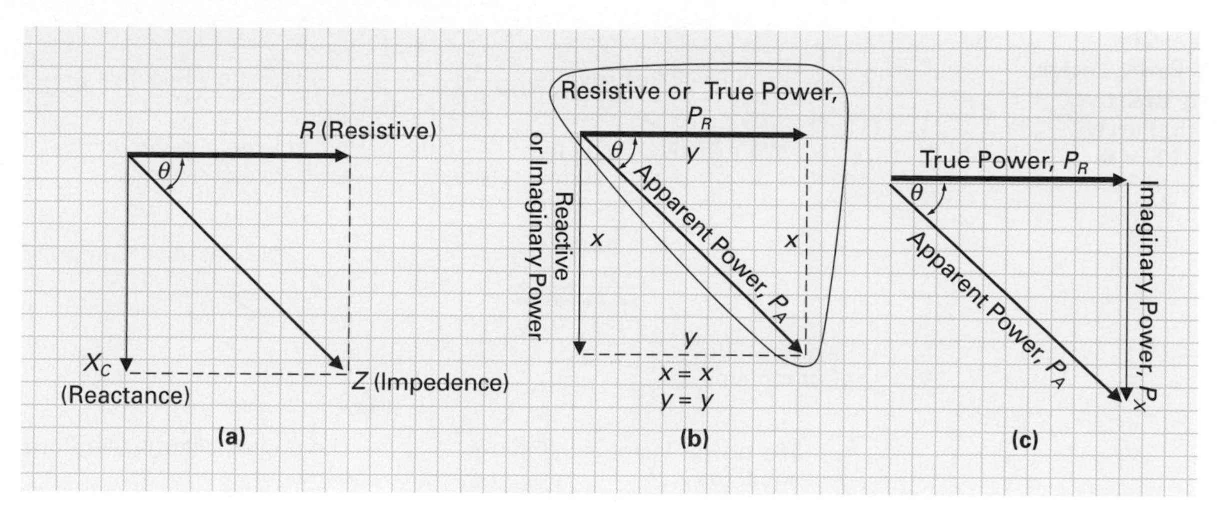

FIGURE 9-43 Apparent Power.

load and includes both the true power dissipated by the resistance and the imaginary power delivered to the capacitor. Applying the Pythagorean theorem once again, we can calculate apparent power by

$$P_A = \sqrt{P_R^{\,2} + P_X^{\,2}}$$

where P_A = apparent power, in volt-amperes (VA)
P_R = true power, in watts (W)
P_X = reactive power, in volt-amperes reactive (VAR)

The **power factor** is a ratio of true power to apparent power and is a measure of the loss in a circuit. It can be calculated by using the formula

Power Factor
Ratio of actual power to apparent power. A pure resistor has a power factor of 1 or 100% and a capacitor has a power factor of 0 or 0%.

$$\text{PF} = \frac{\text{true power } (P_R)}{\text{apparent power } (P_A)}$$

Figure 9-44 helps explain what the power factor of a circuit actually indicates. In a purely resistive circuit, as shown in Figure 9-44(a), the apparent power will equal the true power ($P_A = \sqrt{P_R^{\,2} + 0^2}$). The power factor will therefore equal 1 ($PF = P_R/P_A$). In a purely capacitive circuit, as shown in Figure 9-44(b), the true power will be zero, and therefore the power factor will equal 0 (PF = $0/P_A$). A power factor of 1 therefore indicates a maximum power loss (circuit is resistive), and a power factor of 0 indicates no power loss (circuit is capacitive). With circuits that contain both resistance and reactance, the power factor will be somewhere between zero (0 = reactive) and one (1 = resistive).

Since true power is determined by resistance and apparent power is dependent on impedance, as shown in Figure 9-43(a), the power factor can also be calculated by using the formula

$$\text{PF} = \frac{R}{Z}$$

As the ratio of true power (adjacent) to apparent power (hypotenuse) determines the angle θ, the power factor can also be determined by the cosine of angle θ.

$$\text{PF} = \cos\theta$$

FIGURE 9-44
Circuit's Power Factor. (a) Purely Resistive, PF = 1. (b) Purely Reactive, PF = 0.

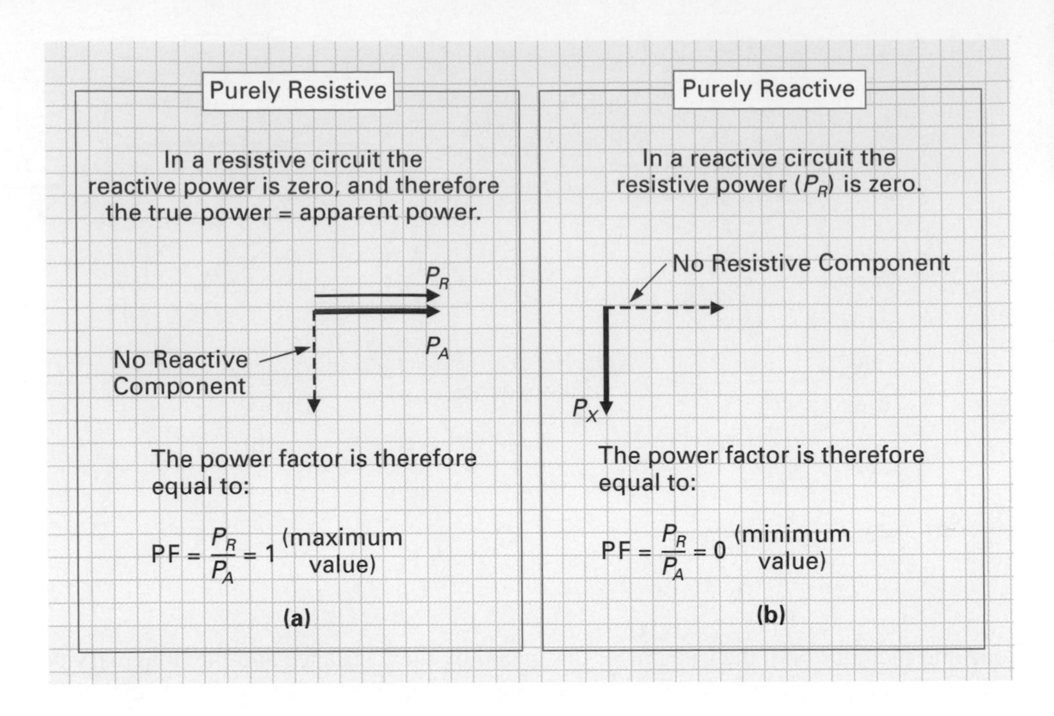

EXAMPLE:

Calculate the following for a series *RC* circuit if $R = 2.2\ \text{k}\Omega$, $X_C = 3.3\ \text{k}\Omega$, and $V_S = 5\ \text{V}$:

a. Z
b. I
c. θ
d. P_R
e. P_X
f. P_A
g. PF

Solution:

a. $Z = \sqrt{R^2 + X_C^2}$
$= \sqrt{(2.2\ \text{k}\Omega)^2 + (3.3\ \text{k}\Omega)^2}$
$= 3.96\ \text{k}\Omega$

b. $I = \frac{V_S}{Z} = \frac{5\ \text{V}}{3.96\ \text{k}\Omega} = 1.26\ \text{mA}$

c. $\theta = \text{invtan}\,\frac{X_C}{R} = \text{invtan}\,\frac{3.3\ \text{k}\Omega}{2.2\ \text{k}\Omega}$
$= \text{invtan}\ 1.5 = 56.3°$

Calculator Sequence

STEP	KEYPAD ENTRY	DISPLAY RESPONSE
1.	3 . 3 E 3	3.3E3
2.	÷	
3.	2 . 2 E 3	2.2E3
4.	=	1.5
5.	inv tan	56.309932

d. True power $= I^2 \times R$
$= (1.26\ \text{mA})^2 \times 2.2\ \text{k}\Omega$
$= 3.49\ \text{mW}$

e. Reactive power $= I^2 \times X_C$
$= (1.26\ \text{mA})^2 \times 3.3\ \text{k}\Omega$
$= 5.24 \times 10^{-3}$ or $5.24\ \text{mVAR}$

f. Apparent power $= \sqrt{P_R^2 + P_X^2}$

$$= \sqrt{(3.49 \text{ mW})^2 + (5.24 \text{ mW})^2}$$

$$= 6.29 \times 10^{-3} \text{ or } 6.29 \text{ mVA}$$

g. Power factor $= \frac{R}{Z} = \frac{2.2 \text{ k}\Omega}{3.96 \text{ k}\Omega} = 0.55$

or

$$= \frac{P_R}{P_A} = \frac{3.49 \text{ mW}}{6.29 \text{ mW}} = 0.55$$

or

$$= \cos\theta = \cos 56.3° = 0.55$$

9-4-4 *Parallel* RC *Circuit*

Now that we have analyzed the characteristics of a series *RC* circuit, let us connect a resistor and capacitor in parallel.

1. Voltage

As with any parallel circuit, the voltage across all components in parallel is equal to the source voltage; therefore,

$$V_R = V_C = V_S$$

2. Current

In Figure 9-45(a), you will see a parallel circuit containing a resistor and a capacitor. The current through the resistor and capacitor is simply calculated by applying Ohm's law:

$$(\text{resistor current})\ I_R = \frac{V_S}{R}$$

$$(\text{capacitor current})\ I_C = \frac{V_S}{X_C}$$

Total current (I_T), however, is not as simply calculated. As expected, resistor current (I_R) is in phase with the applied voltage (V_S), as shown in the vector diagram in Figure 9-45(b). Capacitor current will always lead the applied voltage by 90°, and as the applied voltage is being used as our reference at 0° on the vector diagram, the capacitor current will have to be drawn at +90° in order to lead the applied voltage by 90°, since vector diagrams rotate in a counterclockwise direction. Total current is therefore the vector sum of both the resistor and capacitor currents. Using the Pythagorean theorem, total current can be calculated by

$$I_T = \sqrt{I_R^2 + I_C^2}$$

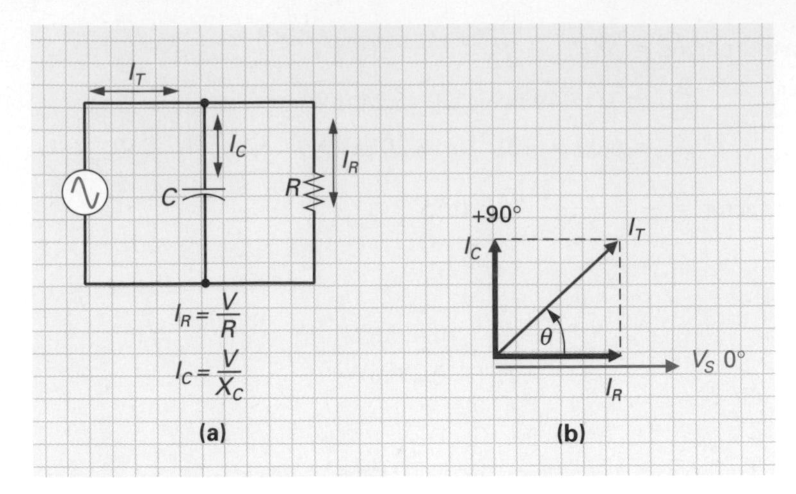

FIGURE 9-45 **Parallel *RC* Circuit.**

3. Phase Angle

The angle by which the total current (I_T) leads the source voltage (V_S) can be determined with either of the following formulas:

$$\theta = \text{invtan}\,\frac{I_C\,(\text{opposite})}{I_R\,(\text{adjacent})} \qquad \theta = \text{invtan}\,\frac{R}{X_C}$$

4. Impedance

Since the circuit is both capacitive and resistive, the total opposition or impedance of the parallel *RC* circuit can be calculated by

$$Z = \frac{V_S}{I_T}$$

The impedance of a parallel *RC* circuit is equal to the total voltage divided by the total current. Using basic algebra, this basic formula can be rearranged to express impedance in terms of reactance and resistance.

$$Z = \frac{R \times X_C}{\sqrt{R^2 + X_C^2}}$$

5. Power

With respect to power, there is no difference between a series circuit and a parallel circuit. The true power or resistive power (P_R) dissipated by an *RC* circuit is calculated with the formula

$$P_R = I_R^2 \times R$$

The imaginary power or reactive power (P_X) of the circuit can be calculated with the formula

$$P_X = I_C^2 \times X_C$$

The apparent power is equal to the vector sum of the true power and the reactive power.

$$P_A = \sqrt{P_R^2 + P_X^2}$$

As with series *RC* circuits the power factor is calculated as

$$\text{PF} = \frac{P_R\,(\text{resistive power})}{P_A\,(\text{apparent power})}$$

A power factor of 1 indicates a purely resistive circuit, and a power factor of 0 indicates a purely reactive circuit.

EXAMPLE:

Calculate the following for a parallel *RC* circuit in which $R = 24\ \Omega$, $X_C = 14\ \Omega$, and $V_S = 10\text{ V}$:

a. I_R
b. I_C
c. I_T
d. Z
e. θ

Solution:

a. $I_R = \dfrac{V_S}{R} = \dfrac{10\text{ V}}{24\ \Omega} = 416.66\text{ mA}$

b. $I_C = \dfrac{V_S}{X_C} = \dfrac{10\text{ V}}{14\ \Omega} = 714.28\text{ mA}$

c.
$$\begin{aligned} I_T &= \sqrt{I_R^2 + I_C^2} \\ &= \sqrt{(416.66\text{ mA})^2 + (714.28\text{ mA})^2} \\ &= \sqrt{0.173 + 0.510} \\ &= 826.5\text{ mA} \end{aligned}$$

d. $Z = \dfrac{V_S}{I_T} = \dfrac{10\text{ V}}{826.5\text{ mA}} = 12\ \Omega$

or

$$\begin{aligned} &= \frac{R \times X_C}{\sqrt{R^2 + X_C^2}} = \frac{24 \times 14}{\sqrt{24^2 + 14^2}} \\ &= 12\ \Omega \end{aligned}$$

e.
$$\begin{aligned} \theta &= \arctan\frac{I_C}{I_R} = \arctan\frac{714.28\text{ mA}}{416.66\text{ mA}} \\ &= \arctan 1.714 = 59.7° \end{aligned}$$

or

$$\begin{aligned} \theta &= \arctan\frac{R}{X_C} = \arctan\frac{24\ \Omega}{14\ \Omega} \\ &= \arctan 1.714 = 59.7° \end{aligned}$$

Calculator Sequence

STEP	KEYPAD ENTRY	DISPLAY RESPONSE
1.	7 1 4 . 2 8 E 3 +/−	714.28E–3
2.	÷	
3.	4 1 6 . 6 6 E 3 +/−	416.66E–3
4.	=	1.7142994
5.	inv tan	59.743762

SELF-TEST EVALUATION POINT FOR SECTION 9-4

Now that you have completed this section, you should be able to:

Objective 13. Explain how the capacitor charges and discharges when ac is applied.

Objective 14. Describe why there is a phase difference between circuit current and capacitor voltage.

Objective 15. Define and explain capacitive reactance.

Objective 16. Describe impedance, phase angle, power, and power factor as they relate to a series and parallel *RC* circuit.

Use the following questions to test your understanding of Section 9-4:

1. What differences occur when ac is applied to a capacitor rather than dc?
2. True or false: A capacitor's reaction to ac and dc accounts for why it is known as an ac short and dc block.
3. What does a voltage-current phase shift mean?
4. True or false: In a resistive circuit, the current leads the voltage by exactly 90°.
5. True or false: In a purely capacitive circuit, a 90° phase difference occurs between current and voltage.
6. True or false: Phase differences between voltage and current only occur when ac is applied.
7. Define *capacitive reactance*.
8. State the formula for capacitive reactance.
9. Why is capacitive reactance inversely proportional to frequency and capacitance?
10. If $C = 4\ \mu F$ and $f = 4$ kHz, calculate X_C.
11. What is the phase relationship between current and voltage in a series *RC* circuit?
12. What is a phasor diagram?
13. Define and state the formula for impedance.
14. What is the phase angle or shift in
    a. A purely resistive circuit?
    b. A purely capacitive circuit?
    c. A series circuit consisting of *R* and *C*?
15. What is the phase relationship between current and voltage in a parallel *RC* circuit?
16. Could a parallel *RC* circuit be called a voltage divider?
17. State the formula used to calculate
    a. I_T
    b. Z
18. Will capacitor current lead or lag resistor current in a parallel *RC* circuit?

9-5 APPLICATIONS OF CAPACITORS

There are many applications of capacitors, some of which will be discussed now; others will be presented later and in your course of electronic studies.

9-5-1 *Combining AC and DC*

Figure 9-46(a) and (b) shows how the capacitor can be used to combine ac and dc. The capacitor is large in value (typically electrolytic) and can be thought of as a very large bucket

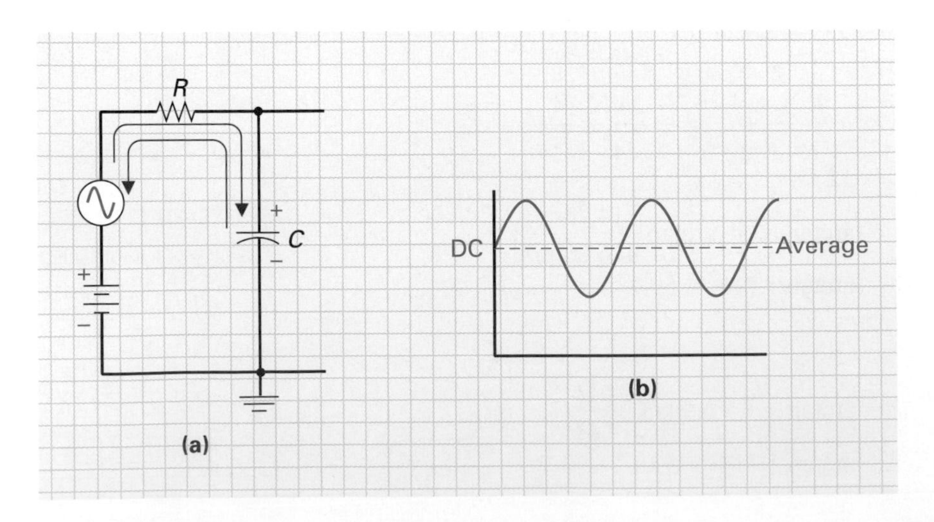

FIGURE 9-46
Superimposing AC on a DC Level.

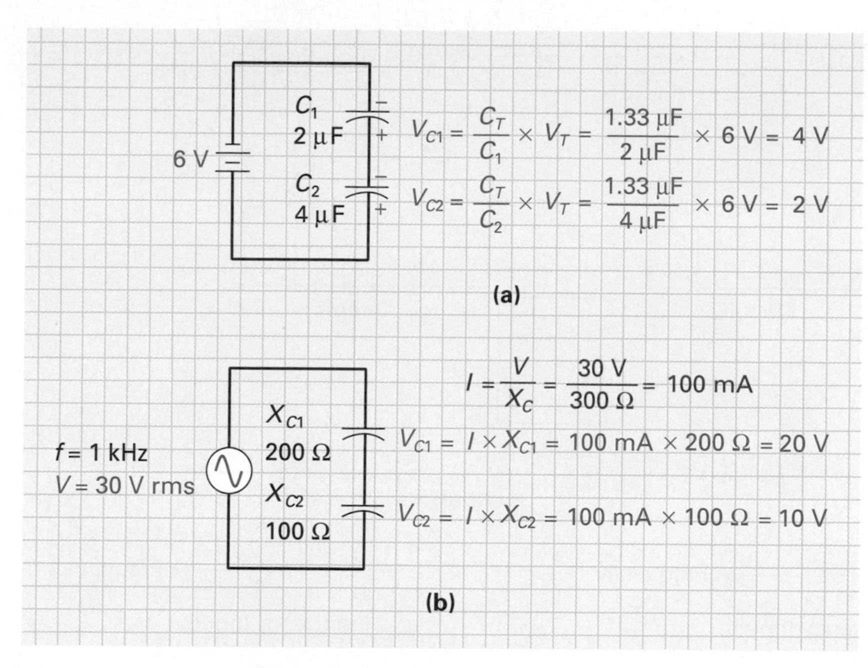

FIGURE 9-47 Capacitive Voltage Divider. (a) DC Circuit. (b) AC Circuit.

that will fill or charge to a dc voltage level. The ac voltage will charge and discharge the capacitor, which is similar to pouring in and pulling out (alternating) more and less water. The resulting waveform is a combination of ac and dc that varies above and below an average dc level. In this instance, the ac is said to be superimposed on a dc level.

9-5-2 *The Capacitive Voltage Divider*

Figure 9-47(a) shows how two capacitors can divide up a dc voltage of 6 V. Figure 9-47(b) shows how two capacitors can be used to divide up an ac voltage source of 30 V rms. Like a resistive voltage divider, voltage drop is proportional to current opposition, which with a capacitor is called *capacitive reactance* (X_C). The larger the capacitive reactance ($X_C\uparrow$), the larger the voltage drop ($V\uparrow$). Since capacitive reactance is inversely proportional to the capacitor value and the frequency of the ac source, a change in input frequency will cause a change in the capacitor's capacitive reactance, and this will change the voltage drop across the capacitor. Although a 90° phase shift is present between voltage and current in a purely capacitive circuit (I leads V by 90°), there is no phase difference between the input voltage and the output voltages.

Filter
Network composed of resistor, capacitor, and inductors used to pass certain frequencies yet block others through heavy attenuation.

Low-Pass Filter
Network or circuit designed to pass any frequencies below a critical or cutoff frequency and reject or heavily attenuate all frequencies above.

9-5-3 RC *Filters*

A **filter** is a circuit that allows certain frequencies to pass but blocks other frequencies. In other words, it filters out the unwanted frequencies but passes the wanted or selected ones. There are two basic *RC* filters:

1. The **low-pass filter**, which can be seen in Figure 9-48(a); as its name implies, it passes the low frequencies but heavily **attenuates** the higher frequencies.
2. The **high-pass filter**, which can be seen in Figure 9-48(b); as its name implies, it allows the high frequencies to pass but heavily attenuates the lower frequencies.

With either the low- or high-pass filter, it is important to remember that capacitive reactance is inversely proportional to frequency ($X_C \propto 1/f$) as stated in the center of Figure 9-48.

Attenuate
To reduce in amplitude an action or signal.

High-Pass Filter
Network or circuit designed to pass any frequencies above a critical or cutoff frequency and reject or heavily attenuate all frequencies below.

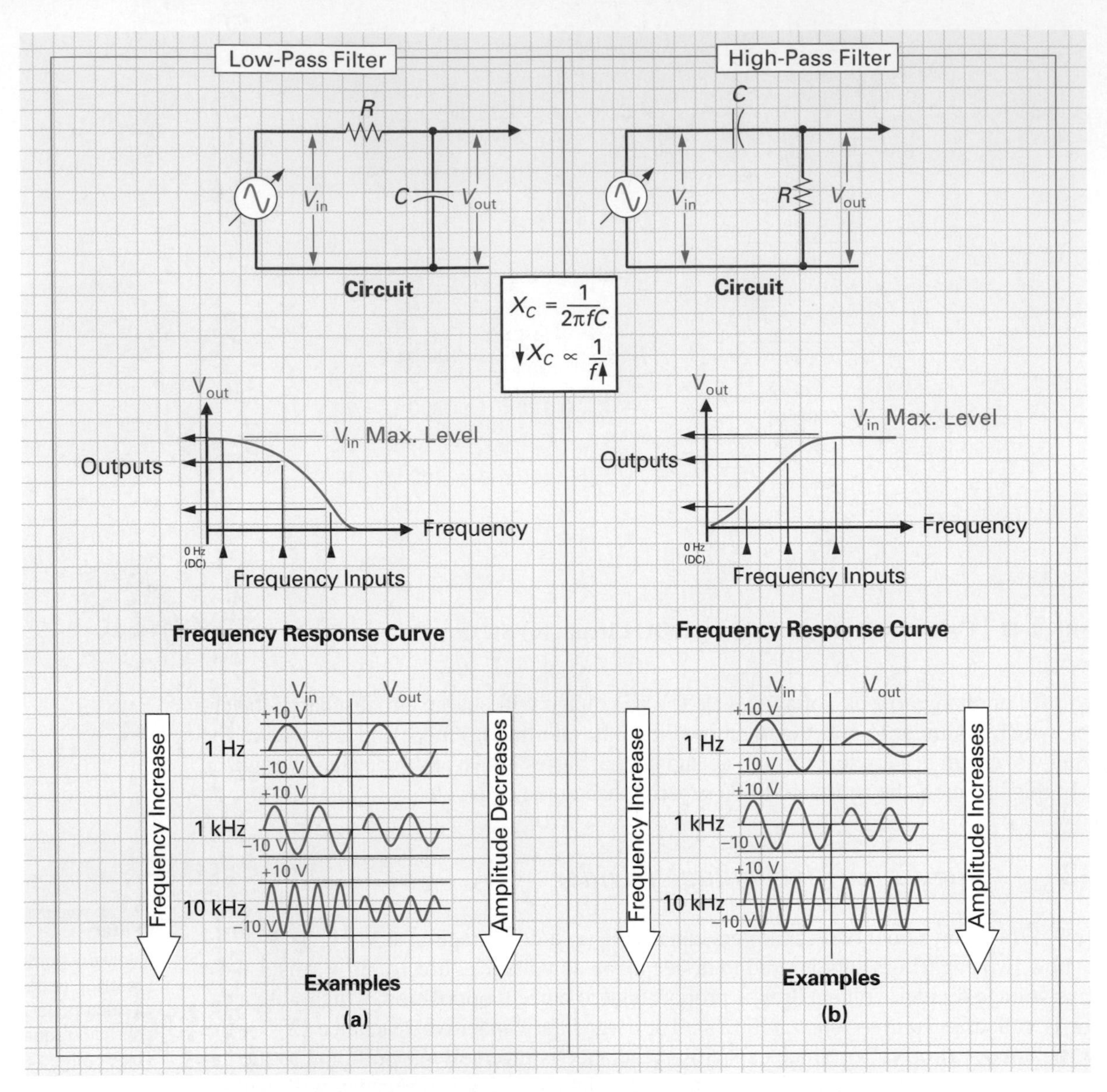

FIGURE 9-48 ***RC*** **Filters. (a) Low-Pass Filter. (b) High-Pass Filter.**

With the low-pass filter shown in Figure 9-48(a), the output is connected across the capacitor. As the frequency of the input increases, the amplitude of the output decreases. At dc (0 Hz) and low frequencies, the capacitive reactance is very large ($X_C\uparrow = 1/f\downarrow$) with respect to the resistor. All the input will appear across the capacitor, because the capacitor and resistor form a voltage divider, as shown in Figure 9-49(a). As with any voltage divider, the larger opposition to current flow will drop the larger voltage. Since the output voltage is determined by the voltage drop across the capacitor, almost all the input will appear across the capacitor and therefore be present at the output.

If the frequency of the input increases, the reactance of the capacitor will decrease ($X_C\downarrow = 1/f\uparrow$) and a larger amount of the signal will be dropped across the resistor. As frequency increases, the capacitor becomes more of a short circuit (lower reactance) and the output, which is across the capacitor, decreases, as shown in Figure 9-49(b).

Frequency Response Curve

A graph indicating how effectively a circuit or device responds to the frequency spectrum.

Below the circuit of the low-pass filter in Figure 9-48(a), you will see a graph known as the **frequency response curve** for the low-pass filter. This curve illustrates that as the frequency of the input increases the voltage at the output will decrease.

With the high-pass filter, seen in Figure 9-48(b), the capacitor and resistor have traded positions to show how the opposite effect for the low-pass filter occurs. At low frequencies, the reactance will be high and almost all of the signal in will be dropped across the capaci-

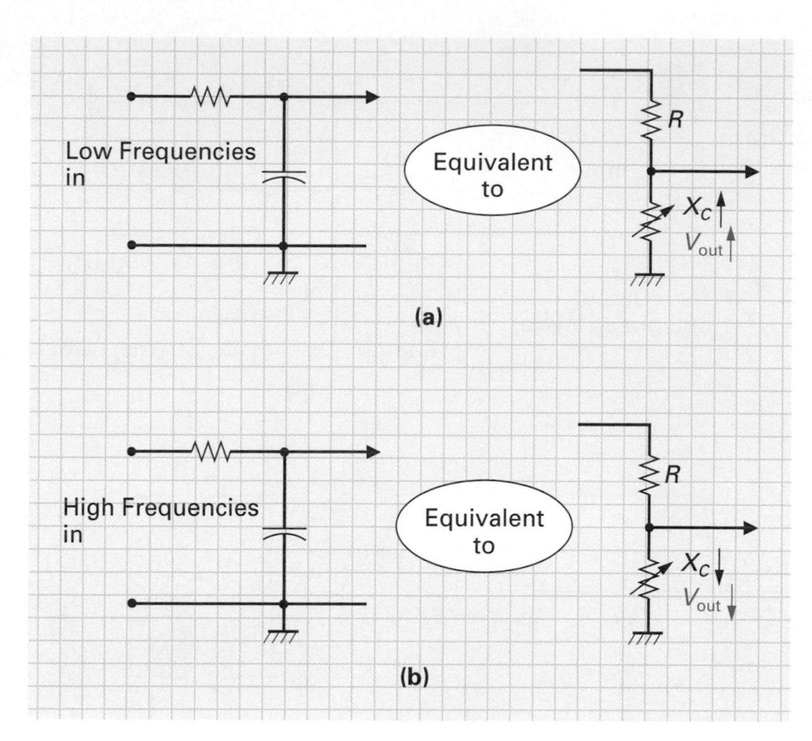

FIGURE 9-49 Low-Pass Filter. (a) Low-Pass Filter, High X_C. (b) Low-Pass Filter, Low X_C.

tor. Very little signal appears across the resistor, and, consequently, the output, as shown in Figure 9-50(a). As the frequency of the input increases, the reactance of the capacitor decreases, allowing more of the input signal to appear across the resistor and therefore appear at the output, as shown in Figure 9-50(b).

Below the circuit of the high-pass filter in Figure 9-48(b), you will see the frequency response curve for the high-pass filter. This curve illustrates that as the frequency of the input increases the voltage at the output increases.

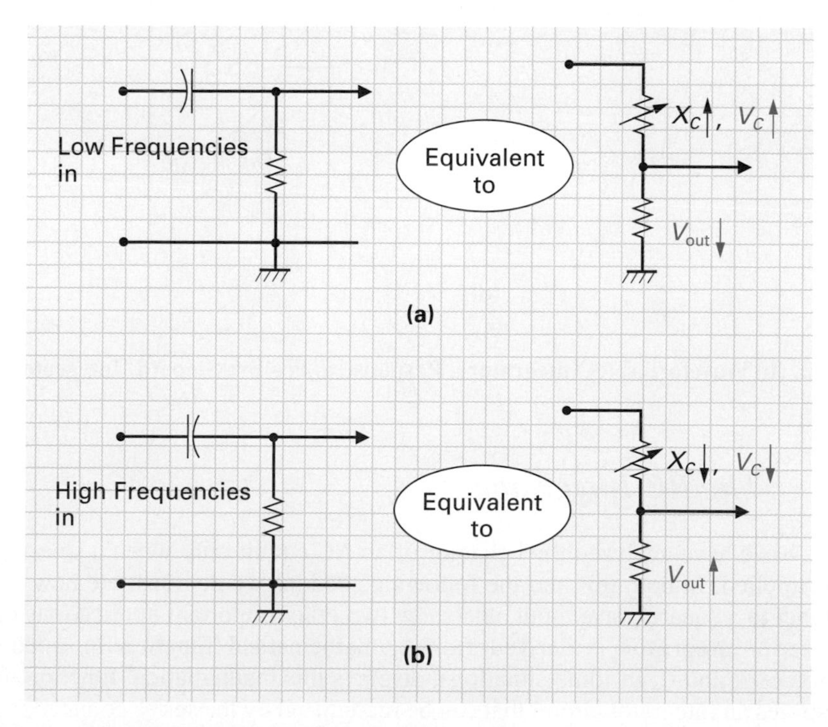

FIGURE 9-50 High-Pass Filter. (a) High X_C. (b) Low X_C.

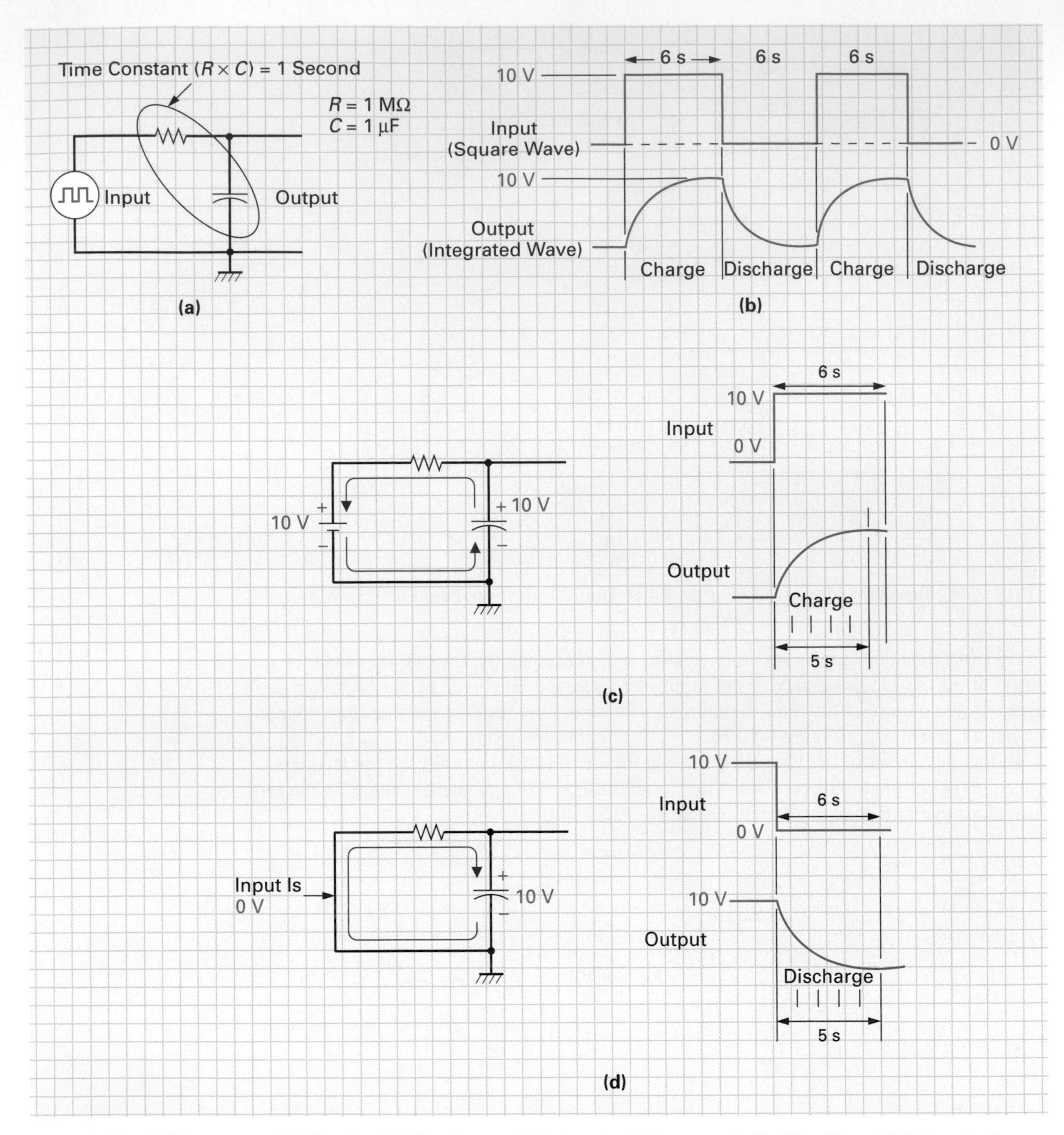

FIGURE 9-51 ***RC* Integrator. (a) Circuit. (b) Waveforms. (c) Integrator's Response to Positive Step. (d) Integrator's Response to Negative Step.**

9-5-4 *The RC Integrator*

Integrator

Device that approximates and whose output is proportional to an integral of the input signal.

Up until now we have analyzed the behavior of an *RC* circuit only when a sine-wave input signal was applied. In both this and the following sections we demonstrate how an *RC* circuit will react to a square-wave input, and show two other important applications of capacitors. The term **integrator** is derived from a mathematical function in calculus. This combination of *R* and *C*, in some situations, displays this mathematical function. Figure 9-51(a) illustrates an integrator circuit that can be recognized by the series connection of *R* and *C*, but mainly from the fact that the output is taken across the capacitor.

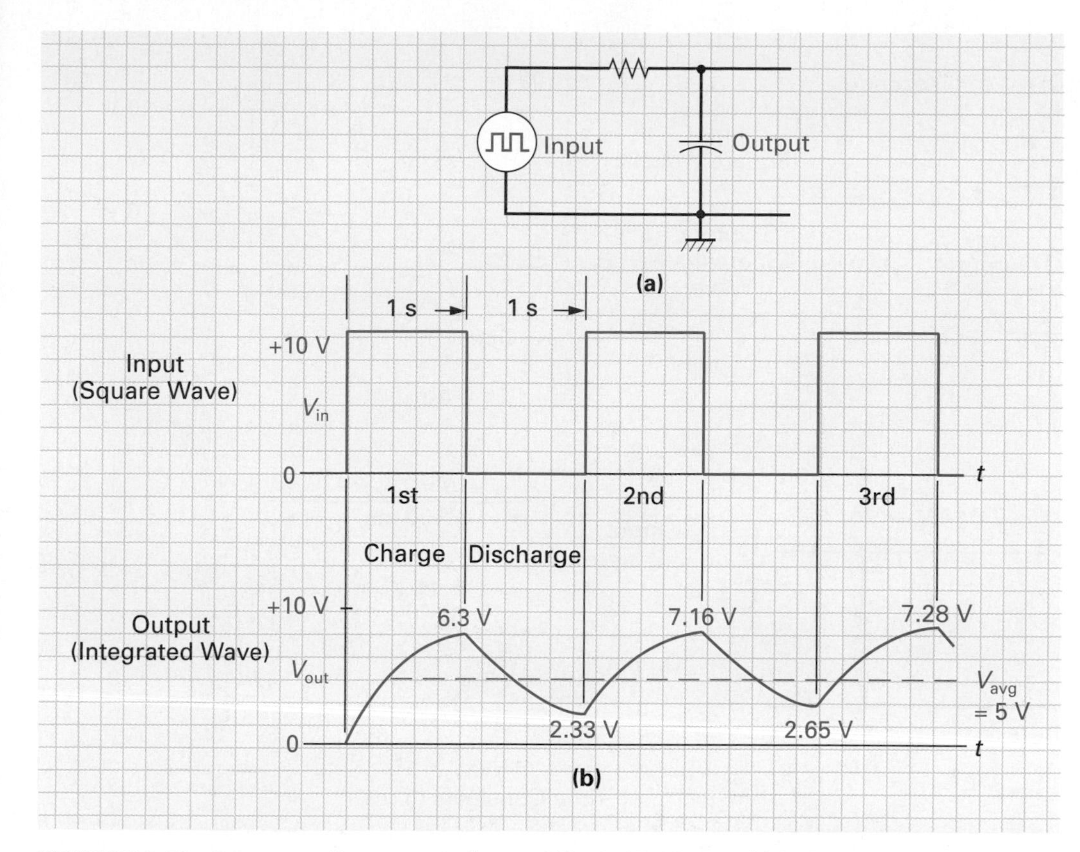

FIGURE 9-52 Integrator Response to Square Wave. (a) Circuit. (b) Waveforms.

If a 10 V square wave is applied across the circuit, as shown in the waveforms in Figure 9-51(b), and the time constant of the *RC* combination calculates out to be 1 second, the capacitor will charge when the square-wave input is positive toward the applied voltage (10 V) and reaches it in five time constants (5 seconds). Since the positive alternation of the square wave lasts for 6 seconds, the capacitor will be fully charged 1 second before the positive alternation ends, as shown in Figure 9-51(c).

When the positive half-cycle of the square-wave input ends after 6 seconds, the input falls to 0 V and the circuit is equivalent to that shown in Figure 9-51(d). The 10 V charged capacitor now has a path to discharge and in five time constants (5 seconds) is fully discharged, as shown in Figure 9-51(d).

If the same *RC* integrator circuit was connected to a square wave that has a 1 second positive half-cycle, the capacitor will not be able to charge fully toward 10 V. In fact, during the positive alternation of 1 second (one time constant), the capacitor will reach 63.2% of the applied voltage (6.32 V), and then during the 0 V half-cycle it will discharge to 63.2% of 6.32 V, to 2.33 V, as shown by the waveform in Figure 9-52. The voltage across the capacitor, and therefore the output voltage, will gradually build up and eventually level off to an average value of 5 V in about five time constants (5 seconds).

In summary, if the period or time of the square wave is decreased, or if the time constant is increased, the same effect results. The capacitor has less time to charge and discharge and reaches an average value of the input voltage, which for a square wave is half of its amplitude.

Differentiator
A circuit whose output voltage is proportional to the rate of change of the input voltage. The output waveform is then the time derivative of the input waveform, and the phase of the output waveform leads that of the input by 90°.

9-5-5 *The RC Differentiator*

Figure 9-53(a) illustrates the **differentiator** circuit, which is the integrator's opposite. In this case the output is taken across the resistor instead of the capacitor, and the time constant is always short with respect to the input square-wave period.

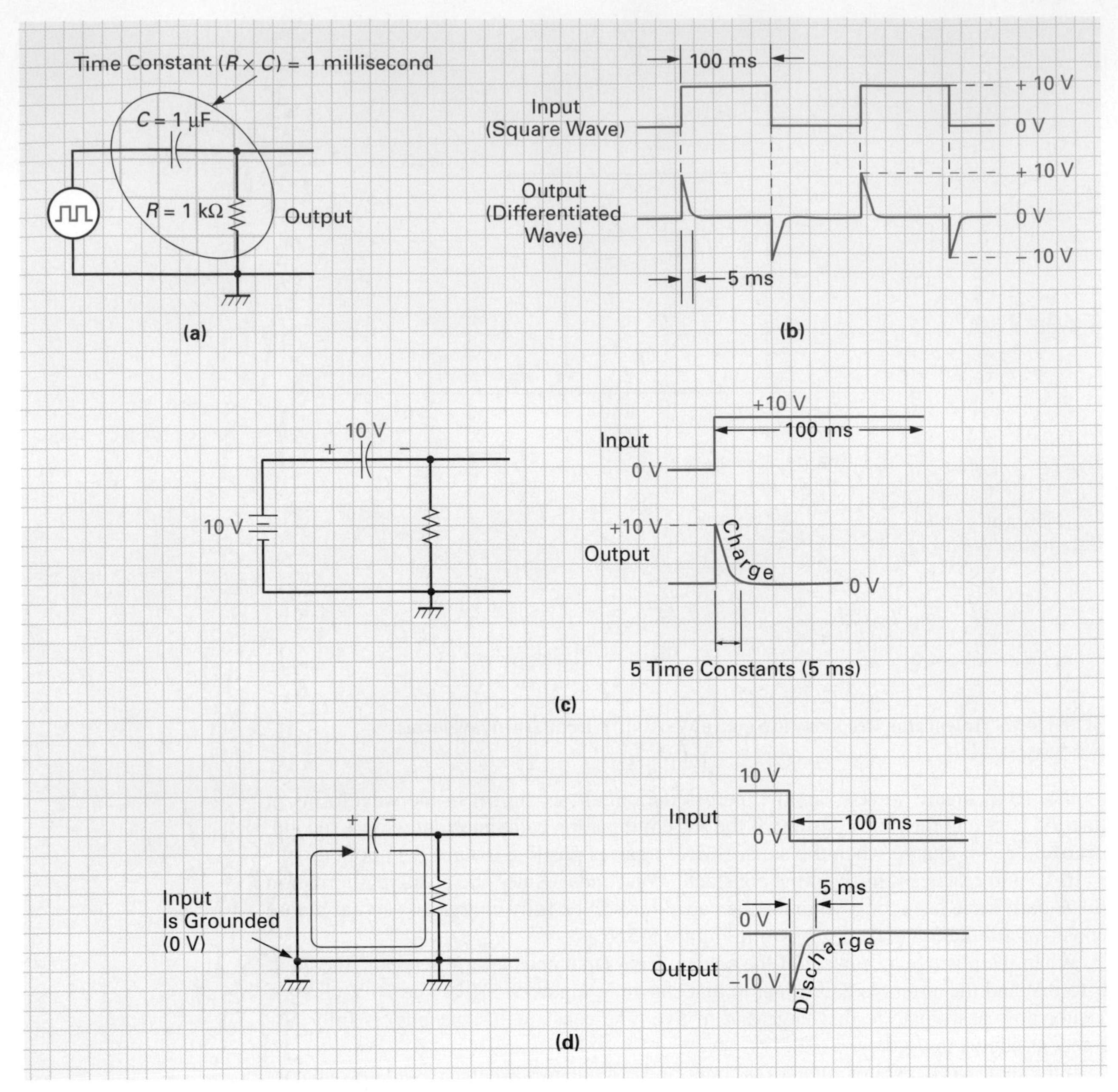

FIGURE 9-53 *RC* **Differentiator. (a) Circuit. (b) Waveforms. (c) Differentiator's Response to Positive Step. (d) Differentiator's Response to Negative Step.**

The differentiator output waveform, shown in Figure 9-53(b), is taken across the resistor and is the result of the capacitor's charge and discharge. When the square wave swings positive, the equivalent circuit is that shown in Figure 9-53(c). When the 10 V is initially applied (positive step of the square wave), all the voltage is across the resistor, and therefore at the output, as the capacitor cannot charge instantly. As the capacitor begins to charge, more of the voltage is developed across the capacitor and less across the resistor. The voltage across the capacitor exponentially increases and reaches 10 V in five time constants (5×1 ms $= 5$ ms), while the voltage across the resistor, and therefore the output, exponentially falls from its initial 10 V to 0 V in five time constants as shown in the waveforms in Figure 9-53(c), at which time all the voltage is across the capacitor and no voltage will be across the resistor.

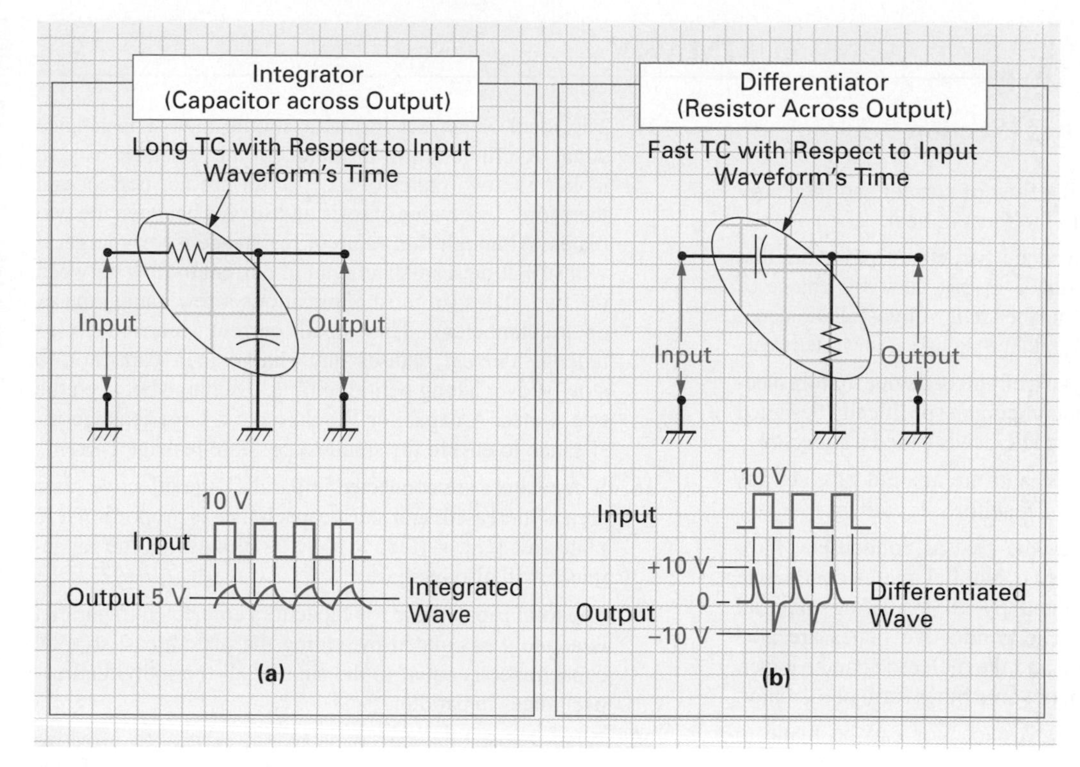

FIGURE 9-54 Summary of Integrator and Differentiator.

When the positive half-cycle of the square wave ends and the input falls to zero, the circuit is equivalent to that shown in Figure 9-53(d). The negative plate of the capacitor is now applied directly to the output. Since the capacitor cannot instantly discharge, the output drops suddenly down to −10 V as shown in the waveforms in Figure 9-53(d). This is the voltage across the resistor, and therefore the output. The capacitor is now in series with the resistor and therefore has a path through the resistor to discharge, which it does in five time constants to 0 V.

Figure 9-54(a) and (b) summarizes the integrator and differentiator circuits and waveforms, which are used extensively in many system applications, such as computers, robots, lasers, and communications.

SELF-TEST EVALUATION POINT FOR SECTION 9-5

Now that you have completed this section, you should be able to:

Objective 17. Describe how the capacitor and resistor in combination can be used

a. To combine ac and dc.
b. To act as a voltage divider.
c. As a filter.
d. As an integrator.
e. As a differentiator.

Use the following questions to test your understanding of Section 9–5:

1. Give three circuit applications for a capacitor.
2. With an *RC* low-pass filter, the ___________ is connected across the output, whereas with an *RC* high-pass filter, the ___________ is connected across the output.
3. An integrator has a ___________ time constant compared to the period of the input square wave.
4. A ___________ circuit produces positive and negative spikes at the output when a square wave is applied at the input.

SUMMARY

Capacitor Characteristics (Section 9-1)

1. Resistance will oppose the flow of current and then convert or dissipate power in the form of heat.
2. Capacitance and inductance are two circuit properties that act differently from resistance in that they will charge or store the supplied energy and then return almost all the stored energy back to the circuit, rather than lose it in wasted heat.
3. Capacitance is the ability of a circuit or device to store electrical charge. A device or component specifically designed to have this capacity or capacitance is called a *capacitor*.
4. Capacitors store electrons, and the amount of electrons stored is a measure of the capacitor's capacitance.
5. Capacitors, referred to in the past as condensers, have two leads connected to two parallel metal conductive plates, which are separated by an insulating material known as a *dielectric*. The conductive plates are normally made of metal foil, and the dielectric can be paper, air, glass, ceramic, mica, or some other form of insulator.
6. Like a secondary battery, a capacitor can be made to charge or discharge, and when discharging it will return all of the energy it consumed during charge.
7. Just as a magnetic field is produced by the flow of current, an electric field is produced by voltage.
8. A charged capacitor has an electric or electrostatic field existing between the positively charged and negatively charged plates. The strength of the electrostatic field is proportional to the charge or potential difference on the plates and inversely proportional to the distance between the plates.
9. The energy in a capacitor is actually stored in the electric or electrostatic field within the dielectric.
10. *Capacitance* is the ability of a capacitor to store an electrical charge, and the unit of capacitance is the farad (F), named in honor of Michael Faraday's work in 1831 in the field of capacitance.
11. A capacitor with the capacity of 1 farad (1 F) can store 1 coulomb of electrical charge (6.24×10^{18} electrons) if 1 volt is applied across the capacitor's plates.
12. The dielectric constant (K) is the ease with which an insulating material can establish an electrostatic (electric) field.
13. The capacitance of a capacitor is directly proportional to the dielectric constant (K) and the area (A) of the plates and is inversely proportional to the dielectric thickness or distance between the plates (d).
14. The amount of charge stored by a capacitor is proportional to the capacitor's capacitance and the voltage applied across the capacitor ($Q = C \times V$).
15. If the voltage across a capacitor is increased, the charge held by the capacitor will increase until the dielectric between the two plates of the capacitor breaks down and a spark jumps or arcs between the plates.
16. The breakdown voltage of a capacitor is determined by the strength of the dielectric used.
17. The ideal or perfect insulator should have a resistance equal to infinite ohms. Insulators or the dielectric used to isolate the two plates of a capacitor are not perfect and therefore some value of resistance exists between the two plates. Although this value of resistance is very large, it will still allow a small amount of current to flow between the two plates (in most applications a few nanoamperes or picoamperes). This small current, referred to as *leakage current*, causes any charge in a capacitor to slowly, over a long period of time, discharge between the two plates. A capacitor should have a large leakage resistance to ensure the smallest possible leakage current.
18. If capacitors are connected in parallel, the effective plate area is increased; and since capacitance is proportional to plate area [$C\uparrow = (8.85 \times 10^{-12}) \times K \times A\uparrow/d$], the capacitance will also increase.
19. The total capacitance for capacitors connected in parallel is actually calculated by adding the plate areas, so total capacitance is equal to the sum of all the individual capacitances in parallel.
20. When capacitors are connected in series the effective plate area is decreased ($A\downarrow$) and the dielectric thickness increased ($d\uparrow$), and both of these effects result in an overall capacitance decrease [$C\downarrow\downarrow = (8.85 \times 10^{-12}) \times K \times A\downarrow/d\uparrow$]. The plate area is actually decreased to the smallest individual capacitance connected in series. If the plate area were the only factor, then capacitance would always equal the smallest capacitor value. However, the dielectric thickness is always equal to the sum of all the capacitor dielectrics, and this factor always causes the total capacitance (C_T) to be less than the smallest individual capacitance when capacitors are connected in series.
21. The total capacitance of two or more capacitors in series therefore is calculated by using the product-over-sum formula or the reciprocal formula.
22. As with series-connected resistors, the sum of all of the voltage drops across the series-connected capacitors will equal the voltage applied (Kirchhoff's voltage law). With capacitors connected in series, the charged capacitors act as a voltage divider, and therefore the voltage-divider formula can be applied to capacitors in series.
23. Capacitors in series will have the same charging current throughout the circuit, and because of this, two or more capacitors in series will always have equal amounts of coulomb charge. If the charge (Q) is equal, the voltage across the capacitor is determined by the value of the capacitor. A small capacitance will charge to a larger voltage ($V\uparrow = Q/C\downarrow$), whereas a large value of capacitance will charge to a smaller voltage ($V\downarrow = Q/C\uparrow$).

Capacitor Types, Coding, and Testing (Section 9-2)

24. A *fixed-value capacitor* is a capacitor whose capacitance value remains constant and cannot be altered. Fixed capacitors normally come in a disk or a tubular package and

consist of metal foil plates separated by one of the following types of insulators (dielectric), which is the means by which we classify them:

a. Mica **d.** Plastic
b. Ceramic **e.** Electrolytic
c. Paper

25. A *variable-value capacitor* is a capacitor whose capacitance value can be changed by rotating a shaft. The variable capacitor normally consists of one electrically connected movable plate and one electrically connected stationary plate. There are basically four types of variable-value capacitors, which are also classified by the dielectric used:

a. Air **c.** Ceramic
b. Mica **d.** Plastic

26. Ceramic chip surface-mount capacitors are now available on the market for both discrete and integrated types of circuits.

27. The electrolytic capacitor must always have a positive charge applied to its positive plate and a negative charge to its negative plate. If this rule is not followed, the electrolytic capacitor becomes a safety hazard, because it can, in the worst-case condition, explode violently.

28. Tantalum instead of aluminum is used in some electrolytic capacitors for the plates and has advantages over the aluminum type, which are

a. Higher capacitance per volt for a given unit volume
b. Longer life and excellent shelf life (storage)
c. Ability to operate in a wider temperature range
d. Better temperature stability
e. More rugged construction features

The cost of tantulum electrolytics, however, is almost four to five times that of aluminum electrolytics, and their operating voltages are much lower.

29. The variable-value-type capacitors use a hand-rotated shaft to vary the effective plate area, and so capacitance.

30. The adjustable-type capacitors use screw-in or screw-out mechanical adjustment to vary the distance between the plates, and so capacitance.

31. Capacitors should only be adjusted with a plastic or nonmetallic alignment tool, because a metal screwdriver may affect the capacitance of the capacitor when nearby, making it very difficult to adjust for a specific value of capacitance.

32. The capacitor value, tolerance, and voltage rating need to be shown in some way on the capacitor's exterior. Presently, two methods are used: alphanumeric labels and color coding.

33. A faulty capacitor may have one of three basic problems:

a. A short, which is easy to detect and is caused by a contact from plate to plate.
b. An open, which is again quite easy to detect and is normally caused by one of the leads becoming disconnected from its respective plate.
c. A leaky dielectric or capacitor breakdown, which is quite difficult to detect, as it may only short at a certain voltage. This problem is usually caused by the deterioration of the dielectric, which starts displaying a much lower dielectric resistance than it was designed for. The capacitor with this type of problem is referred to as a *leaky capacitor*.

34. Capacitors of 0.5 μF and larger can be checked by using a digital multimeter with a bar graph display.

35. The ohmmeter check tests the capacitor under a low-voltage condition. This may be adequate for some capacitor malfunctions; however, a problem that often occurs with capacitors is that they short or leak at a high voltage. A capacitance meter or analyzer can totally check all aspects of a capacitor in a range of values from approximately 1 pF to 20 F. The tests that are generally carried out include

a. Capacitor value change
b. Capacitor leakage
c. Dielectric absorption
d. Equivalent series resistance

DC Capacitive Circuits (Section 9-3)

36. When a capacitor is connected across a dc voltage source, it will charge to a value equal to the voltage applied. If the charged capacitor was then connected across a load, the capacitor would then discharge through the load. The time it takes a capacitor to charge or discharge can be calculated if the circuit's resistance and capacitance are known.

37. Time constant is the time needed for either a voltage or current to rise to 63.2% of the maximum or fall to 36.8% of the initial value. The time constant of an *RC* circuit is equal to the product of *R* and *C*.

38. The voltage waveform produced by the capacitor acquiring a charge is known as an *exponential* waveform, and the voltage across the capacitor is said to rise exponentially. An exponential rise is also referred to as a *natural increase*.

39. In a pure resistive circuit, the current flow through a resistor would be in step with the voltage across that same resistor, in that an increased current would cause a corresponding increase in voltage drop across the resistor. Voltage and current are consequently said to be *in step* or *in phase* with one another.

40. With the capacitive circuit, the current flow in the circuit and voltage across the capacitor are not in step or in phase with one another. The circuit current flow is out of phase with the capacitor voltage, and this difference is referred to as a *phase shift*. In any circuit containing capacitance, current will lead voltage.

41. When a dc voltage is applied across a capacitive circuit, the capacitor will charge and then oppose the dc source, preventing current flow. Since no current can flow when the capacitor is charged, the capacitor acts like an open circuit, and it is this action that accounts for why capacitors are referred to as an "open to dc."

AC Capacitive Circuits (Section 9-4)

42. When an ac voltage is applied across a capacitive circuit, the continual reversal of the applied voltage will cause a continual charging and discharging of the capacitor, and therefore circuit current will always be present. Any type of ac source or in fact any fluctuating or changing dc source will cause a circuit current, and if current is present, the opposition is low. It is this action that accounts for why capacitors are referred to as a "short to ac."

43. This ability of a capacitor to block dc and seem to pass ac will be exploited in applications of capacitors.

44. Thinking of the applied alternating voltage as a dc source that is being continually reversed sometimes makes it easier to understand how a capacitor reacts to ac. Whether the applied voltage is dc or ac, the rule holds true, in that a 90° phase shift or difference exists between circuit current and capacitor voltage.

45. In a capacitive circuit, circuit current leads the capacitor voltage by 90°, and the 90° leading phase shift (current leading voltage) will occur only in a purely capacitive circuit.

46. *Resistance* (R), by definition, is the opposition to current flow with the dissipation of energy and is measured in ohms. Capacitors oppose current flow like a resistor, but a resistor dissipates energy, whereas a capacitor stores energy (when it charges) and then gives back its energy into the circuit (when it discharges). *Capacitive reactance* (X_C), by definition, is the opposition to current without the dissipation of energy and is also measured in ohms.

47. If capacitive reactance is basically opposition, it is inversely proportional to the amount of current flow.

48. When a dc source is connected across a capacitor, current will flow only for a short period of time ($5RC$ seconds) to charge the capacitor. After this time, there is no further current flow. Consequently, the capacitive reactance or opposition offered by a capacitor to dc is infinite (maximum).

49. Alternating current is continuously reversing in polarity, resulting in the capacitor continuously charging and discharging. This means that charge and discharge currents are always flowing around the circuit, and if we have a certain value of current, we must also have a certain value of reactance or opposition.

50. If the applied alternating current is at a high frequency, it is switching polarity more rapidly than a lower frequency and there is very little time between the start of charge and discharge. As the charge and discharge currents are largest at the beginning of the charge and discharge of the capacitor, the reactance has very little time to build up and oppose the current, which is why the current is a high value and the capacitive reactance is small at higher frequencies. With lower frequencies, the applied alternating current is switching at a slower rate, and therefore the reactance, which is low at the beginning, has more time to build up and oppose the current. Capacitive reactance is therefore inversely proportional to frequency.

51. Capacitive reactance is also inversely proportional to the value of capacitance. If a larger capacitor value is used a longer time is required to charge the capacitor ($t\uparrow = C\uparrow R$), which means that current will be present for a longer period of time, so the overall current will be large ($I\uparrow$); consequently, the reactance must be small ($X_C\downarrow$). On the other hand, a small capacitance value will charge in a small amount of time ($t\downarrow = C\downarrow R$) and the current is present for only a short period of time. The overall current will therefore be small ($I\downarrow$), indicating a large reactance ($X_C\uparrow$). Capacitive reactance (X_C) is therefore inversely proportional to both frequency and capacitance.

52. In a purely resistive circuit, the current flowing within the circuit and the voltage across the resistor are in phase with one another.

53. In a purely capacitive circuit, the current flowing in the circuit leads the voltage across the capacitor by 90°.

54. The voltage across the resistor (V_R) is always in phase with the circuit current (I), because maximum points and zero crossover points occur at the same time. The voltage across the capacitor (V_C) lags the circuit current by 90°.

55. The voltage drop across the resistor (V_R) and the voltage drop across the capacitor (V_C) are out of phase with one another, which means that their peaks occur at different times. The signal for the applied voltage (V_S) is therefore obtained by adding the values of V_C and V_R at each instant in time, plotting the results, and then connecting the points with a line.

56. A vector (or phasor) is a quantity that has both magnitude and direction and is represented by a line terminated at the end by an arrowhead.

57. Vectors are generally always used to represent a physical quantity that has two properties.

58. A vector diagram is an arrangement of vectors to illustrate the magnitude and phase relationships between two or more quantities of the same frequency within an ac circuit.

59. By vectorially adding the two voltages V_R and V_C, we obtain a resultant V_S vector that has both magnitude and phase. The angle theta (θ) formed between circuit current (I) and source voltage (V_S) will always be less than 90°.

60. Impedance (designated Z) is also measured in ohms and is the total circuit opposition to current flow. It is a combination of resistance (R) and reactance (X_C).

61. By vectorially adding R and X_C, we have a resulting impedance (Z) vector.

62. In a purely resistive circuit, the total opposition (Z) is equal to the resistance of the resistor, so the phase shift (θ) is equal to 0°.

63. In a purely capacitive circuit, the total opposition (Z) is equal to the capacitive reactance (X_C) of the capacitor, so the phase shift (θ) is equal to $-90°$.

64. When a circuit contains both resistance and capacitive reactance, the total opposition or impedance has a phase shift that is between 0 and 90°.

65. Voltage and current are in phase with one another in a resistive circuit, and instantaneous power is calculated by multiplying voltage by current at every instant through 360° ($P = V \times I$). The sinusoidal power waveform is totally positive, because a positive voltage multiplied by a positive current gives a positive value of power, and a negative voltage multiplied by a negative current will also produce a positive value of power.

66. A resistor is said to generate a positive power waveform, which is twice the frequency of the voltage and current waveforms; two power cycles occur in the same time as one voltage and current cycle.

The power waveform has been split in half, and the line that exists between the maximum point and zero point is the average value of power that is being dissipated by the resistor.

67. The current leads the voltage by 90° in a purely capacitive circuit, and the power wave is calculated by multiplying voltage by current, as before, at every instant through 360°. The resulting power curve is both positive and negative. During the positive alternation of the power curve, the capacitor is taking power as the capacitor charges. When the power alternation is negative, the capacitor is giving back the power it took as it discharges back into the circuit.

    The average power dissipated calculates out to be zero, which means that no power is dissipated in a purely capacitive circuit.

68. When an ac voltage source is applied across a series-connected *RC* circuit, the current leads the voltage by some phase angle less than 90°, and the power waveform is once again determined by the product of voltage and current. The negative alternation of the power cycle indicates that the capacitor is discharging and giving back the power that it consumed during the charge.

69. The positive alternation of the power cycle is much larger than the negative alternation because it is the combination of both the capacitor taking power during charge and the resistor consuming and dissipating power in the form of heat. The average power being dissipated will be some positive value, due to the heat being generated by the resistor.

70. In a purely resistive circuit, all the energy supplied to the resistor from the source is dissipated in the form of heat. This form of power is referred to as *resistive power* (P_R) or *true power*.

71. In a purely capacitive circuit, all the energy supplied to the capacitor is stored from the source and then returned to the source, without energy loss. This form of power is referred to as *reactive power* (P_X) or *imaginary power*.

72. When a circuit contains both capacitance and resistance, some of the supply is stored and returned by the capacitor and some of the energy is dissipated and lost by the resistor.

73. Apparent power is the power that appears to be supplied to the load and includes both the true power dissipated by the resistance and the imaginary power delivered to the capacitor.

74. The power factor is a ratio of true power to apparent power and is a measure of the loss in a circuit.

75. As with any parallel circuit, the voltage across all components in parallel is equal to the source voltage.

76. The current through the resistor and capacitor is simply calculated by applying Ohm's law.

77. Capacitor current will always lead the applied voltage by 90°, and as the applied voltage is being used as our reference at 0° on the vector diagram, the capacitor current will have to be drawn at + 90° in order to lead the applied voltage by 90°, since vector diagrams rotate in a counterclockwise direction.

78. Total current is therefore the vector sum of both the resistor and capacitor currents.

79. The angle by which the total current (I_T) leads the source voltage (V_S) is called the *phase angle*.

80. The impedance of a parallel *RC* circuit is equal to the total voltage divided by the total current.

81. With respect to power, there is no difference between a series circuit and a parallel circuit.

Applications of Capacitors (Section 9-5)

82. The capacitor can be used to combine ac and dc.

83. Capacitors could be connected in series across a dc voltage source to form a voltage divider.

84. A filter is a circuit that allows certain frequencies to pass but blocks other frequencies. In other words, it filters out the unwanted frequencies but passes the wanted or selected ones.

    There are two basic *RC* filters:

    a. The low-pass filter, which as its name implies passes the low frequencies but heavily attenuates the higher frequencies
    b. The high-pass filter, which as its name implies allows the high frequencies to pass but heavily attenuates the lower frequencies

85. The frequency response curve for the low-pass filter illustrates that as the frequency of the input increases, the voltage at the output will decrease. The frequency response curve for the high-pass filter illustrates that as the frequency of the input increases, the voltage at the output increases.

86. The term *integrator* is derived from a mathematical function in calculus. An integrator circuit can be recognized by the series connection of *R* and *C*, but mainly from the fact that the output is taken across the capacitor. The voltage across the capacitor of an integrator, and therefore the output voltage, will gradually build up and eventually level off to an average value in about five time constants.

87. With the differentiator circuit, the output is taken across the resistor instead of the capacitor, and the time constant is always short with respect to the input square-wave period. The differentiator output waveform is taken across the resistor and is the result of the capacitor's charge and discharge.

REVIEW QUESTIONS

Multiple-Choice Questions

1. Capacitors were originally referred to as
- **a.** Vacuum tubes **c.** Inductors
- **b.** Condensers **d.** Suppressors

2. When a capacitor charges,
- **a.** The voltage across the plates rises exponentially.
- **b.** The circuit current falls exponentially.
- **c.** The capacitor charges to the source voltage in 5*RC* seconds.
- **d.** All of the above.

3. A/an __________ field is generated by the flow of current and a/an __________ field is generated by voltage.
- **a.** Magnetic, electrostatic
- **b.** Electric, electrostatic
- **c.** Electric, magnetic
- **d.** All of the above may be considered true.

4. The strength of an electric field in a capacitor is proportional to the __________ and inversely proportional to the __________.
- **a.** Plate separation, charge
- **b.** Plate separation, potential plate difference
- **c.** Plate potential difference, plate separation
- **d.** Both (a) and (b)

5. The plates of a capacitor are generally made out of a
- **a.** Resistive material.
- **b.** Semiconductor material.
- **c.** Conductive material.
- **d.** Two of the above could be true.

6. The energy of a capacitor is stored in
- **a.** The magnetic field within the dielectric.
- **b.** The magnetic field around the capacitor leads.
- **c.** The electric field within the plates.
- **d.** The electric field within the dielectric.

7. What is the capacitance of a capacitor if it can store 24 C of charge when 6 V is applied across the plates?
a. 2 μF **b.** 3 μF **c.** 4.7 μF **d.** None of the above

8. The capacitance of a capacitor is directly proportional to
- **a.** The plate area.
- **b.** The distance between the plates.
- **c.** The constant of the dielectric used.
- **d.** Both (a) and (c).
- **e.** Both (a) and (b).

9. The capacitance of a capacitor is inversely proportional to
- **a.** The plate area.
- **b.** The distance between the plates.
- **c.** The dielectric used.
- **d.** Both (a) and (c).
- **e.** Both (a) and (b).

10. The breakdown voltage of a capacitor is determined by
- **a.** The type of dielectric used.
- **b.** The size of the capacitor plates.
- **c.** The wire gauge of the connecting leads.
- **d.** Both (a) and (b).

11. A capacitor should have a __________ leakage resistance to ensure a __________ leakage current.
- **a.** Large, small **c.** Small, large
- **b.** Large, medium **d.** Small, small

12. Total series capacitance of two capacitors is calculated by
- **a.** Using the product-over-sum formula.
- **b.** Using the voltage-divider formula.
- **c.** Using the series resistance formula on the capacitors.
- **d.** Adding all the individual values.

13. Total parallel capacitance is calculated by
- **a.** Using the product-over-sum formula.
- **b.** Using the voltage-divider formula.
- **c.** Using the parallel resistance formula on capacitors.
- **d.** Adding all the individual values.

14. The mica and ceramic fixed capacitors have
- **a.** An arrangement of stacked plates.
- **b.** An electrolyte substance between the plates.
- **c.** An adjustable range.
- **d.** All of the above.

15. An electrolytic capacitor
- **a.** Is the most popular large-value capacitor.
- **b.** Is polarized.
- **c.** Can have either aluminum or tantalum plates.
- **d.** All of the above.

16. Variable capacitors normally achieve a large variation in capacitance by varying __________, while adjustable trimmer capacitors only achieve a small capacitance range by varying __________.
- **a.** Dielectric constant, plate area
- **b.** Plate area, plate separation
- **c.** Plate separation, dielectric constant
- **d.** Plate separation, plate area

17. In one time constant, a capacitor will charge to __________ of the source voltage.
a. 86.5% **b.** 63.2% **c.** 99.3% **d.** 98.2%

18. When ac is applied across a capacitor, a __________ phase shift exists between circuit current and capacitor voltage.
a. 45° **b.** 60° **c.** 90° **d.** 63.2°

19. A capacitor consists of
- **a.** Two insulated plates separated by a conductor.
- **b.** Two conductive plates separated by a conductor.
- **c.** Two conductive plates separated by an insulator.
- **d.** Two conductive plates separated by a conductive spacer.

20. A 47 μF capacitor charged to 6.3 V will have a stored charge of
- **a.** 296.1 μC **d.** All of the above
- **b.** 2.96×10^{-4} C **e.** Both (a) and (b)
- **c.** 0.296 mC

21. Capacitive reactance is inversely proportional to
- **a.** Capacitance and resistance.
- **b.** Frequency and capacitance.
- **c.** Capacitance and impedance.
- **d.** Both (a) and (c).

22. The impedance of an *RC* series circuit is equal to
    a. The sum of R and X_C.
    b. The square root of the sum of R^2 and X_C^2.
    c. The square of the sum of R and X_C.
    d. The sum of the square root of R and X_C.
23. In a purely resistive circuit,
    a. The current flowing in the circuit leads the voltage across the capacitor by 90°.
    b. The circuit current and resistor voltage are in phase with one another.
    c. The current leads the voltage by 45°.
    d. The current leads the voltage by a phase angle between 0 and 90°.
24. In a purely capacitive circuit,
    a. The current flowing in the circuit leads the voltage across the capacitor by 90°.
    b. The circuit current and resistor voltage are in phase with one another.
    c. The current leads the voltage by 45°.
    d. The current leads the voltage by a phase angle between 0 and 90°.
25. In a series circuit containing both capacitance and resistance,
    a. The current flowing in the circuit leads the voltage across the capacitor by 90°.
    b. The circuit current and resistor voltage are in phase with one another.
    c. The current leads the voltage by 45°.
    d. Both (a) and (b).
26. In a series *RC* circuit, the source voltage is equal to
    a. The sum of V_R and V_C.
    b. The difference between V_R and V_C.
    c. The vectoral sum of V_R and V_C.
    d. The sum of V_R and V_C squared.
27. As the source frequency is increased, the capacitive reactance will
    a. Increase.
    b. Decrease.
    c. Be unaffected.
    d. Increase, depending on harmonic content.
28. The phase angle of a series *RC* circuit indicates by what angle V_S __________ V_R.
    a. Lags   c. Leads or lags
    b. Leads   d. None of the above
29. In a series *RC* circuit, the vector combination of R and X_C is the circuit's __________.
    a. Phase angle   c. Source voltage
    b. Apparent power   d. Impedance
30. In a parallel *RC* circuit, the total current is equal to
    a. The sum of I_R and I_C.
    b. The difference between I_R and I_C.
    c. The vectoral sum of I_R and I_C.
    d. The sum of I_R and I_C squared.
31. __________ is the opposition offered by a capacitor to current flow without the dissipation of energy.
    a. Capacitive reactance   d. Phase angle
    b. Resistance   e. The power factor
    c. Impedance
32. __________ is the total reactive and resistive circuit opposition to current flow.
    a. Capacitive reactance   d. Phase angle
    b. Resistance   e. The power factor
    c. Impedance
33. __________ is the ratio of true (resistive) power to apparent power and is therefore a measure of the loss in a circuit.
    a. Capacitive reactance   d. Phase angle
    b. Resistance   e. The power factor
    c. Impedance
34. In a series *RC* circuit, the leading voltage will be measured across the
    a. Resistor.   c. Source.
    b. Capacitor.   d. Any of the choices are true.
35. In a series *RC* circuit, the lagging voltage will be measured across the
    a. Resistor.   c. Source.
    b. Capacitor.   d. Any of the choices are true.

Communication Skill Questions

36. Describe the construction and main parts of a capacitor. (9-1)
37. What happens to a capacitor during
    a. Charge. (9-1-1)   b. Discharge. (9-1-2)
38. Describe how electrostatics relates to capacitance and give the formula for electric field strength. (9-1-3)
39. Briefly explain the relationship between capacitance, charge, and voltage. (9-1-4)
40. Describe the three factors affecting the capacitance of a capacitor. (9-1-5)
41. Briefly describe
    a. The term *capacitor breakdown*: (9-1-7)
    b. The term *capacitor leakage:* (9-1-7)
    c. Why the smaller-value capacitor in a two-capacitor series circuit has the larger voltage developed across it. (9-1-6)
42. List the formula(s) used to calculate total capacitance when capacitors are connected in
    a. Parallel: (9-1-6)   b. Series: (9-1-6)
43. Describe the following types of fixed-value capacitors (9-2):
    a. Mica   d. Plastic
    b. Ceramic   e. Electrolytic
    c. Paper
44. Briefly describe the following types of variable-value capacitors (9-2):
    a. Air   b. Mica, ceramic, and plastic
45. Explain how the constant 63.2 is used in relation to the charge and discharge of a capacitor. (9-3)
46. How is the coding of a capacitor's value, tolerance, and voltage rating indicated on a capacitor? (9-2)
47. Describe the differences between dc and ac capacitor charging and discharging. (9-4)
48. Calculate the total capacitance of the circuits illustrated in Figure 9-55, and describe which capacitors are in parallel and which are in series. (9-1-6)

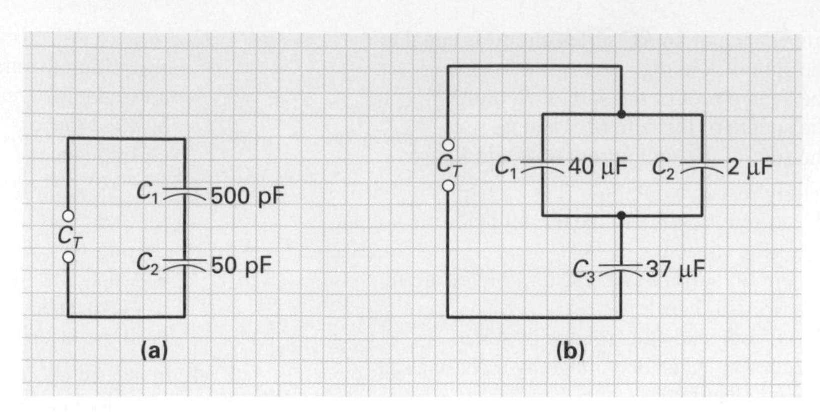

FIGURE 9-55

49. What advantages and disadvantages do tantalum electrolytic capacitors have over the aluminum type? (9-2-1)
50. Which of the fixed-value capacitor types is polarity sensitive, and what does this mean? (9-2-1)
51. Briefly describe why series-connected capacitors are treated like parallel-connected resistors, and why parallel-connected capacitors are treated like series-connected resistors when calculating total capacitance. (9-1-6)
52. Give the formula and define the term *capacitive reactance*. (9-4-2)
53. In a series *RC* circuit, give the formulas for calculating (9-4-3)
    a. V_S, when V_R and V_C are known.
    b. Z, when R and X_C are known.
    c. Z, when I and V are known.
    d. θ, when X_C and R are known.
    e. θ, when V_C and V_R are known.
    f. Power factor, when R and Z are known.
    g. Power factor, when P_R and P_A are known.
54. In a parallel *RC* circuit, give the formulas for calculating (9-4-4)
    a. I_R, when V and R are known.
    b. I_C, when V and X_C are known.
    c. I, when I_R and I_C are known.
    d. Z, when V and I are known.
    e. Z, when R and X_C are known.
55. What is meant by *long* or *short time constant*, and do large or small values of *RC* produce a long or a short time constant? (9-3)
56. Describe how the inverse relationship between frequency and capacitive reactance can be used for the application of filtering. (9-4-2)
57. Describe and illustrate how the capacitor can be used in the following applications (9-5):
    a. Combining ac and dc
    b. A voltage divider
    c. Filtering high and low frequencies
    d. Integrating a square wave
    e. Differentiating a square wave
58. Describe the difference among reactance, resistance, and impedance. (9-4-3)
59. Sketch the phase relationships between (9-4-3)
    a. V_R and I in a purely resistive circuit.
    b. V_C and I in a purely capacitive circuit.
60. What are positive power and negative power? (9-4)

Practice Problems

61. If a 10 μF capacitor is charged to 10 V, how many coulombs of charge has it stored?
62. Calculate the electric field strength within the dielectric of a capacitor that is charged to 6 V and that has a dielectric thickness of 32 μm (32 millionths of a meter).
63. If a 0.006 μF capacitor has stored 125×10^{-6} C of charge, what potential difference would appear across the plates?
64. Calculate the capacitance of the capacitor that has the following parameter values: $A = 0.008$ m^2; $d = 0.00095$ m; the dielectric used is paper.
65. Calculate the total capacitance if the following are connected in
    a. Parallel: 1.7 μF, 2.6 μF, 0.03 μF, 1200 pF.
    b. Series: 1.6 μF, 1.4 μF, 4 μF.
66. If three capacitors of 0.025 μF, 0.04 μF, and 0.037 μF are connected in series across a 12 V source, as shown in Figure 9-56, what would be the voltage drop across each?
67. Give the value of the following alphanumeric capacitor value codes:
    a. 104 c. 0.01
    b. 125 d. 220

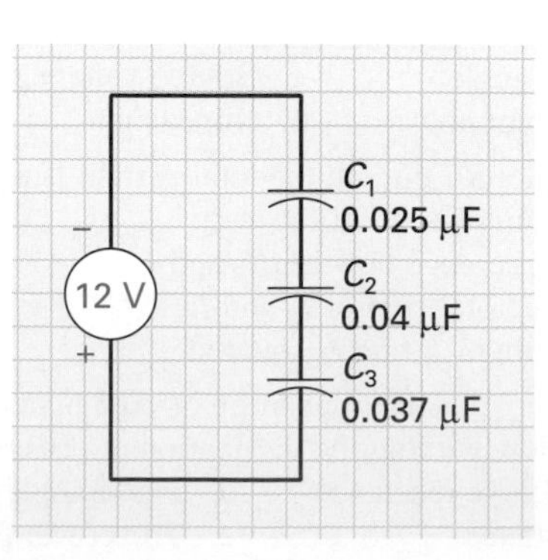

FIGURE 9-56

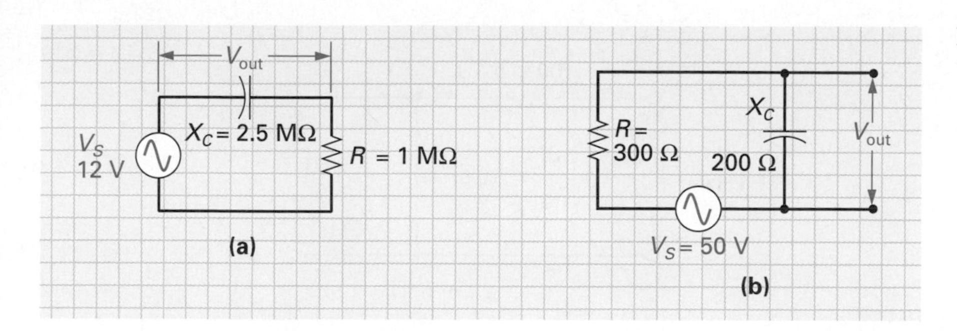

FIGURE 9-57

68. What would be the time constant of the following *RC* circuits?
   a. $R = 6\ \text{k}\Omega$, $C = 14\ \mu\text{F}$
   b. $R = 12\ \text{M}\Omega$, $C = 1400\ \text{pF}$
   c. $R = 170\ \Omega$, $C = 24\ \mu\text{F}$
   d. $R = 140\ \text{k}\Omega$, $C = 0.007\ \mu\text{F}$

69. If 10 V were applied across all the *RC* circuits in Question 67, what would be the voltage across each capacitor after one time constant, and how much time would it take each capacitor to fully charge?

70. In one application, a capacitor is needed to store 25 μC of charge and will always have 125 V applied across its terminals. Calculate the capacitance value needed.

71. Calculate the capacitive reactance of the capacitor circuits with the following parameters:
   a. $f = 1\ \text{kHz}$, $C = 2\ \mu\text{F}$
   b. $f = 100\ \text{Hz}$, $C = 0.01\ \mu\text{F}$
   c. $f = 17.3\ \text{MHz}$, $C = 47\ \mu\text{F}$

72. In a series *RC* circuit, the voltage across the capacitor is 12 V and the voltage across the resistor is 6 V. Calculate the source voltage.

73. Calculate the impedance for the following series *RC* circuits:
   a. 2.7 MΩ, 3.7 μF, 20 kHz
   b. 350 Ω, 0.005 μF, 3 MHz
   c. $R = 8.6\ \text{k}\Omega$, $X_C = 2.4\ \Omega$
   d. $R = 4700\ \Omega$, $X_C = 2\ \text{k}\Omega$

74. In a parallel *RC* circuit with parameters of $V_S = 12$ V, $R = 4\ \text{M}\Omega$, and $X_C = 1.3\ \text{k}\Omega$, calculate:
   a. I_R
   b. I_C
   c. I_T
   d. Z
   e. θ

75. Calculate the total reactance in
   a. A series circuit where $X_{C1} = 200\ \Omega$, $X_{C2} = 300\ \Omega$, $X_{C3} = 400\ \Omega$.
   b. A parallel circuit where $X_{C1} = 3.3\ \text{k}\Omega$, $X_{C2} = 2.7\ \text{k}\Omega$.

76. Calculate the capacitance needed to produce 10 kΩ of reactance at 20 kHz.

77. At what frequency will a 4.7 μF capacitor have a reactance of 2000 Ω?

78. A series *RC* circuit contains a resistance of 40 Ω and a capacitive reactance of 33 Ω across a 24 V source.
   a. Sketch the schematic diagram.
   b. Calculate Z, I, V_R, V_C, I_R, I_C, and θ.

79. A parallel *RC* circuit contains a resistance of 10 kΩ and a capacitive reactance of 5 kΩ across a 100 V source.
   a. Sketch the schematic diagram.
   b. Calculate I_R, I_C, I_T, Z, V_R, V_C, and θ.

80. Calculate V_R and V_C for the circuits seen in Figure 9-57(a) and (b).

81. Calculate the impedance of the four circuits shown in Figure 9-58.

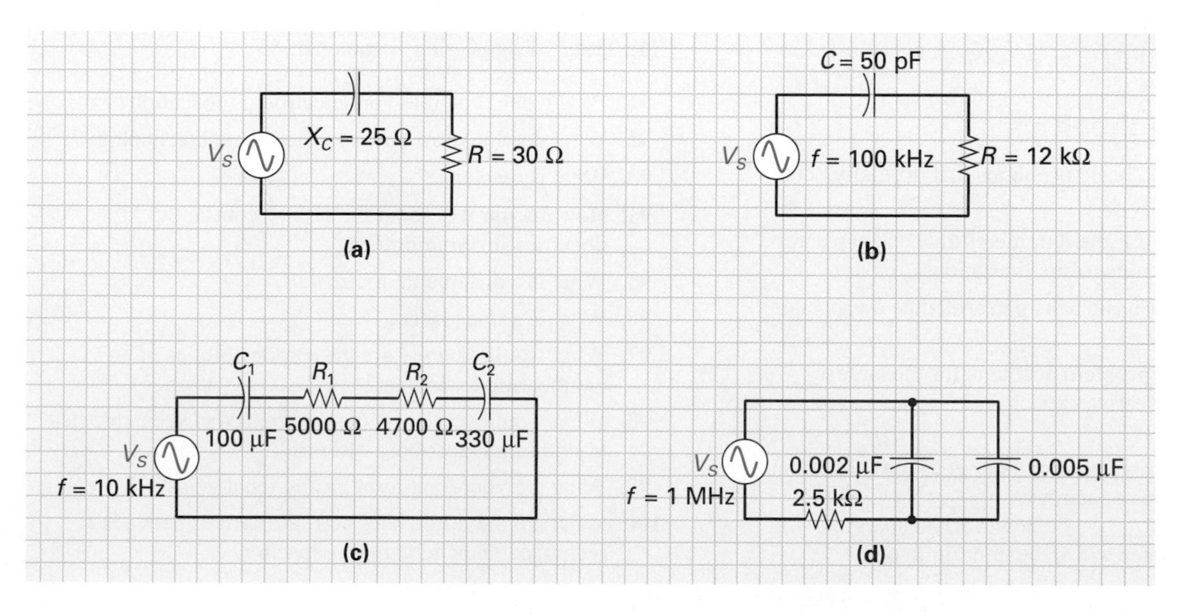

FIGURE 9-58

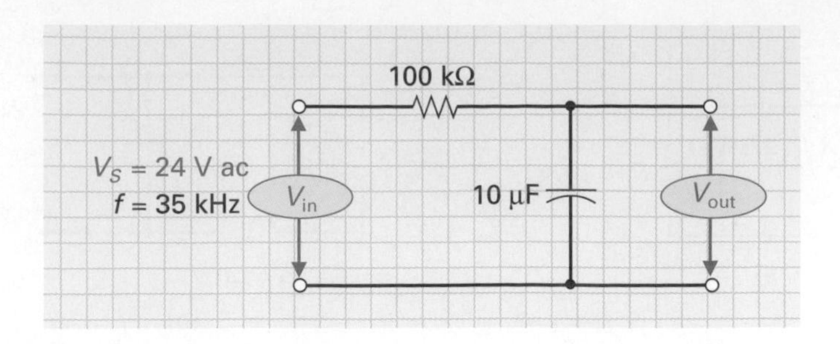

FIGURE 9-59

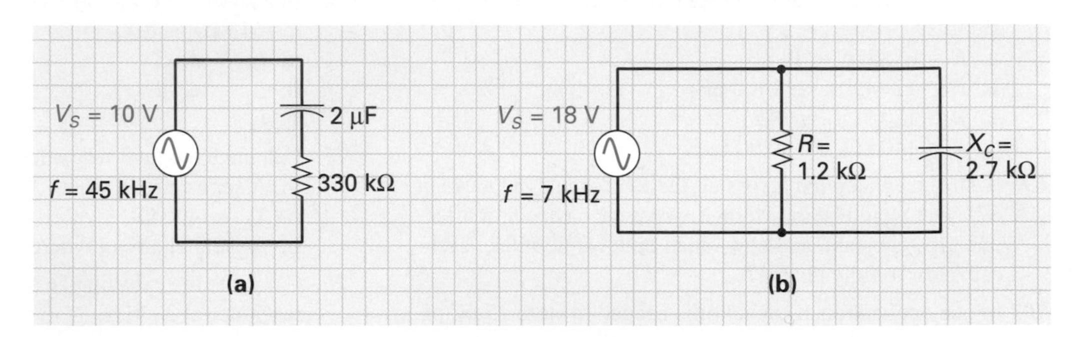

FIGURE 9-60

82. In Figure 9-59, the output voltage, since it is taken across the capacitor, will ___________ the voltage across the resistor by ___________ degrees.

83. If the positions of the capacitor and resistor in Figure 9-59 are reversed, the output voltage, since it is now taken across the resistor, will ___________ the voltage across the capacitor by ___________ degrees.

84. Calculate the resistive power, reactive power, apparent power, and power factor for the circuit seen in Figure 9-59; $V_{in} = 24$ V and $f = 35$ kHz.

85. Refer to Figure 9-60 and calculate the following:
  a. Figure 9-60(a): X_C, I, Z, I_R, θ, V_R, V_C
  b. Figure 9-60(b): V_R, V_C, I_R, I_C, I_T, Z, θ

Troubleshooting Problems

86. Describe the three basic problems that normally occur with faulty capacitors.

87. Describe how to use the ohmmeter to check capacitors and also explain some of its limitations.

88. Describe the four basic tests performed by a capacitor meter or analyzer.

89. Which capacitor problem accounts for the largest percentage of defective capacitors? Explain exactly what this malfunction is.

90. If the bar graph display of a meter goes to 0 ohms and remains there, the capacitor is ___________. If the capacitor is ___________, however, no charging will occur and the meter will indicate infinite ohms.

Job Interview Test

These tests at the end of each chapter will challenge your knowledge up to this point and give you the practice you need for a job interview. To make this more realistic, the test will comprise both technical and personal questions. In order to take full advantage of this exercise, you may want to set up a simulation of the interview environment, have a friend read the questions to you, and record your responses for later analysis.

Company Name: S.A.F.E., Inc.
Industry Branch: Industrial.
Function: Design and Manufacture Security Systems.
Job Title: Engineering Technician.

91. How would you rate your electronics education?

92. What do you know about surface-mount technology?

93. What is reactance?

94. How do you feel about the task of constantly documenting your duties?

95. What is a differentiator circuit?

96. What is a transducer?

97. What would you say are the responsibilities of an engineering technician?

98. What would you say is the difference between electrical systems and electronic systems?

99. What are your long-term professional goals?

100. What are the differences and similiarities between a rechargeable battery and a capacitor?

Answers

91. Answer must be, "very good up to this point," but go on to explain that you see your educaton as an ongoing process, and like industry, you will always be in a learning mode.

92. Chapter 3.

93. Section 9-4-2.

94. Documenting the engineering process is vitally important. As an engineering technician you will be called on to document design successes, and failures, in explicit detail. They are testing here to see if you have any attitude toward a process that most engineers and technicians shy away from. Discuss any lab or theory reports you have completed and how your education has been centered around technical reading and writing.

95. Section 9-5-5.

96. Chapter 4.

97. Quote the details described in the introduction (Engineering technician) and the responsibilities listed in the job advertisement.

98. Chapter 8.

99. This is an individual choice; however, most people discuss furthering their education and finding a job that is both stimulating and rewarding. Be sure to carefully consider your choices—you want to appear confident but not arrogant, show a desire to advance but not be overambitious.

100. Expect the interviewer to sometimes ask somewhat obscure, basic questions that can often set you back on your heels. Take a moment before you answer, and then start. "Well, they both have plates, can be charged and discharged, but their applications are quite different. . . . "

Web Site Questions

Go to the web site http://www.prenhall.com/cook, select the textbook *Introductory DC/AC Electronics* or *Introductory DC/AC Circuits*, this chapter, and then follow the instructions when answering the multiple-choice practice problems.

Electromagnetic Devices

The Great Experimenter

Michael Faraday was born to James and Margaret Faraday on September 22, 1791. At twenty-two, the gifted and engaging Faraday was at the right place at the right time, and with the right talents. He impressed the brilliant professor Humphry Davvy, who made him his research assistant at Count Rumford's Royal Institution. After only two years, Faraday was given a promotion and an apartment at the Royal Institution, and in 1821 he married. He lived at the Royal Institution for the rest of his active life, working in his laboratory, giving very dynamic lectures, and publishing over 450 scientific papers. Unlike many scientific papers, Faraday's papers would never use a calculus formula to explain a concept or idea. Instead, Faraday would explain all of his findings using logic and reasoning, so that a person trying to understand science did not need to be a scientist. It was this gift of being able to make even the most complex areas of science easily accessible to the student, coupled with his motivational teaching style, that made him so popular.

In 1855 he had written three volumes of papers on electromagnetism, the first dynamo, the first transformer, and the foundations of electrochemistry, a large amount on dielectrics and even some papers on plasma. The unit of capacitance is measured in *farads,* in honor of his work in these areas of science.

Faraday began two series of lectures at the Royal Institution, which have continued to this day. Michael and Sarah Faraday were childless, but they both loved children, and in 1860 Faraday began a series of Christmas lectures expressly for children, the most popular of which was called "The Chemical History of a Candle." The other lecture series was the "Friday Evening Discourses," of which he himself delivered over a hundred. These very dynamic, enlightening, and entertaining lectures covered areas of science or technology for the layperson and were filled with demonstrations. On one evening in 1846, an extremely nervous and shy speaker ran off just moments before he was scheduled to give the Friday Evening Discourse. Faraday had to fill in, and to this day a tradition is still enforced whereby the lecturer for the Friday Evening Discourse is locked up for half an hour before the presentation with a single glass of whiskey.

Faraday was often referred to as "the great experimenter," and it was this consistent experimentation that led to many of his findings. He was fascinated by science and technology and was always exploring new and sometimes dangerous horizons. In fact, in one of his reports he states, "I have escaped, not quite unhurt, from four explosions." When asked to comment on experimentation, his advice was to "let your imagination go, guiding it by judgment and principle, but holding it in and directing it by experiment. Nothing is so good as an experiment which, while it sets an error right, gives you as a reward for your humility an absolute advance in knowledge."

10

Outline and Objectives

THE GREAT EXPERIMENTER
INTRODUCTION

10-1 ELECTROMAGNETISM

Objective 1: Describe the left-hand rule of electromagnetism.

10-1-1 DC Versus AC Electromagnetism

Objective 2: Describe the difference between dc and ac electromagnetism.

10-1-2 Magnetic Terms

Objective 3: Explain the following magnetic terms:

a. Magnetic flux
b. Flux density
c. Magnetizing force
d. Magnetomotive force
e. Reluctance
f. Permeability (relative and absolute)

10-1-3 Flux Density *(B)* Versus Magnetizing Force *(H)*

Objective 4: Explain the relationship between flux density and magnetizing force.

Objective 5: Describe the cycle known as the hysteresis loop.

10-1-4 Application of AC Electromagnetism

10-2 ELECTROMAGNETIC INDUCTION

Objective 6: Define *electromagnetic induction.*

10-2-1 Faraday's Law

10-2-2 Lenz's Law

Objective 7: State Faraday's and Lenz's laws relating to eletromagnetic induction.

10-2-3 Applications of Electromagnetic Induction

Objective 8: Describe the following applications of electromagnetic induction:

a. AC generator
b. Moving coil microphone

TIME LINE

In 1819, a Danish physicist, Hans C. Oersted (1777–1851), accidentally discovered an interesting phenomenon. Placing a compass near a current-carrying conductor, he noticed that the needle of the compass pointed to the conductor rather than north. He was quick to realize that electricity and magnetism were related, and in honor of his work, the unit oersted was adopted for the unit of magnetic field strength.

Left-Hand Rule

If the fingers of the left hand are placed around the wire so that the thumb points in the direction of the electron flow, the fingers will be pointing in the direction of the magnetic field being produced by the conductor.

Introduction

It was during a classroom lecture in 1820 that Danish physicist Hans Christian Oersted accidentally stumbled on an interesting reaction. As he laid a compass down on a bench he noticed that the compass needle pointed to an adjacent conductor that was carrying a current, instead of pointing to the earth's north pole. It was this discovery that first proved that magnetism and electricity were very closely related to one another. This phenomenon is now called *electromagnetism* since it is now known that any conductor carrying an electro or electrical current will produce a magnetic field.

In 1831, the English physicist Michael Faraday explored further Oersted's discovery of eletromagnetism and found that the process could be reversed. Faraday observed that if a conductor was passed through a magnetic field, a voltage would be induced in the conductor and cause a current to flow. This phenomenon is referred to as *electromagnetic induction.*

In this chapter, we examine the terms and characteristics of electromagnetism (electricity to magnetism) and electromagnetic induction (magnetism to electricity).

10-1 ELECTROMAGNETISM

Up until now, we have been discussing current flow through a coil in only one direction (dc). To review our coverage of dc electromagnetism in Chapter 4, a dc voltage produces a fixed current in one direction and therefore generates a magnetic field in a coil of fixed polarity, as shown in Figure 10-1(a). The **left-hand rule** can be applied to electromagnets to determine which of the poles will be the north end of the electromagnet, as shown in the inset in Figure 10-1(a). This rule states that if you wrap the fingers of your left hand around the coil so that your fingers are pointing in the direction of current flow, your thumb will be pointing to the north end of the electromagnet.

10-1-1 *DC Versus AC Eletromagnetism*

Alternating current (ac) is continually varying, and as the polarity of the magnetic field is dependent on the direction of current flow (left-hand rule), the magnetic field will also be alternating in polarity, as shown in Figure 10-1(b). Let us look at times 1 through 4 in Figure 10-1 in more detail.

Time 1: The alternating voltage has risen to a maximum positive level and causes current to flow in the direction seen in the circuit. This will cause a magnetic field with a south pole above and a north pole below.
Time 2: Between positions 1 and 2, the voltage, and therefore current, will decrease from a maximum positive value to zero. This will cause a corresponding collapse of the magnetic field from maximum (time 1) to zero (time 2).
Time 3: Voltage and current increase from zero to maximum negative between positions 2 and 3. The increase in current flow causes a similar increase or buildup of magnetic flux, producing a north pole above and south pole below.
Time 4: From time 3 to time 4, the current within the circuit diminishes to zero, and the magnetic field once again collapses. The cycle then repeats.

In regard to current flow, therefore, we can say that

1. A direct current (dc) produces a constant magnetic field of a fixed polarity, for example, north–south.
2. An alternating current (ac) produces an alternating magnetic field, which continuously switches polarity, for example, north–south, south–north, north–south, and so on.

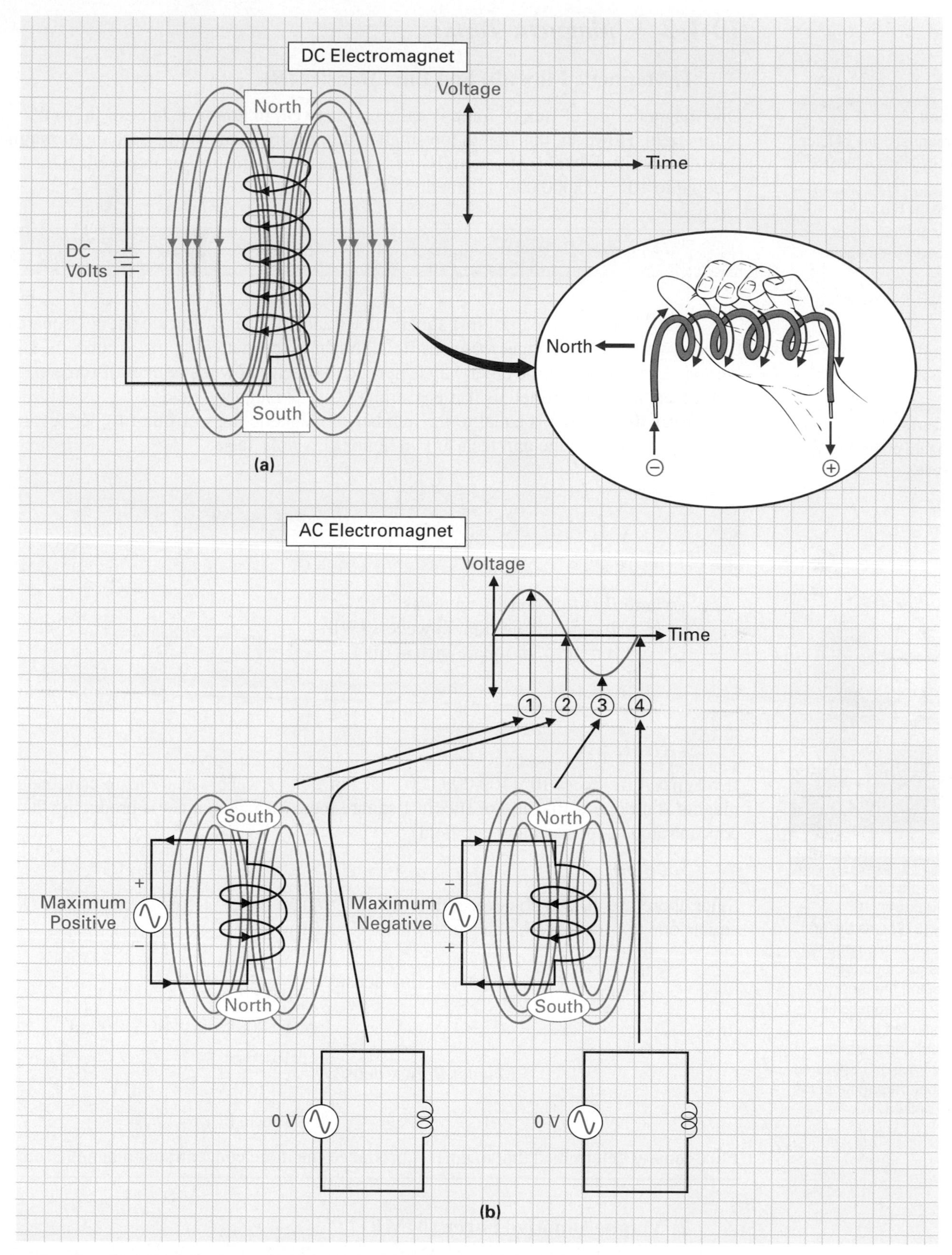

FIGURE 10-1 **Electromagnets. (a) DC Electromagnet. (b) AC Electromagnet.**

10-1-2 *Magnetic Terms*

1. Magnetic Flux (Φ) and Flux Density (*B*)

Maxwell
One magnetic line of force or flux is called a maxwell.

Magnetic Flux
The magnetic lines of force produced by a magnet.

One magnetic line of flux or force is called a **maxwell,** in honor of James Maxwell's work in this field. However, this unit is too small and impractical and so the amount of **magnetic flux** (symbolized Φ, phi) is generally measured in webers instead of maxwells. One *weber* is equal to 10^8 (100,000,000) magnetic lines of force or maxwells.

$$\text{Magnetic flux } (\Phi) = \text{number of lines of force (or maxwells), in webers}$$

If 10^8 lines of force exist in 1 square meter and 10^8 lines of force exist in 1 square centimeter, which is the stronger magnetic field? As far as magnetic flux (Φ) is concerned, 1 weber exists in both. So some other way is needed to specify how many lines of force exist in a given area, and this will tell us if it is a strong or weak magnetic field.

Flux Density
A measure of the strength of a wave.

Flux density (*B*) is equal to the number of magnetic lines of flux (Φ) per square meter, and it is given in the unit tesla (*T*), in honor of Nicola Tesla.

$$\text{Flux density } (B) = \frac{\text{magnetic flux } (\Phi)}{\text{area } (A)}$$

where B = flux density, in teslas (T)
Φ = magnetic flux, in webers (Wb)
A = area, in square meters (m^2)

TIME LINE
Scottish physicist James Clerk Maxwell (1831–1879) was the first professor of experimental physics at Cambridge. Developing the theory of the electromagnetic field, he concluded that electrical and magnetic energy travel in transverse ways that propagate at the speed of light.

EXAMPLE:

If magnet *A* produces 10^8 (100,000,000) lines of flux (1 weber) in 1 square centimeter and magnet *B* produces 10^8 lines of flux in 1 square meter, which magnet is producing the more concentrated or intense magnetic field?

Solution:

Since there are 10,000 square centimeters in 1 square meter, 1 square centimeter (1 cm^2) equals one ten-thousandth of a square meter (0.0001 m^2).

$$\text{Magnet } A\text{: flux density} = \frac{1 \text{ weber}}{0.0001\ (m^2)} = 10{,}000 \text{ tesla}$$

$$\text{Magnet } B\text{: flux density} = \frac{1 \text{ weber}}{1\ (m^2)} = 1 \text{ tesla (T)}$$

Flux density determined that magnet *A* produced the more concentrated magnetic force.

2. Magnetomotive Force (MMF)

Magnetomotive Force
Force that produces a magnetic field.

The magnetic flux produced by an electromagnet is generated by current flowing through a coil of wire. As discussed previously, electromotive force (emf) is the pressure or voltage that forces electrons to move. **Magnetomotive force** (mmf) is the magnetic pressure that produces the magnetic field. The formula for mmf is

$$\text{mmf} = I \times N \text{ (ampere-turns)}$$

where mmf = magnetomotive force, in ampere-turns (At)
I = current, in amperes
N = number of turns in the coil

The formula basically says that if you increase the current through the coil or increase the number of turns in the coil, you will increase the magnetic pressure (magnetomotive force), and by increasing magnetic pressure, you will increase the magnetic field produced, as shown in Figure 10-2.

TIME LINE

In 1600, William Gilbert (1544–1603), an English physician, documented many years of research and experiments on magnets and magnetic bodies such as amber and lodestones. Probably his most important discovery was that when rubbed with a cloth, amber would attract lightweight objects. In 1601, he was appointed physician to Queen Elizabeth I at a salary of $150 a year. He was the first to believe that the earth was nothing but a large magnet and that its magnetic field causes a needle to align itself between north and south.

EXAMPLE:

What is the magnetomotive force produced when 3 A of current flows through five turns in a coil?

Solution:

$$\begin{aligned} \text{mmf} &= I \text{ (current)} \times N \text{ (number of turns)} \\ &= 3 \text{ A} \times 5 \text{ turns} \\ &= 15 \text{ At (ampere-turns)} \end{aligned}$$

3. Magnetizing Force *(H)*

The previous example showed that magnetomotive force (mmf) is equal to the product of current and the number of turns. But if a 20 turn coil is stretched out to twice its length,

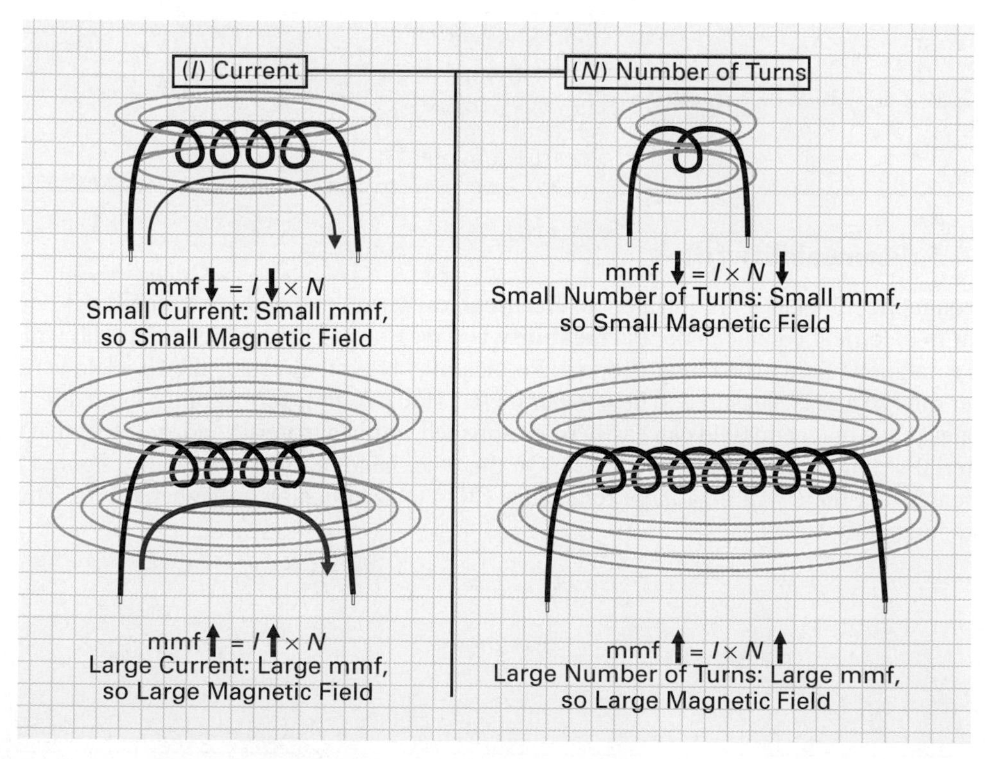

FIGURE 10-2 Magnetomotive Force.

the magnetic field or force will be half as strong, because there would be less of a reinforcing effect between the coils due to the greater distance between them. The length of the coil is therefore a factor that also determines the field intensity, and this new term, **magnetizing force** (H), is the reference we commonly use to describe the magnetic field intensity.

Magnetizing Force
Also called *magnetic field strength,* it is the magnetomotive force per unit length at any given point in a magnetic circuit.

$$\text{Magnetizing force } (H) = \frac{I \times N}{l} \quad \text{or} \quad H = \frac{\text{mmf}}{l}$$

where H = magnetizing force in ampere-turns/meter (At/m)
I = current, in amperes (A)
N = number of coil turns
l = length of coil, in meters (m)

4. Reluctance ($\Re$)

Reluctance
Resistance to the flow of magnetic lines of force.

Reluctance, in reference to magnetic energy, is equivalent to resistance in electrical energy. Reluctance is the opposition or resistance to the establishment of a magnetic field in an electromagnet. The formula for calculating reluctance is the magnetic energy equivalent of the electrical Ohm's law, as shown:

Magnetic

$$\text{Reluctance } (\Re) = \frac{\text{mmf (magnetomotive force)}}{\Phi \text{ (magnetic flux)}}$$

(measured in ampere-turns/weber)

Electric

$$\text{Resistance } (R) = \frac{V \text{ (electromotive force)}}{I \text{ (current)}}$$

(measured in ohms)

5. Permeability (μ)

Permeability
Measure of how much better a material is as a path for magnetic lines of force with respect to air, which has a permeability of 1 (symbolized by the Greek lowercase letter mu, μ).

As magnetic reluctance is equivalent to electrical resistance, magnetic permeability is equivalent to electrical conductance. **Permeability** is a measure of how easily a material will allow a magnetic field to be set up within it. Permeability is symbolized by the Greek lowercase letter mu (μ) and is measured in henrys per meter (H/m). A high permeability figure ($\mu\uparrow$) indicates that a magnetic field can easily be established within a material and therefore that this material's reluctance must be low ($\Re\downarrow$). On the other hand, a low permeability figure ($\mu\downarrow$) indicates that there will be a large reluctance ($\Re\uparrow$) to establishing a magnetic field. This is mathematically stated as

Magnetic

$$\text{Permeability } (\mu) = \frac{1}{\Re \text{ (reluctance)}}$$

(Permeability is inversely proportional to reluctance and is measured in henrys per meter.)

TABLE 10-1 Permeabilities of Various Materials

μ / μ_r | μ_0

MATERIAL	RELATIVE PERMEABILITY (μ_r)	PERMEABILITY
Air or vacuum	1	1.26×10^{-6}
Nickel	50	6.28×10^{-5}
Cobalt	60	7.56×10^{-5}
Cast iron	90	1.1×10^{-4}
Machine steel	450	5.65×10^{-4}
Transformer iron	5,500	6.9×10^{-3}
Silicon iron	7,000	8.8×10^{-3}
Permalloy	100,000	0.126
Supermalloy	1,000,000	1.26

Electric

$$\text{Conductance } (G) = \frac{1}{R \text{ (resistance)}}$$

(Conductance is inversely proportional to resistance and is measured in siemens.)

The permeability figures in henrys/meter for different materials are listed in Table 10-1. **Relative** (with respect to) is a word that means that a comparison has to be made. Relative permeability (μ_r) is the measure of how well another given material will conduct magnetic lines of force with respect to, or relative to, our reference material, air, which has a relative permeability value of 1. Referring to the relative permeability column in Table 10-1, you can see that magnetic lines of flux will pass through nickel 50 times easier than through air. The relative permeability (μ_r) of air, which is equal to 1, should not be confused with the absolute permeability (μ_0) of free space or air, which is equal to $4\pi \times 10^{-7}$ or 1.26×10^{-6}.

Relative
Not independent; compared or with respect to some other value of a measured quantity.

$$\mu = \mu_r \times \mu_0$$

Permeability = relative permeability × absolute permeability of air

Whether permeability (μ), relative permeability (μ_r), or absolute permeability (μ_0) is used as a standard makes little difference. A high permeability figure indicates a low reluctance ($\mu\uparrow$, $\Re\downarrow$), and vice versa.

EXAMPLE:

If the magnetic flux produced by a material is equal to 335 μWb and the mmf equals 15 At,

a. What is the reluctance of the material?

b. What is the material's permeability?

■ *Solution:*

Calculator Sequence

STEP	KEYPAD ENTRY	DISPLAY RESPONSE
1.	1 5	15
2.	÷	
3.	3 3 5 E 6 +/−	335E–6
4.	=	44776.1
5.	1/X	22.3E–6

a. $\text{Reluctance } (\Re) = \dfrac{\text{mmf (magnetomotive force)}}{\Phi \text{ (magnetic flux)}}$

$= \dfrac{15 \text{ At (ampere-turns)}}{335\ \mu\text{Wb}}$

$= 44.8 \times 10^3 \text{ At/Wb}$

b. $\text{Permeability } (\mu) = \dfrac{1}{44.8 \times 10^3 \text{ (reluctance)}}$

$= 22.3 \times 10^{-6} \text{ henrys/meter (H/m)}$

6. Summary

The magnetic field strength of an electromagnet can be increased by increasing the number of turns in its coil, increasing the current through the electromagnet, or decreasing the length of the coil ($H = I \times N/l$). The field strength can be increased further if an iron core is placed within the electromagnet, as illustrated in Figure 10-3(a). This is because an iron core has less reluctance or opposition to the magnetic lines of force than air, so the flux density (B) is increased. Another way of saying this is that the permeability or conductance of magnetic lines of force within iron is greater than that of air, and if the permeability of iron is large, its reluctance must be small. The symbol for an iron-core electromagnet is seen in Figure 10-3(b).

10-1-3 *Flux Density (B) Versus Magnetizing Force (H)*

***B–H* Curve**
Curve plotted on a graph to show successive states during magnetization of a ferromagnetic material.

The **B–H curve** in Figure 10-4 illustrates the relationship between the two most important magnetic properties: flux density (B) and magnetizing force (H). Figure 10-4(a) illustrates the *B–H* curve; Figure 10-4(b) illustrates the positive rising portion of the ac current that is being applied to the iron-core electromagnetic circuit in Figure 10-4(c).

The magnetizing force is actually equal to $H = I$ (current) $\times N$ (number of turns) / l (length of coil), but since the number of turns and length of coil are fixed for the coil being used, the magnetizing force (H) is proportional to the current (I) applied, which is shown in Figure 10-4(b). The positive rise of the current from zero to maximum positive is applied through the electromagnet and will produce a corresponding bloom or buildup in magnetic flux at a rate indicated by the *B–H* curve shape in Figure 10-4(a).

It is important to note that as the magnetizing force (current) is increased, there are three distinct stages in the change of flux density or magnetic flux out.

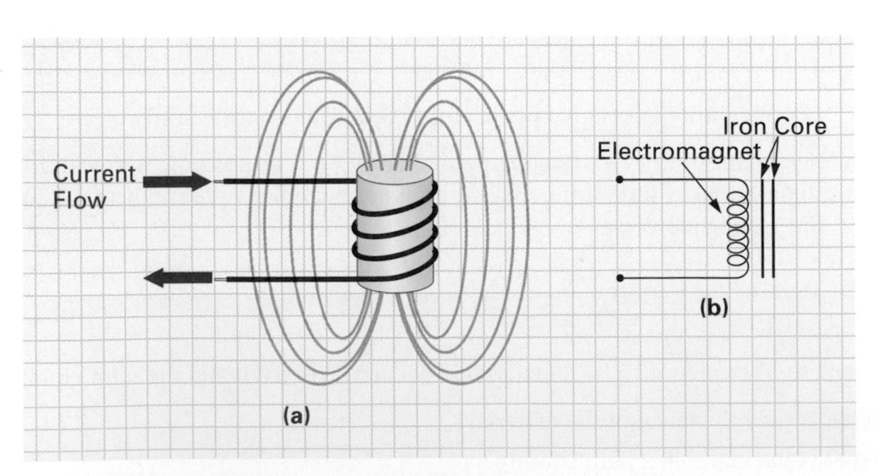

FIGURE 10-3 **Iron-Core Electromagnet. (a) Operation. (b) Symbol.**

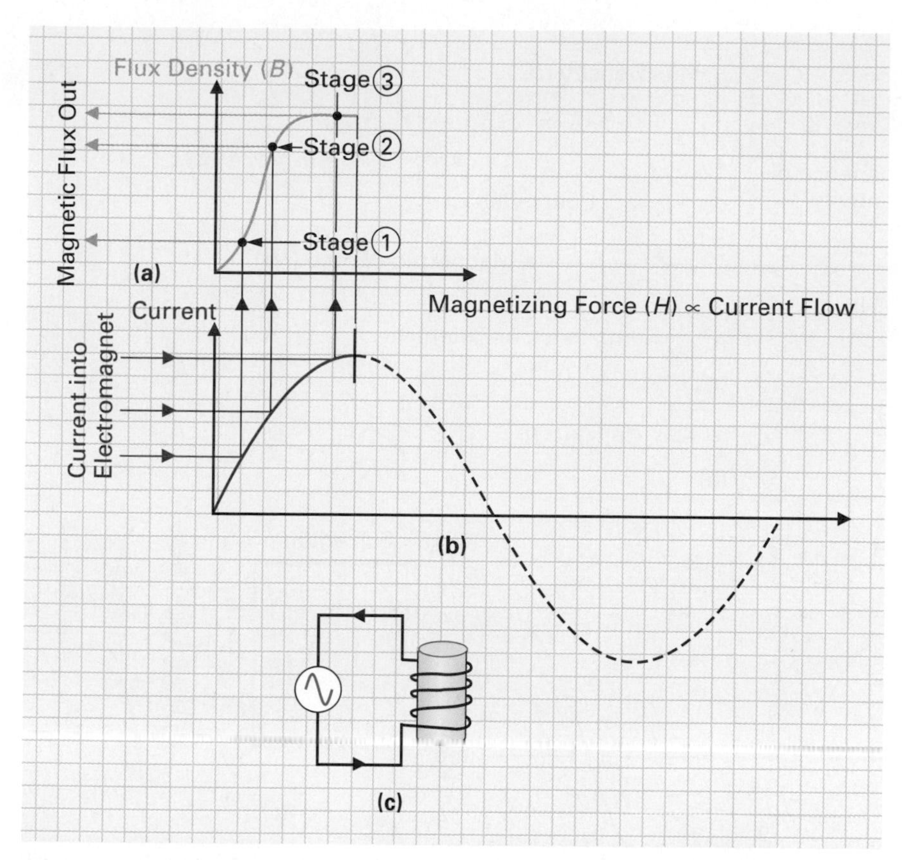

FIGURE 10-4 (a) Flux Density Versus Magnetizing Force Curve (*B–H* Curve). (b) Current Applied to Circuit. (c) Circuit.

Stage 1: Up to this point, the increase in flux density is slow, as a large amount of force is required to begin alignment of the molecule magnets.

Stage 2: Increase in flux density is now rapid and almost linear as the molecule magnets are aligning easily.

Stage 3: In this state the molecule magnets cannot be magnetized any further because they are all fully aligned and no more flux density can be easily obtained. This is called the **saturation point,** and it is the state of magnetism beyond which an electromagnet is incapable of further magnetic strength. It is the point beyond which the *B–H* curve is a straight, horizontal line, indicating no change.

Saturation Point
The point beyond which an increase in one of two quantities produces no increase in the other.

Saturation can easily be described by the simple analogy of a sponge. A dry sponge can only soak up a certain amount of water. As it continues to absorb water, a point will be reached where it will have soaked up the maximum amount of water possible. At this point, the sponge is said to be saturated with water, and no matter how much extra water you supply, it cannot hold any more. The electromagnet is saturated at stage 3 and cannot produce any more magnetic flux, even though more magnetizing force is supplied as the sine wave continues on to its maximum positive level.

Looking at these three stages and the *B–H* curve that is produced, you can see that, in fact, the magnetization (setting up of the magnetic field, *B*) lags the magnetizing force (*H*) because of molecular friction. This lag or time difference between magnetizing force (*H*) and flux density (*B*) is known as **hysteresis.**

Hysteresis
A lag between cause and effect. With magnetism, it is the amount that the magnetization of a material lags the magnetizing force due to molecular friction.

Figure 10-5(a) illustrates what is known as a *hysteresis loop,* which is formed when you plot magnetizing force (*H*) against flux density (*B*) through a complete cycle of alternating current, as seen in Figure 10-5(b). Initially, when the electric circuit switch is open, the iron core is unmagnetized. Therefore, both *H* and *B* are zero at point *a*. When the switch

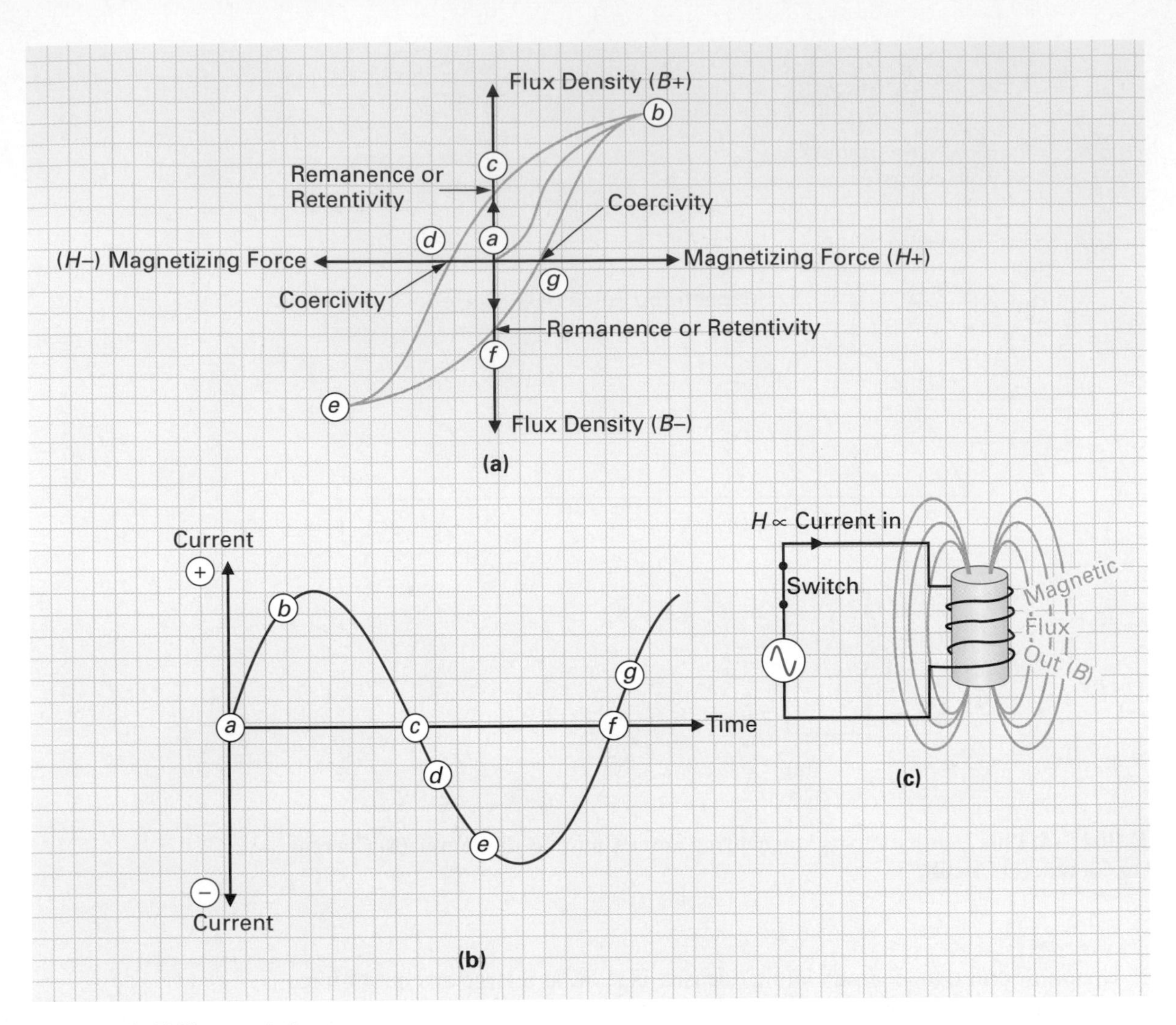

FIGURE 10-5 ***B–H* Hysteresis Loop.**

is closed, as seen in Figure 10-5(c), the current [Figure 10-5(b)] is increased and flux density [Figure 10-5(a)] increases until saturation point *b* is reached. This part of the waveform (from point *a* to *b*) is exactly the same as the *B–H* curve, discussed previously. The current continues on beyond saturation point *b*, but the flux density cannot increase beyond saturation.

At point *c*, the magnetizing force (current) is zero and *B* (flux density) falls to a value *c* that is the positive magnetic flux remaining after the removal of the magnetizing force (*H*). This particular value of flux density is termed **remanence or retentivity.**

Remanence or Retentivity
Amount a material remains magnetized after the magnetizing force has been removed.

The current or magnetizing force now reverses, and the amount of current in the reverse direction that causes flux density to be brought down from *c* to zero (point *d*) after the core has been saturated is termed the **coercive force.** The current, and therefore magnetizing force, continues on toward a maximum negative until saturation in the opposite magnetic polarity occurs at point *e*.

Coercive Force
Magnetizing force needed to reduce the residual magnetism within a material to zero.

At point *f*, the magnetizing force (current) is zero and *B* falls to remanence, which is the negative magnetic flux remaining after the removal of the magnetizing force *H*. The value of current between *f* and *g* is the coercive force needed in the reverse direction to bring flux density down to zero (point *g*).

10-1-4 *Application of AC Electromagnetism*

The speaker detailed in Figure 10-6 uses an electromagnet to move a flexible cone, which generates a sound wave that varies in magnitude and rate in exactly the same manner as the

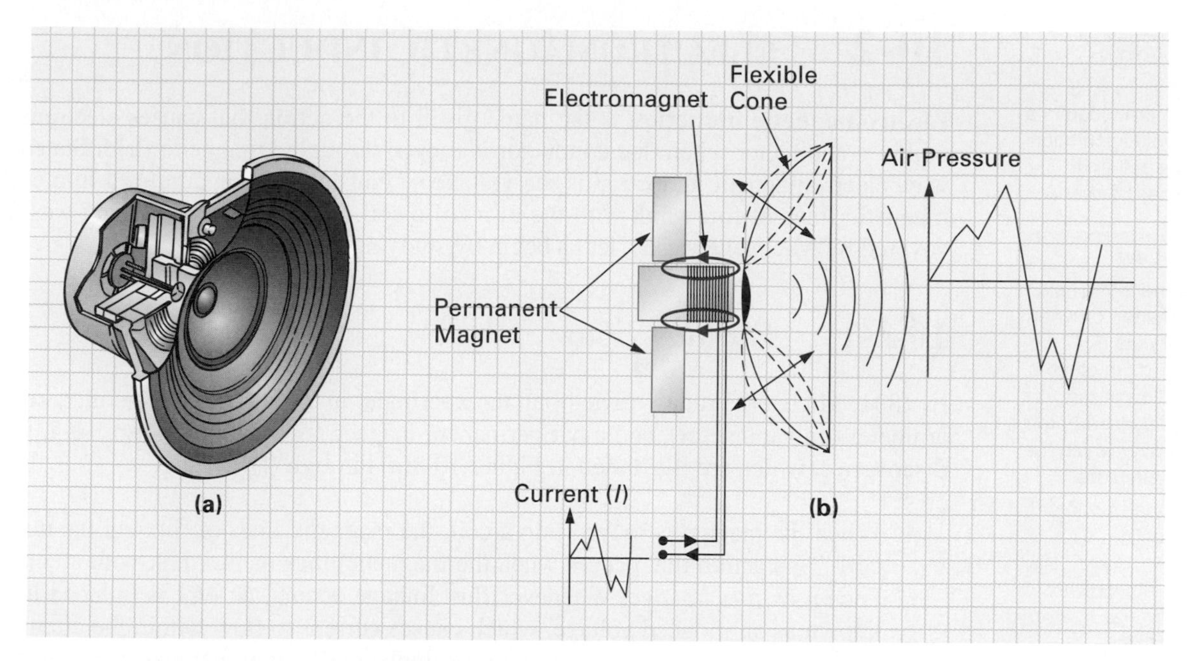

FIGURE 10-6 **Speaker. (a) Construction. (b) Operation.**

amplitude and frequency of the electric current applied to the coil. The electromagnet is a movable core, suspended within a fixed, all-encapsulating permanent magnet. As the signal current changes, so does the magnetic flux pattern generated by the electromagnet, and it is this interaction, between this changing field and that of the permanent magnets' fixed magnetic field, that causes the core to vibrate.

SELF-TEST EVALUATION POINT FOR SECTION 10-1

Now that you have completed this section, you should be able to

Objective 1. Describe the left-hand rule of electromagnetism.

Objective 2. Describe the difference between dc and ac electromagnetism.

Objective 3. Explain the following magnetic terms:

a. Magnetic flux
b. Flux density
c. Magnetizing force
d. Magnetomotive force
e. Reluctance
f. Permeability (relative and absolute)

Objective 4. Explain the relationship between flux density and magnetizing force.

Objective 5. Describe the cycle known as the hysteresis loop.

Use the following questions to test your understanding of section 10-1:

1. What name could be used instead of *electromagnet*?
2. True or false: The greater the number of loops in an electromagnet, the greater or stronger the magnetic field.
3. What form of current flow produces a magnetic field that maintains a fixed magnetic polarity?
4. True or false: A magnetic field is produced only when ac is passed through a conductor.
5. Describe how the left-hand rule applies to electromagnetism.
6. What form of current flow produces a magnetic field that continuously switches in polarity?

Define the following:

7. Magnetic flux
8. Flux density
9. Magnetomotive force
10. Magnetizing force
11. Reluctance
12. Permeability
13. True or false: The hysteresis loop is formed when you plot H against B through a complete cycle of ac.
14. What is magnetic saturation?

Electromagnetic Induction

The voltage produced in a coil due to relative motion between the coil and magnetic lines of force.

Faraday's Law

1. When a magnetic field cuts a conductor, or when a conductor cuts a magnetic filed, an electric current will flow in the conductor if a closed path is provided over which the current can circulate.

2. Two other laws relate to electrolytic cells.

TIME LINE

Eduard W. Weber (1804–1891), a German physicist, made enduring contributions to the modern system of electric units, and magnetic flux is measured in webers in honor of his work.

Weber (Wb)

Unit of magnetic flux. One weber is the amount of flux that, when linked with a single turn of wire for an interval of 1 second, will induce an electromotive force of 1 V.

10-2 ELECTROMAGNETIC INDUCTION

Electromagnetic induction is the name given to the action that causes electrons to flow within a conductor when that conductor is moved through a magnetic field. Stated another way, electromagnetic induction is the voltage or emf induced or produced in a coil as the magnetic lines of force link with the turns of a coil. Since this phenomenon was first discovered by Michael Faraday, let us begin by studying **Faraday's law.**

10-2-1 *Faraday's Law*

In 1831, Michael Faraday carried out an experiment in which he used a coil, a zero center ammeter (galvanometer), and a bar permanent magnet, as shown in Figure 10-7. Faraday's discoveries, which are illustrated in Figure 10-8(a) through (f), are that

a. When the magnet is moved into a coil, the magnetic lines of flux cut the turns of the coil. This action that occurs when the magnetic lines of flux link with a conductor is known as *flux linkage.* Whenever flux linkage occurs, an emf is induced in the coil known as an induced voltage, which causes current to flow within the circuit and the meter to deflect in one direction, for example, to the right. Faraday discovered, in fact, that if the magnet was moved into the coil, or if the coil was moved over the magnet, an emf or voltage was induced within the coil. What actually occurs is that the electrons within the coil are pushed to one end of the coil by the magnetic field, creating an abundance of electrons at one end of the coil (negative charge) and an absence of electrons at the other end of the coil (positive charge). This potential difference across the coil will produce a current flow if a complete path for current (closed circuit) exists.

b. When the magnet is stationary within the coil, the magnetic lines are no longer cutting the turns of the coil, and so there is no induced voltage and the meter returns to zero.

c. When the magnet is pulled back out of the coil, a voltage is induced that causes current to flow in the opposite direction to that of (a) and the meter deflects in the opposite direction, for example, to the left.

d. If the magnet is moved into or out of the coil at a greater speed, the voltage induced also increases, and therefore so does current.

e. If the size of the magnet and therefore the magnetic flux strength are increased, the induced voltage also increases.

f. If the number of turns in the coil are increased, the induced voltage also increases.

Consideration of Faraday's law enables us to take a closer look at the unit of flux (Φ), which is named in honor of Eduard Weber. The **weber** is equal to 10^8 magnetic lines of force,

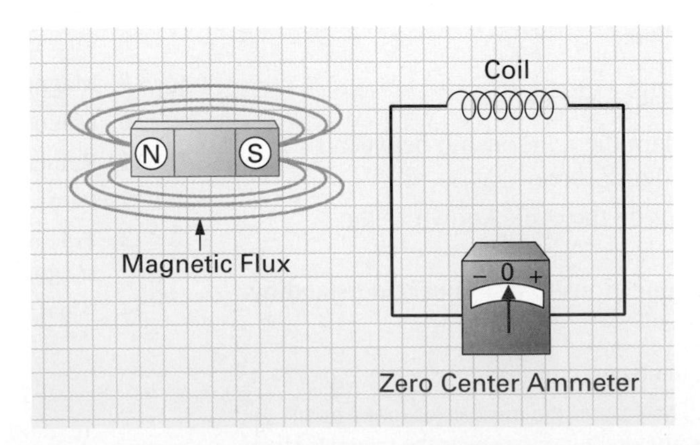

FIGURE 10-7 **Faraday's Electromagnetic Induction Experiment Components.**

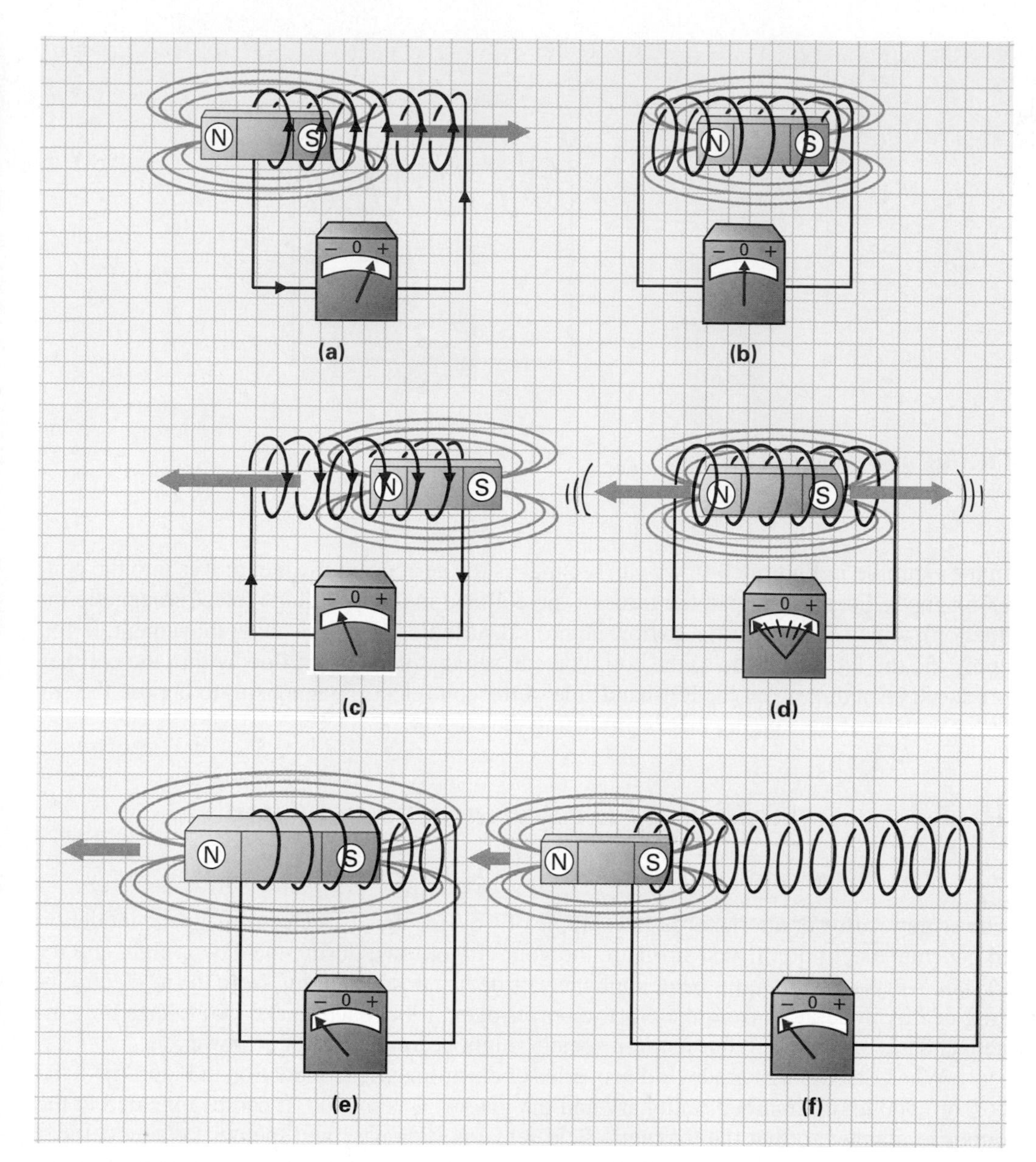

FIGURE 10-8 Faraday's Electromagnetic Induction Discoveries.

and from the electromagnetic induction point of view, if 1 weber of magnetic flux cuts a conductor for a period of 1 second, a voltage of 1 volt will be induced.

In summary, whenever there is relative motion or movement between the coil of a conductor and the magnetic lines of flux, a potential difference will be induced and this action is called *electromagnetic induction.* The magnitude of the induced voltage depends on the number of turns in the coil, rate of change of flux linkage, and flux density.

10-2-2 *Lenz's Law*

At about the same time a German physicist, Heinrich Lenz, performed a similar experiment along the same lines as Faraday's. His law states that the current induced in a coil due to the change in the magnetic flux is such as to oppose the cause producing it.

To explain this law further, refer to Figure 10–9. When the magnet moves into the coil, a voltage is induced in the coil and the current flow resulting from this induced voltage will produce a pole at the face of the coil (left-hand rule), which opposes the entry of the magnet.

TIME LINE

Baltic German physicist Heinrich Friedrich Emil Lenz (1804–1865) deduced his electromagnetic induction theory in 1834, and in honor of this achievement, it was later named Lenz's Law.

Lenz's Law

The current induced in a circuit due to a change in the magnetic field is so directed as to oppose the change in flux, or to exert a mechanical force opposing the motion.

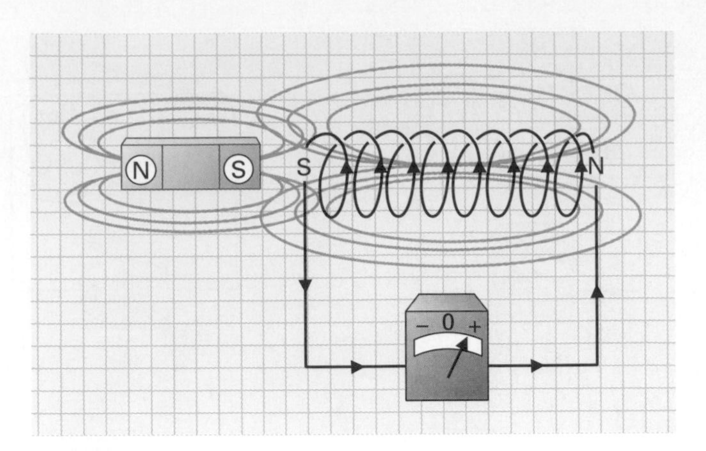

FIGURE 10-9 Lenz's Law.

In the example in Figure 10-9, we can see that as the permanent magnet moves into the coil, its magnetic lines of flux cut the turns of the coil and induce a voltage (electromagnetic induction), which causes current to flow in the coil, as indicated by the meter movement. If you apply the the left-hand rule to the coil, you can see that the current flow within the coil has produced a south pole on the left-hand side of the coil (electromagnetism) to oppose the entry or motion of the magnet that is producing the current.

10-2-3 *Applications of Electromagnetic Induction*

1. AC Generator

Generator

Device used to convert a mechanical energy input into an electrical energy output.

The ac **generator** or alternator is an example of a device that uses electromagnetic induction to generate electricity. If you stroll around your city or town during the day or night and try to spot every piece of equipment, appliance, or device that is running from the ac electricity supplied by generators from the electric power plant, it begins to make you realize how inactive, dark, and difficult our modern society might become without ac power.

When discussing Faraday's discoveries of electromagnetic induction in Figure 10-8 the coil or conductor remained stationary and the magnet was moved. It can be operated in the opposite manner so that the magnetic field remains stationary and the coil or conductor is moved. As long as the magnetic lines of force have relative motion with respect to the conductor, an emf will be induced in the coil. Figure 10-10(a) illustrates a piece of wire wound to form a coil and attached to a galvanometer; its needle rests in the center position, indicating zero current. If the conductor remains stationary within the magnetic lines of flux being generated by the permanent magnet, there is no emf or voltage induced into the wire and so no current flow through the circuit and meter.

If the conductor is moved past the permanent magnet so that it cuts the magnetic lines of flux, as seen in Figure 10-10(b), an emf is generated within the conductor, which is known as an induced voltage, and this will cause current to flow through the wire in one direction and be indicated on the meter by the deflection of the needle to the left. If the conductor is moved in the opposite direction past the magnetic field, it will induce a voltage of the opposite polarity and cause the meter to deflect to the right, as shown in Figure 10-10(c).

The value or amount of induced voltage is indicated by how far the meter deflects; this voltage is dependent on three factors:

1. The speed at which the conductor passes through the magnetic field
2. The strength or flux density of the magnetic field
3. The number of turns in the coil

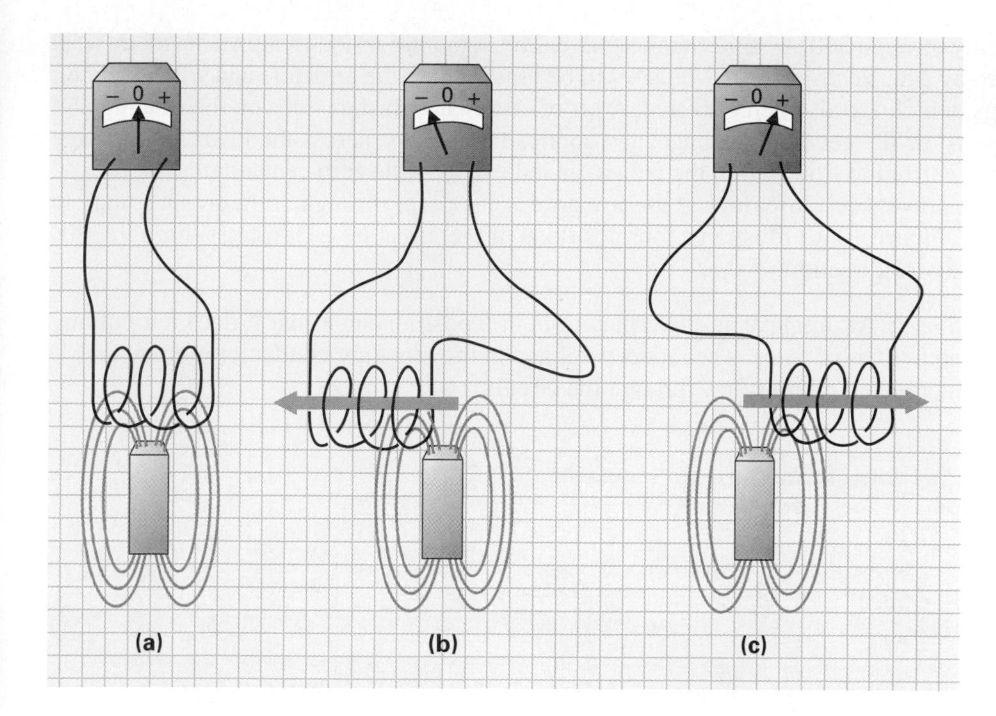

FIGURE 10-10 (a) Stationary Conductor, No Induced Voltage. (b) Moving Conductor, Induced Voltage. (c) Moving Conductor, Induced Voltage.

If the speed at which the conductor passes through the magnetic field or the strength of the magnetic field or the number of turns of the coil is increased, then the induced voltage will also increase. This is merely a repetition of Faraday's law, but in this case we moved the conductor instead of the magnetic field, but the results were the same.

Basic Generator. Basic generator operation makes use of Faraday's discoveries of electromagnetic induction. In Chapter 8 a large, 700,000 kW power plant generator was shown. Figure 10-11(b) illustrates a smaller, mobile generator that can therefore be used in remote locations.

Figure 10-11(a) illustrates the basic generator's construction. The basic principle of operation is that the mechanical drive energy input will produce ac electrical energy out by means of electromagnetic induction. A loop of conductor, known as an **armature,** is rotated

Armature
Rotating or moving component of a magnetic circuit.

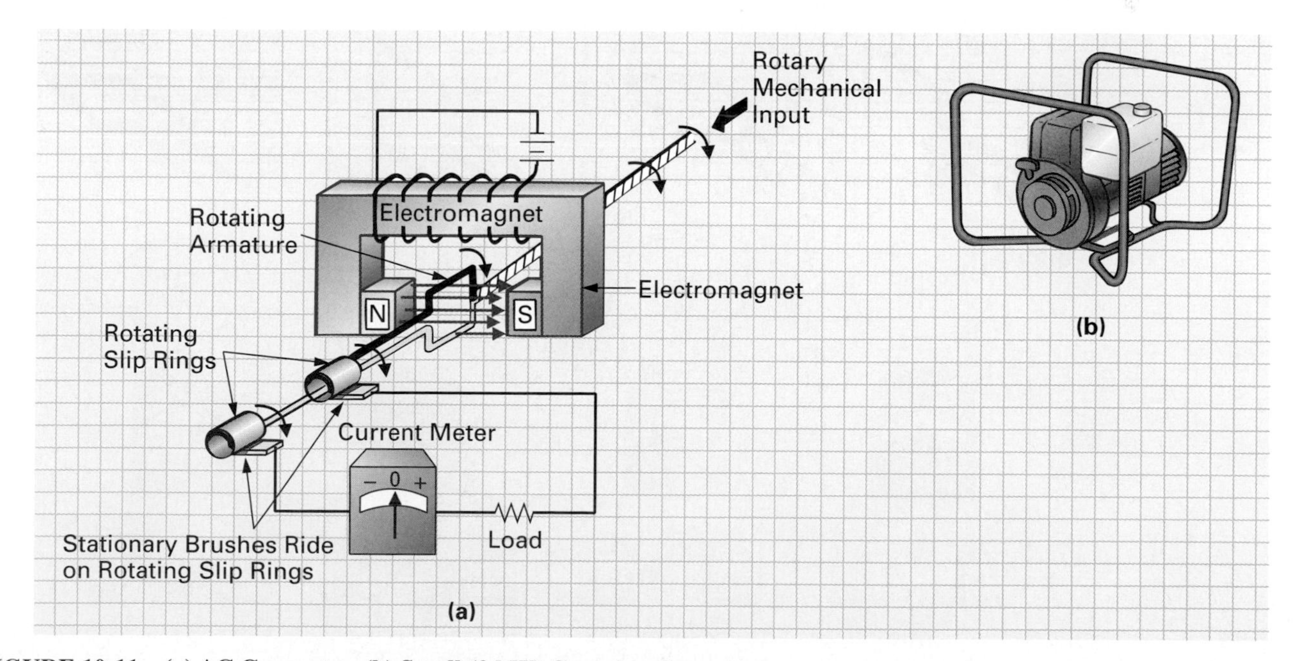

FIGURE 10-11 (a) AC Generator. (b) Small (2 kW) Camping Generator.

continually through 360° by a mechanical drive. This armature resides within a magnetic field produced by a dc electromagnet. Voltage will be induced into the armature and will appear on slip rings, which are also being rotated. A set of stationary brushes rides on the rotating slip rings, picks off the generated voltage, and applies this voltage across the load. This voltage will cause current to flow within the circuit, as indicated by the zero center ammeter.

Let's now take a closer look at the armature as it sweeps through 360°, or one complete revolution. Figure 10-12 illustrates four positions of the armature as it rotates through 360° in the clockwise direction.

> *Position 1:* At this instant, the armature is in a position such that it does not cut any magnetic lines of force. The induced voltage in the armature conductor is equal to 0 V and there is no current flow through the circuit.

FIGURE 10-12 360° Generator Operation.

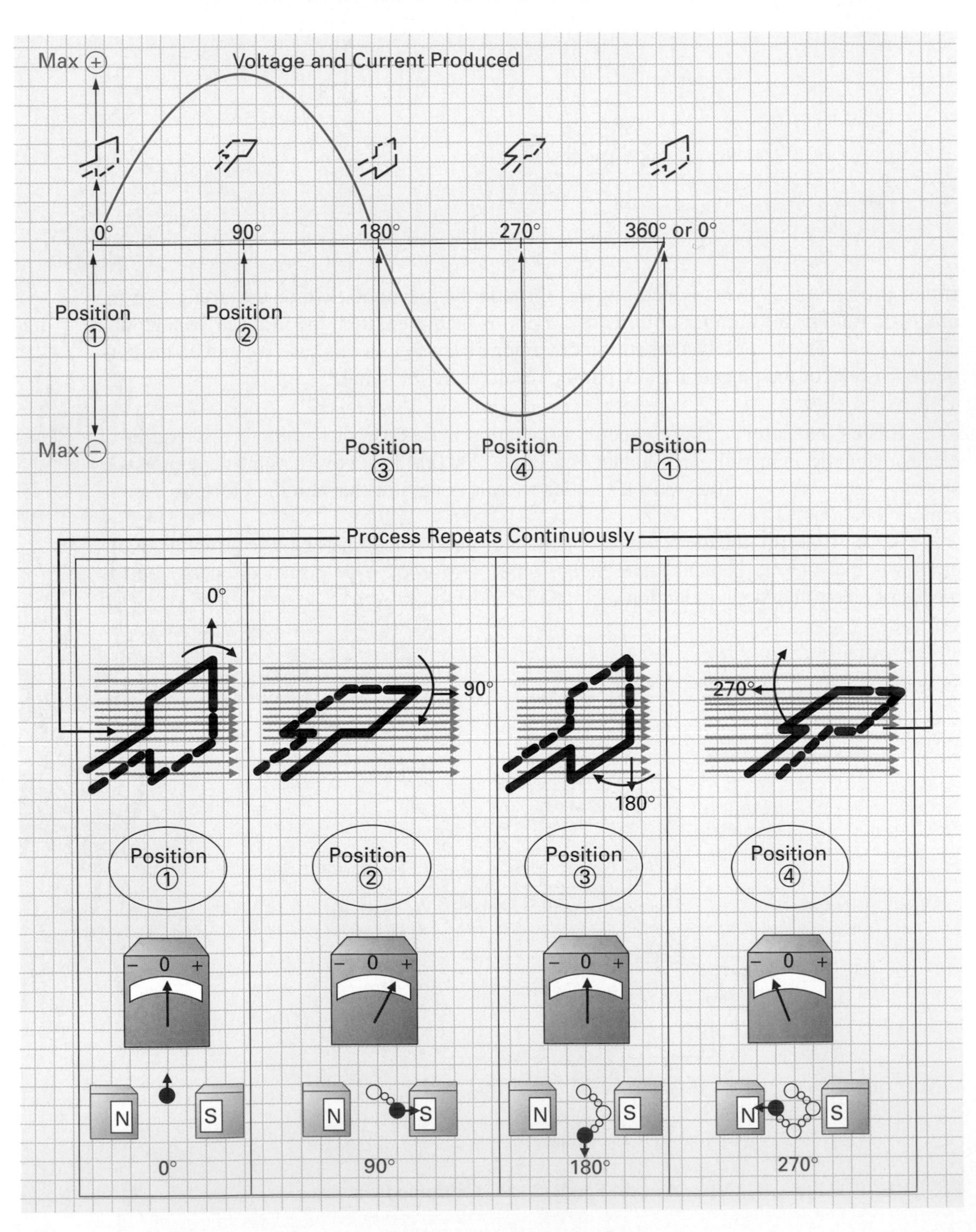

Position 2: As the conducting armature moves from position 1 to position 2, you can see that more and more magnetic lines of flux will be cut, and the induced emf in the armature (being coupled off by the brushes from the slip rings) will also increase to a maximum value. The current flow throughout the circuit will rise to a maximum as the voltage increases, as can be seen by the zero center ammeter deflection to the right. From the ac waveform in Figure 10–12, you can see the sinusoidal increase from zero to a maximum positive as the armature is rotated from 0 to 90°.

Position 3: The armature continues its rotation from 90 to 180°, cutting first through a maximum quantity and then fewer magnetic lines of force. The induced voltage decreases from the maximum positive at 90° to 0 V as 180°. At the 180° position, as with the 0° position, the armature is once again perpendicular to the magnetic field, and so no lines are cut and the induced voltage is equal to 0 V.

Position 4: From 180 to 270°, the armature is still moving in a clockwise direction, and as it travels toward 270°, it cuts more and more magnetic lines of force. The direction of the cutting action between 0 and 90° causes a positive induced voltage; and since the cutting position between 180 and 270° is the reverse, a negative induced voltage will result in the armature causing current flow in the opposite direction, as indicated by the deflection of the zero center ammeter to the left. The voltage induced when the armature is at 270° will be equal but opposite in polarity to the voltage generated when the armature was at the 90° position. The current will therefore also be equal in value but opposite in its direction of flow.

From position 4 (270°), the armature turns to the 360° or 0° position, which is equivalent to position 1, and the induced voltage decreases from maximum negative to zero. The cycle then repeats.

To summarize, in Figure 10-12, one complete revolution of the mechanical energy input causes one complete cycle of the ac electrical energy output, with a sine wave being generated as a result of circular motion.

2. Moving Coil Microphone

The moving coil **microphone** is an example of how electromagnetic induction is used to convert information-carrying sound waves to information-carrying electrical waves. Sound is the movement of pressure waves in the air. To create these pressure waves, a device such as a string, reed, or stretched membrane or the human vocal cords must be vibrated to compress and expand the nearby air molecules. Figure 10-13 illustrates a taut string that is vibrating back and forth and generating maximum (*A*) and minimum (*C*) pressure regions.

Microphone
Electroacoustic transducer that responds and converts a sound wave input into an equivalent electrical wave out.

The frequency or pitch of the sound wave is determined by the number of complete vibrations per second (hertz or cycles/second), and the amplitude or intensity of the sound wave is determined by the amount the string shifts from left to right from its normal position (*B*).

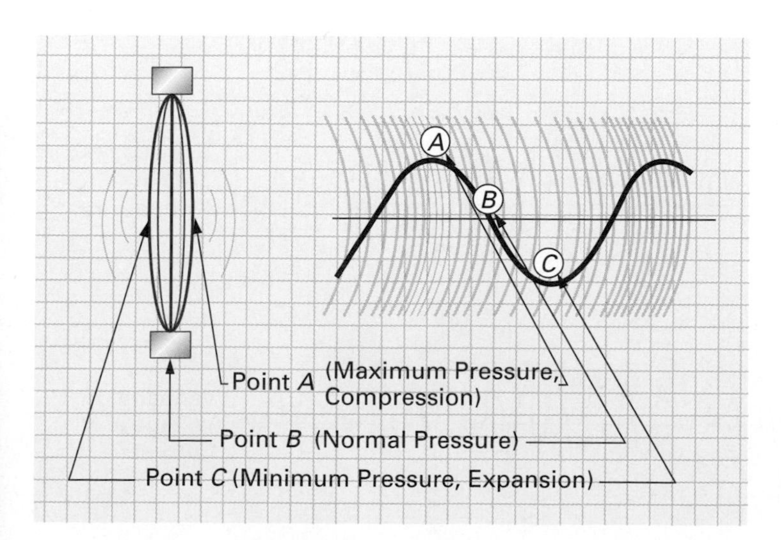

FIGURE 10-13 Sound Wave.

A moving coil microphone converts mechanical sound waves into an electrical replica by use of electromagnetic induction. Figure 10-14 illustrates the physical appearance and construction of a moving coil type of microphone. A coil of wire is suspended in an air gap between magnetic poles and attached to a delicate diaphragm (flexible membrane). A strong magnetic field from a permanent magnet surrounds the coil, and a perforated protecting cover or shield is included to protect the delicate diaphragm.

Sound waves strike the diaphragm, causing it to vibrate back and forth. Since the coil is attached to the diaphragm, it will also be moved back and forth. This movement will cause the coil to cut the magnetic lines of force from the permanent magnet, and a resulting alternating voltage will be induced in the coil (electromagnetic induction). The electrical voltage produces an alternating current, which will have the same waveform shape (and consequently information) as the sound wave that generated it, as seen in Figure 10-14(c). This electrical signal is often referred to as an *analog signal*. *Analog* is a term meaning "similar to," and as you can see in Figure 10-14(c), the electrical wave output of the microphone is an analog of the sound wave input. This microphone is sometimes called a dynamic or moving (dynamic is the opposite of static) coil microphone since it has a moving coil within it.

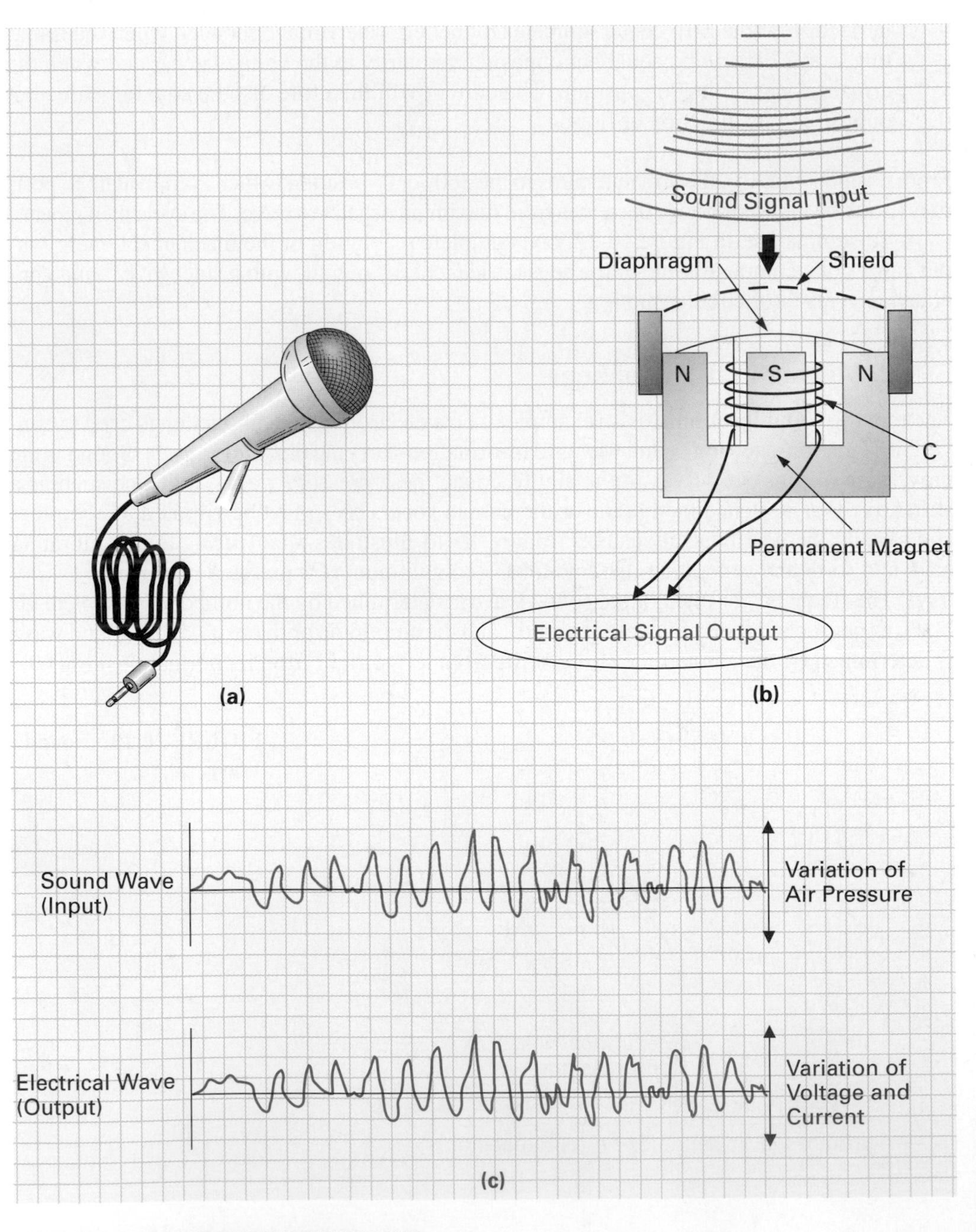

FIGURE 10-14 Moving Coil Microphone. (a) Physical Appearance. (b) Construction. (c) Input/Output Waveforms.

SELF-TEST EVALUATION POINT FOR SECTION 10-2

Now that you have completed this section, you should be able to:

Objective 6. Define electromagnetic induction.

Objective 7. State Faraday's and Lenz's laws relating to electromagnetic induction.

Objective 8. Describe the following applications of electromagnetic induction:

a. AC generator
b. Moving coil microphone

Use the following questions to test your understanding of Section 10-2:

1. Define *electromagnetic induction.*
2. Briefly describe Faraday's and Lenz's laws in relation to electromagnetic induction.
3. What waveform shape does the ac generator produce?
4. Why do you think a microphone is called an *electroacoustical transducer?*

SUMMARY

Electromagnetism (Section 10-1)

1. During a classroom lecture in 1820, Danish physicist Hans Christian Oersted discovered that magnetism and electricity were very closely related to one another. This phenomenon is now called *electromagnetism* since it is now known that any conductor carrying an electrical current will produce a magnetic field.
2. In 1831, the English physicist Michael Faraday explored further Oersted's discovery of electromagnetism and found that the process could be reversed. Faraday observed that if a conductor was passed through a magnetic field, a voltage would be induced in the conductor and cause a current to flow. This phenomenon is referred to as *electromagnetic induction.*
3. The left-hand rule can be applied to electromagnets to determine which of the poles will be the north end of the electromagnet.
4. A dc voltage produces a fixed current in one direction and therefore generates a magnetic field in a coil of fixed polarity and as determined by the left-hand rule.
5. Alternating current (ac) is continually varying, and as the polarity of the magnetic field is dependent on the direction of current flow (left-hand rule), the magnetic field will also be alternating in polarity.
6. One magnetic line of flux or force is called a maxwell, in honor of James Maxwell's work in this field. One *weber* is equal to 10^8 (100,000,000) magnetic lines of force or maxwells.
7. Flux density (B) is equal to the number of magnetic lines of flux (Φ) per square meter, and it is given in the unit of tesla (T).
8. Magnetomotive force (mmf) is the magnetic pressure that produces the magnetic field.
9. The length of the coil determines the field intensity, and magnetizing force (H) is the reference we commonly use to describe the magnetic field intensity.
10. Reluctance is the opposition or resistance to the establishment of a magnetic field in an electromagnet.
11. Permeability is a measure of how easily a material will allow a magnetic field to be set up within it. Permeability is symbolized by the Greek lowercase letter mu (μ) and is measured in henrys per meter (H/m).
12. Relative permeability (μ_r) is the measure of how well another given material will conduct magnetic lines of force with respect to, or relative to, our reference material, air, which has a relative permeability value of 1.
13. The relative permeability (μ_r) of air, which is equal to 1, should not be confused with the absolute permeability (μ_0) of free space or air, which is equal to $4\pi \times 10^{-7}$ or 1.26×10^{-6}.
14. The magnetic field strength of an electromagnet can be increased by increasing the number of turns in its coil, increasing the current through the electromagnet, or decreasing the length of the coil ($H = I \times N/l$). The field strength can be increased further by placing an iron core within the electromagnet. An iron core has less reluctance or opposition to the magnetic lines of force than air, so the flux density (B) is increased.
15. The B-H curve illustrates the relationship between the two most important magnetic properties: flux density (B) and magnetizing force (H).
16. The saturation point is the state of magnetism beyond which an electromagnet is incapable of further magnetic strength.
17. Hysteresis is a lag between cause and effect. With magnetism, it is the amount that the magnetization of a material lags the magnetizing force due to molecular friction.
18. Remanence or retentivity is the amount a material remains magnetized after the magnetizing force has been removed. Coercive force is the magnetizing force needed to reduce the residual magnetism within a material to zero.
19. With a speaker, a movable electromagnet is suspended within a fixed permanent magnet. Signal current changes alter the electromagnet's magnetic flux pattern, and it is the interaction between this changing magnetic field and the permanent magnet's fixed magnetic field that causes the core to vibrate.

Electromagnetic Induction (Section 10-2)

20. Electromagnetic induction is the voltage or emf induced or produced in a coil as the magnetic lines of force link with the turns of a coil.

21. In 1831, Michael Faraday carried out an experiment including the use of a coil, a zero center ammeter (galvanometer), and a bar permanent magnet. Faraday discovered that whenever there is relative motion between a coil and magnetic lines of flux, a voltage will be induced.
22. Heinrich Lenz's law states that the current induced in a coil due to the change in the magnetic flux is such as to oppose the cause producing it.
23. The generator's basic principle of operation is that the mechanical-drive energy input will produce ac electrical energy out by means of electromagnetic induction.
24. One complete revolution of the mechanical energy input to the generator causes one complete cycle of the ac electrical energy output, with a sine wave being generated as a result of circular motion.
25. The moving coil microphone is an example of how electromagnetic induction is used to convert information-carrying sound waves to information-carrying electrical waves.

REVIEW QUESTIONS

Multiple-Choice Questions

1. Electromagnetism was first discovered by
a. Hans Christian Oersted. **c.** James Watt.
b. Heinrich Hertz. **d.** James Clerk Maxwell.

2. One magnetic line of flux is known as a __________.
a. Weber **b.** Tesla **c.** Maxwell **d.** Oersted

3. 10^8 magnetic lines of flux are referred to as 1 __________.
a. Weber **b.** Tesla **c.** Maxwell **d.** Oersted

4. Flux density is equal to the magnetic flux divided by area and is measured in __________.
a. Webers **b.** Teslas **c.** Maxwells **d.** Oersteds

5. By increasing either current or the number of turns in a coil, the mmf, which is an abbreviation for __________, will increase.
a. Multiple magnetic formulas **c.** Electromotive force
b. Magnetomotive force **d.** None of the above

6. __________ is a term used to describe the magnetic field intensity and is equal to the mmf divided by the length of the coil.
a. Flux density **c.** Magnetizing force
b. Magnetomotive force **d.** Reluctance

7. Reluctance is equivalent to electrical
a. Current. **b.** Voltage. **c.** Resistance. **d.** Power.

8. Permeability is equivalent to electrical
a. Current. **c.** Resistance.
b. Conductance. **d.** Voltage.

9. The absolute permeability of air is equal to
a. 1 **c.** 4π
b. 1.26×10^{-6} **d.** None of the above.

10. An electromagnet can be used in the application of
a. A relay. **c.** A circuit breaker.
b. A solenoid. **d.** All of the above.

11. The relative permeability of air or a vacuum is equal to
a. 4π **c.** 1
b. 6.26×10^{-6} **d.** None of the above.

12. An electromagnet is also known as a
a. Coil. **d.** Both (a) and (c).
b. Solenoid. **e.** Both (a) and (b).
c. Resistor.

13. Direct current produces a __________ magnetic field of __________ polarity.
a. Alternating, unchanging **c.** Constant, unchanging
b. Constant, alternating **d.** Both (a) and (c)

14. Magnetizing force (H) is equal to
a. $I \times N \times l$ **b.** $I \times N + l$ **c.** $I \times N/l$ **d.** $N \times l/I$

15. Electromagnetic induction
a. Is the magnetism resulting from electrical current flow.
b. Is the electrical voltage resulting in a coil from the relative motion of a magnetic field.
c. Both (a) and (b).
d. None of the above.

16. The __________ plots magnetizing force against flux density through a complete cycle of alternating current.
a. *B-H* curve **c.** Power curve
b. Coercive force **d.** Hysteresis loop

17. When the magnetic flux linking a conductor is changing, an emf is induced, the magnitude of which depends on the number of coil turns, rate of change of flux linkage change, and flux density. This law was discovered by
a. Heinrich Lenz. **c.** Michael Faraday.
b. Guglielmo Marconi. **d.** Joseph Henry.

18. The current induced in a coil due to the change in the magnetic flux is such as to oppose the cause producing it. This law was discovered by
a. Heinrich Lenz. **c.** Michael Faraday.
b. Guglielmo Marconi. **d.** Joseph Henry.

19. An ac generator uses __________ to generate ac electricity.
a. Electromagnetism **c.** Magnetism
b. Electromagnetic induction **d.** None of the above

20. The generator converts __________ energy into __________ energy.
a. Electrical, electrical **c.** Chemical, electrical
b. Mechanical, electrical **d.** None of the above

21. The loop of conductor rotated through 360° in a generator is known as a/an
a. Electromagnet. **c.** Armature.
b. Field coil. **d.** Both (a) and (b).

22. Sound waves are a form of
a. Electrical energy. **c.** Magnetic energy.
b. Chemical energy. **d.** Mechanical energy.

23. The moving coil microphone converts __________ waves into __________ waves.
a. Sound, electrical **c.** Electromagnetic, sound
b. Electrical, sound **d.** Sound, radio

24. There is a reciprocal relationship between permeability and __________

a. Flux density
b. Magnetizing force
c. Remanence
d. Reluctance

25. The amount of current in the reverse direction needed to reduce the flux density (B) to zero is termed the
a. Coercive force.
b. Magnetizing force.
c. Electromotive force.
d. All of the above.

Communication Skill Questions

26. Describe what is meant by the word *electromagnetism.* (10-1)
27. How many maxwells make up 1 weber? (10-1-2):
28. Give the formulas for the following (10-1-2)
a. Flux density
b. Magnetomotive force
c. Magnetizing force
d. Reluctance
e. Permeability
29. From which pole does the magnetic flux emerge and into which pole end does the flux return? (10-2)
30. What are the differences between magnetic flux and flux density? (10-1-2)
31. What effect does current have when it is passed through a coil of conductor? (10-1)
32. What effect does a magnet have when it is moved into and out of a coil? (10-2)
33. State Faraday's law. (10-2-1)
34. State Lenz's law. (10-2-2)
35. Describe the different effects when ac and dc are passed though a coil. (10-1-1)
36. Illustrate and describe all the different points on a hysteresis curve. (10-1-3)
37. Describe the meaning of the following terms (Chapter 10):
a. Flux density
b. Magnetizing force
c. Remanence
d. Coercive force
e. Electromagnetism
f. Electromagnetic induction
38. Briefly describe the operation of the generator. (10-2-3)
39. Briefly describe the operation of a speaker and a moving coil microphone. (10-1-4 and 10-2-3)
40. Illustrate and describe the operation of the generator through 360°. (10-2-3)

Practice Problems

41. If a magnet has a pole area of 6.4×10^{-3} m^2 and a 1200 μWb total flux, what flux density would the pole produce?
42. Calculate the magnetomotive force produced when 760 mA flows through 25 turns.
43. Calculate the magnetizing force (H) or field intensity if a 15 cm, 40 turn coil has a current of 1.2 A flowing through it.
44. Calculate the permeability (μ) of the following materials ($\mu_0 = 4\pi \times 10^{-7}$):
a. Cast iron b. Nickel c. Machine steel
45. Calculate the reluctance of a magnetic circuit if the magnetic flux produced is equal to 2.3×10^{-4} Wb and is produced by 3 A flowing through a solenoid of 36 turns.
46. If a coil of 50 turns is passing a current of 4 A, calculate the mmf.
47. Calculate the reluctance of an iron core when mmf = 150 At and Φ = 360 μWb.
48. Calculate mmf when a 9 V battery is connected across a 50 turn, 23 Ω coil.
49. Calculate the permeability of a permalloy core.
50. Calculate the magnetizing force (H) of the coil in Question 48 if it were 0.7 m long.

Job Interview Test

These tests at the end of each chapter will challenge your knowledge up to this point and give you the practice you need for a job interview. To make this more realistic, the test will comprise both technical and personal questions. In order to take full advantage of this exercise, you may want to set up a simulation of the interview environment, have a friend read the questions to you, and record your responses for later analysis.

Company Name: POS, Inc.

Industry Branch: Computers.

Function: Service store point-of-sale terminals.

Job Appling For: Field Service Technician.

51. Do you have a PC?
52. What is Ohm's law?
53. What first attracted you to electronics?
54. What application software are you familiar with?
55. How would you deal with an irate customer?
56. What is the power formula?
57. Would you say that you have a likeable personality?
58. Could you define the expression *saturation point*?
59. What is a relay?
60. Would you say that you work better in the morning or the afternoon?

Answers

51. Yes. Go on to explain what computer system you have at home and what was available at school. Describe how you have used them, both personally and professionally.
52. Chapter 2.
53. Individual response.
54. Discuss EWB's Multisim, and all other application software you are familiar with.
55. Work as hard as possible to solve their problem as soon as possible.
56. Chapter 2.
57. They're asking if you can get along well with customers and colleagues. Mention any work experience in which you had direct contact with customers, and discuss how you worked with fellow students in your class.
58. Section 10-1-2
59. Section Chapter 4.
60. Choose neither of the options, but instead answer, "I find that I work consistently well all day."

Web Site Questions

Go to the web site http://www.prenhall. com/cook, select the textbook *Introductory DC/AC Electronics* or *Introductory DC/AC Circuits,* this chapter, and then follow the instructions when answering the multiple-choice practice problems.

Inductors

The Wizard of Menlo Park

Thomas Alva Edison was born to Samuel and Nancy Edison on February 11, 1847. As a young boy he had a keen and inquisitive mind, yet he did not do well at school, so his mother, a former schoolteacher, withdrew him from school and tutored him at home. In later life he said that his mother taught him to read well and instilled a love for books that lasted the rest of his life. In fact, the inventor's personal library of more than 10,000 volumes is preserved at the Edison Laboratory National Monument in West Orange, New Jersey.

At the age of twenty-nine, after several successful inventions, Edison put into effect what is probably his greatest idea—the first industrial research laboratory. Choosing Menlo Park in New Jersey, which was then a small rural village, Edison had a small building converted into a laboratory for his 15-member staff and a house built for his wife and two small daughters. When asked to explain the point of this lab, Edison boldly stated that it would produce "a minor invention every ten days and a big thing every six months or so." At the time, most of the scientific community viewed his prediction as preposterous; however, in the next 10 years Edison would be granted 420 patents, including those for the electric lightbulb, the motion picture, the phonograph, the universal electric motor, the fluorescent lamp, and the medical fluoroscope.

Over 1000 patents were granted to Edison during his lifetime; his achievements at what he called his "invention factory" earned him the nickname "the wizard of Menlo Park." When asked about his genius he said, "Genius is two percent inspiration and ninety-eight percent perspiration."

11

Outline and Objectives

THE WIZARD OF MENLO PARK

INTRODUCTION

11-1 INDUCTOR CHARACTERISTICS

Objective 1: Describe self-inductance.

11-1-1 Inductance

11-1-2 The Inductor

11-1-3 Inductance Formula

Objective 2: List and explain the factors affecting inductance.

Objective 3: Give the formula for inductance.

11-1-4 Inductors in Combination

Objective 4: Identify inductors in series and parallel and understand how to calculate total inductance when inductors are in combination.

11-2 INDUCTOR TYPES AND TESTING

11-2-1 Inductor Types

Objective 5: List and explain the fixed and variable types of inductors.

11-2-2 Inductor Testing

Objective 6: State the three typical malfunctions of inductors and explain how they can be recognized.

11-3 DC INDUCTIVE CIRCUITS

11-3-1 Current Rise

Objective 7: Explain the inductive time constant.

11-3-2 Current Fall

11-4 AC INDUCTIVE CIRCUITS

11-4-1 Inductive Reactance

Objective 8: Give the formula for inductive reactance.

11-4-2 Series *RL* Circuit

11-4-3 Parallel *RL* Circuit

Objective 9: Describe all aspects relating to a series and parallel RL circuit.

11-5 APPLICATIONS OF INDUCTORS

11-5-1 *RL* Filters

11-5-2 *RL* Integrator

11-5-3 *RL* Differentiator

Objective 10: Describe how inductors can be used for the following applications:

a. *RL* filter
b. *RL* integrator
c. *RL* differentiator

Introduction

In Chapter 10 two important rules were discussed that relate to this chapter on inductance. The first was that a magnetic field will build up around any current-carrying conductor (electromagnetism), and the second was that a voltage will be induced into a conductor when it is subjected to a moving magnetic field. These two rules form the basis for a phenomenon covered in this chapter called *self-inductance* or, more commonly, *inductance.* Inductance, by definition, is the ability of a device to oppose a change in current flow, and the device designed specifically to achieve this function is called the *inductor.* There is, in fact, no physical difference between an inductor and an electromagnet, since they are both coils. The two devices are given different names even though their constuction and principle of operation are the same because they are used in different applications. An electromagnet is used to generate a magnetic field in response to current, and an inductor is used to oppose any change in current.

In this chapter, we examine all aspects of inductance and inductors, including inductive circuits, types, reactance, testing, and applications.

11-1 INDUCTOR CHARACTERISTICS

In Figure 11-1 an external voltage source has been connected across a coil, forcing a current through the coil. This current will generate a magnetic field that will increase in field strength in a short time from zero to maximum, expanding from the center of the conductor (electromagnetism). The expanding magnetic lines of force have relative motion with respect to the stationary conductor, so an induced voltage results (electromagnetic induction). The blooming magnetic field generated by the conductor is actually causing a voltage to be induced in the conductor that is generating the magnetic field. This effect of a current-carrying coil of conductor inducing a voltage within itself is known as **self-inductance.** This phenomenon

Self-Inductance

The property that causes a counter-electromotive force to be produced in a conductor when the magnetic field expands or collapses with a change in current.

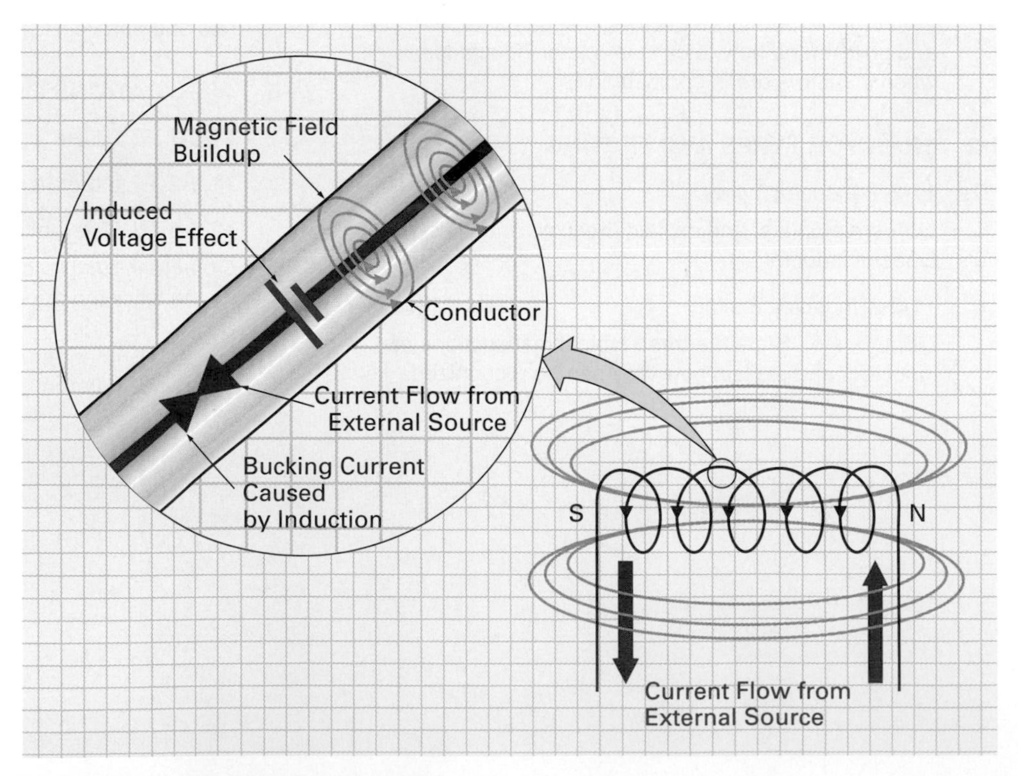

FIGURE 11-1 Self-Induction of a Coil.

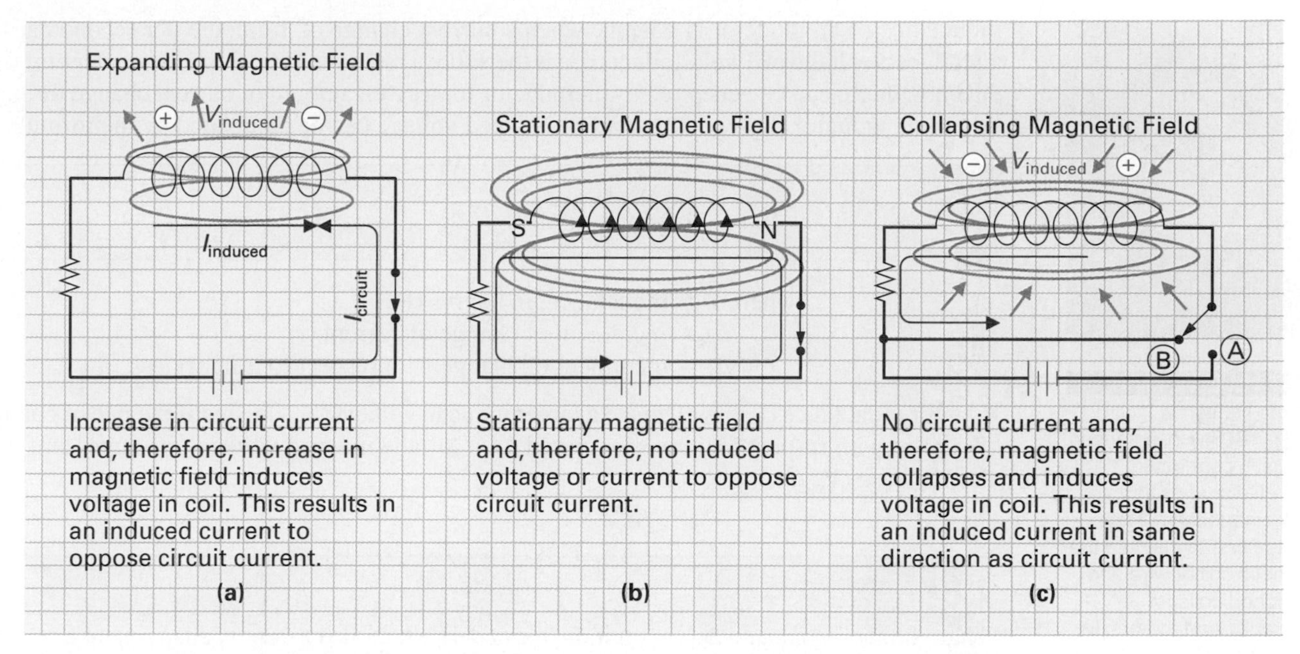

FIGURE 11-2 Self-Inductance. (a) Switch Closed. (b) Constant Circuit Current. (c) Switch Opened.

was first discovered by Heinrich Lenz, who observed that the induced voltage causes an induced (bucking) current to flow in the coil, which opposes the source current producing it.

11-1-1 *Inductance*

Figure 11-2 details the reaction of an inductor to a dc source. When the switch is closed, as seen in Figure 11-2(a), a circuit current will exist through the inductor and the resistor. As the current rises toward its maximum value, the magnetic field expands, and throughout this time of relative motion between field and conductor an induced voltage will be present. This induced voltage will produce an induced current to oppose the change in the circuit current.

When the current reaches its maximum, the magnetic field, which is dependent on current, will also reach a maximum value and then no longer expand but remain stationary. When the current remains constant, no change will occur in the magnetic field and therefore no relative motion will exist between the conductor and magnetic field, resulting in no induced voltage or current to oppose circuit current, as shown in Figure 11-2(b). The coil has accepted electrical energy and is storing it in the form of a magnetic energy field, just as the capacitor stored electrical energy in the form of an electric field.

If the switch is put in position *B,* as shown in Figure 11-2(c), the current from the battery will be zero and the magnetic field will collapse, as it no longer has circuit current to support it. As the magnetic lines of force collapse, they cut the conducting coils, causing relative motion between the conductor and magnetic field. A voltage is induced in the coil, which will produce an induced current to flow in the same direction as the circuit current was flowing before the switch was opened. The coil is now converting the magnetic field energy into electrical energy and returning the original energy that it stored.

After a short period of time, the magnetic field will have totally collapsed, the induced voltage will be zero, and the induced current within the circuit will therefore also no longer be present.

This induced voltage is called a **counter emf** or *back emf.* It opposes the applied emf (or battery voltage). The ability of a coil or conductor to induce or produce a counter emf within itself as a result of a change in current is called *self-inductance* or, more commonly, **inductance** (symbolized *L*). The unit of inductance is the **henry** (H), named in honor of Joseph Henry, an American physicist, for his experimentation within this area of science. The

Counter emf (Counter Electromotive Force)
Abbreviated "counter emf," or "back emf," it is the voltage generated in an inductor due to an alternating or pulsating current and is always of opposite polarity to that of the applied voltage.

Inductance
Property of a circuit or component to oppose any change in current as the magnetic field produced by the change in current causes an induced countercurrent to oppose the original change.

Henry
Unit of inductance.

inductance of an inductor is 1 henry when a current change of 1 ampere per second causes an induced voltage of 1 volt. Inductance is therefore a measure of how much counter emf (induced voltage) can be generated by an inductor for a given amount of current change through that same inductor. This counter emf or induced voltage can be calculated by the formula

$$V_{ind} = L \times \frac{\Delta I}{\Delta t}$$

where L = inductance, in henrys (H)
ΔI = increment of change of current (I)
Δt = increment of change with respect to time (t)

TIME LINE
Joseph Henry (1797–1878), an American physicist, conducted extensive studies into electromagnetism. Henry was the first to insulate the magnetic coil of wire and developed coils for telegraphy and motors. In recognition of his discovery of self-induction in 1832, the unit of inductance is called the henry.

A larger inductance ($L\uparrow$) will create a larger induced voltage ($V_{ind}\uparrow$), and if the rate of change of current with respect to time is increased $\Delta I\uparrow/\Delta t$), the induced voltage or counter emf will also increase ($V_{ind}\uparrow$).

EXAMPLE:

What voltage is induced across an inductor of 4 H when the current is changing at a rate of

a. 1 A/s?
b. 4 A/s?

Solution:

a. a. $V_{ind} = L \times \frac{\Delta I}{\Delta t} = 4\text{ H} \times 1\text{ A/s} = 4\text{ V}$

b. b. $V_{ind} = L \times \frac{\Delta I}{\Delta t} = 4\text{ H} \times 4\text{ A/s} = 16\text{ V}$

The faster the coil current changes, the larger the induced voltage.

11-1-2 *The Inductor*

Inductor
Coil of conductor used to introduce inductance into a circuit.

An **inductor** is basically an electromagnet, as its construction and principle of operation are the same. We use the two different names because they have different applications. The purpose of the electromagnet is to generate a magnetic field; the purpose of an inductor or coil is to oppose any change of circuit current.

In Figure 11-3 a steady value of direct current is present within the circuit and the inductor is creating a steady or stationary magnetic field. If the current in the circuit is suddenly increased (by lowering the circuit resistance), the change in the expanding magnetic field will

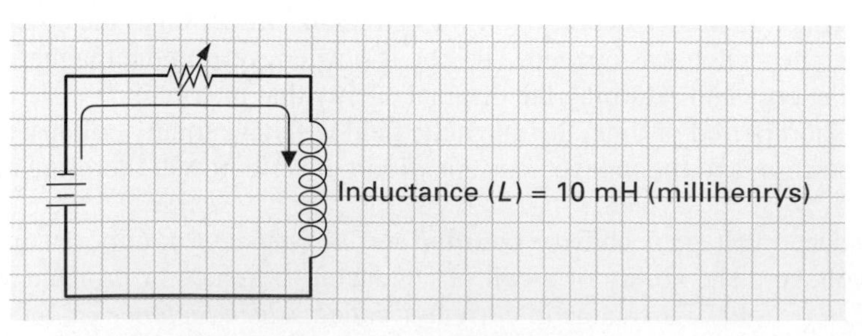

FIGURE 11-3 Inductor's Ability to Oppose Current Change.

induce a counter emf within the inductor. This induced voltage will oppose the source voltage from the battery and attempt to hold current at its previous low level.

The counter emf cannot completely oppose the current increase, for if it did, the lack of current change would reduce the counter emf to zero. Current therefore incrementally increases up to a new maximum, which is determined by the applied voltage and the circuit resistance ($I = V/R$). Once the new higher level of current has been reached and remains constant, there will no longer be a change. This lack of relative motion between field and conductor will no longer generate a counter emf, so the current will remain at its new higher constant value.

This effect also happens in the opposite respect. If current decreases (by increasing circuit resistance), the magnetic lines of force will collapse because of the reduction of current and induce a voltage in the inductor, which will produce an induced current in the same direction as the circuit current. These two combine and tend to maintain the current at the higher previous constant level. Circuit current will fall, however, as the induced voltage and current are only present during the change (in this case, the decrease from the higher current level to the lower); and once the new lower level of current has been reached and remains constant, the lack of change will no longer induce a voltage or current. So the current will then remain at its new lower constant value.

The inductor is therefore an electronic component that will oppose any changes in circuit current, and this ability or behavior is referred to as *inductance*. Since the current change is opposed by a counter emf, inductance may also be defined as the ability of a device to induce a counter emf within itself for a change in current.

11-1-3 *Inductance Formula*

The inductance of an inductor is determined by four factors:

1. Number of turns
2. Area of the coil
3. Length of the coil
4. Core material used within the coil

Let's now discuss how these four factors can affect inductance, beginning with the number of turns.

1. Number of Turns (N) (Figure 11-4)

If an inductor has a greater number of turns, the magnetic field produced by passing current through the coil will have more magnetic force than an inductor with fewer turns. A greater magnetic field will cause a larger counter emf, because more magnetic lines of flux will cut more coils of the conductor, producing a larger inductance value. Inductance (L) is therefore proportional to the number of turns (N):

$$L \propto N$$

2. Area of Coil (A) (Figure 11-5)

If the area of the coil is increased for a given number of turns, more magnetic lines of force will be produced, and if the magnetic field is increased, the inductance will also increase. Inductance (L) is therefore proportional to the cross-sectional area of the coil (A):

$$L \propto A$$

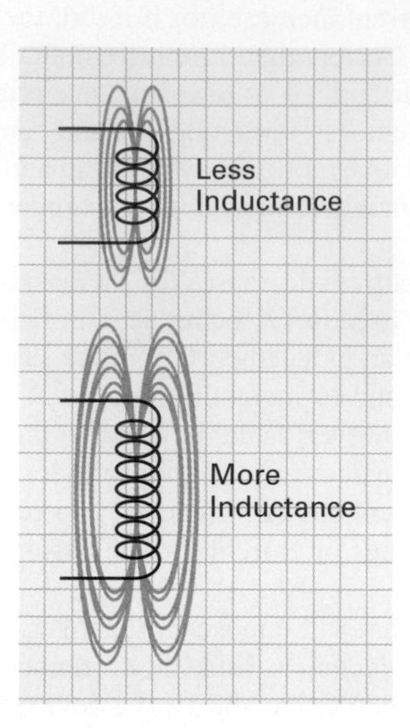

FIGURE 11-4 Inductance Is Proportional to the Number of Turns ($L \propto N$) in a Coil.

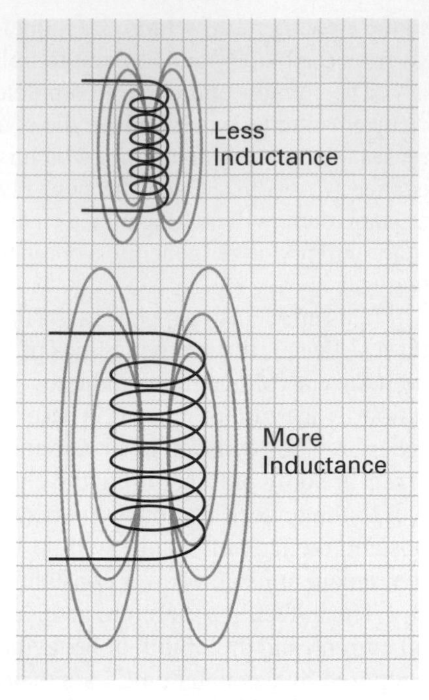

FIGURE 11-5 Inductance Is Proportional to the Area ($L \propto A$) in a Coil.

3. Length of Coil (l) (Figure 11–6)

If, for example, four turns are spaced out (long-length coil), the summation that occurs between all the individual coil magnetic fields will be small. On the other hand, if four turns are wound close to one another (short-length coil), all the individual coil magnetic fields will easily interact and add together to produce a larger magnetic field and, therefore, greater inductance. Inductance is therefore inversely proportional to the length of the coil, in that a longer coil, for a given number of turns, produces a smaller inductance, and vice versa.

$$L \propto \frac{1}{l}$$

4. Core Material (μ) (Figure 11-7)

Most inductors have core materials such as nickel, cobalt, iron, steel, ferrite, or an alloy. These cores have magnetic properties that concentrate or intensify the magnetic field. Permeability is another factor that is proportional to inductance; the figures for various materials are shown in Table 11-1. The greater the permeability of the core material, the greater the inductance.

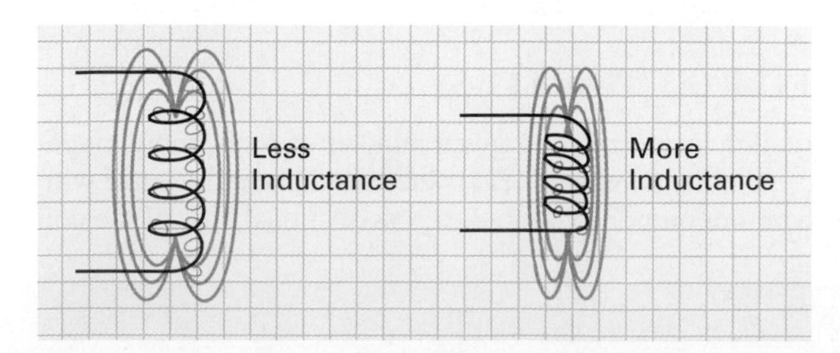

FIGURE 11-6 Inductance Is Inversely Proportional to the Length ($L \propto 1/l$) of a Coil.

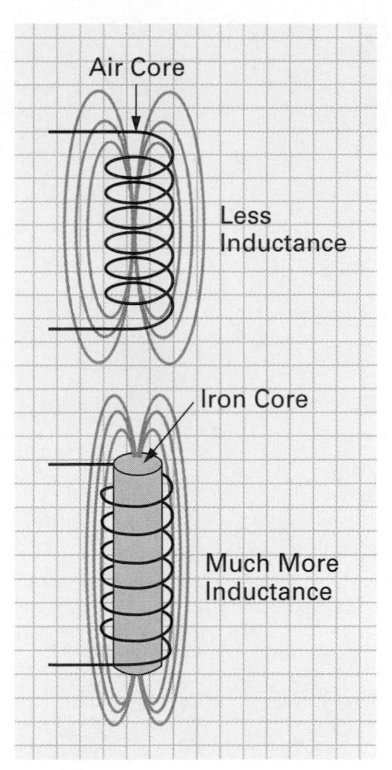

FIGURE 11-7 Inductance Is Proportional to the Permeability of the Coil's Material ($L \propto \mu$).

TABLE 11-1 Permeabilities of Various Materials

MATERIAL	PERMEABILITY
Air or vacuum	1.26×10^{-6}
Nickel	6.28×10^{-5}
Cobalt	7.56×10^{-5}
Cast iron	1.1×10^{-4}
Machine steel	5.65×10^{-4}
Transformer iron	6.9×10^{-3}
Silicon iron	8.8×10^{-3}
Permalloy	0.126
Supermalloy	1.26

$$L \propto \mu$$

5. Formula for Inductance

All the four factors described can be placed in a formula to calculate inductance.

$$L = \frac{N^2 \times A \times \mu}{l}$$

where L = inductance, in henrys (H)
N = number of turns
A = cross-sectional area, in square meters (m^2)
μ = permeability
l = length of core, in meters (m)

EXAMPLE:

Refer to Figure 11-8(a) and (b) and calculate the inductance of each.

Solution:

a. $$L = \frac{5^2 \times 0.01 \times (6.28 \times 10^{-5})}{0.001} = 15.7 \text{ mH}$$

b. $$L = \frac{10^2 \times 0.1 \times (1.1 \times 10^{-4})}{0.1} = 11 \text{ mH}$$

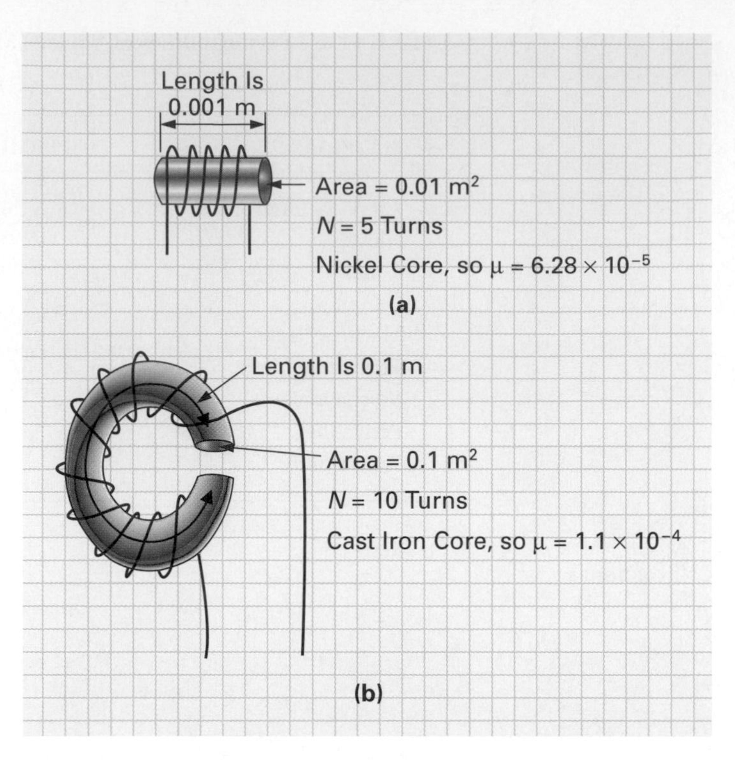

FIGURE 11-8 **Inductor Examples.**

11-1-4 *Inductors in Combination*

Inductors oppose the change of current in a circuit and so are treated in a manner similar to resistors connected in combination. Two or more inductors in series merely extend the coil length and increase inductance. Inductors in parallel are treated in the same way as resistors, with the total inductance being less than that of the smallest inductor's value.

1. Inductors in Series

When inductors are connected in series with one another, the total inductance is calculated by summing all the individual inductances.

$$L_T = L_1 + L_2 + L_3 + \cdots$$

EXAMPLE:

Calculate the total inductance of the circuit shown in Figure 11-9.

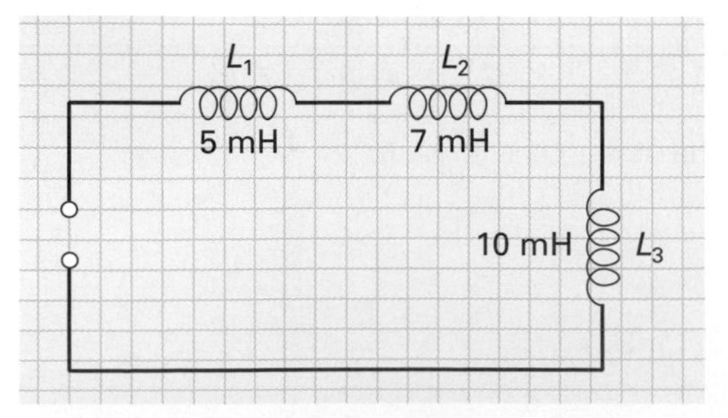

FIGURE 11-9 **Inductors in Series.**

Solution:

$$\begin{aligned} L_T &= L_1 + L_2 + L_3 \\ &= 5 \text{ mH} + 7 \text{ mH} + 10 \text{ mH} \\ &= 22 \text{ mH} \end{aligned}$$

2. Inductors in Parallel

When inductors are connected in parallel with one another, the reciprocal (two or more inductors) or product-over-sum (two inductors) formula can be used to find total inductance, which will always be less than the smallest inductor's value.

$$L_T = \frac{1}{(1/L_1) + (1/L_2) + (1/L_3) + \cdots}$$

$$L_T = \frac{L_1 \times L_2}{L_1 + L_2}$$

EXAMPLE:

Determine L_T for the circuits in Figure 11-10(a) and (b).

Solution:

a. Reciprocal formula:

$$\begin{aligned} L_T &= \frac{1}{(1/L_1) + (1/L_2) + (1/L_3)} \\ &= \frac{1}{(1/10\text{ mH}) + (1/5\text{ mH}) + (1/20\text{ mH})} \\ &= 2.9\text{ mH} \end{aligned}$$

b. Product over sum:

$$\begin{aligned} L_T &= \frac{L_1 \times L_2}{L_1 + L_2} \\ &= \frac{10\ \mu\text{H} \times 2\ \mu\text{H}}{10\ \mu\text{H} + 2\ \mu\text{H}} \\ &= \frac{20 \times 10^{-12}\text{H}^2}{12\ \mu\text{H}} = 1.67\ \mu\text{H} \end{aligned}$$

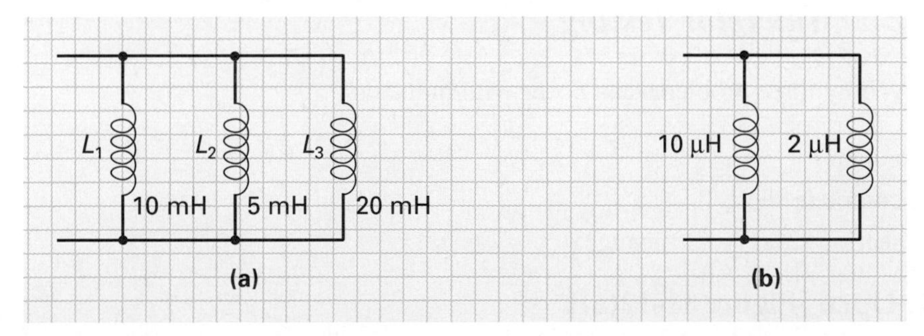

FIGURE 11-10 Inductors in Parallel.

SELF-TEST EVALUATION POINT FOR SECTION 11-1

Now that you have completed this section, you should be able to:

Objective 1. Describe self-inductance.

Objective 2. List and explain the factors affecting inductance.

Objective 3. Give the formula for inductance.

Objective 4. Identify inductors in series and parallel and understand how to calculate total inductance when inductors are in combination.

Use the following questions to test your understanding of Section 11-1:

1. Define self-induction.
2. What is counter emf, and how can it be calculated?
3. Calculate the voltage induced in a 2 mH inductor if the current is increasing at a rate of 4 kA/s.
4. What is the difference between an electromagnet and an inductor?
5. True or false: The inductor will oppose any changes in circuit current.
6. List the four factors that determine the inductance of an inductor.
7. State the formula for inductance.
8. True or false: To calculate total inductance, inductors can be treated in the same manner as capacitors.
9. State the formula for calculating total inductance in
   a. A series circuit.
   b. A parallel circuit.
10. Calculate the total circuit inductance if 4 mH and 2 mH are connected
    a. In series.
    b. In parallel.

11-2 INDUCTOR TYPES AND TESTING

11-2-1 *Inductor Types*

Fixed-value inductor
An inductor whose value cannot be altered.

As with resistors and capacitors, inductors are basically divided into the two categories of fixed-value and variable-value, as shown in Figure 11-11. With **fixed-value inductors,** the inductance value remains constant and cannot be altered. They are usually made of solid copper wire with an insulating enamel around the conductor. This enamel or varnish on the conductor is needed since the turns of a coil are generally wound on top of one another. The three major types of fixed-value inductors on the market today are

1. Air core
2. Iron core
3. Ferrite core

Variable-value inductor
An inductor whose value can be varied.

With **variable-value inductors,** the inductance value can be changed by adjusting the position of the movable core with respect to the stationary coil. There is basically only one type of variable-value inductor in wide use today—the ferrite core.

11-2-2 *Inductor Testing*

Basically, only three problems can occur with inductors:

1. An open
2. A complete short
3. A section short (value change)

1. Open [Figure 11-12(a)]

This problem can be isolated with an ohmmeter. Depending on the winding's resistance, the coil should be in the range of zero to a few hundred ohms. An open accounts for 75% of all defective inductors.

2. Complete or Section Short [Figure 11-12(b)]

A coil with one or more shorted turns or a complete short can be checked with an ohmmeter and thought to be perfectly good because of the normally low resistance of a coil, as it is just a piece of wire. But if it is placed in a circuit with a complete or section short present, it will

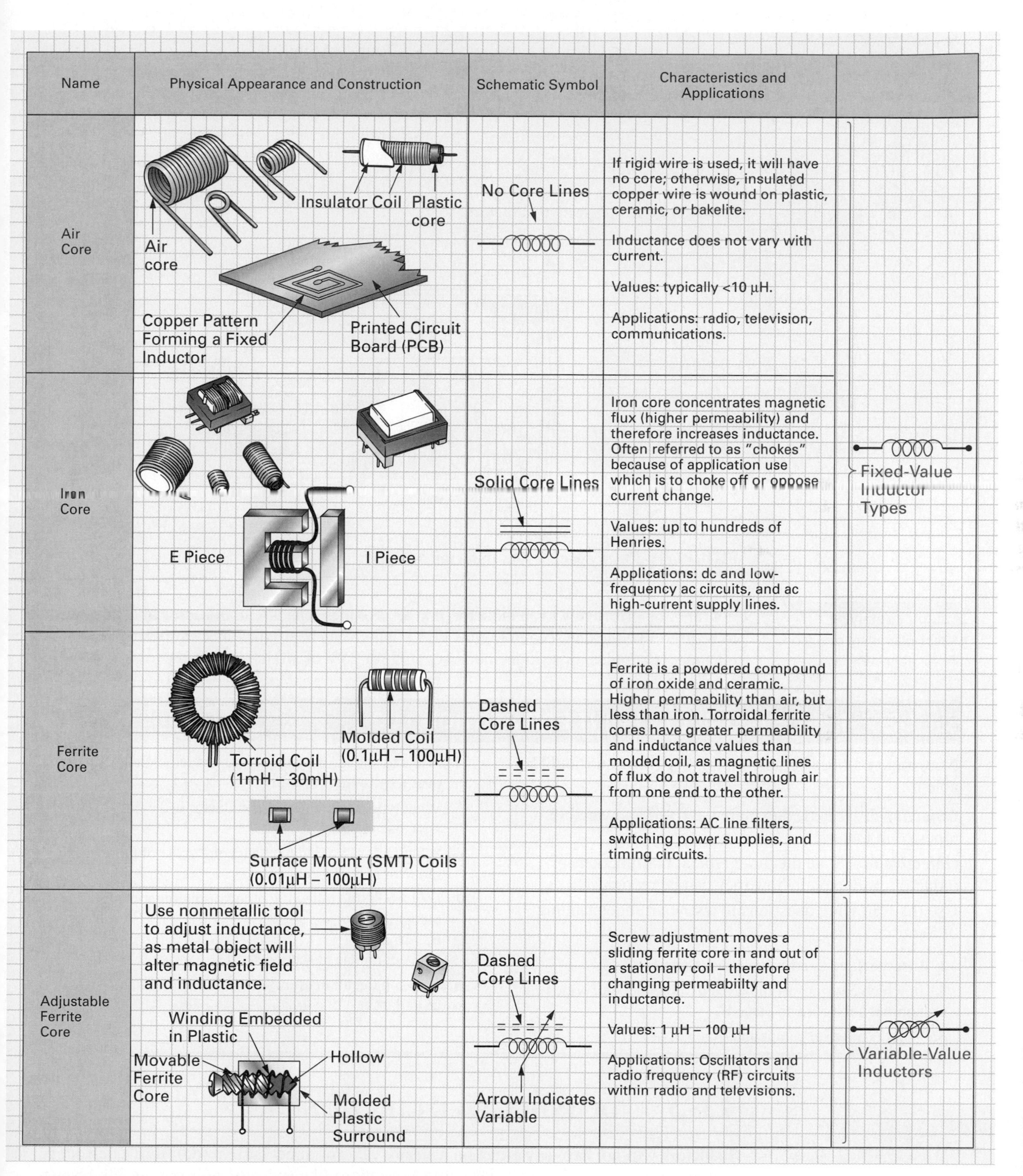

Name	Physical Appearance and Construction	Schematic Symbol	Characteristics and Applications	
Air Core	Air core; Insulator Coil; Plastic core; Copper Pattern Forming a Fixed Inductor; Printed Circuit Board (PCB)	No Core Lines	If rigid wire is used, it will have no core; otherwise, insulated copper wire is wound on plastic, ceramic, or bakelite. Inductance does not vary with current. Values: typically <10 μH. Applications: radio, television, communications.	Fixed-Value Inductor Types
Iron Core	E Piece; I Piece	Solid Core Lines	Iron core concentrates magnetic flux (higher permeability) and therefore increases inductance. Often referred to as "chokes" because of application use which is to choke off or oppose current change. Values: up to hundreds of Henries. Applications: dc and low-frequency ac circuits, and ac high-current supply lines.	
Ferrite Core	Torroid Coil (1mH – 30mH); Molded Coil (0.1μH – 100μH); Surface Mount (SMT) Coils (0.01μH – 100μH)	Dashed Core Lines	Ferrite is a powdered compound of iron oxide and ceramic. Higher permeability than air, but less than iron. Toroidal ferrite cores have greater permeability and inductance values than molded coil, as magnetic lines of flux do not travel through air from one end to the other. Applications: AC line filters, switching power supplies, and timing circuits.	
Adjustable Ferrite Core	Use nonmetallic tool to adjust inductance, as metal object will alter magnetic field and inductance.; Winding Embedded in Plastic; Movable Ferrite Core; Hollow; Molded Plastic Surround	Dashed Core Lines; Arrow Indicates Variable	Screw adjustment moves a sliding ferrite core in and out of a stationary coil – therefore changing permeabiilty and inductance. Values: 1 μH – 100 μH Applications: Oscillators and radio frequency (RF) circuits within radio and televisions.	Variable-Value Inductors

FIGURE 11-11 Fixed-Value and Variable-Value Inductor Types.

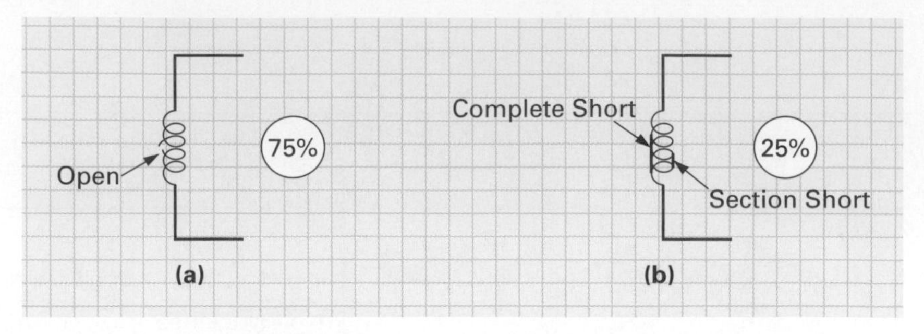

FIGURE 11-12 **Defective Inductors.**

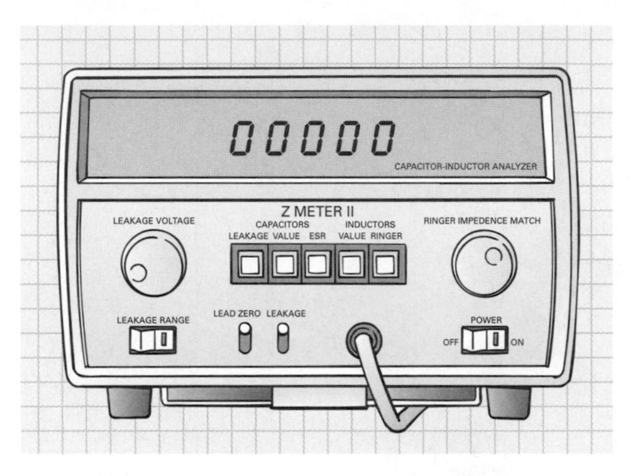

FIGURE 11-13 **Capacitor and Inductor Analyzer.**

Inductor Analyzer
A test instrument designed to test inductors.

not function effectively as an inductor, if at all. For these checks, an **inductor analyzer** needs to be used like the one seen in Figure 11-13, which can be used to check capacitance and inductance. Complete or section shorts account for 25% of all defective inductors.

SELF-TEST EVALUATION POINT FOR SECTION 11-2

Now that you have completed this section, you should be able to:

Objective 5. List and explain the fixed and variable types of inductors.

Objective 6. State the three typical malfunctions of inductors and explain how they can be recognized.

Use the following questions to test your understanding of Section 11-2:

1. List the three fixed-value inductor types.
2. In which category would a fixed-value inductor with a nonmagnetic core be placed?
3. What is a ferrite compound?
4. What advantage does the toroid-shaped inductor have over a cylindrical inductor?
5. What type of variable-value inductor is the only one in wide use today?
6. What factor is varied to change the value of the variable inductor?
7. How could the following inductor malfunctions be recognized?
   a. An open
   b. A complete or section short
8. Which inductor malfunction accounts for almost 75% of all failures?

11-3 DC INDUCTIVE CIRCUITS

Inductors will not have any effect on a steady value of direct current (dc) from a dc voltage source. If, however, the dc is changing (pulsating), the inductor will oppose the change whether it is an increase or decrease in direct current, because a change in current causes the

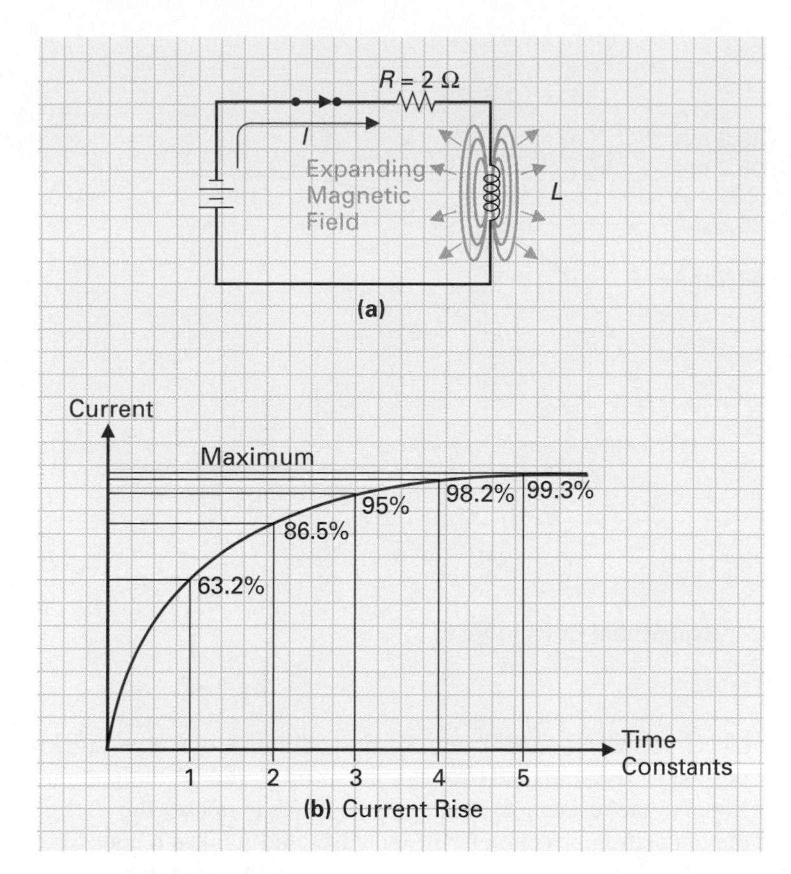

FIGURE 11-14 DC Inductor Current Rise.

magnetic field to expand or contract, and in so doing it will cut the coil of the inductor and induce a voltage that will counter the applied emf.

11-3-1 *Current Rise*

Figure 11-14(a) illustrates an inductor (L) connected across a dc source (battery) through a switch and series-connected resistor. When the switch is closed, current will flow and the magnetic field will begin to expand around the inductor. This field cuts the coils of the inductor and induces a counter emf to oppose the rise in current. Current in an inductive circuit, therefore, cannot rise instantly to its maximum value, which is determined by Ohm's law ($I = V/R$). Current will in fact take time to rise to maximum, as graphed in Figure 11-14(b), due to the inductor's ability to oppose change.

It will actually take five time constants (5τ) for the current in an inductive circuit to reach maximum value. This time can be calculated by using the formula

$$\tau = \frac{L}{R} \quad \text{seconds}$$

The constant to remember is the same as before: 63.2%. In one time constant ($1 \times L/R$) the current in the RL circuit will have reached 63.2% of its maximum value. In two time constants ($2 \times L/R$), the current will have increased 63.2% of the remaining current, and so on, through five time constants. For example, if the maximum possible circuit current is 100 mA and an inductor of 4 H is connected in series with a resistor of 2 Ω, the current will increase as shown in Figure 11-15.

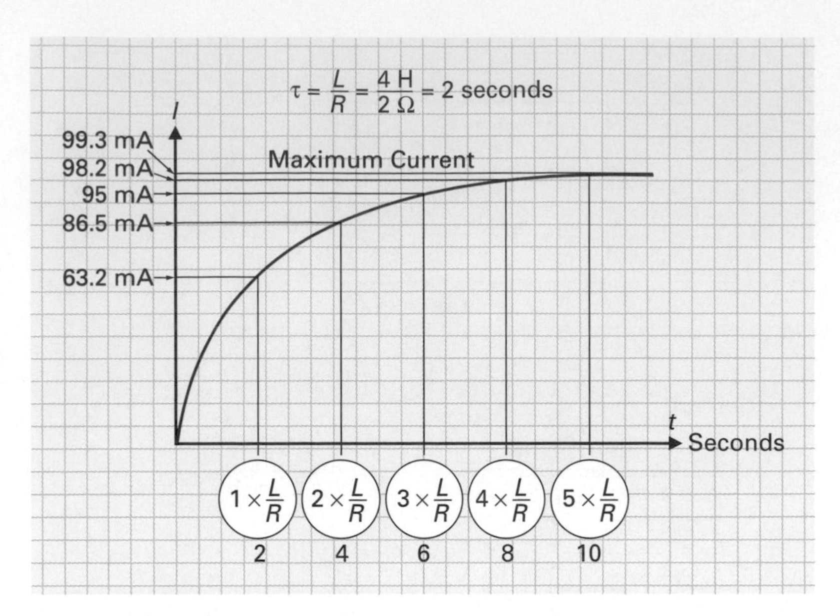

FIGURE 11-15 Exponential Current Rise.

Referring back to the $\tau = L/R$ formula, you will notice that how quickly an inductor will allow the current to rise to its maximum value is proportional to the inductance and inversely proportional to the resistance. A larger inductance increases the strength of the magnetic field, so the opposition or counter emf increases and it takes longer for current to rise to a maximum ($\tau\uparrow = L\uparrow/R$). If the circuit resistance is increased, the maximum current will be smaller, and a smaller maximum is reached more quickly than a higher one ($\tau\downarrow = L/R\uparrow$).

11-3-2 *Current Fall*

When the inductor's dc source of current is removed, as shown in Figure 11-16(a) by placing the switch in position *B,* the magnetic field will collapse and cut the coils of the inductor, inducing a voltage and causing a current to flow in the same direction as the original source current. This current will exponentially decay, or fall, from the maximum to zero level in five time constants ($5 \times L/R = 5 \times 4/2 = 10$ seconds), as shown in Figure 11-16(b).

EXAMPLE

Calculate the circuit current at each of the five time constants if a 12 V dc source is connected across a series *RL* circuit where $R = 60\ \Omega$ and $L = 24$ mH. Plot the results on a graph showing current against time.

Solution:

$$\text{Maximum current, } I_{\text{max}} = \frac{V_S}{R} = \frac{12\text{ V}}{60\ \Omega} = 200\text{ mA}$$

$$\text{Time constant, } \tau = \frac{L}{R} = \frac{24\text{ mH}}{60\ \Omega} = 400\ \mu\text{s}$$

At one time constant (400 μs after source voltage is applied), the current will be

$$\begin{aligned} I &= 63.2\% \text{ of } I_{\text{max}} \\ &= 0.632 \times 200\text{ mA} = 126.4\text{ mA} \end{aligned}$$

FIGURE 11-16
Exponential Current Fall.

At two time constants (800 μs after source voltage is applied):

$$I = 86.5\% \text{ of } I_{\text{max}}$$
$$= 0.865 \times 200 \text{ mA} = 173 \text{ mA}$$

At three time constants (1200 μs or 1.2 ms):

$$I = 95\% \text{ of } I_{\text{max}}$$
$$= 0.95 \times 200 \text{ mA} = 190 \text{ mA}$$

At four time constants (1.6 ms):

$$I = 98.2\% \text{ of } I_{\text{max}}$$
$$= 0.982 \times 200 \text{ mA} = 196.4 \text{ mA}$$

At five time constants (2 ms):

$$I = 99.3\% \text{ of } I_{\text{max}}$$
$$= 0.993 \times 200 \text{ mA} = 198.6 \text{ mA, approximately maximum (200 mA)}$$

See Figure 11-17.

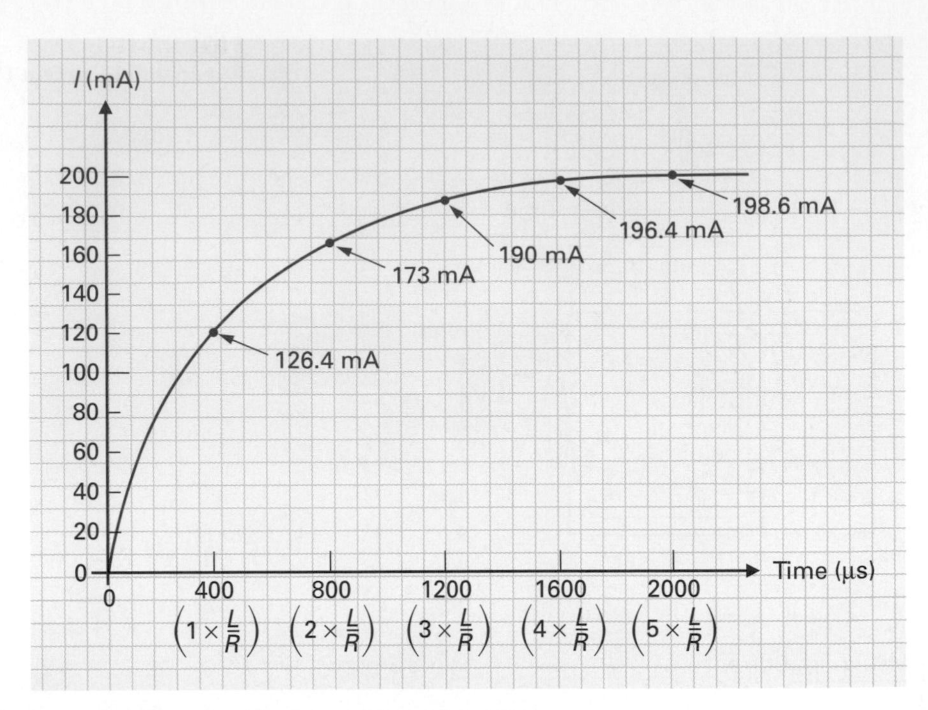

FIGURE 11-17 Exponential Current Rise Example.

SELF-TEST EVALUATION POINT FOR SECTION 11-3

Now that you have completed this section, you should be able to:

Objective 7: Explain the inductive time constant.

Use the following questions to test your understanding of Section 11-3:

1. How does the inductive time constant relate to the capacitive time constant?
2. True or false: The greater the value of the inductor, the longer it would take for current to rise to a maximum.
3. True or false: A constant dc level is opposed continuously by an inductor.

11-4 AC INDUCTIVE CIRCUITS

If an alternating (ac) voltage is applied across an inductor, as shown in Figure 11-18(a), the inductor will continuously oppose the alternating current because it is always changing. Figure 11-18(b) shows the phase relationship between the voltage across an inductor or counter emf and the circuit current.

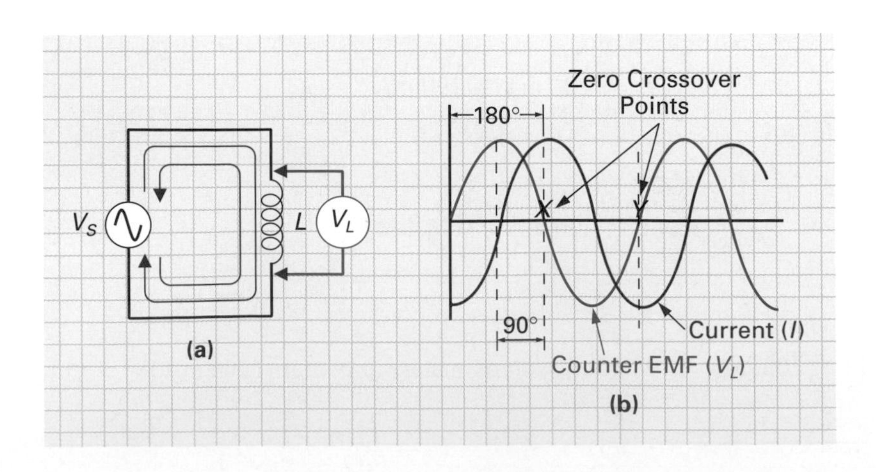

FIGURE 11-18 The Voltage across an Inductor Leads the Circuit Current by 90° in an Inductive Circuit.

The current in the circuit causes the magnetic field to expand and collapse and cut the conducting coils, resulting in an induced counter emf. At points X and Y, the steepness of the current waveform indicates that the current will be changing at its maximum rate, and therefore the opposition or counter emf will also be maximum. When the current is at its maximum positive or negative value, it has a very small or no rate of change (flat peaks). Therefore, the opposition or counter emf should be very small or zero, as can be seen by the waveforms. The counter emf is, therefore, said to be 90° out of phase with the circuit current. To summarize, we can say that the voltage across the inductor (V_L) or counter emf leads the circuit current (I) by 90°.

11-4-1 *Inductive Reactance*

Reactance is the opposition to current flow without the dissipation of energy, as opposed to resistance, which is the opposition to current flow with the dissipation of energy. **Inductive reactance** (X_L) is the opposition to current flow offered by an inductor without the dissipation of energy. It is measured in ohms and can be calculated by using the formula

Inductive Reactance
Measured in ohms, it is the opposition to alternating or pulsating current flow without the dissipation of energy.

$$X_L = 2\pi \times f \times L$$

where X_L = inductive reactance, in ohms (Ω)
2π = 2π radians, 360° or 1 cycle
f = frequency, in hertz (Hz)
L = inductance, in henrys (H)

Inductive reactance is proportional to frequency ($X_L \propto f$) because a higher frequency (fast-switching current) will cause a greater amount of current change, and a greater change will generate a larger counter emf, which is an opposition or reactance against current flow. When 0 Hz is applied to a coil (dc), there exists no change, so the inductive reactance of an inductor to dc is zero ($X_L = 2\pi \times 0 \times L = 0$). Inductive reactance is also proportional to inductance because a larger inductance will generate a greater magnetic field and subsequent counter emf, which is the opposition to current flow.

Ohm's law can be applied to inductive circuits just as it can be applied to resistive and capacitive circuits. The current flow in an inductive circuit (I) is proportional to the voltage applied (V), and inversely proportional to the inductive reactance (X_L). Expressed mathematically,

$$I = \frac{V}{X_L}$$

EXAMPLE:

Calculate the current flowing in the circuit illustrated in Figure 11-19.

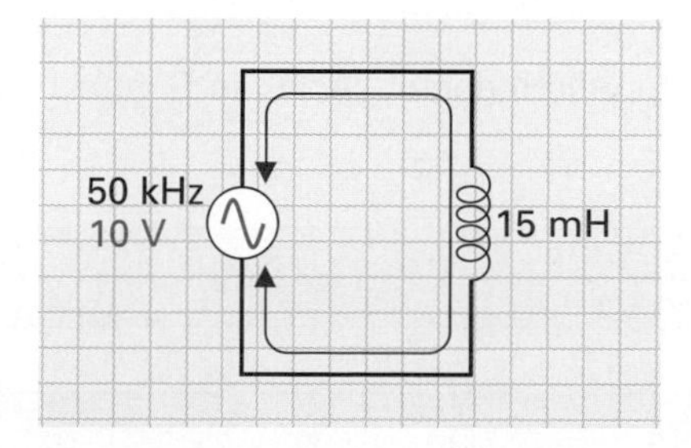

FIGURE 11-19

TIME LINE

In 1950, Kenneth Olsen started work at MIT's Digital Computer Laboratory. For the next seven years, Olsen worked on a new computer system designed to store data obtained from tracking long-range enemy bombers capable of penetrating American air space. IBM won the contract to build the network and Olsen traveled to the plant in New York to supervise production. On completion of the project, Olsen and fellow MIT colleague Harlan Anderson founded a new company, Digital Equipment Corporation or DEC. After three years, they produced their first computer, the PDP 1 (programmed data processor). In 1965, the desktop PDP 8 was launched, and in 1969, the PDP 11 was unveiled. Like its predecessors, this data processor was accepted in many applications, such as tracking telephone calls received on the 911 emergency line, directing welding machines in automobile plants, recording experimental results in laboratories, and processing all types of data in offices, banks, and department stores. Just 20 years after introducing low-cost computing in the 1960s, DEC was second only to IBM as a manufacturer of computers in the United States.

Solution:

The current can be calculated by Ohm's law and is a function of the voltage and opposition, which in this case is inductive reactance.

$$I = \frac{V}{X_L}$$

However, we must first calculate X_L:

$$\begin{aligned} X_L &= 2\pi \times f \times L \\ &= 6.28 \times 50 \text{ kHz} \times 15 \text{ mH} \\ &= 4710\ \Omega, \text{ or } 4.71\ \text{k}\Omega \end{aligned}$$

Current is therefore equal to

$$I = \frac{V}{X_L} = \frac{10\text{ V}}{4.71\ \text{k}\Omega} = 2.12\text{ mA}$$

EXAMPLE:

What opposition or inductive reactance will a motor winding or coil offer if $V = 12$ V and $I = 4.5$ A?

Solution:

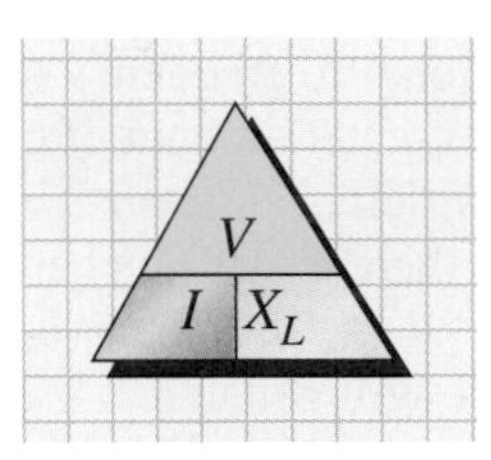

$$X_L = \frac{V}{I} = \frac{12\text{ V}}{4.5\text{ A}} = 2.66\ \Omega$$

11-4-2 *Series* RL *Circuit*

In a purely resistive circuit, as seen in Figure 11-20, the current flowing within the circuit and the voltage across the resistor are in phase with one another. In a purely inductive circuit, as shown in Figure 11-21, the current will lag the applied voltage by 90°. If we connect a resistor and inductor in series, as shown in Figure 11-22(a), we will have the most common combination of R and L used in electronic equipment.

1. Voltage

The voltage across the resistor and inductor shown in Figure 11-22(a) can be calculated by using Ohm's law:

$$V_R = I \times R$$

$$V_L = I \times X_L$$

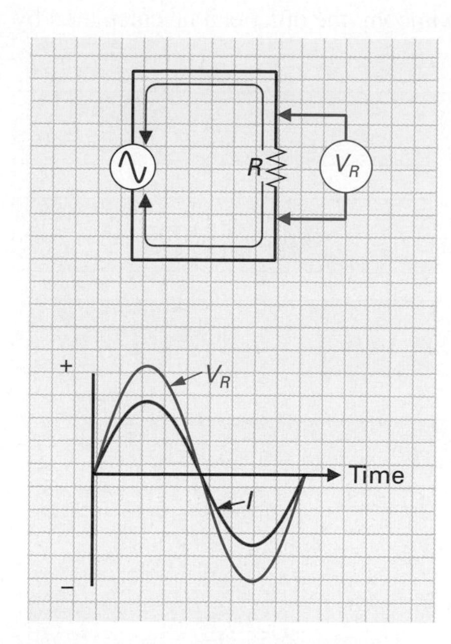

FIGURE 11-20 Purely Resistive Circuit: Current and Voltage Are in Phase.

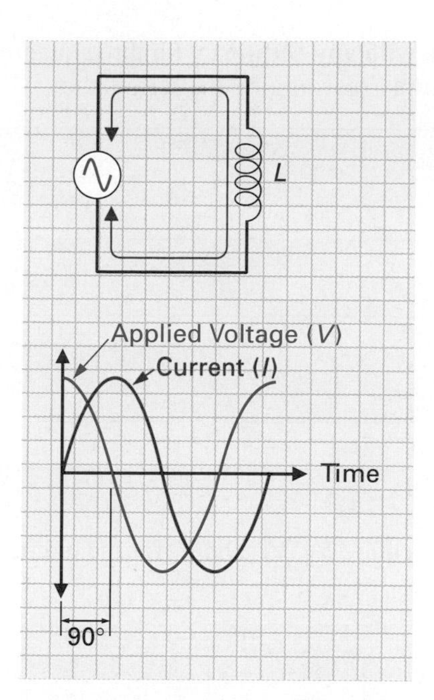

FIGURE 11-21 Purely Inductive Circuit: Current Lags Applied Voltage by 90°.

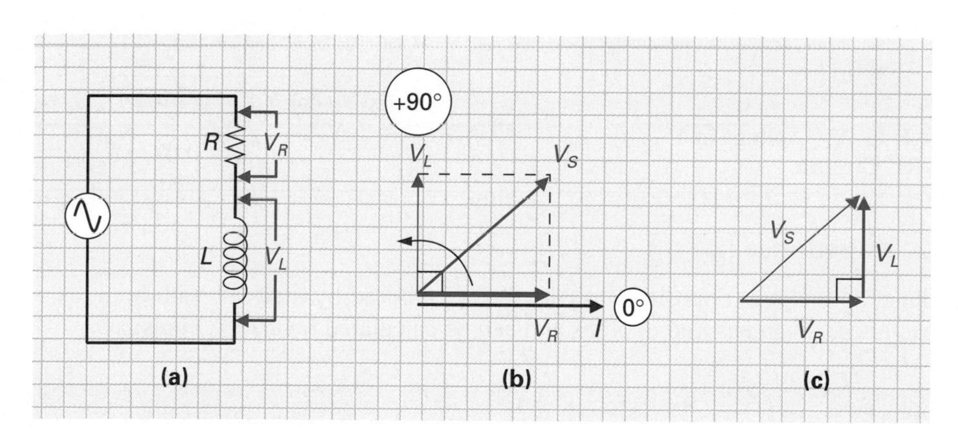

FIGURE 11-22 Series RL Circuit.

The vector diagram in Figure 11-22(b) illustrates current (I) as the reference at 0°, and, as expected, the voltage across the resistor (V_R) is in phase with the circuit current. The voltage across the inductor (V_L) leads the circuit current and the voltage across the resistor (V_R) by 90°, or the circuit current vector lags the voltage across the inductor by 90°.

As with any series circuit, we have to apply Kirchhoff's voltage law when calculating the value of applied or source voltage (V_S), which, due to the phase difference between V_R and V_L, is the vector sum of all the voltage drops. By creating a right triangle from the three quantities, as shown in Figure 11-22(c), and applying the Pythagorean theorem, we arrive at a formula for source voltage.

$$V_S = \sqrt{V_R^2 + V_L^2}$$

As with any formula with three quantities, if two are known, the other can be calculated by simply rearranging the formula to

$$V_S = \sqrt{V_R^2 + V_L^2}$$

$$V_R = \sqrt{V_S^2 - V_L^2}$$

$$V_L = \sqrt{V_S^2 - V_R^2}$$

EXAMPLE:

Calculate V_R, V_L, and V_s for the circuit shown in Figure 11-23.

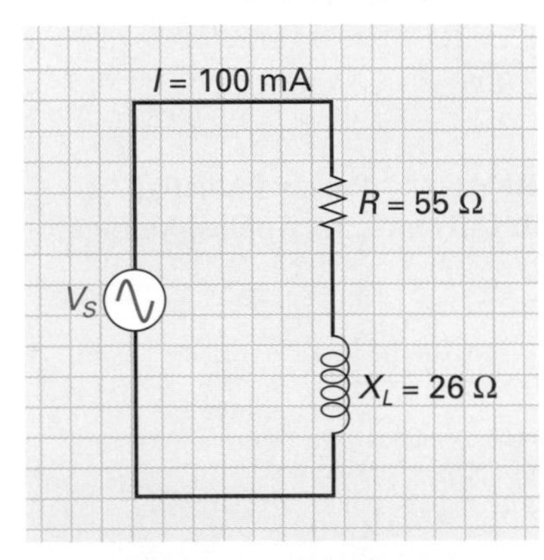

FIGURE 11-23 Voltage in a Series RL Circuit.

Solution:

$$\begin{aligned} V_R &= I \times R \\ &= 100 \text{ mA} \times 55 \ \Omega \\ &= 5.5 \text{ V} \\ V_L &= I \times X_L \\ &= 100 \text{ mA} \times 26 \ \Omega \\ &= 2.6 \text{ V} \\ V_S &= \sqrt{V_R^2 + V_L^2} \\ &= \sqrt{(5.5 \text{ V})^2 + (2.6 \text{ V})^2} \\ &= 6 \text{ V} \end{aligned}$$

2. Impedance (Z)

Impedance is the total opposition to current flow offered by a circuit with both resis-tance and reactance. It is measured in ohms and can be calculated by using Ohm's law.

$$Z = \frac{V}{I}$$

Just as a phase shift or difference exists between V_R and V_L and they cannot be added to find applied voltage, the same phase difference exists between R and X_L, so impedance or total opposition cannot be simply the sum of the two, as shown in Figure 11-24. The impedance of a series *RL* circuit is equal to the square root of the sum of the squares of resistance and reactance, and by rearrangement, X_L and R can also be calculated if the other two values are known.

$$Z = \sqrt{R^2 + X_L^2}$$

$$R = \sqrt{Z^2 - X_L^2}$$

$$X_L = \sqrt{Z^2 - R^2}$$

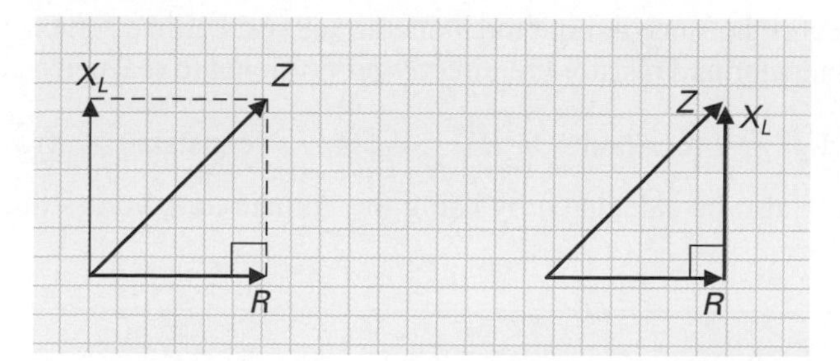

FIGURE 11-24 Impedance in a Series *RL* Circuit.

EXAMPLE:

Referring back to Figure 11-23, calculate *Z*.

Solution:

$$Z = \sqrt{R^2 + X_L{}^2}$$
$$= \sqrt{55^2 + 26^2}$$
$$= 60.8\ \Omega$$

3. Phase Shift

If a circuit is purely resistive, the phase shift (θ) between the source voltage and circuit current is zero. If a circuit is purely inductive, voltage leads current by 90°; therefore, the phase shift is +90°. If the resistance and inductive reactance are equal, the phase shift will equal +45°, as shown in Figure 11-25.

The phase shift in an inductive and resistive circuit is the degrees of lead between the source voltage (V_S) and current (I), and by looking at the examples in Figure 11-25, you can see that the phase angle is proportional to reactance and inversely proportional to resistance. Mathematically, it can be expressed as

$$\theta = \arctan \frac{X_L}{R}$$

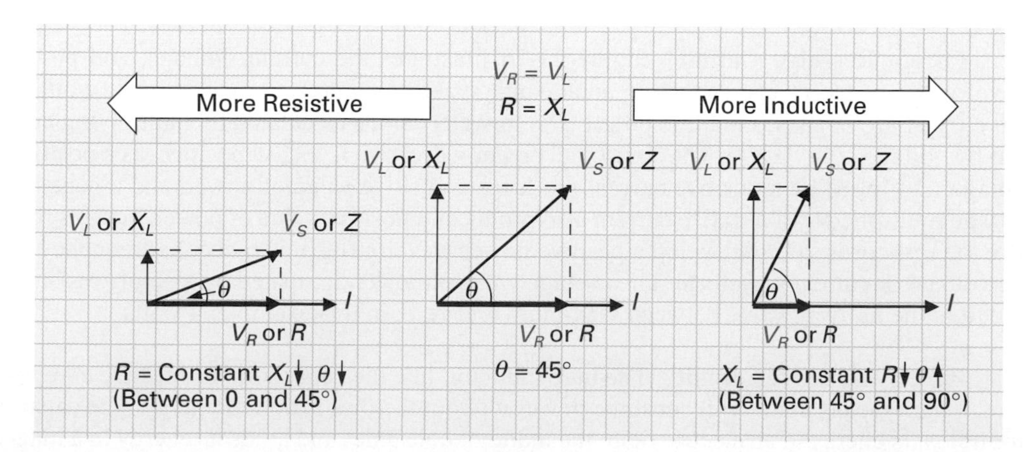

FIGURE 11-25 Phase Shift (θ) in a Series *RL* Circuit.

TIME LINE

Most of the 8-bit computers of the late 1970s used an operating system developed by Gary Kildall in 1974 called CP/M, standing for "Control Program for Microcomputers." Just before IBM officially entered into the personal computer market in August of 1981, two IBM executives went to a prearranged meeting with Kildall, who was at the time flying his private plane. After being contacted by radio, Kildall declined to come down and discuss an operating system for their proposed personal computer or "PC."

Rescheduling the meeting, the executives were surprised to discover that Kildall had flown off to the Caribbean for a vacation. When he came back, IBM had made a deal with Bill Gates of Microsoft for an operating system. Called MS-DOS, which stood for "Microsoft—Disk Operating System" (DOS means an operating system that is loaded into the computer from a disk), it became an instant success largely due to its blessing from IBM. Hundreds of computer manufacturers started producing computers that were compatible with the IBM PC, and therefore used Gates's MS-DOS. Even in an expanding market, Kildall's CP/M could not hold on to any business and is today extinct.

As the current is the same in both the inductor and resistor in a series circuit, the voltage drops across the inductor and resistor are directly proportional to reactance and resistance:

$$V_R\updownarrow = I\,(\text{constant}) \times R\updownarrow \qquad V_L\updownarrow = I\updownarrow\,(\text{constant}) \times X_L\updownarrow$$

The phase shift can also be calculated by using the voltage drop across the inductor and resistor.

$$\theta = \arctan \frac{V_L}{V_R}$$

EXAMPLE:

Referring back to Figure 11-23, calculate the phase shift between source voltage and circuit current.

Solution:

Since the ratio of X_L/R and V_L/V_R are known for this example circuit, either of the phase shift formulas can be used.

$$\begin{aligned}\theta &= \arctan \frac{X_L}{R} \\ &= \arctan \frac{26\ \Omega}{55\ \Omega} \\ &= \arctan 0.4727 \\ &= 25.3°\end{aligned}$$

or

$$\begin{aligned}\theta &= \arctan \frac{V_L}{V_R} \\ &= \arctan \frac{2.6\ \text{V}}{5.5\ \text{V}} \\ &= \arctan 0.4727 \\ &= 25.3°\end{aligned}$$

The source voltage in this example circuit leads the circuit current by 25.3°.

4. Power

Purely Resistive Circuit Figure 11-26 illustrates the current, voltage, and power waveforms produced when applying an ac voltage across a purely resistive circuit. Voltage and current are in phase, and true power (P_R) in watts can be calculated by multiplying current by voltage ($P = V \times I$). The sinusoidal power waveform is totally positive, as a positive voltage multiplied by a positive current produces a positive value of power, and a negative voltage multiplied by a negative current also produces a positive value of power. For this reason, the resistor is said to develop a positive power waveform that is twice the frequency of the voltage or current waveform. The average value of power dissipated by a purely resistive circuit is the halfway value between maximum and zero, in this example, 4 watts.

Purely Inductive Circuit The pure inductor, like the capacitor, is a reactive component, which means that it will consume power without the dissipation of energy. The capacitor holds its energy in an electric field; the inductor consumes and holds its energy in a magnetic field and then releases it back into the circuit.

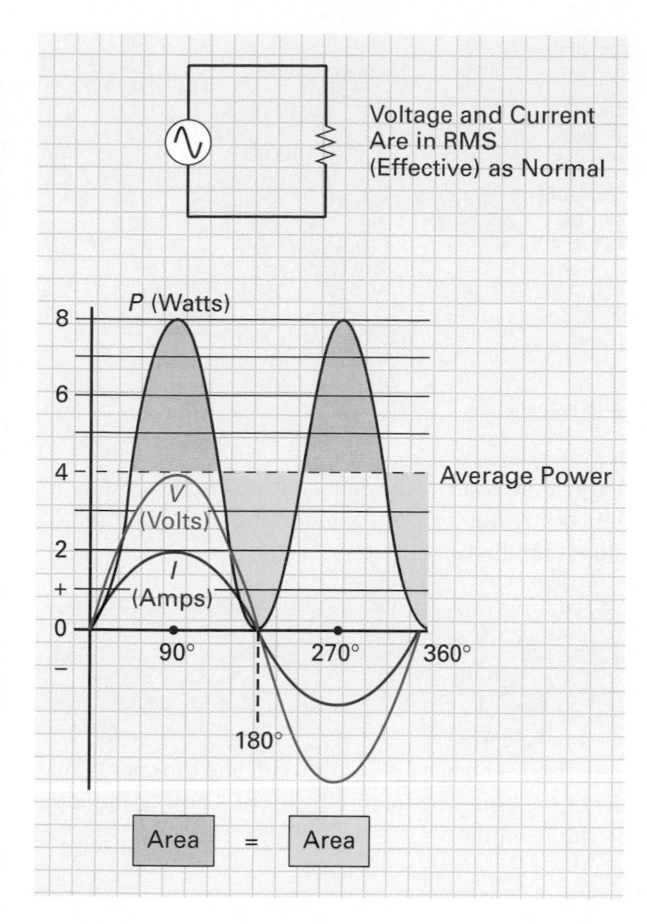

FIGURE 11-26 **Purely Resistive Circuit.**

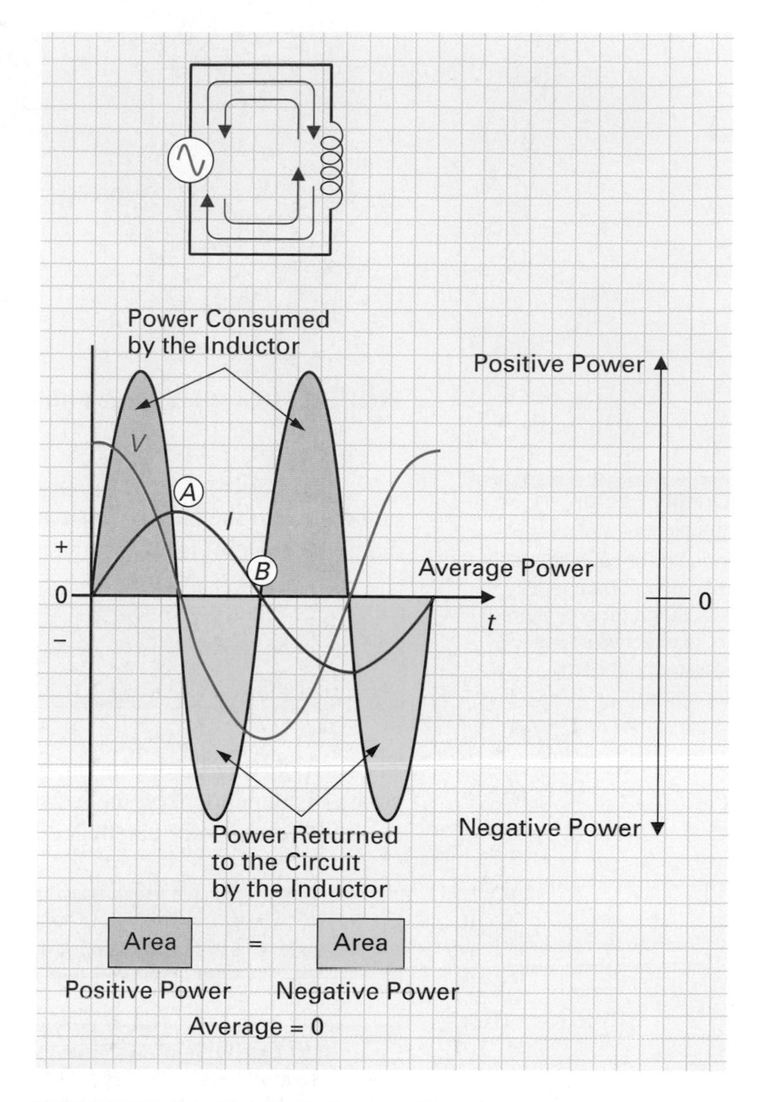

FIGURE 11-27 **Purely Inductive Circuit.**

The power curve alternates equally above and below the zero line, as seen in Figure 11-27. During the first positive power half-cycle, the circuit current is on the increase, to maximum (point A), the magnetic field is building up, and the inductor is storing electrical energy. When the circuit current is on the decline between A and B, the magnetic field begins to collapse and self-induction occurs and returns electrical energy back into the circuit. The power alternation is both positive when the inductor is consuming power and negative when the inductor is returning the power back into the circuit. As the positive and negative power alternations are equal but opposite, the average power dissipated is zero.

Resistive and Inductive Circuit An inductor is different from a capacitor in that it has a small amount of resistance no matter how pure the inductor. For this reason, inductors will never have an average power figure of zero, because even the best inductor will have some value of inductance and resistance within it, as seen in Figure 11-28. The reason an inductor has resistance is that it is simply a piece of wire, and any piece of wire has a certain value of resistance as well as inductance. This coil resistance should be, and normally is, very small and can usually be ignored; however, in some applications even this small resistance can prevent the correct operation of a circuit, so a value or term had to be created to specify the differences in the quality of inductor available.

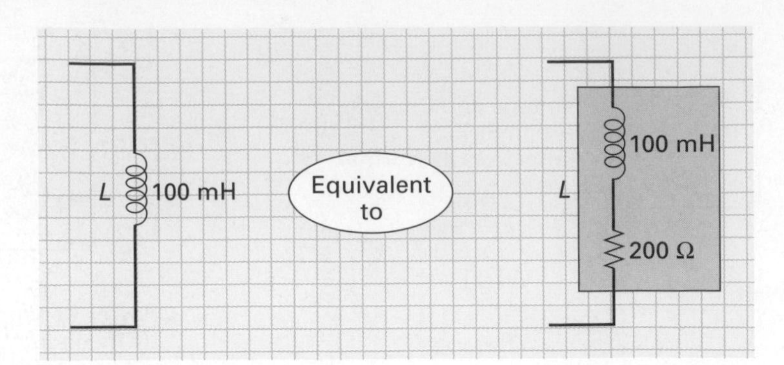

FIGURE 11-28 Resistance within an Inductor.

Quality Factor
Quality factor of an inductor or capacitor is the ratio of a component's reactance (energy stored) to its effective series resistance (energy dissipated).

The **quality factor** (Q) of an inductor is the ratio of the energy stored in the coil by its inductance to the energy dissipated in the coil by the resistance; therefore, the higher the Q, the better the coil is at storing energy rather than dissipating it.

$$\text{Quality } (Q) = \frac{\text{energy stored}}{\text{energy dissipated}}$$

The energy stored is dependent on the inductive reactance (X_L) of the coil, and the energy dissipated is dependent on the resistance (R) of the coil. The quality factor of a coil or inductor can therefore also be calculated by using the formula

$$Q = \frac{X_L}{R}$$

EXAMPLE:

Calculate the quality factor Q of a 22 mH coil connected across a 2 kHz, 10 V source if its internal coil resistance is 27 Ω.

Solution:

$$Q = \frac{X_L}{R}$$

The reactance of the coil is not known but can be calculated by the formula

$$\begin{aligned} X_L &= 2\pi f L \\ &= 2\pi \times 2 \text{ kHz} \times 22 \text{ mH} = 276.5 \ \Omega \end{aligned}$$

Therefore,

$$Q = \frac{X_L}{R} = \frac{276.5 \ \Omega}{27 \ \Omega} = 10.24$$

Inductors, therefore, will never appear as pure inductance, but rather as an inductive and resistive (RL) circuit, and the resistance within the inductor will dissipate *true power.* Figure 11-29 illustrates a circuit containing R and L and the power waveforms produced when $R = X_L$; the phase shift (θ) is equal to 45°. The positive power alternation, which is above the zero line, is the combination of the power dissipated by the resistor and the power consumed by the inductor while circuit current was on the rise. The negative power alterna-

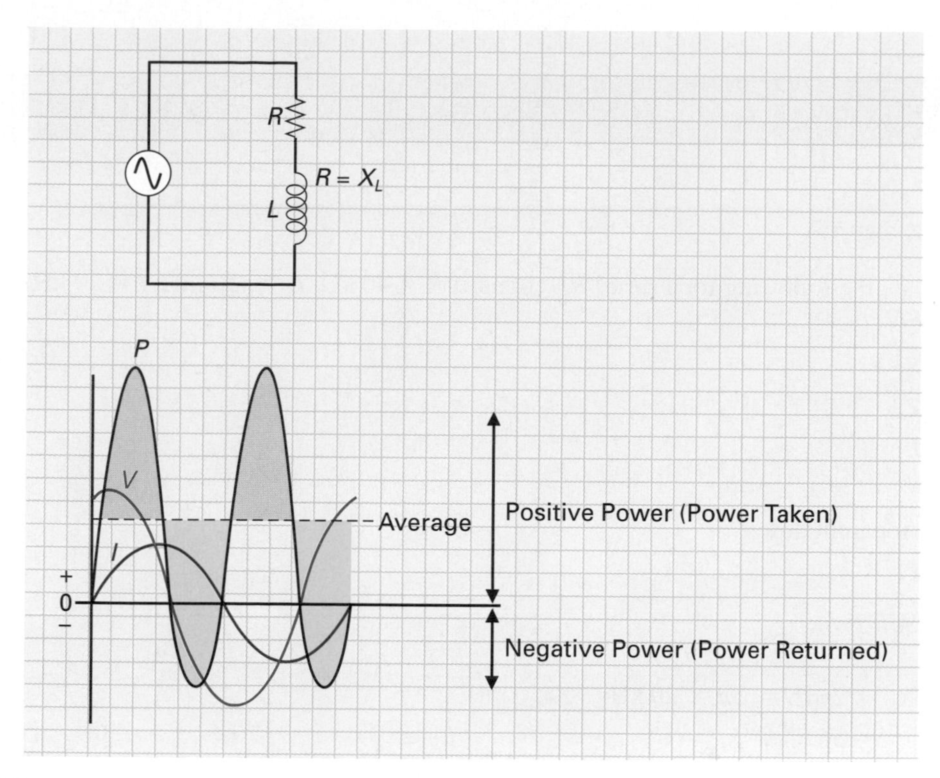

FIGURE 11-29 Power in a Resistive and Inductive Circuit.

tion is the power that was given back to the circuit by the inductor while the inductor's magnetic field was collapsing and returning the energy that was consumed.

Power Factor When a circuit contains both inductance and resistance, some of the energy is consumed and then returned by the inductor (reactive or imaginary power), and some of the energy is dissipated and lost by the resistor (resistive or true power). Apparent power is the power that appears to be supplied to the load and is the vector sum of both the reactive and true power; it can be calculated by using the formula

$$P_A = \sqrt{P_R{}^2 + P_X{}^2}$$

where P_A = apparent power, in volt-amperes (VA)
P_R = true power, in watts (W)
P_X = reactive power, in volt-amperes reactive (VAR)

The power factor is a ratio of the true power to the apparent power and is therefore a measure of the loss in a circuit.

$$\text{PF} = \frac{\text{true power } (P_R)}{\text{apparent power } (P_A)}$$

or

$$\text{PF} = \frac{R}{Z}$$

or

$$\text{PF} = \cos\theta$$

EXAMPLE:

Calculate the following for a series *RL* circuit if $R = 40\ \text{k}\Omega$, $L = 450\ \text{mH}$, $f = 20\ \text{kHz}$, and $V_S = 6\ \text{V}$:

a. X_L
b. Z
c. I
d. θ
e. Apparent power
f. PF

Solution:

a. $X_L = 2\pi fL = 2\pi \times 20\ \text{kHz} \times 450\ \text{mH}$
$= 56.5\ \text{k}\Omega$

b. $Z = \sqrt{R^2 + X_L{}^2}$
$= \sqrt{(40\ \text{k}\Omega)^2 + (56.5\ \text{k}\Omega)^2} = 69.23\ \text{k}\Omega$

c. $I = \dfrac{V_S}{Z} = \dfrac{6\ \text{V}}{69.23\ \text{k}\Omega} = 86.6\ \mu\text{A}$

d. $\theta = \arctan\dfrac{X_L}{R} = \arctan\dfrac{56.5\ \text{k}\Omega}{40\ \text{k}\Omega} = 54.7°$

e. Apparent power $= \sqrt{(\text{true power})^2 + (\text{reactive power})^2}$

$P_R = I^2 \times R = (86.6\ \mu\text{A})^2 \times 40\ \text{k}\Omega = 300\ \mu\text{W}$

$P_X = I^2 \times X_L = (86.6\ \mu\text{A})^2 \times 56.5\ \text{k}\Omega = 423.7\ \mu\text{VAR}$

$P_A = \sqrt{P_R{}^2 + P_X{}^2}$
$= \sqrt{(300\ \mu\text{W})^2 + (423.7\ \mu\text{W})^2} = 519.2\ \mu\text{VA}$

f. $\text{PF} = \dfrac{P_R}{P_A} = \dfrac{300\ \mu\text{W}}{519.2\ \mu\text{W}} = 0.57$
$= \dfrac{R}{Z} = \dfrac{40\ \text{k}\Omega}{69.23\ \text{k}\Omega} = 0.57$
$= \cos\theta = \cos 54.7° = 0.57$

11-4-3 *Parallel* RL *Circuit*

Now that we have seen the behavior of resistors and inductors in series, let us analyze the parallel *RL* circuit.

1. Current

In Figure 11-30(a) you will see a parallel combination of a resistor and inductor. The voltage across both components is equal because of the parallel connection. and the current through each branch is calculated by applying Ohm's law.

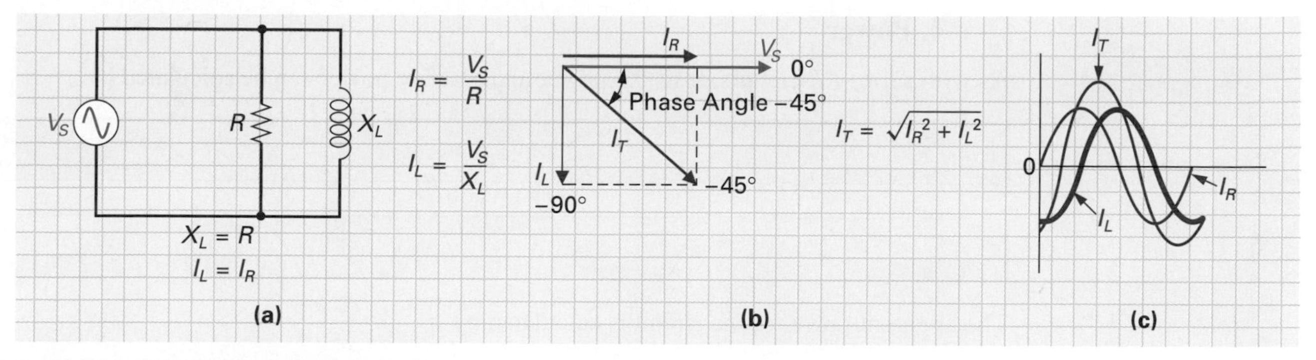

FIGURE 11-30 Parallel *RL* Circuit.

$$\text{(Resistor current) } I_R = \frac{V_S}{R}$$

$$\text{(Inductor current) } I_L = \frac{V_S}{X_L}$$

Total current (I_T) is equal to the vector combination of the resistor current and inductor current, as shown in Figure 11-30(b). Figure 11-30(c) illustrates the current waveforms.

$$I_T = \sqrt{I_R^2 + I_L^2}$$

2. Phase Angle

The angle by which the total current (I_T) leads the source voltage (V_S) can be determined with either of the following formulas:

$$\theta = \arctan \frac{I_L}{I_R}$$

or

$$\theta = \arctan \frac{R}{X_L}$$

3. Impedance

The total opposition or impedance of a parallel *RL* circuit can be calculated by

$$Z = \frac{V_S}{I_T}$$

Using basic algebra, this formula can be rearranged to express impedance in terms of reactance and resistance.

$$Z = \frac{R \times X_L}{\sqrt{R^2 + X_L^2}}$$

4. Power

As with series *RL* circuits, resistive power and reactive power can be calculated by

$$P_R = I_R^2 \times R$$

$$P_X = I_L^2 \times X_L$$

The apparent power of the circuit is calculated by

$$P_A = \sqrt{P_R^2 + P_X^2}$$

and, finally, the power factor is equal to

$$\text{PF} = \cos \theta = \frac{P_R}{P_A}$$

EXAMPLE:

Calculate the following for a parallel *RL* circuit if $R = 45\ \Omega$, $X_L = 1100\ \Omega$, and $V_S = 24$ V:

a. I_R
b. I_L
c. I_T
d. Z
e. θ

Solution:

a. $I_R = \frac{V_S}{R} = \frac{24\text{ V}}{45\ \Omega} = 533.3\text{ mA}$

b. $I_L = \frac{V_S}{X_L} = \frac{24\text{ V}}{1100\ \Omega} = 21.8\text{ mA}$

c. $I_T = \sqrt{I_R^2 + I_L^2}$

$= \sqrt{(533.3\text{ mA})^2 + (21.8\text{ mA})^2} = 533.7\text{ mA}$

d. $Z = \frac{R \times X_L}{\sqrt{R^2 + X_L^2}} = \frac{45\ \Omega \times 1100\ \Omega}{\sqrt{(45\ \Omega)^2 + (1100\ \Omega)^2}}$

$= \frac{49.5\text{ k}\Omega}{1100.9\ \Omega} = 44.96\ \Omega$

e. $\theta = \arctan \frac{R}{X_L} = \arctan \frac{45\ \Omega}{1100\ \Omega} = 2.34°$

Therefore, I_T lags V_S by 2.34°.

SELF-TEST EVALUATION POINT FOR SECTION 11-4

Now that you have completed this section, you should be able to:

Objective 8. Give the formula for inductive reactance.

Objective 9. Describe all aspects relating to a series and parallel *RL* circuit.

Use the following questions to test your understanding of Section 11-4:

1. Define and state the formula for inductive reactance.
2. Why is inductive reactance proportional to frequency and the inductance value?
3. How does inductive reactance relate to Ohm's law?
4. True or false: Inductive reactance is measured in henrys.
5. True or false: In a purely inductive circuit, the current will lead the applied voltage by 90°.
6. Calculate the applied source voltage V_S in an R_L circuit where $V_R = 4$ V and $V_L = 2$ V.
7. Define and state the formula for impedance when R and X_L are known.
8. If $R = X_L$, the phase shift will equal ___________ .
9. What is positive power?
10. Define Q and state the formula when X_L and R are known.
11. What is the difference between true and reactive power?
12. State the power factor formula.
13. True or false: When R and L are connected in parallel with one another, the voltage across each will be different and dependent on the values of R and X_L.
14. State the formula for calculating total current (I_T) when I_R and I_L are known.

11-5 APPLICATIONS OF INDUCTORS

11-5-1 RL *Filters*

The ***RL* filter** will achieve results similar to the *RC* filter in that it will pass some frequencies and block others, as seen in Figure 11–31. The inductive reactance of the coil and the resistance of the resistor form a voltage divider. Since inductive reactance is proportional to frequency ($X_L \propto f$), the inductor will drop less voltage at lower frequencies ($f\downarrow, X_L\downarrow, V_L\downarrow$) and more voltage at higher frequencies ($f\uparrow, X_L\uparrow, V_L\uparrow$).

***RL* Filter**
A selective circuit of resistors and inductors that offers little or no opposition to certain frequencies while blocking or attenuating other frequencies.

With the low-pass filter shown in Figure 11-31(a), the output is developed across the resistor. If the frequency of the input is low, the inductive reactance will be low, so almost all the input will be developed across the resistor and applied to the output. If the frequency of the input increases, the inductor's reactance will increase, resulting in almost all the input being dropped across the inductor and none across the resistor and therefore the output.

With the high-pass filter, shown in Figure 11-31(b), the inductor and resistor have been placed in opposite positions. If the frequency of the input is low, the inductive reactance will be low, so almost all the input will be developed across the resistor and very little will appear across the inductor and therefore at the output. If the frequency of the input is high, the inductive reactance will be high, resulting in almost all of the input being developed across the inductor and therefore appearing at the output.

11-5-2 RL *Integrator*

In the preceding section on *RL* filters we saw how an *RL* circuit reacted to a sine-wave input of different frequencies. In this section and the following we will see how an *RL* circuit will react to a square-wave input and show two other important applications of inductors. In the ***RL* integrator,** the output is taken across the resistor, as seen in the circuit in Figure 11-32(a). The output shown in Figure 11-32(b) is the same as the previously described *RC* integrator's output. When the input rises from 0 to 10 V at the leading positive edge of the square-wave input, the situation is as seen in Figure 11-33(a).

***RL* Integrator**
An RL circuit with an output proportionate to the integral of the input signal.

The inductor's 10 V counter emf opposes the sudden input change from 0 to 10 V, and if 10 V is across the inductor, 0 V must be across the resistor (Kirchhoff's voltage law)

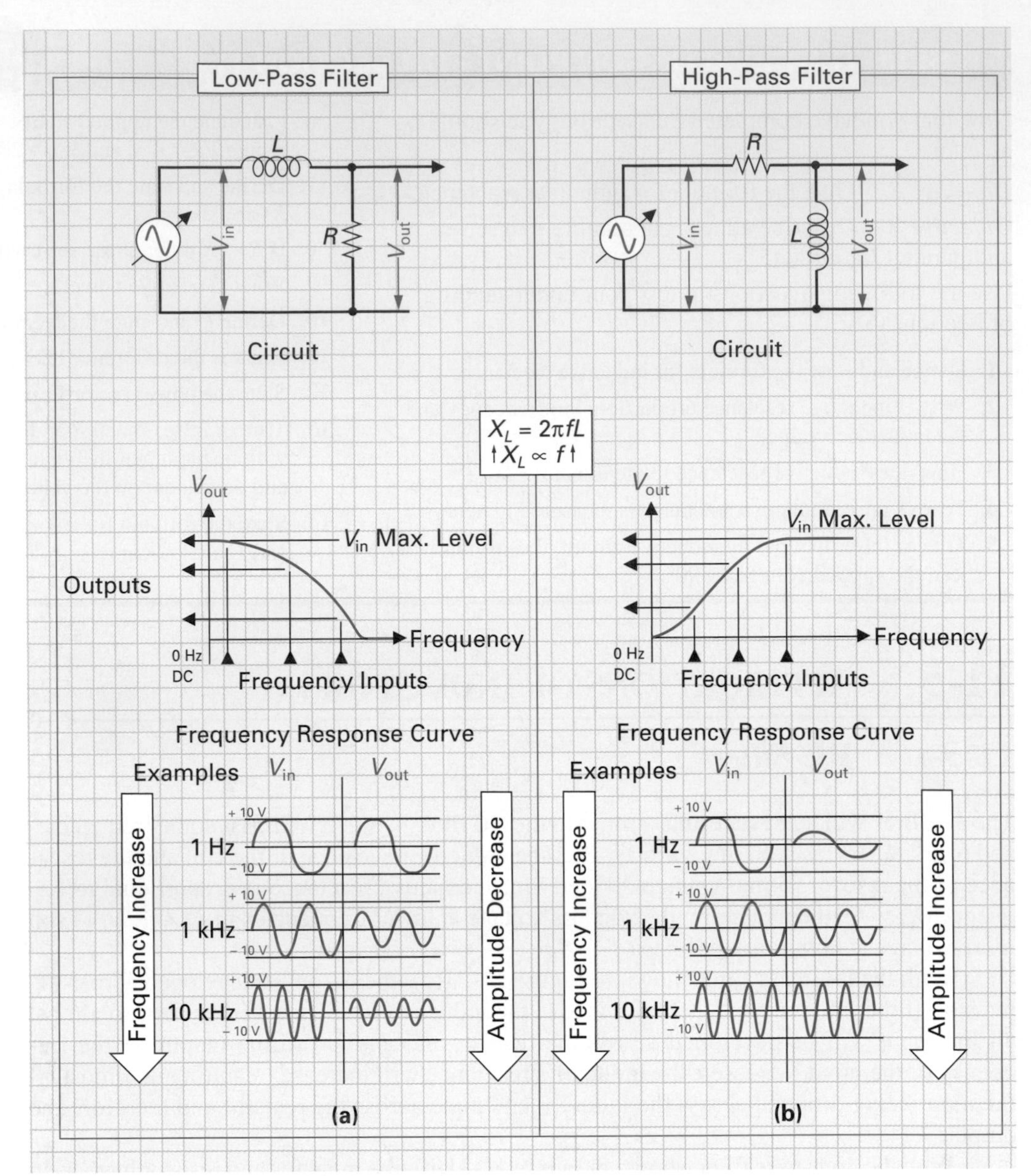

FIGURE 11-31 ***RL*** **Filter. (a) Low-Pass Filter. (b) High-Pass Filter.**

and therefore appear at the output. After five time constants ($5 \times L/R$), the inductor's current in the circuit will have built up to maximum (V_{in}/R), and the inductor will be an equivalent short circuit, because no change and consequently no back emf exist, as shown in Figure 11-33(b). All the input voltage will now be across the resistor and therefore at the output.

When the square-wave input drops to zero, the circuit is equivalent to that seen in Figure 11-33(c), and the collapsing magnetic field will cause an induced voltage within the conductor, which will cause current to flow within the circuit for five time constants, whereupon it will reach 0.

As with the *RC* integrator, if the period of the square wave is decreased or if the time constant is increased, the output will reach an average value of half the input square wave's amplitude, as shown in Figure 11-34.

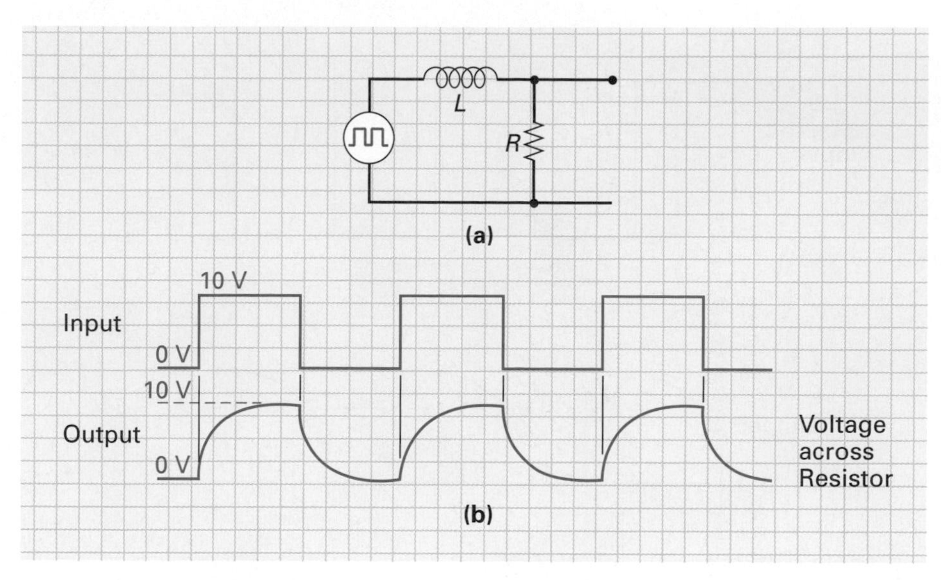

FIGURE 11-32 ***RL* Integrator.**

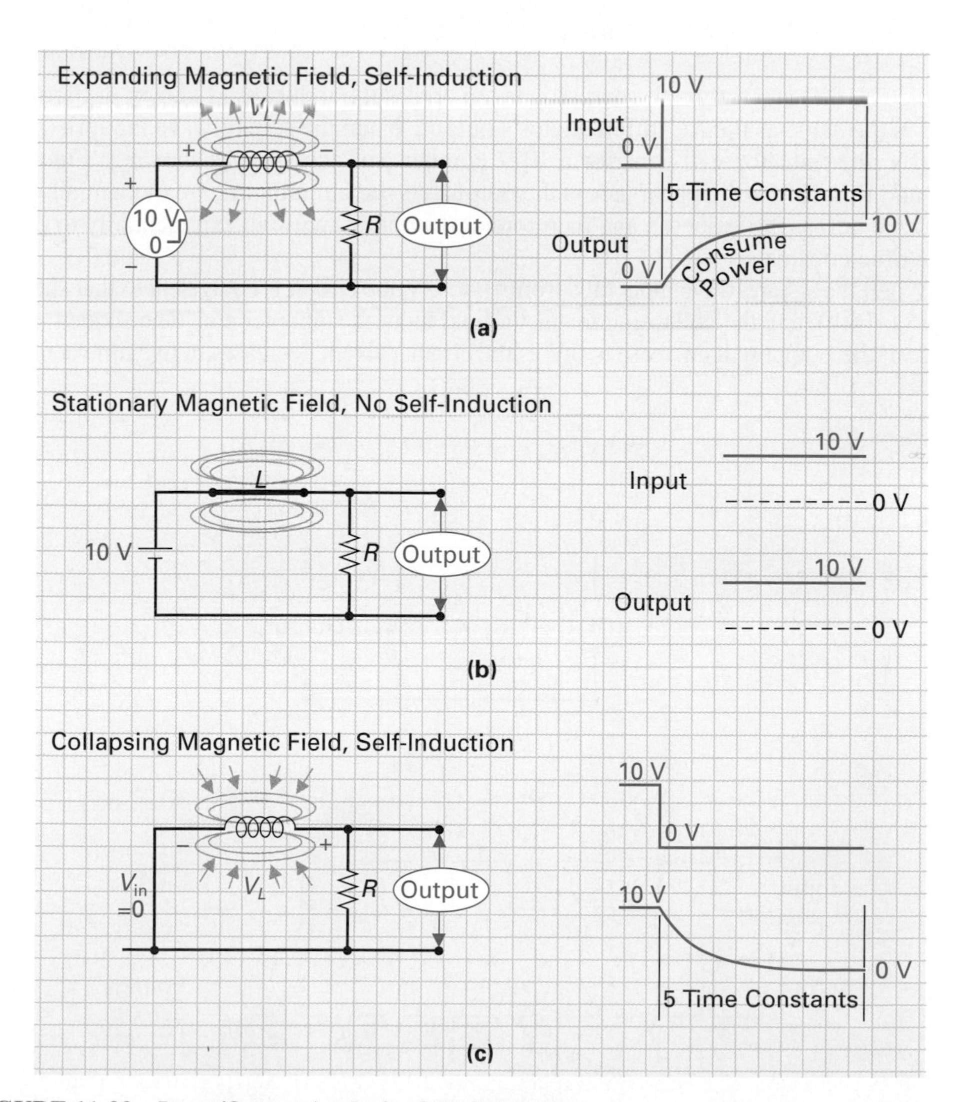

FIGURE 11-33 **Input/Output Analysis of *RL* Integrator.**

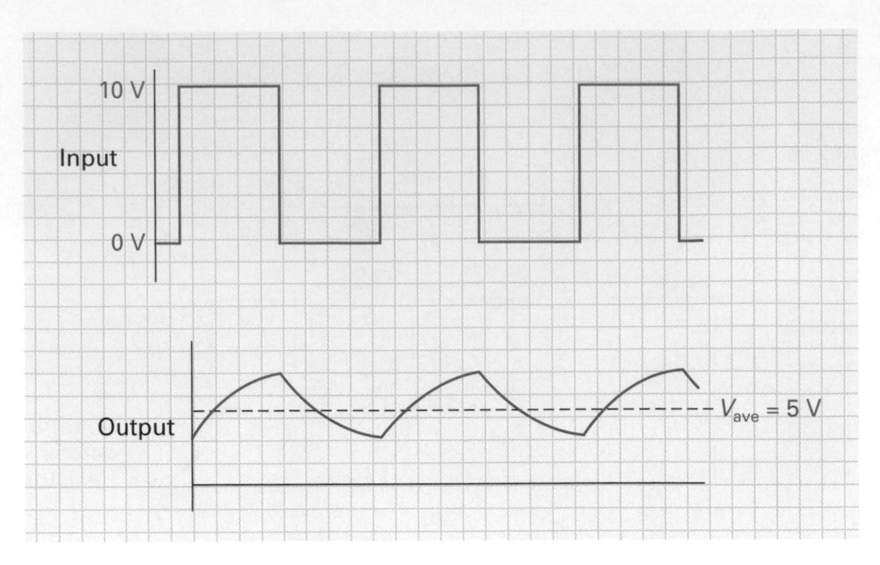

FIGURE 11-34

11-5-3 RL *Differentiator*

***RL* Differentiator**
An *RL* circuit whose output voltage is proportional to the rate of change of the input voltage.

With the ***RL* differentiator,** the output is taken across the inductor, as shown in Figure 11-35 and is the same as the *RC* differentiator's output. When the square-wave input rises from 0 to 10 V, the inductor will generate a 10 V counter emf across it, as shown in Figure 11-36(a), and this 10 V will appear across the output. As the circuit current exponentially rises, the voltage across the inductor, and therefore at the output, will fall exponentially to 0 V after five time constants.

When the square-wave input falls from 10 to 0 V, the circuit is equivalent to that shown in Figure 11-36(b), and the collapsing magnetic field induces a counter emf. The sudden −10 V impulse at the output will decrease to 0 V as the circuit current decreases in five time constants.

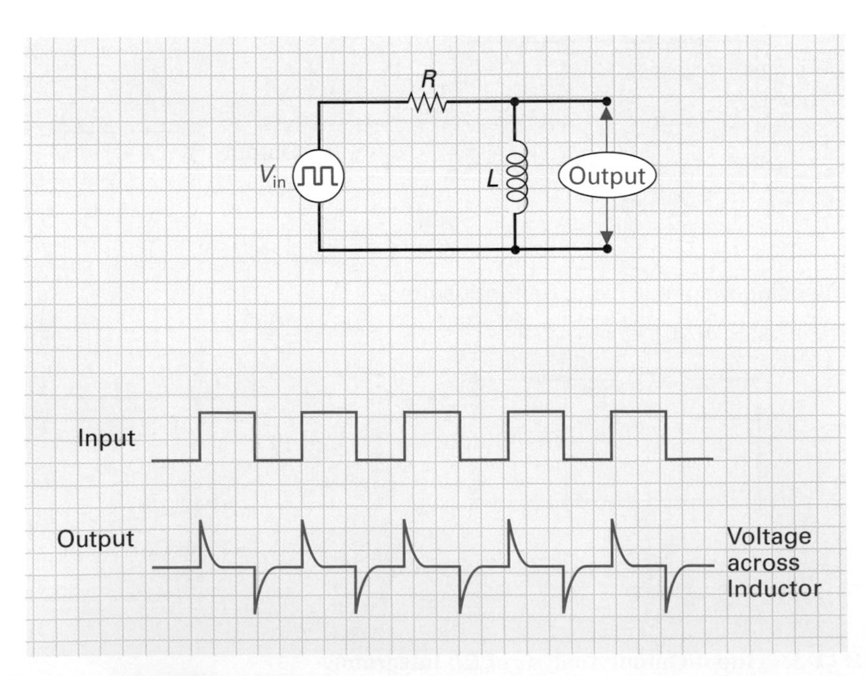

FIGURE 11-35 ***RL* Differentiator.**

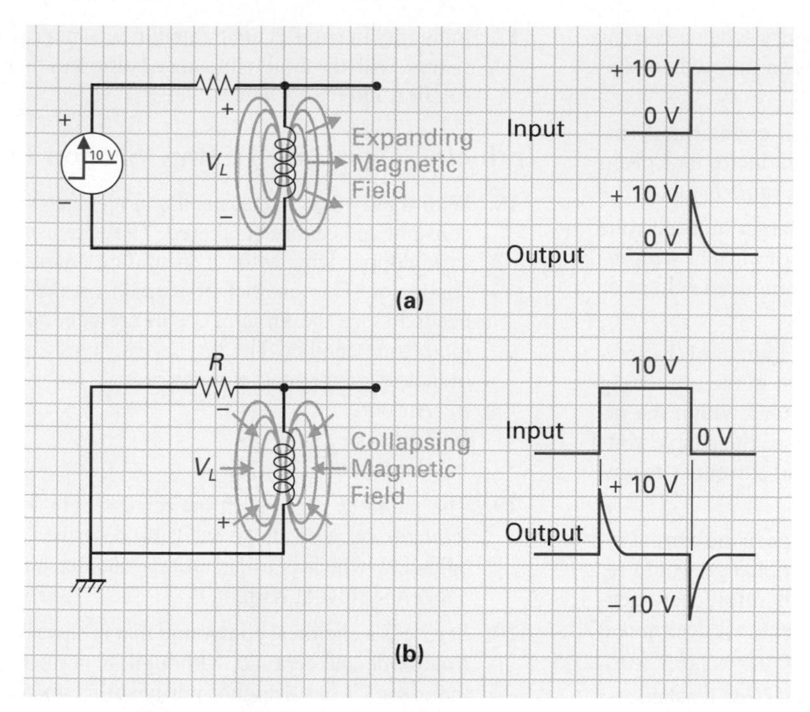

FIGURE 11-36 Input/Output Analysis of *RL* Differentiator.

SELF-TEST EVALUATION POINT FOR SECTION 11-5

Now that you have completed this section, you should be able to:

Objective 10. Describe how inductors can be used for the following applications:

a. *RL* filter
b. *RL* integrator
c. *RL* differentiator

Use the following questions to test your understanding of Section 11-5:

1. List three circuit applications of the inductor.
2. Will an *RL* filter act as a low-pass filter or high-pass filter, or can it be made to function as either?
3. What are the waveform differences between an integrator and differentiator?

SUMMARY

Inductor Characteristics (Section 11-1)

1. Inductance, by definition, is the ability of a device to oppose a change in current flow, and the device designed specifically to achieve this function is called an *inductor*.
2. There is, in fact, no physical difference between an inductor and an electromagnet, since they are both coils. Even though their construction and principle of operation are the same, the two devices are given different names because they are used in different applications. An electromagnet is used to generate a magnetic field in response to current, whereas an inductor is used to oppose any change in current.
3. When an external voltage source is connected across a coil, it will force a current through the coil. This current will generate a magnetic field that will increase in field strength in a short time from zero to maximum, expanding from the center of the conductor (electromagnetism). The expanding magnetic lines of force have relative motion with respect to the stationary conductor, so an induced voltage results (electromagnetic induction). The blooming magnetic field generated by the conductor is actually causing a voltage to be induced in the conductor that is generating the magnetic field. This effect of a current-carrying coil of conductor inducing a voltage within itself is known as self-inductance.
4. The induced voltage is called a *counter emf* or *back emf.* It opposes the applied emf (or battery voltage). The ability of a coil or conductor to induce or produce a counter emf within itself as a result of a change in current is called *self-inductance,* or more commonly inductance (symbolized *L*).
5. The unit of inductance is the henry (H), named in honor of Joseph Henry, an American physicist, for his experimentation within this area of science.

6. The inductance of an inductor is 1 henry when a current change of 1 ampere per second causes an induced voltage of 1 volt.
7. Inductance is a measure of how much counter emf (induced voltage) can be generated by an inductor for a given amount of current change through that same inductor.
8. A larger inductance ($L\uparrow$) will create a larger induced voltage ($V_{ind}\uparrow$), and if the rate of change of current with respect to time is increased ($\Delta I/\Delta t\uparrow$), the induced voltage or counter emf will also increase ($V_{ind}\uparrow$).
9. If the current in the circuit is suddenly increased (by lowering the circuit resistance), the change in the expanding magnetic field will induce a counter emf within the inductor. This induced voltage will oppose the source voltage from the battery and attempt to hold current at its previous low level.
10. The counter emf cannot completely oppose the current increase, for if it did, the lack of current change would reduce the counter emf to zero. Current therefore incrementally increases up to a new maximum, which is determined by the applied voltage and the circuit resistance ($I = V/R$).
11. Once the new higher level of current has been reached and remains constant, there will no longer be a change. This lack of relative motion between field and conductor will no longer generate a counter emf, so the current will remain at its new higher constant value.
12. This effect also happens in the opposite respect. If current decreases (by increasing circuit resistance), the magnetic lines of force will collapse because of the reduction of current and induce a voltage in the inductor, which will produce an induced current in the same direction as the circuit current. These two combine and tend to maintain the current at the higher previous constant level. Circuit current will fall, however, as the induced voltage and current are only present during the change. Once the new lower level of current has been reached and remains constant, the lack of change will no longer induce a voltage or current. So the current will then remain at its new lower constant value.
13. The inductor is an electronic component that will oppose any changes in circuit current, and this ability or behavior is referred to as *inductance.* Since the current change is opposed by a counter emf, inductance may also be defined as the ability of a device to induce a counter emf within itself for a change in current.
14. The inductance of an inductor is determined by four factors:
    a. Number of turns
    b. Area of the coil
    c. Length of the coil
    d. Core material used within the coil

    Inductance (L) is proportional to the number of turns (N), proportional to the cross-sectional area of the coil (A), inversely proportional to the length of the coil (l), and proportional to the permeability (μ) of the core material used.
15. Inductors oppose the change of current in a circuit and so are treated in a manner similar to resistors connected in combination.
16. Two or more inductors in series merely extend the coil length and increase inductance.
17. Inductors in parallel are treated in a manner similar to resistors, with the total inductance being less than that of the smallest inductor's value.

Inductor Types and Testing (Section 11-2)

18. With fixed-value inductors, the inductance value remains constant and cannot be altered. They are usually made of solid copper wire with an insulating enamel around the conductor. This enamel or varnish on the conductor is needed since the turns of a coil are generally wound on top of one another. The three major types of fixed-value inductors on the market today are
    a. Air core
    b. Iron core
    c. Ferrite core
19. The name *choke* is used interchangeably with the word *inductor,* because choke basically describes an inductors' behavior, which is to generate a counter emf to limit or choke the flow of current.
20. Ferrite is a chemical compound that is basically powdered iron oxide and ceramic. Although its permeability is less than iron, it has a higher permeability than air and can therefore obtain a higher inductance value than air for the same number of turns.
21. The toroidal (doughnut-shaped) inductor has an advantage over the cylinder-shaped core in that the magnetic lines of force do not have to pass through air. With a cylindrical core, the magnetic lines of force merge out of one end of the cylinder, propagate or travel through air, and then enter into the other end of the cylinder.
22. With variable-value inductors, the inductance value can be changed by adjusting the position of the movable core with respect to the stationary coil.
23. Variable inductors should only be adjusted with a plastic or nonmetallic alignment tool since any metal object in the vicinity of the inductor will interfere with the magnetic field and change the inductance of the inductor.
24. Basically, only three problems can occur with inductors:
    a. An open
    b. A complete short
    c. A section short (value change)
25. Depending on the winding's resistance, the coil should be in the range of zero to a few hundred ohms.
26. A coil with one or more shorted turns or a complete short can be checked with an ohmmeter and thought to be perfectly good because of the normally low resistance of a coil, as it is just a piece of wire.
27. An inductor analyzer can be used to check capacitance and inductance.

DC Inductive Circuits (Section 11-3)

28. Inductors will not have any effect on a steady value of direct current (dc) from a dc voltage source. If, however, the dc is changing (pulsating), the inductor will oppose the change whether it is an increase or decrease in direct current, because a change in current causes the magnetic

field to expand or contract, and in so doing it will cut the coil of the inductor and induce a voltage that will counter the applied emf.

29. Current in an inductive circuit cannot rise instantly to its maximum value, which is determined by Ohm's law ($I = V/R$). Current will in fact take time to rise to maximum, due to the inductor's ability to oppose change.
30. It will actually take five time constants (5τ) for the current in an inductive circuit to reach maximum value.
31. The constant to remember is 63.2%. In one time constant ($1 \times L/R$) the current in the RL circuit will have reached 63.2% of its maximum value. In two time constants ($2 \times L/R$), the current will have increased 63.2% of the remaining current, and so on, through five time constants.
32. When the inductor's dc source of current is removed, the magnetic field will collapse and cut the coils of the inductor, inducing a voltage and causing a current to flow in the same direction as the original source current. This current will exponentially decay, or fall from the maximum to zero level, in five time constants.

AC Inductive Circuits (Section 11-4)

33. If an alternating (ac) voltage is applied across an inductor, the inductor will continuously oppose the alternating current because it is always changing. The current in the circuit causes the magnetic field to expand and collapse and cut the conducting coils, resulting in an induced counter emf.
34. The voltage across the inductor (V_L) or counter emf leads the circuit current (I) by 90°.
35. Inductive reactance (X_L) is the opposition to current flow offered by an inductor without the dissipation of energy.
36. Inductive reactance is proportional to frequency ($X_L \propto f$) because a higher frequency (fast-switching current) will cause a greater amount of current change, and a greater change will generate a larger counter emf, which is an opposition or reactance against current flow. When 0 Hz is applied to a coil (dc), there exists no change, so the inductive reactance of an inductor to dc is zero ($X_L = 2\pi \times 0 \times L = 0$).
37. Inductive reactance is also proportional to inductance because a larger inductance will generate a greater magnetic field and subsequent counter emf, which is the opposition to current flow.
38. Ohm's law can be applied to inductive circuits just as it can be applied to resistive and capacitive circuits. The current flow in an inductive circuit (I) is proportional to the voltage applied (V) and inversely proportional to the inductive reactance (X_L).
39. In a purely resistive circuit, the current flowing within the circuit and the voltage across the resistor are in phase with one another.
40. In a purely inductive circuit, the current will lag the applied voltage by 90°.
41. The voltage across the resistor (V_R) is in phase with the circuit current. The voltage across the inductor (V_L) leads the circuit current and the voltage across the resistor (V_R) by 90°, or, the circuit current vector lags the voltage across the inductor by 90°.
42. As with any series circuit, we have to apply Kirchhoff's voltage law when calculating the value of applied or source voltage (V_S), which, due to the phase difference between V_R and V_L, is the vector sum of all the voltage drops.
43. Impedance is the total opposition to current flow offered by a circuit with both resistance and reactance. It is measured in ohms and can be calculated by using Ohm's law. The impedance of a series RL circuit is equal to the square root of the sum of the squares of resistance and reactance, and by rearrangement, X_L and R can also be calculated if the other two values are known.
44. If a circuit is purely resistive, the phase shift (θ) between the source voltage and circuit current is zero. If a circuit is purely inductive, voltage leads current by 90°; therefore, the phase shift is +90°. If the resistance and inductive reactance are equal, the phase shift will equal +45°.
45. The phase shift in an inductive and resistive circuit is the degrees of lead between the source voltage (V_S) and current (I). The phase angle is proportional to reactance and inversely proportional to resistance.
46. As the current is the same in both the inductor and resistor in a series circuit, the voltage drops across the inductor and resistor are directly proportional to reactance and resistance.
47. When applying an ac voltage across a purely resistive circuit, voltage and current are in phase, and true power (P_R) in watts can be calculated by multiplying current by voltage ($P = V \times I$).
48. The pure inductor, like the capacitor, is a reactive component, which means that it will consume power without the dissipation of energy. The capacitor holds its energy in an electric field; the inductor consumes and holds its energy in a magnetic field and then releases it back into the circuit. The power alternation is both positive when the inductor is consuming power and negative when the inductor is returning the power back into the circuit. As the positive and negative power alternations are equal but opposite, the average power dissipated is zero.
49. An inductor is different from a capacitor in that it has a small amount of resistance no matter how pure the inductor. For this reason, inductors will never have an average power figure of zero, because even the best inductor will have some value of inductance and resistance within it.
50. The reason an inductor has resistance is that it is simply a piece of wire, and any piece of wire has a certain value of resistance, as well as inductance.
51. The quality factor (Q) of an inductor is the ratio of the energy stored in the coil by its inductance to the energy dissipated in the coil by the resistance.
52. When a circuit contains both inductance and resistance, some of the energy is consumed and then returned by the inductor (reactive or imaginary power), and some of the energy is dissipated and lost by the resistor (resistive or true power).

53. Apparent power is the power that appears to be supplied to the load and is the vector sum of both the reactive and true power.

54. With a parallel combination of a resistor and inductor, the voltage across both components is equal because of the parallel connection, and the current through each branch is calculated by applying Ohm's law.

55. Total current (I_T) is equal to the vector combination of the resistor current and inductor current.

Applications of Inductors (Section 11–5)

56. The *RL* filter will achieve results similar to the *RC* filter in that it will pass some frequencies and block others.

57. In the *RL* integrator, the output is taken across the resistor.

58. With the *RL* differentiator, the output is taken across the inductor.

REVIEW QUESTIONS

Multiple-Choice Questions

1. Self-inductance is a process by which a coil will induce a voltage within __________.
a. Another inductor
b. Two or more inductors in close proximity
c. Itself
d. Both (a) and (b)

2. Inductance is a process by which a coil will induce a voltage within __________.
a. Another inductor
b. Two or more inductors in close proximity
c. Itself
d. Both (a) and (b)

3. The inductor stores electrical energy in the form of a __________ field, just as a capacitor stores electrical energy in the form of a __________ field.
a. Electric, magnetic
b. Magnetic, electric

4. The inductor is basically
a. An electromagnet.
b. A coil of wire.
c. A coil of conductor formed around a core material.
d. All of the above.
e. None of the above.

5. The inductance. of an inductor is proportional to __________ and inversely proportional to __________.
a. $N, A, \mu; l$ **c.** $\mu, l, N; A$
b. $A, \mu\, l; N$ **d.** $N, A, l; \mu$

6. The total inductance of a series circuit is
a. Less than the value of the smallest inductor.
b. Equal to the sum of all the inductance values.
c. Equal to the product over sum.
d. All of the above.

7. The total inductance of a parallel circuit can be calculated by
a. Using the product-over-sum formula.
b. Using *L* divided by *N* for equal-value inductors.
c. Using the reciprocal resistance formula.
d. All of the above.

8. Air-core fixed-value inductors can use air or nonmagnetic forms, such as
a. Iron, cardboard. **c.** Ceramic, cardboard.
b. Ceramic, copper. **d.** Silicon, germanium.

9. Ferrite is a chemical compound that is basically powdered
a. Iron oxide and ceramic. **c.** Mylar and iron.
b. Iron and steel. **d.** Gauze and electrolyte.

10. The ferrite-core variable inductor varies inductance by changing
a. μ **c.** N
b. l **d.** A

11. It will actually take __________ time constants for the current in an inductive circuit to reach a maximum value.
a. 63.2 **c.** 1.414
b. 1 **d.** 5

12. The time constant for a series inductive/resistive circuit is equal to
a. $L \times R$ **c.** V/R
b. L/R **d.** $2\pi \times f \times L$

13. Inductive reactance (X_L) is proportional to
a. Time or period of the ac applied.
b. Frequency of the ac applied.
c. The stray capacitance that occurs due to the air acting as a dielectric between two turns of a coil.
d. The value of inductance.
e. Two of the above are true.

14. In a series *RL* circuit, the source voltage (V_S) is equal to
a. The square root of the sum of V_R^2 and V_L^2.
b. The vector sum of V_R and V_L.
c. $I \times Z$.
d. Two of the above are partially true.
e. Answers (a), (b), and (c) are correct.

15. In a purely resistive circuit, the phase shift is equal to __________, whereas in a purely inductive or capacitive circuit, the phase shift is __________ degrees.
a. 45, 0 **c.** 45, 90
b. 90, 0 **d.** None of the above

16. With an *RL* integrator, the output is taken across the
a. Inductor. **c.** Resistor.
b. Capacitor. **d.** Transformer's secondary.

17. With an *RL* differentiator, the output is taken across the
a. Inductor. **c.** Resistor.
b. Capacitor. **d.** Transformer's secondary.

18. An inductor or choke between the input and output forms a
a. High-pass filter. **b.** Low-pass filter.

19. An inductor or choke connected to ground or in shunt forms a
a. Low-pass filter. **b.** High-pass filter.

20. Lenz's law states that when current is passed through a conductor a self-induced voltage in a coil will
  a. Aid the applied source voltage.
  b. Aid the increasing current from the source.
  c. Produce an opposing current.
  d. Both (a) and (c).

21. When tested with an ohmmeter, an open coil would show
  a. Zero resistance.
  b. An infinite resistance.
  c. A 100 Ω to 200 Ω resistance.
  d. Both (b) and (c).

22. Inductive reactance
  a. Increases with frequency.
  b. Is proportional to inductance.
  c. Reduces the amplitude of alternating current.
  d. All of the above.

23. The current through an inductor __________ the voltage across the same inductor by __________.
  a. Lags, 90° c. Leads, 90°
  b. Lags, 45° d. Leads, 45°

24. The phasor combination of X_L and R is the circuit's
  a. Reactance. c. Power factor.
  b. Total resistance. d. Impedance.

25. In a series RL circuit, where $V_R = 200$ mV and $V_L = 0.2$ V, $\theta =$ __________.
  a. 45° c. 0°
  b. 90° d. 1°

Communication Skill Questions

26. Briefly describe the terms: (Chapter 10, introduction)
  a. Electromagnetism
  b. Electromagnetic induction

27. What is self-induction, and how does it relate to an inductor? (11-1)

28. Give the formula for inductance, and explain the four factors that determine inductance. (11-1-3)

29. List all the formulas for calculating total inductance when inductors are connected in: (11-1-4)
  a. Series b. Parallel

30. What are a fixed- and a variable-value inductor? (11-2)

31. List the important factors of the: (11-2-1)
  a. Air-, iron-, and ferrite-core fixed-value inductors
  b. Ferrite-core variable-value inductor

32. Describe the current rise and fall through an inductor when: (11-3)
  a. DC is applied b. AC is applied

33. Define the following terms:
  a. Inductive reactance (11-4-1)
  b. Impedance (11-4-2)
  c. Phase shift (11-4-2)
  d. Q factor (11-4-2)
  e. Power factor (11-4-2)

34. With illustrations, describe how an inductor and resistor could be used for the following applications: (11-5)
  a. Filtering high and low frequencies
  b. Integration
  c. Differentiation

35. Explain briefly why an inductor acts as an open to an instantaneous change. Why does an inductor act like a short to dc? (11-1)

Practice Problems

36. Convert the following:
  a. 0.037 H to mH c. 862 mH to H
  b. 1760 μH to mH d. 0.256 mH to μH

37. Calculate the impedance (Z) of the following series RL combinations:
  a. 22 MΩ, 25 μH, $f = 1$ MHz
  b. 4 kΩ, 125 mH, $f = 100$ kHz
  c. 60 Ω, 0.05 H, $f = 1$ MHz

38. Calculate the voltage across a coil if ($d = \Delta$ or delta):
  a. $d_i/d_t = 120$ mA/ms and $L = 2$ μH
  b. $d_i/d_t = 62$ μA/μs and $L = 463$ mH
  c. $d_i/d_t = 4$ A/s and $L = 25$ mH

39. Calculate the total inductance of the following series circuits:
  a. 75 μH, 61 μH, 50 mH b. 8 mH, 4 mH, 22 mH

40. Calculate the total inductance of the following parallel circuits:
  a. 12 mH, 8 mH b. 75 μH, 34 μH, 27 μH

41. Calculate the total inductance of the following series–parallel circuits:
  a. 12 mH in series with 4 mH, and both in parallel with 6 mH
  b. A two-branch parallel arrangement made up of 6 μH and 2 μH in series with one another, and 8 μH and 4 μH in series with one another
  c. Two parallel arrangements in series with one another, made up of 1 μH and 2 μH in parallel and 4 μH and 15 μH in parallel

42. Determine the time constant for each of the examples in Question 41, and state how long it will take in each example for current to build up to maximum.

43. In a series RL circuit, if $V_L = 12$ V and $V_R = 6$ V, calculate:
  a. V_S d. Q
  b. I if $Z = 14$ kΩ e. Power factor
  c. Phase angle

44. What value of inductance is needed to produce 3.3 kΩ of reactance at 15 kHz?

45. At what frequency will a 330 μH inductor have a reactance of 27 kΩ?

46. Calculate the impedance of the circuits seen in Figure 11-37.

47. Referring to Figure 11-38, calculate the voltage across the inductor for all five time constants after the switch has been closed.

48. Referring to Figure 11-39, calculate:
  a. L_T
  b. X_L
  c. Z
  d. I_{R1}, I_{L1}, and I_{L2}
  e. θ
  f. True power, reactive power, and apparent power
  g. Power factor

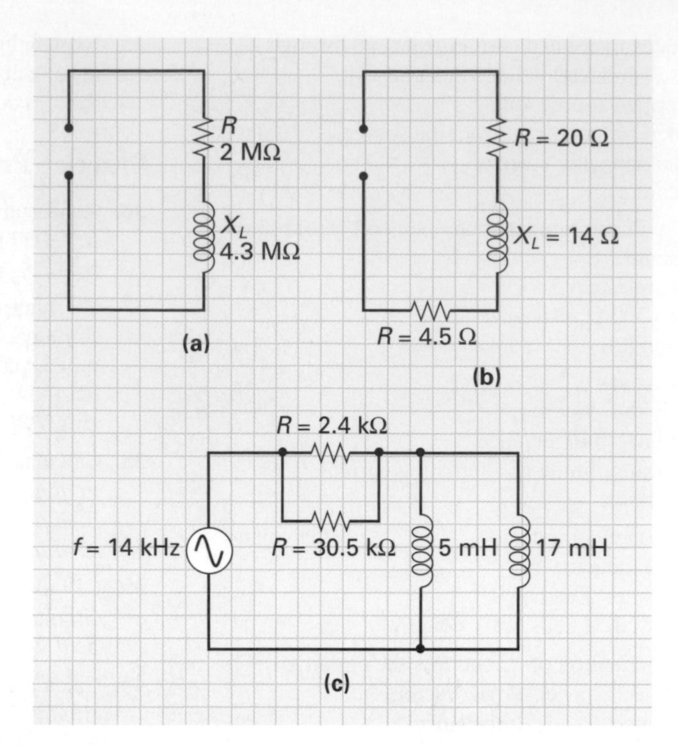

FIGURE 11-37 Calculating Impedance.

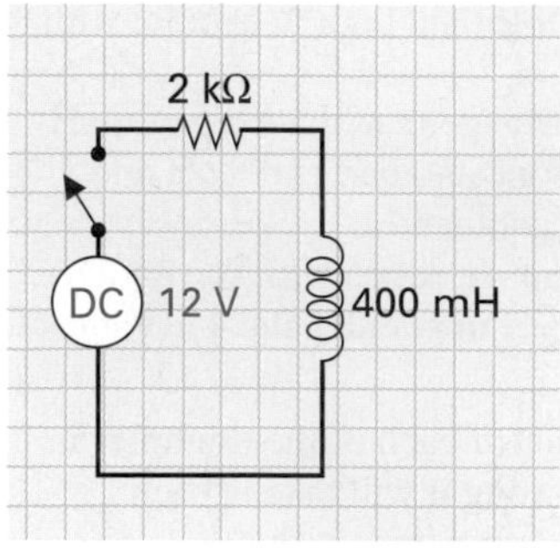

FIGURE 11-38 Inductive Time Constant Example.

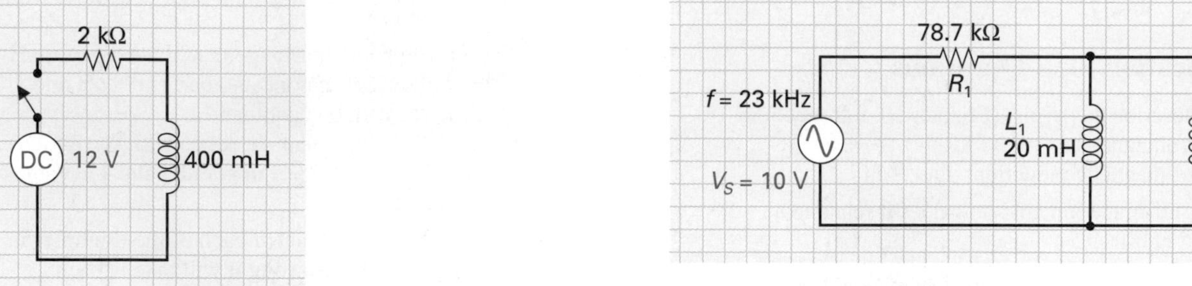

FIGURE 11-39 *RL* Circuit.

49. Referring to Figure 11-40, calculate:

a. R_T
b. L_T
c. X_L
d. Z
e. V_R, V_L
f. I_R, I_L
g. I_T
h. θ
i. Apparent power
j. PF

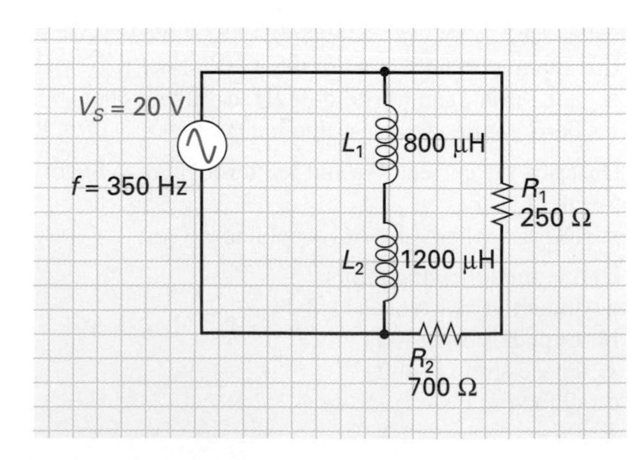

FIGURE 11-40 Parallel *RL* Circuit.

Troubleshooting Problem

50. What problems can occur with inductors, and how can the ohmmeter be used to determine these problems?

Job Interview Test

These tests at the end of each chapter will challenge your knowledge up to this point and give you the practice you need for a job interview. To make this more realistic, the test will comprise both technical and personal questions. In order to take full advantage of this exercise, you may like to set up a simulation of the interview environment, have a friend pose the questions to you, and record your responses for later analysis.

Company Name: Konekt, Inc.

Industry Branch: Consumer Products.

Function: Cell phone repair.

Job Appling For: Customer Service Technician.

51. What do you know about our phones?

52. What made you choose our company?

53. What can a multimeter be used for?

54. Would you enjoy an out-of-state training course?

55. Do you mind working different shifts?

56. What is an oscilloscope?

57. Do you need constant supervision to keep you on task?

58. What is the difference between an inductor and a capacitor?

59. What do you think you will be doing on a day-in, day-out basis if we hire you as a customer service technician?

60. Do you get along well with most people?

Answers

51. Visit the company's web site before the interview to get an overall understanding of the company's ownership, product line, service, and support.

52. Describe why you were first attracted to the company's advertisement, why it felt compatible with your career goals, and how the company's positive attributes listed in its web site matched your expectations.

53. Chapters 1 and 2.

54. Additional training only serves to increase your skills and marketability. If you are unsure, however, don't object to anything at the time. Later, if you decide that it is something you do not want to do, you can call and decline the offer.

55. Individual choice. Once again, if you are unsure, don't object to anything at the time.

56. Chapter 8.

57. No. Discuss how you have disciplined yourself to manage school, studying, and a job and how well you performed in all. Also, describe the lab environment, which was more than likely self-paced.

58. Chapter 14 introduction.

59. Quote intro. section in text and job description listed in newspaper.

60. They're asking if you can get along well with customers and colleagues. Mention any work experience in which you did direct contact with customers, and discuss how you worked with fellow students in your class.

Web Site Questions

Go to the web site http://www.prenhall.com/cook, select the textbook *Introductory DC/AC Electronics* or *Introductory DC/AC Circuits,* this chapter, and then follow the instructions when answering the multiple-choice practice problems.

Transformers

Let's Toss for It!

David Packard

In 1938, Bill Hewlett and Dave Packard, close friends and engineering graduates at Stanford University, set up shop in the one-car garage behind the Packards' rented home in Palo Alto, California. In the garage, the two worked on what was to be the first product of their lifetime business together, an electronic oscillating instrument that represented a breakthrough in technology and was specifically designed to test sound equipment.

Late in the year, the oscillator (designated the 200A "because the number sounded big") was presented at a West Coast meeting of the Institute of Radio Engineers (now the Institute of Electrical and Electronics Engineers—IEEE) and the orders began to roll in. Along with the first orders for the 200A was a letter from Walt Disney Studios asking the two to build a similar oscillator covering a different frequency range. The Model 200B was born shortly thereafter, and Disney purchased eight to help develop the unique sound system for the classic film *Fantasia.* By 1940, the young company had outgrown the garage and moved into a small rented building nearby.

Over the years, the company continued a steady growth, expanding its product line to more than 10,000, including computer systems and peripheral products, test and measuring instruments, hand-held calculators, medical electronic equipment, and systems for chemical analysis. Employees have increased from 2 to almost 82,000, and the company is one of the 100 largest industrial corporations in America, with a net revenue of $8.1 billion. Whose name should go first was decided by the toss of a coin on January 1, 1939, and the outcome was Hewlett-Packard or HP.

12

Outline and Objectives

LET'S TOSS FOR IT!
INTRODUCTION

12-1 TRANSFORMER CHARACTERISTICS

Objective 1: Define mutual inductance and how it relates to transformers.

12-1-1 Basic Transformer Action

Objective 2: Describe the basic operation of a transformer.

12-1-2 Transformer Loading

Objective 3: Explain the differences between a loaded and unloaded transformer.

12-1-3 Windings and Phase

Objective 4: Describe the transformer dot convention as it relates to windings and phase.

12-1-4 Coefficient of Coupling

Objective 5: State the meaning of the coefficient of coupling.

12-1-5 Transformer Losses

Objective 6: Describe the three basic transformer power losses.

12-2 TRANSFORMER RATIOS AND APPLICATIONS

Objective 7: List the three basic applications of transformers.

12-2-1 Turns Ratio

12-2-2 Voltage Ratio

12-2-3 Power and Current Ratio

12-2-4 Impedance Ratio

Objective 8: Describe how a transformer's turns ratio can be used to step up or step down voltage or current, or match impedance.

12-3 TRANSFORMER TYPES, RATINGS, AND TESTING

12-3-1 Transformer Types

Objective 9: List and explain the operation and application of fixed and variable transformer types.

12-3-2 Transformer Ratings

Objective 10: Describe how transformers are rated.

12-3-3 Transformer Testing

Objective 11: Explain how to test the windings of a transformer for opens, partial shorts, or complete shorts.

Introduction

Transformer
Device consisting of two or more coils that are used to couple electric energy from one circuit to another, yet maintain electrical isolation between the two.

Primary Winding
First winding of a transformer, which is connected to the source.

Secondary Winding
Output winding of a transformer, which is connected across the load.

When alternating current was introduced in Chapter 8, it was mentioned that an ac voltage could be stepped up to a larger voltage or stepped down to a smaller voltage by a device called a **transformer.** The transformer is an electrical device that makes use of electromagnetic induction to transfer alternating current from one circuit to another. The transformer consists of two inductors that are placed in very close proximity to one another. When an alternating current flows through the first coil or **primary winding** the inductor sets up a magnetic field. The expanding and contracting magnetic field produced by the primary cuts across the windings of the second inductor or **secondary winding** and induces a voltage in this coil.

By changing the ratio between the number of turns in the secondary winding to the number of turns in the primary winding, some characteristics of the ac signal can be changed or transformed as it passes from primary to secondary. For example, a low ac voltage could be stepped up to a higher ac voltage, or a high ac voltage could be stepped down to a lower ac voltage.

In this chapter we will examine the operation, characteristics, types, testing, and applications of transformers.

12-1 TRANSFORMER CHARACTERISTICS

Mutual Inductance
Ability of one inductor's magnetic lines of force to link with another inductor.

As was discussed in the previous chapter, self-inductance is the process by which a coil induces a voltage within itself. The principle on which a transformer is based is an inductive effect known as **mutual inductance,** the process by which an inductor induces a voltage in another inductor.

Figure 12-1 illustrates two inductors that are magnetically linked, yet electrically isolated from one another. As the alternating current continually rises, falls, and then rises in the opposite direction, a magnetic field will build up, collapse, and then build up in the opposite direction. If a second inductor, or secondary coil (L_2), is in close proximity to the first inductor or primary coil (L_1), which is producing the alternating magnetic field, a voltage will be induced into the nearby inductor, which causes current to flow in the secondary circuit through the load resistor. This phenomenon is known as mutual inductance or transformer action.

As with self-inductance, mutual inductance is dependent on change. As discussed previously, direct current (dc) is a constant current and produces a constant or stationary magnetic field that does not change. Alternating current, however, is continually varying, and as the polarity of the magnetic field is dependent on the direction of current flow (left-hand rule), the magnetic field will also be alternating in polarity, and it is this continual building up and collapsing of the magnetic field that cuts the adjacent inductor's conducting coils and

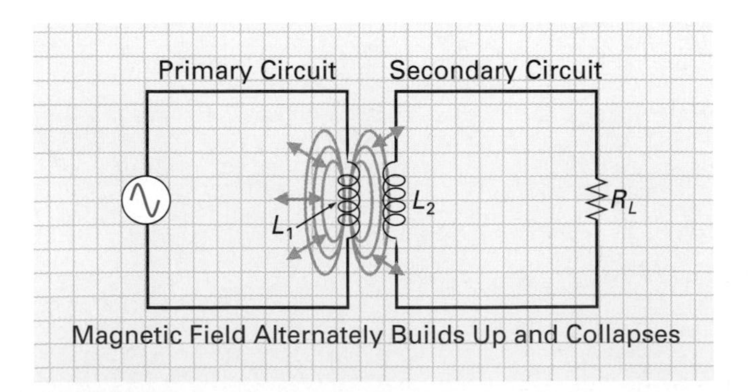

FIGURE 12-1 **Mutual Inductance.**

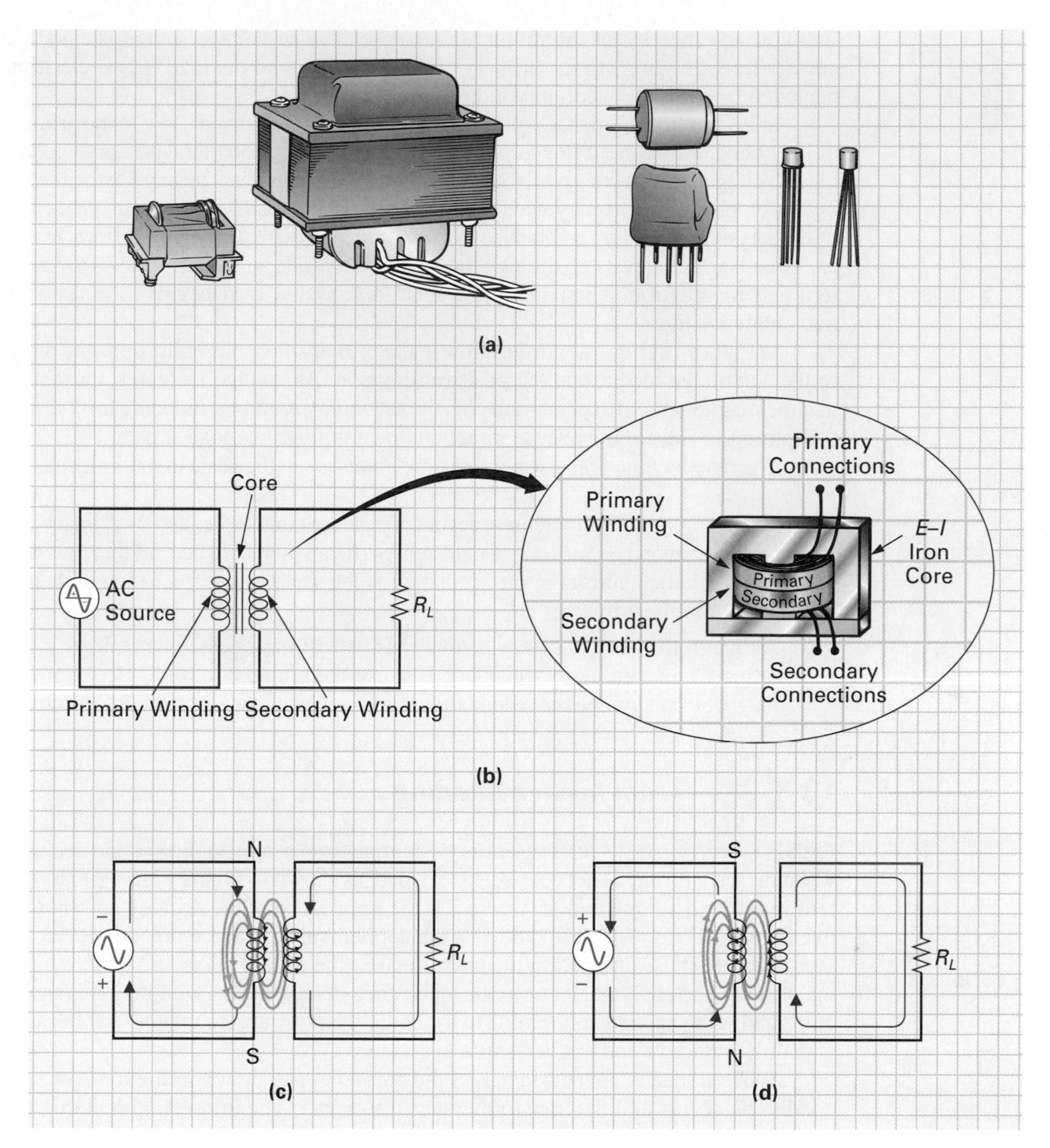

FIGURE 12-2 Transformers and Transformer Action. (a) Physical Appearance. (b) Construction. (c) and (d) Alternating Action.

induces a voltage in the secondary circuit. Mutual induction is possible only with ac and cannot be achieved with dc because of the lack of change.

Self-induction is a measure of how much voltage an inductor can induce within itself. Mutual inductance is a measure of how much voltage is induced in the secondary coil due to the change in current in the primary coil.

12-1-1 *Basic Transformer Action*

Figure 12-2(a) shows the physical appearance of several transformer types. Figure 12-2(b) illustrates a basic transformer, which consists of two coils within close proximity to one another to ensure that the second coil will be cut by the magnetic flux lines produced by the first coil and thereby ensure mutual inductance. The ac voltage source is electrically connected (through wires) to the primary coil or winding, and the load (R_L) is electrically connected to the secondary coil or winding.

In Figure 12-2(c), the ac voltage source has produced current flow in the primary circuit, as illustrated. This current flow produces a north pole at the top of the primary winding and, as the ac voltage input swings more negative, the current increases causing the magnetic field being developed by the primary winding to increase. This expanding magnetic field cuts the coils of the secondary winding and induces a voltage, and a subsequent current flows in the secondary circuit, which travels up through the load resistor. The ac voltage follows a sinusoidal pattern and moves from a maximum negative to zero and then begins to build up toward a maximum positive.

In Figure 12-2(d), the current flow in the primary circuit is in the opposite direction due to the ac voltage increase in the positive direction. As voltage increases, current increases, and the magnetic field expands and cuts the secondary winding, inducing a voltage and causing current to flow in the reverse direction down through the load resistor.

You may have noticed a few interesting points about the basic transformer just discussed.

1. As primary current increases, secondary current increases, and as primary current decreases, secondary current also decreases. It can therefore be said that the frequency of the alternating current in the secondary is the same as the frequency of the alternating current in the primary.
2. Although the two coils are electrically isolated from one another, energy can be transferred from primary to secondary, because the primary converts electrical energy into magnetic energy, and the secondary converts magnetic energy back into electrical energy.

12-1-2 *Transformer Loading*

Let's now carry our discussion of the basic transformer a little further and see what occurs when the transformer is not connected to a load, as shown in Figure 12-3. Primary circuit current is determined by $I = V/Z$, where Z is the impedance of the primary coil (both its inductive reactance and resistance) and V is the applied voltage. Since no current can flow in the secondary, because an open in the circuit exists, the primary acts as a simple inductor, and the primary current is small due to the inductance of the primary winding. This small primary current lags the applied voltage due to the counter emf by approximately 90° because the coil is mainly inductive and has very little resistance.

When a load is connected across the secondary, as shown in Figure 12-4, a change in conditions occurs and the transformer acts differently. The important point that will be observed is that as we go from a no-load to a load condition, the primary current will increase due to mutual inductance. Let's follow the steps one by one.

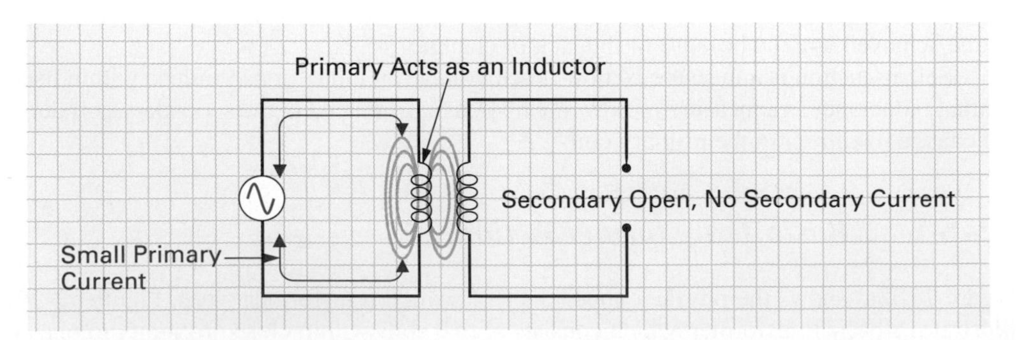

FIGURE 12-3 Unloaded Transformer.

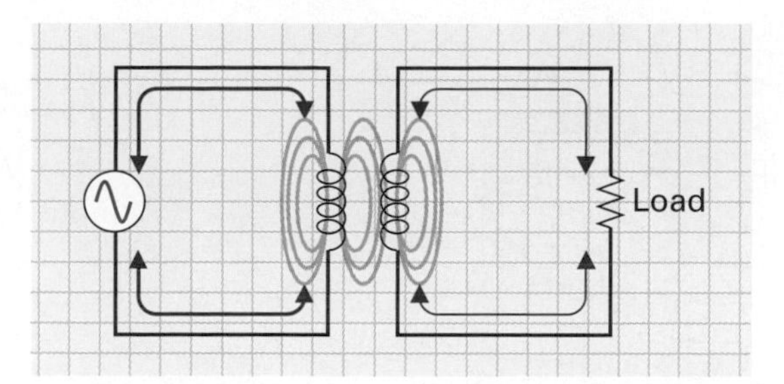

FIGURE 12-4 Loaded Transformer.

1. The ac applied voltage sets up an alternating magnetic field in the primary winding.
2. The continually changing flux of this primary field induces and produces a counter emf in the primary to oppose the applied voltage.
3. The primary's magnetic field also induces a voltage in the secondary winding, which causes current to flow in the secondary circuit through the load.
4. The current in the secondary winding produces another magnetic field that is opposite to the field being produced by the primary.
5. This secondary magnetic field feeds back to the primary and induces a voltage that tends to cancel or weaken the counter emf that was set up in the primary by the primary current.
6. The primary's counter emf is therefore reduced, so primary current can now increase.
7. This increase in primary current is caused by the secondary's magnetic field; consequently, the greater the secondary current, the stronger the secondary magnetic field, which causes a reduction in the primary's counter emf, and therefore a primary current increase.

In summary, an increase in secondary current ($I_s\uparrow$) causes an increase in primary current ($I_p\uparrow$), and this effect in which the primary induces a voltage in the secondary (V_s) and the secondary induces a voltage into the primary (V_p) is known as mutual inductance.

12-1-3 *Windings and Phase*

The way in which the primary and secondary coils are wound around the core determines the polarity of the voltage induced into the secondary relative to the polarity of the primary. In Figure 12-5(a), you can see that if the primary and secondary windings are both wound in a clockwise direction around the core, the voltage induced in the primary will be in phase with the voltage induced in the secondary. Both the input and output will also be in phase with one another if the primary and secondary are both wound in a counterclockwise direction.

In Figure 12-5(b), you will notice that in this case we have wound the primary and secondary in opposite directions, the primary being wound in a clockwise direction and the secondary in a counterclockwise direction. In this situation, the output ac sine-wave voltage is 180° out of phase with respect to the input ac voltage.

In a schematic diagram, there has to be a way of indicating to the technician that the secondary voltage will be in phase or 180° out of phase with the input. The **dot convention** is a standard used with transformers and is illustrated in Figure 12-6(a) and (b).

Dot Convention
A standard used with transformer symbols to indicate whether the secondary voltage will be in phase or out of phase with the primary voltage.

A positive on the primary dot causes a positive on the secondary dot, and similarly, since ac is applied, a negative on the primary dot will cause a negative on the secondary dot. In Figure 12-6(a), you can see that when the top of the primary swings positive or negative, the top of the secondary will also follow suit and swing positive or negative, respectively, and

FIGURE 12-5 Winding and Phase. (a) Primary and Secondary in Phase. (b) Primary and Secondary out of Phase.

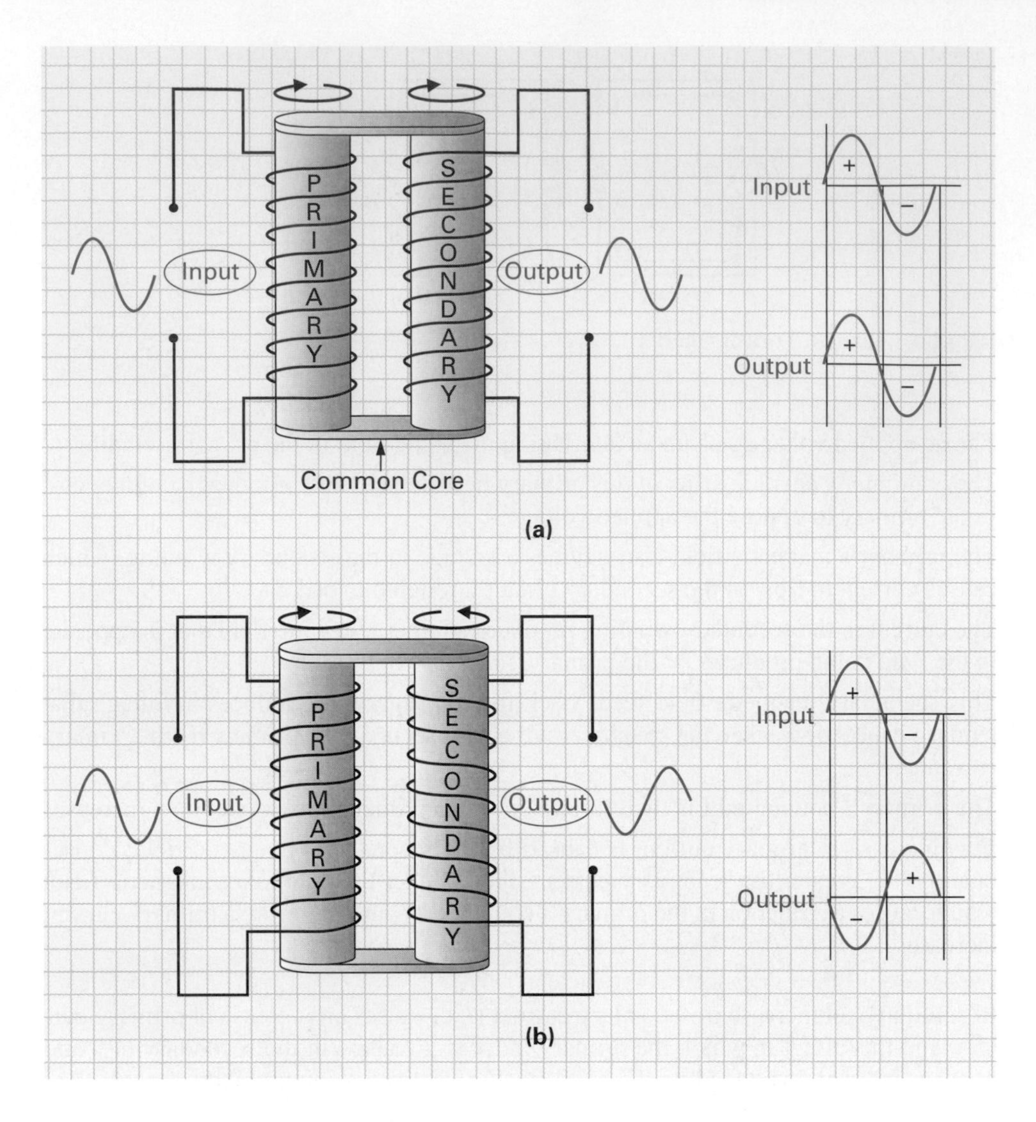

FIGURE 12-6 Dot Convention. (a) In-Phase Dots. (b) Out-of-Phase Dots.

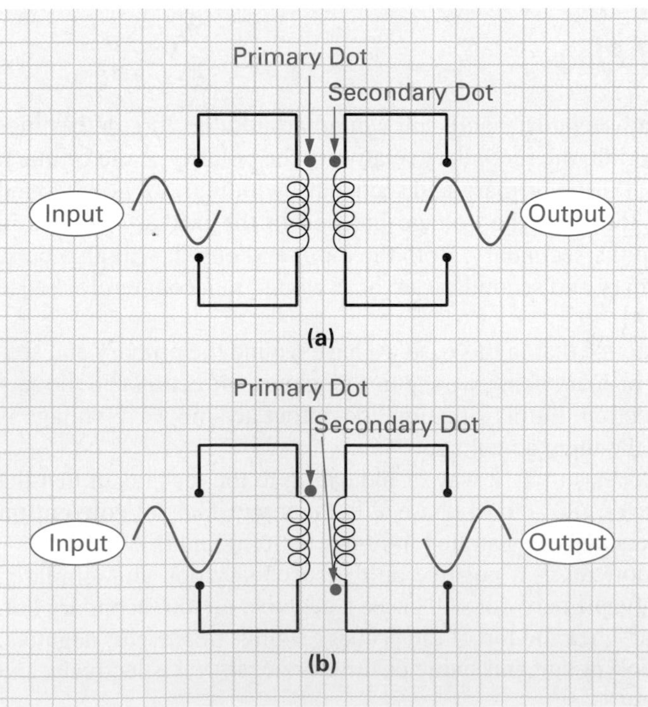

you will now be able to determine that the transformer secondary voltage is in phase with the primary voltage.

In Figure 12-6(b), however, the dots are on top and bottom, which means that as the top of the primary winding swings positive, the bottom of the secondary will go positive, which means the top of the secondary will actually go negative, resulting in the secondary ac voltage being out of phase with the primary ac voltage.

12-1-4 *Coefficient of Coupling*

The voltage induced into the secondary winding is dependent on the mutual inductance between the primary and secondary, which is determined by how much of the magnetic flux produced by the primary actually cuts the secondary winding. The **coefficient of coupling** (k) is a ratio of the number of magnetic lines of force that cut the secondary compared to the total number of magnetic flux lines being produced by the primary and is a figure between 0 and 1.

Coefficient of Coupling
The degree of coupling that exists between two circuits.

$$k = \frac{\text{flux linking secondary coil}}{\text{total flux produced by primary}}$$

If, for example, all the primary flux lines cut the secondary winding, the coefficient of coupling will equal 1. If only half of the total flux lines being produced cut the secondary winding, k will equal 0.5.

EXAMPLE:

A primary coil is producing 65 μWb (microwebers) of magnetic flux. Calculate the coefficient of coupling if 52 μWb links with the secondary coil.

Solution:

$$k = \frac{52\ \mu\text{Wb}}{65\ \mu\text{Wb}} = 0.8$$

This means that 80% of the magnetic flux lines being generated by the primary coil are linking with the secondary coil.

The coefficient of coupling depends on two key factors:

1. How close together the primary and secondary are to one another.
2. The type of core material used.

Figure 12-7(a) illustrates how the primary and secondary can be wound on the same core to achieve a high coefficient of coupling. Figure 12-7(b) shows that a large distance between the primary and secondary will result in a low coefficient of coupling.

Copper Loss
Also called I^2R loss, it is the power lost in transformers, generators, connecting wires, and other parts of a circuit because of the current flow (I) through the resistance (R) of the conductors.

12-1-5 *Transformer Losses*

Internal transformer losses cause the power delivered from the secondary winding of a transformer in reality to be less than the power fed into the primary winding.

The first of these, **copper loss,** is due to the ohmic resistance of the windings, which generates heat ($I^2 \times R$). In addition, during each ac cycle the core is taken through a cycle of magnetization and so energy is lost due to **hysteresis** and appears as heat in the core. Hysteresis loss is proportional to frequency and is minimized by using a core of soft iron or an alloy such as stalloy or permalloy.

Hysteresis
Amount that the magnetization of a material lags the magnetizing force due to molecular friction.

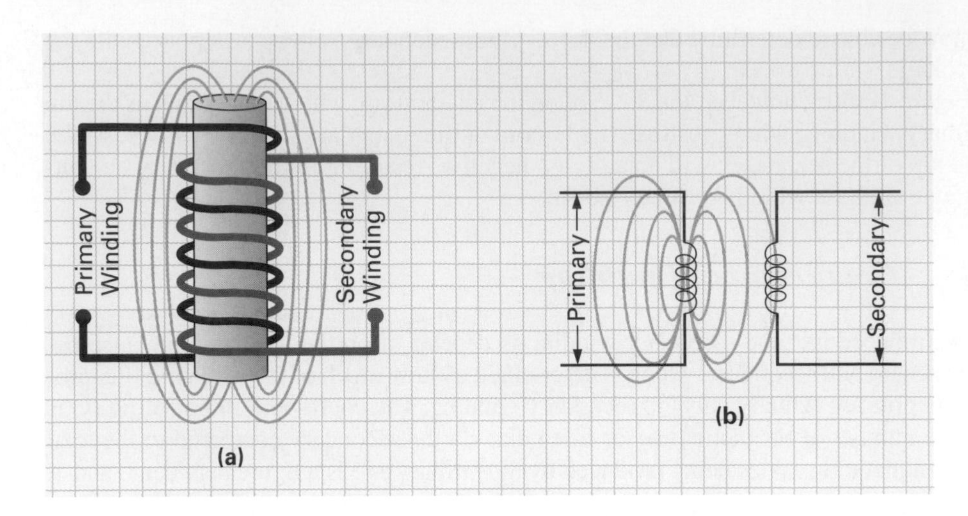

FIGURE 12-7 Coefficient of Coupling. (a) Wound on Same Core, Small Distance Between Primary and Secondary, High *k*. (b) Winding Far Apart, Small *k*.

Eddy Currents
Small currents Induced in a conducting core due to the variations in alternating magnetic flux.

Magnetic Leakage
The passage of magnetic flux outside the path along which it can do useful work.

The other core loss is caused by the continuously changing flux that induces voltages in the conducting core. These small voltages create local **eddy currents** in the core that combine to produce a large circulating current, as seen in Figure 12-8(a). Eddy currents oppose the main flux, generate heat, and are proportional to frequency. To minimize this loss, manufacturers use a laminated core, as described in Figure 12-8(b).

The final loss, **magnetic leakage,** occurs because not all of the flux lines produced in the primary will cut the secondary (the coefficient of coupling is less than perfect).

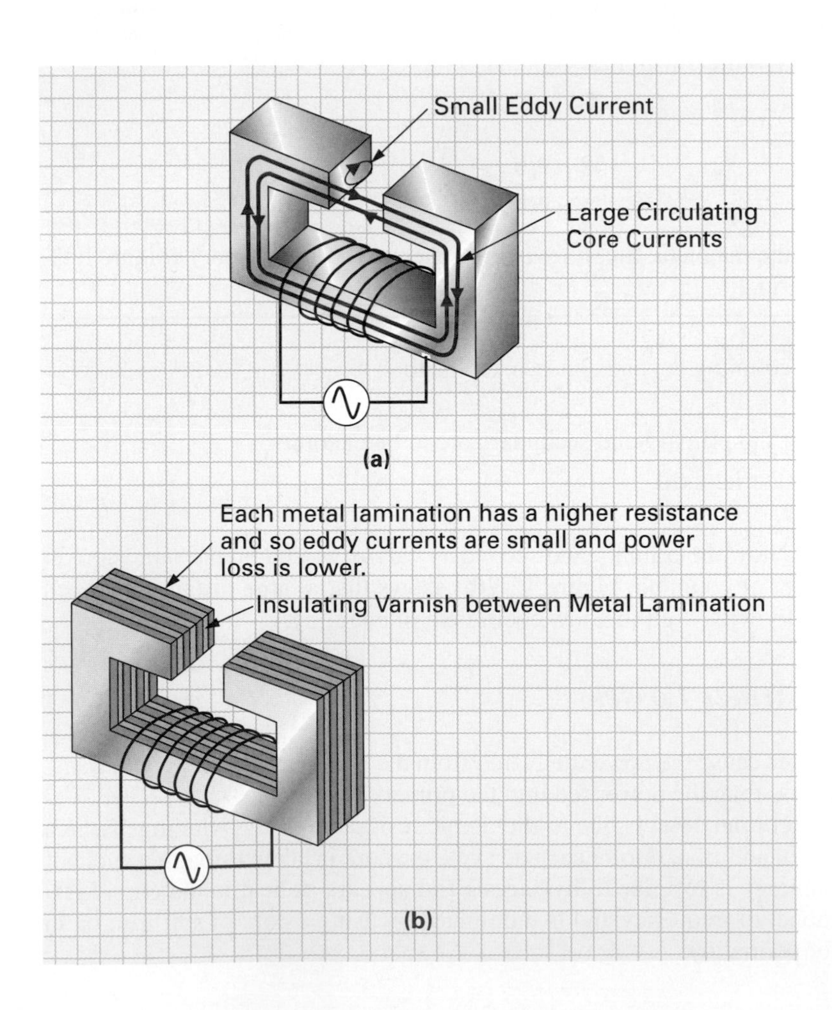

FIGURE 12-8 Eddy Currents. (a) Solid Core, Large Eddy Currents. (b) Laminated Core, Small Eddy Currents.

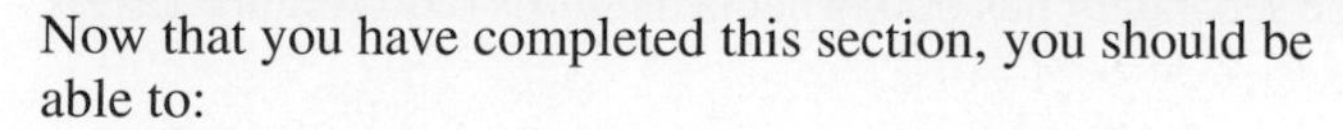

SELF-TEST EVALUATION POINT FOR SECTION 12-1

Now that you have completed this section, you should be able to:

Objective 1. Define mutual inductance and how it relates to transformers.

Objective 2. Describe the basic operation of a transformer.

Objective 3. Explain the differences between a loaded and an unloaded transformer.

Objective 4. Describe the transformer dot convention as it relates to windings and phase.

Objective 5. State the meaning of the coefficient of coupling.

Objective 6. Describe the three basic transformer power losses.

Use the following questions to test your understanding of Section 12-1:

1. Define mutual inductance and explain how it differs from self-inductance.
2. True or false: Mutual induction is possible only with direct current flow.
3. True or false: A transformer achieves electrical isolation.
4. True or false: The transformer converts electrical energy to magnetic, and then from magnetic back to electrical.
5. What are the names given to the windings of a transformer?
6. True or false: An increase in secondary current causes an increase in primary current.
7. True or false: The greater the secondary current, the greater the primary's counter emf.
8. How does the winding of the primary and secondary affect the phase of the output with respect to the input?
9. Describe the transformer dot convention used on schematics.
10. Define and state the formula for the coefficient of coupling.
11. True or false: If $k = 0.75$, then 75% of the total flux lines produced by the primary are cutting the coils of the secondary.
12. List the two factors that determine k.
13. List the three types of transformer losses.
14. Why will a laminated transformer core reduce eddy-current loss?

12-2 TRANSFORMER RATIOS AND APPLICATIONS

Basically, transformers are used for one of three applications:

1. To step up (increase) or step down (decrease) voltage
2. To step up (increase) or step down (decrease) current
3. To match impedances

In all three cases, any of the applications can be achieved by changing the ratio of the number of turns in the primary winding compared to the number of turns in the secondary winding. This ratio is appropriately called the *turns ratio.*

12-2-1 *Turns Ratio*

The **turns ratio** is the ratio between the number of turns in the secondary winding (N_s) and the number of turns in the primary winding (N_p).

Turns Ratio
Ratio of the number of turns in the secondary winding to the number of turns in the primary winding of a transformer.

$$\text{Turns ratio} = \frac{N_s}{N_p}$$

Let us use a few examples to see how the turns ratio can be calculated.

EXAMPLE:

If the primary has 200 turns and the secondary has 600, what is the turns ratio (Figure 12-9)?

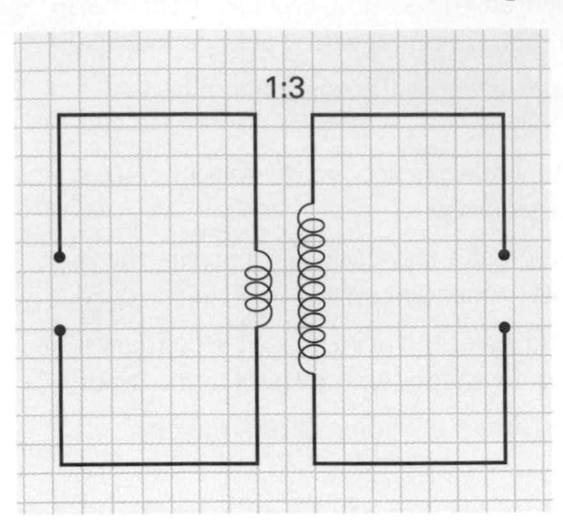

FIGURE 12-9 Step-Up Transformer Example.

Solution:

$$\text{Turns ratio} = \frac{N_s}{N_p} = \frac{600}{200} = \frac{3(\text{secondary})}{1(\text{primary})} = 3$$

This simply means that there are three windings in the secondary to every one winding in the primary. Moving from a small number (1) to a larger number (3) means that we *stepped up* in value. Stepping up always results in a turns ratio figure greater than 1, in this case, 3.

EXAMPLE:

If the primary has 120 turns and the secondary has 30 turns, what is the turns ratio (Figure 12-10)?

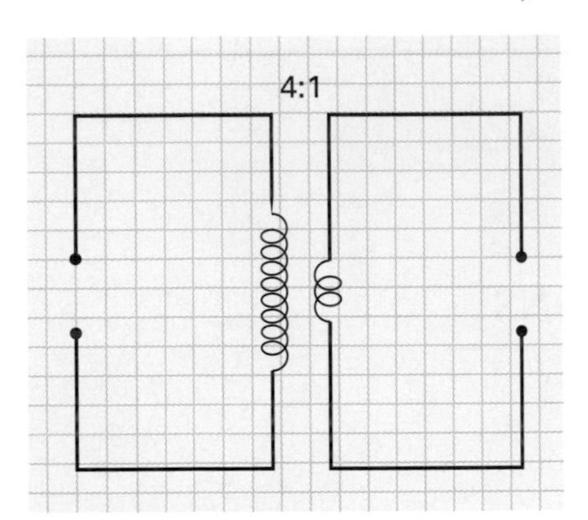

FIGURE 12-10 Step-Down Transformer Example.

Solution:

$$\text{Turns ratio} = \frac{N_s}{N_p} = \frac{30}{120} = \frac{1(\text{secondary})}{4(\text{primary})} = 0.25$$

Said simply, there are four primary windings to every one secondary winding. Moving from a larger number (4) to a smaller number (1) means that we *stepped down* in value. Stepping down always results in a turns ratio figure of less than 1, in this case 0.25.

12-2-2 *Voltage Ratio*

Transformers are used within the power supply unit of almost every piece of electronic equipment to step up or step down the 115 V ac from the outlet. Some electronic circuits require lower-power supply voltages, while other devices may require higher-power supply voltages. The transformer is used in both instances to convert the 115 V ac to the required value of voltage.

Step-Up Transformer
Transformer in which the ac voltage induced in the secondary is greater (due to more secondary windings) than the ac voltage applied to the primary.

Step Up

If the secondary voltage (V_s) is greater than the primary voltage (V_p), the transformer is called a **step-up transformer** ($V_s > V_p$), as shown in Figure 12–11. The voltage is stepped up or increased in much the same way as a generator voltage can be increased by increasing the number of turns.

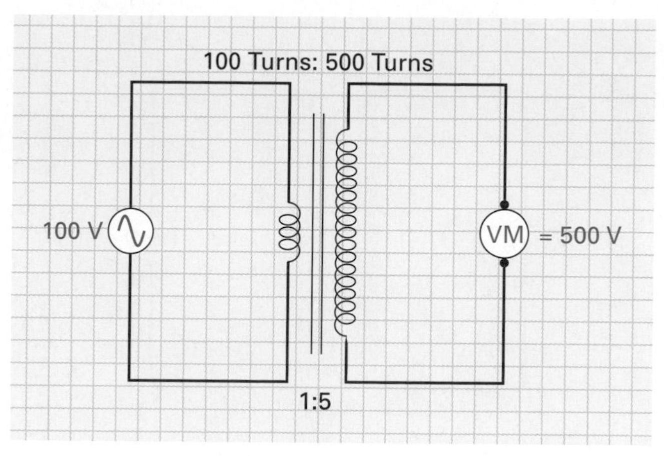

FIGURE 12-11 Step-Up Transformer.

If the ac primary voltage is 100 V and the turns ratio is a 1 : 5 step up, the secondary voltage will be five times that of the primary voltage, or 500 V, because the magnetic flux established by the primary cuts more turns in the secondary and therefore induces a larger voltage. In this example, you can see that the ratio of the secondary voltage to the primary voltage is equal to the turns ratio; in other words,

$$\frac{V_s}{V_p} = \frac{N_s}{N_p}$$

or

$$\frac{500}{100} = \frac{500}{100}$$

To calculate V_s, therefore, we can rearrange the formula and arrive at

$$V_s = \frac{N_s}{N_p} \times V_p$$

In our example, this is

$$V_s = \frac{500}{100} \times 100\text{ V}$$
$$= 500\text{ V}$$

Step Down

If the secondary voltage (V_s) is smaller than the primary voltage (V_p), the transformer is called a **step-down transformer** ($V_s < V_p$), as shown in Figure 12-12. The secondary voltage will be equal to

Step-Down Transformer
Transformer in which the ac voltage induced in the secondary is less (due to fewer secondary windings) than the ac voltage applied to the primary.

$$V_s = \frac{N_s}{N_p} \times V_p$$
$$= \frac{10}{100} \times 1000\text{ V} = 100\text{ V}$$

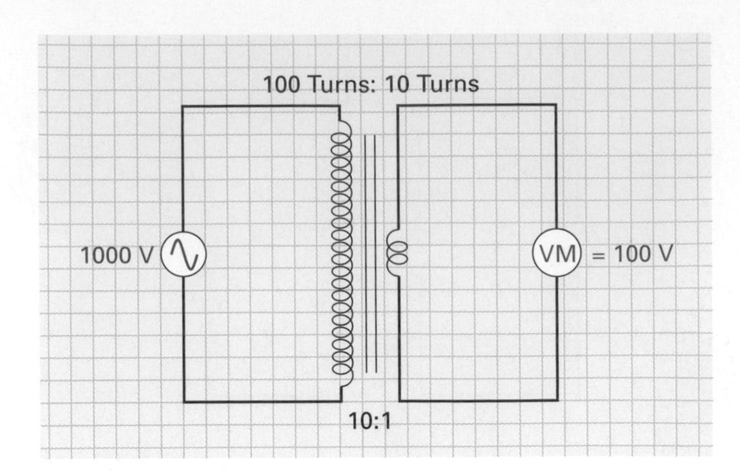

FIGURE 12-12 **Step-Down Transformer.**

TIME LINE

Werner von Siemens (1816–1892) was a German electrical engineer who played an important role in the development of the telegraph industry.

EXAMPLE:

Calculate the secondary voltage (V_s) if a 1 : 6 step-up transformer has 24 V ac applied to the primary.

Solution:

$$V_s = \frac{N_s}{N_p} \times V_p$$

$$= \frac{6}{1} \times 24 \text{ V} = 144 \text{ V}$$

The coupling coefficient (k) in this formula is always assumed to be 1, which for most iron-core transformers is almost always the case. This means that all the primary magnetic flux is linking the secondary, and the secondary voltage is dependent on the number of secondary turns that are being cut by the primary magnetic flux.

The transformer can be used to transform the primary ac voltage into any other voltage, either up or down, merely by changing the transformer's turns ratio.

12-2-3 *Power and Current Ratio*

The power in the secondary of the transformer is equal to the power in the primary ($P_p = P_s$). Power, as we know, is equal to $P = V \times I$, and if voltage is stepped up or down, the current automatically is stepped down or up, respectively, in the opposite direction to voltage to maintain the power constant.

For example, if the secondary voltage is stepped up ($V_s\uparrow$), the secondary current is stepped down ($I_s\downarrow$), so the output power is the same as the input power.

$$P_s = V_s\uparrow \times I_s\downarrow$$

This is an equal but opposite change. Therefore, $P_s = P_p$, and you cannot get more power out than you put in. The current ratio is therefore inversely proportional to the voltage ratio:

$$\frac{V_s}{V_p} = \frac{I_p}{I_s}$$

If the secondary voltage is stepped up, the secondary current goes down:

$$\frac{V_s\uparrow}{V_p} = \frac{I_p}{I_s\downarrow}$$

If the secondary voltage is stepped down, the secondary current goes up:

$$\frac{V_s\downarrow}{V_p} = \frac{I_p}{I_s\uparrow}$$

If the current ratio is inversely proportional to the voltage ratio, it is also inversely proportional to the turns ratio:

$$\frac{I_p}{I_s} = \frac{V_s}{V_p} = \frac{N_s}{N_p}$$

By rearranging the current and turns ratio, we can arrive at a formula for secondary current, which is

$$I_s = \frac{N_p}{N_s} \times I_p$$

EXAMPLE:

The step-up transformer in Figure 12-13 has a turns ratio of 1 to 5. Calculate:

a. Secondary voltage (V_s)
b. Secondary current (I_s)
c. Primary power (P_p)
d. Secondary power (P_s)

Solution:

The secondary has five times as many windings as the primary and, consequently, the voltage will be stepped up by 5 between primary and secondary. If the secondary voltage is going to be five times that of the primary, the secondary current is going to decrease to one-fifth of the primary current.

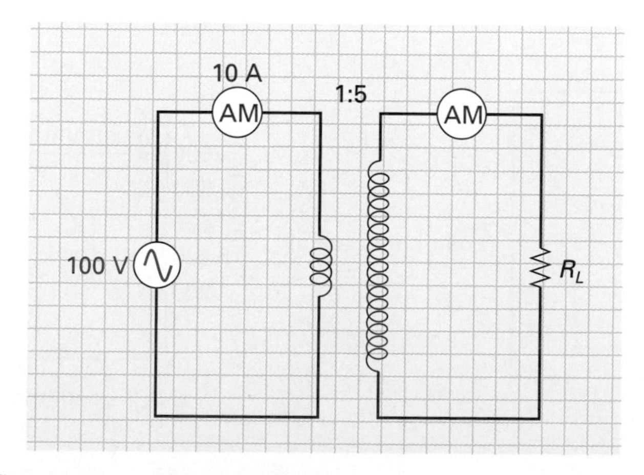

FIGURE 12-13 Step-Up Transformer Example.

a. $V_s = \frac{N_s}{N_p} \times V_p$

$= \frac{5}{1} \times 100 \text{ V}$

$= 500 \text{ V}$

b. $I_s = \frac{N_p}{N_s} \times I_p$

$= \frac{1}{5} \times 10 \text{ A}$

$= 2 \text{ A}$

c. $P_p = V_p \times I_p = 100 \text{ V} \times 10 \text{ A} = 1000 \text{ VA}$

d. $P_s = V_s \times I_s = 500 \text{ V} \times 2 \text{ A} = 1000 \text{ VA}$

Therefore, $P_p = P_s$.

12-2-4 *Impedance Ratio*

Maximum Power Transfer Theorem

The maximum power will be absorbed by the load from the source when the impedance of the load is equal to the impedance of the source.

The **maximum power transfer theorem,** which has been discussed previously and is summarized in Figure 12-14, states that maximum power is transferred from source (ac generator) to load (equipment) when the impedance of the load is equal to the internal impedance of the source. If these impedances are different, a large amount of power could be wasted.

In most cases maximum power transfer is required from a source that has an internal impedance (Z_s) that is not equal to the load impedance (Z_L). In this situation, a transformer can be inserted between the source and the load to make the load impedance appear to equal the source's internal impedance.

For example, let's imagine that your car stereo system (source) has an internal impedance of 100 Ω and is driving a speaker (load) of 4 Ω impedance, as seen in Figure 12-15. By choosing the correct turns ratio, the 4 Ω speaker can be made to appear as a 100 Ω load impedance, which will match the 100 Ω internal source impedance of the stereo system, resulting in maximum power transfer. The turns ratio can be calculated by using the formula

$$\text{Turns ratio} = \sqrt{\frac{Z_L}{Z_S}}$$

where Z_L = Load impedance in ohms
Z_S = Source impedance in ohms

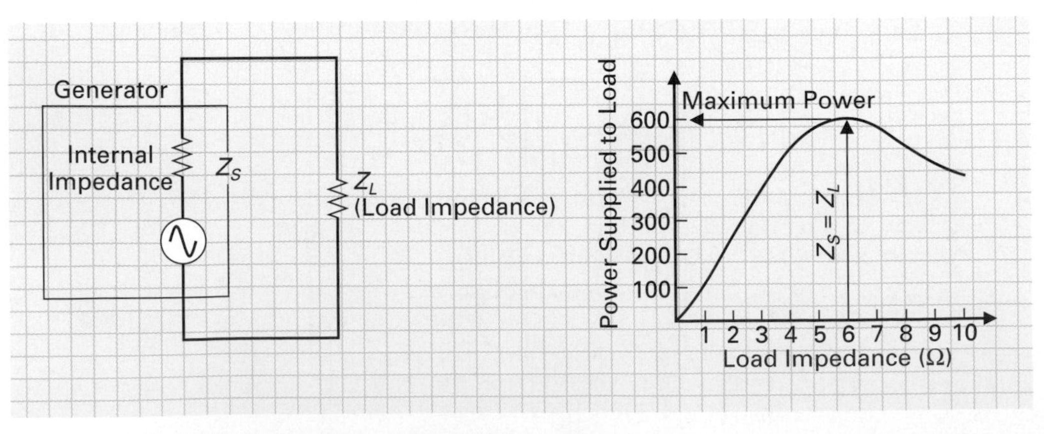

FIGURE 12-14 **Maximum Power Transfer Theorem.**

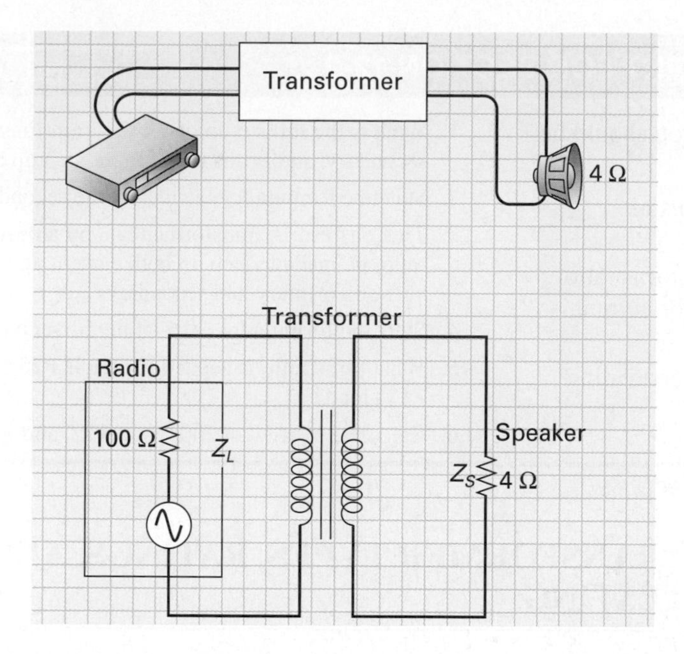

FIGURE 12-15 **Impedance Matching.**

In our example, this will calculate out to be

$$\text{Turns ratio} = \sqrt{\frac{Z_L}{Z_S}}$$

$$= \sqrt{\frac{4}{100}} = \frac{\sqrt{4}}{\sqrt{100}}$$

$$= \frac{2}{10} = \frac{1}{5}$$

$$= 0.2$$

If the turns ratio is less than 1, a step-down transformer is required. A turns ratio of 0.2 means a step-down transformer is needed with a turns ratio of 5 : 1 (⅕ = 0.2).

EXAMPLE:

Calculate the turns ratio needed to match the 22.2 Ω output impedance of an amplifier to two 16 Ω speakers connected in parallel.

Solution:

The total load impedance of two 16 Ω speakers in parallel will be

$$Z_L = \frac{\text{product}}{\text{sum}} = \frac{16 \times 16}{16 + 16} = 8\ \Omega$$

The turns ratio will be

$$\text{Turns ratio} = \sqrt{\frac{Z_L}{Z_S}}$$

$$= \sqrt{\frac{8\ \Omega}{22.2\ \Omega}}$$

$$= \sqrt{0.36} = 0.6$$

Therefore, a step-down transformer is needed with a turns ratio of 1.67:1.

SELF-TEST EVALUATION POINT FOR SECTION 12-2

Now that you have completed this section, you should be able to:

Objective 7. List the three basic applications of transformers.

Objective 8. Describe how a transformer's turns ratio can be used to step up or step down voltage or current, or match impedance.

Use the following questions to test your understanding of Section 12-2:

1. What is the turns ratio of a 402 turn primary and 1608 turn secondary, and is this transformer step up or step down?
2. State the formula for calculating the secondary voltage (V_s).
3. True or false: A transformer can, by adjusting the turns ratio, be made to step up both current and voltage between primary and secondary.
4. State the formula for calculating the secondary current (I_s).
5. What turns ratio is needed to match a 25 Ω source to a 75 Ω load?
6. Calculate V_s if $N_s = 200$, $N_p = 112$, and $V_p = 115$ V.

12-3 TRANSFORMER TYPES, RATINGS, AND TESTING

12-3-1 *Transformer Types*

We will treat transformers the same as every other electronic component and classify them as having either a fixed or variable turns ratio.

1. Fixed Turns Ratio Transformers

The fixed turns ratio transformers, shown in Figure 12-16(a), have a turns ratio that cannot be varied. They are generally wound on a common core to ensure a high k and can be classified by the type of core material used: air core, iron core, or ferrite core.

The air-core transformers typically have a nonmagnetic core, such as ceramic or a carboard hollow shell, and are used in high-frequency applications. The electronic circuit symbol just shows the primary and secondary coils. The more common iron- or ferrite-core transformers concentrate the magnetic lines of force, resulting in improved transformer performance; they are symbolized by two lines running between the primary and secondary. The iron-core transformer's lines are solid; the ferrite-core transformer's lines are dashed.

2. Variable Turns Ratio Transformers

Variable turns ratio transformers, shown in Figure 12-16(b) through (f), have a turns ratio that can be varied.

Center-Tapped Transformer

A transformer that has a connection at the electrical center of the secondary winding.

Neutral Wire

The conductor of a polyphase circuit or of a single-phase three-wire circuit that is intended to have a ground potential. The potential differences between the neutral and each of the other conductors are approximately equal in magnitude and are also equally spaced in phase.

Center-Tapped Secondary If the tapped lead is in the exact center of the secondary, the transformer is said to have a center-tapped secondary, as shown in Figure 12-16(b). With the **center-tapped transformer,** the two secondary voltages are each half of the total secondary voltage. If we assume a 1:1 turns ratio (primary turns = secondary turns) and a 20 V ac primary voltage and therefore secondary voltage, each of the output voltages between either end of the secondary and the center tap will be 10 V waveforms, as shown in Figure 12-16(b). The two secondary outputs will be 180° out of phase with one another from the center tap. When the top of the secondary swings positive (+), the bottom of the secondary will be negative, and vice versa.

Figure 12-17 illustrates a typical power company utility pole, which as you can see makes use of a center-tapped secondary transformer. These transformers are designed to step down the high 4.8 kV from the power line to 120 V/240 V for commercial and residential customers. Within the United States, the actual secondary voltage may be anywhere between 225 and 245 V, depending on the demand. The demand is actually higher during the day and in the winter months (source voltage is pulled down by smaller customer load resistance).

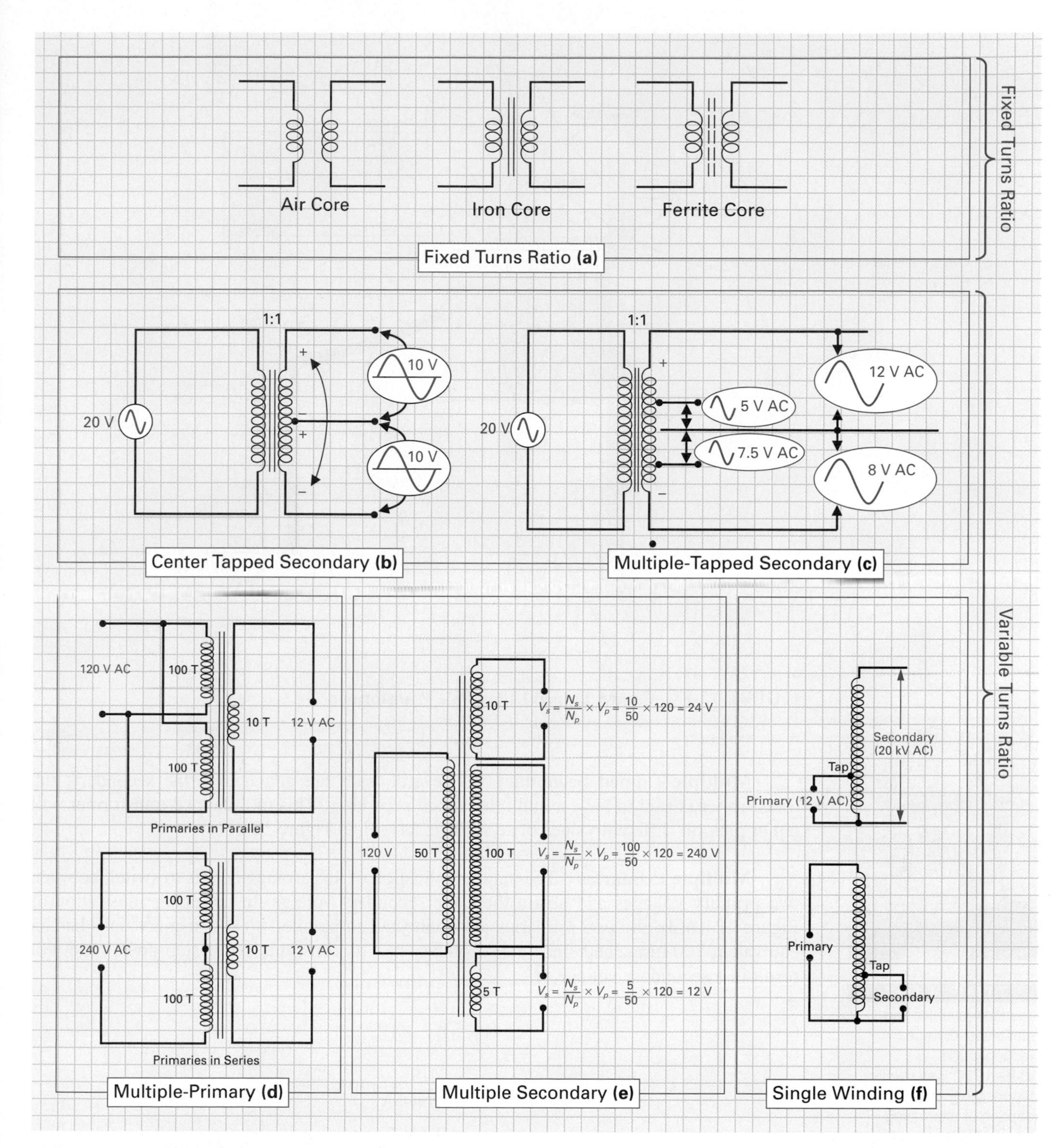

FIGURE 12-16 Transformer Types.

The multiple taps on the primary are used to change the turns ratio and compensate for regional power line voltage differences.

The center tap of the secondary winding is connected to a copper earth grounding rod. This center tap wire is called the **neutral wire,** and within the building it is color coded with a white insulation. Residential or commercial appliances or loads that are designed to operate at 240 V are connected between the two black wires, while loads that are designed to operate at 120 V are connected between a black wire and the white wire (neutral).

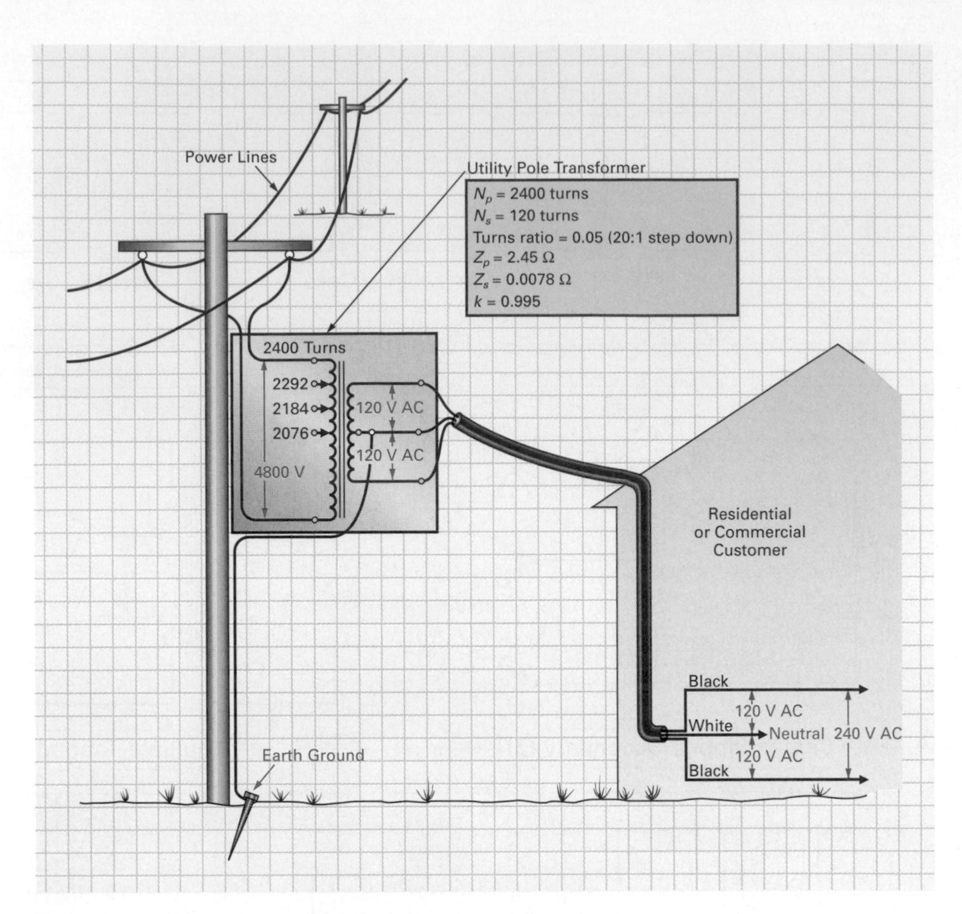

FIGURE 12-17 **Center-Tapped Utility Pole Transformer.**

TIME LINE

American inventor and industrialist George Westinghouse (1846–1914) was chiefly responsible for the adoption of alternating current (ac) for electric power transmission in the United States. Westinghouse won the coveted contract to harness Niagara Falls, bidding half of what Edison bid for a direct current (dc) system, and in 1895 the Niagara Power System transmitted electricity 20 miles to Buffalo, New York.

Multiple-Tapped Secondary If several leads are attached to the secondary, the transformer is said to have a multiple-tapped secondary. This multitapped transformer will tap or pick off different values of ac voltage, as seen in Figure 12-16(c).

Multiple Primaries The major application of this arrangement is to switch between two primary voltages and obtain the same secondary voltage. In the upper illustration of Figure 12-16(d), the two primaries are connected in parallel, so the primary only has an equivalent 100 turns; and since the secondary has 10 turns, this 10 : 1 step-down ratio will result in a secondary voltage of

$$V_s = \frac{N_s}{N_p} \times V_p$$

$$= \frac{10}{100} \times 120 \text{ V}$$

$$= 12 \text{ V}$$

If the two primaries are connected in series, as shown in the lower illustration of Figure 12-16(d), the now 200 turn primary will result in a secondary voltage of

$$V_s = \frac{N_s}{N_p} \times V_p$$

$$= \frac{10}{200} \times 240 \text{ V}$$

$$= 12 \text{ V}$$

Some portable electrical or electronic equipment, such as radios or shavers, have a switch that allows you to switch between 120 V ac (U.S. wall outlet voltage) and 240 V ac (European wall outlet voltage). By activating the switch, you can always obtain the correct voltage to operate the equipment, whether the wall socket is supplying 240 or 120 V.

Multiple Secondaries In some applications, more than one secondary is wound onto a common primary and core. The advantage of this arrangement can be seen in Figure 12-16(e), where many larger and smaller voltages can be acquired from one primary voltage.

Single Winding (Autotransformer) The single winding transformer or autotransformer is used in the automobile ignition system (switched dc pulse system) to raise the low 12 V battery voltage up to about 20 kV for the spark plugs, as shown in the upper illustration in Figure 12-16(f). Autotransformers are constructed by winding one continuous coil onto a core, which acts as both the primary and secondary. This yields three advantages; they are smaller, cheaper, and lighter than the normal separated primary/secondary transformer types. Its disadvantage, however, is that no electrical isolation exists between primary and secondary. The single winding transformer is normally used as a step-up transformer for the automobile or in televisions to obtain 20 kV. As with all transformers, the primary-to-secondary turns ratio can also be arranged to step down voltage, as shown in the lower illustration in Figure 12-16(f).

12-3-2 *Transformer Ratings*

A typical transformer rating could read 1 kVA, 500/100, 60 Hz. The 500 normally specifies the maximum primary voltage, the 100 normally specifies the maximum secondary voltage, and the 1 kVA is the apparent power rating. In this example, the maximum load current will equal

$$I_s = \frac{\text{apparent power } (P_A)}{\text{secondary voltage } (V_s)} \qquad \left(P = V \times I; \text{ therefore, } I = \frac{P}{V} \right)$$

$$= \frac{1 \text{ kVA}}{100 \text{ V}} = 10 \text{ A}$$

With this secondary voltage at 100 V and a maximum current of 10 A, the smallest load resistor that can be connected across the output of the secondary is

$$R_L = \frac{V_s}{I_s}$$

$$= \frac{100 \text{ V}}{10 \text{ A}}$$

$$= 10 \ \Omega$$

Exceeding the rating of the transformer will cause overheating and even burning out of the windings.

EXAMPLE:

Calculate the smallest value of load resistance that can be connected across a 3 kVA, 600/200, 60 Hz step-down transformer.

■ ***Solution:***

$$I = \frac{\text{apparent power } (P_A)}{\text{secondary voltage } (V_s)}$$

$$= \frac{3 \text{ kVA}}{200 \text{ V}} = 15 \text{ A}$$

$$R_L = \frac{V_s}{I_s} = \frac{200 \text{ V}}{15 \text{ A}} = 13.3 \ \Omega$$

12-3-3 *Transformer Testing*

Transformers, which are basically a single device containing two or more coils, can develop one of three problems:

1. An open winding
2. A complete short in a winding
3. A short in a section of a winding

1. Open Primary or Secondary Winding

An open in the primary winding will prevent any primary current, and therefore there will be no induced voltage in the secondary and so no voltage will be present across the load. An open secondary winding will prevent the flow of secondary current, and once again, no voltage will be present across the load.

An open in the primary or secondary winding is easily detected by disconnecting the transformer from the circuit and testing the resistance of the windings with an ohmmeter. Like an inductor, a transformer winding should have a low resistance, in the tens to hundreds range. An open will easily be recognized because of its infinite (maximum) resistance.

2. Complete Short or Section Short in Primary or Secondary Winding

A partial or complete short in the primary winding of a transformer will result in an excessive source current that will probably blow the circuit's fuse or trip the circuit's breaker. A partial or complete short in the secondary winding will cause an excessive secondary current, which in turn will result in an excessive primary current.

In both instances, a short in the primary or secondary will generally burn out the primary winding unless the fuse or circuit breaker opens the excessive current path. An ohmmeter can be used to test for partial or complete shorts in transformer windings; however, all coils have a naturally low resistance, which can be mistaken for a short. An inductive or reactive analyzer can accurately test transformer windings, checking the inductance value of each coil.

It is virtually impossible to repair an open, partially shorted, or completely shorted transformer winding, and therefore defective transformers are always replaced with a transformer that has an identical rating.

SELF-TEST EVALUATION POINT FOR SECTION 12-3

Now that you have completed this section, you should be able to:

Objective 9. List and explain the operation and application of fixed and variable transformer types.

Objective 10. Describe how transformers are rated.

Objective 11. Explain how to test the windings of a transformer for opens, partial shorts, or complete shorts.

Use the following questions to test your understanding of Section 12-3:

1. List the three types of fixed transformers.

2. List the five basic classifications for variable transformers.
3. True or false: The two outputs of a center-tapped secondary transformer will always be 90° out of phase with one another.
4. In what application would a multiple-primary transformer be used?
5. In what application would a single-winding transformer be used?
6. What are the advantages of a single-winding transformer?
7. What do each of the values mean when a transformer is rated as a 10 kVA, 200/100, 60 Hz?
8. If a 100 Ω resistor is connected across the secondary of a transformer that is to supply 1 kV and is rated at a maximum current of 8 A, will the transformer overheat and possibly burn out?
9. Which test instrument could you use to test for a suspected open primary winding?
10. If a secondary winding had a resistance of 50 Ω, would this value indicate a problem?

SUMMARY

Transformer Characteristics (Section 12-1)

1. The transformer is an electrical device that makes use of electromagnetic induction to transfer alternating current from one circuit to another.
2. The transformer consists of two inductors that are placed in very close proximity to one another. When an alternating current flows through the first coil or primary winding, the inductor sets up a magnetic field. The expanding and contracting magnetic field produced by the primary cuts across the windings of the second inductor or secondary winding and induces a voltage in this coil.
3. By changing the ratio between the number of turns in the secondary winding to the number of turns in the primary winding, some characteristics of the ac signal can be changed or transformed as it passes from primary to secondary.
4. The principle on which a transformer is based is an inductive effect known as mutual inductance, which is the process by which an inductor induces a voltage in another inductor.
5. As with self-inductance, mutual inductance is dependent on change. Direct current (dc) is a constant current and produces a constant or stationary magnetic field that does not change. Alternating current, however, is continually varying, and as the polarity of the magnetic field is dependent on the direction of current flow (left-hand rule), the magnetic field will also be alternating in polarity.
6. Self-induction is a measure of how much voltage an inductor can induce within itself. Mutual inductance is a measure of how much voltage is induced in the secondary coil due to the change in current in the primary coil.
7. The basic transformer consists of two coils within close proximity to one another to ensure that the second coil will be cut by the magnetic flux lines produced by the first coil and thereby ensure mutual inductance. The ac voltage source is electrically connected (through wires) to the primary coil or winding, and the load (*RL*) is electrically connected to the secondary coil or winding.
8. As primary current increases, secondary current increases, and as primary current decreases, secondary current also decreases. It can be said that the frequency of the alternating current in the secondary is the same as the frequency of the alternating current in the primary.
9. Although the two coils are electrically isolated from one another, energy can be transferred from primary to secondary because the primary converts electrical energy into magnetic energy, and the secondary converts magnetic energy back into electrical energy.
10. When the transformer is not connected to a load, primary circuit current is determined by $I = V/Z$, where Z is the impedance of the primary coil (both its inductive reactance and resistance) and V is the applied voltage. Primary current lags the applied voltage due to the counter emf by approximately 90° because the coil is mainly inductive and has very little resistance.
11. When a load is connected across the secondary, an increase in secondary current ($I_s\uparrow$) causes an increase in primary current ($I_p\uparrow$), and this effect in which the primary induces a voltage in the secondary (V_s) and the secondary induces a voltage into the primary (V_p) is known as mutual inductance.
12. The dot convention is a standard used with transformer symbols to indicate whether the secondary voltage will be in phase or out of phase with the primary voltage.
13. The coefficient of coupling (k) is a ratio of the number of magnetic lines of force that cut the secondary compared to the total number of magnetic flux lines being produced by the primary and is a figure between 0 and 1.
14. Internal losses cause the power delivered from the secondary winding of a transformer in reality to be less than the power fed into the primary winding. These losses include ohmic resistance of the windings, hysteresis, local eddy currents in the core, and magnetic leakage.

Transformer Ratios and Applications (Section 12-2)

15. Basically, transformers are used for one of three applications:
    a. To step up (increase) or step down (decrease) voltage
    b. To step up (increase) or step down (decrease) current
    c. To match impedances

In all three cases, any of the applications can be achieved by changing the ratio of the number of turns in the primary winding compared to the number of turns in the secondary winding. This ratio is appropriately called the *turns ratio.*

16. The turns ratio is the ratio between the number of turns in the secondary winding (N_s) and the number of turns in the primary winding (N_p).
17. If the secondary voltage (V_s) is greater than the primary voltage (V_p), the transformer is called a step-up transformer ($V_s > V_p$).
18. If the secondary voltage (V_s) is smaller than the primary voltage (V_p), the transformer is called a step-down transformer ($V_s < V_p$).
19. The power in the secondary of the transformer is equal to the power in the primary ($P_p = P_s$).
20. The maximum power transfer theorem states that maximum power is transferred from source (ac generator) to load (equipment) when the impedance of the load is equal to the internal impedance of the source. If these impedances are different, a large amount of power could be wasted.
21. In most cases a source has an internal impedance (Z_S) that is not equal to the load impedance (Z_L). In this situation, a transformer can be inserted between the source and the load to make the load impedance appear to equal the source's internal impedance.

Transformer Types, Ratings, and Testing (Section 12-3)

22. Fixed turns ratio transformers have a turns ratio that cannot be varied.
23. The air-core transformers typically have a nonmagnetic core, such as ceramic or a cardboard hollow shell, and are used in high-frequency applications.
24. The more common iron- or ferrite-core transformers concentrate the magnetic lines of force, resulting in improved transformer performance.
25. Variable turns ratio transformers have a turns ratio that can be varied.
26. If the tapped lead is in the exact center of the secondary, the transformer is said to have a center-tapped secondary. With the center-tapped transformer, the two secondary voltages are each half of the total secondary voltage.
27. A typical transformer rating could read 1 kVA, 500/100, 60 Hz. The 500 normally specifies the maximum primary voltage, the 100 normally specifies the maximum secondary voltage, and the 1 kVA is the apparent power rating.
28. An open in the primary winding will prevent any primary current, and therefore there will be no induced voltage in the secondary and no voltage will be present across the load.
29. An open secondary winding will prevent the flow of secondary current, and once again, no voltage will be present across the load.
30. An open in the primary or secondary winding is easily detected by disconnecting the transformer from the circuit and testing the resistance of the windings with an ohmmeter. Like an inductor, a transformer winding should have a low resistance, in the tens to hundreds range. An open will easily be recognized because of its infinite (maximum) resistance.
31. A partial or complete short in the primary winding of a transformer will result in an excessive source current that will probably blow the circuit's fuse or trip the circuit's breaker.
32. A partial or complete short in the secondary winding will cause an excessive secondary current, which will result in an excessive primary current.
33. In both instances, a short in the primary or secondary will generally burn out the primary winding unless the fuse or circuit breaker opens the excessive current path.
34. An ohmmeter can be used to test for partial or complete shorts in transformer windings. All coils have a naturally low resistance, which can be mistaken for a short.
35. An inductive or reactive analyzer can accurately test transformer windings, checking the inductance value of each coil.
36. It is virtually impossible to repair an open, partially shorted, or completely shorted transformer winding, and therefore defective transformers are always replaced with a transformer that has an identical rating.

REVIEW QUESTIONS

Multiple-Choice Questions

1. Transformer action is based on
   a. Self-inductance.
   b. Air between the coils.
   c. Mutual capacitance.
   d. Mutual inductance.
2. An increase in transformer secondary current will cause a/an __________ in primary current.
   a. Decrease
   b. Increase
3. If 50% of the magnetic lines of force produced by the primary were to cut the secondary coil, the coefficient of coupling would be
   a. 75 **c.** 0.5
   b. 50 **d.** 0.005
4. A step-up transformer will always have a turns ratio __________, whereas a step-down transformer has a turns ratio __________.
   a. $< 1, > 1$ **c.** $> 1, < 1$
   b. $> 1, > 1$ **d.** $< 1, < 1$

5. With an 80 V ac secondary voltage center-tapped transformer, what would be the voltage at each output, and what would be the phase relationship between the two secondary voltages?
   a. 20 V, in phase with one another
   b. 30 V, 180° out of phase
   c. 40 V, in phase
   d. 40 V, 180° out of phase
6. One application of the autotransformer would be
   a. To obtain the final anode high-voltage supply for the cathode-ray tube in a television.
   b. To obtain two outputs, 180° out of phase with one another.
   c. To tap several different voltages from the secondary.
   d. To obtain the same secondary voltage for different voltages.
7. Transformers can only be used with alternating current because
   a. It produces an alternating magnetic field.
   b. It produces a fixed magnetic field.
   c. Its magnetic field is greater than that of dc.
   d. Its rms is 0.707 of the peak.
8. Eddy-current losses are reduced with laminated iron cores because
   a. The air gap is always kept to a minimum.
   b. The resistance of iron is always low.
   c. The laminations are all insulated from one another.
   d. Current cannot flow in iron.
9. Assuming 100% efficiency, the output power, P_s, is always equal to
   a. P_p
   b. $V_s \times I_s$
   c. $0.5 \times P_p$
   d. Both (a) and (c)
   e. Both (a) and (b)
10. If the primary winding of a transformer were open, the result would be
    a. No flux linkage between primary and secondary.
    b. No primary or secondary current.
    c. No induced voltage in the secondary.
    d. All of the above.

Communication Skill Questions

11. What is a transformer? (Introduction)
12. Describe mutual inductance and how it relates to transformers. (12-1)
13. Why does loading the transformer's secondary circuit affect primary transformer current? (12-1-2)
14. Describe what a coefficient of coupling figure of 0.9 means. (12-1-4)
15. List the three basic applications of transformers, and then describe how each of these applications can be achieved by merely changing the transformer's turns ratio. (12-2)
16. Briefly describe how the dot convention is used in schematics to describe phase. (12-1-3)
17. Illustrate the schematic symbol and main points relating to the following transformers (12-3):
    a. Fixed air core
    b. Fixed iron core
    c. Fixed ferrite core
    d. Center-tapped secondary
    e. Four-output tapped secondary
    f. Multiple primary and multiple secondary
    g. Autotransformer
18. Illustrate and explain the following transformer applications (12-3-1):
    a. 120/240 V primary voltage to constant secondary voltage
    b. 20 kV secondary voltage using the autotransformer
19. Briefly describe what is meant by copper losses with transformers. (12-1-5)
20. Briefly describe the following iron and core losses and how they can be reduced (12-1-5):
    a. Hysteresis
    b. Eddy current
21. Briefly describe the transformer loss known as magnetic leakage. (12-1-5)
22. What meter could be used to check a suspected open primary coil? (12-3-3)
23. Could a transformer be considered a dc block? (12-1)
24. Would a step-up voltage transformer step up or step down current? What would happen to secondary power? (12-2)
25. Briefly describe how transformers are used to reduce I^2R losses in ac power distribution. (Chapter 8)

Practice Problems

26. Calculate the turns ratio of the following transformers and state whether they are step up or step down:
    a. P = 12 T, S = 24 T
    b. P = 3 T, S = 250 T
    c. P = 24 T, S = 5 T
    d. P = 240 T, S = 120 T
27. Calculate the secondary ac voltage for all the examples in Question 26 if the primary voltage equals 100 V.
28. Calculate the secondary ac current for all the examples in Question 26 if the primary current equals 100 mA.
29. What turns ratio would be needed to match a source impedance of 24 Ω to a load impedance of 8 Ω?
30. What turns ratio would be needed to step
    a. 120 V to 240 V.
    b. 240 V to 720 V.
    c. 30 V to 14 V.
    d. 24 V to 6 V.
31. For a 24 V, 12 turn primary with a 16, 2, 1, and 4 turn multiple secondary transformer, calculate each of the secondary voltages.
32. If a 2 : 1 step-down transformer has a primary input voltage of 120 V, 60 Hz, and a 2 kVA rating, calculate the maximum secondary current and smallest load resistor that can be connected across the output.
33. Indicate the polarity of the secondary voltages in Figure 12-18(a) and (b).
34. Referring to Figure 12-19(a) and (b), sketch the outputs, showing polarity and amplitude with respect to the inputs.

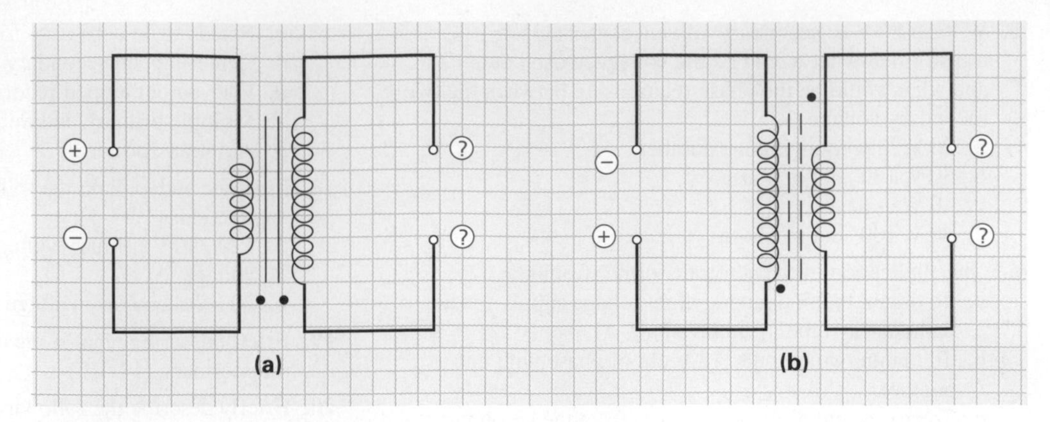

FIGURE 12-18 **Dot Convention Examples.**

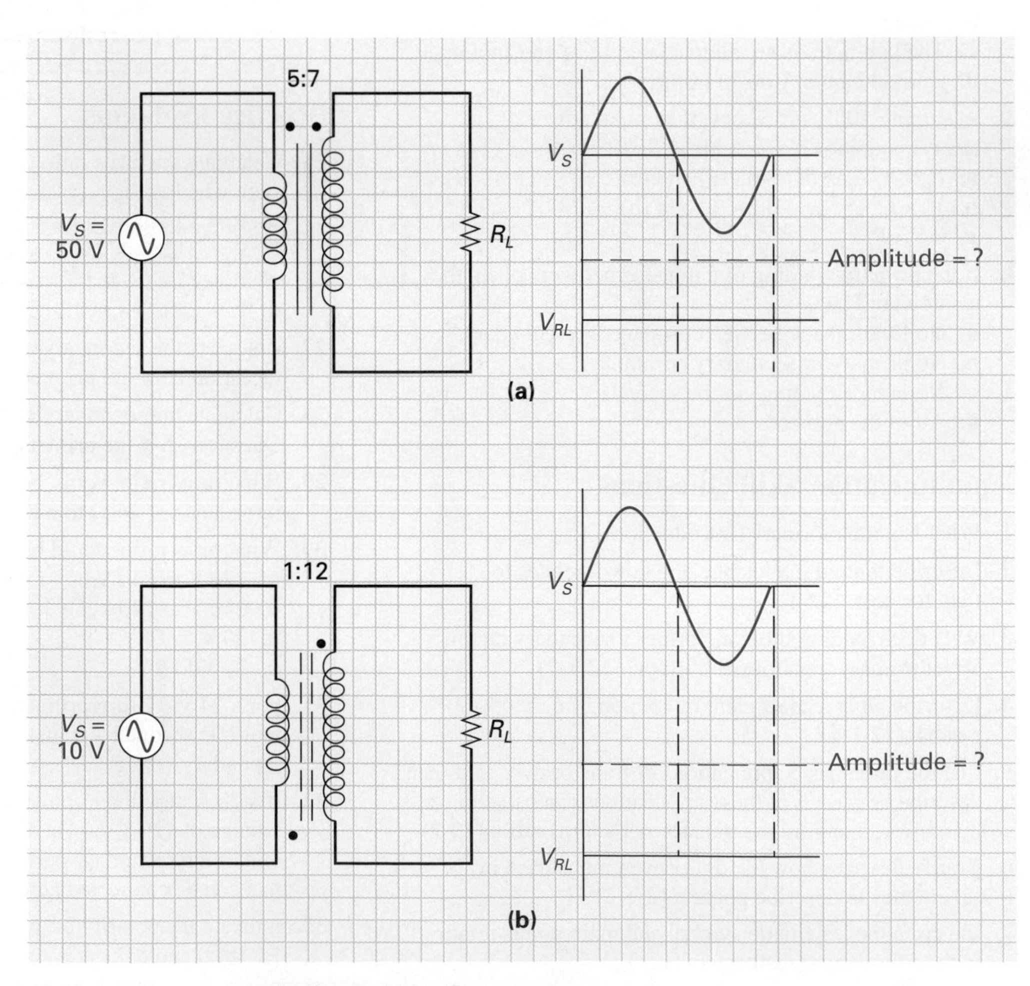

FIGURE 12-19 **Input/Output Polarity Examples.**

35. If a transformer is rated at 500 VA, 60 Hz, the primary voltage is 240 V ac, and the secondary voltage is 600 V ac, calculate:
   a. Maximum load current
   b. Smallest value of R_L

Job Interview Test

These tests at the end of each chapter will challenge your knowledge up to this point and give you the practice you need for a job interview. To make this more realistic, the test will comprise both technical and personal questions. In order to take full advantage of this exercise, you may like to set up a simulation of the interview environment, have a friend read the questions to you, and record your responses for later analysis.

Company Name: SOS, Inc.

Industry Branch: Marine Electronics.

Function: Manufacture Communications Equipment.

Job Title: Calibration Technician.

36. Who was your last employer?
37. What does the term *calibration* mean?
38. What is a potentiometer?
39. Could you describe the steps involved in developing a product from start to finish?
40. Why would the load resistance connected to the secondary of a transformer determine the primary current of the same transformer?
41. What test equipment are you familiar with?
42. What is the job function of a production test technician?
43. What level of supervision would you need for this job?
44. What is the maximum power transfer theorem?
45. During a heavy production cycle we may require that you work every other Saturday for a short space of time. Would this be a problem for you?

Answers

36. Discuss your previous employer and give a copy of your letter of recommendation. If this job was not technical, refer back to how the lab work during training was structured similar to a work environment and discuss other related topics.
37. Chapter 8.
38. Chapter 3.
39. Describe the steps detailed in the Introduction section.
40. Section 12-1-2.
41. Up to this time, the multimeter, oscilloscope, function generator, and frequency counter.
42. Quote intro. section in text and job description listed in newspaper.
43. Say that you expect that in the beginning you would have many questions, but as you became more familiar with company procedures and the product line, you would need less and less supervision.
44. Section 12-2-4.
45. If you are unsure, don't object to anything at the time. Later, if you decide that it is something you do not want to do, you can call and decline the offer.

Web Site Questions

Go to the web site http://www.prenhall.com/cook, select the textbook *Introductory DC/AC Electronics* or *Introductory DC/AC Circuits,* this chapter, and then follow the instructions when answering the multiple-choice practice problems.

Resistive, Inductive, and Capacitive (*RLC*) Circuits

There's No Sleeping When He's Around!

Carl Friedrich Gauss was born April 30, 1777, to poor, uneducated parents in Brunswick, Germany. He was a child of precocious abilities, particularly in mental computation. In elementary school he soon impressed his teachers, who said that mathematical ability came easier to Gauss than speech.

In secondary school he rapidly distinguished himself in ancient languages and mathematics. At 14, Gauss was presented to the court of the duke of Brunswick, where he displayed his computing skill. Until his death in 1806, the duke generously supported Gauss and his family, encouraging the boy with textbooks and a laboratory.

In the early years of the nineteenth century, Gauss met physicist Wilhelm Weber, who would eventually become famous for his work on electricity. They worked together for many years and became close friends, investigating electromagnetism and the use of a magnetic needle for current measurement. In 1833 they constructed an electric telegraph system that could communicate across Göttingen from Gauss's observatory to Weber's physics laboratory. (This telegraph system of communication was later developed independently by U.S. inventor Samuel Morse.)

Gauss conceived almost all of his fundamental mathematical discoveries between the ages of 14 and 17. There are many stories of his genius in his early years, one of which involved a sarcastic teacher who liked giving his students long-winded problems and then resting, or on some occasions sleeping, in class. On his first day with Gauss, who was 8 years old, the teacher began, as usual, by telling the students to find the sum of all the numbers from 1 to 100. The teacher barely had a chance to sit down before Gauss raised his hand and said "5050." The dumbfounded teacher, who believed Gauss must have heard the problem before and memorized the answer, asked Gauss to explain how he had solved the problem. He replied: "The numbers 1, 2, 3, 4, 5, and so on the 100 can be paired as 1 and 100, 2 and 99, 3 and 98, and so on. Since each pair has a sum of 101, and there are 50 pairs, the total is 5050."

13

Outline and Objectives

Introduction

In this chapter we will combine resistors (R), inductors (L), and capacitors (C) into series and parallel ac circuits. Resistors, as we have discovered, operate and react to voltage and current in a very straightforward way, in that the voltage across a resistor is in phase with the resistor current.

Inductors and capacitors operate in essentially the same way, in that they both store energy and then return it back to the circuit. However, they have completely opposite reactions to voltage and current. To help you remember the phase relationships between voltage and current for capacitors and inductors, you may wish to use the following memory phrase:

ELI the ICE man

This phrase states that voltage (symbolized E) leads current (I) in an inductive (L) circuit (abbreviated by the word "ELI"), and current (I) leads voltage (E) in a capacitive (C) circuit (abbreviated by the word "ICE").

In this chapter we will study the relationships among voltage, current, impedance, and power in both series and parallel *RLC* circuits. We will also examine the important *RLC* circuit characteristic called *resonance*, and see how *RLC* circuits can be made to operate as filters. In the final section we will discuss how complex numbers can be used to analyze series and parallel ac circuits containing resistors, inductors, and capacitors.

13-1 SERIES *RLC* CIRCUIT

Figure 13-1 begins our analysis of series *RLC* circuits by illustrating the current and voltage relationships. The circuit current is always the same throughout a series circuit and can therefore be used as a reference. Studying the waveforms and vector diagrams shown alongside the components, you can see that the voltage across a resistor is always in phase with the current, while the voltage across the inductor leads the current by 90° and the voltage across the capacitor lags the current by 90°.

Now let's analyze the impedance, current, voltage, and power distribution of this circuit in a little more detail.

13-1-1 *Impedance*

Impedance is the total opposition to current flow and is a combination of both reactance (X_L, X_C) and resistance (R). An example circuit is illustrated in Figure 13-2(a).

Capacitive reactance can be calculated by using the formula

$$X_C = \frac{1}{2\pi fC}$$

In the example,

$$X_C = \frac{1}{2\pi \times 60 \times 10\ \mu\text{F}} = 265.3\ \Omega$$

Inductive reactance is calculated by using the formula

$$X_L = 2\pi fL$$

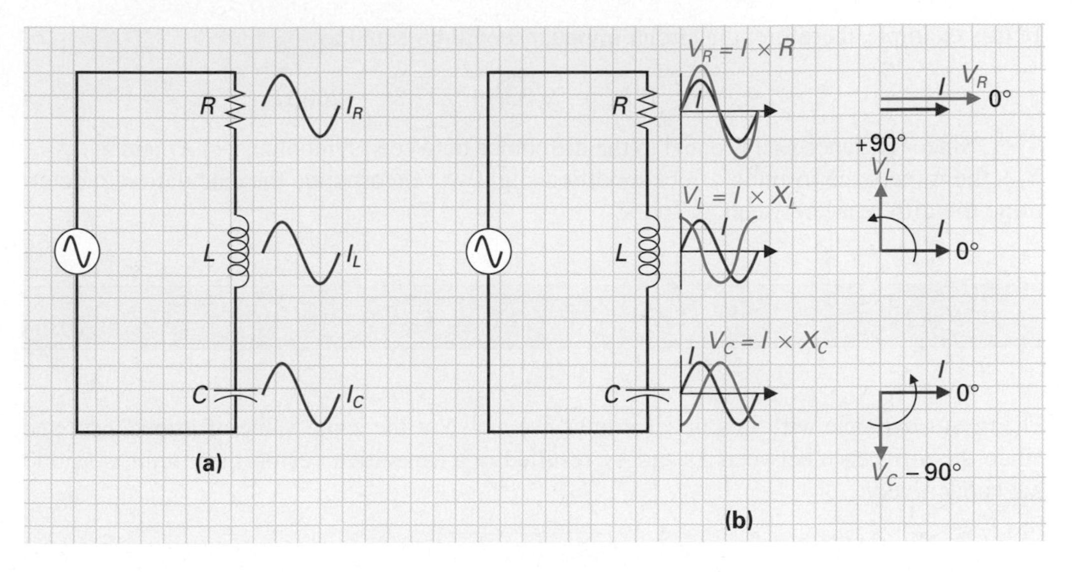

FIGURE 13-1 Series *RLC* Circuit. (a) *RLC* Series Circuit Current: Current Flow Is Always the Same in All Parts of a Series Circuit. (b) *RLC* Series Circuit Voltages: *I* Is in Phase with V_R, *I* Lags V_L by 90°, and *I* Leads V_C by 90°.

In the example,

$$X_L = 2\pi \times 60 \times 20 \text{ Mh} = 7.5\ \Omega$$

Resistance in the example is equal to $R = 33\ \Omega$. Figure 13-2(b) illustrates these values of resistance and reactance in a vector diagram. In this vector diagram you can see that X_L is drawn 90° ahead of R, and X_C is drawn 90° behind R. The capacitive and inductive reactances are 180° out of phase with one another and counteract to produce the vector diagram shown in Figure 13-2(c). The difference between X_L and X_C is equal to 257.8, and since X_C is greater than X_L, the resultant reactive vector is capacitive. Reactance, however, is not in phase with resistance, and impedance is the vector sum of the reactive (X) and resistive (R) vectors. The formula, based on the Pythagorean theorem as illustrated in Figure 13-2(d), is

$$Z = \sqrt{R^2 + X^2}$$

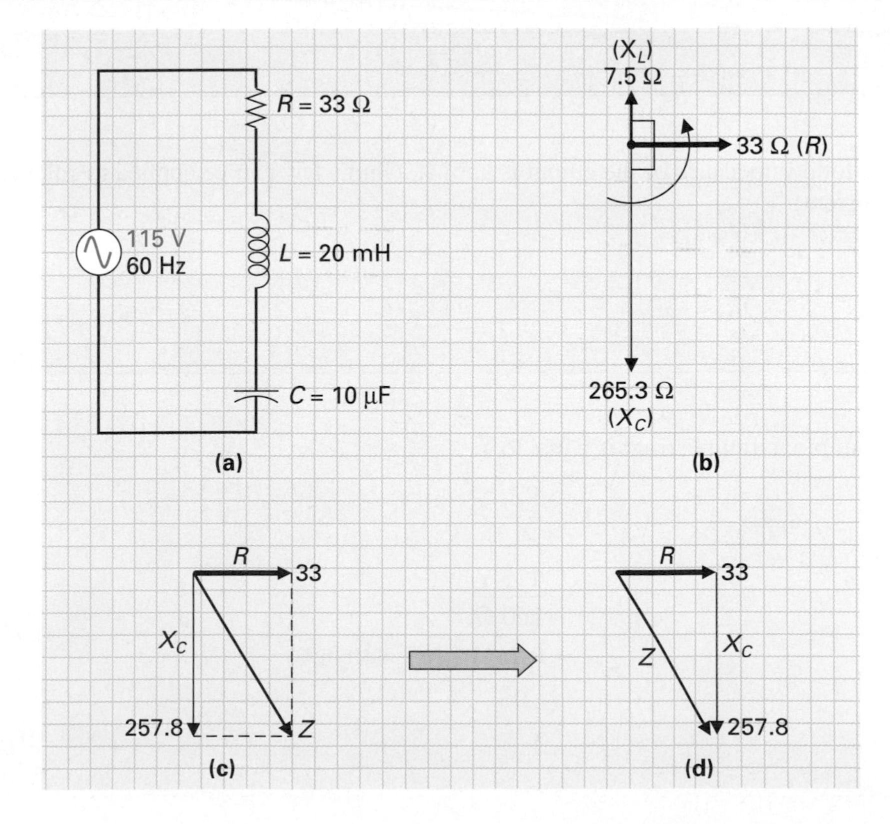

FIGURE 13-2 Series Circuit Impedance.

In this example, therefore, the circuit impedance will be equal to

$$Z = \sqrt{R^2 + X^2} = \sqrt{33^2 + 257.8^2} = 260\ \Omega$$

Since reactance (X) is equal to the difference between (symbolized ~) X_L and X_C ($X_L \sim X_C$), the impedance formula can be modified slightly to incorporate the calculation to determine the difference between X_L and X_C.

$$Z = \sqrt{R^2 + (X_L \sim X_C)^2}$$

Using our example with this new formula, we arrive at the same value of impedance, and since the difference between X_L and X_C resulted in a capacitive vector, the circuit is said to act capacitively.

$$\begin{aligned} Z &= \sqrt{R^2 + (X_L \sim X_C)^2} \\ &= \sqrt{33^2 + (7.5 \sim 265.3)^2} \\ &= \sqrt{33^2 + 257.8^2} \\ &= 260\ \Omega \end{aligned}$$

If, on the other hand, the component values were such that the difference was an inductive vector, then the circuit would be said to act inductively.

13-1-2 *Current*

The current in a series circuit is the same at all points throughout the circuit, and therefore

$$I = I_R = I_L = I_C$$

Once the total impedance of the circuit is known, Ohm's law can be applied to calculate the circuit current:

$$I = \frac{V_S}{Z}$$

In the example, circuit current is equal to

$$\begin{aligned} I &= \frac{V_S}{Z} \\ &= \frac{115\ \text{V}}{260\ \Omega} \\ &= 0.44\ \text{A} \quad \text{or} \quad 440\ \text{mA} \end{aligned}$$

13-1-3 *Voltage*

Now that you know the value of current flowing in the series circuit, you can calculate the voltage drops across each component, as shown in Figure 13-3(a).

$$V_R = I \times R$$

$$V_L = I \times X_L$$

$$V_C = I \times X_C$$

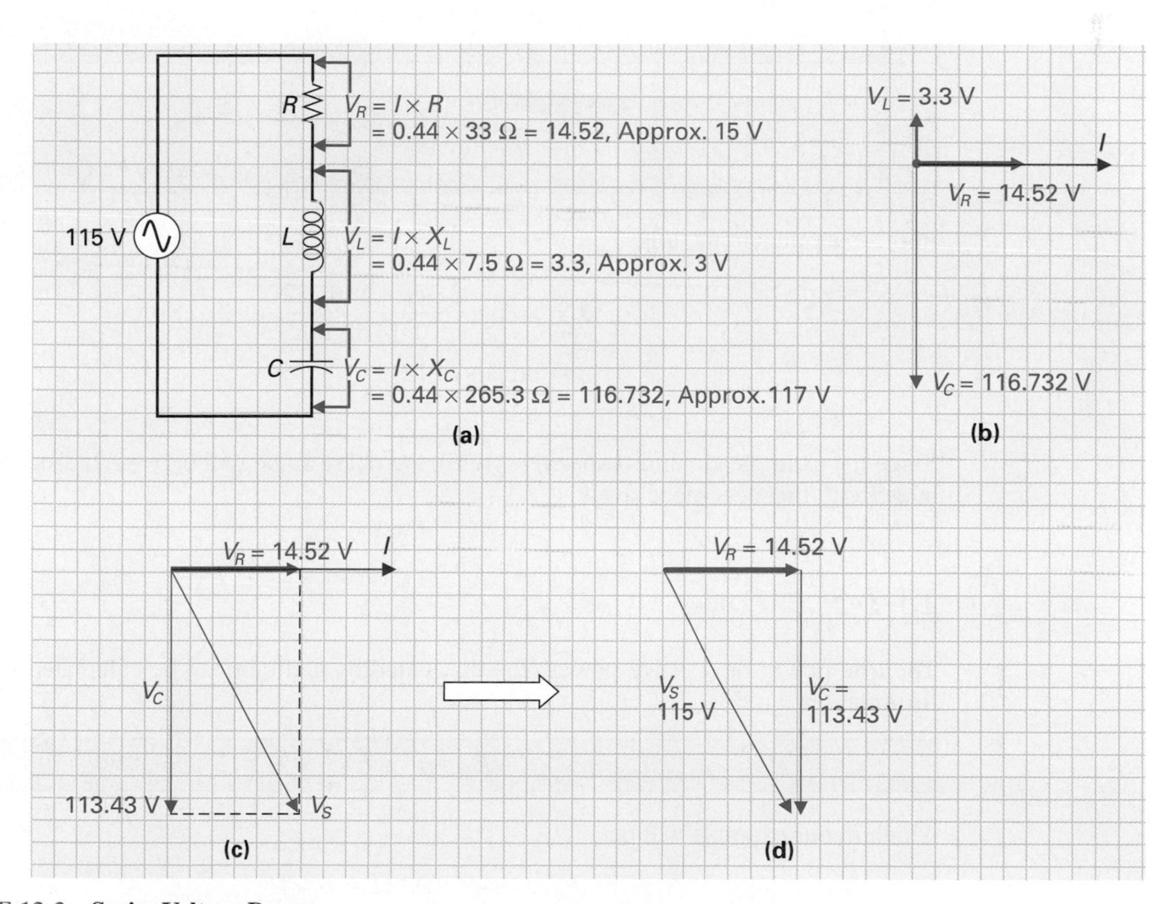

FIGURE 13-3 Series Voltage Drops.

Since none of these voltages are in phase with one another as shown in Figure 13-3(b), they must be added vectorially to obtain the applied voltage. The formula, based on the Pythagorean theorem and illustrated in Figure 13-3(c) and (d), can be used to calculate V_S.

$$V_S = \sqrt{V_R^2 + (V_L \sim V_C)^2}$$

In the example circuit, the applied voltage is, as we already know, 115 V.

$$\begin{aligned} V_S &= \sqrt{15^2 + (3 \sim 117)^2} \\ &= \sqrt{225 + 12{,}996} \\ &= \sqrt{13{,}221} \\ &= 115 \text{ V} \end{aligned}$$

13-1-4 *Phase Angle*

As can be seen in Figure 13-3(c), there is a phase difference between the source voltage (V_S) and the circuit current (I). This phase difference can be calculated with either of the following formulas:

$$\theta = \arctan \frac{V_L \sim V_C}{V_R}$$

$$\theta = \arctan \frac{X_L \sim X_C}{R}$$

In the example circuit, θ is

$$\begin{aligned} \theta &= \arctan \frac{3.3 \text{ V} \sim 116.732 \text{ V}}{14.52 \text{ V}} \\ &= \arctan 7.812 \\ &= 82.7^\circ \end{aligned}$$

Since the example circuit is capacitive (ICE), the phase angle will be -82.7° since V_S lags I in a circuit that acts capacitively.

13-1-5 *Power*

The true power or resistive power (P_R) dissipated by a circuit can be calculated using the formula

$$P_R = I^2 \times R$$

which in our example will be

$$P_R = 0.44^2 \times 33\ \Omega = 6.4 \text{ W}$$

The apparent power (P_A) consumed by the circuit is calculated by

$$P_A = V_S \times I$$

which in our example will be

$$P_A = 115\text{ V} \times 0.44 = 50.6\text{ volt-amperes (VA)}$$

The true or actual power dissipated by the resistor is, as expected, smaller than the apparent power that appears to be in use.

The power factor can be calculated, as usual, by

$$\text{PF} = \cos\theta = \frac{R}{Z} = \frac{P_R}{P_A}$$

PF of 0 = reactive circuit
PF of 1 = resistive circuit

In the example circuit, PF = 0.126, indicating that the circuit is mainly reactive.

EXAMPLE:

For a series circuit where $R = 10\ \Omega$, $L = 5$ mH, $C = 0.05\ \mu$F, and $V_S = 100$ V/2 kHz, calculate:

a. X_C
b. X_L
c. Z
d. I
e. V_R, V_C, and V_L
f. Apparent power
g. True power
h. Power factor
i. Phase angle

Solution:

a. $X_C = \dfrac{1}{2\pi fC} = 1.6\text{ k}\Omega$

b. $X_L = 2\pi fL = 62.8\ \Omega$

c. $Z = \sqrt{R^2 + (X_L \sim X_C)^2}$
$= \sqrt{(10\ \Omega)^2 + (1.6\text{ k}\Omega \sim 62.8\ \Omega)^2}$
$= \sqrt{100\ \Omega^2 + (1.54\text{ k}\Omega)^2}$
$= 1.54\text{ k}\Omega$ (capacitive circuit due to high X_C)

d. $I = \dfrac{V_S}{Z} = \dfrac{100\text{ V}}{1.54\text{ k}\Omega} = 64.9\text{ mA}$

e. $V_R = I \times R = 64.9\text{ mA} \times 10\ \Omega = 0.65\text{ V}$
$V_C = I \times X_C = 64.9\text{ mA} \times 1.6\text{ k}\Omega = 103.9\text{ V}$
$V_L = I \times X_L = 64.9\text{ mA} \times 62.8\ \Omega = 4.1\text{ V}$

f. Apparent power $= V_S \times I = 100\text{ V} \times 64.9\text{ mA} = 6.49\text{ VA}$

g. True power $= I^2 \times R = 64.9^2 \times 10\ \Omega = 42.17\text{ mW}$

h. $\text{PF} = \dfrac{R}{Z} = \dfrac{10\ \Omega}{1.5\text{ k}\Omega} = 0.006$ (reactive circuit)

i. $\theta = \arctan\dfrac{V_L \sim V_C}{V_R}$
$= \arctan\dfrac{4.1\text{ V} \sim 103.9\text{ V}}{0.65\text{ V}}$
$= \arctan 153.54$
$= 89.63°$

Capacitive circuit (ICE); therefore, V_S lags I by $-89.63°$.

SELF-TEST EVALUATION POINT FOR SECTION 13-1

Now that you have completed this section, you should be able to:

Objective 1. Identify the difference between a series and a parallel RCL circuit.

Objective 2. Explain the following as they relate to series RLC circuits:

a. Impedance
b. Current
c. Voltage
d. Phase angle
e. Power

Use the following questions to test your understanding of Section 13-1:

1. List in order the procedure that should be followed to fully analyze a series *RLC* circuit.
2. State the formulas for calculating the following in relation to a series *RLC* circuit:
   a. Impedance
   b. Current
   c. Apparent power
   d. V_S
   e. True power
   f. V_R
   g. V_L
   h. V_C
   i. Phase angle (θ)
   j. Power factor (PF)

13-2 PARALLEL *RLC* CIRCUIT

Now that the characteristics of a series circuit are understood, let us connect a resistor, inductor, and capacitor in parallel with one another. Figure 13-4(a) and (b) show the current and voltage relationships of a parallel *RLC* circuit.

13-2-1 *Voltage*

As can be seen in Figure 13-4(a), the voltage across any parallel circuit will be equal and in phase. Therefore,

$$V_R = V_L = V_C = V_S$$

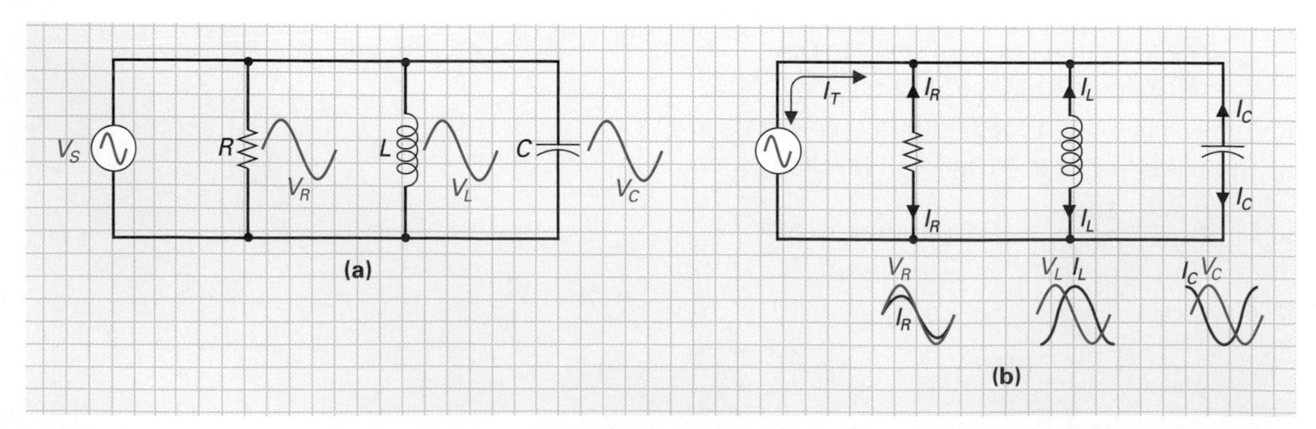

FIGURE 13-4 Parallel *RLC* Circuit. (a) *RLC* Parallel Circuit Voltage: Voltages across Each Component Are All Equal and in Phase with One Another in a Parallel Circuit. (b) *RLC* Parallel Circuit Currents: I_R Is in Phase with V_R, I_L Lags V_L by 90°, and I_C Leads V_C by 90°.

13-2-2 *Current*

With current, we must first calculate the individual branch currents (I_R, I_L, and I_C) and then calculate the total circuit current (I_T). An example circuit is illustrated in Figure 13-5(a), and the branch currents can be calculated by using the formulas

$$I_R = \frac{V}{R}$$

$$I_L = \frac{V}{X_L}$$

$$I_C = \frac{V}{X_C}$$

Figure 13-5(b) illustrates these branch currents vectorially, with I_R in phase with V_S, I_L lagging by 90°, and I_C leading I_R by 90°. The 180° phase difference between I_C and I_L results in a cancellation, as shown in Figure 13-5(c).

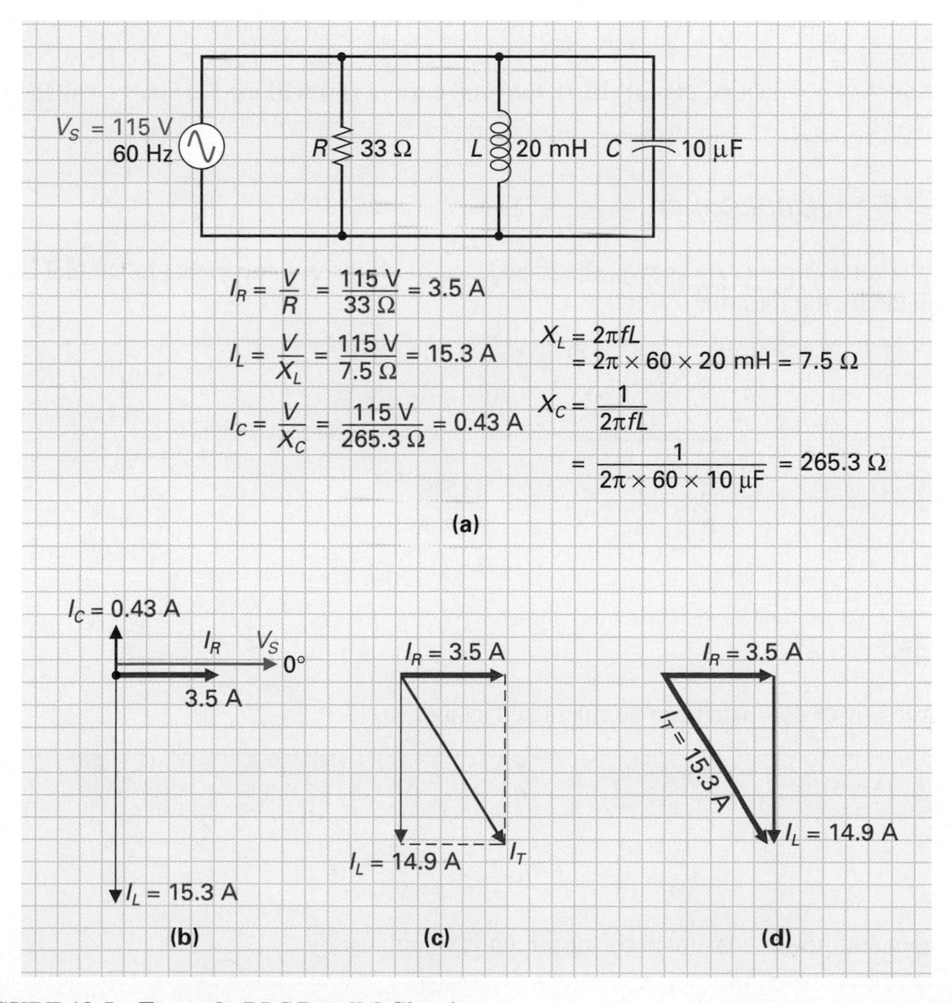

FIGURE 13-5 Example *RLC* Parallel Circuit.

The total current (I_T) can be calculated by using the Pythagorean theorem on the right triangle, as illustrated in Figure 13-5(d).

$$I_T = \sqrt{I_R^2 + I_X^2}$$

$$(I_X = I_L \sim I_C)$$

$$= \sqrt{3.5^2 + 14.9^2}$$
$$= 15.3 \text{ A}$$

13-2-3 *Phase Angle*

As shown in Figure 13-5(b) and (c), there is a phase difference between the source voltage (V_S) and the circuit current (I_T). This phase difference can be calculated using the formula

$$\theta = \arctan \frac{I_L \sim I_C}{I_R}$$

$$= \arctan \frac{15.3 \text{ A} - 0.43 \text{ A}}{3.5 \text{ A}}$$
$$= \arctan 4.25$$
$$= 76.7^\circ$$

Since this is an inductive circuit (ELI), the total current (I_T) will lag the source voltage (V_S) by -76.7°.

13-2-4 *Impedance*

With the total current (I_T) known, the impedance of all three components in parallel can be calculated by the formula

$$Z = \frac{V}{I_T}$$

$$= \frac{115 \text{ V}}{15.3 \text{ A}}$$
$$= 7.5 \ \Omega$$

13-2-5 *Power*

The true power dissipated can be calculated using

$$P_R = I_R^2 \times R$$

$$= 3.5^2 \times 33 \ \Omega$$
$$= 404.3 \text{ W}$$

The apparent power consumed by the circuit is calculated by

$$P_A = V_S \times I_T$$

$= 115\ \text{V} \times 15.3\ \text{A}$
$= 1759.5$ volt-amperes (VA)

Finally, the power factor can be calculated, as usual, with

$$\text{PF} = \cos\theta = \frac{P_R}{P_A}$$

In the example circuit, PF = 0.23.

SELF-TEST EVALUATION POINT FOR SECTION 13-2

Now that you have completed this section, you should be able to:

Objective 3. Explain the following as they relate to parallel *RLC* circuits:

a. Voltage
b. Current
c. Impedance
d. Phase angle
e. Power

Use the following questions to test your understanding of Section 13-2:

1. State the formulas for calculating the following in relation to a parallel *RLC* circuit:
   a. I_R **b.** I_T **c.** I_C **d.** I_L **e.** θ **f.** Z
2. State the formulas for
   a. P_R **b.** P_X **c.** P_A **d.** PF

13-3 RESONANCE

Resonance is a circuit condition that occurs when the inductive reactance (X_L) and the capacitive reactance (X_C) have been balanced. Figure 13-6 illustrates a parallel- and a series-connected *LC* circuit. If a dc voltage is applied to the input of either circuit, the capacitor will act as an open (X_C = infinite Ω) and the inductor will act as a short ($X_L = 0\ \Omega$).

Resonance
Circuit condition that occurs when the inductive reactance (X_L) is equal to the capacitive reactance (X_C).

If a low-frequency ac is now applied to the input, X_C will decrease from maximum and X_L will increase from zero. As the ac frequency is increased further, the capacitive reactance will continue to fall ($X_C\downarrow \propto 1/f\uparrow$) and the inductive reactance to rise ($X_L\uparrow \propto f\uparrow$), as shown in Figure 13-7.

As the input ac frequency is increased further, a point will be reached where X_L will equal X_C, and this condition is known as *resonance*. The frequency at which $X_L = X_C$ in either a parallel or a series *LC* circuit is known as the *resonant frequency* (f_0) and can be

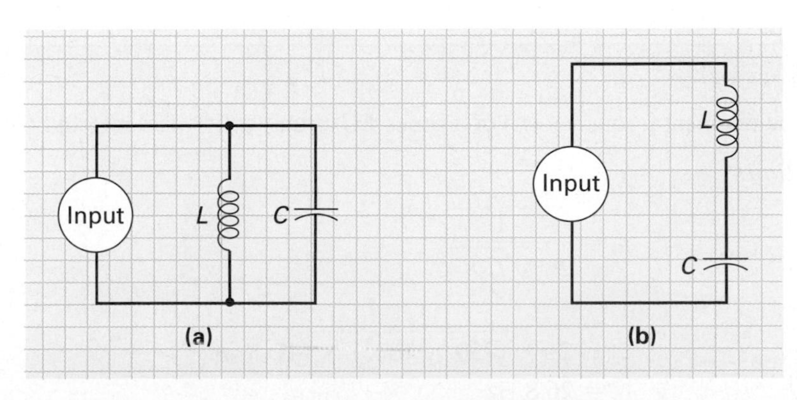

FIGURE 13-6 **Resonance. (a) Parallel *LC* Circuit. (b) Series *LC* Circuit.**

FIGURE 13-7
Frequency Versus Reactance.

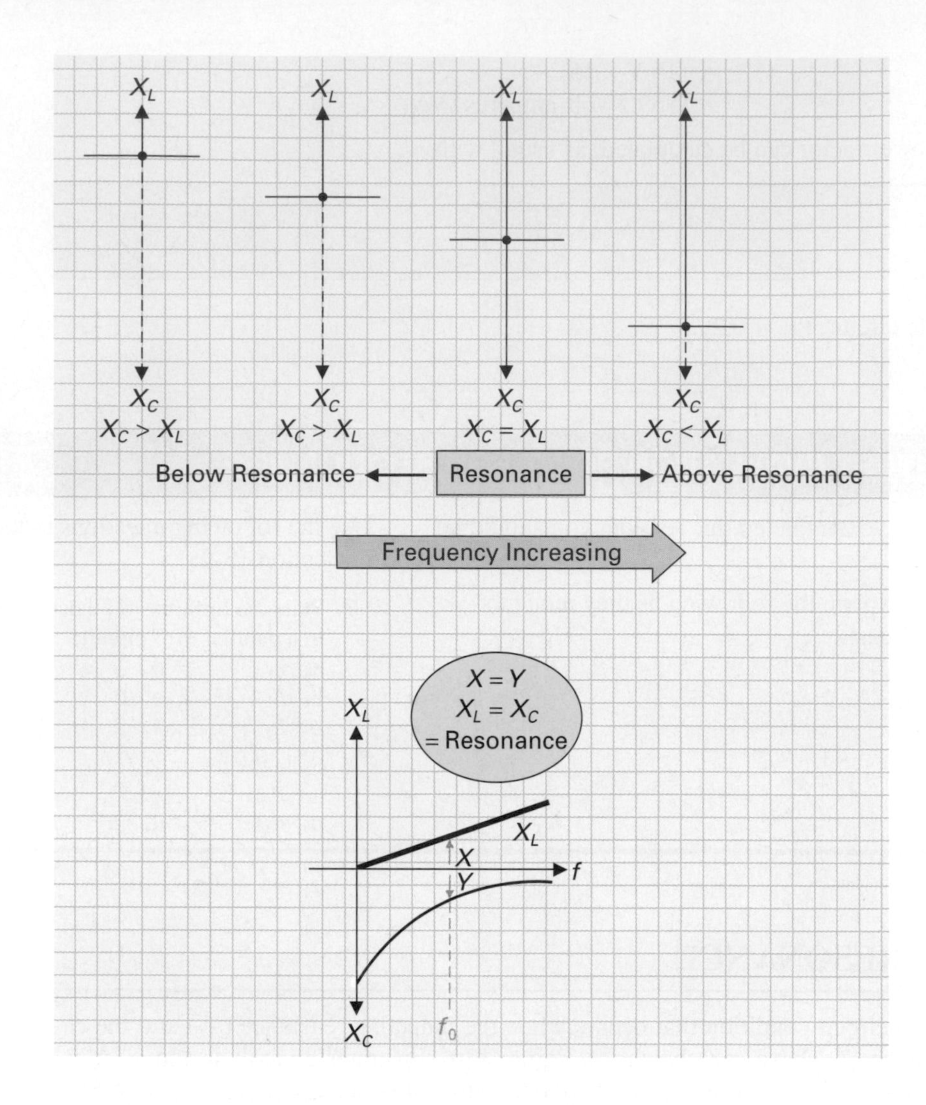

calculated by the following formula, which has been derived from the capacitive and inductive reactance formulas:

$$f_0 = \frac{1}{2\pi\sqrt{LC}}$$

where f_0 = resonant frequency, in hertz (Hz)
L = inductance, in henrys (H)
C = capacitance, in farads (F)

EXAMPLE:

Calculate the resonant frequency (f_0) of a series LC circuit if $L = 750$ mH and $C = 47\ \mu$F.

Solution:

$$\begin{aligned} f_0 &= \frac{1}{2\pi\sqrt{L \times C}} \\ &= \frac{1}{2\pi\sqrt{(750 \times 10^{-3}) \times (47 \times 10^{-6})}} \\ &= 26.8 \text{ Hz} \end{aligned}$$

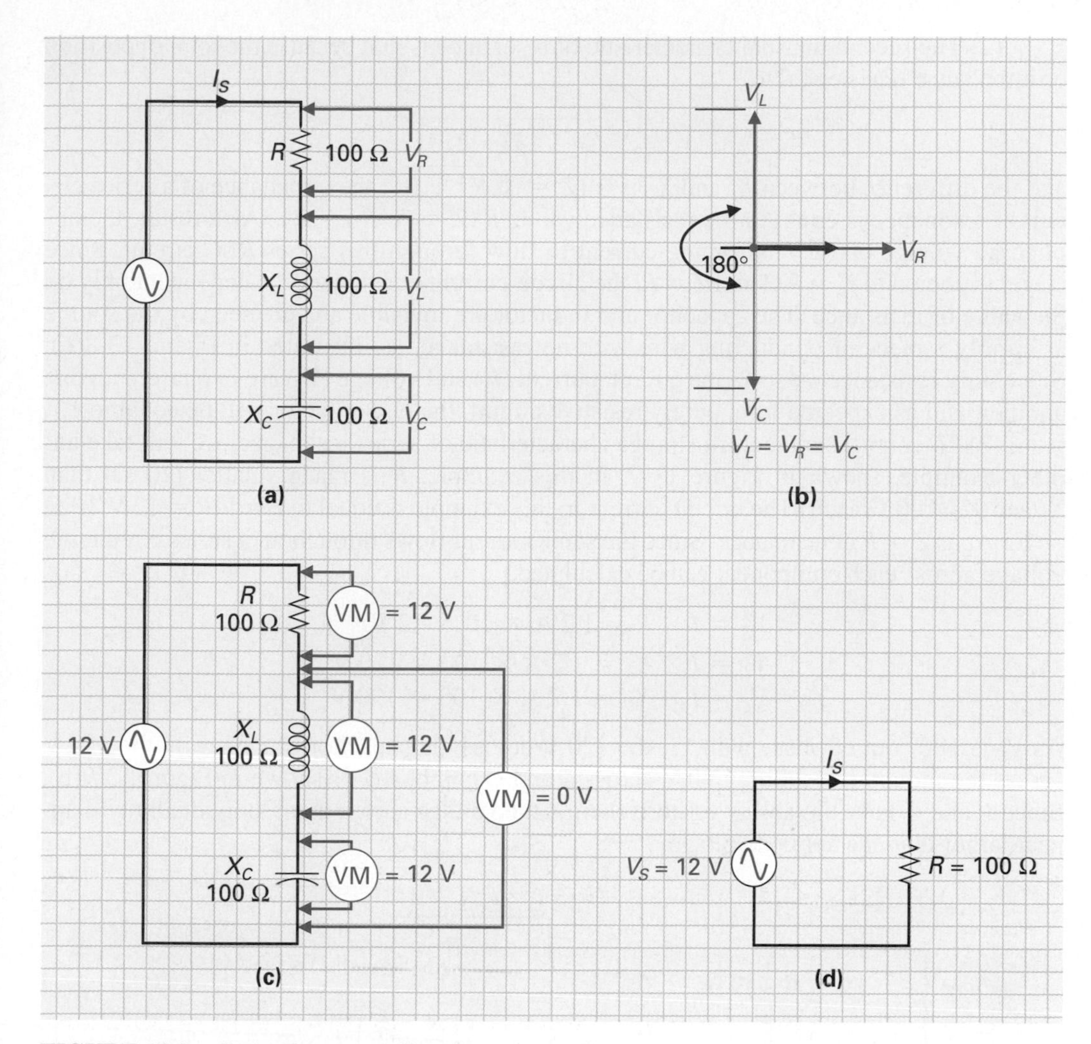

FIGURE 13-8 Series Resonant Circuit.

13-3-1 *Series Resonance*

Figure 13-8(a) illustrates a series *RLC* circuit at resonance ($X_L = X_C$), or **series resonant circuit.** The ac input voltage causes current to flow around the circuit, and since all the components are connected in series, the same value of current (I_S) will flow through all the components. Since R, X_L, and X_C are all equal to 100 Ω and the current flow is the same throughout, the voltage dropped across each component will be equal, as illustrated vectorially in Figure 13-8(b).

Series Resonant Circuit
A resonant circuit in which the capacitor and coil are in series with the applied ac voltage.

The voltage across the resistor is in phase with the series circuit current (I_S); however, since the voltage across the inductor (V_L) is 180° out of phase with the voltage across the capacitor (V_C), and both are equal to one another, V_L cancels V_C, when both are measured in series.

Three unusual characteristics occur when a circuit is at resonance, which do not occur at any other frequency.

(1) The first is that if V_L and V_C cancel, the voltage across L and C will measure 0 V on a voltmeter. Since there is effectively no voltage being dropped across these two components, all the voltage must be across the resistor (V_R = 12 V). This is true; however, since the same current flows throughout the series circuit, a voltmeter will measure 12 V across C, 12 V across L, and 12 V across R, as shown in Figure 13-8(c). It now appears that the voltage drops around the series circuit (36 V) do not equal the voltage applied (12 V). This is not true, as V_L and V_C cancel, because they are out of phase with one another, so Kirchhoff's voltage law is still valid.

TIME LINE

One of the best known and most admired men in the latter half of the eighteenth century was the American Benjamin Franklin (1706–1790). He is renowned for his kite-in-a-storm experiment, which proved that lightning is electricity. Franklin also discovered that there was only one type of electricity, and that the two previously believed types were simply two characteristics of electricity. Vitreous was renamed *positive charge* and resinous was called *negative charge,* terms that were invented by Franklin, along with the terms *battery* and *conductor*, which are still used today.

(2) The second unusual characteristic of resonance is that because the total opposition or impedance (Z) is equal to

$$Z = \sqrt{R^2 + (X_L \sim X_C)^2}$$

and the difference between X_L and X_C is 0 ($Z = \sqrt{R^2 + 0}$), the impedance of a series circuit at resonance is equal to the resistance value R ($Z = \sqrt{R^2} = R$). As a result, the applied ac voltage of 12 V is forcing current to flow through this series *RLC* circuit. Since current is equal to $I_S = V/Z$ and $Z = R$, the circuit current at resonance is dependent only on the value of resistance. The capacitor and inductor are invisible and are seen by the source as simply a piece of conducting wire with no resistance, as illustrated in Figure 13-8(d). Since only resistance exists in the circuit, current (I_S) and voltage (V_S) are in phase with one another, and as expected for a purely resistive circuit, the power factor will be equal to 1.

(3) To emphasize the third strange characteristic of series resonance, we will take another example, shown in Figure 13-9. In this example, R is made smaller (10 Ω) than X_L and X_C (100 Ω each).The circuit current in this example is equal to $I = V/R = 12\text{ V}/10\ \Omega = 1.2\text{ A}$, as $Z = R$ at resonance. Since the same current flows throughout a series circuit, the voltage across each component can be calculated.

$$V_R = I \times R = 1.2\text{ A} \times 10 = 12\text{ V}$$
$$V_L = I \times X_L = 1.2\text{ A} \times 100 = 120\text{ V}$$
$$V_C = I \times X_C = 1.2\text{ A} \times 100 = 120\text{ V}$$

As V_L is 180° out of phase with V_C, the 120 V across the capacitor cancels with the 120 V across the inductor, resulting in 0 V across L and C combined, as shown in Figure 13-9(b). Since L and C have the ability to store energy, the voltage across them individually will appear larger than the applied voltage.

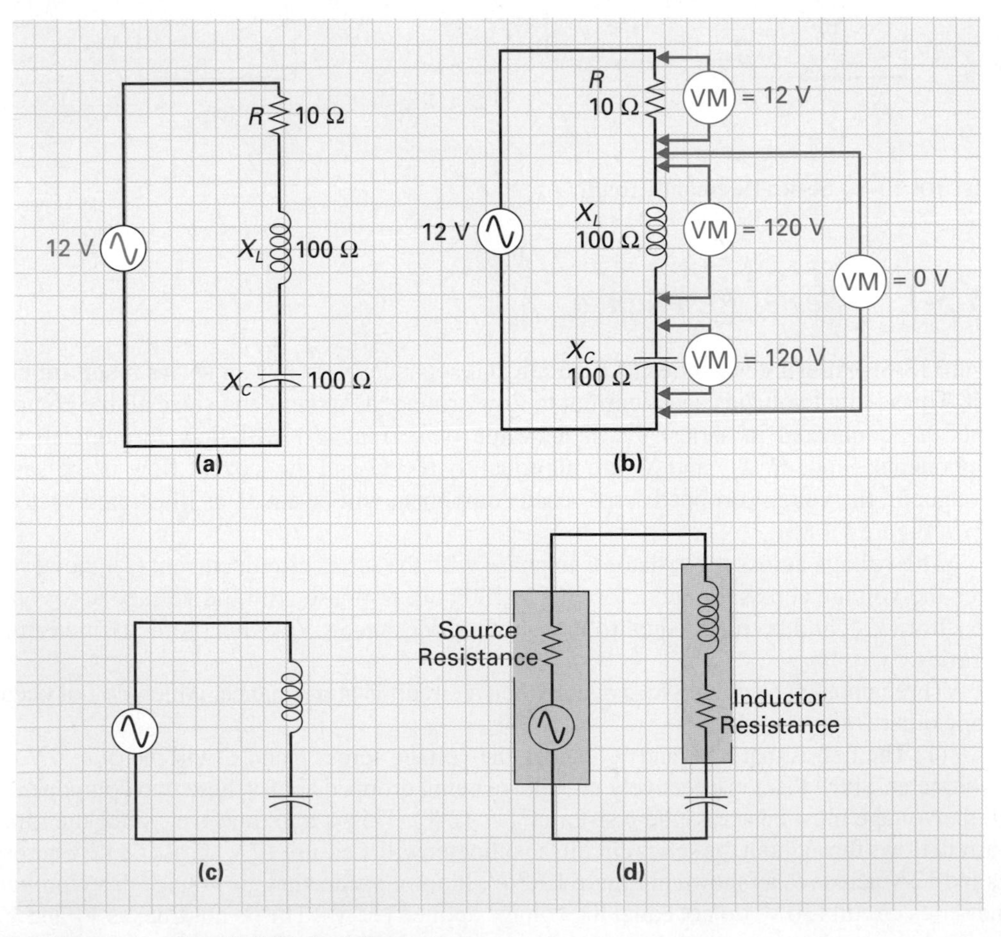

FIGURE 13-9 **Circuit Effects at Resonance.**

If the resistance in the circuit is removed completely, as shown in Figure 13-9(c), the circuit current, which is determined by the resistance only, will increase to a maximum ($I\uparrow = V/R\downarrow$) and, consequently, cause an infinitely high voltage across the inductor and capacitor ($V\uparrow = I\uparrow \times R$). In reality, the ac source will have some value of internal resistance, and the inductor, which is a long length of thin wire ($R\uparrow$), will have some value of resistance, as shown in Figure 13-9(d), which limits the series resonant circuit current,

In summary, we can say that in a series resonant circuit,

1. The inductor and capacitor electrically disappear due to their equal but opposite effect, resulting in a 0 V drop across the series combination, and the circuit seems purely resistive.
2. The current flow is large because the impedance of the circuit is low and equal to the series resistance (R), which has the source voltage developed across it.
3. The individual voltage drops across the inductor or capacitor can be larger than the source voltage if R is smaller than X_L and X_C.

Quality Factor

As discussed previously in the inductance chapter, the Q factor is a ratio of inductive reactance to resistance and is used to express how efficiently an inductor will store rather than dissipate energy. In a series resonant circuit, the Q factor indicates the quality of the series resonant circuit, or is the ratio of the reactance to the resistance.

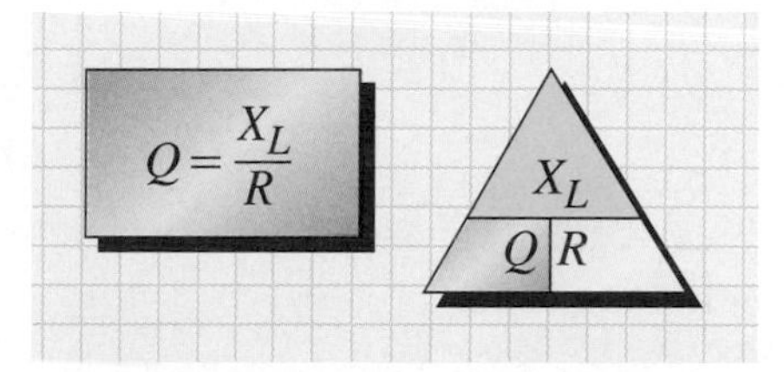

or, since $X_L = X_C$,

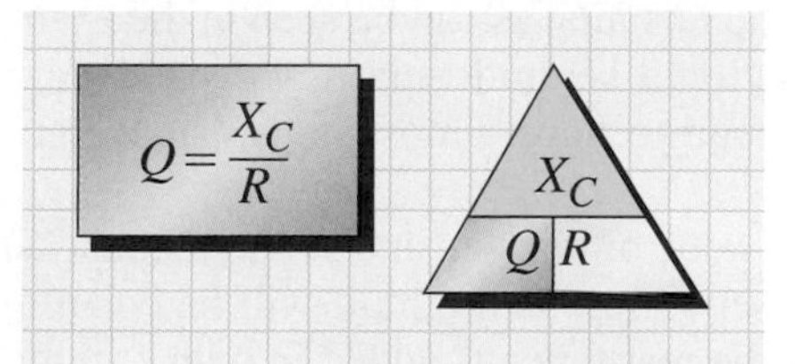

Another way to calculate the Q of a series resonant circuit is by using the formula

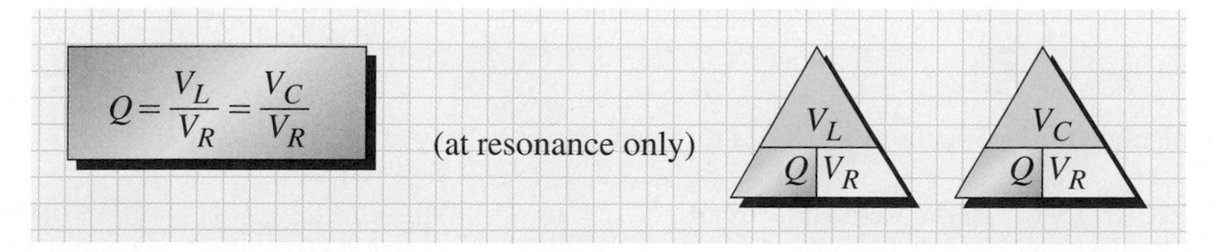

or, since $V_R = V_S$,

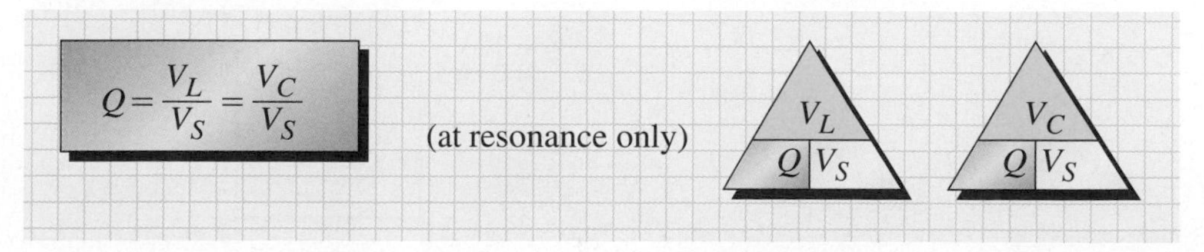

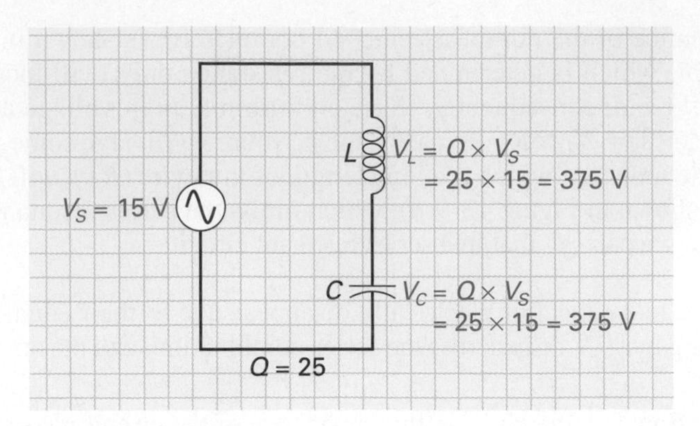

FIGURE 13-10 Quality Factor at Resonance.

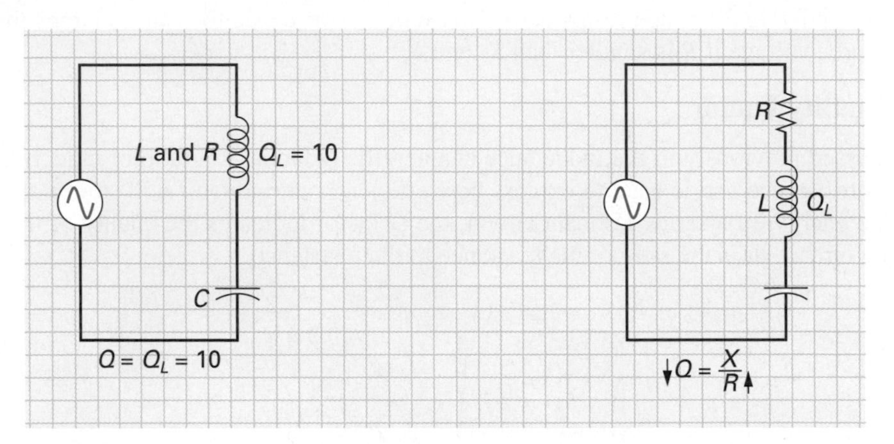

FIGURE 13-11 Resistance within Inductor.

If the Q and source voltage are known, the voltage across the inductor or capacitor can be found by transposition of the formula, as can be seen in the example in Figure 13-10.

The Q of a resonant circuit is almost entirely dependent on the inductor's coil resistance, because capacitors tend to have almost no resistance at all, only reactance, which makes them very efficient.

The inductor has a Q value of its own, and if only L and C are connected in series with one another, the Q of the series resonant circuit will be equal to the Q of the inductor, as shown in Figure 13-11. If the resistance is added in with L and C, the Q of the series resonant circuit will be less than that of the inductor's Q.

EXAMPLE:

Calculate the resistance of the series resonant circuit illustrated in Figure 13-12.

FIGURE 13-12 Series Resonant Circuit Example.

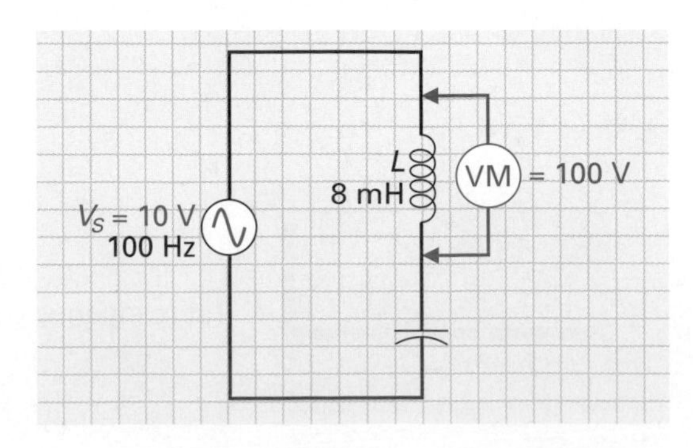

■ *Solution:*

$$Q = \frac{V_L}{V_S} = \frac{100 \text{ V}}{10 \text{ V}} = 10$$

Since $Q = X_L/R$ and $R = X_L/Q$, if the inductive reactance can be found, the R can be determined.

$$X_L = 2\pi \times f \times L$$
$$= 2\pi \times 100 \times 8 \text{ mH}$$
$$= 5 \ \Omega$$

R will therefore equal

$$R = \frac{X_L}{Q}$$
$$= \frac{5}{10}$$
$$= 0.5 \ \Omega$$

Bandwidth

A series resonant circuit is selective in that frequencies at resonance or slightly above or below will cause a larger current than frequencies well above or below the circuit's resonant frequency. The group or band of frequencies that causes the larger current is called the circuit's **bandwidth.**

Bandwidth
Width of the group or band of frequencies between the half-power points.

Figure 13-13 illustrates a series resonant circuit and its bandwidth. The × marks on the curve illustrate where different frequencies were applied to the circuit and the resulting value of current measured in the circuit. The resulting curve produced is called a **frequency response curve,** as it illustrates the circuit's response to different frequencies. At resonance, $X_L = X_C$ and the two cancel, which is why maximum current was present in the circuit (100 mA) when the resonant frequency (100 Hz) was applied.

Frequency Response Curve
A graph indicating a circuit's response to different frequencies.

The bandwidth includes the group or band of frequencies that cause 70.7% or more of the maximum current to flow within the series resonant circuit. In this example, frequencies

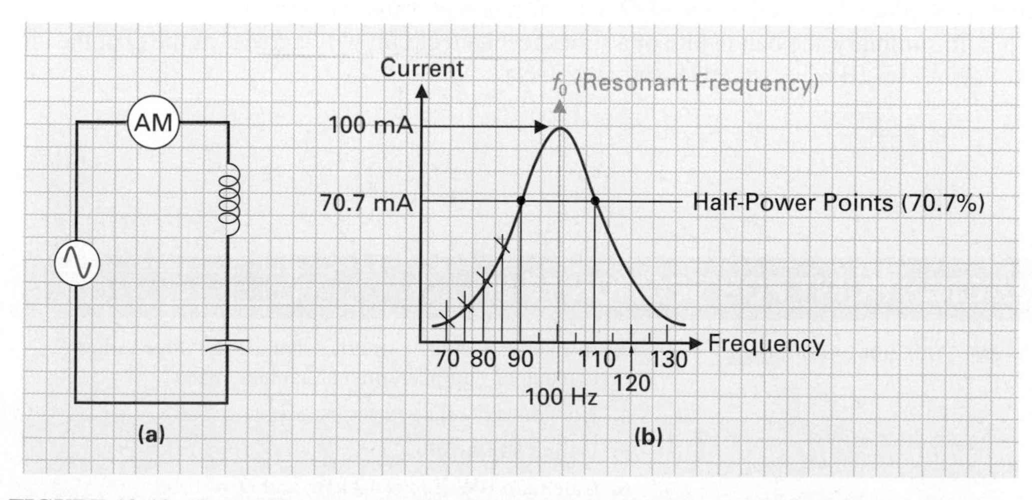

FIGURE 13-13 Series Resonant Circuit Bandwidth. (a) Circuit. (b) Frequency Response Curve.

from 90 to 110 Hz cause 70.7 mA or more, which is 70.7% of maximum (100 mA), to flow. The bandwidth in this example is equal to

$$\text{BW} = 110 - 90 = 20 \text{ Hz}$$

Cutoff Frequency
Frequency at which the gain of the circuit falls below 0.707 of the maximum current or half-power.

Half-Power Point
A point at which power is 50%. This half-power point corresponds to 70.7% of the total current.

(110 Hz and 90 Hz are known as **cutoff frequencies**).

Referring to the bandwidth curve in Figure 13-13(b), you may notice that 70.7% is also called the **half-power point,** although it does not exist halfway between 0 and maximum. This value of 70.7% is not the half-current point but the half-power point, as we can prove with a simple example.

EXAMPLE:

$R = 2\ \text{k}\Omega$ and $I = 100$ mA; therefore, power $= I^2 \times R = (100\ \text{mA})^2 \times 2\ \text{k}\Omega = 20$ W. If the current is now reduced so that it is 70.7% of its original value, calculate the power dissipated.

Solution:

$$P = I^2 \times R = (70.7\ \text{mA})^2 \times 2\ \text{k}\Omega = 10\ \text{W}$$

In summary, the 70.7% current points are equal to the 50% or half-power points. A circuit's bandwidth is the band of frequencies that exists between the 70.7% current points or half-power points.

The bandwidth of a series resonant circuit can also be calculated by use of the formula

$$\text{BW} = \frac{f_0}{Q_{f_0}}$$

f_0 = resonant frequency
Q_{f_0} = quality factor at resonance

This formula states that the BW is proportional to the resonant frequency of the circuit and inversely proportional to the Q of the circuit.

Figure 13-14 illustrates three example response curves. In these three examples, the value of R is changed from 100 Ω to 200 Ω to 400 Ω. This does not vary the resonant frequency, but simply alters the Q and therefore the BW. The resistance value will determine the Q of the circuit, and since Q is inversely proportional to resistance, Q is proportional to current; consequently, a high value of Q will cause a high value of current.

In summary, the bandwidth of a series resonant circuit will increase as the Q of the circuit decreases ($\text{BW}\uparrow = f_0/Q\downarrow$), and vice versa.

SELF-TEST EVALUATION POINT FOR SECTION 13-3-1

Use the following questions to test your understanding of Section 13-3-1:

1. Define *resonance.*
2. What is series resonance?
3. In a series resonant circuit, what are the three rather unusual circuit phenomena that take place?
4. How does Q relate to series resonance?
5. Define *bandwidth.*
6. Calculate BW if $f_0 = 12$ kHz and $Q = 1000$.

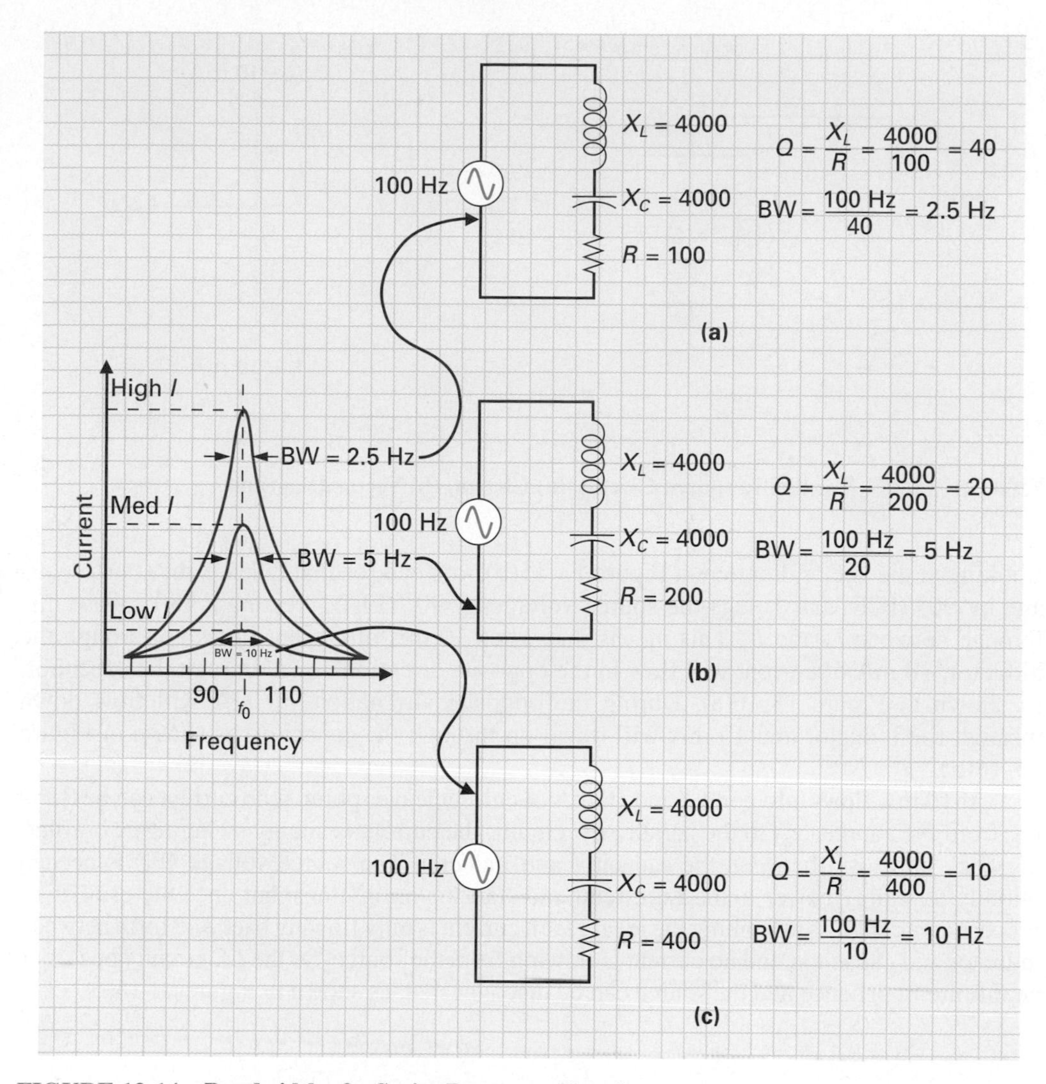

FIGURE 13-14 **Bandwidth of a Series Resonant Circuit.**

13-3-2 *Parallel Resonance*

Parallel Resonant Circuit
Circuit having an inductor and capacitor in parallel with one another, offering a high impedance at the frequency of resonance.

The **parallel resonant circuit** acts differently from the series resonant circuit, and these different characteristics need to be analyzed and discussed. Figure 13-15 illustrates a parallel resonant circuit. The inductive current could be calculated by using the formula

$$\begin{aligned} I_L &= \frac{V_L}{X_L} \\ &= \frac{10 \text{ V}}{1 \text{ k}\Omega} \\ &= 10 \text{ mA} \end{aligned}$$

The capacitive current could be calculated by using the formula

$$\begin{aligned} I_C &= \frac{V_C}{X_C} \\ &= \frac{10 \text{ V}}{1 \text{ k}\Omega} \\ &= 10 \text{ mA} \end{aligned}$$

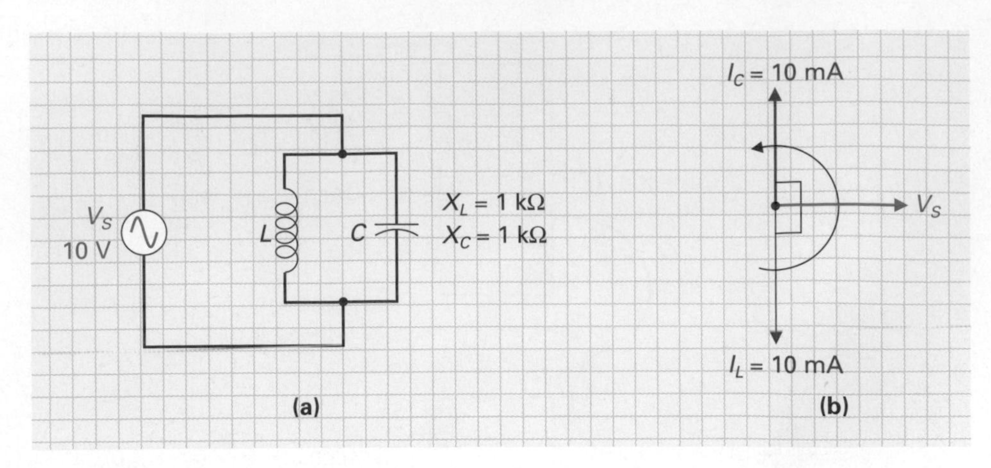

FIGURE 13-15 **Parallel Resonant Circuit. (a) Circuit. (b) Vector Diagram.**

Looking at the vector diagram in Figure 13-15(b), you can see that I_C leads the source voltage by 90° (ICE) and I_L lags the source voltage by 90° (ELI), creating a 180° phase difference between I_C and I_L. This means that when 10 mA of current flows up through the inductor, 10 mA of current will flow in the opposite direction down through the capacitor, as shown in Figure 13-16(a). During the opposite alternation, 10 mA will flow down through the inductor and 10 mA will travel up through the capacitor, as shown in Figure 13-16(b).

If 10 mA flows into point *X* and 10 mA of current leaves point *X*, no current can be flowing from the source (V_S) to the parallel *LC* circuit; the current is simply swinging or oscillating back and forth between the capacitor and inductor. The source voltage (V_S) is needed initially to supply power to the *LC* circuit and start the oscillations; but once the oscillating process is in progress (assuming the ideal case), current is only flowing back and forth between inductor and capacitor, and no current is flowing from the source. So the *LC* circuit appears as an infinite impedance and the source can be disconnected, as shown in Figure 13-16(c).

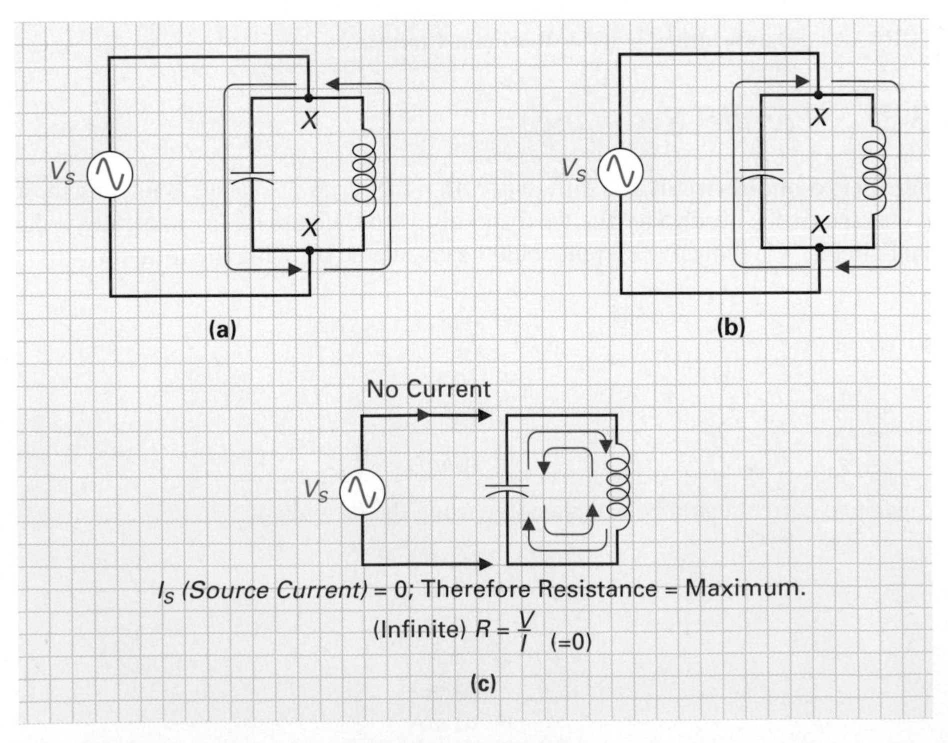

FIGURE 13-16 **Current in a Parallel Resonant Circuit.**

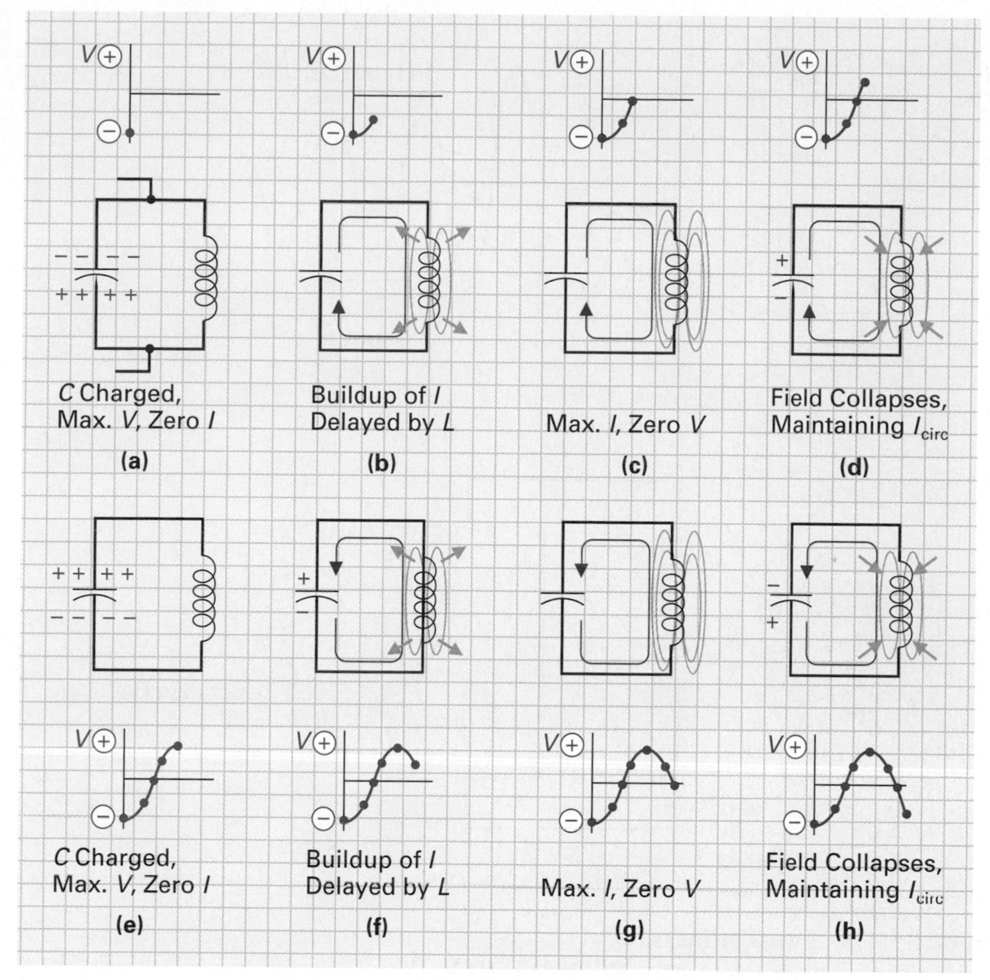

FIGURE 13-17 Energy and Current in an *LC* Parallel Circuit at Resonance.

Flywheel Action

Let's discuss this oscillating effect, called **flywheel action,** in a little more detail. The name is derived from the fact that it resembles a mechanical flywheel, which, once started, will keep going continually until friction reduces the magnitude of the rotations to zero.

The electronic equivalent of the mechanical flywheel is a resonant parallel-connected *LC* circuit. Figure 13-17(a) through (h) illustrate the continual energy transfer between capacitor and inductor, and vice versa. The direction of the circulating current reverses each half-cycle at the frequency of resonance. Energy is stored in the capacitor in the form of an electric field between the plates on one half-cycle, and then the capacitor discharges, supplying current to build up a magnetic field on the other half-cycle. The inductor stores its energy in the form of a magnetic field, which will collapse, supplying a current to charge the capacitor, which will then discharge, supplying a current back to the inductor, and so on. Due to the "storing action" of this circuit, it is sometimes related to the fluid analogy and referred to as a **tank circuit.**

Flywheel Action
Sustaining effect of oscillation in an *LC* circuit due to the charging and discharging of the capacitor and the expansion and contraction of the magnetic field around the inductor.

Tank Circuit
Circuit made up of a coil and capacitor that is capable of storing electric energy.

The Reality of Tanks

Under ideal conditions, a tank circuit should oscillate indefinitely if no losses occur within the circuit. In reality, the resistance of the coil reduces that 100% efficiency, as does friction with the mechanical flywheel. This coil resistance is illustrated in Figure 13-18(a). Unlike reactance, resistance is the opposition to current flow, with the dissipation of energy in the

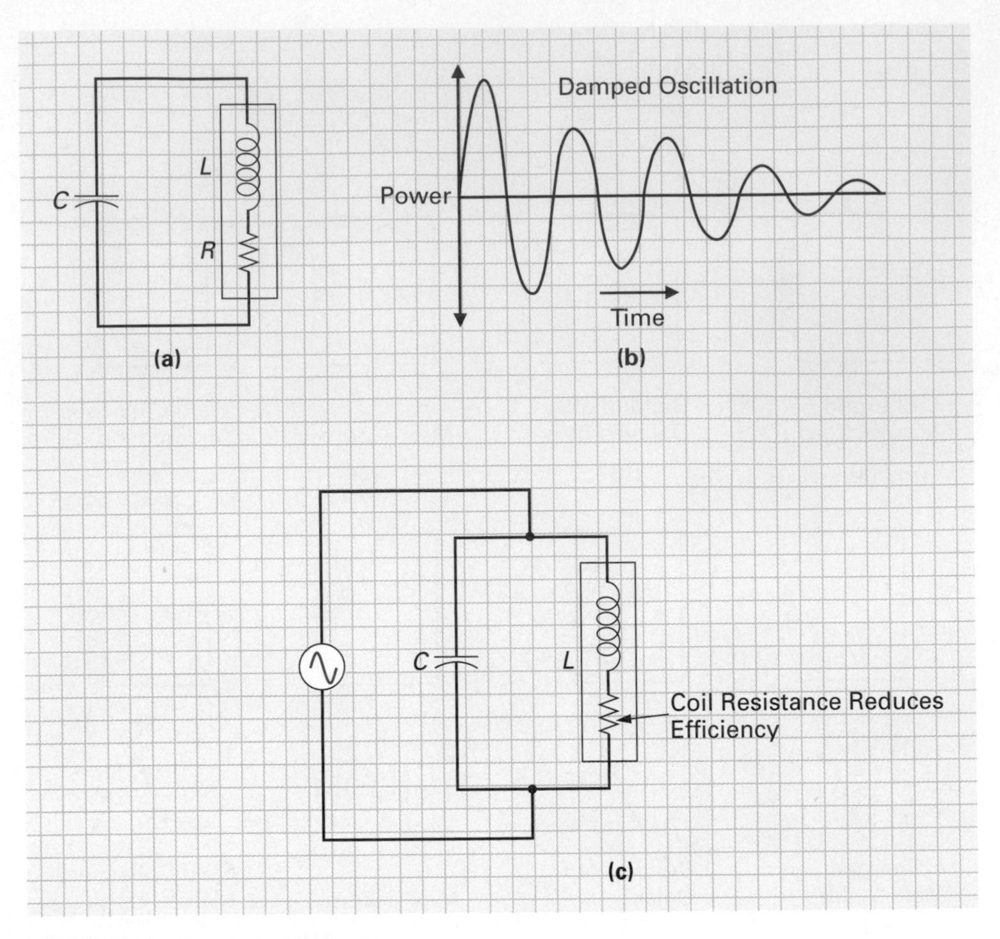

FIGURE 13-18 Losses in Tanks.

form of heat. As a small part of the energy is dissipated with each cycle, the oscillations will be reduced in size and eventually fall to zero, as shown in Figure 13-18(b). If the ac source is reconnected to the tank, as shown in Figure 13-18(c), a small amount of current will flow from the source to the tank to top up the tank or replace the dissipated power. The higher the coil resistance is, the higher the loss and the larger the current flow from source to tank to replace the loss.

Quality Factor

In the series resonant circuit, we were concerned with voltage drops since current remains the same throughout a series circuit, so

$$Q = \frac{V_C \text{ or } V_L}{V_S} \quad \text{(at resonance only)}$$

In a parallel resonant circuit, we are concerned with circuit currents rather than voltage, so

$$Q = \frac{I_{\text{tank}}}{I_S}$$

(at resonance only)

The quality factor, Q, can also be expressed as the ratio between reactance and resistance:

$$Q = \frac{X_L}{R} \quad \text{(at any frequency)}$$

Another formula, which is the most frequently used when discussing and using parallel resonant circuits, is

$$Q = \frac{Z_{tank}}{X_L}$$

(at resonance only)

This formula states that the Q of the tank is proportional to the tank impedance. A higher tank impedance results in a smaller current flow from source to tank. This assures less power is dissipated, and that means a higher-quality tank.

Of all the three Q formulas for parallel resonant circuits, $Q = I_{tank}/I_S$, $Q = X_L/R$, and $Q = Z_{tank}/X_L$, the last is the easiest to use as both X_L and the tank impedance can easily be determined in most cases where C, L, and R internal for the inductor are known.

Bandwidth

Figure 13-19 illustrates a parallel resonant circuit and two typical response curves. These response curves summarize what we have described previously, in that a parallel resonant circuit has maximum impedance [Figure 13-19(b)] and minimum current [Figure 13-19(c)] at resonance. The current versus frequency response curve shown in Figure 13-19(c) is the complete opposite to the series resonant response curve. At frequencies below resonance (< 10 kHz), X_L is low and X_C is high and the inductor offers a low reactance, producing a high current path and low impedance. On the other hand, at frequencies above resonance (> 10 kHz), the capacitor displays a low reactance, producing a high current path and low impedance. The parallel resonant circuit is like the series resonant circuit in that it responds to a band of frequencies close to its resonant frequency.

The bandwidth (BW) can be calculated by use of the formula

$$\text{BW} = \frac{f_0}{Q_{f_0}}$$

f_0 = resonant frequency
Q_{f_0} = quality factor at resonance

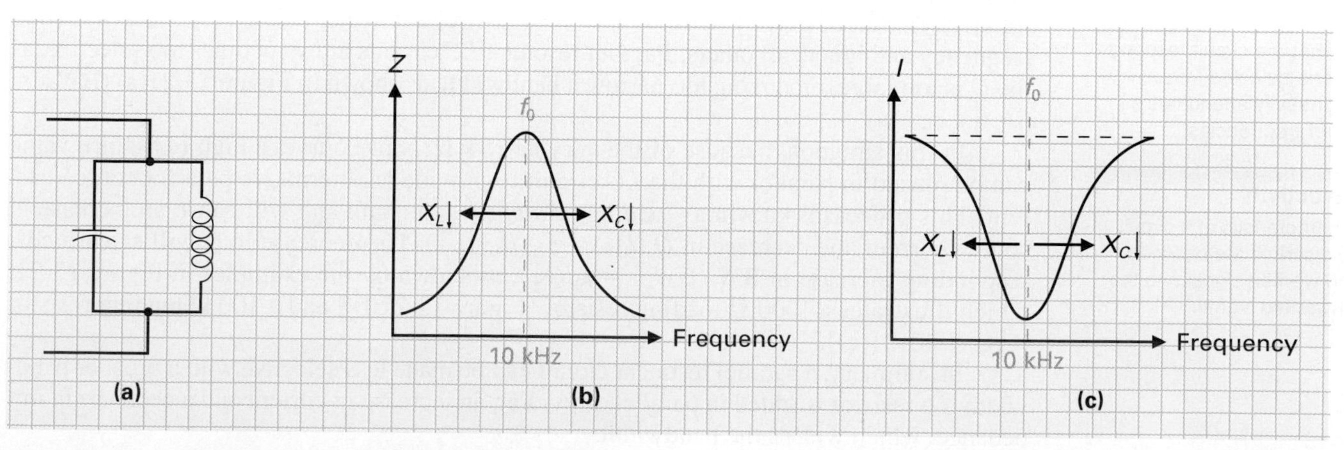

FIGURE 13-19 Parallel Resonant Circuit Bandwidth.

EXAMPLE:

Calculate the bandwidth of the circuit illustrated in Figure 13-20.

FIGURE 13-20 Bandwidth Example.

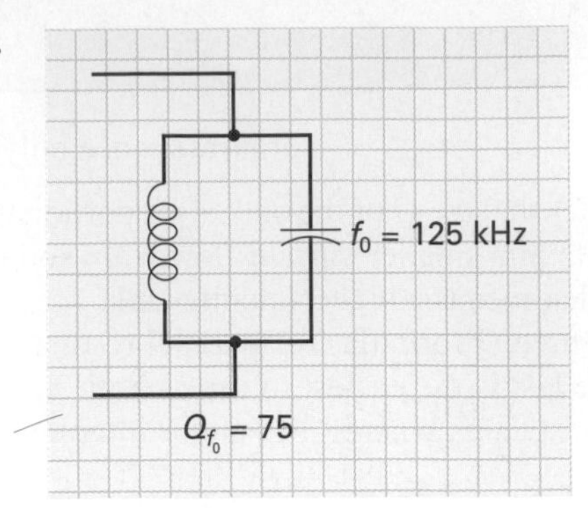

■ *Solution:*

$$\text{BW} = \frac{f_0}{Q_{f_0}}$$

$$= \frac{125 \text{ kHz}}{75}$$

$$= 1.7 \text{ kHz}$$

$$\frac{1.7 \text{ kHz}}{2} = 0.85 \text{ kHz}$$

therefore the bandwidth extends from

$$f_0 + 0.85 \text{ kHz} = 125.85 \text{ kHz}$$

$$f_0 - 0.85 \text{ kHz} = 124.15 \text{ kHz}$$

$$\text{BW} = 124.15 \text{ kHz to } 125.85 \text{ kHz}$$

Selectivity

Tuned Circuit
Circuit that can have its components' values varied so that the circuit responds to one selected frequency yet heavily attenuates all other frequencies.

Selectivity
Characteristic of a circuit to discriminate between the wanted signal and the unwanted signal.

Circuits containing inductance and capacitance are often referred to as **tuned circuits** since they can be adjusted to make the circuit responsive to a particular frequency (the resonant frequency). **Selectivity,** by definition, is the ability of a tuned circuit to respond to a desired frequency and ignore all others. Parallel resonant *LC* circuits are sometimes too selective, as the Q is too large, producing too narrow a bandwidth, as shown in Figure 13-21(a) ($\text{BW}\downarrow = f_0/Q\uparrow$).

In this situation, because of the very narrow response curve, a high resistance value can be placed in parallel with the *LC* circuit to provide an alternative path for source current. This process is known as *loading* or *damping* the tank and will cause an increase in source current and decrease in Q ($Q\downarrow = I_{\text{tank}}/I_{\text{source}}\uparrow$). The decrease in Q will cause a corresponding increase in BW ($\text{BW}\uparrow = f_0/Q\downarrow$), as shown by the examples in Figure 13-21, which illustrates a 1000 Ω loading resistor [Figure 13-21(b)] and a 100 Ω loading resistor [Figure 13-21(c)].

In summary, a parallel resonant circuit can be made less selective with a broader bandwidth if a resistor is added in parallel, providing an increase in current and a decrease in impedance, which widens the bandwidth.

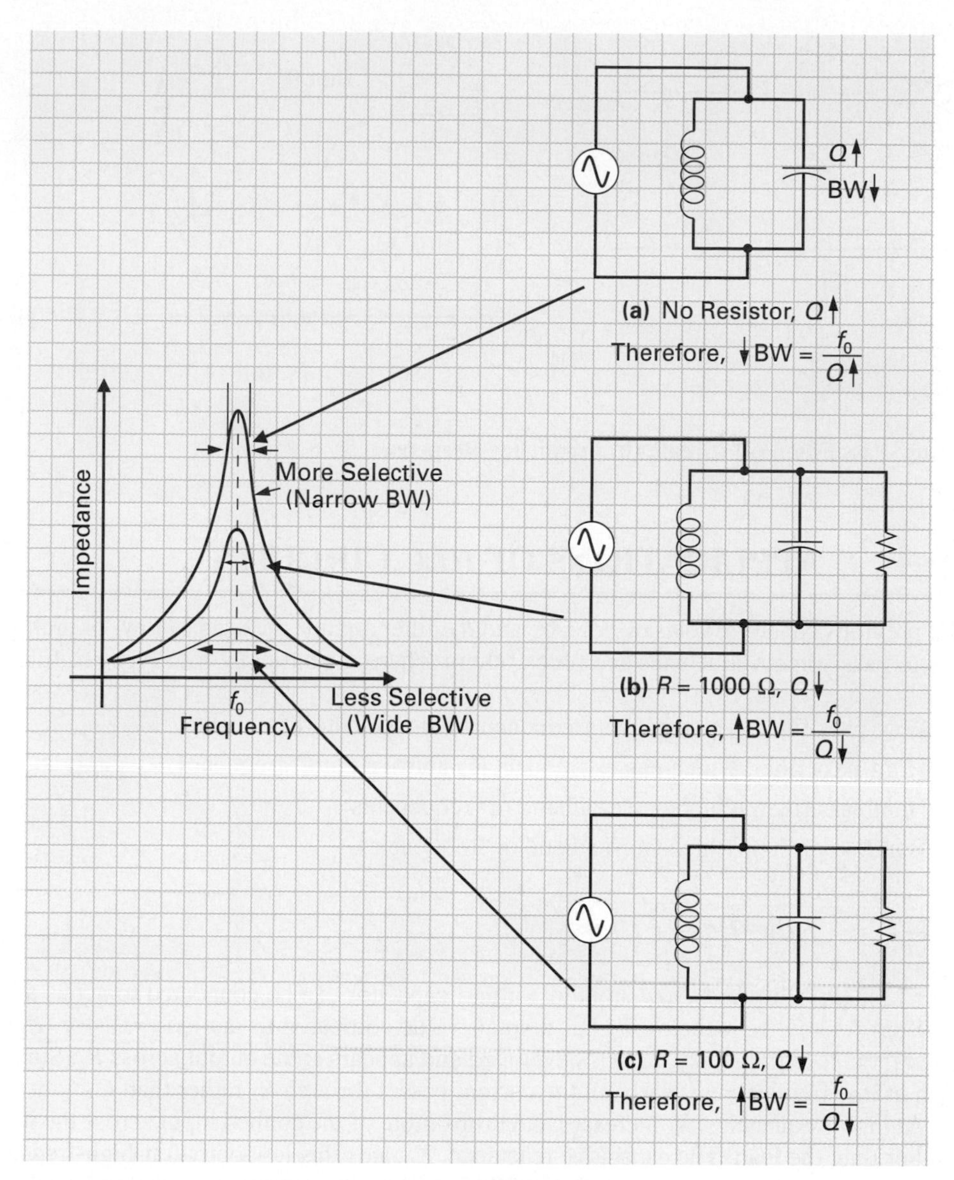

FIGURE 13-21 Varying Bandwidth by Loading the Tank Circuit.

SELF-TEST EVALUATION POINT FOR SECTION 13-3-2

Now that you have completed this section, you should be able to:

Objective 4. Define resonance and explain the characteristics of

a. Series resonance
b. Parallel resonance

Use the following questions to test your understanding of Section 13-3-2.

1. What are the differences between a series and a parallel resonant circuit?
2. Describe flywheel action.
3. Calculate the value of Q of a tank if $X_L = 50\ \Omega$ and $R = 25\ \Omega$.
4. When calculating bandwidth for a parallel resonant circuit, can the series resonant bandwidth formula be used?
5. What is selectivity?

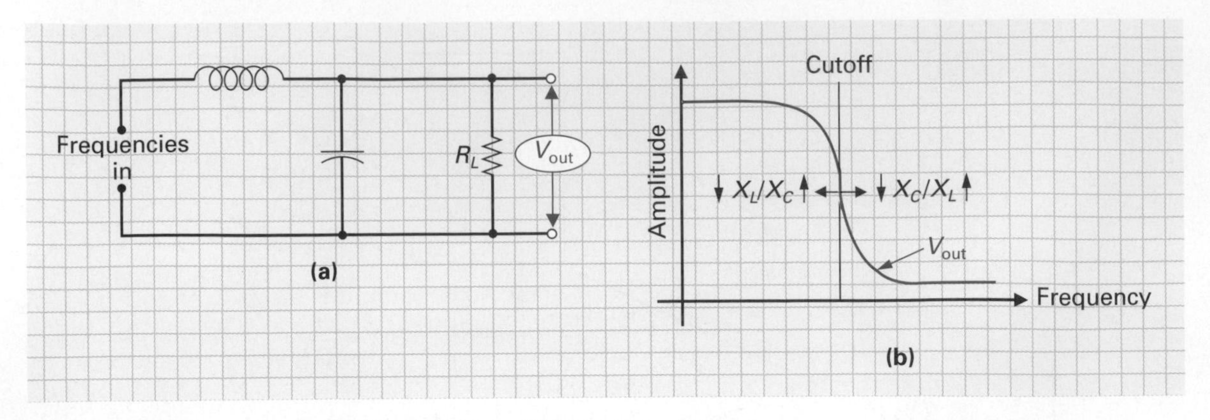

FIGURE 13-22 Low-Pass Filter. (a) Circuit. (b) Frequency Response.

13-4 APPLICATIONS OF *RLC* CIRCUITS

In the previous chapters, you saw how *RC* and *RL* filter circuits are used as low- or high-pass filters to pass some frequencies and block others. There are basically four types of filters:

1. Low-pass filter, which passes frequencies below a cutoff frequency
2. High-pass filter, which passes frequencies above a cutoff frequency
3. Bandpass filter, which passes a band of frequencies
4. Band-stop filter, which stops a band of frequencies

13-4-1 *Low-Pass Filter*

Figure 13-22(a) illustrates how an inductor and capacitor can be connected to act as a low-pass filter. At low frequencies, X_L has a small value compared to the load resistor (R_L), so nearly all the low-frequency input is developed and appears at the output across R_L. Since X_C is high at low frequencies, nearly all the current passes through R_L rather than C.

At high frequencies, X_L increases and drops more of the applied input across the inductor rather than the load. The capacitive reactance, X_C, aids this low-output-at-high-frequency effect by decreasing its reactance and providing an alternative path for current to flow.

Since the inductor basically blocks alternating current and the capacitor shunts alternating current, the net result is to prevent high-frequency signals from reaching the load. The way in which this low-pass filter responds to frequencies is graphically illustrated in Figure 13-22(b).

13-4-2 *High-Pass Filter*

Figure 13-23(a) illustrates how an inductor and capacitor can be connected to act as a high-pass filter. At high frequencies, the reactance of the capacitor (X_C) is low while the reactance of the inductor (X_L) is high, so all the high frequencies are easily passed by the capacitor and blocked by the inductor and all are routed through to the output and load.

At low frequencies, the reverse condition exists, resulting in a low X_L and a high X_C. The capacitor drops nearly all the input, and the inductor shunts the signal current away from the output load.

Bandpass Filter

Filter circuit that passes a group or band of frequencies between a lower and an upper cutoff frequency, while heavily attenuating any other frequency outside this band.

13-4-3 *Bandpass Filter*

Figure 13-24(a) illustrates a series resonant **bandpass filter,** and Figure 13-24(b) shows a parallel resonant bandpass filter. Figure 13-24(c) shows the frequency response curve

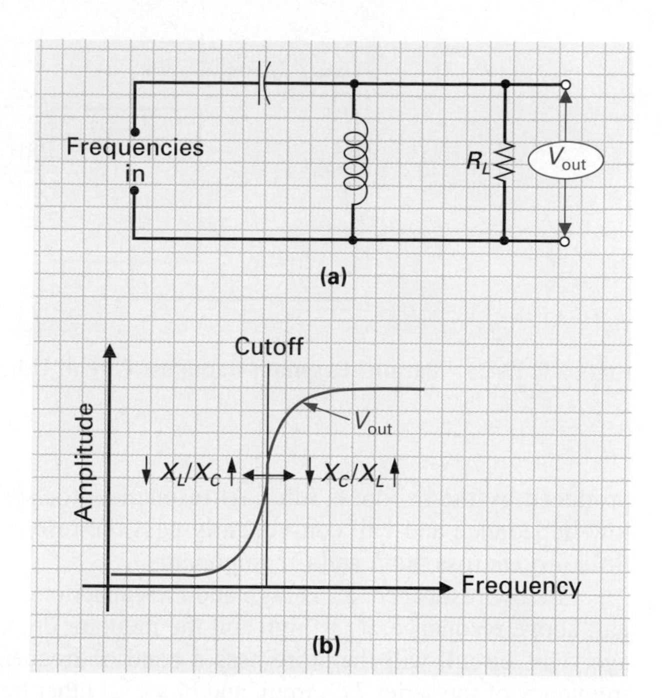

FIGURE 13-23 High-Pass Filter. (a) Circuit. (b) Frequency Response.

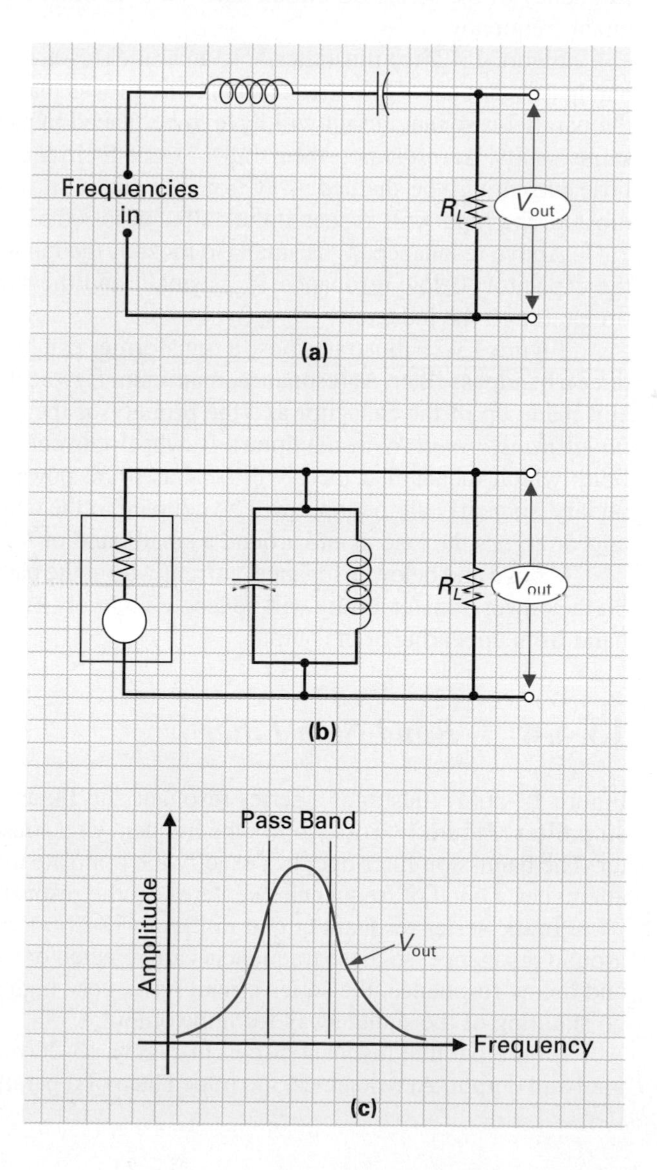

FIGURE 13-24 Bandpass Filter. (a) Series Resonant Bandpass Filter. (b) Parallel Resonant Bandpass Filter. (c) Frequency Response.

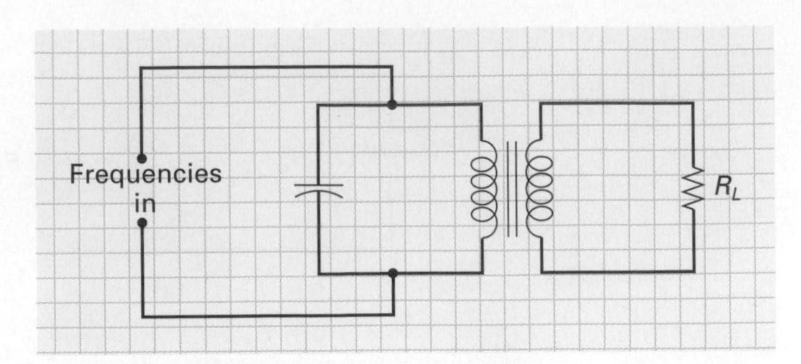

FIGURE 13-25 **Parallel Resonant Bandpass Circuit Using a Transformer.**

produced by the bandpass filter. At resonance, the series resonant *LC* circuit has a very low impedance and will consequently pass the resonant frequency to the load with very little drop across the *L* and *C* components.

Below resonance, X_C is high and the capacitor drops a large amount of the input signal; above resonance, X_L is high and the inductor drops most of the input frequency voltage. This circuit will therefore pass a band of frequencies centered around the resonant frequency of the series *LC* circuit and block all other frequencies above and below this resonant frequency.

Figure 13-24(b) illustrates how a parallel resonant *LC* circuit can be used to provide a bandpass response. The series resonant circuit was placed in series with the output, whereas the parallel resonant circuit will have to be placed in parallel with the output to provide the same results. At resonance, the parallel resonant circuit or tank has a high impedance, so very little current will be shunted away from the output; it will be passed on to the output, and almost all the input will appear at the output across the load.

Above resonance, X_C is small, so most of the input is shunted away from the output by the capacitor; below resonance, X_L is small, and the shunting action occurs again, but this time through the inductor.

Figure 13-25 illustrates how a transformer can be used to replace the inductor to produce a bandpass filter. At resonance, maximum flywheel current flows within the parallel circuit made up of the capacitor and the primary of the transformer (*L*), which is known as a *tuned transformer*. With maximum flywheel current, there will be a maximum magnetic field, which means that there will be maximum power transfer between primary and secondary, so nearly all the input will be coupled to the output (coupling coefficient $k = 1$) and appear across the load at and around a small band of frequencies centered on resonance.

Above and below resonance, current within the parallel resonant circuit will be smaller. So the power transfer ability will be less, effectively keeping the frequencies outside the pass band from appearing at the output.

13-4-4 *Band-Stop Filter*

Band-Stop Filter
A filter that attenuates alternating currents whose frequencies are between given upper and lower cutoff values while passing frequencies above and below this band.

Figure 13-26(a) illustrates a series resonant and Figure 13-26(b) a parallel resonant **band-stop filter.** Figure 13-26(c) shows the frequency response curve produced by a band-stop filter. The band-stop filter operates exactly the opposite to a bandpass filter in that it blocks or attenuates a band of frequencies centered on the resonant frequency of the *LC* circuit.

In the series resonant circuit in Figure 13-26(a), the *LC* impedance is very low at and around resonance, so these frequencies are rejected or shunted away from the output. Above and below resonance, the series circuit has a very high impedance, which results in almost no shunting of the signal away from the output.

In the parallel resonant circuit in Figure 13-26(b), the *LC* circuit is in series with the load and output. At resonance, the impedance of a parallel resonant circuit will be very high

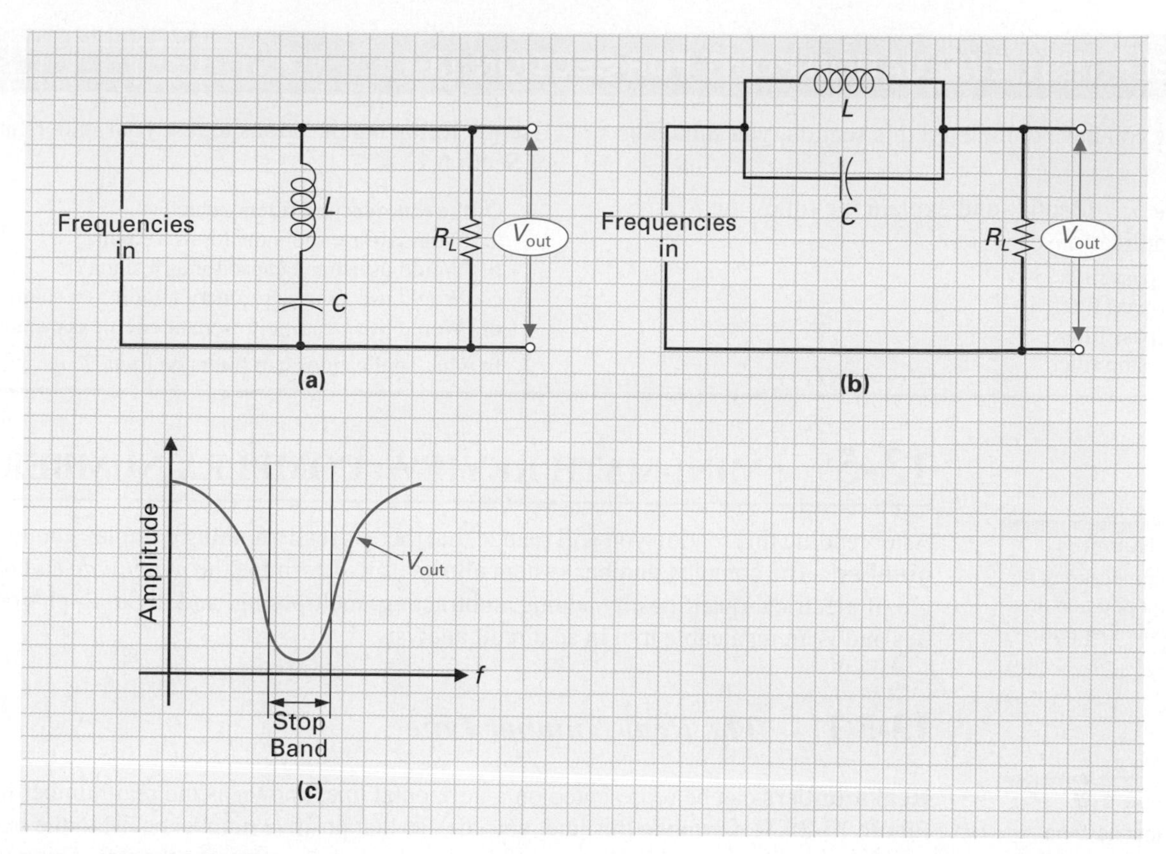

FIGURE 13-26 Band-Stop Filter. (a) Series Resonant Band-Stop Filter. (b) Parallel Resonant Band-Stop Filter. (c) Frequency Response.

and the band of frequencies centered around resonance will be blocked. Above and below resonance, the impedance of the tank is very low, so nearly all the input is developed across the output.

Filters are necessary in applications such as television or radio, where we need to tune in (select or pass) one frequency that contains the information we desire, yet block all the millions of other frequencies that are also carrying information, as shown in Figure 13-27.

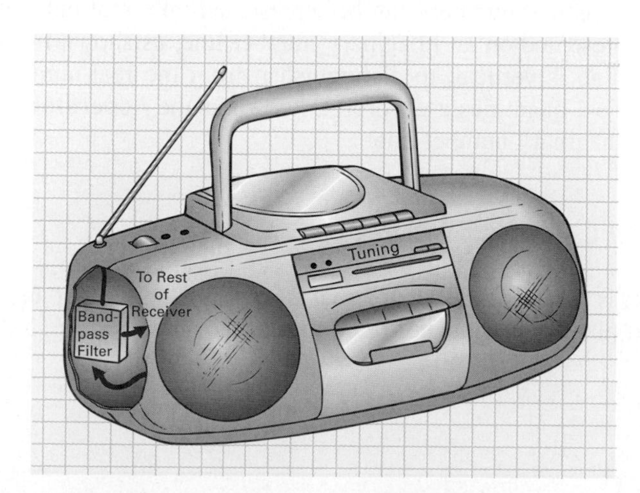

FIGURE 13-27 Tuning in of Station by Use of a Bandpass Filter.

SELF-TEST EVALUATION POINT FOR SECTION 13-4

Now that you have completed this section, you should be able to:

Objective 5. Identify and explain the following *RLC* circuit applications:

a. Low-pass filter
b. High-pass filter
c. Bandpass filter
d. Band-stop filter

Use the following questions to test your understanding of Section 13-4:

1. Of the four types of filters, which
  a. Would utilize the inductor as a shunt?
  b. Would utilize the capacitor as a shunt?
  c. Would use a series resonant circuit as a shunt?
  d. Would use a parallel resonant circuit as a shunt?

2. In what applications can filters be found?

13-5 MINI-MATH REVIEW: COMPLEX NUMBERS

Complex Numbers
Numbers composed of a real-number part and an imaginary-number part.

After reading this section you will realize that there is really nothing complex about **complex numbers.** The complex number system allows us to determine the *magnitude* and *phase angle* of electrical quantities by adding, subtracting, multiplying, and dividing phasor quantities and is an invaluable tool in ac circuit analysis.

13-5-1 *The Real-Number Line*

Real Numbers
Numbers that have no imaginary parts.

Real numbers can be represented on a horizontal line, known as the real-number line, as in Figure 13-28. Referring to this line, you can see that positive numbers exist to the right of the center point corresponding to zero, while negative numbers exist to the left. This representation satisfied most mathematicians for a short time, as they could indicate numbers such as 2 or 5 as a point on the line. Numbers corresponding to $\sqrt{9}$ could also be represented, as three points to the right of zero $(\sqrt{9} = +3)$. However, a problem arose if they wished to indicate a point corresponding to $\sqrt{-9}$. The $\sqrt{-9}$ is not $+3$ [since $(+3) \times (+3) = +9$], and it is not -3 [since $(-3) \times (-3) = +9$]. So it was eventually realized that the square root of a negative number could not be indicated on the real-number line, as it is not a real number.

13-5-2 *The Imaginary-Number Line*

Imaginary Number
A complex number whose imaginary part is not zero.

***j* Operator**
A prefix used to indicate an imaginary number.

Mathematicians decided to call the square root of a negative number, such as $\sqrt{-4}$ or $\sqrt{-9}$, **imaginary numbers,** which are not fictitious or imaginary, but simply a particular type of number. Just as real numbers can be represented on a real-number line, imaginary numbers can be represented on an imaginary-number line, as shown in Figure 13-29. The imaginary-number line is vertical, to distinguish it from the real-number line, and when working with electrical quantities a $\pm j$ prefix, known as the ***j* operator,** is used for values that appear on the imaginary-number line.

13-5-3 *The Complex Plane*

Complex Plane
A plane whose points are identified by means of complex numbers.

A complex number is the combination of a real and imaginary number and is represented on a two-dimensional plane called the **complex plane,** shown in Figure 13-30. Generally, the

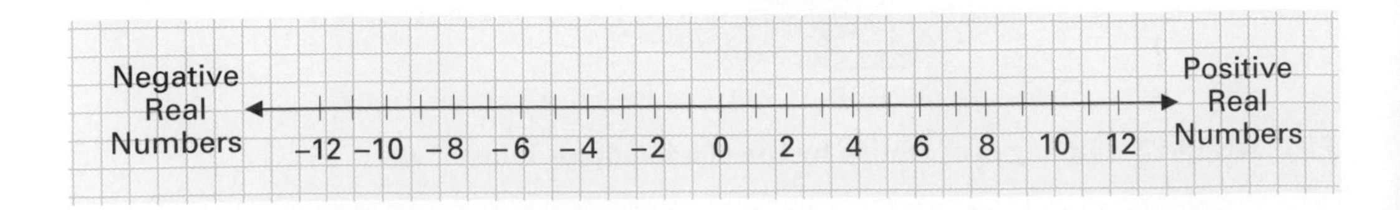

FIGURE 13-28 Real-Number Line.

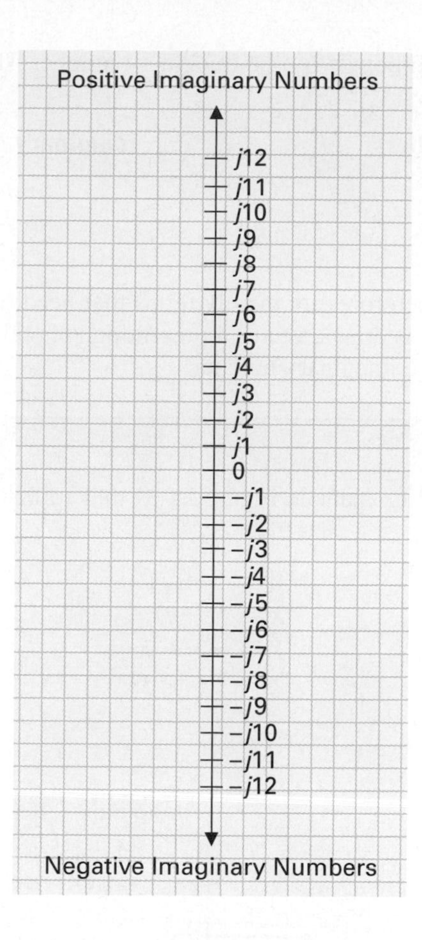

FIGURE 13-29 Imaginary-Number Line.

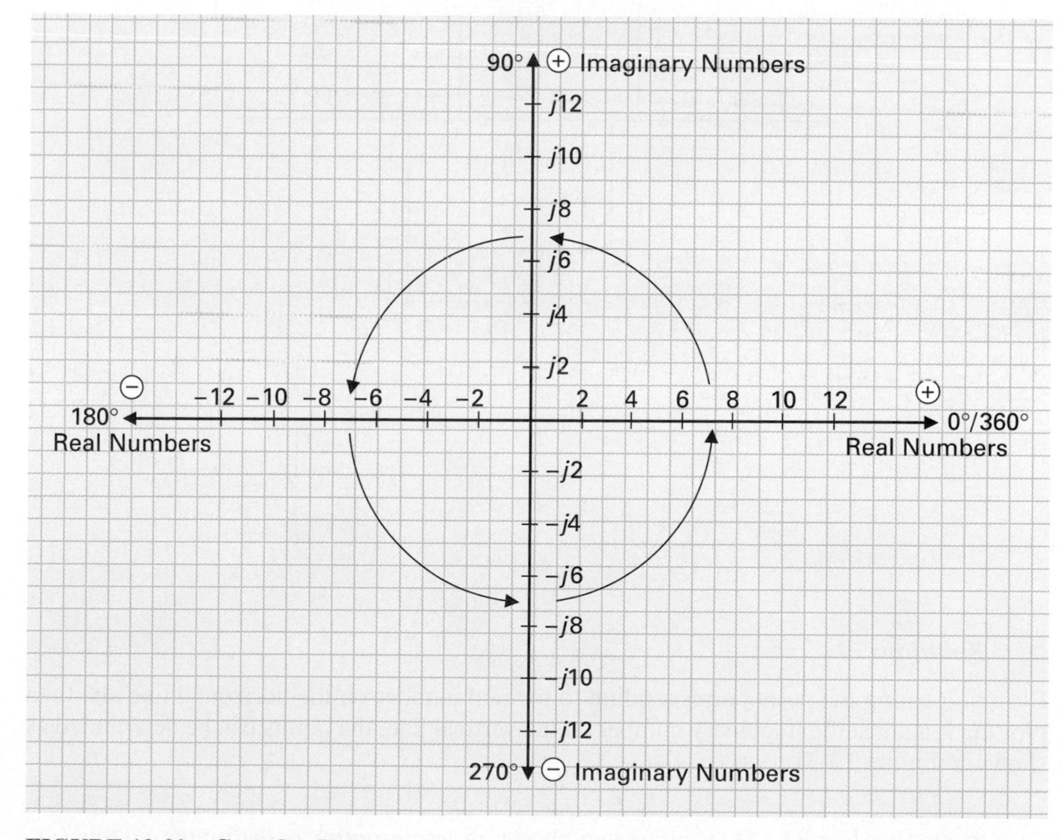

FIGURE 13-30 Complex Plane.

real number appears first, followed by the imaginary number. Here are some examples of complex numbers.

Real Numbers	Imaginary Numbers
3	$+ j4$
−2	$+ j4$
−3	$- j2$

Complex numbers, therefore, are merely terms that need to be added as phasors, and all you have to do basically is draw a vector representing the real number and then draw another vector representing the imaginary number.

EXAMPLE:

Find the points in the complex plane in Figure 13-31 that correspond to the following complex numbers.

$$W = 3 + j4$$
$$X = 5 - j7$$
$$Y = -4 + j6$$
$$Z = -3 - j5$$

FIGURE 13-31 Complex Numbers Examples.

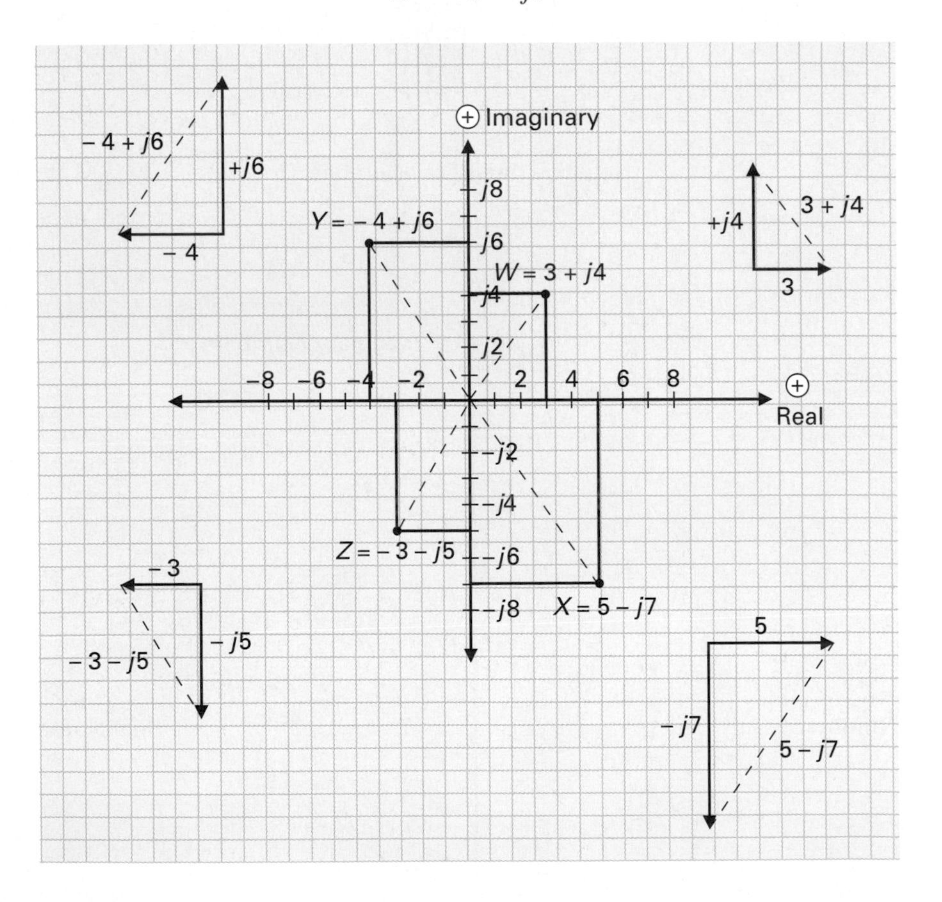

Solution:

By first locating the point corresponding to the real number on the horizontal line and then plotting it against the imaginary number on the vertical line, the points can be determined as shown in Figure 13-31.

Rectangular Coordinates

A Cartesian coordinate of a Cartesian coordinate system whose straight-line axes or coordinate planes are perpendicular.

A number like $3 + j4$ specifies two phasors in **rectangular coordinates,** so this system is the *rectangular representation of a complex number*. There are several other ways to

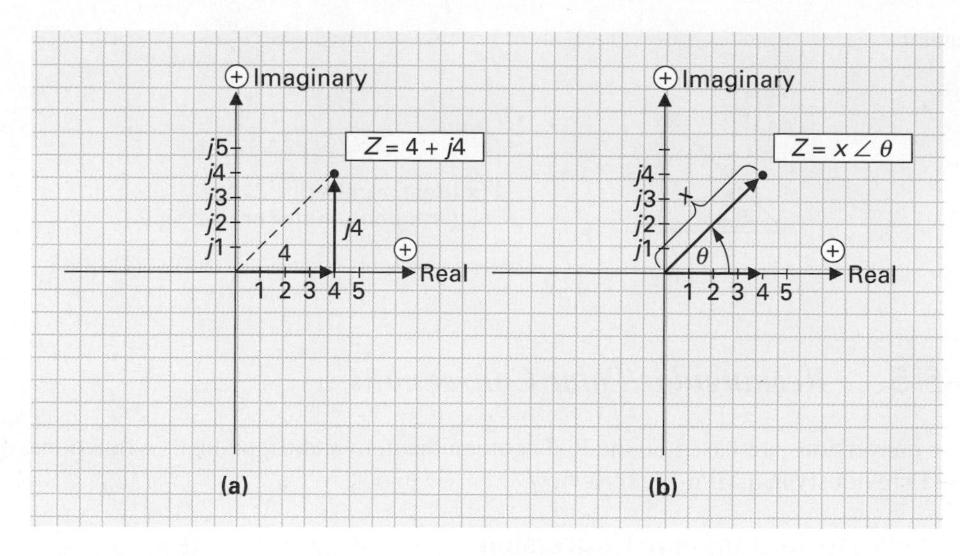

FIGURE 13-32 **Representing Phasors. (a) Rectangular Notation. (b) Polar Notation.**

describe a complex number, one of which is the polar representation of a complex number, using **polar coordinates,** which will be discussed next.

Polar Coordinates
Either of two numbers that locate a point in a plane by its distance from a fixed point on a line and the angle this line makes with a fixed line.

13-5-4 *Polar Complex Numbers*

Phasors can also be expressed in polar form, as shown in Figure 13-32, which compares rectangular and polar notation. With the rectangular notation in Figure 13-32(a), the horizontal coordinate is the real part and the vertical coordinate is the imaginary part of the complex number. With the polar notation shown in Figure 13-32(b), the magnitude of the phasor (x, or size) and the angle ($\angle \theta$, meaning "angle theta") relative to the positive real axis (measured in a counterclockwise direction) are stated.

EXAMPLE:

Sketch the following polar numbers:

a. $5 \angle 60°$
b. $3 \angle 220°$

Solution:

As you can see in Figure 13-33, an equivalent negative angle, which is calculated by subtracting the given positive angle from 360°, can also be used.

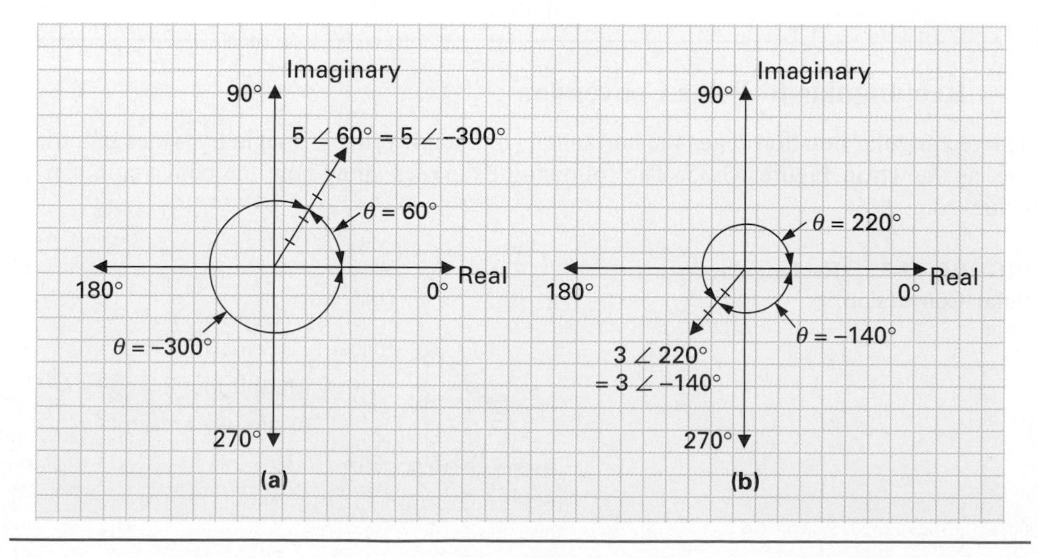

FIGURE 13-33 **Polar Number Examples.**

FIGURE 13-34 Polar-to-Rectangular Conversion.

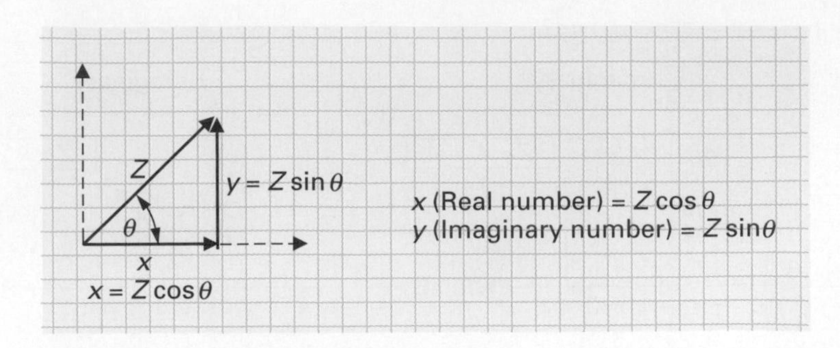

TIME LINE

French mathematician Pierre de Fermat (1601–1665) is often called the founder of the modern theory of numbers.

13-5-5 *Rectangular/Polar Conversions*

These conversions are based on the Pythagorean theorem and trigonometric functions, discussed previously in a Mini-Math Review.

Polar-to-Rectangular Conversion

The polar notation states the magnitude and angle, as shown in Figure 13-34. The following examples show how this conversion can be achieved.

EXAMPLE:

Convert the following polar numbers to rectangular form:

a. $5 \angle 30°$

b. $18 \angle -35°$

c. $44 \angle 220°$

Solution:

a. Real number $= 5 \cos 30° = 4.33$
Imaginary number $= 5 \sin 30° = j2.5$
Polar number, $5 \angle 30°$ = rectangular number, $4.33 + j2.5$

b. Real number $= 18 \cos(-35°) = 14.74$
Imaginary number $= 18 \sin(-35°) = -j10.32$
Polar number, $18 \angle -35°$ = rectangular number, $14.74 - j10.32$

c. Real number $= 44 \cos 220° = -33.7$
Imaginary number $= 44 \sin 220° = -j28.3$

Polar number, $44 \angle 220°$ = rectangular number, $-33.7 - j28.3$

Rectangular-to-Polar Conversion

The rectangular notation states the horizontal (real) and vertical (imaginary) sides of a triangle, as shown in Figure 13-35. The following examples show how the conversion can be achieved.

FIGURE 13-35 Rectangular-to-Polar Conversion.

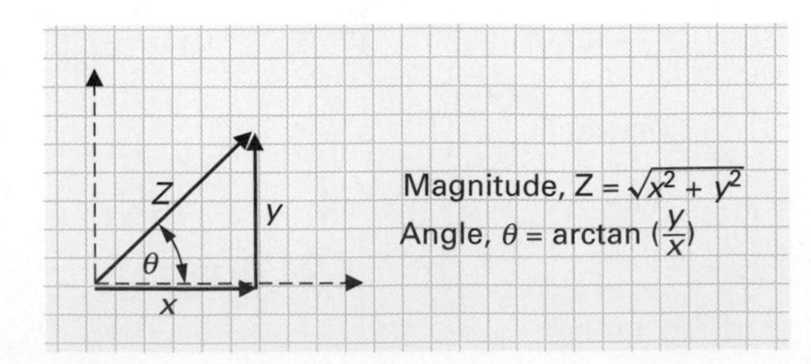

EXAMPLE:

Convert the following rectangular numbers to polar form:

a. $4 + j3$ **b.** $16 - j14$

Solution:

a. Magnitude $= \sqrt{4^2 + 3^2} = 5$
Angle = arctan (3/4) = 36.9°
Rectangular number, $4 + j3$ = polar number, $5 \angle 36.9°$

b. Magnitude $= \sqrt{16^2 + (-14)^2} = 21.3$
Angle = arctan (−14/16) = − 41.2°
Rectangular number, $16 - j14$ = polar number, $21.3 \angle -41.2°$

Many scientific calculators have a feature that allows you to convert rectangular numbers to polar numbers, and vice versa.

CALCULATOR KEYS

Name: Complex numbers

Function: Most scientific calculators have the ability to perform conversions between rectangular and polar numbers and to perform arithmetic operations on either. (Be sure that the calculator is configured for degree angle mode, not radian angle mode).

Example: To enter a complex number in rectangular form, enter the value of *a* (*real component*), press [+] or [−], enter the value of *b* (*imaginary component*), and press [2nd] [*i*] (constant).

real component (+ or −) *imaginary component i*

Rectangular Arithmetic Operation Example
$(12 + 6i) - (6 - 3i)$ ENTER
Answer: $6 + 9i$

To enter a complex number in polar form, enter the value of *r* (*magnitude*), press [2nd] [e^x] (exponential function), enter the value of θ (*angle*), press [2nd] [i] (constant), and then press [i].

magnitude e^(angle i)

Polar Arithmetic Operation Example
$(5\angle 35) - (7\angle 70)$ ENTER
Answer: $-9.058 + 33.80i$ (rectangular)
$35\angle 105$ (Polar)

Conversions
▶**Rect** (display as rectangular) displays a complex result in rectangular form. It is valid only at the end of an expression. It is not valid if the result is real.

Example: $(35 \angle 105)$ ▶rect ENTER
Answer: $-9.058 + 33.80i$

▶**Polar** (display as polar) displays a complex result in polar form. It is valid only at the end of an expression. It is not valid if the result is real.

Example: $(-9.058 + 33.80i)$ ▶polar ENTER
Answer: $35 \angle 105$

13-5-6 *Complex-Number Arithmetic*

Since a phase difference exists between real and imaginary (j) numbers, certain rules should be applied when adding, subtracting, multiplying, or dividing complex numbers.

Addition

The sum of two complex numbers is equal to the sum of their separate real and imaginary parts.

EXAMPLE:

Add the following complex numbers:

a. $(3 + j4) + (2 + j5)$
b. $(4 + j5) + (2 - j3)$

Solution:

a. $(3 + j4) + (2 + j5) = (3 + 2) + (j4 + j5) = 5 + j9$
b. $(4 + j5) + (2 - j3) = (4 + 2) + (j5 - j3) = 6 + j2$

Subtraction

The difference of two complex numbers is equal to the difference between their separate real and imaginary parts.

EXAMPLE:

Subtract the following complex numbers:

a. $(4 + j3) - (2 + j2)$
b. $(12 + j6) - (6 - j3)$

Solution:

a. $(4 + j3) - (2 + j2) = (4 - 2) + j(3 - 2) = 2 + j1$
b. $(12 + j6) - (6 - j3) = (12 - 6) + j[6 - (-3)] = 6 + j9$

Multiplication

Multiplication of two complex numbers is achieved more easily if they are in polar form. The simple rule to remember is to multiply the magnitudes and then add the angles algebraically.

EXAMPLE:

Multiply the following complex numbers:

a. $5 \angle 35° \times 7 \angle 70°$
b. $4 \angle 53° \times 12 \angle -44°$

Solution:

a. Multiply the magnitudes: $5 \times 7 = 35$
Algebraically add the angles: $\angle (35° + 70°) = \angle 105° = 35 \angle 105°$
b. Multiply the magnitudes: $4 \times 12 = 48$
Algebraically add the angles: $\angle [53° + (-44°)] = \angle 9° = 48 \angle 9°$

Division

Division is also more easily carried out in polar form. The rule to remember is to divide the magnitudes and then subtract the denominator angle from the numerator angle.

EXAMPLE:

Divide the following complex numbers:

a. $60 \angle 30°$ by $30 \angle 15°$
b. $100 \angle 20°$ by $5 \angle -7°$

Solution:

a. Divide the magnitudes: $60/30 = 2$
Subtract the denominator angle from the numerator angle: $\angle (30° - 15°) = \angle 15° = 2 \angle 15°$

b. Divide the magnitudes: $100/5 = 20$
Subtract the angles: $\angle [20° - (-7°)] = \angle 27° = 20 \angle 27°$

13-5-7 *How Complex Numbers Apply to AC Circuits*

Complex numbers find an excellent application in ac circuits due to all the phase differences that occur between different electrical quantities, such as X_L, X_C, R, and Z, as shown in Figure 13-36. The positive real-number line, at an angle of 0°, is used for resistance, which in this example is 3 Ω, as shown in Figure 13-36(a) and (b).

On the positive imaginary-number line, at an angle of 90° ($+j$), inductive reactance (X_L) is represented, which in this example is $j4$ Ω ($X_L = 4$ Ω). The voltage drop across an inductor (V_L) is proportional to its inductive reactance (X_L), and both are represented on the $+j$ imaginary-number line, since the voltage drop across an inductor will always lead the current (which in a series circuit is always in phase with the resistance) by 90°.

On the negative imaginary-number line, at an angle of $-90°$ or 270° ($-j$), capacitive reactance (X_C) is represented, which in this example is $-j2$ ($X_C = 2$ Ω). The voltage drop across a capacitor (V_C) is proportional to its capacitive reactance (X_C), and both are represented on the $-j$ imaginary-number line since the voltage drop across a capacitor always lags current (charge and discharge) by $-90°$.

Series AC Circuits

Referring to Figure 13-36(a) and (b) once again, we can calculate total impedance simply by adding the phasors.

Z_T (Rectangular). Total series impedance is equal to the sum of all the resistances and reactances:

$$\begin{aligned} Z &= R + (jX_L \sim jX_C) \\ &= 3 + (+j4 \sim -j2) \\ &= 3 + j2 \end{aligned}$$

Z_T (Polar). The total series impedance can be converted from rectangular to polar form:

$$\text{Magnitude} = \sqrt{3^2 + 2^2} = 3.61\ \Omega$$

$$\text{Angle} = \arctan \frac{2}{3} = 33.7°$$

$$= 3.61 \angle 33.7°$$

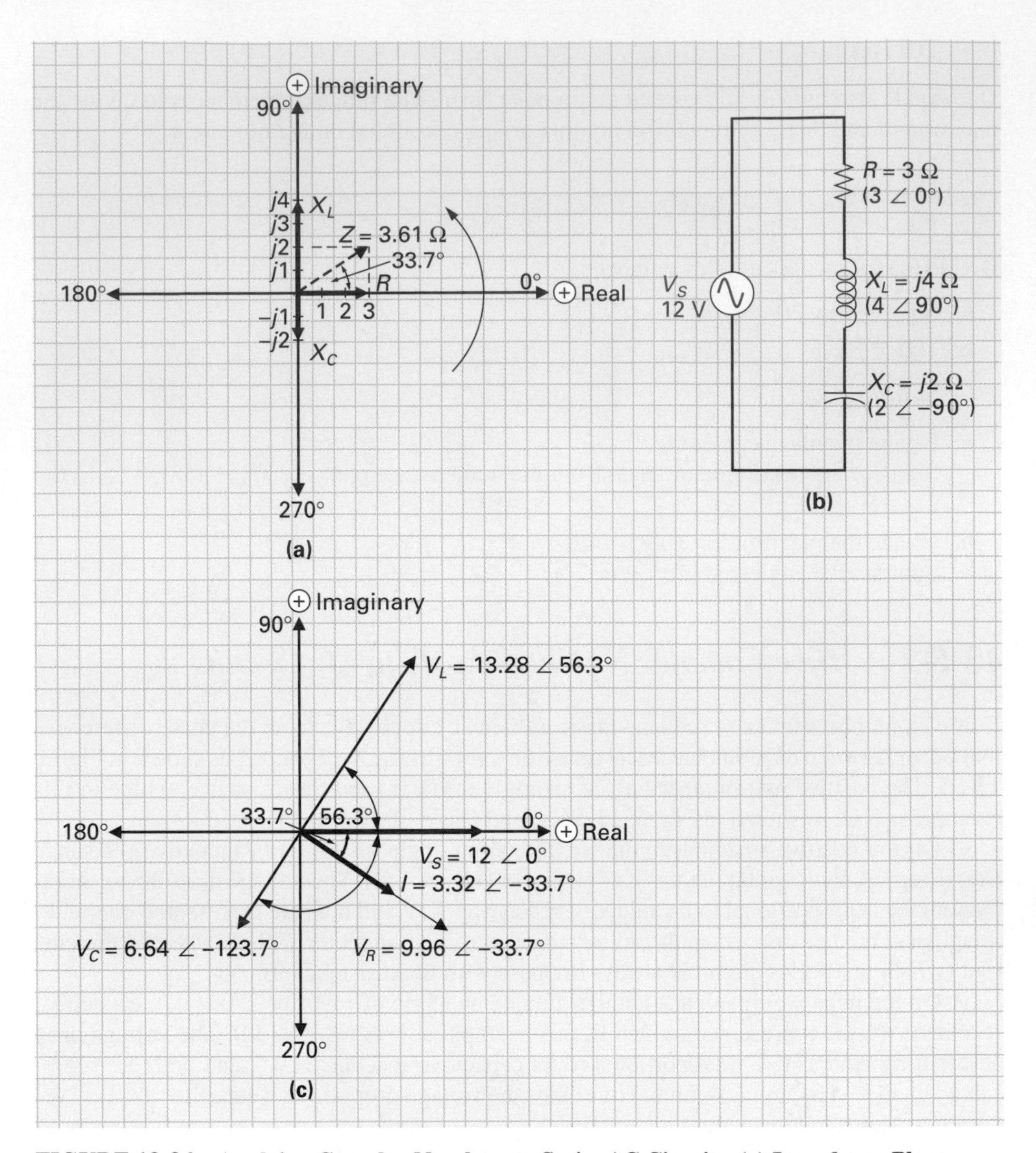

FIGURE 13-36 Applying Complex Numbers to Series AC Circuits. (a) Impedance Phasors. (b) Series Circuit. (c) Voltage and Current Phasors.

Current. Once the magnitude of Z_T is known (3.61 Ω), I can be calculated. The source voltage of 12 V is a real positive number (0°) and is therefore represented as 12 ∠ 0°. Current is equal to

$$I = \frac{V_S}{Z_T} = \frac{12 \angle 0°}{3.61 \angle 33.7°}$$

Polar division Divide the magnitudes: $\frac{12}{3.61} = 3.32 \text{ A}$

Subtract the angles: $0° - 33.7° = -33.7°$

$I = 3.32 \angle -33.7°$

Phase angle. The circuit current has an angle of −33.7°, which means that it lags V_T (inductive circuit, therefore ELI). This negative phase angle is expected since this series circuit is inductive ($X_L > X_C$), in which case current should lag voltage by some phase angle.

This phase angle is less than 45° because the net reactance is less than the circuit resistance; it is shown in Figure 13-36(c).

Voltage drops. The component voltage drops are calculated with the following formulas and are shown in Figure 13-36(c):

$V_R = I \times R = (3.32 \angle -33.7°) \times (3 \angle 0°)$
Multiply the magnitudes: $3.32 \times 3 = 9.96$ V
Algebraically add the angles: $\angle(-33.7° + 0°) = -33.7°$
$V_R = 9.96 \angle -33.7°$
$V_L = I \times X_L = (3.32 \angle -33.7°) \times (4 \angle 90°)$
Multiply the magnitudes: $3.32 \times 4 = 13.28$ V
Algebraically add the angles: $\angle(-33.7° + 90°) = 56.3°$
$V_L = 13.28 \angle 56.3°$
$V_C = I \times X_C = (3.32 \angle -33.7°) \times (2 \angle -90°)$
Multiply the magnitudes: $3.32 \times 2 = 6.64$ V
Algebraically add the angles: $\angle(-33.7° + -90°) = -123.7°$
$V_C = 6.64 \angle -123.7°$

Phase relationships. As shown in Figure 13-36(c), the voltage across an inductor leads the circuit current by +90°, whereas the voltage across a capacitor lags the circuit current by −90°. The source voltage acts as the zero reference phase and leads the circuit current and the voltage across the resistor (V_R) by 33.7°.

Source voltage. Although the source voltage is known, it can be checked to verify all the previous calculations, since the sum of all the individual voltage drops should equal the source voltage.

Polar ⟶		**Rectangular**
$V_L = 13.28 \angle 56.3°$	=	$7.37 + j11.05$
$V_R = 9.96 \angle -33.7°$	=	$8.29 - j5.53$
$V_C = 6.64 \angle -123.7°$	=	$-3.68 - j5.52$
		$11.98 + j0$

Series-Parallel AC Circuits

Figure 13-37 illustrates a series–parallel ac circuit containing an *RL* branch, an *RC* branch, and an *RLC* branch.

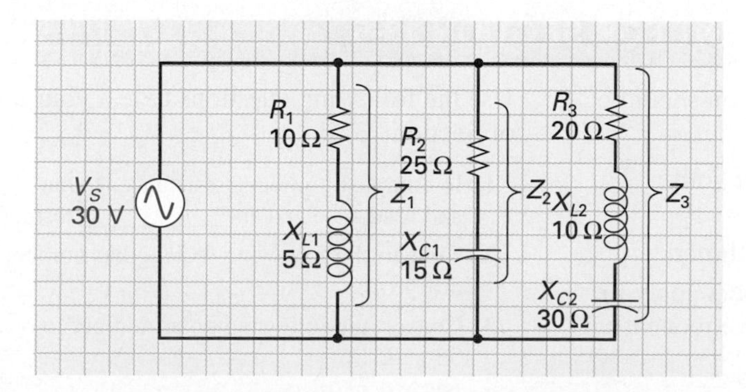

FIGURE 13-37 Applying Complex Numbers to Series–Parallel AC.

Impedance of each branch. The three branches will each have a value of impedance that will be equal to

Rectangular ⟶ Polar

$$Z_1 = 10 + j5 = 11.2 \angle 26.6^\circ\ \Omega$$
$$Z_2 = 25 - j15 = 29.2 \angle -31.0^\circ\ \Omega$$

The third branch is capacitive since the difference between $-j30\ (X_{C2})$ and $+j10\ (X_{L2})$ is $-j20$.

$$Z_3 = 20 - j20 = 28.3 \angle -45^\circ\ \Omega$$

Branch currents. The three branch currents, I_1, I_2, and I_3, are calculated by dividing the source voltage (V_S) by the individual branch impedances.

$$I = \frac{V_S}{Z_1} = \frac{30 \angle 0^\circ}{11.2 \angle 26.6^\circ}$$

Divide magnitudes: $30 \div 11.2 = 2.68$

Subtract angles: $\angle (0^\circ - 26.6^\circ) = -26.6^\circ$

$$I_1 = 2.68 \angle -26.6^\circ = 2.4 - j1.2\ \text{A}$$

$$I_2 = \frac{V_S}{Z_2} = \frac{30 \angle 0^\circ}{29.2 \angle -31^\circ} = 1.03 \angle +31^\circ = 0.88 + j0.5\ \text{A}$$

$$I_3 = \frac{V_S}{Z_3} = \frac{30 \angle 0^\circ}{28.3 \angle -45^\circ} = 1.06 \angle +45^\circ = 0.75 + j0.7\ \text{A}$$

Total current.

$$\begin{aligned} I_T &= I_1 + I_2 + I_3 \\ &= (2.4 - j1.2) + (0.88 + j0.5) + (0.75 + j0.7) \\ &= (2.4 + 0.88 + 0.75) + [-j1.2 + (+j0.5) + (+j0.7)] \\ &= 4.03\ \text{A} \end{aligned}$$

In polar form, this will equal $4.03 \angle 0^\circ$ A.

Total impedance.

$$Z_T = \frac{V_S}{I_T} = \frac{30 \angle 0^\circ}{4.03 \angle 0^\circ} = 7.44 \angle 0^\circ\ \Omega$$

Polar ⟶ Rectangular

$$7.44 \angle 0^\circ = 7.44 + j0$$

The complex ac circuit seen in Figure 13-37 is therefore equivalent to a 7.44 Ω resistor in series with no reactance.

SELF-TEST EVALUATION POINT FOR SECTION 13-5

Now that you have completed this section, you should be able to:

Objective 6. Describe complex numbers in both rectangular and polar form.

Objective 7. Perform complex-number arithmetic.

Objective 8. Describe how complex numbers apply to ac circuits containing series–parallel RLC components.

Use the following questions to test your understanding of Section 13-5:

1. In complex numbers, resistance is a/an __________ term and reactance is a/an __________ term. (imaginary/real)
2. Convert the following rectangular number to polar form: $5 + j6$.
3. Convert the following polar number to rectangular form: $33 \angle 25^\circ$.
4. What is a complex number?

SUMMARY

1. Resistors, as we have discovered, operate and react to voltage and current in a very straightforward way; the voltage across a resistor is in phase with the resistor current.
2. Inductors and capacitors operate in essentially the same way, in that they both store energy and then return it back to the circuit. However, they have completely opposite reactions to voltage and current.
3. The phrase "ELI the ICE man" states that voltage (symbolized *E*) leads current (*I*) in an inductive (*L*) circuit (abbreviated by the word "ELI"), and current (*I*) leads voltage (*E*) in a capacitive (*C*) circuit (abbreviated by the word "ICE").

Series *RLC* Circuit (Section 13-1)

4. The circuit current is always the same throughout a series circuit and can therefore be used as a reference.
5. The voltage across a resistor is always in phase with the current, whereas the voltage across the inductor leads the current by 90° and the voltage across the capacitor lags the current by 90°.
6. Impedance is the total opposition to current flow and is a combination of both reactance (X_L, X_C) and resistance (*R*).
7. The capacitive and inductive reactances are 180° out of phase with one another and counteract.
8. Impedance is the vector sum of the reactive (*X*) and resistance (*R*) vectors.
9. Since none of these voltages are in phase with one another in a series *RLC* circuit, they must be added vectorially to obtain the applied voltage.

Parallel *RLC* Circuit (Section 13-2)

10. The voltage across any parallel circuit will be equal and in phase.
11. With current, we must first calculate the individual branch currents (I_R, I_L, and I_C) and then calculate the total circuit current (I_T).
12. There is a phase difference between the source voltage (V_S) and the circuit current (I_T).
13. Once the total current (I_T) is known, the impedance of all three components in parallel can be calculated.

Resonance (Section 13-3)

14. Resonance is a circuit condition that occurs when the inductive reactance (X_L) and the capacitive reactance (X_C) have been balanced.
15. As the input ac frequency is increased, a point will be reached where X_L will equal X_C, and this condition is known as *resonance*. The frequency at which $X_L = X_C$ in either a parallel or a series *LC* circuit is known as the *resonant frequency* (f_0).
16. A series resonant circuit is a resonant circuit in which the capacitor and coil are in series with the applied ac voltage.
17. In summary, we can say that in a series resonant circuit:
    a. The inductor and capacitor electrically disappear due to their equal but opposite effect, resulting in 0 V drops across the series combination, and the circuit consequently seems purely resistive.
    b. The current flow is large because the impedance of the circuit is low and equal to the series resistance (*R*), which has the source voltage developed across it.
    c. The individual voltage drops across the inductor or capacitor can be larger than the source voltage if *R* is smaller than X_L and X_C.
18. The *Q* factor is a ratio of inductive reactance to resistance and is used to express how efficiently an inductor will store rather than dissipate energy. In a series resonant circuit, the *Q* factor indicates the quality of the series resonant circuit, or is the ratio of the reactance to the resistance.
19. The *Q* of a resonant circuit is almost entirely dependent on the inductor's coil resistance, because capacitors tend to have almost no resistance figure at all, only reactance, which makes them very efficient.
20. A series resonant circuit is selective in that frequencies at resonance or slightly above or below will cause a larger current than frequencies well above or below the circuit's resonant frequency.
21. The group or band of frequencies that causes the larger current is called the circuit's *bandwidth*. A frequency response curve is a graph indicating a circuit's response to different frequencies.
22. The bandwidth includes the group or band of frequencies that cause 70.7% or more of the maximum current to flow within the series resonant circuit.
23. The cutoff frequency is the frequency at which the gain of the circuit falls below 0.707 of the maximum current or half-power (−3 dB).
24. The half-power point is a point at which power is 50%. This half-power point corresponds to 70.7% of the total current.
25. Bandwidth is proportional to the resonant frequency of the circuit and inversely proportional to the *Q* of the circuit. In summary, the bandwidth of a series resonant circuit will increase as the *Q* of the circuit decreases ($\text{BW}\uparrow = f_0 / Q\downarrow$), and vice versa.
26. A parallel resonant circuit is a circuit having an inductor and capacitor in parallel with one another, offering a high impedance at the frequency of resonance.
27. The source voltage (V_S) is needed initially to supply power to the *LC* circuit and start the oscillations; but once the oscillating process is in progress (assuming the ideal case), current is only flowing back and forth between inductor and capacitor, and no current is flowing from the source.
28. Flywheel action is a sustaining effect of oscillation in an *LC* circuit due to the charging and discharging of the capacitor and the expansion and contraction of the magnetic field around the inductor.
29. The electronic equivalent of the mechanical flywheel is a resonant parallel-connected *LC* circuit.

30. The direction of the circulating current reverses each half-cycle at the frequency of resonance. Energy is stored in the capacitor in the form of an electric field between the plates on one half-cycle, and then the capacitor discharges, supplying current to build up a magnetic field on the other half-cycle. The inductor stores its energy in the form of a magnetic field, which will collapse, supplying a current to charge the capacitor, which will then discharge, supplying a current back to the inductor, and so on. Due to the "storing action" of this circuit, it is sometimes related to the fluid analogy and referred to as a tank circuit.

31. Under ideal conditions, a tank circuit should oscillate indefinitely if no losses occur within the circuit. In reality, the resistance of the coil reduces that 100% efficiency, as does friction with the mechanical flywheel.

32. As a small part of the energy is dissipated with each cycle, the oscillations will be reduced in size and eventually fall to zero.

33. If the ac source is reconnected to the tank, a small amount of current will flow from the source to the tank to top up the tank or replace the dissipated power. The higher the coil resistance is, the higher the loss and the larger the current flow from source to tank to replace the loss.

34. In the series resonant circuit, we were concerned with voltage drops since current remains the same throughout a series circuit. In a parallel resonant circuit, we are concerned with circuit currents rather than voltage.

35. Of all the three Q formulas for parallel resonant circuits, $Q = I_{\text{tank}}/I_S$, $Q = X_L/R$, or $Q = Z_{\text{tank}}/X_L$, the last is the easiest to use as both X_L and the tank impedance can easily be determined in most cases where C, L, and R internal for the inductor are known.

36. The current versus frequency response curve in a parallel resonant circuit is the complete opposite of the series resonant response curve. At frequencies below resonance, X_L is low and X_C is high, and the inductor offers a low reactance, producing a high current path and low impedance. On the other hand, at frequencies above resonance, the capacitor displays a low reactance, producing a high current path and low impedance.

37. The parallel resonant circuit is like the series resonant circuit in that it responds to a band of frequencies close to its resonant frequency.

38. Circuits containing inductance and capacitance are often referred to as *tuned circuits* since they can be adjusted to make the circuit responsive to a particular frequency (the resonant frequency).

39. Selectivity, by definition, is the ability of a tuned circuit to respond to a desired frequency and ignore all others.

40. Parallel resonant *LC* circuits are sometimes too selective, as the Q is too large, producing too narrow a bandwidth. In this situation, because of the very narrow response curve, a high resistance value can be placed in parallel with the *LC* circuit to provide an alternative path for line current. This process is known as *loading* or *damping* the tank and will cause an increase in line current and decrease in Q ($Q\downarrow = I_{\text{tank}}/I_{\text{line}}\uparrow$). The decrease in Q will cause a corresponding increase in BW ($\text{BW}\uparrow = f_0/Q\downarrow$).

Applications of *RLC* Circuits (Section 13-4)

41. There are basically four types of filters:
    - **a.** Low-pass filter, which passes frequencies below a cutoff frequency
    - **b.** High-pass filter, which passes frequencies above a cutoff frequency
    - **c.** Bandpass filter, which passes a band of frequencies
    - **d.** Band-stop filter, which stops a band of frequencies

42. An inductor and capacitor can be connected to act as a low-pass filter. Since the inductor basically blocks alternating current and the capacitor shunts alternating current, the net result is to prevent high-frequency signals from reaching the load.

43. An inductor and capacitor can also be connected to act as a high-pass filter.

44. A bandpass filter circuit passes a group or band of frequencies between a lower and an upper cutoff frequency, while heavily attenuating any other frequency outside this band.

45. A band-stop filter attenuates alternating currents whose frequencies are between given upper and lower cutoff values while passing frequencies above and below this band.

46. Filters are necessary in applications such as television or radio, where we need to tune in (select or pass) one frequency that contains the information we desire, yet block all the millions of other frequencies that are also carrying information.

Mini-Math Review: Complex Numbers (Section 13-5)

47. The complex-number system allows us to determine the *magnitude* and *phase angle* of electrical quantities by adding, subtracting, multiplying, and dividing phasor quantities and is an invaluable tool in ac circuit analysis.

48. Real numbers can be represented on a horizontal line, known as the *real-number line*. Positive numbers exist to the right of the center point corresponding to zero, while negative numbers exist to the left. This representation satisfied most mathematicians for a short time, as they could indicate numbers such as 2 or 5 as a point on the line. However, a problem arose if they wished to indicate a point corresponding to $\sqrt{-9}$. $\sqrt{-9}$ is not $+3$ [since $(+3) \times (+3) = +9$], and it is not -3 [since $(-3) \times (-3) = +9$]. So it was eventually realized that the square root of a negative number could not be indicated on the real-number line, as it is not a real number.

49. Mathematicians decided to call the square root of a negative number, such as $\sqrt{-4}$ or $\sqrt{-9}$, *imaginary numbers*, which are not fictitious or imaginary, but simply a particular type of number. Just as real numbers can be represented on a real-number line, imaginary numbers can be represented on an imaginary-number line. The imaginary-number line is vertical, to distinguish it from the real-number line. When working with electrical quantities, a $\pm j$ prefix, known as the *j operator*, is used for values that appear on the imaginary-number line.

50. A complex number is the combination of a real and imaginary number and is represented on a two-dimensional plane called the *complex plane*. Generally, the real number appears first, followed by the imaginary number.
51. Complex numbers are merely terms that need to be added as phasors, and all you have to do basically is draw a vector representing the real number and then draw another vector representing the imaginary number.
52. A number like $3 + j4$ specifies two phasors in rectangular coordinates, so this system is the *rectangular representation of a complex number.*
53. There are several other ways to describe a complex number, one of which is the polar representation of a complex number, using polar coordinates.
54. With the rectangular notation, the horizontal coordinate is the real part and the vertical coordinate is the imaginary part of the complex number. With the polar notation, the magnitude of the phasor (x, or size) and the angle ($\angle \theta$, meaning "angle theta") relative to the positive real axis (measured in a counterclockwise direction) are stated.
55. Since a phase difference exists between real and imaginary (j) numbers, certain rules should be applied when adding, subtracting, multiplying, or dividing complex numbers.
56. The sum of two complex numbers is equal to the sum of their separate real and imaginary parts.
57. The difference of two complex numbers is equal to the difference between the separate real and imaginary parts.
58. Multiplication of two complex numbers is achieved more easily if they are in polar form. The simple rule to remember is to multiply the magnitudes and then add the angles algebraically.
59. Division is also more easily carried out in polar form. The rule to remember is to divide the magnitudes and then subtract the denominator angle from the numerator angle.
60. Complex numbers find an excellent application in ac circuits due to all the phase differences that occur between different electrical quantities, such as X_L, X_C, R, and Z. The positive real-number line, at an angle of 0°, is used for resistance.
61. On the positive imaginary-number line, at an angle of 90° ($+j$), inductive reactance (X_L) is represented.
62. The voltage drop across an inductor (V_L) is proportional to its inductive reactance (X_L), and both are represented on the $+j$ imaginary-number line, since the voltage drop across an inductor will always lead the current (which in a series circuit is always in phase with the resistance) by 90°.
63. On the negative imaginary-number line, at an angle of $-90°$ or 270° ($-j$), capacitive reactance (X_C) is represented.
64. The voltage drop across a capacitor (V_C) is proportional to its capacitive reactance (X_C), and both are represented on the $-j$ imaginary-number line since the voltage drop across a capacitor always lags current (charge and discharge) by $-90°$.

REVIEW QUESTIONS

Multiple-Choice Questions

1. Capacitive reactance is __________ to frequency and capacitance, whereas inductive reactance is __________ to frequency and inductance.
   a. Proportional, inversely proportional
   b. Inversely proportional, proportional
   c. Proportional, proportional
   d. Inversely proportional, inversely proportional
2. Resonance is a circuit condition that occurs when
   a. V_L equals V_C.
   b. X_L equals X_C.
   c. L equals C.
   d. Both (a) and (c).
   e. Both (a) and (b).
3. As frequency is increased, X_L will __________, while X_C will __________.
   a. Decrease, increase
   b. Increase, decrease
   c. Remain the same, decrease
   d. Increase, remain the same
4. In an RLC series resonant circuit, with $R = 500\ \Omega$ and $X_L = 250\ \Omega$, what would be the value of X_C?
   a. 2 Ω c. 250 Ω
   b. 125 Ω d. 500 Ω
5. At resonance, the voltage drop across both a series-connected inductor and capacitor will equal
   a. 70.7 V. c. 10 V.
   b. 50% of the source. d. Zero.
6. In a series resonant circuit the current flow is __________, as the impedance is __________ and equal to __________.
   a. Large, small, R
   b. Small, large, X
   c. Large, small, X
   d. Small, large, R
7. A circuit's bandwidth includes a group or band of frequencies that cause __________ or more of the maximum current, or more than __________ of the maximum power, to appear at the output.
   a. 110, 90
   b. 50%, 70.7%
   c. 70.7%, 50%
   d. Both (a) and (c)
8. The bandwidth of a circuit is proportional to the
   a. Frequency of resonance.
   b. Q of the tank.
   c. Tank current.
   d. Two of the above.

9. Series or parallel resonant circuits can be used to create
   a. Low-pass filters.
   b. Low-pass and high-pass filters.
   c. Bandpass and band-stop filters.
   d. All of the above.
10. Flywheel action occurs in
   a. A tank circuit.
   b. A parallel *LC* circuit.
   c. A series *LC* circuit.
   d. Two of the above.
   e. None of the above.
11. $25 \angle 39°$ is an example of a complex number in
   a. Polar form.
   b. Rectangular form.
   c. Algebraic form.
   d. None of the above.
12. $3 + j10$ is an example of a complex number in
   a. Polar form.
   b. Rectangular form.
   c. Algebraic form.
   d. None of the above.
13. Which complex number form is usually more convenient for addition and subtraction?
   a. Rectangular
   b. Polar
14. Which complex number form is usually more convenient for multiplication and division?
   a. Rectangular
   b. Polar
15. In complex numbers, *resistance* is a real term, while *reactance* is a/an ___________.
   a. *j* term
   b. Imaginary term
   c. Value appearing on the vertical axis
   d. All of the above

Communication Skill Questions

16. Illustrate with phasors and describe the current and voltage relationships in a series *RLC* circuit. (13-1)
17. Describe the procedure for the analysis of a series *RLC* circuit. (13-1)
18. Define *resonance* and give the formula for calculating the frequency of resonance. (13-3)
19. Describe the three unusual characteristics of a circuit that is at resonance. (13-3)
20. Define the following: (13-3)
   a. Flywheel action
   b. Quality factor
   c. Bandwidth
   d. Selectivity
21. Describe a frequency response curve. (13-3)
22. Illustrate with phasors and describe the current and voltage relationships in a parallel *RLC* circuit. (13-2)
23. Describe the differences between a series and a parallel resonant circuit. (13-3)
24. Explain how loading a tank affects bandwidth and selectivity. (13-3-2)
25. Illustrate the circuit and explain the operation of the following, with their corresponding response curves. (13-4)
   a. Low-pass filter
   b. High-pass filter
   c. Bandpass filter
   d. Band-stop filter
26. Describe why capacitive reactance is written as $-jX_C$ and inductive reactance is written as jX_L. (13-5)
27. How are capacitive and inductive reactances written in polar form? (13-5-7)
28. List the rules used to perform complex number (13-5-6)
   a. Addition (rectangular)
   b. Subtraction (rectangular)
   c. Multiplication (polar)
   d. Division (polar)
29. Describe briefly how the real-number and imaginary-number lines are used for ac circuit analysis and what electrical phasors are represented at 0°, 90°, and −90°. (13-5-7)
30. Referring to Figure 13-36, describe why the series circuit current (I) is not in phase with the source voltage (V_S). (13-5-7)

Practice Problems

31. Calculate the values of capacitive or inductive reactance for the following when connected across a 60 Hz source:
   a. 0.02 μF
   b. 18 μF
   c. 360 pF
   d. 2700 nF
   e. 4 mH
   f. 8.18 H
   g. 150 mH
   h. 2 H
32. If a 1.2 kΩ resistor, a 4 mH inductor, and an 8 μF capacitor are connected in series across a 120 V/60 Hz source, calculate:
   a. X_C
   b. X_L
   c. Z
   d. I
   e. V_R
   f. V_L
   g. V_C
   h. Apparent power
   i. True power
   j. Resonant frequency
   k. Circuit quality factor
   l. Bandwidth

33. If a 270 Ω resistor, a 150 mH inductor, and a 20 μF capacitor are all connected in parallel with one another across a 120 V/60 Hz source, calculate:
   a. X_L
   b. X_C
   c. I_R
   d. I_L
   e. I_C
   f. I_T
   g. Z
   h. Resonant frequency
   i. Q factor
   j. Bandwidth

34. Calculate the impedance of a series circuit if $R = 750\ \Omega$, $X_L = 25\ \Omega$, and $X_C = 160\ \Omega$.

35. Calculate the impedance of a parallel circuit with the same values as those of Question 34 when a 1 V source voltage is applied.

36. State the following series circuit impedances in rectangular and polar form:
   a. $R = 33\ \Omega, X_C = 24\ \Omega$
   b. $R = 47\ \Omega, X_L = 17\ \Omega$

37. Convert the following impedances to rectangular form:
   a. $25 \angle 37°$ **c.** $114 \angle -114°$
   b. $19 \angle -20°$ **d.** $59 \angle 99°$

38. Convert the following impedances to polar form:
   a. $-14 + j14$ **c.** $-33 - j18$
   b. $27 + j17$ **d.** $7 + j4$

39. Add the following complex numbers:
   a. $(4 + j3) + (3 + j2)$
   b. $(100 - j50) + (12 + j9)$

40. Perform the following mathematical operations:
   a. $(35 \angle -24°) \times (13 \angle 50°)$
   b. $(100 - j25) - (25 + j5)$
   c. $(98 \angle 80°) \div (40 \angle 17°)$

41. State the impedances of the circuits seen in Figure 13-38 in rectangular and polar form. What is Z_T in ohms and its phase angle?

42. Calculate in polar form the impedances of both circuits shown in Figure 13-39. Then combine the two impedances as if the circuits were parallel connected, using the product-over-sum method. Express the combined impedance in polar form.

43. Referring to Figure 13-40, calculate:
   a. Z_T (rectangular and polar)
   b. Circuit current and phase angle
   c. Voltage drops
   d. V_C, V_L, and V_R phase relationships

44. Sketch an impedance and voltage phasor diagram for the circuits in Figure 13-38.

45. Referring to Figure 13-41, calculate:
   a. Impedance of the two branches
   b. Branch currents
   c. Total current
   d. Total impedance

46. Sketch an impedance and current phasor diagram for the circuit shown in Figure 13-39.

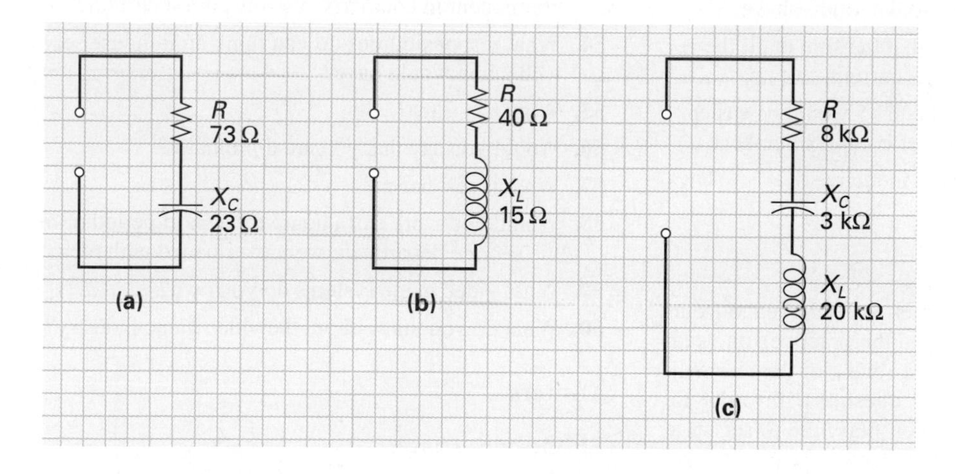

FIGURE 13-38

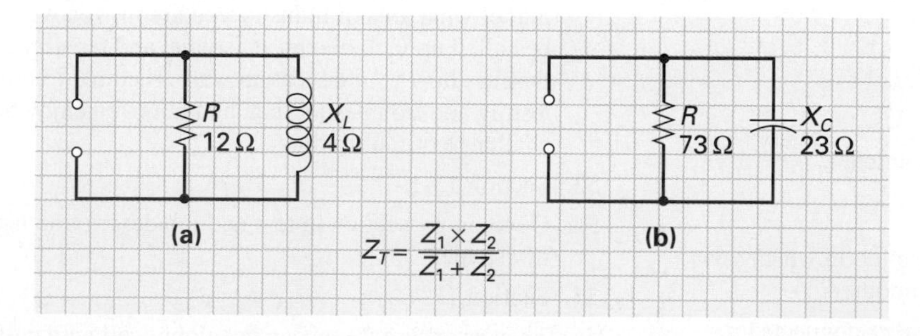

FIGURE 13-39

FIGURE 13-40

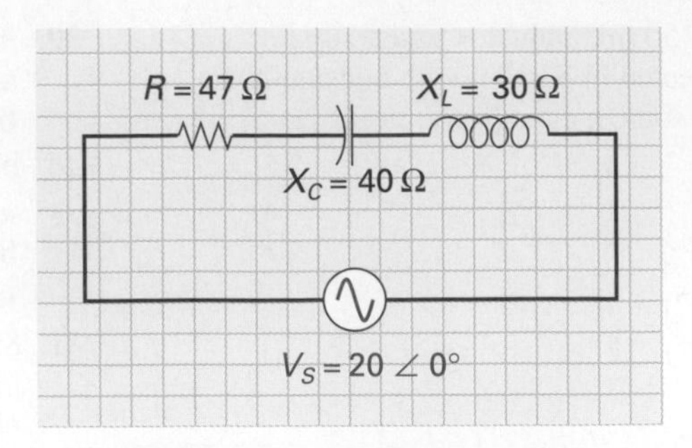

FIGURE 13-41

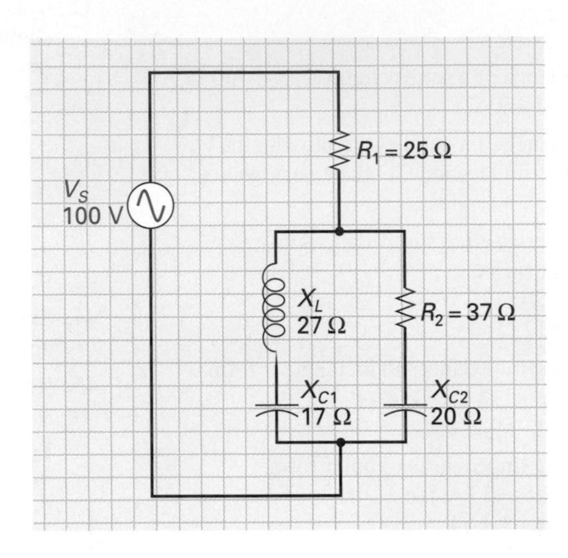

47. Referring to Figure 13-38, verify that the sum of all the individual voltage drops is equal to the total voltage.

48. Referring to Figure 13-39, verify that the sum of all the branch currents is equal to the total current.

49. Determine whether Questions 43 and 45 would be easier to answer with or without the use of complex numbers.

50. Is the circuit in Figure 13-39 more inductive or capacitive?

Job Interview Test

These tests at the end of each chapter will challenge your knowledge up to this point and give you the practice you need for a job interview. To make this more realistic, the test will comprise both technical and personal questions. In order to take full advantage of this exercise, you may like to set up a simulation of the interview environment, have a friend read the questions to you, and record your responses for later analysis.

Company Name: PLAY, Inc.

Industry Branch: Consumer Electronics.

Function: Repair of Video Game Consoles.

Job Appling For: In-house Service Technician.

51. How did you find out about this job?

52. How do you think you would handle a job in which you are besieged with one problem after another?

53. If a resistor, capacitor, and inductor were connected in series across an ac source, what would be the phase relationship between the voltage drop across each component in comparison to the circuit current?

54. What responsibilities do you think an in-house service technician would have?

55. What is a filter?

56. Would you say that you are diplomatic?

57. What is resonance?

58. Would you work extra hours for a short period of time if the service department were overloaded with repairs?

59. What video game systems do you own?

60. What do you know about the video game industry?

Answers

51. Describe source.

52. This has always been the job function of a technician. No matter what area of industry you are in, you will always be called upon to diagnose, isolate, and repair problems. Discuss how your education was structured to train you in testing and troubleshooting, and how you enjoy the challenge of faultfinding.

53. Section 13-1.

54. Quote intro. section in text and job description listed in newspaper.

55. Section 13-4.

56. They're asking if you can get along well with customers and colleagues and if you have the communication skills to tactfully handle daily interaction. Mention any work experience in which you had direct contact with customers and discuss how you worked with fellow students in your class.

57. Section 13-3.

58. If you are unsure, don't object to anything at the time. Later, if you decide that it is something you do not want to do, you can call and decline the offer.

59. Individual answer.

60. If you visit the company's web site before the interview, there will generally be a section describing the industry branch in which they do business—remember, the more informed you are, the better.

Web Site Questions

Go to the web site http://www.prenhall.com/cook, select the textbook *Introductory DC/AC Electronics* or *Introductory DC/AC Circuits*, this chapter, and then follow the instructions when answering the multiple-choice practice problems.

APPENDIX

Electronics Dictionary

Absorption Loss or dissipation of energy as it travels through a medium. For example, radio waves lose some of their electromagnetic energy as they travel through the atmosphere.

AC Abbreviation for "alternating current."

AC alpha (α_{ac}) The ratio of input emitter current change to output collector current change.

AC beta (β_{ac}) The ratio of a transistor's ac output current to input current.

Accelerate To go faster.

AC coupling Circuit or component that couples or passes the ac signal yet blocks any dc level.

AC/DC If indicated on a piece of equipment, it means that the equipment will operate from either an ac or dc supply.

AC generator Device that transforms or converts a mechanical input into an ac electrical power output.

Acoustic Relating to sound or the science of sound.

AC power supply Power supply that delivers one or more sources of ac voltage.

Activate To put ac voltage to work or make active. The application of an enabling signal.

Active bandpass filter A circuit that uses an amplifier with passive filter elements to pass only a band of input frequencies.

Active band-stop filter A circuit that uses an amplifier with passive filter elements to block a band of input frequencies.

Active component Component that amplifies a signal or achieves some sort of gain between input and output.

Active equipment Equipment that will transmit and receive, as opposed to passive equipment, which only receives.

Active filter A filter that uses an amplifier with passive filter elements to provide pass or rejection characteristics.

Active high-pass filter A circuit that uses an amplifier with passive filter elements to pass all frequencies above a cutoff frequency.

Active low-pass filter A circuit that uses an amplifier with passive filter elements to pass all frequencies below a cutoff frequency.

Active operation or in the active region When the base-emitter junction is forward biased and the base-collector junction is reverse biased. In this mode, the transistor is equivalent to a variable resistor between collector and emitter.

Active region Flat part of the collector characteristic curve. A transistor is normally operated in this region, where it is equivalent to a variable resistor between the collector and emitter.

AC voltage Alternating voltage.

ADC Abbreviation for "analog-to-digital converter."

Adjacent The side of a right-angled triangle that has a common endpoint, in that it extends between the hypotenuse and the vertical side.

Adjustable resistor Resistor whose value can be changed.

Admittance (symbolized *Y*) Measure of how easily ac will flow through a circuit. It is equal to the reciprocal of impedance and is measured in siemens.

Aerial Another term used for antenna.

AF Abbreviation for "audio frequency."

AFC Abbreviation for "automatic frequency control."

AGC Abbreviation for "automatic gain control."

Air-core inductor An inductor that has no metal core.

Algebra A generalization of arithmetic in which letters representing numbers are combined according to the rules of arithmetic.

Aligning tool Small, nonconductive, nonmagnetic screwdriver used to adjust receiver or transmitter tuned circuits.

Alkaline cell Also called an alkaline manganese cell, it is a primary cell that delivers more current than the carbon–zinc cell.

Alligator clip Spring clip normally found on the end of test leads and used to make temporary connections.

Alphanumeric Having numerals, symbols, and letters of the alphabet.

Alpha wave Also called alpha rhythm, it is a brain wave between 9 and 14 Hz.

Alternating current Electric current that rises from zero to a maximum in one direction, falls to zero, and then rises to a maximum in the opposite direction, and then repeats another cycle, the positive and negative alternations being equal.

Alternation One ac cycle consists of a positive and negative alternation.

Alternator Another term used to describe an ac generator.

AM Abbreviation for "amplitude modulation."

Ambient temperature Temperature of the air surrounding the components.

American wire gauge American wire gauge (AWG) is a system of numerical designations of wire sizes, with the first being 0000 (the largest size) and then going to 000, 00, 0, 1, 2, 3, and so on up to the smallest sizes of 40 and above.

Ammeter Meter placed in the circuit path to measure the amount of current flow.

Ampere (A) Unit of electric current.

Ampere-hours Multiplying the amount of current, in amperes, that can be supplied for a given time frame in hours, gives a value of ampere-hours.

Ampere-hour meter Meter that measures the amount of current drawn in a unit of time (hours).

Ampere-turn Unit of magnetomotive force.

Ampere-turn per meter Base unit of magnetic field strength.

Amplification Process of making bigger or increasing the voltage or current, thus increasing the power of a signal.

Amplifier Circuit or device that achieves amplification.

Amplitude Magnitude or size an alternation varies from zero.

Amplitude distortion Changing of a wave shape so that it does not match its original form.

Analog Relating to devices or circuits in which the output varies in direct proportion to the input.

Analog data Information that is represented in a continuously varying form, as opposed to digital data, which have two distinct and discrete values.

Analog display Display in which a moving pointer or some other analog device is used to display the data.

Analog electrical circuits A power circuit designed to control power in a continuously varying form.

Analog electronic circuits Electronic circuits that represent information in a continuously varying form.

Analog multimeter Electronic test instrument that can perform multiple tasks in that it can be used to measure voltage, current, or resistance.

Anode Positive electrode or terminal.

Antenna Device that converts an electrical wave into an electromagnetic wave that radiates away from the antenna.

Antenna transmission line System of conductors connecting the transmitter or receiver to the antenna.

Apparent power The power value obtained in an ac circuit by multiplying together effective values of voltage and current, which reach their peaks at different times.

Arc Discharge of electricity through a gas. For example, lightning is a discharge of static electricity buildup through air.

Armature Rotating or moving component of a magnetic circuit.

Armstrong oscillator A tuned transformer oscillator developed by E. H. Armstrong.

Astable multivibrator A device commonly used as a clock oscillator.

Atom Smallest particle of an element.

Atomic number Number of positive charges or protons in the nucleus of an atom.

Atomic weight The relative weight of a neutral atom of an element, based on a neutral oxygen atom having an atomic weight of 16.

Attenuate To reduce in amplitude an action or signal.

Attenuation Loss or decrease in energy of a signal.

Audio Relating to all the frequencies that can be heard by the human ear—from the Latin word meaning "I hear."

Audio frequency Frequency that can be detected by a human ear—from approximately 20 Hz to 20 kHz.

Audio-frequency amplifier An amplifier that has one or more transistor stages designed to amplify audio-frequency signals.

Audio-frequency generator A signal generator that can be set to generate a sinusoidal AF signal voltage at any desired frequency in the audio spectrum.

Autotransformer Single-winding tapped transformer.

Average A single value that summarizes or represents the general significance of a set of unequal values.

Average value Mean value found when the area of a wave above a line is equal to the area of the wave below the line.

Avionics Field of aviation electronics.

AWG Abbreviation for "American wire gauge."

Axial lead Component that has its connecting leads extending from either end of its body.

Balanced bridge Condition that occurs when a bridge is adjusted to produce a zero output.

Balancing Setting the output of an op-amp to zero volts when both inverting and noninverting inputs are at zero volts.

Ballast resistor A resistor that increases in resistance when current increases. It can therefore maintain a constant current despite variations in line voltage.

Bandpass filter Filter circuit that passes a group or band of frequencies between a lower and an upper cutoff frequency while heavily attenuating any other frequency outside this band.

Band-stop filter A filter that attenuates alternating currents whose frequencies are between given upper and lower cutoff values while passing frequencies above and below this band.

Bandwidth Width of the group or band of frequencies between the half-power points.

Bar graph or bar chart A graphic means of quantitative comparison by rectangles with lengths proportional to the measure of the data being compared.

Bar graph display A left-to-right or low-to-high set of bars that are turned on in successive order based on the magnitude they are meant to represent.

Barometer Meter used to measure atmospheric pressure.

Barretter Temperature-sensitive device having a positive temperature coefficient; that is, as temperature increases, resistance increases.

Barrier potential or voltage The potential difference, or voltage, that exists across the junction.

Base The region that lies between an emitter and a collector of a transistor and into which minority carriers are injected.

Base biasing A transistor biasing method in which the dc supply voltage is applied to the base of the transistor via a base-bias resistor.

Base current (I_B) The relatively small current at the transistor's base terminal.

Bass Low audio frequencies normally handled by a woofer in a sound system.

Battery DC voltage source containing two or more cells that converts chemical energy into electrical energy.

Battery charger Piece of electrical equipment that converts ac input power to a pulsating dc output, which is then used to charge a battery.

Baud Unit of signaling speed describing the number of elements per second.

Beta β The transistor's current gain in a common-emitter configuration.

***B-H* curve** Curve plotted on a graph to show successive states during magnetization of a ferromagnetic material.

Bias voltage The dc voltage applied to a semiconductor device to control its operation.

Bidirectional device A device that will conduct current in either direction.

Binary Number system having only two levels and used in digital electronics.

Binary word A numerical value expressed as a group of binary digits.

Bipolar device A device in which there is a change in semiconductor material between the output terminals (NPN or PNP between emitter and collector) so that the charge carriers can be one of two polarities (bipolar).

Bipolar family A group of digital logic circuits that make use of the bipolar junction transistor.

Bipolar transistor A bipolar junction transistor in which excess minority carriers are injected from an emitter region into a base region and from there pass into a collector region.

Bistable multivibrator A digital control device that can be either set or reset.

Bits An abbreviation for "binary digits."

Bleeder current Current drawn continuously from a voltage source. A bleeder resistor is generally added to lessen the effect of load changes or provide a voltage drop across a resistor.

Body resistance The resistance of the human body.

Bolometer Device whose resistance changes when heated.

Branch current A portion of the total current that is present in one path of a parallel circuit.

Breakdown region The point at which the collector supply voltage will cause a damaging value of current through the transistor.

Breakdown voltage The voltage at which breakdown of a dielectric or an insulator occurs. Also, the voltage (V_{BR}) at which a damaging value of I will pass through the JFET.

Bridge rectifier circuit A full-wave rectifier circuit using four diodes that will convert an alternating voltage input into a direct voltage output.

Buffer current amplifier The *C-C* circuit that can ensure that power is efficiently transferred from source to load.

BW Abbreviation for "bandwidth."

Bypass capacitor Capacitor that is connected to provide a low-impedance path for ac.

Byte Group of 8 binary digits or bits.

Cable Group of two or more insulated wires.

CAD Abbreviation for "computer-aided design."

Calculator Device that can come in a pocket (battery-operated) or desktop (ac power) size and is used to achieve arithmetic operations.

Calibrate A procedure in which a meter or other device is adjusted to ensure its accuracy. The unit being calibrated is normally placed in a certain condition and then adjusted until a measured value is equal to an expected standard value.

Calibration To determine by measurement or comparison with a standard the correct value of each scale reading.

Capacitance (*C*) Measured in farads, it is the ability of a capacitor to store an electrical charge.

Capacitance meter Instrument used to measure the capacitance of a capacitor or circuit.

Capacitive filter A capacitor used in a power supply filter system to supress ripple currents while not affecting direct currents.

Capacitive reactance (X_C) Measured in ohms, it is the ability of a capacitor to oppose current flow without the dissipation of energy.

Capacitor Device that stores electric energy in the form of an electric field that exists within a dielectric (insulator) between two conducting plates, each connected to a connecting lead. This device was originally called a condenser.

Capacitor-input filter Filter in which a capacitor is connected to shunt away low-frequency signals such as ripple from a power supply.

Capacitor microphone Microphone that contains a stationary and movable plate separated by air. The arriving sound waves cause movements in the movable plate, which affects the distance between the plates and therefore the capacitance, generating a varying AF electrical wave. Also called an electrostatic microphone.

Capacitor speaker Also called the electrostatic speaker, its operation depends on mechanical forces generated by an electrostatic field.

Capacity Output current capabilities of a device or circuit.

Capture The act of gaining control of a signal.

Carbon composition resistor Fixed resistor consisting of carbon particles mixed with a binder, which is molded and then baked. Also called a composition resistor.

Carbon film resistor Resistor formed by depositing a thin carbon film on a ceramic form to create resistance.

Carbon microphone Microphone whose operation depends on the pressure variation in the carbon granules changing their resistance.

Cardiac pacemaker Device used to control the frequency or rhythm of the heart by stimulating it through electrodes.

Carrier Wave that has one of its characteristics varied by an information signal. The carrier will then transport this information between two points.

Cascode amplifier circuit An amplifier circuit consisting of a self-biased common-source amplifier in series with a voltage-divider-biased common-gate amplifier.

Cascoded amplifier An amplifier that contains two or more stages arranged in a series manner.

Cassette Thin, flat, rectangular device containing a length of magnetic tape that can be recorded onto or played back.

Cathode Term used to describe a negative electrode or terminal.

Cathode-ray tube A vacuum tube in which electrons emitted by a hot cathode are focused into a narrow beam by an electron gun and then applied to a fluorescent screen. The beam can be varied in intensity and position to produce a pattern or picture on the screen.

Cell Single unit having only two plates that convert chemical energy into a dc electrical voltage.

Celsius temperature scale Scale that defines the freezing point of water as 0°C and the boiling point as 100°C (named after Anders Celsius).

Center tap Midway connection between the two ends of a winding or resistor.

Center-tapped rectifier Rectifier that makes use of a center-tapped transformer and two diodes to achieve full-wave rectification.

Center-tapped transformer A transformer that has a connection at the electrical center of the secondary winding.

Ceramic capacitor Capacitor in which the dielectric material used between the plates is ceramic.

Cermet Ceramic-metal mixture used in film resistors.

Channel A path for a signal.

Charge Quantity of electrical energy stored in a battery or capacitor.

Charging current Current that flows to charge a capacitor or battery when a voltage is applied.

Chassis Metal box or frame in which components, boards, and units are mounted.

Chassis ground Connection to the metal box or frame that houses the components and associated circuitry.

Choke Inductor used to impede the flow of alternating or pulsating current.

Circuit Interconnection of many components to provide an electrical path between two or more points.

Circuit breaker Reusable fuse. This device will open a current-carrying path without damaging itself once the current value exceeds its maximum current rating.

Circuit failure A malfunction or breakdown of a component in the circuit.

Circular mil (cmil) A unit of area equal to the area of a circle whose diameter is 1 mil.

Circumference The perimeter of a circle.

Clapp oscillator A series tuned colpitts oscillator circuit.

Class A amplifier An amplifier in which the transistor is in its active region for the entire signal cycle.

Class AB amplifier An amplifier in which the transistor is in its active region for slightly more than half the signal cycle.

Class B amplifier An amplifier in which the transistor is in its active region for approximately half the signal cycle.

Class C amplifier An amplifier in which the transistor is in its active region for less than half the signal cycle.

Clock Generally, a square waveform used for the synchronization and timing of several circuits.

Clock input (*C*) An accurate timing signal used for synchronizing the operation of digital circuits.

Clock oscillator A device for generating a clock signal.

Clock signal Generally a square wave used for the synchronization and timing of several circuits.

Closed circuit Circuit having a complete path for current to flow.

Closed-loop mode A control system containing one or more feedback control loops, in which functions of the controlled signals are combined with functions of the commands that tend to maintain prescribed relationships between the commands and the controlled signals.

Closed-looped voltage gain (A_{CL}) The voltage gain of an amplifier when it is operated in the closed-loop mode.

Coaxial cable Transmission line in which a center signal-carrying conductor is completely covered by a solid dielectric, another conductor, and then an outer insulating sleeve.

Coefficient This is related to the ratio of change under certain conditions.

Coefficient of coupling The degree of coupling that exists between two circuits.

Coercive force Magnetizing force needed to reduce the residual magnetism within a material to zero.

Coil Number of turns of wire wound around a core to produce magnetic flux (an electromagnet) or to react to a changing magnetic flux (an inductor).

Cold resistance The resistance of a device when cold.

Collector A semiconductor region through which a flow of charge carriers leaves the base of the transistor.

Collector characteristic curve A set of characteristic curves of collector voltage versus collector current for a fixed value of transistor base current.

Collector current (I_C) The current emerging out of the transistor's collector.

Color code Set of colors used to indicate a component's value, tolerance, and rating.

Colpitts oscillator An *LC* tuned oscillator circuit in which two tank capacitors are used instead of a tapped coil.

Common Shared by two or more services, circuits, or devices. Although the term *common ground* is frequently used to describe two or more connections sharing a common ground, the term *common* alone does not indicate a ground connection, only a shared connection.

Common-base (*C-B*) circuit Configuration in which the input signal is applied between the transistor's emitter and base, while the output is developed across the transistor's collector and base.

Common-collector (*C-C*) circuit Configuration in which the input signal is applied between the transistor's base and collector, while the output is developed across the transistor's collector and emitter.

Common-drain configuration A FET configuration in which the drain is grounded and common to the input and output signal.

Common-emitter (*C-E*) circuit Configuration in which the input signal is applied between the base and the emitter, while the output signal appears between the transistor's collector and emitter.

Common-gate configuration A FET configuration in which the gate is grounded and common to the input and output signal.

Common-mode gain (A_{CM}) The amplification of common-mode input.

Common-mode input An input signal applied equally to both ungrounded inputs of a balanced amplifier. Also called in-phase input.

Common-mode rejection The ability of a device to reject a voltage signal applied simultaneously to both input terminals.

Common-mode rejection ratio Abbreviated CMRR, it is the ratio of an operational amplifier's differential gain to common-mode gain.

Common-source configuration A FET configuration in which the source is grounded and common to the input and output signal.

Communication Transmission of information between two points.

Comparator An operational amplifier used without feedback to detect changes in voltage level.

Complementary dc amplifier An amplifier in which NPN and PNP transistors are used in an alternating sequence.

Complementary latch circuit A circuit containing a NPN and PNP transistor that once triggered ON will remain latched ON.

Complex numbers Numbers composed of a real-number part and an imaginary-number part.

Complex plane A plane whose points are identified by means of complex numbers.

Component (1) Device or part in a circuit or piece of equipment. (2) In vector diagrams, it can mean a part of a wave, voltage, or current.

Compound A material composed of united separate elements.

Computer Piece of equipment used to process information or data.

Conductance (*G*) Measure of how well a circuit or path conducts or passes current. It is equal to the reciprocal of resistance.

Conduction Use of a material to pass electricity or heat.

Conduction band An energy band in which electrons can move freely within a solid.

Conductivity Reciprocal of resistivity; it is the ability of a material to conduct current.

Conductors Materials that have a very low resistance and pass current easily.

Configurations Different circuit interconnections.

Connection When two or more devices are attached so that a path exists between them, a connection is said to exist between the two.

Connector Conductive device that makes a connection between two points.

Constant Fixed value.

Constant current Current that remains at a fixed, unvarying level despite variations in load resistance.

Constant-current region The flat portion of the drain characteristic curve. In this region I_D remains constant despite changes in V_{DS}.

Contact Current-carrying part of a switch, relay, or connector.

Continuity Occurs when a complete path for electric current exists.

Continuity test Resistance test to determine whether a path for electric current exists.

Contrast The difference between the light and dark areas in a video picture. High-contrast pictures have dark blacks and brilliant whites. Low-contrast pictures have an overall gray appearance.

Control voltage A voltage signal that starts, stops, or adjusts the operation of a device, circuit, or system.

Conventional current flow A current produced by the movement of positive charges toward a negative terminal.

Copper loss Also called I^2R loss, it is the power lost in transformers, generators, connecting wires, and other parts of a circuit because of the current flow (*I*) through the resistance (*R*) of the conductors.

Core Magnetic material within a coil used to intensify or concentrate the magnetic field.

Cosine The trigonometric function that for an acute angle is the ratio between the leg adjacent to the angle when it is considered part of a right triangle, and the hypotenuse.

Coulomb Unit of electric charge. One coulomb equals 6.24×10^{18} electrons.

Counter electromotive force Abbreviated counter emf, or sometimes called back emf, it is the voltage generated in an inductor due to an alternating or pulsating current and is always of opposite polarity to that of the applied voltage.

Covalent bond A pair of electrons shared by two neighboring atoms.

Crossover distortion Distortion of a signal at the zero crossover point.

CRT Abbreviation for "cathode-ray tube."

Crystal A natural or synthetic piezoelectric or semiconductor material whose atoms are arranged with some degree of geometric regularity.

Cube root A number that when multiplied by itself three times gives the number under consideration.

Current (*I*) Measured in amperes or amps, it is the flow of electrons through a conductor.

Current amplifier An amplifier designed to build up the current of a signal.

Current clamp A device used in conjunction with an ac ammeter, containing a magnetic core in the form of hinged jaws that can be snapped around the current-carrying wire.

Current-controlled device A device in which the input junction is normally forward biased and the input current controls the output current.

Current divider A parallel network designed to proportionally divide the circuit's total current.

Current gain The increase in current produced by the transistor circuit.

Current-limiting resistor Resistor that is inserted into the path of current flow to limit the amount of current to some desired level.

Current sink A circuit or device that absorbs current from a load circuit.

Current source A circuit or device that supplies current to a load circuit.

Current-to-voltage converter An amplifier circuit used for converting a low dc current into a proportional low dc voltage.

Cutoff The minimum value of bias voltage that stops output current in a transistor.

Cutoff frequency Frequency at which the gain of the circuit falls below 0.707 of the maximum current or half-power (− 3 dB).

Cutoff region The point at which the collector supply voltage has the transistor operating in cutoff.

Cycle When a repeating wave rises from zero to some positive value, back to zero, and then to a maximum negative value before returning back to zero, it is said to have completed one cycle. The number of cycles occurring in 1 second is the frequency measured in hertz (cycles per second).

DAC Abbreviation for "digital-to-analog converter."

Damping Reduction in the magnitude of oscillation due to absorption.

Darlington pair A current amplifier consisting of two separate transistors with connected collectors and the emitter of one connected to the base of the other.

D'Arsonval movement When a direct current is passed through a small, lightweight, movable coil, the magnetic field produced interacts with a fixed permanent magnetic field that rotates while moving an attached pointer positioned over a scale.

DC Abbreviation for "direct current."

DC alpha Term for the dc current amplification factor of a transistor.

DC beta (β_{dc}) The ratio of a transistor's dc output current to its input current.

DC block Component used to prevent the passage of direct current, normally a capacitor.

DC generator Device used to convert a mechanical energy input into a dc electrical output.

DC load line A line representing all the dc operating points of the transistor for a given load resistance.

DC offsets The change in input voltage that is required to produce an output voltage of zero when no input signal is present.

DC power supply A power line, generator, battery, power pack, or other dc source of power for electrical equipment.

Dead short Short circuit having almost no resistance.

Decibel (dB) One-tenth of a bel (1×10^{-1} bel), the logarithmic unit for the difference in power at the output of an amplifier compared to the input.

Degerative effect An effect that causes a reduction in amplification due to negative feedback.

Degenerative (negative) feedback Feedback in which a portion of the output signal is fed back 180° out of phase with the input signal. Also called degenerative feedback, inverse feedback, or stabilized feedback.

Degree A unit of measure for angles equal to an angle with its vertex at the center of a circle and its sides cutting off $\frac{1}{360}$ of the circumference.

Depletion region A small layer on either side of the junction that becomes empty, or depleted, of free electrons or holes.

Depletion-type MOSFET A field effect transistor with an insulated gate (MOSFET) that can be operated in either the depletion or enhancement mode.

Design engineer Engineer responsible for the design of a product for a specific application.

Desoldering The process of removing solder from a connection.

Device Component or part.

DIAC A bidirectional diode that has a symmetrical switching mode.

Diagnostic Related to the detection and isolation of a problem or malfunction.

Diameter The length of a straight line passing through the center of an object.

Dielectric Insulating material between two (*di*) plates in which the *electric* field exists.

Dielectric constant (*K*) The property of a material that determines how much electrostatic energy can be stored per unit volume when unit voltage is applied. Also called permittivity.

Dielectric strength The maximum potential a material can withstand without rupture.

Difference amplifier An amplifier whose output is proportional to the difference between the voltages applied to the two inputs.

Differential amplifier An amplifier whose output is proportional to the difference between the voltages applied to its two inputs.

Differential gain (A_{VD}) The amplification of differential-mode input.

Differential-mode input signals Input signals to an op-amp that are out of phase with one another.

Differentiator A circuit whose output voltage is proportional to the rate of change of the input voltage. The output waveform is then the time derivative of the input waveform, and the phase of the output waveform leads that of the input by 90°.

Differentiator amplifier A circuit whose output voltage is proportional to the rate of change of the input voltage. The output waveform is the time derivative of the input waveform.

Diffusion A method of producing a junction by diffusing an impurity metal into a semiconductor at a high temperature.

Diffusion current The current that is present when the depletion layer is expanding.

Digital Relating to devices or circuits that have outputs of only two distinct levels or steps, for example, on–off, 0–1, open–closed, and so on.

Digital computer Piece of electronic equipment used to process digital data.

Digital data Data represented in digital form.

Digital display Display in which either light-emitting diodes (LEDs) or liquid crystal diodes (LCDs) are used and each of the segments can be either turned on or off.

Digital electrical (power) circuit An electrical circuit designed to turn power either ON or OFF (two-state).

Digital electronic circuits Electronic circuits that encode information into a group of pulses consisting of HIGH or LOW voltages.

Digital multimeter Multimeter used to measure amperes, volts, and ohms and indicate the results on a digital readout display.

Diode ac semiconductor switch or DIAC A bidirectional diode that has a symmetrical switching mode.

DIP Abbreviation for "dual in-line package."

Direct coupling The coupling of two circuits by means of a wire.

Direct current (dc) Current flow in only one direction.

Direct-current amplifier An amplifier capable of amplifying dc voltages and slowly varying voltages.

Direct voltage (dc voltage) Voltage that causes electrons to flow in only one direction.

Directly proportional The relation of one part to one or more other parts in which a change in one causes a similar change in the other.

Discharge Release of energy from either a battery or capacitor.

Disconnect Breaking or opening of an electric circuit.
Discrete components Separate active and passive devices that were manufactured before being used in a circuit.
Display Visual presentation of information or data.
Display unit Unit designed to display digital data.
Dissipation Release of electrical energy in the form of heat.
Distortion An undesired change in the waveform of a signal.
Distributed capacitance Also known as self-capacitance, it is any capacitance other than that within a capacitor, for example, the capacitance between the coil of an inductor or between two conductors (lead capacitance).
Distributed inductance Any inductance other than that within an inductor, for example, the inductance of a length of wire (line inductance).
Domain Also known as a magnetic domain, it is a moveable magnetized area in a magnetic material.
Doping The process wherein impurities are added to the intrinsic semiconductor material either to increase the number of free electrons or to increase the number of holes.
Dot convention A standard used with transformer symbols to indicate whether the secondary voltage will be in phase or out of phase with the primary voltage.
Double-pole, double-throw switch (DPDT) Switch having two movable contacts (double pole) that can be thrown or positioned in one of two positions (double throw).
Double-pole, single-throw switch (DPST) Switch having two movable contacts (double pole) that can be thrown in only one position (single throw) to either make a dual connection or opened to disconnect both poles.
Drain One of the field effect transistor's electrodes.
Drain characteristic curve A plot of the drain current (I_D) versus the drain-to-source voltage (V_{DS}).
Drain current (I_D) A JFET's source-to-drain current.
Drain-feedback bias A configuration in which the gate receives a bias voltage fed back from the drain.
Drain supply voltage ($+V_{DD}$) The bias voltage connected between the drain and source of the JFET, which causes current to flow.
Drain-to-source current with shorted gate (I_{DSS}) The maximum value of drain current, achieved by holding V_{GS} at 0 V.
Dropping resistor Resistor whose value has been chosen to drop or develop a given voltage across it.
Dry cell DC voltage-generating chemical cell using a nonliquid (paste) type of electrolyte.
Dual-gate D-MOSFET A metal-oxide semiconductor FET having two separate gate electrodes.
Dual in-line package (DIP) Package that has two (dual) sets or lines of connecting pins.
Dual-trace oscilloscope Oscilloscope that can simultaneously display two signals.
Duty cycle A term used to describe the amount of ON time versus OFF time. ON time is usually expressed as a percentage.
Dynamic Related to conditions or parameters that change, i.e., motion.
E-core Laminated form, in the shape of the letter E, onto which inductors and transformers are wound.
Eddy currents Small currents induced in a conducting core due to the variations in alternating magnetic flux.
Efficiency Ratio of output power to input power, normally expressed in percent.
Electrical components Components, circuits, and systems that manage the flow of power.
Electrical equipment Equipment designed to manage or control the flow of power.
Electrical wave Traveling wave propagated in a conductive medium that is a variation in voltage or current and travels at slightly less than the speed of light.
Electric charge Electric energy stored on the surface of a material.
Electric current (*I*) Electron movement or motion.
Electric field Also called a voltage field, it is a field or force that exists in the space between two different potentials or voltages.
Electrician Person involved with the design, repair, and assembly of electrical equipment.
Electricity Science that states certain particles possess a force field, which with electrons is negative and with protons is positive. Electricity can be characterized as static and dynamic. Static electricity deals with charges at rest; dynamic electricity deals with charges in motion.
Electric motor Device that converts an electrical energy input into a mechanical energy output.
Electric polarization A displacement of bound charges in a dielectric when placed in an electric field.
Electroacoustic transducer Device that achieves an energy transfer from electric to acoustic (sound), and vice versa. Examples include a microphone and a loudspeaker.
Electroluminescence Transmission or conversion of electrical energy into light energy.
Electrolyte Electrically conducting liquid (wet) or paste (dry).
Electrolytic capacitor Capacitor having an electrolyte between the two plates; due to chemical action, a very thin layer of oxide is deposited on only the positive plate, which accounts for why this type of capacitor is polarized.
Electromagnet A magnet consisting of a coil wound on a soft iron or steel core. When current is passed through the coil, a magnetic field is generated and the core is strongly magnetized to concentrate the magnetic field.
Electromagnetic communication Use of an electromagnetic wave to pass information between two points. Also called wireless communication.
Electromagnetic energy Radiant electromagnetic energy, such as radio and light waves.
Electromagnetic field Field having both an electric (voltage) and magnetic (current) field.
Electromagnetic induction The voltage produced in a coil due to relative motion between the coil and magnetic lines of force.
Electromagnetic spectrum List or diagram showing the entire range of electromagnetic radiation.
Electromagnetic wave Wave that consists of both an electric and magnetic variation.
Electromagnetism Relates to the magnetic field generated around a conductor when current is passed through it.
Electromechanical transducer Device that transforms electrical energy into mechanical energy (motor), and vice versa (generator).
Electromotive force (emf) Force that causes the motion of electrons due to a potential difference between two points.
Electron Smallest subatomic particle of negative charge that orbits the nucleus of the atom.
Electron flow A current produced by the movement of free electrons toward a positive terminal.
Electron–hole pair When an electron jumps from the valence shell or band to the conduction band, it leaves a gap in the covalent bond called a hole. This action creates an electron–hole pair.
Electronic components Components, circuits, and systems that manage the flow of information.

Electronic equipment Equipment designed to control the flow of information.
Electronics Science related to the behavior of electrons in devices.
Electron–pair bond or covalent bond A pair of electrons shared by two neighboring atoms.
Electron-volt (eV) A unit of energy equal to the energy acquired by an electron when it passes through a potential difference of 1 V in a vacuum.
Electrostatic Related to static electric charge.
Electrostatic field Force field produced by static electrical charges.
Element There are 107 different natural chemical substances, or elements, that exist on earth. These can be categorized as gas, solid, or liquid.
Emitter A transistor region from which charge carriers are injected into the base.
Emitter current (I_E) The current at the transistor's emitter terminal.
Emitter feedback The coupling from the emitter output to the base input in a transistor amplifier.
Emitter follower or voltage-follower The common-collector circuit in which the emitter output voltage seems to track or follow the phase and amplitude of the input voltage.
Emitter modulator A modulating circuit in which the modulating signal is applied to the emitter of a bipolar transistor.
Encoder circuit A circuit that produces different output voltage codes, depending on the position of a rotary switch.
Energized Being electrically connected to a voltage source so that the device is activated.
Energy Capacity to do work.
Energy gap The space between two orbital shells.
Engineer Person who designs and develops materials to achieve desired results.
Engineering notation A floating-point system in which numbers are expressed as products consisting of a number that is greater than 1 multiplied by an appropriate power of 10 that is some multiple of 3.
Enhancement-mode MOSFET A field effect transistor in which there are no charge carriers in the channel when the gate-source voltage is zero.
Equation A formal statement of the equality or equivalence of mathematical or logical expressions.
Equipment Term used to describe electrical or electronic units.
Equivalent resistance (R_{eq}) Total resistance of all the individual resistances in a circuit.
Error voltage A voltage that is proportional to the error that exists between input and output.
Excited state An energy level in which an electron may exist if given sufficient energy to reach this state from a lower state.
Exponent A symbol written above and to the right of a mathematical expression to indicate the operation of raising to a power.
Extrinsic semiconductor A semiconductor whose electrical properties are dependent on impurities added to the semiconductor crystal.
Facsimile Electronic process whereby pictures or images are scanned and the graphical information is converted into electrical signals that can be reproduced locally or transmitted to a remote point, where a likeness or facsimile of the original can be produced.
Fahrenheit temperature scale Temperature scale that indicates the freezing point of water at 32°F and boiling point at 212°F.
Fall time Time it takes a negative edge of a pulse to fall from 90% to 10% of its peak value.
Farad (F) Unit of capacitance.
Faraday's law 1. When a magnetic field cuts a conductor, or when a conductor cuts a magnetic field, an electric current will flow in the conductor if a closed path is provided over which the current can circulate. 2. Two other laws related to electrolytic cells.
FCC Abbreviation for "Federal Communications Commission."
Ferrite A powdered, compressed, and sintered magnetic material having high resistivity. The high resistance makes eddy-current losses low at high frequencies.
Ferrite bead Ferrite composition in the form of a bead.
Ferrite core Ferrite core normally shaped like a doughnut.
Ferrite-core inductor An inductor containing a ferrite core.
Ferrites Compound composed of iron oxide, a metallic oxide, and ceramic. The metal oxides include zinc, nickel, cobalt, or iron.
Ferrous Composed of and/or containing iron. A ferrous metal exhibits magnetic characteristics, as opposed to nonferrous metals.
Fiber optics The carrying of information via laser light output between two points by thin glass optical fibers.
Field effect transistor Abbreviated FET, it is a transistor in which the resistance of the source-to-drain current path is changed by applying a transverse electric field to the gate.
Field strength The strength of an electric, magnetic, or electromagnetic field at a given point.
Filament Thin thread of, for example, carbon or tungsten, which when heated by the passage of electric current will emit light.
Filament resistor The resistor in a lightbulb or electron tube.
Filter Network composed of resistor, capacitor, and inductors used to pass certain frequencies yet block others through heavy attenuation.
Fission The process of splitting the nucleus of heavy elements such as uranium and plutonium into two parts, which results in large releases of radioactivity and heat (fission or division of a nuclei).
Fixed biasing A constant value of bias for a FET in which the voltage is independent of the input signal strength.
Fixed component Component whose value or characteristics cannot be varied or changed.
Fixed-value capacitor A capacitor whose value is fixed and cannot be varied.
Fixed-value resistor A resistor whose value cannot be changed.
Floating ground Ground potential that is not tied or in reference to earth.
Flow soldering Soldering technique used in large-scale electronic assembly to solder all the connections on a printed circuit board by moving the board over a wave or flowing bath of molten solder.
Fluorescent lamp Gas-filled glass tube in which the flow of electric current causes the gas to ionize and then release light.
Flux A material used to remove oxide films from the surfaces of metals in preparation for soldering.
Flux density A measure of the strength of a wave.
Flywheel action Sustaining effect of oscillation in an *LC* circuit due to the charging and discharging of the capacitor and the expansion and contraction of the magnetic field around the inductor.
Force Physical action capable of moving a body or modifying its movement.
Formula A general fact, rule, or principle expressed usually in mathematical symbols.
Forward bias The condition in which a small depletion region at the junction will offer a small resistance and permit a large current. Such a junction is forward biased.
Forward breakover voltage Voltage needed to turn ON an SCR.
Forward voltage drop (V_F) The forward voltage drop is equal to the junction's barrier voltage.

Free electrons Electrons that are not in any orbit around a nucleus.

Free-running Operating without any external control.

Free-running multivibrator A circuit that requires no input signal to start its operation, but simply begins to oscillate the moment the dc supply voltage is applied.

Frequency Rate or recurrences of a periodic wave normally within a unit of 1 second, measured in hertz (cycles/second).

Frequency counter Meter used to measure the frequency or cycles per second of a periodic wave.

Frequency-division multiplex (FDM) Transmission of two or more signals over a common path by using a different frequency band for each signal.

Frequency-domain analysis A method of representing a waveform by plotting its amplitude versus frequency.

Frequency drift A slow change in the frequency of a circuit due to temperature or frequency-determining component value changes.

Frequency multiplier A harmonic conversion circuit in which the frequency of the output signal is an exact multiple of the input signal.

Frequency response Indication of how well a device or circuit responds to the different frequencies applied to it.

Frequency response curve A graph indicating a circuit's response to different frequencies.

Friction The rubbing or resistance to relative motion between two bodies in contact.

Frictional electricity Generation of electric charges by rubbing one material against another.

Full-scale deflection (FSD) Deflection of a meter's pointer to the farthest position on the scale.

Full-wave center-tapped rectifier A rectifier circuit that makes use of a center-tapped transformer to cause an output current to flow in the same direction during both half-cycles of the ac input.

Full-wave rectifier Rectifier that makes use of the full ac wave (in both the positive and negative cycle) when converting ac to dc.

Full-wave voltage-doubler circuit A rectifier circuit that doubles the output voltage by charging capacitors during both alternations of the ac input—making use of the full ac input wave.

Function generator Signal generator that can function as a sine, square, rectangular, triangular, or sawtooth waveform generator.

Fundamental frequency A sine wave that is always the lowest-frequency and largest-amplitude component of any waveform shape and is used as a reference.

Fuse A circuit- or equipment-protecting device consisting of a short, thin piece of wire that melts and breaks the current path if the current exceeds a rated, damaging level.

Fuse holder Housing used to support a fuse with two connections.

Fusion The process of melting the nuclei of two light atoms together to create a heavier nucleus, which results in a large release of energy (fusion or combining of nuclei).

Gain Increase in power from one point to another. Normally expressed in decibels.

Gamma rays High-frequency electromagnetic radiation from radioactive particles.

Ganged Mechanical coupling of two or more capacitors, switches, potentiometers, or any other components so that the activation of one control will operate all.

Gas Any aeriform or completely elastic fluid that is not a solid or liquid. All gases are produced by the heating of a liquid beyond its boiling point.

Gate One of the field effect transistor's electrodes (also used for thyristor devices).

Gate-source bias voltage (V_{GS}) The bias voltage applied between the gate and source of a field effect transistor.

Gate-to-source cutoff voltage or $V_{GS(off)}$ The negative V_{GS} bias voltage that causes I_D to drop to approximately zero.

Geiger counter Device used to detect nuclear particles.

Generator Device used to convert a mechanical energy input into an electrical energy output.

Giga Prefix for 1 billion (10^9).

Graph A diagram (as a series of one or more points, lines, line segments, curves, or areas) that represents the variation of a variable in comparison with that of one or more other variables.

Graph origin Center of the graph where the horizontal axis and vertical axis cross.

Greenwich mean time (GMT) Also known as universal time, it is a standard based on the earth's rotation with respect to the sun's position. The solar time at the meridian of Greenwich, England, which is at zero longitude.

Ground An intentional or accidental conducting path between an electrical circuit or system and the earth or some conducting body acting in the place of the earth.

Gunn diode A semiconductor diode that utilizes the Gunn effect to produce microwave frequency oscillation or to amplify a microwave frequency signal.

Half-power point A point at which power is 50%. This corresponds to 70.7% of the total current.

Half-split method A troubleshooting technique used to isolate a faulty block in a circuit or system. In this method a midpoint is chosen and tested to determine which half is malfunctioning.

Half-wave rectifier circuit A circuit that converts ac to dc by only allowing current to flow during one-half of the ac input cycle.

Hall effect sensor A sensor that generates a voltage in response to a magnetic field.

Hardware Electrical, electronic, mechanical, or magnetic devices or components; the physical equipment.

Harmonic frequency Sine wave that is smaller in amplitude and is some multiple of the fundamental frequency.

Hartley oscillator An *LC* tuned oscillator circuit in which the tank coil has an intermediate tap.

Henry (H) Unit of inductance.

Hertz (Hz) Unit of frequency; 1 hertz is equal to one cycle per second.

Hexadecimal number system A base 16 number system, using the digits 0 through 9 and the six additional digits represented by A, B, C, D, E, and F.

High fidelity (hi-fi) Sound reproduction equipment that is as near to the original sound as possible.

High-pass filter Network or circuit designed to pass any frequencies above a critical or cutoff frequency and reject or heavily attenuate all frequencies below.

High *Q* Abbreviation for quality and generally related to inductors that have a high value of inductance and very little coil resistance.

High tension Lethal voltage in the kilovolt range and above.

High-voltage probe Accessory to the voltmeter that has added multiplier resistors within the probe to divide up the large potential being measured by the probe.

***H*-lines** Invisible lines of magnetic flux.

Hole A mobile vacancy in the valance structure of a semiconductor. This hole exists when an atom has less than its normal number of electrons, and is equivalent to a positive charge.

Hole flow Conduction in a semiconductor when electrons move into holes when a voltage is applied.

Hologram Three-dimensional picture created with a laser.

Holography Science dealing with three-dimensional optical recording.

Horizontal Parallel to the horizon or perpendicular to the force of gravity.

Horizontally polarized wave Electromagnetic wave that has the electric field lying in the horizontal plane.

Hot resistance The resistance of a device when hot due to the generation of heat by electric current.

Hybrid circuit Circuit that combines two technologies (passive and active or discrete and integrated components) on one microelectronic circuit. Passive components are generally made by thin-film techniques; active components are made utilizing semiconductor techniques. Integrated circuits can be mounted on the microelectronic circuit and connected to discrete components also on the small postage-stamp-size boards.

Hydroelectric Generation of electric power by the use of water in motion.

Hypotenuse The side of a right-angled triangle that is opposite the right angle.

Hysteresis Amount that the magnetization of a material lags the magnetizing force due to molecular friction.

IC Abbreviation for "integrated circuit."

Imaginary number A complex number whose imaginary part is not zero.

IMPATT diode A semiconductor diode that has a negative resistance characteristic produced by a combination of impact avalanche breakdown and charge carrier transmit time effects in a thin semiconductor chip.

Impedance (*Z*) Measured in ohms, it is the total opposition a circuit offers to current flow (reactive and resistive).

Impedance coupling The coupling of two signal amplifier circuits through the use of an impedance, such as a choke.

Impedance matching circuit A circuit that can match, or isolate, a high-resistance (low-current) source. Matching of the source impedance to the load impedance causes maximum power to be transferred.

Incandescence State of a material when it is heated to such a high temperature that it emits light.

Incandescent lamp An electric lamp that generates light when an electric current is passed through its filament of resistance, causing it to heat to incandescence.

Induced current Current that flows due to an induced voltage.

Induced voltage Voltage generated in a conductor when it is moved through a magnetic field.

Inductance Opposition of a circuit or component to any change in current, as the magnetic field produced by the change in current causes an induced countercurrent to oppose the original change.

Inductive circuit Circuit that has a greater inductive reactance figure than capacitive reactance figure.

Inductive reactance (X_L) Measured in ohms, it is the opposition to alternating or pulsating current flow without the dissipation of energy.

Inductor Length of conductor used to introduce inductance into a circuit.

Inductor analyzer A test instrument designed to test inductors.

Infinite Having no limits.

Infinity Amount larger than any number can indicate.

Information Data or meaningful signals.

Infrared Electromagnetic heat radiation whose frequencies are above the microwave frequency band and below red in the visible band.

Ingot A mass of metal cast into a convenient shape.

Inhibit To stop an action or block data from passing.

Injection laser diode or ILD A semiconductor *p-n* junction diode that uses a lasing action to increase and concentrate the light output.

In phase Two or more waves of the same frequency whose maximum positive and negative peaks occur at the same time.

Input impedance (Z_{in}) The total opposition offered by the transistor to an input signal.

Input resistance (R_{in}) The amount of opposition offered to an input signal by the input base-emitter junction (emitter diode).

Input signal voltage change The input voltage change that causes a corresponding change in the output voltage.

Input transducers A transducer that generates input control signals.

Insulated The condition produced when a nonconductive material is used to isolate conducting materials from one another.

Insulating material Material that will in nearly all cases prevent the flow of current due to its chemical composition.

Insulation resistance Resistance of the insulating material. The greater the insulation resistance, the better the insulator.

Insulator Material that has few electrons per atom and whose electrons are close to the nucleus and cannot be easily removed.

Integrated When two or more components are combined (into a circuit) and then incorporated in one package or housing.

Integrator Device that approximates and whose output is proportional to an integral of the input signal.

Integrator amplifier An amplifier whose output is the integral of its input with respect to time.

Integrator circuit A circuit with an output that is the integral of its input with respect to time.

Intermediate-frequency amplifier An amplifier that has one or more transistor stages designed to amplify intermediate frequency signals.

Intermittent Occurring at random intervals of time. An intermittent component, circuit, or equipment problem is undesirable and difficult to troubleshoot, as the problem needs to occur before isolation of the fault can begin.

Internal resistance A property that represents the inefficiency of a voltage source. No voltage source is 100% efficient in that not all the energy in is converted to electrical energy out; some is wasted in the form of heat dissipation.

Intrinsic semiconductor materials Pure semiconductor materials.

Inversely proportional The relation of one part to one or more other parts in which a change in one causes an opposite change in the other.

Inverting amplifier circuit An op-amp circuit that produces an amplified output signal that is 180° out of phase with the input signal.

Inverting input The inverting or negative input of an operational amplifier.

Ion The form of an atom that has either more electrons in orbit than protons (negative ion) or fewer electrons in orbit or than protons (positive ion).

Jack Socket or connector into which a plug may be inserted.

JFET A field effect transistor made up of a gate region diffused into a channel region. When a control voltage is applied to the gate, the channel is depleted or enhanced, and the current between source and drain is thereby controlled.

***j* operator** A prefix used to indicate an imaginary number.

Joule The unit of work and energy.

Junction Contact or connection between two or more wires or cables.

Junction diode A semiconductor diode in which the rectifying characteristics occur at a junction between the *n*-type and *p*-type semiconductor materials.

Kilovolt-ampere 1000 volts at 1 ampere.

Kilowatt-hour 1000 watts for 1 hour.

Kilowatt-hour meter A meter used by electric companies to measure a customer's electric power use in kilowatt-hours.

Kinetic energy Energy associated with motion.

Kirchhoff's current law The sum of the currents flowing into a point in a circuit is equal to the sum of the currents flowing out of that same point.

Kirchhoff's voltage law The algebraic sum of the voltage drops in a closed-path circuit is equal to the algebraic sum of the source voltage applied.

Knee voltage The voltage at which a curve joins two relatively straight portions of a characteristic curve.

Lag Difference in time between two waves of the same frequency expressed in degrees, i.e., one waveform lags another by a certain number of degrees.

Laminated core Core made up of sheets of magnetic material insulated from one another by an oxide or varnish.

Lamp Device that produces light.

Laser Device that produces a very narrow, intense beam of light. The name is an acronym for "light amplification by stimulated emission of radiation."

Latched A bistable circuit action that causes the circuit to remain held or locked in its last activated state.

***LC* filter** A selective circuit that makes use of an inductance-capacitance network.

Lead The angle by which one alternating signal leads another in time, or a wire that connects two points in a circuit.

Lead–acid cell Cell made up of lead plates immersed in a sulfuric acid electrolyte.

Leakage current or reverse current (I_R) The extremely small current present at the junction.

Least significant bit (LSB) The rightmost binary digit.

LED Abbreviation for "light-emitting diode."

Left-hand rule If the fingers of the left hand are placed around a wire so that the thumb points in the direction of the electron flow, the fingers will be pointing in the direction of the magnetic field being produced by the conductor.

Lenz's law The current induced in a circuit due to a change in the magnetic field is so directed as to oppose the change in flux, or to exert a mechanical force opposing the motion.

Lie detector Piece of electronic equipment, also called a polygraph, that determines whether a person is telling the truth by looking for dramatic changes in blood pressure, body temperature, breathing rate, heart rate, and skin moisture in response to certain questions.

Lifetime The time difference between an electron jumping into the conduction band and then falling back into a hole.

Light Electromagnetic radiation in a band of frequencies that can be received by the human eye.

Light-emitting diode A semiconductor diode that converts electric energy into electromagnetic radiation at visible and near-infrared frequencies when its *p-n* junction is forward biased.

Light-emitting semiconductor devices Semiconductor devices that will emit light when an electrical signal is applied.

Light-sensitive semiconductor devices Semiconductor devices that change their characteristics in response to light.

Limiter Circuit or device that prevents some portion of its input from reaching the output.

Linear Relationship between input and output in which the output varies in direct proportion to the input.

Linear scale A scale whose divisions are uniformly spaced.

Line graph A graph in which points representing values of a variable are connected by a broken line.

Liquid crystal displays (LCDs) A digital display having two sheets of glass separated by a sealed quantity of liquid crystal material. When a voltage is applied across the front and back electrodes, the liquid crystals' molecules become disorganized, causing the liquid to darken.

Literal number A number expressed as a letter.

Live Term used to describe a circuit or piece of equipment that is on and has current flow within it.

Load A component, circuit, or piece of equipment connected to a source can be called a load and will have a certain load resistance, which will consequently determine the load current.

Load current The current that is present in the load.

Load impedance Total reactive and resistive opposition of a load.

Loading The adding of a load to a source.

Loading effect Large load resistance will cause a small load current to flow, and so the loading down of the source or loading effect will be small (light load), whereas a small load resistance will cause a large load current to flow from the source, which will load down the source (heavy load).

Load resistance The resistance of the load.

Locked To automatically follow a signal.

Lodestone A magnetite stone possessing magnetic polarity.

Logarithms The exponent that indicates the power to which a number is raised to produce a given number.

Logic Science dealing with the principles and applications of gates, relays, and switches.

Logic gate circuits Two-state (ON/OFF) circuits used for decision-making functions in digital logic circuits.

Loss Term used to describe a decrease in power.

Lower sideband A group of frequencies that are equal to the differences between the carrier and modulation frequencies.

Low-pass filter Network or circuit designed to pass any frequencies below a critical or cutoff frequency and reject or heavily attenuate all frequencies above.

Magnet Body that can be used to attract or repel magnetic materials.

Magnetic circuit breaker Circuit breaker that is tripped or activated by use of an electromagnet.

Magnetic coil Spiral of a conductor, which is called an electromagnet.

Magnetic core Material that exists in the center of the magnetic coil to either support the windings (nonmagnetic material) or intensify the magnetic flux (magnetic material).

Magnetic field Magnetic lines of force traveling from the north to the south pole of a magnet.

Magnetic flux The magnetic lines of force produced by a magnet.

Magnetic leakage The passage of magnetic flux outside the path along which it can do useful work.

Magnetic poles Points of a magnet from which magnetic lines of force leave (north pole) and arrive (south pole).

Magnetism Property of some materials to attract and repel others.

Magnetizing force Also called magnetic field strength, it is the magnetomotive force per unit length at any given point in a magnetic circuit.

Magnetomotive force Force that produces a magnetic field.

Mainframe Large computers that initially were only affordable for medium-sized and large businesses who had the space for them. Minicomputers came after mainframes and were affordable for any business, and now we have microcomputers, which are much more affordable.

Majority carriers The type of carrier that constitutes more than half the total number of carriers in a semiconductor material. In *n*-type materials electrons are the majority carriers, whereas in *p*-type materials holes are the majority carriers.

Matched impedance Condition that occurs when the source impedance is equal to the load impedance, resulting in maximum power being transferred.

Matching Connection of two components or circuits so that maximum energy is transferred or coupled between the two.

Maximum power transfer theorem A theorem that states maximum power will be transferred from source to load when the source resistance is equal to the load resistance.

Maxwell One magnetic line of force or flux is called a maxwell.

Measurement Determining the presence and magnitude of variables.

Medical electronics Branch of electronics involved with therapeutic or diagnostic practices in medicine.
Mercury cell Primary cell that has a mercuric oxide cathode, a zinc anode, and a potassium hydroxide electrolyte.
Metal film resistor A resistor in which a film of metal, metal oxide, or alloy is deposited on an insulating substrate.
Metal-oxide resistor A metal film resistor in which an oxide of a metal (such as tin) is deposited as a film onto the substrate.
Metal-oxide varistors (MOVs) Devices that are replacing zener-diode and transient-diode suppressors because they are able to shunt a much higher current surge and are cheaper.
Meter (1) Any electrical or electronic measuring device. (2) In the metric system, it is a unit of length equal to 39.37 inches, or 3.28 feet.
Meter FSD current The value of current needed to cause the meter movement to deflect the needle to its full-scale deflection (FSD) position.
Meter resistance The resistance of a meter's armature coil.
Metric system A decimal system of weights and measures based on the meter and the kilogram.
Mica capacitor Fixed capacitor that uses mica as the dielectric between its plates.
Microphone Electroacoustic transducer that responds and converts a sound wave input into an equivalent electrical wave output.
Microwave Term used to describe a band of very small-wavelength radio waves within the UHF, SHF, and EHF bands.
Mil One-thousandth of an inch (0.001 in.).
Miller-effect capacitance (C_M) An undesirable inherent capacitance that exists between the junctions of transistors.
Minority carriers The type of carrier that constitutes less than half the total number of carriers in a semiconductor material. In *n*-type materials holes are the minority carriers, whereas in *p*-type materials electrons are the minority carriers.
Mismatch Term used to describe a difference between the source impedance and load impedance, which will prevent maximum power transfer.
Modulation Process whereby an information signal is used to modify some characteristic of another, higher-frequency wave known as a carrier.
Molecule Smallest particle of a compound that still retains its chemical characteristics.
Monostable multivibrator A device that when triggered will generate a rectangular pulse of fixed duration.
MOS family A group of digital logic circuits that make use of metal-oxide semiconductor field effect transistors (MOSFETs).
MOSFET A field effect transistor in which the insulating layer between the gate electrode and the channel is a metal-oxide layer. Either a *p* or *n* substrate.
Most significant bit (MSB) The leftmost binary digit.
Moving-coil microphone Microphone that makes use of a moving coil between a fixed magnetic field. Also called a dynamic microphone.
Moving-coil pickup Dynamic phonograph pickup that uses a coil between a fixed magnetic field, which is moved back and forth by the needle or stylus.
Moving-coil speaker Dynamic speaker that uses a coil placed between a fixed magnetic field and converts the electrical wave input into sound waves.
Multimeter Piece of electronic test equipment that can perform multiple tasks in that it can be used to measure voltage, current, or resistance.
Multiple-emitter input transistor A transistor specially constructed to have more than one emitter.
Multiplier resistor A resistor connected in series with the meter movement of a voltmeter.
Mutual inductance Ability of one inductor's magnetic lines of force to link with another inductor.
Navigation equipment Electronic equipment designed to aid in the direction of aircraft and ships to their destination.
***n*-Channel D-type MOSFET** A depletion-type MOSFET having an *n*-type channel between its source and drain terminals.
***n*-Channel E-type MOSFET** An enhancement-type MOSFET having an *n*-type channel between its source and drain terminals.
***n*-Channel JFET** A junction field effect transistor having an *n*-type channel between source and drain.
Negative (neg.) (1) Some value less than zero. (2) Terminal that has an excess of electrons.
Negative charge An electric charge that has more electrons than protons.
Negative feedback Feedback in which a portion of the output signal is fed back 180° out of phase with the input signal.
Negative ground A system whereby the negative terminal of the voltage source is connected to the system's conducting chassis or body.
Negative ion Atom that has more than the normal neutral amount of electrons.
Negative resistance A resistance such that when the current through it increases, the voltage drop across the resistance decreases.
Negative temperature coefficient Effect that if temperature increases, resistance or capacitance will decrease.
Negative temperature coefficient (NTC) thermistor A thermistor in which a temperature increase causes the resistance to decrease.
Neon bulb Glass envelope filled with neon gas, which when ionized by an applied voltage will glow red.
Network Combination and interconnection of components, circuits, or systems.
Neutral The state of being neither positive nor negative.
Neutral atom An atom in which the number of positive charges in the nucleus (protons) is equal to the number of negative charges (electrons) that surround the nucleus.
Neutral wire The conductor of a polyphase circuit or a single-phase three-wire circuit that is intended to have a ground potential. The potential differences between the neutral and each of the other conductors are approximately equal in magnitude and are also equally spaced in phase.
Neutron Subatomic particle residing within the nucleus and having no electrical charge.
Nickel-cadmium cell Most popular secondary cell; it uses a nickel oxide positive electrode and cadmium negative electrode.
No-change or latch condition The condition of the *S-R* flip-flop when both inputs are LOW, where there will be no change in the output.
Node Junction or branch point.
Noise Unwanted electromagnetic radiation within an electrical or mechanical system.
Noninverting amplifier An operational amplifier in which the input signal is applied to the ungrounded positive input terminal to give a gain that is greater than 1 and make the output change in phase with the input voltage.
Noninverting input The noninverting or positive input of an operational amplifier.
Nonlinear scale A scale whose divisions are not uniformly spaced.
Normally closed (N.C.) Designation which states that the contacts of a switch or relay are connected normally; however, when activated, these contacts will open.
Normally open (N.O.) Designation which states that the contacts of a switch or relay are normally not

connected; however, when activated, these contacts will close.

North pole Pole of a magnet out of which magnetic lines of force are assumed to originate.

Norton's theorem Any network of voltage sources and resistors can be replaced by a single equivalent current source (I_N) in parallel with a single equivalent resistance (R_N).

NPN transistor Negative-positive-negative transistor in which a layer of *p*-type conductive semiconductor is located between two *n*-type regions.

***n*-Type semiconductor** A material that has more conduction-band electrons than valence-band holes.

Nuclear energy Atomic energy or power released in a nuclear reaction when either a neutron is used to split an atom into smaller atoms (fission) or when two smaller nuclei are joined together (fusion).

Nuclear reactor Unit that maintains a continuous self-supporting nuclear reaction (fission).

Nucleus Core of an atom; it contains both positive (protons) and neutral (neutrons) subatomic particles.

Octave Interval between two sounds whose fundamental frequencies differ by a ratio of 2 to 1.

Offset null inputs The two balancing inputs used to balance an op-amp.

Ohm Unit of resistance, symbolized by the Greek capital letter omega (Ω).

Ohmmeter Measurement device used to measure electric resistance.

Ohm's law Relationship between the three electrical properties of voltage, current, and resistance. Ohm's law states that the current flow within a circuit is directly proportional to the voltage applied across the circuit and inversely proportional to its resistance.

Ohms per volt Value that indicates the sensitivity of a voltmeter. The higher the ohms per volt rating, the more sensitive the meter.

One-shot multivibrator Produces one output pulse or shot for each input trigger.

Open circuit Break in the path of current flow.

Open collector output A type of output structure found in certain bipolar logic families. Resistive pull-ups are generally added to provide the high-level output voltage.

Open-loop mode A control system that has no means of comparing the output with the input for control purposes.

Operational amplifier Special type of high-gain amplifier; also called an op-amp.

Operator error An incorrect use of the controls of a circuit or system.

Opposite The side of a right-angle triangle that is opposite the angle theta.

Optically coupled isolator DIP module An optocoupler that contains an infrared-emitting diode and a silicon photo-transistor and is generally used to transfer switching information between two electrically isolated points.

Optically coupled isolator interrupter module A device that consists of a matched and aligned emitter and detector and that is used to detect opaque or nontransparent targets.

Optically coupled isolator reflector module A device that consists of a matched and aligned emitter and detector and that is used to detect targets.

Optically coupled isolators Devices that contain a light-emitting and light-sensing device in one package. They are used to optically couple two electrically isolated points.

Optimum power transfer Since the ideal maximum power transfer conditions cannot always be achieved, most designers try to achieve maximum power transfer by having the source resistance and load resistance as close in value as possible.

Ordinary ground A connection in the circuit that is said to be at ground potential or zero volts and, because of its connection to earth, has the ability to conduct electrical current to and from the earth.

OR gate A gate for which when either input *A* OR *B* is HIGH, the output will be HIGH.

Oscillate Continual repetition of or passing through a cycle.

Oscillator Electronic circuit that converts dc to a continuous alternating current out.

Oscilloscope Instrument used to view signal amplitude, period, and shape at different points throughout a circuit.

Out of phase Term that describes when the maximum and minimum points of two or more waveforms do not occur at the same time.

Output Terminals at which a component, circuit, or piece of equipment delivers current, voltage, or power.

Output impedance (Z_{out}) The total opposition offered by the transistor to the output signal.

Output power Amount of power a component, circuit, or system can deliver to its load.

Output resistance (R_{out}) The amount of opposition offered to an output signal by the output base-collector junction (collector diode).

Output signal voltage change Change in output signal voltage in response to a change in the input signal voltage.

Output transducers Transducers that convert output electrical signals to some other energy form.

Overload Situation that occurs when the load is greater than the component, circuit, or system was designated to handle (load resistance too small, load current too high), resulting in waveform distortion and/or overheating.

Overload protection Protective device such as a fuse or circuit breaker that automatically disconnects or opens a current path when it exceeds an excessive value.

Paper capacitor Fixed capacitor using oiled or waxed paper as a dielectric.

Parallax error The apparent displacement of an object's position caused by a shift in the point of observation of the object.

Parallel Also called shunt; circuit having two or more paths for current flow.

Parallel circuit Also called shunt; circuit having two or more paths for current flow.

Parallel data transmission The transfer of information simultaneously over a set of parallel paths or channels.

Parallel resonant circuit Circuit having an inductor and capacitor in parallel with one another, offering a high impedance at the frequency of resonance.

Parallel-to-serial converter An IC that converts the parallel word applied to its inputs into a serial data stream.

Parasitic oscillations An unwanted self-sustaining oscillation or self-generated random pulse.

Pass band Band or range of frequencies that will be passed by a filter circuit.

Passive component Component that does not amplify a signal, such as a resistor or capacitor.

Passive filter A filter that contains only passive (nonamplifying) components and provides pass or rejection characteristics.

Passive system System that emits no energy. In other words, it only receives; it does not transmit and consequently reveal its position.

***p*-Channel D-type MOSFET** A depletion-type MOSFET having a *p*-type channel between its source and drain terminals.

***p*-Channel E-type MOSFET** An enhancement-type MOSFET having a *p*-type channel between its source and drain terminals.

***p*-Channel JFET** A junction field effect transistor having a *p*-type channel between source and drain.

Peak Maximum or highest-amplitude level.

Peak inverse voltage Abbreviated PIV; the maximum rated value of an ac voltage acting in the direction opposite to that in which a device is designed to pass current.

Peak to peak Difference between the maximum positive and maximum negative values.

Peak voltage (V_p) The maximum value of voltage.

Pentode A five-electrode electron tube that has an anode, cathode, control grid, and two additional electrodes.

Percent In the hundred; of each hundred.

Percent of regulation The change in output voltage that occurs between no-load and full-load in a dc voltage source. Dividing this change by the rated full-load value and multiplying the result by 100 gives percent regulation.

Percent of ripple The ratio of the effective or rms value of ripple voltage to the average value of the total voltage. Expressed as a percentage.

Period Time taken to complete one complete cycle of a periodic or repeating waveform.

Permanence Magnetic equivalent of electrical conductance and consequently equal to the reciprocal of reluctance, just as conductance is equal to the reciprocal of resistance.

Permanent magnet Magnet, normally made of hardened steel, that retains its magnetism indefinitely.

Permeability Measure of how much better a material is as a path for magnetic lines of force with respect to air, which has a permeability of 1 (symbolized by the Greek lowercase letter mu, μ).

Phase Angular relationship between two waves, normally between current and voltage in an ac circuit.

Phase angle Phase difference between two waves, normally expressed in degrees.

Phase-locked loop (PLL) circuit A circuit consisting of a phase comparator that compares the output frequency of a voltage-controlled oscillator with an input frequency. The error voltage out of the phase comparator is then coupled via an amplifier and low-pass filter to the control input of the voltage controlled oscillator to keep it in phase, and therefore at exactly the same frequency as the input.

Phase shift Change in phase of a waveform between two points, given in degrees of lead or lag.

Phase-shift oscillator An *RC* oscillator circuit in which the 180° phase shift is achieved with several *RC* networks.

Phase splitter A circuit that takes a single input signal and produces two output signals that are 180° apart in phase.

Phonograph Piece of equipment used to reproduce sound.

Phosphor Luminescent material applied to the inner surface of a cathode-ray tube that when bombarded with electrons will emit light.

Photoconduction A process by which the conductance of a material is changed by incident electromagnetic radiation in the light spectrum.

Photoconductive cell Material whose resistance decreases or conductance increases when light strikes it.

Photoconductive cell or light-dependent resistor (LDR) A two-terminal device that changes its resistance when light is applied.

Photodetector Component used to detect or sense light.

Photodiode A semiconductor diode that changes its electrical characteristics in response to illumination.

Photometer Meter used to measure light intensity.

Photon Discrete portion of electromagnetic energy. A small packet of light.

Photoresistor Also known as a photoconductive cell or light-dependent resistor, a device whose resistance varies with the illumination of the cell.

Photo-transistor, photo-darlington, and photo-SCR Examples of light-reactive devices.

Photovoltaic action A process by which a device generates a voltage as a result of exposure to radiation.

Photovoltaic cell Component, commonly called a solar cell, used to convert light energy into electric energy (voltage).

Photovoltaic cell or solar cell A device that generates a voltage across its terminal that will increase as the light level increases.

Pi Value representing the ratio between the circumference and diameter of a circle and equal to approximately 3.142 (symbolized by the lowercase Greek letter π).

Pie chart or circle graph A circular chart cut by radii into segments illustrating relative magnitudes.

Pierce crystal oscillator An oscillator circuit in which a piezoelectric crystal is connected in a tank between output and input.

Piezoelectric crystal Crystal material that will generate a voltage when mechanical pressure is applied and conversely will undergo mechanical stress when subjected to a voltage.

Piezoelectric effect The operation of a voltage between the opposite sides of a piezoelectric crystal as a result of pressure or twisting. Also, the reverse effect in which the application of a voltage to opposite sides causes deformation to occur at the frequency of the applied voltage.

Piezoresistive diaphragm pressure sensor A sensor that changes its resistance in response to pressure.

Pinch-off region A region on the characteristic curve of a FET in which the gate bias causes the depletion region to extend completely across the channel.

Pinch-off voltage (V_P) The value of V_{DS} at which further increases in V_{DS} will cause no further increase in I_D.

PIN diode A semiconductor diode that has a high-resistance intrinsic region between its low-resistance *p*-type and *n*-type regions.

Pitch Term used to describe the inflection or frequency scale of sounds. When the pitch is increased by one octave, twice the original frequency will be the result.

Plastic film capacitor A capacitor in which alternate layers of metal aluminum foil are separated by thin films of plastic dielectric.

Plate Conductive electrode in either a capacitor or battery.

Plug Movable connector that is normally inserted into a socket.

***p-n* junction** The point at which two opposite doped materials come in contact with one another.

PNP transistor Positive-negative-positive transistor in which a layer of *n*-type conductive semiconductor is located between two *p*-type regions.

Polar coordinates Either of two numbers that locate a point in a plane by its distance from a fixed point on a line and the angle this line makes with a fixed line.

Polarity Term used to describe positive and negative charges.

Polarized electrolytic capacitor An electrolytic capacitor in which the dielectric is formed adjacent to one of the metal plates, creating a greater opposition to current in one direction only.

Positive Point that attracts electrons, as opposed to negative, which supplies electrons.

Positive charge The charge that exists in a body that has fewer electrons than normal.

Positive ground A system whereby the positive terminal of the voltage source is connected to the system's conducting chassis or body.

Positive ion Atom that has lost one or more of its electrons and therefore has more protons than electrons, resulting in a net positive charge.

Positive shunt clipper A circuit that has a diode connected in shunt with the lead or output: its orientation is such that it will clip off the positive alternation of the ac input.

Positive temperature coefficient of resistance The rate of increase in resistance relative to an increase in temperature.

Positive temperature coefficient (PTC) thermistors A thermistor in which a temperature increase causes the resistance to increase.

Potential difference (PD) Voltage difference between two points, which will cause current to flow in a closed circuit.

Potential energy Energy that has the potential to do work because of its position relative to others.

Potentiometer Three-lead variable resistor that through mechanical turning of a shaft can be used to produce a variable voltage or potential.

Power Amount of energy converted by a component or circuit in a unit of time, normally seconds. It is measured in units of watts (joules/second).

Power amplifier An amplifier designed to deliver maximum output power to a load.

Power dissipation Amount of heat energy generated by a device in 1 second when current flows through it.

Power factor Ratio of actual power to apparent power. A pure resistor has a power factor of 1, or 100%, whereas a capacitor has a power factor of 0 or 0%.

Power gain (A_P) The ratio of the output signal power to the input signal power.

Power loss Ratio of power absorbed to power delivered.

Power supply Piece of electrical equipment used to deliver either ac or dc voltage.

Prefix Name used to designate a factor or multiplier.

Pressure The application of force to something by something else in direct contact with it.

Primary First winding of a transformer that is connected to the source, as opposed to the secondary that is connected to the load.

Primary cell Cell that produces electrical energy through an internal electrochemical action; once discharged, it cannot be reused.

Printed circuit board (PCB) Insulating board that has conductive tracks printed onto the board to make the circuit.

Programmable unijunction transistor or PUT A unijunction transistor whose peak voltage can be controlled.

Propagation Traveling of electromagnetic, electrical, or sound waves through a medium.

Propagation time Time it takes for a wave to travel between two points.

Proportional A term used to describe the relationship between two quantities that have the same ratio.

Protoboard An experimental arrangement of a circuit on a board. Also called a breadboard.

Proton Subatomic particle within the nucleus that has a positive charge.

***p*-Type semiconductor** A material that has more valence-band holes than conduction-band electrons.

Pull-up resistor A resistor connected between a signal output and positive V_{CC} to pull the signal line HIGH when it is not being driven LOW.

Pulse Rise and fall of some quantity for a period of time.

Pulse fall time Time it takes for a pulse to decrease from 90% to 10% of its maximum value.

Pulse repetition frequency The number of times per second that a pulse is transmitted.

Pulse repetition time The time interval between the start of two consecutive pulses.

Pulse rise time Time it takes for a pulse to increase from 10% to 90% of its maximum value.

Pulse width, pulse length, or pulse duration The time interval between the leading edge and trailing edge of a pulse at which the amplitude reaches 50% of the peak pulse amplitude.

Push-pull amplifier A balanced amplifier that uses two similar equivalent amplifying transistors working in phase opposition.

Pythagorean theorem A theorem in geometry: The square of the length of the hypotenuse of a right triangle equals the sum of the squares of the lengths of the other two sides.

Q Quality factor of an inductor or capacitor; it is the ratio of a component's reactance (energy stored) to its effective series resistance (energy dissipated).

Quiescent point The voltage or current value that sets up the no-signal input or operating point bias voltage.

Race condition An unpredictable bistable circuit condition.

Radar Acronym for "radio detection and ranging"; a system that measures the distance and direction of objects.

Radioastronomy Branch of astronomy that studies the radio waves generated by celestial bodies and uses these emissions to obtain more information about them.

Radio broadcast Transmission of music, voice, and other information on radio carrier waves that can be received by the general public.

Radio communication Term used to describe the transfer of information between two or more points by use of radio or electromagnetic waves.

Radio-frequency (RF) amplifier An amplifier that has one or more transistor stages designed to amplify radio-frequency signals.

Radio-frequency generator A generator capable of supplying RF energy at any desired frequency in the radio spectrum.

Radio-frequency probe A probe used in conjunction with an ac meter to measure high-frequency RF signals.

Radius A line segment extending from the center of a circle or sphere to the circumference.

Ratio The relationship in quantity, amount, or size between two or more things.

RC Abbreviation for "resistance-capacitance."

***RC* circuit** Circuit containing both a resistor and capacitor.

***RC* coupling** A coupling method in which resistors are used as the input and output impedances of the two stages. A coupling capacitor is generally used between the stages to couple the ac signal and block the dc supply bias voltages.

***RC* filter** A selective circuit that makes use of a resistance-capacitance network.

***RC* time constant** A measurement equal to the product of resistance and capacitance in seconds; in one time constant a capacitor will have charged or discharged 63.2% of the maximum applied voltage.

Reactance (*X*) Opposition to current flow without the dissipation of energy.

Reactive power Also called imaginary power or wattless power, the power value obtained by multiplying the effective value of current by the effective value of voltage and the sine of the angular phase difference between current and voltage.

Real number A number that has no imaginary part.

Receiver Unit or piece of equipment used for the reception of information.

Recombination The combination and resultant neutralization of particles or objects having unlike charges; for example, a hole and an electron or a positive ion and negative ion.

Rectangular coordinates A Cartesian coordinate of a Cartesian coordinate system whose straight-line axes or coordinate planes are perpendicular.

Rectangular wave Also known as a pulse wave, a repeating wave that only alternates between two levels or values and remains at one of these values for a small amount of time relative to the other.

Rectification Process that converts alternating current (ac) into direct current (dc).

Rectifier A device that converts alternating current into a unidirectional or dc current.

Rectifier circuit A circuit that achieves rectification.

Rectifier diodes or rectifiers Junction diodes that achieve rectification.

Reed relay Relay that consists of two thin magnetic strips within a glass envelope with a coil wrapped around the envelope so that when it is energized the relay's contacts or strips will snap together, making a connection between the leads attached to each of the reed strips.

Regulator A device that maintains a desired quantity at a predetermined voltage.

Relative Not independent; compared with or with respect to some other value of a measured quantity.

Relaxation oscillator An oscillator circuit whose frequency is determined by an *RL* or *RC* network, producing a rectangular or sawtooth output waveform.

Relay Electromechanical device that opens or closes contacts when a current is passed through a coil.

Reluctance Resistance to the flow of magnetic lines of force.

Remanence Amount a material remains magnetized after the magnetizing force has been removed.

Reset and carry An action that occurs in any column that reaches its maximum count.

Reset state A circuit condition in which the output is reset to binary 0.

Residual magnetism Magnetism remaining in the core of an electromagnet after the coil current has been removed.

Resistance Symbolized *R* and measured in ohms (Ω); the opposition to current flow with the dissipation of energy in the form of heat.

Resistive power (true power) The average power consumed by a circuit during one complete cycle of alternating current.

Resistive temperature detector (RTD) A temperature detector consisting of a fine coil of conducting wire (such as platinum) that will produce a relatively linear increase in resistance as temperature increases.

Resistivity Measure of a material's resistance to current flow.

Resistor Component made of a material that opposes the flow of current and therefore has some value of resistance.

Resistor color code Coding system of colored stripes on a resistor that indicates the resistor's value and tolerance.

Resonance Circuit condition that occurs when the inductive reactance (X_L) is equal to the capacitive reactance (X_C).

Resonant circuit Circuit containing an inductor and capacitor tuned to resonate at a certain frequency.

Resonant frequency Frequency at which a circuit or object will produce a maximum amplitude output.

Resultant vector A vector derived from or resulting from two or more other vectors.

Reverse bias The condition in which a large depletion region at the junction will offer a large resistance and permit only a small current. Such a junction is reverse biased.

Reverse leakage current (I_R) The undesirable flow of current through a device in the reverse direction.

Reverse voltage drop (V_R) The reverse voltage drop is equal to the source voltage (applied voltage).

RF Abbreviation for "radio frequency."

RF amplifier/frequency tripler circuit A radio-frequency amplifier circuit in which the frequency of the output is three times that of the input frequency.

RF amplifier/multiplier circuit A radio-frequency amplifier circuit in which the frequency of the output is an exact multiple of the input frequency.

RF local oscillator circuit The radio-frequency oscillator in a superheterodyne receiver.

Rheostat Two-terminal variable resistor used to control current.

Right-angle triangle A triangle having a 90° or square corner.

Ripple frequency The frequency of the ripple present in the output of a dc source.

Rise time Time it takes a positive edge of a pulse to rise from 10% to 90% of its high value.

***RL* differentiator** An *RL* circuit whose output voltage is proportional to the rate of change of the input voltage.

***RL* filter** A selective circuit of resistors and inductors that offers little or no opposition to certain frequencies while blocking or attenuating other frequencies.

***RL* integrator** An *RL* circuit with an output proportional to the integral of the input signal.

RMS Abbreviation for "root mean square."

RMS value The rms value of an ac voltage, current, or power waveform equal to 0.707 times the peak value. The rms value is the effective or dc value equivalent of the ac wave.

Rotary switch Electromechanical device that has a rotating shaft connected to one terminal that is capable of making or breaking a connection.

***R–2R* ladder circuit** A network or circuit composed of a sequence of L networks connected in tandem. This *R–2R* circuit is used in digital-to-analog converters.

Rounding off An operation in which a value is abbreviated by applying the following rule: When the first digit to be dropped is a 6 or more, or a 5 followed by a digit that is more than zero, increase the previous digit by 1. When the first digit to be dropped is a 4 or less, or a 5 followed by a zero, do not change the previous digit.

Saturation The condition in which a further increase in one variable produces no further increase in the resultant effect.

Saturation point The point beyond which an increase in one of two quantities produces no increase in the other.

Saturation region The point at which the collector supply voltage has the transistor operating at saturation.

Sawtooth wave Repeating waveform that rises from zero to a maximum value linearly and then falls to zero and repeats.

Scale Set of markings used for measurement.

Schematic diagram Illustration of the electrical or electronic scheme of a circuit, with all the components represented by their respective symbols.

Schottky diode A semiconductor diode formed by contact between a semiconductor layer and a metal coating. Hot carriers (electrons for *n*-type material and holes for *p*-type material) are emitted from the Schottky barrier of the semiconductor and move to the metal coating. Since majority carriers predominate, there is essentially no injection or storage of minority carriers to limit switching speed.

Scientific notation System in which numbers are entered and displayed in terms of a power of 10. For example:

Number	Scientific Notation
7642	7.642×10^3
0.000096	96×10^{-6}
0.0012	1.2×10^{-3}
64,000,000	64×10^6

Scopemeter A handheld, battery-operated instrument that combines a multimeter, oscilloscope, frequency counter, and signal generator in one.

Secondary Output winding of a transformer that is connected across the load.

Secondary cells Electrolytic cells for generating electricity. Once discharged the cell may be restored or recharged by sending an electric current through the cell in the opposite direction to that of the discharge current.

Selectivity Property of a circuit to discriminate between the wanted signal and the unwanted signal.

Self-biasing Gate bias for a FET in which a resistor is used to drop the supply voltage and provide gate bias.

Self-inductance The property that causes a counter electromotive force to be produced in a conductor when the magnetic field expands or collapses with a change in current.

Semiconductor transducer An electronic device that converts one form of energy to another.

Semiconductors Materials that have properties that lie between insulators and conductors.

Serial data transmission The transfer of information sequentially through a single path or channel.

Serial-to-parallel converter An IC that converts a serial data stream applied to its input into a parallel output word.

Series circuit Circuit in which the components are connected end to end so that current has only one path to follow throughout the circuit.

Series clipper A circuit that will clip off part of the input signal. Also known as a limitor since the circuit will limit the ac input. A series clipper circuit has a clipping or limiting device in series with the load.

Series–parallel circuit Network or circuit that contains components that are connected in both series and parallel.

Series resonance Condition that occurs when the inductive and capacitive reactances are equal and both components are connected in series with one another, and therefore the impedance is minimum.

Series resonant circuit A resonant circuit in which the capacitor and coil are in series with the applied ac voltage.

Series switching regulators A regulator circuit containing a power transistor in series with the load that is switched ON and OFF to regulate the dc output voltage delivered to the load.

Set-reset flip-flop A bistable multivibrator circuit that has two inputs that are used to either SET or RESET the output.

Set state A circuit condition in which the output is set to binary 1.

Seven-segment display Component that normally has eight LEDs, seven of which are mounted into segments or bars that make up the number 8, with the eighth LED used as a decimal point.

Shells or bands An orbital path containing a group of electrons that have a common energy level.

Shield Metal grounded cover that is used to protect a wire, component, or piece of equipment from stray magnetic and/or electric fields.

Shock The sudden pain, convulsion, unconsciousness, or death produced by the passage of electric current through the body.

Short circuit Also called a short; a low-resistance connection between two points in a circuit, typically causing a large amount of current flow.

Shorted out Term used to describe a component that has either internally malfunctioned, resulting in a low-resistance path through the component, or a component that has been bypassed by a low-resistance path.

Shunt clipper A circuit that will clip off part of the input signal. Also known as a limitor since the circuit will limit the ac input. A shunt clipper circuit has a clipping or limiting device in shunt with the load.

Shunt resistor A resistor connected in parallel or shunt with the meter movement of an ammeter.

Side bands A band of frequencies produced by modulation on both sides of the carrier frequency of a modulated signal.

Signal Conveyor of information.

Signal-to-noise ratio Ratio of the magnitude of the signal to the magnitude of the noise, normally expressed in decibels.

Signal voltage Effective voltage or rms value of a signal.

Silicon (Si) Nonmetallic element (atomic number 14) used in pure form as a semiconductor.

Silicon controlled rectifier Abbreviated SCR; a three-junction, three-terminal, unidirectional P-N-P-N thyristor that is normally an open circuit. When triggered with the proper gate signal it switches to a conducting state and allows current to flow in one direction.

Silicon transistor Transistor using silicon as the semiconducting material.

Silver (Ag) Precious metal that does not easily corrode and is more conductive than copper.

Silvered mica capacitor Mica capacitor with silver instead of a conducting metal foil deposited directly onto the mica sheets.

Silver solder Solder composed of silver, copper, and zinc with a melting point lower than silver but higher than the standard lead-tin solder.

Simplex Communication in only one direction at a time, for example, facsimile and television.

Simulcast Broadcasting of a program simultaneously in two different forms, for example, a program on both AM and FM.

Sine The trigonometric function of an angle equal to the side opposite divided by the hypotenuse of a right-angle triangle.

Sine wave Wave whose amplitude is the sine of a linear function of time. It is drawn on a graph that plots amplitude against time or radial degrees relative to the angular rotation of an alternator.

Single in-line package (SIP) Package containing several electronic components (generally resistors) with a single row of external connecting pins.

Single-pole, double-throw (SPDT) Three-terminal switch or relay in which one terminal can be thrown in one of two positions.

Single-pole, single-throw (SPST) Two-terminal switch or relay that can either open or close one circuit.

Single sideband (SSB) AM radio communication technique in which the transmitter suppresses one sideband and therefore only transmits a single sideband.

Single-throw switch Switch containing only one set of contacts, which can be either opened or closed.

Sink Device, such as a load, that consumes power or conducts away heat.

Sintering Process of bonding either a metal or powder by cold-pressing it into a desired shape and then heating to form a strong, cohesive body.

Sinusoidal Varying in proportion to the sine of an angle or time function; for example, alternating current (ac) is sinusoidal.

SIP Abbreviation for "single in-line package."

Skin effect Tendency of high-frequency (RF) currents to flow near the surface layer of a conductor.

Slide switch Switch having a sliding bar, button, or knob.

Slow-acting relay Slow-operating relay that when energized may not pull up the armature for several seconds.

Slow-blow fuse Fuse that can withstand a heavy current (up to ten times its rated value) for a small period of time without opening.

Snap switch Switch containing a spring under tension or compression that causes the contacts to come together suddenly when activated.

SNR Abbreviation for "signal-to-noise ratio."

Soft magnetic material Ferromagnetic material that is easily demagnetized.

Software Program of instructions that directs the operation of a computer.

Solar cell Photovoltaic cell that converts light into electric energy. They are especially useful as a power source for space vehicles.

Solder Metallic alloy that is used to join two metal surfaces.

Soldering Process of joining two metallic surfaces to make an electrical contact by melting solder (usually tin and lead) across them.

Soldering gun Soldering tool having a trigger switch and pistol shape that at its tip has a fast-heating resistive element for soldering.

Soldering iron Soldering tool having an internal heating element that is used for heating a connection to melt solder.

Solenoid Coil and movable iron core that when energized by an alternating or direct current will pull the core into a central position.

Solid conductor Conductor having a single solid wire, as opposed to strands.

Solid state Pertaining to circuits and devices that use solid semiconductors such as silicon. Solid-state electronic devices have a solid material between their input and output pins (transistors, diodes), whereas vacuum tube electronics uses tubes, which have a vacuum between input and output.

Sonar Acronym for "sound navigation and ranging." A system using sound waves to determine a target's direction and distance.

Sonic Pertaining to the speed of sound waves.

Sound wave Traveling wave propagated in an elastic medium that travels at a speed of approximately 1133 ft/s.

Source Device that supplies the signal power or electric energy to a load. Also, one of the field effect transistor's electrodes.

Source-follower A FET amplifier in which the signal is applied between the gate and drain and the output is taken between the source and drain. Used to handle large-input signals and applications requiring a low-input capacitance.

Source impedance Impedance a source presents to a load.

South pole Pole of a magnet into which magnetic lines of force are assumed to enter.

Spark Momentary discharge of electric energy due to the breakdown of air or some other dielectric material separating two terminals.

SPDT Abbreviation for "single-pole, double-throw."

Speaker Also called a loudspeaker; an electroacoustic transducer that converts an electrical wave input into a mechanical sound wave (acoustic) output into the air.

Specification sheet or data sheet Details the characteristics and maximum and minimum values of operation of a device.

Spectrum The frequency spectrum displays all the frequencies and their applications.

Spectrum analyzer Instrument that can display the frequency domain of a waveform, plotting amplitude against frequency of the signals present.

Speed of light Physical constant—the speed at which light travels through a vacuum—equal to 186,282.397 miles/s, 2.997925×10^8 m/s, 161,870 nautical miles/s, or 328 yards/μs.

Speed of sound Speed at which a sound wave travels through a medium. In air it is equal to about 1133 ft/s or 334 m/s, while in water it is equal to approximately 4800 ft/s or 1463 m/s. Also known as sonic speed.

Speedup capacitor Capacitor connected in a circuit to speed up an action due to its inherent behavior.

SPST Abbreviation for "single-pole, single-throw."

Square The product of a number multiplied by itself.

Square root A factor of a number that when squared gives the number.

Square wave Wave that alternates between two fixed values for an equal amount of time.

Square-wave generator A circuit that generates a continuously repeating square wave.

***S-R* latch** Another name for *S-R* flip-flop, so called because the output remains latched in the set or reset state even though the input is removed.

***S-R* (set-reset) flip-flop** A multivibrator circuit in which a pulse on the SET input will "flip" the circuit into the set state and a pulse on the RESET input will "flop" the circuit into its reset state.

Staircase-wave generator A signal generator circuit that produces an output signal voltage that increases in steps.

Standard Exact value used as a basis for comparison or calibration.

Static Crackling noise heard on radio receivers, caused by electric storms or electric devices in the vicinity.

Static electricity Electricity at rest or stationary.

Statistics A branch of mathematics dealing with the collection, analysis, interpretation, and presentation of masses of numerical data.

Stator Stationary part of some rotating device.

Statute mile Distance unit equal to 5280 ft or 1.61 km.

Step-down transformer Transformer in which the ac voltage induced in the secondary is less (due to fewer secondary windings) than the ac voltage applied to the primary.

Stepper motor A motor that rotates in small angular steps.

Step-up transformer Transformer in which the ac voltage induced in the secondary is greater (due to more secondary windings) than the ac voltage applied to the primary.

Stereo sound Sound system in which the sound is delivered through at least two channels and loudspeakers, arranged to give the listener a replica of the original performance.

Stranded conductor Conductor composed of a group of twisted wires.

Stray capacitance Undesirable capacitance that exists between two conductors, such as two leads or a lead and a metal chassis.

Subassembly Components contained in a unit for convenience in assembling or servicing the equipment.

Subatomic Particles such as electrons, protons, and neutrons that are smaller than atoms.

Substrate The mechanical insulating support on which a device is fabricated.

Summing amplifier circuit or adder circuit An op-amp circuit that will sum or add all of the input voltages.

Superconductor Metal such as lead or niobium that, when cooled to within a few degrees of absolute zero, can conduct current with no electrical resistance.

Superheterodyne receiver Radio-frequency receiver that converts all RF inputs to a common intermediate frequency (IF) before demodulation.

Superhigh frequency (SHF) Frequency band between 3 and 30 GHz, so designated by the Federal Communications Commission (FCC).

Superposition theorem Theorem designed to simplify networks containing two or more sources. It states: In a network containing more than one source, the current at any point is equal to the algebraic sum of the currents produced by each source acting separately.

Supersonic Faster than the speed or velocity of sound (Mach 1).

Supply voltage Voltage produced by a power source or supply.

Surface mount technology A method of installing tiny electronic components on the same side of a circuit board as the printed wiring pattern that interconnects them.

SW Abbreviation for "shortwave."

Sweep generator Test instrument designed to generate a radio-frequency voltage that continually and automatically varies in frequency within a selected frequency range.

Swing Amount a frequency or amplitude varies.

Switch Manual, mechanical, or electrical device used for making or breaking an electric circuit.

Switching power supply A dc power supply that makes use of a series switching regulator controlled by a pulse-width modulator to regulate the output voltage.

Switching transistor Transistor designed to switch either on or off.

Symmetrical bidirectional switch A device that has the same value of breakover voltage in both the forward and reverse direction.

Synchronization Also called sync; the precise matching or keeping in step of two waves or functions.

Synchronous Two or more circuits or devices in step or in phase.

Sync pulse or signal Pulse waveform generated to synchronize two processes.

System Combination or linking of several parts or pieces of equipment to perform a particular function.

Tachometer Instrument that produces an output voltage indicating the angular speed of the input in revolutions per minute.

Tangent The trigonometric function that for an acute angle is the ratio between the leg opposite the angle when it is considered part of a right triangle and the leg adjacent.

Tank circuit Circuit made up of a coil and capacitor that is capable of storing electric energy.

Tantalum capacitor Electrolytic capacitor having a tantalum foil anode.

Tap Electrical connection to some point, other than the ends, on the element of a resistor or coil.

Tapered Nonuniform distribution of resistance per unit length throughout the element.

Technician Expert in troubleshooting circuit and system malfunctions. Along with a thorough knowledge of all test equipment and how to use it to diagnose problems, the technician is also familiar with how to repair or replace faulty components. Technicians basically translate theory into action.

Telegraphy Communication between two points by sending a series of coded current pulses either through wires or by radio.

Telemetry Transmission of instrument readings to a remote location either through wires or by radio waves.

Telephone Apparatus designed to convert sound waves into electrical waves, which are then sent to and reproduced at a distant point.

Telephone line Wires existing between subscribers and central stations.

Telephony Telecommunications system involving the transmission of speech information, therefore allowing two or more persons to converse verbally.

Teletypewriter Electric typewriter that like a teleprinter can produce coded signals corresponding to the keys pressed or print characters corresponding to the coded signals received.

Television (TV) System that converts both audio and visual information into corresponding electric signals that are then transmitted through wires or by radio to a receiver, which reproduces the original information.

Telex Teletypewriter exchange service.

Temperature coefficient of frequency Rate frequency changes with temperature.

Temperature coefficient of resistance Rate resistance changes with temperature.

Temperature stability The ability of a resistor to maintain its value of resistance despite changes in temperature.

Tera (T) Prefix that represents 10^{12}.

Terminal Connecting point for making electric connections.

Tesla (T) SI unit of magnetic flux density (1 tesla = 1 Wb/m^2).

Test Sequence of operations designed to verify the correct operation or the malfunction of a system.

Tetrode A four-electrode electron tube that has an anode, cathode, control grid, and an additional electrode.

Thermal relay Relay activated by a heating element.

Thermal stability The ability of a circuit to maintain stable characteristics despite changes in the ambient temperature.

Thermistor Temperature-sensitive semiconductor that has a negative temperature coefficient of resistance (as temperature increases, resistance decreases).

Thermocouple Temperature transducer consisting of two dissimilar metals welded together at one end to form a junction that when heated will generate a voltage.

Thermometry Relating to the measurement of temperature.

Thermostat Temperature-sensitive device that opens or closes a circuit.

Theta The eighth letter of the Greek alphabet, used to represent an angle.

Thévenin's theorem Theorem that replaces any complex network with a single voltage source in series with a single resistance. It states: Any network of resistors can be replaced with an equivalent voltage source (V_{TH}) and an equivalent series resistance (R_{TH}).

Thick-film capacitor Capacitor consisting of two thick-film layers of conductive film separated by a deposited thick-layer dielectric film.

Thick-film resistor Fixed-value resistor consisting of a thick-film resistive element made from metal particles and glass powder.

Thin-film capacitor Capacitor in which both the electrodes and the dielectric are deposited in layers on a substrate.

Thin-film detector (TFD) A temperature detector containing a thin layer of platinum and used for very precise temperature readings.

Three-phase supply A supply that consists of three ac voltages that are 120° out of phase with one another.

Threshold Minimum point at which an effect is produced or indicated.

Thyristor A semiconductor switching device in which bistable action depends on P-N-P-N regenerative feedback. A thyristor can be bidirectional or unidirectional and have from two to four terminals.

Time constant Time needed for either a voltage or current to rise to 63.2% of the maximum or fall to 36.8% of the initial value. The time constant of an *RC* circuit is equal to the product of *R* and *C*; the time constant of an *RL* circuit is equal to the inductance divided by the resistance.

Time-division multiplex (TDM) Transmission of two or more signals on the same path but at different times.

Time-domain analysis A method of representing a waveform by plotting its amplitude versus time.

Tinned Coated with a layer of tin or solder to prevent corrosion and simplify the soldering of connections.

Toggle switch Spring-loaded switch that is put in one of two positions, either on or off.

Tolerance Permissible deviation from a specified value, normally expressed as a percentage.

Tone A term describing both the bass and treble of a sound signal.

TO package Cylindrical, metal can type of package for some semiconductor components.

Toroidal coil Coil wound on a doughnut-shaped core.

Torque Moving force.

Totem pole circuit A transistor circuit containing two transistors connected one on top of the other with two inputs and one output.

Transconductance Also called mutual conductance; the ratio of a change in output current to the initiating change in input voltage.

Transducer Any device that converts energy from one form to another.

Transformer Device consisting of two or more coils that are used to couple electric energy from one circuit to another, yet maintain electrical isolation between the two.

Transformer coupling Also called inductive coupling; the coupling of two circuits by means of mutual inductance provided by a transformer.

Transient suppressor diode A device used to protect voltage-sensitive electronic devices in danger of destruction by high-energy voltage transients.

Transistance The characteristic achieved by a transistor that makes possible the control of voltages and currents so as to achieve gain or switching action.

Transistor (TRANSfer resISTOR) Semiconductor device having three main electrodes, called the emitter, base, and collector, that can be made to either amplify or rectify.

Transistor tester Special test instrument that can be used to test both NPN and PNP bipolar transistors.

Transmission Sending of information.

Transmission line Conducting line used to couple signal energy between two points.

Transmitter Equipment used to achieve transmission.

Transorb (AbsORB TRANSients) Another name for transient suppressor diode.

Transposition The transfer of any term of an equation from one side to the other side with a corresponding change of sign.

Treble High audio frequencies normally handled by a tweeter in a sound system.

TRIAC A bidirectional, gate-controlled thyristor that provides full-wave control of ac power.

Triangular wave A repeating wave with equal positive and negative ramps that have linear rates of change with time.

Triangular-wave generator A signal generator circuit that produces a continuously repeating triangular wave output.

Trigger Pulse used to initiate a circuit action.

Triggering Initiation of an action in a circuit, that then functions for a predetermined time, for example, the duration of one sweep in a cathode-ray tube.

Trigger input pulse used to initiate a circuit action.

Trigonometry The study of the properties of triangles and trigonometric functions and of their applications.

Trimmer Small-value variable resistor, capacitor, or inductor.

Triode A three-electrode vacuum tube that has an anode, cathode, and control grid.

Troubleshooting The process of locating and diagnosing malfunctions or breakdowns in equipment by means of systematic checking or analysis.

Tune To adjust the resonance of a circuit so that it will select the desired frequency.

Tuned circuit Circuit whose components' values can be varied so that the circuit responds to one selected frequency and heavily attenuates all other frequencies.

Tunnel diode A heavily doped junction diode that has negative resistance in the forward direction of its operating range due to quantum mechanical tunnelling.

Turns ratio Ratio of the number of turns in the secondary winding to the number of turns in the primary winding of a transformer.

Twin-T sine-wave oscillator An oscillator circuit that makes use of two T-shaped feedback networks.

Two phase Refers to two repeating waveforms having a phase difference of 90°.

UHF Abbreviation for "ultrahigh frequency."

Ultrasonic Signals that are just above the range of human hearing, approximately 20 kHz.

Uncharged Having a normal number of electrons and therefore no electrical charge.

Unidirectional device A device that will conduct current in only one direction.

Unijunction transistor Abbreviated UJT; a *p-n* device that has an emitter connected to the *p-n* junction on one side of the bar and two bases at either end of the bar. Used primarily as a switching device.

Unipolar device A device in which only one type of semiconductor material exists between the output terminals and therefore the charge carriers have only one polarity (unipolar).

Unit A determinate quantity adopted as a standard of measurement.

Upper sideband A group of frequencies that are equal to the sums of the carrier and modulation frequencies.

VA Abbreviation for "volt-ampere."

Vacuum tube Electron tube evacuated to such a degree that its electrical characteristics are essentially unaffected by the presence of residual gas or vapor. Eventually replaced by the transistor for amplification and rectification.

Valence shell or ring Outermost shell formed by electrons.

Valley voltage (V_V) The voltage at the dip or valley in the characteristic curve.

Varactor diode A *p-n* semiconductor diode that is reverse biased to increase or decrease its depletion region width and vary device capacitance.

Variable Quantity that can be altered or controlled to assume a number of distinct values.

Variable capacitor Capacitor whose capacitance can be changed by varying the effective area of the plates or the distance between the plates.

Variable resistor A resistor whose value can be changed. *See also* rheostat *and* potentiometer.

Variable-tuned RF amplifier A tuned radio-frequency amplifier in which the tuned circuit(s) can be adjusted to select the desired station carrier frequency.

Variable-value capacitor A capacitor whose value can be varied.

Variable-value inductor An inductor whose value can be varied.

Varistor Voltage-dependent resistor.

VCR Abbreviation for "videocassette recorder."

Vector or phasor Quantity that has both magnitude and direction. They are normally represented as a line, the length of which indicates magnitude and the orientation of which, due to the arrowhead on one end, indicates direction.

Vector addition Determination of the sum of two out-of-phase vectors using the Pythagorean theorem.

Vector diagram Arrangement of vectors showing the phase relationships between two or more ac quantities of the same frequency.

Vertical-channel E-MOSFET An enhancement-type MOSFET that, when turned ON, forms a vertical channel between source and drain.

Very high frequency (VHF) Electromagnetic frequency band from 30 to 300 MHz as set by the FCC.

Very low frequency (VLF) Frequency band from 3 to 30 kHz as set by the FCC.

Video Relating to any picture or visual information, from the Latin word meaning "I see."

Video amplifier An amplifier that has one or more transistor stages designed to amplify video signals.

Video signals A signal that contains visual information for television or radar systems.

Virtual ground A ground for voltage but not for current.

Voice coil Coil attached to the diaphragm of a moving-coil speaker, which is moved through an air gap between the pole pieces of a permanent magnet.

Voice synthesizer Synthesizer that can simulate speech in any language by stringing together phonemes.

Volt (V) Unit of voltage, potential difference, or electromotive force. One volt is the force needed to produce 1 ampere of current in a circuit containing 1 ohm of resistance.

Voltage (*V* or *E*) Term used to designate electrical pressure, or the force that causes current to flow.
Voltage amplifier An amplifier designed to build up the voltage of a signal.
Voltage-controlled device A device in which the input junction is normally reverse biased and the input voltage controls the output current.
Voltage divider Fixed or variable series resistor network that is connected across a voltage to obtain a desired fraction of the total voltage.
Voltage-divider biasing A biasing method used with amplifiers in which a series arrangement of two fixed-value resistors is connected across the voltage source. The result is that a desired fraction of the total voltage is obtained at the center of the two resistors and is used to bias the amplifier.
Voltage drop Voltage or difference in potential developed across a component or conductor due to the loss of electric pressure as a result of current flow.
Voltage-follower An operational amplifier that has a direct feedback to give unity gain so that the output voltage follows the input voltage. Used in applications where a very high input impedance and very low output impedance are desired.
Voltage gain Also called voltage amplification; the difference between the output voltage level and the input signal voltage level. This value is normally expressed in decibels, which are equal to 20 times the logarithm of the ratio of the output voltage to the input voltage.
Voltage rating Maximum voltage a component can safely withstand without breaking down.
Voltage regulator A device that maintains the output voltage of a voltage source within required limits despite variations in input voltage and load resistance.
Voltage source A circuit or device that supplies voltage to a load circuit.
Voltaic cell A battery cell having two unlike metal electrodes immersed in a solution that chemically interacts with the plates to produce an emf.
Volt-ampere (VA) Unit of apparent power in an ac circuit containing reactance. Apparent power is equal to the product of voltage and current.
Voltmeter Instrument designed to measure the voltage or potential difference. Its scale can be graduated in kilovolts, volts, or millivolts.
Volume Magnitude or power level of a complex audio-frequency (AF) wave, expressed in volume units (VU).
VOM meter Abbreviation for volt-ohm-milliamp meter.
VRMS Abbreviation for "volts root-mean-square."
Wall outlet Spring-contact outlet mounted on the wall to which a portable appliance is connected to obtain electric power.
Watt (W) Unit of electric power required to do work at a rate of 1 joule/second. One watt of power is expended when 1 ampere direct current flows through a resistance of 1 ohm. In an ac circuit, the true power is effective volts multiplied by effective amperes, multiplied by the power factor.
Wattage rating Maximum power a device can safely handle continuously.
Watt-hour (Wh) Unit of electrical work equal to a power of 1 watt being absorbed continuously for 1 hour.
Wattmeter A meter used to measure electric power in watts.
Wave Electric, electromagnetic, acoustic, mechanical, or other form whose physical activity rises and falls or advances and retreats periodically as it travels through some medium.
Waveform Shape of a wave.
Waveguide Rectangular or circular metal pipe used to guide electromagnetic waves at microwave frequencies.
Wavelength (λ) Distance between two points of corresponding phase and equal to waveform velocity or speed divided by frequency.
Weber (Wb) Unit of magnetic flux. One weber is the amount of flux that when linked with a single turn of wire for an interval of 1 second will induce an electromotive force of 1 V.
Wet cell Cell using a liquid electrolyte.
Wetting The coating of a contact surface.
Wheatstone bridge A four-arm, generally resistive bridge that is used to measure resistance.
Wideband amplifier Also called broadband amplifier; an amplifier that has a flat response over a wide range of frequencies.
Wien-bridge oscillator An *RC* oscillator circuit in which a Wien bridge determines frequency.
Winding One or more turns of a conductor wound to form a coil.
Wire Single solid or stranded group of conductors having a low resistance to current flow.
Wire gauge American wire gauge (AWG) is a system of numerical designations of wire sizes, with the first being 0000 (the largest size) and then going to 000, 00, 0, 1, 2, 3, and so on up to the smallest sizes of 40 and above.
Wireless Term describing radio communication that requires no wires between the two communicating points.
Wirewound resistor Resistor in which the resistive element is a length of high-resistance wire or ribbon, usually nichrome, wound onto an insulating form.
Wire wrapping Method of prototyping in which solderless connections are made by wrapping wire around a rectangular terminal.
Woofer Large loudspeaker designed primarily to reproduce low audio-frequency signals at large power levels.
Work The transformation of energy from one type to another. The amount of work done is dependent on the amount of energy transformed.
X Symbol for reactance.
x-axis Horizontal axis.
Y Symbol for admittance.
y-axis Vertical axis.
z-axis Axis perpendicular to both the x- and y axes.
Zener diode A semiconductor diode in which a reverse breakdown voltage current causes the diode to develop a constant voltage.
Zener voltage (V_Z) The voltage drop across the zener when it is being operated in the reverse zener breakdown region.
Zero biasing A configuration in which no bias voltage is applied at all.
Zeroing Calibrating a meter so that it shows a value of zero when zero is being measured.

APPENDIX

Electronic Schematic Symbols

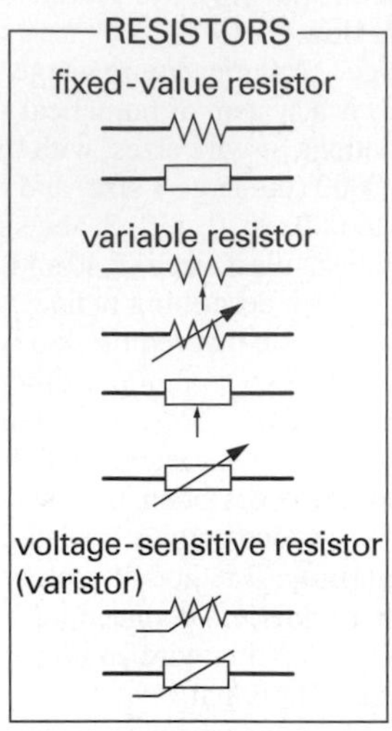

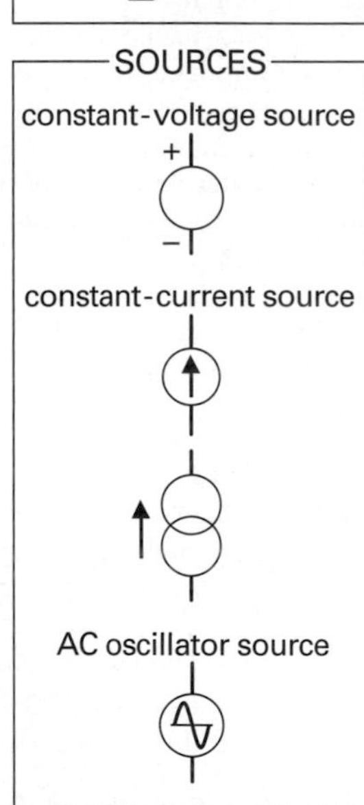

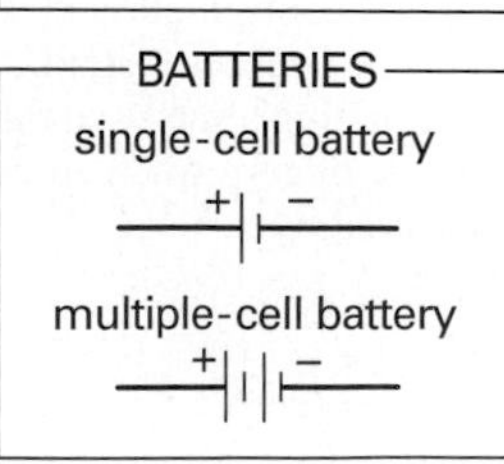

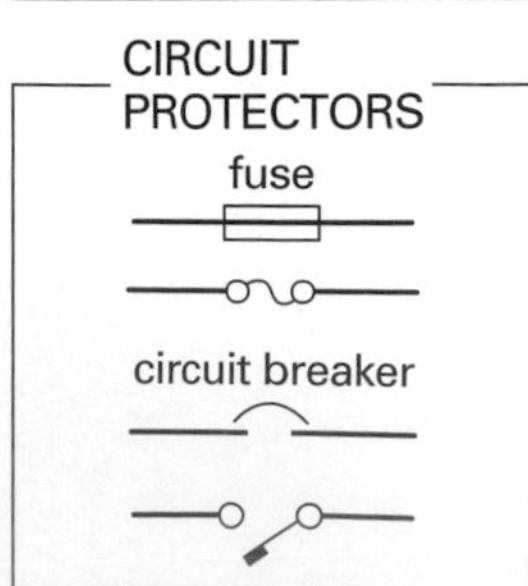

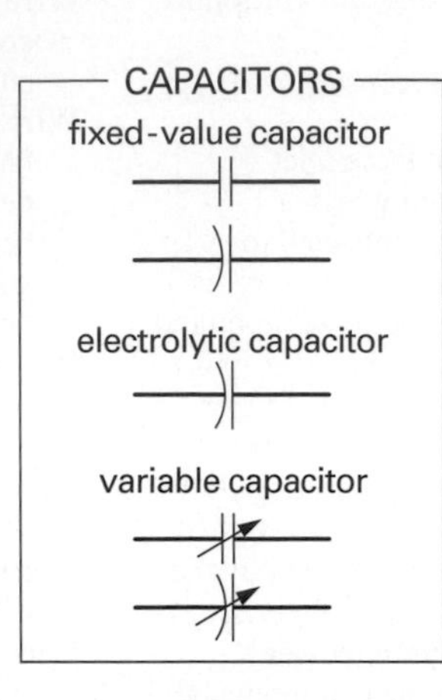

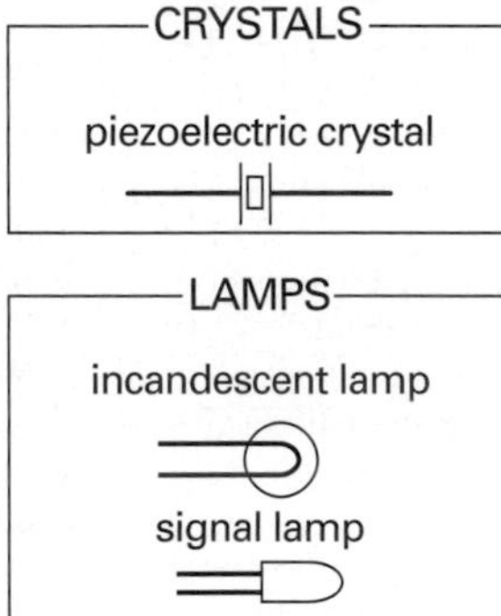

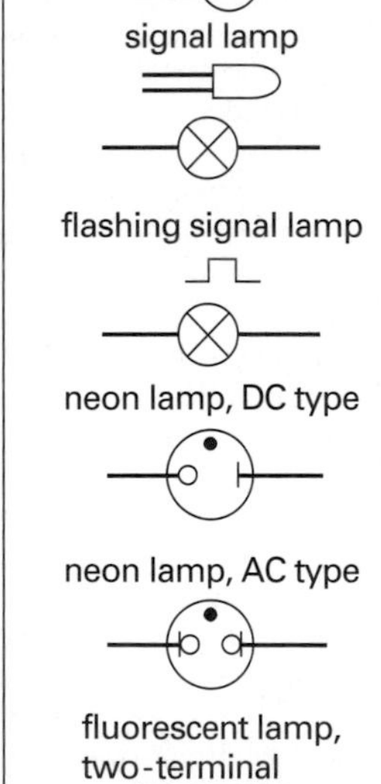

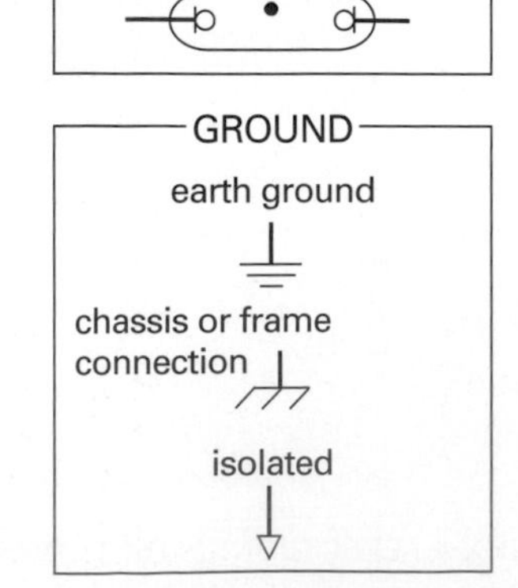

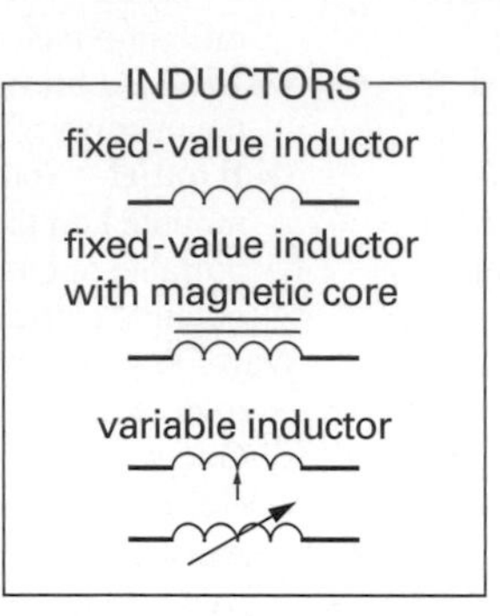

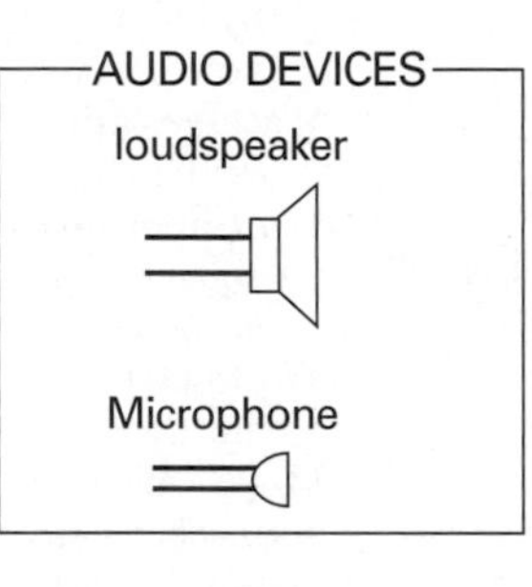

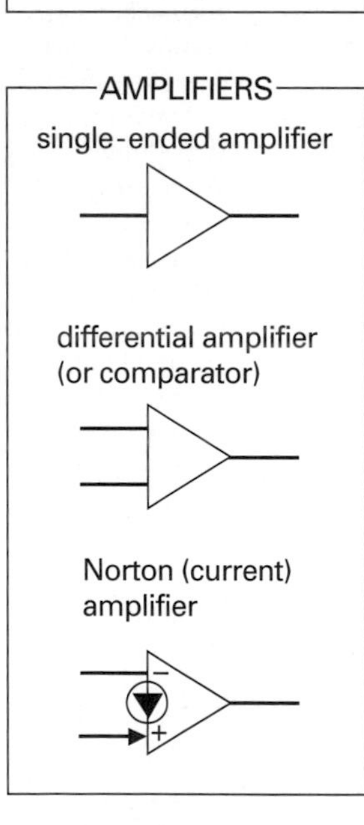

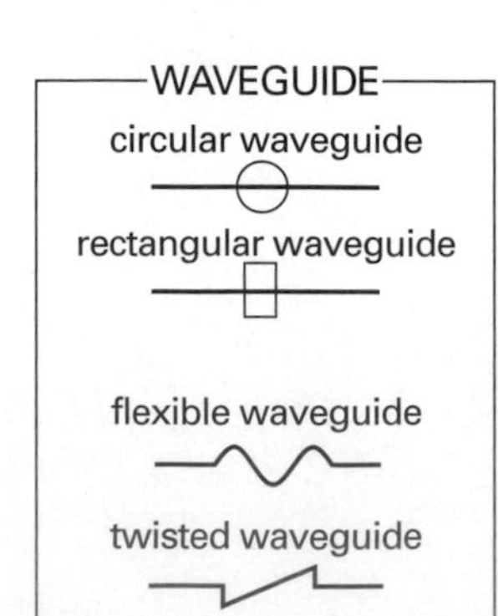

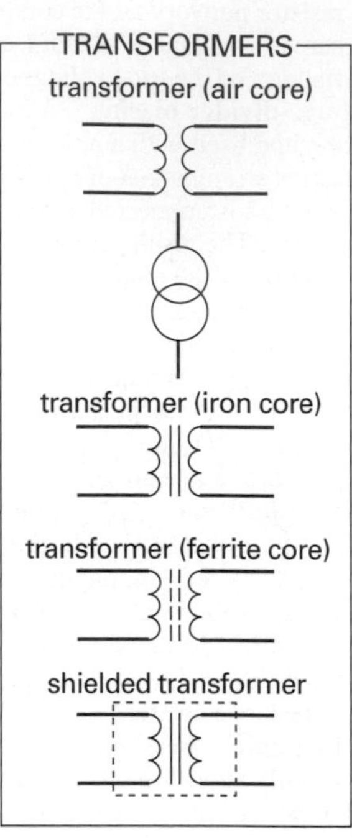

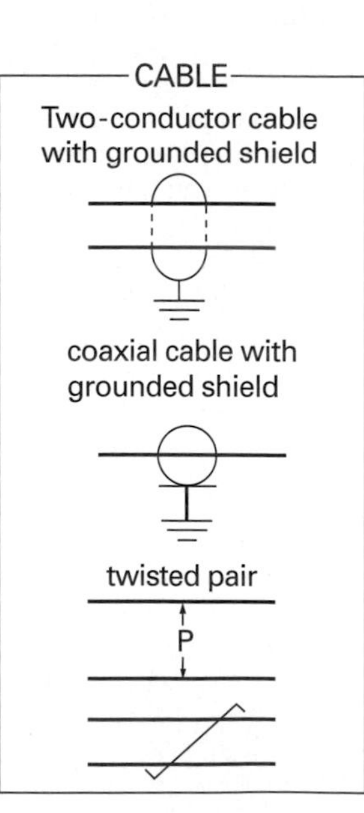

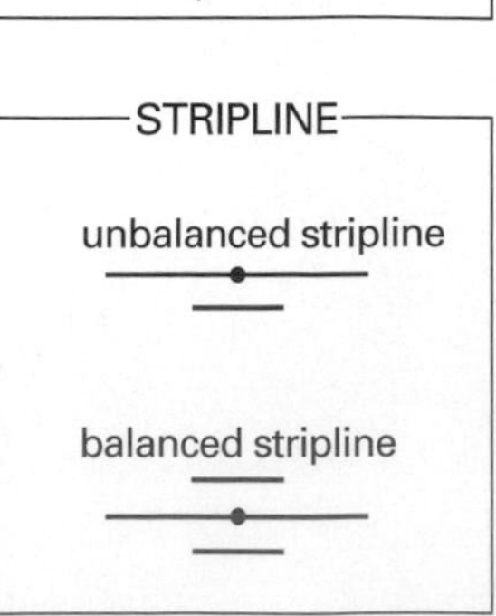

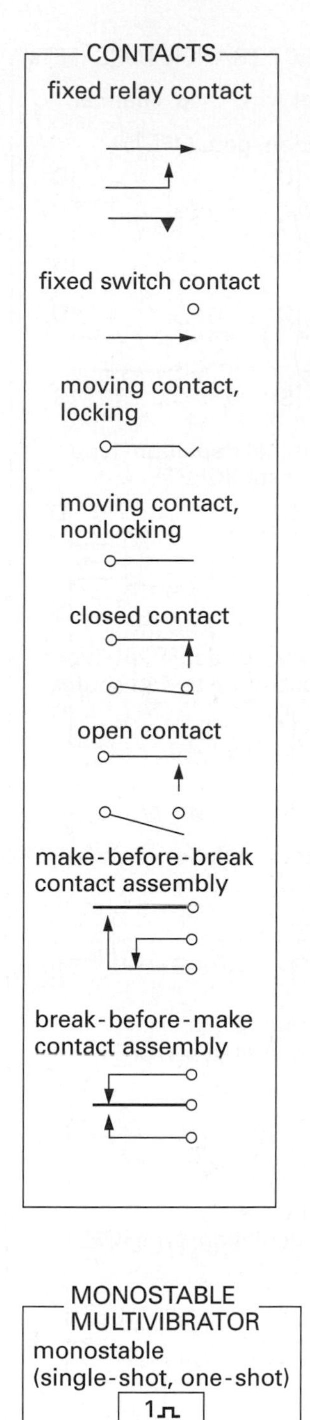

MONOSTABLE MULTIVIBRATOR

monostable (single-shot, one-shot)

1⎍

SS

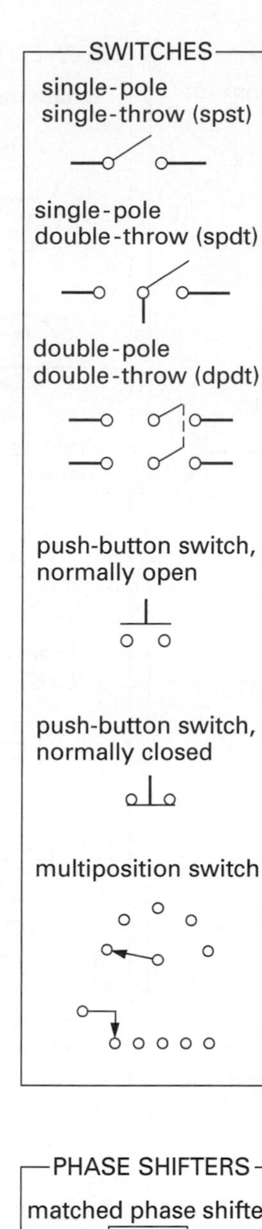

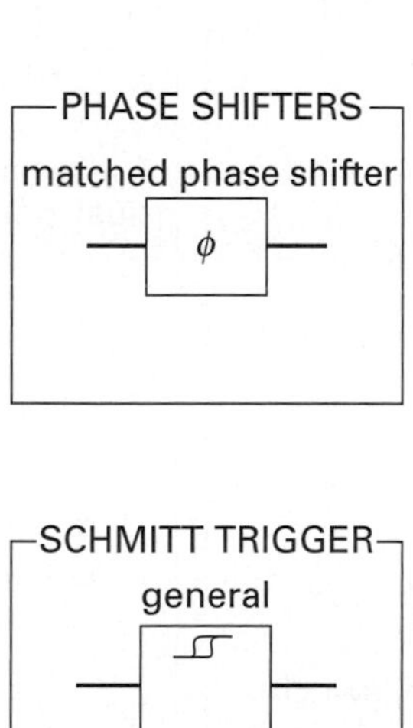

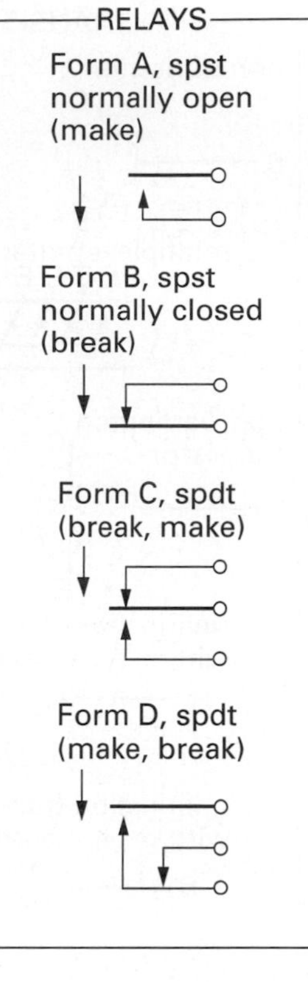

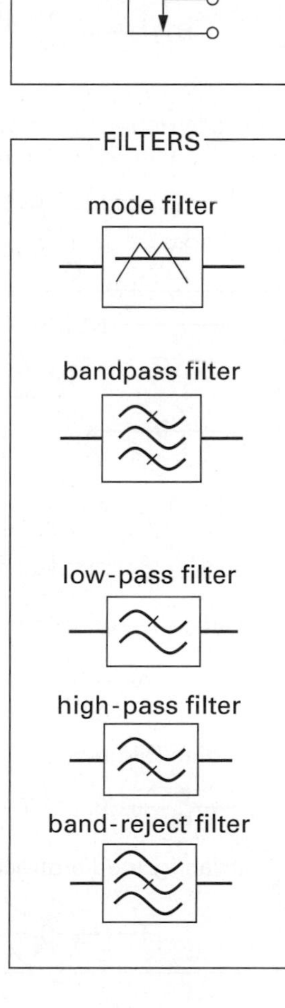

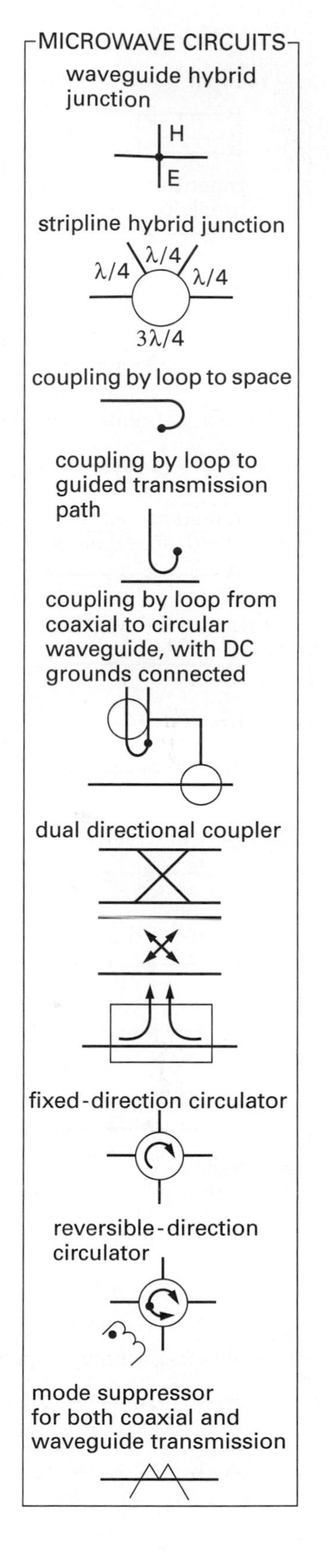

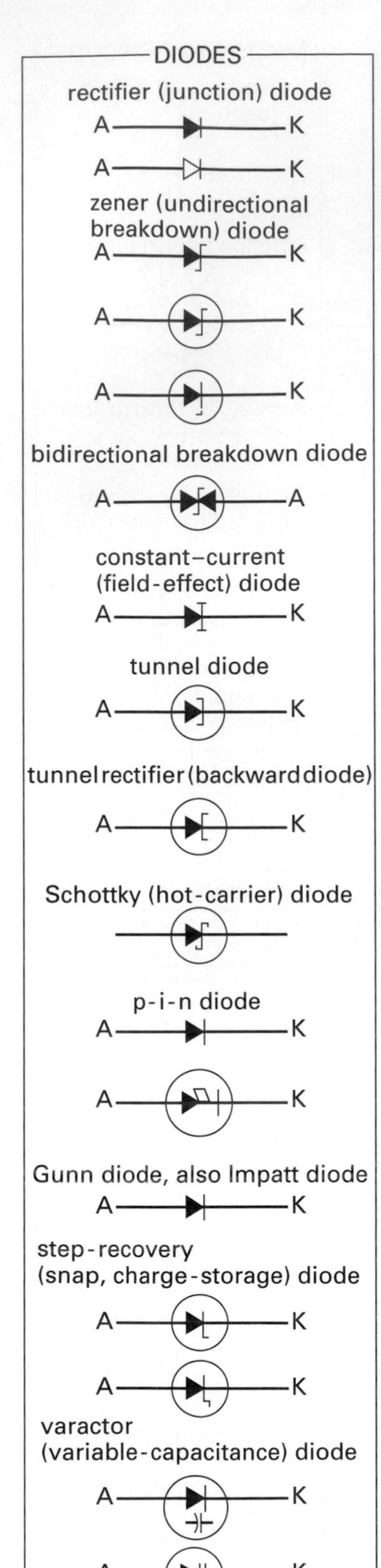
DIODES
rectifier (junction) diode
A
K
zener (undirectional breakdown) diode
bidirectional breakdown diode
constant–current (field-effect) diode
tunnel diode
tunnel rectifier (backward diode)
Schottky (hot-carrier) diode
p-i-n diode
Gunn diode, also Impatt diode
step-recovery (snap, charge-storage) diode
varactor (variable-capacitance) diode

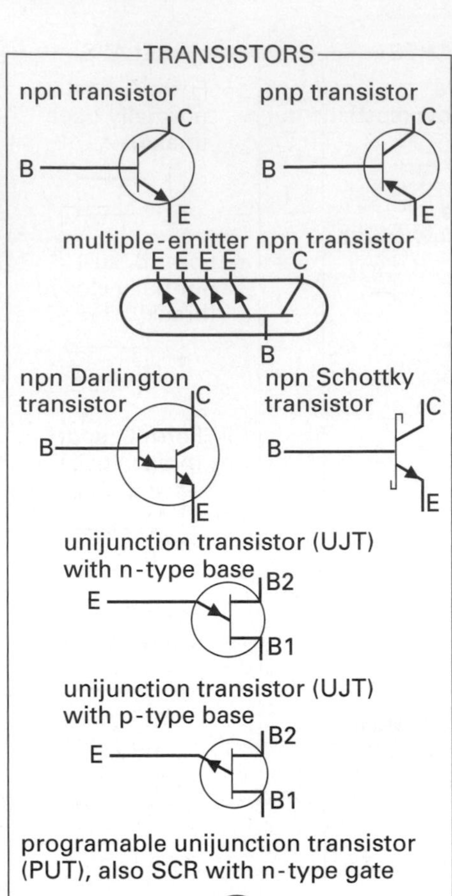
TRANSISTORS
npn transistor
pnp transistor
B
C
E
multiple-emitter npn transistor
E E E E
npn Darlington transistor
npn Schottky transistor
unijunction transistor (UJT) with n-type base
B2
B1
unijunction transistor (UJT) with p-type base
programable unijunction transistor (PUT), also SCR with n-type gate
A
K
G

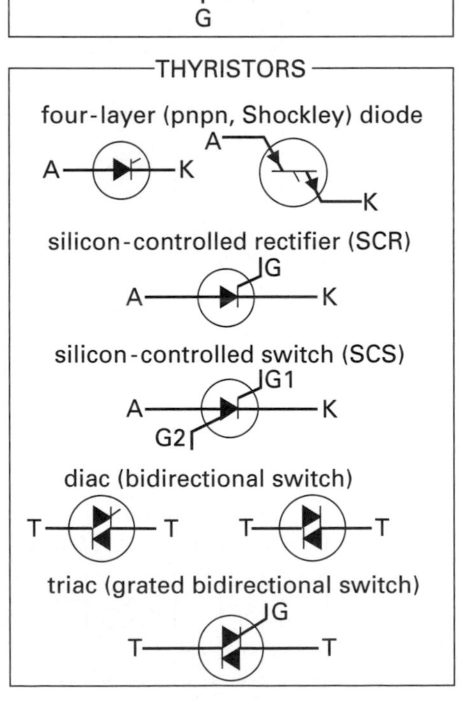
THYRISTORS
four-layer (pnpn, Shockley) diode
A
K
silicon-controlled rectifier (SCR)
G
silicon-controlled switch (SCS)
G1
G2
diac (bidirectional switch)
T
triac (grated bidirectional switch)

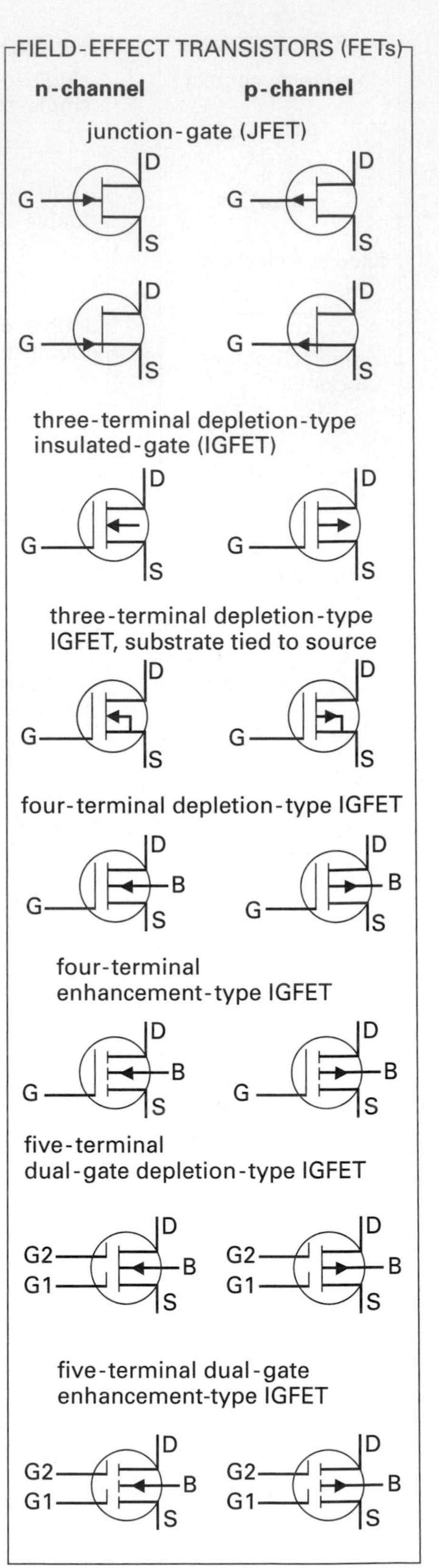
FIELD-EFFECT TRANSISTORS (FETs)
n-channel
p-channel
junction-gate (JFET)
G
D
S
three-terminal depletion-type insulated-gate (IGFET)
three-terminal depletion-type IGFET, substrate tied to source
four-terminal depletion-type IGFET
B
four-terminal enhancement-type IGFET
five-terminal dual-gate depletion-type IGFET
G2
G1
five-terminal dual-gate enhancement-type IGFET

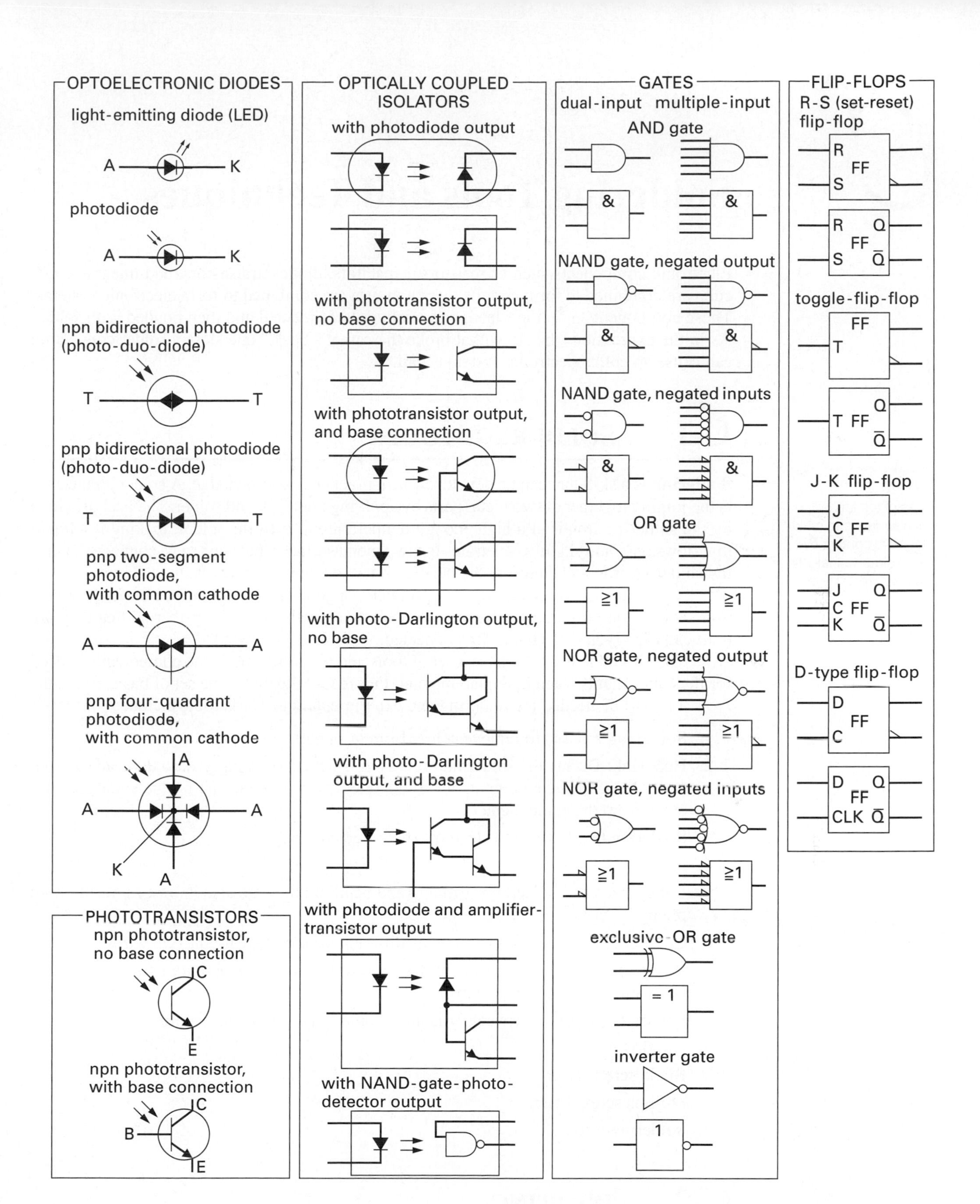
OPTOELECTRONIC DIODES
light-emitting diode (LED)
A
K
photodiode
A
K
npn bidirectional photodiode (photo-duo-diode)
T
T
pnp bidirectional photodiode (photo-duo-diode)
T
T
pnp two-segment photodiode, with common cathode
A
A
pnp four-quadrant photodiode, with common cathode
A
A
A
K
A
PHOTOTRANSISTORS
npn phototransistor, no base connection
C
E
npn phototransistor, with base connection
C
B
E
OPTICALLY COUPLED ISOLATORS
with photodiode output
with phototransistor output, no base connection
with phototransistor output, and base connection
with photo-Darlington output, no base
with photo-Darlington output, and base
with photodiode and amplifier-transistor output
with NAND-gate-photo-detector output
GATES
dual-input
multiple-input
AND gate
&
&
NAND gate, negated output
&
&
NAND gate, negated inputs
&
&
OR gate
≧1
≧1
NOR gate, negated output
≧1
≧1
NOR gate, negated inputs
≧1
≧1
exclusivc-OR gate
= 1
inverter gate
1
FLIP-FLOPS
R-S (set-reset) flip-flop
R
FF
S
R
Q
FF
S
Q̄
toggle-flip-flop
FF
T
Q
T FF
Q̄
J-K flip-flop
J
C FF
K
J
Q
C FF
K
Q̄
D-type flip-flop
D
FF
C
D
Q
FF
CLK
Q̄

APPENDIX

C Soldering Tools and Techniques

Electronic components such as resistors, capacitors, diodes, transistors, and integrated circuits are combined to form circuits, which in turn are combined to form electronic systems. These components have their leads physically interconnected and then bonded with solder. As in the expression "the straw that broke the camel's back," one sloppy solder connection can cause an entire electronic system to fail.

C-1 SOLDERING TOOLS

Soldering
Process of joining two metallic surfaces to make an electrical contact by melting solder (usually tin and lead) across them.

Soldering is a skill that can be developed with practice and knowledge. A solder connection is the joining together of two metal parts by applying both heat and solder. The heat provided by the soldering iron is at a high enough temperature to melt the solder, making it a liquid that flows onto and slightly penetrates the two metal surfaces that need to be connected. Once the soldering iron and therefore the heat are removed, the solder cools. In so doing, it solidifies and bonds the two metal surfaces, producing a good *electrical and mechanical connection.* This is the purpose of soldering: to bond two metal surfaces together so that they are both electrically and mechanically connected.

You should at this time have a set of tools and toolbox with all the components needed for the experiments listed in the lab manual. Figure C-1 illustrates the set of basic electronic tools that will be needed for soldering and experimentation. These tools include:

1. Solder wick: Used to remove solder from terminals.
2. Heat sink: During the soldering process, this device is clamped onto the lead of especially heat-sensitive components to conduct the heat generated by the soldering iron away from the component.
3. Soldering brush: Used to clean off flux after soldering.
4. Pliers
5. Long-nose pliers: Used for gripping and bending; they are also known as needle-nose pliers.
6. Cutters: These are available in many different sizes and are used to cut wires or cables.
7. Wire strippers: The strippers are designed to be adjustable and are used for stripping the insulating sleeve off a wire.
8. Solder: 60/40 rosin-core solder is most commonly used in electronics.
9. Nut driver
10. Blade screwdriver
11. Phillips screwdriver
12. Soldering iron

C-2 WETTING

Every time you make a good electrical and mechanical connection, the solder will flow, when heated to its melting temperature, over the lead and terminal to be connected, as shown in Figure C-2. The solder actually penetrates the metals, and this embedding of the solder

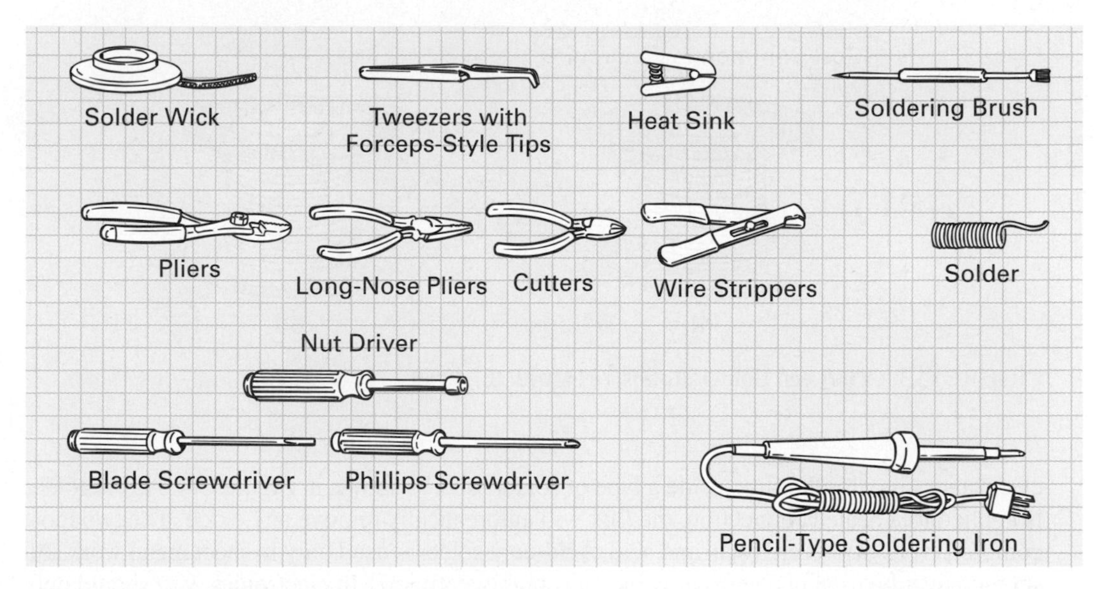

FIGURE C-1 Basic Electronic Tools.

into the metal is called **wetting.** If the solder feathers out to a thin edge, good wetting is said to have occurred, and it is this that gives the connection its physical strength and electrical connection.

Wetting
The coating of a contact surface.

C-3 SOLDER AND FLUX

Solder is a mixture of tin and lead, both of which have a low melting temperature with respect to other metals. This is necessary so that the soldering iron can melt the solder and not the terminals or leads.

Solder
Metallic alloy that is used to join two metal surfaces.

Different proportions of tin and lead are available to produce solder with different characteristics. For example, Figure C-3(a) shows a 60/40 solder, which is a mixture of 60% tin and 40% lead; Figure C-3(b) shows a 40/60 solder that consists of 40% tin and 60% lead. The proportions of tin to lead determine the melting temperature of the solder; for example, 60/40 solder has a lower melting temperature than 40/60 solder. In electronics, 60/40 is most commonly used because of its low melting temperature, which means that a component lead or terminal will not have to be heated to a high temperature in order to make a connection. A 63/37 solder has an even lower melting temperature and is therefore even safer than 60/40 solder. It is, however, more expensive.

Any lead or terminal is always exposed to the air, which forms an invisible insulating layer on the surface of leads, pins, terminals, and any other surface. A chemical substance is needed to remove this layer, otherwise the solder would not be able to flow and stick to the metal contacts. **Flux** removes this invisible insulating oxide layer, and nearly all solder used in

Flux
A material used to remove oxide films from the surfaces of metals in preparation for soldering.

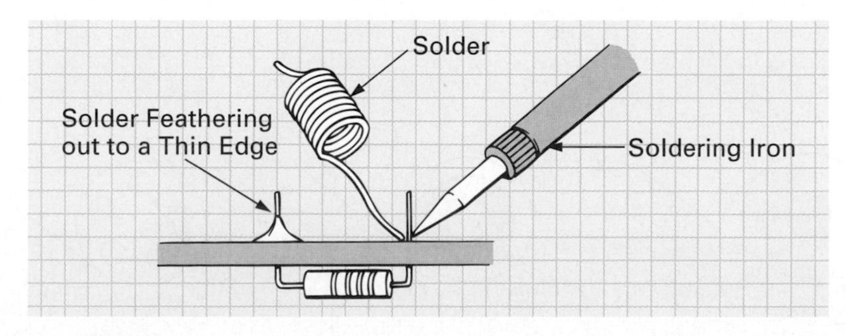

FIGURE C-2 Wetting.

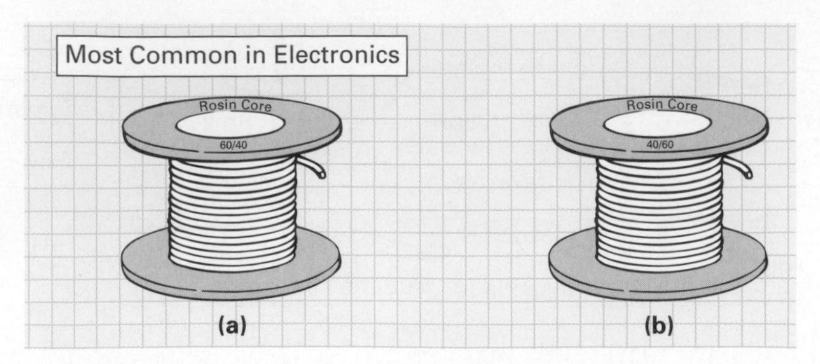

FIGURE C-3 **Tin/Lead Solder Ratios. (a) 60/40. (b) 40/60.**

electronics contains the flux inside a type of solder tube, as shown in Figure C-4. As the solder is applied to a heated connection, the flux will automatically remove any oxide. The two most common types of flux are rosin and acid. Acid-core solder is used only in sheet metal work and should never be used in electronics since it is highly corrosive. In electronics, you should only use rosin-core solder, and this is normally indicated as shown in Figure C-3.

A variety of solder diameters are available. The larger-diameter solder is used for terminals and large component leads and connections, whereas the smaller-diameter solder is used for soldering terminals that are very close to one another on a printed circuit board, where the amount of solder being applied to the connection needs to be carefully controlled.

C-4 SOLDERING IRONS

Soldering Iron
Tool having an internal heating element that is used for heating a connection to melt solder.

Figure C-5 illustrates the two basic types of **soldering irons.** They are rated in terms of wattage, which indicates the amount of power consumed. More important, the wattage rating of a soldering iron indicates the amount of heat it produces. When an iron is applied to a connection, heat transfer takes place and heat is drained away from the iron to the metals to be connected, so a larger connection will need a larger-wattage soldering iron. A 25 to 60 W pencil iron is ideal for most electronic work.

For your safety and to protect voltage- and current-sensitive components, you should always use a soldering iron with a three-pin plug and three-wire ac power cord. The third ground wire will ground the iron's exposed metal areas and the tip to prevent electrical shock. This grounded tip will protect delicate MOS integrated circuits from leakage electricity and static charges. On the subject of safety, always turn off the equipment before soldering, as the grounded tip may cause a short circuit in the equipment and possible damage.

The tip of a soldering iron can sometimes be changed, and a different temperature tip will allow your iron to be used for different applications. The temperature of the tip selected should be governed by the size of the connection and the temperature sensitivity of the components. In general, a 700°F tip is ideal for most electronic applications; however, for delicate printed circuit boards, use a 600°F tip.

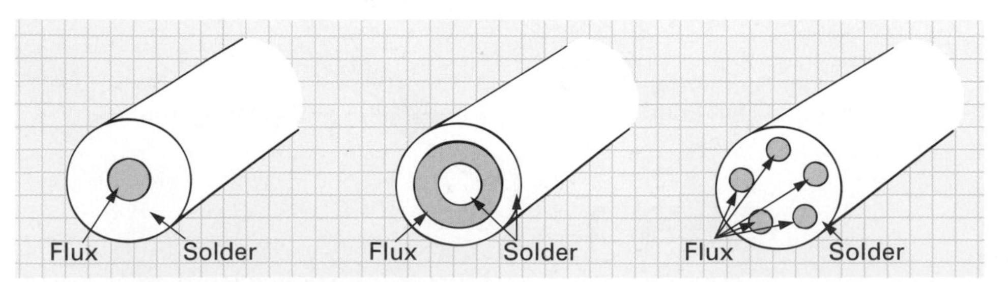

FIGURE C-4 **Rosin-Core Solder.**

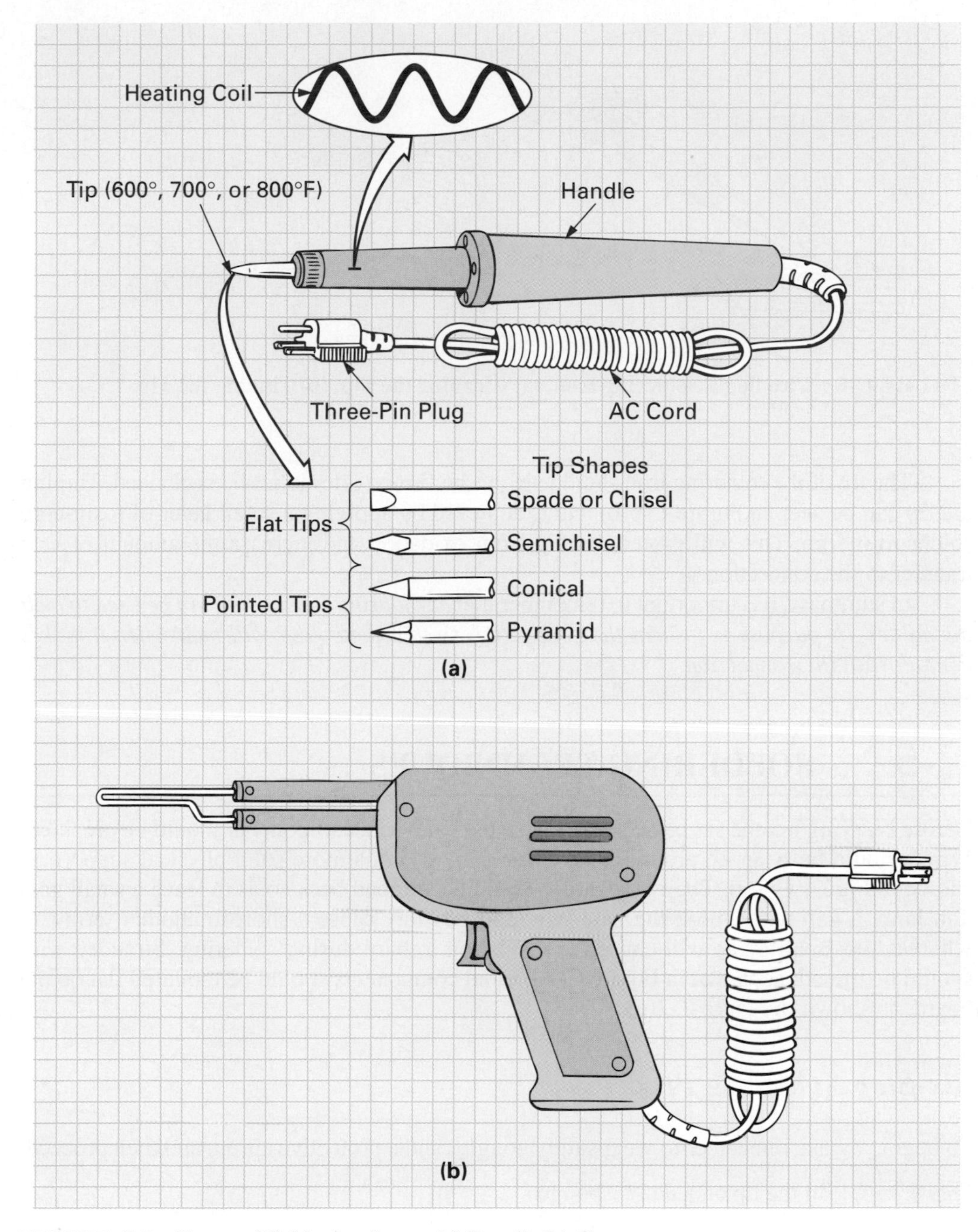

FIGURE C-5 Types of Soldering Irons. (a) Pencil. (b) Gun.

The tip of a soldering iron comes in a variety of shapes and sizes, as shown in Figure C-5(a). The heat produced by the soldering iron heats the connection and melts the solder so that it flows over the connection. A tip shape should be chosen that will conduct the heat to the connection as quickly as possible, so as not to damage the component or PCB terminal. The tip shape that makes the best contact with the connection, and will therefore conduct the most heat, should be used.

As you regularly use your soldering iron, the tip will naturally accumulate dirt and oxide, and this contamination will reduce its effectiveness. The tip should be regularly cleaned by wiping it across a damp sponge, as shown in Figure C-6(a). Soldering iron tips are generally made of copper, plated with either iron or nickel, so you should never clean a tip by filing the end, as this will probably remove the plating.

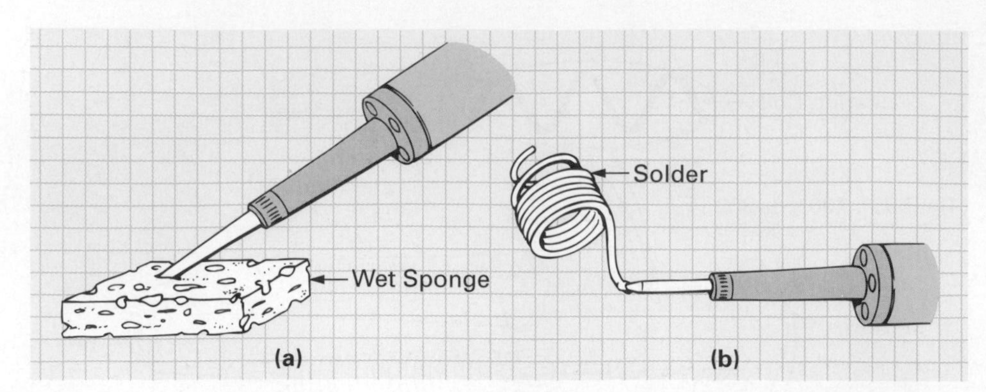

FIGURE C-6 **Cleaning a Soldering Iron. (a) Cleaning the Tip. (b) Tinning the Tip.**

Tinned
Coated with a layer of tin or solder to prevent corrosion and simplify the soldering of connections.

The tip of the soldering iron should always be **tinned** after it has been cleaned. Tinning the tip can be seen in Figure C-6(b) and is achieved by applying a small layer of rosin-core solder to the tip. This will protect the tip from oxidation and increase the amount of heat transfer to the connection.

In summary, it is important to remember that *a soldering iron should not be used to melt the solder. Its purpose is to heat the connection so that the solder will melt when it makes contact with the connection.*

C-5 SOLDERING TECHNIQUES

Before carrying out the six basic soldering steps, wires and components should be prepared. Wires should be wrapped around a terminal to give them a more solid physical support, as shown in Figure C-7(a). Their insulation should be stripped back so as to leave a small gap. Too large a gap will expose the bare wire and possibly cause a short to another terminal, whereas too small a gap will cause the insulation to burn during soldering. Stranded wire should be tinned, as shown in Figure C-7(b), and components should be mounted flat on the board, as shown in Figure C-7(c).

Six-Step Soldering Procedure

To begin, always remember to wear safety goggles and a protective apron, and then proceed.

Step 1: Clean the tip on a damp sponge.

Step 2: Tin the tip if necessary.

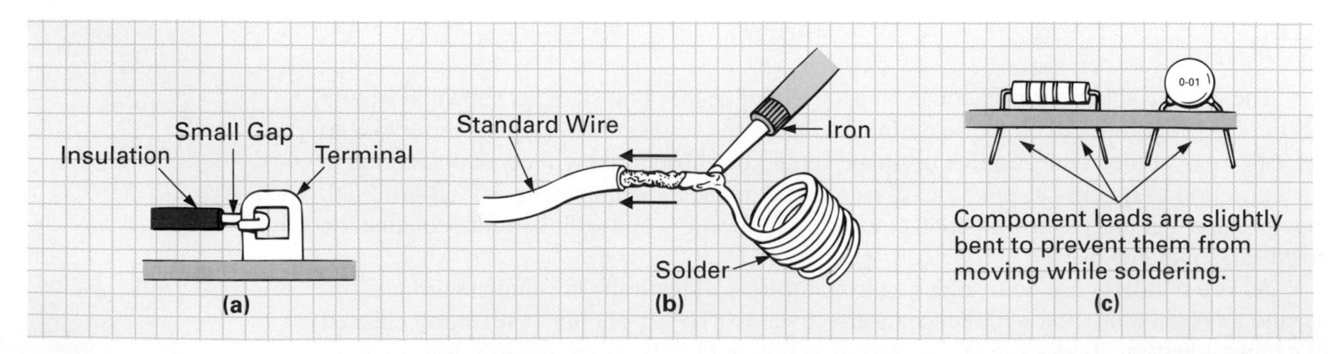

FIGURE C-7 **Wire and Component Preparation.**

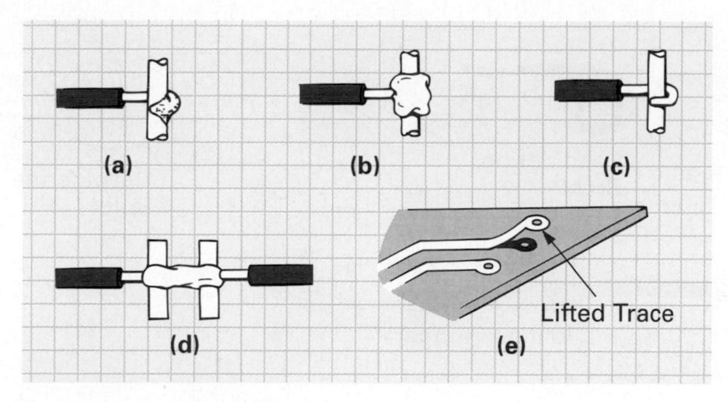

FIGURE C-8 **Poor Solder Connections.**

Step 3: Heat the connection.

Step 4: Apply the solder to the opposite side of the connection.

Step 5: Leave the tip of the soldering iron on the connection only long enough to melt the solder.

Step 6: Remove the solder and then the iron; then let the melted solder cool and solidify undisturbed.

Once this procedure is complete, you should remove the excess leads with the cutters and remove the flux residue, because it collects dust and dirt that could produce electrical leakage paths. When inspecting your work, you should notice the following:

1. The connection is *smooth and shiny.*
2. The solder should *feather out to a thin edge.*

Poor Solder Connections

1. *Cold solder joint:* This connection has a dull gray appearance and it makes a poor mechanical and electrical connection. It is normally caused due to connector lead movement while the solder was cooling or to insufficient heat [Figure C-8(a)]. The cure is to resolder the connection.
2. *Excessive solder joint:* This is caused by too much solder being applied to the connection or by using solder of too large diameter [Figure C-8(b)]. The cure is to desolder the excess with a wick.
3. *Insufficient solder joint:* Not enough solder was applied, so a poor bond exists between the lead and the terminal [Figure C-8(c)]. The cure is to resolder as if it were a new joint.
4. *Solder bridge:* This occurs when adjacent terminals or traces on printed circuit boards are connected accidentally as a result of using too much solder [Figure C-8(d)]. Most can be removed by desoldering the excess with a wick.
5. *Excessive heat:* On printed circuit boards, too much heat can lift the traces or track and ruin the entire board [Figure C-8(e)].

C-6 DESOLDERING

If a faulty component has to be replaced or if a component or wire was soldered in an incorrect place, it will have to be desoldered. Figure C-9 illustrates some of the tools used to

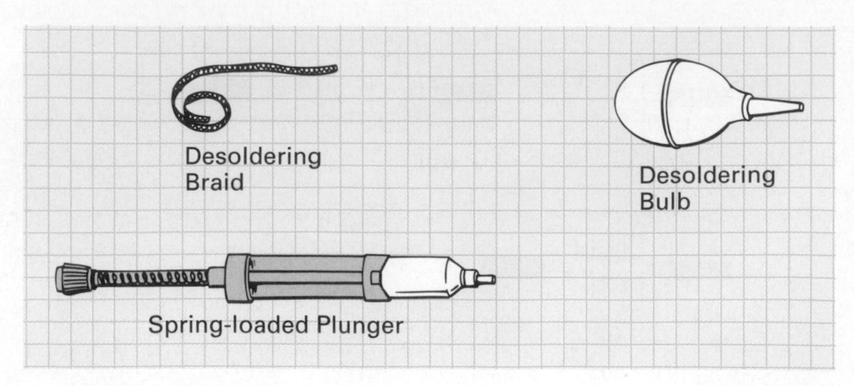

FIGURE C-9 **Desoldering Tools.**

desolder. The easiest way to remove a component is to remove all the solder and then disconnect the leads. All three tools shown in Figure C-9 are designed to remove solder.

Use of the braided wick to remove solder can be seen in Figure C-10(a). The braid contains many small wire strands, and when it is placed between the solder and the iron, the melted solder flows into the braid by capillary action. The solder-filled braid is then cut off and discarded.

Desoldering
The process of removing solder from a connection.

The **desoldering** bulb or spring-loaded plunger relies on suction rather than capillary action to remove the solder, as shown in Figure C-10(b). The steps to follow are

Step 1: Squeeze the desoldering bulb or cock the spring-loaded plunger.

Step 2: Melt the solder with the soldering iron.

Step 3: Remove the iron's tip.

Step 4: Quickly insert the tip of the bulb or plunger into the molten solder and then activate the suction.

When as much of the solder as possible has been removed, use the long-nose pliers to hold the lead and remove the component. It may be necessary to use the iron to heat the lead slightly to loosen up the residual solder so that the component lead can be removed with the pliers. Always be careful not to apply an excessive amount of heat or stress to either the component or the board.

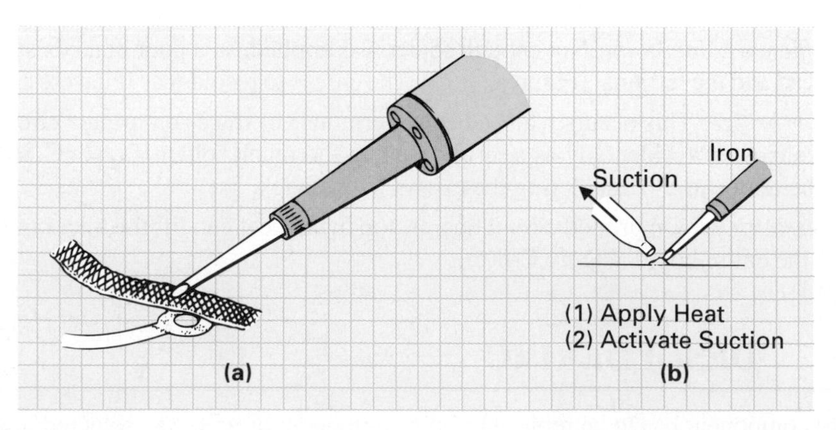

FIGURE C-10

C-7 HOW TO SOLDER AND DESOLDER SURFACE MOUNT COMPONENTS (SMCs)

In this section, we will discuss the soldering and desoldering of surface mount components (SMCs). One of the key differences between SMC soldering and through-hole component soldering is that through-hole components, like the resistor shown in Figure C-11(a), have wire leads that hold the device in place. Surface mount components, on the other hand, have nothing to hold them in place while you apply solder to their terminals.

One of the easiest ways to hold the SMC in place is to apply a small spot of liquid flux to the pretinned circuit board footprints, as shown in the upper diagram in Figure C-11(b), and then place the SMC on the sticky footprints. To solder an SMC device onto a printed circuit board (PCB) you should use a 10 watt to 15 watt soldering iron and 15 mil solder. When soldering a chip SMC, like the one shown in Figure C-11(b), follow this procedure:

Step 1: Place the chip SMC on its flux-coated footprint using tweezers.

Step 2: Touch the soldering iron tip to the footprint (not the SMC).

Step 3: Apply solder to the junction until a neat fillet, about half the height of the end terminal, is formed.

Step 4: Remove solder and iron.

When soldering a leaded SMC, like the one shown in Figure C-11(c), follow this procedure:

Step 1: Place the leaded SMC on its flux-coated footprint using tweezers.

Step 2: Touch the soldering iron tip to the SMC pin.

Step 3: Apply solder to the junction, and allow solder to flow around the footprint and pin.

Step 4: Remove solder and iron.

If you have a U-shaped tip available, like the one shown in Figure C-11(d), the soldering of SMCs will be a lot easier. In this instance the procedure is as follows:

Step 1: With iron off, install U-shaped tip to the end of iron.

Step 2: Melt some solder over the tip, and then clean off tip by wiping it across damp sponge.

Step 3: Pre-tin the footprint on the printed circuit board with a thin layer of solder.

Step 4: Holding SMC in place with tweezers, place U-shaped soldering iron tip over the SMC so that both forks touch the pretinned footprint.

Step 5: As pretinned solder on footprint melts, use tweezers to gently press SMC on its footprint pads.

Step 6: Remove U-shaped soldering tip, wait a few seconds until solder has solidified and bonded SMC to its footprint, and then remove tweezers.

Step 7: Apply additional solder to footprint as shown in Figure C-11(b).

Several U-shaped soldering iron tips are available for all sizes of SMCs. These specialized tips are designed specifically for soldering and desoldering. To desolder and remove an SMC using a U-shaped tip, follow this procedure:

Step 1: Place the hot U-shaped iron tip over the SMC so that both forks touch the footprint of the SMC.

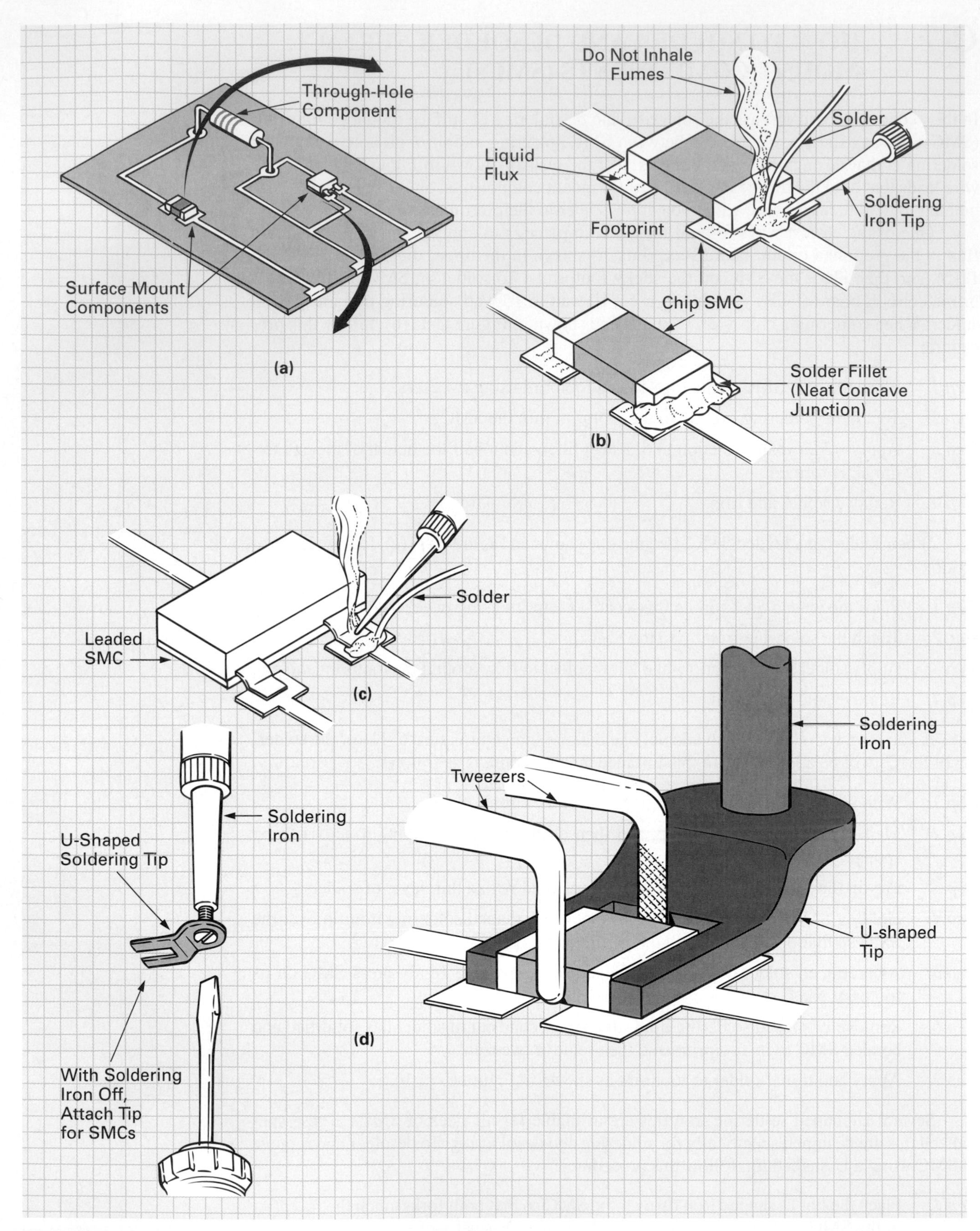

FIGURE C-11 Soldering Surface Mount Components (SMCs).

Step 2: When the solder liquefies, twist the tip of the iron to free the SMC from its footprint.

Step 3: Use tweezers to remove the SMC.

To remove an SMC using a wedge-shaped soldering iron tip, follow the standard desoldering procedure described in Section C–6.

C-8 SAFETY

Molten solder, like any other liquid, can splash or spill, and soldering irons are even hotter than the solder. Consequently, you should always wear protective safety glasses and an apron to protect yourself from burns. NEVER fling hot solder off an iron; always use the sponge.

APPENDIX

D

Safety When Troubleshooting

The following procedures and precautions have been acquired from experienced technicians and should be applied whenever possible.

1. Except when absolutely necessary, do not work on electrical or electronic circuits or equipment when power is on.
2. When troubleshooting inside equipment, people tend to lean one hand on the chassis (the metal framework of the equipment) and hold the test lead or probe in the other hand. If the probing hand comes in contact with a high voltage, current will flow from one hand, through your chest (heart and lungs), and finally through the other hand to the chassis ground. To avoid this dangerous situation, always place the free hand in a pocket or behind your back while testing a piece of equipment with power on.
3. When troubleshooting equipment with power on, try to insulate yourself by wearing rubber-soled electrical safety shoes or standing on a rubber mat.
4. Electrolytic and other large-value capacitors (which are covered in Chapters 13 and 14) can hold a dangerous voltage charge even after the equipment has been turned off.
5. All tools should be well insulated. If not, not only is it dangerous to you but probes that are not well insulated right down to the tip can cause short circuits between two points, resulting in additional circuit and equipment problems.
6. Always switch off the equipment and disconnect the power (since some equipment has power present even when off) before replacing any components. When removing components that get hot, such as resistors, allow enough time for cooling after the equipment has been turned off.
7. Necklaces, rings, and wristwatches have a low resistance and should be removed when working on equipment.
8. Inspect the equipment before working on it, and if it is in poor condition (frayed, cracked, or burned power cords, chipped plugs), turn off the equipment and replace these hazards.
9. Make sure that someone is present who can render assistance in the event of an emergency.
10. Make it a point to know the location of the power-off switch.
11. Cathode-ray tubes, which are the picture tubes within televisions and computer monitors, are highly evacuated and should therefore be handled with extreme care. If broken, the relatively high external pressure will cause an implosion (burst inward), which will result in the inner metal parts and glass fragments being expelled violently outward.

D-1 FIRST AID, TREATMENT, RESUSCITATION

Shock. Electric shock is the effect produced on the body and in particular on the nerves by an electric current passing through it. Its magnitude depends on the strength of the current, which, in turn, depends on the voltage value present. Its effect varies according to the ohmic resistance of the body, which varies in different persons, and also according to the parts of the body between which the current flows (contact in the cardiac region can be particularly dangerous). It also depends on the current flow and on the surface resistance of the skin, which is much reduced when the skin is wet and is reduced to zero if the skin is penetrated.

Shock can be felt from voltages as low as 15 V. At 20 to 25 V most people will experience pain, and the victim may be unable to let go of the conductor. It is believed that under certain conditions, death can be caused by voltages as low as 70 V, but generally the danger below 120 V ac is believed to be small (although not entirely negligible). Most serious and fatal accidents occur at the industrial 200 to 240 V ac when current is greater than 25 mA.

Injury can also be caused by a minor shock, not serious in itself but which has the effect of contracting the muscles sufficiently to result in a fall or other reaction.

Burns. Burns can be caused by the passage of a heavy current through the body or by direct contact with an electrically heated surface. Burns can also be caused by the intense heat generated by the arcing from a short circuit. All cases of burns require immediate medical attention.

Explosion. An explosion can be caused by the ignition of flammable gases by a spark from an electric contact.

Eye injuries. These can be caused by exposure to the strong ultraviolet rays of an electric arc. In these cases, the eyes may become inflamed and painful after a lapse of several hours, and there may be temporary loss of sight. Although very painful, the condition usually passes off within 24 hours.

Lasers are also dangerous to the eyes due to their intensely concentrated beam, and therefore specially filtered protective glasses should be worn.

Precautions to protect the eyes must always be taken by wearing protective goggles when clipping leads or soldering.

Body injuries from microwave and radio-frequency equipment. The energy in microwave and radio-frequency equipment can damage the body, especially those parts with a low blood supply. The eyes are particularly vulnerable. The highest energy level to which operators should be subject is 1.0 mW/cm^2, and intensities exceeding 10 mW/cm^2 should always be avoided.

Resuscitation. You should familiarize yourself with the various methods of artificial respiration by contacting your local Red Cross for complete instruction. The mouth-to-mouth method of artificial respiration is the most effective of the resuscitation techniques. It is comparatively simple and produces the best and quickest results when done correctly.

It is essential to begin artificial respiration without delay. *Do not touch the victim with your bare hands until the circuit is broken.* If this is not possible, *protect yourself* with dry insulating material and pull the victim clear of the conductor.

Step 1: Lay the patient on his or her back and, if on a slope, have the stomach slightly lower than the chest.

Step 2: Make a brief inspection of the mouth and throat to ensure that they are clear of obvious obstructions.

Step 3: Give the patient's head the maximum backward tilt so that the chin is prominent and the neck stretched to give a clear airway, as shown in Figure D-1(a).

Step 4: Sealing off the patient's nose with your thumb and finger, open your mouth wide and make an airtight seal over the patient's open mouth and then blow, as shown in Figure D-1(b). (New steps indicate using a plastic bag airway).

Step 5: After exhaling, turn your head to watch for chest movement, while inhaling deeply in readiness for blowing again, as shown in Figure D-1(c).

Step 6: If the chest does not rise, check that the patient's mouth and throat are free of obstruction and that the head is tilted back as far as possible, and then blow again.

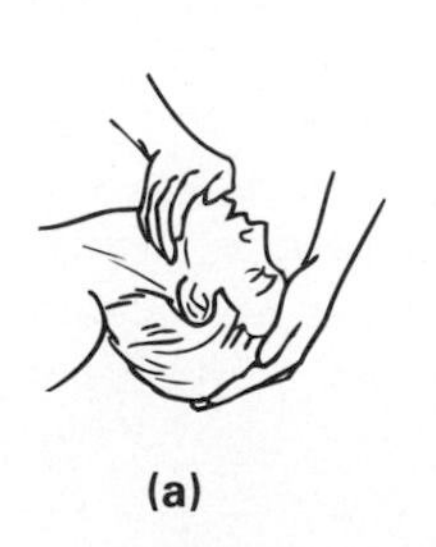

(a)

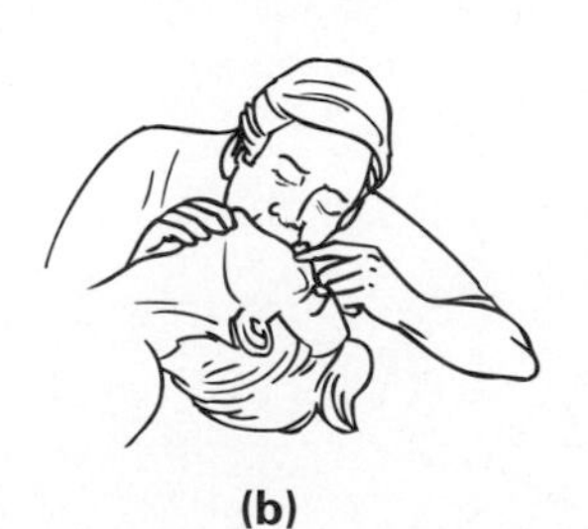

(b)

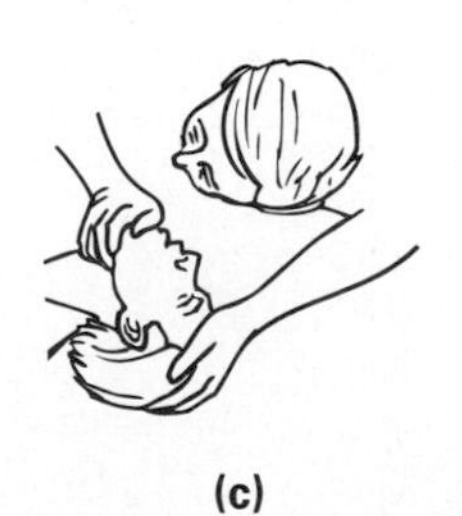

(c)

FIGURE D-1 Mouth-to-Mouth Resuscitation.

APPENDIX

E

Frequency Spectrum

The term *frequency* describes the number of alternations occurring in 1 second. Direct current (dc) is a steady or constant current that does not alternate and is therefore listed as zero cycles per second or 0 Hz. This appendix illustrates in detail the entire range of frequencies from the lowest subaudible frequency to the highest cosmic rays, along with their applications.

The following page is an overall summary of the complete range of frequencies or spectrum, as it is normally called. The subsequent pages cover each band or section of frequencies in a lot more detail and list the different frequency applications.

This and all of the other appendixes will be useful as references throughout your course of electronic study.

Overview of Frequency Spectrum

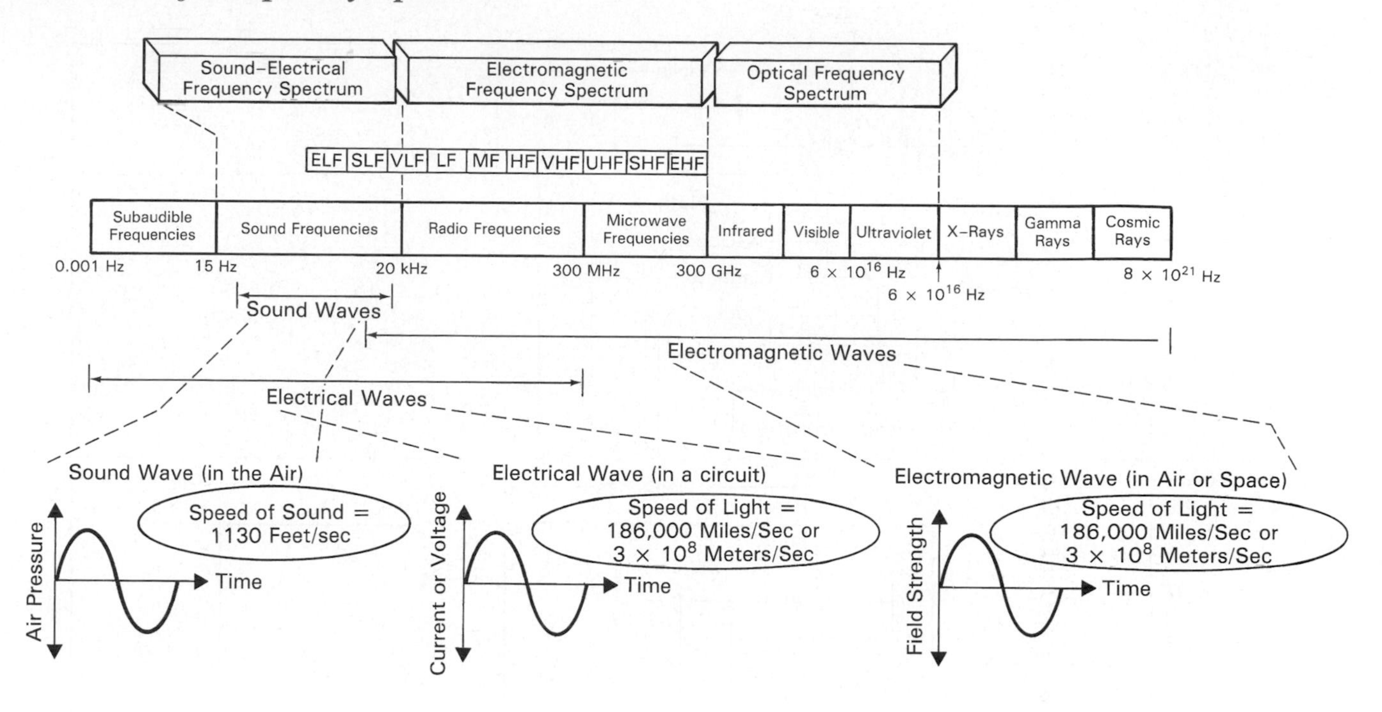

0 Hz	16 → 16 kHz	16 → 30 kHz	10 kHz → 300 GHz
Direct Current (DC) Motors, Relays, Supply Voltages	Audio Frequencies (AC) Motors, Amplifiers, Music Equipment, Speakers, Microphones, Oscillators	Sound (Ultrasonic) Frequencies Sonar, Music, Speech	Electromagnetic (Radio) Frequencies Voice Communications Television Navigation Medical, Scientific, and Military

300 GHz → 4×10^{14}	4×10^{14} → 7.69×10^{14}	4×10^{14} → 6×10^{16}	9.375×10^{15} → 3×10^{19}
Infrared (R) Heating, Photography, Sensing, Military	Visible Color, Photography, Movies, TV	Ultraviolet (UV) Sterilizing, Medical	X–Rays Medical, Gauge Thickness, Inspection

3×10^{19} → 5×10^{20}	5×10^{20} →
Gamma Rays Deeper Penetrating Than X–Rays, Detection of Radiation	Cosmic Rays Present in Outer Space

The Sound Spectrum

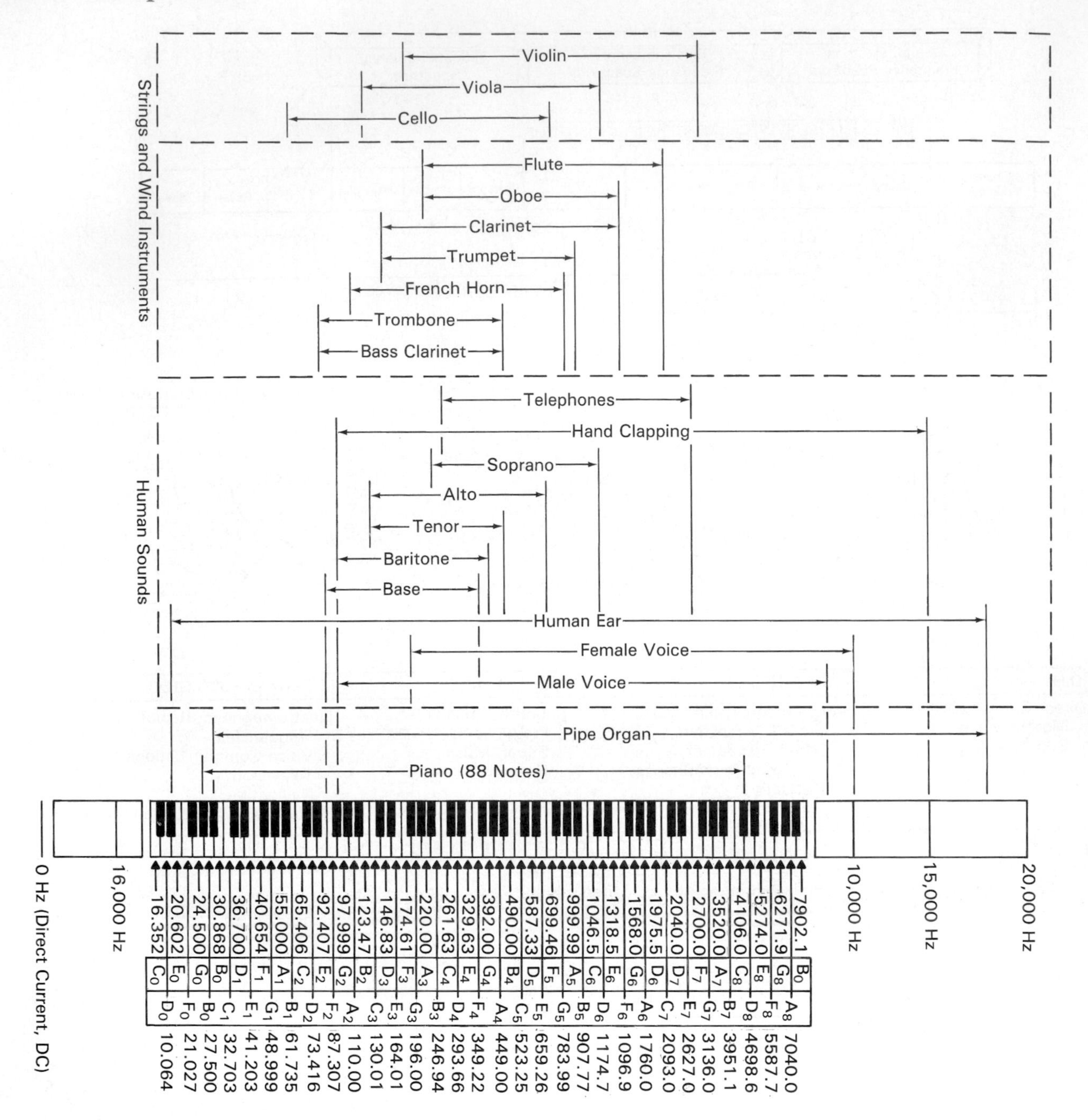

VLF (Very Low–Frequency Band)

3

Not Allocated

3 kHz

Not Allocated

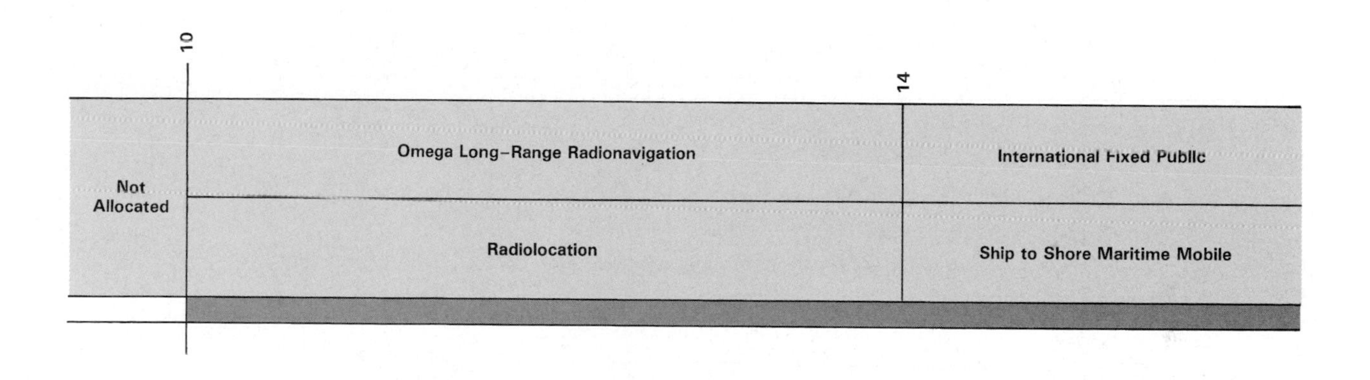

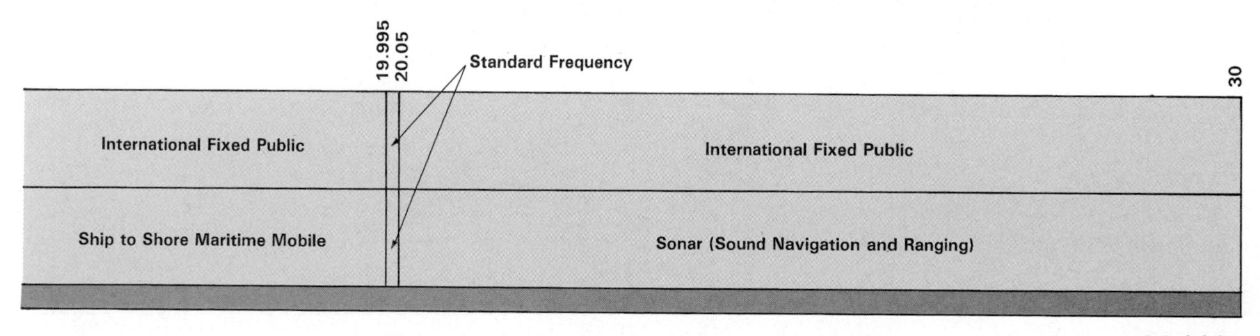

30 kHz

LF (Low–Frequency Band)

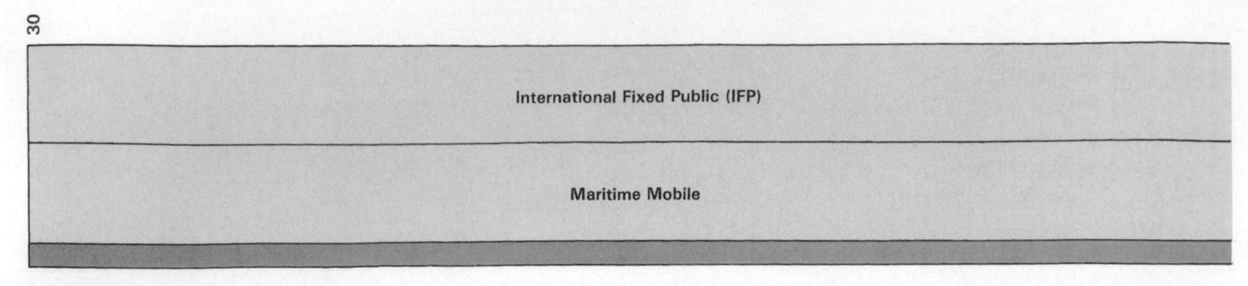

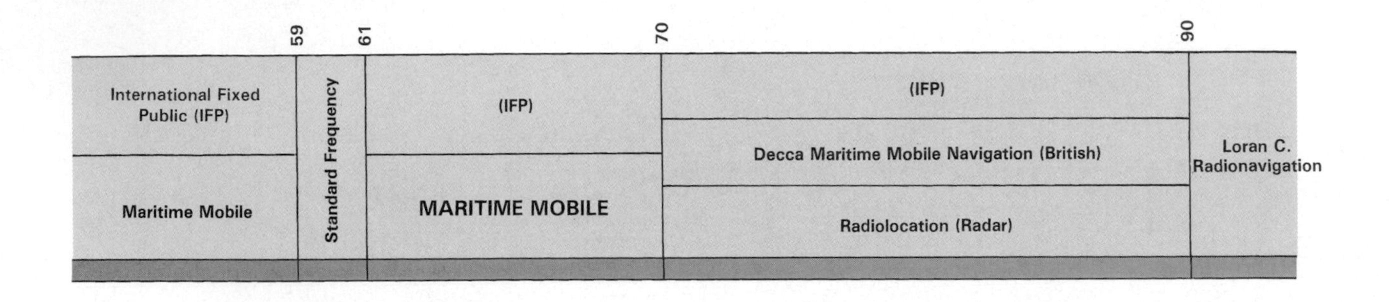

110
130
160
Loran C. Radionavigation
(IFP)
Maritime Mobile
Radar (Radio Detection and Ranging)
(IFP)
Maritime Mobile (MM)
(IFP)
Maritime Mobile (MM)

200
285
300
(IFP)
Maritime Mobile (MM)
Aeronautical Mobile Communications
Maritime Radionavigation
Maritime Radio–navigation
Aeronautical Radio–navigation

300 kHz

MF (Medium–Frequency Band)

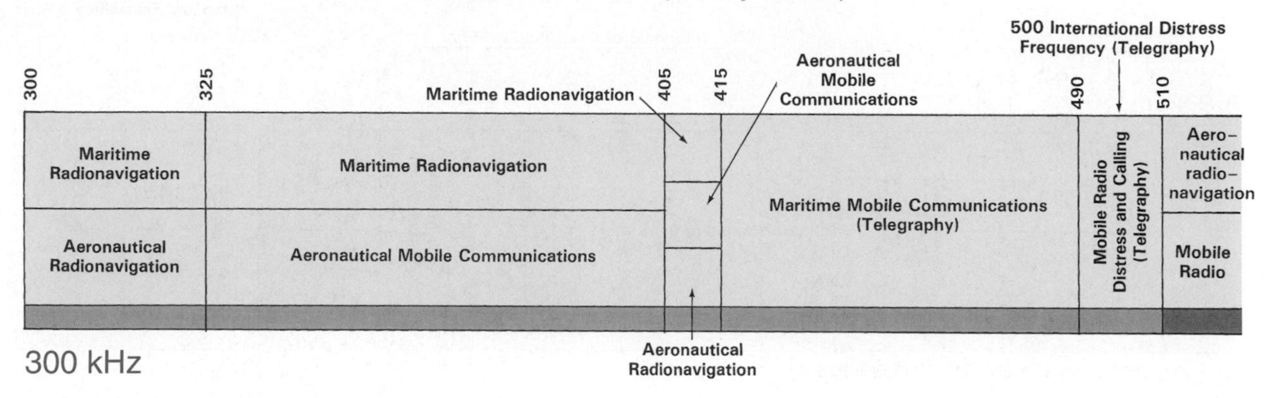

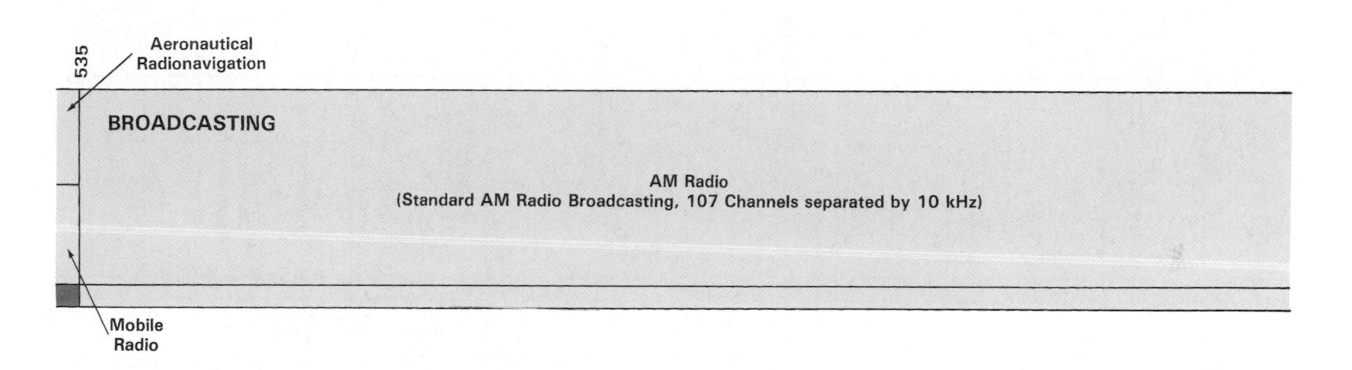

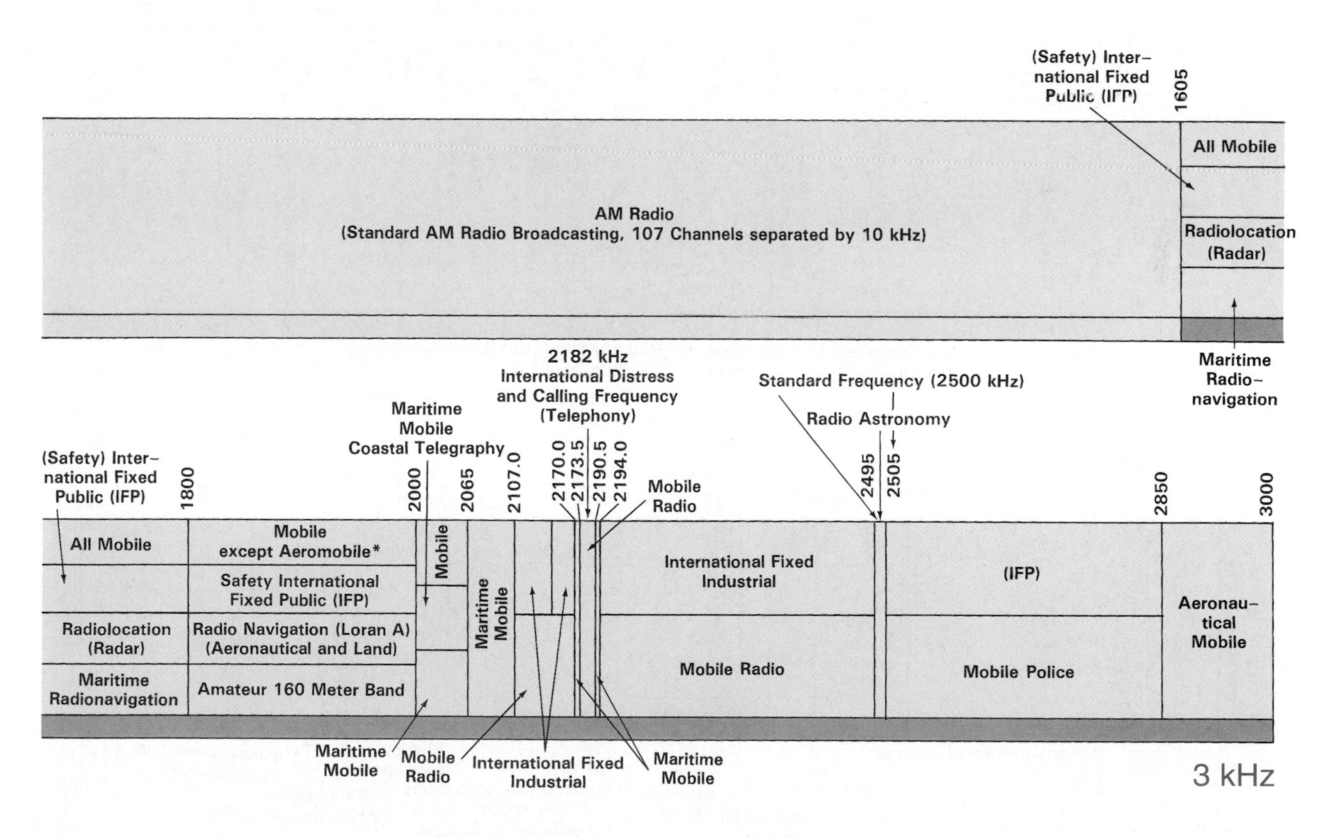

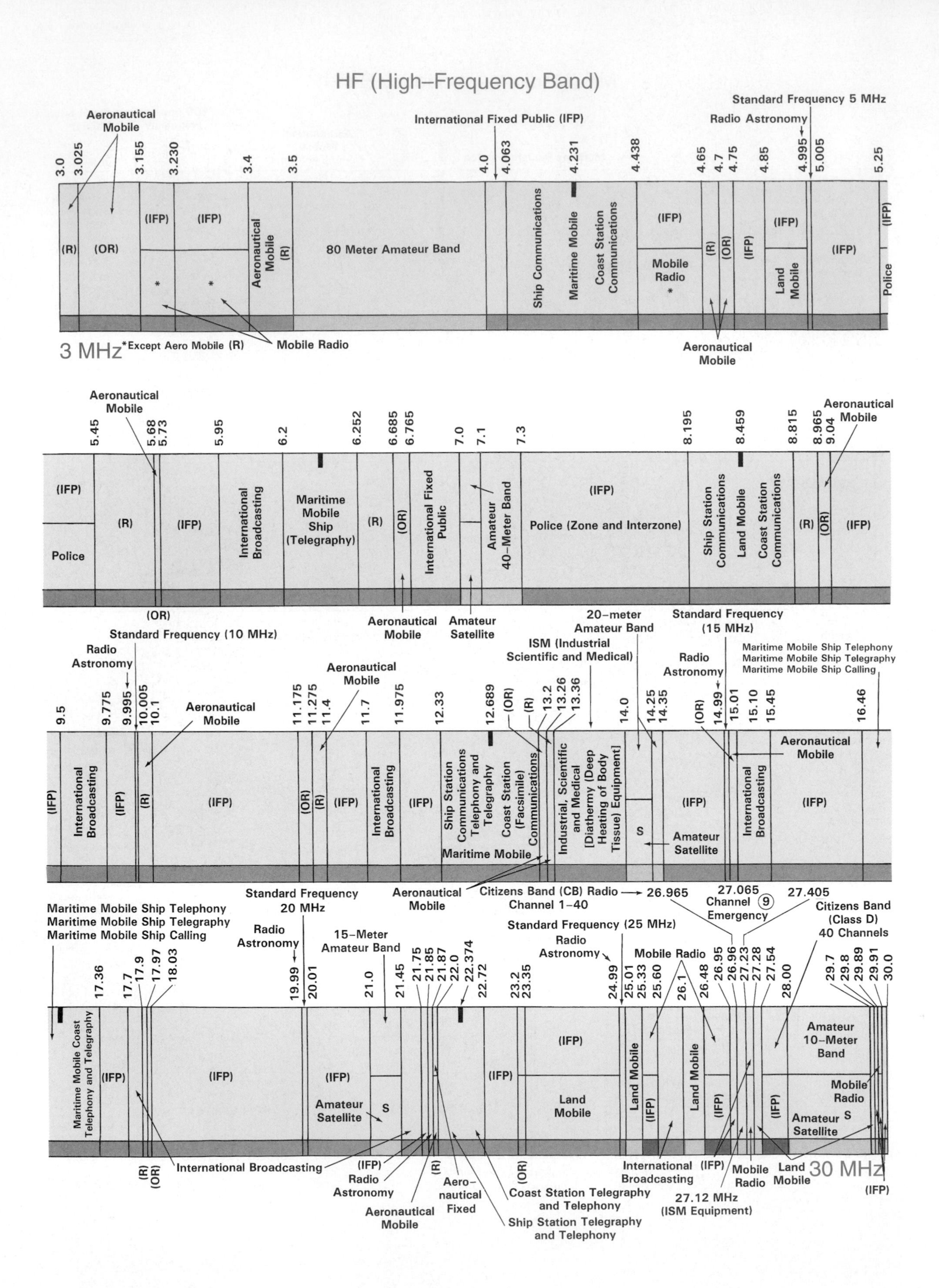
HF (High–Frequency Band)
3 MHz
*Except Aero Mobile (R)
Mobile Radio
Aeronautical Mobile
International Fixed Public (IFP)
Standard Frequency 5 MHz
Radio Astronomy
80 Meter Amateur Band
Ship Communications
Maritime Mobile
Coast Station Communications
Mobile Radio
Land Mobile
Police
International Broadcasting
Maritime Mobile Ship (Telegraphy)
International Fixed Public
Amateur 40–Meter Band
Police (Zone and Interzone)
Ship Station Communications
Land Mobile
Coast Station Communications
Amateur Satellite
Standard Frequency (10 MHz)
20–meter Amateur Band
ISM (Industrial Scientific and Medical)
Standard Frequency (15 MHz)
Maritime Mobile Ship Telephony
Maritime Mobile Ship Telegraphy
Maritime Mobile Ship Calling
Ship Station Communications Telephony and Telegraphy
Coast Station (Facsimile) Communications
Maritime Mobile
Industrial, Scientific and Medical [Diathermy (Deep Heating of Body Tissue) Equipment]
Standard Frequency 20 MHz
15–Meter Amateur Band
Citizens Band (CB) Radio → 26.965
Channel 1–40
Standard Frequency (25 MHz)
Channel 9 Emergency
Citizens Band (Class D) 40 Channels
Maritime Mobile Coast Telephony and Telegraphy
Amateur 10–Meter Band
International Broadcasting
Aeronautical Fixed
Coast Station Telegraphy and Telephony
Ship Station Telegraphy and Telephony
27.12 MHz (ISM Equipment)
30 MHz

VHF (Very High–Frequency Band)

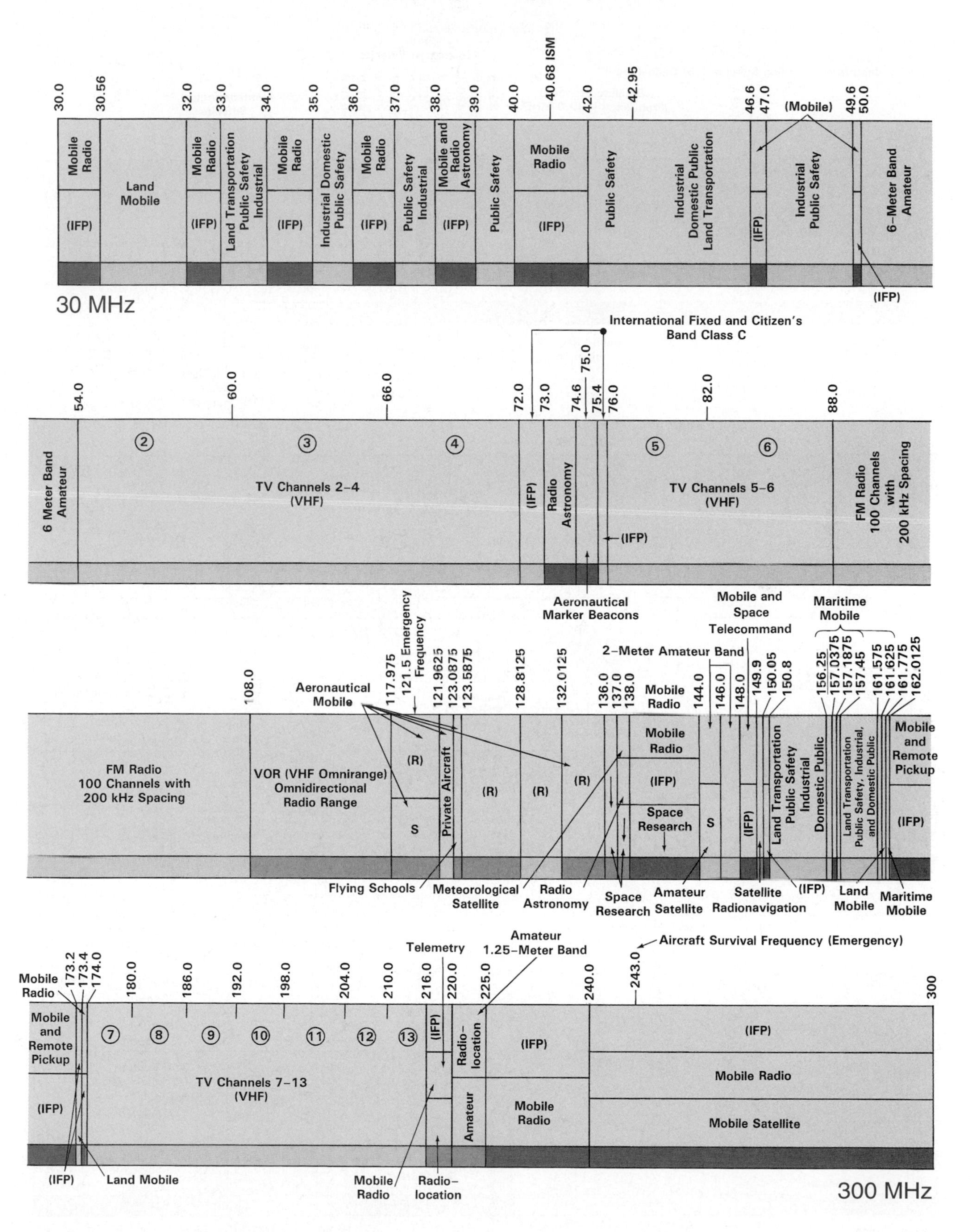

UHF (Ultra High–Frequency Band)

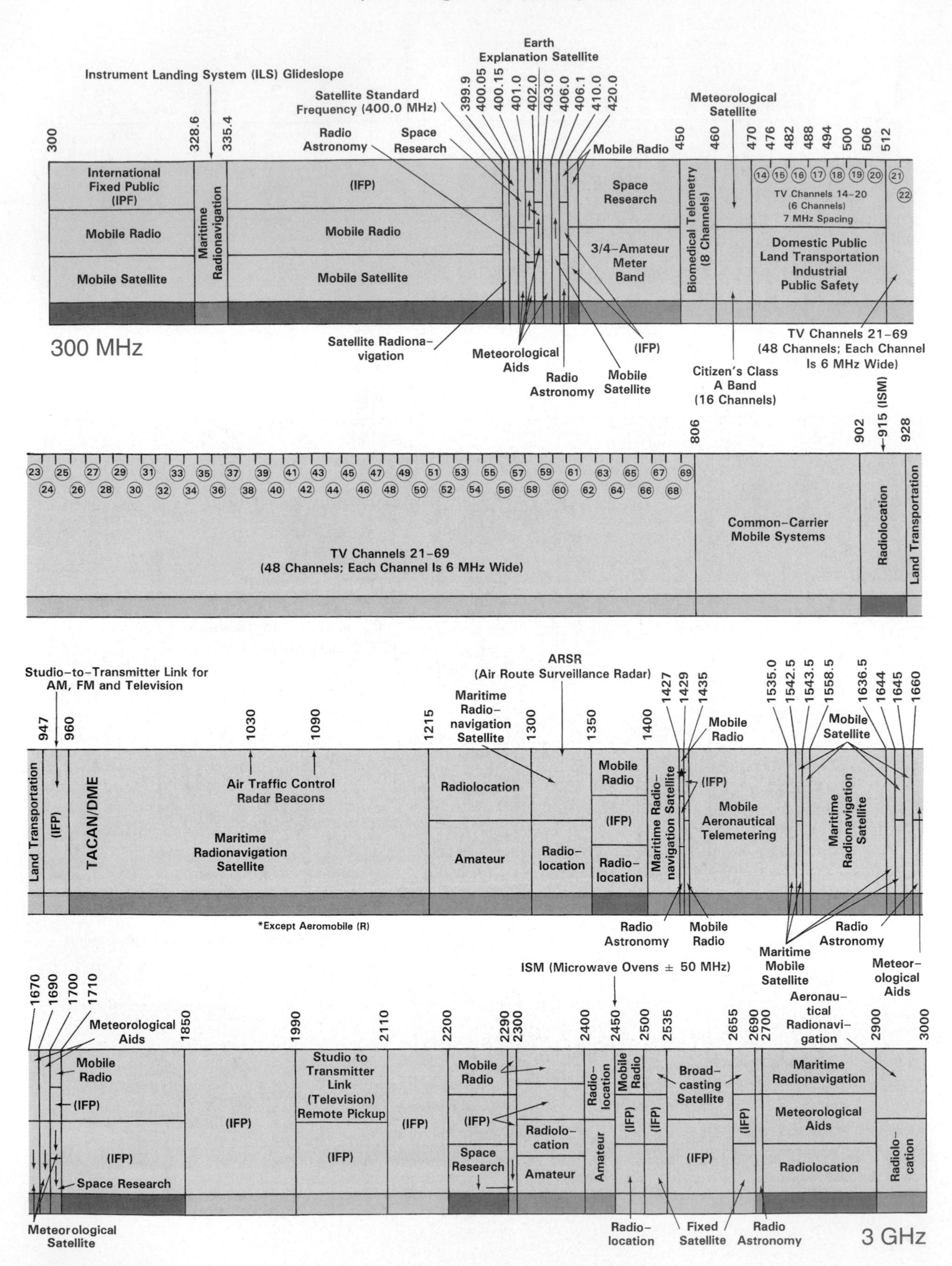

SHF (Super High–Frequency Band)

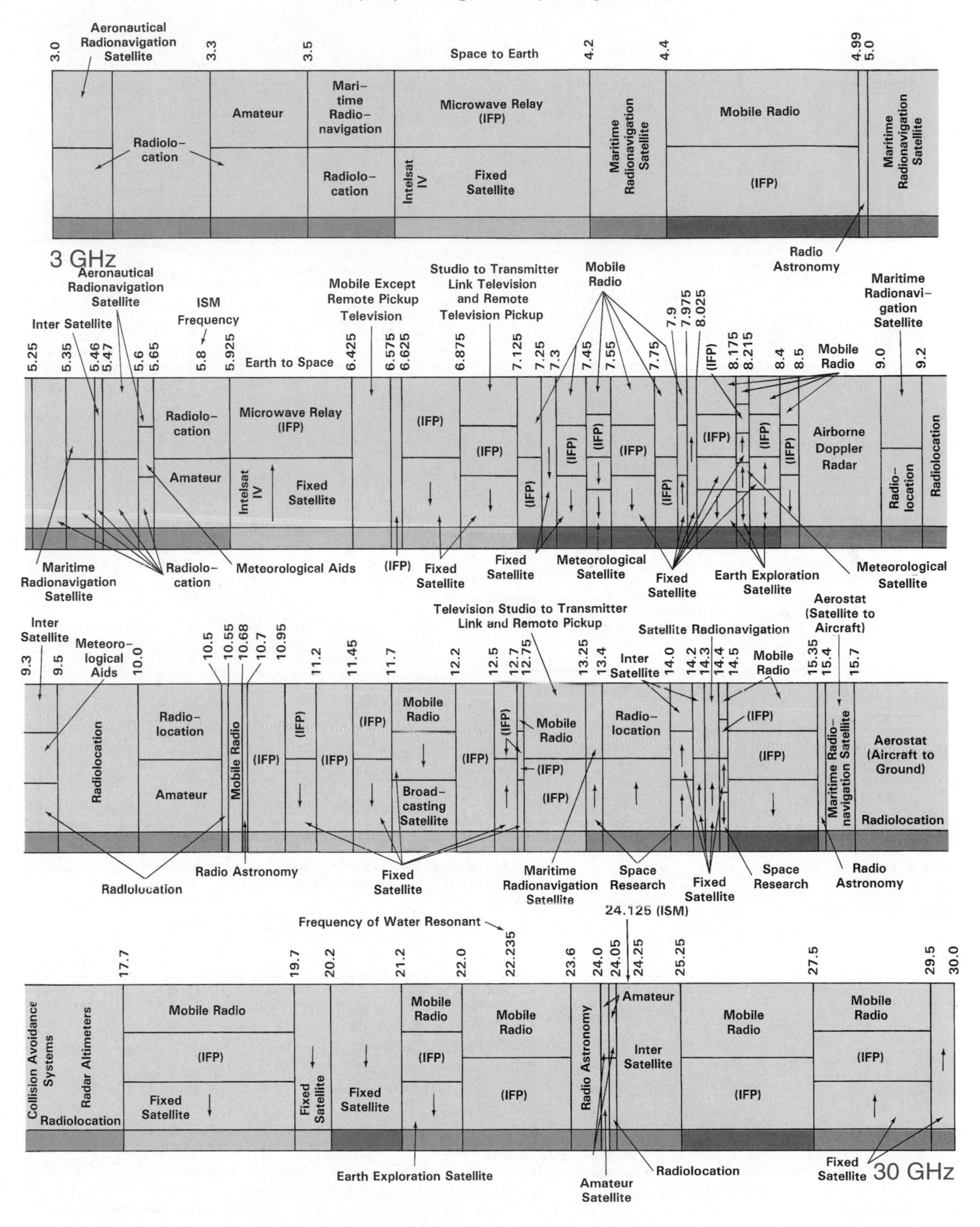

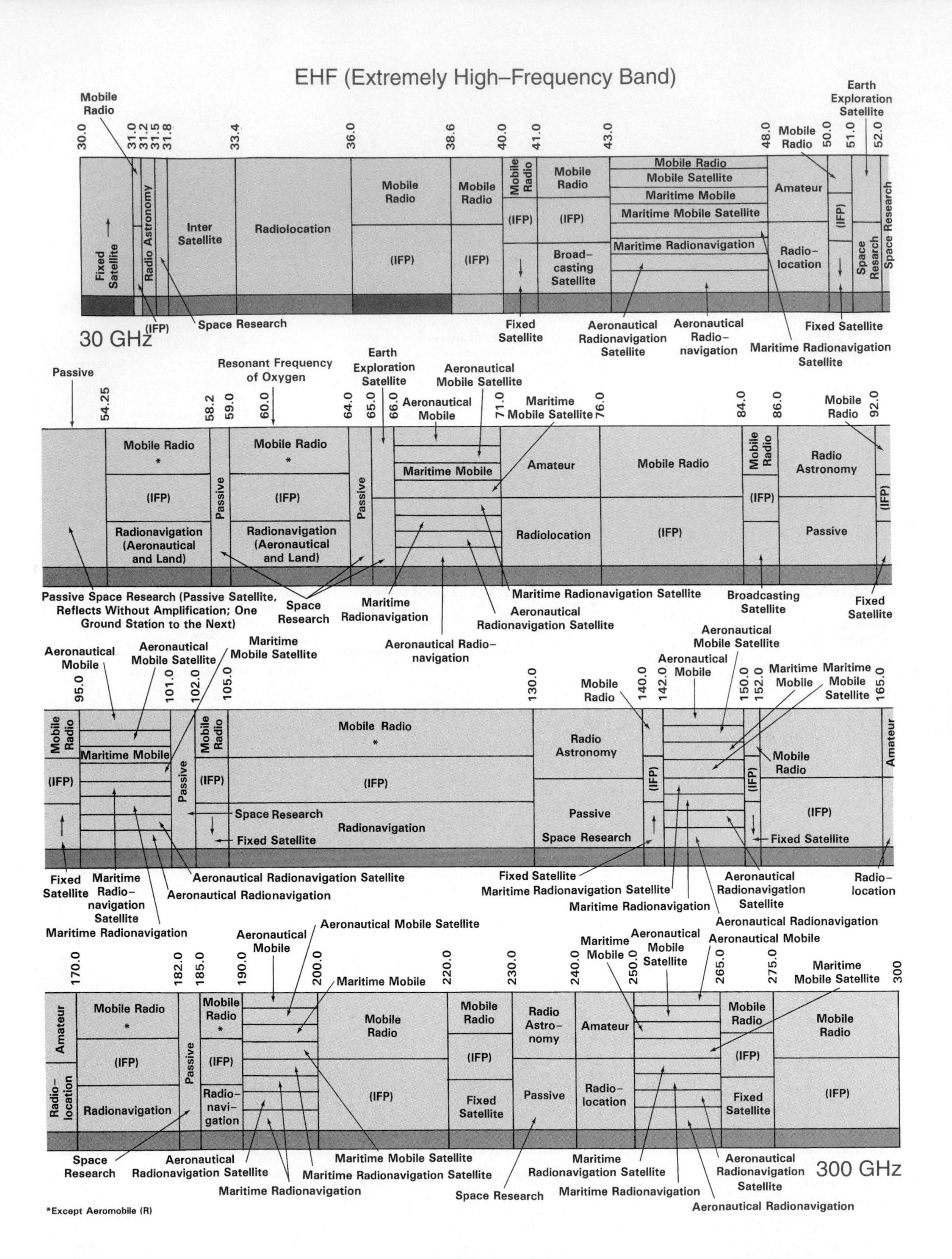
EHF (Extremely High–Frequency Band)
30 GHz
300 GHz
Mobile Radio
Fixed Satellite
Radio Astronomy
Inter Satellite
Radiolocation
(IFP)
Space Research
Fixed Satellite
Broadcasting Satellite
Mobile Satellite
Maritime Mobile
Maritime Mobile Satellite
Maritime Radionavigation
Aeronautical Radionavigation Satellite
Aeronautical Radionavigation
Maritime Radionavigation Satellite
Amateur
Earth Exploration Satellite
Passive
Resonant Frequency of Oxygen
Radionavigation (Aeronautical and Land)
Aeronautical Mobile
Aeronautical Mobile Satellite
Passive Space Research (Passive Satellite, Reflects Without Amplification; One Ground Station to the Next)
Radio Astronomy
Radionavigation
Radio–location
Radio Astro–nomy
*Except Aeromobile (R)

OPTICAL SPECTRUM

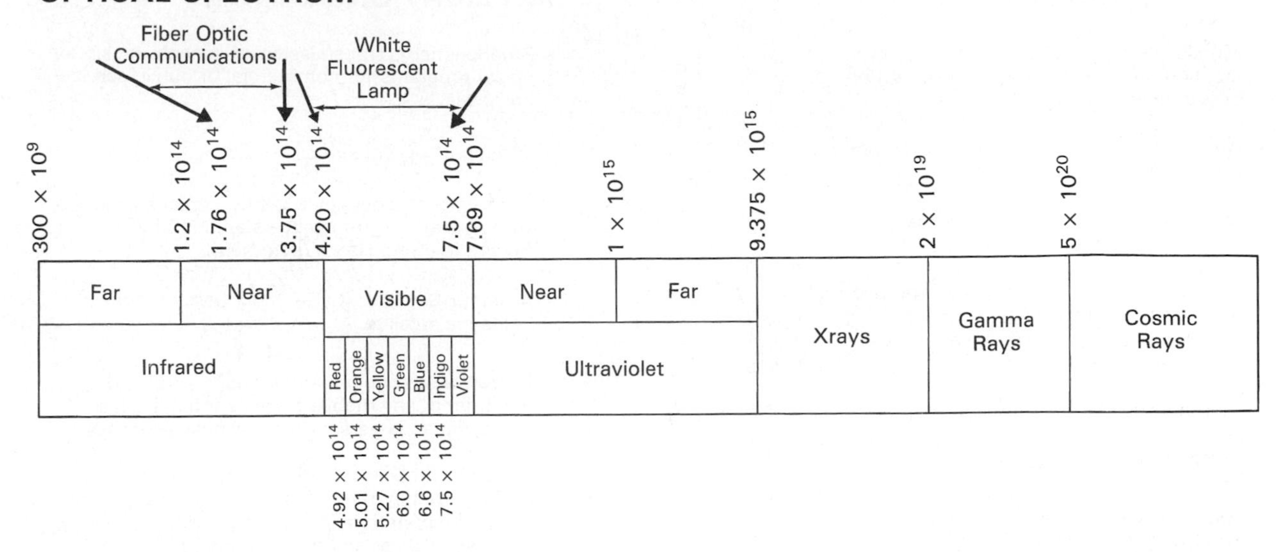

DEFINITIONS AND CODES

USAGE

Exclusively used by government, state, and federal agencies

Shared by government and nongovernment agencies

Nongovernment; publicly used frequencies

TYPES OF EMISSION

Amplitude (AM;Amplitude Modulation)
A0–Steady Unmodulated Pure Carrier
A1–Telegraphy on Pure Continuous Waves
A2–Amplitude Tone–modulated Telegraphy
A3–AM Telephony Including Single and Double Sideband with full, Reduced, or Suppressed Carrier
A4–Facsimile (Pictures)
A5–Television

Frequency FM;Frequency Modulation)
F0–Steady Unmodulated Pure Carrier
F1–Carrier–Shift Telegraphy
F2–Audio–frequency–shift Telegraphy
F3–Frequency or Phase–modulated Telephony
F4–Facsimile
F5–Television

Pulse (PM;Pulse Modulation)
P0–Pulsed Radar
P1–Telegraphy: On/Off Keying of Pulsed Carrier
P2–Telegraphy: Pulse Modulation of Pulse Carrier, Pulse Width, Phase or Position Tone Modulated
P3–Telephony: Amplitude (PAM), Width (PWM), Phase or Position (PPM) Modulated Pulses

ABBREVIATED TERMS

(R), International major air route (air traffic control)
(OR), Off route military (air traffic control)
S Assigned satellite frequency
↑ Earth to space communication (uplink)
↓ Space to earth communication (downlink)

ISM: industrial, scientific, medical
Passive communication equipment: generates no electromagnetic radiation (receive only)
Active communication equipment: generates electromagnetic radiation (transmits)

APPLICATIONS

- International Fixed: public radiocommunication service. Transmitters designed for one frequency.
- Aeronautical Fixed: a service intended for the transmission of air navigation and preparation for safety of flight.
- Fixed Satellite: satellites that maintain a geostationary orbit (same position above the earth)
- Radiolocation: radio waves used to detect an object's direction, position, or motion.
- Radio Astronomy: radio waves emitted by celestial bodies that are used to obtain data about them.
- Space Operations
- Space Research

APPLICATIONS (continued)

- Mobile: Radio service between a fixed location and one or more mobile stations or between mobile stations.
- Aeronautical Mobile: As above, except mobile stations are aircraft.
- Maritime Mobile: As above, except mobile stations are marine.
- Land Mobile: As above, except mobile stations are automobiles.
- Mobile Satellite
- Aeronautical Mobile Satellite
- Maritime Mobile Satellite
- Radionavigation (Aeronautical and Land): Navigational use of radiolocation equipment such as direction finders, radio compass, radio homing beacons, etc.
- Aeronautical Radionavigation
- Maritime Radionavigation
- Satellite Radionavigation
- Aeronautical Radionavigation Satellite
- Maritime Radionavigation Satellite
- Inter Satellite

APPLICATIONS (continued)

- Broadcasting: Transmission of speech, music, or visual programs for commercial or public service purposes.
- Broadcasting Satellite
- Amateur: a frequency used by persons licensed to operate radio transmitters as a hobby. Person is also called a radio ham.
- Amateur Satellite: Communication by radio hams via satellite.
- Citizen's: a radiocommunications service of fixed, land, and mobile stations intended for short-distance personal or business purposes.
- Standard Frequency: Highly accurate signal broadcasted by the national bureau of standards (NBS) radio station (WWV) to provide frequency, time, solar flare, and other standards.
- Standard Frequency Satellite: NBS broadcast via satellite.
- Meteorological Aids: a radio service in which the emission consists of signals used solely for meteorological use.
- Meteorological Satellite: Meteorological broadcast via satellite.
- Earth Exploration Satellite: Radio frequencies used for earth exploration.

APPENDIX

F

Answers to Self-Test Evaluation Points

STEP 1-1

1. **a.** $16^4 = 16 \times 16 \times 16 \times 16 = 65{,}536$
   b. $32^3 = 32 \times 32 \times 32 = 32{,}768$
   c. $112^2 = 112 \times 112 = 12{,}544$
   d. $15^6 = 15 \times 15 \times 15 \times 15 \times 15 \times 15 = 11{,}390{,}625$
   e. $2^3 = 2 \times 2 \times 2 = 8$
   f. $3^{12} = 3 \times 3 \times 3 \times 3 \times 3 \times 3 \times 3 \times 3 \times 3 \times 3 \times 3 \times 3 =$ 531,441
2. **a.** $\sqrt[2]{144} = 12$
   b. $\sqrt[3]{3375} = 15$
   c. $\sqrt[2]{20} = 4.47$
   d. $\sqrt[3]{9} = 2.08$
3. **a.** 10^2
   b. 10^0
   c. 10^1
   d. 10^6
   e. 10^{-3}
   f. 10^{-6}
4. **a.** $6.3 \times 10^3 = 6.300. = 6300.0$ or 6300
   b. $114{,}000 \times 10^{-3} = 114.000. = 114.0$ or 114
   c. $7{,}114{,}632 \times 10^{-6} = 7.114632. = 7.114632$
   d. $6624 \times 10^6 = 6624.000000. = 6{,}624{,}000{,}000.0$
5. **a.** $\sqrt{3 \times 10^6} - \sqrt{3{,}000{,}000} = 1732.05$
   b. $(2.6 \times 10^{-6}) - (9.7 \times 10^{-9}) = 0.0000025$ or 2.5×10^{-6}
   c. $\frac{(4.7 \times 10^3)^2}{3.6 \times 10^6} = (4.7 \times 10^3)^2 \div (3.6 \times 10^6) = 6.14$
6. **a.** $47{,}000 = 47000. = 47 \times 10^3$
   b. $0.00000025 = 0.000000250. = 250 \times 10^{-9}$
   c. $250{,}000{,}000 = 250.000000. = 250 \times 10^6$
   d. $0.0042 = 0.004.2 = 4.2 \times 10^{-3}$
7. **a.** 10^3
   b. 10^{-2}
   c. 10^{-3}
   d. 10^6
   e. 10^{-6}

STEP 1-2

1. Elements are made up of similar atoms; compounds are made up of similar molecules.
2. Protons, neutrons, electrons
3. Copper
4. Like charges repel; unlike charges attract.

STEP 1-3

1. Amp
2. Current = Q/t (number of coulombs divided by time in seconds).
3. The direction of flow is different: negative to positive is termed electron current flow; positive to negative is known as conventional current flow.
4. The ammeter

STEP 1-4

1. Volts
2. 3000 kV
3. The voltmeter
4. They are directly proportional.
5. No
6. Yes
7. A short circuit provides an unintentional path for current to flow. A closed circuit has a complete path for current.

STEP 2-1

1. Yes
2. $\frac{144}{12} \times \square = \frac{36}{6} \times 2 \times \square = 60$

   $12 \times \square = 6 \times 2 \times \square = 60$

   $12 \times 5 = 12 \times 5 = 60$

   $\square = 5$
3. Yes
4. **a.**
   $$x + 14 = 30$$
   $$x + 14 - 14 = 30 - 14 \quad (-\ 14 \text{ from both sides})$$
   $$x = 30 - 14$$
   $$x = 16$$
   b.
   $$8 \times x = \frac{80 - 40}{10} \times 12$$
   $$8 \times x = 4 \times 12$$
   $$8 \times x = 48$$
   $$\frac{8 \times x}{8} = \frac{48}{8} \quad (\div\ 8)$$
   $$x = \frac{48}{8}$$
   $$x = 6$$
   c.
   $$y - 4 = 8$$
   $$y - 4 + 4 = 8 + 4 \quad (+\ 4)$$
   $$y = 8 + 4$$
   $$y = 12$$
   d.
   $$(x \times 3) - 2 = \frac{26}{2}$$
   $$(x \times 3) - 2 + 2 = \frac{26}{2} + 2 \quad (+\ 2)$$
   $$x \times 3 = \frac{26}{2} + 2$$
   $$x \times 3 = 15$$
   $$\frac{x \times 3}{3} = \frac{15}{3} \quad (\div\ 3)$$
   $$x = \frac{15}{3}$$
   $$x = 5$$

e. $x^2 + 5 = 14$

$x^2 + 5 - 5 = 14 - 5 \quad (-5)$

$x^2 = 14 - 5$

$\sqrt{x^2} = \sqrt{14 - 5} \quad (\sqrt{\ })$

$x = \sqrt{14 - 5}$

$x = \sqrt{9}$

$x = 3$

f. $2(3 + 4x) = 2(x + 13)$

$6 + 8x = 2x + 26 \quad$ (remove parentheses)

$6 + 8x - 2x = \cancel{2x} + 26 - \cancel{2x} \quad (-2x)$

$6 + (8x - 2x) = 26$

$6 + 6x = 26$

$6 + 6x - 6 = 26 - 6 \quad (-6)$

$6x = 26 - 6$

$6x = 20$

$\frac{\cancel{6} \times x}{\cancel{6}} = \frac{20}{6} \quad (\div 6)$

$x = \frac{20}{6}$

$x = 3.3333$

5. a. $x + y = z, y = ?$

$x + y - x = z - x \quad (-x)$

$y = z - x$

b. $Q = C \times V, C = ?$

$\frac{Q}{V} = \frac{C \times \cancel{V}}{\cancel{V}} \quad (\div V)$

$\frac{Q}{V} = C$

$C = \frac{Q}{V}$

c. $X_L = 2 \times \pi \times f \times L, L = ?$

$\frac{X_L}{2 \times \pi \times f} = \frac{\cancel{2} \times \cancel{\pi} \times \cancel{f} \times L}{\cancel{2} \times \cancel{\pi} \times \cancel{f}} \quad (\div 2 \times \pi \times f)$

$\frac{X_L}{2 \times \pi \times f} = L$

$L = \frac{X_L}{2 \times \pi \times f}$

d. $V = I \times R, R = ?$

$\frac{V}{I} = \frac{\cancel{I} \times R}{\cancel{I}} \quad (\div I)$

$\frac{V}{I} = R$

$R = \frac{V}{I}$

6. a. $I^2 = 9$

$\sqrt{I^2} = \sqrt{9} \quad (\sqrt{\ })$

$I = \sqrt{9}$

$I = 3$

b. $\sqrt{Z} = 8$

$\sqrt{Z^2} = 8^2$

$Z = 8^2$

$Z = 64$

7. $x = y \times z$ and $a = x \times y, y = 14, z = 5, a = ?$

$a = x \times y$

$a = y \times z \times y \quad$ (substitute $y \times z$ for x)

$a = y^2 \times z \quad (y \times y = y^2)$

$a = 14^2 \times 5$

$a = 980$

STEP 2-2

1. A circuit is said to have a resistance of 1 ohm when 1 volt produces a current of 1 ampere.
2. $I = V/R = 24\text{ V}/6\ \Omega = 4\text{ A}$
3. A memory aid to help remember Ohm's law:

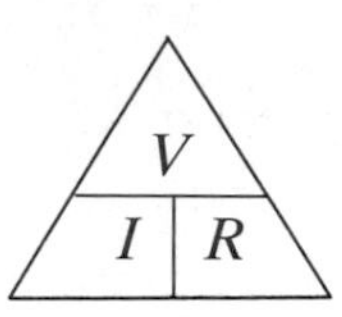

4. Current is proportional to voltage and inversely proportional to resistance.
5. $V = I \times R = 25\text{ mA} \times 1\text{ k}\Omega = (25 \times 10^{-3}) \times (1 \times 10^3) = 25\text{ V}$
6. $R = V/I = 12\text{ V}/100\ \mu\text{A} = 12\text{ V}/100 \times 10^{-6} = 120\text{ k}\Omega$
7. Ohmmeter
8. (Section 2-2-3)

STEP 2-3

1. Light, heat, magnetic, chemical, electrical, mechanical
2. When energy is transformed, work is done. Power is the rate at which work is done or energy is transformed.
3. $W = Q \times V, P = V \times I$
4. When 1 kW of power is used in 1 hour

STEP 3-1

1. $\frac{176 \div 8}{8 \div 8} = \frac{22}{1}$ or 22:1 (ratio of 22 to 1)
2. **a.** 86.44
   b. 12,263,415.01
   c. 0.18
3. 86.43760 **a.** 7 **b.** 6
   12,263,415.00510 **a.** 13 **b.** 12
   0.176600 **a.** 6 **b.** 4
4. **a.** 0.075 **b.** 220 **c.** 0.235 **d.** 19.2
5. **a.** 35 **b.** 4004
6.

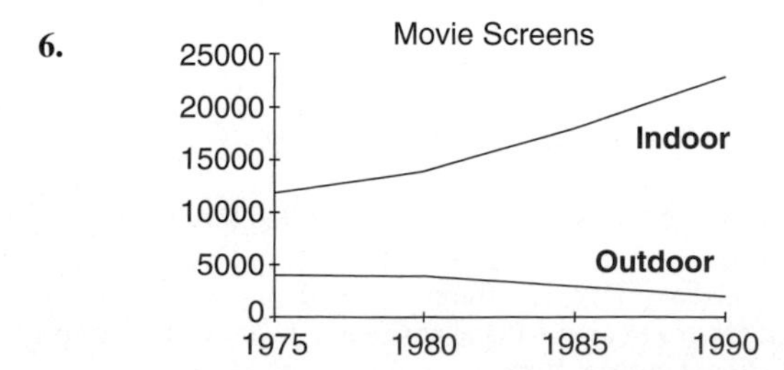

STEP 3-2

1. Type of conducting material used, cross-sectional area, length, temperature
2. The resistance of a conductor is proportional to its length and resistivity, and inversely proportional to its cross-sectional area.
3. True
4. True
5. False
6. **a.** Electrons are far away from nucleus.
   b. Many electrons per atom
   c. Incomplete valence shell
7. Copper
8. 28.6 millisiemens

STEP 3-3

1. True
2. Mica
3. The voltage needed to cause current to flow through a material
4. Small

STEP 3-4

1. Carbon composition, carbon film, metal film, wirewound, metal oxide, thick film
2. SIPs have one row of connecting pins; DIPs have two rows of connecting pins.
3. Rheostat has two terminals; potentiometer has three terminals.
4. Linear means that the resistance changes in direct proportion to the amount of change of the input, while a tapered potentiometer varies nonuniformly.
5. General purpose are ±5% or greater; precision are ±2% or less.
6. 222 × 0.1, ±0.25% = 22.2 Ω, ±0.25%
7. Yellow, violet, red, silver
8. True (general purpose, ±20% tolerance)

STEP 4-1

1. A device that generates a voltage output when heat is applied to its two-metal junction
2. By using a photovoltaic cell
3. Generator
4. Negative plate, positive plate, electrolyte
5. False

STEP 4-2

1. b
2. Decrease
3. c
4. b
5. A piezoresistive device changes its resistance in response to pressure, whereas a piezoelectric device generates a voltage output in response to a pressure input.

STEP 4-3

1. Indicate the maximum current that can flow without blowing the fuse and the maximum voltage value that will not cause an arc.
2. Fast-blow fuses open almost instantly after an excessive current occurs, whereas a slow-blow fuse will have a certain time delay until it opens.
3. Thermal, magnetic, thermomagnetic
4. A circuit breaker can be reset.
5. Section 4-3-3
6. Section 4-3-3

STEPS 5-1 AND 5-2

1. A circuit in which current has only one path
2. 8 A

STEP 5-3

1. $R_T = R_1 + R_2 + R_3 + \cdots$
2. $R_T = R_1 + R_2 + R_3 = 2\text{ k}\Omega + 3\text{ k}\Omega + 4700\ \Omega = 9.7\text{ k}\Omega$

STEP 5-4

1. True
2. True
3. $R_T = R_1 + R_2 = 6 + 12 = 18\Omega$; $I_T = V_S/R_T = 18/18 = 1\text{ A}$
   $V_{R_1} = 1\text{ A} \times 6\ \Omega = 6\text{ V}$
   $V_{R_2} = 1\text{ A} \times 12\ \Omega = 12\text{ V}$
4. $V_X = (R_X/R_T) = V_S$
5. Potentiometer
6. No

STEP 5-5

1. $P = I \times V$ or $P = V^2/R$ or $P = I^2 \times R$
2. $P = V^2/R = 12^2/12 = 144/12 = 12\text{ W}$
3. Wirewound, at least 12 W, ideally a 15 W
4. $P_T = P_1 + P_2 = 25\text{ W} + 3800\text{ mW} = 28.8\text{ W}$

STEP 5-6

1. Component will open, component's value will change, and component will short.
2. No current will flow; source voltage dropped across it.
3. False
4. No voltage drop across component, yet current still flows in circuit; resistance equals zero for component.

STEPS 6-1 AND 6-2

1. When two or more components are connected across the same voltage source so that current can branch out over two or more paths
2. False
3. $V_{R_1} = V_S = 12\text{ V}$
4. No

STEP 6-3

1. The sum of all currents entering a junction is equal to sum of all currents leaving that junction.
2. $I_2 = I_T - I_1 = 4\text{ A} - 2.7\text{ A} = 1.3\text{ A}$
3. $I_X = (R_T/R_X) \times I_T$
4. $I_T = V_T/R_T = 12/1\text{ k}\Omega = 12\text{ mA}$; $I_1 = 1\text{ k}\Omega/2\text{ k}\Omega \times 12\text{ mA} = 6\text{ mA}$

STEP 6-4

1. $R_T = \dfrac{R_1 \times R_2}{R_1 + R_2}$
2. $R_T = \dfrac{1}{(1/R_1) + (1/R_2) + (1/R_3)} + \cdots$
3. $R_T = \dfrac{\text{common value of resistors } (R)}{\text{number of parallel resistors } (n)}$
4. $R_T = \dfrac{1}{(1/2.7\text{ k}\Omega) + (1/24\text{ k}\Omega) + (1/1\text{ M}\Omega)} = 2.421\text{ k}\Omega$

STEP 6-5

1. True
2. $P_1 = I_1 \times V = 2\text{ mA} \times 24\text{ V} = 48\text{ mW}$
3. $P_T = P_1 + P_2 = 22\text{ mW} + 6400\ \mu\text{W} = 28.4\text{ mW}$
4. Yes

STEP 6-6

1. No current will flow in the open branch; total current will decrease.
2. Maximum current is through shorted branch; total current will increase.
3. Will cause a corresponding opposite change in branch current and total current
4. False

STEPS 7-1 AND 7-2

1. By tracing current to see if it has one path (series connection) or more than one path (parallel connection)
2. $R_{1,2} = R_1 + R_2 = 12\text{ k}\Omega + 12\text{ k}\Omega = 24\text{ k}\Omega$

$$R_{1,2,3} = \frac{R_{1,2} \times R_3}{R_{1,2} + R_3} = \frac{24\text{ k}\Omega \times 6\text{ k}\Omega}{24\text{ k}\Omega + 6\text{ k}\Omega} = \frac{144\text{ k}\Omega^2}{30\text{ k}\Omega} = 4.8\text{ k}\Omega$$

3. STEP A: Find equivalent resistances of series-connected resistors. STEP B: Find equivalent resistances of parallel-connected combinations. STEP C: Find equivalent resistances of remaining series-connected resistances.
4. $R_{1,2} = R_1 + R_2 = 470 + 330 = 800\ \Omega$

$$R_{1,2,3} = \frac{R_{1,2} \times R_3}{R_{1,2} + R_3} = \frac{800 \times 270}{800 + 270} = \frac{216\text{ k}\Omega^2}{1.07\text{ k}\Omega} = 201.9\ \Omega$$

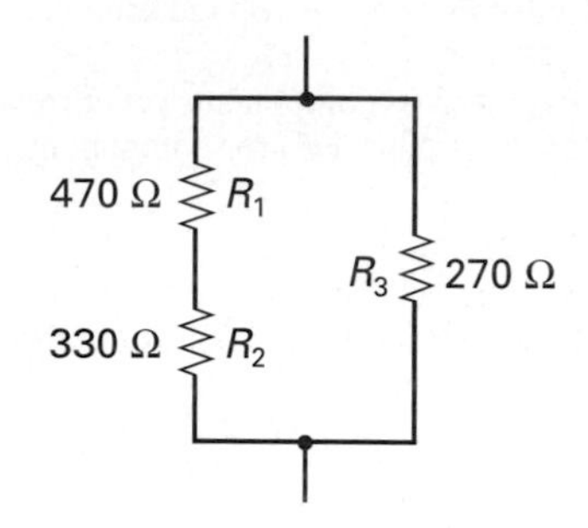

Figure 7-1, 4

STEP 7-3

1. STEP 1: Find total resistance (STEPs A, B and C). STEP 2: Find total current. STEP 3: Find voltage drop with $I_T \times R_X$.
2. The voltage drops previously calculated would not change since the ratio of the series resistor to the series equivalent resistors remains the same, and therefore the voltage division will remain the same.

STEPS 7-4, 7-5, AND 7-6

1. Find total resistance; find total current; find the voltage across each series and parallel combination resistors; find the current through each branch of parallel resistors; find the total and individual power dissipated.
2. This is a do-it-yourself question; each answer will vary.

STEP 7-7

1. When a load resistance changes the circuit and lowers output voltage

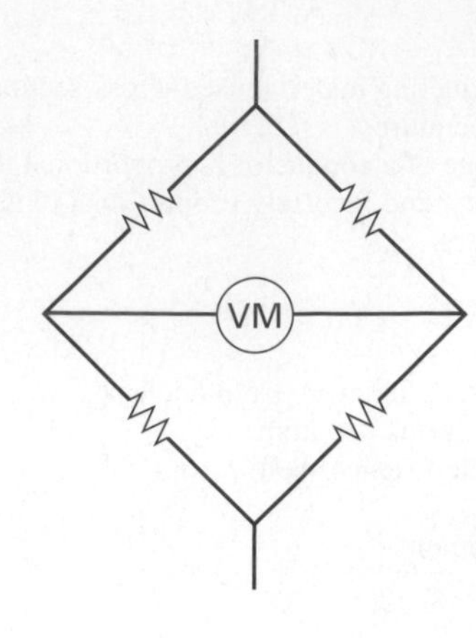

Figure 7-7, 2

2. Used to check for an unknown resistor's resistance
3. *R*
4. Current divider for a digital-to-analog converter

STEP 7-8

1. If component is in series with circuit, no current will flow and source voltage will be across bad component. If component is parallel, no current will flow in that branch.
2. Total resistance will decrease and bad component will have 0 V dropped across it.
3. Will cause the circuit's behavior to vary

STEP 7-9

1. Small, large
2. Voltage source
3. Voltage, resistor
4. Current

STEP 8-1

1. $C = \sqrt{A^2 + B^2} = \sqrt{15^2 + 22^2} = \sqrt{225 + 484} = \sqrt{709} = 26.63\text{ m}$

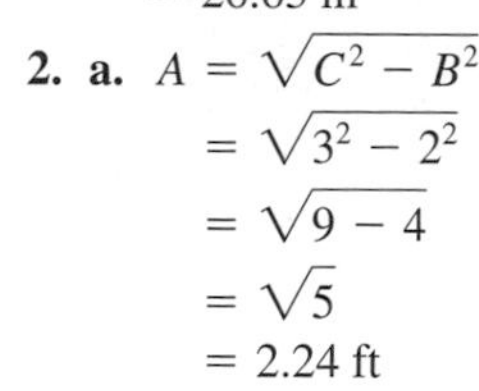

2. a. $A = \sqrt{C^2 - B^2} = \sqrt{3^2 - 2^2} = \sqrt{9 - 4} = \sqrt{5} = 2.24\text{ ft}$

   b. $B = \sqrt{C^2 - A^2} = \sqrt{160^2 - 80^2} = \sqrt{25{,}000 - 6400} = \sqrt{19{,}200} = 138.56\text{ km}$

   c. $C = \sqrt{A^2 + B^2} = \sqrt{112^2 + 25^2} = \sqrt{12{,}544 + 625} = \sqrt{13{,}169} = 114.76\text{ mm}$

3\. a. $(O) = 35$ mm $\theta = 36°$
$(H) = ?$
↓

SOH or $\sin\theta = \dfrac{O}{H}$

$$\sin 36° = \frac{35\text{ mm}}{H}$$
$$0.59 = \frac{35}{H}$$
$$0.59 \times H = \frac{35}{\cancel{H}} \times \cancel{H} \quad (\times H)$$
$$\frac{0.59 \times H}{0.59} = \frac{35}{0.59} \quad (\div 0.59)$$
$$H = \frac{35}{0.59} = 59.32\text{ mm}$$

b. $(H) = 160$ km $\theta = 38°$
$(A) = ?$
↓

CAH or $\cos\theta = \dfrac{A}{H}$

$$\cos 38° = \frac{A}{160\text{ km}}$$
$$0.79 = \frac{A}{160}$$
$$0.79 \times 160 = \frac{A}{\cancel{160}} \times \cancel{160} \quad (\times 160)$$
$$0.79 \times 160 = A$$
$$A = 126.4\text{ km}$$

c. $(O) = 163$ cm $\theta = 72°$
$(A) = ?$
↓

TOA or $\tan\theta = \dfrac{O}{A}$

$$\tan 72° = \frac{163\text{ cm}}{A}$$
$$3.08 = \frac{163}{A}$$
$$3.08 \times A = \frac{163 \times \cancel{A}}{\cancel{A}} \quad (\times A)$$
$$\frac{3.08 \times A}{3.08} = \frac{163}{3.08} \quad (\div 3.08)$$
$$A = \frac{163}{3.08} = 52.92\text{ cm}$$

4\. a. $(H) = 120$ miles $\theta = ?$
$(O) = 38$ miles
↓

SOH or $\sin\theta = \dfrac{O}{H}$

$$\sin\theta = \frac{38\text{ miles}}{120\text{ miles}}$$
$$\sin\theta = 0.32$$
$$\cancel{\text{invsin}} \times \cancel{\sin}\,\theta = \text{invsin } 0.32 \quad (\times \text{ invsin})$$
$$\theta = \text{invsin } 0.32$$
$$\theta = 18.66°$$

b. $(H) = 25$ ft $\theta = ?$
$(A) = 17$ ft
↓

CAH or $\cos\theta = \dfrac{A}{H}$

$$\cos\theta = \frac{17\text{ ft}}{25\text{ ft}}$$
$$\cos\theta = 0.68$$
$$\cancel{\text{invcos}} \times \cancel{\cos}\,\theta = \text{invcos } 0.68$$
$$\theta = \text{invcos } 0.68$$
$$\theta = 47.16°$$

c. $(A) = 69$ cm $\theta = ?$
$(O) = 51$ cm
↓

TOA or $\tan\theta = \dfrac{O}{A}$

$$\tan\theta = \frac{51\text{ cm}}{69\text{ cm}}$$
$$\tan\theta = 0.74$$
$$\cancel{\text{invtan}} \times \cancel{\tan}\,\theta = \text{invtan } 0.74$$
$$\theta = \text{invtan } 0.74$$
$$\theta = 36.5°$$

5\. a.
$$x = \sqrt{A^2 + B^2} = \sqrt{40^2 + 30^2} = \sqrt{1600 + 900} = \sqrt{2500} = 50\text{ V}$$

b.
$$x = \sqrt{A^2 + B^2} = \sqrt{75^2 + 26^2} = \sqrt{5625 + 676} = \sqrt{6301} = 79.38\text{ W}$$

c.
$$x = \sqrt{A^2 + B^2} = \sqrt{93^2 + 36^2} = \sqrt{8649 + 1296} = \sqrt{9945} = 99.72\text{ mm}$$
$$y = \sqrt{A^2 + B^2} = \sqrt{48^2 + 96^2} = \sqrt{2304 + 9216} = \sqrt{11{,}520} = 107.33\text{ mm}$$

STEP 8-2

1. **a.** Alternating current
   b. Direct current
2. DC; current only flows in one direction
3. AC
4. DC
5. Power transfer, information transfer
6. DC flows in one direction, whereas ac first flows in one direction and then in the opposite direction.

STEP 8-3-1

1. AC generators can be larger, less complex, and cheaper to run; transformers can be used with ac to step up/down, so low-current power lines can be used; easy to change ac to dc, but hard the other way around.
2. False
3. $P = I^2 \times R$
4. A device that can step up or down ac voltages
5. 120 V ac
6. AC, DC

STEP 8-3-2

1. The property of a signal or message that conveys something meaningful to the recipient; the transfer of information between two points
2. Sound, electromagnetic, electrical
3. **a.** Sound wave
   b. Electromagnetic wave
   c. Electrical wave
4. 1133 feet per second; 186,000 miles per second
5. Sound, electromagnetic, electrical
6. **a.** Microphone
   b. Speaker
   c. Antenna
   d. Human ear
7. Electronic
8. Electrical

STEP 8-4

1.

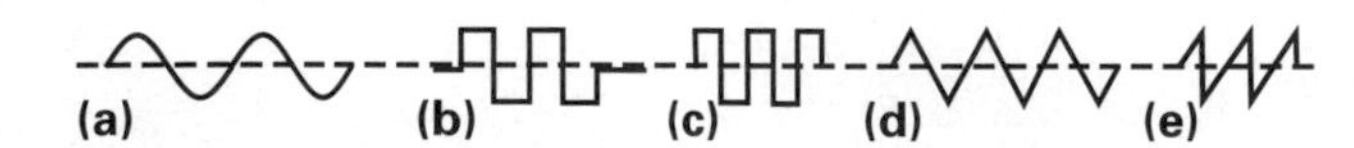

2. Time, frequency
3. (Sound wave); $\lambda(\text{mm}) = \dfrac{344.4 \text{ m/s}}{f(\text{Hz})}$;

   (Electromagnetic wave): $\lambda(\text{m}) = \dfrac{3 \times 10^8 \text{ m/s}}{f(\text{Hz})}$;

   different because sound waves travel at a different speed than do electromagnetic waves.
4. Odd

STEP 8-5-1

1. True; no
2. RF probe is used to measure high-frequency electrical waves and uses a special high-frequency rectifier. HV probe is used to measure high voltages and uses multiplier resistors.

STEP 8-5-2

1. Oscilloscope
2. Cathode-ray tube (CRT)
3. $t = 80\ \mu s$
   $f = 12.5$ kHz
4. $V_p = 4$ V
   $V_{p\text{-}p} = 8$ V
5. Waveforms can be compared.

STEP 8-5-3

1. Sine, square, triangular
2. Measure the frequency of a periodic wave

STEP 8-5-4

1. A handheld test instrument that combines an oscilloscope and multimeter
2. Portability, ease of setup, and ease of use

STEP 9-1

1. True
2. None
3. The farad
4. Capacitance, C (farads) = charge, Q (coulombs)/voltage, V (volts)
5. $30{,}000 \times 10^{-6} = 0.03 \times 10^0 = 0.03$ F
6. $C = Q/V = 17.5/9 = 1.94$ F
7. $C = \dfrac{(8.85 \times 10^{-12}) \times K \times A}{d}$
8. $C_T = C_1 + C_2 + C_3 = 7 \text{ pF} + 2 \text{ pF} + 14 \text{ pF} = 23 \text{ pF}$
9. $V_{CX} = (C_T/C_X) \times V_T$
10. True

STEP 9-2

1. Mica, ceramic, paper, plastic, electrolytic
2. Electrolytic
3. 470 pF, 2% tolerance
4. 0.47 μF or 470 nF, 5% tolerance
5. 0.5 μF
6. Capacitance meter or analyzer

STEP 9-3

1. The time it takes a capacitor to charge to 63.2%
2. 63.2%
3. 36.8%
4. False

STEP 9-4

1. Capacitive opposition is low for ac (short) and high for dc (open).
2. True
3. It means that the voltage and current maximums and minimums occur at different times.
4. False
5. True
6. False
7. Opposition to current flow without the dissipation of energy
8. $X_C = 1/2\pi fC$
9. When frequency or capacitance goes up, there is more charge and discharge current, so X_C is lower.
10. $X_C = 1/2\pi fC = 1/2\pi \times 4 \text{ kHz} \times 4\ \mu\text{F} = 9.95\ \Omega$
11. Current leads voltage by some phase angle less than 90°.
12. An arrangement of vectors to illustrate the magnitude and phase relationships between two or more quantities of the same frequency
13. Z = total opposition to current flow; $Z = \sqrt{R^2 + X_C^2}$
14. **a.** 0°
    b. 90°
    c. Between 0 and 90°
15. Resistor current is in phase with voltage; capacitor current is 90° out of phase (leading) with voltage.
16. No
17. **a.** $I_T = \sqrt{I_R^2 + I_C^2}$
    b. $Z = V_S/I_T$;

    also $Z = (R \times X_C)/\sqrt{R^2 + X_C^2}$
18. Lead

STEP 9-5

1. **a.** Filter
   b. Voltage divider
   c. Differentiator
2. Capacitor, resistor
3. Long
4. Differentiator

STEP 10-1

1. Coil
2. True
3. DC
4. False
5. Used to determine the magnetic polarity; placing your left-hand fingers in the direction of the current, your thumb points in the direction of the north magnetic pole.
6. AC
7. Number of lines of force (or maxwells) in webers
8. Number of magnetic lines of flux per square meter
9. The magnetic pressure that produces the magnetic field
10. Magnetomotive force divided by length of coil
11. Opposition or resistance to the establishment of a magnetic field
12. A measure of how easily a material will allow a magnetic field to be set up within it
13. True
14. The state beyond which an electromagnet is incapable of further magnetic strength

STEP 10-2

1. The voltage induced or produced in a coil as the magnetic lines of force link with the turns of a coil
2. Faraday's: When the magnetic flux linking a coil is changing, an emf is induced.
   Lenz's: The current induced in a coil due to the change in the magnetic flux is such as to oppose the cause producing it.
3. Sine wave
4. Because it converts sound waves of air pressure (acoustical) to electrical waves (electro)

STEP 11-1

1. When the current-carrying coil of a conductor induces a voltage within itself
2. The induced voltage, which opposes the applied emf; $V_{\text{ind}} = L \times (\Delta i/\Delta t)$
3. $V_{\text{ind}} = L \times (\Delta i/\Delta t) = 2 \text{ mH} \times 4 \text{ kA/s} = 8 \text{ V}$
4. Different applications: An electromagnet is used to generate a magnetic field; an inductor is used to oppose any changes of circuit current.
5. True
6. Number of turns, area of coil, length of coil, core material used
7. $L = \dfrac{N^2 \times A \times \mu}{l}$
8. False
9. **a.** $L_T = L_1 + L_2 + L_3 + \cdots$
   b. $L_T = 1/(1/L_1) + (1/L_2) + (1/L_3) + \cdots$
10. **a.** $L_T = L_1 + L_2 = 4 \text{ mH} + 2 \text{ mH} = 6 \text{ mH}$
    b. $L_T = \dfrac{L_1 \times L_2}{L_1 + L_2}$ (using product over sum)
    $= \dfrac{4 \text{ mH} \times 2 \text{ mH}}{4 \text{ mH} + 2 \text{ mH}} = \dfrac{8 \text{ mH}^2}{6 \text{ mH}} = 1.33 \text{ mH}$

STEP 11-2

1. Air core, iron core, ferrite core
2. Air core
3. Chemical compound, basically powdered iron oxide and ceramic
4. Toroidal-type inductors have a greater inductance.
5. Ferrite-core variable inductor
6. Core permeability
7. **a.** With an ohmmeter; resistance will be infinite instead of low.
   b. With an inductor analyzer to check its inductance value
8. An open

STEP 11-3

1. Current in an inductive circuit builds up in the same way that voltage does in a capacitive circuit, but the capacitive time constant is proportional to resistance, whereas the inductive time constant is inversely proportional to resistance.
2. True
3. False

STEP 11-4

1. The opposition to current flow offered by an inductor without the dissipation of energy: $X_L = 2 \times \pi \times f \times L$
2. The higher the frequency (and therefore the faster the change in current) or the larger the inductance of the inductor, the larger the magnetic field created, the larger the counter emf will be to oppose applied emf; so the inductive reactance (opposition) will be large also.
3. Inductive reactance is an opposition, so it can be used in Ohm's law in place of resistance.
4. False
5. False
6. $V_S = \sqrt{V_R^2 + V_L^2} = \sqrt{4^2 + 2^2} = \sqrt{16 + 4} = \sqrt{20}$
   $= 4.47 \text{ V}$
7. The total opposition to current flow offered by a circuit with both resistance and reactance: $Z = \sqrt{R^2 + X_L^2}$
8. $+45°$
9. Power consumption above the zero line, caused by positive current and voltage or negative current and voltage
10. Quality factor of an inductor that is the ratio of the energy stored in the coil by its inductance to the energy dissipated in the coil by the resistance: $Q = X_L/R$
11. True power is energy dissipated and lost by resistance; reactive power is energy consumed and then returned by a reactive device.
12. PF = true power (P_R)/apparent power (P_A) or PF = R/Z or PF = $\cos \theta$
13. False
14. $I_T = \sqrt{I_R^2 + I_L^2}$

STEP 11-5

1. *RL* integrator, *RL* differentiator, *RL* filter
2. Can be made to function as either a high- or a low-pass filter
3. The integrator will approximate a dc level; the differentiator will output spikes of current.

STEP 12-1

1. Mutual inductance is the process by which an inductor induces a voltage in another inductor, whereas self-inductance is the process by which a coil induces a voltage within itself.

2. False
3. True
4. True
5. Primary and secondary
6. True
7. False
8. If both primary and secondary are wound in the same direction, output is in phase with input; but if primary and secondary are wound in different directions, the output will be 180° out of phase with the input.
9. When there is a positive on the primary dot, there will be a positive on the secondary dot. If there is a negative on the primary dot, there will be a negative on the secondary dot.
10. It is the ratio of the number of magnetic lines of force that cut the secondary compared to the total number of magnetic flux lines being produced by the primary: k = flux linking secondary coil/total flux produced by primary.
11. True
12. How close together the primary and secondary are to one another and the type of core material used
13. Copper losses, hysteresis, eddy-current loss, magnetic leakage
14. Eddy currents cannot cross the laminated parts of the core, so reduced eddy currents mean reduced opposition to main flux and reduced heat, resulting in a reduced loss.

STEP 12-2

1. Turns ratio $= N_s/N_p = 1608/402 = 4$; step up
2. $V_s = N_s/N_p \times V_p$
3. False
4. $I_s = N_p/N_s \times I_p$
5. Turns ratio $= \sqrt{Z_L/Z_S} = \sqrt{75\ \Omega/25\ \Omega} = \sqrt{3} = 1.732$
6. $V_s = (N_s/N_p) \times V_p = (200/112) \times 115 = 205.4$ V

STEP 12-3

1. Air core, iron or ferrite core
2. Center-tapped secondary, multiple-tapped secondary, multiple winding, single winding
3. False
4. To be able to switch between two primary voltages and obtain the same secondary voltage
5. Automobile ignition system
6. Smaller, cheaper, lighter-than-normal, separated primary/secondary transformer types
7. 10 kVA is the apparent power rating, 200 V the maximum primary voltage, 100 V the maximum secondary voltage, at 60 cycles per second (Hz).
8. $R_L = V_s/I_s = 1$ kV/8 A $= 125\ \Omega$; $125 > 100$, so the transformer will overheat and possibly burn out.
9. An ohmmeter
10. Probably not since 50 Ω is a normal coil resistance

STEP 13-1

1. Calculate the inductive and capacitive reactance (X_L and X_C), the circuit impedance (Z), the circuit current (I), the component voltage drops (V_R, V_L, and V_C), and the power distribution and power factor (PF).
2. **a.** $Z = \sqrt{R^2 + (X_L \sim X_C)^2}$
   b. $I = V_s/Z$
   c. Apparent power $= V_s \times I$ (volt-amperes)
   d. $V_S = \sqrt{V_R^2 + (V_L \sim V_C)^2}$
   e. True power $= I^2 \times R$(watts)
   f. $V_R = I \times R$
   g. $V_L = I \times X_L$
   h. $V_C = I \times X_C$
   i. $\theta = \arctan \dfrac{V_L \sim V_C}{V_R}$
   j. PF $= \cos \theta$

STEP 13-2

1. **a.** $I_R = V/R$ **c.** $I_C = V/X_C$
   b. $I_T = \sqrt{I_R^2 + I_X^2}$ **d.** $I_L = V/X_L$
2. **a.** $P_R = I^2 \times R$ **c.** $P_A = V_S \times I_T$
   b. $P_X = I^2 \times X_L$ **d.** PF $= \cos \theta$

STEP 13-3-1

1. A circuit condition that occurs when the inductive reactance (X_L) and the capacitive reactance (X_C) have been balanced
2. A series *RLC* circuit that at resonance X_L equals X_C, so V_L and V_C will cancel, and $Z = R$
3. Voltage across L and C will measure 0; impedance only equals R; voltage drops across inductor or capacitor can be higher than source voltage.
4. Q factor indicates the quality of the series resonant circuit, or is the ratio of the reactance to the resistance.
5. Group or band of frequencies that causes the larger current flow
6. BW $= f_0/Q = 12$ kHz/1000 $= 12$ Hz

STEP 13-3-2

1. In a series resonant *RLC* circuit, source current is maximum; in a parallel resonant circuit, source current is minimum at resonance.
2. Oscillating effect with continual energy transfer between capacitor and inductor
3. $Q = X_L/R = 50\ \Omega/25\ \Omega = 2$
4. Yes
5. Ability of a tuned circuit to respond to a desired frequency and ignore all others

STEP 13-4

1. **a.** High-pass **c.** Band-stop
   b. Low-pass **d.** Bandpass
2. Television, radio, and other communications equipment

STEP 13-5

1. Real, imaginary
2. Magnitude $= \sqrt{5^2 + 6^2} = 7.81$
   Angle $= \arctan (6/5) = 50.2°$
   Rectangular number $65 + j6$ = polar number $7.81 \angle 50.2°$
3. Real number $= 33 \cos 25° = 29.9$
   Imaginary number $= 33 \sin 25° = 13.9$
   Polar number $33 \angle 25°$ = rectangular number $29.9 + j13.9$
4. Combination of a real and imaginary number

APPENDIX

Answers to Odd-Numbered Problems

CHAPTER 1

1. a	**7.** c	**13.** b	**17.** a
3. d	**9.** b	**15.** a	**19.** d
5. a	**11.** d		

(The anwsers to Communication Skill Questions 21 through 35 can be found in the sections indicated that follow the questions.)

37. a. $\sqrt{81} = 9$
b. $\sqrt{4} = 2$
c. $\sqrt{0} = 0$
d. $\sqrt{36} = 6$
e. $\sqrt{144} = 12$
f. $\sqrt{1} = 1$

39. a. $\sqrt[3]{343} = 7$
b. $\sqrt{1296} = 36$
c. $\sqrt[4]{760} = 5.25$
d. $\sqrt[6]{46{,}656} = 6$

41. a. $10^{-4} = \frac{1}{10{,}000}$ or 0.0001
b. $10^{-2} = \frac{1}{100}$ or 0.01
c. $10^{-6} = \frac{1}{1{,}000{,}000}$ or 0.000001
d. $10^{-3} = \frac{1}{1000}$ or 0.001

43. a. 0.005 ampere = 0∘005. = 5×10^{-3} ampere = 5 milliamperes
b. 8000 meters = 8.000∘ = 8×10^{3} meters = 8 kilometers
c. 15,000,000 ohms = 15,000000∘ = 15×10^{6} ohms = 15 megaohms
d. 0.000016 second = 0∘000016. = 16×10^{-6} second = 16 microseconds.

45. $I = Q/t = 3/4 = 0.75$ amps or 750 milliamps

47. a. 1.473 V **c.** 139 V
b. 7.143 kV **d.** 390 kV

49. 0.47 mA

CHAPTER 2

1. a	**5.** a	**9.** c	**13.** d	**17.** d
3. c	**7.** b	**11.** b	**15.** b	**19.** a

(The answers to Communication Skill Questions 21 through 35 can be found in the sections indicated that follow the questions.)

37. a.
$$4x = 11$$
$$4 \times x = 11$$
$$\frac{\cancel{4} \times x}{\cancel{4}} = \frac{11}{4} \quad (\div 4)$$
$$x = \frac{11}{4}$$
$$x = 2.75$$

b.
$$6a + 4a = 70$$
$$10a = 70$$
$$10 \times a = 70$$
$$\frac{\cancel{10} \times a}{\cancel{10}} = \frac{70}{10} \quad (\div 10)$$
$$a = \frac{70}{10}$$
$$a = 7$$

c.
$$5b - 4b = \frac{7.5}{1.25}$$
$$1b = \frac{7.5}{1.25} \quad (1b = 1 \times b = b)$$
$$b = \frac{7.5}{1.25}$$
$$b = 6$$

d.
$$\frac{2z \times 3z}{4.5} = 2z \qquad \left[\begin{aligned} 2z \times 3z &= 2 \times z \times 3 \times z \\ &= (2 \times 3) \times (z \times z) \\ &= 6 \times z^2 \\ &= 6z^2 \end{aligned}\right]$$
$$\frac{6z^2}{4.5} = 2z$$
$$\frac{6z^2}{\cancel{4.5}} \times \cancel{4.5} = 2z \times 4.5 \quad (\times 4.5)$$
$$6z^2 = 2z \times 4.5$$
$$\frac{\cancel{6z} \times \cancel{z} \times z}{2 \times \cancel{z}} = \frac{(2 \times z) \times 4.5}{2 \times z} \quad (\div 2z)$$
$$3z = 4.5$$
$$\frac{\cancel{3} \times z}{\cancel{3}} = \frac{4.5}{3} \quad (\div 3)$$
$$z = \frac{4.5}{3}$$
$$z = 1.5$$

39. a. Power (P) = voltage (V) × current (I)
1500 watts = 120 volts × ?
Transpose formula to solve for I.
$$P = V \times I$$
$$\frac{P}{V} = \frac{\cancel{V} \times I}{\cancel{V}} \quad (\div V)$$
$$\frac{P}{V} = I$$
$$I = \frac{P}{V}$$
$$\text{Current } (I) \text{ in amperes} = \frac{\text{power } (P) \text{ in watts}}{\text{voltage } (V) \text{ in volts}}$$
$$I = \frac{1500 \text{ W}}{120 \text{ V}}$$
Current (I) = 12.5 amperes

b. Voltage (V) in volts = current (I) in amperes × resistance (R) in ohms

$$120 \text{ volts} = 12.5 \text{ amperes} \times ?$$

Transpose formula to solve for R.

$$V = I \times R$$

$$\frac{V}{I} = \frac{\not{I} \times R}{\not{I}}$$

$$\frac{V}{I} = R$$

$$R = \frac{V}{I}$$

$$\text{Resistance } (R) \text{ in ohms} = \frac{\text{voltage } (V) \text{ in volts}}{\text{current } (I) \text{ in amperes}}$$

$$R = \frac{120 \text{ V}}{12.5 \text{ A}}$$

$$\text{Resistance } (R) = 9.6 \text{ ohms}$$

41. $V = I \times R = 8 \text{ mA} \times 16 \text{ k}\Omega = 128 \text{ V}$

43. $P = I \times V$, 2.4 kW, 1.024 W, 1.2 kW

45. Total cost = power in (kW) × time × cost per hour
- **a.** 0.3 × 10 × 9 = 27 cents
- **b.** 0.1 × 10 × 9 = 9 cents
- **c.** 0.06 × 10 × 9 = 5 cents
- **d.** 0.025 × 10 × 9 = 2 cents

47. $R = V/I = 120 \text{ V}/500 \text{ mA} = 240\ \Omega$
$P = I \times V = 500 \text{ mA} \times 120 \text{ V} = 60 \text{ W}$

49. $P = V \times I$
- **a.** 120 V × 20 mA = 2.4 W
- **b.** 12 V × 2 A = 24 W
- **c.** 9 V × 100 μA = 900 μW
- **d.** 1.5 V × 4 mA = 6 mW

51. **a.** 1 kW **b.** 345 mW **c.** 1.25 MW **d.** 1250 μW

53. **a.** 7500 = 7.5 kW
Energy Consumed = 7.5 kW × 1hr = 7.5 kWh
b. 25 W = 0.025 kW
Energy Consumed = 0.025 kW × 6 hrs = 0.15 kWh
c. 127,000 = 127 kW
Energy Consumed = 127 kW × 0.5 hr = 63.5 kWh

55. (see following)

57. $G = 1/100\ \Omega = 0.01$ or 10 millisiemens

59. 13

61. 0.945/1 = 0.945

63. 10 V/750 kV = 0.0000133 cm or 0.000133 mm

65. 35 × (2000 kV/cm ÷ 10) = 7000 kV

67. **a.** $P = I^2 \times R = 50 \text{ mA}^2 \times 10 \text{ k}\Omega = 25 \text{ W}$; **b.** no

69. In tolerance.

CHAPTER 3

1. b	**7.** d	**13.** a	**19.** a	**23.** b
3. b	**9.** d	**15.** a	**21.** c	**25.** b
5. d	**11.** c	**17.** a		

(The answers to Communication Skill Questions 26 through 50 can be found in the sections indicated that follow the questions.)

51. **a.** $\frac{20 \text{ ft}}{5 \text{ ft}}$ — Reduce to lowest terms

$$\frac{20 \div 5}{5 \div 5} = \frac{4}{1}$$

The ratio of 20 ft to 5 ft is 4 to 1 (4:1).

b. Both quantities must be alike, so we must first convert minutes to seconds.

$$2\frac{1}{2} \text{ min} = 2\frac{1}{2} \times 60 \text{ s} = 150 \text{ s}$$

$$\frac{150 \text{ s} \div}{30 \text{ s} \div} = \frac{5}{1}$$

The ratio of $2\frac{1}{2}$ min to 30 s is 5 to 1 (5:1).

53. 10.9

55. **a.** 6
- **b.** 17.25 or $17\frac{1}{4}$ s
- **c.** 116 V
- **d.** 505.75, or $505\frac{3}{4}\ \Omega$
- **e.** 155 m

CHAPTER 4

1. b	**7.** d	**13.** b	**19.** a	**23.** c
3. d	**9.** a	**15.** b	**21.** c	**25.** b
5. c	**11.** c	**17.** a		

(The answers to Communication Skill Questions 26 through 45 can be found in the sections indicated that follow the questions.)

47.

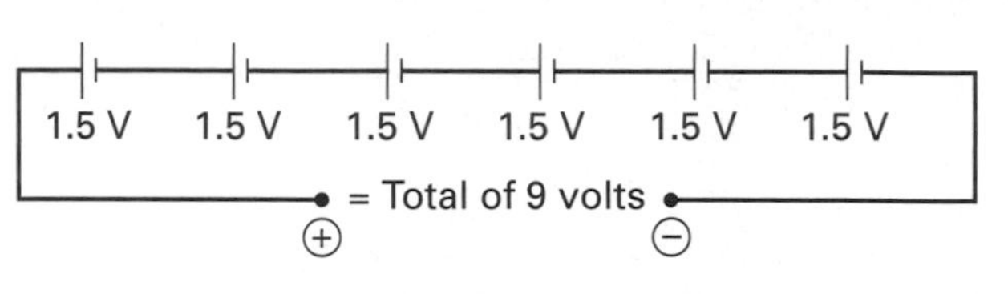

(a)

1.5 V 1.5 V 1.5 V 1.5 V 1.5 V 1.5 V Total = 1.5 V

(b)

49.

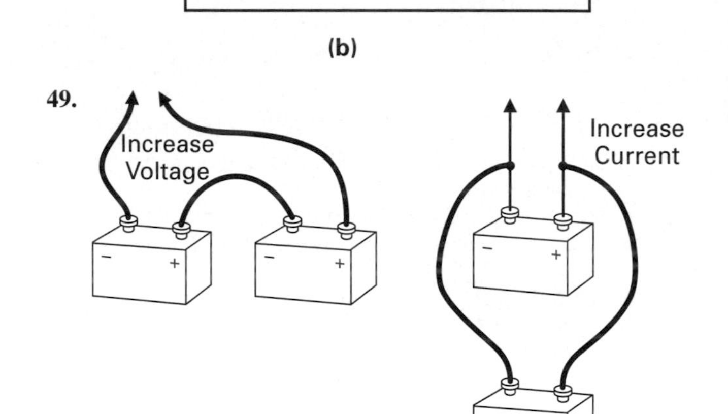

CHAPTER 5

1. d	**5.** a	**9.** a	**13.** d
3. c	**7.** d	**11.** c	**15.** d

(The answers to Communication Skill Questions 16 through 25 can be found in the sections indicated that follow the questions.)

27. $I = \frac{V_S}{R_T}$, $R_T = R_1 + R_2 = 40 + 35 = 75\ \Omega$
$I = 24/75 = 320$ mA; 150 Ω (double 75 Ω) needed to halve current

29. 40 Ω, 20 Ω, 60 Ω. (Any values can be used, as long as the ratio remains the same.)

31. $I_{R_1} = I_T = 6.5 \text{ mA}$

33. $P_T = P_1 + P_2 + P_3 = 120 + 60 + 200 = 380$ W; $I_T = P_T/V_S = 380/120 = 3.17$ A
$V_1 = P_1/I_T = 120$ W/3.17 A = 38 V; $V_2 = P_2/I_T = 60$ W/3.17 A = 18.9 V
$V_3 = P_3/I_T = 200$ W/3.17 A = 63.1 V

35. a. $R_T = R_1 + R_2 + R_3 = 22\text{ k}\Omega + 3.7\text{ k}\Omega + 18\text{ k}\Omega = 43.7\text{ k}\Omega$
$I = V/R = 12\text{ V}/43.7\text{ k}\Omega = 274.6\ \mu\text{A}$

b. $R_T = V/I = 12\text{ V}/10\text{ mA} = 1.2\text{ k}\Omega$
$P_T = V \times I = 12\text{ V} \times 10\text{ mA} = 120\text{ mW}$

c. $R_T = R_1 + R_2 + R_3 + R_4 = 5 + 10 + 6 + 4 = 25\ \Omega$; $V_S = I \times R_T = 100\text{ mA} \times 25\ \Omega = 2.5\text{ V}$
$V_{R_1} = I \times R_1 = 100\text{ mA} \times 5 = 500\text{ mV}$, $V_{R_2} = I \times R_2 = 100\text{ mA} \times 10 = 1\text{ V}$, $V_{R_3} = I \times R_3 = 100\text{ mA} \times 6 = 600\text{ mV}$
$V_{R_4} = I \times R_4 = 100\text{ mA} \times 4 = 400\text{ mV}$, $P_1 = I \times V_1 = 100\text{ mA} \times 500\text{ mV} = 50\text{ mW}$
$P_2 = I \times V_2 = 100\text{ mA} \times 1\text{ V} = 100\text{ mW}$, $P_3 = I \times V_3 = 100\text{ mA} \times 600\text{ mV} = 60\text{ mW}$
$P_4 = I \times V_4 = 100\text{ mA} \times 400\text{ mV} = 40\text{ mW}$

d. $P_T = P_1 + P_2 + P_3 + P_4 = 12\text{ mW} + 7\text{ mW} + 16\text{ mW} + 3\text{ mW} = 38\text{ mW}$
$I = P_T/V_S = 38\text{ mW}/12.5\text{ V} = 3.04\text{ mA}$
$R_1 = P_1/I^2 = 12\text{ mW}/(3.04\text{ mA})^2 = 1.3\text{ k}\Omega$
$R_2 = P_2/I^2 = 7\text{ mW}/(3.04\text{ mA})^2 = 757.4\ \Omega$
$R_3 = P_3/I^2 = 16\text{ mW}/(3.04\text{ mA})^2 = 1.73\text{ k}\Omega$
$R_4 = P_4/I^2 = 3\text{ mW}/(3.04\text{ mA})^2 = 324.6\ \Omega$

37. Zero voltage drop across the shorted component, while there is also an increase in voltage across the others.

39. a. No current at all (zero)
b. Go to infinity (∞)
c. Measure source voltage
d. No voltage across any other component

CHAPTER 6

1. b **3.** d **5.** c **7.** b **9.** a

(The answers to Communication Skill Questions 11 through 20 can be found in the sections indicated that follow the questions.)

21. $R_T = R/\text{no. of } R\text{'s} = 30\text{ k}\Omega/4 = 7.5\text{ k}\Omega$

23. $R_T = R/\text{no. of } R\text{'s} = 25/3 = 8.33\ \Omega$, $I_T = V_S R_T = 10/8.33 = 1.2\text{A}$
$I_1 = I_2 = I_3 = R_T/R_X \times I_T = 8.33/25 \times 1.2 = I_T/\text{no. of } R\text{'s} = 1.2\text{ A}/3 = 400\text{ mA}$

25. $I_T = V_S/R_T = 14/700 = 20\text{ mA}$; $I_X = I_T/\text{no. of } R = 20\text{ mA}/3 = 6.67\text{ mA}$

27. a. $R_T = \dfrac{R_1 \times R_2}{R_1 + R_2} = \dfrac{33 \times 22}{33 + 22} = \dfrac{726\text{ k}\Omega}{55\text{ k}\Omega} = 13.2\text{ k}\Omega$

b. $I_T = V_S/R_T = 20/13.2\text{ k}\Omega = 1.5\text{ mA}$

c. $I_1 = R_T/R_1 \times I_T = (13.2\text{ k}\Omega/33\text{ k}\Omega) \times 1.5\text{ mA} = 600\ \mu\text{A}$
$I_2 = R_T/R_2 \times I_T = 13.2\text{ k}\Omega/22\text{ k}\Omega \times 1.5\text{ mA} = 900\ \mu\text{A}$

d. $P_T = I_T \times V_S = 1.5\text{ mA} \times 20 = 30\text{ mW}$

e. $P_1 = I_1 \times V_1$, $(V_1 = V_S = 20\text{ V})$, $600\ \mu\text{A} \times 20 = 12\text{ mW}$
$P_2 = I_2 \times V_2$, $(V_2 = V_S = 20\text{ V})$, $900\ \mu\text{A} \times 20 = 18\text{ mW}$

29. a. $R_T = \dfrac{R_1 \times R_2}{R_1 + R_2} = \dfrac{22\text{ k}\Omega \times 33\text{ k}\Omega}{22\text{ k}\Omega + 33\text{ k}\Omega} = \dfrac{726\text{ k}\Omega}{55\text{ k}\Omega} = 13.2\text{ k}\Omega$

$$I_T = \frac{V_S}{R_T} = \frac{10}{13.2\text{ k}\Omega} = 757.6\ \mu\text{A}$$

$$I_1 = \frac{R_T}{R_1} \times I_T = \frac{13.2\text{ k}\Omega}{22\text{ k}\Omega} \times 757.6\ \mu\text{A} = 454.56\ \mu\text{A}$$

$$I_2 = \frac{R_T}{R_1} \times I_T = \frac{13.2\text{ k}\Omega}{33\text{ k}\Omega} \times 757.6\ \mu\text{A} = 303.04\ \mu\text{A}$$

b. $$R_T = \frac{1}{(1/R_1) + (1/R_2) + (1/R_3)} = \frac{1}{(1/220\ \Omega) + (1/330\ \Omega) + (1/470\ \Omega)} = 103\ \Omega$$

$I_T = V_S/R_T = 10/103 = 97\text{mA}$, $I_1 = R_T/R_1 \times I_T = 103/220 \times 97\text{ mA} = 45.4\text{ mA}$
$I_2 = (R_T/R_2) \times I_T = (103/330) \times 97\text{ mA} = 30.3\text{ mA}$
$I_3 = (R_T/R_2) \times I_T = (103/470) \times 97\text{ mA} = 21.3\text{ mA}$

31. a. $G_T = \dfrac{1}{R_1} + \dfrac{1}{R_2} + \dfrac{1}{R_3} = \dfrac{1}{5} + \dfrac{1}{5} + \dfrac{1}{5} = 0.6\text{ S}$,

$R_T = \dfrac{1}{G} = \dfrac{1}{.06} = 1.67\ \Omega$

b. $G_T = \dfrac{1}{R_1} + \dfrac{1}{R_2} = \dfrac{1}{200} + \dfrac{1}{200} = 10\text{ mS}$,

$R_T = \dfrac{1}{G} = \dfrac{1}{10\text{ mS}} = 100\ \Omega$

c. $G_T = \dfrac{1}{R_1} + \dfrac{1}{R_2} + \dfrac{1}{R_3} = \dfrac{1}{1\text{ M}\Omega} + \dfrac{1}{500\text{ M}\Omega} + \dfrac{1}{3.3\text{ M}\Omega}$

$= 1.305\ \mu\text{S}$, $R_T = \dfrac{1}{\text{G}} = \dfrac{1}{1.305\ \mu\text{S}} = 766.3\text{ k}\Omega$

d. $G_T = \dfrac{1}{R_1} + \dfrac{1}{R_2} + \dfrac{1}{R_3} = \dfrac{1}{5} + \dfrac{1}{3} + \dfrac{1}{2} = 1.033\text{ S}$,

$R_T = \dfrac{1}{G} = \dfrac{1}{1.033} = 967.7\text{ m}\Omega$

33. a. $R_T = \dfrac{R_1 \times R_2}{R_1 + R_2} = \dfrac{15 \times 7}{15 + 7} = \dfrac{105}{22} = 4.77\ \Omega$

b. $$R_T = \frac{1}{(1/R_1) + (1/R_2) + (1/R_3)} = \frac{1}{(1/26\ \Omega) + (1/15\ \Omega) + (1/30\ \Omega)} = 7.22\ \Omega$$

c. $R_T = \dfrac{R_1 \times R_2}{R_1 + R_2} = \dfrac{5.6\text{ k}\Omega \times 2.2\text{ k}\Omega}{5.6\text{ k}\Omega + 2.2\text{ k}\Omega} = \dfrac{12.32\text{ M}\Omega}{7.8\text{ k}\Omega} = 1.58\ \Omega$

d. $$R_T = \frac{1}{(1/R_1) + (1/R_2) + (1/R_3) + (1/R_4) + (1/R_5)} = \frac{1}{(1/1\text{ M}\Omega) + (1/3\text{ M}\Omega) + (1/4.7\text{ M}\Omega) + (1/10\text{ M}\Omega) + (1/33\text{ M}\Omega)} = 596.5\text{ K}\Omega$$

35. a. $I_2 = I_T - I_1 - I_3 = 6\text{ mA} - 2\text{ mA} - 3.7\text{ mA} = 300\ \mu\text{A}$

b. $I_T = I_1 + I_2 + I_3 = 6\text{ A} + 4\text{ A} + 3\text{ A} = 13\text{ A}$

c. $R_T = \dfrac{R_1 \times R_2}{R_1 + R_2} = \dfrac{5.6\text{ M} \times 3.3\text{ M}}{5.6\text{ M} + 3.3\text{ M}} = \dfrac{18.48\text{ M}\Omega}{8.9\text{ M}\Omega} = 2.08\text{ M}\Omega$

$V_S = I_T \times R_T = 100\text{ mA} \times 2.08\text{ M}\Omega = 208\text{ kV}$

$$I_1 = \frac{R_T}{R_1} \times I_T = \frac{2.08\text{ M}\Omega}{5.6\text{ M}\Omega} \times 100\text{ mA} = 37\text{mA}$$

$$I_2 = \frac{R_T}{R_2} \times I_T = \frac{2.08\text{ m}\Omega}{3.3\text{ m}\Omega} \times 100\text{ mA} = 63\text{ mA}$$

d. $I_1 = V_{R_1}/R_1 = 2/200\text{ k}\Omega = 10\ \mu\text{A}$,
$I_2 = I_T - I_1 = 100\text{ mA} - 10\ \mu\text{A} = 99.99\text{ mA}$

$R_2 = \dfrac{V_{R_2}}{I_2} = \dfrac{2}{99.99\text{ mA}} = 20.002\ \Omega$,

$P_T = I_T \times V_S = 100\text{ mA} \times 2\text{ V} = 200\text{ mW}$

37. a

39. Total current would increase, and the branch current with the shorted resistor would increase to 20 V/1 Ω = 20 A.

CHAPTER 7

1. c	**7.** a	**13.** a	**17.** c
3. b	**9.** c	**15.** a	**19.** c
5. c	**11.** b		

(The answers to Communication Skill Questions 21 through 35 can be found in the sections indicated that follow the questions.)

37. $V_A = V_s = 100$ V, $V_B = V_S - V_R = 100 - 11.125 = 88.875$ V
$V_C = V_S - V_{R_1} - V_{R_2} = 44.3$, $V_D = V_{R_4} \times R_4 = 4.43$ mA $\times$ 2.5 kΩ $= 11.075$ V
$V_E = 0$ V

39. a. $V_{RL} = \dfrac{R_L}{R_T} \times V_S = \dfrac{R_L}{R_L + R_{int}} \times V_S = \dfrac{25}{25 + 15} \times 15 = 9.375$

b. $V_{RL} = \dfrac{R_L}{R_T} \times V_S = \dfrac{R_L}{R_L + R_{int}} \times V_S = \dfrac{2.5\text{ k}\Omega}{2.5\text{ k}\Omega + 15} \times 15 = 14.91$ V

c. $V_{RL} = \dfrac{R_L}{R_T} \times V_S = \dfrac{R_L}{R_L + R_{int}} \times V_S = \dfrac{2.5\text{ M}\Omega}{2.5\text{ M}\Omega + 15} \times 15 = 14.99991$ V

41. a. For V_1: $R_T = R_1 + R_{2,3}$, $R_{2,3} = \dfrac{2\text{ k}\Omega \times 6\text{ k}\Omega}{2\text{ k}\Omega + 6\text{ k}\Omega} = \dfrac{12\text{ M}\Omega}{8\text{ k}\Omega} = 1.5\text{ k}\Omega$, $R_T = 8\text{ k}\Omega + 1.5\text{ k}\Omega = 9.5\text{ k}\Omega$, $I_T = V_S/R_T = \dfrac{28}{9.5\text{ k}\Omega} = 2.95$ mA, $I_{R_2} = R_{2,3}/R_2 \times I_T = \dfrac{1.5\text{ k}\Omega}{2\text{ k}\Omega} \times 2.95\text{ mA} = 2.21$ mA

For V_2: $R_T = R_3 + R_{1,2}$, $R_{1,2} = \dfrac{8\text{ k}\Omega \times 2\text{ k}\Omega}{8\text{ k}\Omega + 2\text{ k}\Omega} = \dfrac{16\text{ M}\Omega}{10\text{ k}\Omega} = 1.6\text{ k}\Omega$, $R_T = 6\text{ k}\Omega + 1.6\text{ k}\Omega = 7.6\text{ k}\Omega$, $I_T = V_S/R_T = 20/7.6\text{ k}\Omega = 2.63$ mA, $I_{R_2} = R_{1,2}/R_2 \times I_T = 1.6\text{ k}\Omega/2\text{ k}\Omega \times 2.63\text{ mA} = 2.1$ mA, I_{R_2} total $= 2.21\text{ mA} + 2.1\text{ mA} = 4.31$ mA

b. For V_1: $R_T = R_1 + R_{2,3} + R_4$, $R_{2,3} = \dfrac{R_2 \times R_3}{R_2 + R_3} = \dfrac{15 \times 75}{15 + 75} = \dfrac{1125}{90} = 12.5\ \Omega = R_T = 10 + 12.5 + 5 = 27.5$, $I_T = V_S/R_T = 3.5/27.5 = 127$ mA, $I_{R_3} = (R_{2,3}/R_3) \times I_T = (12.5/75) \times 127\text{ mA} = 21$ mA

For V_2: $R_T = R_2 + \dfrac{1}{(1/R_{1,4}) + (1/R_3)}$, $R_{1,4} = R_1 + R_4 = 10\ \Omega + 5\ \Omega = 15\ \Omega$, $R_T = 15\ \Omega + \dfrac{1}{(1/15\ \Omega) + (1/75\ \Omega)} = 27.5\ \Omega$, $I_T = V_S/R_T = 1.5\text{ V}/27.5\ \Omega = 54.5$ mA, $I_{R_3} = (R_{1,3,4}/R_3) \times I_T = (12.5\ \Omega/75\ \Omega) \times 54.5\text{ mA} = 9$ mA, I_{R_3} Total $= 21\text{ mA} - 9\text{ mA} = 12$ mA

43. $V_{TH} = 5$ V, $R_{TH} = 3$ kΩ
$I_{RL} = V_{TH}/R_T = 5/(3\text{ k}\Omega + 1\text{ k}\Omega) = 5/4\text{ k}\Omega = 1.25$ mA
$I_N = I_{R_2} = V_S/R_2 = 5/3\text{ k}\Omega = 1.67$ mA, $R_N = 3$ kΩ

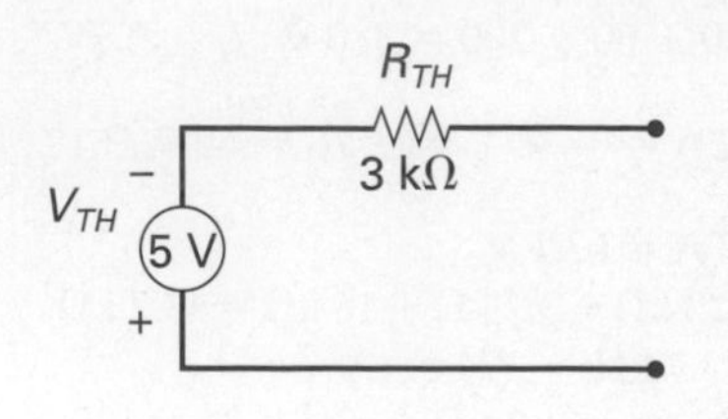

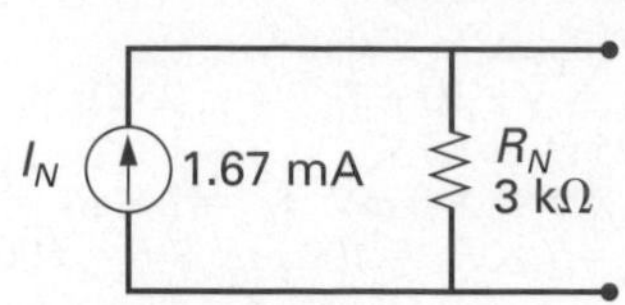

FIGURE Ch. 7, 43

45. a. $V = I \times R = 5\text{ mA} \times 5\text{ M}\Omega = 25$ kV
b. $V = I \times R = 10\text{ A} \times 10\text{ k}\Omega = 100$ kV
c. $V = I \times R = 0.0001\text{ A} \times 2.5\text{ k}\Omega = 250$ mV

47. This answer will vary with each person.

49. $V_{TH} = \dfrac{R_4}{R_T} \times V_S$, $R_T = R_{1,2} + R_4$, $R_{1,2} = \dfrac{R_1 \times R_2}{R_1 + R_2} = \dfrac{2 \times 3}{2 + 3} = \dfrac{6}{5} = 1.2\ \Omega$

$R_T = 1.2 + 7 = 8.2\ \Omega$, $V_{TH} = \dfrac{7}{8.2} \times 20 = 17.1$ V

$R_{TH} = \dfrac{1}{(1/R_1) + (1/R_2) + (1/R_4)} = \dfrac{1}{(1/2) + (1/3) + (1/7)} = 1.024\ \Omega$

a. If R_2 shorts, $V_{TH} = 20$ V, $R_{TH} = 0\ \Omega$

b. If R_2 opens, $V_{TH} = \dfrac{R_4}{R_T} \times V_S$, $R_T = R_1 + R_4 = 2 + 7 = 9\ \Omega$,

$V_{TH} = \dfrac{7}{9} \times 20 = 15.56$ V, $R_{TH} = \dfrac{R_1 \times R_4}{R_1 + R_4} = \dfrac{2 \times 7}{2 + 7} = \dfrac{14}{9} = 1.56\ \Omega$

CHAPTER 8

1. d	**7.** d	**13.** d	**19.** a	**25.** b
3. b	**9.** a	**15.** b	**21.** d	**27.** c
5. c	**11.** c	**17.** b	**23.** b	**29.** d

(The answers to Communication Skill Questions 31 through 50 can be found in the sections indicated that follow the questions.)

51. a. $C = \sqrt{A^2 + B^2} = \sqrt{20^2 + 53^2} = \sqrt{400 + 2809} = \sqrt{3209} = 56.65$ mi

b. $C = \sqrt{2^2 + 3^2} = 3.6$ km

c. $C = \sqrt{4^2 + 3^2} = 5$ inches

d. $C = \sqrt{12^2 + 12^2} = 16.97$ mm

53. a. $\sin 0° = 0$
b. $\sin 30° = 0.5$
c. $\sin 45° = 0.707$
d. $\sin 60° = 0.866$
e. $\sin 90° = 1.0$
f. $\cos 0° = 1.0$
g. $\cos 45° = 0.707$
h. $\cos 60° = 0.5$
i. $\cos 90° = 0$
j. $\tan 0° = 0$
k. $\tan 30° = 0.577$
l. $\tan 45° = 1.0$
m. $\tan 60° = 1.73$
n. $\tan 90° = \infty$ (infinity)

55. **a.**

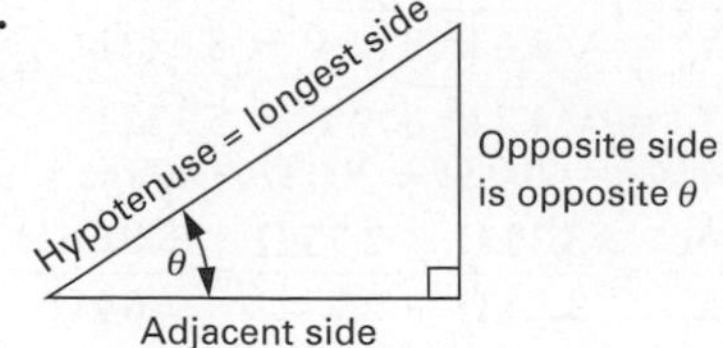

Known values	Unknown values
H = 27 miles	O = ?
θ = 37°	A = ?

We must calculate the length of the opposite and adjacent sides. To achieve this we can use either H or θ to calculate O (SOH), or H and θ to calculate A (CAH).

$$\sin\theta = \frac{O}{H}$$

$$\sin 37^\circ = \frac{O}{27}$$

$$0.6 = \frac{O}{27}$$

$$0.6 \times 27 = \frac{O}{\cancel{27}} \times \cancel{27}$$

$$O = 0.6 \times 27$$

Opposite = 16.2 mi

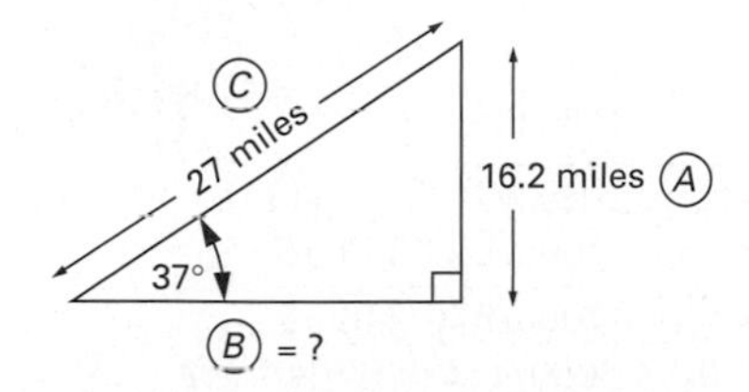

The next step is to use the Pythagorean theorem to calculate the length of the unknown side now that the length of two sides are known.

$$B = \sqrt{C^2 - A^2}$$
$$= \sqrt{27^2 - 16.2^2}$$
$$= \sqrt{729 - 262.44}$$
$$= \sqrt{466.56}$$

Adjacent or B = 21.6 mi

b.

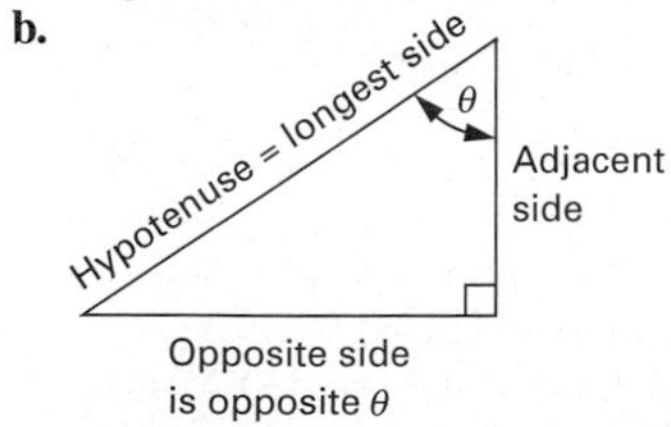

Known values	Unknown values
O = 29 cm	H = ?
T = 21°	A = ?

We can use either O and θ to calculate H (SOH), or O and θ to calculate A (TOA).

$$\tan\theta = \frac{O}{A}$$

$$\tan 21^\circ = \frac{29\text{ cm}}{A}$$

$$0.384 = \frac{29}{A}$$

$$0.384 \times A = \frac{29}{\cancel{A}} \times \cancel{A} (\times A)$$

$$\frac{\cancel{0.384} \times A}{\cancel{0.384}} = \frac{29}{0.384} (\div 0.384)$$

$$A = \frac{29}{0.384}$$

$$A = 75.5\text{ cm}$$

Now that O and A are known, we can calculate H.

$$C\ (\text{or } H) = \sqrt{A^2 + B^2}$$
$$= \sqrt{75.5^2 - 29^2}$$
$$= \sqrt{5700.25 + 841}$$
$$= \sqrt{6541.25}$$
$$= 80.88\text{ cm}$$

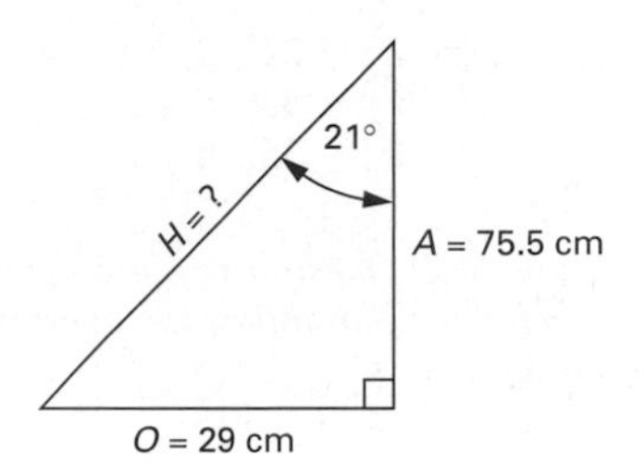

c.

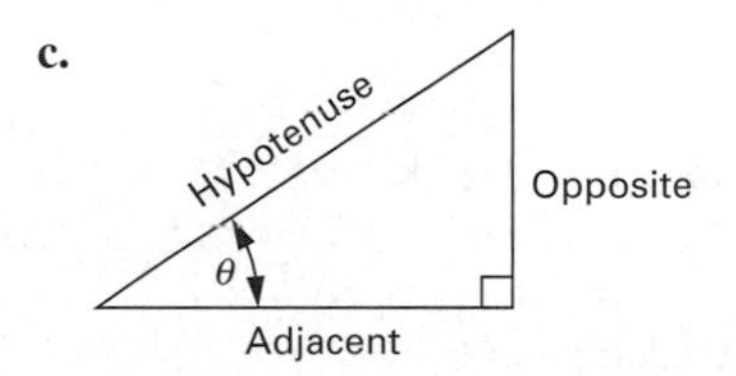

Known values	Unknown values
H = 34 volts	θ = ?
A = 18.5 volts	O = ?

Because both H and A are known, we can use cosine to calculate θ (CAH).

$$\cos\theta = \frac{\cancel{A}}{\cancel{H}}$$

$$\cos\theta = \frac{18.5\text{ V}}{34\text{ V}}$$

$$\cos\theta = 0.544$$

$$\cancel{\cos} \times \cancel{\text{invcos}}\ \theta = \text{invcos } 0.544\ (\times \text{ invcos})$$

$$\theta = \text{invcos } 0.544$$

$$\theta = 57^\circ$$

To calculate the length of the unknown side, we can use the Pythagorean theorem.

$$B = \sqrt{C^2 - A^2}$$
$$= \sqrt{34^2 - 18.5^2}$$
$$= \sqrt{813.75}$$
$$= 28.5\text{ V}$$

57. Frequency = 1/time:
a. 1/16 ms = 62.5 Hz;
b. 1/1 s = 1 Hz;
c. 1/15 μs = 66.67 kHz;
d. 1/0.05s = 20 Hz;
e. 1/200 μs = 5 kHz;
f. 1/350 ms = 2.86 Hz

59. **a.** Peak = 1.414 × rms = 1.414 × 40 mA = 56.56 mA
b. Peak to peak = 2 × peak = 2 × 56.56 mA = 113.12 mA
c. Average = 0.637 × peak = 0.637 × 56.56 mA = 36 mA

61. Duty cycle % = $(P_w/t) \times 100$, $t = 1/f = 1/10$ kHz = 100 μs; duty cycle = (10 μs/100 μs) × 100 = 10%

63. I_{avg} = baseline + (duty cycle × I_p), duty cycle = $P_w/t \times 100$, $t = 1/f = 1/10$ kHz = 100 μs, duty cycle = 10 μs/100 μs × 100 = 10%, $I_{avg} = 0 + (0.1 \times 15\text{ A}) = 1.5$ A

65. **a.** Third harmonic = 3 × fundamental = 3 × 1 kHz = 3 kHz
b. Second harmonic = 2 × fundamental = 2 × 1 kHz = 2 kHz
c. Seventh harmonic = 7 × fundamental = 7 × 1 kHz = 7 kHz

67. $V_{pk\text{-}pk} = 2 \times V_{pk}$, $V_{pk} = V_{rms} \times 1.414 = 6 \times 1.414 = 8.484$; $V_{pk\text{-}pk} = 2 \times 8.484 = 16.968$, 2 V/cm × 8 would only show 16 V, so 5 V/cm is lowest setting. $t = 1/f = 1/350$ kHz = 2.857 μs; 0.2 μs/cm × 10 would only show 2 μs, so 0.5 μs/cm is lowest setting.

69. t = 5.5 cm × 1 μs/cm = 5.5 μs; frequency = $1/t$ = 1/5.5 μs = 181.818 kHz

CHAPTER 9

1. b	**9.** b	**17.** b	**25.** d	**31.** a
3. a	**11.** a	**19.** c	**27.** b	**33.** e
5. c	**13.** d	**21.** b	**29.** d	**35.** b
7. d	**15.** d	**23.** b		

(The answers to Communication Skill Questions 36 through 60 can be found in the sections indicated that follow the questions.)

61. $Q = C \times V = 10\ \mu\text{F} \times 10\text{ V} = 100\ \mu\text{C}$

63. $V = Q/C = 125 \times 10^{-6}/0.006\ \mu\text{F} = 20.83\text{ kV}$

65. **a.** $C_T = C_1 + C_2 + C_3 + C_4 = 1.7\ \mu\text{F} + 2.6\ \mu\text{F} + 0.03\ \mu\text{F} + 1200\text{ pF} = 4.3312\ \mu\text{F}$

b. $C_T = \dfrac{1}{(1/C_1) + (1/C_2) + (1/C_3)} = \dfrac{1}{(1/1.6\ \mu\text{F}) + (1/1.4\ \mu\text{F}) + (1/4\ \mu\text{F})} = 0.629\ \mu\text{F}$

67. **a.** 10 + 4 zeros = 100,000 pF or 0.1 μF
b. 12 + 5 zeros = 1,200,000 pF or 1.2 μF
c. 0.01 μF
d. 220 pF

69. **a.** V_{ITC} = 63.2% of V_S = 0.632 × 10 V = 6.32 V; 5TC = 5 × 84 ms = 420 ms
b. V_{ITC} = 63.2% of V_S = 0.632 × 10 V = 6.32 V; 5TC = 5 × 16.8 ms = 84 ms
c. V_{ITC} = 63.2% of V_S = 0.632 × 10 V = 6.32 V; 5TC = 5 × 4.08 ms = 20.4 ms
d. V_{ITC} = 63.2% of V_S = 0.632 × 10 V = 6.32 V; 5TC = 5 × 980 μs = 4.9 ms

71. $X_C = \dfrac{1}{2\pi FC}$:

a. $\dfrac{1}{2\pi(1\text{ kHz})\,2\mu\text{F}} = 79.6\ \Omega$;

b. $\dfrac{1}{2\pi(100)0.01\ \mu\text{F}} = 159.2\text{ k}\Omega$;

c. $\dfrac{1}{2\pi(17.3\text{ MHz})\,47\ \mu\text{F}} = 195.7\ \mu\Omega$

73. **a.** $Z = \sqrt{R^2 + X_C^2}$, $X_C = \dfrac{1}{2\pi\text{FC}}$

$= \dfrac{1}{2\pi(20\text{ kHz})\,3.7\ \mu\text{F}} = 2.15\ \Omega$,

$Z = \sqrt{2.7\text{ M}^2 + (2.15)^2} = 2.7\text{ M}\Omega$

b. $Z = \sqrt{R^2 + X_C^2}$,

$X_C = \dfrac{1}{2\pi fc} = \dfrac{1}{2\pi(3\text{ MHz})0.005\ \mu\text{F}} = 10.61\ \Omega$,

$Z = \sqrt{350^2 + 10.61^2} = 350.16\ \Omega$

c. $Z = \sqrt{R^2 + X_C^2} = \sqrt{8.6\text{ k}^2 + 2.4^2} = 8.6\text{ k}\Omega$

d. $Z = \sqrt{R^2 + X_C} = \sqrt{4.7\text{ k}^2 + 2\text{ k}^2} = 5.1\text{ k}\Omega$

75. **a.** $X_{C_T} = X_{C_1} X_{C_2} + X_{C_3} = 200\ \Omega + 300\ \Omega + 400\ \Omega = 900\ \Omega$

b. $X_{C_T} = \dfrac{X_{C_1} \times X_{C_2}}{X_{C_1} + X_{C_2}} = \dfrac{3.3\text{ k}\Omega \times 2.7\text{ k}\Omega}{3.3\text{ k}\Omega + 2.7\text{ k}\Omega} = \dfrac{8.91\ \mu\Omega}{6\text{ k}\Omega}$
$= 1.485\text{ k}\Omega$

77. $F = \dfrac{1}{2\pi X_C C} = \dfrac{1}{2\pi(2000\ \Omega)\,4.7\ \mu\text{F}} = 16.93\text{ Hz}$

79. **a.**

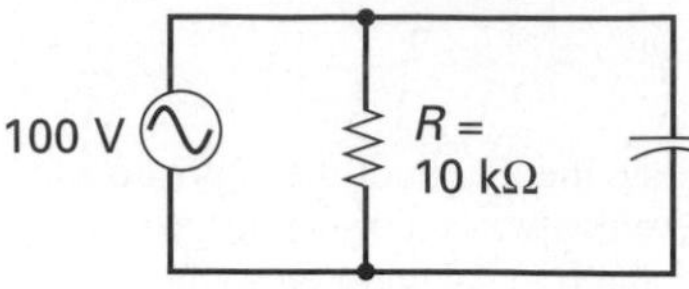

FIGURE Ch. 9, 79(a)

b. $I_R = V/R = \dfrac{100\text{ V}}{10\text{ k}\Omega} = 10\text{ mA}$

$I_C = V/X_C = 100\text{ V}/5\text{ k}\Omega = 20\text{ mA}$

$I_T = \sqrt{I_R^2 + I_C^2} = \sqrt{(10\text{ mA})^2 + (20\text{ mA})^2}$
$= 22.36\text{ mA}$

$Z = V/I_T = 100/22.36\text{ mA} = 4.5\text{ k}\Omega$

$V_R = V_C = V_S = 100\text{ V}$

$\theta = \arctan(R/X_C) = \arctan(10\text{ k}\Omega/5\text{ k}\Omega) = 63.4°$

81. **a.** $Z = \sqrt{R^2 + X_C^2} = \sqrt{30^2 + 25^2} = 39.05\ \Omega$

b. $X_C = \dfrac{1}{2\pi fc} = \dfrac{1}{2\pi(100\text{ kHz})50\text{ PF}} = 31.8\text{ k}\Omega$

$Z = \sqrt{12\text{k }\Omega^2 + 31.8\text{ k}\Omega^2} = 33.94\text{ k}\Omega$

c. $C_T = \dfrac{C_1 \times C_2}{C_1 + C_2} = \dfrac{100\ \mu\text{F} \times 330\ \mu\text{F}}{100\ \mu\text{F} + 330\ \mu\text{F}} = 76.7\ \mu\text{F}$,

$R_T = R_1 + R_2 = 5000 + 4700 = 9.7\text{ k}\Omega$

$X_C = \dfrac{1}{2\mu fC} = \dfrac{1}{2\pi(10\text{ kHz})\,76.7\ \mu\text{F}} = 207.5\text{ m}\Omega$

$Z = \sqrt{R^2 + X_C^2} = \sqrt{9.7\text{ k}\Omega + 207.5\text{ m}\Omega^2} = 9.7\text{ k}\Omega$

d. $C_T = C_1 + C_2 = 0.002\ \mu\text{F} + 0.005\ \mu\text{F} = 0.007\ \mu\text{F}$

$X_C = \dfrac{1}{2\pi fc} = \dfrac{1}{2\pi(1\text{ MHz})\,0.007\ \mu\text{F}} = 22.7\ \Omega$

$Z = \sqrt{R^2 + X_C^2} = \sqrt{2.5\text{ k}\Omega + 22.7\ \Omega^2} = 2.5\text{ k}\Omega$

83. Lead, 90

85. **a.** $X_C = \dfrac{1}{2\pi fc} = \dfrac{1}{2\pi(45\text{ kHz})\,2\ \mu\text{F}} = 1.77\ \Omega$

$Z = \sqrt{R^2 + X_C^2} = \sqrt{330\text{ k}\Omega^2 + 1.77^2} = 330\ \Omega$

$I = V/Z = 10/330 = 30.3\text{ mA}$, $I_R = I = 30.3\text{ mA}$

$\theta = \arctan(X_C/R) = \arctan 1.77/330 = 0.3°$

$V_R = I \times R = 30.3\text{ mA} \times 330 = 9.999\text{ V}$

$V_C = I \times X_C = 30.3\text{ mA} \times 1.77 = 53.631\text{ mV}$

b. $V_R = V_C = V_S = 18\text{ V}$

$I_R = V/R = \dfrac{18\text{V}}{1.2\text{ k}\Omega} = 15\text{ mA}$

$I_C = V/X_C = 18\text{ V}/2.7\text{ k}\Omega = 6.67\text{ mA}$

$I_T = \sqrt{I_R^2 + I_C^2} = \sqrt{15\text{ mA}^2 + 6.67\text{ mA}^2} = 16.4\text{ mA}$

$Z = V/I_T = \dfrac{18\text{ V}}{16.4\text{ mA}} = 1.1\text{ k}\Omega$

$\theta = \arctan(R/X_C) = \arctan \dfrac{1.2\text{ k}\Omega}{2.7\text{ k}\Omega} = 23.96°$

87. Short capacitor leads to discharge it; set meter to highest ohm scale and zero meter. Connect meter to capacitor observing correct polarity. Meter's pointer should go to 0 ohms and then move back to infinity as it charges. Cannot be used to reliably check capacitors with values less than 0.5 μF.

89. Capacitor leakage: imperfection of the dielectric, resistance between plates may become too small, effectively in parallel with the capacitor, which can cause circuit to malfunction

CHAPTER 10

1. a **3.** a **5.** b **7.** c **9.** b **11.** c **13.** c **15.** b **17.** c **19.** b **21.** c **23.** a **25.** a

(The answers to Communication Skill Questions 26 through 40 can be found in the sections indicated that follow the questions.)

41. $\beta = \dfrac{1200\ \mu\text{Wb}}{6.4 \times 10^{-3}} = 0.1875$ teslas

43. $H = (I \times N)/l = (1.2 \times 40)/0.15 = 48/0.15 = 320$ A · t/m

45. R = mmf/ϕ: mmf $= I \times N = 3 \times 36 = 108$
$R = 108/2.3 \times 10^{-4} = 469.57$ k ampere-turns/weber

47. $R = \text{mmf}/\phi = 150/360 \times 10^{-6} = 416.67$ kAt/Wb

49. $\mu = \mu_r \times \mu_0 = 100.000 \times 4\pi \times 10^{-7} = 0.126$ H/m

CHAPTER 11

1. c **3.** b **5.** a **7** d **9.** a **11.** d **13.** c **15.** d **17.** a **19.** b **21.** b **23.** a **25.** a

(The answers to Communication Skill Questions 26 through 35 can be found in the sections indicated that follow the questions.)

37. a. 22 MΩ
b. 78.6 kΩ
c. 314.2 kΩ

39. a. $L_T = L_1 + L_2 + L_3 = 75\ \mu\text{H} + 61\ \mu\text{H} + 50\ \text{mH} = 50.136\ \text{mH}$
b. $L_T = L_1 + L_2 + L_3 = 8\ \text{mH} + 4\ \text{mH} + 22\ \text{mH} = 34\ \text{mH}$

41. a. 4.36 mH
b. 4.7 μH
c. 3.82 μH

43. a. $V_S = \sqrt{V_R^2 + V_L^2} = \sqrt{12^2 + 6^2} = \sqrt{144 + 36} = \sqrt{180} = 13.4$ V
b. $I = V_S/Z = 13.4\ \text{V}/14\ \text{k}\Omega = 957.1\ \mu\text{A}$
c. $\angle = \arctan V_L/V_R = \arctan 2 = 63.4°$.
d. $Q = V_L/V_R = 12/6 = 2$.
e. PF = cos θ = 0.448.

45. $f = X_L/2\pi L = 27\ \text{k}\Omega/2\pi 330\ \mu\text{H} = 13.02$ MHz

47. $\tau = \dfrac{L}{R} = \dfrac{400\ \text{mH}}{2\ \text{k}\Omega} = 200\ \mu\text{S}$

V_L will start at 12 V and then exponentially drop to 0 V

Time	Factor	V_S	V_L
0	1.0	12	12
1 TC	0.365	12	4.416
2 TC	0.135	12	1.62
3 TC	0.05	12	0.6
4 TC	0.018	12	0.216
5 TC	0.007	12	0.084

49. a. $R_T = R_1 + R_2 = 250 + 700 = 950\ \Omega$
b. $L_T = L_1 + L_2 = 800\ \mu\text{H} + 1200\ \mu\text{H} = 2$ mH
c. $X_L = 2\pi fL = 2\pi(350\ \text{Hz})\ 2\ \text{mH} = 4.4\ \Omega$
d. $Z = \dfrac{R \times X_L}{\sqrt{R^2 + X_L^2}} = \dfrac{950 \times 4.4}{\sqrt{950^2 + 4.42^2}} = 4.39$
e. $V_{R_T} = V_{L_T} = V_S = 20$ V
f. $I_{R_T} = \dfrac{V_S}{R_T} = \dfrac{20\ \text{V}}{950\ \Omega} = 21$ mA,
$I_{L_T} = \dfrac{V_{S_1}}{X_{L_T}} = \dfrac{20\ \text{V}}{4.4\ \Omega} = 4.5$ A
g. $I_T = \sqrt{I_R^2 + I_L^2} = \sqrt{21\ \text{mA}^2 + 4.5\ \text{A}^2} = 4.5\ A$
h. $\theta = \arctan (R/X_L) = \arctan 950/4.4 = 89.7°$
i. $P_R = I^2 \times R = 21\text{mA}^2 \times 950 = 418.95$ mW
$P_X = I^2 \times X_L = 4.5\ \text{A}^2 \times 4.4 = 89.1$ VAR
$P_A = \sqrt{P_R^2 + P_X^2} = \sqrt{418.95\ \text{mW}^2 + 89.1\ \text{A}^2} = 89.1$ VA
j. PF $= P_R/P_A = 418.95\ \text{mW}/89.1\ \text{W} = 0.0047$

CHAPTER 12

1. d **3.** c **5.** d **7.** a **9.** a

(The answers to Communication Skill Questions 11 through 25 can be found in the sections indicated that follow the questions.)

27. a. $V_s = \dfrac{N_s}{N_P} \times V_P = 24/12 \times 100\ \text{V} = 200$ V
b. $V_s = \dfrac{N_s}{N_P} \times V_P = 250/3 \times 100\ \text{V} = 8.33$ kV
c. $V_s = \dfrac{N_s}{N_P} \times V_P = 5/24 \times 100\ \text{V} = 20.83$ V
d. $V_s = \dfrac{N_s}{N_P} \times V_P = 120/240 \times 100\ \text{V} = 50$ V

29. Turns ratio $= \sqrt{Z_L/Z_S} = \sqrt{8\ \Omega/24\ \Omega} = \sqrt{1/3} = 0.58$

31. For 16 turns: $V_s = (N_s/N_p) \times V_p = (16/12) \times 24\ \text{V} = 32$ V
For 2 turns: $V_s = (N_s/N_p) \times V_p = (2/12) \times 24\ \text{V} = 4$ V
For 1 turn: $V_s = (N_s/N_p) \times V_p = (1/12) \times 24\ \text{V} = 2$ V
For 4 turns: $V_s = (N_s/N_p) \times V_p = (4/12) \times 24\ \text{V} = 8$ V

33. Follow polarity dots.

35. a. I_s − apparent power/V_s = 500 VA/600 V = 833.3 mA
b. $R_L = V_s/I_s = 600\ \text{V}/833.3\ \text{mA} = 720\ \Omega$

CHAPTER 13

1. b **3.** b **5.** d **7.** c **9.** c **11.** a **13.** a **15.** d

(The answers to Communication Skill Questions 16 through 30 can be found in the sections indicated that follow the questions.)

31. a. $X_C = \dfrac{1}{2\pi fC} = \dfrac{1}{2\pi(60)0.02\mu\text{F}} = 132.6\ \text{k}\Omega$
b. $X_C = \dfrac{1}{2\pi fC} = \dfrac{1}{2\pi(60)18\mu\text{F}} = 147.4\ \Omega$
c. $X_C = \dfrac{1}{2\pi fC} = \dfrac{1}{2\pi(60)360\ \text{pF}} = 7.37\ \text{M}\Omega$
d. $X_C = \dfrac{1}{2\pi fC} = \dfrac{1}{2\pi(60)2700\ \text{nF}} = 982.4\ \Omega$
e. $X_L = 2\pi fL = 2\pi(60)4\ \text{mH} = 1.5\ \Omega$
f. $X_L = 2\pi fL = 2\pi(60)8.18\ \text{H} = 3.08\ \text{k}\Omega$
g. $X_L = 2\pi fL = 2\pi(60)150\ \text{mH} = 56.5\ \Omega$
h. $X_L = 2\pi fL = 2\pi(60)2\ \text{H} = 753.98\ \Omega$

33. a. $X_L = 2\pi fL = 2\pi(60\ \text{Hz})150\ \text{mH} = 56.5\ \Omega$
b. $X_C = \dfrac{1}{2\pi fc} = \dfrac{1}{2\pi(60\ \text{Hz})20\ \mu\text{F}} = 132.6\ \Omega$
c. $I_R = V/R = 120\ \text{V}/270\ \Omega = 444.4$ mA
d. $I_L = V/X_L = 120\ \text{V}/56.5\ \Omega = 2.12$ A
e. $I_C = V/X_C = 120\ \text{V}/132.6\ \Omega = 905$ mA
f. $I_T = \sqrt{I_R^2 + I_X^2} = \sqrt{(444.4\ \text{mA})^2 + (1.215)^2} = 1.29$ A
g. $Z = V/I_T = 120\ \text{V}/1.29\ \text{A} = 93.02\ \Omega$

h. Resonant frequency $= \dfrac{1}{2\pi\sqrt{LC}}$

$= \dfrac{1}{2\pi\sqrt{150 \text{ mH} \times 20\ \mu\text{F}}}$

$= 91.89$ Hz

i. $X_L = 2\pi fL$

$= 6.28 \times 91.89 \text{ Hz} \times 150 \text{ mH}$

$= 86.54\ \Omega$

$Q = \dfrac{X_L}{R} = \dfrac{86.54\ \Omega}{270\ \Omega} = 0.5769$

j. $\text{BW} = \dfrac{f_0}{Q} = \dfrac{91.89 \text{ Hz}}{0.5769} = 159.28$ Hz

35. Using a source voltage of 1 volt:

$Z = V/I_T, I_T = \sqrt{I_R^2 + I_X^2}, I_R = V/R = 1/750 = 1.33$ mA

$I_L = V/X_L = 1/25 = 40$ mA, $I_C = V/X_C = 1/160 = 6.25$ mA

$I_X = I_L - I_C = 40 \text{ mA} - 6.25 \text{ mA} = 33.75$ mA

$I_T = \sqrt{(1.33 \text{ mA})^2 + (33.75 \text{ mA})^2} = 33.78$ mA

$Z = 1/33.78\text{mA} = 29.6\Omega$

37. a. Real number = 25 cos 37° = 19.97; imaginary number = 25 sin 37° = 15, 19.97 + j15

b. Real number = 19 cos − 20° = 17.9; imaginary number = 19 sin − 20° = −6.5, 17.9 − j6.5

c. Real number = 114 cos−114° = −46.4; imaginary number = 114 sin −114° = −104.1, −46.4 −j104.1*

d. Real number = 59 cos 99° = +9.2; imaginary number = 59 sin 99° = 58.3. + 9.2 + j58.3

39. a. $(4 + j3) + (3 + j2) = (4 + 3) + (j3 + j2) = 7 + j5$

b. $(100 - j50) + (12 + j9) = (100 + 12) + (-j50 + j9) = 112 - j41$

*These examples were for practice purposes only; real numbers are always positive when they represent impedances.

41. a. $Z_T = 73 - j23, \sqrt{73^2 + 23^2} = 76.5, \angle = \arctan(23/73) = -17.5°, 76.5 \angle -17.5°; Z_T = 76.5\ \Omega$ at −17.5° phase angle

b. $Z_T = 40 + j15, \sqrt{40^2 + 15^2} = 42.7, \angle = \arctan(15/40) = 20.6°, 42.7 \angle 20.6°; Z_T = 42.\ \Omega$ at 20.6° phase angle

c. $Z_T = 8\text{ k}\Omega - j3\text{ k}\Omega + j20\text{ k}\Omega = 8\text{ k}\Omega + 17\text{ k}\Omega$ $\sqrt{8\text{ k}\Omega^2 + 17\text{ k}\Omega^2} = 18.8\text{ k}\Omega, \angle = \arctan(17\text{ k}\Omega/8\text{ k}\Omega) = 64.8°, 18.8\text{ k}\Omega \angle 64.8°; Z_T = 18.8\text{ k}\Omega$ at 64.8° phase angle

43. a. $Z_T = 47 - j40 + j30 = 47 - j10, \sqrt{47^2 + 10^2} = 48.05, \angle = \arctan(-10/47) = -12°, Z_T = 48.05 \angle -12°$

b. $I = \dfrac{V_S}{Z_T} = \dfrac{20 \angle 0°}{48.05 \angle -12} = \dfrac{20}{48.05\angle 0 - (-12)°} = 416.2 \text{ mA} \angle 12°$

c. $V_R = I \times R = 416.2 \text{ mA}\angle 12° \times 47 \angle 0 = 416.2 \text{ mA} \times 47 \angle (12 + 0) = 19.56 \text{ V} \angle 12°$

$V_C = I \times X_C = 416.2 \text{ mA} \angle 12° \times 40 \angle -90° = 416.2 \text{ mA} \times 40 \angle (12° - 90°) = 16.65 \angle -78°$

$V_L = I \times X_L = 416.2 \text{ mA} \angle 12° \times 30 \angle 90° = 416.2 \text{ mA} \times 30 \angle (12° + 90°) = 12.49 \text{ V} \angle 102°$

d. V_C lags I_T by 90°, V_L leads I by 90°, V_R is in phase with I.

45. a. $Z_1 = 0 + j27 - j17 = j10, \sqrt{0^2 + 10^2} = 10 \angle = 90°, Z_1 = 10 \angle 90°; Z_2 = 37 - j20 = 42.06 \angle -28.39°$; Z combined = Product/Sum; Product = $420.59\angle 61.61°$; Sum = $37 - j10 = 38.33 \angle -15.12°$; Z combined = $10.97 \angle 76.73°$

b. $I_1 = 3.72 \angle -34.48$ A; $I_2 = 884 \angle 83.91$ mA

c. $I_T = 3.39 \angle -21.21°$ A

d. $Z_T = 29.52 \angle 21.21°\ \Omega$

47. $V_R = 19.56 \angle 12° = 19.56 \cos 12° + j19.56 \sin 12° = 19.13 + j4.07$

$V_C = 16.65 \angle -78° = 16.65 \cos -78° + j16.65 \sin -78° = 3.46 - j16.29$

$V_L = 12.49 \angle 102° = 12.49 \cos 102° + j12.49 \sin 102° = = -2.6 + j12.22$

49. Easier with use of complex numbers

Index